William Feller
Selected Papers II

W. Feller (around 1960)
Picture taken by R.W. Collier at the IBM Thomas J. Watson Research Center.
Photograph courtesy of Joanne Elliott

William Feller
Selected Papers II

Edited by

René L. Schilling · Zoran Vondraček
Wojbor A. Woyczyński

Springer

William Feller
(1906 Zagreb – 1970 New York City)

Editors
René L. Schilling
Institut für Mathematische Stochastik
Technische Universität Dresden
Dresden
Germany

Zoran Vondraček
Department of Mathematics
University of Zagreb
Zagreb
Croatia

Wojbor A. Woyczyński
Department of Mathematics
Case Western Reserve University
Cleveland, OH
USA

ISBN 978-3-319-16855-5
Library of Congress Control Number: 2012954381

Math. Subj. Classification (2010): 01A75 60G05, 60F05, 60E07, 60J35, 60K05, 47D07

Springer Cham Heidelberg New York Dordrecht London

Printed on acid-free paper

Springer International Publishing AG Switzerland is part of Springer Science+Business Media
(www.springer.com)

Preface

These volumes contain a selection of William Feller's most important works on probability theory, mathematical biology, analysis and geometry. Feller was a prolific writer, with many groundbreaking contributions which, almost 45 years after his death, are still frequently quoted – both directly and indirectly. This abundance meant that the decision which of his more than 100 research papers should be included in these Selecta was not an easy task. Most of Feller's English-language research contributions have been included, and we also selected a fair share of his pre-1940 publications in German and French, six of which were translated into English for these Selecta. His two most cherished contributions, the definitive Feller–Lindeberg version of the *Central Limit Theorem* [Feller 1935c], and the deep *On the theory of stochastic processes* [Feller 1936c] – a companion to Kolmogorov's seminal work on stochastic processes and PDEs – are here being made accessible in English for the first time. We decided not to reprint Feller's work on measure theory, and the geometry selection contains only the fundamental *Acta Mathematica* contribution by Busemann and Feller; in fact, Feller did not continue to work in these directions, and we felt that his impact on these areas remained rather limited.

Feller is widely known for his brilliant two-volume textbook, *An Introduction to Probability Theory and Its Applications*, on which he continuously worked since the late 1940s. It is one of the few books which changed considerably from edition to edition. The observant reader of these Selecta will notice that in later editions Feller included many results from his (then) recent research papers. Consequently, the *Introduction to Probability Theory* is still a "modern" book, and a treat for every reader as well. For obvious reasons we could not include these books in the present selection.

The papers are arranged in chronological order: Volume 1 covers the years 1928–1950, and Volume 2 the period 1951–1971. Each volume contains additional material, of which the most important are commentaries and essays written by the leading experts in the areas covered by the Selecta. We are grateful to them for their willingness to spend considerable time and effort to put various aspects of Feller's work, and its afterlife, into modern perspective; Ellen Baake & Anton Wakolbinger wrote on Mathematical Biology, Hans Fischer on Foundations and the Central Limit Theorem, Masatoshi Fukushima on Diffusions and Boundaries, Niels Jacob on Functional Analysis, Ross Maller on Limit Theorems, Goran Peskir on Boundary Conditions and Erhard Scholz on Geometry. The volumes would not have the same relevance to modern mathematical research without their incisive analyses.

Thanks are due to Springer-Verlag for undertaking the publication. The team around Catriona Byrne and Marina Reizakis was very supportive and helped the project move forward throughout the past couple of years. We are grateful to all owners of the copyrights for their cooperation, and for generously granting the reprint permissions free of charge. These days this is no longer the rule, but an exception. Only one publisher declined to help us. Thus the absence of the difficult-to-access 1938 survey paper [*Feller 1938a] on the foundations of probability theory leaves a distressing gap in these Selecta.

Many colleagues and friends helped us during the preparation of these volumes. We would like to thank all of them, in particular, Christian Berg, Ulrich Brehm, Jürgen Elstrodt, Tony Knapp, Heinz König, Anders Martin-Löf and Hrvoje Šikić for valuable comments. We are especially grateful to Joanne Elliott, a long-time friend, and neighbour of the Feller family for insightful conversations about Feller's life in America, and for sharing generously her vast trove of historical photographs and other materials; some are reproduced here with her permission. Several of the surviving Feller's PhD students contacted by us also expressed support for the idea of this publication. Our student Ms. Franziska Kühn (TU Dresden) helped us with the intricacies of typesetting this work in Latex and other administrative chores.

Dresden, Zagreb, and Cleveland
September 2014

René L. Schilling
Zoran Vondraček
Wojbor A. Woyczyński

Acknowledgements

The Editors gratefully acknowledge the kindness of these institutions and individuals in granting the following permissions:

Acta Universitatis Szegediensis, Bolyai Institute

W. Feller: Über das Gesetz der großen Zahlen. *Acta Litterarum ac Scientiarum. Regiae Universitatis Hungaricae Francisco-Josephinae. Sectio Scientiarum Mathematicarum. (Acta. Sci. Litt. Szeged)* **8** (1937) 191–201. © Acta Universitatis Szegediensis, Bolyai Institute. Reprinted with permission of Acta Universitatis Szegediensis, Bolyai Institute.

American Mathematical Society

W. Feller: On the integro-differential equations of purely discontinuous Markoff processes. *Transactions of the American Mathematical Society* **48** (1940) 488–515. © American Mathematical Society. Reprinted with permission of the American Mathematical Society.

W. Feller: The general form of the so-called law of the iterated logarithm. *Transactions of the American Mathematical Society* **54** (1943) 373–402. © American Mathematical Society. Reprinted with permission of the American Mathematical Society.

W. Feller: Generalization of a probability limit theorem of Cramér. *Transactions of the American Mathematical Society* **54** (1943) 361–372. © American Mathematical Society. Reprinted with permission of the American Mathematical Society.

W. Feller: The fundamental limit theorems in probability. *Bulletin of the American Mathematical Society* **51** (1945) 800–832. © American Mathematical Society. Reprinted with permission of the American Mathematical Society.

H. Busemann, W. Feller: Regularity properties of a certain class of surfaces. *Bulletin of the American Mathematical Society* **51** (1945) 583–598. © American Mathematical Society. Reprinted with permission of the American Mathematical Society.

P. Erdös, W. Feller, H. Pollard: A property of power series with positive coefficients. *Bulletin of the American Mathematical Society* **55** (1949) 201–204. © American Mathematical Society. Reprinted with permission of the American Mathematical Society.

W. Feller: Fluctuation theory of recurrent events. *Transactions of the American Mathematical Society* **67** (1949) 98–119. © American Mathematical Society. Reprinted with permission of the American Mathematical Society.

W. Feller: Some recent trends in the mathematical theory of diffusion. *Proceedings of the International Congress of Mathematicians, Cambridge (MA) 1950.* Vol. 2. American Mathematical Society, Providence (RI) 1952, pp. 322–339. © American Mathematical Society. Reprinted with permission of the American Mathematical Society.

W. Feller: Diffusion processes in one dimension. *Transactions of the American Mathematical Society* **77** (1954) 1–31. © American Mathematical Society. Reprinted with permission of the American Mathematical Society.

W. Feller: Boundaries induced by non-negative matrices. *Transactions of the American Mathematical Society* **93** (1956) 19–54. © American Mathematical Society. Reprinted with permission of the American Mathematical Society.

J. Elliott, W. Feller: Stochastic processes connected with harmonic functions. *Transactions of the American Mathematical Society* **82** (1956) 392–420. © American Mathematical Society. Reprinted with permission of the American Mathematical Society.

Cambridge University Press

W. Feller: Some new connections between probability and classical analysis. *Proceedings of the International Congress of Mathematicians, Edinburgh 1958.* Cambridge University Press, Cambridge 1960, pp. 69–86. © Cambridge University Press. Reprinted with permission of the Cambridge University Press.

W. Feller: On fitness and the cost of natural selection. *Genetics Research* **9** (1967) 1–15. © Cambridge University Press. Reprinted with permission of the Cambridge University Press.

Croatian Academy of Sciences and Arts

W. Feller: Neuer Beweis für die Kolmogoroff–P. Lévysche Charakterisierung der unbeschränkt teilbaren Verteilungsfunktionen. *Bulletin international de l'académie Yougoslave des sciences et des beaux-arts, Zagreb. Classe des sciences mathématiques et naturelles* **32** (1939) 106–113. © Croatian Academy of Sciences and Arts. Reprinted with permission of the Croatian Academy of Sciences and Arts.

Duke University Press

W. Feller: Completely monotone functions and sequences. *Duke Mathematical Journal* **5** (1939) 661–674. © Duke University Press. Reprinted with permission of the Duke University Press.

W. Feller: Some geometric inequalities. *Duke Mathematical Journal* **9** (1942) 885–892. © Duke University Press. Reprinted with permission of the Duke University Press.

Elsevier

W. Feller: The birth and death processes as diffusion processes. *Journal de Mathématiques pures et appliquées, IX. série* **38** (1959) 301–345. © Elsevier. Reprinted with permission of Elsevier.

Illinois Journal of Mathematics

W. Feller: Generalized second order differential operators and their lateral conditions. *Illinois Journal of Mathematics* **1** (1957) 459–504. Reprinted with permission of the Illinois Journal of Mathematics, published by the University of Illinois at Urbana-Champaign.

W. Feller: On the intrinsic form for second order differential operators. *Illinois Journal of Mathematics* **2** (1958) 1–18. Reprinted with permission of the Illinois Journal of Mathematics, published by the University of Illinois at Urbana-Champaign.

W. Feller: Differential operators with the positive maximum property. *Illinois Journal of Mathematics* **3** (1959) 182–186. Reprinted with permission of the Illinois Journal of Mathematics, published by the University of Illinois at Urbana-Champaign.

Indiana University Mathematics Journal

W. Feller, S. Orey: A renewal theorem. *Journal of Mathematics and Mechanics* **10** (1961) 619–624. © Indiana University Mathematics Journal. Reprinted with permission of the Indiana University Mathematics Journal.

W. Feller: On the Fourier representation for Markov chains and the strong ratio theorem. *Journal of Mathematics and Mechanics* **15** (1966) 273–283. © Indiana University Mathematics Journal. Reprinted with permission of the Indiana University Mathematics Journal.

W. Feller: An extension of the law of the iterated logarithm to variables without variance. *Journal of Mathematics and Mechanics* **18** (1968) 343–355. © Indiana University Mathematics Journal. Reprinted with permission of the Indiana University Mathematics Journal.

Institute of Mathematical Statistics, Baltimore

W. Feller: On the integral equation of renewal theory. *Annals of Mathematical Statistics* **12** (1941) 243–267. © Institute of Mathematical Statistics, Baltimore. Reprinted with permission of the Institute of Mathematical Statistics, Baltimore.

W. Feller: On a general class of "contagious" distributions. *Annals of Mathematical Statistics* **14** (1943) 389–400. © Institute of Mathematical Statistics, Baltimore. Reprinted with permission of the Institute of Mathematical Statistics, Baltimore.

W. Feller: On the normal approximation to the binomial distribution. *Annals of Mathematical Statistics* **16** (1945) 319–329. © Institute of Mathematical Statistics, Baltimore. Reprinted with permission of the Institute of Mathematical Statistics, Baltimore.

W. Feller: Note on the law of large numbers and "fair" games. *Annals of Mathematical Statistics* **16** (1945) 301–304. © Institute of Mathematical Statistics, Baltimore. Reprinted with permission of the Institute of Mathematical Statistics, Baltimore.

W. Feller: On the Kolmogorov–Smirnov limit theorems for empirical distributions. *Annals of Mathematical Statistics* **19** (1948) 177–189. Erratum: *Annals of Mathematical Statistics* **21** (1950) 301—302. Both: © Institute of Mathematical Statistics, Baltimore. Reprinted with permission of the Institute of Mathematical Statistics, Baltimore.

W. Feller: The asymptotic distribution of the range of sums of independent random variables. *Annals of Mathematical Statistics* **22** (1951) 427–432. © Institute of Mathematical Statistics, Baltimore. Reprinted with permission of the Institute of Mathematical Statistics, Baltimore.

W. Feller: Non-Markovian processes with the semigroup property. *Annals of Mathematical Statistics* **30** (1959) 1252–1253. © Institute of Mathematical Statistics, Baltimore. Reprinted with permission of the Institute of Mathematical Statistics, Baltimore.

The Editors of Ann. Math. Statist.: William Feller, 1906–1970. *Annals of Mathematical Statistics* **41** No. 6 (1970), iv–xiii. © Institute of Mathematical Statistics, Baltimore. Reprinted with permission of the Institute of Mathematical Statistics, Baltimore.

Institute of Mathematics Polish Academy of Sciences

H. Busemann, W. Feller: Zur Differentiation der Lebesgueschen Integrale. *Fundamenta Mathematicae* **22** (1934) 226–256. © Institute of Mathematics Polish Academy of Sciences. Reprinted with permission of the Institute of Mathematics Polish Academy of Sciences.

W. Feller: Über die Existenz von sogenannten Kollektiven. *Fundamenta Mathematicae* **32** (1939) 87–96. © Institute of Mathematics Polish Academy of Sciences. Reprinted with permission of the Institute of Mathematics Polish Academy of Sciences.

W. Feller: On positivity preserving semigroups of transformations on $C[r_1, r_2]$. *Annales de la Société Polonaise de Mathématique* **25** (1952) 85–94. © Institute of Mathematics Polish Academy of Sciences. Reprinted with permission of the Institute of Mathematics Polish Academy of Sciences.

John Wiley & Sons, Inc.

W. Feller: On differential operators and boundary conditions. *Communications on Pure and Applied Mathematics* **8**:1 (1955) 203–216. ©1955 John Wiley & Sons, Inc. This material is reproduced with permission of John Wiley & Sons, Inc.

W. Feller: A simple proof for renewal theorems. *Communications on Pure and Applied Mathematics* **14**:3 (1961) 285–293. ©1961 John Wiley & Sons, Inc. This material is reproduced with permission of John Wiley & Sons, Inc.

L'Enseignement Mathématique

W. Feller: One-sided analogues of Karamata's regular variation. *L'Enseignement Mathématique, IIe. Série* **15** (1969) 107–121. © L'Enseignement Mathématique. Reprinted with permission of L'Enseignement Mathématique.

Lund University, Centre for Mathematical Sciences

W. Feller: On a generalization of Marcel Riesz' potentials and the semigroups generated by them.

Meddelanden från Lunds Universitets Matematiska Seminarium. Supplement M. Riesz, Uppsala 1952, pp. 73–81. © Lund University, Centre for Mathematical Sciences. Reprinted with permission of Lund University, Centre for Mathematical Sciences.

Princeton University, Department of Mathematics

W. Feller: The law of the iterated logarithm for identically distributed random variables. *Annals of Mathematics* **47** (1946) 631–638. © Princeton University, Department of Mathematics. Reprinted with permission of Princeton University, Department of Mathematics.

W. Feller: Two singular diffusion problems. *Annals of Mathematics* **54** (1951) 173–182. © Princeton University, Department of Mathematics. Reprinted with permission of Princeton University, Department of Mathematics.

W. Feller: The parabolic differential equations and the associated semigroups of transformation. *Annals of Mathematics* **55** (1952) 468–519. © Princeton University, Department of Mathematics. Reprinted with permission of Princeton University, Department of Mathematics.

W. Feller: Semigroups of transformations in general weak topologies. *Annals of Mathematics* **57** (1953) 287–308. © Princeton University, Department of Mathematics. Reprinted with permission of Princeton University, Department of Mathematics.

W. Feller: On the generation of unbounded semigroups of bounded linear operators. *Annals of Mathematics* **58** (1953) 166–174. © Princeton University, Department of Mathematics. Reprinted with permission of Princeton University, Department of Mathematics.

W. Feller: The general diffusion operator and positivity preserving semigroups in one dimension. *Annals of Mathematics* **60** (1954) 417–436. © Princeton University, Department of Mathematics. Reprinted with permission of Princeton University, Department of Mathematics.

W. Feller: On second order differential operators. *Annals of Mathematics* **61** (1955) 90–105. © Princeton University, Department of Mathematics. Reprinted with permission of Princeton University, Department of Mathematics.

W. Feller: On boundaries and lateral conditions for the Kolmogorov differential equations. *Annals of Mathematics* **65** (1957) 527–570. Additional notes: *Annals of Mathematics* **68** (1958) 735–736. Both: © Princeton University, Department of Mathematics. Reprinted with permission of Princeton University, Department of Mathematics.

W. Feller: On the oscillations of sums of independent random variables. *Annals of Mathematics* **91** (1970) 402–418. © Princeton University, Department of Mathematics. Reprinted with permission of Princeton University, Department of Mathematics.

Society for Industrial and Applied Mathematics

W. Feller: Infinitely divisible distributions and Bessel functions associated with random walks. *SIAM Journal of Applied Mathematics* **14** (1966) 864–875. © Society for Industrial and Applied Mathematics. Reprinted with permission of the Society for Industrial and Applied Mathematics.

Springer

W. Feller: Über algebraisch rektifizierbare transzendente Kurven. *Mathematische Zeitschrift* **27** (1928) 481–495. © Springer. Reprinted with permission of Springer.

W. Feller: Über die Lösungen der linearen partiellen Differentialgleichungen zweiter Ordnung vom elliptischen Typus. *Mathematische Annalen* **102** (1930) 633–649. © Springer. Reprinted with permission of Springer.

H. Busemann, W. Feller: Krümmungseigenschaften konvexer Flächen. *Acta Mathematica* **66** (1936) 1–47. © Springer. Reprinted with permission of Springer.

W. Feller: Über den zentralen Grenzwertsatz der Wahrscheinlichkeitsrechnung. *Mathematische Zeitschrift* **40** (1936) 521–559. © Springer. Reprinted with permission of Springer.

W. Feller: Zur Theorie der stochastischen Prozesse. (Existenz- und Eindeutigkeitssätze). *Mathematische Annalen* **113** (1936) 113–160. © Springer. Reprinted with permission of Springer.

W. Feller: Über den zentralen Grenzwertsatz der Wahrscheinlichkeitsrechnung. II. *Mathematische Zeitschrift* **42** (1937) 301–312. Erratum: *Mathematische Zeitschrift* **44** (1939) 794. Both: © Springer. Reprinted with permission of Springer.

W. Feller: Die Grundlagen der Volterraschen Theorie des Kampfes ums Dasein in wahrscheinlichkeitstheoretischer Behandlung. *Acta Biotheoretica, Leiden* **5** (1939) 11–39. © Springer. Reprinted with permission of Springer.

W. Feller: On the classical Tauberian theorems. *Archiv der Mathematik* **14** (1963) 317–322. © Springer. Reprinted with permission of Springer.

W. Feller: On the Berry-Esseen theorem. *Zeitschrift für Wahrscheinlichkeitstheorie und verwandte Gebiete* **10** (1968) 261–268. © Springer. Reprinted with permission of Springer.

W. Feller: General analogues to the law of the iterated logarithm. *Zeitschrift für Wahrscheinlichkeitstheorie und verwandte Gebiete* **14** (1969) 21–26. © Springer. Reprinted with permission of Springer.

W. Feller: Limit theorems for probabilities of large deviations. *Zeitschrift für Wahrscheinlichkeitstheorie und verwandte Gebiete* **14** (1969) 1–20. © Springer. Reprinted with permission of Springer.

The Johns Hopkins University Press

W. Feller: A limit theorem for random variables with infinite moments. *American Journal of Mathematics* **68**:2 (1946) 257–262. ©1946 The Johns Hopkins Press. Reprinted with permission of The Johns Hopkins University Press.

University of California Press

W. Feller: On regular variation and local limit theorems. In: J. Neyman and L. LeCam (eds.): *Proceedings of the Fifth Berkeley Symposium on Mathematical Statistics and Probability 1965/66*, Vol. 2, Pt. 1, University of California Press, Berkeley, Los Angeles (CA) 1967, pp. 373–388. © University of California Press. Reprinted with permission of the University of California Press.

J.L. Doob: William Feller and Twentieth-Century Probability. In: L. LeCam, J. Neyman, E.L. Scott (eds.): *Proceedings of the Sixth Berkeley Symposium on Mathematical Statistics and Probability 1970/71*, Vol. 2, University of California Press, Berkeley, Los Angeles (CA) 1972, pp. 15–20. © University of California Press. Reprinted with permission of the University of California Press.

M. Kac: William Feller, In Memoriam. In: L. LeCam, J. Neyman, E.L. Scott (eds.): *Proceedings of the Sixth Berkeley Symposium on Mathematical Statistics and Probability 1970/71*, Vol. 2, University of California Press, Berkeley, Los Angeles (CA) 1972, pp. 21–23. © University of California Press. Reprinted with permission of the University of California Press.

For the following publications, the copyright resides with the author(s). Despite our best efforts, we were not able to trace the present copyright owners and we do ask them to come forward if they feel that their rights are infringed or that the material has been used in violation of copyright law.

W. Feller: On probability problems in the theory of counters. In: *Studies, Essays, presented to R. Courant (Courant anniversary volume)*. Interscience Publishers, New York 1948, pp. 105–115.

K.L. Chung, W. Feller: On fluctuations in coin-tossing. *Proceedings of the National Academy of Sciences, USA* **35** (1949) 605–608.

W. Feller, H.P. McKean Jr.: A diffusion equivalent to a countable Markov chain. *Proceedings of the National Academy of Sciences, USA* **42** (1956) 351–354.

W. Feller: On boundaries defined by stochastic matrices. In: *Proceedings of Symposia in Applied Mathematics*. Vol. 7. McGraw-Hill Book Co., New York (for the American Mathematical Society, Providence (RI)) 1957, 35–40.

W. Feller: On semi-Markov processes. *Proceedings of the National Academy of Sciences, USA* **51** (1964) 653–659.

W. Feller: On the influence of natural selection on population size. *Proceedings of the National Academy of Sciences, USA* **55** (1966) 733–738.

The following papers are already in the public domain. We are grateful to the University of California Press for this information.

W. Feller: On the theory of stochastic processes, with particular reference to applications. In: J. Neyman (ed.): *Proceedings of the (First) Berkeley Symposium on Mathematical Statistics and Probability 1945/46*. University of California Press, Berkeley, Los Angeles (CA) 1949, pp. 403–432.

W. Feller: Diffusion processes in genetics. In: J. Neyman (ed.): *Proceedings of the Second Berkeley Symposium on Mathematical Statistics and Probability 1950*. University of California Press, Berkeley, Los Angeles (CA) 1951, pp. 227–246.

Contents of Volume 2

Preface v

Acknowledgements vii

Contents of Volume 2 xiii

Contents of Volume 1 xvii

Curriculum Vitae: William Feller 1906–1970 xxi

Bibliography of William Feller xxv

PhD Students of William Feller xxxv

R. L. Schilling and W. A. Woyczyński: William Feller. A Biography 1

E. Baake and A. Wakolbinger: Feller's Contributions to Mathematical Biology 25

N. Jacob: Feller on Differential Operators and Semi-groups 45

M. Fukushima: Feller's Contributions to the One-Dimensional Diffusion Theory and Beyond 63

G. Peskir: On Boundary Behaviour of One-Dimensional Diffusions: From Brown to Feller and Beyond 77

R. Maller: Feller's Work in Renewal Theory, the Law of the Iterated Logarithm and Karamata Theory 95

[Feller 1951a] The Asymptotic Distribution of the Range of Sums of Independent Random Variables 115

[Feller 1951d] Diffusion Processes in Genetics 121

[Feller 1951e] Two Singular Diffusion Problems 141

[Feller 1952a] The Parabolic Differential Equations and the Associated Semigroups of Transformation 151

[Feller 1952b] On a Generalization of Marcel Riesz' Potentials and the Semigroups Generated by Them 203

[Feller 1952c] Some Recent Trends in the Mathematical Theory of Diffusion 213

[Feller 1952e] On Positivity Preserving Semigroups of Transformations on $C[r_1, r_2]$ 231

[Feller 1953a] Semigroups of Transformations in General Weak Topologies 241

[Feller 1953b] On the Generation of Unbounded Semigroups of Bounded Linear Operators 263

[Feller 1954a] The General Diffusion Operator and Positivity Preserving Semigroups in One Dimension 273

[Feller 1954b] Diffusion Processes in One Dimension 293

[Feller 1955a] On Second Order Differential Operators 325

[Feller 1955b] On Differential Operators and Boundary Conditions 341

[Feller 1956a] Boundaries Induced by Non-negative Matrices 355

[Feller 1956b] A Diffusion Equivalent to a Countable Markov Chain 391

[Feller 1956c] Stochastic Processes Connected with Harmonic Functions 395

[Feller 1957a] Generalized Second Order Differential Operators and Their Lateral Conditions 425

[Feller 1957d] On Boundaries and Lateral Conditions for the Kolmogorov Differential Equations 471

[Feller 1957e] On Boundaries Defined by Stochastic Matrices 517

[Feller 1958] On the Intrinsic Form for Second Order Differential Operators 523

[Feller 1959b] The Birth and Death Processes as Diffusion Processes 541

[Feller 1959c] Non-Markovian Processes with the Semigroup Property 587

[Feller 1959d] Differential Operators with the Positive Maximum Property 589

[Feller 1960] Some New Connections Between Probability and Classical Analysis 595

[Feller 1961a] A Renewal Theorem 613

[Feller 1961b] A Simple Proof for Renewal Theorems 619

[Feller 1963] On the Classical Tauberian Theorems 629

[Feller 1964] On Semi-Markov Processes 635

[Feller 1966a] On the Fourier Representation for Markov Chains and the Strong Ratio Theorem 643

[Feller 1966b] On the Influence of Natural Selection on Population Size 655

[Feller 1966c] Infinitely Divisible Distributions and Bessel Functions Associated with Random Walks 661

[Feller 1967a] On Regular Variation and Local Limit Theorems 673

[Feller 1967c] On Fitness and the Cost of Natural Selection 689

[Feller 1968b] On the Berry-Esseen Theorem 705

[Feller 1968d] An Extension of the Law of the Iterated Logarithm to Variables Without Variance 713

[Feller 1969b] One-sided Analogues of Karamata's Regular Variation 727

[Feller 1969d] General Analogues to the Law of the Iterated Logarithm 743

[Feller 1969c] Limit Theorems for Probabilities of Large Deviations 749

[Feller 1970] On the Oscillations of Sums of Independent Random Variables 769

Contents of Volume 1

Preface v

Acknowledgements vii

Contents of Volume 1 xiii

Contents of Volume 2 xvii

Curriculum Vitae: William Feller 1906–1970 xxi

Bibliography of William Feller xxv

PhD Students of William Feller xxxv

Z. W. Birnbaum (ed.): William Feller, 1906–1970 1

J. L. Doob: William Feller and Twentieth Century Probability 7

M. Kac: William Feller, *In Memoriam* 13

R. L. Schilling and W. A. Woyczyński: William Feller. A Biography 17

H. Fischer: Feller's Early Work on Measure Theory and Mathematical Foundations of Probability 43

H. Fischer: Feller's Early Work on Limit Theorems 69

E. Scholz: Feller and Busemann on Surface Theory — Contributions to Geometry 87

[Feller 1928] Über algebraisch rektifizierbare transzendente Kurven 101

[Feller 1930] Über die Lösungen der linearen partiellen Differentialgleichungen zweiter Ordnung vom elliptischen Typus 117

[Feller 1934a] Zur Differentiation der Lebesgueschen Integrale 135

[Feller 1935c] Über den zentralen Grenzwertsatz der Wahrscheinlichkeitsrechnung 167

[Feller 1935c] (translation) On the Central Limit Theorem of Probability Theory 207

[Feller 1936b] Krümmungseigenschaften konvexer Flächen 245

[Feller 1936c] Zur Theorie der stochastischen Prozesse. (Existenz-und Eindeutigkeitssätze.) 293

[Feller 1936c] (translation) On the Theory of Stochastic Processes. (Existence and Uniqueness.) 341

[Feller 1937a] Über den zentralen Grenzwertsatz der Wahrscheinlichkeitsrechnung. II 389

[Feller 1937a] (translation) On the Central Limit Theorem of Probability Theory. II 403

[Feller 1937b] Über das Gesetz der großen Zahlen 417

[Feller 1937b] (translation) On the Law of Large Numbers 429

[Feller 1939a] Die Grundlagen der Volterraschen Theorie des Kampfes ums Dasein in wahrscheinlichkeitstheoretischer Behandlung 441

[Feller 1939a] (translation) The Foundations of Volterra's Theory of the Struggle for Life in a Probabilistic Treatment 471

[Feller 1939b] Completely Monotone Functions and Sequences 497

[Feller 1939c] Über die Existenz von sogenannten Kollektiven 511

[Feller 1939c] (translation) On the Existence of So-called Kollektivs 521

[Feller 1939d] Neuer Beweis für die Kolmogoroff–P. Lévysche Charakterisierung der unbeschränkt teilbaren Verteilungsfunktionen 531

[Feller 1940c] On the Integro-differential Equations of Purely Discontinuous Markoff Processes 539

[Feller 1941a] On the Integral Equation of Renewal Theory 567

[Feller 1942] Some Geometric Inequalities 593

[Feller 1943b] Generalization of a Probability Limit Theorem of Cramér 601

[Feller 1943c] The General Form of the So-called Law of the Iterated Logarithm 613

[Feller 1943d] On a General Class of "Contagious" Distributions 643

[Feller 1945a] On the Normal Approximation to the Binomial Distribution 655

[Feller 1945b] The Fundamental Limit Theorems in Probability 667

[Feller 1945c] Regularity Properties of a Certain Class of Surfaces 701

[Feller 1945d] Note on the Law of Large Numbers and "Fair" Games 717

[Feller 1946a] A Limit Theorem for Random Variables with Infinite Moments 721

[Feller 1946b] The Law of the Iterated Logarithm for Identically Distributed Random Variables 727

[Feller 1948a] On the Kolmogorov–Smirnov Limit Theorems for Empirical Distributions 735

[Feller 1948c] On Probability Problems in the Theory of Counters 751

[Feller 1949a] A Property of Power Series with Positive Coefficients 761

[Feller 1949b] On Fluctuations in Coin-tossing 765

[Feller 1949c] On the Theory of Stochastic Processes, with Particular Reference to Applications 769

[Feller 1949d] Fluctuation Theory of Recurrent Events 799

Curriculum Vitae: William Feller 1906–1970

7 July 1906	William Feller is born in Zagreb (German: Agram), Croatia (then part of the Austro-Hungarian empire), as son of Ida (née Oehmichen) and Eugen Viktor Feller. Feller's Christian name is Vilibald; during his life Feller uses several translations of his Christian name: Vilim, Willi, Willy, William
1923–1925	Feller studies 4 semesters mathematics at the University of Zagreb (Croatia)
1925–1926	Feller studies 2 semesters mathematics at the Georg-August-University, Göttingen (Germany)
1926	PhD at the Georg-August-University, Göttingen (Germany). Advisor: Richard Courant. Thesis: *Über algebraisch rektifizierbare transzendente Kurven* (*On algebraically rectifiable transcendental curves*), see [*Feller 1926] and [Feller 1928] in Feller's bibliography, p. xxv
1926–1928	Post-doc at the Georg-August-University, Göttingen (Germany)
1928	Habilitation in mathematics at Kiel University, Kiel (Germany)
1928–1933	Lecturer and *Privatdozent* at Kiel University, Kiel (Germany)
1933–1934	Visiting University of Copenhagen (Denmark)
1934–1939	Lecturer at the University of Stockholm (Sweden)
27 July 1938	William Feller marries Clara Maria Nielsen
1939	Editor-in-Chief of the newly founded *Mathematical Reviews*
1939–1945	Associate professor at Brown University, Providence (RI)
1944	US citizenship
1945–1950	Full professor at Cornell University, Ithaca (NY)
1950–1970	Eugene Higgins Professor of Mathematics at Princeton University, Princeton (NJ)

1966	Permanent title of visiting professor of the Rockefeller University, New York (NY)
1969/1970	National medal of science
14 January 1970	William Feller dies of cancer at the age of 63 in New York City (NY)

Lebenslauf.

Ich, Willy Feller, wurde den 7. Juli 1906 zu Zagreb (Agram) geboren; ich bin jugoslavischer Staatangehöriger und römisch katholischer Konfession. In meiner Heimatsstadt besuchte ich das Realgymnasium und durfte mit Erlaubnis der Regierung zwei Klassen kontrahieren, so dass ich schon im Juni 1923 die Reifeprüfung bestehen konnte. Danach widmete ich mich dem Studium der Mathematik und Physik. Zunächst studierte ich vom Oktober 1923 bis Juli 1925 d.h. vier Semester an der Universität zu Zagreb und bin nun seit dem vergangenen Winterhalbjahre in Göttingen immatrikuliert.

Göttingen, den 10. August 1926.

Handwritten CV of Feller's application for the admission to the PhD examination at the University of Göttingen. Reproduction courtesy of Universitätsarchiv Göttingen [Math Nat Prom 0010, 23]. Translation of the text:
Curriculum Vitae. I, Willy Feller, was born on 7 July 1906 in Zagreb (*Agram*)*; I am Yugoslav citizen and of Roman Catholic confession. In my native town I went to secondary school* (Realgymnasium) *and I got the permission of the government to skip two grades so that I could already graduate from school in June 1923. Then I started studying mathematics and physics. First I studied from October 1923 until July 1925, i. e. four semesters, at the University of Zagreb; since the last winter term I am enrolled at* [the University of] *Göttingen. Göttingen, 10 August 1926.*

Bibliography of William Feller

A star "*" indicates that the respective entry is *not contained* in these Selecta.

[*Feller 1926] *Über algebraisch rektifizierbare transzendente Kurven* [*On algebraically rectifiable transcendental curves*]. PhD-Thesis Universität Göttingen, June 1926. Advisor: Richard Courant. Also published as [Feller 1928].

[Feller 1928] Über algebraisch rektifizierbare transzendente Kurven [On algebraically rectifiable transcendental curves]. *Mathematische Zeitschrift* **27** (1928) 481–495.

[*Feller 1929] *Über die Lösungen der linearen partiellen Differentialgleichungen zweiter Ordnung vom elliptischen Typus* [*On the solutions of second-order linear partial differential equations of elliptic type*]. Habilitationsschrift Universität Kiel, February 1929. Also published as [Feller 1930].

[Feller 1930] Über die Lösungen der linearen partiellen Differentialgleichungen zweiter Ordnung vom elliptischen Typus [On the solutions of second-order linear partial differential equations of elliptic type]. *Mathematische Annalen* **102** (1930) 633–649.

[*Feller 1931] (with Erhard Tornier) Mengentheoretische Untersuchung von Eigenschaften der Zahlenreihe [Set-theoretic investigation of some properties of the natural numbers]. *Zentralblatt* **1** (1931) 257–259.

[*Feller 1932a] (with Erhard Tornier) Maß- und Inhaltstheorie des Baireschen Nullraumes [Measure and integration theory of Baire's null space]. *Mathematische Annalen* **107** (1932) 165–187.

[*Feller 1932b] (with Erhard Tornier) Mengentheoretische Untersuchung von Eigenschaften der Zahlenreihe [Set-theoretic investigation of some properties of the natural numbers]. *Mathematische Annalen* **107** (1932) 188–232.

[*Feller 1932c] Allgemeine Maßtheorie und Lebesguesche Integration [General measure theory and Lebesgue integration]. *Sitzungsberichte der Preußischen Akademie der Wissenschaften, Physikalisch–Mathematische Klasse* **27** (1932) 459–472.

[Feller 1934a] (with Herbert Busemann) Zur Differentiation der Lebesgueschen Integrale [On the differentiation of Lebesgue's integrals]. *Fundamenta Mathematicae* **22** (1934) 226–256.

[*Feller 1934b] Bemerkungen zur Maßtheorie in abstrakten Räumen [Remarks on measure theory in abstract spaces]. *Bulletin international de l'académie Yougoslave des sciences et des beaux-arts, Zagreb. Classe des sciences mathématiques et naturelles* **28** (1934) 30–45. Croatian original publication: Dr. Vilim (W.) Feller: Prilog teoriji mjera u apstraktnim prostorima. *RAD* **249** (1934) 204–224.

[*Feller 1935a] (with Herbert Busemann) Bemerkungen zur Differentialgeometrie der konvexen Flächen. I. Kürzeste Linien auf differenzierbaren Flächen [Remarks on the differential geometry of convex surfaces. I. Shortest lines on differentiable surfaces]. *Matematisk Tidsskrift B* (1935) 25–36.

[*Feller 1935b] (with Herbert Busemann) Bemerkungen zur Differentialgeometrie der konvexen Flächen. II. Über die Krümmungsindikatrizen [Remarks on the differential geometry of convex surfaces. II. On curvature indicatrices]. *Matematisk Tidsskrift B* (1935) 87–115.

[Feller 1935c] Über den zentralen Grenzwertsatz der Wahrscheinlichkeitsrechnung [On the central limit theorem of probability theory]. *Mathematische Zeitschrift* **40** (1935/36) 521–559. *This paper has been translated for the present Selecta.*

[*Feller 1936a] (with Herbert Busemann) Bemerkungen zur Differentialgeometrie der konvexen Flächen. III. Über die Gauss'sche Krümmung [Remarks on the differential geometry of convex surfaces. III. On Gauss' curvature]. *Matematisk Tidsskrift B* (1936) 41–70.

[Feller 1936b] (with Herbert Busemann) Krümmungseigenschaften konvexer Flächen [Curvature properties of convex surfaces]. *Acta Mathematica* **66** (1936) 1–47.

[Feller 1936c] Zur Theorie der stochastischen Prozesse. (Existenz- und Eindeutigkeitssätze) [On the theory of stochastic processes. (Existence- and uniqueness theorems)]. *Mathematische Annalen* **113** (1936) 113–160. *This paper has been translated for the present Selecta.*

[Feller 1937a] Über den zentralen Grenzwertsatz der Wahrscheinlichkeitsrechnung. II [On the central limit theorem of probability theory. II]. *Mathematische Zeitschrift* **42** (1937) 301–312.
Erratum: *Mathematische Zeitschrift* **44** (1939) 794. *This paper has been translated for the present Selecta.*

[Feller 1937b] Über das Gesetz der großen Zahlen [On the law of large numbers]. *Acta Litterarum ac Scientiarum. Regiae Universitatis Hungaricae*

Francisco-Josephinae. Sectio Scientiarum Mathematicarum. (Acta. Sci. Litt. Szeged) **8** (1937) 191–201. *This paper has been translated for the present Selecta.*

[*Feller 1937c] Über die Theorie der stochastischen Prozesse [On the theory of stochastic processes]. In: *Comptes Rendus du Congrès International des Mathématiciens. Oslo 1936*. Vol. 2. A. W. Brøggers Boktrykkeri A/S, Oslo 1937, pp. 194–196.

[*Feller 1938a] Sur les axiomatiques du calcul des probabilités et leurs relations avec les expériences [On the axioms of probability theory and their relations with the experience]. In: P. Cantelli, M. Fréchet, R. von Mises, I. F. Steffensen and A. Wald (eds.): *Conférences internationales de sciences mathématiques. Colloque consacré à la théorie des probabilités. II: Les fondements du calcul des probabilités.* Actualités scientifiques et industrielles **735**. Hermann & Cie., Paris 1938, pp. 7–21.

[*Feller 1938b] Note on regions similar to the sample space. *Statistical Research Memoirs, University College London* **2** (1938), 117–125.

[*Feller 1938c] (with J. Runnström and E. Sperber) Die Aufnahme von Glucose durch Bäckerhefe unter aeroben und anaeroben Bedingungen [On the absorption of glucose by baker's yeast under aerobic and anaerobic conditions]. *Naturwissenschaften* **26** (1938) 547–548.

[Feller 1939a] Die Grundlagen der Volterraschen Theorie des Kampfes ums Dasein in wahrscheinlichkeitstheoretischer Behandlung [The foundations of Volterra's theory on the struggle for life in a probabilistic treatment]. *Acta Biotheoretica, Leiden* **5** (1939) 11–39. *This paper has been translated for the present Selecta.*

[Feller 1939b] Completely monotone functions and sequences. *Duke Mathematical Journal* **5** (1939) 661–674.

[Feller 1939c] Über die Existenz von sogenannten Kollektiven [On the existence of so-called Kollektivs] *Fundamenta Mathematicae* **32** (1939) 87–96. *This paper has been translated for the present Selecta.*

[Feller 1939d] Neuer Beweis für die Kolmogoroff–P. Lévysche Charakterisierung der unbeschränkt teilbaren Verteilungsfunktionen [A new proof of Kolmogoroff's and P. Levy's characterization of infinitely divisible distribution functions]. *Bulletin international de l'académie Yougoslave des sciences et des beaux-arts, Zagreb. Classe des sciences mathématiques et naturelles* **32** (1939) 106–113. Croatian original publication: Vilim (W.) Feller: O Kolmogoroff–P. Lévyjevu predočivanju beskonačno djeljivih funkcija repartície. *RAD* **263** (1939) 95–112.

[*Feller 1940a] On the logistic law of growth and its empirical verifications in biology. *Acta Biotheoretica, Leiden* **5** (1940) 51–65.

[*Feller 1940b] On the time distribution of so-called random events. *Physical Reviews, II. Series* **57** (1940) 906–908.

[Feller 1940c] On the integro-differential equations of purely discontinuous Markoff processes. *Transactions of the American Mathematical Society* **48** (1940) 488–515.

[*Feller 1940d] Statistical aspects of ESP. *Journal of Parapsychology* **4** (1940) 271–298.

[Feller 1941a] On the integral equation of renewal theory. *Annals of Mathematical Statistics* **12** (1941) 243–267.

[*Feller 1941b] (with Jacob David Tamarkin) *Partial Differential Equations*. Brown University Summer Session for Advanced Instructions & Research in Mechanics. 6/23- 9/13 1941, Providence (RI) 1941.
Chapters 1–3 by J. D. Tamarkin, Chapters 4–7 by William Feller.
Reprinted by the Lewis Flight Propulsion Laboratory, National Committee for Aeronautics, Cleveland 1956.

[Feller 1942] Some geometric inequalities. *Duke Mathematical Journal* **9** (1942) 885–892.

[*Feller 1943a] On A. C. Aitken's method of interpolation. *Quarterly of Applied Mathematics* **1** (1943) 86–87.

[Feller 1943b] Generalization of a probability limit theorem of Cramér. *Transactions of the American Mathematical Society* **54** (1943) 361–372.

[Feller 1943c] The general form of the so-called law of the iterated logarithm. *Transactions of the American Mathematical Society* **54** (1943) 373–402.

[Feller 1943d] On a general class of "contagious" distributions. *Annals of Mathematical Statistics* **14** (1943) 389–400.

[Feller 1945a] On the normal approximation to the binomial distribution. *Annals of Mathematical Statistics* **16** (1945) 319–329.

[Feller 1945b] The fundamental limit theorems in probability. *Bulletin of the American Mathematical Society* **51** (1945) 800–832.

[Feller 1945c] (with Herbert Busemann) Regularity properties of a certain class of surfaces. *Bulletin of the American Mathematical Society* **51** (1945) 583–598.

[Feller 1945d] Note on the law of large numbers and "fair" games. *Annals of Mathematical Statistics* **16** (1945) 301–304.

[Feller 1946a] A limit theorem for random variables with infinite moments. *American Journal of Mathematics* **68** (1946) 257–262.

[Feller 1946b] The law of the iterated logarithm for identically distributed random variables. *Annals of Mathematics* **47** (1946) 631–638.

[Feller 1948a] On the Kolmogorov–Smirnov limit theorems for empirical distributions. *Annals of Mathematical Statistics* **19** (1948) 177–189.
Erratum: *Annals of Mathematical Statistics* **21** (1950) 301—302.

[*Feller 1948b] Spanish Translation of the paper [Feller 1945b] *Revista Matemática Hispano-Americana, IV. Seria* **8** (1948) 95–132.

[Feller 1948c] On probability problems in the theory of counters. In: *Studies, Essays, Presented to R. Courant (Courant Anniversary Volume)*. Interscience Publishers, New York 1948, pp. 105–115.

[Feller 1949a] (with Paul Erdös and Harry Pollard) A property of power series with positive coefficients. *Bulletin of the American Mathematical Society* **55** (1949) 201–204.

[Feller 1949b] (with Kai Lai Chung) On fluctuations in coin-tossing. *Proceedings of the National Academy of Sciences, USA* **35** (1949) 605–608.

[Feller 1949c] On the theory of stochastic processes, with particular reference to applications. In: J. Neyman (ed.): *Proceedings of the (First) Berkeley Symposium on Mathematical Statistics and Probability 1945/46*. University of California Press, Berkeley, Los Angeles 1949, pp. 403–432.

[Feller 1949d] Fluctuation theory of recurrent events. *Transactions of the American Mathematical Society* **67** (1949) 98–119.

[*Feller 1950] *An Introduction to Probability Theory and Its Applications*. John Wiley & Sons, New York 1950.

[Feller 1951a] The asymptotic distribution of the range of sums of independent random variables. *Annals of Mathematical Statistics* **22** (1951) 427–432.

[*Feller 1951b] (with George E. Forsythe) New matrix transformations for obtaining characteristic vectors. *Quarterly of Applied Mathematics* **8** (1951) 325–331.

[*Feller 1951c] The problem of *n* liars and Markov chains. *American Mathematical Monthly* **58** (1951) 606–608.

[Feller 1951d] Diffusion processes in genetics. In: J. Neyman (ed.): *Proceedings of the Second Berkeley Symposium on Mathematical Statistics and Probability 1950*. University of California Press, Berkeley, Los Angeles (CA) 1951, pp. 227–246.

[Feller 1951e] Two singular diffusion problems. *Annals of Mathematics* **54** (1951) 173–182.

[Feller 1952a] The parabolic differential equations and the associated semigroups of transformation. *Annals of Mathematics* **55** (1952) 468–519.

[Feller 1952b] On a generalization of Marcel Riesz' potentials and the semigroups generated by them. In: *Meddelanden från Lunds Universitets Matematiska Seminarium. Supplement M. Riesz*, Uppsala 1952, pp. 73–81.

[Feller 1952c] Some recent trends in the mathematical theory of diffusion. In: *Proceedings of the International Congress of Mathematicians, Cambridge (MA) 1950*. Vol. 2. American Mathematical Society, Providence (RI) 1952, pp. 322–339.

[*Feller 1952d] (with George E. Forsythe) New matrix transformations for obtaining characteristic vectors. In: *Proceedings of the International Congress of Mathematicians, Cambridge (MA) 1950*. Vol. 1. American Mathematical Society, Providence (RI) 1952, p. 661.

[Feller 1952e] On positivity preserving semigroups of transformations on $C[r_1, r_2]$. *Annales de la Société Polonaise de Mathématique* **25** (1952) 85–94.

[Feller 1953a] Semigroups of transformations in general weak topologies. *Annals of Mathematics* **57** (1953) 287–308.

[Feller 1953b] On the generation of unbounded semigroups of bounded linear operators. *Annals of Mathematics* **58** (1953) 166–174.

[Feller 1954a] The general diffusion operator and positivity preserving semigroups in one dimension. *Annals of Mathematics* **60** (1954) 417–436.

[Feller 1954b] Diffusion processes in one dimension. *Transactions of the American Mathematical Society* **77** (1954) 1–31.

[Feller 1955a] On second order differential operators. *Annals of Mathematics* **61** (1955) 90–105.

[Feller 1955b] On differential operators and boundary conditions. *Communications on Pure and Applied Mathematics* **8** (1955) 203–216.

[Feller 1956a] Boundaries induced by non-negative matrices. *Transactions of the American Mathematical Society* **83** (1956) 19–54.

[Feller 1956b] (with Henry P. McKean Jr.) A diffusion equivalent to a countable Markov chain. *Proceedings of the National Academy of Sciences, USA* **42** (1956) 351–354.

[Feller 1956c] (with Joanne Elliott) Stochastic processes connected with harmonic functions. *Transactions of the American Mathematical Society* **82** (1956) 392–420.

[*Feller 1956d] On generalized Sturm–Liouville operators. In: J. B. Diaz and L. E. Payne (eds.): *Proceedings of the Conference on Differential Equations (Dedicated to A. Weinstein)*, University of Maryland Book Store, College Park (MD) 1956, pp. 251–270.

[Feller 1957a] Generalized second order differential operators and their lateral conditions. *Illinois Journal of Mathematics* **1** (1957) 459–504.

[*Feller 1957b] The numbers of zeros and of changes of sign in a symmetric random walk. *L'Enseignement Mathématique, IIe. Série* **3** (1957) 229–235.

[*Feller 1957c] Sur une forme intrinsèque pour les opérateurs différentiels du second ordre [On an intrinsic form of second-order differential operators]. *Publications de l'Institut de Statistique de l'Université de Paris. Université de Paris VI, Institut de Statistique, Paris* **6** (1957) 291–301.

[Feller 1957d] On boundaries and lateral conditions for the Kolmogorov differential equations. *Annals of Mathematics* **65** (1957) 527–570.
Additional notes: *Annals of Mathematics* **68** (1958) 735–736.

[Feller 1957e] On boundaries defined by stochastic matrices. In: *Proceedings of Symposia in Applied Mathematics*. Vol. 7. McGraw-Hill Book Co., New York (for the American Mathematical Society, Providence (RI)) 1957, 35–40.

[*Feller 1957f] *An Introduction to Probability Theory and Its Applications. Vol. 1.* 2nd edn. of [*Feller 1950]. John Wiley & Sons, New York 1957.

[Feller 1958] On the intrinsic form for second order differential operators. *Illinois Journal of Mathematics* **2** (1958) 1–18.

[*Feller 1959a] On combinatorial methods in fluctuation theory. In: U. Grenander (ed.): *The Harald Cramér Volume*. Almqvist & Wiksell, Stockholm; John Wiley & Sons, New York 1959, pp. 75–91.

[Feller 1959b] The birth and death processes as diffusion processes. *Journal de Mathématiques Pures et Appliquées, IX. série* **38** (1959) 301–345.

[Feller 1959c] Non-Markovian processes with the semigroup property. *Annals of Mathematical Statistics* **30** (1959) 1252–1253.

[Feller 1959d] Differential operators with the positive maximum property. *Illinois Journal of Mathematics* **3** (1959) 182–186.

[*Feller 1959e] On the equation of the vibrating string. *Journal of Mathematics and Mechanics* **8** (1959) 339–348.

[Feller 1960] Some new connections between probability and classical analysis. *Proceedings of the International Congress of Mathematicians, Edinburgh 1958.* Cambridge University Press, Cambridge 1960, pp. 69–86.

[Feller 1961a] (with Steven Orey) A renewal theorem. *Journal of Mathematics and Mechanics* **10** (1961) 619–624.

[Feller 1961b] A simple proof for renewal theorems. *Communications on Pure and Applied Mathematics* **14** (1961) 285–293.

[*Feller 1961c] Chance processes and fluctuations. In: E. F. Beckenbach and M. R. Hestenes (eds.): *Modern Mathematics for the Engineer: Second Series. University of California Engineering Extension Series.* Mc Graw-Hill, New York 1961, pp. 167–181.

[Feller 1963] On the classical Tauberian theorems. *Archiv der Mathematik* **14** (1963) 317–322.

[Feller 1964] On semi-Markov processes. *Proceedings of the National Academy of Sciences, USA* **51** (1964) 653–659.

[Feller 1966a] On the Fourier representation for Markov chains and the strong ratio theorem. *Journal of Mathematics and Mechanics* **15** (1966) 273–283.

[Feller 1966b] On the influence of natural selection on population size. *Proceedings of the National Academy of Sciences, USA* **55** (1966) 733–738.

[Feller 1966c] Infinitely divisible distributions and Bessel functions associated with random walks. *SIAM Journal of Applied Mathematics* **14** (1966) 864–875.

[*Feller 1966d] *An Introduction to Probability Theory and Its Applications. Vol. 2.* John Wiley & Sons, New York 1966.

[Feller 1967a] On regular variation and local limit theorems. In: J. Neyman and L. LeCam (eds.): *Proceedings of the Fifth Berkeley Symposium on Mathematical Statistics and Probability 1965/66*, Vol. 2, Pt. 1, University of California Press, Berkeley, Los Angeles (Ca) 1967, pp. 373–388.

[*Feller 1967b] A direct proof of Stirling's formula. *American Mathematical Monthly* **74** (1967) 1223–1225.
Erratum: *American Mathematical Monthly* **75** (1968) 518.

[Feller 1967c] On fitness and the cost of natural selection. *Genetics Research* **9** (1967) 1–15.

[*Feller 1968a] On Müntz' theorem and completely monotone functions. *American Mathematical Monthly* **75** (1968) 342–350.

[Feller 1968b] On the Berry-Esseen theorem. *Zeitschrift für Wahrscheinlichkeitstheorie und verwandte Gebiete* **10** (1968) 261–268.

[*Feller 1968c] On probabilities of large deviations. *Proceedings of the National Academy of Sciences, USA* **61** (1968) 1224–1227.

[Feller 1968d] An extension of the law of the iterated logarithm to variables without variance. *Journal of Mathematics and Mechanics* **18** (1968) 343–355.

[*Feller 1968e] *An Introduction to Probability Theory and Its Applications. Vol. 1.* 3rd edn. of [*Feller 1950], [*Feller 1957f]. John Wiley & Sons, New York 1968.

[*Feller 1969a] A geometrical analysis of fitness in triply allelic systems. *Mathematical Biosciences* **5** (1969) 19–38.

[Feller 1969b] One-sided analogues of Karamata's regular variation. *L'Enseignement Mathématique, IIe. Série* **15** (1969) 107–121.

[Feller 1969c] Limit theorems for probabilities of large deviations. *Zeitschrift für Wahrscheinlichkeitstheorie und verwandte Gebiete* **14** (1969) 1–20.

[Feller 1969d] General analogues to the law of the iterated logarithm. *Zeitschrift für Wahrscheinlichkeitstheorie und verwandte Gebiete* **14** (1969) 21–26.

[*Feller 1969e] On the fluctuations of sums of independent random variables. *Proceedings of the National Academy of Sciences, USA* **63** (1969) 637–639.

[*Feller 1969f] Are life scientists overawed by Statistics? (Too much faith in statistics). *Scientific Research* **4**.3 (1969) 24–29.

[Feller 1970] On the oscillations of sums of independent random variables. *Annals of Mathematics* **91** (1970) 402–418.

[*Feller 1971] *An Introduction to Probability Theory and Its Applications. Vol. 2.* 2nd edn. of [*Feller 1966d]. John Wiley & Sons, New York 1971. Posthumously published. The final version was published with the help of J. Goldman, A. Grunbaum, H. P. McKean, L. Pitt and A. Pittenger.

Textbooks

A star "*" indicates that the respective entry is *not contained* in the present selection.

[*Feller 1941b] (with Jacob David Tamarkin) *Partial Differential Equations.* Brown University Summer Session for Advanced Instructions & Research in Mechanics. 6/23- 9/13 1941, Providence (RI) 1941.
Chapters 1–3 by J. D. Tamarkin, Chapters 4–7 by William Feller.
Reprinted by the Lewis Flight Propulsion Laboratory, National Committee for Aeronautics, Cleveland 1956.

[*Feller 1950] *An Introduction to Probability Theory and Its Applications.* John Wiley & Sons, New York 1950.

[*Feller 1957f] *An Introduction to Probability Theory and Its Applications. Vol. 1.* 2nd edn. of [*Feller 1950]. John Wiley & Sons, New York 1957.

[*Feller 1966d] *An Introduction to Probability Theory and Its Applications. Vol. 2.* John Wiley & Sons, New York 1966.

[*Feller 1968e] *An Introduction to Probability Theory and Its Applications. Vol. 1.* 3rd edn. of [*Feller 1950], [*Feller 1957f]. John Wiley & Sons, New York 1968.

[*Feller 1971] *An Introduction to Probability Theory and Its Applications. Vol. 2.* 2nd edn. of [*Feller 1966d]. John Wiley & Sons, New York 1971. Posthumously published. The final version was published with the help of J. Goldman, A. Grunbaum, H. P. McKean, L. Pitt and A. Pittenger.

Unpublished Lecture Notes

A star "*" indicates that the respective entry is *not contained* in the present selection.

[*Feller 1929/30] *Partielle Differentialgleichungen der Physik. Kiel W.S. 1929/30* [*Partial Differential Equations of Physics. Kiel Winter Term 1929/30*]. Handwritten lecture notes (in German, approx. 100 pp.). Deposited as a bound manuscript at Princeton University Library.

[*Feller 1942b] *Vectors in a Plane; Notes for Mathematics 5, Summer 1942*. Brown University, Providence (RI) 1942. (Brown University Libary: SCI – Level 7, Aisle 1B 1-SIZE QA261.F45)

[*Feller 1945e] *Lectures on probability. Brown University, Semester II, 1944-1945.* Brown University, Providence (RI) 1945. (Brown University Library: SCI QA273.F34)

PhD Students of William Feller

1. Forsythe, George E.: *Riesz summability methods of order r, for* $R(r) < 0$*, Cesaro summability of independent random variables.* (Brown University 1941)
2. Kincaid, Wilfred MacDonald: *Part I: On non-cut sets of locally connected continua. Part II: An application of orthogonal moments to problems in statistically indeterminate structures. Part III: Numerical methods for finding characteristic roots and vectors of matrices.* (Brown University 1946)
3. Juncosa, Mario L.: *On the asymptotic behavior of the minimum in a sequence of random variables.* (Cornell 1949)
4. Weber, Maria A.: *The solution of a linear differential equation of parabolic type.* (Cornell 1949)
5. Seifert, George H.: *Some third order boundary value problems.* (Cornell 1950)
6. Shapley, Lloyd Stowell: *Additive and non-additive set functions.* (Princeton 1953. Co-advisor: A. W. Tucker)
7. Murdoch, Brian Hughes: *Preharmonic functions.* (Princeton 1955)
8. Billingsley, Patrick Paul: *The invariance principle for dependent random variables.* (Princeton 1955)
9. McKean, Henry P. Jr.: *Sample functions of stable processes.* (Princeton 1955)
10. Trotter, Hale F.: *Convergence of semi-groups of operators.* (Princeton 1956)
11. Gaver, Donald Paul: *Some results in the theory of queues.* (Princeton 1956)
12. Knight, Frank Beardsley: *Construction of diffusion processes by means of random walks.* (Princeton 1959)
13. George, Melvin D.: *The approximation of solutions of nonlinear differential equations.* (Princeton 1959)
14. Freedman, David Amiel: *Mixtures of stochastic processes.* (Princeton 1960)
15. Shepp, Lawrence A.: *Recurrent sums of random variables.* (Princeton 1961)

16. Schay, Geza: *The equations of diffusion in the special theory of relativity.* (Princeton 1961)

17. Goldman, Jay Robert: *Stochastic point processes: limit theorems and infinite divisibility.* (Princeton 1965)

18. Silverstein, Martin Louis: *Many particle processes.* (Princeton 1965)

19. Weiss, Benjamin: *Vibrating systems and positivity preserving semi-groups.* (Princeton 1965)

20. Kurtz, Robert. *On regularly varying functions in analysis and probability.* (Princeton 1966)

21. Pitt, Loren Dallas: *Two problems in Markov processes. Extending the life span of Markov processes. Products of Markovian semi-groups.* (Princeton 1967)

22. Cleveland, William Swain II: *Time series projections, theory and practice.* (Yale 1969. Co-advisor: L. J. Savage)

William Feller. A Biography

by *René L. Schilling* from Dresden
and *Wojbor A. Woyczyński* from Cleveland

William Feller was born on 7 July 1906 in Zagreb, Croatia, which was then a Southern province of the Austro-Hungarian Empire. He was the ninth of twelve children of Eugen Viktor Feller and Ida Feller. Following the Roman-Catholic tradition, the young baby boy was named Vilibald after the saint (St. Willibald) whose feast day fell on his birthday. In the church register of births (city register)[1] he had been registered as Vilibald Srećko[2] Feller. Feller called himself Vilim[3] (which is the Croatian form of William), and throughout his life he would adopt the native versions of his first name depending on the country he lived in: thus in Germany and Scandinavia he called himself Willi/Willy, and William/Will in America.

Feller's father Eugen was born in 1871 in Lemberg, Galicia, in the Northeastern corner of the Austro-Hungarian Empire (now Lviv, in Western Ukraine), and died in 1936 in Zagreb. His wife Ida was born in 1870 (place unknown), and passed away in Zagreb two years after her husband's death. His paternal grandparents were David Feller and Elizabeth (Elsa) Holzer from Lemberg, and his maternal grandparents were Ferdinand Oemichen and Hermina Peerz (or Perc). Feller's parents owned a pharmaceutical company. The then-famous, now-obscure Elsa fluid, named after Eugen's mother, was one of the mainstays of the enterprise. It had been marketed as an elixir capable to cure all kinds of maladies, such as headaches, colds, back pains, etc., and was the source of the family's substantial wealth. Like many upper-class families, the Fellers were bilingual and William spoke both Croatian and German. Young William was raised in the Roman-Catholic tradition even though his paternal

[1]Žubrinić [47, p. 9]. The references [Feller 19*nn*] and [*Feller 19*nn*] (the star indicating that the respective paper is not contained in these Selecta) refer to Feller's bibliography, while [*n*] points to the list of references at the end of this essay.

[2]Srećko is the Croatian equivalent of Felix, 'the lucky'.

[3]The rather anecdotal story in Seneta [40, p. 87] is based on the claim that July 7 is St. Willibald in the "(German) Catholic saint's list" while it is Sv. Vilim's feast day in the "Croatian Catholic list" which cannot be verified (www.namecalendar.net, accessed 3 August 2014, gives for Vilim the dates 6 April, 23 May, 8 June and 25 June, the Croatian Catholic Calendar www.hkr.hr/kalendar, accessed 29 August 2014, gives the dates 10 February, 23 May, 8 June and 29 July). It is more likely that Vili(m) was used as a family nickname instead of the more formal Vilibald; also in German, it is common practice to abbreviate Willibald by Willi or Willy.

R.L. Schilling et al. (eds.), *Selected Papers II*,

Picture 1: William Feller as a young boy, aged 4. The handwritten note in the lower left corner says "Unser Willy" (our Willy). The photograph is taken from the family's picture album, courtesy of Joanne Elliott.

grandfather David was Jewish; he had converted to Catholicism when marrying Elsa.[4] Feller's ancestors probably came to Zagreb during the second half of the nineteenth century. More details on Feller's parents and grandparents can be found in the two studies by Fatović-Ferenčić, and Ferber-Bogdan [10, 11].

Zagreb 1906–1925

We do not know the exact year when the Fellers moved to their modern, spacious and elegantly decorated villa on 31a, Jurjevska street,[5] but it is safe to assume that William spent most of his childhood there. Judging by family photos and personal stories, they lived in a pleasant and productive environment. Several of William's brothers had successful careers of their own: Ferdinand (Ferdo) (1897–1960) was a well-known pharmacist, Miroslav (Fritz) (1901–1961) was a poet, novelist, philosopher, and an art critic, and Marijan (1903–1974) was a pianist with an academic career in Zagreb and Sarajevo.[6]

Not much is known about William Feller's schooldays. Until the end of the first

[4] Žubrinić [47, pp. 2*ff.*]

[5] The Villa was built in 1910/11 and designed by Mathias Feller from Munich, William's uncle. The whole September 1914 issue of the German architectural magazine *Innendekoration* [12] is devoted to its description.

[6] Žubrinić [47, pp. 26*ff.*] and the Croatian Biographical Encyclopedia [22].

Picture 2: The Fellers' Villa at 31a, Jurjevska Street, in Zagreb. Clockwise from the top left: View from the South in 1913/14 and today, from the East (today), and the West (in 1913/14). Photographs on the left side courtesy of Joanne Elliott, on the right side courtesy of Zoran Vondraček.

World War (1918), Croatia and Zagreb (Agram in German) were part of the autonomous "Kingdom of Dalmatia, Croatia and Slavonia", associated with the Kingdom of Hungary, itself part of the Austro-Hungarian monarchy. From 1918 onwards it was part of the Kingdom of Serbs, Croats and Slovenes, which was renamed "Kingdom of Yugoslavia" in 1929. In Feller's days the educational system still followed the Austrian and German tradition, and William (as did five of his brothers) attended the *I. Realgymnasium*[7,8] in Zagreb, graduating in June 1923.[9] During his first years there he was a private pupil[10]; he also skipped two years.[11] One of Feller's private

[7] German for a "secondary school that prepares students for the university, that offers Latin but no Greek, and that typically emphasizes sciences and modern languages" (www.merriam-webster.com/dictionary/realgymnasium, accessed 30/7/2014).

[8] The former I. Realgymnasium today houses the Mimara museum of art, one of Croatia's foremost art collections.

[9] Žubrinić [47, p. 12]

[10] I.e. he had private teachers at home, only the exams to progress to the next year were taken at the Realgymnasium.

[11] "[I]ch [. . .] durfte mit Erlaubnis der Regierung zwei Klassen kontrahieren" [1, handwritten CV in the Promotionsakte].

teachers was Stanko Vlögel, a mathematician and faculty member at the University of Zagreb.[12]

In October 1923 Feller enrolled at the University of Zagreb as a student of Mathematics and Physics. From the matriculation forms, which for the summer terms of 1924, and 1925, and the winter term of 1924/25, are reprinted in Žubrinić [47, p. 14, p. 23], we learn that Feller was then using Vilim (rather than Vilibald) as his first name. The forms also contain information about the courses Feller had been taking: Mathematics I–IV (with M. Kiseljak), Differential and Integral Calculus (V. Varićak), Infinite Series (S. Bohniček), Number Theory 1,2 (S. Bohniček), Theory of Real Functions (V. Vranić), Calculus of Variations (V. Varićak), and two mathematical seminars (V. Varićak), along with a panoply of experimental and theoretical physics[13], some chemistry, and a sample of courses in pedagogy and psychology. The fact that Feller simultaneously studied two subjects was well within the Austro-German tradition. The final exam was either the *Staatsexamen*, which would qualify him to teach at grammar schools (*Gymnasium* and *Realgymnasium*), or the doctorate (the *Diplom*, Master's Diploma, was introduced only later, in the 1930s – 50s). From today's perspective, Feller's educational pathway at the University of Zagreb is comparable to the standard education at an average German or Austrian university of the 1920s. Among Feller's teachers Vladimir Varićak (1865–1942) was the most internationally visible scientist, renowned for his contributions to Einstein's theory of relativity and non-Euclidean geometry. Very likely Varićak's lectures and mathematical seminars made a lasting impact on Feller's early work in geometry.

The German years: Göttingen & Kiel 1925–1933

In Constance Reid's biographies of David Hilbert and Richard Courant we find a confirmation of the fact that

> At the beginning of the twentieth century, mathematics students all over the world were receiveing the same advice: "Pack your suitcase and take yourself to Göttingen!"[14]

Thus, after four semesters of study at Zagreb,[15] Feller enrolled for the winter term of 1925/26 at the University of Göttingen. After the first World War,

> ... in Göttingen there was again a scientific paradise. Gifted students flocked to the university. There was a constant procession of distinguished visitors from all over the world. Sometimes they came merely

[12] Žubrinić [47, p. 12]

[13] Among the courses were: Mechanics & Acoustics, Experiments in Physics, Theoretical Mechanics, Thermodynamics, Optics, Theory of Heat and a Physics Colloquium.

[14] Reid [33, p. 102]

[15] Feller did not obtain a formal degree from Zagreb. The Promotionsakte from Göttingen [1] lists an *Abgangszeugnis* (transcript of marks) of the University in (*sic*!) Agram (Zagreb) dated 13 October 1925. The *Abgangszeugnis*, however, was returned to the candidate after the defence of the thesis and is not part of the Promotionsakte.

for a single talk or a series of talks to the *mathematische Gesellschaft*. Often they lectured for a full term as guest professors. The very air seemed to [...] crackle [...] with scientific electricity.[16]

Feller was only nineteen when he moved to Göttingen. We do not know many details of his student life, but the *Promotionsakte* (dissertation file) [1] contains his address: Nikolausbergerweg 43, Göttingen. He attended the *Anfängerpraktikum*, problem classes for beginners, a revolutionary idea introduced by Richard Courant[17], and very quickly caught the attention of Courant's assistants.[18] Reid [34, p. 111] contains an anecdote about the first contact between Feller and Courant:

> At the beginning of the [winter] term in 1925, [...] [the assistants] promptly alerted Courant to the presence of Willy Feller. After the third calculus lecture, to Feller's amazement, the professor – an unbelievably august personage to a European student of that day – approached. Questioning the boy about his education in his native land, Courant discovered that Feller was already doing mathematics on his own. He told him to bring his work to the next lecture. Even thus instructed, Feller was too bashful to produce his papers on the appointed day. The next morning he was awakened by a commotion on the stairs leading to his attic room. There was a knock on the door. Courant entered and left a few moments later with the desired papers.[19]

Very likely the papers were related to the topics covered in the Mathematics Seminar Feller attended in his Zagreb days. Soon thereafter Feller joined Courant's circle of students and collaborators, which proved to be hugely beneficial to him: on 10 August 1926, after only two semesters at Göttingen, Feller, who had just turned twenty, submitted his Ph.D. thesis *Über algebraisch rektifizierbare transzendente Kurven* [*Feller 1926]. Courant gave the thesis the grade of I, or at least II,[20] remarking

> that [Feller] independently, without outside input, found the topic and worked on it, notably in an environment, (Agram), where no outside input was possible.[21]

Courant was slightly unhappy about the fact that the topic was not completely in the mainstream of the mathematics of the day,[22] but he found this shortcoming balanced

---

[16] Reid [34, p. 119]

[17] Reid [34, p. 99]. In those years, Courant's assistants, O. Neugebauer and K. O. Friedrichs were in charge of the *Anfängerpraktikum*. Feller would meet them again later in his career.

[18] Around the same time quite a few very gifted students were at Göttingen, e.g. Herbert Busemann (1905–1994), Hans Lewy (1904–1988), John von Neumann (1903–1957), and Franz Rellich (1906–1955).

[19] Rota also mentions this as an example of one of Feller's "bombastic stories" [38, p. 230].

[20] Promotionsakte [1]. At German universities (and also at Göttingen) a PhD thesis/examination is marked on the following scale: I is *outstanding, summa cum laude*, II *very good, magna cum laude*, III *good, cum laude*, IV *satisfactory, rite*.

[21] dass er [Feller] selbständig, ohne jede äussere Anregung, sein Problem sich gestellt und bearbeitet hat und zwar in einer Umgebung, (Agram), in welcher er keinerlei Anregung von aussen empfangen konnte.

[22] Courant notes "that the choice of the topic leads away from those questions which nowadays ap-

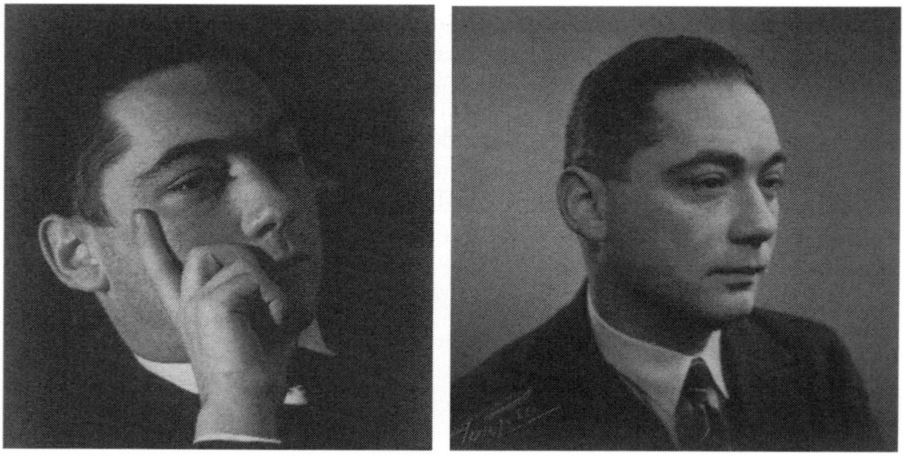

Picture 3: William Feller in his twenties or thirties. Photographs courtesy of Joanne Elliott.

by the complete independence of Feller's effort. Indeed, the topic of Feller's thesis (algebraically rectifiable transcendental curves) was much closer to Varićak's work than to anything done at Göttingen at that time, and Feller would never come back to this line of research.

The 90-minute oral exam took place on 3 November 1926 and the examination subjects were Physics (examiner: J. Franck), Mathematical Analysis (G. Herglotz), and Mathematics – Geometry (R. Courant). Feller passed this part with a I (outstanding). German universities require that the thesis be printed, so the degree was not finally awarded until 18 July 1927, after the thesis had been accepted for publication in *Mathematische Zeitschrift* [Feller 1928].

One year later Abraham Adolf Fraenkel[23] invited the "highly gifted" Feller to join Kiel University. He moved there in 1928 and almost immediately (in February of 1929), with Fraenkel's backing, obtained the habilitation in mathematics,[24] and became *Privatdozent*. The habilitation thesis [*Feller 1929] (published as [Feller 1930]) *Über die Lösungen der linearen partiellen Differentialgleichungen zweiter Ordnung vom elliptischen Typus*[25] was obviously influenced by his time at Göttingen (see Jacob's detailed commentary [24] in these Selecta).

pear to be interesting and important" ["dass die Stellung des Themas etwas aus dem Rahmen der Fragen hin[aus]führt, die uns heute interessant und wichtig erscheinen"].

[23] of Zermelo–Fraenkel fame, worked on number theory, set theory and the foundations of mathematics. He came to Kiel in 1928. His relation to Feller and Tornier is described in his autobiography *Lebenskreise* [17, pp. 154*f*].

[24] Habilitation is a post-doctoral degree that bestows on the recipient the *venia legendi*, the permission to teach, examine and supervise students independently at the university which conferred the degree. Usually a full professor of the faculty has to support one's application for habilitation.

[25] On the solutions of second-order linear partial differential equations of elliptic type

At about the same time, in October 1929, Erhard Tornier started his job as a visiting professor at Kiel,[26] an innocuous event at the time which had momentous repercussions in Feller's life four years later. He became Feller's colleague and it is very likely that they began discussing mathematics right away since Tornier was interested in the foundational questions of probability theory. They were then a very active topic in mathematics and several big open problems remained following introduction of von Mises' Kollektivs (1919), and the appearance of Steinhaus' (1923) *Fundamenta Mathematicae* paper [44] which tried to provide probability theory with a measure-theoretic foundation in the sense of Lebesgue. But Kolmogorov's axiomatization (1933) had not yet been published, and Tornier's strategy could be described as fitting in-between the above approaches.[27] This may have been Feller's first contact with modern probability and measure theory. Having completed his visiting position appointment at Kiel, Tornier left for Halle but at the end of 1931 he asked Fraenkel

> to be allowed to transfer his habilitation [which he obtained at Halle in 1930] to Kiel, in order to be 'in my [Fraenkel's] environment and to gain substantial knowledge from me'. I knew Tornier only superficially,[28] but I did grant his wish, *mainly because of Feller* [our emphasis].[29]

Thus Feller changed his subject yet again, and started a collaboration with Erhard Tornier on measure theory and the foundations of probability theory, (see the essay of Fischer [15] in these Selecta) which resulted in several joint publications [*Feller 1931, *Feller 1932a, *Feller 1932b] and his own papers on Lebesgue integration [*Feller 1932c] appear; the later papers [Feller 1934a] (with Busemann), [*Feller 1938a], and [Feller 1939c], also belong to this circle of ideas.

The other side of the coin was that Tornier turned out to be an ardent Nazi supporter joining the Nazi party (NSDAP) as early as 1932.[30] As soon as Hitler came to power on 30 January 1933, Tornier started denouncing his 'Jewish' and non-Aryan colleagues, including Feller, whose paternal grandfather was Jewish. In Nazi language, Feller was a *Vierteljude* ('Quarter-Jew') and, as such, not fit for civil service. The *Gesetz zur Wiederherstellung des Berufsbeamtentums* (*BBG*)[31] required all staff members to fill out a questionnaire and to prove their Aryan ancestry (*Ariernach-*

[26] Hochkirchen [21, p. 26]

[27] Hochkirchen [21]. Feller's paper [*Feller 1938a] is largely trying to explain why Tornier's theory is in line with Kolmogorov's approach; cf. also Fischer [15] in these Selecta.

[28] When Tornier was a visiting faculty member at Kiel, Fraenkel was in Jerusalem. Both, however, have had the same doctoral advisor: Kurt Hensel at the University of Marburg.

[29] Fraenkel [17, p. 131; our translation from German]

[30] He joined on 1 May 1932, after the philosophical faculty at Kiel agreed to recognize his Halle habilitation on 2 March 1932, cf. [21, p. 29].

[31] Act for the re-instatement of the professional civil service (7 April 1933) a rather euphemistically named Nazi law which allowed to cleanse the civil service from non-Aryan and politically nonconforming members. Its §3 is one of the earliest Aryan paragraphs. The full text is available under www.documentarchiv.de/ns/beamtenges.html (accessed 1 August 2014), excerpts are in Uhlig [45, pp. 143*ff.*].

weis) – and so did Feller.[32] Since Feller was considered to be a *Vierteljude*, the *Freie Kieler Studentenschaft* – a Nazi student organization – demanded on 21 April 1933 (two days after Feller's questionnaire had been evaluated), that Feller be suspended, and this request was supported by Tornier who emphasized Feller's Croatian, hence Slavic, origins. On the basis of the BBG, Feller's right to teach at Kiel university was revoked on 9 September 1933.[33] A month later he left Germany for Denmark.[34]

During Feller's Kiel days a young student, Clara Nielsen, attended one of his classes.[35] Several years later, in 1938, she became his wife. Unfortunately, we do not have much information regarding Clara Nielsen. Most likely she was German, possibly a member of the Danish minority in Schleswig; we know from Feller's letter[36] that "Clara's father (who is 2/3 danish [*sic*!], her mother being dutch [*sic*!]) lives in Flensburg", a city on the German–Danish border. It seems she was born on 15 September 1910,[37] which would have made her twenty-four years old in 1933, the year Feller left Germany.

European Exile: Copenhagen & Stockholm 1933–1939

Feller spent the Academic Year 1933/34 in Copenhagen, where Harald and Niels Bohr had been actively helping German emigrés. There he ran again into Herbert Busemann and Otto Neugebauer who, in the years 1934–38, continued editing the *Zentralblatt für Mathematik und ihre Grenzgebiete*[38] from Denmark. Feller knew

[32] Doob [9] writes of an "Nazi oath" which Feller "refused to sign [. . .] and was forced to leave"; this story is reiterated at several places but it cannot be confirmed. On the contrary, Feller filled out all required forms and he was forced to leave on the basis of his ancestral information. The form, which appeared as an appendix to the BBG, is available online alex.onb.ac.at/cgi-content/alex?apm=0&aid=dra&datum=19330004&zoom=2&seite=00000253&ues=0&x=13&y=7 (accessed 20 August 2014).

[33] See the official web-page of the University of Kiel, www.uni-kiel.de/ns-zeit/bios/feller-willy.shtml (accessed 31 July 2014) and the documentation in Uhlig [45, p. 23].

[34] Uhlig [45, p. 23]

[35] Doob [9, p. xvii], Birnbaum [4, p. iv]

[36] Handwritten letter of Feller to Borge Jessen [13]

[37] U.S. Social Security Death Index, accessed on 1 August 2014 via death-records.findthebest.com/l/107145366/Clara-Feller. This record shows that a Clara Feller from Princeton, Mercer County, was born on 15 September 1910 and died on 1 October 1973. Her Social Security number was also provided.

[38] The following historical remarks on the Zentralblatt are from zbmath.org/about/ (accessed 1 August 2014): "The Zentralblatt für Mathematik und ihre Grenzgebiete was founded in 1931 with the aim to publish reviews of the entire world literature in mathematics and related areas. Zentralblatt became the second comprehensive review journal for mathematics in Germany after the Jahrbuch über die Fortschritte der Mathematik (established in 1868) which has been active until the 1940s. Although the Zentralblatt had, essentially, the same agenda as the Jahrbuch, the latter aimed at maintaining the completeness of the coverage and the classification of all articles in each calendar year, whereas Zentralblatt put more emphasis on the promptness of the reviews and the international aspect.

"The initiative for the foundation of a new mathematical reviewing journal came from mathematicians Otto Neugebauer, Richard Courant, and Harald Bohr, together with the publisher Ferdinand Springer. The rapidly growing number of newly published mathematics works in the 1920s and the scientists' need for obtaining quick information on recent material motivated the decision to create an alternative service to the Jahrbuch. [...]

"Neugebauer directed the new periodical for several years until the political situation in Germany made his position as editor-in-chief unsustainable. In 1933, shortly after the Nazi party rose to power, a law was

both from his Göttingen days and very quickly started collaborating with them.

His work with Busemann focused on Lebesgue integration [Feller 1934a] and on differential geometry resulting in a series of papers [*Feller 1935a, *Feller 1935b, *Feller 1936a, Feller 1936b]. Erhard Scholz discusses their contribution to the surface theory later in these Selecta [39].

Neugebauer recruited Feller to write for Zentralblatt, where we find eight of his contributions written in the years 1933-38. Two reviews, of Kolmogorov's seminal 1933 *Grundbegriffe* [Zbl. 0007.21601], and Khinchine's *Asymptotische Gesetze* [Zbl. 0007.21601] are particularly noteworthy: Both are enthusiastically friendly, in spite of the prevailing neutral tone of Zentralblatt's entries, and show Feller's familiarity with the new measure-theoretic probability theory. It should be emphasized that in 1933 Kolmogorov's work was just one more attempt to axiomatize probability theory. Feller was among the first authors who fully recognized the importance of this approach and adopted it throughout his career.

In 1934, with Harald Bohr serving as an intermediary, Feller obtained a position with Harald Cramér's group in Stockholm,

> and stayed on in Stockholm for five years. He made a great number of Swedish friends, collaborating with economists and biologists as well as with the members of our [Cramér's] probabilistic group. He had studied in Göttingen and was well initiated in the great traditions of this mathematical center. We tried hard to get a permanent position for him in Sweden, but in those years before the war this was next to impossible, and it was with great regret that we saw him leave [in 1939] for the United States, where an outstanding career was awaiting him.[39]

With his already published work in probability theory and analysis, and papers on differential geometry, Feller had choices to make, and the contact with Cramér tipped the scale in favour of probability theory. During his Stockholm years Feller wrote two seminal papers which would establish his prominent position as a probabilist: his work on limit theorems [Feller 1935c], where he found the necessary and sufficient conditions for the validity of the Central Limit Theorem (CLT), and his extension [Feller 1936c] of the results of Kolmogorov's *Analytische Methoden* paper [27], which provided the foundations of the theory of Markov processes. The mathematical methods used in both papers reflected his strong analysis and PDE background which he had acquired in Göttingen. There was more work on limit theorems [Feller 1937a] (influenced by Feller's interaction with Marcel Riesz), the paper on the weak law of

enacted which banned Jews and political enemies from holding jobs as civil servants. A call to dismiss Courant, Neugebauer, Landau, Bernays, and Noether appeared in a local newspaper and, soon afterwards, Courant escaped to the UK and later moved to New York. Due to this pressure, Neugebauer decided to resign from his post at Göttingen University and in 1934 took up a professorship in Copenhagen, from where he continued his work for Zentralblatt.

"The struggle to produce the reviewing journal became more difficult throughout this period, however, for the Nazis tried to influence the editorial policy. Neugebauer eventually gave up his position as editor-in-chief in 1938 after a series of incidents, including Levi-Civita being dismissed from the editorial board without his knowledge."

[39] Cramér [8, p. 519]

large numbers [Feller 1937b] (dedicated to Harald Bohr and written during Feller's February 1937 visit to Lund), and another on infinite divisibility [Feller 1939d]; their mathematical contents is scrutinized in detail in Hans Fischer's essay later in these Selecta [16].

At the memorable Geneva *Colloque sur le Calcul des Probabilités*[40] Cramér and Feller met Neyman who "gave a lecture on his theory of confidence intervals, which was then something quite new", and met with a skeptical reception;[41] Feller, however, immediately took up this topic in [*Feller 1938b].[42] His own contribution to the colloquium [*Feller 1938a][43] was a comparison of the von Mises', Tornier's and Kolmogorov's approaches to the foundations of probability theory; he continued this theme in [Feller 1939c]. Interestingly enough, after all the Tornier's political scheming in Kiel, Feller unflinchingly gave him the mathematical credit due. A critical assessment of this work can be found, again, in Fischer's essay [15]. At the same time, clearly with Kolmogorov [27] and [Feller 1936c], as well as Risser [36], in mind he came up with a probabilistic interpretation of Volterra's theory of biological populations using discrete-time and continuous-time stochastic processes [Feller 1939a] (see the commentary by Baake and Wakolbinger later in these Selecta [2]). During his Stockholm years Feller also started to collaborate with applied scientists (see [*Feller 1938c, *Feller 1940a]). In [Feller 1939b] he solved a problem "proposed [...] by Conny Palm, of the telephone-administration of Stockholm", in connection with studies of telephone traffic; this started a new line of research for Feller and he would return to this topic time and again, see the commentary by Maller [28] in these Selecta.

During Feller's stay in Sweden, important events took place in his private life as well. On 27 July 1938 he married his former Kiel student, Clara Nielsen. Back in Zagreb, his parents passed away, his father in 1936 and his mother in 1938. Most of his siblings still resided in Zagreb, and we know that he remained in contact with his homeland. In particular, his two papers [*Feller 1934b, Feller 1939d] appeared first in Croatian in *Rad JAZU*, the journal of the Yugoslav (now: Croatian) Academy of Arts and Sciences, and a shorter German version ("Auszug", meaning excerpt) with essentially the same mathematical content appeared in the international issue of *Rad*. At least since 1937, Feller had been a corresponding member of the Yugoslav Academy of Arts and Sciences.[44]

---

[40] 11–16 October 1937, the proceedings are [18].

[41] Cramér [8, p. 528]

[42] Also, see Neyman [30, p. 57, Footnote (3)].

[43] Unfortunately, this important contribution cannot be reprinted in these Selecta because of the fees demanded by the publisher (Hermann, Paris).

[44] On the titlepage of the *Rad*-version of [Feller 1939d] there is a note, "Napisao član dopisnik" ["written by corresponding member"], and the acceptance date (22 November 1937) is given. This, curiously, only appears in the original Croatian version of [Feller 1939d], cf. also Seneta [40, p. 87*f.*] and Žubrinić [47, p. 45*f.*; he confirms that Feller became corresponding member in 1937].

Picture 4: Clara and William Feller on the train in the 1930s. Photograph courtesy of Joanne Elliott.

The 1940s: Brown & Cornell

In the summer of 1939 the Fellers were on the move again. Since William could not get a permanent position in Sweden,[45] he accepted the offer of an associate professorship at Brown University in Providence, Rhode Island. At that time R. G. D. Richardson served as dean of the Graduate School at Brown, as well as Secretary of the American Mathematical Society (from 1921 to 1940). Richardson recognized early on the tremendous potential for growth of mathematics in the U.S. created by European emigré mathematicians and he was one of the "principal agents of the mathematicians in aiding emigrés".[46] He mobilized various institutions in his efforts, including the well-known *Emergency Committee in Aid of Displaced German* [later: *European*] *Scholars*, and was instrumental in creating permanent positions at Brown University for W. Feller, O. Neugebauer, W. Prager and J. D. Tamarkin. These actions eventually led to the foundation of a distinguished applied mathematics division at Brown. Wartime Brown was an exciting place for mathematics: Run by Richardson, and with Tamarkin and then Prager as scientific directors, the programme focussing on fluid dynamics, elasticity, and PDEs was set up;

> it included Feller, Prager, and Tamarkin, who had Brown appointments, and at one time or another, Stefan Bergman, K. O. Friedrichs, Witold Hurewicz, Charles Loewner, F. D. Murnaghan, I. S. Solonikoff, Richard von Mises, Stefan E. Warschawski, Antoni Zygmund and myself [Lip-

[45] Possibly, also because of the more or less latent anti-Semitism in Sweden, see the comments and excerpts from Feller's letters from 1934 and 1939 in Siegmund-Schultze [41, pp. 135*f*.].

[46] Reingold [35, pp. 178, 184]

Picture 5: William relaxing on the grass during his Cornell days. Photograph courtesy of Joanne Elliott.

man Bers].[47]

In 1941 Richardson launched the *Brown University Summer Session for Advanced Instruction and Research in Mechanics* (June 23 – September 13, 1941), and Tamarkin and Feller contributed with their lecture notes [*Feller 1941b] on PDEs. Ironically, Richardson's move torpedoed initially a parallel Courant's initiative to set up a center for basic and applied mathematics at New York University.[48] Mathematically, Feller's work fit well into Richardson's plans at Brown. There are several papers [*Feller 1940b, *Feller 1940d] where Feller engages with a non-mathematical (scientific) audience, and one of his notes [*Feller 1943a] appears to be specifically written for the new Brown University journal *Quarterly of Applied Mathematics*. By 1941 Feller's first graduate student (co-mentored with Tamarkin) George E. Forsythe had defended his dissertation on summability methods. Later, Forsythe would become the founder of the Computer Science Department at Stanford. These developments were just a small component in a vast web of strong influences that Feller and other immigrant mathematicians have exerted on the American scientific environment after World War II; a thorough discussion of this historical phenomenon can be found in Siegmund-Schultze [41].

The main thrust of Feller's research continued to involve limit theorems and Markov processes. In [Feller 1943c] he strengthens the results of Kolmogorov's paper on the law of the iterated logarithm (LIL), giving necessary and sufficient criteria for norming functions to be in Lévy's upper and lower classes. Feller would come back to this theme time and again. Mark Kac[49] called this paper a "veritable *tour de force*", and it can be argued that this paper still is unsurpassed in its completeness and its methods. The paper is essentially self-contained, relying only on his own previous work [Feller 1943b] where asymptotic estimates for the tails of sums of (bounded) independent random variables were obtained. In a major survey paper [Feller 1945b] he reviews developments on the LIL, CLT, and its "little brother", the weak law of large numbers (WLLN); he was also interested in applications of the CLT and the

[47]Bers [3, p. 240]

[48]Reingold [35, p. 197], Reid [34, p. 228*f.*]

[49]M. Kac: William Feller, *in memoriam* – reprinted in these Selecta.

WLLN, e.g. in connection with the St. Petersburg paradox [Feller 1945d], and normal approximations of the binomial law [Feller 1945a]. His paper on conditional probability functions [Feller 1940c] which determine a not necessarily diffusive Markov process completed his seminal contribution in [Feller 1936c], and actually introduced the term "Markoff process". Later, Doob[50] would say that

> Feller completely transformed the subject of Markov processes. Going beyond his 1935 paper, he put the analysis into a modern framework, applying semigroup theory [in the 1950s] to the semigroups generated by these processes. He observed that the appropriate boundary conditions for the parabolic differential equations governing the transition probabilities correspond on the one hand to the specification of the domains of the infinitesimal generators of the semigroups and on the other hand to the conduct of the process trajectories at the boundaries of the process state spaces. In particular, he found a beautiful perspicuous canonical form for the infinitesimal generator of a one dimensional diffusion. In this work, he was a pioneer yet frequently obtained definitive results.

The geometry papers [Feller 1942], [Feller 1945c] (jointly with H. Busemann) were the last *contributions to geometry*; both continued his earlier work. One could argue that during his stay at Brown Feller mostly was adding finishing touches to his earlier achievements. The only paper which contained completely new material was his publication on renewal theory [Feller 1941a]. The reason could have been the new environment he found himself in, as well as various not research-related activities he was involved in during those years.

Around the year 1938 several mathematicians started a campaign for an independent American abstracting journal modelled on the *Zentralblatt* (Zbl) which was compromised by Nazi racist policies. By the end of May 1939 the name *Mathematical Reviews* (MR) was already in use and O. Neugebauer (who had been editing Zentralblatt from 1931 to 1934 from Göttingen, and then from 1934 till 1938 from Copenhagen) and J. D. Tamarkin were approached with the request to serve as founding editors; both were employed by Brown University and Brown was also the first location for the editorial office (until 1951). The first issue of MR saw the light in January 1940, covering articles published from July 1939 onwards.

> Dr Willy [*sic*!] Feller [was appointed] to be technical assistant to the editors for a three year term effective 1 July 1939". [...] The name of Feller first appears, with the title Executive Editor in vol. 5 (1944).[51]

In fact, Feller was Executive Editor from 1941 to 1945 and served in the editorial committee from 1948 until 1953.[52] Since there was not much staff besides the editors, working for MR was quite time-consuming. Many reviews were actually written by the editors themselves, and a reviewer search on the MathSciNet database reveals

[50] J. L. Doob: William Feller and twentieth century probability. Reprinted in these Selecta.
[51] Pitcher [32, p. 72]
[52] Pitcher [32]

that Feller wrote 843 abstracts in 14 years (1940–1954). On top of that he served as President of the Institute of Mathematical Statistics during 1947.

Towards the end of World War II, in 1944, William Feller became a US citizen. By 1945 he had quit as Executive Editor of MR, and soon thereafter decided to leave Brown University in 1945. He accepted a (full) professorship at Cornell University, where he stayed until 1950. During those five years at Cornell, Mark Kac and Richard Feynman were his colleagues, and K. L. Chung, G. A. Hunt and G. Elfving were among the many permanent and visiting staff members. Most of his articles at the time dealt with limit theorems [Feller 1946a, Feller 1946b], [Feller 1948a] (containing an interesting foray into statistics) and he developed important ideas regarding fluctuations and recurrent events [Feller 1949a, Feller 1949b, Feller 1949d]. In the paper [Feller 1949c] presented at the first Berkeley Symposium Feller collected and surveyed, for the first time, his research interests in the theory of Markov processes. Towards the end of his stay at Cornell, Feller had completed the first edition of the first volume of his landmark monograph, *An Introduction to Probability Theory and its Applications* [*Feller 1950]. Volume 1 is 'elementary' in the sense that it does not use measure theory. The more advanced parts of the theory would appear later in the second volume, but it is interesting to note that many of Feller's recent research results were directly included. Work on this monograph must have started several years earlier. Indeed, in a letter to Borge Jessen[53] Feller writes "I am writing myself a book which is to incorporate the mathematical theory of probability with many applications and we are just mimeographing the first few chapters." From this period also comes the anecdote which was later told by J. Doob:

> While writing my book [i.e. Doob: Stochastic Processes, 1953] I had an argument with Feller. He asserted that everyone said "random variable" and I asserted that everyone said "chance variable". We obviously had to use the same name in our books, so we decided the issue by a stochastic procedure. That is, we tossed for it and he won.[54]

The Princeton Period 1950–1970

In the summer of 1950 Feller was appointed Eugene Higgins Professor of Mathematics at Princeton University. Shortly thereafter Feller gave a 45-minute address [Feller 1952c] in the *Statistical Mechanics* section at the 1950 International Congress of Mathematicians, which took place between 30 August and 6 September 1950 in Cambridge, Massachusetts. Feller talked about recent trends in the theory of diffusion processes, covering some familiar terrain including his and Kolmogorov's work on Markov processes, but now clad it in the new language of semigroup theory. Moreover, he explicitly mentioned Itô's approach, Bochner's subordination, and Riesz' fractional derivatives. At that time this was truly visionary, mapping out a

[53]Typewritten letter [14] of Feller to Borge Jessen dated 2 April 1947

[54]Snell, interview with Doob [43, p. 307]. Interestingly, this did not prevent Feller to speak of "chance processes" in [*Feller 1961c].

Picture 6: Clara and William Feller around 1950. Photograph courtesy of Joanne Elliott.

programme which would combine probability theory, (partial) differential equations and functional analysis.

The Princeton years, in particular the 1950s, marked the second extremely innovative period in Feller's research. He introduced semigroups into probability theory (following Bochner [6] and Yosida [46])[55] and started discussing processes on subsets of $\mathbb{R}$ (or $\mathbb{R}^d$), leading to parabolic initial value problems with additional boundary conditions. At the same time he understood (but rarely worked himself on)[56] the pathwise interpretation of diffusions offered by Itô's approach. The collaboration of one of his students, H. P. McKean, with K. Itô, culminating in the celebrated monograph [23], was certainly no accident, see also the discussions in Fukushima's and Peskir's essays [19, 31] in these Selecta.

In a series of papers[57] he develops semigroup theory with probabilistic applications in mind which would usually be phrased in terms of the Kolmogorov backward and forward equations. Very important contributions are contained in the 1953 papers where he relaxes the continuity conditions (both in space and time) which are needed for the backward and forward Kolmogorov equations. Even more important is his discovery in [Feller 1952a] that the corresponding adjoint of the backward equation leads to an "essentially new boundary condition"[58] of second order. This line of research led to the general characterization of boundary conditions for one-

[55]Kendall writes in Reuter's obituary: "Now Feller, in the preceding year [1951/52] had given a long course on aspects of the Hille–Yosida theory of semigroups in the context of probability theory" and he goes on how "semigroup analysis" helped him (Kendall) to understand and work (with Harry Reuter) on general Markov chains [25, p.178].

[56]The interpretation of the boundary conditions in terms of the processes in [Feller 1954b] is one notable exception. In general, however "[h]e was one of the first generation who thought probabilistically [...], but when it came to writing down any of his results for publication, he would chicken out and recast the mathematics in purely analytic terms" (Rota [38, p. 227]). The same pattern can already be observed in Feller's early contributions to probability theory, see Fischer [16].

[57]notably in [Feller 1952a, Feller 1952b, Feller 1952e, Feller 1953a, Feller 1953b, Feller 1959c]

[58]Yosida, review MR0047886 (13,948a) of [Feller 1952a].

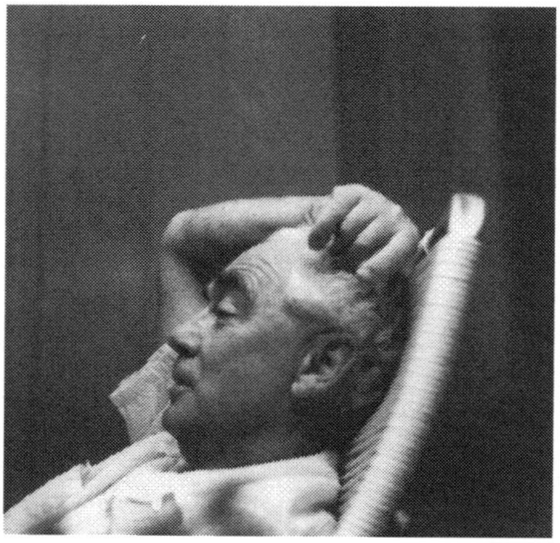

Picture 7: William Feller at his home in Princeton. Photograph courtesy of Joanne Elliott.

dimensional diffusion operators which appeared in a cascade of papers[59]; this body of work certainly is Feller's most cited contribution to pure mathematics, see the essays by Fukushima [19], Jacob [24] and Peskir [31] in these Selecta.

At about the same time, building on his ideas from [Feller 1939a], he presents a groundbreaking paper at the Second Berkeley Symposium on Mathematical Statistics and Probability [Feller 1951d], paving the way for non-trivial applications of stochastic processes, in particular in genetics. At the end of the decade, he connects his diffusion and semigroup theory with birth and death processes [Feller 1959b] – an idea which was already present in the 1939 paper. Feller's pioneering work on mathematical biology is among his most influential contributions to science, see the commentary by Baake and Wakolbinger [2] included in these Selecta.

At Princeton he was able to attract many exceptionally talented graduate students and young researchers to either work with him or to be strongly influenced by his work. Among his Princeton PhD students were Patrick Billingsley (1955), Henry McKean (1955), Hale Trotter (1956), Frank Knight (1956), David Freedman (1960), Lawrence Shepp (1961), Martin Silverstein (1965), Benjamin Weiss (1965) and Loren Pitt (1967); the Mathematics Genealogy Project[60] lists 22 students and 1043 descendants. Many established international visitors were attracted to Princeton as well. For example, Kiyosi Itô visited the Institute for Advanced Study in Princeton from 1954 to 1956, and almost immediately started working with Henry McKean, then one of Feller's graduate students, and soon they both started exchanging their ideas with Feller. Their celebrated work on diffusion processes was in many ways influenced by Feller's programme: In Itô and McKean's words "W. Feller has our best thanks, his ideas run through the whole book."[61]. In his MR book review, S. Watanabe wrote:

> Around 1955, W. Feller's work on linear diffusion [MR0047886(13,948a)

[59]including [Feller 1952a, Feller 1954a, Feller 1954b, Feller 1955a, Feller 1955b, Feller 1957a, Feller 1957d, Feller 1958]

[60]genealogy.math.ndsu.nodak.edu/id.php?id=33019 (accessed 5 August 2014).

[61]Itô and McKean [23, p. XI]

> [Feller 1952a]; MR0068082(16,824g) [Feller 1955a]; and many others], which was primarily of analytic character, spurred some outstanding probabilists (including the present authors [Itô, McKean]) to re-establishing Feller's results by more probabilistic methods, solving some conjectures presented by Feller (in his paper or privately) and, finally, studying the structure of the sample paths of linear diffusion profoundly.[62]

This nicely illustrates how Feller handled his interactions with his students. He never competed with them head-on but carefully led them onto independent, but still related to his own work, paths of research.

The 1950s were not only an exceptionally productive period for Feller but they also marked a breakthrough in the appreciation of his work in a wider mathematical and scientific community. At the age of only 45 he got appointed to a named chair at Princeton University, which was just about to enter top rankings of the mathematical world, taking over the role Göttingen had played before the war. In 1958 he was a plenary lecturer at the International Congress of Mathematicians (ICM) in Edinburgh and in the same year he was named a Fellow of the American Academy of Arts and Sciences. Two years later, William Feller became an elected member of the National Academy of Sciences (USA)[63], and among the many signs of international recognition was his appointment to serve on the *Fields Medal Committee* for the 1966 ICM in Moscow.

Feller continued working on his benchmark monograph. The second, slightly enlarged, but substantially changed, edition of Volume 1 appeared in 1957 and the preparations for the much more advanced Volume 2 (which would appear in 1966) must have been in full swing. The content of both volumes reflects Feller's research interests, for example, the material from the papers on fluctuations and coin tossing [Feller 1949b, *Feller 1951c, *Feller 1957b, *Feller 1959a] have been included in Volume 1, while Volume 2 would establish semigroup theory as a tool in probability and it is still one of the most complete discussions of limit theorems.

The 1958 ICM at Edinburgh showcased Feller at the height of his career. In the one-hour plenary talk [Feller 1960] he summed up his boundary theory, reviewed its connections with Martin boundaries, and explained the intimate link between probability and potential theory. But after the Congress had adjourned he left the field which he had initiated over the past decade, and turned his attention again to limit theorems and applications. As in Stockholm, a quarter-century earlier, he started interacting again with scientists and engineers. The expository paper [*Feller 1961c] was aimed at engineering students and he returned to research in mathematical population biology and genetics: [Feller 1966b, Feller 1967c, *Feller 1969a, *Feller 1969f]. Jointly with S. Orey he studied renewal theorems for random walks [Feller 1961a, Feller 1961b], and strong ratio theorems [Feller 1966a]. In the field of limit theorems he returned to the topics he investigated in the decade starting in 1935, studying in a series of papers fluctuations and oscillations of series of random variables [Feller 1966c, *Feller 1969e, Feller 1970], and limit theorems for large deviations

[62] S. Watanabe, review MR 0199891 (33 #8031) of the monograph [23] by Itô and McKean.

[63] Rosenblatt [37, p. 12]

Picture 8: One of the portrait pictures taken by R.W. Collier at IBM's Thomas J. Watson Research Center, where Feller served as a consultant for many years. These photographs were also used by Monroe Donsker's wife, Mary, to paint Feller's watercolor portrait which is in the possession of Dr. Elliot. Photograph courtesy of Joanne Elliott.

and the LIL [*Feller 1968c, Feller 1968d, Feller 1969d, Feller 1969c]. In connection with limit theorems he contributed to Tauberian theory (for Laplace transforms) and the theory of regular variation [Feller 1963, Feller 1967a, *Feller 1968a] and [Feller 1969b]. A discussion of the mathematical contents of those papers can be found in Maller's commentary [28] in these Selecta.

By the end of 1965 Feller had completed the first edition of the second volume of his exceptional monograph on probability and statistics. He would continue working on the next edition until the end of his life, and it is thanks to many of his former graduate students and collaborators that the last edition was eventually completed after his death. In 1966 Feller was also awarded the title of permanent visiting professor at the Rockefeller University in New York, NY. The university, having recently changed its name from Rockefeller Institute, was expanding its mission and started to hire faculty with expertise in physics and mathematics. Feller's collaborators there included T. Dobzhansky[64], among others, and the paper [Feller 1966b] was the first of Feller's publications with both affiliations, Rockefeller and Princeton University.[65] During that time Feller also worked on the the third edition of Volume 1 of his probability textbook which appeared in 1968 [*Feller 1968e] (with a corrected reprint in 1970); again it differs considerably from the two earlier editions.

William Feller is remembered not only for his scientific achievements and his monographs, but lives on in the memories of many contemporaries as a "man of wide interests, with a profound knowledge of subjects like history and literature, and a

[64]Theodosius Dobzhanski was one of the founding fathers of what is now known as *evolutionary synthesis* within modern evolutionary biology. It is worth pointing out that Feller's publications in mathematical biology were strongly influenced by his work.

[65]Rosenblatt [37, p. 12]

Picture 9: Clara Feller around 1960. Photograph courtesy of Joanne Elliott.

great love of music." [66]

> Feller was an ebullient man, who would rather be wrong than undecided, and who preferred getting a hearing for his views to getting applauded for them. [...] He spoke loudly, very fast, with a strong Yugoslav accent, with wit and charm and understanding.[67]
>
> He was short, compact, with a mop of wooly gray hair, irrepressible. In conversation quick, always ready with an opinion (or two) addicted to exaggeration. If you knew the code, you applied the "Feller factor" (discount by 90%) [...] [he was] so full of fun.[68]

As a lecturer he was "loud and entertaining" and, although his proofs were not always correct or complete, the underlying idea was usually sound.[69] Coming to the 'end' of such an incomplete proof, Feller would stare out at the people in the class "with a set chin and an earnest look on his face, trying to elicit a slight nod from people, and he would often get such a nod."[70] In this context Mark Kac coined the expression "proof by intimidation".[71] Feller did not adopt Landau's Definition – Theorem – Proof style for his lectures and he did not routinely use the headlines "Theorem" or "Proof" on the blackboard.[72]

We do not know when exactly Feller learned about his terminal illness, but we do know that he was aware of it for a while:

> Having accepted the verdict himself he tried to make it easy for all of

[66] Cramér [7, p. 436]

[67] Halmos [20, p. 94]

[68] McKean quoted in Rosenblatt [37, pp. 13*f.*]

[69] Rota [38, p. 227]

[70] Knapp [26]

[71] Rota [38, pp. 227*f.*]. Knapp [26] confirms that the phrase "proof by intimidation" was actually used in Feller's class in the academic year 1962/63, but he does not remember who had introduced it; it may well have been Feller's own joke on various ways to prove things.

[72] Knapp [26]

us to accept it too. He behaved so naturally and he took such interest in things around him that he made us almost forget from time to time that he was mortally ill.[73]

William Feller died of cancer in a New York hospital on 14 January 1970. He did not live long enough to be present at the award ceremony for perhaps the most distinguished of his many recognitions, the *National Medal of Science* for 1969. His widow Clara received the award at the White House on 16 February 1970. After her husband's death, Clara continued her engagement with the mathematical community, and served as the Technical Editor for the Annals of Mathematics taking care of the Volumes 94/95 (1971/72). The second edition of Volume 2 of Feller's textbook [*Feller 1971] appeared posthumously in 1971; "[t]he manuscript had been finished at the time of the author's death but no proofs had been received" and proofreading, indexing and final touches were done by Feller's students.[74]

Acknowledgement. Many colleagues supported us when writing this biographical sketch. In particular we would like to thank Christian Berg for supplying Feller's letters to Borge Jessen, Joanne Elliott for sharing with us her recollections and supplying many photographs of Feller, Hans Fischer, Niels Jacob, Joseph Rosenblatt and Kenīti Sato for critical remarks and stimulating discussions. Tony Knapp kindly told us about Feller's teaching at Princeton and he shared with us his personal lecture notes of some of Feller's classes. Our Croatian colleagues, Nikola Sandrić, Hrvoje Šikić and Zoran Vondraček helped us with many documents and sources in Croatian which would otherwise have been inaccessible to us. Part of this exposition is based on Hrvoje Šikić's article [42] on Feller. We are also indebted to Ulrich Hunger from the *Staats- und Universitätsbibliothek Göttingen* who provided scans of the *Promotionsakte*, to Ms. Christiane Weber of TU Dresden for digitizing the photographs and to Ms. Sarah Hamdouchi of the *Einwohnermeldeamt* Flensburg, Germany, for checking the birth register of the city of Flensburg.

References

All citations of the form [Feller 19*nn*], resp., [*Feller 19*nn*] (if the respective paper is not included in these Selecta) point to Feller's bibliography, pp. xxv–xxxiv.

[1] Promotionsakte [dissertation file] "Willy Feller". Universitätsarchiv Göttingen, Signatur [Math Nat Prom 0010, 23].

[2] Baake, E., Wakolbinger, A.: Feller's Contributions to Mathematical Biology. *These Selecta*, Vol. 2, pp. 25*ff*.

[3] Bers, L.: The migration of European mathematicians to America. In: Duren, P. L. et al.: *A Century of Mathematics in America* Volume 1. Am. Math. Soc., Providence (RI) 1988, pp. 231–243.

[73]Mark Kac in Rosenblatt [37, p. 13].

[74][*Feller 1971, p. xi]; J. Goldman, A. Grunbaum, H. McKean, L. Pitt and A. Pittenger are explicitly

[4] Birnbaum, Z. W. (ed.): William Feller, 1906–1970. *Ann. Math. Statist.* **41**.6 (1970) iv–xiii. Reprinted in these Selecta, Vol. 1.

[5] Brillinger, D. R., Davis, R. A.: A conversation with Murray Rosenblatt. *Statistical Science* **24** (2009) 116–140.

[6] Bochner, S.: Diffusion equation and stochastic processes. *Proc. Natl. Acad. Sci. USA* **35** (1949) 368–370.

[7] Cramér, H.: Obituary Note. William Feller 1906–1970. *Revue de l'Institut International de Statistique / Review of the International Statistical Institute* **38** (1970) 435–436.

[8] Cramér, H.: Half a century with probability theory. Some personal recollections. *Ann. Probab.* **4** (1976) 509–546. Reprinted in Martin-Löf, A.: *Collected Works of Harald Cramér* Vol. 2. Spinger, Berlin 1994, pp. 1352–1389.

[9] Doob, J. L.: William Feller and Twentieth Century Probability. In: Le Cam, L. and Neyman, J. (eds.): *Proceedings of the Sixth Berkeley Symposium on Mathematical Statistics and Probability*. Vol. II, pp. xv–xxi. University of California Press, Berkeley (CA) 1972. Reprinted in these Selecta, Vol. 1.

[10] Fatović-Ferenčić, S. and Ferber-Bogdan, J.: Ljekarnik Eugen Viktor Feller (Pharmacist Eugen Viktor Feller). (Croatian with English summary) *Medicus*, **6** (1997) 277–283.

[11] Fatović-Ferenčić, S. and Ferber-Bogdan, J.: Ljekarna K sv. Trojstvu: izgubljeni sjaj zagrebačke secesije. (Croatian) *Medicus*, **16** (2007) 121–128.

[12] Feller, M.: Das Landhaus E. V. Feller in Agram. Von Architekt Mathias Feller in München. *Innendekoration* **25** (September 1914) 367–398. digi.ub.uni-heidelberg.de/diglit/innendekoration1914/0396 – digi.ub.uni-heidelberg.de/diglit/innendekoration1914/0428

[13] Feller, W.: Handwritten letter to Borge Jessen, dated 1 September 1946. (The letter was kindly provided by Christian Berg, June 2014).

[14] Feller, W.: Typed letter to Borge Jessen, dated 2 April 1947. (The letter was kindly provided by Christian Berg, June 2014).

[15] Fischer, H.: Feller's Early Work on Measure Theory and Mathematical Foundations of Probability. *These Selecta*, Vol. 1, pp. 43*ff*.

[16] Fischer, H.: Feller's Early Work on Limit Theorems. *These Selecta*, Vol. 1, pp. 69*ff*.

[17] Fraenkel, A. A.: *Lebenskreise. Aus den Erinnerungen eines jüdischen Mathematikers*. Deutsche Verlags-Anstalt, Stuttgart 1967.

mentioned.

[18] Fréchet, M. (ed.): *Colloque Consacré à la Théorie des Probabilités* (I–VIII). Actualités Scientifiques et Industrielles, tomes 734–740, 766, Hermann, Paris 1938/39.

[19] Fukushima, M.: Feller's Contributions to the One-Dimensional Diffusion Theory and Beyond. *These Selecta*, Vol. 2, pp. 63*ff.*

[20] Halmos, P.: *I Want to be a Mathematician. An Automatography*. Springer, New York (NY) 1985.

[21] Hochkirchen, T.: Wahrscheinlichkeitsrechnung im Spannungsfeld von Maß- und Häufigkeitstheorie – Leben und Werk des „Deutschen" Mathematikers Erhard Tornier (1894–1982). *NTM International Journal of History & Ethics of Natural Sciences, Technology & Medicine* **6** (1998) 22–41.

[22] Hrvatski Biografski Leksikon, Vol 4 (E–Gm), ed. by Trpimir Macan, Leksikografski Zavod Miroslav Krleža, Zagreb, 1998. URL: hbl.lzmk.hr

[23] Itô, K., McKean, H. P.: *Diffusion Processes and Their Sample Paths*. Springer, Grundlehren Bd. 125, Berlin 1965.

[24] Jacob, N.: Feller on Differential Operators and Semi-groups. *These Selecta*, Vol. 2, pp. 45*ff.*

[25] Kendall, D. G., Bingham, N. H., Sondheimer, E. H.: Obituary. Gerd Edzard Harry Reuter. *Bull. London Math. Soc.* **27** (1995) 177–188.

[26] Knapp, A.: Private communication. E-mail to the editors dated 28 August 2014.

[27] Kolmogoroff, A. N.: Über die analytischen Methoden in der Wahrscheinlichkeitsrechnung. *Math. Ann.* **104** (1931) 415–458. English translation: On analytical methods in probability theory, in: Shiryayev, A.Ṅ.: *Selected Works of A. N. Kolmogorov*. Kluwer, Dordrecht 1992, Vol. 2, pp. 62–108.

[28] Maller, R.: Feller's Work in Renewal Theory, the Law of the Iterated Logarithm, Karamata Theory and Related Topics. *These Selecta*, Vol. 2, pp. 95*ff.*

[29] Mukhopadhyay, N: A conversation with Ulf Grenander. *Statistical Science* **21** (2006) 404–426.

[30] Neyman, J.: L'estimation statistique traitée comme un problème classique de probabilité. Exposé III, pp. 25–57 in Vol. VI of [18].

[31] Peskir, G.: On boundary behaviour of one-dimensional diffusions: From Brown to Feller and beyond. *These Selecta*, Vol. 2, pp. 77*ff.*

[32] Pitcher, E.: Mathematical Reviews. In: Pitcher, E. (ed.): *A History of the Second Fifty Years, American Mathematical Society 1939–1988: Volume I*. American Mathematical Society, Providence (RI) 1988, pp. 69–89.

[33] Reid, C.: *Hilbert*. Springer, New York (NY) 1970.

[34] Reid, C.: *Courant in Göttingen and New York. The Story of an Improbable Mathematician*. Springer, New York (NY) 1976.

[35] Reingold, N.: Refugee mathematicians in the United States of America, 1933–1941: Reception and reaction. In: Duren, P. L. et al. (eds.): *A Century of Mathematics in America* Volume 1. Am. Math. Soc., Providence (RI) 1988, pp. 175–200. Originally published as *Annals of Science* **38** (1981) 313–338.

[36] Risser, R.: *Applications de la statistique à la démographie et à la biologie*. Traité du calcul des probabilités et de ses applications, III.3. Gauthier–Villars, Paris 1932.

[37] Rosenblatt, M.: William Feller. July 7, 1906–January 14, 1970. In: National Academy of Sciences, Office of the Home Secretary (ed.): *Biographical Memoirs*. Volume 90, National Academies Press 2009. (This is mainly based on [4] and [9].)

[38] Rota, G. C.: Fine Hall in its Golden Age: Remembrances of Princeton. In: Duren, P. L. et al. (eds.): *A Century of Mathematics in America. Volume 2*. Am. Math. Soc., Providence (RI) 1988, pp. 195–236.

[39] Scholz, E.: Feller and Busemann on Surface Theory — Contributions to Geometry. *These Selecta*, Vol. 1, pp. 87*ff*.

[40] Seneta, E.: Karamata's characterization theorem, Feller, and regular variation in probability theory. *Publications de l'Institut Mathématique, Nouvelle série*, **71(85)** (2002) 79–89.

[41] Siegmund-Schultze, R.: *Mathematicians Fleeing from Nazi Germany. Individual Fates and Global Impact*. Princeton University Press, Princeton (NJ) 2009.

[42] Šikić, H.: William Feller. In: J. Herak: *Distinguished Croatian Scientists in America*. Croatian–American Society and Matica Hrvatska, Zagreb 1997, pp. 105–115.

[43] Snell, J. L.: A conversation with Joe Doob. *Statistical Science* **12** (1997) 301–311.

[44] Steinhaus, H. : Les probabilités dénombrables et leur rapport à la théorie de la mesure. *Fundamenta Mathematicae* **4** (1923) 286–310.

[45] Uhlig, R. (ed.): *Vertriebene Wissenschaftler der Christian-Albrechts-Universität zu Kiel (CAU) nach 1933*. Kieler Werkstücke Reihe A: Beiträge zur schleswig-holsteinischen und skandinavischen Geschichte Band 2, Peter Lang, Frankfurt am Main 1991.

[46] Yosida, K: An operator-theoretical treatment of temporally homogeneous Markoff process. *J. Math. Soc. Japan* **1** (1949) 244–253.

[47] Žubrinić, D.: *Vilim Feller / William Feller* (Croatian and English). Graphis, Zagreb 2010.

Prof. René L. Schilling
Fachrichtung Mathematik
Institut für Mathematische Stochastik
Technische Universität Dresden
Helmholtzstraße 10
01069 Dresden
Germany
rene.schilling@tu-dresden.de

Prof. Wojbor A. Woyczyński
Department of Mathematics,
Applied Mathematics and Statistics
Case Western Reserve University
10900 Euclid Avenue
Cleveland, Ohio 44106
U.S.A.
waw@case.edu

Feller's Contributions to Mathematical Biology

by *Ellen Baake* from Bielefeld
and *Anton Wakolbinger* from Frankfurt am Main

1 Introduction

Feller had a persistent interest in biology. This is documented in numerous examples from mathematical genetics in his monograph [*Feller 1950, *Feller 1966d],[1] and by a couple of influential research papers at the interface of population biology and probability theory. Looking back at these papers in historical perspective is highly rewarding: They are cornerstones of biomathematics; they mirror the development of probability theory of their time; and at least one of them ([Feller 1951d]) had lasting impact on probability theory.

Feller's important papers on the interface to biology are the articles [Feller 1939a], [Feller 1951d], and [Feller 1967c]. The first one addresses general population dynamics, the other two are mainly concerned with models in population genetics. The area of *population dynamics* is concerned with the growth, stabilisation, decay, or extinction of populations. Models of population dynamics describe how the *size* of populations changes over time under given assumptions on birth and death rates of individuals, which may depend on the current population size since individuals interact (e. g. compete) with each other. In contrast, *population genetics* is concerned with the *genetic composition* of populations under the action of various evolutionary processes, such as mutation and selection. Naturally, there is no sharp boundary between the fields, as we will also see in Feller's contributions. Let us now look at them.

2 Feller and population dynamics

In [Feller 1939a], a paper still in German entitled (in English translation) *The foundations of a probabilistic treatment of Volterra's theory of the struggle for life*, Feller

[1] The references [Feller 19*nn*] and [*Feller 19*nn*] (the star indicating that the respective paper is not contained in these Selecta) refer to Feller's bibliography, while [*n*] points to the list of references at the end of this essay.

R.L. Schilling et al. (eds.), *Selected Papers II*,

presents a synthesis of two fundamental developments that both started in 1931. On the one hand, Volterra presented his book *Lessons about the mathematical theory of the struggle for life* [28]; on the other hand, Kolmogorov published his seminal paper *On analytical methods in probability theory* [17]. Volterra's book laid the foundations for the *deterministic* description of population dynamics in terms of systems of ordinary differential equations that model birth, death, and interaction of individuals. These models imply that populations are so large that random fluctuations can be neglected, and population sizes are measured in units so large that the size can be considered a continuous quantity. Kolmogorov presented the general and systematic formalism for the description of *stochastic* dynamics in terms of Markov chains in continuous time; in particular, he found the description for the evolution of probability weights and the transport of expectations in terms of differential equations, which we know today under the names of *Kolmogorov forward equations* and *Kolmogorov backward equations*.

In his 1939 paper, Feller ties these two fundamental developments together by applying Kolmogorov's new formalism to some examples of Volterra's population dynamics. We see here the birth of the stochastic description of population dynamics, which today has its firm place in mathematical biology, and is highly developed both in analytical terms and in terms of simulations.

Feller's paper is devoted to the description of single populations (except from a small excursion to predator–prey models in the end) and consists of two large parts. The first establishes the Kolmogorov forward equations (KFE) for the Markov jump processes (namely, birth-and-death processes) that describe finite populations (remarkably, there is no mention of the Kolmogorov backward equations in this paper). The second part discusses a continuum analogue of such processes, a special case of which seems to be the first appearance of what today is called *Feller's branching diffusion*.

It is remarkable to see (and a pleasure to read) that Feller notices some of the crucial relationships between corresponding deterministic and stochastic models in this early paper, which appear as a central theme.

For the sake of clarity, let us make explicit here the two fundamental limits of birth-death processes that are addressed in [Feller 1939a]. Consider a birth-death process $K_N(t)$ with birth rate $n\lambda$ and death rate $n\mu$ when in state n, with N being the initial population size. Then, as the initial population size N tends to ∞, the sequence of process $(K_N(t)/N)_{t\geq 0}$, $N = 1, 2, \ldots$, converges in distribution to the solution of the differential equation

$$\dot{x} = (\lambda - \mu)x, \quad x(0) = 1. \tag{1}$$

This reflects a dynamical version of the Law of Large Numbers (see e. g. [19]). (Notably, due to the linearity, the expectation $M(t) := \mathbb{E}(K_1(t))$ satisfies (1) as well.) A different kind of limit emerges if one assumes that the individual split and death rates λ and μ depend on N and the process is *nearly critical* in the sense that

$$\lambda_N = \beta + \theta_1/N \quad \text{and} \quad \mu_N = \beta + \theta_2/N,$$

with $\theta_1 - \theta_2 =: \alpha$. The Law of Large Numbers then says that the limit of the processes

$(K_N(t)/N)_{t\geq 0}$ is the constant 1. However, on a larger time scale the fluctuations become visible: the sequence of processes $(K_N(Nt)/N)_{t\geq 0}$ converges in distribution to the solution of the *stochastic* differential equation (5.1′) stated in paragraph 3.1.1, whose diffusion equation is (4). This is a prototype of a diffusion limit for birth-death processes. In [Feller 1939a], these limiting procedures are not made explicit (but see [Feller 1951d] for a major step in this direction). Feller in 1939 goes rather the other way, in search for stochastic processes that correspond to a given deterministic model. Let us now explain the major lines of his article.

2.1 Markov jump processes for population dynamics

In the first part (Sections 1–4), devoted to the stochastic description of finite populations, Feller explains a variety of birth-and-death processes and sets up the Kolmogorov forward equations for them, i. e. he establishes the system of differential equations that describe how the probability weights for the number of individuals alive at time t evolve over time. He starts with the simple linear death process (where each individual dies at rate λ, independently of all others), proceeds via the corresponding birth process and the linear birth-and-death process and finally arrives at the general birth-and-death process. In an individual-based picture, the latter includes interaction between individuals, so that the birth and/or death rates are no longer linear in the number of individuals. The case of logistic growth, which includes a quadratic competition term, serves as an important example; the case of 'positive interaction' (such as symbiosis) is not treated explicitly here. Let us comment on the major insights of this part.

2.1.1 Kolmogorov equations, their solutions, and relationship with deterministic description

Feller notices that for a given net reproduction rate α per individual, by choosing $\lambda - \mu = \alpha$, one obtains a variety of linear birth-death processes whose expectation value $M(t)$ satisfies one and the same ODE (1), whereas for $\alpha > 0$, there is exactly one linear pure birth process ($\lambda = \alpha, \mu = 0$) with this property. Feller states this ambiguity explicitly when discussing *logistic growth*. Its deterministic version is given by the differential equation

$$\dot{m} = m(\lambda - \gamma m) =: f(m), \tag{2}$$

which Feller also calls the *Pearl-Verhulst* equation. Here m is shorthand for $m(t)$, the 'deterministic version' of the population size at time t, λ denotes the per capita net reproduction rate in the absence of competition, and γ is the competition parameter. Again, Feller notices that there are many possibilities in terms of birth-death processes that correspond to (2). They are parametrised in his Eq. (27), which describes the process with per capita birth at rate $\omega - \nu n$ and per capita death at rate $\tau - \sigma n$ if there are currently n individuals. Here, we have renamed γ in Feller's Eq. (27) by ν in order to achieve compatibility with the notation in (2).

Feller starts out by calculating the explicit solution to the KFE of the pure linear death process, that is, the number of inidividuals alive at time t; he states this as the result of a recursive construction. With a typo corrected ($e^{\lambda t} - 1$ must be replaced by $1 - e^{-\lambda t}$ in his formula (6)), this same formula says that the number of inidividuals alive at time t has a binomial distribution with parameters N and $e^{-\lambda t}$ if there are initially N individuals. (Today, after [*Feller 1950], we would conclude this immediately, without solving systems of Kolmogorov forward equations, via the probabilistic argument that there are initially N independent individuals, each of which dies at rate λ and is therefore alive at time t with probability $e^{-\lambda t}$.) Likewise, the solution of the pure linear *birth* process with per capita birth rate λ, which he gives in his Eq. (17), is the negative binomial distribution with parameters N and $e^{-\lambda t}$, which arises as the distribution of the sum of N independent random variables that are geometrically distributed with parameter $e^{-\lambda t}$. Again, this has a nice interpretation as the offspring of N independently reproducing ancestors.

For the general birth process, with arbitrary birth rates p_n, Feller notes that the KFE define a probability distribution if and only if either only finitely many of the p_n are positive, or if $\sum_n 1/p_n$ diverges; this is a standard textbook result today (usually presented in the generalisation to birth-and-death processes). Under the conditions stated, he also gives the explicit solution in passing.

2.1.2 The moments of the stochastic process, and their relationship with the deterministic equation

Feller is particularly interested in the expectation, variance, and other moments of the (random) number of individuals alive at time t. In a trendsetting way, he does not calculate them from the explicit solution of the KFE, even where this is known; he rather uses the KFE to derive *differential equations for the moments*. Let $M(t) = \sum_k kP_k(t) = \mathbb{E}[K(t)]$ be the expected number of individuals at time t. As stated above, Feller observes that, for the linear birth-and-death process, $M(t)$ follows the differential equation for the deterministic population model, and hence the expectation of the stochastic process coincides with the deterministic solution. In contrast, for the logistic model, he finds from the differential equation relating the first and the second moment that

$$\dot{M} < f(M) \tag{3}$$

with f of Eq. (2). From this he argues that M is always less than the solution of the logistic equation. An alternative way to see (3) would be to observe that the KFE gives

$$\frac{d}{dt}\mathbb{E}[K(t)] = \mathbb{E}[g(K(t)]$$

where $g(k) = \sum_n Q(k,n)n$ and

$$Q(n,n+1) = \lambda n, \quad Q(n,n-1) = \gamma n^2,$$
$$Q(n,n) = -(\lambda n + \gamma n^2), \quad Q(k,n) = 0 \text{ otherwise.}$$

As a matter of fact, it turns out that $g(k) = \lambda k - \gamma k^2$, which is strictly concave, and hence (3) is a consequence of Jensen's inequality. Since Feller does not consider models with positive interaction (such as symbiosis) in this part of the paper, he does not encounter the convex situation.

2.2 Diffusion equations for population dynamics

The second part of the paper (Sections 5–8 and 10) is devoted to the diffusion limit of stochastic population dynamics. We cannot resist to quote Feller's thoughts from the beginning of Section 5, formulated in an almost literary German, about the *substantially more lithesome probabilistic treatment, in which the population size is no longer assumed as integer-valued*, and where he alludes to similarities to the Brownian motion:

> Wir wenden uns nun der anderen von der in der Einleitung erwähnten wahrscheinlichkeitstheoretischen Behandlungsweisen des Wachstumsproblems zu, welche wesentlich geschmeidiger ist, und bei der die Grösse der Population nicht mehr ganzzahlig vorausgesetzt wird. Den Mechanismus des Vorgangs kann man sich hier ähnlich wie bei der Brownschen Molekularbewegung vorstellen. Der Zustand der betrachteten Population, d. h. ihre gesamte Lebensenergie ist einer dauernden Veränderung unterworfen [...]

Starting from the transition density, Feller calculates the infinitesimal drift $a(x)$ and the infinitesimal variance $b(x)$ (provided they exist). With remarkable intuition, and a clear view of the branching property, he states that, in the case of a stochastically independent reproduction of the individuals, $a(x)$ and $b(x)$ must be proportional to x. Again, let us quote in German:

> Nimmt man beispielsweise an, dass die Grösse der Population keinen Einfluss hat auf die durchschnittliche Vermehrungsgeschwindigkeit der Einzelindividuen, d. h. dass diese untereinander stochastisch unabhängig sind [...], so müssen $a(x)$ und $b(x)$ offenbar proportional zu x sein [...]

This gives rise to his equation (38), which reads as

$$\frac{\partial w(t,x)}{\partial t} = \beta \frac{\partial^2 (xw(t,x))}{\partial x^2} - \alpha \frac{\partial (xw(t,x))}{\partial x}. \tag{4}$$

Here $w(t,\cdot)$ is the density of population size at time t, and α and β are positive constants. This seems to be the first appearance of what became famous as *Feller's branching diffusion*. We will come back to this in Section 3.1.

It is interesting to note that the diffusion process in this second part of the paper is *not derived* from the birth-death jump processes which Feller has presented in the first part; maybe, at this early stage, the subtle rescaling required for this limit was not yet at his fingertips. A decade later, however, he had these techniques; see Section 3.1 on *Diffusion processes in genetics*. In 1939, Feller does allude to the birth-death processes, but the connection is not yet clear. For example, he tells us that the α in (4) corresponds to the λ encountered in the pure birth process. This is correct for the expected growth rate, but as a matter of fact a pure birth process cannot have the diffusion limit (4), since the paths of the former can only increase in time, whereas the paths of the latter have fluctuations in both directions.

This issue reappears when Feller discusses the extinction probability of the diffusion process. He notes the important fact that this quantity increases with β, since it is tied to the fluctuations of the process, and at the same time emphasises as a sort of paradox that, even for a positive net growth $\alpha > 0$, the diffusion process may die out with positive probability, while the population described by the deterministic differential equation (1), as well as a pure birth process, cannot die out. (This paradox is resolved when one has in mind the different rescalings that lead to (1) and (4).)

Following these considerations of the linear birth-and-death process, Feller includes dependence between individuals in Sections 8 and 10. He presents two specific examples. The first is his Equation (51), which is the diffusion version of his Equation (7) and known today as *Feller's branching diffusion with logistic growth* [20, 23]. The second is the (two-dimensional) diffusion describing a two-species model with predator-prey interactions, which now is also called *Lotka-Volterra process*, see Eq. (1.2) in [3]. As with the jump processes in the first part of his paper, Feller is concerned with the moments of the diffusion processes and writes down a general recursion for the kth moments M_k. In the two-species model, the interaction is positive (from the point of view of the predator), so we finally encounter the convex case, in which the expectation is greater than the solution of the corresponding ODE.

In Section 9, Feller makes some final remarks concerning the deterministic limit of both the birth-death jump processes and the branching diffusion. These are brief, heuristic calculations, which hint at the convergence of the stochastic models to Volterra's population models in the limit of infinite population size. Today, powerful laws of large numbers are available for large classes of such processes [8, Chap. 11]. They go far beyond the simple linear case alluded to in (1); rather, they include quite general forms of density dependence. This leads us to the present state of population dynamics.

2.3 Afterthoughts

Today, 75 years after [Feller 1939a], stochastic population dynamics constitute a vibrant area of research, so wide that it is impossible to give an overview in a short paragraph. Suffice it to say that major questions raised by Feller continue to be ardent research themes. Above all, this is true of interactions within and between populations. Even simple models for the competition of two populations, whose deterministic limit can be tackled as an easy exercise, turn into hard problems when considered probabilistically. Specifically, diffusion models with interaction have become objects of intense research, see, e.g., [6, 27] and references therein.

In the context of this commentary, it is particularly noteworthy that a class of models known under the name of *adaptive dynamics* brings together ecological aspects (on a short time scale) and genetical aspects (on a longer time scale) and thus builds a bridge between population dynamics and population genetics. A nice overview of this topic and many others may be found in the monograph by Haccou, Jagers, and Vatutin [12]. Let us now turn to Feller's contributions to population genetics.

3 Feller and population genetics

As already laid out in Section 1, important processes in population genetics are those that describe the evolution of type frequencies, or in other words, of proportions of subpopulation sizes within a total population, whose size may vary as well. In this context, we may think of the individuals as *genes*, where each gene is of a certain type, say a or A.

The foundations of mathematical population genetics were laid starting in the 1920s by Fisher, Wright, and Haldane. Their work mirrors the genetics of their time, today known as *classical genetics*. It had to rely on the phenotypic appearance of individuals (colour of flower, surface structure of peas, body weight, milk yield ...). The molecular basis of genetics was still unknown, so genes had to be treated as abstract entities. When *molecular genetics* entered the labs in the 1960s, population genetics changed dramatically, with Kimura as a leading figure, see Section 3.2.1. The next (and, from a 2015 perspective, the last) big leap took place in 1982, when Kingman introduced the genealogical perspective via the *coalescent process*. Comprehensive overviews of population genetics theory are given in the textbooks by Ewens [9] and Durrett [5]; for coalescent theory in particular, we further recommend Berestycki [2] (from a mathematical point of view) and Wakeley [29] (from a more biological perspective).

With his contributions to population genetics, Feller thus was in the midst of an important line of development. We will comment on two of these articles. The first, *Diffusion processes in genetics* [Feller 1951d], is a landmark contribution towards stochastic modelling and analysis via diffusion processes, and, as a matter of fact, reaches far beyond population genetics as such. The second, *On fitness and the cost of natural selection* [Feller 1967c], uses deterministic modelling (and is therefore similar in spirit to the 'Volterra equations').

3.1 Diffusion processes in genetics

Feller's article *Diffusion processes in genetics* [Feller 1951d] appeared in the Proceedings of the 2nd Berkeley Symposium on Mathematical Statistics and Probability, which took place in 1950. The central role of [Feller 1951d] is nicely put into perspective by the following quote from Thomas Nagylaki's review [24] on Gustave Malécot and the transition from classical to modern population genetics:

> Mathematical research in diffusion theory influenced population genetics only gradually. As described in more detail below, Wright was unaware of Kolmogorov's (1931) pioneering paper, and Wright, Malécot, and Kimura were all apparently unacquainted with Khintchine's (1933)

book. [...] Thus, the mutually beneficial cross-fertilization between diffusion theory and population genetics did not start until Feller published his seminal 1951 paper.

In the introduction of that paper, Feller sets the stage by writing:

> Relatively small populations require discrete models, but for large populations it is possible to apply a continuous approximation, and this leads to processes of the diffusion type.

Two diffusion processes are in the focus of the paper. One is what is nowadays called *Feller's branching diffusion*, the other is the so-called *Wright–Fisher diffusion*. Feller describes them by their *diffusion equations* (5.1) and (7.1), which are the Kolmogorov forward equations (or Fokker–Planck equations) for the densities, here called $u(t,x)$, cf. Section 2.2. Feller writes on pp. 228–229:

> It is known that an essential part of Wright's theory is mathematically equivalent to assuming a certain diffusion equation for the *gene frequency* (that is, the proportion of a-genes).

In a footnote on the same page, Feller gives hints to the roots of this knowledge in the work of Kolmogorov, Fisher, Wright, and Malécot.

3.1.1 A foresight: Feller's diffusions as solutions of stochastic differential equations

Nowadays we do not hesitate to write the process described by Feller's equations (5.1) and (7.1) as solutions of stochastic differential equations in the sense of Itô:

$$(5.1') \qquad dZ_t = \sqrt{2\beta Z_t}\, dW_t + \alpha Z_t\, dt,$$

$$(7.1') \qquad dY_t = \sqrt{2\beta Y_t(1-Y_t)}\, dW_t + (\gamma_2(1-Y_t) - \gamma_1 Y_t)\, dt,$$

where W is a standard Brownian motion. Feller legitimately resisted writing the processes in this form. In [Feller 1952c], which grew out of Feller's invited lecture at the International Congress of Mathematicians in the year 1950, he writes about Itô's Stochastic Analysis:

> This approach has the advantage that it permits a direct study of the properties of the path functions, such as their continuity, etc. In principle, we have here a possibility of proving the existence theorems for the partial differential equations [...] directly from the properties of the path functions. However, the method is for the time being restricted to the infinite interval and the conditions on [the diffusion and drift coefficients] a and b are such as to guarantee the uniqueness of the solution. So far, therefore, we cannot obtain any new information concerning the 'pathological' cases.

3.1.2 An emerging theme: What happens at the boundary?

Indeed, the state spaces of (5.1) and (7.1) are not the 'infinite interval' $(-\infty,\infty)$ but $[0,\infty)$ and $[0,1]$, and it took 20 years until T. Yamada and S. Watanabe proved that the coefficients in the above stated SDEs are good enough to guarantee strong uniqueness of the solution, see [32] and also [31]. A coupling argument from Stochastic Analysis then guarantees that the solution of (7.1′) converges in law to the unique equilibrium distribution whose density is the unique invariant probability density of (7.1), which is the Beta($\gamma_2/\beta, \gamma_1/\beta$)-density. Thus, although for $\gamma_1 < \beta$ the random path Y hits 0 with probability one (and similarly for $\gamma_2 < \beta$ it hits 1 with probability one), these visits to the boundary do not lead (as conjectured by Feller on p. 239) to a non-vanishing *accumulation of the masses concentrated at* $x = 0$ *and* $x = 1$ *which is maintained in the steady state*, in other words, the coefficient μ in his equation (7.3) is in fact equal to 1.

Questions like these may have been one source of motivation for Feller to initiate his groundbreaking studies on the boundary classification of diffusion processes, see his footnote on p. 234, where he speaks of *boundary conditions of an altogether new type*, and the one on p. 229 *added in proof*, where he announces that *a systematic theory, including the new boundary condition, is to appear in the Annals of Math.* Feller's *classification of boundaries* is reviewed and commented in Section 2 of Masatoshi Fukushima's essay in these selecta.

3.1.3 The diffusion approximation of the Wright–Fisher chain and beyond

As already indicated, another important aspect that is taken up in Feller's paper is that of the *diffusion approximation*, i. e. the convergence of a sequence of (properly scaled) discrete processes to the solution of (5.1) and (7.1), respectively. In the former case the underlying discrete process is a Galton–Watson process, in the latter it is the Wright–Fisher Markov chain. The transition probabilities of the Wright–Fisher chain are given by (3.2), (3.4) and (3.5). The *diffusive mass-time-scaling* is given by (8.5): a unit of time consists of N generations, and a unit of mass consists of N (or here $2N$) individuals, with N being the total population size. The scaling (8.4) of the individual mutation probabilites α_1, α_2 is that of *weak mutation*, which leads in the scaling limit to the *infinitesimal mean displacement* $a(x)$ and the *infinitesimal variance* $2b(x)$, see (8.6) and (8.7). In the context of (7.1), the *drift coefficient* $a(x)$ is due to the effect of mutation, and the *diffusion coefficient* $b(x)$ describes the strength of the fluctuations that come from the random reproduction. (In order to be consistent with (7.1), β_i should be replaced by γ_i in (8.4), (8.6) and (8.9)). The 'convergence of generators' which emerges from (8.6) and (8.7) can be lifted to the convergence of the corresponding semigroups, see e. g. the chapter on *Genetic Models* in the monograph [8] by Ethier and Kurtz.

The convergence theorems in [8] comply with Feller's programmatic proposal: *It should be proved that our passage to the limit actually leads from* (8.2) *to* (8.6), i. e. from the probability weights of the Wright–Fisher chain to the probability densities the Wright–Fisher diffusion. To achieve this, Feller proposed an expansion into eigenfunctions (in particular he found the eigenvalues of the Wright–Fisher transition semigroup) and checked part of the convergence in Section 8 and Appendix I. Such a representation is not required in the systematic approach presented in [8]. Still, the approach via eigenfunctions is interesting in its own right, and has been extensively used in Mathematical Biology.

At the beginning of Section 9 (entitled *Other possibilities*) Feller writes :

> The described passage to the limit which leads to Wright's diffusion equation (7.1) is different from the familiar similar processes in physical diffusion theory where the ratio $\Delta x/\Delta t$ tends to infinity rather than to a constant. It rests entirely on the assumption (8.4) [of weak mutation]. We shall now see that *any modification of this assumption leads to a non-singular diffusion equation of the familiar type* (*to normal distributions*).

Indeed, for the scaling (9.1), (9.2) $\alpha_i = \gamma_i \varepsilon$, $i = 1,2$, $N\varepsilon \to \infty$, which corresponds to *strong mutation*, Feller states a law of large numbers, i. e. a convergence of the type frequencies to the equilibrium point $\left(\frac{\gamma_2}{\gamma_1+\gamma_2}, \frac{\gamma_1}{\gamma_1+\gamma_2}\right)$, and argues that the (properly scaled) process of fluctuations around this equlibrium point converges to a process whose probability density satisfies the diffusion equation (9.10) (and thus is an Ornstein-Uhlenbeck process).

3.1.4 The diffusion approximation of Galton–Watson processes

The diffusion equation (5.1) appeared already in [Feller 1939a], see Eq. (4) in Section 2.2. However, as we have seen there, certain issues concerning the (scaling) limits of Galton–Watson processes had remained unrevealed in [Feller 1939a]. Towards 1950, Feller was ready to attack this. As to the diffusion approximation of a sequence of 'nearly critical' Galton–Watson processes by (5.1), Feller gives a proof in Appendix II. His idea is to take the iterates f_n of the offspring generating function (which are known to describe the generating functions of the subsequent generation sizes) to their scaling limit. This limit turns out to satisfy the PDE (12.9) (which, in turn, corresponds to (5.1)). Feller writes:

> We effect this passage to the limit formally: it is not difficult to justify these steps, since the necessary regularity properties of the generating functions $f_n(x)$ were establihed by Harris [13].

Again, from today's perspective, an alternative way is provided by the convergence of generators, see [8]. In the very last lines of his Appendix II, Feller remains a bit sketchy when he writes that

> the boundary condition $u(t,0)$ follows from the fact that in the branching process the probability mass flowing out into the origin tends to zero.

In fact, for the solution Z of (5.1$'$) (with $Z_0 = 1$, say), the probability mass flowing out into the origin is non-zero at any fixed time t, and the density of Z_t does not vanish near the origin. Again, the desire to obtain clarity on questions like these may have been a motivation for Feller's then upcoming research on the boundary behaviour of one-dimensional diffusions.

3.1.5 From two-type to continuum-type generalisations of Feller's branching diffusion

In the introduction, Feller points out that *serious difficulties arise if one wishes to construct population models with interactions among the individuals*, and that *the situation grows worse if the population consists of different types of individuals.* He then writes:

> In fact, the bivariate branching process leads to such difficulties that apparently not one single truly bivariate case has been treated in the literature. In the theory of evolution this difficulty is overcome by the assumption of a constant population size [...] In Section 10 the assumptions of constant population size is dropped and a truly bivariate model is constructed which takes into account selective advantages in a more flexible way. [...] The same limiting process which leads [...] to the diffusion equation of Wright's theory can be applied to our new bivariate model and leads to a diffusion equation in two dimensions.

These *two-dimensional Markov processes with branching property* have been taken up and analysed in a broader context in 1969 in the paper [31] which carries that title. Already before, Watanabe had published his seminal paper [30] which established Feller's branching diffusions with a continuum of types. This together with the poineering work of Don Dawson gave rise to a class of processes that were later called Dawson–Watanabe superprocesses ([6, 4]). A good part of Perkins' Saint Flour Lecure Notes (part 2 of [4]) is devoted to superprocesses with interactions, and thus is fully on the line of Feller's program to construct population models with different types of individuals and with interactions among the individuals.

3.1.6 The inner life of Feller's branching diffusion: excursions and continuum trees

This is a good place to mention another fascinating development which is connected with Feller's branching diffusions and is associated with the names of Daniel Ray and Frank Knight (the latter was Feller's doctoral student and Ed Perkins' PhD advisor).

Thanks to the branching property (and the thereby implied infinite divisibility), the random path Z of a Feller branching diffusion is a Poissonian sum of countably many 'Feller excursions' ζ. In fact, each of them has an 'internal life' in the sense that ζ_t is the size at time t of a continuum population originating from one single ancestor. The genealogical tree of this population can be described by a Brownian excursion η reaching level t, which can be imagined as the 'exploration path' of a continuum random tree whose mass alive at level t is ζ_t. The second Ray–Knight theorem says that ζ_t can be represented as the local time spent by η at level t. In a discrete setting of Galton–Watson processes, this is ancticipated in Harris' work [11] with its section on *walks and trees.* The correspondence between a Feller excursion ζ and an Itô excursion η is depicted on the first page of [25], framed by pictures of Feller and Itô, who met in person at Princeton in 1954. See [25] for more explanations, and references to groundbreaking developments that dealt with the genealogical structure behind Feller's branching diffusion, such as Aldous' *Continuum Random Tree* (which plays in the world of random trees a similar role to that of Brownian motion in the classical invariance principle) and Le Gall's *Random Snake*, which provides a representation of the Dawson–Watanabe super-Brownian motion as a continuum-tree-indexed Markov motion.

For more on excursions and excursion point processes in relation with Feller's work, see Section 3.1 of the contribution of M. Fukushima.

3.1.7 Frequencies in multivariate continuum branching: conditioning and time change

Another interesting question which Feller addresses at the end of his introduction concerns the relative frequencies in a bivariate model of branching diffusions. Feller writes:

> [...] it is to be observed that *in no truly bivariate case does the gene frequency satisfy a diffusion equation* (Sections 6 and 10). In fact, if the population size is not constant, then the gene frequency is not a random variable of a Markov process. Thus, conceptually at least, the assumption of a constant population size plays a larger role than would appear on the surface.

Indeed, as it turns out (and Feller may have been well aware of this), one way of passing from (5.1) to (7.1), say with $\alpha = \gamma_1 = \gamma_2 = 0$, is to consider two independent Feller branching diffusions (solutions of (5.1′)) $Z^{(1)}$ and $Z^{(2)}$, conditioned to $Z^{(1)} + Z^{(2)} = 1$. Of course, this must be given a precise meaning, and this has been done in a much more general context in the papers by Etheridge and March [7] and Perkins [26]. The title of Perkins' paper is programmatic: A Dawson–Watanabe superprocess conditioned to have constant mass one is a Fleming–Viot process (which is the continuum-type, and thus measure-valued, generalisation of the Wright–Fisher diffusion). In the context of (5.1) to (7.1) this means that under the conditioning $Z^{(1)} + Z^{(2)} = 1$ the process $Y := Z^{(1)}$ is a Wright–Fisher diffusion. On the level of genealogies, the conditioning to a constant total mass takes the continuum random forest that underlies (5.1′) into Kingman's coalescent.

A second way to get from (5.1′) to (7.1′) (again for $\alpha = \gamma_1 = \gamma_2 = 0$) is to consider the relative frequency $Y := \frac{Z^{(1)}}{Z^{(1)}+Z^{(2)}}$ after a time change $ds = dt/(Z^{(1)} + Z^{(2)})$. In this way, the relative frequencies again become Markovian and, by an easy application of Itô's formula, turn out to solve (7.1′).

The concept of time change is central also in the work of John Lamperti. Lamperti's work can be seen as a direct continuation and extension of Feller's ideas, introducing and analysing the continuum mass limits of Galton–Watson processes also for heavy-tailed offspring distributions [21]. His article [22], which was communicated by H. P. McKean, another former PhD student of Feller, introduced what is now called Lamperti's transform, a time change which establishes the link between Lévy processes and continuous state branching processes.

To conclude: Feller's paper *Diffusion processes in genetics* is a remarkable contribution at the interface of probability theory and population biology, with enduring stimulations in either direction. It takes a central place in the development of mathematical population genetics, and has triggered substantial new directions in the modern theory of stochastic processes.

3.2 The cost of natural selection

Feller's article *On fitness and the cost of natural selection* [Feller 1967c] appeared in *Genetical Research Cambridge*, a renowned biology journal. The introduction contains the disclaimer

> This paper is written by a mathematician, and accordingly no new biological models or hypotheses are advanced.

It may be added that the mathematics is fairly elementary as well, and the strength of the article is the concise conceptual thinking by which Feller dissects the logics of an argument that enormously influenced the genetic thinking of that time, and finds a fundamental weak point in it.

3.2.1 The 1960s and the neutral theory of population genetics

As hinted at already, the 1960s were turbulent times for genetics – and for population genetics in particular. Until then, the variation between individuals could only be observed at the phenotypical level, and much of this was easily explained by selection: Stronger beaks crack harder nuts; webbing eases swimming; fat pads protect against the cold. Then, in the 1960s, the first observations of *variation at the molecular level* became available – not yet via sequencing, but via so-called restriction length polymorphsims (RFLP) of DNA or via gel electrophoresis of proteins. The resolution of these methods is lower than that of sequencing but, nevertheless, the variation was so much larger than expected on phenotypic grounds that researchers were shocked – and were puzzled about the question: *Can this all be explained by selection?*

These considerations were strongly influenced by the concept of the *genetic load*, coined by Haldane [10]; in particular the concept of the *substitutional load*. This is the number of *selective deaths*, that is, the number of individuals 'killed' by selection in the process of substituting one type by a fitter one. Put differently, the substitutional load (or *cost of natural selection*) is the number of excess individuals that must be produced in a population under selection. In 1968, Kimura [15] concluded that, if a large fraction of the observed variability is selective, the load is forbidding. This led to one of the most influential and conflict-prone hypotheses of evolutionary theory, namely, to the so-called *neutral theory*, which claims that the overwhelming proportion of the observed molecular variation is selectively neutral, that is, most mutations do not change fitness.

In what follows, we look more closely into the concept of the substitutional load, and into Feller's criticism of it. We restrict ourselves to the case of haploid populations (i.e., carrying only one copy of the genetic information) that reproduce asexually (Sections 1–5 in Feller's paper). In Sections 6–9, he tackles additional complications that emerge in diploid individuals (with two copies of every gene), but the conceptual issues are more transparent in the haploid case.

3.2.2 Absolute and relative frequencies in population genetics

Feller considers a population of individuals that consists of our two types A and a, large enough to justify deterministic treatment. He assumes discrete generations where every A-individual leaves an average of μ offspring for the next generation, whereas every a-individual produces an average of $\mu' = \mu(1-k)$ descendants where $0 < k < 1$. The quantities μ and μ' are known as (absolute) fitnesses in population genetics. Each of the two subpopulations then grows (or decays) geometrically,

$$(5) \qquad N_n = \mu N_{n-1}, \quad N'_n = \mu(1-k)N'_{n-1},$$

so $N_n = N_0\mu^n$ and $N'_n = N'_0(\mu(1-k))^n$, where N_n and N'_n denote the size of the A- and a-subpopulations, respectively.

Now population genetics is traditionally more concerned with relative frequencies of types than with absolute ones; the main reason is that relative frequencies are easier to measure. One therefore considers

$$p_n := \frac{N_n}{N_n + N'_n}, \qquad q_n := \frac{N'_n}{N_n + N'_n}.$$

Clearly, under (5),

$$(6) \qquad q_n = \frac{q_0(1-k)^n}{p_0 + q_0(1-k)^n},$$

which is Feller's Eq. (3.5). Obviously, the powers of μ cancel out, which is a strength and a weakness at the same time: On the one hand, this means that knowledge only of the ratio of the fitness values is required to predict the behaviour of the population. Actually, (6) holds more generally than the simple derivation may suggest: It continues to hold if (5) is replaced by

$$(7) \qquad N_n = \mu N_{n-1} f(N_n + N'_n), \quad N'_n = \mu(1-k)N'_{n-1} f(N_n + N'_n).$$

Here f is a function that depends on the total population size only and acts on both types in the same way. In typical ecological models, one uses a monotonically decreasing function f with $f(0) = 1$ in order to describe how population size decreases the per capita offspring size due to competition. In particular, for suitable choices of f, both the population size and the relative frequencies will, in the long run, approach stationary values.

The downside of thinking in terms of relative fitnesses is that one loses sight of the absolute population sizes. The latter may lead to absurd conclusions, in particular in cases where one or both subpopulations go extinct. This leads us to Feller's criticism of the genetic load.

3.2.3 The substitutional load and Feller's criticism

Feller now recalls Haldane's definition of the genetic load: In generation n, the mixed population has a loss of $d_n := kq_n$ offspring relative to a population consisting of A individuals only, where d_n must be measured in units of the total population size of generation n. Over M generations, Haldane takes $D := \sum_{n=0}^{M} d_n$ as the total genetic load. Calculating this with reasonable parameters and as $M \to \infty$, he arrives at a representative value of 30 for the cost of substitution of one gene by a fitter one.

Haldane's definition of D makes sense only if population size remains constant over time, or if changes in population size are so small that they can safely be neglected. Haldane does not make this explicit; Feller has an eye on both possibilities.

Feller first considers the case that the population size is not constant; rather, N_n and N_n' behave as in (5). Since $\mu = 1$ and $k < 1$ is assumed, this means that $N_n \equiv N_0$ and $N_n' \to 0$ as $n \to \infty$, so the total population size decreases from $N_n + N_n'$ to N_0. On an absolute scale, the loss of individuals due to selection is $N_n' - N_{n+1}' = kN_n'$ in generation n, and altogether

$$(N_0' - N_1') + (N_1' - N_2') + \cdots = N_0',$$

in agreement with the decrease of the total population size from $N_0 + N_0'$ to N_0. In contrast, Haldane's D, which neglects the size change, can give much larger values; in particular, it can be larger than the number of a individuals ever born. Obviously, this understanding of the load produces severe artifacts.

Feller then discusses how Haldane's argument may be 'rescued'. One possibility is to keep population size constant by immigration from a reservoir population, in exactly the same proportions as in the current population under consideration. Then, Haldane's D gives the correct answer (but it must be kept in mind that the cost of selection is then borne by the reservoir population). The second possibility is to assume an ecological model rather than geometric growth. Feller speaks of μ depending on population size; maybe as in our Eq. (7), where μ is replaced by $\mu f(N_n + N_n')$. But more general models are also possible; for example, each type may be affected by competition in a different way. The genetic load would then be identified with the decrease in the stationary population size due to the reduced reproduction rate of one type. But now there is a wealth of possible models, and the genetic load would depend on the details. Specifying these is a task Feller assigns to the biologists.

3.2.4 Feller's criticism in the general context

With remarkable insight, Feller dissects a conceptual problem of his time: Load arguments can be inconsistent if they blindly rely on relative frequencies. His fellow researchers in biology do, however, not seem to have taken too much notice of his criticism. After all, as already mentioned in Section 3.2.1, a year after Feller's paper (and without citing it), Kimura did put forward his neutral theory of molecular evolution, to a large extent on the basis of load arguments [15]. More precisely, Kimura used an extension of Haldane's argument. He assumed a sequence of numerous genetic loci (rather than Haldane's and Feller's single locus), each of which can be of a favourable or a less favourable type, with fitness assumed as *multiplicative across loci*. As a consequence, there is a multitude of possible genotypes; fitness differences between individuals can become enormous; and the load (if calculated in Haldane's manner) can become astronomical (Kimura and Ohta [16] give a value of $D = 10^{78}$). Here the misconception lies in the assumption of multiplicativity over loci, which is completely unrealistic, but was hardly questioned at that time. For the details, see the insightful presentation in [9, Chapter 2.11].

Let us return to the original question that load arguments were supposed to answer: Can all the variation observed at the molecular level be explained by selection? Indeed, today, a large fraction of the molecular variation is considered selectively neutral or nearly so, although single mutations with spectacular selective effects are well known. But this insight is no longer built on load arguments – rather, the assumption of neutrality has proved extremely successful in describing patterns of genetic variation.

Last but not least, it should be noted that there is a lot of truth in Feller's general warning not to neglect population size in population genetics. Indeed, load arguments are not the only artifacts of this kind. Another example is the famous phenomenon of *Muller's ratchet*, which describes the *ad infinitum* accumulation of deleterious mutations due to stochastic effects in finite populations of constant size. If described in terms of an ecologically more realistic (and conceptually more correct) model with variable population size, the accumulation does not continue forever. Rather, when fitness has declined below a threshold value, the population experiences a *mutational meltdown*, which ultimately leads to extinction (see [1] for a review).

Acknowledgement. The interaction between mathematics and biology, which was greatly advanced by Feller, has a continuing impact on both fields. The authors gratefully acknowledge support from the Priority Programme *Probabilistic Structures in Evolution* (*SPP 1590*) funded by Deutsche Forschungsgemeinschaft (German Research Foundation, DFG).

References

All citations of the form [Feller 19*nn*], resp., [*Feller 19*nn*] (if the respective paper is not included in these Selecta) point to Feller's bibliography, pp. xxv–xxxiv.

[1] Baake, E. and Gabriel, W.: Biological evolution through mutation, selection, and drift: An introductory review. In: *Ann. Rev. Comput. Phys. Vol. VII*, ed. D. Stauffer, World Scientific, Singapore, 2000, 203–264.

[2] Berestycki, N.: Recent progress in coalescent theory. *Ensaios Matematicos* **16** (2009) 1–193.

[3] Cattiaux, P. and Méléard, S.: Competitive or weak cooperative stochastic Lotka–Volterra systems conditioned on non-extinction. *J. Math. Biol.* **60** (2010) 797–829.

[4] Dawson, D. A. and Perkins, E.: *Superprocesses at Saint Flour*. Springer, Berlin 2012.

[5] Durrett, R.: *Probability Models for DNA Sequence Evolution*, 2nd ed. Springer, New York 2008.

[6] Etheridge, A.: *An Introduction to Superprocesses*. American Mathematical Society, Providence (RI) 2000.

[7] Etheridge, A. and March, P.: A note on superprocesses. *Probab. Theory Relat. Fields* **89** (1991) 141–147.

[8] Ethier, S. and Kurtz, T.: *Markov Processes: Characterisation and Convergence*. Wiley, New York 1986.

[9] Ewens, W.: *Mathematical Population Genetics I. Theoretical Introduction*. Springer, New York 2004.

[10] Haldane, J. B. S.: The cost of natural selection. *J. Genet.* **57** (1957) 511–524.

[11] Harris, T.: First passage and recurrence distributions. *Trans. Am. Math. Soc.* **73** (1952) 471–486.

[12] Haccou, P., Jagers, P., and Vatutin, V. A.: *Branching Processes: Variation, Growth, and Extinction of Populations*. Cambridge University Press, Cambridge 2005.

[13] Harris, T.: Branching processes. *Ann. Math. Stat.* **19** (1948) 474–494.

[14] Khintchine, A.: *Asymptotische Gesetze der Wahrscheinlichkeitsrechnung*. Springer, Ergebnisse Math. Grenzgeb. Bd. 2, Teil 4, Berlin 1933.

[15] Kimura, M.: Evolutionary rate at the molecular level. *Nature* **217** (1968) 624–626.

[16] Kimura, M. and Ohta, T.: *Theoretical Aspects of Population Genetics*. Princeton University Press, Princeton 1971.

[17] Kolmogorov, A.: Über die analytischen Methoden in der Wahrscheinlichkeitsrechnung. *Math. Ann.* **49** (1931) 415–458. (An English Translation *On analytical methods in probability theory* is contained in the Selected Works of Kolmogorov [18, pp. 62–108])

[18] Kolmogorov, A. N. and Shiryayev, A. N. (Ed.): *Selected Works of A. N. Kolmogorov Volume II: Probability Theory and Mathematical Statistics*. Kluwer, Mathematics and Its Applications (Soviet Series) vol. 26, Dordrecht 1992.

[19] Kurtz, T. G.: Solutions of ordinary differential equations as limits of pure jump Markov processes. *J. Appl. Probab.* **7** (1970) 49–58.

[20] Lambert, A.: The branching process with logistic growth. *Ann. Appl. Probab.* **15** (2005) 1506–1535.

[21] Lamperti, J.: The limit of a sequence of branching processes. *Z. Wahrscheinlichkeitstheorie verw. Gebiete* **7** (1967) 271–288.

[22] Lamperti, J.: Continuous state branching processes. *Bull. Am. Math. Soc* **73** (1967) 382–386.

[23] Le, V., Pardoux, E. and Wakolbinger, A.: Trees under attack: a Ray–Knight representation of Feller's branching diffusion with logistic growth. *Probab. Theor. Relat. Fields* **155** (2013) 583–619.

[24] Nagylaki,T.: G. Malécot and the transition from classical to modern population genetics. *Genetics* **122** (1989) 253–268.

[25] Pardoux, E. and Wakolbinger, A.: From exploration paths to mass excursions – variations on a theme of Ray and Knight. In: Blath, J., Imkeller, P., Roelly, S.: *Surveys in Stochastic Processes*. European Mathematical Society, Zürich 2011, pp. 87–106.

[26] Perkins, E.: Conditional Dawson–Watanabe processes and Fleming–Viot processes. In: Cinlar, E., Chung. K. L. and Sharpe, M. J.: *Seminar on Stochastic Processes 1991, Progress in Probability 29*. Birkhäuser, Boston 1992, pp. 42–155.

[27] Perkins, E.: Dawson–Watanabe Superprocesses and Measure Valued Diffusion Processes. In: *Lectures on Probability Theory and Statistics, Ecole d'Eté de Probabilités de Saint Flour 1999*, Springer Lecture Notes in Mathematics 1781, Berlin 2002, pp. 125–324.

[28] Volterra, V.: *Leçons sur la théorie mathématique de la lutte pour la vie*. Gauthier–Villars, Paris 1931.

[29] Wakeley, J.: *Coalescent Theory: An Introduction*. Roberts & Company, Greenwood Village (CO) 2009.

[30] Watanabe, S.: A limit theorem of branching processes and continuous state branching processes. *J. Math. Kyoto Univ.* **8** (1968) 141–167.

[31] Watanabe, S.: On two dimensional Markov processes with branching property. *Trans. Am. Math. Soc.* **136** (1969) 447–466.

[32] Yamada, T. and Watanabe, S.: On the uniqueness of solutions of stochastic differential equations, *J. Math. Kyoto Univ.* **11** (1971) 155–167.

Prof. Ellen Baake
Universität Bielefeld
Technische Fakultät
Postfach 10 01 31
33501 Bielefeld, Germany
ebaake@techfak.uni-bielefeld.de

Prof. Anton Wakolbinger
Goethe-Universität Frankfurt
Institut für Mathematik
Robert-Mayer-Straße 10
60325 Frankfurt am Main, Germany
wakolbinger@math.uni-frankfurt.de

Feller on Differential Operators and Semi-groups

by *Niels Jacob* from Swansea

Feller's work on one-dimensional diffusions is one of his most influential contributions to modern probability theory, and the articles of M. Fukushima and G. Peskir are dealing with this part of his work. It is based on "purely" analytical considerations on parabolic differential equations and one-parameter operator semi-groups associated with these equations. In this article we want to have a closer look at the analysis behind his theory of one-dimensional diffusions and to put it into context with the contemporary analysis of that time.

First, however, we want to discuss two papers, [Feller 1930],[1] which is his "Habilitationsschrift" and, hence, *per se* of importance for his career, and [Feller 1952b], which deals with Riesz potentials and – in some anachronistic language – with subordination in the sense of Bochner and fractional powers of the Laplacian.

1

The paper [Feller 1930] is Feller's "Habilitationsschrift" [*Feller 1929] which he submitted to the University of Kiel. It must be seen in the context of extending results of potential theory to general (symmetric) elliptic partial differential equations of second order. Feller made the very interesting observation that for these equations the naturally associated Riemannian volume element can be used to derive representation formulae for the solutions to these equations which are completely analogous to those known for the Laplacian. Once the analogous form for Green's second identity is derived, it is possible to prove the mean-value theorem, hence the maximum principle, or representations for potentials. Furthermore, Harnack's theorems can be extended. For some results symmetry is not even needed.

In its time this paper was seen as a possible extension of potential theory, see J. Lense's short review [32] in *Jahrbuch über die Fortschritte der Mathematik*, but

[1] The references [Feller 19*nn*] and [*Feller 19*nn*] (the star indicating that the respective paper is not contained in these Selecta) refer to Feller's bibliography, while [*n*] points to the list of references at the end of this essay.

R.L. Schilling et al. (eds.), *Selected Papers II*,

it seems that this work was not taken up by contemporary authors. In fact, I could find only a short remark in C. Miranda [37], p. 54. Clearly, the breakthrough in the theory of elliptic partial differential operators of second order in that time was the contribution of J. Schauder. It is, however, a bit surprising that those researchers (for example, S. Bochner) who started to look more closely at Laplace–Beltrami operators never (to my knowledge) referred to this paper of Feller. Feller himself seemed never to have returned to this study. In fact, this paper looks quite isolated in the body of his work devoted to partial differential equations.

2

The paper [Feller 1952b] is dedicated to Marcel Riesz and indeed picks up a topic close to M. Riesz, the Riemann–Liouville integrals and what is now called Riesz potentials. However, Feller put these potentials in quite a different, new context and, in fact, established the connection to what we call nowadays *subordination in the sense of Bochner*. Denote by J_+^α and J_-^α the one-sided Riemann–Liouville integrals, and by I^α the Riesz potential of order α; recall that $\alpha = 2$ is the Newton potential. From Riesz's work [43] it has been known that $(I^\alpha)_{\alpha>0}$ enjoys the semi-group property, $I^\alpha \circ I^\beta = I^{\alpha+\beta}$, and that relations such as

$$\Delta I^{\alpha+2} f = -I^\alpha f \quad \text{or} \quad I^{-2k} f = (-1)^k \Delta^k f$$

(Δ being the Laplacian) hold for suitable values of α, smooth functions f and $k \in \mathbb{N}$. In [2] S. Bochner introduced (in our modern language) subordination, and linked symmetric stable distributions to fractional powers of the Laplacian.

Feller introduces now the two-parameter family of operators

$$I_\delta^\alpha f(x) = \frac{1}{\Gamma(\alpha)\sin(\alpha\pi)} \int_{-\infty}^{\infty} f(y)|y-x|^{\alpha-1} \cdot \sin\left[\alpha(\tfrac{\pi}{2} + \delta\,\mathrm{sgn}(y-x))\right] dy$$

for $\delta \in \mathbb{R}$, $0 < \alpha < 1$ and f smooth enough to make $I_\delta^\alpha f$ meaningful. He first derives relations such as

$$I_\delta^\alpha f = \frac{\sin(\alpha(\delta + \frac{\pi}{2}))}{\sin \alpha\pi} J_+^\alpha f + \frac{\sin(\alpha(-\delta + \frac{\pi}{2}))}{\sin \alpha\pi} J_-^\alpha,$$

$$I_0^\alpha = I^\alpha, \quad I_{\frac{\pi}{2}}^\alpha = J_+^\alpha, \quad I_{-\frac{\pi}{2}}^\alpha = J_-^\alpha,$$

and further

$$I_\delta^0 = \mathrm{id}, \quad I_\delta^{-2} = -\frac{d^2}{dx^2}, \quad I_\delta^{\alpha+\beta} = I_\delta^\alpha \circ I_\delta^\beta,$$

etc. But some care is needed with respect to domains. Now he returns to the equation

$$\frac{\partial}{\partial t} u(t,x) = -I_\delta^{-\alpha} u(t,x), \quad u(0,x) = f(x),\, t > 0,\, x \in \mathbb{R},$$

and seeks a formal solution in the form

$$u(t,x) = \sum_{n=0}^{\infty} \frac{(-t)^n}{n!} I_\delta^{-n} f(x)$$

for a suitable class of functions f, and derives the integral kernel for the solution operator as

$$U_{\alpha,\delta}(x) = \frac{1}{\pi} \operatorname{Re} \int_0^\infty \exp\left\{-ir|x| - tr^\alpha e^{-i\alpha\delta \operatorname{sgn}(x)}\right\} dr.$$

For $\delta = 0$ this is of course the semi-group corresponding to the generator

$$I_0^{-\alpha} = -\left(-\frac{d^2}{dx^2}\right)^\alpha,$$

and hence Bochner's result is covered. Moreover, the power series expansion of $U_{\alpha,\delta}$ allows to derive expansions for (symmetric) stable distributions.

Feller had to work in the frame of classical complex analysis in order to deal with all these "*divergent*" objects. Often he refers to "*formal solutions*" and "*formal calculations*". The paper is very clear in its ideas, but it seems that in analysis it was not much taken up, neither it is quoted in the "bible" on Riesz potentials, the monograph of N. S. Landkof [30], nor in monographs on singular integral operators or pseudo-differential operators. However, the paper had its impact in probability theory, especially the idea of handling stable distributions with the help of the kernel $U_{\alpha,\delta}$ was taken up, see for example V. M. Zolotarev [57].

3

Now we turn to the main topic of this article, the analytical part of Feller's work leading to his theory of one-dimensional diffusions. Of course this distinction is a bit artificial, but helpful as we believe. As a prelude to Feller's work on parabolic equations, we want to mention briefly the paper [Feller 1936c]. Here Feller picks up Kolmogorov's seminal paper [27], and provides solutions for the corresponding parabolic equations (as well as their formal adjoints, i. e. forward and backward equations) under suitable conditions on the coefficients. He can rely on E. E. Levi's work [33] for parametrix-constructions, and he approaches the problem by constructing and investigating fundamental solutions. The paper is well received and, of course, it was written before semi-groups had been introduced.

When looking at the papers he started to publish in the early 1950s, perhaps it is sharpening our view when starting with a surprising observation: Seminal papers such as [Feller 1952a] *The parabolic differential equations and the associated semi-groups of transformation*, or [Feller 1957a] *Second order differential operators and their lateral conditions*, just to mention two, had essentially no impact on the partial differential equations community (PDE community) of that time. Avner Friedman's book [16] *Partial Differential Equations of Parabolic Type* published 1964 – which was most influential to anyone (in the West) working on parabolic partial differential equations – does not quote any of Feller's work, the same applies to A. Friedman [17] which contains a long chapter on semi-groups and parabolic equations, and the "eastern" counterpart to Friedman's book, the monograph [12] of S. D. Eidelman (Russian publication in 1964, English translation in 1969). O. A. Ladyzhenskaya, V. A. Solonnikov, and N. N. Uraltseva [29] mention only [Feller 1936c], and in J. R. Cannon's *Encyclopedia of Mathematics and its Applications* volume on the one-dimensional heat equation [6] we find again a reference to [Feller 1936c], but the more than 680 items in the bibliography covering the period of 1950–1973 do not contain a single item of Feller, despite the fact that 270 items are listed under *initial boundary value problems*; no mentioning either in: L. Hörmander [26], J.-L. Lions and E. Magenes [34], and none of the 9 volumes on partial differential equations in *Encyclopaedia of Mathematical Sciences* makes any reference to Feller, not even the volume on elliptic and parabolic equations, [19].

A hint for this surprising neglect is maybe given in H. O. Fattorini [14, p. 168]:

> [...] Roughly along the same lines Yosida [55] and Hille [21] considered a general non-self adjoint A_0 in relation to applications to unidimensional Markov process. The problem is that of obtaining extensions of the operator A_0 in L^1 and in spaces of continuous functions by means of boundary conditions at finite points and at infinity; these extensions are required to be infinitesimal generators of positive semigroups. The work of Feller [Feller 1954a], [Feller 1954b], [Feller 1955b] and [*Feller 1956d] and other proceed in the same direction but use **extremely weak assumptions on the operator** [my bold].

Feller needed these "*extremely weak assumptions*" because his aim was to find the "*characteristics*" of stochastic processes related to semi-groups, more precisely to semi-groups in spaces of continuous functions and dual objects, i. e. measures (which, however, could be related to functions of bounded variation since he was working in one dimension). This probabilistic need took his contributions more or less out of the view of the PDE community, who at that time just progressed in different directions.

More different was the perception in parts of the functional analysis community and, of course, among probabilists. But even in the semi-group community he had to convince colleagues that results for generators beyond the Hille–Yosida theory are needed.

4

The paper [Feller 1952a] is our starting point, more precisely it was Feller's starting point for *a systematic research* (cf. [56, p. 398]) using semi-group theory to understand Markov processes (with local generators). In the following we use, in general, our modern terminology to describe Feller's work, the reader of the papers will have no problem to switch from one to the other.

Consider the parabolic differential equation

$$(1)\qquad \frac{\partial}{\partial t}u(t,x) = a(x)\frac{\partial^2}{\partial x^2}u(t,x) + b(x)\frac{\partial}{\partial x}u(t,x) + c(x)u(t,x) = Au(t,x),$$

and its formal adjoint

$$(2)\qquad \frac{\partial}{\partial t}v(t,y) = \frac{\partial}{\partial y}\left(\frac{\partial}{\partial y}(a(y)v(t,y)) - b(y)v(t,y)\right) + c(y)v(t,y) = A^*v(t,y),$$

where $t > 0$ and $-\infty \leq r_1 < x, y < r_2 \leq \infty$, and a, b, c are continuous functions on $[r_1, r_2]$. Moreover, subject (1) and (2) to initial conditions, i.e. $u(0,x) = f(x)$ and $v(0,y) = g(y)$. Feller wants to represent the corresponding solutions to (1) and (2) with the help of semi-groups

$$u(t,x) = T_t f(x) \quad \text{and} \quad v(t,y) = T_t^* g(y)$$

where $(T_t)_{t \geq 0}$ and $(T_t^*)_{t \geq 0}$ are in "duality". The problem is, of course, triggered by the work of K. Yosida and E. Hille, compare also their comments on Feller's work in [56] and [22]. Although it is not apparent in this paper, Feller clearly has the construction of Markov processes in mind. With this background he immediately spots a few problems to overcome: Initial data are sought to come from $C([r_1, r_2])$ and $L^1((r_1, r_2))$, respectively, and solutions to (1), (2) shall lead to transition densities or transition functions, but $C([r_1, r_2])$—$L^1((r_1, r_2))$ is not a proper duality pairing; or transition densities or functions need not be good enough to lead to strong continuity at 0; etc. These problems he will pick up in later papers. Here he handles mainly A and A^*, and the programme: Given A, find all semi-groups (with specified properties) whose generator is a restriction (his wording is *contraction*) of A and find the "true" adjoint problem.

He starts with fixing the spaces: $C([r_1, r_2])$ for A and $BV([r_1, r_2])$ (functions of bounded variation) for A^*, but eventually he considers L^1 as embedded into BV, which allows him to return to L^1. The following definition [Feller 1952a, p. 478] is central (modern language used):

> Let $(A, D(A))$ be a linear operator in a Banach space X and $\Sigma \subset D(A)$ a subspace. If the restriction of A to Σ is the infinitesimal generator of strongly continuous contraction semi-group $(T_t)_{t \geq 0}$ on X, then we say that A, *restricted by the lateral condition* Σ, generates $(T_t)_{t \geq 0}$.

Of course $(A|_\Sigma, \Sigma)$ must be a closed operator, but we want to emphasize that Σ "contains" also uniqueness conditions, for example boundary conditions. Now the paper starts to investigate A under the assumption that a, a', b and c are continuous in (r_1, r_2) and $a > 0$. In two steps the problem is reduced to the study of the operator

$$\Omega = \frac{d^2}{dx^2} + b(x)\frac{d}{dx},$$

the important step to arrive at $a = 1$ is to introduce as new variable

$$\xi(x) := \int_0^x \frac{1}{\sqrt{a(s)}}\, ds.$$

Note that this transformation was already used in [Feller 1936c] in order to find the zero-order approximation of the parametrix. Next, the equation $\lambda u - \Omega u = 0$ is considered; of course, this is done with the Hille–Yosida theorem in mind, and here Feller can rely on a well-known theory for eigenvalue problems for second order ordinary differential equations. Then he switches to the adjoint equation $\lambda v - \Omega^* v = 0$ with the formal adjoint

$$\Omega^* = \frac{d}{dx}\left(\frac{d}{dx} - b(x)\right).$$

To proceed further, we need, of course, boundary conditions, and here the famous boundary point classification has its origin. The important observation is that the properties of $B(x) := \int_0^x b(s)\,ds$ determine the behaviour (regular, exit, entrance, natural). Thus, lateral conditions might be given by boundary conditions.

Now Green's functions are constructed, the inhomogeneous equations are investigated, and so are properties of solutions. Eventually, a very detailed study of the corresponding semi-groups (the adjoint is always included, and Feller is very carefully making a distinction between the formal adjoint A^* of A, and the generator of the adjoint semi-group) in dependence of the boundary conditions follows. We see here essentially the whole analytical part of Feller's theory, however still under severe regularity restrictions on a and b, a certain asymmetry between A and A^*, and the conditions of the semi-group being positivity preserving are not yet discussed.

Using semi-group theory, the problem to solve these partial differential equation is reduced to solving eigenvalue problems for certain ordinary differential equations, this is an approach depending on the space dimension, and this might have further contributed to not having much impact in the PDE community of that period.

5

The paper [Feller 1952a] treats "general" semi-groups generated by a second order ordinary differential operator (and the adjoint problem) within the Hille–Yosida setting. However, Feller is coming from a different background: He wants to use the theory of semi-groups to handle stochastic processes in relation to Kolmogorov's results. For this he has to overcome some problems:

- When starting with the operator A as in (1) having continuous coefficients, then the formal adjoint operator A^* in (2) need not be defined due to a lack of regularity of the coefficients.
- The requirement of strong continuity at $t = 0$ can cause problems.
- When are semi-groups generated by A (or A^*) positivity preserving?

In a series of papers Feller addresses these problems. Remarkable and full of ideas and insights is already [Feller 1952e]. Feller starts with a contraction semi-group $(T_t)_{t\geq 0}$ on $C([r_1, r_2])$, and assumes weak measurability of $t \mapsto T_t f$ which implies strong continuity of $(T_t)_{t>0}$ but not strong continuity of $(T_t)_{t\geq 0}$, i. e. strong continuity at 0. Now he asks for conditions for $(T_t)_{t\geq 0}$ being positivity preserving when the generator is a local operator. Today, we would define an operator to be local if it decreases the support, in the sense that $\operatorname{supp} Au \subset \operatorname{supp} u$. Feller uses an approximation scheme for A to define the notion of a local generator: A is said to be *local* if for every $x \in [r_1, r_2]$ and $f \in C([r_1, r_2])$ such that $f(y) = 0$ in some interval $|x - y| < \delta$ it follows that

$$\lim_{t\to 0} \frac{T_t f(x) - f(x)}{t} = 0.$$

(This definition is influenced by the Lindeberg condition discussed in [Feller 1936c].) Local operators in this sense need not be differential operators (as understood in his times). However, adding that for $f \in D(A)$ it follows that f' and f'' exist will imply that A is a second order differential operator. Next, regular and singular points for

$$A = a(x)\frac{\partial^2}{\partial x^2} + b(x)\frac{\partial}{\partial x} + c(x)$$

were introduced, and an example shows that, at singular points, A need not be of the type as above, i. e. of the form (1). The example depends on a variable change which does not preserve regularity. The main result is that if $(T_t)_{t\geq 0}$ is positivity preserving, if A is local and $f \in D(A)$ satisfies differentiability conditions, then at regular points A is of the form (1) with $a \geq 0$ and $c \leq 0$. Feller, again, turns to the adjoint problem and explains why the theory is not yet sufficient to treat the "forward equation" where the operator shall act on transition functions, i. e. set functions $p(t,x;A)$, while for the "backward equation", where the operator acts on densities $p(t,x,y)$, the theory makes already sense. In [50] A. D. Ventcel pointed to an error in the proof of the main result in [Feller 1952a] which Feller is addressing in [Feller 1957a].

In the paper [Feller 1953a] he investigates semi-groups $(T_t)_{t\geq 0}$ on a Banach space X which are not strongly continuous at $t = 0$, i. e. it is only assumed that for $t > 0$ the mapping $t \mapsto T_t f$ is weakly measurable. Feller develops a quite general theory by introducing weak measurability with respect to a subspace of X^*. The paper provides him with tools for investigations to follow, but this full generality is not often taken up. However [Feller 1953b] is a key contribution. Independently of J. Miyadera [38] and R. S. Phillips [41], he established what is nowadays considered as the general form of the Hille–Yosida theorem.

With these results on the general theory of operator semi-groups he returns in the article [Feller 1954a] to positivity preserving semi-groups with local generators. This paper introduces the definition of "*diffusion processes by the property that their infinitesimal generator is of local character*", and by making some regularity assumptions, they are seen to be differential operators. In the paper he links also the continuity of sample paths to the locality of the generator. Moreover, here for the first time (to my knowledge) a minimum principle is used to characterize generators of positivity preserving semi-groups (Lemma 2, p. 425).

Feller further introduces the transformation $f \mapsto \varphi f$ on $C([r_1, r_2])$ with a suitable function φ, and considers the corresponding semi-group

$$T_t^* f := \frac{1}{\varphi} T_t(\varphi f),$$

and its corresponding generator A^*. (Here it is worth staying with Feller's original notation, but note that "$*$" does not denote the adjoint object in this context.) It is shown that properties of $(T_t)_{t\geq 0}$ and A (such as positivity preserving, strong continuity, locality) hold for the new semigroup. This transformation becomes now key in a step towards a representation of A which allows a better treatment of the adjoint, and eventually leads to "Feller's theory" of generalized differential operators.

Null points are introduced as points $x_0 \in (r_1, r_2)$ where $T_t f(x_0) = 0$ for all functions $f \in C([r_1, r_2])$ and $t > 0$, and they are used to study $(T_t^*)_{t\geq 0}$ and A^*, in particular A^*1. Now two sets of "*singular points*" are defined:

absorption points	$Af(x_0) = cf(x_0)$;
right [resp. left] translation points	$A(f(x)g(x_0) - f(x_0)g(x)) > 0$ in $\lvert x - x_0 \rvert < \varepsilon$ implies locally that f/g is strictly increasing [resp. decreasing].

Next, for a monotone function v define

$$(3) \qquad D_v f(x) := \begin{cases} \dfrac{f(x+) - f(x-)}{v(x+) - v(x-)}, & \text{at points of discontinuity of } v. \\[2ex] \lim\limits_{h \to 0} \dfrac{f(x+h) - f(x)}{v(x+h) - v(x)}, & \text{elsewhere.} \end{cases}$$

A main result in the paper states that if A generates a diffusion process and $I \subset [r_1, r_2]$ is an open interval with no absorption and translation points, then there exist a strictly increasing function v, a continuous function φ and a choice of the dependent variable x such that

$$Af = \varphi D_v \left(\frac{f}{\varphi} \right)' + cf.$$

(For simplicity, we have worked with $D(A)$ rather than with local domains $D(A, x_0)$.)

This is a big step towards the "Feller form" of the generators of one-dimensional diffusions. In [Feller 1954b] this result is taken up and discussed from a more probabilistic point of view. We refer to the contributions of M. Fukushima and G. Peskir in these Selecta.

6

In the papers [Feller 1955a], [Feller 1955b] and [*Feller 1956d], Feller approaches linear second order ordinary differential operators with continuous coefficients "only" from the differential operator point of view, i. e. no connection to probability theory or to operator semi-groups is made. In these papers certain notions are clarified and some proofs are simplified, most of all he wants to show the "universality" of his theory.

The topic is to handle

$$(4) \qquad Af(x) = a(x) \frac{d^2}{dx^2} f(x) + b(x) \frac{d}{dx} f(x) + c(x) f(x)$$

with continuous coefficients and $a > 0, c \leq 0$. The basic assumption is that the operators under consideration are local which finally in [Feller 1955b] is defined close to what we would expect nowadays: If f vanishes in a neighbourhood of x_0, then $Af(x_0) = 0$. (Clearly he has to modify his earlier definition of locality since no generic semi-group can be used.) Further, he assumes the *minimum principle*: If $f(x_0) = 0$ and $f(x_0)$ is a minimum of f, then $Af(x_0) \geq 0$. These two assumptions will determine the structure of A. (We do not make here a distinction between A defined on a local domain $D(A,x_0)$ and the "full" domain $D(A)$, the locality of A allows such a shortcut.)

In [Feller 1955a] the aim is to reduce A to the standard (new, generalized) form

$$\Omega f = D_v D_u f, \tag{5}$$

where D_v and D_u are derivatives with respect to monotone increasing functions, and u is also continuous, compare (3) for the definition of D_v and D_u.

Before we have a closer look at this problem, we want to discuss Feller's philosophy. For Feller an operator of type

$$\frac{d^2}{dx^2}(af), \qquad a \text{ continuous and } f \text{ twice differentiable,} \tag{6}$$

is not a well-defined object, in particular, not a differential operator. However such an object will show up when looking at the formal adjoint of $a\frac{d^2}{dx^2}f$. The lack of differentiability of a is the main problem, and Feller's aim is to find a change of variable (in a certain sense) to get rid of this regularity problem.

We must note here that this approach is, of course, strictly limited to one dimension. Moreover, within the theory of L. Schwartz's distributions, there is no problem: If a is continuous and f twice (continuously) differentiable, then af is a distribution, hence arbitrarily often differentiable in the sense of distributions. Thus, when introducing the correct duality pairing respecting the distributional orders of a and f, respectively, we have no problem with the definition of (6). However, Feller's approach avoids distribution theory and by this he obtains much deeper insights into properties of (4). (Calling the derivatives D_u or D_v *generalized* derivatives makes sense, but of course already in that time the general understanding of a *generalized* derivative was a different one.)

In order to find suitable changes of variables we need, as before, the notion of regular and irregular points with respect to A. In an interval consisting only of regular points x_0, the transformations are

$$u(x) = \int_{x_0}^{x} \exp\left(-\int_{x_0}^{t} \frac{b(s)}{a(s)}\,ds\right) dt$$

and

$$v(x) = \int_{x_0}^{x} \frac{1}{a(t)} \exp\left(\int_{x_0}^{t} \frac{b(s)}{a(s)}\,ds\right) dt.$$

In addition, the general form of all transformations is derived, and we understand that the choice $u(x) \equiv x$ is, in principle, possible. Thus, for a regular interval A can be reduced to Ω, and in case of irregular points x_0, certain one-sided derivatives must be included, whose type depends on the type of irregularity of x_0. Given A in the reduced form (5), i.e. Ω, it is possible to find the formal adjoint to be $D_u D_v$, and an intrinsic scalar product and Hilbert space by introducing the norm

$$\|f\|^2 = \int f^2(x)\,dv(x).$$

Moreover, many points of the "standard " theory of ordinary differential equations (Green's function, Wronski determinant, etc.) carry over to the new representation.

In [Feller 1955b] a type of overview is given including a discussion of the problem when the Cauchy problem $u_t = Au$, $u(0,x) = f(x)$, leads to positivity preserving semi-groups. It seems that here, for the first time, a characterization using the minimum principle is explicitly given (in case of local operators).

Both, [Feller 1955a] and [Feller 1955b] include some examples and they illustrate well which type of problem Feller wants to cover: Coefficients with jump discontinuities. Furthermore, he takes up a problem by W. F. Bauer [1], and deals with an extended Sturm–Liouville problem where a second order derivative is given at one boundary point.

A short remark on [*Feller 1956d]. Of course, Sturm–Liouville operators are central objects in spectral theory, and it is natural to expect Feller to study Sturm–Liouville problems for his reduced operators, and to try to reduce given "*classical*" Sturm–Liouville operators to his generalized type. He first derives an "*intrinsic form*" of Ω for which he needs essentially to assume the existence of a solution with no zeroes, and then he discusses spectral properties following the Weyl theory.

7

The influence of *Generalized Second Order Differential Operators and Their Lateral Conditions*, [Feller 1957a], can hardly be overstated when having in mind its impact in the development of the theory of (one-dimensional) diffusion processes and beyond. The reason is that here Feller eventually gives an almost final form for the boundary and lateral conditions that we may add to a (reduced or generalized) operator in order to associate a stochastic process with the operator in the sense of the operator being a generator of a positivity preserving semi-group. In his contribution to these Selecta, M. Fukushima gives a detailed discussion of these issues and their impact. Again, we restrict ourselves to the "pure" analytic content of [Feller 1957a]. The theory so far published by Feller is first of all polished, notational inconsistencies are abolished (see the remark to Ventcel's paper in Section 5), and central ideas are clarified. Starting with $A = a\frac{d^2}{dx^2} + b\frac{d}{dx}$ a first reduction to $A = \alpha\frac{d^2}{dy^2}$ is introduced and then, using the change of variable,

$$m(x) = \int_{x_0}^{x} \frac{1}{\alpha(t)}\,dt,$$

he arrives at the canonical form

$$A = D_m D_x.$$

For certain considerations, using a solution ψ to $A\psi + c\psi = 0$, the operator

$$A_\psi f = \frac{1}{\psi} D_m \left(\psi^2 D_x \frac{f}{\psi}\right)$$

is introduced to treat operators of type $Af + cf$. All these "reductions" and their assumptions are now systematically handled and, in particular, the interpretation of

m as speed measure turns up, see again the essay of M. Fukushima.

From the analytical point of view the first interesting feature of this paper is the discussion of $A = D_m D_x$ in the Hilbert space H_m induced by the norm

$$\|f\|_m^2 := \int_{\beta_1+}^{\beta_2-} |f'|^2 \, dm \tag{7}$$

where (β_1, β_2) is an interval in which $A = D_m D_x$ holds. An extension is given to the case $[\beta_1, \beta_2]$ with the new norm

$$|||f|||_m^2 = \|f\|_m^2 + m_1 f^2(\beta_1) + m_2 f^2(\beta_2), \tag{8}$$

so while (7) incorporates zero boundary data, in (8) non-zero boundary data are taken into account. In the Hilbert space setting self-adjointness is discussed, as well as spectral properties (Weyl and Stone–Weyl theory). These spectral results (limit circle vs. limit point dichotomy) are important for the endpoint classification. But Feller is again probabilist and influenced by Kolmogorov. He needs or wants to consider the operator $A = D_m D_x$ in a space of bounded continuous functions, hence the adjoint will act on measures. In this context he now asks which boundary and lateral condition can be assumed so that the local operator A, satisfying the minimum principle, is associated with a positivity preserving semi-group, hence a stochastic process. This is the central content of Section 12 and 13 of [Feller 1957a] and discussed in the contributions by M. Fukushima and G. Peskir.

The paper [Feller 1958] has, in Feller's own words,

> the purpose [...] to present an improved derivation of the canonical forms of operators with the properties (1) [locality], (2) [non-triviality of A], (3) [strong minimum principle] and to provide additional material on various points.

A short remark on [Feller 1959d]. Here Feller switches from the minimum principle to the *positive maximum principle*: If f attains a positive maximum at x_0, then $Af(x_0) \leq 0$. First he characterizes local operators in one dimension satisfying the positive maximum principle, and secondly he shows (Theorem 2) that a (local) operator generating a strongly continuous positivity preserving contraction semi-group on a space of continuous functions must satisfy the positive maximum principle.

8

We want to indicate some immediate impact or non-impact of Feller's contributions discussed here on the development of analysis. A class of examples in which Feller is rather interested are operators where a has a jump discontinuity. In physics such operators show up when media of different properties, for example heat conductivity, are in contact. On the contact surface (or point) additional matching conditions must be added (transmission conditions, for example). Feller reduces such operators (in one dimension) to his canonical form, and doing so he can avoid these matching conditions. In [*Feller 1959a] a vibrating string is discussed within his framework. It seems that applied mathematicians never took up his approach. Maybe, to change with every problem the notion of differentiability is a price too high for concrete applications.

The impact in probability theory or at the interface of probability theory and analysis is enormous. In the article of M. Fukushima we can learn about these developments following Feller's work, especially in relation to one-dimensional diffusion theory with many contributions of K. Itô and H. P. McKean. Of course, M. Fukushima himself gave central contributions extending Feller's boundary theory to diffusions in higher dimensions, culminating in the modern theory of Dirichlet forms, see [18]. Much more about this part of Feller's impact, i. e. to the theory of Markov processes and their boundary theory can be found in Z.-Q. Chen and M. Fukushima [7].

Feller himself never referred in his original papers to the contributions of E. B. Dynkin who, since 1955, started to publish on Markov processes, their generators and associated semi-groups. (However, Feller give Dynkin a brief mentioning in his ICM paper [Feller 1960].) In the monograph [10] E. B. Dynkin himself gives fair credits to Feller's work and compares the different approaches. Further, Dynkin's contributions are thoroughly discussed in his *Selected Papers* [11], in particular by P. A. Meyer [36].

David Williams, in his groundbreaking monograph (jointly with L. C. G. Rogers) [44] suggests to call strongly continuous, positivity preserving contraction semi-groups in the space of *continuous function vanishing at infinity* Feller–Dynkin semi-groups. A well justified name, but unfortunately this suggestion has not been taken up by the community.

Three "pure" analytical developments need a discussion: The general structure of operators satisfying the positive maximum principle, the general boundary or lateral conditions satisfied by a generator of a Feller–Dynkin semi-group, and Krein–Feller differential operators.

We recall that K. Yosida in [55] already derived for generators of positivity preserving semi-groups on L^1 some integro-differential expression. J. Neveu in his thesis, see [39] and [40], derived for positivity preserving semi-groups on $C(\mathbb{R}^n)$ the nowadays well-known Lévy-type description of the corresponding generator A assuming that $(A\exp(it\cdot))(t,x)$ is well-defined and has certain continuity properties. This work is independent of Feller's. Later, however, Ph. Courrège [8] and [9] combined Feller's work and Neveu's ideas to give a complete characterisation of operators satisfying the positive maximum principle. His work was influenced by earlier work by W. von Waldenfels [53]-[54]. Courrège derives the integro-differential form, i. e. the form of a Lévy-type operator, but using the Lévy–Khinchine formula, he also gives a pseudo-differential operator representation.

Using this pseudo-differential operator representation it is possible on the state space $\mathbb{R}^n$ to construct concrete semi-groups, hence stochastic processes, and to investigate them using properties of the symbol of the generator. Note that this symbol is the natural generalization of the characteristic exponent of a Lévy process. For details we refer to W. Hoh [23]-[25], R. L. Schilling [46]-[49] as well as [5].

We learnt already from Ventcel's paper [50], see also [51] and [52], about the general form of boundary conditions for generators of Feller–Dynkin semi-groups. For years to come the definitive paper in this direction of pure analytical content was [4] written by J.-M. Bony, Ph. Courrège and P. Priouret. On the probabilistic side we must mention the contribution of K. Sato and T. Ueno [45]. It seems only when these approaches were combined with the theory of elliptic pseudo-differential operators, further progress was made. However, it is still an open problem and a challenge to state and to solve for a Lévy-type operator the general Feller-Ventcel boundary value problem and to construct the corresponding process from analytic data.

In one dimension we have a different situation. In the 1950s Rolf Nevanlinna started to investigate indefinite metrics (associated with a bilinear form) on Hilbert spaces and suggested to study differential operators in this setting, we refer to the survey [35] and the monograph [3] by J. Bognar. Nevanlinna's interest was stimulated by the work of the physicist K. Bleuler on photons. Earlier L. S. Pontryagin [42] studied operators in spaces with indefinite metrics influenced by problems in quantum physics. It was M. G. Krein who established in the late 1940s and early 1950s what is now known as theory of *Krein spaces*, see again [3] for details. In I. C. Gohberg and M. G. Krein [20] we can find much about extensions of operators in an L^2-setting related to Krein spaces. H. Langer in the 1960s contributed substantially to the theory of operators in Krein spaces, and in the early 1970s he started to establish a connection to Feller's theory. A highly praised contribution is [31] where the existence of a semi-group, hence of a Markov process, associated with $D_m D_x$ under general (non-local) Feller-Ventcel conditions was proved. Since then, Krein–Feller differential operators (or diffusions) form a special subject in this part of analysis, a more recent discussion can be found in A. Fleige [15].

References

All citations of the form [Feller 19*nn*], resp., [*Feller 19*nn*] (if the respective paper is not included in these Selecta) point to Feller's bibliography, pp. xxv–xxxiv.

[1] Bauer, W. F.: Modified Sturm–Liouville systems. *Quart. Appl. Math.* **11** (1953) 273–284.

[2] Bochner, S.: Diffusion equation and stochastic processes. *Proc. Nat. Acad. Sci. USA* **35** (1949) 368–370.

[3] Bognar, J.: *Indefinite Inner Product Spaces*. Ergebnisse der Mathematik und ihrer Grenzgebiete **78**. Springer, Berlin 1974.

[4] Bony, J.-M., Courrège, Ph. and Priouret, P.: Semi-groupes de Feller sur une variété à bord compacte et problèmes aux limites intégro-différentiels du second ordre donnant lieu au principe du maximum. *Ann. Inst. Fourier Grenoble* **18.2** (1968) 369–521.

[5] Böttcher, B., Schilling, R. L. and Wang, J.: *Lévy-type Processes: Construction, Approximation and Sample Path Properties*. Lecture Notes in Mathematics Vol. 2099. Springer, Berlin 2014.

[6] Cannon, J. R.: *The One-Dimensional Heat Equation*. Encyclopedia of Mathematics and its Applications **23**. Addison-Wesley Publishing Company, Menlo Park (CA) 1984.

[7] Chen, Z.-Q. and Fukushima, M.: *Symmetric Markov Processes, Time Change, and Boundary Theory*. London Mathematical Society Monographs **35**. Princeton University Press, Princeton (NJ) 2012.

[8] Courrège, Ph.: Sur la forme intégro-différentielle des opérateurs de C_K^∞ dans C satisfaisant au principe du maximum. *Séminaire Brelot-Choquet-Deny (Théorie du Potentiel)*. 10e année (1965/66) n^o2, 2-01–2-38.

[9] Courrège, Ph.: Sur la forme intégro-différentielle du générateur infinitésimal d'un semi-groupe de Feller sur une variété. *Séminaire Brelot-Choquet-Deny (Theorie du Potentiel)*. 10e année (1965/66) n^o3, 3-01–3-48.

[10] Dynkin, E. B.: *Markov Processes. Vol I & 2*. Grundlehren der mathematischen Wissenschaften **121**, **122**. Springer, Berlin 1965.

[11] Dynkin, E. B., Yuskevich, A. A., Seitz, G. M., Onishchuk, A. L. (eds.): *Selected Papers of E. B. Dynkin With Commentary*. American Mathematical Society, Providence (RI) 2000.

[12] Eidel'man, S. D.: *Parabolic Systems*. North-Holland Publishing Company, Amsterdam & Wolters-Noordhoff Publishing, Groningen 1969.

[13] Ejdel'man, S. D.: Parabolic Equations. In: Egorov, Yu. V. and Shubin, M. A. (eds.): *Partial Differential Equations VI*. Encyclopedia of Mathematical Sciences **63**, Springer, Berlin 1994, pp. 203–316.

[14] Fattorini, H. O.: *The Cauchy Problem*. Encyclopedia of Mathematics and its Applications **18**. Addison-Wesley Publishing Company, Reading (MA) 1983.

[15] Fleige, A.: *Spectral Theory of Indefinite Krein–Feller Differential Operators*. Mathematical Research **98**. Akademie Verlag, Berlin 1996.

[16] Friedman, A.: *Partial Differential Equations of Parabolic Type*. Prentice-Hall, Englewood Cliffs (NJ) 1964.

[17] Friedman, A.: *Partial Differential Equations*. Holt, Rinehart and Winston, New York (NY) 1969.

[18] Fukushima, M.: *Dirichlet Forms and Markov Processes*. North-Holland Mathematical Library **23**. North-Holland, Amsterdam & Kodansha, Tokyo 1980.

[19] Gamkrelidze, R. V.: *Encyclopaedia of Mathematical Sciences — Partial Differential Equations I–IX*. Springer, Berlin 1992–97.

[20] Gohberg, J. C. and Krein, M. G.: *Introduction to the Theory of Linear Non-Selfadjoint Operators in Hilbert Space*. Translations of Mathematical Monographs **18**. American Mathematical Society, Providence (RI) 1969.

[21] Hille, E.: The abstract Cauchy problem and Cauchy's problem for parabolic differential equations. *J. Analyse Math.* **3** (1954) 81–196.

[22] Hille, E. and Phillips, R. S.: *Functional Analysis and Semigroups.* American Mathematical Society Colloquium Publications **23**, American Mathematical Society, Providence (RI) 1957.

[23] Hoh, W.: The martingale problem for a class of pseudo differential operators. *Math. Ann.* **300** (1994) 121–147.

[24] Hoh, W.: A symbolic calculus for pseudo differential operators generating Feller semi-groups. *Osaka J. Math.* **35** (1998) 798–821.

[25] Hoh, W.: *Pseudo Differential Operators Generating Markov Processes.* Habilitationsschrift, Universität Bielefeld, Bielefeld 1998. www.math.uni-bielefeld.de/~hoh/temp/pdo_mp.pdf

[26] Hörmander, L.: *Linear Partial Differential Operators.* Grundlehren der mathematischen Wissenschaften **116**, Springer, Berlin 1963.

[27] Kolmogorov, A.: Über die analytischen Methoden in der Wahrscheinlichkeitsrechnung. *Math. Ann.* **49** (1931) 415–458. (An English Translation *On analytical methods in probability theory* is contained in the Selected Works of Kolmogorov [28, pp. 62–108])

[28] Kolmogorov, A. N. and Shiryayev, A. N. (ed.): *Selected Works of A. N. Kolmogorov Volume II: Probability Theory and Mathematical Statistics.* Kluwer, Mathematics and Its Applications (Soviet Series) vol. 26, Dordrecht 1992.

[29] Ladyzhenskaya, O. A., Solonnikov, V. A. and Ural'ceva, N. N.: *Linear and Quasilinear Equations of Parabolic Type.* Translations of Mathematical Monographs **23**, American Mathematical Society, Providence (RI) 1968.

[30] Landkof, N. S.: *Foundations of Modern Potential Theory.* Grundlehren der mathematischen Wissenschaften **180**, Springer, Berlin 1972.

[31] Langer, H., Partzsch, L. and Schütze, D.: Über verallgemeinerte gewöhnliche Differentialoperatoren mit nichtlokalen Randbedingungen und die von ihnen erzeugten Markov-Prozesse. *Publ. Res. Inst. Math. Sci.* **7** (1971/72) 659–702.

[32] Lense, J.: Review of [Feller 1930]. *Jahrbuch Fortschritte der Mathematik.* JFM 56.0419.01.

[33] Levi, E. E.: Sulle equazioni lineari totalmente ellittiche alle derivate parziali. *Rend. Circ. Mat. Palermo* **24** (1907) 275–317.

[34] Lions, J.-L. and Magenes, E.: *Non-Homogenous Boundary Value Problems and Applications. Vol I, II & III.* Grundlehren der mathematischen Wissenschaften **181–183**, Springer, Berlin 1972–73.

[35] Louhivaara, I. S.: Über die neuere Entwicklung der Theorie der linearen Räume mit indefiniten Bilinearformen. In: Künzi, H. P. and Pfluger, A. (eds.): *Festband zum 70. Geburtstag von Rolf Nevanlinna*. Springer, Berlin 1966, pp. 66–81.

[36] Meyer, P.-A.: *Dynkin and the Theory of Markov Processes*. In [11], 763–771.

[37] Miranda, C.: *Partial Differential Equations of Elliptic Type*. (2nd revised ed.) Ergebnisse der Mathematik und ihrer Grenzgebiete **2**. Springer, Berlin 1970.

[38] Miyadera, J.: Generation of a strongly continuous semi-group of operators. *Tohôku Math. J.* **4** (1952) 109–121.

[39] Neveu, J.: Semi-groupes généralisés et processus de Markoff. *C. R. Acad. Sci. Paris* **240** (1955) 1046–1047.

[40] Neveu, J.: Théorie des semi-groupes de Markov. *Univ. California Publ. Statist.* **2** (1958) 319–394.

[41] Phillips, R. S.: An inversion formula for Laplace transform and semi-groups of linear operators. *Ann. Math.* **59** (1954) 325–356.

[42] Pontryagin, L. S.: Hermitian operators in spaces with indefinite metric. (Russian) *Izv. Akad. Nauk SSSR Ser. Mat.* **8** (1944) 243–280.

[43] Riesz, M.: Intégrales de Riemann–Liouville et potentiels. *Acta Sci. Math. Szeged* **9** (1938) 1–42.

[44] Rogers, L. C. G. and Williams, D.: *Diffusions, Markov Processes, and Martingales. Vol. 1 (2nd edn.) & 2*. Wiley Series in Probability and Mathematical Statistics. Wiley, Chichester 1994 & 1987. (Reprinted by Cambridge University Press, Cambridge 1990).

[45] Sato, K. and Ueno, T.: Multi-dimensional diffusion and the Markov process on the boundary. *J. Math. Kyoto Univ.* **4** (1965) 529–605.

[46] Schilling, R. L.: On Feller processes with sample paths in Besov Spaces. *Math. Ann.* **309** (1997) 663–675.

[47] Schilling, R. L.: Feller processes generated by pseudo-differential operators: On the Hausdorff dimension of their sample paths. *J. Theor. Probab.* **11** (1998) 303–330.

[48] Schilling, R. L.: Growth and Hölder conditions for the sample paths of Feller processes. *Probab. Theory Relat. Fields* **112** (1998) 565–611.

[49] Schilling, R. L.: Conservativeness and extensions of Feller semi-groups. *Positivity* **2** (1998) 239–256.

[50] Ventcel, A. D.: Semigroups of operators generated by a generalised differential operator of second order. (Russian) *Doklady Akad. Nauk SSSR* **111** (1956) 269–272.

[51] Ventcel, A. D.: On boundary conditions for multidimensional diffusion processes. (Russian) *Teor. Veroyatnost. Primen.* **4** (1959) 172–185.

[52] Ventcel, A. D.: General boundary problems connected with diffusion processes. (Russian) *Uspehi Mat. Nauk* 15.2 (92) (1960) 202–204.

[53] Waldenfels, W. v.: Positive Halbgruppen auf einem n-dimensionalen Torus. *Arch. Math.* **15** (1964) 191–203.

[54] Waldenfels, W. v.: Fast positive Operatoren. *Z. Wahrscheinlichkeitstheor. verw. Geb.* **4** (1965) 159–174.

[55] Yosida, K.: An operator-theoretical treatment of temporally homogeneous Markoff processes. *J. Math. Soc. Japan.* **1** (1949) 244–253.

[56] Yosida, K.: *Functional Analysis.* (6th ed.) Grundlehren der mathematischen Wissenschaften **123**. Springer, Berlin 1980.

[57] Zolotarev, V. M.: *One Dimensional Stable Distributions.* Translations of Mathematical Monographs **65**. American Mathematical Society, Providence (RI) 1986.

Prof. Niels Jacob
Department of Mathematics
Swansea University
Singleton Park
Swansea, SA2 8PP, Wales, UK
n.jacob@swansea.ac.uk

Feller's Contributions to the One-Dimensional Diffusion Theory and Beyond

by *Masatoshi Fukushima* from Osaka

1 Intrinsic forms of infinitesimal generators

For an open interval $I \subset \mathbb{R}$, we denote by $C_b(I)$ the space of bounded continuous functions on I. Consider a second order elliptic differential operator

$$
(1) \qquad \begin{aligned} &\mathscr{L}u(x) = a(x)u''(x) + b(x)u'(x) + c(x)u(x), \quad x \in I, \\ &\text{for} \quad a,\, b,\, c \in C_b(I) \quad \text{with} \quad a > 0,\, c \leq 0 \quad \text{on} \quad I. \end{aligned}
$$

This differential operator $\mathscr{L}$ can be converted by

$$
ds = e^{-B}\,d\xi, \quad dm = \frac{1}{a}e^{B}\,d\xi, \quad dk = -\frac{c}{a}e^{B}\,d\xi
$$

where $B(\xi) = \int_\gamma^\xi \frac{b(\eta)}{a(\eta)}d\eta$, $\gamma \in I$, into the operator $\mathcal{A}$ of the *canonical form*

$$
(2) \qquad \mathcal{A}u(x) = \frac{dD_s u - u\,dk}{dm}(x), \qquad x \in I.
$$

The triplet (s, m, k) appearing in the expression (2) consists of

(3) *canonical scale* s : a strictly increasing continuous function on I,

(4) *speed measure* m : a positive Radon measure on I with full support,

(5) *killing measure* k : a positive Radon measure on I.

We take the domain $\mathscr{D}(\mathcal{A})$ of $\mathcal{A}$ to be

$$
(6) \qquad \mathscr{D}(\mathcal{A}) = \left\{ u \in C_b(I) \;\middle|\; \begin{array}{l} du \text{ is absolutely continuous with respect to } ds, \\ du/ds \text{ has a version } D_s u \text{ of bounded variation,} \\ D_s u - u\,dk \text{ is absolutely continuous w.r.t. } m \\ \text{with a Radon–Nikodým derivative in } C_b(I) \end{array} \right\}.
$$

R.L. Schilling et al. (eds.), *Selected Papers II*,

The operator (2) has an obvious advantage over (1) in that it is a *topological invariant*: if the interval I is mapped onto another interval J by any strictly increasing continuous function, then $(\mathcal{A}, \mathscr{D}(\mathcal{A}))$ is transformed into an operator based on J of the same type as (2), while $\mathscr{L}$ in (1) is not – unless the map is differentiable.

Let us consider a Markov process $X = (X_t, \zeta, \mathbb{P}_x)$ on I whose transition function $P_t(x,A) = \mathbb{P}_x(X_t \in A)$ makes the space $C_b(I)$ invariant. X then induces a semigroup on $C_b(I)$: $P_s P_t u = P_{s+t} u$, $0 < s,t$, $u \in C_b(I)$. Define the *infinitesimal generator* of this transition semigroup by

$$\mathcal{G}u(x) = \lim_{t \downarrow 0} \frac{1}{t}(P_t u(x) - u(x)), \quad x \in I. \tag{7}$$

The domain of the operator $\mathcal{G}$ consists of those functions $u \in C_b(I)$ for which the right hand side converges for every $x \in I$ with the limit function in $C_b(I)$.

In 1931, A. N. Kolmogorov showed that, under certain regularity conditions on the transition function P_t, $\mathcal{G}u$ admits an expression $\mathscr{L}u$ of (1) with $c = 0$ for smooth functions u. The imposed regularity conditions included a *local property of the transition function*:

$$P_t(x, U(x)^c) = o(t), \quad t \downarrow 0, \quad \text{for any } x \in I \text{ and for any neigborhood } U(x) \text{ of } x, \tag{8}$$

that is equivalent to the following *local property of the generator* $\mathcal{G}$:

(9) If $u, v \in \mathscr{D}(\mathcal{G})$ coincide in a neighborhood of $x \in I$, then $\mathcal{G}u(x) = \mathcal{G}v(x)$.

In [Feller 1936c],[1] W. Feller had called a Markov process satisfying condition (8) a *continuous process* and constructed, among other things, the fundamental solution of the parabolic differential equation with (1) on its right-hand side. He allowed (1) to contain a coefficient c of arbitrary sign in order to solve Kolmogorov's backward and forward equations simultaneously.

Soon after moving to Princeton in 1950, Feller resumed his study of the one-dimensional diffusion taking a remarkable lead on the subject. In a series of papers [Feller 1954a, Feller 1955a, Feller 1957a, Feller 1958], he successfully derived the canonical form (2) with $k = 0$ of the infinitesimal generator $\mathcal{G}$ defined by (7) from its local property (9) and its *strong minimum property*:

(10) If u in the domain of $\mathcal{G}$ has a local minimum at $x \in I$, then $\mathcal{G}u(x) \geq 0$.

In view of the definition (7), the last property is obviously satisfied by $\mathcal{G}$ when X is, for instance, conservative ($P_t 1 = 1$).

Thus Feller's method of deriving the canonical form is purely analytic, and yet Feller had clear picture of the sample paths of Markov processes behind his analytic presentations. In [Feller 1954a], he defined a diffusion process to be a Markov process satisfying the local property (8) as in [Feller 1936c], but added a sentence

[1]The references [Feller 19*nn*] and [*Feller 19*nn*] (the star indicating that the respective paper is not

The functions X_t are continuous with probability one if, and only if, the process is of a local character.

by quoting a paper by D. Ray [13] that was not yet published at that time. He also wrote in [Feller 1954b] that

> [...] the continuity of X_t is a topological property, whereas a continuous change of scale may change differentiable functions into nondifferentiable ones. It now appears possible to generalize the differential equation $u_t(t,x) = \mathscr{L}_x u(t,x)$ and to reformulate all the results of [Feller 1952a] in a topological invariant form.

From 1955 to 1956, K. Itô stayed in Princeton as a research fellow where he started his collaboration with H. McKean, a Ph.D. student of Feller. In his lecture at Kyoto in 1985 recorded in a video film, Itô recollected the period as follows.

> When I was in Princeton, Feller calculated repeatedly, for a one-dimensional diffusion with a simple generator $\mathscr{L}u = au'' + bu'$, the quantities like
>
> $$s(x) = \mathbb{P}_x(\sigma_\alpha > \sigma_\beta), \quad e(x) = \mathbb{E}_x[\sigma_\alpha \wedge \sigma_\beta], \quad x \in (\alpha, \beta) \subset I,$$
>
> as a harmonic function and a solution of a Poisson equation for $\mathscr{L}$. At first, I wondered why he was repeating so simple computations as exercise, not for objects in higher dimensions. However, in this way, he was bringing out intrinsic topological invariants. I understood that the one-dimensional diffusion is a topological concept but I was not as thoroughgoing as Feller. When Feller told me about this, he said he once listened to a lecture by Hilbert who mentioned: 'Study a quite simple case very much profoundly, then you will truly understand a general case'.

However, Feller was unable to attain by his analytic method the canonical form (2) in that generality involving the killing measure k. A reason for this seems to be the following. The true *minimum property* for the generator $\mathcal{G}$ of a Markov process X allowing a possible finite life-time ($\mathbb{P}_x(\zeta < \infty) > 0$) is

(11) If u in the domain of $\mathcal{G}$ has a local minimum at $x \in I$ and $u(x) \leq 0$, then $\mathcal{G}u(x) \geq 0$,

which is much weaker than (10). But Feller employed an even weaker principle called a *weak minimum property*:

(12) If u in the domain of $\mathcal{G}$ has a local minimum at $x \in I$ and $u(x) = 0$, then $\mathcal{G}u(x) \geq 0$,

---

contained in these Selecta) refer to Feller's bibliography, while [*n*] points to the list of references at the end of this essay.

allowing not only killing but also creation of paths so that the corresponding measure k, if it exists, could be a signed measure.

The canonical form (2) was eventually established by McKean [12] by a probabilistic method using results due to Ray [13] and E. B. Dynkin [3] on a *strong Markov process* (a Markov process satisfying the strong Markov property) with continuous sample paths, which is nowadays adopted as a definition of a *diffusion process*. Here the so called *Dynkin formula* expressing $\mathcal{G}u(x)$ in terms of exit times of X from neighborhoods of x played a key role.

To be more precise, let $I = (r_1, r_2)$ and let us start with a minimal diffusion X^0 on I. A Markov process $X^0 = (X_t^0, \zeta^0, \mathbb{P}_x^0)$ on I is called a *minimal diffusion* if

(d.1) X^0 is a Hunt process on I,

(d.2) X^0 is a diffusion process: X_t^0 is continuous in $t \in (0, \zeta^0)$ almost surely,

(d.3) X^0 is irreducible: $\mathbb{P}_x^0(\sigma_y < \infty) > 0$ for any $x, y \in I$.

Denote the one-point compactification of I by $I_\partial = I \cup \{\partial\}$. X_t^0 takes values in I_∂. For $B \subset I_\partial$, we define

$$\sigma_B = \inf\left\{t > 0 : X_t^0 \in B\right\}, \quad \inf \emptyset = \infty, \quad \tau_B = \sigma_{I_\partial \setminus B}.$$

We write σ_B as σ_b when $B = \{b\}$ is a one-point set. $\{\partial\}$ plays the role of cemetery for X^0: $\zeta^0 = \sigma_\partial$, $X_t^0 = \partial$ for any $t \geq \zeta^0$. Condition **(d.1)** means that X^0 is a strong Markov process whose sample paths X_t^0 are right continuous and have left limits on $[0, \infty)$, and are absorbed upon approaching $\{\partial\}$: $\lim_{n \to \infty} \tau_{J_n} = \zeta^0$ whenever $\{J_n\}$ are subintervals of I with $\overline{J}_n \subset I$, $J_n \uparrow I$. X^0 is *minimal* in this sense. Under **(d.1)** and **(d.2)**, the condition **(d.3)** is equivalent to the requirement for each point $a \in I$ to be *regular* in the sense that

$$\mathbb{E}_a[e^{-\alpha \sigma_{a+}}] = \mathbb{E}_a[e^{-\alpha \sigma_{a-}}] = 1, \quad \text{for } \alpha > 0,$$

where $E_a[e^{-\alpha \sigma_{a\pm}}] = \lim_{b \to \pm a} E_a[e^{-\alpha \sigma_b}]$.

Let $\{R_\alpha^0; \alpha > 0\}$ be the *resolvent* of a minimal diffusion X^0:

$$R_\alpha^0 f(x) = \mathbb{E}_x^0 \left[\int_0^\infty e^{-\alpha t} f(X_t^0) dt \right].$$

Denote by $\mathscr{B}_b(I)$ the space of all bounded Borel measurable functions in I. Then

$$R_\alpha^0(\mathscr{B}_b(I)) \subset C_b(I)$$

due to the above regularity of each point of I. R_α^0 is a one-to-one map from $C_b(I)$ into itself because of the resolvent equation $R_\alpha^0 - R_\beta^0 + (\alpha - \beta) R_\alpha^0 R_\beta^0 = 0$ and

$$\lim_{\alpha \to \infty} \alpha R_\alpha^0 f(x) = f(x), \quad x \in I, \ f \in C_b(I).$$

Thus, the generator $\mathcal{G}^0$ of X^0 is well defined by

$$(13) \quad \begin{cases} \mathscr{D}(\mathcal{G}^0) = R_\alpha^0(C_b(I)), \\ (\mathcal{G}^0 u)(x) = \alpha u(x) - f(x) \quad \text{for} \quad u = R_\alpha^0 f, \ f \in C_b(I), \ x \in I; \end{cases}$$

$\mathcal{G}^0$ so defined is independent of $\alpha > 0$. Let us call $\mathcal{G}^0$ the *C_b-generator* of X^0.

In [12, Chapter 4], it was proved that, for a given minimal diffusion X^0, there exist a canonical scale s, a speed measure m and a killing measure k such that, if we define $(\mathcal{A}, \mathcal{D}(\mathcal{A}))$ by (2) and (6) using this triplet (s,m,k), then

$$(14) \qquad \mathcal{D}(\mathcal{G}^0) \subset \mathcal{D}(\mathcal{A}) \quad \text{and} \quad \mathcal{G}^0 u = \mathcal{A}u \quad \text{for any} \quad u \in \mathcal{D}(\mathcal{G}^0),$$

namely, $\mathcal{G}^0$ is a restriction of $\mathcal{A}$. In particular, $u = R^0_\alpha f$ for $\alpha > 0$, $f \in C_b(I)$, satisfies an inhomogeneous equation

$$(15) \qquad (\alpha - \mathcal{A})u = f.$$

The triplet (s,m,k) is unique up to a multiplicative constant in the sense that, for another such triplet $(\widetilde{s}, \widetilde{m}, \widetilde{k})$, there exists a constant $c > 0$ such that $d\widetilde{s} = c\,ds$, $d\widetilde{m} = c^{-1}\,dm$ and $d\widetilde{k} = c^{-1}\,dk$.

We call (s,m,k) satisfying (14) a *triplet attached to* the minimal diffusion X^0.

The triplet (s,m,k) attached to the minimal diffusion X^0 was constructed in [12, §4.4] probabilistically as follows: For $r_1 < a < b < r_2$ and $J = (a,b)$, consider the hitting probabilities and mean exit time

$$p_{ab}(\xi) = \mathbb{P}^0_\xi(\sigma_a < \sigma_b), \quad p_{ba}(\xi) = \mathbb{P}^0_\xi(\sigma_b < \sigma_a), \quad e_{ab}(\xi) = \mathbb{E}^0_\xi[\tau_J], \quad \xi \in J,$$

and define

$$\begin{cases} s(d\xi) = s_{ab}(d\xi) = p_{ab}(\xi)p_{ba}(d\xi) - p_{ba}(\xi)p_{ab}(d\xi) \\ k(d\xi) = k_{ab}(d\xi) = D_s p_{ab}(d\xi)/p_{ab}(\xi) \\ m(d\xi) = m_{ab}(d\xi) = -\{D_s e_{ab}(d\xi) - e_{ab}(\xi)k_{ab}(d\xi)\}, \quad a < \xi < b. \end{cases}$$

For another choice of $\widetilde{a}$, $\widetilde{b}$ with $r_1 < \widetilde{a} < a < b < \widetilde{b} < r_2$, we have

$$s_{\widetilde{a}\widetilde{b}}(d\xi) = c s_{ab}(d\xi), \; k_{\widetilde{a}\widetilde{b}}(d\xi) = c^{-1}k_{ab}(d\xi), \; m_{\widetilde{a}\widetilde{b}}(d\xi) = c^{-1}m_{ab}(d\xi), \; a < \xi < b,$$

for a constant $c > 0$ depending on a, b, $\widetilde{a}$, $\widetilde{b}$, so that a universal triplet (s,m,k) can be introduced on I. Here, $f(d\xi)$ denotes the measure induced by a function $f(\xi)$ of bounded variation.

In [4, Chap. 15,16], Dynkin gave similar but slightly different representations of the infinitesimal generators of one-dimensional diffusions.

2 Classification of boundaries and lateral conditions

Let $X^0 = (X^0_t, \zeta^0, \mathbb{P}^0_x)$ be a minimal diffusion on an interval $I = (r_1, r_2)$ with attached triplet (s,m,k). Denote by I^* an interval of the type $[r_1, r_2]$ or $[r_1, r_2)$ or $(r_1, r_2]$, and $X = (X_t, \zeta, \mathbb{P}_x)$ be a Markov process on I^* satisfying conditions **(d.1)**, **(d.3)** with I^* in place of I. We call X a *Markovian extension* of X^0, if the subprocess of X being killed upon hitting $I^* \setminus I$ is identical with X^0 in law. If X further satisfies condition **(d.2)**, then we call it a *diffusion extension* of X^0. The *boundary problem of the one-dimensional diffusion* that Feller first raised and partly solved in [Feller 1952a, Feller 1957a] can be phrased as follows:

$$\text{(16)} \qquad \text{Characterize and construct all possible Markovian extensions of } X^0 \text{ in terms of } (s,m,k) \text{ and other intrinsic quantities.}$$

For any diffusion extension X of X^0, its C_b-generator $\mathcal{G}$ is well defined by (13) with I^* in place of I. Just as $\mathcal{G}^0$, $\mathcal{G}$ can been seen to be a restriction of $\mathcal{A}$ and, therefore, the problem is equivalent to finding a *lateral condition* Σ imposed on the functions $u \in \mathscr{D}(\mathcal{A})$ such that

$$\mathscr{D}(\mathcal{G}) = \mathscr{D}(\mathcal{A}) \cap \Sigma. \tag{17}$$

Since the resolvent R_α of X satisfies (15) along with R_α^0 and the difference $R_\alpha - R_\alpha^0$ is a positive kernel, the problem can be also reduced to the study of behavior of positive solutions u of the homogeneous equation

$$(\alpha - \mathcal{A})u = 0. \tag{18}$$

We write $j = m + k$ and define for $r_1 < c < r_2$

$$\lambda_1 = \int_{r_1}^{c} s(dx) \int_{x}^{c} j(dy), \quad \mu_1 = \int_{r_1}^{c} j(dx) \int_{x}^{c} s(dy), \quad r_1 < c < r_2.$$

The left boundary r_1 of I is called

$$\begin{array}{rlll} \textit{regular} & \text{if} & \lambda_1 < \infty, & \mu_1 < \infty, \\ \textit{exit} & \text{if} & \lambda_1 < \infty, & \mu_1 = \infty, \\ \textit{entrance} & \text{if} & \lambda_1 = \infty, & \mu_1 < \infty, \\ \textit{natural} & \text{if} & \lambda_1 = \infty, & \mu_1 = \infty. \end{array}$$

An analogous classification of r_2 is in force. This classification goes back to the paper [Feller 1952a], where Feller considered a very simple differential operator $\mathscr{L}u = u'' + bu'$ on I so that

$$ds = e^{-B} d\xi, \quad dj = dm = e^{B} d\xi, \quad dk = 0, \tag{19}$$

for $B(\xi) = \int_c^\xi b(\eta) d\eta$, $c \in I$.

In this special case, he made the classification in terms of (19) in accordance with the boundary behavior of positive solutions of the homogeneous equation (18). It was so nice and intrinsic that it was carried over straightforwardly into the case of a general pair (s,m) by [Feller 1957a] and a general triplet (s,m,k) by Itô–McKean [12] where the usage of the terms were somewhat modified, though.

The boundary problem of the one-dimensional diffusion was essentially settled by Itô–McKean [11, 12].

First of all, the generator $\mathcal{G}^0$ of the minimal diffusion X^0 can be characterized simply as follows ([Feller 1952a], [5]):

(20) $\mathscr{D}(\mathcal{G}^0) = \{u \in \mathscr{D}(\mathcal{A}) : u(r_i) = 0 \quad \text{whenever } r_i \text{ is regular or exit}, \ i = 1,2\}.$

Next, suppose r_1 is regular and r_2 is natural. Then the most general diffusion extension X of X^0 is an extension to $I^* = [r_1, r_2)$ whose generator $\mathcal{G}$ is characterized as follows ([12], [5]):

$u \in \mathscr{D}(\mathcal{G})$ if and only if

$$u \in C_b(I^*), \qquad \frac{dD_s u - u\,dk}{dm} \in C_b(I^*), \tag{21}$$

$$D_s u(r_1) - u(r_1)k^*(r_1) = \mathcal{G}u(r_1)m^*(r_1). \tag{22}$$

where $\mathcal{G}u(r_1)$ denotes the value of the function $(dD_s u - u\,dk)/dm$ (which is in $C_b(I^*)$) at r_1. Here, $k^*(r_1)$ and $m^*(r_1)$ are non-negative constants indicating the killing and sojourn at r_1 of the sample paths of X. $D_s u(r_1)$ indicates the reflection of paths at r_1. We may think that the killing measure k and the speed measure m are extended to $[r_1, r_2)$ by assigning the above point masses at r_1. Boundary conditions analogous to (20) and (22) can be also found in Dynkin [4, Chapters 15, 16].

Let X^0 be the standard absorbing Brownian motion on $I = (0,\infty)$. In this case, $ds = dx$, $dm = 2dx$, $dk = 0$, $\mathcal{A}u = \frac{1}{2}u''$ and $\mathscr{D}(\mathcal{A}) = C_b^2(I)$, the boundary 0 is regular and ∞ is natural. The most general Markovian extension X of X^0 is then an extension to $[0,\infty)$ whose generator $\mathcal{G}$ is characterized as follows ([Feller 1952a], [11]):

$u \in \mathscr{D}(\mathcal{G})$ if and only if $u \in C_b^2[0,\infty)$ and

$$p_1 u(0) - p_2 u'(0) + p_3 \mathcal{G}u(0) = \int_I [u(\ell) - u(0)] p_4(d\ell), \tag{23}$$

where $\mathcal{G}u(0) = \frac{1}{2}u''(0+)$. Here p_1, p_2, p_3 are non-negative constants and p_4 is a measure on I with

$$p_1 + p_2 + p_3 + \int_I (\ell \wedge 1) p_4(d\ell) = 1 \tag{24}$$

and

$$p_4(I) = \infty \quad \text{in case} \quad p_2 = p_3 = 0. \tag{25}$$

Itô and McKean ([11, 12]) call X *Feller's Brownian motion.*

The general boundary condition (23) with p_4 being a finite measure on I appeared already in [Feller 1952a], where Feller also mentioned that the reflection rate p_2 should vanish when r_1 is exit and r_2 is natural. The diffusion extension of X^0 corresponding to the boundary condition (23) with $p_3 = 0$, $p_4 = 0$ is called an *elastic barrier process*, while a Markovian extension of X^0 corresponding $p_1 = p_2 = p_3 = 0$ with finite measure p_4 is called an *elementary return process*. In [Feller 1954b], he stated a plan to construct a general Markovian extension X of X^0 as a limit of a sequence of elementary return processes. Later, Itô recollected in his article [10]

> There was once an occasion when McKean tried to explain to Feller my work on the stochastic differential equations along with the above mentioned idea of tangent. It seemed to me that Feller did not fully understand its significance, but when I explained Lévy's local time to Feller, he immediately appreciated its relevance to the study of the one-dimensional diffusion. Indeed, Feller later gave us a conjecture that the Brownian motion on $[0,\infty)$ with an elastic boundary condition could be constructed from the reflecting barrier Brownian motion by killing it at the time when its local time at the origin exceeds an independent exponentially distributed random time, which was eventually substantiated in my joint paper with McKean.

Thus the most general Markovian extension X on $[0,\infty)$ of the absorbing Brownian motion on I subjected to the boundary condition (23) was constructed in [11] purely probabilistically starting with the reflecting Brownian motion X^r on $[0,\infty)$, making a killing at 0 as described above, and making the sample paths to jump in I from 0 according to the infinite measure p_4 by using an increasing Lévy process.

Furthermore, a similar construction was carried out in [11, §17] for the Brownian motion on $\mathbb{R}$ with *two sided barriers* at 0. Let X^0 be the standard absorbing Brownian motion on $\mathbb{R}\setminus\{0\}$. Then the most general Markovian extension X of X^0 to $\mathbb{R}$ ought to have the generator $\mathcal{G}$ characterized as follows:

$u \in \mathscr{D}(\mathcal{G})$ if and only if

$$u \in C_b^2(\mathbb{R}\setminus\{0\}), \quad u''(0-) = u''(0+)$$

and

$$p_1 u(0) + p_{-2}u^-(0) - p_{+2}u^+(0) + p_3\mathcal{G}u(0\pm) = \int_{\mathbb{R}\setminus\{0\}} [u(\ell) - u(0)]p_4(d\ell), \tag{26}$$

where $u^-(0)$ (resp. $u^+(0)$) denotes the left (resp. right) derivative at 0, and $\mathcal{G}u(0\pm) = \frac{1}{2}u''(0\pm)$. Here p_1, $p_{\pm 2}$, p_3 are non-negative constants and p_4 is a measure on $\mathbb{R}\setminus\{0\}$ subject to

$$p_1 + p_{-2} + p_{+2} + p_3 + \int_{\mathbb{R}\setminus\{0\}} (|\ell| \wedge 1)p_4(d\ell) = 1 \tag{27}$$

and

$$p_4(\mathbb{R}\setminus\{0\}) = \infty \quad \text{in case} \quad p_{\pm 2} = p_3 = 0. \tag{28}$$

Thus, what is nowadays called a *skew Brownian motion* was studied already in [11] by allowing a killing and a sojourn at 0 as well as a jump into I from 0.

Actually, a boundary condition with two-sided barriers, similar to (26), had appeared in [Feller 1957a, (12.6a), (12.6b)] in order to describe analytically all possible Markovian extensions of an absorbing Brownian motion on a finite interval (β_1, β_2) on the circle $\mathbb{S}$ that is obtained by identifying β_1 and β_2. The corresponding Markov process on $\mathbb{S}$ was also constructed in [11, §17] by projecting a periodic process on $\mathbb{R}$ onto $\mathbb{S}$.

3 Some later developments and Feller measures

Along with his study of the boundary problem of the one-dimensional diffusion, Feller wrote an influential paper [Feller 1957d] in which a boundary problem for a minimal (time continuous) Markov process X^0 on a discrete countable state space E was formulated in an analogous manner to (16). In particular, he introduced what is called a *Feller measure* U (intuitively speaking, measuring the excursion rate from an entrance boundary point to an exit boundary point) as well as a *supplementary Feller measure* τ (measuring the excursion rate from an entrance boundary point never returning back to exit boundary points in finite time) as global intrinsic quantities for X^0 and indicated their possible roles in the study of the boundary problem for a general Markov process. Here the terms *entrance* and *exit* are used in a similar sense to Itô–McKean [12] but differently from Feller's sense mentioned in §2.

The study of the one-dimensional diffusion initiated by [Feller 1952a] was almost finalized until the middle of 1960's. In what follows, I briefly mention some of closely related developments of general features after 1970. Particularly, we will find that the supplementary Feller measure plays important roles even in the study of the one-dimensional diffusion.

3.1 Excursion point processes

Let X^0 (resp. $X^r = (X^r_t, \mathbb{P}^r_x)$) be the absorbing (resp. reflecting) Brownian motion on $(0,\infty)$ (resp. $[0,\infty)$). The transition function of X^0 is denoted by $\{P^0_t,\ t>0\}$. A σ-finite measure $\{\nu_t,\ t>0\}$ is called an X^0*-entrance law* if $\nu_s P^0_t = \nu_{s+t},\ s,t>0$.

Let $\{\ell_t,\ t\geq 0\}$ be the local time of X^r at the origin 0 normalized in a way that its Revuz measure relative to the (symmetrizing) speed measure $dm = 2dx$ is the Dirac mass at 0. Then its inverse S_t is a subordinator and, by associating with each of its jumping times an excursion of X^r around 0, we can get an excursion-valued point process $(p,\ \mathbb{P}^r_0)$. According to the general theory due to Itô [9], p is a *Poisson point process* whose characteristic measure n called the *excursion law* is a σ-finite measure on the excursion path space

$$W = \Big\{ w : [0,\zeta(w)) \to (0,\infty) \text{ continuous}, \\ 0<\zeta(w),\ w(0+)=0,\ w(\zeta-)\in\{0\}\cup\{\infty\}\Big\},$$

determined by

$$(29)\qquad \int_W f_1(w(t_1))f_2(w(t_2))\cdots f_n(w(t_n))\,n(dw) \\ = \mu_{t_1} f_1 P^0_{t_2-t_1} f_2 \ldots P^0_{t_{n-1}-t_{n-2}} f_{n-1} P^0_{t_n-t_{n-1}} f_n, \qquad 0<t_1<t_2<\cdots<t_n,$$

for some X^0-entrance law $\{\mu_t,\ t>0\}$.

In this case μ_t is known to be

$$(30)\qquad \mu_t(dx) = \frac{1}{(2\pi t^3)^{1/2}}\, x e^{-x^2/(2t)}\, m(dx)$$

so that $\mu_t(dx)/m(dx)$ is the density of the approaching time distribution of X^0 to 0 starting at $x \in (0,\infty)$. Therefore the excursion law n is intrinsically determined by the absorbing Brownian motion X^0, and we can recover from X^0 the reflecting Brownian motion X^r starting at 0 by piecing together the excursions around 0 using the Poisson point process p with characteristic measure n. More than that, we can construct from X^0 Feller's Brownian motion subjected to the boundary condition (23) at once by modifying the excursion law n in a way to allow jumps to the cemetery with rate p_1/p_2 and to $d\ell \subset (0,\infty)$ with rate $p_4(d\ell)/p_2$.

Actually, when X^0 is a minimal diffusion on $(0,\infty)$ with 0 being an exit boundary, Itô [8] constructed and characterized all jump in extensions of X^0 in this way. In this case, no continuous entry from 0 is possible and such extensions of X^0 with finite jump in measures was analytically described in [Feller 1952a] already. A typical example of such a minimal diffusion on $(0,\infty)$ is the celebrated *Feller's branching diffusion* corresponding to $\mathscr{L}u(x) = \beta x u''(x)$, $ds = dx$, $dm = dx/(\beta x)$, $x > 0$, as is presented in Section 3.1 of the essay by Ellen Baake and Anton Wakolbinger in these Selecta.

Next, we consider a general minimal diffusion $X^0 = (X_t^0, \zeta^0, \mathbb{P}_x^0)$ on $(0,\infty)$ with attached triplet (s,m,k) for which 0 is a regular boundary and also a unique diffusion extension $X^r = (X_t^r, \mathbb{P}_x^r)$ of X^0 to $[0,\infty)$ with no killing nor sojourn at 0. Define the approaching probability of X^0 to 0 in finite time by $\varphi(x) = \mathbb{P}_x^0(\zeta^0 < \infty,\ X^0_{\zeta^0-} = 0)$, $x \in (0,\infty)$. When X^0 is the absorbing Brownian motion, $\varphi = 1$ and the entrance law (30) trivially satisfies the equation

$$\int_0^\infty \mu_t\, dt = \varphi \cdot m, \tag{31}$$

but $\varphi \neq 1$ in general. Nevertheless, there exists a unique X^0-entrance law $\{\mu_t,\ t > 0\}$ satisfying (31) due to a general theorem by P. J. Fitzsimmons ([6]). It has been known (see [2, §7.5] and references therein) that the excursion point process $(p, \mathbb{P}_0^r)$ produced by the local time of X^r at 0 is a certain Poisson point process $\widetilde{p}$ being stopped at an independent exponentially distributed random time with rate $L^0(\varphi, 1-\varphi)$ which is the weight of the *supplementary Feller measure* for the one-point set $\{0\}$ intrinsically determined by X^0 that will be explained in the next subsection. Further the characteristic measure n of $\widetilde{p}$ is determined by (29) and (31). Thus, we can recover X^r from X^0 intrinsically just in the same way as above.

3.2 Symmetric Markov processes

Let X^0 be a minimal diffusion on an interval $I = (r_1, r_2)$ with attached triplet (s,m,k). The boundary r_i is called *approachable* if $|s(r_i)| < \infty$, $i = 1,2$. As is shown in [5], X^0 is m-symmetric in the sense that its transition function $\{P_t,\ t > 0\}$ satisfies

$$\int_I P_t f \cdot g\, dm = \int_I g \cdot P_t f\, dm, \quad t > 0,\ f,g \in \mathscr{B}_+(I). \tag{32}$$

Furthermore, the Dirichlet form $(\mathscr{E}^0, \mathscr{F}^0)$ of X^0 on $L^2(I;m)$ can be described in terms of the triplet as follows:

$$\mathscr{F}^0 = \mathscr{F}_0^{s,k}(I) \cap L^2(I;m), \quad \mathscr{E}^0(u,v) = \mathscr{E}_I^{s,k}(u,v), \tag{33}$$

where

$$(34)\qquad \mathscr{F}^{s,k}(I) = \left\{u : du \text{ is absolutely continuous w. r. t. } ds \text{ and } \mathscr{E}_I^{s,k}(u,u) < \infty\right\},$$

$$(35)\qquad \mathscr{E}_I^{s,k}(u,v) = \int_I D_s u(x) D_s v(x)\, ds(x) + \int_I u(x)v(x)\, dk(x),$$

$$(36)\qquad \mathscr{F}_0^{s,k}(I) = \left\{u \in \mathscr{F}^{s,k}(I) : u(r_i) = 0 \text{ whenever } r_i \text{ is approachable}\right\}.$$

For a general topological measure space (E,m), the notion of a *Dirichlet form* $(\mathscr{E},\mathscr{F})$ on the space $L^2(E;m)$ was introduced by A. Beurling and J. Deny [1]. M. L. Silverstein, one of the last Ph.D. students of Feller, then introduced the associated fundamental notions of an *extended Dirichlet space* and *reflected Dirichlet space* as well in [14]. The significance of the extended Dirichlet space $(\mathscr{F}_e,\mathscr{E})$ of $(\mathscr{E},\mathscr{F})$ is in that it is an invariant under time changes. It always holds that

$$(37)\qquad \mathscr{F} = \mathscr{F}_e \cap L^2(E;m).$$

If an m-symmetric Hunt process X associated with a regular Dirichlet form $(\mathscr{E},\mathscr{F})$ on $L^2(E;m)$ is changed into $\widetilde{X}$ by a time substitution with respect to its positive continuous additive functional (PCAF) whose Revuz measure $\widetilde{m}$ is of full quasi-support, then $\widetilde{X}$ becomes $\widetilde{m}$-symmetric while its extended Dirichlet space is unchanged so that we only need to replace m by $\widetilde{m}$ in (37) to obtain the Dirichlet form of $\widetilde{X}$ on $L^2(E;\widetilde{m})$ (cf. [2, Chapter 5]). In the special case of the above minimal diffusion X^0, the extended Dirichlet space of $(\mathscr{E}^0,\mathscr{F}^0)$ equals $(\mathscr{F}_0^{s,k}(I),\mathscr{E}_I^{s,k})$ that depends only on s and k. Concurring with a famous saying of Feller, we may say that

> the symmetric Hunt process X travels according to a road map indicated by the Beurling-Deny formula for the extended Dirichlet space $(\mathscr{F}_e,\mathscr{E})$ and with speed indicated by the symmetrizing measure m.

When the quasi-support F of the Revuz measure $\widetilde{m}$ of a time changing PCAF of X is a proper subset of E, things are not that simple. The time-changed process $\widetilde{X}$ on F is then $\widetilde{m}$-symmetric, and the extended Dirichlet space $(\widetilde{\mathscr{F}_e},\widetilde{\mathscr{E}})$ of the Dirichlet form of $\widetilde{X}$ on $L^2(F;\widehat{m})$ is also changed from the original one $(\mathscr{F}_e,\mathscr{E})$ according to the following rule ([2, Chapter 5]): $\widetilde{\mathscr{F}_e}$ is the restriction of $\mathscr{F}_e$ to F and $\widetilde{\mathscr{E}}(u,v)$, $u,v \in \widetilde{\mathscr{F}_e}$, is the sum of the due restriction of $\mathscr{E}(u,v)$, $u,v \in \mathscr{F}_e$, to F and the integral form

$$(38)\qquad \frac{1}{2}\int_{F\times F\setminus\Delta} (u(\xi)-u(\eta))(v(\xi)-v(\eta))\,U(d\xi d\eta) + \int_F u(\xi)v(\xi)\,V(d\xi).$$

Here U is a measure on $F\times F$ off the diagonal $\Delta \subset F\times F$, and V is a measure on F specified by (39) below. Let $X^0 = (X_t^0,\zeta^0,\mathbb{P}_x^0)$ be the subprocess of X on $E_0 = E\setminus F$ obtained by killing upon the hitting time σ_F of F. Define $H_F f(x) = \mathbb{E}_x^0[f(X_{\zeta^0-}^0);\ \zeta^0<\infty]$, $x\in E_0$, for $f\in\mathscr{B}_+(F)$. Denote by $L^0(f,g)$ the *energy functional* of a purely excessive function f and an excessive function g relative to X^0 (cf.

[6], [2, §5.4]). Then, for $f, g \in \mathscr{B}_+(F)$ with $fg = 0$,

$$
\begin{aligned}
(39) \qquad & \int_{F\times F} f(\xi)g(\eta)U(d\xi d\eta) = L^0(H_F f, H_F g), \\
& \int_F f(\xi)V(d\xi) = L^0(H_F f, 1 - H_F 1_F).
\end{aligned}
$$

We call U and V the *Feller measure* and the *supplementary Feller measure*, respectively, because they are analogous to the quantities U, τ introduced by [Feller 1957d].

Let us consider the case where X is a minimal diffusion on $\mathbb{R}$ with attched triplet (s, m, k), $F = [0, \infty)$ and $\widetilde{m}$ is a positive Radon measure on $\mathbb{R}$ with support F. For simplicity, we assume that ∞ is non-approachable: $s(\infty) = \infty$. The PCAF of X with Revuz measure $\widetilde{m}$ is then given by $A_t = \int_F \ell_t(x)\widetilde{m}(dx)$ where $\ell_t(x)$ is the local time of X with Revuz measure δ_x relative to m. The subprocess X^0 of X on $E_0 = (-\infty, 0)$ is the minimal diffusion on E_0 with attached triplet consisting of the restrictions to E_0 of s, m and k. Then the measure $H_F(x, \cdot)$, $x \in E_0$, is concentrated on the one-point set $\{0\}$ so that the Feller measure U vanishes and the *supplementary Feller measure V* has the expression

$$(40) \qquad V = L^0(\varphi, 1 - \varphi)\delta_0,$$

where L^0 is the energy functional for X^0 and $\varphi(x) = \mathbb{P}^0_x(\zeta^0 < \infty, X^0_{\zeta^0-} = 0)$, $x \in E_0$.

The time-changed process $\widetilde{X}$ of X by means of A is therefore an $\widetilde{m}$-symmetric diffusion on $[0, \infty)$ whose extended Dirichlet space $(\widetilde{\mathscr{F}}_e, \widetilde{\mathscr{E}})$ is given by

$$(41) \qquad \widetilde{\mathscr{F}}_e = \mathscr{F}^{s,k}([0, \infty)),$$

$$(42) \qquad \widetilde{\mathscr{E}}(u, v) = \mathscr{E}^{s,k}_{[0,\infty)}(u, v) + L^0(\varphi, 1 - \varphi)u(0)v(0).$$

Using a similar method as in [5], we can also deduce from this the following characterization of the C_b-generator $\widetilde{\mathfrak{G}}$ of $\widetilde{X}$:

$u \in \mathscr{D}(\widetilde{\mathfrak{G}})$ if and only if

$$(43) \qquad u \in C_b([0, \infty)), \qquad \frac{dD_s u - u\,dk}{d\widetilde{m}} \in C_b([0, \infty)),$$

$$(44) \qquad D_s u(0) - u(0)\left[k(\{0\}) + L^0(\varphi, 1 - \varphi)\right] = \widetilde{\mathfrak{G}}u(0)\widetilde{m}(\{0\}).$$

Itô–McKean [12, §5.3] obtained the above charcterization in the special case that X is a standard Brownian motion on $\mathbb{R}$. In this case, $k = 0$ and also $L^0(\varphi, 1 - \varphi) = 0$ because $\varphi(x) = 1$ for $x < 0$. But, if $|s(-\infty)| < \infty$ and $m(-\infty, c) = \infty$ for $c < 0$, then $\varphi(x) < 1$ for $x < 0$, and accordingly $L^0(\varphi, 1 - \varphi)$ does not vanish. For the higher dimensional Brownian motion, the jump term with the Feller measure U really appears in the Dirichlet form under a time change by a PCAF of non-full support (cf. [2, §5.8]).

References

All citations of the form [Feller 19*nn*], resp., [*Feller 19*nn*] (if the respective paper is not included in these Selecta) point to Feller's bibliography, pp. xxv–xxxiv.

[1] Beurling, A. and Deny, J.: Dirichlet spaces. *Proc. Nat. Acad. Sci. USA* **45** (1959) 208–215.

[2] Chen, Z.-Q. and Fukushima, M.: *Symmetric Markov Processes, Time Changes and Boundary Theory*. Princeton University Press, London Math. Soc. Monographs Series **35**, Princeton (NJ) 2012.

[3] Dynkin, E. B.: Infinitesimal operators of Markov stochastic processes. *Dokl. Akad. Nauk SSSR* **105** (1955) 206–209.

[4] Dynkin, E. B.: *Markov Processes I, II*. Springer, Grundlehren math. Wiss. **121, 122**, Berlin 1965.

[5] Fukushima, M.: On general boundary conditions for one-dimensional diffusions with symmetry. *J. Math. Soc. Japan* **66** (2014) 289–316.

[6] Getoor, R. K.: *Excessive measures*. Birkhäuser, Boston 1990.

[7] Itô, K.: *Essentials of Stochastic Processes*. American Mathematical Society, Transl. Math. Monogr. **231**, Providence (RI) 2006. (Originally published in Japanese, Iwanami Shoten, 1957.)

[8] Itô, K.: Poisson point processes and their application to Markov processes. Lecture Notes of the Mathematics Department, Kyoto University, Kyoto September 1969. (unpublished) Electronic version available from the official web site of Mathematical Society of Japan for the "Centennial Anniversary of the Birth of Kiyosi Itô": mathsoc.jp/publication/ItoArchive/index.html

[9] Itô, K.: Poisson point processes attached to Markov processes. In: Le Cam, L., Neyman, J. and Scott, E. L. (eds.): *Proceedings of the Sixth Berkeley Symposium on Mathematical Statistics and Probability, Vol. 3*, University of California Press, Berkeley (CA) 1970, pp. 225–239.

[10] Itô, K.: Memoirs of my research on stochastic analysis. In: Benth, F. E. et al. (eds.): *Stochastic Analysis and Applications, The Abel Symposium 2005*. Springer, Berlin 2007, pp. 153–196.

[11] Itô, K. and McKean Jr., H. P.: Brownian motions on a half line. *Illinois J. Math.* **7** (1963) 181–231.

[12] Itô, K. and McKean, Jr., H. P.: *Diffusion Processes and Their Sample Paths*. Springer, Grundlehren math. Wiss. **125** 1965. (Reprinted in: Classics in Mathematics, Springer 1996.)

[13] Ray, D.: Stationary Markov processes with continuous paths. *Trans. Am. Math. Soc.* **82** (1956) 452–493.

[14] Silverstein, M. L.: *Symmetric Markov Processes*. Springer, Lecture Notes in Math. **516**, Berlin 1974.

Prof. Masatoshi Fukushima
Professor Emeritus
Osaka University
Department of Mathematical Science
Faculty of Engineering Science
Toyonaka, Osaka 560-8531, Japan
fuku2@mx5.canvas.ne.jp

On Boundary Behaviour of One-Dimensional Diffusions: From Brown to Feller and Beyond

by *Goran Peskir* from Manchester

1 Introduction

Feller's greatest discovery in mathematics was *sticky* (or *slowly reflecting*) boundary behaviour of one-dimensional diffusion processes (characterised by the appearance of the *second derivative* at the boundary point). Before him (Dirichlet[1], Neumann[2], Robin[3]) it was not known that this was possible. A boundary condition containing the second derivative (in one dimension) should therefore be referred to as *Feller boundary condition* (in addition to *Dirichlet condition* containing the function itself, *Neumann condition* containing its first derivative, *Robin condition* containing a linear combination of the function and its first derivative) and extensions of this condition to higher dimensions should likewise contain the name of Feller.

In this brief review I will address the relevance of this discovery and Feller's work on boundary classification of one-dimensional diffusion processes within a general context of mathematics and physics. Feller's motivating aim, often stated explicitly in his papers, was to disclose the "most general" conditions, or to describe the "most general" situations, and this attitude happened to be the key to unlock the mystery of the still unseen boundary behaviour. Although enduring the progress was slow and it took over 10 years (1951-1965) including help and insights of other people (Itô and McKean via Lévy and Volkonskii) to complete it at the level of a sample path. The general picture in one dimension is complete at present.

After reading Feller's papers time and again I wonder whether he ever thought

[1]Peter Gustav Lejeune Dirichlet (1805–1859) was a German mathematician.
[2]Carl Gottfried Neumann (1832–1925) was a German mathematician.
[3]Victor Gustave Robin (1855–1897) was a French mathematician.

R.L. Schilling et al. (eds.), *Selected Papers II*,

that the ultimate goal of the "most general" was achieved. The line had to be always drawn somewhere but only to be stretched further in a new paper. Likewise my review will fail to draw definite lines and at the end I will briefly indicate how to go beyond the "most general" in hope that the suggested possibility would only please Feller in his never ending quest for the "most general" conditions and situations. I will now return to the beginning of the story leaping forward in big jumps to get to Feller's time as quickly as possible.

2 From Brown (1828) to Feller (1936)

A pillar of Newtonian mechanics is Newton's first law[4] (or Galileo's law of inertia[5]). It states that if a body is not acted upon by a force then the body stays at rest or moves at straight line with constant velocity. The observations made by Brown [2][6] in 1828 about the unpredictable "zigzag" motion of spherical particles with a diameter of the order of several $\mu\text{m}\,(=10^{-6}\text{m})$ suspended in a fluid at rest (Brownian particles) were therefore curious. The numerous attempts to explain the phenomenon over the subsequent 75 years have failed until the appearance of the paper by Einstein [5] in 1905. Einstein's aim was to probe the molecular-kinetic theory of heat according to which the fluid consists of atoms/molecules in motion whose average kinetic energy is measured by the fluid temperature. The fluid atoms/molecules have a diameter of the order of several $\text{Å}\,(=10^{-10}\text{m})$ and thus could not be observed by the microscope. Einstein's ingenious idea (one of his "gedanken" experiments) was to take a single atom/molecule of the fluid, enlarge it to the size of the order of several μm that is visible by the microscope, and witness the existence of the other atoms/molecules of the fluid (around 10^{23} of them) through their bombardment and the observed motion of the enlarged atom/molecule (i.e. Brownian particle). In the first (physical) part of his paper Einstein derives a closed form expression for the diffusion coefficient D expressed in terms of other physical quantities (the Einstein relation), and in the second (mathematical) part of his paper Einstein derives the diffusion equation

$$p_t = Dp_{xx} \tag{1}$$

for the transition density $p = p(t,x)$ of the position X_t of a Brownian particle. It means that

$$\mathbb{P}(X_t \in A) = \int_A p(t,x)\,dx \tag{2}$$

for a (measurable) set A at time $t > 0$ upon assuming that $X_0 = 0$. Thus instead of inferring that the particle occupies a particular position at a particular time (Newtonian mechanics) we speak of the particle occupying a particular set at a particular time

[4] Isaac Newton (1642–1727) was an English physicist and mathematician.

[5] Galileo Galilei (1564–1642) was an Italian physicist and mathematician.

[6] The references [Feller 19*nn*] and [*Feller 19*nn*] (the star indicating that the respective paper is not contained in these Selecta) refer to Feller's bibliography, while [*n*] points to the list of references at the end of this essay.

with certain probability. Since (1) can be used to estimate D through the observation of a Brownian particle, the Einstein relation then enables him to estimate the physical quantity of interest (Avogadro's number[7]). In this way Einstein succeeds in connecting the micro scale of atoms/molecules to the macro scale of the observer through the meso scale of the Brownian particle. Experiments made subsequently by Perrin[8] confirmed the predictions and established that the atoms/molecules are real.

In his physical arguments Einstein makes use of Fick's law [7] from 1855 (which is a definition of the diffusion coefficient) that in turn mimics Fourier's law [9] from 1822 (a special case of which is Newton's law of cooling). A simple integration by parts (or the divergence theorem in higher dimensions) transforms Fick's law into the *diffusion equation* and likewise Fourier's law into the *heat equation* (both being transport phenomena). A curious fact is that Einstein appears to be unfamiliar with this derivation of the diffusion equation. Instead he ingeniously postulates that the position process X has stationary and independent increments. This enables him to write down an integral equation for the transition density (later to be named after Chapman and Kolmogorov) which he then expands in powers of time (first order) and space (second order). Passing to the limit (upon violating his own claim on the breakdown of independent increments at small time scales) he then formally obtains the diffusion equation (1). I will refer to this notable derivation as Einstein's argument in the sequel (see [5, p. 95 of the English translation] for details). Wiener [30] then in 1923 proves the existence of a stochastic process satisfying the Einstein's postulates (stationary independent increments plus continuous sample paths which Einstein finds too obvious to state explicitly). The Einstein–Wiener process does not fit the observed data at all time scales (due to the breakdown of independent increments). The Ornstein–Uhlenbeck position process [28] based on the equation of Langevin [21] (representing Newton's second law) does. Nonetheless, due to its canonical form and tractability, the Einstein–Wiener process turns out to the most important stochastic process that is often referred to as Brownian motion itself (in mathematical circles in particular).

Subsequently Smoluchowski [26] in 1916 extends the forward equation (1) to the case when the Brownian particle is under influence of an external force (transforming the zero/constant drift into a function of [time and] space), and Chapman [3] in 1928 studies the case when the Brownian particle is in a fluid with non-constant temperature (transforming the constant diffusion coefficient into a function of [time and] space). Stationary and independent increments in both cases break down and get superseded by the Markov[9] property (formalised by Markov in 1906 although the idea constitutes the backbone of Newtonian mechanics) in that the future probabilistic behaviour of the process depends only on its current position when the entire past is given. Consequently the drift and the diffusion coefficient become non-constant in these more general physical situations. This more general forward equation has also been derived in some particular cases by Fokker [8] in 1914 and Planck [25] in 1917 (and it is often named after them). For further details on the facts reviewed so far see

[7]Lorenzo Romano Amedeo Carlo Avogadro (1776–1856) was an Italian scientist.

[8]Jean Baptiste Perrin (1870–1942) was a French physicist.

[9]Andrey Andreyevich Markov (1856–1922) was a Russian mathematician.

[22] and [23].

Kolmogorov [19] in 1931 undertakes a systematic study of the forward and backward equations for the transition densities starting with the integral equation (of Einstein) arising from the Markov property (named later after Chapman and Kolmogorov) as a new axiom for the stochastic motion. Kolmogorov derives the *backward equation*

$$p_s = -\mu p_x - D p_{xx} \tag{3}$$

by extending the Einstein's argument referred to above (compare incidentally the same page 95 in the English translations of both [19] and [5]) and the *forward equation*

$$p_t = -(\mu p)_y + (Dp)_{yy} \tag{4}$$

integrating by parts (twice) so that the two equations become adjoints to each other (in the sense of a scalar product). In these equations $p = p(s,x,t,y)$ denotes the transition density of the position process X in the sense that

$$\mathbb{P}(X_t \in A \mid X_s = x) = \int_A p(s,x,t,y)\, dy \tag{5}$$

for a (measurable) set A, while $\mu = \mu(t,x)$ is the drift of X and $\sigma = \sigma(t,x)$ is the (mathematical) diffusion coefficient of X (the two are connected by $D = \sigma^2/2$). Kolmogorov establishes some existence and uniqueness results for the backward and forward equations (in [19] when the state space is countable and in [20] when the state space is a Riemannian manifold) but largely leaves the existence and uniqueness question open.

At this point Feller enters the scene with his paper [Feller 1936c] that was written during his early Stockholm years (1934–1939) and published in 1936. Firstly, he establishes more definite existence and uniqueness results for the backward and forward equations (both for continuous and discontinuous Markov processes). Secondly, he also provides a more complete treatment of Markov processes with jumps (this was indicated by Kolmogorov) leading to partial integro-differential equations in place of the backward and forward differential equations (4) and (5). Feller expresses all solutions as series in his paper. His method relies upon an adaptation of the construction of fundamental solutions due to Hadamard[10] and Gevrey[11] (dating back to 1911 and 1913 respectively). Both Kolmogorov [19] and Feller [Feller 1936c] treat the backward and forward equations in a purely analytic fashion starting with the Chapman-Kolmogorov integral equation that is implied by a Markov process but also non-Markov processes as shown later by Lévy[12] using a sequence of pairwise independent random variables that are not mutually independent (Feller's paper [Feller 1959c] deals with that issue). While Kolmogorov retains his vision of a stochastic process as the time evolution of marginal laws, Feller will gradually soften

[10] Jacques Salomon Hadamard (1865–1963) was a French mathematician.
[11] Maurice-Joseph Gevrey (1884–1957) was a French mathematician.
[12] Paul Pierre Lévy (1886–1971) was a French mathematician.

his analytic view and start to appreciate sample paths of the process in later years. Nonetheless, the backward and forward equations will remain Feller's main pillar for stochastic processes and he will never get tired from studying them in new and different settings from different standpoints as well. In line with this intention Feller devotes the entire Section 6 in his paper [Feller 1936c] to purely discontinuous processes (those changing their states only by jumps). He demonstrates that this case of backward and forward equations can be treated under much weaker assumptions than the general case and continues his study in [Feller 1940c] using more general methods and establishing more general existence and uniqueness results. At that time Feller was already at Brown University and after passing through Cornell and settling down at Princeton in 1950 he was to enter the golden era of his research on diffusion processes. The start was again through the backward and forward equations. This time however the solutions are not to be defined on the entire state space but only on its *subspace* of interest. One of the key motivations for Feller was coming from genetics. If the population size is described by a diffusion process then clearly this process is *non-negative* and the question arises what happens after the process hits zero. In line with his incessant search for "most general" Feller aims to disclose all possibilities. It turns out however that the question is not as simple as one could expect at first glance.

3 From Feller (1951) to Itô and McKean (1965)

Motivated by applications in genetics [Feller 1951d] where the diffusion process by its nature takes non-negative values, Feller begins his work on boundary behaviour of one-dimensional diffusions in the paper [Feller 1951e] by studying the forward equation

$$u_t = -((bx+c)\,u)_x + (ax\,u)_{xx} \tag{6}$$

for $x > 0$ where $u = u(t,x)$ is a function of time and space, and a,b,c are constants with $a > 0$. The key feature of this equation is its singularity at the boundary point 0 in the sense that the diffusion coefficient $D = ax$ vanishes at $x = 0$. Feller shows that this forces the equation to change its character completely depending on whether $c \le 0$ (exit), $0 < c < a$ (regular), $c \ge a$ (entrance). Roughly speaking, it means that in the first and third case one has that the desired solution (positivity preserving and norm decreasing/preserving) is uniquely determined by its initial values $u(0,x)$ for $x > 0$, while in the second case one no longer has this uniqueness and the boundary values such as $u(t,0)$ or $u_x(t,0)$ need be imposed to obtain it. In terms of sample paths of the process X to which Feller does not refer explicitly, it means that the singularity of the diffusion coefficient in the first and third case determines the sample path behaviour at the boundary uniquely, while in the second case leaves the sample path behaviour at the boundary unspecified. Singular equations of that kind were not studied before Feller's paper [Feller 1951e] and the natural question that Feller takes up in the 1950s is to understand and classify all possible boundary behaviour of one-dimensional diffusion processes associated with the backward and forward equations (both singular and non-singular i.e. regular). Singularity also refers to non-existence of a finite limit of the drift or diffusion coefficient when the boundary is approached from the interior. Feller's main tool in [Feller 1951e] was Laplace transform but he also states on page 174 that

> Our result is also of some interest for the *theory of semigroups*[13] and is a counterpart to the investigations of Yosida and Hille [10][14] who have studied the semigroups defined over $-\infty < x < \infty$ by Fokker-Planck equations with singularities at infinity.

A year later Feller publishes his fundamental paper [Feller 1952a] in which he discloses a full classification of all possible boundary behaviour for one-dimensional diffusions. In his paper Feller studies the backward and forward equations on a general bounded or unbounded interval of the real line and on pages 468–469 states that "The use of *semigroup methods* for partial differential equations was initiated by Hille [10]" and then continues

> It seems that hitherto only the question has been asked whether the right side [of the forward equation] is the (or possibly the extension of the) infinitesimal generator of a semigroup. This amounts to requiring *uniqueness* of the solution of the initial value problem. Actually, the case of uniqueness for [the forward equation] is most exceptional, and the main problems relate to the case of *non-uniqueness*.

It is exactly this non-uniqueness that is the essence of Feller's problem which enables him to depart from previous research. After addressing the method of proof with reference to semigroup theory Feller concludes on page 470 that

> Curious phenomena are encountered. The general lateral condition for [the backward equation] depends on bounded linear functionals and on differential operators of *second order*. By contrast, the lateral condition of the adjoint problem is always of a local character and depends on differential operators of at most *first order*.

This is the first reference to the *second order* conditions (see equation (19.4) on page 505 for the explicit form). Feller will rarely point this out in his papers again probably because a fuller understanding of this condition will only be obtained in the 1960s as we shall see below. When addressing the problem of uniqueness Feller reveals on pages 472–473 that

> In [Feller 1936c] the general case of time dependent coefficients [of the backward and forward equation] is treated. In the time independent case greater elegance can be achieved by the *semigroup method* and this approach has been used by Yosida [31] and, more specifically, by Hille [10], [11]. Both authors limit themselves to a discussion of conditions which guarantee uniqueness.

[13]All emphases in the quotations throughout are my own (the original emphases are not reproduced).

[14]Citations in the quotations throughout point to the list of references at the end of this essay (the

In the footnote to Hille's paper [10] above Feller acknowledges that "Its manuscript (which was available to [Feller] in the Fall of 1949) drew [Feller's] attention to the possibilities opened by the study of the resolvent". That realisation was crucial for Feller. It opened a plethora of possibilities that he will tirelessly explore. I will now briefly explain the method that Feller applied in his papers.

When the Markov process X is time-homogeneous then one has $p(s,x,t,y) = p(0,x,t-s,y)$ for $s<t$. Setting $p = p(t,x) := p(0,x,t,y)$ this shows that the backward equation (3) reads

$$p_t = \mu p_x + D p_{xx} \tag{7}$$

for any y given and fixed. Imposing the initial condition $u(0,x) = f(x)$ and integrating over y one sees that the solution to (7) can be expressed as

$$u(t,x) = \mathbb{E}_x f(X_t) =: \mathsf{P}_t f(x) \tag{8}$$

where $X_0 = x$ under $\mathbb{P}_x$. When f is the indicator function of a (measurable) set then the solution (8) can be interpreted as the transition probability but (8) also makes sense for other functions f including continuous ones (which approximate others). The Markov property of X then reads

$$\mathsf{P}_{t+s} = \mathsf{P}_t \mathsf{P}_s \tag{9}$$

for $t,s \geq 0$ with $\mathsf{P}_0 = \mathsf{I}$ showing that $(\mathsf{P}_t)_{t\geq 0}$ defines a *semigroup* of operators. To describe its action one considers the *infinitesimal generator* defined by

$$\mathsf{L} f := \lim_{t \downarrow 0} \frac{\mathsf{P}_t f - f}{t} = (\mu \partial_x + D \partial_{xx}) f \tag{10}$$

for f in its domain $D(\mathsf{L})$ where the limit exists (in the sense to be specified). Another action of interest is expressed by the *resolvent* defined by

$$\mathsf{R}_\lambda f := (\lambda \mathsf{I} - \mathsf{L})^{-1} f = \int_0^\infty e^{-\lambda t} \mathsf{P}_t f \, dt \tag{11}$$

for $\lambda > 0$ representing the Laplace transform of the semigroup. Then under some mild regularity conditions one establishes the following identity

$$D(\mathsf{L}) = \text{Range}(\mathsf{R}_\lambda) \tag{12}$$

for any $\lambda > 0$. This is the key fact for Feller's approach: *To describe the domain of the infinitesimal generator* L *it is sufficient (and essentially the same as) to describe the range of the resolvent* R_λ *for any* $\lambda > 0$. This is what Feller does and the description is accomplished in terms of *boundary conditions* when the state space has finite boundaries in particular. It means that f satisfying these boundary conditions

original numbers are rewritten).

belongs to $D(\mathsf{L})$ and hence letting $h \downarrow 0$ in

$$(13)\qquad \frac{u(t+h,x)-u(t,x)}{h} = \left(\frac{\mathsf{P}_{t+h}-\mathsf{P}_t}{h}\right)f(x) = \left(\frac{\mathsf{P}_h-\mathsf{I}}{h}\right)\mathsf{P}_t f(x)$$

$$(14)\qquad \frac{u(t+h,x)-u(t,x)}{h} = \left(\frac{\mathsf{P}_{t+h}-\mathsf{P}_t}{h}\right)f(x) = \mathsf{P}_t\left(\frac{\mathsf{P}_h-\mathsf{I}}{h}\right)f(x)$$

one obtains the backward and forward equations

$$(15)\qquad u_t = \mathsf{L}u\ (= \mathsf{L}\mathsf{P}_t f)$$

$$(16)\qquad u_t = \mathsf{L}^* u\ (= \mathsf{P}_t \mathsf{L} f)$$

as adjoints to each other. Thus the problem reduces to determine the range of the resolvent (11) and this is what Feller does in his papers. Once this is accomplished the domain of the infinitesimal generator is determined and boundary conditions satisfied by functions belonging to the domain can be read off. The infinitesimal generator and its domain generate a unique probability measure on the space of path functions and those functions in a (minimal) set of probability one then form the sample paths of the process. This measure-theoretic aspect of the general picture is mostly implicit in Feller's papers and will only be clarified later in fuller generality. The key point in Feller's approach to boundary classification is to insist that one looks for solutions of the backward and forward equation in the form (8). This assumption is so natural from the standpoint of probability that it is hard to imagine that anything else would be possible in a more canonical form. We shall return to this issue at the end of this review.

Equipped with this machinery Feller classifies all boundary behaviour in his Section 11 of [Feller 1952a] into (i) *regular* (non-singular), (ii) *exit*, (iii) *entrance*, and (iv) *natural*. The conditions are expressed in terms of the infinitesimal characteristics of the process (drift and diffusion coefficient). In the footnote of the same page 487 where this classification is stated Feller acknowledges (with the exposition slightly adapted to the present setting):

> In the case of $r_1 = -\infty$, $r_2 = +\infty$ [i.e. when the state space equals the entire real line] E. Hille [11] has, by different methods, discovered that the (exit) boundary condition (ii) is necessary and sufficient for the uniqueness for the forward equation and the (entrance) boundary condition (iii) is necessary and sufficient for the uniqueness for the backward equation.

Feller discusses the implications of the boundary classification on the backward and forward equations in his Section 23 of [Feller 1952a]. To express the boundary conditions of Section 11 he introduces at the bottom of page 515 ("following Hille") what is later to be termed as the (derivative of the) *scale function s*. Feller's discussion is mainly analytic in this section/paper and he makes no attempt to interpret the results in terms of a sample path of the process. Regular boundaries are briefly addressed at the end of the paper. On page 518 Feller writes "If r_1 is a natural but r_2 a regular boundary, then there exist *infinitely many* solutions for the initial value problem both for [the backward and forward equation]". He then continues

> It is curious that the lateral condition for [the backward equation] involves a differential operator of *second order* and terms of global character, whereas the lateral conditions determining the adjoint problem are always of the simple form [specified by two conditions in the paper].

Clearly these "second order" conditions have been puzzling Feller quite a bit, however he was ahead of his time as early as in 1952. A curious fact in this regard is that Feller allows his diffusion process to jump into the interior of the state space after hitting the boundary. It might be that this was partly motivated by what Feller states on page 474 of his paper [Feller 1952a]:

> Various probability considerations and a longstanding problem in genetics made it increasingly clear that, in addition to the *classical* types of boundary conditions there exist some of a *new* type. Physically speaking, there exist diffusion processes where a particle can be absorbed, and stay absorbed for a finite time, after which it penetrates *slowly* back into the interior. Attempts at a mathematical formulation of such phenomena remained *unsuccessful* for several years. It is now *possible* to identify one of our new boundary conditions as the one describing this phenomenon.

This might be viewed as another hint of sticky (or slowly reflecting) boundary behaviour, however, it is not entirely clear what Feller meant since a strong Markov process must leave a holding point by a jump (as it is well known now). If the jump density is highly concentrated in the interior close to the boundary then this could create the impression that the particle virtually "penetrates *slowly* back into the interior" although strictly speaking jumps off the boundary are still present.

Feller addresses probabilistic interpretations of the analytic results from his paper [Feller 1952a] in [Feller 1954b]. In that paper he also attempts to give a rigorous definition of a *diffusion process* and on pages 1-2 states "Probably it soon will be possible to define diffusion processes as Markov processes with continuous path functions". The adjective "strong" is not present in front of "Markov" as it will only be recognised as a necessity through the work of Dynkin [4] and Hunt [12] a couple of years later. In his paper Feller refers to X_t as "the position of the particle at time t" and states that the sample paths $t \mapsto X_t$ are continuous (except perhaps at the boundary). Feller discusses the existence of a stochastic processes associated with the backward and forward equations. These elements demonstrate Feller's shift from the almost purely analytic view to a more complete probabilistic understanding. His analysis includes absorption at the boundary, reinsertion in the interior, termination (killing), instantaneous reflection, and elastic boundary behaviour. In relation to the latter Feller states on page 4 that his "limiting procedure sheds some new light on the *elastic barrier* processes, but does not give a measure-theoretic description of the path functions near the boundary". Later he will form a conjecture on the latter that will be proved by Itô and McKean (see below). No mention of sticky (or slowly reflecting) boundary behaviour is made in Feller's paper. I am not aware of any conjecture on that either. It seems that Feller had no more definite idea on how to construct sticky processes (we will return to this point in the final section below). Feller concludes on page 4 that "the diffusion processes constructed in Section 12 [of his paper] represent the most general type of diffusion process in the open interval (r_1, r_2). For the closed interval we have *no* such complete result".

In his paper [Feller 1957a] Feller establishes the canonical infinitesimal generator form

$$\mathsf{L} = \frac{d}{dm}\frac{d}{ds} \tag{17}$$

where s is the *scale function* (due to Hille) and m is the *speed measure* (original to Feller) with probabilistic interpretations given in the papers [Feller 1954b, p. 25] and [Feller 1955a, p. 95] combined with [Feller 1954b, p. 12] respectively. Feller writes on page 459 of his paper that "It will be observed that (17) is much more general than the final expression in (10)". He then states that "This generalization introduces a remarkable simplification and unification of the theory". It is interesting that Feller relates the speed measure to a vibrating string. On page 460 he writes "The latter [speed measure] plays an important role in the theory of the vibrating string, m representing the distribution of the actual mass". He then states that "As a matter of curiosity it is possible to treat the motion of a continuous string with the entire mass concentrated at rational points" referring to his paper [Feller 1956b] with McKean. In terms of a diffusion process X it is known that large m corresponds to slow motion of X. This finally provides some clue to the sticky (slowly reflecting) boundary behaviour, however no explicit claim appears in the paper and no further probabilistic discussion is given either. Feller also writes a number of other papers in the 1950s related to his work in [Feller 1952a], [Feller 1954b], [Feller 1957a] however these three papers remain central. Among others in his paper [Feller 1959d] Feller extends the canonical infinitesimal generator form (17) to include a *killing measure* (corresponding to disappearance of Brownian particles due to chemical reactions for instance).

The state of the art on the boundary classification established in Feller's papers (from the standpoint of the semigroup theory) by the end of 1950s was as follows. All possible boundary conditions were known in their analytic form. Some of these were classical (Dirichlet, Neumann, Robin) and some were new (like those involving the second derivative). A natural question considered already by Feller in his paper [Feller 1954b] was to construct the underlying diffusion process corresponding to the given boundary condition and in this way obtain a fuller understanding of its meaning. In fact the entire boundary classification problem considered by Feller could be recast in these purely probabilistic terms. To illustrate the idea consider the simplest case of a standard Brownian motion and the state space $[0, \infty)$. The problem then is to describe/construct all strong Markov continuous processes X with values in $[0, \infty)$ that behave like a standard Brownian motion when off zero in $(0, \infty)$. In this case 0 is a regular boundary point and Feller's result states that the functions belonging to the domain of the infinitesimal generator of X can be characterised by the condition

$$p_1 f(0) - p_2 f'(0) + p_3 f''(0) = 0 \tag{18}$$

for some $p_1, p_2, p_3 \geq 0$ satisfying $p_1 + p_2 + p_3 = 1$. The case $f(0) = 0$ (Dirichlet condition) corresponds to X being killed upon reaching zero (i.e. moved to a coffin state);

the case $f'(0)=0$ (Neumann condition) corresponds to X being instantaneously reflected upon reaching zero; the case $f(0)=(p_2/p_1)f'(0)$ (Robin condition) corresponds to an elastic boundary behaviour of X at zero; the case $f''(0)=0$ (Feller condition) corresponds to X being absorbed upon reaching zero (infinite stickiness); the case $f'(0)=(p_3/p_2)f''(0)$ (Feller condition) corresponds to X being stickily (slowly) reflected upon reaching zero; the case $f(0)=-(p_3/p_2)f''(0)$ (Feller condition) corresponds to X being absorbed upon reaching zero and then killed after some independent exponentially distributed time has elapsed; the general case (18) with $p_1,p_2,p_3\neq 0$ corresponds to X being stickily (slowly) reflected upon reaching zero, then absorbed at zero after some independent exponentially distributed time has elapsed, and then finally killed after another independent exponentially distributed time has elapsed. All these cases can be expressed more systematically in terms of the speed measure and the killing measure (see e.g. [1, Chapter II] for a modern exposition). Elastic and sticky (slowly reflecting) constructions were not known when Feller had finished his work on the paper [Feller 1957a].

Shortly afterwards Feller came up with a conjecture on how to construct a diffusion process exhibiting *elastic* boundary behaviour. This has been recalled by Itô in his memoirs [15, pp 1-2] who spent the years 1954-56 at Princeton:

> Feller had just finished his works on the most general one-dimensional diffusion process, especially representing its local generator as (17) by means of a canonical scale function s and a speed measure m. I learned about these from Henry McKean, a graduate student of Feller, while I explained my previous work to McKean. There was once an occasion when McKean tried to explain to Feller my work on the stochastic differential equations along with the above mentioned idea of tangent. It seemed to me that Feller did not fully understand its significance, but when I explained Lévy's *local time* to Feller, he immediately appreciated its relevance to the study of the one-dimensional diffusion. Indeed, Feller later gave us a *conjecture* that the Brownian motion on $[0,\infty)$ with an *elastic* boundary condition could be constructed from the reflecting barrier Brownian motion by killing its local time ℓ^0 at the origin by an independent exponentially distributed random time, which was eventually substantiated in my joint paper with McKean [16] published in 1963.

Section 10 (pp 200–201) of that paper contains the elastic Brownian motion construction described above. Itô and McKean state on page 184 of their paper that "M. Kac [18] cited the problem of describing the sample paths of the elastic Brownian motion". In that paper of Kac I could not find a direct reference to the elastic boundary behaviour; however, when discussing one drawback of his approach on page 208 in relation to the Neumann condition on Γ he states that "one can again appeal to the theory of Brownian motion except that the *absorbing barrier* Γ must now be replaced by the *reflecting barrier*. Unfortunately, no rigorous treatment of reflecting barriers seems to be available". Feller published his paper [Feller 1951d] in the same proceedings as Kac and most likely was familiar with these claims.

It seems that Feller had no conjecture on how to construct a diffusion process exhibiting *sticky* boundary behaviour. Equipped with the concept of the local time mentioned above (which is an additive functional), and the result by Volkonskii [29] that a Markov process run by the clock obtained as the inverse of an additive functional preserves the Markov property, it was also accomplished by Itô and McKean. In his memoirs [15, p. 3] Itô recalls the background and relevance of this construction:

> My joint paper [16] with McKean in 1963 gave a probabilistic construction of the Brownian motion on $[0,\infty)$ subjected to the most general boundary condition whose analytic study had been established by Feller under some restrictions. Our methods involved the probabilistic idea originated in Lévy about the local time and excursions away from 0.

He then continues by explaining how the idea was extended in his fundamental paper [14] on excursion theory.

The sample-path approach to one-dimensional diffusions (return to Einstein) has been completed in the book by Itô and McKean [17]. In his memoirs [15, p. 2] Itô explains their key insight in relation to Feller's vision (with the mathematical notation slightly adjusted to the present style):

> A popular saying by Feller goes as follows: A one-dimensional diffusion traveler X makes a trip in accordance with the road map indicated by the scale function s and with the speed indicated by the measure m appearing in the generator L of X. This was substantiated in my joint book with McKean in the following fashion. Given a one-dimensional standard Brownian motion X which corresponds to $ds = dx$ and $dm = 2dx$, consider its local time ℓ^x at $x \in \mathbb{R}$ and the additive functional A defined by $A_t = \int_{\mathbb{R}} \ell_t^x \, m(dx)$ for $t \geq 0$. Then the time-changed process X_T by means of the inverse T of A turns out to be the diffusion governed by the generator $d^2/dm\,dx$.

In other words, recalling that the scale function s composed with the diffusion process X produces a new diffusion process (local martingale) on natural scale (for which the scale function is identity), this argument has established that each one-dimensional diffusion process can be obtained as a scale and time changed standard Brownian motion [i.e. a standard Brownian motion run by a new clock (controlled by the speed measure) in a new path space (shaped by the scale function)]. Section 5.7 in Itô–McKean's book [17] entitled "Feller's Brownian motions" provides a review of Feller's boundary classification results for Brownian motion in $(0,\infty)$ including the sample path constructions of elastic and sticky (slowly reflecting) processes. It remains to be a most complete book on the topic to date.

4 Further developments

In this section I will briefly address two subsequent developments among many others present in the literature. The first one deals with stochastic differential equations (in the sense of Itô's integral [13]) for *sticky* reflecting Brownian motion. The second one deals with the initial boundary value of Stroock and Williams [27] to which Feller's *semigroup approach* is *not* applicable in some cases of the parameters involved.

1. Recall that *sticky* boundary behaviour of diffusion processes was discovered by Feller in his paper [Feller 1952a]. The problem considered by Feller was to describe domains of the infinitesimal generators associated with strong Markov processes X in $[0,\infty)$ that behave like a standard Brownian motion B while in $(0,\infty)$. Using a semigroup approach Feller showed that a possible boundary behaviour of X at 0 is described by the following condition

$$f'(0+) = \tfrac{1}{2\mu} f''(0+) \tag{19}$$

where $\mu \in (0,\infty)$ is a given and fixed constant. In addition to the classic boundary conditions of (i) Dirichlet ($f(0+)=0$), (ii) Neumann ($f'(0+)=0$) and (iii) Robin ($f'(0+)=(1/2\mu)f(0+)$) corresponding to (i) killing, (ii) instantaneous reflection and (iii) elastic boundary behaviour of X at 0 respectively, the key novelty of the condition (19) is the appearance of the second derivative $f''(0+)$ measuring the "stickiness" of X at 0.

The problem of how to construct the sample paths of X in a canonical manner was solved by Itô and McKean [16, Section 10]). They showed that X can be obtained from the reflecting Brownian motion $|B|$ by the time change $t \mapsto T_t := A_t^{-1}$ where $A_t = t + (1/\mu)\ell_t^0(B)$ for $t \geq 0$ and $\ell^0(B)$ is the local time of B at 0. When $X = |B_T|$ is away from 0, then both the additive functional A and the "new clock" T run ordinarily (like the "old clock") and hence the motion of X is the same as the motion of $|B|$. When X is at 0 however, then A gets an extra increase from the local time $\ell^0(B)$ causing the new clock T to slow down and forcing X to stay longer at 0 (in comparison with the reflecting Brownian motion $|B|$ under the old clock). Thus the two processes X and $|B|$ have the same sample paths and the two motions differ only by their speeds (the former being slower).

The problem whether the sticky reflecting Brownian motion arising from (19) above can be obtained from a stochastic differential equation (SDE) driven by B has been considered in the literature. Skorokhod[15] conjectured in the 1980s that the stochastic differential equation for the sticky reflecting Brownian motion X has no strong solution. Building on the Itô–McKean construction one finds that the SDE system for X is given as follows:

$$dX_t = \tfrac{1}{2} d\ell_t^0(X) + I(X_t > 0)\, dB_t \tag{20}$$

$$I(X_t = 0)\, dt = \tfrac{1}{2\mu} d\ell_t^0(X) \tag{21}$$

where $X_0 = x$ in $[0,\infty)$ and $\ell^0(X)$ is the local time of X at 0. Moreover one can verify using the Itô-Tanaka formula that the system (20)+(21) is equivalent to the single equation

$$dX_t = \mu I(X_t = 0)\, dt + I(X_t > 0)\, dB_t \tag{22}$$

[15] Anatoliy Volodymyrovych Skorokhod (1930–2011) was a Soviet and Ukrainian mathematician.

obtained by incorporating (21) into (20). Then it can be shown that (i) the system (20)+(21) or equivalently the equation (22) has a jointly unique weak solution, and (ii) the system (20)+(21) or equivalently the equation (22) has no strong solution thus verifying Skorokhod's conjecture. For details see [6] and the references therein.

2. Consider the initial boundary value problem

$$u_t = \mu u_x + \tfrac{1}{2} u_{xx} \quad (t > 0, x \geq 0) \tag{23}$$

$$u(0,x) = f(x) \quad (x \geq 0) \tag{24}$$

$$u_t(t,0) = \nu u_x(t,0) \quad (t > 0) \tag{25}$$

of Stroock and Williams [27] where $\mu, \nu \in \mathbb{R}$ and the boundary condition is not of Feller's type ([Feller 1952a], [Feller 1954b], [Feller 1957a]) when $\nu < 0$. If $\nu > 0$ then it is known that the solution to (23)–(25) with $f \in C_b([0,\infty))$ can be represented as

$$u(t,x) = \mathbb{E}_x f(\tilde{X}_t) \tag{26}$$

where $\tilde{X}$ starts at x under $\mathbb{P}_x$, behaves like Brownian motion with drift μ when in $(0,\infty)$, and exhibits a sticky boundary behaviour at 0. The process $\tilde{X}$ can be constructed by a time change of the reflecting Brownian motion X with drift μ (the inverse of the running time plus the local time of X at 0 divided by ν) forcing it to spend more time at 0 (cf. [16, p. 186]). If $\nu = 0$ then (26) remains valid with $\tilde{X}$ being absorbed at 0 (corresponding to the limiting case of infinite stickiness). If $\nu < 0$ then Feller's semigroup approach ([Feller 1952a], [Feller 1954b], [Feller 1957a]) is no longer applicable since the speed measure of $\tilde{X}$ cannot be negative. Stroock and Williams [27] show that the minimum principle breaks down in this case (non-negative f can produce negative u) so that the solution to (23)–(25) cannot be represented by (26) where $\tilde{X}$ is a strong Markov process which behaves like Brownian motion with drift μ when in $(0,\infty)$.

Inspired by their insights in the recent paper [24] we develop an entirely different approach to solving (23)–(25) probabilistically that applies to smooth initial data f vanishing at ∞. Firstly, exploiting higher degrees of smoothness of the solution u in the interior of the domain (which is a well-known fact from the theory of parabolic PDEs) we reduce the *sticky* boundary behaviour at 0 to (i) a *reflecting* boundary behaviour when $\nu = \mu$ and (ii) an *elastic* boundary behaviour when $\nu \neq \mu$. Secondly, writing down the probabilistic representations of the solutions to the resulting initial boundary value problems expressed in terms of the reflecting Brownian motion with drift μ and its local time at 0, choosing joint realisations of these processes where the initial point is given explicitly so that the needed algebraic manipulations are possible (making use of the extended Lévy's distributional theorem), we find that the following probabilistic representation of the solution is valid

$$u(t,x) = \mathbb{E}_x F(X_t, \ell_t^0(X)) \tag{27}$$

where X is a reflecting Brownian motion with drift μ starting at x under $\mathbb{P}_x$, and $\ell^0(X)$ is the local time of X at 0. The function F is explicitly given by

$$F(x,\ell) = f(x) - f'(x) \int_0^\ell e^{-2(\nu-\mu)s}\, ds \tag{28}$$

for $x \geq 0$ and $\ell \geq 0$. The derivation applies simultaneously to all μ and ν with no restriction on the sign of ν, and the process X (with its local time) plays the role of a fundamental solution in this context (a building block for all other solutions).

Since $(X, \ell^0(X))$ is a Markov process we see that the solution u is generated by the semigroup of transition operators $(\mathsf{P}_t)_{t \geq 0}$ acting on f by means of (27) and (28) (in the reverse order). Moreover, it is clear from (27) and (28) that the solution can be interpreted in terms of X and its *creation* in 0 at rate proportional to $\ell^0(X)$. Note that this also holds when $\nu < 0$ in which case the Feller's semigroup approach based on the probabilistic representation (26) is not applicable. Finally, invoking the law of $(X_t, \ell_t^0(X))$ we derive a closed integral formula for u expressed in terms of μ, ν and f. Integrating further by parts yields a closed formula for u where smoothness of f is no longer needed (see [24] for details).

References

All citations of the form [Feller 19*nn*], resp., [*Feller 19*nn*] (if the respective paper is not included in these Selecta) point to Feller's bibliography, pp. xxv–xxxiv.

[1] Borodin, A. N. and Salminen, P.: *Handbook of Brownian Motion: Facts and Formulae*. Birkhäuser, Basel 2002.

[2] Brown, R.: A brief account of microscopical observations made in the months of June, July, and August, 1827, on the particles contained in the pollen of plants; and on the general existence of active molecules in organic and inorganic bodies. *Edinburgh New Philosophical Journal* (1928) 358–371. [Originally printed as part of a privately circulated pamphlet.]

[3] Chapman, S.: On the Brownian displacements and thermal diffusion of grains suspended in a non-uniform fluid. *Proc. Roy. Soc. London Ser. A* **119** (1928) 34–54.

[4] Dynkin, E. B.: Infinitesimal operators of Markov processes. *Theory Probab. Appl.* **1** (1956) 34–54.

[5] Einstein, A.: Über die von der molekularkinetischen Theorie der Wärme geforderte Bewegung von in ruhenden Flüssigkeiten suspendierten Teilchen. *Ann. Phys.* **17** (1905) 549–560. [English translation "On the motion of small particles suspended in liquids at rest required by the molecular-kinetic theory of heat" in: Stachel, J. (ed.): *Einstein's Miraculous Year*. Princeton Univ. Press, Princeton (NJ) 1998, pp. 85–98.]

[6] Engelbert, H. J. and Peskir, G.: Stochastic differenial equations for sticky Brownian motion. Stochastics 86 (2014) 993-1021.

[7] Fick, A.: Ueber Diffusion. (*Poggendorff's*) *Annalen der Physik und Chemie* **94** (1855) 59–86.

[8] Fokker, A. D.: Die mittlere Energie rotierender elektrischer Dipole im Strahlungsfeld. *Ann. Phys.* **43** (1914) 810–820.

[9] Fourier, J.: *Théorie Analytique de la Chaleur*. Firmin Didot Père et Fils, Paris 1822.

[10] Hille, E.: On the integration problem for Fokker–Planck's equation in the theory of stochastic processes. Den 11te Skandinaviske Matematikerkongress, Trondheim, 1949. Johan Grundt Tanums Forlag, Oslo 1952, pp. 183–194.

[11] Hille, E.: Les probabilités continues en chaîne. *C. R. Acad. Sci. Paris* **230** (1950) 34–35.

[12] Hunt, G. A.: Some theorems concerning Brownian motion. *Trans. Amer. Math. Soc.* **81** (1956) 294–319.

[13] Itô, K.: Stochastic integral. *Proc. Imperial Acad. Tokyo* **20** (1944) 519–524.

[14] Itô, K.: Poisson point processes attached to Markov processes. In: Le Cam, L. et al. (eds.): *Proceedings of the Sixth Berkeley Symposium on Mathematical Statistics and Probability, Vol. 3*. University of California Press, Berkeley (CA) 1970, pp. 225–239.

[15] Itô, K.: Memoirs of my research on stochastic analysis. In: Benth, F. E. et al. (eds.): *Stochastic Analysis and Applications, The Abel Symposium 2005*. Springer, Berlin 2007, pp. 153–196.

[16] Itô, K. and McKean, H. P. Jr.: Brownian motions on a half line. *Illinois J. Math.* **7** (1963) 181–231.

[17] Itô, K. and McKean, H. P. Jr.: *Diffusion Processes and their Sample Paths*. Springer, Berlin 1965.

[18] Kac, M.: On some connections between probability theory and differential and integral equations. In: Neyman, J. (ed.): *Proceedings of the Second Berkely Symposium on Mathematical Statistics and Probability, August 1950*. University of California Press, Berkeley (CA) 1951, pp. 189–225.

[19] Kolomogoroff, A. N.: Über die analytischen Methoden in der Wahrscheinlichkeitsrechnung. *Math. Ann.* **104** (1931) 415–458. [English translation "On analytical methods in probability theory" in: A. N. Shiryayev (ed.): *Selected works of A. N. Kolmogorov* Vol II. Kluwer, Dordrecht 1992, pp. 62–108.]

[20] Kolmogoroff, A. N.: Zur Theorie der stetigen zufälligen Prozesse. *Math. Ann.* **108** (1933) 149–160. [English translation "On the theory of continuous random processes" in: A. N. Shiryayev (ed.): *Selected works of A. N. Kolmogorov* Vol II. Kluwer, Dordrecht 1992, pp. 156–168.]

[21] Langevin, P.: Sur la théorie du mouvement brownien. *C. R. Acad. Sci. Paris* **146** (1908) 530–533.

[22] Nelson, E.: *Dynamical Theories of Brownian Motion.* Princeton Univ. Press, Princeton (NJ) 1967.

[23] Peskir, G.: On the diffusion coefficient: The Einstein relation and beyond. *Stoch. Models* **19** (2003) 383–405.

[24] Peskir, G.: A probabilistic solution to the Stroock–Williams equation. *Ann. Probab.* **42** (2014) 2197–2206.

[25] Planck, M.: Über einen Satz der statistischen Dynamik and seine Erweiterung in der Quantentheorie. *Sitzungsber. Preuß. Akad. Wiss.* **24** (1917) 324–341.

[26] Smoluchowski, M. von: Über Brownsche Molekularbewegung unter Einwirkung äußerer Kräfte und deren Zusammenhang mit der verallgemeinerten Diffusionsgleichung. *Ann. Phys.* **24** (1916) 1103–1112.

[27] Stroock, D. W. and Williams, D.: A simple PDE and Wiener-Hopf Riccati equations. *Commun. Pure Appl. Math.* **58** (2005) 1116–1148.

[28] Uhlenbeck, G. E. and Ornstein, L. S.: On the theory of Brownian motion. *Phys. Rev.* **36** (1930) 823–841.

[29] Volkonskii, V. A.: Random substitution of time in strong Markov processes. *Theory Probab. Appl.* **3** (1958) 310–326.

[30] Wiener, N.: Differential space. *J. Math. and Phys.* **2** (1923) 131–174.

[31] Yosida, K.: An operator-theoretical treatment of temporally homogeneous Markoff process. *J. Math. Soc. Japan* **1** (1949) 244–253.

Prof. Goran Peskir
School of Mathematics
The University of Manchester
Oxford Road
Manchester M13 9PL
United Kingdom
goran@maths.man.ac.uk

Feller's Work in Renewal Theory, the Law of the Iterated Logarithm and Karamata Theory

by *Ross Maller* from Canberra

1 Introduction

In this chapter we review and consider in some detail Feller's work in renewal theory, the iterated logarithm law and its generalizations and extensions, and Feller's systematic use of the theory and properties of regularly varying functions in probability. The range of topics is immense as is the span of time covered. Feller returned again and again to problems that tantalized him, each time making progress, sometimes incrementally, sometimes by leaps, but always with a dedicated view to obtaining the fullest understanding of the problems he set out to solve, and to expressing them in the simplest and most elegant manner, consistent with complete rigour. The advances he made set the stage for an enormous amount of future development as well as solidly establishing the foundations of many subjects.

2 Renewal Theory

[Feller 1941a][1] represents Feller's first substantial foray into renewal theory. He deals with the solution to

$$U(t) = G(t) + \int_0^t U(t-x)\,dF(x),\ t > 0, \tag{1}$$

[1] The references [Feller 19*nn*] and [*Feller 19*nn*] (the star indicating that the respective paper is not contained in these Selecta) refer to Feller's bibliography, while [*n*] points to the list of references at the end of this essay.

R.L. Schilling et al. (eds.), *Selected Papers II*,

where F and G are finite, nondecreasing, right-continuous functions on $[0,\infty)$, with $F(0) = G(0) = 0$, such that the Laplace transforms

$$\phi(s) := \int_0^\infty e^{-st}\, dF(t) \quad \text{and} \quad \gamma(s) := \int_0^\infty e^{-st}\, dG(t) \tag{2}$$

converge at least for s greater than some $\sigma \geq 0$.

He motivates the investigation by mention of *industrial replacement* models studied by Lotka[2] in 1939 and 1940 (just as renewal theory is commonly introduced today with examples of a need to change burnt-out light globes, etc.) and *population theory* – dealing with rates of reproduction of offspring in a population model. Here we see at an early stage Feller's desire to relate theoretical analyses to practical examples – as well as his liking for simple proofs which cut through technicalities as much as possible. Another distinguishing characteristic also appears here: an expressed desire early in the paper to give rigorous (and correct) proofs and precise conditions for validity of results sometimes presented previously by others as being of a *heuristic nature*.

Feller motivates the use of discontinuous F and G and the Stieltjes integral in Eq. (1) by means of the population example (we cannot start with all members being of zero age). The solution of (1) is specified in Theorem 1 as follows:

> *When $\phi(\sigma+) > 1$ let $\sigma' > \sigma$ be the root of the characteristic equation $\phi(s) = 1$; σ' exists and is unique. When $\phi(\sigma+) \leq 1$ let $\sigma' = \sigma$. Then there exists exactly one finite, nondecreasing, function $U(t)$ satisfying Eq. (1). Its Laplace transform $\omega(s) := \int_0^\infty e^{-st}\, dU(t)$ converges for $s > \sigma'$ and satisfies*
>
> $$\omega(s) = \frac{\gamma(s)}{1-\phi(s)}. \tag{3}$$

Proof of this proceeds by showing that $\omega(s)$ is completely monotone, then by reference to [Feller 1939b], which deals with such functions as solutions to (1) (and was apparently written with the present purpose in mind). Theorem 2 gives a density version, namely, a solution to

$$u(t) = g(t) + \int_0^t u(t-x)f(x)\, dx,\ t > 0, \tag{4}$$

where the functions f and g are measurable, nonnegative and bounded in every finite interval $0 \leq t \leq T$. The Laplace transforms

$$\phi(s) := \int_0^\infty e^{-st} f(t)\, dt \quad \text{and} \quad \gamma(s) := \int_0^\infty e^{-st} g(t)\, dt \tag{5}$$

are assumed to converge for s greater than some $\sigma \geq 0$. Then there exists exactly one nonnegative solution $u(t)$ of (4) which is bounded in every finite interval. For

[2] Works not referenced directly in the present review can be found referenced in the corresponding

this function the integral $\omega(s) := \int_0^\infty e^{-st}u(t)\,dt$ converges at least for $s > \sigma'$, where $\sigma' > \sigma$ is the root of the characteristic equation $\phi(s) = 1$ when $\phi(\sigma+) > 1$, and $\sigma' = \sigma$ when $\phi(\sigma+) \leq 1$. For $s > \sigma'$, $\omega(s)$ satisfies (3).

Feller in [Feller 1941a] goes on to study the asymptotic behaviour of $u(t)$, not directly, at first, but via its mean value

$$u^*(t) := \frac{1}{t}\int_0^t u(\tau)\,d\tau, \quad t > 0.$$

Assuming the integrals

$$a := \int_0^\infty f(t)\,dt \quad \text{and} \quad b := \int_0^\infty g(t)\,dt \tag{6}$$

are finite, he shows that a necessary and sufficient condition for $\lim_{t\to\infty} u^*(t)$ to exist and equal C, a finite positive constant, is that $a = 1$ and $m := \int_0^\infty tf(t)\,dt$ is finite. In that case,

$$C = \frac{b}{m}.$$

A corresponding result is also found for the case $a > 1$, and the case $a < 1$ is easily dealt with. The stationary case $a = 1$ is treated in more detail, and with reference to the convergence of $u(t)$, rather than that of $u^*(t)$. Under some further assumptions on moments and the total variation of f and g, $u(t)$ is shown to approach b/m at a polynomial rate, depending on the finiteness of certain moments. Proof of this is longish (more than four pages) and uses Tauberian theory due to Haar from 1927. Further discussion in [Feller 1941a] is concerned with the possible oscillation behaviour of $u(t)$ (in case f and g vanish for large t), and a final section develops in detail computational methods – series solutions – for $u(t)$, expanding on and correcting in certain ways Lotka's methods.

Feller mentions that

> [t]he existence of a solution to (4) can be deduced directly from a well-known result of Paley and Wiener (in 1933) [...] However, the case of nonnegative functions f and g is too simple to justify the deep methods used by Paley and Wiener.

As a followup paper, [Feller 1948c] applies the theory in the context of an instrument counting events in what we now call a Poisson process. Times between events are exponentially distributed and explicit expressions and/or approximations are obtained for the Laplace transforms and other quantities occurring in [Feller 1941a]. Prior to this, [*Feller 1940b] calculates the conditional probability that the interval between two consecutive events in a Poisson process exceeds $T < D$, given that exactly $n \geq 2$ events occurred in a randomly selected interval of length D, thus correcting a previously published result due to Ruark.

[Feller 1949a], joint with P. Erdös and H. Pollard, is purely analytic in nature, concerned with properties of discrete renewal sequences.

---

paper of Feller; e. g., [Feller 1941a] contains references to four papers of Lotka in 1939 and 1940.

Suppose the complex valued function $P(x) := \sum_{k=0}^{\infty} p_k x^k$ *is not a power series in* x^t *for any integer* $t > 1$, *where* $\sum_{k=0}^{\infty} p_k = 1$ *and* $m := \sum_{k=0}^{\infty} k p_k \leq \infty$. *Then* $1 - P(x)$ *has no zeroes in the circle* $|x| < 1$, *and the series*

$$\frac{1}{1-P(x)} = \sum_{k=0}^{\infty} u_k x^k$$

has the property

$$\lim_{n \to \infty} u_n = 1/m.$$

By contrast with [Feller 1941a], [Feller 1949d] adopts a more probabilistic notation and approach. Feller realises that much theory concerned with sequences of repeated trials (which may be independent or dependent) can be subsumed into a study of *recurrent events*. Success runs in a series of coin tosses and their often counter-intuitive properties are used as motivation. For example, an equalisation (the event that the number of heads equals the number of tails) forms a recurrent event, $\mathscr{E}$, and the times at which the event occurs, the durations between occurrences, the number of occurrences of $\mathscr{E}$ in a series of given length, etc., form interesting objects of study. The corresponding results are of *renewal* type. Further examples of a recurrent event are the zeroes of sums $S_n := X_1 + \cdots + X_n$ of i.i.d. random variables; the passage of a system through a specified state of a Markov chain; etc. This material was to form a substantial but accessible part of Feller's first, and famous, book [Feller 1966b, Chapter XIII, p. 303].

Feller in [Feller 1949d] proceeds very carefully, with a formal definition of a recurrent event: given a series of trials labeled by integers $n = 1, 2, \ldots$, event $\mathscr{E}$ is specified by a rule which at each n and outcome in the sample space allows us to decide whether or not $\mathscr{E}$ has occurred. Thus, we can make a classification of each finite subsequence $\{j_1, \ldots, j_n\}$ into an $\mathscr{E}$-class or a non–$\mathscr{E}$-class. A probability measure P is supposed given on the space of infinite sequences $\{j_1, j_2, \ldots\}$ of positive integers, where j_k represents the possible outcomes of the kth trial. Then $\mathscr{E}$ is said to be recurrent if for every $\mathscr{E}$-sequence $\{j_1, \ldots, j_n\}$ and every sequence $\{j_{n+1}, \ldots, j_{n+\nu}\}$, $\nu = 1, 2, \ldots$, the following two conditions hold: first,

$$P(j_1, \ldots, j_n, j_{n+1}, \ldots, j_{n+\nu}) = P(j_1, \ldots, j_n) P(j_{n+1}, \ldots, j_{n+\nu}),$$

and, second, $\{j_1, \ldots, j_n\}$ and $\{j_{n+1}, \ldots, j_{n+\nu}\}$ belong to the same class.

The number of trials up to and including the nth occurrence of $\mathscr{E}$ is denoted by $S_n := X_1 + \cdots + X_n$, where X_k is 1 plus the number of trials between the $(k-1)$st and kth occurrence of $\mathscr{E}$. The X_k ("recurrence times") are independent with common discrete distribution

$$\begin{aligned} p_\nu :=& P(X_k = \nu) \\ =& \text{probability that } \mathscr{E} \text{ occurs for the first time at the } \nu\text{th trial}, \ \nu = 1, 2, \ldots . \end{aligned}$$

This distribution is not necessarily proper; if $\sum_\nu p_\nu < 1$ there is a positive probability for non-occurrence of $\mathscr{E}$, in which case $\mathscr{E}$ is called *uncertain*; (later, *transient*,

[Feller 1966b, p. 310]); otherwise, *certain (persistent)*. $\mathscr{E}$ is *periodic* with period t if it can only occur at times $t, 2t, \ldots,$. The cumulative distribution function corresponding to $(p_\nu)_{\nu=1,2,\ldots}$ is $F(x)$, $x \geq 0$, and (with $u_0 = 1$)

$$\begin{aligned} u_n &= \text{probability that } \mathscr{E} \text{ occurs at the } n\text{th trial} \\ &= u_0 p_n + u_1 p_{n-1} + \cdots + u_{n-1} p_n. \end{aligned}$$

With these definitions, $\mathscr{E}$ is shown to be uncertain if and only if $\sum_n u_n < \infty$; if $\mathscr{E}$ is certain with period t then $\lim_{n\to\infty} u_{nt} = t/\mu$, where $\mu := \sum_\nu \nu p_\nu \in (0, \infty]$. Proofs of these using generating functions are very easy and avoid the use of "*deep tools of Fourier analysis and complex variables*" as employed by Täcklind in 1945. But Feller references a 1948 paper of Blackwell for the case of a continuous time parameter.

In this paper, [Feller 1949d], Feller improves on another theorem of Täcklind to the effect that, when $\mathscr{E}$ is certain and aperiodic, and $M := \sum_\nu (\nu - 1)\nu p_\nu < \infty$, then, as $n \to \infty$,

$$(a) \quad \sum_{i=0}^{n} u_i - \frac{n+1}{\mu} \to \frac{M}{2\mu^2};$$

$$(b) \quad \sum_{n \geq 0} \left| u_n - \frac{1}{\mu} \right| < \infty;$$

$$(c) \quad u_n = \frac{1}{\mu} + o\left(\frac{1}{n}\right).$$

The number of realisations of $\mathscr{E}$ in n trials, N_n, satisfies a central limit theorem (is asymptotically normal as $n \to \infty$ after centering and norming) provided the X_k have finite variance. Feller transcribes a result of Doeblin (announced in [8]) to the effect that, more generally, N_n has a (proper, nondegenerate) limiting distribution as $n \to \infty$ after centering and norming if and only if

$$\lim_{x\to\infty} \frac{1 - F(cx)}{1 - F(x)} \text{ exists and equals } c^{-\alpha} \tag{7}$$

for each $c > 0$ for some $\alpha \in (0, 2)$. This is a very early appearance of Karamata's regular variation condition in probability, though apparently not recognised as such (as yet) by Feller (or Doeblin). The limiting distribution of N_n when this holds is of course the famous arc-sine law due to Lévy in 1937, and attributed accordingly by Feller. Feller derives this result from a lemma (which *amplifies Doeblin's criterion*) on the convergence of S_n as $n \to \infty$ after centering and norming. These calculations later appeared in the second edition of the monograph. Further results on moments of N_n, including strong laws and iterated log laws, and an application of the theory of recurrent events to denumerable Markov chains, complete [Feller 1949d].

Some time after [Feller 1949d], Feller with coauthor Stephen Orey returns to general renewal theory in [Feller 1961a]. The notations now are very much random walk style as we currently use them: $S_n = \sum_{k=0}^{n} X_k$, $n = 0, 1, 2, \ldots$, is a sum of i.i.d. random variables $(X_k)_{k=1,2,\ldots}$ in $\mathbb{R}$ (with $X_0 = S_0 = 0$) whose distribution is the n-fold convolution F^{n*}. The renewal function

(8) $$U(B) := \sum_0^\infty P(S_n \in B) = \sum_0^\infty F^{n*}(B)$$

is the expected number of times S_n visits Borel subset B of $\mathbb{R}$. Chung and Fuchs showed in 1951 that S_n is interval recurrent, i.e., $P(S_n \in \Delta \text{ i.o. as } n \to \infty) = 1$ for some (hence every) bounded interval $\Delta = [0, \delta)$ if and only if $U(\Delta) = \infty$.[3] So interest lies in the alternative, transient, case, when for some (hence every) interval Δ we have $0 < U(\Delta) < \infty$. Feller and Orey deal with this case, using a combination of Fourier analytic (characteristic function) methods, with somewhat similar calculations to those in corresponding parts of Chung and Fuchs, together with probabilistic methods.

Soon after, [Feller 1961b] publishes a simpler proof of the main result in the paper [Feller 1961a] as:

Theorem 1. *For an arbitrary transient F we have $U(\Delta + t) \to 0$ as $t \to \pm\infty$ for each bounded interval Δ except if the moment*

(9) $$\mu := \int_{-\infty}^{\infty} x F(dx)$$

exists. In this case necessarily $\mu \neq 0$ and for non-arithmetic distributions

(10) $$U(\Delta + t) \to \begin{cases} |\mu^{-1}||\Delta| & \text{as } \mu t \to \infty \\ 0 & \text{as } \mu t \to -\infty. \end{cases}$$

(The obvious amendment holds for arithmetic distributions.)

(In Theorem 1, $|\Delta|$ denotes the length of Δ and $\Delta + t$ denotes the translation of Δ by real number t; F is arithmetic if all points of increase occur at multiples of a fixed number $d > 0$). By "existence" of μ in (9) is meant absolute convergence of the integral: nonexistence thus corresponds to the cases

$$\infty = \int_{-\infty}^{0} |x| F(dx) > \int_0^\infty x F(dx),$$

$$\text{or} \quad \int_{-\infty}^{0} |x| F(dx) < \int_0^\infty x F(dx) = \infty,$$

$$\text{or} \quad \infty = \int_{-\infty}^{0} |x| F(dx) = \int_0^\infty x F(dx).$$

Feller's proof in the third case may be "simple" as per the title of his paper but it is also especially instructive; it involves showing that if $\limsup_{t \to \pm\infty} U(\Delta + t)/|\Delta|$ is positive, then one or the other side (S_n taking positive or negative values) must dominate.

[3]"i.o." stands for "infinitely often". In an acknowledgement, Chung and Fuchs attribute their *simple proofs of these (particular) results* to Feller.

The earlier proof in [Feller 1961a] is quite different and also worthy of perusal for the deep probabilistic intuition it displays. The analysis is in terms of quantities

$$W(B) := P\left(\bigcup\{S_n \in B\}\right), \quad \text{for Borel } B, \tag{11}$$

and, for a given interval $I \subset \mathbb{R}$ and Borel $B \subset I$,

$$W_I(B) := P(\text{first entry into } I \text{ occurs at a point } x \in B). \tag{12}$$

In a preliminary lemma, Feller gives a version of (10) for a symmetrised measure $V(I) := (U(I) - U(-I))/2$. This is done by Fourier (characteristic function) methods, with somewhat similar calculations to those in corresponding parts of Chung and Fuchs (but Chung and Fuchs also consider random walks in dimensions $1, 2, 3$). Next, for a fixed $\delta > 0$, let $\Delta = [0, \delta]$ and $\Delta^* = \Delta \cup (-\Delta) = (-\delta, \delta)$. Then if I is any interval of length δ, Feller shows that

$$U(I) = \int_I W_I(dx) U(I - x) \leq U(\Delta^*) W(I), \tag{13}$$

hence, for all intervals of length δ,

$$W(I) \leq U(I) \leq U(\Delta^*) W(I). \tag{14}$$

With these bounds and some further analysis of probabilities the limit behaviour of $U(x + I)$ as $x \to \infty$ is determined.

Feller's specific contributions to renewal theory end[4] with [Feller 1961b], though, as the subject merges at its edges into various other aspects of the theory (and applications) of random walks, Feller's work in a variety of kinds of limit theorems for random walk continues to be relevant. And then there was the major development and synthesis of ideas in the book [*Feller 1971].

Many other major figures of course contributed to the systematic development of renewal ideas, at around the same period as Feller worked, and subsequently; the work in particular, of Spitzer[5] (see Spitzer [55, 56, 57], Blackwell [3], Doney [10, 11], Stone [58, 59, 60], Port and Stone [51], Erickson [16, 17, 18], and Kesten (see for example the review paper Kesten [33], and Kesten [34]) should be mentioned as well as that of Chung and his coworkers, as referenced earlier. But otherwise, it is impossible in the present review to cover the scope and breadth of Feller's influence in the field; literally hundreds of papers, both theoretical and applied, show traces of his presence, as the effects of renewal theory spread far and wide.

3 Iterated Logarithm and Related Results

In [Feller 1943b], Feller generalizes a 1938 result of Cramér. Here $S_n = X_1 + X_2 + \cdots + X_n$ is a sum of independent (not necessarily i.i.d.) random variables with distri-

[4] Though it is worth mentioning that [Feller 1966c] is concerned with hitting times of points by a simple random walk; in particular, the time of the first visit to a point is shown to be infinitely divisible with a quite explicit expression for its generating function.

[5] In his book, Spitzer thanks his teachers, among them Feller, *who introduced me to the theory of*

butions F_n having expectations 0 and variances $\sigma_k^2 = EX_k^2$. The X_k are bounded:

$$\max_{1 \le k \le n} |X_k| \le \lambda_n s_n, \tag{15}$$

where $s_n^2 := \sigma_1^2 + \sigma_2^2 + \cdots + \sigma_n^2 \to \infty$ and $\lambda_n \to 0$ as $n \to \infty$. Assuming in addition that $0 < \lambda_n x < (3 - 5^{1/2})/4$ (Feller enjoys specifying exact numerical values where a generic "c" constant is often otherwise used, though he stresses that these values are usually not in any way sharp), Theorem 1 gives a large deviations result of the form

$$1 - F_n(xs_n) = e^{-x^2 Q_n(x)/2} \left[1 - \Phi(x) + \theta e^{-x^2/2} \right] \tag{16}$$

where the parameter θ is bounded independent of the sequence (X_k) and of n, and $Q_n(x) = \sum_{\nu \ge 1} q_{n,\nu} x^\nu$ is a series whose coefficients $q_{n,\nu}$ depend only on the first ν moments of $X_1, X_2, \ldots, X_n$. The $q_{n,\nu}$ are defined in terms of Thiele's semi-invariants of the distributions of the X_k, and the proof of the theorem proceeds by considering, as Cramér did, an Esscher transform (the X_k have a finite Laplace transform, since bounded), and application of a 1941 bound of Berry for the distance between the distribution of a normed, centred sum and the normal distribution. Interestingly, Esscher's result in 1932 was derived in an early insurance risk scenario (related of course to Cramér's famous work in this area).

Theorems 2 and 3 of [Feller 1943b] give bounds, uniform in n, and, when $x = o(1/\lambda_n)$, a limiting result as $x \to \infty$, for the ratio

$$\frac{1 - F_n((x + \alpha/x)s_n)}{1 - F_n(xs_n)},$$

where $\alpha > 0$. (The limit is $e^{-\alpha}$). Feller mentions briefly the possibility of application of results like these to the law of the iterated logarithm, and this promise is realized in a series of papers which followed.

[Feller 1943c] assumes the same setup and similar notation as for [Feller 1943b], and considers *upper and lower classes*, $\mathscr{U}$ and $\mathscr{L}$, for S_n and a deterministic sequence ϕ_n (assumed nondecreasing). Following Lévy's work in the early 1930s and in Lévy's 1937 book, Feller says that $(\phi_n) \in \mathscr{L}$ if $S_n > s_n \phi_n$ for infinitely many n, almost surely (a.s.); whereas $(\phi_n) \in \mathscr{U}$ if $S_n > s_n \phi_n$ only for finitely many n, a.s. Assuming (15) with $\lambda_n \le (1/200)\phi_n$, his Theorem 1 tells us that

> $(\phi_n) \in \mathscr{U}$ $((\phi_n) \in \mathscr{L})$ *if and only if the series*
>
> $$\sum_{n \ge 1} \frac{\sigma_n^2}{s_n^2} e^{-\phi_n^2 [1 + Q_n(\phi_n)]/2}$$
>
> *converges (diverges).*

Proof of this and a series of corollaries and extensions is by careful and com-

stochastic processes.

plicated choices of sequences and subsequences of integers, splitting of the X_k into independent blocks, and what amount to various versions of Borel–Cantelli type arguments to get the a.s. statements; a procedure much used still today. The large deviation theorems of [Feller 1943b] are crucial but otherwise the paper is self-contained.

In a useful introduction to this paper, Feller motivates interest in the law of the iterated logarithm, originally due to Kolmogorov, as stemming from *a special problem in number theory*, and gives a step-by-step précis of the history of a.s. bounds for S_n, starting with the strong law then tracing its development through results of Hausdorff, Hardy-Littlewood, Steinhaus, Khintchine, Lévy and Erdös.

The survey address [Feller 1945b] to the American Mathematical Society is similarly concerned with an overview of classical results to that date. Feller begins with a reprise of the central limit theorem and its "little brother", the law of large numbers,[6] before moving on to a discussion of the law of the iterated logarithm. He mentions that its present form grew out of a 1914 problem of Hardy and Littlewood in Diophantine approximations.

[Feller 1946a] is a short but important paper, restricted to the i.i.d. case, but motivated by reference to work of Lévy [39] and Marcinkiewicz [46] on independent (not necessarily i.i.d.) random variables. The event "$\mathscr{L} : |S_n| > a_n$ takes place for infinitely many n", where a_n is a nondecreasing nonstochastic sequence, has (by the Hewitt–Savage 0-1 law) probability zero or 1. The same is true for the event "$\mathscr{L}^* : |X_n| > a_n$ takes place for infinitely many n". Are the probabilities of these events the same? Theorems 1 and 2 answer this question affirmatively in cases when X_1 has finite expectation (equal to 0) but $\int_{-\infty}^{\infty} |x|^{1+\delta} dV(x) = \infty$ for some $\delta \in (0,1)$ (for Theorem 1), and when $\int_{-\infty}^{\infty} |x| dV(x) = \infty$ (for Theorem 2). Here $V(x)$ is the distribution function of X_1. Growth restrictions such as $\liminf_{n\to\infty} a_{2n}^2/a_n^2 > 2$ are placed on the a_n. These seemingly relatively simple results (proved of course by relating the probability of event $\mathscr{L}$ to convergence or divergence of the series $\sum_{n\geq 1} P(|X_1| > a_n)$) have received wide application subsequently, not least for example in the spectacular result of Kesten [32] concerning limiting behaviour of S_n/n in the i.i.d. case. Feller as he says in his introduction

> leave(s) the domain of applicability of the central limit theorem (for) practically unknown terrain; [...] no systematic tools have as yet been developed [...].

The works of Lévy and Marcinkiewicz mentioned above are apparently the only precursors.

[Feller 1946b] returns to the concerns of [Feller 1943c], but is again restricted to i.i.d. X_k, where results, definitive in a certain sense, can be obtained. The X_k are assumed to have distribution function F with expectation 0 and variance 1. The novel insights lie in a different way of defining the test event, and seemingly a new application of Kronecker's lemma at a certain point. Define $\eta_k^2 = k(\log\log k)^{-4}$ for

[6] Although not directly relevant, we mention here [Feller 1945d], which gives necessary and sufficient conditions for degenerate convergence of S_n, with application to the St. Petersburg paradox. See also Hans Fischer's chapter in the present volume.

$k > 80$, and in any monotone way for $k < 80$, and set

$$B_n = \sum_{k=1}^{n} EX_1^2 \mathbb{1}_{\{|X_1| < \eta_k\}}.$$

Theorem 2 of the paper is

Theorem 2. *For any monotonic sequence ϕ_n the divergence (convergence) of the series*

$$\sum_{n \geq 1} \frac{\phi_n}{n} e^{-(1/2)\phi_n^2} \tag{17}$$

is necessary and sufficient that, with probability one, the inequality

$$S_n > B_n^{1/2} \phi_n$$

be satisfied for infinitely (only finitely) many n.

A comment to the effect that B_n may be replaced by the simpler expression

$$B_n = nEX_1^2 \mathbb{1}_{\{|X_1| < \eta_n\}}$$

is made, but "*the author was unable to devise a proof simple enough to be justified*" (for inclusion).

There are very subtle and clever choices of subsequences and truncation levels involved in these proofs, which have provided a source of inspiration for later writers. But, ranging widely and deeply over so many areas, as it did, it's not surprising that there were occasional glitches in Feller's work. His proof of Theorem 2 overlooked a mean correction term and is strictly correct only for symmetric X_n. The gap was eventually filled by Einmahl [13] and Theorem 2 is correct as stated.

From Theorem 2 is deduced (Feller's Theorem 1)

Theorem 3. *Suppose, as $x \to \infty$,*

$$\int_{|t|>x} t^2 \, dF(t) = O\left((\log \log x)^{-1} \right);$$

then $(\phi_n) \in \mathscr{U}$ or $(\phi_n) \in \mathscr{L}$ according as the series in (17) converges or diverges.

After a long period during which Feller turns his attention to diffusion processes, among other things, he returns to iterated log laws for random walks in [Feller 1968d]. Here he introduces a unique and important idea. First he notes that a generalized kind of LIL for sums of i.i.d. random variables, of the form

$$\limsup_{n \to \infty} \frac{S_n}{a_n} = 1 \quad \text{a.s.,} \tag{18}$$

for a nonstochastic sequence $a_n > 0$, may hold when the variance of the X_k is infinite. The X_k in [Feller 1968d] are restricted to be symmetric about 0, so there is no need for centering constants in (18). In this case the a_n will not be proportional to $\sqrt{n\log\log n}$, as shown by Strassen [61]. His Theorem 1 states that (18) holds with a_n taken as $\rho_n(2\log\log\rho_n)^{1/2}$, ρ_n being the largest root of $\rho_n^2 = nH(\rho_n)$, provided $a_n/\sqrt{n}$ increases, a_{2n}/a_n is bounded for all n, and

$$\int_0^\infty \frac{H(dx)}{H(x)\log\log x} < \infty,$$

where $H(x) = \int_{|t|\le x} t^2 F(dt)$. Such random walks need not be in the domain of attraction of the normal distribution. Feller's result is not correct as stated for the sequence a_n as defined; see Einmahl's papers [12] and [14] for detailed discussion and explication of this. Feller's proof of (18) is actually worked in terms of a sequence $\widetilde{a}_n$ chosen to be the largest root of $\widetilde{a}_n^2 = nH(\widetilde{a}_n)\log\log\widetilde{a}_n$. Kesten [33] (see his Theorem 7 and Remark 9 at the end of his paper) discusses this and gives a slightly more general result relating to similar choices of norming sequence. For another nice addendum to [Feller 1968d], see Einmahl and Li [15].

In Theorem 2 of [Feller 1968d], Feller shows that in some circumstances a version of (18) will hold with S_n replaced by ${}^{(1)}S_n := S_n - \max_{1\le k\le n} X_k$. Thus the generalized LIL (18) holds or fails *due to the influence of one single (large) component* X_k. This fertile idea of "trimming" a large value from the sample sum and thereby "improving" its behaviour, familiar to statisticians as an inferential or robustness tool, has subsequently developed a substantial literature elucidating both almost sure and weak convergence behaviour of S_n. See for example, Maller [43], Mori [48, 49], Griffin [22, 23], Griffin and Pruitt [25, 26, 27], Griffin and Qazi [29], Hahn, Mason and Weiner [30], and many others. Precursors to [Feller 1968d], but dealing with convergence in distribution or in probability, are in Darling [7] and Arov and Bobrov [1], for example. (There is no evidence that Feller knew about these particular results, at this time.)

[Feller 1969c, Feller 1969d] are a kind of reprise of the papers [Feller 1943b] and [Feller 1943c]. [Feller 1969c] gives a large deviations result and [Feller 1969d] applies it to give some very general LIL theorems. The methods of the later papers, likewise, reprise those of the earlier ones. The setting now is of a triangular array $(X_{k,n})$ of independent random variables with row sums S_n, and again the Esscher transform is used for estimation purposes. Results of the form

$$P(S_n > z_n) = e^{-r_n + o(r_n)}, \quad r_n \to \infty,$$

with $z_n \to \infty$, somewhat cruder than those in [Feller 1943b], as he says, are derived with the intention of applying them to get generalized LILs of the form (18). This is accomplished in [Feller 1969c] with the aid of these estimates. [Feller 1969c] returns to normed sums of independent (not necessarily i.i.d.) random variables and obtains (18) for certain a_n defined implicitly in terms of the Laplace transform of S_n.

Feller points out that the a_n may bear no resemblance to the classical norming sequence, $(2s_n^2\log\log s_n)^{1/2}$, but makes no attempt to ascertain the domain of applicability of a relation like (18). This direction was explored initially by Rogozin [52] and Heyde [31] at much the same time, and completed by Kesten [33], at least, for a centered version of (18). Feller's intuition was perhaps a little amiss in this matter; in [Feller 1969c] he observes that

> [...] the very form of the (LIL) is closely related to the central limit theorem, and no essential generalization can be expected by means of limit theorems involving the normal distribution.

A natural enough deduction, given the history so far; but the Rogozin–Heyde–Kesten result is that a centered version of (18) holds if and only if (in the case of i.i.d. summands) S_n is in the domain of *partial* attraction of the normal – a very remarkable insight and an elegant, complete, resolution of the LIL question (for i.i.d. summands, with a general norming sequence). These results spawned a further constellation of beautiful subsequent results which we cannot go into here.

[Feller 1970] (results also announced in [*Feller 1969e]) is Feller's last attempt to settle the LIL question for sums of independent (not necessarily i.i.d.) random variables. The aim is to give conditions for

$$p(I) := P\left(\frac{S_n - a_n}{b_n} \in I \text{ i.o.}\right)$$

for intervals $I = (L_1, L_2)$, $-\infty \leq L_1 < L_2 \leq \infty$, of the real line, and nonstochastic sequences a_n and b_n. The test function is

$$\Sigma(I) := \sum_{n=1}^{\infty} \min\left[1, \frac{a_{n+1} - a_n}{b_n}\right] P\left(\frac{S_n - a_n}{b_n} \in I\right),$$

and his Theorem 1 gives:

> $\Sigma(I) < \infty$ *implies* $p(I_\varepsilon) = 1$ *for every* $\varepsilon > 0$*; alternatively, if* $0 \leq L_1 < L_2 < \infty$ *and* $\Sigma(I) = \infty$ *then* $p(I_\varepsilon) = 1$ *for every* $\varepsilon > 0$.

Here $I_\varepsilon = (L_1 - \varepsilon, L_2 + \varepsilon)$. The restriction to bounded intervals in this statement is shown to be necessary. But as a corollary:

> *If for some* $\rho > 0$
>
> $$\sum_{n=1}^{\infty} \min\left[1, \frac{a_{n+1} - a_n}{a_n}\right] P(S_n > xa_n)$$
>
> *converges for* $x > \rho$ *and diverges for* $x < \rho$*, then*
>
> $$\limsup_{n\to\infty} \frac{S_n}{a_n} = \rho, \text{ a.s.} \qquad (19)$$
>
> *If the series converges for all* $x > 0$ *then* (19) *holds with* $\rho = 0$.

Further Theorems 2,3,4 in [Feller 1970] give *more pleasing* versions of these results when the ratios

$$\frac{P(S_n > xa_n + b_n u)}{P(S_n > xa_n + b_n v)}$$

remain bounded away from 1 and ∞ for fixed $u < v$; Theorems 3 and 4 give versions written in terms of convergence of series of subsequences of $(S_n - a_n)/b_n$. Theorem 5 gives a nice application to the case of normally distributed summands.

4 Regular Variation, Applications and Variants

An early foray of Feller into regularly varying functions is [Feller 1963], which gives an elementary proof of a Tauberian theorem due to Karamata (together with the Abelian version). The

> emphasis of the paper is not on new results, but on the extreme simplicity of the proof.

The property of being "regularly varying" (as $t \to \infty$) refers to the existence of the limit

$$\lim_{t\to\infty} \frac{L(tx)}{L(t)} = x^{\rho}, \quad x > 0, \tag{20}$$

and of the index $\rho \in (-\infty, \infty)$, for a real valued, positive function $L(\cdot)$. If "$t \to \infty$" is replaced by "$t \downarrow 0$" we have regular variation at 0, and when $\rho = 0$, L is said to be "slowly varying" in the respective cases.

Let U be a measure on $[0, \infty)$, finite on bounded sets, with Laplace transform

$$\omega(\lambda) = \int_0^{\infty} e^{-\lambda x} U(dx),$$

assumed to converge for all $\lambda > 0$. Feller proves:

Theorem 4. *If $L(t)$ is slowly varying as $t \to \infty$ ($t \downarrow 0$) then each of the relations*

$$U(t) \sim \frac{1}{\Gamma(\alpha - 1)} t^{\alpha} L(t),\ t \to \infty\ (t \downarrow 0),$$

and

$$\omega(\lambda) \sim \lambda^{-\alpha} L(\lambda^{-1}),\ \lambda \downarrow 0\ (\lambda \to \infty),$$

implies the other. (Here $\alpha \geq 0$.)

The simplification in Feller's approach is via application of the Helly method to choose a convergent subsequence in a way which is now commonly done in the literature. An extension to signed measures is formulated as follows:

Theorem 5. *Suppose that U is of bounded variation in every finite interval and that*

$$V(x) := \int_0^x y U(dy) = o(x) \quad \text{as } x \to \infty.$$

If the Laplace transform ω is slowly varying at 0, then

$$U(t) - \omega(t^{-1}) = o(1) \quad \text{as } t \to \infty.$$

In particular, $\omega(\lambda) \to 1$ as $\lambda \to 0$ implies $U(t) \to 1$ as $t \to \infty$.

[Feller 1963] is purely analytical, with no mention of probability applications. But [Feller 1967a] is introduced with mention of Doeblin's work in the 1940s characterizing domains of attraction of stable distributions, of Dynkin and others for renewal theory, and of Darling [7] for relations between a random walk and its maximum summand. Feller identifies Karamata's far-reaching concept of regular variation as the unifying theme in these results, and in following sections gives applications of it to (a) domains of attraction of stable distributions (with detailed attention to special cases such as $\alpha = 1$); (b) asymptotic distributions of maxima of i.i.d. random variables; (c) local limit theorems for random walks in a domain of attraction of a stable distribution.[7]

In the final sections of [Feller 1967a], Feller introduces a new and very powerful concept: that of one-sided, or *dominated variation.* This relates to the *stochastic compactness* of a sequence of distributions of random walks in a similar way as regular variation characterizes the domain of attraction of a stable distribution. To explain: Feller calls a sequence of distribution functions $(F_n)_{n=1,2,...}$ stochastically compact if each sequence of integers $n_i \to \infty$ contains a subsequence for which F_{n_i} converges weakly to a proper, nondegenerate distribution. When applied to random walks, he proves:

Theorem 6. *Let $(S_n)_{n=1,2,...}$ be a random walk with summands $(X_i)_{i=1,2,...}$, having cumulative distribution function F, assumed symmetric around 0. Then there exist $a_n > 0$ such that $F^{n*}(a_n x)$ is stochastically compact if and only if*

$$\limsup_{x\to\infty} \frac{x^2(1-F(x)+F(x-))}{\int_{|y|\le x} y^2\,dF(y)} < \infty. \tag{21}$$

If this holds there exist constants $\alpha > 0$, C and τ such that

$$\frac{\int_{|y|\le tx} y^2\,dF(y)}{\int_{|y|\le t} y^2\,dF(y)} \le Cx^{2-\alpha} \quad \textit{for } x > 1, t > \tau. \tag{22}$$

The assumed symmetry of F is inessential, as is easily shown; for general, i.i.d. X_i, (21) is equivalent to the statement:

> *There are nonstochastic sequences $a_n > 0$, b_n, such that each sequence of integers $n_i \to \infty$ contains a subsequence through which $(S_{n_i} - b_{n_i})/a_{n_i}$ converges weakly to a proper, nondegenerate distribution.*

(22) is stated in a uniform way; the one-sided generalization of regular variation is perhaps clearer if the equivalence with (21) is written as:

$$\limsup_{t\to\infty} \frac{\int_{|y|\le tx} y^2\,dF(y)}{\int_{|y|\le t} y^2\,dF(y)} \le Cx^{2-\alpha} \quad \text{for } x > 1, \tag{23}$$

for some $\alpha \in (0,2]$ (not the same α as in (22)). If $C = 1$ and the "limsup" in (23) were replaced by "lim", we would have regular variation with index $2-\alpha$ of the trun-

[7] See [53] for a detailed discussion of the Feller-Karamata connection.

cated second moment integral $\int_{|y|\le t} y^2\,dF(y)$ – compare with (20) – and a necessary condition for S_n to be in the domain of attraction of a stable law of index α.

[Feller 1969b] directly addresses one-sided analogues of Karamata's regular variation, expanding on (22). Thus a nondecreasing function U on $[0,\infty)$ varies dominatedly if, for some $C>0$,

$$\limsup_{t\to\infty}\frac{U(tx)}{U(t)}\le Cx^{\rho} \quad \text{for } x>1, \rho\ge 0. \tag{24}$$

A nonincreasing function varies dominatedly if (24) holds with $x>1$ replaced by $x<1$ and $\rho\le 0$. Auxiliary functions U_p and Z_q are defined by

$$U_p(x)=\int_x^{\infty} y^{-p-q} Z_q(dy) \quad \text{and} \quad Z_q(x)=\int_0^x y^q\,U(dy)$$

for values of p and q for which these integrals converge. The ratio

$$R_U(t)=\frac{t^p U_p(t)}{U(t)}$$

plays an important role. Existence of a limit for $R_U(t)$ as $t\to\infty$ is related to regular variation of U and U_p. Also: let $\underline{r}=\liminf_{t\to\infty} R_U(t)$ and $\overline{r}=\limsup_{t\to\infty} R_U(t)$. Then

Theorem 7. *U varies dominatedly if and only if $\overline{r}<\infty$. Similarly U_p varies dominatedly if and only if $\underline{r}>0$.*

Feller also specifies what he calls *ratio limit theorems*. Let U and V be nondecreasing unbounded functions on $[0,\infty)$, and suppose L is a slowly varying function. Then U and V are said to be L–equivalent if the ratio

$$\lim_{x\to\infty}\frac{U(x)L(x)}{V(x)}=1. \tag{25}$$

Define R_V analogously as for R_U.

Theorem 8. *Let U be of dominated variation. Then there is a slowly varying function L such that* (25) *holds if and only if*

$$\lim_{x\to\infty}(R_U(x)-R_V(x))=0, \ \textit{boundedly}. \tag{26}$$

Section 6 of [Feller 1969b] deals with Tauberian theorems, to the effect that U varies regularly (dominatedly) at infinity with index $\alpha\ge 0$ if and only if its Laplace transform ω varies regularly (dominatedly) at 0 with index $-\alpha$. Further, U varies dominatedly at infinity if and only if ω varies dominatedly at 0. Feller gives an application of (25) in this context.

The final Section 7 of [Feller 1969b] gives an application of dominated variation to infinitely divisible laws. Let H be an infinitely divisible distribution with Lévy measure $M(dx)$. If M varies dominatedly at infinity then

$$1-H(x)\sim M\{(x,\infty)\} \quad \text{as } x\to\infty. \tag{27}$$

Feller's concepts of dominated variation and stochastic compactness opened up a wide field of applications and generalizations and have given rise to a large literature, not only in the asymptotic theory of random walks but in the asymptotics of what we now call Lévy processes as well; and, for the latter, not only large time but small time asymptotics are explored. We cannot do justice to this topic here but mention a few key developments selected preferentially, of course, according to the present writer's areas of interest.

Early concepts of the non-normal stable laws, as a significant subclass of infinitely divisible distributions, and their domains of attraction, were developed as far back as the 1920s and 1930s; see, e.g., Lévy [39, 40], Khintchine [37, 36]. Lévy also introduced the notion of a domain of partial attraction (convergence of a normed, centered, random walk to an infinitely divisible random variable through a subsequence of integers), characterising, in particular, the domain of partial attraction of the normal distribution. Consult Fischer [19], Ch. 5.2, for a fuller exposition of this part of the history.

Lévy's student Doeblin developed the domain of attraction ideas in a variety of directions, giving another characterisation of the domain of attraction of the normal distribution, introducing *universal laws* (those partially attracted to every infinitely divisible random variable), and defining another concept: *strong compactness* (e. g., Doeblin [9]). Some discussion of these ideas is in Maller [41, 42]. Mason [47] gives an extensive discussion of Doeblin's [9] paper and related developments. Substantial later contributions were made also by Griffin [21] (multivariate formulations); Griffin, Jain and Pruitt [24] (local limit theorems); and Mason and his coworkers; the latter using a quantile approach to the domain of attraction question (e. g., Csörgő, Haeusler and Mason [4, 5, 6]. These authors coined the very appropriate description *Feller class* for the stochastically compact random walks. Griffin and Maller [28] extended the stochastic compactness idea in other directions, relating it to dominance conditions of the sum compared to the maximum, thus completing a circle from Feller back to Lévy in another way.

More recently, attention has turned to Lévy processes, as a kind of continuous time generalisation of random walks, and concepts of domains of attraction, the Feller class, etc., have been transferred to them sometimes in a straightforward, sometimes not so routine, way. Again, a large literature is being developed; see, e. g., Maller and Mason [44, 45].

Acknowledgement. I am grateful to Uwe Einmahl, Bruce Erickson, Hans Fischer, Phil Griffin and David Mason for feedback on earlier drafts of this chapter which have added greatly to it.

References

All citations of the form [Feller 19*nn*], resp., [*Feller 19*nn*] (if the respective paper is not included in these Selecta) point to Feller's bibliography, pp. xxv–xxxiv.

[1] Arov, D. Z. and Bobrov, A. A.: The extreme terms of a sample and their role in the sum of independent variables. *Theory Probab. Appl.* **5** (1960) 377–396.

[2] Berkes, I., Horváth, L. and Schauer, J.: Asymptotic behavior of trimmed sums. *Stochastics and Dynamics* **12** (2012) 1150002–1–1150002–14.

[3] Blackwell, D. H.: A renewal theorem. *Duke Math. J.* **15** (1948) 145–150.

[4] Csörgő, S., Haeusler, E. and Mason, D. M.: A probabilistic approach to the asymptotic distribution of sums of independent, identically distributed random variables. *Adv. Appl. Math.* **9** (1988) 259–333.

[5] Csörgő, S., Haeusler, E. and Mason, D. M.: The asymptotic distribution of trimmed sums. *Ann. Probab.* **16** (1988) 672–699.

[6] Csörgő, S., Haeusler, E. and Mason, D. M.: The quantile-transform-empirical-process approach to limit theorems for sums of order statistics. In: Hahn, M. G., Mason, D. M. and Weiner, D. C.: *Sums, Trimmed Sums and Extremes*. Birkhäuser, Boston 1991, pp. 215–267.

[7] Darling, D. A.: The influence of the maximum term in the sum of independent random variables. *Trans. Am. Math. Soc.* **73** (1952) 95–107.

[8] Doeblin, W.: Étude de l'ensemble de puissances d'une loi de probabilité. *C. R. Acad. Sci.* **206** (1938) 718–720.

[9] Doeblin, W.: Sur l'ensemble de puissances d'une loi de probabilité. *Studia Math.* **9** (1940) 71–96.

[10] Doney, R. A.: Recurrent and transient sets for 3-dimensional random walks. *Z. Wahrscheinlichkeitstheorie Verw. Gebiete* **4** (1965) 253–259.

[11] Doney, R. A.: An analogue of the renewal theorem in higher dimensions. *Proc. London Math. Soc.* **16** (1966) 669–684.

[12] Einmahl, U.: Strong approximations for partial sums of iid B-valued R.V.s in the domain of attraction of a Gaussian law. *Probab. Theor. Relat. Fields* **77** (1988) 65–85.

[13] Einmahl, U.: The Darling-Erdös theorem for sums of i.i.d. random variables. *Probab. Theor. Relat. Fields* **82** (1989) 241–257.

[14] Einmahl, U.: Stability results and strong invariance principles for partial sums of Banach space valued random variables. *Ann. Probab.* **17** (1989) 333–352.

[15] Einmahl, U. and Li, D.: Some results on two-sided LIL behavior. *Ann. Probab.* **33** (2005) 1601–1624.

[16] Erickson, K. B.: Strong renewal theorems with infinite mean. *Trans. Am. Math. Soc.* **151** (1970) 263–291.

[17] Erickson, K. B.: A renewal theorem for distributions on $\mathbb{R}^1$ without expectation. *Bull. Am. Math. Soc.* **77** (1971) 406–410.

[18] Erickson, K. B.: The strong law of large numbers when the mean is undefined. *Trans. Am. Math. Soc.* **185** (1973) 371–381.

[19] Fischer, H.: *A History of the Central Limit Theorem: From Classical to Modern Probability Theory*. (Sources and Studies in the History of Mathematics and Physical Sciences) Springer, New York.

[20] Gnedenko, B. V. and Kolmogorov, A. N.: *Limit Distributions for Sums of Independent Random Variables*. Addison–Wesley Co., Reading (MA) 1968 2nd ed.

[21] Griffin, P. S.: Matrix normalized sums of independent identically distributed random vectors. *Ann. Probab.* **14** (1986) 224–246.

[22] Griffin, P. S.: The influence of extremes on the law of the iterated logarithm. *Probab. Theor. Relat. Fields* **77** (1988) 241–270.

[23] Griffin, P. S.: Non-classical law of the iterated logarithm behavior for trimmed sums. *Probab. Theor. Relat. Fields* **78** (1988) 293–319.

[24] Griffin, P. S., Jain, N. C. and Pruitt, W. E.: Approximate local limit theorems for laws outside domains of attraction. *Ann. Probab.* **12** (1984) 45–63.

[25] Griffin, P. S. and Pruitt, W. E.: The central limit problem for trimmed sums. *Math. Proc. Cambridge Philos. Soc.* **102** (1987) 329–349.

[26] Griffin, P. S. and Pruitt, W. E.: Asymptotic normality and subsequential limits of trimmed sums. *Ann. Probab.* **17** (1989) 1186–1219.

[27] Griffin, P. S. and Pruitt, W. E.: Weak convergence of trimmed sums. In: Hahn, M. G., Mason, D. M. and Weiner, D. C.: *Sums, Trimmed Sums and Extremes*. Birkhäuser, Boston 1991, pp. 55–80.

[28] Griffin, P. S. and Maller, R. A.: Dominance of the sum over the maximum and some new classes of stochastic compactness. In: Bramson, M. and Durrett, R.: *Perplexing Problems in Probability, Festschrift in Honor of Harry Kesten*. Birkhäuser, Boston 1999, pp. 219–246.

[29] Griffin, P. S. and Qazi, F. S.: Limit laws of modulus trimmed sums. *Ann. Probab.* **30** (2002) 1466–1485.

[30] Hahn, M. G., Mason, D. M. and Weiner, D. C.: *Sums, Trimmed Sums and Extremes*. Birkhäuser, Boston 1991.

[31] Heyde, C. C.: A note concerning behaviour of iterated logarithm type. *Proc. Am. Math. Soc.* **23** (1969) 85–90.

[32] Kesten, H.: The limit points of a normalized random walk. *Ann. Math. Statist.* **41** (1970) 1173–1205.

[33] Kesten, H.: The 1971 Rietz Lecture: Sums of independent random variables – without moment conditions. *Ann. Math. Statist.* **43** (1972) 701–732.

[34] Kesten, H.: Random difference equations and renewal theory for products of random matrices. *Acta Math.* **131** (1973) 207–248.

[35] Kesten, H. and Maller, R. A.: Some effects of trimming on the law of the iterated logarithm. *J. Appl. Probab.* **41A** (2004) 253–271.

[36] Khintchine, A. Ya.: Sul dominio di attrazione della legge di Gauss. *Giorn. 1st. Ital. Attuary* **6** (1935) 371–393.

[37] Khintchine, A. Ya, and Lévy, P.: Sur les lois stables. *C. R. Acad. Sci.* **202** (1936) 374–376.

[38] Lévy, P.: Sur les séries dont les termes sont des variables éventualles indépendantes. *Studia Math.* **3** (1931) 119–155.

[39] Lévy, P.: Proprietés asymptotiques des sommes de variables aléatoires indépendantes ou enchainées. *J. Math. Pure Appl. (VII)* **14** (1935) 347–402.

[40] Lévy, P.: *Théorie de l'Addition des Variables Aléatoires*. Gauthier-Villars, Paris 1937.

[41] Maller, R. A.: A note on domains of partial attraction. *Ann. Probab.* **8** (1980) 576–583.

[42] Maller, R. A.: Some properties of stochastic compactness. *J. Austral. Math. Soc. (Series A)* **30** (1981) 264–277.

[43] Maller, R. A.: Asymptotic normality of lightly trimmed means – a converse. *Math. Proc. Cambridge Phil. Soc.* **92** (1982) 535–545.

[44] Maller, R. A. and Mason, D. M.: Stochastic compactness of Lévy processes. In: Houdré, C., Koltchinskii, V., Mason, D. M. and Peligrad, M.: *High Dimensional Probability V: The Luminy Volume*. Institute of Mathematical Statistics, Beachwood (OH) 2009, pp. 239–257.

[45] Maller, R. A. and Mason, D. M.: Small-time compactness and convergence behavior of deterministically and self-normalised Lévy processes. *Trans. Am. Math. Soc.* **362** (2010) 2205–2248.

[46] Marcinkiewicz, J.: Quelques théorèmes de la théorie des probabilités. *Travaux de la Société des Sciences et des Lettres de Wilno, Classes des Sciences Mathématiques et Naturelles* **13** (1939) 1–13.

[47] Mason, D. M.: Notes on the 1940 Wolfgang Doeblin paper: L'ensemble de puissances d'une loi de probabilité. To appear.

[48] Mori, T.: The strong law of large numbers when extreme terms are excluded from sums. *Probab. Theor. Relat. Fields* **36** (1976) 189–194.

[49] Mori, T.: Stability for sums of i.i.d. random variables when extreme terms are excluded. *Probab. Theor. Relat. Fields* **40** (1977) 159–167.

[50] Mori, T.: On the limit distributions of lightly trimmed sums. *Math. Proc. Cambridge Philos. Soc.* **96** (1984) 507–516.

[51] Port, S. C. and Stone, C. J.: Hitting times for transient random walks on groups. *J. Math. Mech.* **17** (1968) 1117–1130.

[52] Rogozin, B. A.: On the existence of exact upper sequences. *Theory Probab. Appl.* **13** (1968) 667–672.

[53] Seneta, E.: Karamata's characterisation theorem, Feller, and regular variation in probability theory. Publications de l'Institut Mathématique, nouvelle series, 71(85) (2002), 79-89.

[54] Silvestrov, D. S. and Teugels, J. L.: Limit theorems for mixed max-sum processes with renewal stopping. *Ann. Appl. Probab.* **14** (2004) 1838–1868.

[55] Spitzer, F.: A combinatorial lemma and its application to probability theory. *Trans. Am. Math. Soc.* **82** (1956) 323–339.

[56] Spitzer, F.: A Tauberian theorem and its probability interpretation. *Trans. Am. Math. Soc.* **94** (1960) 150–169.

[57] Spitzer, F.: *Principles of Random Walk.* Springer, New York 1976 2nd ed.

[58] Stone, C.: On characteristic functions and renewal theory. *Trans. Am. Math. Soc.* **120** (1965) 327–342.

[59] Stone, C.: A local limit theorem for nonlattice multi-dimensional distribution functions. *Ann. Math. Statist.* **36** (1965) 546–551.

[60] Stone, C.: The growth of a recurrent random walk. *Ann. Math. Statist.* **37** (1966) 1040–1041.

[61] Strassen, V.: A converse to the law of the iterated logarithm. *Z. Wahrscheinlichkeitstheorie Verw. Gebiete* **4** (1965) 265–268.

Prof. Ross Maller
Mathematical Sciences Institute
Australian National University
Canberra, ACT 2601
Australia
Ross.Maller@anu.edu.au

THE ASYMPTOTIC DISTRIBUTION OF THE RANGE OF SUMS OF INDEPENDENT RANDOM VARIABLES

By William Feller

Princeton University

Summary. The asymptotic distribution of the range and normalized range of the sum of n independent variables is derived using the theory of Brownian motion.

1. Introduction. Let $[X_k]$ be a sequence of mutually independent random variables with a common distribution $V(x)$, and suppose that $E(X_k) = 0$, $\text{Var}(X_k) = 1$. Put $S_n = X_1 + \cdots + X_n$ and let

$$(1.1)\qquad \begin{aligned} M_n &= \max [0, S_1, S_2, \cdots, S_n], \\ m_n &= \min [0, S_1, S_2, \cdots, S_n]. \end{aligned}$$

The random variable

$$(1.2)\qquad R_n = M_n - m_n$$

will be called *the range of the cumulative sums* S_n.

In applications[1] it is advantageous to modify this definition. One considers instead of the values of the sums S_k their deviations from the straight line joining the origin to the point (n, S_n). Thus we replace the random variables S_k by

$$(1.3)\qquad S_k^* = S_k - kS_n/n \qquad (k = 1, \cdots, n)$$

and define the corresponding variables M_n^*, m_n^*, R_n^* in analogy with (1.1) and (1.2). The variable R_n^* will be called the *adjusted range of the cumulative sums* S_n.

The adjusted range has a greater sampling stability, but its main advantage is probably due to the fact that it eliminates the trend when $E(X_k) \neq 0$, so that it can be used even when the means do not vanish.

It is practically impossible to calculate the exact distribution of the ranges even for $n = 3$ and simple forms of the underlying distribution $V(x)$. Now the sums S_n are obviously asymptotically normally distributed, and therefore the asymptotic distribution of the ranges is independent of the form of $V(x)$. It suffices accordingly to consider the case where the variables X_k are normal. The sum S_n can then be considered as the value at time $t = n$ of a continuously changing normal variable $S(t)$ which is subject to a Bachelier-Wiener process (or ordinary diffusion). Since the sequence $[S_k]$ is a subsequence of the values assumed by $S(t)$, the range R_n is certainly not smaller than the range at time

[1] Cf. in particular Hurst [4]. A surprising statistical phenomenon discovered by Hurst is discussed at the end of Section 2. The author is indebted to Mr. G. W. Alexander of the State Rivers and Water Supply Commission, Melbourne, for drawing his attention to Hurst's paper and the interesting statistical problems connected with it.

R.L. Schilling et al. (eds.), *Selected Papers II*,

$t = n$ of the variable $S(t)$, and it is clear that for large n the two ranges will be practically the same.

In Section 3 we shall find *the exact distribution of the range* $R(t)$ *of the continuous variable* $S(t)$ [cf. (3.7)]. *One gets in particular*

$$(1.4)\qquad \begin{aligned} E(R(n)) &= 2(2n/\pi)^{\frac{1}{2}} = 1.5958 \cdots n^{\frac{1}{2}}, \\ \mathrm{Var}(R(n)) &= 4n(\log 2 - 2/\pi) = 0.2181 \cdots n. \end{aligned}$$

These quantities are, asymptotically, the mean and varance of the range R_n.

For the *adjusted range* R_n^* we have to introduce the corresponding continuously changing variable

$$(1.5)\qquad S^*(t) = S(t) - tS(T)/T \qquad (0 < t < T).$$

This variable appears more complicated than $S(t)$. Fortunately the stochastic process defined by (1.5) happens to be equivalent to a process which has been studied in an exceedingly elegant and simple manner by Doob in connection with his heuristic approach to the Kolmogorov-Smirnov theorems. Using Doob's results it is easy to obtain *the exact distribution of the adjusted range* $R^*(T)$ *for the continuously changing variable* $S(t)$. It is given in (4.3) and represents the desired asymptotic distribution of the adjusted range R_n^* for $n = T$. One gets in particular

$$(1.6)\qquad \begin{aligned} E(R^*(T)) &= (T\pi/2)^{\frac{1}{2}} = 1.2533 \cdots T^{\frac{1}{2}}, \\ \mathrm{Var}(R^*(T)) &= \left(\frac{\pi^2}{6} - \frac{\pi}{2}\right) T = 0.07414 \cdots T. \end{aligned}$$

2. Discussion. A comparison of (1.4) and (1.6) shows that the adjusted range has the advantage of greater sampling stability.

In order to get an idea about the goodness of the approximations (1.4) and (1.6) we compare them with the exact values in the perhaps most unfavorable case, namely where each variable X_k assumes only the values ± 1, each with probability $\frac{1}{2}$. For $n = 6, 10, 12$ we get

(2.1)

	Exact value	*Approximation* (1.4)
$E(R_6)$	3.0625	3.909···
$\mathrm{Var}(R_6)$	1.18360	1.309···
$E(R_{10})$	4.1523···	5.046···
$\mathrm{Var}(R_{10})$	2.0872···	2.181···
$E(R_{12})$	4.6377···	5.528···
$\mathrm{Var}(R_{12})$	2.545 ···	2.617···

For the adjusted ranges (and the same Bernoulli variables) the corresponding figures are

		Exact value	*Approximation* (1.6)
(2.2)	$E(R_6^*)$	2	$3.070\cdots$
	$\mathrm{Var}(R_6^*)$	$0.4396\cdots$	$0.445\cdots$
	$E(R_{10}^*)$	$2.954\cdots$	$3.963\cdots$
	$\mathrm{Var}(R_{10}^*)$	$0.5822\cdots$	$0.7414\cdots$

Considering the smallness of our n and the fact that the assumed distribution of the X_k is most unfavorable for our approximation, the above results appear surprisingly good. They also bear out the expectation that the ranges of the sums S_n should be smaller than those of the corresponding continuously varying variables $S(t)$.

If the model of cumulative sums of independent random variables applies to a particular type of empirical phenomena, then the observed ranges should, on the average, increase with the square root of the length T of the observational period. Now there is available a huge body of statistics concerning annual water levels of rivers and lakes all over the world. It has naturally been assumed that such levels could reasonably be treated as the cumulative effect of sums of random variables, but in an interesting paper [4] H. E. Hurst discovered puzzling systematic departures. In fact, Hurst has collected an impressively large statistical material relating to water levels and other phenomena which seems to bear out the contention that *the observed adjusted ranges do not increase, as expected, like the square root of the observational period T, but like a higher power T^c.* The most surprising feature is the stability of the observed values of the exponent c: it varies only from 0.69 to 0.80, with a mean of 0.729 and standard deviation 0.092. Within the several separate groups of phenonema the stability of c is even greater. Hurst himself has not attempted an explanation of his interesting discovery.

It is conceivable that the phenomenon can be explained probabilistically, starting from the assumption that the variables X_k are not independent, but that X_{n+1} depends only on the actual value of S_n. For example, a high lake level creates additional outlets for the outflow and this in practice means a restoring force towards the average size. Mathematically this would require treating the variables X_k as a Markov process. In theory the method presented in this paper applies to this more general case, but the simple ordinary diffusion equation would have to be replaced by a general Fokker-Planck equation, and the solution of the corresponding boundary value problem is not explicitly known. We are here confronted with a problem which is interesting from both a statistical and a mathematical point of view.

3. The range. We have to deal with the variable $S(t)$ of a Bachelier-Wiener process; this means that $S(t)$ is a normal variable with mean 0 and variance t

$(t > 0)$, and the increment $S(t + h) - S(t)$ is a normal variable with mean 0 and variance h which is independent of $S(t)$ (and the values $S(\tau)$ for $\tau < t$). For fixed $u > 0$ and $v > 0$ we require the probability $F(T; u, v)$ of the event

$$M(T) \leq v, m(T) \geq -u, \tag{3.1}$$

where $M(T) > 0$ and $m(T) < 0$ denote, respectively, the maximum and minimum of $S(t)$ for $0 \leq t \leq T$. The corresponding probability density is given by the mixed derivative

$$f(T; u, v) = F_{uv}(T; u, v), \tag{3.2}$$

and it is easily seen that the density function $\delta(T; r)$ of the range $R(T) = M(T) + | m(T) |$ is

$$\delta(T; r) = \int_0^r f(T; u, r - u)\, du. \tag{3.3}$$

To calculate $F(T; u, v)$ we start from the density function $w(t, x; u, v)$ of the event that simultaneously $S(t) = x$, $M(t) \leq v$, and $m(t) \geqq -u$. By the definition of these functions we have

$$F(T; u, v) = \int_{-u}^{v} w(T, x; u, v)\, dx, \tag{3.4}$$

so that the required density $\delta(T; r)$ follows from $w(T, x; u, v)$ by routine calculations. Now it is easily seen[2] that $w(t, x; u, v)$ is simply the fundamental solution of the ordinary diffusion equation $w_t = \frac{1}{2}w_{xx}$ for the interval $-u < x < v$ with the boundary conditions $w(t, x; u, v) = 0$ when $x = -u$ or $x = v$. One gets by the so-called method of images[3]

$$\begin{aligned} t^{\frac{1}{2}} w(t, x; u, v) = \sum_{k=-\infty}^{\infty} \phi\left(\frac{2ku + 2kv - x}{t^{\frac{1}{2}}}\right) \\ - \sum_{k=-\infty}^{\infty} \phi\left(\frac{2ku + 2(k-1)v + x}{t^{\frac{1}{2}}}\right), \end{aligned} \tag{3.5}$$

where $\phi(x)$ stands for the normal density function with zero mean and unit variance. Carrying out the indicated operations we find finally for *the density function of the range* $R(t)$

$$\delta(t; r) = 8 \sum_{k=1}^{\infty} (-1)^{k-1} k^2 \phi\left(\frac{kr}{t^{\frac{1}{2}}}\right). \tag{3.6}$$

In this form it is not even obvious that the function is positive, and it is readily seen that the mean can not be obtained by termwise integration of

[2] The reasoning is substantially the same as in the case of discrete random walks (cf. [3], chap. 14).

[3] Cf. problem 5 on p. 304 of [3]. Formula (3.5) can be derived from the formula given there by the passage to the limit described in section 6 of chapter 14. Cf. also [5], p. 213.

(3.6). Fortunately this function is closely related to the distribution function $L(z)$ which occurs in the Kolmogorov-Smirnov theorem on empirical distribution functions. The distribution function $L(z)$ can be written in two equivalent forms[4]

$$L(z) = 1 - 2\sum_{k=1}^{\infty} (-1)^{k-1} \exp(-2k^2 z^2) = (2\pi)^{\frac{1}{2}} z^{-1} \sum_{k=1}^{\infty} \exp(-(2k-1)^2 \pi^2/8z^2). \tag{3.7}$$

Clearly

$$\delta(t; r) = (2/\pi)^{\frac{1}{2}} r^{-1} L'(r/(2t^{\frac{1}{2}})). \tag{3.8}$$

The second representation in (3.7) shows that $z^{-a}L(z) \to 0$ as $z \to 0$ for any a, and hence an integration by parts shows that

$$\int_0^{\infty} \delta(t; r)\, dr = 8\pi^{-2} \sum_{k=1}^{\infty} (2k-1)^{-2} = 1. \tag{3.9}$$

A similar procedure then leads to the formulas (1.4).

4. The adjusted range. We have now to find the range of the continuously changing variable defined by (1.5). It is clear that $S^*(t)$ is normally distributed with mean 0, variance $t(T - t)$, and $\mathrm{Cov}[S^*(s), S^*(t)] = s(T - t)/T$, for $0 < s < t < T$. Thus the stochastic process defined by (1.5) is, for the particular value $T = 1$, the process studied by Doob [1]. According to Doob a simple transformation permits one to reduce (1.5) to the ordinary Bachelier-Wiener process with the interval $0 < t < T$ going over into the entire interval $0 < t < \infty$. This actually simplifies matters inasmuch as the probabilities corresponding to (3.4) and (3.5) are no longer time dependent, so that the preceding boundary value problem for a partial differential equation is replaced by a simpler functional equation. At any rate, Doob's last equation furnishes us with the probability $F(T; u, v)$ that $S^*(t)$ is for $0 < t < T$ contained in the interval $(-u, v)$. We have[5]

$$\begin{aligned} F(T; u, v) &= 1 + e(u + v) \\ &\quad - \sum_{k=1}^{\infty} \{e(ku + (k-1)v) + e((k-1)u + kv) - e(ku + kv) \\ &\qquad - e((k-1)u + (k-1)v)\}, \end{aligned} \tag{4.1}$$

[4] Cf. formula (1.4) of [2], where however a factor 2 is missing in the exponent.

[5] Doob's formula looks simpler than (4.1), but the rearrangement (4.1) was necessary to make it possible to perform the required differentiations and integrations in the routine manner. (In the original form each term of the series contains a singular probability distribution along the axes, and formal manipulations lead into apparent contradictions. Also, Doob has $T = 1$.)

where we put for abbreviation

$$e(x) = \exp(-2x^2/T). \tag{4.2}$$

Formula (4.1) corresponds to (3.4), and it remains to perform the calculations indicated in (3.2) and (3.3). In this way we get for *the density function of the range $R^*(T)$ of the variable $S^*(t)$*

$$\begin{aligned}\delta(T; r) = re''(r) + \sum_{k=2}^{\infty} \{2k(k-1)\,[e'((k-1)r) - e'(kr)] \\ + (k-1)^2 re''((k-1)r) + k^2 re''(kr)\}.\end{aligned} \tag{4.3}$$

To see how the moments are calculated note, for example, that

$$\begin{aligned}\int_0^\infty r^2 \delta(T; r)\, dr = \int_0^\infty r^3 e''(r)\, dr \\ + \sum_{k=2}^{\infty} \left\{\left[\frac{2k}{(k-1)^2} - \frac{2k-2}{k^2}\right] \int_0^\infty r^2 e'(r)\, dr \right. \\ \left. + \left[\frac{1}{(k-1)^2} + \frac{1}{k^2}\right] \int_0^\infty r^3 e'(r)\, dr\right\}.\end{aligned} \tag{4.4}$$

But

$$\int_0^\infty r^3 e''(r)\, dr = -3 \int_0^\infty r^2 e'(r)\, dr = 6 \int_0^\infty re(r)\, dr = 3T/2,$$

and therefore

$$\begin{aligned}\int_0^\infty r^2 \delta(T; r)\, dr = \frac{3}{2}T - \frac{1}{2}T \sum_{k=2}^{\infty} \left\{\frac{2k}{(k-1)^2} - \frac{(2k-2)}{k^2} - \frac{3}{(k-1)^2} - \frac{3}{k^2}\right\} \\ = T \sum_{k=1}^{\infty} k^{-2} = \pi^2 T/6.\end{aligned} \tag{4.5}$$

In this way formulas (1.6) are obtained.

REFERENCES

[1] J. L. Doob, "Heuristic approach to the Kolmogorov-Smirnov theorems," *Annals of Math. Stat.*, Vol. 20 (1949), pp. 393–403.

[2] W. Feller, "On the Kolmogorov-Smirnov limit theorems for empirical distributions," *Annals of Math. Stat.*, Vol. 19 (1948), pp. 177–189.

[3] W. Feller, *An Introduction to Probability Theory and Its Applications*, Vol. 1, John Wiley and Sons, 1950.

[4] H. E. Hurst, "Long-term storage capacity of reservoirs," *Proc. Amer. Soc. Civil Engrs.*, Vol. 76 (1950), separate no. 11, 30 pp.

[5] P. Lévy, *Processus stochastiques*, Gauthier-Villars, Paris, 1948.

DIFFUSION PROCESSES IN GENETICS

WILLIAM FELLER
PRINCETON UNIVERSITY

1. Introduction

The theory of evolution provides examples of stochastic processes which have not yet been treated systematically. We find here many open problems whose solution promises new insights into the general theory.

There exists a huge literature on the mathematical theory of evolution and statistical genetics, but existing methods and results are due almost entirely to R. A. Fisher and Sewall Wright.[1] They have attacked individual problems with great ingenuity and an admirable resourcefulness, and had in some instances to discover for themselves isolated facts of the general theory of stochastic processes. However, as is only natural with such pioneer work, it is not easy to penetrate to the mathematical core of the arguments or to discover the explicit and implicit assumptions underlying the theory. In the following an attempt is made to formulate the basic mathematical problems and to discuss their connection with other stochastic processes. Such a systematic approach leads automatically to more general formulations which may be useful at least for a better understanding of the underlying assumptions.

We are concerned with mathematical models of population growth. Relatively small populations require discrete models, but for large populations it is possible to apply a continuous approximation, and this leads to processes of the diffusion type.

By way of introduction we start with the simple *branching process* which became popular in connection with its application to nuclear chain reactions, but which had been previously used by Galton in a discussion of the survival of family names, and by R. A. Fisher in his treatment of the survival of mutant genes. This branching process describes the simplest possible populations: the individuals are of like kind, and there is absolutely no interaction among them so that they are statistically independent. Thus, contrary to a widespread belief, the branching process does not represent an isolated type of stochastic process and is remarkable mainly because of its simplicity. For example, in the case of a growing population the process in its later stages converges to a diffusion process regulated by the Fokker-Planck equation describing the simplest growth (compare appendix II, and end of section 5).

Serious difficulties arise if one wishes to construct population models with in-

This research was done under an Office of Naval Research contract at Cornell University for developing probability theory.

[1] See references [5], [12], and [13]. It is difficult to give useful references to original papers, since these are mostly highly technical and inaccessible to nonspecialists.

teractions among the individuals.[2] The situation grows worse if the population consists of different types of individuals. Explicit examples of such populations with various types of interactions between individuals are provided by various breeding systems. Their treatment leads to Markov chains which are in general finite. [4, chaps. 15, 16].

The populations with which we would like to deal in genetics are far too complicated for a mathematical treatment, and it proved necessary to introduce artificial simplifications. Thus, almost the entire theory is devoted to the case where the population consists of only two types of individuals, called a- and A-genes. Each individual may have a certain number of descendants of either kind. We have here a bivariate branching process with the added difficulty that the individuals are not statistically independent. A mathematical theory of such bivariate branching processes is extremely desirable, but even a beginning is hard to get. For example, there are three possibilities of extinction (only a-genes, only A-genes, or both). If the elementary transition probabilities of the process are known, the extinction probabilities can be calculated from an infinite system of linear equations, but these in general do not have unique solutions.

In fact, the bivariate branching process leads to such difficulties that apparently not one single truly bivariate case has been treated in the literature. In the theory of evolution this difficulty is overcome by the assumption of *a constant population size:* if there are j a-genes, the number of A-genes is $2N - j$, and thus it suffices to study the univariate population of a-genes. Presumably this is a justifiable approximation in specific applications, but it is of interest to note that essential features of the whole mathematical theory depend on this assumption. Dropping it will lead us to an entirely new theoretical model (the practical applicability of which is not discussed).

In section 3 we discuss a discrete model of a population of fixed size of a- and A-genes. The probability relations which connect the numbers of a-genes in two succeeding generations reflect random mating and take into account mutation pressure and selective advantages in a way which seems to correspond to the basic assumptions of the accepted theories.[3]

In section 10 the assumption of constant population size is dropped and a truly bivariate model is constructed which takes into account selective advantages in a more flexible way.

It is known[4] that an essential part of Wright's theory is mathematically equiva-

[2] A special problem arises in connection with differentiation according to age. If there are m possible age groups then the population may be classified into m classes and the process becomes an m-variate process of the Markov type to which the remarks apply. This is an artificial device and often it is preferable to insist in discussing only the total population size. However, in this case the past history has after effects, and the process is univariate but not of the Markov type. In practice, this is a compelling reason to neglect effects of age differences, and we shall adhere to this tacit convention throughout the paper.

[3] However, our formalization leads to a 'sampling variance' which is somewhat larger than the usually accepted value.

[4] Apparently [14] this was first observed by Kolmogorov who had previously treated a particular problem of genetics as a diffusion problem. However, as early as 1922 R. A. Fisher was led to an equation of the heat conduction type, and this was before Kolmogorov's famous paper [7] on stochastic processes. For Fisher's method and references to the original papers compare [5, p. 88]. A variation of Wright's arguments which leads directly to the diffusion equation was given by Malécot [11].

lent to assuming a certain diffusion equation for the *gene frequency* (that is, the proportion of a-genes). This equation is discussed in section 7 and obtained in section 8 from the model of section 2 by a limiting process which is actually implicitly implied in Wright's theory. Now this diffusion equation is of a peculiar type and it should be realized that the limiting process in question is but one in a family of possible processes (depending on the ratios of mutation rates to population size).

In general (section 9) we obtain in the limit *ordinary* diffusion equations with drift, and thus *normal distributions* for the deviation of the gene frequency from the equilibrium point. Presumably this corresponds to actual phenomena in nature. However this may be, it is of mathematical interest that the singular diffusion equation of Wright's theory appears as a degenerate case of ordinary diffusion.

This diffusion equation (as well as others occurring in population theory) are of a *singular type* and lead to new types of boundary conditions and mathematical problems which have not yet been investigated.* Of significance is the fact that (even in the case of the diffusion equation of Wright's theory) a *stable solution* $u(x)$ *can exist but nevertheless the actual solution may approach, not* $u(x)$, *but a certain fraction* $qu(x)$. The possibility of such an occurrence seems to have been completely overlooked in the literature (section 7).

The same limiting process which leads from the model of section 3 to the diffusion equation of Wright's theory can be applied to our new bivariate model and leads to a diffusion equation in two dimensions. It is very different from the univariate diffusion equation, but the latter is contained in it (section 10). The solutions of this new equation are not discussed, but it is to be observed that *in no truly bivariate case does the gene frequency satisfy a diffusion equation* (sections 6 and 10). In fact, if the population size is not constant, then the gene frequency is not a random variable of a Markov process. Thus, conceptually at least, the assumption of a constant population size plays a larger role than would appear on the surface.

2. The branching process and Markov chains

The classical branching process is too well known to require treatment here,[5] and we shall not use its special properties. However, we start from it to tie it in with our notations, and to discuss its relation with the diffusion equation of section 5.

In each succeeding generation the population of the classical branching process consists of like individuals, each of which may give birth to 0, 1, 2, . . . direct descendants (that is, elements of the succeeding generation). It is assumed that the corresponding probabilities $p_0, p_1, p_2, \ldots$ $\left(\text{with} \sum p_k = 1\right)$ are the same for all individuals and, furthermore, that the individuals of any generations are statistically independent of each other. Denote the number of individuals in the n-th generation by $Z^{(n)}$. This is a random variable of a Markov chain, and the whole theory hinges on the *transition probabilities* p_{jk} (the conditional probability that $Z^{(n+1)} = k$ given that $Z^{(n)} = j$).

* *Added in proof*. A systematic theory, including the new boundary condition, is to appear in the *Annals of Math.*

[5] Compare, for example, [4, chapter 11].

An explicit expression for p_{jk} is cumbersome, but (as for many Markov chains) the use of generating functions makes the required calculations possible. Putting

$$f(x) = \sum_{k=0}^{\infty} p_k x^k , \tag{2.1}$$

the classical theory depends on the easily verified remark that (due to the assumption of statistical independence)

$$p_{jk} = \text{coefficient of } x^k \text{ in } f^j(x) . \tag{2.2}$$

From this it follows that if the original population size is r, then the generating function of the population size $Z^{(n)}$ in the n-th generation is given by $f_n^r(x)$ where

$$f_1(x) = f(x) , \qquad f_2(x) = f[f(x)] , \ldots , f_{n+1}(x) = f_n[f(x)] . \tag{2.3}$$

The main point of interest to us is the fact that the definition of the process permits us to calculate the transition probabilities p_{jk}; these, together with the initial population size, completely determine all probability relations of the process. In particular, the *n-step transition probabilities* $p_{jk}^{(n)}$ (giving the probability that if the population size $Z^{(m)}$ is at any time j, it will n generations later amount to k), can be calculated recursively from $p_{jk}^{(1)} = p_{jk}$,

$$p_{jk}^{(n+1)} = \sum_{\nu} p_{j\nu}^{(n)} p_{\nu k} . \tag{2.4}$$

As a further illustration consider the *extinction probability* ϵ_r that the population size will even be reduced to zero if the initial population size is r. This case can occur only if $p_0 > 0$. In the terminology of Markov chains the population size 0 is an 'absorbing state' (or 'trap'), which once it is reached becomes a permanent state. It follows then easily [4, p. 334] that the extinction probabilities ϵ_r are a solution of the infinite system of equations

$$\epsilon_r = \sum_{\nu=1}^{\infty} p_{r\nu}\epsilon_\nu + p_{r0} . \tag{2.5}$$

The fact that $p_{r\nu}$ is the ν-th coefficient in the power series for $f^r(x)$ suggests investigating the possibility of a solution of the form $\epsilon_r = \lambda^r$. With it the right hand member reduces to $f^r(\lambda)$, and hence (2.5) requires that $\lambda^r = f^r(\lambda)$, or

$$\lambda = f(\lambda) . \tag{2.6}$$

Conversely, if λ is a solution of (2.6), then $\epsilon_r = \lambda^r$ is a solution of (2.5). Now $\lambda = 1$ is always a root of (2.6), but if $f'(1) > 1$ there exists another root $\lambda < 1$. In this case, therefore, the solution of the system (2.5) is not unique. It follows from the general theory (or the recursion formula for the probability of extinction within n generations) that we are interested in the smallest solution of (2.5), and this determines our extinction probabilities uniquely. We have thus derived the well known basic theorem concerning branching processes from the general theory of Markov chains.[6]

[6] As a further example of the applicability of the general theory to the branching process let us note that all states except 0 are transient so that $p_{jk}^{(n)} \to 0$ as $n \to \infty$ $(k \neq 0)$. A deeper theorem of

Leaving the branching process we arrive at more general population models by assuming appropriate transition probabilities p_{jk}. The simplest generalization consists in permitting the fertility to depend on the actual population size; in our notation this means that the probability distribution $\{p_k\}$ changes from generation to generation and depends on the instantaneous population size.

In theory the methods apply also to *bivariate populations*, but to describe the actual state of the population we then require two integers. Accordingly the transition probabilities will now be of the form $p(j_1, j_2; k_1, k_2)$: this is the (conditional) probability that a generation will consist of k_1 individuals of the first kind and k_2 individuals of the second kind, given that the corresponding numbers for the preceding generation are j_1 and j_2. Everything becomes rather involved. For example, the three types of extinction probabilities can be determined from infinite systems of linear equations analogous to (2.5), but the solutions are not unique and are hard to get even in the simplest cases. As was mentioned in the introduction, apparently no truly bivariate model has ever been treated, and this should be a challenge to mathematicians.

3. Discrete model with constant population size

With respect to a particular pair of genes, say a and A, each individual belongs to one of the three genotypes (a, a), (a, A), or (A, A), and we should actually consider a trivariate population of N individuals. However, an accepted standard simplification consists in introducing the genes themselves as elements of the population so that we deal with a population of $2N$ elements which are either a or A. Here N is a fixed number.

For a first orientation let us begin with the simple *random mating* case disregarding mutation pressures and selective forces. Mathematically speaking, the assumption of random mating amounts to saying that the $2N$ genes of any generation are formed in $2N$ independent trials: if the parent population consists of j a-genes and $2N - j$ A-genes, then each trial results in a or A with probabilities

$$(3.1) \qquad p_j = \frac{j}{2N}, \qquad q_j = 1 - \frac{j}{2N},$$

respectively. In other words, we have now a Markov chain with transition probabilities given by the binomial distribution

$$(3.2) \qquad p_{jk} = \binom{2N}{k} p_j^k q_j^{2N-k}.$$

These, together with the initial population size, determine the entire process. We have now the two 'absorbing states' 0 and $2N$ which, once reached, are perpetuated: in biological language these occurrences are called *fixation*. The probability ϵ_r of a fixation at 0 (extinction of a-genes) if the initial number of a-genes is r, is again given by the system of equations (2.5), which is now finite. Its solution is easily found to be $\epsilon_r = 1 - r/2N$. The speed with which this fixation is reached can be

Doeblin [1] states that there exists a sequence $\lambda_n \to \infty$ such that $\lambda_n^{-1} p_{jk}^{(n)}$ converges to a limit $a_k > 0$ which is independent of j. It is easily verified that in our case the sequence λ_n can be chosen as $\lambda_n = f'^n(1)$. It follows that *the probability distribution of the normed variable Z_n/λ_n converges.* This interesting fact is contained in stronger results of T. E. Harris [6].

judged from the largest non-trivial characteristic value of the matrix (p_{jk}). This turns out to be $1 - 1/2N$. This means that if a_n is the probability of no fixation within n generations, a_{n+1}/a_n is approximately $1 - 1/2N$. This is an agreement with estimates of the ratio of decay obtained by quite different methods by Fisher and Wright.[7] It also agrees with a result to be obtained from an approximating diffusion equation (compare section 8).

We now modify the model so as to take into account mutations. We assume that at the formation of each new generation each gene has the possibility to mutate, that is, to change into a gene of the other kind. More precisely we assume that the mutation

$$a \to A \text{ occurs with probability } \alpha_1 \qquad (3.3)$$

$$A \to a \text{ occurs with probability } \alpha_2 .$$

Again we assume that the $2N$ genes are statistically independent. This means that if a particular generation consists of j a-genes and $2N - j$ A-genes, the $2N$ genes of the next generation are formed by $2N$ independent trials each of which results in a with probability

$$p_j = \frac{j}{2N}(1 - \alpha_1) + \left(1 - \frac{j}{2N}\right)\alpha_2 \qquad (3.4)$$

and in A with probability

$$q_j = \frac{j}{2N}\alpha_1 + \left(1 - \frac{j}{2N}\right)(1 - \alpha_2) . \qquad (3.5)$$

With this new definition of p_j and q_j the transition probabilities p_{jk} are still given by (3.2).

We still have a simple Markov chain, but the process exhibits an entirely new character. Thus, if $\alpha_1 > 0$, $\alpha_2 > 0$ there is no possibility of fixations. Instead, the n-step transition probabilities $p_{jk}^{(n)}$ [cf. (2.4)] approach a *steady state probability distribution* $\{x_k\}$ which is given by the solution of the system of equations [obtained from (2.4) letting $n \to \infty$]

$$x_k = \sum_{\nu=1}^{2N} x_\nu p_{\nu k} . \qquad (3.6)$$

Whatever the initial composition of the population, the probability that the n-th generation is composed of k a-genes and $2N - k$ A-genes tends to x_k. The distribution $\{x_k\}$ is called the *steady state gene frequency distribution* and is of greatest interest. An explicit formula for it is not known, and most of the work consists in deriving suitable approximations.

Selective forces in nature convey advantages to one kind of gene and disadvantages to the other. Qualitatively they act in the same direction as mutations and it is therefore possible, in a first approximation, to account for them in the same way. At any rate, the usual models of genetic theories amount to assuming

[7] Compare, for example, [5, p. 87] or [12, p. 16]. The Markov chain (3.2) was considered by Malécot [9]. He tried to determine the first characteristic value approximately from the consideration of an integral equation resembling (3.2). His approximation is $1 - 1/(2N + 1)$. Formula (8.1) gives the characteristic values of the more general chain of the text, and reduces to the present case if $\alpha_1 = \alpha_2 = 0$. That $1 - 1/2N$ is a characteristic value is easily verified by noting that $x_k = k^2 - k$ is a characteristic vector.

that the coefficients α_i in (3.4) and (3.5) represent the combined influence of mutation pressure and selective forces. Accordingly, we shall introduce no further improvement in our scheme.

It must be emphasized, however, that with a constant population size any decrease say of a-genes implies an equal increase of A-genes. If the population size changes then the same ratio of the numbers of a- and A-genes may represent different compositions of the population and, strictly speaking, *selective pressure can not be a function of the gene ratios*. Accordingly, any theory derived from the present model (or an equivalent to it) *depends essentially on the assumption of a constant population* size and the number of a-genes must not be identified with the a-gene frequency in a variable population. Only a bivariate model permits to treat the latter, and such models lead to new types of equations (section 10).

Let the random variable $Z^{(n)}$ represent the number of a-genes in the n-th generation, and $\phi_n = Z^{(n)}/2N$ the corresponding *gene frequency*. At a time when $Z^{(n)} = j$ the *expected change of the gene frequency in the next generation is*

$$E(\Delta\phi_n) = \alpha_2 - (\alpha_1 + \alpha_2)\frac{j}{2N} \tag{3.7}$$

and the corresponding variance

$$\operatorname{var}(\Delta\phi_n) = \frac{1}{2N} p_j q_j . \tag{3.8}$$

This quantity corresponds to the '*sampling variance*' of S. Wright. His point of departure is the expression (3.7), but the sampling variance is usually identified with the expression obtained from (3.8) by putting $\alpha_1 = \alpha_2 = 0$. The larger variance (3.8) takes into account that chance fluctuations are due not only to the random mating process, but also to mutations and selective forces.

4. The general diffusion equation

Let $Z(t)$ be a random variable (such as gene frequency, population size, etc.) depending on the time t which we now agree to treat as a continuous parameter. The stochastic process described by $Z(t)$ is said to be of the Markov type if, roughly speaking, future changes depend on the present state, but not on the past history which led to this present state. All processes which we shall consider are of the Markov type.[8]

The variable $Z(t)$ may change in jumps: such is the case, for example, if $Z(t)$ represents the number of incoming telephone calls, or the size of a population as long as we count individuals and do not introduce the usual continuous approximation. With discontinuous processes the probability of a change during a small time interval $(t, t+h)$ is small (of the order of magnitude h), but *if* a change occurs, it is of finite magnitude. Opposed to such processes are processes of the diffusion type where $Z(t)$ changes continually. In this case there is certainty that during any time interval, however small, $Z(t)$ undergoes some change, but for small time intervals this change is practically sure to be small. More precisely, the probability that $|Z(t+h) - Z(t)| > \epsilon$ is of smaller order of magnitude than h.[9]

[8] Compare footnote 1.

[9] For this definition compare [2] where the equation is derived under more general conditions than stated in the text. The whole theory, of course, is due to Kolmogorov [7]. It should be noted

As in the case of finite chains, the whole process is determined by the initial value of $Z(t)$ at time 0 and by the *transition probability density*. By this we mean the function $p(t; z_1, x)$ which gives the (conditional) probability density that $Z(t_0 + t) = x$ given that $Z(t_0) = z$ at some fixed time t_0. This is the continuous analogue of the transition probabilities $p_{jk}^{(n)}$ of section 2. In the discrete case all the $p_{jk}^{(n)}$ could be expressed in terms of the one step transition probabilities p_{jk}. It is the most remarkable fact of diffusion theory that $p(t; z, x)$ can be calculated for all t if only the *infinitesimal mean displacement* $a(x)$ *and the infinitesimal variance* $2b(x)$ are known. These quantities are defined in analogy with (3.7) and (3.8) as follows.

Suppose it is known that at some fixed time t_0 we have $Z(t_0) = z$. The expected value of $Z(t)$ at a later time $t_0 + h$ is then

$$E[Z(t_0+h)] = \int x p(h; z, x)\, dx \tag{4.1}$$

and the expected value of the increase (the so called mean displacement) is

$$E[\Delta Z(t_0)] = \int (x - z)\, p(h; z, x)\, dx . \tag{4.2}$$

Similarly

$$\operatorname{var}[\Delta Z(t_0)] = \int (x - z)^2 p(h; z, x)\, dx . \tag{4.3}$$

These quantities are functions of h and z, and for small h they must be of the order of magnitude h. It is only a mild regularity assumption that as $h \to 0$

$$\frac{1}{h} E[\Delta Z(t_0)] \to a(z) \tag{4.4}$$

$$\frac{1}{h} \operatorname{var}[\Delta Z(t_0)] \to 2b(z) .$$

These are the infinitesimal mean displacement and variance.

It was shown by Kolmogorov [7] *that the probability density* $u(t, x)$ *of* $Z(t)$ *satisfies the diffusion equation* (or 'Fokker-Planck equation')

$$u_t(t, x) = \{b(x)\, u(t, x)\}_{xx} - \{a(x)\, u(t, x)\}_x \tag{4.5}$$

(where subscripts indicate partial derivatives). Given the coefficients $a(x)$ and $b(x)$ and the initial distribution $u(0, x)$ of $Z(0)$, the solution $u(t, x)$ of (4.5) is in general[10] uniquely determined. In particular, if initially $Z(0) = z$, then the distribution $u(t, x)$ of $Z(t)$ coincides with $p(t; z, x)$ so that, for every fixed z, the transition

that there exist other types of Markov processes. Furthermore, it is in the text tacitly assumed that the transition probability depends only on the length of the time interval, but not on its position; in the general case the coefficients of the diffusion equation depend on x and t.

[10] We are here slurring over a delicate point of greatest interest from the point of view of differential equations. In physical diffusion theory the solution depends not only on the initial distribution, but also on boundary conditions. By contrast, our equations are of a *singular type* with the coefficient $b(x)$ vanishing on the boundaries. This leads to a new phenomenon, namely *natural boundaries* where no conditions need, or can, be imposed. In some cases there exists only one solution, in others there are many solutions, but only one corresponding to our problem. The theory of such singular equations is practically *terra incognita* and presents many interesting problems. It will be seen that the theory of evolution leads to boundary conditions of an altogether new type (compare section 7).

probability $p(t; z, x)$ is a solution of (4.5). In other words, the entire process is determined by the two coefficients $a(x)$ and $b(x)$.

Population growth is, strictly speaking, a discontinuous process, but for *large populations* the continuous model becomes a reasonable approximation, provided an appropriate time scale is used. In fact, it seems that this approximation comes closer to realities than many other simplifying assumptions. Most models used in evolution theory assume continuous growth, although this assumption may be more or less hidden. It follows then that in the univariate case such a model is mathematically equivalent to a particular choice of the coefficients $a(x)$ and $b(x)$.[11] In particular, as has been noted in the introduction, an essential part of the results of S. Wright can be derived from a certain diffusion equation. We shall discuss this aspect in section 7, but a few preparations seem desirable.

5. Exponential population growth and the branching process

The simplest illustration of the possibilities of the method is provided by a model which is the continuous analogue of the classical branching process and is actually an approximation to it when the population is large, that is, in its later phases.

We wish to describe a continuous population growth in which the individuals multiply and are statistically independent of each other with a fertility function independent of the total population size. In this case the infinitestimal mean displacement and variance are necessarily proportional to the instantaneous population size, and the diffusion equation (4.5) takes on the form

$$u_t(t, x) = \beta\{xu(t, x)\}_{xx} - \alpha\{xu(t, x)\}_x, \qquad 0 < x < \infty, \tag{5.1}$$

where α and β are constants. Their numerical values depend on the choice of units, and using appropriate scales on the t- and x-axes one can always achieve that $\beta = 1$, $\alpha = 1, -1$, or 0. The coefficient α measures the *drift* and is positive or negative according as the population average increases or decreases.

The equation (5.1) contains implicitly all probability relations governing our process and we proceed to verify that this model leads to sensible results and to compare it to the branching process.

The following explicit solution of (5.1) is given for the sake of completeness only. If the population size at time $t = 0$ is ξ, and $\alpha \neq 0$, then[12]

$$\begin{aligned} u(t, x) &= \frac{\alpha}{\beta(e^{\alpha t}-1)}\left(\frac{\xi e^{\alpha t}}{x}\right)^{1/2} I_1\left(\frac{2\alpha(\xi x e^{\alpha t})^{1/2}}{\beta(e^{\alpha t}-1)}\right) e^{-\alpha(\xi e^{\alpha t}+x)/\beta(e^{\alpha t}-1)} \\ &= \frac{\xi\alpha^2 e^{\alpha t}}{\beta^2(e^{\alpha t}-1)^2} e^{-\alpha(\xi e^{\alpha t}+x)/\beta(e^{\alpha t}-1)} \sum_{\nu=0}^{\infty} \frac{1}{\nu!(\nu+1)!}\left\{\frac{\alpha(\xi x e^{\alpha t})^{1/2}}{\beta(e^{\alpha t}-1)}\right\}^{2\nu}. \end{aligned} \tag{5.2}$$

[11] This is, of course, true only if one disregards after effects due to variations in age etc., which would make the process non-Markovian.

[12] Compare [3] where more general models are treated, but no proofs are given. Equation (5.1) is an example of a singular diffusion equation with a *unique* solution, where no boundary conditions can be imposed. (Compare footnote 10.) More curious is the equation (4.5) with $b(x) = \beta x$, $a(x) = \alpha x + \nu$ [which for $\nu = 0$ reduces to (5.1)]. Here uniqueness and various types of boundary effects depend on whether $\nu < 0$, $0 < \nu < 1$, or $\nu > 1$ (unpublished results).

The *expected population size*

$$M(t) = E[Z(t)] = \int_0^\infty xu(t, x)\,dx \tag{5.3}$$

can be obtained from (5.2), but it is simpler to note that easy manipulations on (5.1) show that $M'(t) = aM(t)$, and hence

$$M(t) = \xi e^{at}. \tag{5.4}$$

Thus the population average increases exponentially, as should be expected since the rate of increase is proportional to the instantaneous population size. In a similar way one sees that

$$\operatorname{var}[Z(t)] = 2\frac{\beta}{a} e^{at}(e^{at} - 1). \tag{5.5}$$

Finally, the probability that the population dies out before time t is

$$\delta(t) = 1 - \int_0^\infty u(t, x)\,dx = e^{-a\xi e^{at}/\beta(e^{at}-1)} \tag{5.6}$$

and hence we find for the probability of *ultimate extinction*

$$\lim_{t\to\infty} \delta(t) = \begin{cases} e^{-a\xi/\beta} & \text{if} \quad a > 0, \\ 1 & \text{if} \quad a \leq 0. \end{cases} \tag{5.7}$$

These results agree closely with the corresponding results for the simple *branching process* of section 2. There too the extinction probability is of the form γ^ξ. The generating function of the population size $Z^{(n)}$ in the n-th generation is $f^\xi(x)$, where ξ is the initial population size. As is well known (compare, for example, [4] or [6]),

$$E(Z^{(n)}) = \xi\mu^n \tag{5.8}$$

and

$$\operatorname{var}(Z^{(n)}) = \xi \frac{\lambda}{\mu(\mu - 1)} \mu^n(\mu^n - 1) \tag{5.9}$$

where $\mu = f'(1)$ is the expectation of the number of direct descendants and $\lambda = f''(1) + f'(1) - f'^2(1)$ the variance.

These formulas are not only of the same form as (5.4) and (5.5), but the latter follow from (5.8) and (5.9) as a limiting case for large populations.

To see this note that if in the branching process the population size is at any time x, then the size of the next generation has mean μx and variance λx. This means that the quantities $\mu - 1$ and λ have the significance of a and β in the continuous process.

Consider now a branching process in which the population is very large, and let us agree to measure the population in units of N individuals. In the new units each individual has weight $1/N$ and therefore the quantities $\mu - 1$ and λ are of the order of magnitude $1/N$. Accordingly we write

$$\mu - 1 = \frac{a}{N}, \qquad \lambda = \frac{\beta}{N}. \tag{5.10}$$

We consider the process after a number of generations which is of the order of magnitude N. This amounts to introducing a time scale in which one generation corresponds to a time interval $\Delta t = 1/N$, and an interval of length t contains some $n = tN$ generations. With these units (5.8) becomes

$$\xi\left(1+\frac{a}{N}\right)^{tN} \sim \xi e^{at} \tag{5.11}$$

and (5.9)

$$\xi \frac{\beta}{a\left(1+\frac{a}{N}\right)}\left(1+\frac{a}{N}\right)^{tN}\left\{\left(1+\frac{a}{N}\right)^{tN}-1\right\} \sim \xi \frac{\beta}{a} e^{at}(e^{at}-1). \tag{5.12}$$

In other words, *as $N \to \infty$ the formulas for the branching process go over into those for the continuous process.* That the same is true for the probability distributions will be shown in appendix II.

6. Gene frequency in the bivariate case

As we shall see in the next section, the theory of S. Wright assumes a diffusion equation for the gene frequency, that is, the proportion of a-genes in the population. With a population of constant size this frequency is proportional to the number of a-genes. This is not so in general. In any realistic bivariate model the numbers of a- and A-genes will be negatively correlated, but if the population size is not constant, the correlation coefficient will not be -1. In this case the *gene frequency cannot depend on a diffusion equation, and indeed, is not subject to a Markovian process.* This point will be helpful for a general understanding of the theory and may be illustrated by means of a model which is the extreme counterpart to the case of a constant population size: when the numbers of a- and A-genes are uncorrelated.

Consider two statistically independent populations each subject to the process described in the preceding section. Let the corresponding sizes be $X(t)$ and $Y(t)$, and the corresponding densities $u(t, x)$ and $v(t, y)$ respectively. The bivariate density of the pair $[X(t), Y(t)]$ has then the density $w(t, x, y) = u(t, x)v(t, y)$. Since $u(t, x)$ and $v(t, y)$ are solutions of equations of the form (5.1), it is easily seen that $w(t, x, y)$ satisfies the equation,

$$w_t(t, x, y) = \beta_1 \{xw(t, x, y)\}_{xx} + \beta_2 \{yw(t, x, y)\}_{yy} - a_1 \{xw(t, x, y)\}_x - a_2 \{yw(t, x, y)\}_y, \tag{6.1}$$

which would have been derived directly from the general theory for the bivariate case.

From $w(t, x, y)$ it is in theory possible to calculate the probability distributions of other random variables connected with the combined population. For example, the total population size is $S(t) = X(t) + Y(t)$, and the quantity corresponding to the gene frequency is $Z(t) = X(t)/S(t)$. Their probability densities are

$$\int_0^x w(t, \xi, x-\xi)\, d\xi \tag{6.2}$$

and

$$\frac{1}{x^2}\int_0^\infty \xi w\left(t, \xi, \frac{1-x}{x}\xi\right) d\xi, \qquad 0 < x < 1, \tag{6.3}$$

respectively. It is readily seen that neither is a solution of an equation of a diffusion type. The reason is that neither $S(t)$ nor $Z(t)$ are Markovian. In fact, it is intuitively clear that a prolonged series of observations on $S(t)$ or $Z(t)$ will permit better predictions as to future developments than would the sole knowledge of the present state; this, however, is but another way of saying that the past history has an after effect.

7. The diffusion equation of Wright's theory

As has been mentioned before, the theory of S. Wright is essentially equivalent to the assumption[13] that the gene frequency satisfies a diffusion equation of the form

$$u_t(t, x) = \{\beta x(1-x)u(t, x)\}_{xx} - \{[\gamma_2 - (\gamma_1 + \gamma_2)x]u(t, x)\}_x . \tag{7.1}$$

The theory hinges on the assumption that the gene frequency satisfies some diffusion equation, and on the particular choice of the coefficients $b(x)$ and $a(x)$. We have seen that in a bivariate population the gene frequency does not satisfy a diffusion equation unless *the population size is constant*. This, then, is theoretically an essential assumption of the theory. Presumably, however, it is in practice satisfied to a sufficient degree.

As for the coefficients, we shall derive them (section 8) by a passage to the limit from the Markov chain of section 2. This method reveals the assumptions underlying the theory and shows also what happens under other assumptions concerning the ratio of mutation pressure to population size.

First, however, let us describe a curious phenomenon concerning the *steady state*. If, as $t \to \infty$, the solution $u(t, x)$ of (4.5) tends to a limit $u(x)$, the latter must be a solution of the ordinary differential equation[14]

$$\{b(x)u(x)\}_{xx} - \{a(x)u(x)\}_x = 0 . \tag{7.2}$$

Various forms of solutions were discussed by S. Wright.

If our diffusion were of the regular type, we would be assured that whenever (7.2) admits of a solution, the limiting relation $u(t, x) \to u(x)$ takes place, so that $u(x)$ describes the later stages of the process. Unfortunately this is *not* so for our problems, and mathematical problems of a new kind confront us.

Consider (7.1) in the simple case $\gamma_1 = \gamma_2 = 0$ (absence of mutation and selection pressures). We know that the discrete model in this case implies certainty of ultimate fixation, and it would be unfortunate for the continuous model if the results did not agree. Actually in our case $u(t, x) \to 0$, as desired. However, (7.2) has the solution $u(x) = C\{x(1-x)\}^{-1}$, which is in reality not related to our problem.[15]

[13] It should be understood that even for a specific biological population the coefficients in (7.1) vary from place to place, that is, are functions of two additional geographic parameters. Geographic *migrations* constitute a new diffusion process so that Wright's theory actually envisages a compound diffusion process in a three dimensional phase space. In practice this process is simplified by discretizing it partially: this is done by means of the notion of population *isolates*, for which the coefficients are fixed. Even in this case, however, additional terms are introduced to account for migration effects, so that the theory is much more complicated than presented in the text.

[14] That the integration constant occurring on the right side of (7.2) vanishes follows from the fact that the left side represents the flow of probability mass at the origin.

[15] It is true that this is not a probability density, but this fact was erroneously attributed to the limited applicability of the continuous model near the boundaries, and it was assumed that $u(x)$ gives a reasonable approximation at least in the central part.

Even if the solution of (7.2) is a probability density it is not legitimate to conclude that $u(t, x) \to u(x)$. In general, there exists a fraction μ such that

$$u(t, x) \to \mu u(t, x), \qquad 0 < \mu < 1. \tag{7.3}$$

The difference $1 - \mu$ gives the probability of 'pure' populations in the steady state. At the beginning of the process probability mass flows out to the boundaries $x = 0$ and $x = 1$, but constant mutation lets part of this mass flow back into the interval where it is once more subject to the diffusion process. The net effect is that the masses concentrated at $x = 0$ and $x = 1$ increase steadily to a saturation point which is maintained in the steady state.

The actual determination of μ presents a difficult problem and what we have just described is a stochastic process of a type not yet studied and which deserves attention.

8. Passage to the limit

We proceed to show how the coefficients in (7.1) can be obtained by a limiting process from the discrete model of section 3.

We know that the probability distribution $\{x_k\}$ of the population size $Z^{(n)}$ in this model approaches a steady state distribution given by (3.6). The speed with which $p_{jk}^{(n)} \to x_k$ depends on the characteristic values $\lambda_0, \lambda_1, \ldots, \lambda_{2N}$ of the matrix (p_{jk}). It will be shown in appendix I that

$$\lambda_r = (1 - \alpha_1 - \alpha_2)^r \binom{2N}{r} \frac{r!}{(2N)^r}, \qquad r = 0, 1, \ldots, 2N. \tag{8.1}$$

[It is remarkable that λ_r depends only on the sum $\alpha_1 + \alpha_2$, but a similar statement is true also for the diffusion equation (7.1)]. Roughly speaking, $\lambda_0 = 1$ determines the steady state distribution $\{x_0\}$, and the next largest value determines the speed of the convergence. In fact, it is known from matrix theory that $p_{jk}^{(n)}$ can be written in the form

$$p_{jk}^{(n)} = x_k + a_1\lambda_1^n + a_2\lambda_2^n + \ldots + a_{2N}\lambda_{2N}^n, \tag{8.2}$$

where the coefficients a_ν depend on j and k, but not on n. Since $\lambda_1 > \lambda_2 > \ldots > \lambda_{2N}$ it follows that $p_{jk}^{(n)} - x_k$ decreases roughly as

$$\lambda_1^n = (1 - \alpha_1 - \alpha_2)^n \left(1 - \frac{1}{2N}\right)^n. \tag{8.3}$$

Here the first factor represents the influence of the mutation and selection pressures, the second the 'sampling variance.' Wright's theory considers mainly the case where the α_i are of the order of magnitude of N^{-1}, and we put

$$\alpha_1 = \frac{\beta_1}{N}, \qquad \alpha_2 = \frac{\beta_2}{N}. \tag{8.4}$$

We consider the β_i constant and study the asymptotic behavior of our model as $N \to \infty$. (Other possibilities are described in section 9.)

To observe a tendency to equilibrium we require a number of generations which is of the order of magnitude N. We, therefore, choose the time scale so that the time required for one generation becomes $\Delta t = 1/N$: a time interval of duration t then corresponds to some $t/\Delta t = tN$ generations. For large populations we treat both time and the gene frequency as continuous variables and put

$$t = \frac{n}{N}, \qquad x = \frac{k}{2N}. \tag{8.5}$$

In the new notations the mean (3.7) and variance (3.8) of the change in population size correspond to the quantities occurring in equation (4.4). Hence,

$$a(x) = \lim_{N \to \infty} \frac{1}{\Delta t} E(\Delta \phi_n) = \beta_2 - (\beta_1 + \beta_2) x \tag{8.6}$$

and similarly

$$2b(x) = \lim_{N \to \infty} \frac{1}{\Delta t} \operatorname{var}(\Delta \phi_n) = \tfrac{1}{2} x (1 - x). \tag{8.7}$$

We get thus the coefficients in (7.1) with $\gamma = \frac{1}{4}$ (this numerical value may be changed at will by a proper choice of scales).

It is noteworthy that[16] the solutions of (7.1) can be written in the form

$$u(t, x) = \sum e^{-\nu_r t} a_r f_r(x), \tag{8.8}$$

analogous to (8.2). Here

$$\nu_r = r\{\beta_1 + \beta_2 + \tfrac{1}{4}(r - 1)\}. \tag{8.9}$$

It should be proved that our passage to the limit actually leads from (8.2) to (8.8). This has not yet been done, but at least it is easily seen that

$$\lim_{N \to \infty} \lambda_r^n = e^{-\nu_r t}, \qquad t = n\Delta t = \frac{n}{N}. \tag{8.10}$$

9. Other possibilities

The described passage to the limit which led to Wright's diffusion equation (7.1) is different from the familiar similar process in physical diffusion theory where the ratio $\Delta x/\Delta t$ tends to infinity rather than to a constant. It rests entirely on the assumption (8.4). We shall now show that *any modification of this assumption leads to a nonsingular diffusion equation of the familiar type (to normal distributions).*

A continuous approximation to the discrete model of section 3 is sensible only if α_1 and α_2 are small. Accordingly we shall put

$$\alpha_1 = \beta_1 \epsilon, \qquad \alpha_2 = \beta_2 \epsilon. \tag{9.1}$$

We shall keep β_1 and β_2 fixed and perform the passage to the limit $\epsilon \to 0$, $N \to \infty$. In the preceding section we have solved the problem for the special case $\epsilon = 1/N$. Since the units of measurement are arbitrary, there remain essentially only the cases $N\epsilon \to 0$ and $N\epsilon \to \infty$.

A glance at (8.1) shows that if $N\epsilon \to 0$ the influence of the coefficients α_i becomes, in the limit, negligible; the process is asymptotically equivalent to the case of a pure random mating without mutation and selection. We assume therefore that

$$N\epsilon \to \infty. \tag{9.2}$$

[16] Compare the thesis by S. Goldberg (not yet published) giving explicit solutions of (7.1).

In this case the influence of the mutation pressure is predominant. Now this pressure is directed towards the *equilibrium point*

$$k = \frac{\alpha_2}{\alpha_1 + \alpha_2} = \frac{\beta_2}{\beta_1 + \beta_2}. \tag{9.3}$$

It is to be expected that once this equilibrium point is reached, large deviation from it will not occur. As a matter of fact, the gene frequency now tends in probability to the equilibrium point, and the fluctuation is concerned with the deviations from the equilibrium point measured in a scale in which they remain finite. In other words, we shall replace the population size $Z^{(n)}$ by the variable

$$Y^{(n)} = \frac{1}{\delta}\left\{\frac{1}{2N} Z^{(n)} - \frac{\beta_2}{\beta_1 + \beta_2}\right\}, \tag{9.4}$$

where $\delta > 0$ is a scale unit still to be determined.

Since the speed of the convergence towards the steady state depends essentially on α_1 and α_2, the number of generations required for continuous changes will be of the order of magnitude $1/\epsilon$. Accordingly, we introduce a time scale in which each generation requires time

$$\Delta t = \epsilon\,. \tag{9.5}$$

If in any generation $Y^{(n)} = x$ then the expectation of the change $Y^{(n+1)} - Y^{(n)} = Y^{(n+1)} - x$ turns out to be

$$-x\,(\alpha_1 + \alpha_2) = -\,x\,(\beta_1 + \beta_2)\,\Delta t\,. \tag{9.6}$$

In accordance with the definition (4.4) we get therefore for the linear coefficient

$$a\,(x) = -x\,(\beta_1 + \beta_2)\,. \tag{9.7}$$

Similarly

$$\operatorname{var}(\Delta Y^{(n)}) = \frac{1}{2N\,\delta^2}\, p_j q_j \sim \frac{1}{2N\,\delta^2}\,\frac{\beta_1\beta_2}{(\beta_1+\beta_2)^2}. \tag{9.8}$$

Hence we get $b(x) = 1$ provided we choose δ so that the last expression becomes $2\Delta t$. Accordingly we put

$$\delta^2 = \frac{1}{4N\epsilon}\,\frac{\beta_1\beta_2}{(\beta_1+\beta_2)^2}. \tag{9.9}$$

In view of (9.2) this quantity tends to 0. We thus are led to a process $Y(t)$ with a probability density satisfying *the diffusion equation*

$$u_t\,(t, x) = u_{xx}\,(t, x) + (\beta_1 + \beta_2)\,\{xu\,(t, x)\}_x\,. \tag{9.10}$$

Here the range of x is the *entire axis*. For an initial value $Y(0) = \eta$ (corresponding to a gene frequency $\beta_2/(\beta_1 + \beta_2) + \delta\eta$), the solution of (9.10) is

$$u\,(t, x) = \left\{\frac{\beta_1+\beta_2}{2\pi\,(1-e^{-2t(\beta_1+\beta_2)})}\right\}^{1/2} \exp\left\{-\frac{(\beta_1+\beta_2)\,(x+\eta\, e^{-(\beta_1+\beta_2)t})}{2\,(1-e^{-2(\beta_1+\beta_2)t})}\right\}. \tag{9.11}$$

Equation (9.10) is an ordinary diffusion equation with a *drift toward the origin*, and (9.11) is a normal probability density with mean and variance tending 0 and $1/(\beta_1 + \beta_2)$, respectively.

10. A bivariate model

The purpose of this section is to generalize the discrete model of section 3 by dropping the assumption that the population size is constant. This permits a more flexible treatment of selection forces. Many schemes can be devised, but for illustrative purposes it is preferable to describe a particular model rather than to go into vague generalities.

The composition of the population in the n-th generation is described by a pair of integers (j_1, j_2) indicating the numbers of a- and A-genes, respectively. These numbers are random variables to be denoted by $X_1^{(n)}$ and $X_2^{(n)}$. We wish to describe a natural scheme leading to explicit expressions for the transition probabilities $p(j_1, j_2; k_1, k_2)$ (that is, the probability that $X_1^{(n+1)} = k_1$ and $X_2^{(n+1)} = k_2$ given that $X_1^{(n)} = j_1$, $X_2^{(n)} = j_2$).

In the model of section 3 the new generation is formed in $2N$ independent trials. We want to take into account the possibility that, say, an a-gene dies without contributing to the succeeding generation and without thus indirectly increasing the (absolute) number of A-genes. Instead of keeping the population size rigidly constant we can adjust the scheme so that only the expected population size remains fixed. For that purpose we accept the following model.

If in a generation the totals of a- and A-genes are j_1 and j_2, then *the succeeding generation is formed in* $N = \sigma_1 j_1 + \sigma_2 j_2$ *independent multiple trials as follows. Each trial consists in first selecting an a- or an A-gene, with corresponding probabilities* $\sigma_1 j_1/N$ *and* $\sigma_2 j_2/N$. *If the result is an a-gene it has probability* ν_1 *to die without further influencing the next generation; if it survives, it has probability* α_1 *to mutate into an A-gene.* These act in a similar way with corresponding probabilities ν_2 and α_2.

In other words, each of the $N = \sigma_1 j_1 + \sigma_2 j_2$ trials has three possible outcomes. It results in an a-gene with probability

$$p_1 = \frac{\sigma_1 j_1}{N}(1 - \nu_1)(1 - \alpha_1) + \frac{\sigma_2 j_2}{N}(1 - \nu_2)\alpha_2; \tag{10.1}$$

the outcome is A with probability

$$p_2 = \frac{\sigma_1 j_1}{N}(1 - \nu_1)\alpha_1 + \frac{\sigma_2 j_2}{N}(1 - \nu_2)\alpha_2; \tag{10.2}$$

finally, the outcome is 'no contribution to the new generation' with probability $1 - p_1 - p_2$. Hence the transition probability is given by the trinomial distribution

$$p(j_1, j_2; k_1, k_2) = \frac{N!}{k_1!k_2!(N - k_1 - k_2)!} p_1^{k_1} p_2^{k_2} (1 - p_1 - p_2)^{N-k_1-k_2}. \tag{10.3}$$

The essential difference between this scheme and that of section 3 is that in the present case the population can increase indefinitely or die out. No special results have been worked out for this new scheme, and we shall be satisfied with deriving the diffusion equation to which it leads in the same way as the model of section 3 leads to Wright's equation (7.1).

If in the n-th generation $X_1^{(n)} = j_1$, $X_2^{(n)} = j_2$, then we have for the increments $\Delta X_i = X_i^{(n+1)} - X_i^{(n)}$

$$\begin{aligned} E(\Delta X_i) &= N p_i - j_i, \\ \operatorname{var}(X_i) &= N p_i (1 - p_i). \end{aligned} \tag{10.4}$$

New to the present case is only the covariance between the two increments,

$$\text{cov}\,(X_1, X_2) = -Np_1p_2 \tag{10.5}$$

(or a correlation coefficient $-\{p_1p_2/(1-p_1)(1-p_2)\}^{1/2}$ instead of -1 as in section 3).

Since we are in the bivariate case, the diffusion equation will be of the form

$$u_t(t, x, y) = \sum_{i,j=1}^{2} \{ b_{ij}(x_1, x_2)\, u(t, x_1, x_2) \}_{x_i x_j} - \sum_{i=1}^{2} \{ a_i(x_1, x_2)\, u(t, x_1, x_2) \}_{x_i x_j}\,. \tag{10.6}$$

A special case is equation (6.1), where $b_{12} = 0$ because of the assumed independence of $X(t)$ and $Y(t)$. The coefficients in (10.6) (the infinitesimal mean displacements and variances) are now to be obtained by a limiting process from (10.4) and (10.5).

Again we denote the time between two succeeding generations by $\Delta t = \epsilon$ and measure the population in units of ϵ^{-1}. In section 8 we had $\epsilon = 1/2N$, but now N is not a constant. As there we put

$$\alpha_i = \beta_i \Delta t\,, \qquad i = 1, 2\,. \tag{10.7}$$

Fluctuations in population size will be slow only if the survival chances $1 - \nu_i$ are close to the inverses of σ_i, and we put in analogy with (10.7)

$$1 - \nu_i = \frac{1 + \mu_i \Delta t}{\sigma_i}\,, \qquad i = 1, 2\,. \tag{10.8}$$

In the new units the expected change of the number of a-genes during a time interval of length Δt is

$$\begin{aligned} \epsilon E\,(\Delta X_1) &= \epsilon\,(Np_1 - j_1) \\ &= \Delta t\,\{-x_1(\beta_1 + \mu_1 - \beta_1\mu_1\epsilon) + x_2\beta_2\mu_2\epsilon\} \end{aligned} \tag{10.9}$$

and hence

$$a_i(x_1, x_2) = -x_i(\beta_i + \mu_i)\,, \qquad i = 1, 2\,. \tag{10.10}$$

A similar calculation for the variances leads to

$$b_{ii}(x_1, x_2) = \tfrac{1}{2} x_i \left(1 - \frac{x_i}{\sigma_1 x_1 + \sigma_2 x_2}\right), \tag{10.11}$$

and to

$$b_{12}(x_1, x_2) = \tfrac{1}{2}\, \frac{x_1 x_2}{\sigma_1 x_1 + \sigma_2 x_2}\,. \tag{10.12}$$

We have thus a bivariate diffusion equation in which the coefficients are not linear functions. If $\sigma_1 = \sigma_2 = 1$ the expressions (10.10) and (10.11) reduce to those in Wright's theory except that in the latter $x_1 + x_2$ is replaced by 1. Even if we put $\sigma_1 = \sigma_2 = 1$, our model is more general than the diffusion equation (7.1). In fact, with $\sigma_1 = \sigma_2 = 1$ the determinant $b_{11}b_{22} - b_{12}^2$ vanishes, and the introduction of new variables $\xi = x_1 + x_2$, $\eta = x_1 - x_2$ will reduce the second order terms in

(10.6) to a single second derivative. However, both variables still appear in the linear terms. If $\sigma_1 = \sigma_2 = 1$, $\mu_1 = \mu_2$, then (10.6) reduces to the equation (7.1) of Wright's theory.

APPENDIX I. THE CHARACTERISTIC VALUES OF THE MATRIX (3.2)

We want to verify formula (8.1) for λ_r and, incidentally, calculate the characteristic column vectors of the general matrix (p_{jk}) defined by (3.2) with p_j, q_j given in (3.4) and (3.5). For this purpose we have to show that for $r = 0, 1, \ldots, 2N$ the system of linear equations

$$\sum_{k=0}^{2N} p_{jk} x_k = \lambda_r x_j \tag{11.1}$$

admits of a nontrivial solution $\{x_0, \ldots, x_{2N}\}$.

Put for abbreviation

$$k_{(\nu)} = k(k-1)\ldots(k-\nu+1) \tag{11.2}$$

and note that

$$\sum_{k=0}^{2N} p_{jk} k_{(\nu)} = \frac{d^\nu}{dx^\nu}(q_j + p_j x)^{2N}\Big|_{x=1} = (2N)_{(\nu)} p_j^\nu . \tag{11.3}$$

We now prove that it is always possible to determine constants $a_0, a_1, \ldots, a_r$ (not all of them zero) so that

$$x_k = a_r k_{(r)} + a_{r-1} k_{(r-1)} + \ldots + a_1 k + a_0 \tag{11.4}$$

is a solution of (11.1). This means that *for the characteristic vector belonging to* λ_r *the component* x_k *is a polynomial in* k *of degree at most* r.

The assertion to be proved is that the a_i can be chosen so as to satisfy the $2N + 1$ equations

$$\sum_{\nu=0}^{r} a_\nu (2N)_{(\nu)} p_j^\nu = \lambda_r \sum_{\nu=0}^{r} a_\nu j_{(\nu)} . \tag{11.5}$$

Now both p_j^ν and $j_{(\nu)}$ are polynomials of degree ν in j, and it is possible to write

$$p_j^\nu = \sum_{s=0}^{\nu} c_{s,\nu} j_{(s)} , \tag{11.6}$$

where the $c_{s,\nu}$ are independent of j. Equating the coefficients in (11.5) we get

$$\lambda_r a_t = \sum_{\nu=t}^{r} a_\nu (2N)_{(\nu)} c_{t,\nu} , \qquad t = 0, 1, \ldots, r . \tag{11.7}$$

But

$$c_{\nu,\nu} = (1 - \alpha_1 - \alpha_2)^\nu N^{-\nu} \tag{11.8}$$

so that $(2N)_\nu c_{\nu,\nu} = \lambda_\nu$, and hence (11.7) is satisfied for $\nu = r$ and arbitrary a_r. Put $a_r = 1$. Then (11.7) permits us to calculate in succession $a_{r-1}, a_{r-2}, \ldots$ and this process breaks down only if for some $\nu < r$ one has $\lambda_\nu = \lambda_r$. In this case one can put $a_\nu = 0$, and thus we get $2N + 1$ independent characteristic vectors.

APPENDIX II. THE PASSAGE FROM THE BRANCHING PROCESS TO THE DIFFUSION EQUATION (5.1)

The results of section 5 show that a direct passage from the simple branching process of section 2 to the equation (5.1) must be possible. We effect this passage to the limit formally: it is not difficult to justify the steps, since the necessary regularity properties of the generating functions $f_n(x)$ were established by Harris [6].

We consider the case of large populations, that is, the later stages of a branching process in which the mean number of descendants $\mu = f'(1) > 1$. We introduce units for measuring time and population size, such that during 'small' time intervals the fluctuations will be 'small.' This requires that in the new units the quantity

$$\mu = 1 + \epsilon \tag{12.1}$$

must be small. Accordingly we introduce new units in such a way that an individual in the old counting and the time of one generation correspond to

$$\Delta x = \Delta t = \epsilon\,. \tag{12.2}$$

Since the population size will no longer be an integer, we pass from the generating function $f(x)$ to the characteristic function $f(e^{iz})$ or, in the new units, $f(e^{iz\epsilon})$.

Put

$$f''(1) = \lambda\,, \tag{12.3}$$

so that the variance of the number of direct descendants becomes

$$\sigma^2 = \lambda + \mu - \mu^2\,. \tag{12.4}$$

Then, as $\epsilon \to 0$

$$\begin{aligned} 1 - f(e^{iz\epsilon}) &= (1 - e^{iz\epsilon})\,\mu + \tfrac{1}{2}(1 - e^{iz\epsilon})^2\,\lambda + O(\epsilon^3) \\ &= -iz\mu\epsilon - \tfrac{1}{2}z^2(\lambda + \mu)\,\epsilon^2 + O(\epsilon^3) \end{aligned} \tag{12.5}$$

and hence

$$\log f(e^{iz\epsilon}) = i\epsilon z\left\{z + \epsilon\left(z + \frac{i}{2}\,z^2\sigma^2\right) + O(\epsilon^2)\right\}. \tag{12.6}$$

Now put

$$\phi(t, z) = f_n(e^{iz\epsilon})\,, \tag{12.7}$$

where $t = n\epsilon$. Then

$$\begin{aligned} \phi(t+\epsilon, z) &= f_n[f(e^{iz\epsilon})] \\ &= \phi\left(t, \frac{1}{i\epsilon}\log f(e^{iz\epsilon})\right) \\ &= \phi\left[t,\ z + \epsilon\left(z + \frac{i}{2}\,z^2\sigma^2\right) + O(\epsilon^2)\right]. \end{aligned} \tag{12.8}$$

In the limit we are led to the differential equation

$$\phi_t(t, z) = \left(z + \frac{i}{2}\,z^2\sigma^2\right)\phi_z(t, z)\,. \tag{12.9}$$

It follows that $\phi(t, z)$ is the characteristic function of a density $u(t, x)$ which satisfies the diffusion equation

$$(12.10) \qquad u_t(t, x) = \frac{\sigma^2}{2}\{xu(t, x)\}_{xx} - \{xu(t, x)\}_x$$

and the boundary condition $u(t, 0) = 0$. Now (12.10) is a special case of (5.1), and the boundary condition follows from the fact that in the branching process the probability mass flowing out into the origin tends to zero.

Harris' remark that in the branching process the distribution of $Z^{(n)}\mu - n$ converges holds also for our differential equation.

REFERENCES

[1] W. Doeblin, "Sur deux problèmes de M. Kolmogoroff concernant les chaines dénombrables," *Bull. Soc. Math. France*, Vol. 66 (1939), pp. 1–11.

[2] W. Feller, "Zur Theorie der stochastischen Prozesse," *Math. Annalen*, Vol. 113 (1936), pp. 113–160.

[3] ———, "Die Grundlagen der Volterraschen Theorie des Kampfes ums Dasein in wahrscheinlichkeitstheoretischer Behandlung," *Acta Biotheoretica*, Vol. 5 (1939), pp. 11–40.

[4] ———, *An Introduction to Probability Theory and Its Applications*, Wiley, New York, 1950.

[5] R. A. Fisher, *The Genetical Theory of Natural Selection*, Oxford University Press, Oxford, 1930.

[6] T. E. Harris, "Branching Processes," *Annals of Math. Stat.*, Vol. 19 (1948), pp. 474–494.

[7] A. Kolmogorov, "Über die analytischen Methoden der Wahrscheinlichkeitsrechnung," *Math. Annalen*, Vol. 94 (1931), pp. 415–458.

[8] ———, "Deviations from Hardy's formula in partial isolation," *C. R. Acad. Sci. URSS*, Vol. 3, No. 8 (1935), pp. 129–132.

[9] G. Malécot, "Sur un problème de probabilités en chaine que pose la génétique," *C. R. Acad. Sci. Paris*, Vol. 219 (1944), pp. 379–381.

[10] ———, *Les Mathématiques de l'Hérédité*, Masson et Cie, Paris, 1948.

[11] ———, "Les processus stochastiques de la génétique," to be published.

[12] S. Wright, "Statistical genetics in relation to evolution," *Actualités Scientifiques et Industrielles*, No. 802, Hermann et Cie, Paris, 1939.

[13] ———, "Statistical genetics and evolution. (Gibbs Lecture)," *Bull. Amer. Math. Soc.*, Vol. 48 (1942), pp. 223–246.

[14] ———, "The differential equation of the distribution of gene frequencies," *Proc. Nat. Acad. Sci.*, Vol. 31 (1945), pp. 382–389.

ANNALS OF MATHEMATICS
Vol. 54, No. 1, July, 1951

TWO SINGULAR DIFFUSION PROBLEMS*

BY WILLIAM FELLER

(Received August 28, 1950)

1. Introduction

The main object of this paper is the study of the parabolic equation

$$u_t = (axu)_{xx} - ((bx + c)u)_x, \qquad 0 < x < \infty \tag{1.1}$$

where $u = u(t, x)$, and a, b, c are constants, $a > 0$. This can be interpreted as the Fokker-Planck equation of a diffusion problem in which $bx + c$ represents the drift, and ax the diffusion coefficient. Its adjoint is of the same type except that it contains an additional term bu. This can be disposed of in the familiar way by introducing $v(t, x) = e^{bt}u(t, x)$.

Equation (1.1) is singular both at $x = 0$ and $x = \infty$, but no exciting phenomena occur at infinity. However, the vanishing of the leading coefficient at $x = 0$ has the effect that the equation completely changes its character according as $c \leqq 0$, $0 < c < a$, or $c > a$. Moreover, the most important solutions are unbounded near $x = 0$.

If (1.1) were regular, it would be possible to prescribe for a solution its initial values $u(0, x) = \phi(x)$ and certain boundary conditions for $x = 0$. The latter could be homogeneous, such as $u(t, 0) = 0$ ('absorbing barrier'), flux zero ('reflecting barrier'), $u_x(t, 0) = 0$, etc., or inhomogeneous, such as $u(t, 0) = g(t)$, etc. The solutions corresponding to the homogeneous boundary conditions would preserve positivity and norm, or be norm decreasing; that is, if the initial value $\phi(x) \in L$ and $\phi(x) \geqq 0$, then $u(t, x) \geqq 0$ for all t and

$$\int_0^\infty u(t, x)\, dx \leqq \int_0^\infty \phi(x)\, dx. \tag{1.2}$$

The purpose of this paper is to prove the following properties of (1.1)

(i) *If $c \leqq 0$, then the initial values $u(0, x)$ can be prescribed arbitrarily, and they uniquely determine a solution (so that no boundary conditions at $x = 0$ can be imposed). This solution is positivity preserving and norm decreasing.*

(ii) *If $0 < c < a$ the situation is almost the same as with regular equations. There exists now a 'reflecting barrier' solution of the initial value problem (that is a positivity and norm preserving solution defined by the condition that the flux at $x = 0$ vanishes). There exist infinitely many other positivity preserving norm decreasing solutions, among which one and only one satisfies the condition $u(t, o) < \infty$ (which now defines an 'absorbing barrier' and corresponds to $u(t, 0) = 0$ in the regular case). Moreover, there exist a unique solution with vanishing initial values and prescribed flux at $x = 0$; it is integrable, but in general unbounded.*

---

* Research under an ONR contract for development of probability theory at Cornell University.

(iii) *If $c > a$ there exists a positivity and norm preserving solution of the initial value problem such that both it and its flux vanish at $x = 0$. This means that $x = 0$ acts both as absorbing and reflecting barrier and that no homogeneous boundary conditions can be imposed. There exist solutions with an arbitrarily prescribed flux at $x = 0$, but any solution other than the one mentioned either assumes negative values or is, for some values of t, norm increasing.*

The proofs are based on the use of the Laplace transform with respect to x. Incidentally, *explicit expressions* are obtained (section 6) for the unique solution in the case $c \leqq 0$ and the absorbing barrier solution for $0 < c < a$.

Singular diffusion equations are of great interest for the theory of stochastic processes and have also applications in biology (cf. [1]). However, it appears that they have not been studied and that the phenomenon here discussed is new.[1] Our result is also of some interest for the theory of semigroups and is a counterpart to the investigations of Yosida and of Hille [4] who have studied the semigroups defined over $-\infty < x < \infty$ by Fokker-Planck equations with singularities at infinity.

Our method depends essentially on the linearity of the coefficients, and is therefore admittedly of limited applicability. However, it does apply to some other problems.

As an example we treat as a second problem the *Ehrenfest model of diffusion*. This appears in the physical literature in the form of a finite Markov chain.[2] We shall modify the model by the use of a continuous time parameter (section 7). In this way one is led to a system of ordinary differential equations and the problem simplifies without losing its physical significance.

2. Definitions and preliminaries

We are interested in solutions $u(t, x)$ which for every fixed t belong to $L(0, \infty)$, but the only regularity property which will actually be used is that the Laplace transformation with respect to x can be applied. We shall say that $u(t, x)$ *is a solution of* (1.1) *if for $x > 0$ it has continuous partial derivatives satisfying* (1.1) *and if for every fixed $s > 0$ and $t > 0$ the functions $e^{-sx}u(t, x)$ and $e^{-sx}u_t(t, x)$ are integrable over $0 < x < \infty$, and this uniformly in every interval $0 < t_0 \leqq t \leqq t_1 < \infty$.*

We put

$$(2.1) \qquad \omega(t, s) = \int_0^\infty e^{-sx}u(t, x)\, dx.$$

[1] An exception is a paper by Kepinski [6] who obtained part of our results for the special case $b = 0$. His results appear to imply that the behavior is the same for $0 < c < a$ as for $c \leqq 0$. This formal contradiction to our results is apparently due to the restrictive definition of solution used by Kepinski. He excludes the possibility that $u(t, x) \to \infty$ as $x \to 0$, but even the solutions used in similar cases in biological diffusion problems exhibit this behavior. An example of a solution rejected by Kepinski is

$$u = (t/x)^{\frac{1}{2}}e^{-x/t} - 2\pi^{\frac{1}{2}}\{1 - \Phi((2x/t)^{\frac{1}{2}})\},$$

where $\Phi(x)$ is the normal distribution function. This solution corresponds to $a = 1$, $c = \frac{1}{2}$, and vanishing initial values.

[2] For the definition cf. [2]. For more complete discussion cf. Friedman [3] and, in particular, Kac [5].

The Laplace transform of $u_t(t, x)$ exists and can be obtained by formal differentiation in (2.1). Note, however, that in general $u(t, x) \to \infty$ as $x \to 0$, so that the Laplace transform of $u_x(t, x)$ need not exist.

Integrating (1.1) from x to 1 and letting $x \to 0$ it is seen that in view of $u_t \in L(0, 1)$

$$\lim_{x\to 0} \{(axu)_x - (bx + c)u\} = -f(t) \tag{2.2}$$

exists and is bounded in every finite t-interval. *The function $f(t)$ will be called the flux of u at the origin.*[3]

For *initial values* it is most convenient to admit also discontinuous functions, for otherwise special considerations are required for the fundamental solutions.

Let $P(x)$ be a function of bounded variation in $(0, \infty)$ for which $P(x) \to 0$ as $x \to 0$, and the Laplace transform

$$\pi(s) = \int_0^\infty e^{-sx}\, dP(x) \tag{2.3}$$

exists.

We shall say that the solution $u(t, x)$ has the initial values $P(x)$ if

$$\lim_{t\to 0} \int_0^x u(t, x)\, dx = P(x) \tag{2.4}$$

at every continuity point of $P(x)$.[4] This amounts to requiring that $\omega(t, s) \to \pi(s)$.

A solution $u(t, x; \xi)$ depending on the parameter $\xi > 0$ will be called a fundamental solution if it assumes the initial values 0 *when $x < \xi$ and* 1 *when $x > \xi$, that is, if*

$$\omega(t, s;\xi) \to e^{-s\xi}. \tag{2.5}$$

In this case

$$u(t, x) = \int_0^\infty u(t, x; \xi)\, dP(\xi) \tag{2.6}$$

assumes the initial values $P(x)$, provided that the regularity conditions imposed on our solutions are satisfied by $u(t, x; \xi)$ uniformly with respect to ξ. (The last condition will be trivially satisfied by the fundamental solutions constructed in the sequel.)

---

[3] It follows from (1.1) that the time derivative of $\int_\alpha^\beta u(t, x)\, dx$ equals the flux at α minus flux at β.

[4] The solution $u(t, x)$ is not necessarily finite for $x = 0$. It is important to realize that (2.4) nevertheless implies a restriction on the behavior of $u(t, x)$ at the origin. For example

$$u(t, x) = \frac{b}{e^{bt} - 1}\left(\frac{b}{1 - e^{-bt}}\right)^{1-c/a} \exp\left\{\frac{-bx}{a(e^{bt} - 1)}\right\}$$

is a solution of (1.1) which, if $0 < c < a$, tends to 0 as $t \to 0$ for every fixed $x > 0$. However, the limit (2.4) is infinite, and therefore $u(t, x)$ does *not* assume vanishing initial values in the sense of our definition. An example of a solution with zero initial values is given in the preceding footnote.

3. Laplace transforms

In taking Laplace transforms of (1.1) some caution is necessary inasmuch as the individual terms on the right are not necessarily integrable. However, by assumption the Laplace transform of the left side converges for $s > 0$. Treating then the right side as a unit and using (2.2) we get

$$\omega_t = f(t) + s \int_0^\infty e^{-sx}\{a(xu)_x - (bx + c)u\}\, dx.$$

Since the Laplace transforms of u and xu converge for $s > 0$ the same is true of the transforms of $(xu)_x$. It follows that $(xu)_x$ is absolutely integrable near $x = 0$ and that $xu(t, x)$ approaches a limit as $x \to 0$. This limit is necessarily 0 since otherwise $u(t, x)$ would not be integrable. Hence, by the familiar manipulations

$$\omega_t + s(as - b)\omega_s = -cs\omega + f(t). \tag{3.1}$$

We require solutions with

$$\lim_{t \to 0} \omega(t, s) = \pi(s). \tag{3.2}$$

To avoid ambiguities we suppose, during the calculations, that $b \neq 0$. The final results retain a sense also for $b = 0$.

The characteristic equations of (3.1) are

$$dt = \frac{ds}{s(as - b)} = \frac{d\omega}{f(t) - cs\omega}. \tag{3.3}$$

The first equation has the general solution

$$e^{-bt}\,\frac{as - b}{s} = C_1. \tag{3.4}$$

Using it we find for ω as function of t

$$\omega = \left| C_1 - ae^{-bt} \right|^{c/a} \left\{ C_2 + \int_0^t \frac{f(\tau)\, d\tau}{|C_1 - ae^{-b\tau}|^{c/a}} \right\}. \tag{3.5}$$

The general solution $\omega(t, s)$ of (3.1) is obtained by putting

$$C_2 = A(C_1) \tag{3.6}$$

where $A(y)$ is an arbitrary function, and substituting the values C_1 and C_2 from (3.4) and (3.6) into (3.5). We require in particular a solution satisfying (3.2). Now for $t = 0$ we get $C_1 = a - b/s$ and therefore

$$\omega(0, s) = \left| \frac{b}{s} \right|^{c/a} A\left(a - \frac{b}{s}\right). \tag{3.7}$$

Equating this to $\pi(s)$ we find

$$A(y) = |a - y|^{-c/a}\, \pi\left(\frac{b}{a - y}\right). \tag{3.8}$$

Using this form we find $\omega(t, s)$ from (3.4)–(3.6). An easy calculation leads to

LEMMA 1. *The solutions of the initial value problem* (3.1)–(3.2) *are given by*

$$\omega(t, s) = \left(\frac{b}{sa(e^{bt}-1)+b}\right)^{c/a} \pi\left(\frac{sbe^{bt}}{sa(e^{bt}-1)+b}\right) + \int_0^t f(\tau)\left\{\frac{b}{sa(e^{b(t-\tau)}-1)+b}\right\}^{c/a} d\tau, \tag{3.9}$$

(with an obvious interpretation of the fractions if $b = 0$).

4. General form of the solutions

We now know that the Laplace transform $\omega(t, s)$ of any solution $u(t, x)$ of (1.1) is necessarily of the form (3.9). Our problem consists in finding the condition on $f(t)$ which will ensure that (3.9) is the Laplace transform of a solution (if it is, then it is obvious that $u(t, x)$ solves the initial value problem for $P(x)$).

LEMMA 2. *If* $c \leqq 0$ *then* (3.9) *can represent the Laplace transform of a solution* $u(t, x)$ *only if* $f(t)$ *is given by*

$$\pi\left(\frac{b}{a(1-e^{-bt})}\right) + \int_0^t f(\tau)\left\{\frac{e^{bt}-1}{e^{b(t-\tau)}-1}\right\}^{c/a} d\tau = 0. \tag{4.1}$$

PROOF. By definition a solution $u(t, x)$ is, for fixed t, integrable near $x = 0$, and therefore $\omega(t, s) \to 0$ as $s \to \infty$. For $c \leqq 0$ we have a fortiori

$$| sa(e^{bt} - 1) + b |^{c/a}\omega(t, s) \to 0,$$

and the lemma follows by bounded convergence.

LEMMA 3. *If* $c > 0$ *and* $f(t)$ *is continuous*[5] *for* $t \geqq 0$, *then* (3.9) *is the Laplace transform of a solution* $u(t, x)$ *with initial values* $P(x)$. *The solution is positivity preserving at least whenever* $f(t) \geqq 0$.

PROOF. It is clear that for fixed t and τ the first factor in (3.9) as well as the factor of $f(\tau)$ under the integral represent Laplace transforms of positive functions which have all regularity properties imposed on our solutions. If $P(x)$ is non-decreasing then $\pi(s)$ is completely monotonic (i.e. $\pi^{(n)}(s)\ (-1)^n \geqq 0$). The argument of $\pi(\ )$ in (3.9) is absolutely monotonic (i.e. $\psi^{(n)}(s) \geqq 0$), and hence the factor $\pi(\ )$ is completely monotonic or the difference of two such functions, according as $P(x)$ is non-decreasing or only of bounded variation. The product of two Laplace transforms corresponds to the convolution of the original functions, and this operation preserves the regularity properties imposed on our solutions.

LEMMA 4. *If* $c \geqq 0$ *and* $P(x)$ *is non-decreasing, then* (3.9) *with* $f(t) \equiv 0$ *defines a non-negative solution* $u(t, x)$ *of* (1.1) *with initial values* $P(x)$ *and*

$$\int_0^\infty u(t, x)\, dx \equiv P(\infty). \tag{4.2}$$

[5] This condition could easily be replaced by boundedness, which we know to be a necessary condition.

(*Norm preserving or 'reflecting barrier' solution.*) *For this solution*

$$(4.3)\qquad \lim_{x\to 0} u(t, x) = \begin{cases} \infty & \text{if } 0 < c < a \\ 0 & \text{if } c > a \\ \dfrac{b}{a(e^{bt}-1)}\,\pi\left(\dfrac{be^{bt}}{a(e^{bt}-1)}\right) & \text{if } c = a \end{cases}$$

PROOF. The existence of a non-negative solution is guaranteed by Lemma 3. Equation (4.2) is equivalent to $\omega(t, 0) = \pi(0)$ and follows trivially from (3.9) (also for $b = 0$). To prove (4.3) note that as $s \to \infty$ the argument of $\pi(\;\;)$ tends to the finite value $5e^{bt}/a(e^{bt} - 1)$. For the function $P^*(x)$ whose Laplace transform is given by expression $\pi(\;\;)$ in (3.9) this means that a finite mass is concentrated at $x = 0$. Near $x = 0$ therefore the solution $u(t, x)$ behaves like the function belonging to the factor of $\pi(\;\;)$, and the lemma follows.

LEMMA 5. *Let $c > a$. If $f(t) \geqq 0$ but $f(t)$ does not vanish identically, then the solution $u(t, x)$ of (3.9) is norm increasing. If $f(t_o) < 0$, then $u(t_o, x)$ assumes negative values.*

PROOF. Write for the second term in (3.9)

$$(4.4)\qquad \omega_o(t, s) = \int_o^t f(\tau)\left\{\frac{b}{sa(e^{b(t-\tau)} - 1) + b}\right\}^{c/a} d\tau.$$

If $f(t) \geqq 0$ then, by the proof of Lemma 3, $\omega_o(t, s)$ is the Laplace transform of a positive solution $u_o(t, x)$ with zero initial values. The first assertion now follows from the fact that the first term in (3.9) is norm preserving.

Next, if $u_o(t_o, x) \to 0$ as $x \to 0$, then

$$(4.5)\qquad s\omega_o(t_o, s) \to 0 \qquad s \to \infty.$$

Now if the interval of integration in (4.4) is restricted to $(0, t - \varepsilon)$, then the integral becomes $O(s^{-c/a}) = o(s^{-1})$. For $f(t_o) < 0$ we get then because of the continuity of $f(t)$

$$(4.6)\qquad \begin{aligned} s\omega(t_o, s) &= \{f(t_o) + o(1)\}s\int_o^{t_0}\left\{\frac{b}{sa(e^{b(t-\tau)} - 1) + b}\right\}^{c/a} d\tau \\ &= \{f(t_o) + o(1)\}s\int_o^{t_0}\left\{\frac{b}{sa(e^{b(t-\tau)} - 1) + b}\right\}^{c/a} e^{b(t-\tau)}\, d\tau \\ &= \{f(t_o) + o(1)\}\frac{1}{c - a}. \end{aligned}$$

This contradicts (4.5) and the Lemma is proved.

5. The integral equation (4.1)

The preceding lemmas contain the existence and uniqueness theorems for the case $c > a$; they show that in the case $0 < c \leqq a$ a norm and positivity preserving solution exists: and they show that when $c \leqq 0$ at most one solution of the initial value problem exists.

The next step consists in showing that when $c < a$ the integral equation (4.1) has a unique solution which, when substituted into (3.9) leads to a solution of our differential equation which is positivity preserving.

In order to simplify the formulas we shall consider only the *fundamental solution*, that is, we now suppose that $\pi(s) = e^{-s}$. The solution in the general case then follows by a simple integration (cf. (2.6)).

LEMMA 6. *Let* $c < a$. *(The integral equation*

$$\int_0^t f(\tau)\left(\frac{1-e^{-bt}}{e^{-b\tau}-e^{-bt}}\right)^{c/a} d\tau = -\exp\left\{-\frac{\xi b}{a(1-e^{-bt})}\right\} \tag{5.1}$$

has the unique solution

$$f(t) = \frac{-b}{\Gamma\left(-\frac{c}{a}+1\right)} \cdot \frac{e^{-bt}}{1-e^{-bt}}\left\{\frac{\xi b}{a(1-e^{-bt})}\right\}^{(-c+a)/a} \cdot \exp\left\{-\frac{\xi b}{a(1-e^{-bt})}\right\}. \tag{5.2}$$

PROOF. The substitutions

$$1-e^{-bt} = z^{-1}, \qquad 1-e^{-b\tau} = \zeta^{-1} \tag{5.3}$$

reduce (5.1) to

$$\int_z^\infty g(\zeta)(\zeta-z)^{-c/a}\, d\zeta = -\exp\left\{-\frac{\xi b z}{a}\right\} \tag{5.4}$$

where

$$g(\xi) = \frac{1}{b} f(\tau) \frac{\xi^{c/a}}{\xi(\xi-1)}. \tag{5.5}$$

It follows that

$$g(z) = -\frac{1}{\Gamma\left(1-\frac{c}{a}\right)}\left(\frac{\xi b}{a}\right)^{(-c+a)/a} \exp\left\{-\frac{\xi b z}{a}\right\}. \tag{5.6}$$

This proves the lemma.

LEMMA 7. *If* $f(t)$ *is the function of Lemma* 6 *and* $\pi(s) = e^{-s\xi}$, *then the expression* (3.9) *takes on the form*

$$\omega(t, s) = \left\{\frac{b}{sa(e^{bt}-1)+b}\right\}^{c/a} \exp\left\{\frac{-sb\xi e^{bt}}{sa(e^{bt}-1)+b}\right\} \cdot \Gamma\left(1-\frac{c}{a}; \frac{b\xi e^{bt}}{a(e^{bt}-1)} \cdot \frac{b}{sa(e^{bt}-1)+b}\right) \tag{5.7}$$

where for abbreviation

$$\Gamma(n; z) = \frac{1}{\Gamma(n)} \int_0^z e^{-x} x^{n-1}\, dx. \tag{5.8}$$

PROOF. The lengthy formal manipulations required for the proof are best performed using the variables defined in (5.3) and (5.5).

6. Fundamental solutions

LEMMA 8. *If $c \leqq 0$, then the equation* (1.1) *has a unique fundamental solution. If $0 < c < a$, then there exists a unique fundamental solution satisfying the boundary condition $u(t, 0; \xi) < \infty$. In either case the fundamental solution is non-negative and*

$$\int_0^\infty u(t, x)\, dx = \Gamma\left(1 - \frac{c}{a}; \frac{b\xi e^{bt}}{a(e^{bt} - 1)}\right); \tag{6.1}$$

its Laplace transform is given by (5.7).

PROOF. We have to prove the completely monotonic character of the solution (5.7). However, an explicit expression of $u(t, x)$ showing the positivity and the required regularity conditions will be given in the next lemma. The assertion $u(t, 0) = 0$ for $0 < c < a$ follows from $s\omega(t, s) < \infty$ as $s \to \infty$. Similarly, (6.1) follows from (5.7) for $s = 0$.

LEMMA 9. *The explicit form of the fundamental solution of Lemma 8 is given by*

$$\begin{aligned} u(t, x; \xi) = {} & \frac{b}{a(e^{bt} - 1)} \cdot \exp\left\{\frac{-b(x + \xi e^{bt})}{a(e^{bt} - 1)}\right\} \cdot (4b^2 e^{-bt} x/\xi)^{(c-a)/2a} \\ & \cdot I_{1-c/a}\left(\frac{2b}{a(1 - e^{-bt})} (e^{-bt} x\xi)^{1/2}\right) \end{aligned} \tag{6.2}$$

where $I_k(x)$ is the Bessel function

$$I_k(x) = \sum_{r=0}^{\infty} \frac{(x/2)^{2r+k}}{r!\Gamma(r + 1 + k)}. \tag{6.3}$$

(Note that $u(t, x; \xi)$ represents a power series in x.)

PROOF. Put

$$A = \frac{b\xi}{a(1 - e^{-bt})} \qquad s = \frac{1}{b}\{sa(e^{bt} - 1) + b\}. \tag{6.4}$$

Then an easy calculation shows that

$$\omega(t, s) = \frac{e^{-A} A^{1-c/a}}{(1 - c/a)} \int_0^1 (1 - v)^{-c/a} e^{Av/z \cdot z^{-1}}\, dv. \tag{6.5}$$

It is well known that $e^{Av/z} \cdot z^{-1}$ is the Laplace transform of $I_0(2(Avx)^{1/2})$. Hence we get by routine manipulations

$$\begin{aligned} u(t, x; \xi) = {} & \frac{1}{\Gamma(1 - c/a)} \cdot \frac{b}{a(e^{bt} - 1)} \cdot \left\{\frac{b\xi}{a(1 - e^{-bt})}\right\}^{1-c/a} \\ & \cdot \exp\left\{\frac{-b(x + \xi e^{bt})}{a(e^{bt} - 1)}\right\} \cdot \int_0^1 (1 - v)^{-c/a} I_0\left(\frac{2b}{a(1 - e^{-bt})} (\xi e^{-bt} vx)^{1/2}\right) dv. \end{aligned} \tag{6.6}$$

This is an alternative form for $u(t, x; \xi)$. The lemma now follows by expanding (6.6) into a power series in x.

7. The Ehrenfest model of diffusion

In the original formulation, N molecules are placed in two containers. At regular intervals one among them is chosen at random and placed into the other urn. We modify this by assuming that the changes can occur at any time and follow a Poisson law. In the usual way (cf. [2], Chapter 17) we get for the probability $p_j(t)$ of finding exactly j molecules in the first urn the system of differential equations

$$p_j' = -p_j + \left(1 - \frac{j-1}{N}\right) p_{j-1} + \frac{j+1}{N} p_{j+1} \tag{7.1}$$

where $0 \leqq j \leqq N$ and we define $p_{-1}(t) = p_{N+1}(t) = 0$. We seek a solution with $p_r(0) = 1$, $p_j(0) = 0$ when $j \neq r$, where r is the number of molecules initially in the first container. Put

$$P(t, x) = \sum_0^N p_j(t)x^j, \qquad P(0, x) = x^r. \tag{7.2}$$

Then

$$P_t = (x-1)P - \frac{x^2-1}{N} P_x. \tag{7.3}$$

The characteristic equations of this equation are

$$dt = \frac{N\,dx}{x^2-1} = \frac{dP}{(x-1)P}. \tag{7.4}$$

Their solutions are

$$e^{-2t/N} \frac{1-x}{1+x} = C_1, \qquad P\cdot(x+1)^{-N} = C_2. \tag{7.5}$$

The general solution of (7.3) is, therefore,

$$P = (x+1)^N A\left(e^{-2t/N} \frac{1-x}{1+x}\right) \tag{7.6}$$

where $A(y)$ is an arbitrary function. But $P(0, x) = x^r$, and hence

$$x^r = (x+1)^N A\left(\frac{1-x}{1+x}\right). \tag{7.7}$$

Therefore

$$P(t, x) = 2^{-N}\{1 + x - (1-x)e^{-2t/N}\}^r\{1 + x + (1-x)e^{-2t/N}\}^{N-r}, \tag{7.8}$$

and hence

$$p_j(t) = 2^{-N} \sum_{\nu=0}^{j} \binom{r}{\nu}\binom{N-r}{j-\nu} (1 - e^{-2t/N})^{r+j-2\nu}(1 + e^{-2t/N})^{N-r-j+2\nu}. \tag{7.9}$$

PRINCETON UNIVERSITY

REFERENCES

[1] W. FELLER: Diffusion Problems in Genetics. Second Symposium on Probability and Statistics. Berkeley.

[2] W. FELLER: An Introduction to Probability Theory and its Applications. New York (Wiley) 1950.

[3] B. FRIEDMAN: *A Simple Urn Model*. Communications on Pure and Applied Mathematics. Vol. 2, pp. 59–70 (1949).

[4] E. HILLE: On the Integration Problem for Fokker-Planck's Equation in the Theory of Stochastic Processes (To appear).

[5] M. KAC: *Random Walk and the Theory of Brownian Motion*. Amer. Math. Monthly, vol. 54, pp. 369–391 (1947).

[6] S. KEPINSKI: *Über die Differentialgleichung* $\frac{\partial^2 z}{\partial x^2} + \frac{m+1}{x}\frac{\partial z}{\partial x} - \frac{n}{x}\frac{\partial z}{\partial t} = 0$. Math. Ann., vol. 61, pp. 397–405 (1905).

ANNALS OF MATHEMATICS
Vol. 55, No. 3, May, 1952
Printed in U.S.A.

THE PARABOLIC DIFFERENTIAL EQUATIONS AND THE ASSOCIATED SEMI-GROUPS OF TRANSFORMATIONS

BY WILLIAM FELLER

(Received July 26, 1951)

Dedicated to the cherished
memory of HARALD BOHR

1. Introduction

We are concerned primarily with the pair of parabolic differential equations

$$u_t(t, x) = a(x)u_{xx}(t, x) + b(x)u_x(t, x) + c(x)u(t, x) \tag{1.1}$$

$$v_t(t, y) = \frac{\partial}{\partial y}\left\{\frac{\partial}{\partial y}(a(y)v(t, y)) - b(y)v(t, y)\right\} + c(y)v(t, y), \tag{1.2}$$

$(-\infty \leqq r_1 < x, y < r_2 \leqq \infty)$ and the more general functional equations which appear as their true adjoints.

The general parabolic equation can be written in either of these forms and, accordingly, no distinction is usually made between them. Nevertheless, the two equations are supposed to be adjoints to each other[1] and much of the theory depends on the interplay between them. It is therefore not surprising that clean results are obtainable only if they are treated in conjugate spaces. This is true even of the "self-adjoint equation" (cf. (2.4)) (and the neglect of this principle has contributed to confusion in the theory of partial differential equations).

The appropriate spaces are suggested by the fact that in diffusion theory (1.2) is satisfied by probability densities and (1.1) by transition probabilities as functions of the initial positions. This leads us to consider (1.2) in the space $\mathfrak{L}(r_1, r_2)$ of integrable functions and (1.1) in the space $\mathfrak{C}[r_1, r_2]$ of continuous functions. Neither of these spaces is the full conjugate of the other, but the possibility of using them as starting point will be explained in section 3. For finer points we shall have to extend the spaces.

For each fixed $t > 0$ we shall consider the solutions as linear transforms of the initial values, $u(t, x) = T_t f(x)$ and $v(t, y) = T_t^* g(y)$, where $f \in \mathfrak{C}[r_1, r_2]$ and $g \in \mathfrak{L}(r_1, r_2)$. The salient feature is that in each case these transformations will have the semi-group property $T_{t+s} = T_t(T_s)$ and $T_{t+s}^* = T_t^*(T_s^*)$. In our spaces it is possible to construct positivity preserving and norm preserving (or decreasing) transformations,[2] and we take particular interest in them.

The use of semi-group methods for partial differential equations was initiated by Hille [6]. More specifically, the integration problem for (1.2) has been taken up by Yosida [11] and Hille [6], [7]. It seems that hitherto only the question has

[1] Strictly speaking, the sign of u_t should be reversed, but this is a matter of formal convenience. We are actually referring to the differential operators on the right sides.

[2] This is not the case if a Hilbert space metric is used. Note that $u_t = u_{xx}$ is not self-adjoint even with a Hilbert space interpretation.

been asked whether the right side is the (or possibly the extension of the) infinitesimal generator of a semi-group. This amounts to requiring uniqueness of the solution of the initial value problem. Actually, the case of uniqueness for (1.2) is most exceptional, and the main problems relate to the case of non-uniqueness.

The infinitesimal generator of a semi-group is a closed operator. A given differential operator Ω (say the right side of (1.1)) can be the extension of infinitely many closed operators, and among them there may be infinitely many which generate semi-groups.

The general problem connected with a huge class of functional equations can now be formulated thus:

Given an operator Ω, find all semi-groups (or semi-groups with a specified property) whose infinitesimal generator is a contraction of Ω (that is, in its domain of existence it coincides with Ω).

From this point of view there is little difference between our parabolic differential equations and certain integrodifferential equations and infinite systems of ordinary differential equations. Also, our main results have close analogues in these cases.

Now the domain of the infinitesimal generator is essentially the same as the range of the resolvent. Knowledge of this range and of Ω determines uniquely the infinitesimal generator. Thus:

Each semi-group of transformations is completely determined by the operator Ω together with the range Σ of the resolvent.

The requirement that the range coincide with a prescribed set Σ will be called a *lateral condition.* It goes without saying that not every set can serve as range of a resolvent, and *our first problem consists in finding all possible lateral conditions.* In practically all cases our lateral conditions will involve unbounded operators, and frequently they will be of a *global character.* The sets Σ are almost always everywhere dense.

The notion of lateral condition has thus acquired the algebraic precision of functional analysis and fortunately retains it when applied to differential equations. Instead of imposing artificial boundary conditions which happen to work in the simplest cases, we are able to construct all possible lateral conditions specifying solutions (which, in turn, are defined in terms of semi-groups of transformations).

We shall encounter the *classical boundary conditions* but in a new form. Indeed, the expressions entering the classical boundary conditions are meaningless except in exceptionally smooth cases. They will be replaced by meaningful expressions and thus the solution of one long outstanding problem of diffusion theory is solved, namely to formulate *the analogue of the known boundary conditions for the case where the coefficients are singular at a boundary.* (A singularity arises if the coefficient $a(x)$ tends to zero, or if a coefficient does not approach a finite limit. For example, for the infinite interval an equation with periodic coefficients is singular.)

In addition to the generalized classical boundary conditions we shall find *lateral conditions of a new (and rather unexpected) type.* They are here derived by pure methods from a discussion of the resolvent equation, but it will be shown in another paper that they admit of probabilistic and physical interpretations. This makes it possible to state the appropriate boundary conditions for certain peculiar diffusion processes (cf. section 2d) which had for years resisted all attempts at a mathematical formulation.[3] This is the second instance of a concrete problem with physical significance where physical intuition failed, but our abstract methods provided a clue.

A matter of great concern to us is *the true form of the adjoint equations.* If the operators T_t form a semi-group of transformations, then the same is true of their formal adjoints T_t^*. Many finer points of the theory depend on the intimate interplay between a semi-group and its adjoint, and it is therefore surprising that the adjoint semi-group is hardly mentioned in the literature.

In each case treated in the present paper it will turn out that the formal adjoint of the resolvent is the resolvent of the adjoint semi-group. In each case we shall find the differential operator and the lateral conditions determining the semi-group. In this way we arrive at a rigorous formal theory of adjoint equations and adjoint lateral conditions.

The classical boundary conditions (in the generalized form found in the sequel) *can now be characterized as the lateral conditions which make* (1.1) *and* (1.2) *adjoints to each other. In all other cases the true adjoints contain additional terms.* The relationship is symmetrical and to illustrate the principle we shall construct all semi-groups related to (1.1) and thus find the most general true adjoint to (1.1). This leads to generalizations of (1.2). We could as well have started from (1.2) and constructed its adjoints.

Curious phenomena are encountered. *The general lateral condition for* (1.1) *depends on bounded linear functionals and on differential operators of second order. By contrast, the lateral condition of the adjoint problem is always of a local character and depends on differential operators of at most first order.*

In the simplest case the complete adjoint to (1.1) is defined by (1.2) and we have to deal with transformations from $\mathfrak{L}(r_1, r_2)$ to itself. In the next simplest case this space still suffices for the description, but the adjoint is no longer a differential equation. As more refined boundary conditions are imposed, we have to enlarge the space $\mathfrak{L}(r_1, r_2)$. Curiously enough the new terms occurring can be interpreted physically as representing masses at the boundary. One can now get a step further and consider the full adjoint to the adjoint, that is, *the second adjoint to* (1.1), etc.

In this way one is led to successive extensions of the same semi-group and also to bigger and better semi-groups. It provides again satisfaction that this formal play of functional analysis actually permits a probabilistic interpretation and that to each enlarged space there corresponds a larger class of stochastic processes. We shall not dwell on these successive conjugate spaces and only the principle is made clear.

[3] The first application will appear in a forthcoming paper by S. Goldberg on diffusion processes in genetics.

For an orientation as far as our differential equations are concerned the reader is referred to sections 2 and 23, where the main problems and results are described. It is hoped that they are understandable independently of the remainder.[4]

Sections 3–6 contain various preparations. Our constructions depend on a theorem due to Hille and Yosida which is restated in a form useful for our purposes. Much of the possible elegance of formulations is lost by the regrettable restriction to semi-groups which are strongly continuous at the origin.[5]

In section 7 a transformation suggested by the Sturm-Liouville theory is introduced which permits us in the sequel to put $a \equiv 1$ at the expense of changing the fundamental interval. By this device we achieve a uniform treatment in which there is no difference between finite or infinite intervals, or between regular and singular equations. For example, $u_t = u_{xx}$ is equivalent to a singular equation in a finite interval.

The completeness of our results is due to the lucky circumstance that the study of the resolvents in our case depends on ordinary differential equations. In section 9 a preparatory study of the solutions of $\lambda u - (u'' + bu') = 0$ is undertaken (following lines suggested by Hille, cf. footnote to section 2a) and in section 10 a duality principle between these solutions and those of $\lambda v - (v' - bv)' = 0$ is established. This leads to the classification of boundaries in section 11 (or, in the general form, in section 23).

There follows a study of the inhomogeneous equations for a fixed λ. Although we pursue a different purpose, many points of contact exist between our results and Stone's treatment [10] of the Sturm-Liouville operator in Hilbert space. Further points of contact (in particular, a counterpart to Weyl's theory and the spectral properties in Hilbert space) are now under investigation.

The actual construction begins when we let λ vary and try to adjust the solving kernel so that our transformations have ranges independent of λ.

In the simplest case (two natural boundaries) there is complete uniqueness and no further problem arises. The criterion for this case generalizes conditions for the uniqueness of the initial value problem in $(-\infty, +\infty)$ previously found by Hille.

The next simplest case is that of an *exit* boundary, paired with a natural boundary (where all non-uniqueness is due to one boundary). Here (1.2) has a unique solution, but not (1.1). In section 16 the general lateral condition is given, in section 17 it is proved that no other lateral conditions are possible, and in section 18 the true adjoints are constructed. They are not differential equations in the usual sense. Their solutions are unique.

[4] Explicit expressions for the lateral conditions and resolvents are given in the text only for the reduced form with $a = 1$. In section 23 the simplest of these equations are restated in their general form. Pressure of space necessitates that the translation of the remaining equations into their general form be left to the reader.

[5] In our applications this restriction causes only a clumsiness of formulation and in the rare cases where the range is not everywhere dense, the gap can be filled in a trite fashion. However, the elegance and power of the theory of semi-groups suffers from this artificial restriction.

If one boundary is *regular*, all solutions acquire additional regularity properties, and a more general lateral condition becomes possible. Curiously, the adjoint equation does not change, but its solution is no longer unique, and lateral conditions must be imposed. This is shown in section 19.

Finally, if *neither* boundary is natural, we get two lateral conditions and a compact matrix notation is required to express them and the resolvent. (Sections 20–22.)[6]

2. The main problems

(a) *Uniqueness.* The general form of the parabolic equation seems to have been studied only in the case of an *infinite interval.* The so-called integration problem then consists in finding solutions $u(t, x)$ and $v(t, y)$ of (1.1) and (1.2) which as $t \to 0$ converge in some sense to given functions $f(x)$ and $g(y)$. Analogy with the simple diffusion equation and also the physical meaning of the two equations led to the assumption that (under sufficient regularity conditions) these initial value problems should have unique solutions. Furthermore, it was expected that these solutions are expressible by means of fundamental solutions (or Green functions) in the form

$$u(t, x) = \int_{-\infty}^{+\infty} U(t; x, s)f(s)\, ds \tag{2.1}$$

$$v(t, y) = \int_{-\infty}^{+\infty} V(t; y, s)g(s)\, ds, \tag{2.2}$$

with the familiar symmetry relation

$$U(t; x, y) = V(t; y, x). \tag{2.3}$$

This would make the transformations $f \to u(t, x)$ and $g \to v(t, y)$ (with t fixed) adjoints to each other.

It has been shown [2] that this conjecture holds generally, but there are exceptions. Even for the "self-adjoint" equation

$$u_t = \alpha(\alpha u_x)_x \qquad -\infty < x < \infty \tag{2.4}$$

(which is at the same time of the form (1.1) and (1.2)) there will be uniqueness if and only if the integral of $\alpha^{-\frac{1}{2}}$ diverges both at $+\infty$ and $-\infty$. This unexplained phenomenon was particularly disturbing in view of the role of the equation in Kolmogorov's theory (cf. below under e). In [2] the general case of time dependent coefficients is treated. In the time independent case greater elegance can be achieved by the semi-group method and this approach has been used

[6] Preliminary results were presented to the International Congress of Mathematicians in September 1950, and in lectures and seminars. Thanks is due to Joanne Elliott for help, in particular for the removal of redundant conditions. [*Added in proof*] It has been shown in the meantime that the semi-groups of transformations are best treated in weak topologies. This removes the condition of strong continuity and leads to a more general theory of the resolvent and removes the defects mentioned in the text.

by Yosida [11] and, more specifically, by Hille [6],[7] [7]. Both authors limit themselves to a discussion of conditions which guarantee uniqueness.

From the point of view of the present paper there is no difference between infinite and finite intervals, and the uniqueness problem is solved together with the following problems.

(b) *The classical boundary conditions.* For the diffusion equation $u_t = u_{xx}$ in a *finite* interval (r_1, r_2) it is customary to add to the initial value condition $u(0, x) = f(x)$ two boundary conditions of the form

$$p_i u(t, r_i) + (-1)^i q_i u_x(t, r_i) = 0. \tag{2.5}$$

(This is the "elastic barrier" condition. For $p_i = 0$ one gets a "reflecting", for $q_i = 0$ an "absorbing" boundary.) Now (2.5) is derived merely by analogy with random walks, and in more complicated cases the formulation of appropriate boundary conditions apparently escapes even the best physical intuition.

The trouble with (2.5) is that in general it can not be postulated that u_x possesses boundary values. It will be shown that actually u_x must be replaced by another operator and that the correct form of (2.5) is given by (23.7) in the case of (1.1), and by (23.8) in the case of (1.2). It is remarkable that these conditions can be imposed only for "regular" boundaries, where the limits automatically exist for *all* solutions. In no case can special regularity properties be obtained by imposing boundary conditions.

In addition to the generalized classical boundary conditions we shall get new types of lateral conditions, partly of a global character.

(c) *Singular equations.* Of greatest theoretical, and also practical,[8] interest are equations where $a(x)$ vanishes at one boundary or where one of the coefficients has no finite limit. Practically nothing is known about such equations. Their curious properties may be guessed from the typical special case (connected with a problem in biology)

$$v_t = (\alpha y v)_{yy} - ((\beta y + \gamma)v)_y \qquad 0 < y < \infty \tag{2.6}$$

where α, β, γ are constants, $\alpha > 0$. The following facts were established in [3]:

If $\gamma \leqq 0$ the initial value problem has a unique solution (for integrable functions) and no boundary conditions whatever can be imposed. If $0 < \gamma < \alpha$, the equation behaves essentially like a regular equation, that is, an arbitrary boundary condition of type (23.8) can be imposed.[9] As $\gamma \to \alpha$ these solutions

[7] In this paper (1.2) is treated. Its manuscript (which was available to the author in the Fall of 1949) drew the author's attention to the possibilities opened by the study of the resolvent. The author is indebted to Hille also for many discussions and for the communication of further results. In particular, in [7] Hille found necessary and sufficient conditions for the uniqueness. These conditions will be here derived by a different method and with a different significance.

[8] For singular diffusion problems in the mathematical theory of evolution cf. [4]. Note that also the introduction of polar coordinates leads to singular equations.

[9] Actually this is a new result since the establishment of precise boundary conditions seemed unobtainable at the time when [3] was written. It was then only established that in addition to the known unique *bounded* solution there exist infinitely many integrable solutions among which exactly one is norm preserving.

converge to the same solution and for $\gamma \geqq \alpha$ we have a case of quasi-uniqueness with exactly one norm preserving solution but additional norm increasing solutions (whose existence is rather disturbing in diffusion theory).

From the present point of view there is no difference between regular and singular equations since a transformation of variables changes one into another.

(d) *New boundary value problems.* Various probability considerations and a long outstanding problem in genetics made it increasingly clear that, in addition to the classical types of boundary conditions there exist some of a new type. Physically speaking, there exist diffusion processes where a particle can be absorbed, and stay absorbed for a finite time, after which it penetrates slowly back into the interior. Attempts at a mathematical formulation of such phenomena remained unsuccessful for several years. It is now possible to identify one of our new boundary conditions as the one describing this phenomenon.

(e) *Stochastic processes.* Kolmogorov's theory [9] depends implicitly on the assumption that the relationship between (1.1) and (1.2) is such as described under (a). The theory has been developed only for infinite intervals assuming the uniqueness of the initial value problems. The non-uniqueness discovered in [2] was therefore disturbing.

It should be understood that even though the physicists use only (1.2) (under the name of Fokker-Planck equation), actually only (1.1) can be derived directly under fairly general conditions (cf. [2]),[10] and (1.2) must be derived as a consequence of (1.1). The derivation of (1.1) works also for finite intervals. It is for this reason that we start with (1.1), determine its most general solution, and then construct the adjoint equation. The latter, then, is the true form of the Fokker Planck equation.

That (1.2) is not the general form was suggested by a discovery of Doob [1] in the denumerable case (where instead of (1.1) and (1.2) we have infinite systems of ordinary differential equations). This suggested by analogy that only (1.1) is true and that the equality sign in (1.2) should be replaced by an inequality. We now see that this is so and we are able to specify the missing term. This result can readily be generalized to other stochastic processes.

3. The underlying spaces

As explained in the introduction we shall study (1.1) in the space $\mathfrak{C}[r_1, r_2]$ of functions continuous in the *closed* interval with the usual norm $\| u(x)\| = \sup | u(x)|$. The adjoint semi-group should then consistently be studied in the conjugate space, that is the space $\mathfrak{BV}[r_1, r_2]$ of functions of bounded variation. However, in each case where this adjoint semi-group is directly connected with (1.2), its range lies entirely within the space $\mathfrak{L}(r_1, r_2)$ of integrable functions, and it is therefore possible to study (1.2) in this space. (We consider the functions as defining set functions, and in this sense $\mathfrak{L}(r_1, r_2)$ is imbedded in $\mathfrak{BV}[r_1, r_2]$.)[11]

[10] [Added in proof]. For a more general abstract derivation cf. [2a].

[11] Actually it is possible to extend our semi-groups to the full space $\mathfrak{BV}[r_1, r_2]$ without increasing the range space. For $t = 0$ the semi-group is strongly continuous only

Even in the construction of the full adjoint to (1.1), we shall have to deal only with a subspace of $\mathfrak{BB}[r_1, r_2]$. In fact, the ranges of our semi-groups will contain only functions which are absolutely continuous within (r_1, r_2), but which attribute finite masses to the two endpoints. For unity of exposition and in order to avoid clumsy integrals we now introduce a notation which separates the three components.

DEFINITION. *By* $\bar{\mathfrak{L}}[r_1, r_2]$ *we shall denote the Banach space of triples* $[g_1, g(x), g_2]$ *with the norm*

$$(3.1) \qquad \| [g_1, g(x), g_2] \| = | g_1 | + | g_2 | + \int_{r_1}^{r_2} | g(x) | \, dx,$$

and the obvious addition rule. Here $g(x) \in \mathfrak{L}(r_1, r_2)$ and the g_i are real numbers. The space $\mathfrak{L}(r_1, r_2)$ is then the subspace of elements with $g_1 = g_2 = 0$.

We shall use the same sign for the norm in the several spaces, but no confusion can result from this.

4. Semi-groups

Let $\mathfrak{X}$ be a real Banach space in which positive elements are defined. A set $\mathfrak{S}$ of transformations T_t (with $t > 0$) is a semi-group if to each $x \in \mathfrak{X}$ and each $t > 0$ there corresponds an element $T_t x \in \mathfrak{X}$ such that $T_{t+s}x = T_t(T_s x)$. We shall deal exclusively with semi-groups which *satisfy the strong continuity condition*

$$(4.1) \qquad \| T_{t+h}x - T_t x \| \to 0 \quad \textit{as} \quad h \downarrow 0.$$

Theoretically, this is a restriction, but in the spaces with which we shall deal (4.1) is a consequence of weak measurability and thus really indispensable (cf. Hille [8], p. 183). To what extent (4.1) holds for $t = 0$ is stated in Lemma 4.

Following Hille [6] *we shall say that* $\mathfrak{S}$ *is a contraction semi-group if it satisfies the conditions* (4.1), *and*

$$(4.2) \qquad \| T_t x \| \leqq \| x \|,$$

and if

$$(4.3) \qquad x \geqq 0 \quad \textit{implies} \quad T_t x \geqq 0.$$

In particular, $\mathfrak{S}$ *is norm preserving if*

$$(4.4) \qquad \| T_t x \| = \| x \| \quad \textit{whenever} \quad x \geqq 0.$$

If $\{T_t\}$ is a contraction semi-group, then the operators $T'_t = e^{mt}T_t$ form a semi-group with properties (4.1), (4.3), and (4.4), but with (4.2) replaced by

$$(4.5) \qquad \| T_t x \| \leqq e^{mt} \| x \|.$$

Conditions (4.2) and (4.5) are therefore essentially equivalent.

We shall concentrate our attention to contraction semi-groups, but in each

in $\mathfrak{L}(r_1, r_2)$. Actually we shall be compelled to extend the resolvent to $\mathfrak{BB}[r_1, r_2]$ and shall for that purpose generalize (1.2); cf. Theorem 13.3.

case it will be obvious which conditions have to be dropped in order to arrive at a statement applicable to semi-groups with properties (4.1) and (4.5).

The *range* $\mathfrak{R}(\mathfrak{S})$ *is defined as the set of elements* $x \in \mathfrak{X}$ *such that* $x = T_t y$ for some $y \in \mathfrak{X}$ and $t > 0$. In practically all cases with which we shall deal $\mathfrak{R}(\mathfrak{S})$ will be dense in $\mathfrak{X}$, but there are exceptional cases. Of great importance is

LEMMA 4. *One has*[12]

$$(4.6) \qquad \| T_h x - x \| \to 0 \qquad \text{if, and only if,} \qquad x \in \overline{\mathfrak{R}(\mathfrak{S})}$$

(assuming, of course, (4.1) and (4.2) or (4.5)).

PROOF. The "only if" is obvious since $T_t x \to x$ implies $x \in \overline{\mathfrak{R}(\mathfrak{S})}$. Next assume that $x \in \mathfrak{R}(\mathfrak{S})$. Then $x = T_t y$ for some y and $t > 0$ and hence $\| T_h x - x \| = \| T_{t+h} y - T_t y \| \to 0$ by (4.1). Finally, to each $x \in \overline{\mathfrak{R}(\mathfrak{S})}$ and each $\delta > 0$ we can find a $\xi \in \mathfrak{R}(\mathfrak{S})$ such that $\| x - \xi \| < \delta$. Then $\| T_h x - x \| \leqq \| T_h \xi - \xi \| + \| T_h \xi - T_h x \| + \| x - \xi \| \leqq 2\delta + \| T_h \xi - \xi \|$, by (4.2).

We can always introduce $\overline{\mathfrak{R}(\mathfrak{S})}$ *as a new Banach space and consider* $\mathfrak{S}$ *as a semi-group on it*, that is, consider $T_t x$ only for $x \in \overline{\mathfrak{R}(\mathfrak{S})}$. The range of the thus contracted semi-group may be smaller, but it is everywhere dense as evidenced by (4.6). Therefore, *the contracted semi-group is strongly continuous for* $t = 0$.

The fact that we study the semi-groups connected with parabolic differential equations in $\mathfrak{C}[r_1, r_2]$ (instead of, say, $\mathfrak{C}(r_1, r_2)$) and $\mathfrak{L}(r_1, r_2)$ (instead of $\mathfrak{B}\mathfrak{V}(r_1, r_2)$) amounts essentially to studying contractions of the type just described (cf. the last footnote).

The *infinitesimal generator* A of $\mathfrak{S}$ is the operator defined by

$$(4.7) \qquad Ax = \lim_{h \to 0+} \frac{T_h x - x}{h}$$

for all those x for which the limit exists; they form the *domain* $\mathfrak{D}(A)$ of A (cf. Hille [8], p. 185). Obviously $\mathfrak{D}(A) \subset \overline{\mathfrak{R}(\mathfrak{S})}$, and hence *the infinitesimal generator is the same for a semi-group and its contraction to* $\overline{\mathfrak{R}(\mathfrak{S})}$.

Our constructions depend on the characterization of infinitesimal generators given in the

Hille-Yosida Theorem.[13] Let $\mathfrak{S}$ be a contraction semi-group with range dense

---

[12] This means that the strong continuity condition (4.1) extends to the origin $t = 0$ only within the closure of the range of $\mathfrak{S}$. For an understanding one should think of the initial value problem for $u_t = u_{xx}(-\infty < x < \infty)$ in the space $\mathfrak{B}\mathfrak{V}$. Here each $\phi \in \mathfrak{B}\mathfrak{V}$ has a transform which is absolutely continuous. For it

$$\frac{d}{dx} T_t \phi = (4\pi t)^{-\frac{1}{2}} \int_{-\infty}^{+\infty} \exp\{-(x-s)^2/4t\}\, d\phi(s).$$

As $t \to 0$ one has $T_t \phi \to \phi$ (in norm) if and only if ϕ is absolutely continuous. In all other cases one has only *pointwise* convergence of $T_t \phi$ to ϕ. Similarly most of our semi-groups can be extended to larger spaces but without *strong* continuity for $t = 0$.

[13] Hille [8] p. 238 and (with a simpler proof) Yosida [12]. We use the latter formulation. The theorem remains true when all references to positivity are omitted and (4.2) is replaced by (4.5).

in $\mathfrak{X}$. The infinitesimal generator A is an additive operator whose domain is dense in $\mathfrak{X}$ and such that to each $x \in \mathfrak{X}$ and to each $\lambda > 0$ there exists a unique $y_\lambda \in \mathfrak{D}(A)$ with

$$\lambda y_\lambda - A y_\lambda = x, \tag{4.8}$$

$$\lambda \| y_\lambda \| \leqq \| x \|, \tag{4.9}$$

and

$$y_\lambda \geqq 0 \qquad \textit{whenever} \qquad x \geqq 0. \tag{4.10}$$

Conversely, an additive operator with these properties[14] is the infinitesimal generator[15] of a contraction semi-group with range dense in X. It is norm preserving if and only if

$$\lambda \| y_\lambda \| = \| x \| \qquad \textit{whenever} \qquad x \geqq 0. \tag{4.11}$$

The inverse

$$y_\lambda = (\lambda I - A)^{-1} x \tag{4.12}$$

is called the resolvent of $\mathfrak{S}$. In view of the last statement in the Hille-Yosida theorem we shall say that *the resolvent is norm preserving if* (4.11) *holds.*

Applying the Hille-Yosida theorem to the contraction of $\mathfrak{S}$ to $\overline{\mathfrak{R}(\mathfrak{S})}$ we conclude: *for every contraction semi-group*

$$\overline{\mathfrak{D}(A)} = \overline{\mathfrak{R}(\mathfrak{S})} \tag{4.13}$$

5. Lateral conditions for semi-groups

The existing terminology is not adapted for the actual construction of semi-groups and for applications in general. For a given operator A it permits only the question whether or not A is the infinitesimal generator and it has led to the erroneous impression that no semi-group is associated with A if the solution of (4.8) is not unique. This in turn suggests introducing artificial spaces in order to achieve a uniqueness.

In general a given operator A will be associated with many semi-groups, that is, different semi-groups will have infinitesimal generators which are contractions of A. Usually then ranges will be dense in $\mathfrak{X}$. The semi-group is, of course, completely determined by its infinitesimal generator, but this generator is defined only by giving A and the subset Σ of $\mathfrak{D}(A)$ on which it coincides with A.[16]

---

[14] Yosida introduces the further condition that $\lambda A y_\lambda \to Ax$ for $x \in \mathfrak{D}(A)$. However, this condition is superfluous. In fact, $\lambda A y_\lambda - A A y_\lambda = Ax$ and hence $\lambda \| A y_\lambda \| <$ const. by (4.9). It follows that $\lambda y_\lambda \to x$ whenever $x \in \mathfrak{D}(A)$. Now $\mathfrak{D}(A)$ is dense in $\mathfrak{X}$, and hence $\lambda y_\lambda \to x$ for *every* x.

[15] Yosida's wording states explicitly only that A is a contraction of the true infinitesimal generator A_1. However, that $A_1 \equiv A$ can be proved as follows. Let $\eta \in \mathfrak{D}(A_1)$ and put $\lambda\eta - A_1\eta = x$, where λ is arbitrary but fixed. By assumption there exists a unique solution y_λ of (4.8). Now A_1 is an extension of A and hence $y_\lambda \in \mathfrak{D}(A_1)$ and $A_1 y_\lambda = A y_\lambda$. Since the solution of $\lambda\eta - A_1\eta = x$ is unique we have $\eta = y_\lambda$. Thus $\eta \in \mathfrak{D}(A)$.

[16] The infinitesimal generator is always a closed operator, and in all our cases A will be a closed extension of it.

In order to change the existing terminology as little as possible we reserve to the term infinitesimal generator its traditional meaning, but amend it by the following

DEFINITION. *Let A be an additive operator and let Σ be a linear set, $\Sigma \subset \mathfrak{D}(A)$. If the contraction of A to Σ is the infinitesimal generator of a semi-group $\mathfrak{S}$, then we shall say that A, restricted by the lateral condition Σ, generates $\mathfrak{S}$.*

Thus the letter Σ refers both to the set and to the condition that A is restricted (contracted) to Σ.

THEOREM 5. *Suppose that to each $\lambda > 0$ there corresponds a linear operator Γ_λ such that* (1) *$y_\lambda = \Gamma_\lambda x$ is defined for all $x \in \mathfrak{X}$,* (2) *$\lambda y_\lambda - Ay_\lambda = x$,* (3) *$\lambda \| y_\lambda \| \leqq \| x \|$,* (4) *$y_\lambda \geqq 0$ if $x \geqq 0$,* (5) *the range Σ of the transformation $y_\lambda = \Gamma_\lambda x$ is independent of λ.*

Then A, restricted by the lateral condition Σ, generates a contraction semi-group $\mathfrak{S}$.

Conversely, if $\mathfrak{S}$ is a contraction semi-group generated by A with range dense in $\mathfrak{X}$, then there exists a transformation Γ_λ with properties (1)–(5) *and such that Σ is the conditioning set.*

PROOF. The second part is a restatement of the first part of the Hille-Yosida theorem. The first part follows from the same theorem applied to the Banach space $\bar{\Sigma}$. In fact for $x \in \bar{\Sigma}$ equation (4.8) has by assumption (5) a unique solution $y_\lambda \in \Sigma$ and it satisfies (4.9) and (4.10). It suffices therefore to prove that the set $\{\Gamma_\lambda x \mid x \in \Sigma\}$ (the second range of Γ_λ) is dense in $\bar{\Sigma}$. This, however, is true since for $x \in \Sigma$ we have $\lambda y_\lambda \to x$ (cf. the footnote to the Hille-Yosida theorem).

NOTE. The Hille-Yosida theorem is tailored for semi-groups with ranges dense in $\mathfrak{X}$. Practically all of our semi-groups will have this property, but there are important cases where the range lies in a closed manifold. The Hille-Yosida theorem says nothing about the resolvent for x outside this manifold. It is for this reason that the unnecessary restriction appears in the second part, and that Γ_λ is not referred to as resolvent in the first part. These are aesthetic defects without consequences for our particular cases.

6. Solutions as semi-groups

We use now the notion of semi-groups to give a precise definition for the notion of solving an initial value problem and for boundary conditions.

For an intuitive justification consider the equation (1.1) in the simplest case where the solution $u(t, x)$ is uniquely determined by its initial values $f(x)$. By this is meant that to every $f(x)$ of a certain function class there corresponds uniquely a $u(t, x)$ which in some sense tends to $f(x)$ as $t \to 0$. For fixed t each $u(t, x)$ is in the same function class and may be considered as the result of a linear transformation on f; that is, $u(t, x) = T_t f$. Now, $u(t + s, x) = T_{t+s} f$ but, because of the assumed uniqueness, we can also consider $u(s, x)$ as the given initial function so that $u(t + s, x) = T_t u(s, x)$ or $T_{t+s} = T_t(T_s)$. The importance of this semi-group property has been emphasized by Hadamard and by Hille.

We encounter the same situation in the case of non-uniqueness for all classical solutions with stationary (or time independent) boundary conditions (such as those in section 2(b)). In fact, the semi-group property $T_{t+s} = T_t(T_s)$ can be taken as definition of stationarity in linear problems, and this point of view is adopted in the present paper.[17]

Accordingly, we relate each differential equation to a certain function space, and in particular (1.1) to $\mathfrak{C}[r_1, r_2]$. We impose the semi-group property $T_{t+s} = T_t(T_s)$ and also the strong continuity condition (4.1), (which is weaker than the assumptions which are usual in the theory of differential equations). The connection between a particular semi-group and a differential equation is established through the infinitesimal generator A. Thus, if $u(t, x) = T_t f$ then, formally from (1.1)

$$\text{(6.1)} \qquad \frac{u(t+h, x) - u(t, x)}{h} = \frac{T_{t+h} - T_t}{h} f = \frac{T_h - I}{h} u(t, x),$$

and a comparison with (4.7) shows that for each fixed $t > 0$ we must have $Au = au_{xx} + bu_x + cu$. We are now led to the

DEFINITION. *By a solution of the initial value problem for* (1.1) *we mean a linear transformation* $u(t, x) = T_t f$, *where* $\{T_t\}$ *is a semi-group generated by the operator* $A = a(x)(d^2/dx^2) + b(x)(d/dx) + c(x)$. *The lateral condition* Σ *defining the semi-group will be referred to as lateral (or boundary) condition for the differential equation.*

It is remarkable that this definition contains no explicit reference to the initial values $f(x)$ and the way in which they are assumed. However, we know that the strong continuity condition (4.6) holds if, and only if, f is in the closure of Σ. Hence:

If $u(t, x)$ *is a solution of* (1.1) *with the lateral condition* Σ, *then*

$$\text{(6.2)} \qquad \| u(t, x) - f(x) \| \to 0 \quad \text{as} \quad t \to 0$$

if and only if $f \in \bar{\Sigma}$.

In principle, $\bar{\Sigma}$ can be a closed linear subspace of $\mathfrak{C}[r_1, r_2]$ and the solution $u(t, x)$ of the initial value problem may be defined for *all* $f \in \mathfrak{C}[r_1, r_2]$. Then the initial values $f(x)$ are assumed *uniformly* in $[r_1, r_2]$ only if $f \in \bar{\Sigma}$. The existing theory of semi-groups tells us nothing about the type of convergence (if any) for initial values outside $\bar{\Sigma}$. In most cases with which we shall be concerned Σ will be everywhere dense and (6.2) will hold for *all* f. In the exceptional cases Σ will lie in the manifold characterized by the vanishing of one or two linear functionals and it is easy to see that outside $\bar{\Sigma}$ we have *pointwise convergence* of $u(t, x)$ to $f(x)$. However, it is always possible to restrict f to $\bar{\Sigma}$.

We shall concentrate our attention to contraction semi-groups, that is, solutions which are positivity preserving and norm not increasing. They are of particular interest. In each case it will be rather obvious which conditions of

[17] This excludes boundary conditions, such as $u(t, 0) = t^2$ into which time enters explicitly. It seems that in each known instance such solutions are obtained from stationary solutions by elementary methods.

our existence theorems are to be omitted to get more general semi-groups (whereas the conditions themselves are far from obvious).

It goes without saying that these considerations are valid for differential equations other than (1.1). The *resolvent* of the semi-group is, essentially, the Laplace transform of a resolving kernel. In all our cases *the adjoint transformation to the resolvent will again be the resolvent of a semi-group. We shall call it the adjoint semi-group and establish that it is related to a differential or other functional equation which will be called the adjoint to* (1.1).

7. Reduction to $a \equiv 1$

We consider the equations (1.1) and (1.2) in the open interval $-\infty \leqq r_1 < x < r_2 \leqq \infty$ *assuming that* $a(x)$, $a'(x)$, $b(x)$, $c(x)$ *are continuous in the interior and that* $a(x) > 0$. *The right side of* (1.2) *is to be interpreted as indicated, so that* v_x *need not exist.* The coefficients need not be bounded.

For definiteness assume that $r_1 < 0$, $r_2 > 0$. Let $h(x)$ be positive and differentiable in the *open* interval (r_1, r_2). The functions

$$(7.1) \qquad \xi(x) = \int_0^x h(s)\, ds, \qquad \eta(y) = \int_0^y h(s)\, ds$$

are monotonic and map the interval (r_1, r_2) onto an interval (ρ_1, ρ_2). For $\rho_1 < \xi < \rho_2$ we put

$$(7.2) \quad a_1(\xi) = a(x)h^2(x), \qquad b_1(\xi) = a(x)h'(x) + b(x)h(x), \qquad c_1(\xi) = c(x),$$

and consider the pair of equations

$$(7.3) \qquad \frac{\partial u_1(t, \xi)}{\partial t} = a_1(\xi)\frac{\partial^2 u_1(t, \xi)}{\partial \xi^2} + b_1(\xi)\frac{\partial u_1(t, \xi)}{\partial \xi} + c_1(\xi)u_1(t, \xi),$$

$$(7.4) \qquad \frac{\partial v_1(t, \eta)}{\partial t} = \frac{\partial}{\partial \eta}\left\{\frac{\partial}{\partial \eta}(a_1(\eta)v_1(t, \eta)) - b_1(\eta)v_1(t, \eta)\right\} + c_1(\eta)v_1(t, \eta).$$

These equations are of the same form as (1.1) and (1.2), and the following theorem (the verification of which is left to the reader) shows that the two pairs are equivalent.

THEOREM 7. *If* $u(t, x) \in \mathfrak{C}[r_1, r_2]$ *and* $v(t, y) \in \mathfrak{L}(r_1, r_2)$ *are solutions of* (1.1) *and* (1.2), *respectively, then*

$$(7.5) \qquad u_1(t, \xi) = u(t, x) \qquad \text{and} \qquad v_1(t, \eta) = \frac{v(t, y)}{h(y)}$$

are solutions of (7.3) *and* (7.4). *Furthermore,* $u_1 \in \mathfrak{C}[\rho_1, \rho_2]$ *and* $v_1 \in \mathfrak{L}(\rho_1, \rho_2)$ *and*

$$(7.6) \qquad \| u_1 \| = \| u \|, \qquad \| v_1 \| = \| v \|.$$

(This theorem implies in particular that if the original equations have a common solving kernel $U(t; x, y)$ in the sense of (2.1)–(2.3), then

$$U_1(t;\xi,\eta) = \frac{U(t;x,y)}{h(y)} \tag{7.7}$$

is a common solving kernel for (7.3) – (7.4).)

We shall use the transformation for the particular choice

$$h(x) = a^{-\frac{1}{2}}(x) \tag{7.8}$$

which leads to

$$a_1 \equiv 1, \qquad b_1 = (b - \tfrac{1}{2}a')a^{-\frac{1}{2}}, \qquad c_1 = c. \tag{7.9}$$

Accordingly *we can (and shall from Section 9 on) without loss of generality assume that* $a \equiv 1$, but this convention excludes any further restrictions on the interval and the remaining coefficients.

8. Reduction to $c \equiv 0$

This reduction is not always possible and the corresponding transformation changes the nature of the underlying spaces.

We continue to use the assumptions and notations of section 7.

THEOREM 8.1. *Suppose that* $p(x) > 0$ *satisfies in* (r_1, r_2) *the equation*

$$a(x)p'' + b(x)p' + c(x)p = kp \tag{8.1}$$

where k *is a positive constant. Put*

$$U(t,x) = e^{kt}p(x)u(t,x), \qquad V(t,y) = \frac{e^{kt}v(t,y)}{p(y)} \tag{8.2}$$

where $u(t,x)$ *and* $v(t,y)$ *are solutions of* (1.1) *and* (1.2), *respectively. Then*

$$\frac{\partial U}{\partial t} = a(x)U_{xx} + \beta(x)U_x \tag{8.3}$$

$$\frac{\partial V}{\partial t} = \frac{\partial}{\partial y}\left\{\frac{\partial}{\partial y}(a(y)V) - \beta(y)V\right\} \tag{8.4}$$

with

$$\beta(x) = b(x) + 2a(x)p'(x)/p(x). \tag{8.5}$$

The essential point is, of course, that (8.3) and (8.4) stand in the same relation to each other as (1.1) and (1.2).

LEMMA 8.1. *A sufficient condition for the existence of a positive solution of* (8.1) *is that*

$$c(x) \leq k. \tag{8.6}$$

PROOF. If (8.6) holds then no solution of (8.1) can have a positive maximum or a negative minimum. It follows that no solution can have two zeros, and the assertion now follows from Lemma 8.2(b).

The next theorem shows that the existence of a positive solution of (8.1) is *necessary* for the existence of a positivity preserving solution of the initial value problem of (1.1). This result will not be used in the sequel, however.

THEOREM 8.2. *In order that the right side of* (1.1) *generate a contraction semi-group in* $\mathfrak{C}[r_1, r_2]$ *it is necessary that* (8.1) *has a positive solution for each k sufficiently large.*

(An analogous theorem holds for (1.2).) The proof will be based on

LEMMA 8.2. (a) *If there exists a solution* $p_1(x)$ *of* (8.1) *with* $p_1(x_0) = p_1(x_1) = 0$, *then every solution of* (8.1) *has a zero somewhere in* $[x_0, x_1]$.

(b) *If no solution of* (8.1) *has two zeros, then there exists at least one positive solution.*

PROOF. (a) This assertion is known. To prove it, assume that $p_1(x) > 0$ in (x_0, x_1) and $p(x) > 0$ in $[x_0, x_1]$. The solution $p(x) + \tau p_1(x)$ has no zero in $[x_0, x_1]$ if τ is small, and two zeros if τ is large enough. For some τ we therefore get a non-negative solution tangent to the x-axis which is impossible.

(b) Let $p_1(x)$ be a solution with $p_1(0) = 1$. Assume that $p_1(\xi) = 0$, since otherwise nothing is to be proved. Let $p_0(x)$ be the solution with $p_0(0) = 0$, $p_0'(0) = 1$. Then $p_1(x) \gtrless 0$ for $x \lessgtr \xi$ and $p_0(x) \gtrless 0$ for $x \gtrless 0$. It is now clear that to every $s \neq 0 (r_1 < s < r_2)$ there corresponds a n_s such that the solution $p_1(x) + n_s p_0(x)$ vanishes at s (and nowhere else). Assume $\xi > 0$. As s varies from ξ to r_2 the function n_s varies monotonically from 0 to $\alpha (0 < \alpha \leqq \infty)$. Letting s decrease from ξ to 0, and then from 0 to r_1, we find n_s going from 0 to $-\infty$, and then from $+\infty$ to β. Any solution with $\beta < n_s < \alpha$ would have a positive and a negative zero. Therefore $\alpha \leqq \beta < \infty$, and it is obvious that the solutions $p_1(x) + \tau p_0(x)$ with $\alpha \leqq \tau \leqq \beta$ are positive.

PROOF OF THEOREM 8.2. If a contraction semi-group existed, then by the Hille-Yosida theorem (section 4) the equation

$$\lambda F - \{aF'' + bF' + cF\} = f \tag{8.7}$$

would have a positive solution $F \in \mathfrak{C}[r_1, r_2]$ for each positive $f \in \mathfrak{C}[r_1, r_2]$. Now if (8.1) has no positive solution, then Lemma 8.2 asserts the existence of an interval $[x_0, x_1]$ such that each solution has a zero in it. Choose $f(x)$ so that $f(x) = 0$ for $x_0 \leqq x \leqq x_1$. Then for $x_0 \leqq x \leqq x_1$ any solution of (8.7) coincides with a multiple of a solution of (8.1), and therefore can not be positive.

9. Solutions of $\lambda u - \Omega u = 0$

From now on we consider only the case $a \equiv 1$, $c \equiv 0$ and *suppose* $b(x)$ *continuous in the* <u>*open*</u> *interval* (r_1, r_2) $(-\infty \leqq r_1 < 0 < r_2 \leqq \infty)$. We introduce the operator

$$\Omega = \frac{d^2}{dx^2} + b(x) \frac{d}{dx} \tag{9.1}$$

and study solutions of the ordinary differential equation

$$\lambda u - \Omega u = 0, \qquad \lambda > 0. \tag{9.2}$$

For abbreviation we put

$$B(x) = \int_0^x b(s)\, ds. \tag{9.3}$$

We shall formulate lemmas involving the boundary r_2 ; analogous assertions hold for r_1 with "decreasing" replaced by "increasing".

Clearly no solution of (9.2) can have a positive maximum or a negative minimum. It follows that for each $s > 0$ there exists exactly one solution $u_s(x)$ joining $(s, 0)$ to $(0, 1)$, and this $u_s(x)$ is non-increasing throughout (r_1, r_2). As s increases to r_2 the slope $u_s'(0)$ increases to a limit γ which must be negative since the solution with $u(0) = 1$, $u'(0) = 0$ has a minimum at $x = 0$ and therefore can not be the limit of non-increasing solutions. The solution with $u(0) = 1$, $u'(0) = \gamma$ is positive non-increasing. This proves

LEMMA 9.1.[18] *There exists at least one positive non-increasing solution of* (9.2).

We proceed to analyze the conditions that this solution be unique and that it tend to 0.

LEMMA 9.2. *The solutions of* (9.2) *passing through* (0, 0) *are bounded near* r_2 *if and only if*

$$e^{-B(x)} \in \mathfrak{L}(0, r_2) \qquad e^{-B(x)} \int_0^x e^{B(s)}\, ds \in \mathfrak{L}(0, r_2). \tag{9.4}$$

The second condition implies the first.

PROOF. From (9.2)

$$u'(x) = e^{-B(x)} \left\{ u'(0) + \lambda \int_0^x e^{B(s)} u(s)\, ds \right\}. \tag{9.5}$$

The solutions through the origin differ only by a multiplicative constant, and we may assume $u'(0) > 0$. Since $u(x)$ can have no positive maximum all terms in (9.5) are positive for $x > 0$. In order that $u(x)$ be bounded at r_2 it is necessary and sufficient that $u'(x) \in \mathfrak{L}(0, r_2)$. This proves the necessity of (9.4). Conversely,

$$0 \leqq u'(x) \leqq u'(0)e^{-B(x)} + \lambda u(x) e^{-B(x)} \int_0^x e^{B(s)}\, ds = B_1(x) + u(x) B_2(x) \tag{9.6}$$

and (9.4) implies that $B_j(x) \in \mathfrak{L}(0, r_2)$. The solution of $w' = B_1 + wB_2$ with $w(0) = 0$ is bounded for $x > 0$, and since $u < w$ this proves the sufficiency of (9.4).

[18] Due to E. Hille [6].

LEMMA 9.3. *In order that there exist a non-increasing solution of* (9.2) *with*

$$\lim_{x \to r_2} u(x) = \omega > 0 \tag{9.7}$$

it is necessary and sufficient that

$$e^{B(x)} \in \mathfrak{L}(0, r_2) \qquad e^{-B(x)} \int_x^{r_2} e^{B(s)}\, ds \in \mathfrak{L}(0, r_2). \tag{9.8}$$

For this solution

$$e^{B(x)} u'(x) \to 0 \tag{9.9}$$

except perhaps, if

$$e^{B(x)} \in \mathfrak{L}(0, r_2) \qquad e^{-B(x)} \in \mathfrak{L}(0, r_2). \tag{9.10}$$

PROOF. Suppose that $u(x) \downarrow \omega > 0$. For this solution the right side in (9.5) is never positive so that the second term in the braces is bounded by $| u'(0) |$. In view of (9.7) this implies the first of the conditions (9.8). For $x > 0$ we can write (9.5) in the form

$$-u'(x) = e^{-B(x)} \left\{ k + \lambda \int_x^{r_2} e^{B(s)} u(s)\, ds \right\} \tag{9.11}$$

whence

$$-u'(x) > e^{-B(x)} \left\{ k + \lambda\omega \int_x^{r_2} e^{B(s)}\, ds \right\}, \tag{9.12}$$

where $x > 0$ and $k \geqq 0$ is a constant. Since $-u'(x) \in \mathfrak{L}(0, r_2)$ this implies the second of the conditions (9.8).

Next assume (9.8) and let $u(x)$ be a positive non-increasing solution (cf. Lemma 9.1). For it (9.11) holds for some $k \geqq 0$. If $k = 0$, then

$$-u'(x) < u(x) \lambda e^{-B(x)} \int_x^{r_2} e^{B(s)}\, ds \tag{9.13}$$

and therefore $-u'/u$ is bounded by a function assumed to be integrable over $(0, r_2)$. It follows that $-\log u(x)$ is bounded. Hence (9.7) and (9.9) hold. On the other hand if $k > 0$ then (9.11) implies (9.10) and lemma (9.4) (c) shows that in this case there exist infinitely many solutions satisfying (9.7).

LEMMA 9.4. *Suppose that* (9.10) *holds. Then:* (a) *Each solution of* (9.2) *is bounded for* $x > 0$. (b) *The limits*

$$\lim_{x \to r_2} u(x) \quad \textit{and} \quad \lim_{x \to r_2} e^{B(x)} u'(x) \tag{9.14}$$

exist. (c) *There exist two positive monotonic solutions* $u_j(x)$ *with*

$$\begin{aligned} u_0(x) \downarrow 0 \qquad & e^{B(x)} u_0'(x) \to -1 \\ u_1(x) \downarrow 1 \qquad & e^{B(x)} u_1'(x) \to \ \ 0, \end{aligned} \tag{9.15}$$

and a solution $u(x)$ is positive non-increasing if and only if

$$u(x) = p_0u_0(x) + p_1u_1(x), \qquad p_j \geqq 0. \tag{9.16}$$

PROOF. Assertion (a) is implied by Lemmas 9.1 and 9.2. That $u(x)$ tends to a limit follows from the non-existence of positive maxima and negative minima. The existence of the second limit in (9.14) follows from (9.5). Finally, it is clear that each limit is attained by exactly one solution $u(x)$ passing through (0, 1) and that a positive solution is non-increasing if, and only if,

$$\lim e^{B(x)}u'(x) \leqq 0. \tag{9.17}$$

Now the (unique) solution $u(x)$ with $u(0) = 1$, $u(x) \to 0$ satisfies (9.11) with $k > 0$ (since $k = 0$ implies $u(x) \downarrow \omega > 0$ as was shown in the second part of the proof to Lemma 9.3). From $k > 0$ it follows that the limit (9.17) is negative, say $-\gamma$. The solution $u_0(x) = \gamma^{-1}u(x)$ then satisfies the conditions stated in (c).

Next consider the solution $u_2(x)$ with $u_2(0) = 0$, $u_2'(0) = 1$. Since it is not decreasing we have $\lim e^{B(x)}u_2'(x) > 0$, and it is therefore possible to determine $\tau > 0$ so that for the solution $u(x) = u_0(x) + \tau u_2(x)$ the limit (9.17) will be zero. This solution is then positive and non-increasing for $x > 0$, and hence for all x. The solution $u_1(x) = u(x)/u(r_2)$ then satisfies the conditions of the Lemma. Finally, the assertion concerning (9.16) follows from the fact that (9.17) is necessary and sufficient for a positive solution to be non-increasing.

LEMMA 9.5. *For every non-increasing positive solution $e^{B(x)}u_2'(x)$ tends to a limit as $x \to r_2$.*

PROOF. This follows from (9.5) since the expression within braces is non-negative and bounded by $| u'(0) |$.

10. Dual results

We continue the initial assumptions and notations from section 9 and introduce the operator

$$\Omega^* = \frac{d}{dx}\left\{\frac{d}{dx} - b(x)\right\}. \tag{10.1}$$

It is understood that for $g(x)$ to belong to the domain of Ω^* it is necessary that $g(x)$ and $g'(x) - b(x)g(x)$ be differentiable, but *neither $g''(x)$ nor $b'(x)$ need exist.*

We shall study solutions (in this generalized sense) of the differential equation

$$\lambda v - \Omega^* v = 0, \qquad \lambda > 0. \tag{10.2}$$

We want to show that to each statement of the preceding section there corresponds an analogous assertion for solutions of (10.2) with the condition of boundedness replaced by integrability.

For this purpose we introduce the differential equation

$$\lambda w - (w'' - bw') = 0, \tag{10.3}$$

which is of the form (9.2) but with $b(x)$ replaced by $-b(x)$. Clearly the Lemmas of section 9 apply to (10.3).

LEMMA 10.1. *The transformation*

$$v(x) = w'(x) \tag{10.4}$$

establishes a one-to-one correspondence between solutions $w(x)$ *of* (10.3) *which are bounded at* r_2 *and solutions* $v(x)$ *of* (10.2) *which are integrable over* $(0, r_2)$. *To non-decreasing* $w(x)$ *there correspond non-negative* $v(x)$; *furthermore*

$$\lim_{x \to r_2} (v' - bv) = -\lambda \lim_{x \to r_2} w(x) \tag{10.5}$$

exists, and if $v > 0$ *also*

$$\lim_{x \to r_2} e^{-B(x)} v(x) = -\lim_{x \to r_2} e^{-B(x)} w'(x). \tag{10.6}$$

PROOF. If $w(x) \in \mathfrak{C}[0, r_2]$ is a solution of (10.3) then $w'(x) \in \mathfrak{L}(0, r_2)$, and it is clear that w' satisfies (10.2). Conversely, if $v(x) \in \mathfrak{L}(0, r_2)$ satisfies (10.2), then the limit on the left side of (10.5) exists, and

$$w(x) = \int_x^{r_2} v(s)\, ds - \lambda^{-1} \lim_{x \to r_2} (v' - bv) \tag{10.7}$$

is a solution of (10.3) with $w \in \mathfrak{C}[0, r_2]$. The existence of the limit (10.6) is implied by Lemma 9.5 (where $B(x)$ is to be replaced by $-B(x)$).

This lemma enables us to transpose all results of section 9, in each case replacing $b(x)$ by $-b(x)$. Now we have encountered three different sets of conditions, namely (9.4), (9.8), and (9.10). When $B(x)$ is replaced by $-B(x)$ the condition (9.10) remains unchanged. Of greatest importance is

LEMMA 10.2. *When* $B(x)$ *is replaced by* $-B(x)$ *the condition* (9.4) *goes over into* (9.8) *and vice versa.*

PROOF. Formally (9.8) goes over into

$$e^{-B(x)} \in \mathfrak{L}(0, r_2), \qquad e^{B(x)} \int_x^{r_2} e^{-B(s)}\, ds \in \mathfrak{L}(0, r_2). \tag{10.8}$$

It suffices therefore to prove

LEMMA 10.3. *The conditions* (9.4) *and* (10.8) *are equivalent.*

PROOF. Integrate the second function in (9.4) from 0 to r_2 and interchange the order of integration.

The duality expressed in Lemma 10.1 permits us to establish a direct relationship between certain types of solutions of (10.2) and (9.2).

LEMMA 10.4. *The transformation*

$$v(x) = e^{B(x)} u(x) \tag{10.9}$$

sets up a one-to-one correspondence between positive non-increasing solutions of $\lambda u - \Omega u = 0$ *and solutions* $v(x)$ *of* $\lambda v - \Omega^* v = 0$ *satisfying the conditions*

$$v(x) \geqq 0, \qquad v(x) \in \mathfrak{L}(0, r_2)$$

and

$$\lim_{x \to r_2} \{v' - bv\} \leqq 0. \tag{10.10}$$

Proof. That (10.9) transforms solutions into solutions is obvious. If $u(x)$ is non-increasing, then (9.5) and Lemma 9.5 show that $v \in \mathfrak{L}(0, r_2)$ and that (10.10) is equivalent to $\lim e^B u' \leqq 0$. To prove the converse one uses Lemma 10.1 and the results of section 9 to enumerate all solutions $v(x)$ satisfying our condition. It is easily verified that all such solutions are obtained by (10.9).

11. Types and properties of boundaries

We use the notations and assumptions introduced in the opening passages of sections 9 and 10. In the following $u(x)$ and $v(x)$ will always denote solutions of $\lambda u - \Omega u = 0$ and $\lambda v - \Omega^* v = 0$, respectively.

The purpose of this section is to summarize the main findings of section 9 and add the corresponding results for $\lambda v - \Omega^* v = 0$ to be obtained by the method of section 10. We state the criteria for the right boundary. When restating them for r_1 it is clear that $u'(x)$ and $v' - bv$ are to be replaced by

$$-u'(x) \quad \text{and} \quad -(v' - bv).$$

For examples cf. section 23.

Definition. *The boundary* r_2 *is: REGULAR if*

$$e^{B(x)} \in \mathfrak{L}(0, r_2) \qquad e^{-B(x)} \in \mathfrak{L}(0, r_2); \tag{11.1}$$

an EXIT boundary if it is not regular and[19]

$$e^{-B(x)} \int_0^x e^{B(s)}\, ds \in \mathfrak{L}(0, r_2); \tag{11.2}$$

an ENTRANCE boundary if it is not regular and

$$e^{B(x)} \int_0^x e^{-B(s)}\, ds \in \mathfrak{L}(0, r_2); \tag{11.3}$$

a NATURAL boundary in all other cases.

[19] An alternative form is given in (10.8), and (11.2) can be rewritten in a similar way. In the case $r_1 = -\infty$, $r_2 = +\infty$ E. Hille [7] has, by different methods, discovered that (11.2) is necessary and sufficient for the uniqueness for (1.2) and (11.3) for (1.1). Cf. the footnote to section 2a.

Types of solutions

I. *If r_2 is REGULAR there exist two non-increasing solutions $u_j(x)$ with*

$$u_0 \downarrow 0, \qquad e^{B}u_0' \to -1 \text{ and } u_1 \downarrow 1, \; e^{B}u_1' \to 0.$$

Each $u(x)$ is bounded at r_2 but it is positive and non-increasing if and only if $u = p_0u_0 + p_1u_1$ with $p_j \geqq 0$.

The solutions $v_j = e^{B}u_j$ are positive, $v_j \in \mathfrak{L}(0, r_2)$, and $e^{-B}v_0 \to 0$, $v_0' - bv_0 \to -1$, while $e^{-B}v_1 \to 1$, $v_1' - bv_1 \to 0$. Each solutiln $v(x)$ belongs to $\mathfrak{L}(0, r_2)$; one has

$$v(x) \geqq 0 \qquad \lim_{x\to r_2} (v' - bv) \leqq 0 \tag{11.4}$$

if and only if $v = p_0v_0 + p_1v_1$ with $p_j \geqq 0$.

II. *If r_2 is an EXIT boundary there exists a non-increasing*

$$u_1(x) \downarrow 0, \qquad e^{B}u_1' \to -\alpha \leqq 0.$$

Each $u(x)$ is bounded at r_2; there exist non-increasing solutions independent of u_1, but they tend to negative limits.

The solution $v_1 = e^{B}u_1 \in \mathfrak{L}(0, r_2)$, and $e^{-B}v_1 \to 0$, $v_1' - bv_1 \to -\alpha$. No solution independent of v_1 belongs to $\mathfrak{L}(0, r_2)$.

III. *If r_2 is an ENTRANCE boundary, there exists a non-increasing $u_1(x) \downarrow 1$, $e^{B}u_1' \to 0$. No independent solution is bounded at r_2.*

The solution $v_1 = e^{B}u_1 \in \mathfrak{L}(0, r_2)$, and $e^{-B}v_1 \to 1$, $v_1' - bv_1 \to 0$. Each $v(x)$ belongs to $\mathfrak{L}(0, r_2)$, but for positive $v(x)$ (independent of v_1) one has lim $(v' - bv) > 0$.

IV. *If r_2 is NATURAL there exists a non-increasing u_1 with $u_1 \downarrow 0$, $e^{B}u_1' \to 0$; no independent solution is bounded at r_2.*

For $v_1 = e^{B}u_1$ one has $e^{-B}v_1 \to 0$, $v_1' - bv_1 \to 0$. This is the only solution in $\mathfrak{L}(0, r_2)$.

12. Regular Green functions

Let $u_1(x)$ and $u_2(x)$ be two solutions of $\lambda u - \Omega u = 0$ (cf. (9.1)) and define

$$K(x, s) = \begin{cases} \dfrac{u_1(x)u_2(s)}{W(s)} & \text{for } x \leqq s \\[2ex] \dfrac{u_2(x)u_1(s)}{W(s)} & \text{for } x \geqq s \end{cases} \tag{12.1}$$

where

$$W(s) = u_1'(s)u_2(s) - u_1(s)u_2'(s) \tag{12.2}$$

is the Wronskian. Then $K(x, s)$ is called a Green function of $\lambda u - \Omega u = 0$. This definition applies to all differential equations and also to the generalized differential equation $\lambda v - \Omega^* v = 0$ (cf. (10.1)). In order to utilize the interplay between the two equations we have to restrict the class of Green functions.

Definition. *The function $K(x, s)$ will be called a regular Green function of Ω if $u_1(x)$ is positive non-decreasing, and $u_2(x)$ positive non-increasing.*

Similarly, a Green function for Ω^* *will be called regular if it is formed by means of two positive solutions* $v_j(x)$ *where* $v_1(x) \in \mathfrak{L}(r_1, 0)$, $v_2(x) \in \mathfrak{L}(0, r_2)$, *and*

(12.3) $$\lim_{x\to r_1} (v_1' - bv_1) \geqq 0, \qquad \lim_{x\to r_2} (v_2' - bv_2) \leqq 0.$$

Clearly *there exists exactly one regular Green function for each equation unless one of the boundaries is regular*, in which case there exist infinitely many.

It is readily verified that

(12.4) $$W(s) = W_0\, e^{-B(s)}$$

where W_0 is a positive constant. Lemma 10.3 then shows the truth of the following theorem (which is false for arbitrary Green functions).

THEOREM 12. *If* $K(x, s)$ *is a regular Green function for* Ω, *then* $K(s, x)$ *is a regular Green function for* Ω^* *and vice versa. Moreover*

(12.5) $$K(x, s) = \begin{cases} u_1(x)v_2(s) & \text{for } x \leqq s \\ u_2(x)v_1(s) & \text{for } x \geqq s, \end{cases}$$

where the u_j and v_j are solutions with properties described in the preceding definition.

13. The inhomogeneous equations

THEOREM 13.1. *Let*[20] $f(x) \in \mathfrak{C}[r_1, r_2]$ *and let* $K(x, s)$ *be a regular Green function of* Ω. *Put*

(13.1) $$F(x) = \int_{r_1}^{r_2} K(x, s)f(s)\, ds.$$

Then $F(x) \in \mathfrak{C}[r_1, r_2]$,

(13.2) $$\lambda F - \Omega F = f$$

and[21]

(13.3) $$\lambda \| F \| \leqq \| f \| .$$

Furthermore $f \geqq 0$ *implies* $F \geqq 0$.

PROOF. That the integral in (13.1) converges follows from Theorem 12, and (13.2) can be verified by differentiation using the representation (12.1). By definition $K(x, s)$ is positive, and hence F varies monotonically with f. It suffices therefore to prove (13.3) for f vanishing outside an interval $[r_1', r_2']$ contained in (r_1, r_2), and $f \geqq 0$. In this case F coincides in (r_j', r_j) with a non-negative multiple

[20] The theorem can be extended to essentially bounded f and leads to a corresponding extension of the semi-group. It will be noticed that the proof applies without change if f is permitted to have discontinuities of the first kind. This remark will be used in the proof of Theorem 13.3.

[21] We use the same sign for the conventional norms in the several spaces.

of $u_j(x)$, which is by assumption monotonic. Hence the maximum of F is assumed at an interior point x_0. However, at a point of maximum

$$\lambda F(x_0) = f(x_0) + \Omega F(x_0) = f(x_0) + F''(x_0) \leqq f(x_0), \tag{13.4}$$

which proves (13.3). It remains to prove that $F(x)$ approaches a limit as $x \to r_j$. Our argument shows that $F(x) \to 0$ if $f(x) \to 0$, and it suffices therefore to show that $F(x) \to \alpha$ if $f \equiv 1$. Now in this case $F = \lambda^{-1}$ plus a bounded solution of $\lambda u - \Omega u = 0$, if any. However, we know that each such solution approaches a limit since it has neither positive maxima nor negative minima.

This concludes the proof. For the case where each bounded solution necessarily tends to zero we found the sharper result that $\lambda F(r_j) = f(r_j)$. We note this for later purposes as a

COROLLARY. *If r_j is a natural boundary, then $\lambda F(r_j) = f(r_j)$.*

THEOREM 13.2.[22] *Let $g(x) \in \mathfrak{L}(r_1, r_2)$ and let $K(x, s)$ be a regular Green function of Ω. Put*

$$G(x) = \int_{r_1}^{r_2} K(s, x)g(s)\, ds. \tag{13.5}$$

Then $G(x) \in \mathfrak{L}(r_1, r_2)$, and[23]

$$\lambda G - \Omega^* G = g. \tag{13.6}$$

Furthermore

$$\lambda \| G \| \leqq \| g \| \tag{13.7}$$

and $g \geqq 0$ implies $G \geqq 0$.

PROOF. The preceding proof carries over, except that instead of considering the maximum of F we now use the relation

$$\lambda \int_{r_1'}^{r_2'} G(x)\, dx = \int_{r_1'}^{r_2'} g(x)\, dx + \{G' - bG\} \Big|_{r_1'}^{r_2'} \tag{13.8}$$

obtained by integrating (13.6). If $g(x) = 0$ for $r_2' < x < r_2$ then in this interval $G = pv_2$, where p is a non-negative constant. Because of (12.3) the contribution at r_2 of the last term in (13.8) is never positive. This argument not only proves (13.7) but also the

COROLLARY. *In order that the transformation* (13.5) *be norm preserving*[24] *it is necessary and sufficient that $K(x, s)$ is formed by means of solutions such that $v_j' - bv_j \to 0$ as $x \to r_j$. In this case also*

$$G' - bG \to 0 \quad as \quad x \to r_j. \tag{13.9}$$

[22] Essentially contained in Hille [6].
[23] For the definition of a solution of (13.6) cf. section 10.
[24] That is, $\lambda \| G \| = \| g \|$ *whenever* $g \geqq 0$.

To construct the general semi-group generated by Ω in the case of non-uniqueness we shall have to consider the full adjoint space to $C[r_1, r_2]$, namely the functions of bounded variation. We shall then require the following extension of the last Theorem.

THEOREM 13.3. *Let* $p(x) \in \mathfrak{BB}(r_1, r_2)$ *and*[25]

$$(13.10) \qquad P(x) = \int_{r_1}^{r_2} K(s, x)\, dp(s).$$

Then $P(x) \in \mathfrak{L}(r_1, r_2)$*, the derivative* $P'(x)$ *exists, and at any two points of continuity* x_j *of* $p(x)$ *one has*

$$(13.11) \qquad \lambda \int_{x_1}^{x_2} P(x)\, dx - \{P'(x) - b(x)P(x)\}\Big|_{x_1}^{x_2} = p(x_2) - p(x_1).$$

Furthermore

$$(13.12) \qquad \lambda \,\|P\| = \lambda \int_{r_1}^{r_2} |P(x)|\, dx \leqq \mathrm{Var}\,(p)$$

and $P \geqq 0$ *if* $p(x)$ *is non-decreasing.*

When $p'(x)$ exists it is possible to differentiate (13.11) and obtain $\lambda P - \Omega^* P = p'$. In order to avoid the clumsy expression (13.11) and to be reminded of the more familiar form we introduce the

Notational Convention. The equation

$$(13.13) \qquad \lambda P - \Omega^* P \approxeq \tilde{p}$$

shall mean the same as (13.11).

Thus (13.13) is valid only in the integrated form, and this is indicated by the sign $\approxeq$. The tilde indicates that in the integral $dp(x)$ occurs rather than $p(x)\, dx$.

PROOF. One method of proof consists in approximating $p(x)$ by functions in $\mathfrak{L}(r_1, r_2)$. It is more instructive to utilize the duality principle. Putting

$$v_2 = -w_2'$$

and $v_1 = w_1'$ (so that $\lambda w_j - (w_j'' - bw_j') = 0$), (13.10) takes on the form

$$(13.14) \quad W_0 P(x) = -w_1'(x) \int_x^{r_2} w_2'(s) e^{-B(s)}\, dp(s) - w_2'(x) \int_{r_1}^{x} w_1'(s) e^{-B(s)}\, dp(s),$$

where W_0 is a positive constant. Integrating by parts and noticing that $(w_j' e^{-B})' = \lambda w_j e^{-B}$ we find

$$(13.15) \qquad P(x) = \frac{d}{dx} \int_{r_1}^{r_2} \bar{K}(x, s) p(s)\, ds$$

[25] Note that $p(x)$ is not permitted to attach masses to the boundaries and that the integral in (13.10) extends over the open interval. Actually masses at the endpoints will play an essential role in the sequel. Instead of a transform $P(x) \in \mathfrak{L}(r_1, r_2)$ we get then $[P_1, P(x), P_2] \in \mathfrak{L}[r_1, r_2]$ (cf. section 3). However, whether or not $P_j = 0$ depends on the type of boundary, and (13.10) can not be generalized without considering separately the different cases.

where $\bar{K}(x, s)$ is a regular Green function of $\lambda w - (w'' - bw') = 0$ (considered in $\mathfrak{C}[r_1, r_2]$). Now $p(x)$ is continuous except for jumps, and according to the footnote to Theorem 13.1 the latter applies to such functions. Hence, denoting the integral in (13.15) by $\bar{P}(x)$ we have $\lambda\bar{P} - (\bar{P}'' - b\bar{P}') = p$ at each point of continuity of p, and this is the same as (13.11).

14. Properties of the solutions

To avoid interruptions we collect here a few properties of the solutions (13.1) and (13.5).

LEMMA 14.1. *If r_2 is an exit boundary, then*

$$\lim_{x \to r_2} F(x) = 0. \tag{14.1}$$

The same is true if r_2 is regular and the Green function is formed by means of the solution $u_2(x)$ which tends to 0 as $x \to r_2$.

PROOF. In view of (12.1) and (12.4) we have

$$W_0 F(x) = u_1(x) \int_x^{r_2} f(s)u_2(s)e^{B(s)}\, ds + u_2(x) \int_{r_1}^{x} f(s)u_1(s)e^{B(s)}\, ds. \tag{14.2}$$

Under our assumptions $u_1(x)$ is bounded at r_2 and $u_2 \downarrow 0$ as $x \to r_2$. It suffices therefore to show that

$$u_2(x) \int_0^x e^{B(s)}\, ds \to 0. \tag{14.3}$$

This is obvious in the case of a regular boundary. Otherwise we recall that by (9.5)

$$\lambda \int_0^{r_2} u_2(s)e^{B(s)}\, ds \leqq |\, u_1'(0)\, |, \tag{14.4}$$

so that for every $\varepsilon > 0$, y sufficiently large, and $y < x < r_2$

$$\varepsilon > \int_y^x u_2(s)e^{B(s)}\, ds \geqq u_2(x) \int_y^x e^{B(s)}\, ds = u_2(x)\left\{\int_0^x e^{B(s)}\, ds - M(\varepsilon)\right\}. \tag{14.5}$$

From this (14.3) follows easily.

The corresponding statement for Ω^* can be put into a sharper form which has no counterpart for Ω. We use the term norm preserving as explained in the footnote to the Corollary to Theorem 13.2. This corollary leads directly to

LEMMA 14.2. *If each of the boundaries is either a natural or an entrance boundary, then the transformation* (13.5) *is norm preserving.*

LEMMA 14.3. *If r_j is a regular boundary then the limit*

$$\lim_{x \to r_j} (-1)^{j-1} e^{B(x)} F'(x) \tag{14.6}$$

exists and is non-negative whenever $f \geqq 0$. *It is possible to choose the Green function* $K(x, s)$ *so that*

$$p_j(-1)^j \lim_{x \to r_j} e^B F' + q_j \lim_{x \to r_j} F = 0, \qquad \frac{p_j}{q_j} = \alpha_j \tag{14.7}$$

where $0 \leqq \alpha_j \leqq \infty$ *is arbitrarily prescribed and this boundary condition (or conditions in the case of two regular boundaries) uniquely determine* $K(x, s)$.

PROOF. If r_2 is regular then $u_1(x)$ is bounded and $e^B u_1'$ approaches a finite limit as $x \to r_2$. It follows then from (14.2) easily that

$$\begin{aligned} W_0 \lim e^B F' &= (\lim e^B u_2') \int_{r_1}^{r_2} f(s) u_1(s) e^{B(s)}\, ds \\ &= (\lim e^B u_2') \cdot W_0 F(r_2)/u_2(r_2) \end{aligned} \tag{14.8}$$

(with $F(r_2) = 0$ if $u_2(r_2) = 0$). Now there exists exactly one positive non-increasing solution $u_2(x)$ for which $-e^B u_2'/u_2$ tends to a prescribed non-negative limit, and this concludes the proof.

The corresponding assertion for Ω^* takes on the form

LEMMA 14.4. *If* r_j *is a regular boundary* $e^{-B}G$ *and* $G' - bG$ *tend to finite limits as* $x \to r_j$. *If the Green function is so chosen that* (14.7) *holds, then*

$$-p_j(-1)^j \lim_{x \to r_j} \{G' - bG\} + q_j \lim e^{-B} G = 0. \tag{14.9}$$

It is clear from the proof that this lemma remains true of the generalized solutions in the sense of Theorem 13.3.

LEMMA 14.5. *The domains of the operators* Ω *and* Ω^* *are dense in* $\mathfrak{C}[r_1, r_2]$ *and* $\mathfrak{L}(r_1, r_2)$, *respectively.*

PROOF. If $f \in \mathfrak{C}[r_1, r_2]$ and $r_1 < r_1' < r_2' < r_2$, then it is possible to find a function $\phi(x)$ such that $\phi''(x) \in \mathfrak{C}[r_1', r_2']$, $\phi''(r_j') = \phi'(r_j') = 0$, $\phi(r_j') = f(r_j')$ and $|f(x) - \phi(x)| < \varepsilon$. Now define $\phi(x) = \phi(r_2')$ for $x > r_2'$ and $\phi(x) = \phi(r_1')$ for $x < r_1'$. Then $\phi(x) \in \mathfrak{D}(\Omega)$ in $[r_1, r_2]$ and $\| \phi - f \| < \varepsilon$ if r_j' is chosen close enough to r_j.

LEMMA 14.6. *If* r_2 *is an exit boundary then there exists a* $w(x)$ *such that*

$$\Omega w \in \mathfrak{C}[r_1, r_2], \qquad \| w \| < \varepsilon,$$

and $\Omega w \to 1$ *as* $x \to r_2$.

PROOF. In the case of an exit boundary the integral

$$w(x) = \int_r^x e^{-B(t)}\, dt \int_r^t e^{B(s)} h(s)\, ds \tag{14.10}$$

is meaningful for $r < x < r_2$ and $h(x) \in \mathfrak{C}[r_1, r_2]$. Also $w'' + bw' = h$ for $x > r$, and $w(r) = w'(r) = 0$. Now choose $h(x)$ so that $h(r) = 0$, $h(r_2) = 1$ and $0 < h < 1$ for $r < x < r_2$. Define w by (14.10) for $x \geqq r$ and put $w = 0$ for $x \leqq r$. Then w satisfies the conditions of the lemma provided r is chosen large enough.

LEMMA 14.7. *If r_2 is a regular boundary then there exists a $w(x)$ such that*

$$\Omega w \in \mathfrak{C}[r_1, r_2], \qquad \| w \| < \varepsilon$$

and $w(x) \to 0$, $e^{B(x)}w'(x) \to 1$ as $x \to r_2$.

PROOF. In the case of a regular boundary one can extend the first integral in (14.10) from x to r_2, and the result follows.

LEMMA 14.8. *If r_1 and/or r_2 are exit boundaries then*

$$\int_{r_1}^{r_2} K(x, s)\, ds = \lambda^{-1} - \xi(x) \tag{14.11}$$

where $\xi(x)$ is a strictly positive solution of $\lambda\xi - \Omega\xi = 0$ with $\xi(x) \to 1/\lambda$ as $x \to r_j$. (The same is true for regular boundaries if the corresponding solution u_j used in forming $K(x, s)$ tends to 0 as $x \to r_j$.)

PROOF. Applying Theorem 13.1 to $f \equiv 1$ we see that the left side of (14.11) is a solution of $\lambda F - \Omega F = 1$, or $F = \lambda^{-1} - \xi(x)$ where $\xi(x)$ is bounded and $\lambda\xi - \Omega\xi = 0$. Furthermore $\xi(x) \geqq 0$ since $\lambda \| F \| \leqq 1$. Finally, if r_2 is an exit boundary then $F \to 0$ as $x \to r_2$ by Lemma 14.1. Therefore $\xi(x) \to \lambda^{-1}$ and $\xi(x)$ can not be identically zero. However, $\xi(x)$ can not be tangent to the x-axis, and therefore $\xi(x) > 0$.

15. Semi-groups defined by regular Green functions

When none of the boundaries is regular there exists exactly one *regular* Green function, even when the solution of (13.2) is not unique. The latter occurs if and only if $\lambda u - \Omega u = 0$ has a bounded solution, that is, if one of the boundaries is an exit boundary or regular.

Accordingly, if the boundaries are either natural or entrance boundaries, the solution of (13.2) is unique and the range of the transformation (13.1) coincides with the domain of Ω. According to Lemma 14.5 this range is dense in $\mathfrak{C}[r_1, r_2]$ and the Hille-Yosida theorem (section 4) shows that the transformation (13.1) is the resolvent of a contraction semi-group with range space dense in $\mathfrak{C}[r_1, r_2]$. If r_j is an exit boundary ($j = 1$ or 2 or both), then by Lemma 14.1 the range of (13.1) lies entirely in the closed manifold $F(r_j) = 0$, and the above argument shows that the range consists of the intersection of $\mathfrak{D}(\Omega)$ and this manifold. In this case we get a semi-group for which the strong continuity condition

$$\| T_t f - f \| \to 0 \qquad t \to 0 \tag{15.1}$$

holds only if $f(r_j) = 0$. The semi-group can actually be extended to the entire space $\mathfrak{C}[r_1, r_2]$, but for f with $f(r_j) \neq 0$ we shall have only pointwise convergence instead of (15.1). We shall not discuss this detail and state only

THEOREM 15.1. *If none of the boundaries is regular then the (unique) regular Green function $K(x, s)$ of Ω is the resolvent kernel of a contraction semi-group defined and satisfying* (15.1) *for all $f \in \mathfrak{C}[r_1, r_2]$ except if a boundary r_j is an exit boundary; in the latter case the domain of definition $\mathfrak{C}[r_1, r_2]$ is to be replaced by the subspace of those f with $f(r_j) = 0$.*

The corresponding theorem for Ω^* requires no special cases and is more precise. Non-uniqueness arises now in connection with entrance boundaries, and the range (13.5) is in this case characterized by the boundary condition (13.9) (cf. Lemma 14.2 and the Corollary to Theorem 13.2). It follows that this range is dense in $\mathfrak{L}(r_1, r_2)$. Moreover, by Lemma 14.2 the semi-group is norm-preserving if the boundaries are natural or of the entrance type. If r_j is an exit boundary, a comparison of (14.11) and (13.5) shows the latter transformation to be strictly norm decreasing (i.e. $\lambda \| G \| < \| g \|$ for each $g > 0$). Thus we have

THEOREM 15.2. *If none of the boundaries is regular then the regular Green function $K^*(x, s)$ of Ω^* is the resolving kernel of a contraction semi-group $\mathfrak{S}^*$ defined and satisfying*

$$(15.2) \qquad \| T_t^* g - g \| \to 0 \qquad t \to 0$$

for all $g \in \mathfrak{L}(r_1, r_2)$. Furthermore, $\mathfrak{S}^*$ is norm preserving unless one of the boundaries is of the exit type, in which case $\mathfrak{S}^*$ is strictly norm decreasing.

COROLLARY. *The semi-groups $\mathfrak{S}$ and $\mathfrak{S}^*$ are adjoints of each other, that is, $K^*(x, s) = K(s, x)$.*

We now turn to the case where r_j is a regular boundary. For each fixed λ and a given $K(x, s)$ every F in the range of the transformation (13.1) satisfies a relation of the form (14.7) and it is obvious that conversely each F in the domain of Ω satisfying (14.7) can be represented in the form (13.1). The range $\mathfrak{R}$ of (13.1) is therefore the intersection of $\mathfrak{D}(\Omega)$ and the set of functions $F \in \mathfrak{C}[r_1, r_2]$ satisfying (14.7). Using Lemmas 14.5 and 14.7 one sees easily that $\mathfrak{R}$ is dense in $\mathfrak{C}[r_1, r_2]$ except if $p_j = 0$, in which case the closure $\bar{\mathfrak{R}}$ coincides with the subset defined by $F(r_j) = 0$ ($j = 1$ or 2 or both). The range $\mathfrak{R}$ is independent of λ if and only if the ratio, or ratios, p_j/q_j are constants. A similar argument applies to Ω^* and we have thus

THEOREM 15.3. (1) *To each regular boundary one can choose a fixed*

$$\alpha \ (0 \leqq \alpha \leqq \infty)$$

and determine a regular Green function $K(x, s)$ of Ω so that it will be the resolvent kernel of a contraction semi-group $\mathfrak{S}$, determined by the lateral condition, or conditions (14.7).[26] *This semi-group is defined and satisfies the strong continuity condition* (15.1) *for all $f \in \mathfrak{C}[r_1, r_2]$, except if $p_j = 0$ or if one of the boundaries is of the exit type, in which case the statement applies only to $f \in \mathfrak{C}[r_1, r_2]$ vanishing at that boundary.*

(2) *The transposed kernel $K^*(x, s) = K(s, x)$ is the resolving kernel for a contraction semi-group $\mathfrak{S}^*$ satisfying* (15.2) *for all $g \in \mathfrak{L}(r_1, r_2)$. Its range is determined by the lateral condition* (14.9). *It is norm preserving if and only if $p_j = 0$ and none of the boundaries is of the exit type.*

[26] More precisely, the lateral condition is defined by the set Σ of all $F \in \mathfrak{D}(\Omega)$ satisfying (14.7). It has been shown that in the case of a natural boundary the limits occurring in (14.7) exist for each $F \in \mathfrak{D}(\Omega)$.

(3) *In this way one obtains all semi-groups generated by Ω and Ω^* whose resolvents are defined by regular Green functions.*

It will be seen that both Ω and Ω^* generate other contraction semi-groups. However, in no other case will we find that they are adjoints to each other.

16. The general semi-group for an exit boundary

For definiteness we now assume that r_1 is a natural and r_2 an exit boundary. There exists then a unique regular Green function $K(x, s)$ of Ω and (by Theorem 15.1) (13.1) defines the resolvent of one semi-group, but there are others. To construct them, we require the general solution of (13.2) which is determined only up to a multiple of a bounded solution of the homogeneous equation.

DEFINITION 16.1. *If r_1 is a natural and r_2 an exit boundary we denote by $\xi(x)$ the solution of $\lambda\xi - \Omega\xi = 0$ for which*

$$\xi(r_1) = 0, \qquad \xi(r_2) = 1/\lambda. \tag{16.1}$$

Note that every bounded solution is a multiple of $\xi(x)$. Furthermore, by Lemma 14.8 this $\xi(x)$ is positive (therefore non-decreasing), and satisfies (14.11).

For any fixed λ we can write the most general linear transformation taking $f \in \mathfrak{C}[r_1, r_2]$ into a solution of $\lambda F - \Omega F = f$ in the form

$$F(x) = \int_{r_1}^{r_2} K(x, s)f(s)\, ds + \xi(x)\cdot Q^*(f) \tag{16.2}$$

where Q^* is an arbitrary linear functional. For this transformation to be the resolvent of a contraction semi-group it is necessary that its range be independent of λ. Our first aim is therefore *to define the functional Q^* in its dependence on λ in such a way that the totality of $F \in \mathfrak{C}[r_1, r_2]$ representable in the form* (16.2) *becomes independent of λ.*

The functional Q^* can be represented by means of a function of bounded variation in $[r_1, r_2]$. It will turn out that this function is absolutely continuous in (r_1, r_2), so that Q^* is given by an element $[q_1, Q(x), q_2] \in \bar{\mathfrak{L}}[r_1, r_2]$ (cf. section 3).

DEFINITION 16.2. *We denote by Σ the set of all $F(x) \in \mathfrak{C}[r_1, r_2]$ which are in the domain of Ω and satisfy the condition*

$$p_2 F(r_2) = p_1 F(r_1) + \tau \int_{r_1}^{r_2} F(x)\, dp(x) - \sigma \cdot \lim_{x \to r_2} \Omega F(x); \tag{16.3}$$

here $p_j \geq 0$, $\tau \geq 0$, $\sigma \geq 0$ are constants,

$$p_1 + \tau \leq p_2, \tag{16.4}$$

and $p(x)$ is a non-decreasing function with[27]

$$p(r_1+) = 0, \quad p(r_2-) = 1. \tag{16.5}$$

[27] Note that $p(x)$ is defined only in (r_1, r_2). It is possible to define it over $[r_1, r_2]$ and thus to absorb the first two terms in (16.3) in the integral, but this is inconvenient for other purposes and complicates the enunciation of the conditions on $p(x)$.

LEMMA 16.1. *If $\sigma > 0$ the set Σ is everywhere dense in $\mathfrak{C}[r_1, r_2]$. If $\sigma = 0$, then Σ is dense in the closed linear manifold defined by*

$$p_2 f(r_2) = p_1 f(r_1) + \tau \int_{r_1}^{r_2} f(x)\, dp(x). \tag{16.6}$$

PROOF. When $\sigma > 0$, Lemma 14.6 assures us that to each $h \in \mathfrak{D}(\Omega)$ we can construct a $w \in \mathfrak{D}(\Omega)$ such that $\|w\| < \varepsilon$ and $h + w \in \Sigma$. The proof is now completed using Lemma 14.5. For $\sigma = 0$ the assertion is contained in Lemma 14.5.

THEOREM 16. *Let r_1 be a natural boundary and r_2 an exit boundary. Then Ω, restricted by the lateral condition Σ, generates a contraction semi-group $\mathfrak{S}$; its range is dense either in $\mathfrak{C}[r_1, r_2]$ or in the linear manifold (16.6) according as $\sigma > 0$ or $\sigma = 0$. Its resolvent is given by the linear transformation*

$$F = \int_{r_1}^{r_2} K(x, s) f(s)\, ds + \xi(x)\left\{q_1 f(r_1) + q_2 f(r_2) + \int_{r_1}^{r_2} f(s) Q(s)\, ds\right\} \tag{16.7}$$

with

$$q_1 = p_1 A/\lambda, \qquad q_2 = A\sigma, \tag{16.8}$$

$$Q(s) = A\tau \int_{r_1}^{r_2} K(x, s)\, dp(x), \tag{16.9}$$

where we put for abbreviation[28]

$$A = \frac{\lambda}{\sigma\lambda + p_2 - \tau\lambda \int_{r_1}^{r_2} \xi(x)\, dp(x)}. \tag{16.10}$$

PROOF. It is clear that $\lambda F - \Omega F = f$. Our first aim is to show that for a fixed λ a function $F \in \mathfrak{D}(\Omega)$ is representable in the form (16.7) if and only if $F \in \Sigma$ (i.e. that the range of (16.7) is independent of λ). Accordingly (keeping λ fixed) we choose $F \in \mathfrak{D}(\Omega)$ and define $f = \lambda F - \Omega F$. Then this F can differ from the right side of (16.7) only by a multiple of $\xi(x)$. On the other hand, the first term on the right tends to 0 as $x \to r_2$ by Lemma 14.1, and $\xi(x) \to \lambda^{-1}$ by (16.1). It follows that F is in the range of (16.7) if, and only if,

$$F(r_2) = \lambda^{-1}\left\{q_1 f(r_1) + q_2 f(r_2) + \int_{r_1}^{r_2} f(s) Q(s)\, ds\right\}. \tag{16.11}$$

We now prove the equivalence of the conditions (16.11) *and* (16.3).

First we observe that $\lambda F - \Omega F = f$ implies

$$\lim_{x \to r_2} \Omega F = \lambda F(r_2) - f(r_2). \tag{16.12}$$

[28] It will be shown that $A > 0$ (cf. Lemma 16.2).

Moreover,

$$\lambda F(r_1) = f(r_1); \tag{16.13}$$

in fact, by the Corollary to Theorem 13.1 this is true of the regular Green function solution of $\lambda F - \Omega F = f$, and any other solution differs from it by a multiple of $\xi(x)$. The latter tends to 0 as $x \to r_1$ by (16.1). We note also that by (16.9)

$$\int_{r_1}^{r_2} f(s)Q(s)\, ds = A\tau \int_{r_1}^{r_2} dp(x) \int_{r_1}^{r_2} K(x, s)f(s)\, ds. \tag{16.14}$$

Suppose now that (16.3) holds. For every $F \in \mathfrak{D}(\Omega)$ we can obviously write

$$F = \int_{r_1}^{r_2} K(x, s)f(s)\, ds + \lambda F(r_2)\cdot\xi(x). \tag{16.15}$$

Hence by (16.14)

$$\tau \int_{r_1}^{r_2} F(x)\, dp(x) = A^{-1} \int_{r_1}^{r_2} f(s)Q(s)\, ds + \lambda\tau F(r_2) \int_{r_1}^{r_2} \xi(x)\, dp(x). \tag{16.16}$$

On the other hand, by (16.3), (16.8), (16.12) and (16.13)

$$\tau \int_{r_1}^{r_2} F(x)\, dp(x) = (p_2 + \sigma\lambda)F(r_2) - A^{-1}\{q_1 f(r_1) + q_2 f(r_2)\}. \tag{16.17}$$

Equating (16.16) to (16.17) and using (16.10) we get (16.11). Thus (16.3) implies (16.11), and the converse is proved in similar fashion, by retracing the steps.

It remains to show that (16.7) is positivity preserving and that $\lambda \| F \| \leqq \| f \|$. The first remark is to be used again in section 17 and we formulate it therefore as

LEMMA 16.2. *The transformation* (16.2) *is positivity preserving and*

$$\lambda \| F \| \leqq \| f \|$$

if, and only if, Q^* *is a positive functional with* $\| Q^* \| \leqq 1$.

In fact, the first integral in (16.2) is positivity preserving and tends to 0 as $x \to r_2$ (Lemma 14.1). On the other hand, $\xi(x)$ is positive, non-decreasing, and $\xi(r_2) = \lambda^{-1}$. The Lemma now follows in view of (14.11)

To complete the proof of the Theorem we have therefore to show that

$$q_j \geqq 0, \qquad Q(s) \geqq 0, \qquad q_1 + q_2 + \int_{r_1}^{r_2} Q(s)\, ds \leqq 1. \tag{16.18}$$

First, it is clear from (16.4) and (16.5) that the denominator in (16.10) can not vanish. Hence $A \geqq 0$. That $Q(s) \geqq 0$ follows now from Theorem 13.3, and that $q_j \geqq 0$ from (16.8). Finally, integrating (16.9) and using (14.11) we find

$$\int_{r_1}^{r_2} Q(s) = A\tau\left\{\lambda^{-1} - \int_{r_1}^{r_2} \xi(x)\, dp(x)\right\} \tag{16.19}$$

and hence

$$(16.20)\qquad q_1 + q_2 + \int_{r_1}^{r_2} Q(s)\,ds = \lambda^{-1} A\left\{p_1 + \tau + \sigma\lambda - \lambda \int_{r_1}^{r_2} \xi(x)\,dp(x)\right\}.$$

The right side is less than or equal to unity according as there is inequality or equality in (16.4).

17. Uniqueness

THEOREM 17. (1) *Let r_1 be a natural and r_2 an exit boundary. Then Definition* 16.2 *with $\sigma > 0$ contains the most general lateral condition for a contraction semi-group generated by Ω with range dense*[29] *in $\mathfrak{C}[r_1, r_2]$.* (2) *If the boundary r_1 is natural and r_2 regular, the same is true if a term $-\gamma \lim e^{B(x)}F'(x)$ is added to the right side, where $\gamma \geqq 0$ is a constant* (cf. (19.4)).

PROOF. We saw that the resolvent of the assumed semi-group is necessarily of the form (16.2) where Q^* is a positive functional of norm not exceeding 1 (cf. Lemma 16.2). We imagine Q^* represented by a function $Q(x; \lambda)$ of bounded variation over the closed interval $[r_1, r_2]$. For any given F in the range of the resolvent and for any fixed λ we shall then have

$$(17.1)\qquad F(r_2) = \int_{r_1-}^{r_2+} F(x)\,dQ(x;\lambda) - \lambda^{-1}\int_{r_1-}^{r_2+} \Omega F(x)\,dQ(x;\lambda).$$

Since the variation of $Q(x; \lambda)$ is uniformly bounded by 1 we can pick out a sequence $\lambda \to \infty$ such that $Q(x; \lambda)$ converges. Two cases arise.

First, the limiting function does not center the entire mass at the right end-point. In this case the second integral in (17.1) is asymptotically negligible, and we get in the limit a relation of the form (16.6). This is the degenerate case where the range lies within a closed linear submanifold, and we have excluded this case by requiring that the range be everywhere dense. (Alternatively, we can consider semi-groups in the manifold and extend them to the entire space.)

Second, with increasing λ the total mass may tend to be concentrated near or at $x = r_2$. In this case we get obviously

$$(17.2)\qquad -\,\Omega F(x)\,|_{x=r_2} = \lim_{\lambda\to\infty} \lambda\left\{F(r_2) - \int_{r_1-}^{r_2+} F(x)\,dQ(x;\lambda)\right\}.$$

Since the range of the resolvent is dense in $\mathfrak{C}[r_1, r_2]$ we can pick $F(x)$ so that $F(r_2) = 0$ and that within any subinterval $r_1 \leqq x \leqq r < r_2$ the function $F(x)$ comes arbitrarily near unity. It then follows that

$$(17.3)\qquad \lim_{\lambda\to\infty} \lambda Q(x;\lambda) = Q(x) \geqq 0$$

[29] If $\sigma = 0$ we get a contraction semi-group whose range is contained in a closed linear subspace, but the group itself is defined on the whole space. Outside the range we have only pointwise convergence for $t = 0$. The restricted formulation of the text is due to the desire to keep within the familiar theory of semi-groups where (unfortunately) strong continuity at the origin is usually postulated.

exists for every fixed $x < r_2$. However, we can not conclude that $Q(x)$ is of bounded variation since the convergence of the integral in (17.2) may be due to the assumption $F(r_2) = 0$. Now, for fixed $r < r_2$ we get from (17.2)

$$(17.4) \qquad -\Omega F(x)\,|_{x=r_2} + \int_{r_1-}^{r} F(x)\,dQ(x) = \lim_{\lambda\to\infty} \lambda\left\{F(r_2) - \int_{r}^{r_2+} F(x)\,dQ(x;\lambda)\right\}$$

at every continuity point r of $Q(x;\lambda)$.

An integration by parts changes the right side into

$$(17.5) \qquad F(r)\lambda Q(r;\lambda) + F(r_2)\lambda\,\{1 - Q(r_{2+}\,;\lambda)\} + \lambda \int_{r}^{r_2+} F'(x)Q(x;\lambda)\,dx.$$

The first term tends to $F(r)Q(r)$, by (17.3). We can choose (as will be seen in a moment) $F(x)$ so that $F(r_2) > 0$ and $F'(x) > 0$ in $r < x < r_2$. This choice implies that the second and third term must converge separately. The second gives a contribution of the form $pF(r_2)$, where $p \geqq 0$. There remains the last term.

For fixed x we have (17.3), and we have to show that $Q(x)$ is bounded; moreover, we have to investigate whether an additional term can arise from the upper limit, in case the functions $\lambda Q(x;\lambda)$ converge to a limit which is discontinuous at $x = r_2$. This is the case if r_2 is *regular*. For then (as is easily verified and will be seen in section 19), $e^{B(x)}F'(x)$ always converges to a limit as $x \to r_2$, and therefore the last term in (17.5) converges to a non-negative multiple of this limit plus a contribution from the interior of the interval (r, r_2). As $r \to r_2$ the last contribution tends to zero. Thus, letting $\lambda \to \infty$, $r \to r_2$ we are in the limit led to an equation of the form (16.3) augmented on the right by a term $\gamma\cdot\lim e^{B(x)}F'(x)$.

In the case of an exit boundary the last limit does not exist. To see that in this case (16.3) represents the most general lateral condition we must investigate the behavior of $F'(x)$ as $x \to r_2$.

Now the integral in (16.2) is of the form indicated in (14.2) where $u_1(x)$ is the same as $\xi(x)$ in our present notations. Accordingly, as $x \to r_2$

$$(17.6) \qquad F'(x) \sim u_2'(x) \int_{r_1}^{x} f(s)\,\xi_1(s)e^{B(s)}\,ds + \xi'(s)\,\{\lambda F(r_2) + o(1)\}.$$

We know that $e^B u_2'$ tends to a finite limit, while

$$(17.7) \qquad e^{B(x)}\,\xi'(x) \sim \int_0^x e^{B(s)}\,ds,$$

by (9.5). The last integral diverges since r_2 is an exit boundary. From (17.6) and (17.7) we conclude

$$(17.8) \qquad e^{B(x)}\,F'(x) \sim \{A_0 f(r_2) + A_1 F(r_2)\} \int_0^x e^{B(s)}\,ds,$$

where A_0 and A_1 are constants independent of the choice of $F(x)$ but dependent on λ. Now $f(r_2) = \lambda F(r_2) - \lim \Omega F(x)$, and thus (17.8) shows that if the last term in (17.5) led to a contribution, this would be a linear combination of terms appearing in (16.3) so that the form of the latter relation remains unaltered.[30]

18. The adjoint

The adjoint transformation of the resolvent (16.7) is, strictly speaking, a linear transformation from $\mathfrak{BB}[r_1, r_2]$ to itself. However, this transformation takes any function which is absolutely continuous in (r_1, r_2) into a function of the same type. That is to say, the subspace $\bar{\mathfrak{L}}[r_1, r_2]$ (cf. section 4) is mapped into itself. Instead of introducing the full adjoint operating on $\mathfrak{BB}[r_1, r_2]$ we can therefore restrict our attention to its contraction to $\bar{\mathfrak{L}}[r_1, r_2]$. It is defined as the linear transformation taking an arbitrary $[g_1, g(x), g_2] \in \bar{\mathfrak{L}}[r_1, r_2]$ into an element $[G_1, G(x), G_2]$ such that for each $f \in \mathfrak{C}[r_1, r_2]$

$$(18.1)\qquad g_1 F(r_1) + g_2 F(r_2) + \int_{r_1}^{r_2} F(x)g(x)\,dx = G_1 f(r_1) + G_2 f(r_2) + \int_{r_1}^{r_2} f(x)G(x)\,dx.$$

Now write for abbreviation

$$(18.2)\qquad G^*(x) = \int_{r_1}^{r_2} K(s, x)g(x)\,dx,$$

where $K(s, x)$ is (as function of x) the unique Green function of Ω^*. Then (18.2) is the unique solution of $\lambda G^* - \Omega^* G^* = g$ (cf. Theorem 13.2).

With this notation we find from (16.7) easily

$$(18.3)\qquad G(x) = G^*(x) + Q(x)\left\{\lambda^{-1}g_2 + \int_{r_1}^{r_2} \xi(s)g(s)\,ds\right\}$$

$$(18.4)\qquad \begin{aligned} G_1 &= \lambda^{-1}g_1 + q_1\left\{\lambda^{-1}g_2 + \int_{r_1}^{r_2} \xi(s)g(s)\,ds\right\} \\ G_2 &= q_2\left\{\lambda^{-1}g_2 + \int_{r_1}^{r_2} \xi(s)g(s)\,ds\right\}. \end{aligned}$$

Keeping the notations and assumptions of Definition 16.2 and of Theorem 16 we now prove

THEOREM 18.1. *If $\sigma > 0$ then* (18.3) – (18.4) *defines the resolvent of a contraction semi-group $\mathfrak{S}^*$ with range dense in $\bar{\mathfrak{L}}[r_1, r_2]$. It is norm preserving if, and only if, there is equality in* (16.4). *The resolvent can be written in the form*

$$(18.5)\qquad [G_1, G(x), G_2] = (\lambda I - \overline{\Omega^*})^{-1}[g_1, g(x), g_2]$$

[30] A closer (but rather painful) study reveals that actually no such contributions are possible at all. In the case of a regular boundary the new contribution is due to a bounded solution of $\lambda v - \Omega^* v = 0$.

where the generator $\overline{\Omega^*}$ *is defined by*[31]

(18.6) $$\lambda G(x) - \Omega^* G(x) - \frac{\tau}{\sigma} G_2 \tilde{p}(x) \approxeq g(x)$$

(18.7) $$\lambda G_1 - \frac{p_1}{\sigma} G_2 = g_1$$

(18.8) $$\lambda G_2 + \frac{p_2}{\sigma} G_2 + \lim_{x \to r_2} \{G'(x) - b(x)G(x)\} = g_2 .$$

This is the only contraction semi-group generated by $\overline{\Omega^*}$.

Note that the existence of the limit occurring in (18.8) is proved and not assumed.

PROOF. We first verify that the range of the transformation (18.3)–(18.4) is independent of λ. In view of the arbitrariness of g_1 and g_2 it suffices to consider (18.3). For a fixed λ a function $G \in \mathfrak{L}(r_1, r_2)$ is representable in the form (18.3) if, and only if, it is a linear combination of the (unique) solution $G^* \in \mathfrak{L}(r_1, r_2)$ of $\lambda G^* - \Omega^* G^* = g$ for some $g \in \mathfrak{L}(r_1, r_2)$ and of the (unique) solution $H \in \mathfrak{L}(r_1, r_2)$

(18.9) $$\lambda H(x) - \Omega^* H(x) \approxeq \tilde{p}(x),$$

where $p(x)$ is the fixed function given in Definition 16.2. Now, if a particular function $H_0(x)$ satisfies (18.9) for $\lambda = \lambda_0$, then

(18.10) $$\lambda H_0(x) - \Omega^* H_0(x) \approxeq \tilde{p}(x) + (\lambda - \lambda_0) H_0(x).$$

This shows that H_0 is a linear combination of the solution of (18.9) and a solution of $\lambda G_0^* - \Omega^* G_0^* = H_0(x)$, and hence the range of (18.3) is independent of λ.

That this range is dense in $\mathfrak{L}(r_1, r_2)$ is clear from the preceding (cf. Lemma 14.5).

Next note that (18.6) follows directly from (18.3) using (18.9) and (16.8)–(16.9). Similarly, (18.7) is obvious. To prove (18.8) we remember that $G^*(x)$ is the solution of Theorem 13.2. Since r_1 is a natural boundary, $G^{*\prime} - bG^* \to 0$ as $x \to r_1$ by the Corollary to that theorem. Hence, integrating $\lambda G^* - \Omega^* G^* = g$ we get

(18.11) $$\lambda \int_{r_1}^{r_2} G^*(x)\, dx - \int_{r_1}^{r_2} g(x)\, dx = \lim_{x \to r_2} \left\{ \frac{d}{dx} G^* - bG^* \right\}.$$

On the other hand, from (18.2) and (14.11)

(18.12) $$\lambda \int_{r_1}^{r_2} G^*(x)\, dx = \int_{r_1}^{r_2} g(x)\, dx - \lambda \int_{r_1}^{r_2} \xi(x) g(x)\, dx,$$

whence

(18.13) $$\lambda \int_{r_1}^{r_2} \xi(x) g(x)\, dx = -\lim_{x \to r_2} \left\{ \frac{d}{dx} G^* - bG^* \right\}.$$

---

[31] Cf. the notational convention (13.13). The operator $\overline{\Omega^*}$ is the infinitesimal generator since no lateral condition is imposed.

By the same argument applied to (16.9)

$$\lambda \tau A \int_{r_1}^{r_2} \xi(x)\, dp(x) = -\lim_{x \to r_2} \{Q' - bQ\}. \tag{18.14}$$

Therefore from (18.3)

$$\begin{aligned} -\lim_{x \to r_2} \{G' - bG\} &= \lambda \int_{r_1}^{r_2} \xi(x) g(x)\, dx \\ &\quad + \lambda \tau A \cdot \int_{r_1}^{r_2} \xi(x)\, dp(x) \cdot \left\{\lambda^{-1} g_2 + \int_{r_1}^{r_2} \xi(x) g(x)\, dx\right\} \\ &= -g_2 + \left\{1 + \tau A \int_{r_1}^{r_2} \xi(x)\, dp(x)\right\} \left\{g_2 + \lambda \int_{r_1}^{r_2} \xi(x) g(x)\, dx\right\}. \end{aligned} \tag{18.15}$$

Introducing the expression (16.10) for A we get

$$-\lim_{x \to r_2} \{G' - bG\} = -g_2 + A(\lambda\sigma + p_2)\left\{\lambda^{-1} g_2 + \int_{r_1}^{r_2} \xi(x)\, dp(x)\right\}. \tag{18.16}$$

This equation is equivalent to (18.8) in view of the second equation in (18.4) and of (16.8).

Next we show that $\mathfrak{S}^*$ is a contraction semi-group. The defining equations reveal that it is positivity preserving.

Integrating (18.6) we get in view of (16.5)

$$\lambda \int_{r_1}^{r_2} G(x)\, dx = \int_{r_1}^{r_2} g(x)\, dx + \frac{\tau}{\sigma} G_2 + \lim_{x \to r_2} \{G' - bG\} \tag{18.17}$$

and hence

$$\lambda \left\{G_1 + G_2 + \int_{r_1}^{r_2} G(x)\, dx\right\} = g_1 + g_2 + \int_{r_1}^{r_2} g(x)\, dx + G_2 \frac{\tau + p_1 - p_2}{\sigma}. \tag{18.18}$$

Thus the norm decreases unless $\tau + p_1 = p_2$ in which case the semi-group is norm preserving.

To prove the *uniqueness* note that the difference of two solutions satisfies (18.6)–(18.8) with $g_1 = g_2 = g(x) = 0$. Assume that a nontrivial solution exists. If $G_2 = 0$ then $G(x) = 0$ since $\lambda G - \Omega^* G = 0$ has no integrable solution. We can therefore assume $G_2 = 1$. Then there exists a unique integrable solution $G(x)$ of (18.6) (cf. Theorem 13.3). For it we get in view of (16.5)

$$-\lim_{x \to r_2} (G' - bG) = -\lambda \int_{r_1}^{r_2} G(x)\, dx + \frac{\tau}{\sigma}. \tag{18.19}$$

By Theorem 13.3 the right side remains bounded as $\lambda \to \infty$. On the other hand (18.8) requires that this right side equal $\lambda + p_2/\sigma$. This contradiction proves the uniqueness.

We now turn to the case $\sigma = 0$ excluded in the last theorem. The set Σ of

Definition 16.2 now lies within the closed manifold

$$p_2 F(r_2) = p_1 F(r_1) + \tau \int_{r_1}^{r_2} F(x)\, dp(x) \tag{18.20}$$

where $p_2 > 0$ in view of (16.4). The closure of the range of the semi-group $\mathfrak{S}$ (and of its resolvent (16.7)) lies entirely within the manifold (18.20). The conjugate space to it is isomorphic to the space $\mathfrak{BB}[r_1, r_2)$. The adjoint to the resolvent can be represented as a transformation taking the subspace $G_2 = 0$ of $\bar{\mathfrak{L}}[r_1, r_2]$ into itself. It is interesting to note that in physical applications this amounts to an identification, or joining, of the two boundaries rather than to assuming zero mass at r_2.

The adjoint to the resolvent of $\mathfrak{S}$ is given by (18.3) and the first equation (18.4) with $g_2 = \sigma = 0$. (The last equation (18.4) states $G_2 = 0$.) Now (18.3) reduces to

$$G(x) = G^*(x) + Q(x) \int_{r_1}^{r_2} \xi(s) g(s)\, ds, \tag{18.21}$$

so that $G(x)$ in no way depends on G_1. We can therefore disregard G_1 (which in turn can be calculated in terms of g_1 and $G(x)$) and study the contraction of the full adjoint semi-group on $\mathfrak{L}(r_1, r_2)$. In this way we get

THEOREM 18.2. If $\sigma = 0$ *then the semi-group* $\mathfrak{S}$ *of Theorem* 16 *has an adjoint contraction semi-group* $\mathfrak{S}^*$ *operating on* $\mathfrak{L}(r_1, r_2)$ *with range dense in* $\mathfrak{L}(r_1, r_2)$. *Its resolvent is given by* (18.21) *and is the inverse of*

$$\lambda G - \Omega^* G + \frac{\tau}{p_2} \cdot \lim_{x \to r_2} \{G' - bG\} \cdot \tilde{p}(x) \approx g(x). \tag{18.22}$$

For $\mathfrak{S}^*$ *to be norm preserving it is necessary and sufficient that* $\tau = p_2$.

The last statement requires an explanation. The condition $\tau = p_2$ implies $p_1 = 0$ and hence $\lambda G_1 = g_1$. If $\tau < p_2$ but $p_1 + \tau = p_2$ the norm of $G(x)$ will decrease, but we shall have

$$\lambda \left\{ G_1 + \int_{r_1}^{r_2} G(x)\, dx \right\} = g_1 + \int_{r_1}^{r_2} g(x)\, dx \tag{18.23}$$

whenever $g(x) \geqq 0$ and $g_1 \geqq 0$.

19. Regular boundaries

In the preceding constructions the function $Q(x)$ defined in (16.9) played an important role. Up to a norming factor dependent on λ this function coincides with the unique solution of (18.9). When r_2 is a regular boundary, the complication arises that (18.9) has infinitely many solutions. This gives us a greater freedom in imposing lateral conditions, and they will now contain an additional arbitrary constant. On the other hand, we require a new quantity to express the resolvent. This quantity is furnished by

$$\lim_{x \to r_2} e^{B(x)} \xi'(x) = \omega \tag{19.1}$$

which is finite in the case of a natural boundary while it was infinite in the case of an exit boundary.

We put

$$\eta(x) = \lambda e^{B(x)} \xi(x) \tag{19.2}$$

so that $\eta(x) \in \mathfrak{L}(r_1, r_2)$ is a solution of $\lambda\eta - \Omega^*\eta = 0$. The condition $\eta \in \mathfrak{L}(r_1, r_2)$ determines this solution up to a multiplicative constant. It satisfies

$$\lim_{x\to r_2} e^{-B(x)}\eta(x) = 1, \qquad \lim_{x\to r_2} (\eta' - b\eta) = \lambda\omega. \tag{19.3}$$

The Green function is no longer unique and we introduce

DEFINITION 19.1. *In the following, $K_0(x, s)$ denotes that Green function of Theorem 15.3 which corresponds to $\alpha = 0$.*

In classical terminology this is the Green function corresponding to an absorbing barrier at r_2, that is, it is determined by the lateral condition $F(r_2) = 0$ for Ω and $e^{-B}G \to 0$ for Ω^*.

The uniqueness theorem of section 17 shows that the most general lateral condition for a contraction semi-group generated by Ω is given in

DEFINITION 19.2. *We denote by Σ' the set of all $F(x) \in \mathfrak{C}[r_1, r_2]$ which are in the domain of Ω and for which*[32]

$$\begin{aligned} p_2 F(r_2) = p_1 F(r_1) + \tau \int_{r_1}^{r_2} F(x)\, dp(x) &- \sigma \lim_{x\to r_2} \Omega F(x) \\ &- \gamma \lim_{x\to r_2} e^{B(x)}F'(x). \end{aligned} \tag{19.4}$$

Here $\gamma \geqq 0$ is a constant and the remaining symbols are defined in Definition 16.2.

LEMMA 19.1. *If $\sigma + \gamma > 0$ the set Σ' is dense in $\mathfrak{C}[r_1, r_2]$. If $\sigma = \gamma = 0$ then Σ' is dense in the closed manifold* (16.6).

The *proof* follows the pattern of Lemma 16.1 using Lemma 14.7.

THEOREM 19.1. *Let r_1 be a natural boundary and r_2 regular. The most general lateral condition under which Ω generates a contraction semi-group $\mathfrak{S}$ with range dense in $\mathfrak{C}[r_1, r_2]$ is given by* (19.4) *with $\sigma + \gamma > 0$. (If $\sigma = \gamma = 0$, then Ω generates a semi-group with range dense in a closed linear manifold.) The resolvent of $\mathfrak{S}$ is given by*

$$\begin{aligned} F = &\int_{r_1}^{r_2} K_0(x, s)f(s)\, ds \\ &+ \xi(s)A_r\left\{\lambda^{-1}p_1 f(r_1) + \sigma f(r_2) + \gamma \int_{r_1}^{r_2} \eta(s)f(s)\, ds + \tau \int_{r_1}^{r_2} Q_0(s)f(s)\, ds\right\} \end{aligned} \tag{19.5}$$

[32] In the limit $\gamma \to 0$ the equation passes into $\lim e^{B(x)}F'(x) = 0$. This is a permissible lateral condition but is omitted here since it is covered by a Green function solution of Theorem 15.3.

where

$$A_r = \frac{\lambda}{(\sigma + \gamma\omega)\lambda + p_2 - \tau\lambda \int_{r_1}^{r_2} \xi(x)\, dp(x)} \tag{19.6}$$

and

$$Q_0(s) = \int_{r_1}^{r_2} K_0(x, s)\, dp(x) \tag{19.7}$$

(and ω is defined in (19.1)). Note again that the existence of the limit of $e^B F'$ is not postulated but proved).

PROOF. That Σ' represents the most general lateral condition was proved in section 17. Otherwise the proof follows the pattern of the proof of Theorem 16, except that we require means to express the integral of ηf in terms of F. Now putting

$$F_0(x) = \int_{r_1}^{r_2} K_0(x, s) f(s)\, ds \tag{19.8}$$

one verifies easily that

$$\lim_{x \to r_2} e^{B(x)} F_0'(x) = -\int_{r_1}^{r_2} \eta(s) f(s)\, ds \tag{19.9}$$

(and $F_0(r_2) = 0$ by definition). It suffices then to calculate $F(r_2)$ and $\lim e^B F'$ from (19.5) to verify that

$$\int_{r_1}^{r_2} \eta(s) f(s)\, ds = \lambda\omega F(r_2) - \lim_{x \to r_2} e^{B(x)} F'(x). \tag{19.10}$$

The remaining part of the proof can be left to the reader.

The most interesting fact concerning the adjoint semi-group is that *it is generated by the same operator* $\bar{\Omega}^*$ as in the case of an exit boundary, but the generator now requires a *lateral* condition to determine a semi-group.

The adjoint to the resolvent (19.5) is the transformation taking $[g_1, g(x), g_2] \in \bar{\mathfrak{L}}[r_1, r_2]$ into $[G_1, G(x), G_2]$ where

$$G(x) = \int_{r_1}^{r_2} K_0(s, x) g(s)\, ds + A_r \left\{ \tau Q_0(x) + \gamma\eta(x) \right\} \left\{ \lambda^{-1} g_2 + \int_{r_1}^{r_2} \xi(s) g(s)\, ds \right\}, \tag{19.11}$$

$$G_1 = \lambda^{-1} g_1 + p_1 \lambda^{-1} A_r \left\{ \lambda^{-1} g_2 + \int_{r_1}^{r_2} \xi(s) g(s)\, ds \right\} \tag{19.12}$$

$$G_2 = \sigma A_r \left\{ \lambda^{-1} g_2 + \int_{r_1}^{r_2} \xi(s) g(s)\, ds \right\}. \tag{19.13}$$

THEOREM 19.2. *Let* $\sigma > 0$. *Then* (19.11)–(19.13) *define the resolvent of a contraction semi-group* $\mathfrak{S}^*$ *with range dense in* $\bar{\mathfrak{L}}[r_1, r_2]$. *The infinitesimal generator is* *given by the operator $\bar{\Omega}^*$ defined in Theorem* 18.1, *but restricted by the lateral condition*

$$\gamma G_2 = \sigma \lim_{x \to r_2} e^{-B(x)} G(x). \tag{19.14}$$

This condition determines the solution of (18.6)–(18.8) *uniquely.*

PROOF. The proof of Theorem 18.1 applies without change and it suffices to prove that (19.14) holds. To $Q_0(x)$ and to the first integral in (19.11) Lemma 14.4 applies for $j = 2, p_j = 0$. Hence

$$\lim e^{-B(x)} G(x) = \gamma A_r \left\{ \lambda^{-1} g_2 + \int_{r_1}^{r_2} \xi(s) g(s)\, ds \right\} \lim_{x \to r_2} e^{-B(x)} \eta(x). \tag{19.15}$$

Using first (19.3) and then (19.13) we get (19.14).

In a similar way we can treat the case $\sigma = 0$. As in section 18 the resolvent reduces to a transformation in $\mathfrak{L}(r_1, r_2)$, namely

$$G(x) = \int_{r_1}^{r_2} K_0(s, x) g(s)\, ds + A_r \{\tau Q_0(x) + \gamma\eta(x)\} \int_{r_1}^{r_2} \xi(x) g(x)\, dx. \tag{19.16}$$

The final result is embodied in

THEOREM 19.3. *If $\sigma = 0$ the semi-group of Theorem* 19.1 *has an adjoint contraction semi-group $\mathfrak{S}^*$ with range dense in $\mathfrak{L}(r_1, r_2)$. Its resolvent is given by* (19.16) *and is the inverse of* (18.22) *when this equation is restricted by the lateral condition*[33]

$$\gamma \lim_{x \to r_2} (G' - bG) = \lim_{x \to r_2} e^{-B} G(x). \tag{19.17}$$

This lateral condition determines the solution uniquely. The semi-group is norm preserving if and only if $\tau = p_2$.

20. Preparations

When neither of the boundaries is natural, *two* lateral conditions are required. The pattern of the construction is clear from the foregoing, but this time the requirement of preservation of positivity produces special difficulties. Also, the fundamental quantities in terms of which the resolvent is expressed can be chosen in several ways. We introduce those which lead to reasonably simple expressions.

DEFINITION 20.1. *If both boundaries are regular we denote by $\xi_j(x)$ the (positive, monotonic) solutions of $\lambda\xi - \Omega\xi = 0$ defined by*

$$\xi_j(r_j) = \lambda^{-1}, \qquad \xi_1(r_2) = \xi_2(r_1) = 0. \tag{20.1}$$

We put

$$\lim_{x \to r_j} e^{B(x)} \xi_i'(x) = (-1)^j \omega_{ij}, \tag{20.2}$$

[33] The case $\gamma = \infty$ corresponds to the "reflecting barrier" condition $G' - bG \to 0$ covered by a Green function solution of Theorem 15.3.

and

(20.3) $$\eta_i(x) = \lambda e^{B(x)} \xi_i(x).$$

Then η_j is a solution of $\lambda\eta - \Omega^*\eta = 0$, $\eta_j \in \mathfrak{L}(r_r, r_2)$, and

(20.4) $$\lim_{x \to r_j} (\eta_i' - b\eta_i) = (-1)^j \lambda\omega_{ij}.$$

LEMMA 20.1. *We have*

(20.5) $$(-1)^{i+j} \omega_{ij} > 0, \qquad \omega_{12} = \omega_{21}$$

and

(20.6) $$|\omega_{12}| < \omega_{ii}.$$

PROOF. The first relation in (20.5) for $j = 2$ (including the assertion that the ω_{ij} are finite) follows from Lemma 9.5 when ξ_1 and ξ_2 are expressed as combinations of two non-increasing solutions. Next calculate the Wronskian (12.2) of the pair ξ_1, ξ_2 and apply (12.4). Letting $x \to r_1$ and $x \to r_2$ one gets $\omega_{12} = \omega_{21}$. Finally, integrating (20.3) we get

(20.7) $$\lambda \int_{r_1}^{r_2} \eta_j \, dx = \eta_j' - b\eta_j \Big|_{r_1}^{r_2} = \omega_{i2} + \omega_{i1} > 0.$$

DEFINITION 20.2. *The Green function of Ω formed by ξ_1, ξ_2 will be denoted*[34] *by $K_{abs}(x, s)$. We put*

(20.8) $$F_{abs}(x) = \int_{r_1}^{r_2} K_{abs}(x, s) f(s) \, ds, \qquad G_{abs}(x) = \int_{r_1}^{r_2} K_{abs}(s, x) g(s) \, ds.$$

Clearly

(20.9) $$K_{abs}(x, s) = \begin{cases} -\omega_{12}^{-1} \xi_2(x)\eta_1(s) & \text{for} \quad x \leq s \\ -\omega_{12}^{-1} \xi_1(x)\eta_2(s) & \text{for} \quad x \geq s. \end{cases}$$

LEMMA 20.2. *We have*

(20.10) $$\int_{r_1}^{r_2} K_{abs}(x, s) \, ds = \lambda^{-1} - \xi_1(x) - \xi_2(x).$$

PROOF. Equation (14.11) holds and $\xi(x)$ is a positive solution of $\lambda u - \Omega u = 0$ with $\xi(r_j) = \lambda^{-1}$. This shows that $\xi = \xi_1 + \xi_2$.

LEMMA 20.3. *We have*

(20.11) $$F_{abs}(r_j) = 0, \qquad \lim_{x \to r_j} e^{-B(x)} G_{abs}(r_j) = 0$$

(20.12) $$(-1)^{j-1} \lim_{x \to r_j} e^{B(x)} F_{abs}'(x) = \int_{r_1}^{r_2} \eta_j(s) f(s) \, ds$$

(20.13) $$(-1)^{j-1} \lim_{x \to r_j} (G_{abs}' - bG_{abs}) = \lambda \int_{r_1}^{r_2} \xi_j(s) g(s) \, ds.$$

[34] The subscript abs stands for *absorbing* boundaries.

PROOF. Equation (20.11) follows from Lemmas 14.1 and 14.3. From (20.9)

$$F'_{\mathrm{abs}}(x) = -\omega_{12}^{-1}\xi_1'(x)\int_{r_1}^{x}\eta_2(s)f(s)\,ds - \omega_{12}^{-1}\xi_2'(x)\int_{x}^{r_2}\eta_1(s)f(s)\,ds. \tag{20.14}$$

Since $f \in \mathfrak{C}[r_1, r_1]$ and $\eta_j \in \mathfrak{L}(r_1, r_2)$ we get (20.12) using (20.2) and (20.5). A similar calculation leads to (20.13).

LEMMA 20.4. *For every $F \in \mathfrak{C}[r_1, r_2]$ in the domain of Ω the limits*

$$\lim_{x \to r_j} e^{B(x)}F'(x) \tag{20.15}$$

exist.

PROOF. For $F \in \mathfrak{D}(\Omega)$ we have $\lambda F - \Omega F = f$ with $f \in \mathfrak{C}[r_1, r_1]$. Using this f we define F_{abs} as above and find that $F = F_{\mathrm{abs}} + \alpha\xi_1 + \beta\xi_2$ where α and β are constants depending on λ. The assertion now follows from Lemma 14.3 and (20.2).

21. Two regular boundaries

Notations. In the following we use brackets to denote column vectors; thus

$$[a_i] = \begin{pmatrix} a_1 \\ a_2 \end{pmatrix}. \tag{21.1}$$

The transposed row vector is denoted by

$$[a_i]' = (a_1, a_2). \tag{21.2}$$

For $F \in \mathfrak{D}(\Omega)$ we put for abbreviation[35]

$$(-1)^j \lim_{x \to r_j} e^{B(x)}F'(x) = \delta_j F \tag{21.3}$$

$$\lim_{x \to r_j} \Omega F(x) = \Omega_j F \tag{21.4}$$

For $f \in \mathfrak{C}[r_1, r_2]$ and $\phi(x) \in \mathfrak{L}(r_1, r_2)$ we write

$$\int_{r_1}^{r_2} f(x)\phi(x)\,dx = (f, \phi\,dx) \tag{21.5}$$

and for $p(x) \in \mathfrak{BV}(r_1, r_2)$

$$\int_{r_1}^{r_2} f(x)\,dp(x) = (f, dp), \tag{21.6}$$

the integral extending over the *open* interval. Heavy print will be used to indicate matrices of order two.

DEFINITION 21.1. *By Σ_2 we denote the set of all $F \in \mathfrak{D}(\Omega)$ in $\mathfrak{C}[r_1, r_2]$ which satisfy the two conditions expressed in the matrix equation*

$$(\mathbf{I} - \boldsymbol{\pi} - \mathbf{p})[F(r_i)] = [\tau_i(F, dp_i)] - \boldsymbol{\pi}[\delta_i F] - [\sigma_i\,\Omega_i F], \tag{21.7}$$

[35] Cf. Lemma 20.4.

where **I** *is the identity matrix,* $\boldsymbol{\pi} = (\pi_{ij})$, $\mathbf{p} = (p_{ij})$, *the* π_{ij} *etc. are constants and* $p_i(x) \in \mathfrak{BB}(r_1, r_2)$.

(Note that the existence of $\delta_i F$ is guaranteed by Lemma 20.4.) Obviously (21.7) represents the most general form of two linear relations between $\delta_i F$, $\Omega_i F$, and two functionals of $F(x)$.

DEFINITION 21.2. *We say that* Σ_2 *satisfies condition* (p) *if*

$$\boldsymbol{\pi} = \begin{pmatrix} \pi_1 & 0 \\ 0 & \pi_2 \end{pmatrix} \tag{21.8}$$

$$p_{ij} \geqq 0, \qquad \pi_i \geqq 0, \qquad \sigma_i \geqq 0, \qquad \tau_i \geqq 0, \tag{21.9}$$

$$P_{i1} + p_{i2} + \pi_i + \tau_i \leqq 1, \qquad (i = 1, 2) \tag{21.10}$$

and the $p_i(x)$ *are non-decreasing with*

$$p_j(r_1+) = 0, \qquad p_j(r_2-) = 1. \tag{21.11}$$

LEMMA 21.1 (1) *If* $\pi_{ij} = 0$, $\sigma_i = 0$ $(i, j = 1, 2)$ *the closure* $\bar{\Sigma}_2$ *coincides with the intersection of two linear manifolds each characterized by the vanishing of a linear functional.* (2) *If both vectors* $[\pi_{i1}]$ *and* $[\pi_{i2}]$ *are parallel to* $[\sigma_i]$ *then* $\bar{\Sigma}_2$ *coincides with one such manifold.* (3) *In every other case* Σ_2 *is dense in* $\mathfrak{C}[r_1, r_2]$.

PROOF. The density assertions follow easily from Lemmas 14.5–14.7. In case (1) equation (21.7) states that two functionals vanish, and in case (2) the functional is obtained on premultiplication by the row vector $(\sigma_2, -\sigma_1)$.

THEOREM 21. *Let the set* Σ_2 *satisfy condition* (p).[36] *Then* Ω, *restricted by the lateral condition* Σ_2 *generates a contraction semi-group* $\mathfrak{S}$; *its norm is unity if and only if equality signs hold in* (21.10). *Conversely, all contraction semi-groups generated by* Ω *with range dense in* $\mathfrak{C}[r_1, r_2]$ *are obtained in this way.*[37]

The resolvent of $\mathfrak{S}$ *is given by the matrix equation*

$$F = F_{\text{abs}} + \lambda[\xi_i]' \, \mathbf{R} \left\{ \boldsymbol{\pi}[(f, \eta_i \, dx)] + [(f, Q_i \, dx] + \boldsymbol{\sigma}[f(r_i)] \right\}. \tag{21.12}$$

Here

$$\boldsymbol{\sigma} = \begin{pmatrix} \sigma_1 & 0 \\ 0 & \sigma_2 \end{pmatrix}, \tag{21.13}$$

$$Q_i(x) = \tau_i \int_{r_1}^{r_2} K_{\text{abs}}(s, x) \, dp_i(x) \tag{21.14}$$

and **R** *is the inverse of the matrix*

$$\mathbf{R}^{-1} = \mathbf{I} + \lambda\boldsymbol{\sigma} + \lambda\boldsymbol{\pi}\boldsymbol{\omega} - \mathbf{p} - \boldsymbol{\pi} - \lambda\boldsymbol{\tau}, \tag{21.15}$$

---

[36] Without (p) we get semi-groups which are not positivity preserving or norm increasing. The first contingency can arise even if all conditions are satisfied except $\pi_{12} = \pi_{21} = 0$ (and with $\pi_{ij} \geqq 0$).

[37] The same is true for the case of ranges dense in a manifold defined by the vanishing of one or two functionals. It appears that we obtain all contraction semi-groups operating on the entire space without projections (i.e. $T_t f = 0$ implies $f = 0$).

where $\boldsymbol{\tau} = (\tau_{ij})$ *is the matrix of*

$$\tau_{ij} = \tau_i \int_{r_1}^{r_2} \xi_j(x)\, dp_i(x) = \tau_i(\xi_j, dp_i); \tag{21.16}$$

$\boldsymbol{\omega} = (\omega_{ij})$, *etc. are defined in Definition* 20.1.

PROOF. *We first show that* (21.12) *implies* (21.7) *and vice versa*. For this purpose the condition (p) is not used so that this part of the argument applies also to semi-groups which are not of the contraction type. It follows from the assertion that the range of the transformation (21.12) is independent of λ whence that (21.12) is a resolvent. Conversely, all resolvents are of the form (21.12) provided that the lateral conditions are of the form (21.7). That they are necessarily so is proved by the methods of section 18 and the proof will not be repeated here.[38]

In the following λ is fixed. For any given $F \in \mathfrak{D}(\Omega)$ we define f by $\lambda F - \Omega F = f$. We have then

$$\begin{aligned} F &= F_{\text{abs}} + [\xi_i]'[A_i] \\ &= F_{\text{abs}} + A_1\xi_1 + A_2\xi_2 \end{aligned} \tag{21.17}$$

where the A_j are constants (or functionals of f if (21.17) is a resolvent).

Now in view of (20.1) and (20.11)

$$[F(r_i)] = \lambda^{-1}[A_i]. \tag{21.18}$$

Next, using (20.12) and (20.2)

$$[\delta_i F] = -[(f, \eta_i\, dx)] + \boldsymbol{\omega}[A_i]. \tag{21.19}$$

Again, because of (20.11) we have $\Omega_i F_{\text{abs}} = -f(r_i)$ and since $\Omega_i \xi_j(r_i) = \lambda^{-1}$ we get

$$[\Omega_i F] = -[f(r_i)] + [A_i]. \tag{21.20}$$

Finally, multiplying (21.17) by $\tau_i dp_i(x)$ and integrating we get, using in succession (20.6), (21.14) and (21.16)

$$[\tau_i(F, dp_i)] = [(f, Q_i\, dx)] + \boldsymbol{\tau}[A_i]. \tag{21.21}$$

We use these equations in two ways. First, assume that $F \in \Sigma_2$, that is, assume that (21.7) holds. Substituting from (21.18)–(21.21) into (21.7) we get

$$\begin{aligned} &\{\lambda^{-1}(\mathbf{I} - \boldsymbol{\pi} - \mathbf{p}) - \boldsymbol{\tau} + \boldsymbol{\pi\omega} + \sigma\}[A_i] \\ &\qquad = [(f, Q_i\, dx)] + \boldsymbol{\pi}[(f, \eta_i\, dx)] + \boldsymbol{\sigma}[f(r_1)]. \end{aligned} \tag{21.22}$$

---

[38] It is of some interest to analyze how much the following proof yields without using the argument of section 18. The general form of the resolvent involves two linear functionals of f, and they can be represented by means of two functions in $\mathfrak{BB}[r_1, r_2]$. Now (21.14) shows that in *the open* interval (r_1, r_2) these functions are absolutely continuous and their derivatives Q_i are in the domain of Ω^* (in the generalized sense of Theorem 13.3). If this latter statement is *assumed*, then the following proof shows that the resolvent is of the form (21.12). The argument of section 18 merely restricts the form of the two functionals.

The matrix on the left is $\lambda^{-1}R^{-1}$ by (21.15), and hence (21.17) reduces to (21.12). This means that (21.7) imples (21.12).

Conversely, *define* A_i by (21.19). Then the calculations just performed show that (21.7) is satisfied identically, so that (21.12) in turn implies (21.7).

We complete the proof by proving

LEMMA 21.2. *If* Σ_2 *satisfies the condition* (p), *the transformation* (21.12) *is positivity preserving and* $\lambda \| F \| \leqq \| f \|$. *The norm of the transformation is unity if and only if the equality signs hold in* (21.10). (In particular, $\mathbf{R}^{-1}$ exists.)

PROOF. We know that $f > 0$ implies $F_{\text{abs}} > 0$ and that all vectors occurring within the braces in (21.12) are non-negative. To prove that (21.12) is positivity preserving it suffices therefore to prove that the matrix $\mathbf{R}$ is non-negative. For that purpose denote by ρ_{ij} the elements of $\mathbf{R}^{-1}$ as defined in (21.15). It suffices to prove that $\rho_{ii} > 0$, $\rho_{12} \leqq 0$, $\rho_{21} \leqq 0$, and $| \rho_{ij} | < \rho_{ii}$.

Now $\xi_1(x) + \xi_2(x)$ is a positive solution of $\lambda\xi - \Omega\xi = 0$ and can therefore have no maximum. Thus, by (20.1), $0 < \xi_1 + \xi_2 < \lambda^{-1}$, and hence by (21.16) and (21.11)

$$\tau_{ij} \geqq 0, \qquad \tau_{i1} + \tau_{i2} < \lambda^{-1}\tau_i . \tag{21.23}$$

It follows that for $i \neq j$

$$\rho_{ii} = (1 - p_{ii} - \pi_i - \lambda\tau_{ii}) + \lambda\sigma_i + \lambda\pi_i\omega_{ii} \geqq \lambda(\tau_{ij} + \sigma_i + \pi_i\omega_{ii}) > 0 \tag{21.24}$$

while

$$\rho_{ij} = -\lambda\tau_{ij} + \pi_i\omega_{ij} \leqq 0 \tag{21.25}$$

by (20.5). Finally $| \rho_{ij} | < \rho_{ii}$ by virtue of (20.6) and (21.23).

Having proved that (21.12) is positivity preserving it suffices to consider $f \equiv 1$. Then from (20.7)

$$\boldsymbol{\pi}[(1, \eta_i\, dx)] = \boldsymbol{\pi\omega} \begin{pmatrix} 1 \\ 1 \end{pmatrix}. \tag{21.26}$$

Similarly we get from (21.14) using (20.10), (21.11), and (21.16)

$$(1, Q_i\, dx) = \lambda^{-1}\tau_i - \tau_{i1} - \tau_{i2} . \tag{21.27}$$

Thus we get for the expression within the braces in (21.12)

$$\begin{aligned} &\boldsymbol{\pi}[(1, \eta_i\, dx)] + [(1, Q_i\, dx)] + [\sigma_i] \\ &\qquad = \left\{ \boldsymbol{\pi\omega} - \boldsymbol{\tau} + \boldsymbol{\sigma} + \lambda^{-1}\begin{pmatrix} \tau_1 & 0 \\ 0 & \tau_2 \end{pmatrix} \right\} \begin{pmatrix} 1 \\ 1 \end{pmatrix}. \end{aligned} \tag{21.28}$$

However, by (21.15) this can be written in the form

$$\lambda^{-1}\mathbf{R}^{-1}\begin{pmatrix} 1 \\ 1 \end{pmatrix} - \left\{ \mathbf{I} - \mathbf{p} - \boldsymbol{\pi} - \begin{pmatrix} \tau_1 & 0 \\ 0 & \tau_2 \end{pmatrix} \right\} \begin{pmatrix} 1 \\ 1 \end{pmatrix}. \tag{21.29}$$

By (21.10) the last vector is non-negative; it vanishes if and only if the equality signs hold in (21.10).

Substituting back into (21.12) we find using (20.10) that the transform of 1 is

$$\lambda^{-1} - \lambda \mathbf{R}\left\{\mathbf{I} - \mathbf{p} - \boldsymbol{\pi} - \begin{pmatrix} \tau_1 & 0 \\ 0 & \tau_2 \end{pmatrix}\right\}\begin{pmatrix} 1 \\ 1 \end{pmatrix},$$

and this accomplishes the proof.

22. The adjoint

As in section 18, the full adjoint to the resolvent (21.12) is a transformation from $\mathfrak{BV}[r_1, r_2]$ to itself. In the exceptional case where the set Σ_2 lies within a submanifold of $\mathfrak{C}[r_1, r_2]$ the adjoint operates on a smaller space, and its range is automatically restricted to the same subspace. To avoid trivial complications we rule out the exceptional case. The adjoint then transforms functions which are absolutely continuous in (r_1, r_2) into functions with the same property. It is therefore entirely appropriate to consider the adjoint as an operation from $\bar{\mathfrak{L}}[r_1, r_2]$ to itself (that is, consider only a contraction).

The adjoint to (21.12) in this sense takes a triple $[g_1, g(x), g_2] \in \bar{\mathfrak{L}}[r_1, r_2]$ into $[G_1, G(x), G_2]$ where, assuming $\sigma_i \neq 0$,

$$\left[\frac{1}{\sigma_i} G_i\right]' = [\lambda(\xi_i, g\,dx) + g_i]'\,\mathbf{R} \tag{22.1}$$

$$G = G_{\text{abs}} + [\lambda(\xi_i, g\,dx) + g_i]'\mathbf{R}\{\boldsymbol{\pi}[\eta_i] + [Q_i]\} \tag{22.2}$$

(for notations cf. (20.3), (20.8), and section 21).

To express the inverse we introduce the *notation*

$$(-1)^j \lim_{x \to r_j} (G'(x) - b(x)G(x)) = \Phi_j G. \tag{22.3}$$

The existence of the limits is guaranteed by Lemma 14.4.

THEOREM 22. *Suppose $\sigma_i > 0$ $(i = 1, 2)$ and that the vectors $[\pi_i]$ and $[\sigma_i]$ are not parallel.*[39] *Suppose that the conditions (p) (Definition 21.2) are satisfied. Then the transformation (22.1)–(22.2) is the resolvent of a contraction semi-group $\mathfrak{S}^*$ with range dense in $\bar{\mathfrak{L}}[r_1, r_2]$. It is norm preserving if and only if the equality signs hold in (21.10).*

The resolvent (22.1)–(22.2) is the inverse of a transformation

$$(\lambda I - \bar{\Omega}^*)[G_1, G(x), G_2] = [g_1, g(x), g_2]$$

given by[40]

$$\lambda G(x) - \Omega^* G(x) - \sum_{i=1}^{2} \frac{\tau_i}{\sigma_i} G_i \tilde{p}_i(x) \approxeq g(x) \tag{22.4}$$

$$\lambda[G_i]' + \left[\frac{1}{\sigma_i} G_i\right]'(\mathbf{I} - \boldsymbol{\pi} - \mathbf{p}) + [\Phi_i G]' = [g_i]'. \tag{22.5}$$

---

[39] This condition guarantees that the set Σ_2 is dense in $\mathfrak{C}[r_1, r_2]$, cf. Lemma 21.1.

[40] Cf. the convention (13.13).

The semi-group is generated by $\overline{\Omega^*}$ *restricted by the lateral conditions*

$$\pi_i G_i = \sigma_i \lim_{x \to r_i} e^{-B(x)} G(x). \tag{22.6}$$

PROOF. Combining (22.1) and (22.2) we get (22.4) directly by the very definitions of G_{abs}, Q_i and η_i. Next from (20.13)

$$\Phi_i G_{\text{abs}} = -\lambda \int_{r_1}^{r_2} \xi_i(x) g(x)\, dx = -\lambda(\xi_i, g\, dx). \tag{22.7}$$

Combining this with (22.1) we can write

$$[\Phi_i G_{\text{abs}}]' = [g_i]' - \left[\frac{1}{\sigma_i} G_i\right]' \mathbf{R}^{-1}. \tag{22.8}$$

By analogy with (22.7) we have from (21.14) and (21.16)

$$\Phi_i Q_j = -\lambda \tau_i(\xi_i, dp_j) = -\lambda \tau_{ij}. \tag{22.9}$$

Finally, by (20.4)

$$\Phi_i \eta_j = \lambda \omega_{ji}. \tag{22.10}$$

Using (22.8)–(22.10) we get from (22.2) and (22.1)

$$[\Phi_i G]' = [g_i]' - \left[\frac{1}{\sigma_i} G_i\right]' \{\mathbf{R}^{-1} - \lambda \boldsymbol{\pi\omega} + \lambda \boldsymbol{\tau}\}. \tag{22.11}$$

When the definition (21.15) of $\mathbf{R}^{-1}$ is introduced into this expression, (22.11) reduces to (22.5), which is thus proved.

The lateral condition (22.6) follows directly from (22.2) since by Lemma 14.4

$$\lim_{x \to r_j} e^{-B} G_{\text{abs}} = 0, \qquad \lim_{x \to r_j} e^{-B} Q_i = 0 \tag{22.12}$$

and by (20.1), (20.3)

$$\lim_{x \to r_j} e^{-B} \eta_i = \begin{matrix} 0 & i \neq j \\ 1 & i = j \end{matrix}. \tag{22.13}$$

Finally, that the resolvent (22.1)–(22.2) is a contraction operator follows from the properties of the adjoint. However, it also can be seen directly that the transformation is positivity preserving. Integrating (22.4) (remembering (22.11)) and adding to (22.5) one finds then

$$\begin{aligned} \lambda\left\{G_1 + G_2 + \int_{r_1}^{r_2} G(x)\, dx\right\} &= g_1 + g_2 + \int_{r_1}^{r_2} g(x)\, dx \\ &\quad - \left[\frac{1}{\sigma_i} G_i\right]' \left\{\mathbf{I} - \boldsymbol{\pi} - \mathbf{p} - \begin{pmatrix} \tau_1 & 0 \\ 0 & \tau_2 \end{pmatrix}\right\} \begin{pmatrix} 1 \\ 1 \end{pmatrix}. \end{aligned} \tag{22.14}$$

The last expression is non-negative or zero, according as (21.10) holds with inequalities or equalities. This completes the proof.

It remains to prove that *the equations* (22.4)–(22.6) *determine the solution uniquely*. It suffices to prove that there exists no non-trivial solution when $g_1 = g_2 = g(x)$. First we note that in this case

$$G(x) = \sum_{i=1}^{2} \frac{1}{\sigma_i} G_i\{Q_i(x) + \pi_i\eta_i(x)\}, \tag{22.15}$$

with $Q_i(x)$ as defined in (21.14). In fact, by Theorem 13.3 this represents a solution of (22.4) (with $g(x) = 0$) and any other solution differs from it by a linear combination of $\eta_i(x)$. However, the boundary conditions (22.6) determine the solution uniquely, and that (22.15) satisfies (22.6) is seen using the fact that as $x \to r_j$

$$e^{-B}Q_i(x) \to 0, \qquad e^{-B}\eta_j \to 1, \qquad e^{-B}\eta_i \to 0 \tag{22.16}$$

$(i \neq j)$ in view of (20.11) and (20.3).

Next, using in succession (20.13) with g replaced by p_i, then (21.16), and finally (20.4) one finds from (22.15)

$$[\Phi_i G]' = \left[\frac{1}{\sigma_i} G_i\right]' (-\lambda\boldsymbol{\tau} + \lambda\boldsymbol{\pi}\boldsymbol{\omega}). \tag{22.17}$$

Inserting this into (22.5) we get (assuming $g_1 = g_2 = 0$) by virtue of (21.15)

$$\left[\frac{1}{\sigma_i} G_i\right]' \mathbf{R}^{-1} = 0,$$

and hence $G_i \equiv 0$.

23. Results for differential equations

The purpose of this section is to survey the meaning and the implications of the main results for the differential equations[41]

$$u_t(t, x) = a(x)u_{xx}(t, x) + b(x)u_x(t, x) \tag{23.1}$$

and

$$v_t(t, x) = \frac{\partial}{\partial x}\left\{\frac{\partial}{\partial x}\{a(x)v(t, x)\} - b(x)v(t, x)\right\}. \tag{23.2}$$

It is assumed that $a'(x)$ and $b(x)$ are continuous, but not necessarily bounded, in the open interval (r_1, r_2); furthermore $a > 0$. (Note that $a''(x)$ need not exist.)

To express the criteria of section 11 for the case $a \not\equiv 1$ we introduce, following Hille,[42] the function

$$W(x) = \exp\left\{-\int_{x_0}^{x} b(s)a^{-1}(s)\, ds\right\} \tag{23.3}$$

where $x_0 \in (r_1, r_2)$ is fixed.

[41] If $c < M$ is *bounded above*, the more general equations (1.1)–(1.2) can be reduced to (23.1)–(23.2) by the transformation of section 8.

[42] Cf. the footnote to section 11.

The boundary r_j is REGULAR if

$$W(x) \in \mathfrak{L}(x_0, r_j), \qquad a^{-1}(x)W^{-1}(x) \in \mathfrak{L}(x_0, r_j); \tag{23.4}$$

it is an EXIT boundary if[43]

$$a^{-1}W^{-1} \notin \mathfrak{L}(x_0, r_j), \qquad W(x) \int_{x_0}^{x} a^{-1}(s)W^{-1}(s)\, ds \in \mathfrak{L}(x_0, r_j); \tag{23.5}$$

it is an ENTRANCE boundary if[44]

$$a^{-1}W^{-1} \in \mathfrak{L}(x_0, r_j), \qquad a^{-1}(x)W^{-1}(x) \int_{x_0}^{x} W(s)\, ds \in \mathfrak{L}(x_0, r_j); \tag{23.6}$$

it is NATURAL in all other cases.

EXAMPLES. (1) For $u_t = u_{xx}$ a boundary r_j is natural if $r_j = \pm\infty$, regular if r_j is finite. (2) For the singular equation (2.6) the *origin* is an exit boundary if $\gamma \leqq 0$, regular if $0 < \gamma < \alpha$, and an entrance boundary if $\alpha \geqq \gamma$. The boundary at infinity is natural. (3) For $u_t = u_{rr} + r^{-1}u_r$ (the diffusion equation in polar coordinates) both $r = 0$ and $r = \infty$ are natural boundaries.

Using the term solution as defined in section 6 the situation is roughly as follows.

(a) *Common fundamental solutions.* When none of the boundaries is regular then there exists *exactly one fundamental solution common to* (23.1) *and* (23.2) even though the initial value problem as such may have many solutions. When the boundary r_j, or both boundaries, are regular, there exist infinitely many common fundamental solutions. If $u(t, x)$ is a solution of (23.1) generated by such a fundamental solution, then

$$q_j \lim_{x \to r_j} u(t, x) + p_j(-1)^j \lim_{x \to r_j} W^{-1}(x)u_x(t, x) = 0 \tag{23.7}$$

for some constants p_j, q_j. Conversely, to arbitrarily prescribed constants p_j, q_j there corresponds uniquely a fundamental solution common to (23.1) and (23.2). The corresponding solutions of (23.2) satisfy the boundary relation

$$q_j \lim_{x \to r_j} W(x)a(x)v(t, x) + p_j(-1)^j \lim_{x \to r_j} \{(a(x)v(t, x))_x - b(x)v(t, x)\} = 0 \tag{23.8}$$

and this condition determines the solution of (23.2). The limits occurring in (23.7) and (23.8) are finite (but only in the case of regular boundaries).

The relations (23.7)–(23.8) are the *generalized classical boundary conditions*, now applicable also for singular equations. They represent the most general lateral conditions leading to a fundamental solution common to (23.1) and (23.2).

In order that our solutions be positivity preserving and norm not increasing it is necessary and sufficient that $p_j \geqq 0$, $q_j \geqq 0$. The solution $v(t, x)$ is strictly

[43] For alternative forms cf. (10.8) and (11.2).

[44] For alternative forms cf. (10.8) and (11.2).

norm decreasing if $q_j > 0$. It is norm preserving if $q_1 = q_2 = 0$ in the case of two regular boundaries; if only r_2 is regular, $v(t, x)$ is norm preserving if $q_2 = 0$ and r_1 is either a natural or an entrance boundary. The solutions $u(t, x)$ in no case preserve the norm of *all* initial data.

(b) *Two natural boundaries.* In this case the initial value problem both for (23.1) and (23.2) is uniquely determined and the solutions are generated by the common fundamental solution described under (a). No lateral conditions can be imposed.

(c) *One exit boundary.* Let r_1 be a natural and r_2 an exit boundary. *The initial value problem for* (23.1) *has infinitely many solutions, but that for* (23.2) *is uniquely determined.* The unique common fundamental solution described under (a) generates this unique solution for (23.2) and solutions of (23.1) characterized by the boundary relation $u(t, r_2) = 0$.

The most general solution of the initial value problem for (23.1) *is determined by a lateral condition of the form* (16.3) (where $F(x)$ is to be replaced by $u(t, x)$ and Ω by the right side of (23.1)). This solution is positivity preserving and norm not increasing if and only if the conditions of Definition 16.2 are satisfied.

The adjoint is no longer the differential equation (23.2). To express it we require in addition to $v(t, x)$ two functions $V_1(t)$ and $V_2(t)$ which in diffusion theory can be interpreted as representing masses at the two boundaries. If $\sigma > 0$, *the fundamental solution of the above described initial value problem for* (23.1) *generates a corresponding solution for the adjoint problem*[45]

$$(23.9)\qquad v_t(t, x) \approx \frac{\partial}{\partial x} \{(a(x)v(t, x))_x - b(x)v(t, x)\} + \frac{\tau}{\sigma} V_2(t)\tilde{p}(x)$$

$$(23.10)\qquad V_1'(t) = \frac{p_1}{\sigma} V_2(t)$$

$$(23.11)\qquad V_2'(t) = -\frac{p_2}{\sigma} V_2(t) - \lim_{x \to r_2} \{(a(x)v(t, x))_x - b(x)v(t, x)\}.$$

Here the initial values of $v(t, x)$, $V_1(t)$ and $V_2(t)$ are given. If the norm is defined in the obvious way (cf. section 3) then the solution of (23.9)–(23.11) (given in section 18) is positivity preserving. It is norm decreasing unless the equality sign holds in (16.4) in which case it is norm-preserving.

It is interesting that *in the special case* $\sigma = 0$ *the adjoint to* (23.1) *can be ex-*

[45] If $p(x)$ is absolutely continuous then $\tilde{p}(x)$ in (23.9) stands for $p'(x)$ and the sign $\approx$ can be replaced by an equality sign. If $p(x)$ is only known to be of bounded variation, then (23.9) indicates that

$$\int_x^y v_t(t, s)\, ds = \{(a(s)v(t, s))_s - b(s)v(t, s)\} \Big|_x^y + \frac{\tau}{\sigma} V_2(t)\{p(y) - p(x)\}$$

for all continuity points x, y of $p(x)$. (Cf. (13.13).)

pressed by the single equation

$$v_t(t, x) \simeq \frac{\partial}{\partial x}\{(a(x)v(t, x))_x - b(x)v(t, x)\} \tag{23.12}$$
$$- p_2\tau\tilde{p}(x) \lim_{x\to r_2} \{(a(x)v(t, x))_x - b(x)v(t, x)\}.$$

This is the simplest example of a differential equation in $\mathfrak{C}[r_1, r_2]$ *having an adjoint in* $\mathfrak{L}(r_1, r_2)$ *which is not a differential equation.*

(d) *Physical interpretation.* If $v(t, x)$ is interpreted as mass density in the interior and $V_j(t)$ as mass at r_j then (23.9)–(23.11) express that mass flows out of the interior at r_2, but not at r_1. The mass concentrates at r_2 from where it flows out at the constant rate p_2/σ. The outflowing mass goes in part to r_1, in part to the interior of (r_1, r_2) and in part it disappears. The three parts are in the ratios $p_1 : \tau : (p_2 - p_1 - \tau)$. The part going to r_1 remains forever inactive (because r_1 is a natural boundary). The part going to the interior is distributed according to the distribution function $p(x)$, and starts an ordinary diffusion process.

It is seen that (23.9)–(23.11) *represent the complete Fokker-Planck equation* replacing the over-simplified diffusion equation (23.2). In the case of a regular boundary an additional reflection process may take place, and if neither boundary is natural all sorts of interactions enter into the play. The situation is completely described in the following equations, but we desist from interpreting them physically.

(e) *Regular boundary.* If r_1 is a natural but r_2 a regular boundary, then there exist infinitely many solutions for the initial value problem both for (23.1) and for (23.2). In addition to the solutions generated by the common fundamental solutions each equation admits of solutions such that the adjoint operation is not given by the other equation. We describe the situation for (23.1), but one could also take (23.2) as starting point.

For every solution of (23.1) the limits appearing in (23.7) (for $j = 2$) automatically exist. *The most general solution of the initial value problem for* (23.1) *is now given by* (19.4) (where the last limit is to be interpreted as the second limit appearing in (23.7)). *If* $\sigma \neq 0$, *the adjoint equations are again given by* (23.9)–(23.11), *but the corresponding initial value problem is now uniquely determined only by means of the lateral condition*

$$\gamma V_2(t) = \sigma \lim_{x\to r_2} W(x)a(x)v(t, x) \tag{23.13}$$

(the existence of the limit being proved). In the case $\sigma = 0$ we get again (23.12) with a lateral condition of the form (23.8).

It is curious that the lateral condition for (23.1) involves a differential operator of second order and terms of global character, whereas the lateral conditions determining the adjoint problem are always of the simple form (23.13) or (23.8).

Both initial value problems are uniquely determined. If the lateral condition is of the type specified in Definition 19.2 both solutions are positivity preserving and norm not increasing. If equality signs hold in (16.4) the solution of the adjoint problem is norm preserving (but this is never the case for the solutions of the initial value problem for (23.1)).

(g) *The general case.* If neither boundary is natural then there are two sources for non-uniqueness and, accordingly, there will be two lateral conditions. A complication arises from the fact that each lateral condition may depend on both boundaries and on global conditions. Fortunately the most general lateral conditions can be expressed in compact matrix form. The most general lateral condition for the case of two regular boundaries (and $a = 1$) is given in definition 21.1; the adjoint is treated in section 22. The modification required for the simpler case where one boundary is an exit or an entrance boundary should now be obvious. Also, it was shown in the foregoing how equations (22.4)–(22.5) are to be interpreted as the adjoint to (23.1).

PRINCETON UNIVERSITY

REFERENCES

[1] J. L. DOOB. *Markoff chains—Denumerable case.* Trans. Am. Math. Soc. **58** (1945), 455–473.
[2] W. FELLER.[46] *Zur Theorie der Stochastischen Prozesse (Existenz und Eindeutigkeitssätze).* Math. Ann. **113** (1936) 113–160.
[3] W. FELLER. *Two singular diffusion problems.* Ann. of Math. **54** (1951), 173–182.
[4] W. FELLER. Diffusion processes in the mathematical theory of evolution. Proc. 2nd. Symposium on Probability and Statistics, Berkeley, 1950.
[5] W. FELLER. Some recent trends in the mathematical theory of diffusion. Intern. Congress of Math. Cambridge, Sept. 1950.
[6] E. HILLE. On the integration problem for Fokker-Planck's equation. 10th Congr. of Scandin. Math., Trondheim 1950.
[7] E. HILLE. *Les probabilités continues en chaine.* C. R. Acad. Sci. Paris **230** (1950), 34–35.
[8] E. HILLE. Functional Analysis and Semi-groups. Am. Math. Soc. Colloquium Publ., vol. 31. New York 1948.
[9] A. KOLMOGOROFF. *Über die analytischen Methoden in der Wahrscheinlichkeitsrechnung.* Math. Ann. **104** (1931) 415–458.
[10] M. H. STONE. Linear Transformations in Hilbert Space. Am. Math. Soc. Colloquium Publ., vol. 15, New York 1932.
[11] K. YOSIDA. *An operator-theoretical treatment of temporally homogeneous Markoff process.* J. Math. Soc. Japan **1** (1949), 244–253.
[12] K. YOSIDA. *On the differentiability and the representation of one-parameter semi-group of linear operators.* J. Math. Soc. Japan **1** (1948), 15–21.

[46] [2a; *added in proof.*] On positivity preserving semi-groups of transformations on $\mathfrak{C}[r_1, r_2]$. Ann. Soc. Polonaise Math. 25 (1952).

On a Generalization of Marcel Riesz' Potentials and the Semi-Groups generated by them.

By

William Feller.

To Marcel Riesz in appreciation of his mathematical work and of what I learned from it and from him.

---

1. **Introduction.** The Riemann-Liouville integrals have always been a favorite and successful tool in the hands of MARCEL RIESZ. For our purpose it is most convenient to fix the arbitrary limit at either $+\infty$ or $-\infty$, and accordingly we consider the two forms

$$(1.1)\quad J_+^a f = \frac{1}{\Gamma(a)} \int_x^\infty f(y)\,(y-x)^{a-1}\,dy, \qquad J_-^a f = \frac{1}{\Gamma(a)} \int_{-\infty}^x f(y)\,(x-y)^{a-1}\,dy.$$

The more familiar form with the integral extending over the finite interval $(0, x)$ is obtained from the second integral in (1.1) when the domain of the operator is restricted to functions $f(x)$ which vanish for $x<0$. In his latest work M. Riesz has introduced a modified form of such transforms with the domain of integration extending over the entire n-dimensional Euclidean or Lorentzian space. For $n=1$ the new Riesz potential takes on the form

$$(1.2)\quad I^a f = \frac{1}{2\Gamma(a)\cos\frac{\pi a}{2}} \int_{-\infty}^{+\infty} f(y)\,|y-x|^{a-1}\,dy.$$

Here, as in (1.1), the function $f(x)$ must satisfy certain obvious regularity requirements, but even then the integral converges only for $0<a<1$. However, Marcel Riesz has always excelled in using divergent algorithms and giving them a meaning by original methods. In the present case he considers the above integrals for a fixed function $f(x)$ as analytic functions of a. He shows that a process of analytic continuation

R.L. Schilling et al. (eds.), *Selected Papers II*,

permits us to attach a meaning to the transforms (1.1) and (1.2) also for values of a for which the integrals diverge. In this way he creates a powerful analytic apparatus by which many results can be derived by surprisingly simple formal operations.

It will be shown in the sequel that the two transforms (1.1) and (1.2) are not so far apart as it appears at first glance. In fact, *we shall introduce a two-parametric family of potentials which contains as special cases both the Riesz potentials and the Riemann-Liouville integrals.* These generalized potentials will preserve the main properties of the Riesz potentials, namely (1) the semi-group property

$$I^a I^\beta = I^{a+\beta}, \tag{1.3}$$

(2) the fact that I^0 reduces to the identity, and (3) that within its domain of definition

$$I^{-2} = -\frac{d^2}{dx^2}. \tag{1.4}$$

Equation (1.3) expresses the fact that the potentials I^a form a semi-group of operators in which a is the group parameter. This semi-group must be distinguished from the semi-group $[T_t]$ of operators whose infinitesimal generator is represented by the operator I^a with a fixed a. For example, when $a=-1$, the latter semi-group can be written in the form

$$T_t f = \frac{1}{\pi} \int_{-\infty}^{+\infty} f(y) \frac{t}{(x-y)^2+t^2}\, dy. \tag{1.5}$$

Putting $T_t f = u(x, t)$ we see that $u(x, t)$ is the harmonic function with $u(x, 0) = f(x)$ defined by the Poisson integral.

We shall briefly discuss the semi-groups generated by our generalized potentials. When $0<a<1$ an explicit kernel representation is easily obtained, and it can be extended by the method of analytic continuation. In particular, *we obtain in this way power-series representations for all stable distribution functions.* These formal developements were suggested by a discovery of S. Bochner linking the stable distributions (and in particular the harmonic function (1.5)) to the familiar diffusion equation. This connection is discussed in section 3. Special attention is paid to harmonic functions in section 5.

2. Generalized Riesz potentials. For simplicity let us consider only a function $f(x)$ which is continuous and vanishes outside a finite interval. Let δ be a fixed real number and define an operator I_δ^a by

$$(2.1) \qquad I_\delta^a f = \frac{1}{\Gamma(a)\sin a\pi} \int_{-\infty}^{+\infty} f(y)\,|y-x|^{a-1}\cdot \sin a\left(\frac{\pi}{2}+\frac{y-x}{|y-x|}\delta\right) dy.$$

The integral converges for all complex a in the half-plane $\Re(a)>0$. For these values (2.1) represents an analytic function of a with a pole at $a=1$. (In exceptional cases the function may be analytic also at $a=1$, cf. [3] p. 22). If $f(x)$ is sufficiently regular, the process of analytic continuation can be used to define I_δ^a in a bigger half-plane or in the whole plane.

It will be observed that for $\delta=0$ the equations (2.1) and (1.2) coincide. For $\delta=\pi/2$ the integrand in (2.1) vanishes in the half-line $x<y$ and our potential reduces to a Riemann-Liouville integral. Thus

$$(2.2) \qquad I_0^a = I^a; \qquad I_{\pi/2}^a = J_+^a; \qquad I_{-\pi/2}^a = J_-^a.$$

More generally, it is easily seen that

$$(2.3) \qquad I_\delta^a = \frac{\sin a\left(\frac{\pi}{2}+\delta\right)}{\sin a\pi} I_{\pi/2}^a + \frac{\sin a\left(\frac{\pi}{2}-\delta\right)}{\sin a\pi} I_{-\pi/2}^a.$$

In view of (2.2) this shows that *each potential* (2.1) *is a linear combination of two Riemann-Liouville integrals.* From the well-known properties of the latter it follows at once that I_δ^0 reduces to the identity for every fixed δ, and also that

$$(2.4) \qquad I_\delta^{-2} = -\frac{d^2}{dx^2}$$

We next prove the semi-group property

$$(2.5) \qquad I_\delta^{a+\beta} = I_\delta^a\, I_\delta^\beta.$$

Following Riesz it suffices to consider parameter values with $0<a,\beta<1$, $a+\beta<1$. We know already that (2.5) is true when $\delta=\pi/2$ and when $\delta=-\pi/2$. An easy calculation shows that

$$(2.6) \qquad \frac{\Gamma(a)\,\Gamma(\beta)}{\Gamma(a+\beta)} I_{\pi/2}^a\, I_{-\pi/2}^\beta f =$$

$$= \left(I_{-\pi/2}^{a+\beta} f\right) \int_0^a \frac{ds}{s^{1-a}(1+s)^{1-\beta}} + \left(I_{\pi/2}^{a+\beta} f\right) \int_0^\infty \frac{ds}{s^{1-\beta}(1+s)^{1-a}}.$$

However,

$$(2.7)\qquad \int_0^\infty \frac{ds}{s^{1-\alpha}\,(1+s)^{-\beta}}=\frac{\Gamma(\alpha)\,\Gamma(1-\alpha-\beta)}{\Gamma(1-\beta)}=\frac{\Gamma(\alpha)\,\Gamma(\beta)}{\Gamma(\alpha+\beta)}\cdot\frac{\sin\beta\pi}{\sin(\alpha+\beta)\pi}$$

whence,

$$(2.8)\qquad I^{\alpha}_{\pi/2}\,I^{\beta}_{-\pi/2}=\frac{\sin\beta\pi}{\sin(\alpha+\beta)\pi}\,I^{\alpha+\beta}_{-\pi/2}+\frac{\sin\alpha\pi}{\sin(\alpha+\beta)\pi}\,I^{\alpha+\beta}_{\pi/2}.$$

It follows that either side of (2.5) is a linear combination of $I^{\alpha+\beta}_{\pi/2}$ and $I^{\alpha+\beta}_{-\pi/2}$, and it remains to show that the coefficients are the same. This is not quite obvious but can be achieved by routine manipulations with trigonometric identities.

3. **Bochner's interpretation of stable distributions.** The solution of the initial value problem for the classical diffusion equation

$$(3.1)\qquad u_t(t,x)=u_{xx}(x,t)\qquad u(t,0)=f(x)$$

is given by

$$(3.2)\qquad u(t,x)=\frac{1}{2(\pi t)^{1/2}}\int_{-\infty}^{+\infty} f(y)\exp\{-(x-y)^2/4t\}\,dy.$$

This is of the same form as (1.5) except that the Poisson kernel is replaced by the normal density. As is well known, both the normal density and the Poisson kernel (known in probability theory as the Cauchy density) are special cases of the *stable densities* introduced by Paul Lévy. The latter name applies to all density functions whose Fourier transforms are of the form

$$(3.3)\qquad \Phi(z)=\exp\left\{-\tau\,|z|^{\alpha}\left(1+i\gamma\frac{z}{|z|}\tan\frac{\pi\alpha}{2}\right)\right\}$$

with $\tau>0$. (This amounts to saying that a density function $g(x)$ is stable if the convolution of $g(ax)$ and $g(bx)$ for any positive a and b is of the form $g(cx)$). Paul Lévy has shown (cf. [2]) that (3.3) defines a stable density if and only if $0<\alpha\leq 2$ and[1] $|\gamma|\leq 1$. In particular, for $\alpha=2$, $\gamma=0$ we obtain the normal density, for $\alpha=1$, $\gamma=0$ the Poisson kernel.

---

[1] When $\alpha=1$ the expression $\gamma\tan\pi\alpha/2$ should be replaced by an arbitrary real number.

Let $K_{a,\gamma}(x)$ be the density function whose Fourier transform is given by (3.3) and put

$$u(t, x) = T_t f(x) = t^{-1/a} \int_{-\infty}^{+\infty} f(y)\, K_{a,\gamma}\left(t^{-1/a}(x-y)\right) dy. \tag{3.4}$$

This includes as special cases both (1.5) and (3.2). It is obvious that in each case the semi-group relation $T_{s+t} f = T_s T_t f$ is true and that $T_t f \to f$ as $t \to 0$ (at least when f has certain regularity properties). Thus each stable density defines a semi-group of transformations. In the particular case of the normal density this semi-group provides the solution of the initial value problem (3.1), and this can be expressed by saying that the semi-group is generated by the differential operator d^2/dx^2.

Bochner's important observation [1] can now be formulated to the effect that the semi-group corresponding to the symmetrical density $K_{a,0}(x)$ is generated by an operator which can symbolically be written in the form

$$-\left(-\frac{d^2}{dx^2}\right)^{a/2}. \tag{3.5}$$

This was derived by considering spectra. In view of (1.4) Riesz would write the diffusion equation in the form $u_t = -I^{-2} u$, and this leads us tentatively to identify Bochner's operator (3.5) with $-I^{-a}$.

More generally *we introduce the functional equation* (3.6)

$$\frac{\partial u(t, x)}{\partial t} = -I_\delta^{-a}\, u(t, x) \tag{3.6}$$

where a and δ are fixed (but not necessarily restricted to values corresponding to a stable density). We seek to solve *the initial value problem*, that is, we seek a solution $u(t, x)$ of (3.6) which in some sense tends to a prescribed $f(x)$ as $t \to 0$. (The type of convergence depends on the underlying metric and is made precise by the theory of semi-groups since the transformation $u(t, x) = T_t f$ automatically has the semi-group property).

For $a=2$, $\delta=0$ equation (3.6) reduces to the classical diffusion equation $u_t = u_{xx}$. When $0 < a < 1$ it can be written in the form of a singular integral equation to which the Carleman theory applies (the iterated kernels of sufficiently high order are continuous). *For other values of a we have in* (3.6) *a functional equation providing a link between partial differential equations and integral equations.* Great progress will be achieved when it will be possible to generalize (3.6) by letting the potentials on the right side vary with x. Such an equation would be related to (3.6) in the same way as the general parabolic differential equation

$$u_t = a(x)\, u_{xx} + b(x)\, u_x + c(x)\, u \tag{3.7}$$

is to the simple diffusion equation $u_t = u_{xx}$.

We now proceed to show how the Riesz theory leads in a simple way to formal solutions of (3.6). As a byproduct we shall obtain explicit formulas for all stable density functions.

4. **Formal solutions.** The operational calculus suggests writing the solution of the initial value problem for (3.6) in the form

$$u(t, x) = e^{-t I_\delta^{-a}} f, \tag{4.1}$$

and all depends on giving a precise meaning to the operator on the right. A formal expansion takes on the form

$$\sum \frac{(-t)^n}{n!} I_\delta^{-na} \tag{4.2}$$

In order that f belong to the domain of existence of each term separately it is necessary that f be an entire function.

On the other hand, using in a purely formal way the definition (2.1) for each term in (4.2), the latter operator can be written as an integral transform with the kernel $U_{a,\delta}(x-y)$ where

$$U_{a,\delta}(x) = \sum \frac{(-t)^n}{n!} \frac{|x|^{-1-na}}{\Gamma(-na)\sin na\pi} \sin na\left(\frac{\pi}{2} + \frac{x}{|x|}\delta\right). \tag{4.3}$$

Now the series converges for every $x \neq 0$ and every δ provided that $0 < a < 1$. We can rewrite it in the form

$$U_{a,\delta}(x) = \frac{-1}{\pi|x|} \sum \left(\frac{-t}{|x|^a}\right)^n \frac{\Gamma(1+na)}{n!} \sin na\left(\frac{\pi}{2} + \frac{x}{|x|}\delta\right). \tag{4.4}$$

A formal solution of (3.6) for $0 < a < 1$ is now obtained in the form (1.5) or (3.2) with the kernels replaced by $U_{a,\delta}(x)$. For other values of a a process of analytic continuation has to be used.

We proceed to justify the expansion (4.4) and to show how the analytic continuation can be effected.

We have (for $0 < a < 1$)

$$U_{a,\delta}(x) = \frac{1}{\pi|x|} \Im \sum \left\{\frac{-t}{|x|^a} e^{-ia\left(\frac{\pi}{2} + \frac{x}{|x|}\delta\right)}\right\}^n \frac{\Gamma(1+na)}{n!}. \tag{4.5}$$

Here we express $\Gamma(1+na)$ by means of the familiar integral to obtain

$$(4.6)\qquad U_{a,\delta}(x)=\frac{1}{\pi}\Im\int_0^\infty \exp\left\{-z\,|x|-t\,z^a\,e^{-ia\left(\frac{\pi}{2}+\frac{x}{|x|}\delta\right)}\right\}dz.$$

If $a|\delta|<\frac{\pi}{2}$ Cauchy's formula permits us to change the path of integration to the imaginary axis. Then

$$(4.7)\qquad U_{a,\delta}(x)=\frac{1}{\pi}\Re\int_0^\infty \exp\left\{-ir\,|x|-tr^a\,e^{-ia\delta x/|x|}\right\}dr.$$

On the right we recognize the inverse Fourier transform of (3.3) with

$$(4.8)\qquad \tau=t\cos a\delta,\qquad \gamma=-\frac{\tan a\delta}{\tan\frac{\pi a}{2}}.$$

Hence, *when* $0<a<1$ *and* $a|\delta|<\frac{\pi}{2}$ *the equation* (4.5) *represents a stable density function, and all stable densities with* $0<a<1$ *are obtained in this way.*

Of course, the series in (4.5) converges also when $a|\delta|\geq\pi/2$ but the representation (4.7) is then no longer valid. Thus (4.5) provides an analytic continuation of the stable densities. For the latter we know that they generate a semi-group, and therefore we know that (4.5) actually represents a solution of (3.6).

When $a>1$ the series (4.5) diverges, but now the integral (4.7) retains a sense and can be used to effect an analytic continuation of (4.5). In fact, when $a>1$ and $a|\delta|<\pi/2$ we get from (4.7) for $x>0$

$$(4.9)\qquad U_{a,\delta}(x)=\frac{1}{\pi}\Re\sum\frac{(-ix)^n}{n!}\int_0^\infty z^n\exp\{-z^a\,e^{-ia\delta}\}\,dz=$$

$$=\frac{1}{\pi}\Re\,e^{i\delta}\sum\frac{(-ix\,e^{i\delta})^n}{n!}\int_0^\infty r^n\,e^{-r^a}\,dr=$$

$$=\frac{1}{a\pi}\Re\,e^{i\delta}\sum\frac{(-ix\,e^{i\delta})^n}{n!}\,\Gamma\left(\frac{n+1}{a}\right).$$

Hence finally

$$U_{a,\delta}(x) = \frac{-1}{\pi|x|}\,\Im \sum \frac{(-i|x|\,e^{i\delta x/|x|})^{n+1}}{(n+1)!}\,\Gamma\left(\frac{n+1}{a}+1\right). \tag{4.10}$$

This formula gives all stable densities with $1 < a \leq 2$.

5. **Harmonic functions.** The case $a=1$, $\delta=0$ is of particular interest. Both series representations break down but lead nevertheless to the Poisson kernel. The series in (4.5) reduces to a geometric series converging for $|x|>t$. Thus for no value of t do we have convergence over the entire axis. However, for $|x|>t$ we get formally the Poisson kernel

$$\frac{1}{\pi}\,\frac{t}{t^2+x^2}. \tag{5.1}$$

The same result can be obtained also from (4.10).

In this way a correct result is derived by a process of doubtful legality. It seems more in Riesz' spirit to start directly from the formal expansion (4.2) and to give it a meaning.

Suppose, then, that $f(x)$ is an entire function and consider the formal expansion

$$\sum \frac{(-t)^n}{n!}\,I^{-n} f. \tag{5.2}$$

According to (1.4) the contribution of the *even* terms must be interpreted as

$$\sum \frac{t^{2k}\,f^{(2k)}(x)}{(2k)!} = \Re f(x+it) = u(x,t) \tag{5.3}$$

where we put for abbreviation

$$f(x+it) = u(x,t) + iv(x,t). \tag{5.4}$$

Since $u(x,0)=f(x)$ and $v(x,0)=0$ it is clear that $u(x,t)$ is not given by the Poisson integral. To interpret the odd terms we require the analytic function

$$F(x+it) = U(x,t) + i\,V(x,t) \tag{5.5}$$

where

$$U(x,t) = \frac{1}{\pi}\int_{-\infty}^{+\infty} \frac{f(y)\,t}{(x-y)^2+t^2}\,dy, \tag{5.6}$$

and $V(x,t)$ is the conjugate harmonic function.

It is then easily seen that for every fixed $t>0$ the difference $v(x,t)-$ $-V(x,t)$ equals the *Hilbert transform* of $u(x,t)$.

We now write the contribution of the odd terms in (5.2) in the form

$$\text{(5.7)}\qquad -\sum \frac{t^{2n+1}}{(2n+1)!} I^{-(2n+1)} f = -\int_0^t I^{-1} \sum \left(\frac{\tau^{2n}}{(2n)!} I^{-2n} f \right) d\tau =$$

$$= -\int_0^t I^{-1} u(x,\tau)\, d\tau$$

But

$$\text{(5.8)}\qquad I^{-1} = I^{-2} I = -\frac{d^2}{dx^2} I = -\frac{d}{dx} (\text{Hilbert transform})$$

In view of the preceding results and using properties of conjugate harmonic functions we get therefore for the last expression in (5.7)

$$\text{(5.9)}\qquad \int_0^t \frac{\partial}{\partial x} \big(v(x,\tau) - V(x,\tau)\big)\, d\tau = -\int_0^t \frac{\partial}{\partial \tau} \big(u(x,\tau) - U(x,\tau)\big)\, d\tau =$$

$$= -u(x,t) + U(x,t).$$

It follows that with this interpretation (5.2) reduces to $U(x,t)$. Thus the formal solution of

$$\frac{\partial \Phi}{\partial t} = -I^{-1} \Phi$$

leads to the harmonic functions in the upper half plane. This agrees with the result which Bochner has obtained by different methods.

References.

[1] S. Bochner: Diffusion equation and stochastic processes. Proc. Nat. Acad. Sci. USA. *35* (1949), 368—370.

[2] P. Lévy: Théorie de l'addition des variables aléatoires. Paris (Gauthier-Villars) 1937).

[3] M. Riesz: L'intégrale de Riemann-Liouville et le problème de Cauchy. Acta Math. *81* (1948), 1—223.

SOME RECENT TRENDS IN THE MATHEMATICAL THEORY OF DIFFUSION

WILLIAM FELLER

1. Introduction. This lecture may seem to sail under a false flag since it appears under the general heading of Applied Mathematics. Only a small part of the mathematical diffusion theory is connected with immediate applications, and much of the recent literature on the subject has an abstract character. The situation is here as in the theory of harmonic functions which likewise stems from practical problems and has many applications, but which has spread into and influenced many fields: complex variable theory, boundary value problems for general differential equations, subharmonic functions, generalized potentials, etc.

Similarly, the diffusion theory has outgrown its origin as a special topic in partial differential equations. Nevertheless, not even this particular chapter is closed: recent applications to the theory of genetic evolution lead us to singular diffusion equations and confront us with a new type of boundary problems. We find in this connection unsolved problems, but of even greater interest is the fact that in diffusion theory we find for the first time an intimate interplay between differential equations and measure theory in function spaces. The latter throws new light even on the classical parts of diffusion theory; it leads to new types of solutions, and opens a new avenue of attack on problems connected with boundary conditions which were left open for a long time. It turns out that the adjoint of a differential equation is not necessarily itself a differential equation, and this new information is of value for the general theory of semigroups of operators. This is underlined by the fact that the probability approach leads us to consider diffusion equations which are not of local character and which can be treated by the methods developed, under different forms, by Bochner, M. Riesz, and L. Schwartz. In this way the theory of differential equations is tied in with other theories and we shall find unexpected connections between harmonic functions and diffusion processes.

It is the purpose of the present address to outline some of the new results and open problems. It is believed that the new methods promise to be fruitful also in other fields, in particular for other types of differential equations. [*Added in proof*: Since this was written, new results were obtained and some of the problems are no longer open; cf. [8].]

2. Classical diffusion and the Wiener space. For a first orientation consider the classical diffusion (or heat-conduction) equation

$$u_t = u_{xx} \tag{2.1}$$

over the interval $-\infty < x < \infty$. Its fundamental solution (or Green function) is given by the normal density function

R.L. Schilling et al. (eds.), *Selected Papers II*,

(2.2) $$u(y;\, t,\, x) = \frac{1}{2(\pi t)^{1/2}} \exp \{-(x - y)^2/4t\}.$$

Under very general conditions on $f(x)$ the function

(2.3) $$u(t,\, x) = \int_{-\infty}^{+\infty} f(y)u(y;\, t,\, x)\, dy$$

represents the unique solution of (2.1) which, for fixed $t > 0$, is integrable with respect to x and which tends to $f(x)$ as $t \to 0$.

Now equation (2.1) is supposed to be connected with Brownian movement, the description of which involves a function space. In fact, the position of a Brownian particle is a function $X(t)$, and "observing a particle" means observing $X(t)$. Thus each $X(t)$ represents a sample point in our experiment, and the problem is to describe the properties of the possible paths $X(t)$ in probabilistic terms. This was first done by N. Wiener [25]. He starts from the assumption that (2.2) defines the transition probability density of our process, that is, the conditional probability density of the relation $X(t_0 + t) = x$ if it is known that $X(t_0) = y$; it is furthermore assumed that the increments $\Delta X(t) = X(t + h) - X(t)$ over nonoverlapping time intervals are statistically independent (or that the corresponding probabilities multiply). Wiener (and by other methods P. Lévy [20]) shows that with probability one the path function is continuous but of unbounded variation in every t-interval; the set of t-values for which $X(t) = 0$ has the structure of a Cantor set; with probability one $X(t)$ satisfies a Lipschitz condition of order 1/2 but no Lipschitz condition of order $1/2 + \epsilon$ (more precise results are given by the local law of the iterated logarithm).

Obviously statements of this type describe the hypothetic diffusion process more directly than does the differential equation (2.1). It is less obvious that even the formal theory of (2.1) can profit from the function space approach. Consider, for example, the boundary value problem for (2.1) for a region bounded by the part $x > a$ of the x-axis and a curve $x = \varphi(t)$ with $\varphi(0) = a$. We are here confronted with a situation similar to that in the Dirichlet problem for $\Delta u = 0$. In general, an essentially unique solution exists, but it does not assume the prescribed boundary values at all points of the curve. As Fortet [10] noticed, the question whether a particular point x_0 of the boundary curve is an exceptional point can be solved by a comparision of the local properties of the boundary with those of the path curves $X(t)$. A final criterion is obtained from the local law of the iterated logarithm. Thus the measure-theoretical approach leads on one hand to an existence theorem which is more general than those obtained by purely analytical methods, and on the other hand we obtain a regularity criterion which admits of a direct interpretation in terms of path functions.

We want to show that even the simpler problem of diffusion in a finite fixed interval is intimately related to the theory of path functions and apparently cannot be treated in a satisfactory manner without the function space approach. It is well known that in the case of diffusion in a finite interval the solution is not determined by the initial values alone, but that different boundary conditions can be imposed. Perhaps the most familiar boundary condition is that of an *absorbing barrier*, which is usually defined formally by the condition $u(t, x_0) = 0$. It must be emphasized that this condition is obtained only heuristically by a passage to the limit from a discrete random walk and is justified by the results. This could be considered satisfactory were it not that the heuristic method completely breaks down in other cases. In fact, a perusal of the literature shows that even in comparatively simple physical problems the appropriate boundary conditions remain unknown. We shall see later on that in a certain case the absorbing barrier is completely described by the condition $u(t, x_0) < \infty$, but again this was found only from the properties of the solutions to which this condition leads. In other cases we shall see that no boundary conditions can be imposed. In short, the appropriate analytical formulation of various boundary conditions is an unsolved problem. Apparently it is unavoidable to revert to the original definition of the various conditions, and this definition usually refers directly to the path functions. Thus, diffusion with absorbing barriers can be described by saying that the laws of free diffusion prevail in the interior but that the process stops when the particle for the first time reaches a boundary, that is, when $X(t)$ equals x_0 or x_1.

3. The adjoint diffusion equations. Consider now a more general diffusion process in the infinite line with transition probability $u(y; t, x)$. By this we mean that if at any time t_0 the position of the particle is $X(t_0) = y$, then

$$\Pr\{a < X(t + t_0) < b\} = \int_a^b u(y; t, x)\, dx. \tag{3.1}$$

If the initial position $X(0)$ of the particle is a random variable with probability density $f(x)$, then the probability density of $X(t)$ is given by the integral (2.3).

A general stochastic process cannot be described solely in terms of the initial distribution and the transition probabilities. A diffusion process, however, is of the Markov type which means, very roughly, that the future depends statistically on the present state, but not on the past history which led to it.[1] This implies in particular the fundamental identity

$$u(y; s + t, x) = \int_{-\infty}^{+\infty} u(y; s, \xi)u(\xi; t, x)\, d\xi, \tag{3.2}$$

known as the *Chapman-Kolmogorov equation*. By virtue of the Markov property the transition probability $u(y; t, x)$ determines all probability relation in the space of path functions $X(t)$. In particular, the joint probability density of $(X(t_1), X(t_2), \cdots, X(t_n))$ for $0 < t_1 < t_2 < \cdots < t_n$ is given by

$$u(t_1, x_1)u(x_1; t_2 - t_1, x_2) \cdots u(x_{n-1}; t_n - t_{n-1}, x_n), \tag{3.3}$$

where $u(t, x)$ is defined by (2.3). Thus the properties of the path functions $X(t)$ are implicitly given by $u(y; t, x)$, and could be derived exactly as Wiener ob-

[1] Cf. the discussion in §8.

tained the properties of the paths in Brownian motion from (2.2). Actually, Wiener's results have been only partially generalized; for example, general conditions which guarantee continuity of almost all path functions seem to be unknown.

Among all stochastic processes for which the Chapman-Kolmogorov equation (3.2) holds, the diffusion processes are characterized by the requirement that the probability of a change exceeding ϵ during a short time interval Δt is small as compared to Δt; more precisely, we assume[2] that

$$\frac{1}{\Delta t}\int_{|x-y|>\epsilon} u(y; \Delta t, x)\, dx \to 0 \tag{3.4}$$

for every fixed y.

If at some time t the position is $X(t) = y$, then the mean and the variance of the displacement $\Delta X(t)$ during the following time interval of duration Δt are

$$\int_{-\infty}^{+\infty} (x - y)u(y; \Delta t, x)\, dx \qquad \text{and} \qquad \int_{-\infty}^{+\infty} (x - y)^2 u(y; \Delta t, x)\, dx. \tag{3.5}$$

Actually these integrals may diverge,[3] and we introduce therefore the truncated moments

$$\int_{|x-y|<\epsilon} (x - y)u(y; \Delta t, x)\, dx = a(\Delta t; y), \tag{3.6}$$

$$\int_{|x-y|<\epsilon} (x - y)^2 u(y; \Delta t, x)\, dx = 2b(\Delta t; y). \tag{3.7}$$

In view of (3.4) the asymptotic behavior of $a(\Delta t; y)$ and $b(\Delta t; y)$ as $t \to 0$ is independent of ϵ, and the physical significance of these quantities is essentially the same as that of the moments (3.5), except that less weight is attributed to large displacements. It is therefore natural to suppose that the limits

$$a(y) = \lim \frac{a(\Delta t; y)}{\Delta t}, \qquad b(y) = \lim \frac{b(\Delta t; y)}{\Delta t} \tag{3.8}$$

exist.[4]

[2] This definition was given in [5] where diffusion processes are called Markov processes of the "purely continuous" type. It seems to be an open problem whether (3.4) implies that the path functions $X(t)$ are continuous with probability one.

[3] An (unpublished) example is given by the density $u(y; t, x) = \{2\pi^{1/2}t^{1/2}\}^{-1}\phi'(x) \exp\{-[\phi(x) - \phi(y)]^2/4t\}$ where $\phi(x) = \log^{1/4}(100 + x/(1 + e^{-x}))$. This is a solution of the equations (3.12)–(3.13) with $b(x) = \{\phi'(x)\}^{-2}$, $a(x) = (1/2)\, b'(x)$.

[4] It will be seen from the following derivation (which follows [5]) that our assumptions actually imply the existence of the partial derivative $u_t(y; t, x)$. Conversely, if it is assumed that u is differentiable and that (3.4) holds, then it suffices to assume that *one* of the ratios occurring in (3.8) has a finite point of accumulation as $\Delta t \to 0$. Letting then Δt approach zero through an appropriate sequence of values it follows from (3.11) that for this particular sequence also the other limit in (3.8) exists, and this leads to (3.12). The first derivation of (3.12) is due to Kolmogorov [17] who, however, uses strong uniformity conditions and assumes the existence of the moments (3.5).

It is then easy to derive a differential equation for $u(y; t, x)$ as a function of t and y. From (3.2) and (3.4) we have

$$(3.9)\qquad \begin{aligned} u(y; t + \Delta t, x) &= \int_{-\infty}^{+\infty} u(y; \Delta t, \xi)u(\xi; t, x)\, d\xi \\ &= \int_{|\xi - y| < \epsilon} u(y; \Delta t, \xi)u(\xi; t, x)\, d\xi + o(\Delta t), \end{aligned}$$

provided $u(\xi; t, x)$ is uniformly bounded for every fixed $t > 0$. If we now assume that for fixed t, x the function $u(\xi; t, x)$ has two continuous derivatives with respect to ξ, then the Taylor formula leads to

$$(3.10)\qquad \begin{aligned} u(y; t + \Delta t, x) &= u(y; t, x) \int_{|\xi - y| < \epsilon} u(y; \Delta t, \xi)\, d\xi \\ &\quad + u_y(y; t, x) \int_{|\xi - y| < \epsilon} (\xi - y)u(y; \Delta t, \xi)\, d\xi \\ &\quad + \frac{1}{2} u_{yy}(y; t, x) \int_{|\xi - y| < \epsilon} (\xi - y)^2 u(y; \Delta t, \xi)\, d\xi + o(\Delta t). \end{aligned}$$

Now the integral of $u(y; \Delta t, \xi)$ should be identically unity.[5] Using, then, (3.4) we can write (3.10) as

$$(3.11)\qquad \begin{aligned} &u(y; t + \Delta t, x) - u(y; t, x) \\ &\qquad = b(\Delta t; y)u_{yy}(y; t, x) + a(\Delta t; y)u_y(y; t, x) + o(\Delta t). \end{aligned}$$

Hence, as a function of t and y, *the transition probability $u(y; t, x)$ satisfies the first (or backward) diffusion equation*

$$(3.12)\qquad z_t(t, y) = b(y)z_{yy}(t, y) + a(y)z_y(t, y).$$

The obvious initial conditions on transition probabilities show that $u(y; t, x)$ is, in fact, a fundamental solution of (3.12). One should therefore expect that for fixed y the function $u(y; t, x)$ should satisfy the adjoint equation

$$(3.13)\qquad w_t(t, x) = \{b(x)w(t, x)\}_{xx} - \{a(x)w(t, x)\}_x ,$$

known as *second (or forward) diffusion equation*, and also as *Fokker-Planck equation.*[6]

We shall see (in §5) that this surmise is true in general, but that there exist cases where $u(y; t, x)$ satisfies only the first diffusion equation; instead of (3.13) we get in these cases an inequality, and a positive operator is added to the right side. This phenomenon is connected with the existence of a new type of solutions which has not been encountered before. It changes our concept of adjoint operators and throws new light on the theory of semigroups in general.

[5] It is possible to relax this condition so as to take account of more general situations.

[6] It is possible to introduce additional terms in (3.12) and (3.13) which will account for the possibility of particles disappearing or being created.

4. Singular equations. The theory of the preceding section applies also if the infinite x-axis is replaced by a finite or semi-infinite interval, except that now boundary conditions have to be specified. As has been mentioned before, the derivation of the familiar boundary conditions (such as absorbing, reflecting, and elastic barriers) is essentially of a heuristic nature, and it would be desirable to derive them from the basic assumptions in the same way as the diffusion equation itself has been obtained. We want now to discuss some new aspects of the boundary problems for a type of equations which is of both theoretical and practical interest.

In physical applications the coefficient $b(y)$ is essentially positive. Now modern diffusion theory has found a new and interesting field of applications in biology, in particular in the mathematical theory of evolution (cf. [6]). Here the particle under diffusion is to be interpreted symbolically and stands for a population size, gene frequency, etc. For example, the frequency of a particular gene in a population is a random variable $X(t)$ which by definition is restricted to the interval $0 < x < 1$, and Sewall Wright's theory of evolution assumes essentially that $X(t)$ is subject to a diffusion process with

$$b(x) = \beta x(1 - x), \qquad a(x) = -\gamma x + \delta(1 - x) \tag{4.1}$$

where β, γ, δ are positive constants. Similarly, in the simplest growth process $X(t)$ stands for a population size and can assume all positive values; the underlying stochastic process is described by our diffusion equations with

$$b(x) = \beta x, \qquad a(x) = \alpha x + \gamma. \tag{4.2}$$

The fact that $b(x)$ vanishes on the boundaries represents a singularity and has important and curious consequences.

Consider, for example, the diffusion equations with the coefficients (4.2), which are discussed in [7]. If $\gamma \leq 0$, then there exists a unique solution so that no boundary conditions can be imposed. For this solution the probability

$$\int_0^\infty u(y; t, x)\, dx = \pi(t, y) \tag{4.3}$$

decreases steadily in time. The difference $1 - \pi(t, y)$ represents the probability that a population of initial size y dies out before time t. (This is the absorption or extinction probability.)

The situation changes radically when $0 < \gamma < \beta$, and resembles the more familiar circumstances of ordinary diffusion. There exist infinitely many solutions, among which there is one for which $\pi(t, y) \equiv 1$; this corresponds to reflecting barriers. For all other solutions $\pi(t, y)$ decreases, and the solution for which the rate of decrease is greatest obviously corresponds to absorbing barriers. It is the only solution for which $u(y; t, 0) < \infty$, so that in this particular case an absorbing barrier is described by the boundary condition $u < \infty$.

For $\gamma > \beta$ the situation changes once more. In passing to the limit $\gamma \to \beta-$, the absorbing and the reflecting barrier solutions become identical, and for

$\gamma > \beta$ we are once more in the situation that no boundary conditions can be imposed and that our problem is completely determined by the differential equation itself.[7]

Accordingly, we can say that for $\gamma \leqq 0$ and $\gamma > \beta$ the origin acts as *natural boundary* where no artificial conditions can be imposed, while for $0 < \gamma < \beta$ we are confronted with conditions similar to those in physical diffusion theory.

It is possible (cf. [8]) to give criteria to determine whether for a particular singular equation the boundary acts as a natural one or whether there exist infinitely many solutions, but we are still unable to formulate the appropriate boundary conditions in all cases. Moreover, the biological applications lead us to a new type of boundary problems.

The classical concept of an elastic barrier at $x = 0$ may be interpreted by saying that whenever the particle reaches $x = 0$ it has a probability p to be absorbed and probability $q = 1 - p$ to be reflected. The limiting cases $p = 0$ and $p = 1$ represent reflecting and absorbing barriers, respectively. Now biological problems compel us to consider the more general case where the particle is only temporarily absorbed; that is, if the particle is absorbed, it remains fixed at the boundary for a finite time, and reverts then to the diffusion process. The sojourn time at the boundary is a random variable with an exponential distribution. In the familiar terminology of a probability mass spread over the x-axis this means that in addition to the solution $u(t, x)$ we have to consider a finite mass $m(t)$ concentrated at the origin. This mass flows at a rate proportional to $m(t)$ back into the interval $x > 0$, but it is partly replaced by new mass being absorbed at the origin. If the diffusion starts with all the initial mass distributed over $x > 0$, then $m(0) = 0$, and one would expect that $m(t)$ will increase to a certain saturation value.

Unfortunately the appropriate boundary conditions have not been formulated and the precise conditions on $a(y)$ and $b(y)$ under which such a process can take place are not known. For an interesting special case which is of some importance in mathematical genetics cf. [6].

5. Existence problems. Pathological solutions. We now return to the case of an infinite interval and the two diffusion equations (3.12) and (3.13). As was mentioned at the end of §3, the whole setup as well as the analogy with the classical diffusion equation $u_y = u_{xx}$ leads one to assume that the transition probability $u(y; t, x)$ will satisfy the two equations (3.12) and (3.13) and, conversely, that each of these equations should have a fundamental solution which is essentially uniquely determined and can serve as $u(y; t, x)$. However, already the first attempt to prove this conjecture showed that it cannot be true under

[7] However, for $\gamma > \beta$ there exist solutions for which $\pi(t, y)$ increases. These solutions have no probability significance and correspond to the solutions of the theory of heat with heat flowing into the medium. The existence of analogous solutions for the whole axis was discovered by Hille [13]; his "explosive" solutions are of this kind with $\pi(t, y) \to \infty$ as $t \to t_0 > 0$.

all circumstances. In fact, it was shown in [5] that whenever one of the two integrals

$$(5.1) \qquad \int^{\infty} a^{-1/2}(x)\, dx, \qquad \int_{-\infty} a^{-1/2}(x)\, dx$$

converges, an appropriate $b(x)$ can be found such that (3.12) admits of a solution which is non-negative, integrable, and such that $w(0, x) = 0$. Thus in this case the equation (3.12) does not suffice to describe our process. However, assuming that both integrals (5.1) diverge, the construction of [5] shows that our conjecture is valid under very mild regularity restrictions on $a(x)$ and $b(x)$. In other words, in case of divergence of (5.1) the whole classical theory carries over to our diffusion equations.[8] The problem was taken up by Yosida [26] and Hille [11, 12] who obtained similar results under slightly different conditions. In particular, Hille showed that (3.13) has a unique, non-negative solution $u(t, x)$ satisfying the conditions

$$(5.2) \qquad u(0, x) = f(x) \geqq 0, \qquad \int_{-\infty}^{+\infty} u(t, x)\, dx = \int_{-\infty}^{+\infty} f(x)\, dx$$

provided that

$$(5.3) \qquad | a'(x) - b(x) | \leqq K\{ | x | + 1\}$$

and that the integral

$$(5.4) \qquad \int \frac{x}{a(x)}\, dx$$

diverges both at $+\infty$ and $-\infty$; conversely, if (5.3) holds, the divergence of (5.4) is a necessary condition for the theorem.[9]

These unexpected results were at first rather disturbing since they seemed to indicate that the diffusion equations are, after all, an inadequate description of the actual process. Fortunately a satisfactory explanation can be found which makes the theory more harmonious than it would be if our diffusion equations always behaved as the classical equation $u_t = u_{xx}$. At the same time we shall be led to a new type of solution of the backward equation and to an interesting phenomenon concerning the forward equation.

The singular diffusion equations discussed in the preceding section point in

[8] In [5] the more general case is treated where the coefficients a and b depend on t. The later investigations depend on the theory of semigroups and apply therefore only to the temporally homogeneous case treated in the text.

[9] The apparent discrepancy between Hille's conditions and Feller's condition (5.1) is readily explained. If one of the integrals in (5.1) converges, then it is impossible that $| a'(x) | < K (| x | + 1)$, and hence (5.3) can hold only if the growth of $b(x)$ is adjusted in a special way to that of $a(x)$. On the other hand, if $| a'(x) | < K (| x | + 1)$, then $a(x) = O(x^2)$ and the divergence of (5.1) implies that of (5.4). [*Added in proof.* Unfortunately the text takes account only of Hille's paper [12] in which (3.13) is treated. The author is indebted to Hille for the manuscript and for many inspiring discussions. Subsequently Hille treated also (3.12) and obtained conclusive results concerning the uniqueness problems for the two equations; they are announced in [11].]

the proper direction. The first point to be noted is that there is no essential difference between diffusion in a finite or infinite interval. In fact, let $\gamma(s)$ be an arbitrary positive function and consider the transformation of variables

$$\xi = \int_0^x \gamma(s)\,ds, \qquad \eta = \int_0^y \gamma(s)\,ds. \tag{5.5}$$

If $u(y; t, x)$ satisfies the diffusion equations (3.12) and (3.13), then $u^*(\eta; t, \xi)/\gamma(\xi)$ (as a function of t, η and of t, ξ, respectively) satisfies similar equations with $a(y)$ and $b(y)$ replaced by $a(y)\cdot\gamma^2(y)$ and $a(y)\cdot\gamma'(y) + b(y)\cdot\gamma(y)$. Moreover, if $u(y; t, x)$ is, as a function of x, a probability density, then the same is true of u^* as a function of ξ. Thus we have a whole group of transformations which change the pair of diffusion equations into equivalent equations. We can make use of this fact either to simplify the equations or to change the interval. In particular, it is always possible to transform a finite interval into the entire real axis and vice versa. However, even if the coefficients are regular in the interior, the transformed equations will in general be of a singular type, with the coefficients unbounded at the endpoints, or $a(x)$ vanishing at an endpoint.

In particular, we may take the ordinary diffusion equation $u_t = u_{xx}$ for the finite interval $-1 < x < 1$, and transform the latter into $-\infty < \xi < \infty$. We obtain in this way a pair of diffusion equations for the infinite interval which exhibits all the anomalies which at first appeared so perplexing. There is no uniqueness, since we have the transforms of the classical absorbing and reflecting barrier solutions, etc. Furthermore, there exist positive solutions (well known from the theory of heat conduction) with zero initial values and so on.

In probabilistic language we can describe the situation as follows. The coefficients a and b can be such that with probability one the particle never reaches a boundary. In this case no boundary conditions need (or can) be imposed, and we have the phenomenon of *natural boundaries*. The typical example is the homogeneous diffusion on the entire axis, since Wiener's result that almost all path functions are continuous implies that the particle cannot reach infinity.

If the coefficients a and b are such that the particle has positive probability to reach a boundary, several contingencies may arise. For example, the drift toward the barrier can be so strong that the boundary automatically acts as an absorbing barrier, and again no boundary conditions can be imposed. A typical example is the origin in the case of the singular equation with coefficients given by (4.2) when $\gamma \leqq 0$ (cf. §4). Alternatively, the situation may be as with homogeneous diffusion in a finite interval, where various boundary conditions can be imposed.

In particular, it is possible that there exists a unique transition probability $u(y; t, x)$ which satisfies the two equations (3.12) and (3.13), but for which the total probability mass $\pi(t, y)$ (defined in (4.3)) is steadily decreasing. Here the boundaries are natural in the sense that no boundary conditions in the classical sense can be imposed. Nevertheless, we are led to a new type of solutions at least for the backward equation (3.12) and a curious phenomenon regarding the forward equation.

Following a remark of Doob[10] we can modify our diffusion process by stipulating that whenever the particle reaches a boundary, it is instantaneously transferred back to the interior. More precisely, let $\phi_i(x)$ $(i = 1, 2)$ be two density functions and let us assume that if the particle reaches the right or left boundary, it jumps to a position Y_1 or Y_2, where the Y_i are random variables with probability densities $\phi_i(x)$. It is easy to calculate the transition probability $u^*(y; t, x)$ of the so modified process, but for our purposes the explicit form is of no particular interest. It is rather obvious that $u^*(y; t, x)$ behaves asymptotically for $t \to 0$ as the transition probability $u(y; t, x)$ of the original process, so that according to the theory of §3 also $u^*(y; t, x)$ must satisfy the *backward equation* (3.12). It follows that $u^* - u$ is a non-negative solution of (3.12) with identically vanishing initial values, so that in this case *the initial value problem for* (3.12) *cannot be determined.* The interesting fact is that our phenomenon can occur even when the initial value problem for (3.13) is uniquely determined. In this case $u^*(y; t, x)$ cannot satisfy (3.13) and we have thus a diffusion process where *the transition probability satisfies the backward equation but not the forward equation.* Instead, the inequality sign $\geqq$ holds in (3.13) and the right side is to be modified by adding a positive operator.[11] We are thus led to the conclusion that the adjoint of the backward equation (3.12) is not always given by (3.13), and is in general not even a differential equation. This is a new phenomenon which seems of interest for the general theory of semigroups and of differential operators.

6. Ito's approach. Before passing to more general processes a word should be said on a new way of describing our diffusion processes which is due to Ito [14, 15]. He does not make direct use of the differential equations or the transition probabilities, but expresses (at least in certain cases) the path functions of a general process by means of those of the Wiener process.

Let $X(t)$ represent the path functions of the latter. We have seen that with probability one $X(t)$ is not of bounded variation so that the classical definition of the integral

$$\int_0^t f(\tau)\, dX(\tau) \tag{6.1}$$

[10] Doob discussed in detail a similar phenomenon for discrete Markov chains, cf. [2].

[11] For details cf. [8]. In the case of discrete Markov chains one is led to two infinite systems of ordinary differential equations playing the role of our diffusion equations. As Doob has shown, the analogue of (3.12) holds always, but the analogue of (3.13) holds only with the equality replaced by the sign $\geqq$. We now see why a direct derivation of (3.13) is impossible. In attempting a derivation along the lines which led to (3.12) one would introduce the function $F(t, y) = \int_{-\infty}^{+\infty} u(y; t, x)f(x)\, dx$ where $f(x)$ vanishes for $|x| > A$ and has, say, three continuous derivatives. Then (*) $F(t+\Delta t, y) = \int_{-\infty}^{+\infty} u(y; \Delta t, z)F(t, z)\, dz \geqq \int_{-\infty}^{+\infty} u(y; t, z)\, dz \cdot \int_{|x-z|<\epsilon} f(x)u(z; \Delta t, x)\, dx$. If one had here the equality sign, it would be easy to derive (3.13) under fairly mild conditions (essentially supposing that the limits in (3.4) and (3.5) are attained boundedly in every finite interval). However, in order to replace the inequality sign in (*) by the equality sign one has to introduce strong uniformity conditions which are actually not justified even in relatively simple cases.

breaks down, even if $f(t)$ is continuous: in fact, for almost every $X(t)$ we can find a sequence of subdivisions $t_k^{(n)}$ such that the corresponding Riemann sums will diverge. Nevertheless, it is possible to give a meaning to the integral (6.1). In fact, consider a fixed sequence of subdivisions, say $t_k^{(n)} = k/2^n$. The corresponding Riemann sum

$$\sum_k f(t_k^{(n)}) \Delta X(t_k^{(n)}) \tag{6.2}$$

is a random variable, and it has been shown by various authors that it converges in probability to a random variable which has all the desired properties of an integral.

Using this notion Ito shows that under certain restrictions on a and b the stochastic integral equation

$$Y(t) = c + \int_0^t b(Y(\tau))\,d\tau + \int_0^t a(Y(\tau))\,dX(\tau) \tag{6.3}$$

admits of an essentially unique solution $Y(t)$. The latter is defined as a random variable on the Wiener space $\{X(t)\}$. It is fairly obvious from (6.3) that the infinitesimal transition probabilities of $Y(t)$ have the desired properties (3.4) and (3.8), and thus we have arrived at a representation of a stochastic process of the required type.

This approach has the advantage that it permits a direct study of the properties of the path functions $Y(t)$, such as their continuity, etc. In principle, we have here a possibility of proving existence theorems for the partial differential equations (3.12)–(3.13) directly from the properties of the path functions. However, the method is for the time being restricted to the infinite interval and the conditions on a and b are such as to guarantee the uniqueness of the solution. So far, therefore, we cannot obtain any new information concerning the "pathological" cases.

7. Diffusion in phase space. The method of §3 works also in two and more dimensions. In two dimensions the transition probability is of the form $u(y_1, y_2; t, x_1, x_2)$, and the equations corresponding to (3.12) and (3.13) are

$$u_t = \sum a_{ij} u_{y_i y_j} - \sum b_i u_{y_i}, \qquad i, j = 1, 2 \tag{7.1}$$

and

$$u_t = \sum (a_{ij} u)_{x_i x_j} + \sum (b_i u)_{x_i}, \qquad i, j = 1, 2 \tag{7.2}$$

where the coefficients may depend on the two space variables and the matrix (a_{ij}) is symmetric and positive definite. The meaning of the coefficients a_{ii} and b_i is the same as before, and the mixed coefficient $a_{1,2}$ gives the infinitesimal covariance:

$$\begin{aligned} &a_{12}(y_1, y_2) \\ &= \lim_{\Delta t \to 0} \frac{1}{\Delta t} \iint_{|x_i - y_i| < \epsilon} (x_1 - y_1)(x_2 - y_2) u(y_1, y_2; \Delta t, x_1, x_2)\, dx_1\, dx_2. \end{aligned} \tag{7.3}$$

These equations were first derived by Kolmogorov [17]. Feller's constructions of the transition probability was generalized to two dimensions by Dressel [4]. In the time homogeneous case the equations were integrated by Yosida [26] using the theory of semigroups. However, nothing is known for the case of singular equations or the case where no uniqueness exists, and practically nothing is known about the appropriate boundary conditions in various problems involving bounded domains.

We shall not pursue this line of investigation but shall be content to indicate how our equations lead to a refined model of diffusion in which the path functions have derivatives.

The fact that the original model of diffusion leads to the conclusion that the particles have no velocities has been pointed out by many authors. Actually this fact is neither perturbing nor surprising: the whole theory is based on the assumption that the process is Markovian, that is, that the particle has no memory. Now if the particle had a finite velocity, shocks in the direction of motion would be less probable, so that any change of velocity would affect the chances of further changes. In other words, finite velocities would imply that the position of the particle is a random variable of a stochastic process with aftereffects.

In a Section meeting at this Congress two proposals were made to modify the basic assumptions so as to endow the particles with finite velocities. In both cases the diffusion equations would be replaced by equations of the hyperbolic type. Now the general solution of the initial value problem for such equations does not preserve positivity or the integral mean. Moreover, it depends not only on the initial value but also on their derivatives, which have no apparent probabilistic meaning. In short, the hyperbolic equations in question cannot serve as appropriate descriptions of a stochastic process, and if a particular transition probability is a solution, this is a lucky coincidence.

However, a refined model of diffusion in which the particles have finite velocities (but no accelerations) is due to Ornstein and Uhlenbeck.[12] This theory assumes that the particle has a velocity $X_2(t)$ which is the subject of an ordinary diffusion process regulated by (2.1). The position $X_1(t)$ of the particle is then obtained from

$$X_1(t) = \int_0^t X_2(\tau)\, d\tau, \tag{7.4}$$

the integral having a meaning since $X_2(t)$ is continuous with probability one. As shown by Doob, $X_1(t)$ is the variable of a Gaussian (non-Markovian) process whose probability relations approach for large t those of the ordinary Brownian motion process.

An alternative way of reaching the same conclusion consists in studying the vector $(X_1(t), X_2(t))$ as the variable of a two-dimensional diffusion process (so

[12] Cf. [23]. A thorough discussion of this process which is more appropriate for our purposes is contained in Doob [3].

that the plane is really the phase space of a particle in one-dimensional motion). Under the above assumptions this two-dimensional process is Markovian, and the corresponding transition probability $u(y_1, y_2; t, x_1, x_2)$ should satisfy a pair of equations of the form (7.1), (7.2). Now according to Ornstein and Uhlenbeck the variable $X_2(t)$ is subject to a symmetric homogeneous diffusion process, so that $b_2 = 0$ and a_{22} is a constant which we can assume as unity. From (7.4) it follows that if at any time $X_1(t) = y_1$, $X_2(t) = y_2$, then

$$(7.5)\qquad \begin{aligned} E(X_1(t+\Delta t) - X_1(t)) &= y_2\Delta t + o(\Delta t) \\ E((X_1(t+\Delta t) - X_1(t))^2) &= o(\Delta t), \end{aligned}$$

and hence $a_{11} = 0$, $b_1 = y_2$. From the second equation (7.5) and Schwarz' inequality it follows that also $a_{12} = 0$. Hence (7.1) reduces to

$$(7.6)\qquad u_t = u_{y_2y_2} - y_2u_{y_1}.$$

This equation has first been given by Kolmogorov [19]. If the condition of homogeneity is dropped, the same consideration leads to the more general equation

$$(7.7)\qquad u_t = b(y_1, y_2)u_{y_2y_2} + a(y_1, y_2)u_{y_2} - y_2u_{y_1}.$$

This equation is of a degenerate type and practically nothing is known about the appropriate boundary conditions for finite domains. A fundamental solution for the infinite plane was constructed by M. Weber [24].

To prove the equivalence of the two approaches one should prove that the solution of (7.7) satisfies the condition (7.4). This is intuitively rather obvious, but a satisfactory proof has not been given. Otherwise the same method could be applied to the calculation of various Wiener functionals, which were the object of investigation by Cameron and Martin, and by Kac (cf. [16]). Let again $X_2(t)$ be the random variable of the simplest diffusion process, and put

$$(7.8)\qquad X_1(t) = \int_0^t V(X_2(\tau))\,d\tau,$$

where $V(x)$ is a given function, say x^2. Then the pair (X_1, X_2) is the variable of a two-dimensional diffusion process, and should satisfy an equation of the form (7.6) with the coefficient y_2 replaced by $V(y_2)$. The difficulty of this approach (as well as in Ito's approach) seems to lie in a direct verification that the transition probability of the pair (X_1, X_2) has all the properties assumed in the derivation of (7.6). Formally, however, one can start with our diffusion equation and derive Kac's results from it.

In the Ornstein-Uhlenbeck model of diffusion the position $X_1(t)$ is not the variable of a Markovian process, but the process becomes Markovian when the state of the particle is defined so as to include position and velocity. Thus the same physical process can be described as a Markovian or non-Markovian process depending on the parameters used for its description. For example, in cosmic ray showers the most interesting variable is the total number of particles, but it is not the variable of a Markov process. However, we may use the conceptual simplicity of and the tools available for Markov processes at the expense of describing the state of the system by the positions, masses, and velocities of each particle. In this way we are led to a Markov process in a rather complicated phase space. If the process is continuous (which excludes instantaneous changes like splitting of particles), we shall again be led to partial differential equations of the diffusion type.

The most radical step in this direction was taken by Feynman [9] in his new approach to quantum physics. The starting point of Feynman's theory is the remark that the phase-time space can always be split into two parts involving the "past" and "future" in such a way that the physical process becomes what we would call a stochastic process of the Markov type. As a physicist Feynman does not mind generalizations and abstractions which would make mere mathematicians shudder, and thus his "past" is permitted even to include future events (thus introducing advance effects instead of aftereffects). Feynman then introduces an assumption equivalent to continuity and shows heuristically that the diffusion equation to which one is led is essentially the Schrödinger equation. It has complex coefficients since Feynman deals with complex probability amplitudes rather than with classical transition probabilities. It is not clear at present whether this is only an excellent analytical tool or whether we have here an essential generalization of the classical concept of a stochastic process.

8. Generalizations by means of M. Riesz' potentials. Usually partial differential equations appear to play a special role among functional equations, but from a probabilistic point of view the diffusion processes are only a particular type of a Markov process. Starting from the Chapman-Kolmogorov equation (3.2) we have derived a diffusion equation by imposing certain conditions on the asymptotic behavior of the transition probability $u(y; t, x)$ as $t \to 0$. Probability considerations suggest to us also more general conditions which then lead to a generalization of the diffusion equation in which certain integrals involving u are added to the right-hand member (cf. [5] and Yosida [27]). However, even the integro-differential equations thus obtained describe only a special class of Markov processes and represent from a certain point of view an unnatural type of functional equations. Our real problem is *to find a linear functional equation equivalent to the Chapman-Kolmogorov equation* (3.2). In the language of the theory of semigroups this amounts to saying that we desire to find *the infinitesimal generators to the most general semigroup* corresponding to (3.2).

Before proceeding it should be emphasized that the Chapman-Kolmogorov equation (3.2) is by no means restricted to probability problems, but that an analogous relation holds for all linear differential equations. In fact, suppose that the solution $u(t, x)$ of an equation (whose coefficients do not depend on t) is determined by its initial values. Then $u(t, x)$ will be given by an integral of the form (2.3), and the Chapman-Kolmogorov equation merely expresses the fact that the solution $u(s + t, x)$ can be calculated either directly in terms of its initial values at time 0 or, alternatively, in terms of its values at time t which, in turn, are expressible in terms of the initial values. Equating the two expressions we obtain (3.2). It is clear that a similar reasoning holds, for example, for a hyperbolic differential equation except that a vector notation must be used since now the problem is determined by the initial values of the solution and its time derivative.

In short, a kernel $u(y; t, x)$ induces two associated linear transformations

$$T_t f(x) = \int f(y)u(y; t, x)\, dy, \tag{8.1}$$

$$S_t g(y) = \int g(x)u(y; t, x)\, dx$$

and the Chapman-Kolmogorov equation (3.2) merely expresses the fact that as t varies each of these transformations forms a semigroup. In operator theoretical language the equation $u_y = u_{xx}$ can be interpreted as stating that the infinitesimal generator (time derivative) of T_t is given by the operator d^2/dx^2, or symbolically

$$\frac{d}{dt} T_t = T \frac{d^2}{dx^2}. \tag{8.2}$$

A formal integration leads to the symbolic equation

$$T_t = e^{td^2/dx^2}. \tag{8.3}$$

If $f(x)$ is an entire function, then (8.3) gives

$$T_t f(x) = e^{td^2/dx^2} f(x) = \sum_{k=0}^{\infty} t^k/k!\, f^{(2k)}(x), \tag{8.4}$$

which agrees with the classical solution (2.2) and (2.3). The more general diffusion equations merely replace the operator d^2/dx^2 on the right by other differential operators. In the case of the integro-differential equations mentioned above we have on the right a combination of a differential and integral operator. However, for a general solution we shall have to consider much more general operators.

The stable distributions are the best known class of solutions of the Chapman-Kolmogorov equation which are not connected with either differential or integro-differential equations. Bochner [1] was the first to notice that the symmetric stable distributions can be interpreted as solutions of an operator equation

$$u_t = Au,$$

where the operator A is a certain fractional power of d^2/dx^2. A better understanding of this important observation may be derived from an application of a technique developed by M. Riesz [22] which also points to various generalizations.

A *fractional potential of order* α is defined, according to Riesz, by

$$I^\alpha f(x) = \frac{1}{2\Gamma(\alpha)\cos \pi\alpha/2}\int_{-\infty}^{+\infty} f(y)\,|\,y - x\,|^{\alpha-1}\,dy. \tag{8.5}$$

This integral diverges in general, but Riesz showed that a meaning can be attached to it for a large class of functions $f(x)$. In particular, $I^0 f(x) = f(x)$, and $I^{-2n}f(x) = (-1)^n f^{(2n)}(x)$, whenever n is a positive integer. This relation is interesting for two reasons. First, it is clear that in general I^α is a global operator, that is, $I^\alpha f(x)$ depends on all values of $f(x)$. For the particular values $\alpha = 0, -2, -4, \cdots$, however, the operator takes on a purely local character. Secondly, since I^{-2} is the same as $-d^2/dx^2$, we can interpret I^{-1} as a square root of $-d^2/dx^2$, and similarly for other powers. This enables us to pursue in more detail Bochner's remark. We are thus led to consider the *general functional equation*

$$u_t = -I^{-\alpha}u. \tag{8.6}$$

If $\alpha = 2$, this reduces to the classical diffusion equation $u_t = u_{xx}$, but for other values of α the functional equation (8.6) has quite different a character, since the operator on the right is of a global character.

From Bochner's remark one should expect that for $0 < \alpha \leqq 1$ the equation (8.6) should lead to *symmetric stable distributions*. This is so, and we proceed to verify it in a purely formal fashion for $\alpha = 1$.

The symmetric stable distribution of order $\alpha = 1$ is characterized by the density

$$u(y;\, t,\, x) = \frac{t}{(x - y)^2 + t^2}, \tag{8.7}$$

which is known as *Cauchy distribution*. It plays for the Cauchy process the role of the normal density (2.2), and the solution (2.3) reduces in the present case to the function $u(t, x)$ *which is harmonic in the half plane* $t > 0$, *vanishes at infinity, and reduces for* $t = 0$ *to* $f(x)$.

If we integrate (8.6) for $\alpha = 1$ formally by analogy with (8.4), we are led to an expansion

$$u(t, x) = \sum_{k=0}^{\infty} \frac{(-t)^k}{k!} I^{-k} f(x). \tag{8.8}$$

Substituting from the definition (8.5) it is easily verified that the right side reduces to the anticipated solution

$$u(t, x) = \frac{1}{\pi}\int_{-\infty}^{+\infty} f(y)\,\frac{t}{(y - x)^2 + t^2}\,dy. \tag{8.9}$$

The formal steps leading to this formula require justification. However, by similar calculations all symmetric stable distributions can be obtained, and by a slight generalization of Riesz' definitions of potentials we can derive also the unsymmetric stable distribution. It appears, in fact, that all probabilistic solutions of the Chapman-Kolmogorov equation can be obtained in this way, except that the parameters of the potentials on the right side will depend on x.

It is of particular interest that we have obtained the solution of a *Dirichlet problem* for the Laplace equation $\Delta u = 0$ as the solution of a generalized diffusion problem. The explanation lies in the semigroup property of the Green function (8.7), and a similar statement applies also to the Dirichlet problem for closed curves. The curve itself corresponds to the x-axis, and certain closed curves in the interior to the lines t = const. These contract to a single point which corresponds to $t = \infty$. From the values of u along the boundary curve the values along the curves in the interior can be calculated by the methods just described. The fundamental fact that as $t \to \infty$ all the values tend to the same limit follows then from the ergodic principle for Markov processes on closed manifolds. We have thus a direct connection of the Dirichlet problem for elliptic equations with certain generalized diffusion equations, which in turn are closely connected with measure in function spaces where almost no function is continuous.

References

1. S. Bochner, *Diffusion equation and stochastic processes*, Proc. Nat. Acad. Sci. U.S.A. vol. 35 (1949) pp. 368–370.
2. J. L. Doob, *Markoff chains—Denumerable case*, Trans. Amer. Math. Soc. vol. 58 (1945) pp. 455–473.
3. ———, *The Brownian movement and stochastic equations*, Ann. of Math. vol. 43 (1942) pp. 351–369.
4. F. G. Dressel, *The fundamental solution of the parabolic differential equation*, Duke Math. J. vol. 7 (1940) pp. 186–203.
5. W. Feller, *Zur Theorie der Stochastischen Prozesse (Existenz und Eindeutigkeitssätze)*, Math. Ann. vol. 113 (1936) pp. 113–160.
6. ———, *Diffusion processes in the mathematical theory of Evolution*, Proceedings of the Second Symposium on Probability and Mathematical Statistics, Berkeley [to appear].
7. ———, *Two singular diffusion problems*, Ann. of Math. vol. 52 (1951), to appear.
8. ———, *The parabolic differential equations and the associated semi-groups of transformations*, to appear, with a companion paper, in Ann. of Math.
9. R. P. Feynman, *Space-time approach to non-relativistic quantum mechanics*, Reviews of Modern Physics vol. 20 (1948) pp. 367–387.
10. R. Fortet, *Les fonctions aléatoires du type de Markoff associées à certaines équations linéaires aux derivées partielles du type parabolique*, J. Math. Pures Appl. vol. 22 (1943) pp. 177–243.
11. E. Hille, *Les probabilités continues en chaîne*, C. R. Acad. Sci. Paris vol. 230 (1950) pp. 34–35.
12. ———, *On the integration problem for Fokker-Planck's equation in the theory of stochastic processes*, [To appear in the proceedings of the Tenth Congress of Scandinavian Mathematicians in Trondheim, 1950].
13. ———, *"Explosive" solutions of Fokker-Planck's equation*, Proceedings of the International Congress of Mathematicians, Cambridge, Mass., 1950.
14. Kiyosi Ito, *On a stochastic integral equation*, Proceedings of the Japan Academy of Sciences vol. 22 (1946) pp. 32–35.
15. ———, *On stochastic differential equations*, Memoirs of the American Mathematical Society, no. 4 (1951).
16. M. Kac, *On the distribution of certain Wiener functionals*, Trans. Amer. Math. Soc. vol. 65 (1949) pp. 1–13.

17. A. KOLMOGOROFF, *Ueber die analytischen Methoden in der Wahrscheinlichkeitsrechnung*, Math. Ann. vol. 104 (1931) pp. 415–458.

18. ———, *Zur Theorie der stetigen zufälligen Prozesse*, Math. Ann. vol. 108 (1933) pp. 149–160.

19. ———, *Zufällige Bewegungen*, Ann. of Math. vol. 35 (1934) pp. 116–117.

20. P. LÉVY, *Théorie de l'addition des variables aléatoires*, Paris, Gauthier-Villars, 1937.

21. ———, *Processus stochastiques et mouvement Brownien*, Paris, Gauthier-Villars, 1948.

22. M. RIESZ, *L'intégrale de Riemann-Liouville et le problème de Cauchy*, Acta Math. vol. 81 (1948) pp. 1–223.

23. G. E. UHLENBECK and L. S. ORNSTEIN, *On the theory of the Brownian motion*, Physical Reviews vol. 36 (1930) pp. 823–841.

24. M. WEBER, *The fundamental solution of the diffusion equation in phase space* [To appear].

25. N. WIENER, *Generalized harmonic analysis*, Acta Math. vol. 55 (1930) pp. 117–258.

26. K. YOSIDA, *An operator-theoretical treatment of temporally homogeneous Markoff process*, Journal of the Mathematical Society of Japan vol. 1 (1949) pp. 244–253.

27. ———, *An extension of Fokker-Planck's equation*, Proceedings of the Japan Academy of Sciences vol. 25 (1949) pp. 1–3.

28. ———, *On the differentiability and the representation of one-parameter semi-group of linear operators*, Journal of the Mathematical Society of Japan vol. 1 (1948) pp. 15–21.

PRINCETON UNIVERSITY,
PRINCETON, N. J., U. S. A.

ON POSITIVITY PRESERVING SEMIGROUPS OF TRANFORMATIONS ON $C[r_1, r_2]$.

By W. FELLER (Princeton)

1. Introduction. In the sequel $\mathfrak{C} = \mathfrak{C}[r_1, r_2]$ denotes the Banach space of functions $x = x(s)$ continuous in $-\infty \leq r_1 \leq s \leq \leq r_2 \leq \infty$ with the usual norm $\|x\| = \sup |x(s)|$. We speak of a ***semigroup*** of transformations $\{T_t\}$ if to each $t > 0$ there corresponds a linear transformation $T_t x$ defined for all $x \in \mathfrak{C}$ such that $T_t x \in \mathfrak{C}$ and $T_{t+\tau} = T_t T_\tau$. Moreover, we shall always tacitly assume that $T_t x$ is weakly measurable and that

$$\|T_t\| \leq 1. \tag{1.1}$$

Since for each real a the transformations $e^{at} T_t$ form a new semigroup it is obvious that (1.1) is not more restrictive than

$$\|T_t\| \leq e^{at} \quad \text{for some } a. \tag{1.2}$$

It is known (cf. Hille [4], p. 183) that the weak measurability (together with the separability of $\mathfrak{C}$) implies the *strong continuity property*

$$\|(T_{t+h} - T_t)x\| \to 0 \qquad h \downarrow 0 \tag{1.3}$$

for each $t > 0$ (but not necessarily for $t = 0$).

The semigroup $\{T_t\}$ is called *positivity preserving* if $x \geq 0$ implies $T_t x \geq 0$ for all $t > 0$.

The purpose of this paper is to give *a characterization of all positivity preserving semigroups on $\mathfrak{C}$ which are generated by an operator of a local character* in the sense of Definition 1. It will be seen that whenever a mild differentiability condition is satisfied, the infinitesimal generator Ω of the semi-

R.L. Schilling et al. (eds.), *Selected Papers II*,

group in its domain of existence coincides for almost all s with the differential operator (4.1). In section 5 this result is reformulated in a way as to provide *a characterization of the parabolic differential equation*

$$\frac{\partial u}{\partial t}=a(s)\frac{\partial^2 u}{\partial s^2}+b(s)\frac{\partial u}{\partial s}+c(s)u. \tag{1.4}$$

This is the basic equation of the theory of stochastic processes of the diffusion type and is essentially equivalent to the so-called *Fokker-Planck equation.* This point is discussed in section 7. It will be seen that our theorem leads to a *new derivation of the Fokker-Planck equation* under less stringent conditions.

The semigroup $\{T_t\}$ possesses an adjoint semigroup $\{T_t^*\}$ of transformations operating on the conjugate space to $\mathfrak{C}$, namely the space $\mathfrak{BB}[r_1,r_2]$ of functions of bounded variation in $[r_1,r_2]$. Our results therefore apply to semigroups in this space (cf. section 6).

2. The Infinitesimal Generator. As is well-known, the semigroup $\{T_t\}$ is determined by its infinitesimal generator Ω. This is an operator defined only on a set $\mathfrak{D}\subset\mathfrak{C}$ consisting of those $x\in\mathfrak{C}$ for which $h^{-1}(T_h-I)x$ approaches a limit

$$\Omega x=\lim_{h\downarrow 0}\frac{T_h-I}{h}x \qquad x\in\mathfrak{D}. \tag{2.1}$$

Here I is the identity operator and the limit is taken in the strong sense (uniform convergence). The set $\mathfrak{D}$ is dense in the closure of the range space of $T_t x$, $x\in\mathfrak{C}$.

Definition 1. *The infinitesimal generator will be said to be of a local character if for each point* $\xi\in[r_1,r_2]$ *and any* $x(s)\in\mathfrak{C}$ *which vanishes in some interval* $|s-\xi|<\delta$ *one has*

$$\left.\frac{T_h-I}{h}x(s)\right|_{s=\xi}\to 0. \tag{2.2}$$

(The left side is the value of the function $(T_h-I)x(s)h^{-1}$ at the point ξ).

Ann. Polon. Math. **25** (1952) 85–94

The fact that the infinitesimal generator Ω is of a local character does not guarantee that it is a differential operator[1]). To see this it suffices to consider a transformation of the independent variable $\sigma = f(s)$, where $f(s)$ is a strictly increasing but not differentiable function. Such a transformation does not affect the space $\mathfrak{C}$ nor the character of the semigroup. However, it changes a differential operator Ω in s into an operator of a different kind in σ.

To restrict the consideration to differential operators it is therefore necessary to introduce the following

Differentiability condition. *For each $x(s) \in \mathfrak{D}$ and each $s \in [r_1, r_2]$ the derivatives $x'(s)$ and $x''(s)$ exist.*

(It is not necessary to require that $x'(s) \in \mathfrak{C}$).

3. Singular Points. In this section we assume that the differentiability condition is satisfied. The set $\mathfrak{D}$ is usually so large that to each $\xi \in [r_1, r_2]$ and to any three real numbers p_0, p_1, p_2 it is possible to find an $x(s) \in \mathfrak{D}$ such that

$$x(\xi) = p_0, \quad x'(\xi) = p_1, \quad x''(\xi) = p_2. \tag{3.1}$$

If this is the case we shall say that ξ is a ***regular*** point.

Definition 2. *A point $\xi \in [r_1, r_2]$ will be called singular if there exists at least one linear relation*

$$q_0 x(\xi) + q_1 x'(\xi) + q_2 x''(\xi) = 0 \tag{3.2}$$

satisfied by each $x(s)$ in the domain $\mathfrak{D}$ of the infinitesimal generator.

Since $\mathfrak{D}$ is a linear set, each point is either regular or singular. The existence of singular points is proved, and their nature illustrated, by the following

Example. Let $-\infty < s < \infty$ and consider the semigroup

$$T_t x(s) = 2\pi^{-1/2} \int_{-\infty}^{+\infty} x(\xi) \cdot \exp\left\{- \frac{(s^3 - \xi^3)^2}{4t}\right\} \cdot 3\xi^2 \cdot d\xi. \tag{3.3}$$

[1]) If one considers semigroups of transformations in the space of essentially bounded functions, then the local character of the infinitesimal generator does not guarantee that continuous functions are taken into continuous functions.

Introducing the substitutions $s^3=y$, $\xi^3=\eta$ it is seen that the function $u(t,y)=T_t x(s)$ is the well-known solution of the initial value problem for the diffusion equation $u_t=u_{yy}$. Accordingly, $x(s)$ will belong to the domain $\mathfrak{D}$ of the infinitesimal generator of (3.3) if and only if

$$\frac{d^2x(y^{1/3})}{dy^2}\in\mathfrak{C}[-\infty,\infty]. \tag{3.4}$$

In this case we have for $s\to 0$

$$x(s)=A+Bs^3+Cs^6+o(s^6). \tag{3.5}$$

Hence $x(0)=x'(0)=x''(0)=0$ for each $x(s)\in\mathfrak{C}$, and $s=0$ is triply singular. For $s\neq 0$

$$\Omega x(s)=\frac{d^2x(y^{1/3})}{dy^2}=\frac{1}{9}s^{-4}\frac{d^2x(s)}{ds^2}-\frac{2}{9}s^{-5}\frac{dx(s)}{ds}, \tag{3.6}$$

so that the infinitesimal generator coincides with a differential operator of second order. However, for $s=0$

$$\Omega x(s)\Big|_{s=0}=2C=\frac{2}{6!}\frac{d^6x(s)}{ds^6}\Big|_{s=0} \tag{3.7}$$

which shows that at a singular point the infinitesimal generator may be of a form different from the one given in the following theorem.

4. Theorem. *Suppose that the infinitesimal generator Ω of the semigroup $\{T_t\}$ is of local character and that it satisfies the differentiability condition of section 2. Then at each regular point Ω coincides with a differential operator*

$$\Lambda=a(s)\frac{d^2}{ds^2}+b(s)\frac{d}{ds}+c(s), \tag{4.1}$$

where $a(s)\geq 0$, $c(s)\leq 0$.

Note. Even in the absence of singular points Ω is in general not identical with Λ but only a contraction of Λ.

Proof. Let ξ be a regular point and consider an $x_0(s)\in\mathfrak{D}$ such that

$$x_0(\xi)=x_0'(\xi)=0,\quad x_0''(\xi)=1. \tag{4.2}$$

Construct a function $\bar{x}_0(s) \in \mathfrak{C}$ such that $\bar{x}_0(s) \geq 0$ for all s and $\bar{x}_0(s) = x_0(s)$ in some interval $|s-\xi| < \delta$, Since $x_0(s) \in \mathfrak{D}$ and since Ω is of a local character, there exists a finite limit

$$a = \lim_{h \to 0} \frac{T_h - I}{h} x_0(s) \bigg|_{s=\xi} = \lim_{h \to 0} \frac{T_h - I}{h} \bar{x}_0(s) \bigg|_{s=\xi} = \lim_{h=0} h^{-1} T_h \bar{x}_0(s) \bigg|_{s=\xi}. \tag{4.3}$$

We have assumed that T_t is positivity preserving, and therefore $a \geq 0$.

Let $x(s)$ be any function in $\mathfrak{C}$ such that $x''(\xi)$ exists and

$$x(\xi) = x'(\xi) = 0, \quad x''(\xi) = \varrho > 0. \tag{4.4}$$

Then for ε sufficiently small there exists a neighbourhood of ξ such that in it

$$0 \leq (\varrho - \varepsilon) \bar{x}_0(s) \leq x(s) \leq (\varrho + \varepsilon) \bar{x}_0(s). \tag{4.5}$$

We now construct a function $\bar{x}(s) \in \mathfrak{C}$ which coincides with $x(s)$ in that neighbourhood but satisfies (4.5) for all s. Then

$$(\varrho - \varepsilon) T_t \bar{x}_0(s) \leq T_t \bar{x}(s) \leq (\varrho + \varepsilon) T_t \bar{x}_0(s) \tag{4.6}$$

for all $t > 0$. Now $\bar{x}_0(\xi) = \bar{x}(\xi) = 0$, and therefore (4.6) and (4.3) together imply that

$$\lim \frac{T_h - I}{h} \bar{x}(s) \bigg|_{s=\xi} = \varrho \lim h^{-1} T_h \bar{x}_0(s) = a\varrho. \tag{4.7}$$

Again, $\bar{x}(s)$ and $x(s)$ coincide in some interval about ξ and hence

$$\lim \frac{T_h - I}{h} x(s) \bigg|_{s=\xi} = a\varrho = ax''(\xi) \tag{4.8}$$

for every function $x(s)$ satisfying (4.4) (even if $x(s) \in \mathfrak{D}$).

The same argument goes through if $\varrho < 0$, and also if $\varrho = 0$ provided that $x(s)$ does not change sign in some neighbourhood of ξ. It $x''(\xi) = 0$ and $x(s)$ does change sign, one has to consider the function $x_+(s)$ which vanishes for $s \leq \xi$ and equals $x(s)$ for $s \geq \xi$, and separately $x_-(s) = x(s) - x_+(s)$. In this way, one establishes the validity of (4.8) for all functions for which $x''(\xi)$ exists and $x(\xi) = x'(\xi) = 0$.

Next consider an arbitrary $x(s) \in \mathfrak{C}$ for which $x''(\xi)$ exists. Define the functions $\Phi_\xi(s)$ and $\Psi_\xi(s)$ so that in some neighbourhood of

$$\begin{aligned} \Phi_\xi(s) &= x(\xi)+(s-\xi)x'(\xi) \\ \Psi_\xi(s) &= x(s)-\Phi_\xi(s), \end{aligned} \tag{4.9}$$

and that they are continuous and bounded. Then $\Psi_\xi(\xi)= = \Psi'_\xi(\xi)=0$ and hence

$$\frac{T_h-I}{h}\Psi_\xi(s)\Big|_{s=\xi} \to a x''(\xi). \tag{4.10}$$

Therefore for each $x(s) \in \mathfrak{D}$ the limit of

$$\frac{T_h-I}{h}\Phi_\xi(s)\Big|_{s=\xi} \tag{4.11}$$

exists. Now choose first $x(s)$ so that $x(\xi)=0$, $x'(\xi)=1$ and next so that $x(\xi)=1$ and $x'(\xi)=0$. Call the corresponding limits b and c. Then, in the general case, the expression (4.11) tends to $bx'(\xi)+cx(\xi)$.

That $c\leq 0$ follows from the fact that for $x(s)\equiv 1$ we have $0\leq T_h x\leq 1$ because of (1.1).

We have thus proved not only the theorem but also the

Corollary. *If ξ is a regular point then for any function $x(s) \in \mathfrak{C}$ for which $x''(\xi)$ exists*

$$\lim \frac{T_h-I}{h}x(s)\Big|_{s=\xi} = a(\xi)x''(\xi)+b(\xi)x'(\xi)+c(\xi)x(\xi)$$

exists (even if $x(s)$ does not belong to the domain $\mathfrak{D}$ of Ω).

5. Reformulation for Differential Equations. Let Λ be an arbitrary differential operator in the variable s and consider the partial differential equation

$$\frac{\partial u}{\partial t} = \Lambda u. \tag{5.1}$$

Suppose that the initial value problem has a unique solution in the sense that to each $x(s) \in \mathfrak{C}$ there exists one and only one solution $u(t,s)$ defined and bounded for each fixed $t>0$ and such that $u(t,s)\to x(s)$ uniformly as $t\to 0$. Then, putting $u(t,s)=T_t x(s)$ we get a semigroup in $\mathfrak{C}$.

If there is no uniqueness, lateral conditions must be imposed on the solution, and to each determining homogeneous lateral condition there corresponds a semigroup of transformations (cf. [2]). In each case the infinitesimal generator is a contraction of the operator Λ. Accordingly:

The solutions of the initial value problem of a partial differential equation of form (5.1) *(with or without homogeneous lateral conditions) will be positivity preserving and norm not increasing if and only if the right side is given by* (4.1).

6. Measure Preserving Transformations. The conjugate space to $\mathfrak{C}$ is isomorphic to the functions of bounded variation in $[r_1,r_2]$ or, what amounts to the same, to the Banach space of completely additive set functions in $[r_1,r_2]$. The adjoint transformations T_t^* to T_t transforms each such set function into another. If T_t is positivity preserving so is T_t^*, and if (1.1) holds then also $\|T_t^*\| \leq 1$. Finally, T_t^* transforms a probability measure into a probability measure if and only if T_t transforms the function $x(s) \equiv 1$ into itself. Needless to say that the transformations $\{T_t^*\}$ form a semigroup.

Our theorem is therefore equivalent to a theorem on semigroups of transformations of measures. The new semigroup T_t^* is generated by an operator which is adjoint to Ω. In differential equation theory it is customary to consider the following equation as adjoint to (5.1)

$$\frac{\partial v}{\partial t} = \frac{\partial^2}{\partial s^2}(a(s)v) - \frac{\partial}{\partial s}(b(s)v) + c(s)v. \tag{6.1}$$

The truth is more complicated inasmuch as the complete adjoint to Λ is not necessarily a differential operator and (6.1) must be replaced by a more general equation. A detailed discussion of the relation between (5.1) and (6.1) is given in [2] together with the complete adjoint to Ω. This paper contains also a construction of all possible lateral conditions determining semigroups T_t and T_t^*.

7. Diffusion Theory and the Fokker-Planck Equation. A stationary stochastic process of the Markov type can be defined abstractly as a semigroup ot transformations T_t^* taking probability measures into probability measures. It is customary to assume that this semigroup can be represented as a kernel transformation

$$(7.1) \qquad T_t^* m(A) = \int P(t, s, A) m(ds)$$

where for fixed $t > 0$, $s \in [r_1, r_2]$ the kernel represents a completely additive set function, and the integration extends over the whole interval. Probabilistically P represents the transition probability of finding the system at time t within the set A if it was originally at the point s. Such a transition probability must satisfy the Chapman-Kolmogorov equation which is an analytic formulation of the assertion that the transformations (7.1) form a semigroup.

Perhaps the best known and most interesting type of such processes is represented by diffusion processes. Kolmogorov [5] was first to treat them systematically, and his treatment has been improved in [1].

To derive the fundamental equations of diffusion theory it is necessary to assume that the set function $P(t,s,A)$ has a density function $p(t,s,y)$. It is the first aim of the theory to show that for fixed y the density $p(t,s,y)$ must satisfy an equation of form

$$(7.2) \qquad \frac{\partial p}{\partial t} = a(s) \frac{\partial^2 p}{\partial s^2} + b(s) \frac{\partial p}{\partial s}$$

and for fixed s

$$(7.3) \qquad \frac{\partial p}{\partial t} = \frac{\partial^2}{\partial y^2}(a(y)p) - \frac{\partial}{\partial y}(b(y)p).$$

These equations are special cases of (5.1) and (6.1) and are known as *Fokker-Planck equations* or as the *backward and forward equations of diffusion theory*. The second equation is really a consequence of the first.

Our theorem represents a new derivation of (7.2) *stating necessary and sufficient conditions* [2]). We do not require the existence of the transition density $p(t,s,y)$, and the differentiability condition of section 2 is much weaker than the regu-

[2]) The argument applies also to $c \leq 0$, but when $c < 0$ the measure of the whole space will decrease with increasing t.

larity restrictions previously used to derive (7.2). The local character of the infinitesimal generator replaces the so-called *Lindeberg condition* used in [1] which requires that for each $\varepsilon > 0$ and each fixed s

$$\lim_{t\to 0} t^{-1} \int_{|y-s|>\varepsilon} p(t,s,y)\,dy = 0, \tag{7.4}$$
$$\lim_{t\to 0} t^{-1} \int_{|y-s|<\varepsilon} p(t,s,y)\,dy = 1.$$

This conditions obviously implies the local character of the infinitesimal generator.

In additions to the Lindeberg condition one had to assume that the limits

$$\lim_{t\to 0} t^{-1} \int_{|y-s|<\varepsilon} (y-s)\,p(t,s,y)\,dy = b(s) \tag{7.5}$$

and

$$\lim_{t\to 0} t^{-1} \int_{|y-s|<\varepsilon} (y-s)^2 p(t,s,y)\,dy = 2a(s) \tag{7.6}$$

exist [3]). The corollary to our theorem shows that whenever a kernel $p(t,s,y)$ exists

$$t^{-1} \int_{r_1}^{r_2} \frac{y-s}{1+(y-s)^2}\, p(t,s,y)\,dy \to b(s) \tag{7.7}$$

and

$$\frac{1}{2} t^{-1} \int_{r_1}^{r_2} \frac{(y-s)^2}{1+(y-s)^2}\, p(t,s,y)\,dy \to a(s). \tag{7.8}$$

In view of the Lindeberg condition (7.4) these relations are obviously equivalent to (7.5) and (7.6). It should be noted, however, that the abstract formulation gives a meaning to these relations without assuming the existence of a kernel $p(t,s,y)$.

---

[3]) For a slightly weaker formulation cf. [3].

REFERENCES

[1] W. Feller, *Zur Theorie der stochastischen Prozesse (Existenz und Eindeutigkeitssätze)*, Math. Annalen **113** (1936), p. 113-160.

[2] W. Feller, *The parabolic differential equation and the associated semigroups of transformations*, Annals of Math. **55** (1952), p. 468-519.

[3] W. Feller, *Some recent trends in the mathematical theory of diffusion*, Proc. Internat. Congress Math., Cambridge 1950.

[4] E. Hille, *Functional Analysis and Semi-groups*, Am. Math. Soc. Colloquium Publ. **31**, New York 1948.

[5] A. Kolmogoroff, *Über die analytischen Methoden in der Wahrscheinlichkeitsrechnung*, Math. Annalen **104** (1931), p. 415-458.

Halifax, N. S.

August 1951.

Annals of Mathematics
Vol. 57, No. 2, March, 1953
Printed in U.S.A.

SEMI-GROUPS OF TRANSFORMATIONS IN GENERAL WEAK TOPOLOGIES

By William Feller[1]

(Received July 15, 1952)

1. Definitions and Summary

1.1. *Introduction.* Our point of departure is a real Banach space $\mathfrak{X}$ and a family of bounded linear operators $\{T_t\}$ from $\mathfrak{X}$ to $\mathfrak{X}$, $(0 < t < \infty)$, with the semi-group property $T_{t+s} = T_t T_s$.

Usually one assumes that $\{T_t\}$ is strongly measurable. From this one derives[2] the basic fact that $\| T_t \|$ is bounded in each closed subinterval of $(0, \infty)$, and then that $\{T_t\}$ is strongly continuous for $t > 0$.

Now let $\mathfrak{X}$ stand for the familiar space of signed measures (functions of bounded variation) on the real line. Then *the group of translations is neither strongly nor weakly measurable* (cf. the end of this section). Similarly, if $\mathfrak{X}$ is the space of bounded *continuous* functions $\mathfrak{C}(-\infty, +\infty)$ the translations are obviously not strongly continuous and therefore (cf. [3], p. 183) *not strongly measurable.*

We see that the assumption of strong measurability is a severe restriction. It turns out to be a particular handicap when one tries to formulate, in a general and rigorous way, the problem of integration and of lateral conditions for certain partial differential equations[3] (for preliminary results cf. [2]).

R. S. Phillips [4] has constructed an example to show that mere *weak* measurability does *not* imply the boundedness of $\| T_t \|$ in any interval and that it does not imply the measurability of $\| T_t \|$.

At first sight this observation would seem to imply that we require more stringent hypotheses than weak measurability. Actually we can do with considerably less. When talking about weak measurability one considers the real valued functions $x^*(T_t x)$, where $x \in \mathfrak{X}$ and x^* is a functional on $\mathfrak{X}$. Now when one studies real measurable functions it is natural to adopt the point of view of the Lebesgue theory and not to distinguish between equivalent functions.

It will be shown that this simple principle leads us surprisingly far. First of all, we have to abandon the ordinary norm $\| T_t \|$ in favor of what we shall call the *essential norm.* It always behaves well. We shall see that integrations can always be performed, that certain continuity properties hold, that there is

---

[1] Research connected with an Office of Ordnance Research Project at Princeton University, OOR Contract No. DA-36-034-ORD-640.

[2] This result is due to R. S. Phillips [4]. Hille ([3], p. 183) introduces the property as hypothesis.

[3] In the theory of stochastic processes we are concerned with semigroups of transformations of measures. The translations of measures are a degenerate simple special case of the general diffusion process. Thus the simplest applications to analysis compel one to eliminate the strong measurability requirement.

a strongly continuous center, and that the theory of the infinitesimal generator can be generalized.

For that we shall not require weak measurability, but the theory begins to apply if *there exists a single functional* x^* *such that* $x^*(T_t x)$ *is Lebesgue measurable for all* $x \in \mathfrak{X}$.

Before proceeding with the summary we introduce the basic concepts and notations.

1.2. *Weak* ($\mathfrak{Y}$) *Measurability*. As usual, $\mathfrak{X}^*$ shall denote the conjugate (or adjoint) space to $\mathfrak{X}$. The real number obtained by applying a functional $x^* \in \mathfrak{X}^*$ to $x \in \mathfrak{X}$ will be denoted by x^*x (instead of the usual $x^*(x)$), and referred to as the inner product.

For each t there is an adjoint of T_t operating from $\mathfrak{X}^*$ to $\mathfrak{X}^*$. It is usually denoted by T_t^*, but for our purposes it is preferable to follow the symbolism of matrix algebra. Accordingly, for each $x^* \in \mathfrak{X}^*$ and $t > 0$ *we define* x^*T_t *as the functional for which*

$$(1.1) \qquad (x^*T_t)x = x^*(T_t x).$$

The common value will be denoted by $x^*T_t\, x$: for fixed x and x^* it is a real function of t.

DEFINITION 1. *Let* $\mathfrak{Y}$ *be a subspace*[4] *of* $\mathfrak{X}^*$ *which is invariant under* $\{T_t\}$ *(that is,* $y \in \mathfrak{Y}$ *implies* $yT_t \in \mathfrak{Y}$*). The semi-group* $\{T_t\}$*, operating on* $\mathfrak{X}$*, will be called weakly* ($\mathfrak{Y}$) *measurable if for each* $x \in \mathfrak{X}$ *and* $y \in \mathfrak{Y}$ *the function* $yT_t x$ *is Lebesgue measurable.*

This definition allows for degrees of measurability. Ordinary weak measurability is ($\mathfrak{X}^*$) measurability. On the other extreme, if there exist *one* functional x_0^* such that $x_0^* T_t x$ is measurable for all x, then we can take for $\mathfrak{Y}$ *the subspace generated by the elements* $x_0^* T_t$. The advantage of this relativization will be seen on considering once more the translations of measures: this semi-group is not weakly measurable in the ordinary sense. However, one can choose for $\mathfrak{Y}$, say, the space $\mathfrak{C}(-\infty, +\infty)$ of bounded continuous functions and the group is ($\mathfrak{Y}$) measurable.

It goes without saying that the larger $\mathfrak{Y}$ is, the more information we can get from the relations between $\mathfrak{X}$ and $\mathfrak{Y}$. However, remarkably small spaces $\mathfrak{Y}$ will reveal the structure of the semi-group. One desideratum on the size of $\mathfrak{Y}$ follows from a symmetry consideration.

The relationship between $\mathfrak{X}$ and $\mathfrak{Y}$ is not quite symmetric inasmuch as, by the very definition,

$$(1.2) \qquad \| y \| = \sup_{x \in \mathfrak{X}} \frac{| yx |}{\| x \|}$$

whereas we can only assert that

$$(1.3) \qquad \| x \| \geqq \sup_{y \in \mathfrak{Y}} \frac{| yx |}{\| x \|}.$$

[4] The word "subspace" refers to a closed linear set.

This has the consequence that T_t may have two different norms, namely

$$(1.4)\qquad \| T_t \| = \sup_{x \in \mathfrak{X}} \frac{\| T_t x \|}{\| x \|} \qquad \text{and} \qquad \| T_t \|_{\mathfrak{Y}} = \sup_{y \in \mathfrak{Y}} \frac{\| y T_t \|}{\| y \|},$$

according as T_t is considered as operator in $\mathfrak{X}$ or in $\mathfrak{Y}$.

DEFINITION 2. *The spaces $\mathfrak{X}$ and $\mathfrak{Y} \subset \mathfrak{X}^*$ will be called reciprocal if for each $x \in \mathfrak{X}$*

$$(1.5)\qquad \| x \| = \sup_{y \in \mathfrak{Y}} \frac{| yx |}{\| y \|}.$$

Even though, for convenience, this definition is formulated in an unsymmetric way, it is clear that *the relationship between two reciprocal spaces is in every respect symmetrical.* In particular, $\mathfrak{X}$ admits of the canonical imbedding in the adjoint space $\mathfrak{Y}^*$. If $\mathfrak{X}$ and $\mathfrak{Y}$ are reciprocal and if $\{T_t\}$, is weakly ($\mathfrak{Y}$) measurable, then the same semi-group considered in $\mathfrak{Y}$ will be weakly ($\mathfrak{X}$) measurable. In this case it makes no difference whether we take $\mathfrak{X}$ or $\mathfrak{Y}$ as primary space. We have also

LEMMA 1.1. *If $\mathfrak{X}$ and $\mathfrak{Y}$ are reciprocal, then*

$$(1.6)\qquad \| T_t \|_y = \| T_t \|.$$

(whereas in general) one can only assert that

$$(1.7)\qquad \| T_t \|_y \leqq \| T_t \|.$$

For conciseness of exposition we defer the proof to Section 1.4.

The theory becomes more elegant if $\mathfrak{X}$ and $\mathfrak{Y}$ are reciprocal, and for some results this assumption is necessary. It is of some interest to note that it is always possible to *refine* the metric in $\mathfrak{X}$ (without altering the norms in $\mathfrak{Y}$), the inner products, or the semi-group) so as to achieve that $\mathfrak{X}$ and $\mathfrak{Y}$ become reciprocal. (Cf. Lemma 1.2.)

1.3. *Summary of results.* Our point of view is that with a given semi-group each functional $y \in \mathfrak{Y}$ becomes an operator mapping $\mathfrak{X}$ onto the collection of real functions $yT_t x$. If these functions are treated in the sense of Lebesgue, then the true norm $\| T_t \|$ ceases to be of interest. Instead, we introduce the *essential norm* $\| T_t \|_e$. It suffices here to describe this notion roughly.[5]

If $\mathfrak{X}$ and $\mathfrak{Y}$ are reciprocal, the true norm could be defined by

$$(1.8)\qquad \| T_t \| = \sup \frac{| yT_t x |}{\| y \| \cdot \| x \|}$$

for all $x \in \mathfrak{X}$, $y \in \mathfrak{Y}$. We get an idea of the notion of *essential norm* $\| T_t \|_e$ if we replace the numerator in (1.8) by the *essential* upper limit[6] of $| yT_{t+h} x |$ as

[5] The rigorous definition is given in Section 2. It appears less elegant than the present approach, but is better adapted for the proofs. Incidentally, the two definitions do not necessarily lead to the same numerical value *at each point t*.

[6] That is the smallest upper limit that can be obtained by neglecting a set of measure zero.

$h \downarrow 0$. (An essential norm is also defined if $\mathfrak{X}$ and $\mathfrak{Y}$ are not reciprocal.) It is, of course, not required in advance that $\| T_t \|_e$ be finite.

Section 2 is devoted to the basic theorem that for a ($\mathfrak{Y}$) measurable semi-group $\| T_t \|_e$ *is Lebesgue measurable and bounded in every closed subinterval* of $(0, \infty)$.

It may be noted that $\| T_t \|_e$ can vanish identically, but only if for each $x \in \mathfrak{X}$ and $y \in \mathfrak{Y}$ the function of $yT_t x$ vanishes for almost all t. (This is the case in the example of Phillips mentioned above). Since we wish to neglect sets of measure zero such a semi-group is from our point of view equivalent to annihilation.

Armed with the stated theorem, we are able fully to utilize the interplay between $\mathfrak{X}$ and $\mathfrak{Y}$. First (Section 3) *it is possible to define integrals*

$$\int_a^b (yT_t)\, dt \tag{1.9}$$

for $y \in \mathfrak{Y}$, $0 < a < b < \infty$. They are integrals in the sense of Dunford and Pettis and are functionals in $\mathfrak{X}^*$ (not necessarily in $\mathfrak{Y}$). We can now *enlarge* $\mathfrak{Y}$ *to a larger space* $\bar{\mathfrak{Y}} \subset \mathfrak{X}^*$ *which is closed under integration and* $\{T_t\}$ *will be* ($\bar{\mathfrak{Y}}$) *measurable*. Furthermore, the closure of the set of elements of the form (1.9) is *a space* $\mathfrak{Y}_1 \subset \bar{\mathfrak{Y}}$ in which $\{T_t\}$ is *strongly continuous* for $t > 0$. We call Y_1 the *center*.

If $\mathfrak{X}$ and $\mathfrak{Y}$ are reciprocal, the same applies to $\mathfrak{X}$. We can then without losing the weak measurability property *extend* $\{T_t\}$ *to two reciprocal spaces* $\bar{\mathfrak{X}}$ and $\bar{\mathfrak{Y}}$ *which are closed under integration and each of these spaces contains a subspace in which* $\{T_t\}$ *is strongly continuous for* $t > 0$ (not necessarily for $t = 0$).

The fact that the existence of a strongly continuous center can be demonstrated, instead of postulated, is of importance for the proper formulation of the Cauchy problem. For the group of translations of measures the strongly continuous center is formed by absolutely continuous measures.

From our point of view it is natural to identify two elements y_1 and y_2 if for each $x \in \mathfrak{X}$ one has $y_1T_t x = y_2T_t x$ almost everywhere. This leads (Section 4) to the introduction of *the space* $\mathfrak{Y}'$ *of equivalence classes*. In truth, the semi-group $\{T_t\}$ is, from the point of view of our weak topology, almost indistinguishable from *the semi-group introduced on* $\mathfrak{Y}'$. One would expect that $\| T_t \|_e$ coincides with the norm of this induced semi-group, but this is not necessarily so.

Properties of, and relations between, the various norms are studied in Section 4. It turns out that elegance and precise results can be obtained only if one adjoins to $\mathfrak{Y}$ all those functionals in $\mathfrak{X}^*$ which are equivalent to some $y \in \mathfrak{Y}$.

In Section 4 we also define *Banach limits* and show that our semi-groups have a *weak continuity property* expressible in terms of such limits.

It is well-known that in the theory of semi-groups difficulties arise from the fact that $\| T_t \|$ may be unbounded near $t = 0$. It is therefore of some interest that *by a refinement of the topology in* $\mathfrak{X}$ *one can always achieve that* $\{T_t\}$ *receives a true norm bounded in every finite interval*. Cf. the summary at the end of Section 5.

Finally, in Section 6 we show how the definition of *infinitesimal generator* and *resolvent* can be generalized. The latter may be an *unbounded* operator. This extension is designed for applications to partial differential equations.

The author is thankful to Solomon Leader and R. S. Phillips for reading the manuscript.

1.4. PROOF OF LEMMA 1.1. From

$$| yT_t x | \leqq \| y \| \cdot \| T_t \| \cdot \| x \| \tag{1.10}$$

we have

$$\| yT_t \| \leqq \| y \| \cdot \| T_t \|, \tag{1.11}$$

which proves (1.7). If $\mathfrak{X}$ and $\mathfrak{Y}$ are reciprocal, then to given x, t, and ε we can find a $y_0 \in Y$ such that

$$| y_0 T_t x | > \| y_0 \| \cdot \| T_t x \| \cdot (1 - \varepsilon). \tag{1.12}$$

Now choose x_0 so that

$$| T_t x_0 | > \| T_t \| \cdot \| x_0 \| \cdot (1 - \varepsilon) \tag{1.13}$$

and apply (1.12) to $x = x_0$. Then

$$| y_0 T_t x_0 | > \| y_0 \| \cdot \| T_t \| \cdot \| x_0 \| \cdot (1 - \varepsilon)^2 \tag{1.14}$$

while by definition

$$| y_0 T_t x_0 | \leqq \| y_0 T_t \| \cdot \| x_0 \| \leqq \| y_0 \| \cdot \| T_t \|_y \cdot \| x_0 \|. \tag{1.15}$$

We conclude that

$$\| T_t \|_y \geqq \| T_t \| \tag{1.16}$$

which together with the reversed inequality proves (1.6).

The following lemma bears out the last assertion in subsection 1.2. It stands by itself and will *not* be used in the sequel.

LEMMA 1.2. *If $\mathfrak{X}$ and $\mathfrak{Y}$ are not reciprocal but $\{T_t\}$ is ($\mathfrak{Y}$) measurable, define for $x \in \mathfrak{X}$*

$$\| x \|^1 = \sup_{y \in \mathfrak{Y}} \frac{| yx |}{\| y \|}. \tag{1.17}$$

Let $\mathfrak{X}'$ be the Banach space obtained by completing $\mathfrak{X}$ under this norm. For $x' \varepsilon \mathfrak{X}'$ define yx' and $T_t x'$, if necessary, as limit of corresponding Cauchy sequences. Then: (a) *$\mathfrak{X}'$ and $\mathfrak{Y}$ are reciprocal spaces.* (b) *The norm of T_t in either space equals $\| T_t \|_{\mathfrak{Y}}$.* (c) *The semi-group in $\mathfrak{X}'$ is weakly ($\mathfrak{Y}$) measurable.*

PROOF. The assertions (a) and (c) are obvious and (b) follows from Lemma 1.1 and the fact that the norm of T_t in $\mathfrak{Y}$ remains unchanged.

The assertion that the group of translations in the space of signed measures is not weakly measurable is presumably known. We prove it by means of a simplified example modifications of which will be used in the sequel.

An example. Take for $\mathfrak{X}$ the space of real functions which vanish except at denumerably many points, and put $\| x \| = \sum_z | x(z) |$. For T_t we take *translations*, that is, $T_t x(z) = x(z + t)$.

Here X^* is represented by the space of bounded functions with $\| x^* \| = \sup | x^*(z) |$, and $x^* x = \sum_z x^*(z)x(z)$. Clearly $x^* T_t x$ is measurable for all x if and only if x^* represents a measurable function. Thus $\{T_t\}$ *is not weakly measurable, but is* $(\mathfrak{Y})$*measurable* if one chooses for $\mathfrak{Y}$ the subspace of measurable functions.

Note that $\mathfrak{X}$ is isomorphic to the subspace of discrete (atomic) measures, or of step-functions considered as subspace of $\mathfrak{BB}(-\infty, +\infty)$. Therefore translations in the bigger spaces can not be measurable.

2. The Essential Norm

Note. In the following, the word density is used as an abbreviation for "*right density.*" Thus, if Γ is a set in $0 < t < \infty$, the upper and lower density of Γ at t_0 are defined as the upper and lower limit, respectively, of the ratio $m(\Gamma_h)/h$, where Γ_h is the intersection of Γ with the interval $(t_0, t_0 + h)$, m denotes the Lebesgue measure, and $h \downarrow 0$. They are the upper and lower right derivatives of the integral of the characteristic function of Γ.

DEFINITION. *Let* $\{T_t\}$ *be weakly* $(\mathfrak{Y})$ *measurable. For fixed* t_0 *(with* $0 \leqq t_0 \leqq \infty$*) and* $y \in \mathfrak{Y}$ *we define the essential norm* $\| yT_{t_0} \|_e$ *to be the infinum of all numbers* α *with the following property: for each* $x \in \mathfrak{X}$ *and each* $\varepsilon > 0$ *the t-set*

$$\{t \,\|\, yT_t x \,| > (\alpha + \varepsilon) \cdot \| x \|\} \tag{2.1}$$

has density 0 *at* t_0. *The essential norm* $\| T_t \|_e$ *is defined by*

$$\| T_t \|_e = \sup_{y \in \mathfrak{Y}} \frac{\| yT_t \|_e}{| y |}. \tag{2.2}$$

Since we shall have occasion to consider different spaces $\mathfrak{Y}$, we shall occasionally refer to $\| T_t \|_e$ as the *essential norm with respect to* $(\mathfrak{X}, \mathfrak{Y})$.

It goes without saying that we could have defined $\| T_t x \|_e$ in the same way. However, one must not expect that $\sup \| T_t x \|_e / \| x \|$ equals $\| T_t \|_e$ unless $\mathfrak{X}$ and $\mathfrak{Y}$ are reciprocal spaces. That *in the case of reciprocal spaces all relations are symmetrical* is shown by the following lemma, which is essentially a restatement of the definition.

LEMMA 2.1. *Let* $t_0 \geqq 0$ *and* $\varepsilon > 0$ *be fixed. For each* $x \in \mathfrak{X}$ *and each* $y \in \mathfrak{Y}$ *the t-set*

$$\{t \,\|\, yT_t x \,| > (1 + \varepsilon) \cdot \| y \| \cdot \| T_{t_0} \|_e \cdot \| x \|\} \tag{2.3}$$

has density zero at t_0. *However, there exists a pair* $x_0 \in \mathfrak{X}$, $y_0 \in \mathfrak{Y}$ *such that*

$$\{t \,\|\, y_0 T_t x_0 \,| > (1 - \varepsilon) \cdot \| y \| \cdot \| T_{t_0} \|_e \cdot \| x \|\} \tag{2.4}$$

has positive upper density at t_0.

THEOREM 2.1. *The function* $\| T_t \|_e$ *is Lebesgue measurable and bounded in every closed subinterval of* $(0, \infty)$.

The proof is accomplished by Lemmas 2.2–2.8.

LEMMA 2.2. *Let I be an interval on the positive t-axis. There exists a constant M and a closed set $\Gamma \subset I$ of positive measure such that*

$$\| T_t \|_e < M \qquad \textit{for} \quad t \in \Gamma. \tag{2.5}$$

PROOF. Assume the contrary. Then there exists a point $t_1 \in I$ with $\| T_{t_1} \|_e > 1$. By Lemma 2.1 it is possible to find a pair $x_1 \in \mathfrak{X}$, $y_1 \in \mathfrak{Y}$ such that

$$| y_1 T_t x_1 | > \| y_1 \| \cdot \| x_1 \| \tag{2.6}$$

for $t \in \Gamma_1$ where Γ_1 has *positive* measure. We shall suppose Γ_1 *closed* (taking a subset, if necessary).

Now Γ_1 has density 1 at almost all of its points, and we can therefore find a closed subset $\Gamma_1' \subset \Gamma_1$ of positive measure such that for each $t \in \Gamma_1'$ the set Γ_1 has density 1 at t.

By assumption, $\| T_t \|_e$ is unbounded on Γ_1' and we can therefore find a point $t_2 \varepsilon \Gamma_1'$ with $\| T_{t_2} \|_e > 2$. Then there exists a pair $x_2 \in \mathfrak{X}$, $y_2 \in \mathfrak{Y}$ such that

$$| y_2 T_t x_2 | > 2 \| y_2 \| \cdot \| x_2 \| \qquad \text{for } t \in A_2 \tag{2.7}$$

where A_2 is a set of positive upper density at t_2. Now the complement $I - \Gamma_1$ has density 0 at t_2 and it follows that $A_2 \cap \Gamma_1$ has positive upper density at t_2. Accordingly $A_2 \cap \Gamma_1$ has positive measure, and we can choose a *closed* set $\Gamma_2 \subset A_2$ such that

$$m(\Gamma_1 \cap \Gamma_2) > 0. \tag{2.8}$$

We now proceed by induction. Suppose that for $k = 1, \cdots, n - 1$ we have closed sets Γ_k with

$$m(\Gamma_1 \cap \cdots \cap \Gamma_k) > 0 \tag{2.9}$$

and that to each Γ_k there corresponds a pair x_k, y_k such that

$$| y_k T_t x_k | > k \cdot \| y_k \| \cdot \| x_k \| \qquad \text{for } t \in \Gamma_k. \tag{2.10}$$

Let Γ_{n-1}' be a closed subset of $\Gamma_1 \cap \cdots \cap \Gamma_{n-1}$ with positive measure and such that $\Gamma_1 \cap \cdots \cap \Gamma_{n-1}$ has density 1 at each point of Γ_{n-1}'. By assumption we can find $t_n \in \Gamma_n'$ with $\| T_{t_n} \|_e > n$. There exists then a pair x_n, y_n such that (2.10) holds with $k = n$ for $t \in A_n$, where A_n has positive upper density at t_n. Again we conclude that $A_n \cap \Gamma_{n-1}'$ has positive measure and we can therefore choose a closed subset Γ_n of A_n to which (2.9) and (2.10) will apply.

Because of compactness the sets Γ_n have a non-empty intersection. Now for any $t \in \cap \Gamma_n$ all the inequalities (2.10) hold and hence $\| T_t \| > n$ for each n. This contradicts the assumption that T_t is finite for each t.

LEMMA 2.3.

$$\| T_{\tau+s} \|_e \leqq \| T_\tau \| \cdot \| T_s \|_e. \tag{2.11}$$

PROOF. Assume, on the contrary, that for some $\varepsilon > 0$ and some τ, s

$$\| T_{\tau+s} \|_e > \| T_\tau \| \cdot \{ \| T_s \|_e + \varepsilon \} + \varepsilon. \tag{2.12}$$

We can find a pair x, y such that

$$(2.13)\qquad |\, y T_{\tau+t} x \,| > \{\| T_{\tau+s} \|_s - \varepsilon\}\cdot \| y \| \cdot \| x \|$$

for all $t \in A$, where the set A has positive upper density at s. Therefore (2.12) would imply that for $t \in A$

$$(2.14)\qquad \begin{aligned} |\, yT_{\tau+t}x \,| &\geqq \{\| T_s \|_e + \varepsilon\}\cdot \| T_\tau \| \cdot \| y \| \cdot \| x \| \\ &\geqq \{\| T_s \|_s + \varepsilon\}\cdot \| yT_\tau \| \cdot \| x \|. \end{aligned}$$

But we know that the inequality

$$(2.15)\qquad |\, yT_{\tau+t}x \,| = |\, (yT_\tau)T_t x \,| \geqq \| yT_\tau \| \cdot \{\| T_s \|_s + \varepsilon\}\cdot \| x \|$$

takes place only on a t-set of density 0 at s, so that (2.12) leads to a contradiction.

LEMMA 2.4. *To any interval $0 < a < t < b < \infty$ there corresponds a set $\{t_n\}$ dense in it and a constant C such that*

$$(2.16)\qquad \| T_{t_n} \| < C.$$

Note. The assertion holds for any (not necessarily measurable) non-negative $f(t)$ for which $f(t + s) \leqq f(t)f(s)$.

PROOF. Let $A_N = \{t \mid \| T_t \| < N\}$. The union $\bigcup A_N$ covers the interval $(0, a)$. By Baire's category theorem (cf. [3], p. 5, or [1], p. 14) there exists an N such that the closure $\overline{A_N}$ contains some interval $I \subset (0, a)$. Finitely many translates of I, say $I + s_1, I + s_2, \cdots, I + s_n$ will cover the given interval. Because of $\| T_{t+s_k} \| \leqq \| T_t \| \cdot \| T_{s_k} \|$ the set A_M contains $A_N + s_k$ and is therefore dense in $I + s_k$ at least when $M \geqq N \cdot \|T_{s_k}\|$. Then (2.16) is true with $C = N \cdot \max \| T_{s_k} \|$.

LEMMA 2.5. *Let I denote the interval (a, b) and let Γ be a set of positive measure contained in $(\frac{1}{3}a, \frac{2}{3}a)$. Let $\{t_n\}$ be a denumerable set which is dense in $(\frac{1}{3}a, b)$. Then almost every point of I is contained in $\bigcup (\Gamma + t_n)$.*

PROOF. Let s be a point at which Γ has density 1. The set $\{s + t_n\}$ is dense in I. Let $t \in I$ be arbitrary, but fixed, and let I_h denote the interval $(t, t + h)$. To given $\varepsilon > 0$, $\delta > 0$ we can find $0 < h < \delta$ and an index n such that

$$(2.17)\qquad m(I_h \cap (\Gamma + t_n)) > (1 - \varepsilon)h.$$

It follows that t cannot be a point of density 0 of $\bigcup (\Gamma + t_n)$, whence $m(I \cap (\bigcup(\Gamma + t_n))) = m(I)$.

LEMMA 2.6. *There exists a constant K such that*

$$(2.18)\qquad \| T_t \|_s < K$$

for almost all $t \in (a, b)$.

PROOF. Apply Lemma 2.2 to the interval $(\frac{1}{3}a, \frac{2}{3}a)$, and Lemma 2.4 to $(\frac{1}{3}a, b)$. By Lemma 2.5 the sets $\Gamma + t_n$ cover almost every point of (a, b). But for $t \in \Gamma + t_n$ we have $\| T_t \|_s < CM$, by Lemma 2.3.

LEMMA 2.7. For each $t \in (a, b)$

$$(2.19)\qquad \| T_t \|_s \leqq 2K.$$

PROOF. If (2.19) is not true for $t = t_0$, we can find a pair x, y such that

$$| yT_t x | > \tfrac{3}{2}K \cdot \| y \| \cdot \| x \| \tag{2.20}$$

for $t \in A$, where A is a t-set of positive measure. At almost all its points A has density 1. By Lemma 2.6 we can therefore find a point τ at which A has density 1 but for which $\| T_\tau \|_e < K$. This in turn requires that the set on which (2.20) holds, has density 0 at τ, and this contradiction proves the Lemma.

LEMMA 2.8. *$\| T_t \|_e$ is Lebesgue measurable.*

PROOF. For fixed a, b put

$$A = \{t \mid \| T_t \|_e > a, t < b\}. \tag{2.21}$$

We have to prove that A is measurable.

Under any circumstances there exists an F_σ-set $A_i \subset A$ whose measure equals the *inner measure* $m_*(A)$. If A is not measurable then

$$m_*(A - A_i) = 0 \qquad m^*(A - A_i) = \eta > 0. \tag{2.22}$$

Let $\tau \in A - A_i$ be a point at which A_i has density 0. Since $\tau \in A$, we can find a pair x, y such that

$$| yT_t x | > a \cdot \| y \| \cdot \| x \| \tag{2.23}$$

for $t \in \Gamma$, where Γ has positive upper density δ at τ. Let $I_h = (\tau, \tau + h)$. The last assertion states that there exists a sequence $h_n \to 0$ such that

$$m(\Gamma \cap I_{h_n}) > \delta h_n . \tag{2.24}$$

Now, since A_i has density 0 at τ,

$$m(A_i \cap I_{h_n}) = o(h_n). \tag{2.25}$$

Therefore $\Gamma - \Gamma \cap A_i = \Gamma_1$ has positive measure. For $t \in \Gamma_1$ the inequality (2.23) holds and therefore at each point t_1 at which Γ_1 has density 1 we have $\| T_{t_1} \|_e > a$. We conclude that $\| T_t \|_e > a$ on a set of positive measure $\Gamma_1 \subset A - A_i$, which contradicts the first relation in (2.22).

THEOREM 2.2. *For each $y \in \mathfrak{Y}$ the function $\| yT_t \|_e$ is Lebesgue measurable. For each fixed $x \in \mathfrak{X}$ one has almost everywhere*

$$\| yT_t x \| \leqq \| yT_t \|_e \cdot \| x \|. \tag{2.26}$$

PROOF. The boundedness of $\| yT_t \|_e$ follows from $\| yT_t \|_e \leqq \| y \| \cdot \| T_t \|_e$. The measurability is now proved by repeating the argument of Lemma 2.8. If (2.26) were not true almost everywhere, one could find two rationals $r_1 < r_2$ and a set of positive measure Γ such that for $t \in \Gamma$ simultaneously

$$| yT_t x | > r_2, \qquad \| yT_t \|_e \cdot \| x \| < r_1 . \tag{2.27}$$

Now choose for t a point at which Γ has density 1, and the contradiction between the two statements of (2.27) becomes manifest.

3. The Strongly Continuous Center of $\{T_t\}$

Let $\{T_t\}$ be weakly ($\mathfrak{Y}$) measurable. For any a $\epsilon\, \mathfrak{X}$, $y \,\epsilon\, \mathfrak{Y}$, and $0 < a < b < \infty$ Theorem 2.2 permits us to write

$$(3.1)\quad \left| \int_a^b (yT_t x)\, dt \right| \leqq \| x \| \cdot \int_a^b \| yT_t \|_e \cdot dt \leqq \| x \| \cdot \| y \| \cdot \int_a^b \| T_t \|_e \cdot dt < \infty.$$

For fixed y the integral on the left side therefore represents a bounded linear functional. This proves the

THEOREM 3.1. *For fixed $y \,\epsilon\, \mathfrak{Y}$ and $0 < a < b < \infty$ there exists a unique functional $y_{a,b} \,\epsilon\, \mathfrak{X}^*$ such that for each $x \,\epsilon\, \mathfrak{X}$*

$$(3.2)\quad y_{a,b}x = \int_a^b (yT_t x) \cdot dt.$$

This functional will be denoted by

$$(3.3)\quad y_{a,b} = \int_a^b (yT_t)\, dt.$$

This is a special case of the Dunford-Pettis integral, and only its existence was to be proved. When $a = 0$ or $b = \infty$ the integral need not exist, but the same notation will be used whenever meaningful.

We recall that the adjoint to T_t is defined for all $\mathfrak{X}^*$. Obviously

$$(3.4)\quad y_{a,b} T_s = y_{a+s,b+s} = \int_a^b (yT_{t+s})\, dt.$$

It follows in particular that $y_{a,b}T_t x$ is a *continuous* function of t for each $x \,\epsilon\, \mathfrak{X}$.

NOTATION. *The subspace of $\mathfrak{X}^*$ spanned by all elements of the form* (3.3) *with $0 < a < b < \infty$ will be denoted by $\mathfrak{Y}_1$. The subspace spanned by $\mathfrak{Y}$ and $\mathfrak{Y}_1$ will be denoted by $\tilde{\mathfrak{Y}}$.*

THEOREM 3.2. (a) *The space $\tilde{\mathfrak{Y}}$ is invariant, that is, $\tilde{\mathfrak{Y}}T_t \subset \tilde{\mathfrak{Y}}$.* (b) *The semi-group $\{T_t\}$ is weakly ($\tilde{\mathfrak{Y}}$) measurable.* (c) *If $\mathfrak{X}$ and $\mathfrak{Y}$ are reciprocal, then so are $\mathfrak{X}$ and $\tilde{\mathfrak{Y}}$.* (d) *The space $\tilde{\mathfrak{Y}}$ is closed under the operation of integration introduced above.*

PROOF. (a) follows from (3.4) and (b) from the sentence following (3.4). The assertion (c) is obvious since $\mathfrak{Y} \subset \tilde{\mathfrak{Y}}$. To prove (d) note that for h small enough

$$(3.5)\quad y_{a,b}(T_{s+h} - T_s) = \int_{b+s}^{b+s+h} (yT_t)\, dt - \int_{a+s}^{a+s+h} (yT_t)\, dt.$$

From this it follows that as $h \to 0$

$$(3.6)\quad \| y_{a,b}(T_{s+h} - T_s) \| \to 0.$$

It is then easily seen that the integral

$$\int_\alpha^\beta (y_{a,b}\,T_s)\,ds \tag{3.7}$$

can be obtained as the *strong limit* of Riemann sums

$$\sum (y_{a,b}\,T_{s_n})\cdot \Delta s_n\,, \tag{3.8}$$

that is, (3.7) can be defined as a Bochner integral.

We now turn to the space $\mathfrak{Y}_1$. It is clear that it is an invariant subspace of $\mathfrak{X}^*$, and we can therefore study the contraction of our semi-group to $\mathfrak{Y}_1$. For all elements of the form (3.3) we have (3.6). We shall show that more generally for each $y \in \mathfrak{Y}_1$ and $s > 0$

$$yT_{s+h} \to yT_s \tag{3.9}$$

as $h \downarrow 0$. This means that our new semi-group is strongly continuous. To find a bound for its norm note that because of continuity

$$y_{a,b}\,T_t x = \lim_{h\to 0} \frac{1}{h}\int_0^h (y_{a,b}\,T_{t+s}\,x)\,ds. \tag{3.10}$$

We now require the fact that for h sufficiently small

$$\left| \frac{1}{h}\int_0^h (y_{a,b}\,T_{t+s}\,x)\,ds \right| < (1+\epsilon)\cdot \| y_{a,b} \|\cdot\| T_t \|_e\cdot\|x\| \tag{3.11}$$

(where, as before, $\| T_t \|_e$ is the essential norm with respect to $\mathfrak{Y}$). The inequality (3.11) is an immediate consequence of (4.33). However, (4.33) is best proved using Banach limits and the proof is therefore deferred to Section 4.6. This is legitimate since the Theorem 3.3 will not be used in the sequel. It follows from (3.11) that

$$\| y_{a,b}\,T_t \| \leqq \| y_{a,b} \|\cdot\| T_t \|_e\,. \tag{3.12}$$

Using this and Theorem 2.1 one obtains the strong continuity property (3.9) as an immediate consequence of (3.6) and the fact that the elements of the form $y_{a,b}$ are dense in $\mathfrak{Y}_1$.

We have thus

THEOREM 3.3. *Our semi-group induces a semi-group of transformations on the space $\mathfrak{Y}_1$ which is strongly continuous for $t > 0$ (but not necessarily for $t = 0$). Its norm is bounded by $\| T_t \|_e$, and may be unbounded near $t = 0$.*

The last statement can be substantiated by a simple example.

Since the relation between $\mathfrak{X}$ and $\mathfrak{Y}$ is symmetric, we could as well have obtained a strongly continuous semi-group in a space $\mathfrak{X}_1$ imbedded in the adjoint $\mathfrak{Y}^*$.

THEOREM 3.4. *If*

$$\limsup_{t\to 0} \| T_t \|_e \leqq 1 \tag{3.13}$$

and $\mathfrak{X}$ and $\mathfrak{Y}$ are reciprocal spaces, then so are $\mathfrak{X}_1$ and $\mathfrak{Y}_1$. The induced strongly continuous semi-groups are adjoints to each other.

PROOF. It is obvious that $\mathfrak{X}_1$ and $\mathfrak{Y}_1$ are dual spaces in the sense that the inner product y_1x_1 is defined for every pair $x_1 \in \mathfrak{X}_1$, $y_1 \in \mathfrak{Y}_1$. We have to verify that the norms in $\mathfrak{X}_1$ and $\mathfrak{Y}_1$ have the required properties. Because of the symmetry between $\mathfrak{X}$ and $\mathfrak{Y}$ it suffices to prove that to each element $y_{a,b}$ of the form (3.3) there corresponds an element $x_1 \in \mathfrak{X}_1$ such that

$$\| y_{a,b}x_1 \| \geqq (1 - \varepsilon)\cdot\| y_{a,b} \|\cdot\| x_1 \|. \tag{3.14}$$

By definition there exists $x \in \mathfrak{X}$ such that

$$\| y_{a,b}x \| \geqq (1 - \tfrac{1}{3}\varepsilon)\cdot\| y_{a,b} \|\cdot\| x \|. \tag{3.15}$$

Put

$$x_1 = \frac{1}{h}\int_0^h (T_s x)\, ds \tag{3.16}$$

where h is chosen small enough to have

$$\| x_1 \| \leqq \left(1 + \frac{\varepsilon}{3}\right)\cdot\| x \|. \tag{3.17}$$

In view of the strong continuity in $\mathfrak{Y}_1$ we know that as $h \to 0$

$$\frac{1}{h}\int_0^h (y_{a,b}\, T_s x)\, ds \to y_{a,b}x. \tag{3.18}$$

Therefore, if h is small enough, we get from (3.15)

$$| y_{a,b}x_1 | > \left(1 - \frac{\varepsilon}{3}\right)\cdot| y_{a,b}x | > \left(1 - \frac{\varepsilon}{3}\right)^2\cdot\| y_{a,b} \|\cdot\| x \|. \tag{3.19}$$

This combined with (3.17) proves (3.14).

4. Equivalence Classes. Relations for the Norms

4.1. *Basic definitions and notations. Two functionals x_1^* and x_2^* will be called equivalent, or in symbols,*

$$x_1^* \sim x_2^*, \tag{4.1}$$

if for each $x \in \mathfrak{X}$

$$x_1^*T_tx = x_2^*T_tx \qquad \textit{for almost all } t. \tag{4.2}$$

The set of all functionals equivalent to $y \in \mathfrak{Y}$ will be called the equivalence class of y and denoted by y'. The prime will be used consistently to denote the operation of mapping an element onto its equivalence class.

The subspace of $\mathfrak{X}^*$ spanned by $\mathfrak{Y}$ and all functionals in the equivalence classes of elements $y \in \mathfrak{Y}$ will be denoted by $\mathfrak{Y}_{enl}$ and referred to as the enlarged space.

Clearly, $\{T_t\}$ is weakly $(\mathfrak{Y}_{enl})$ measurable. *The essential norm with respect to*

$\mathfrak{Y}_{enl}$ *will be denoted by* $\| T_t \|_E$ to distinguish it from $\| T_t \|_e$. The relation between $\| T_t \|_e$ and $\| T_t \|_E$ is studied in subsection 4.4.

The justification for passing from $\mathfrak{Y}$ to $\mathfrak{Y}_{enl}$ lies in the greater elegance and the more precise results obtained in this way. Of course, we could have from the beginning substituted $\mathfrak{Y}_{enl}$ for $\mathfrak{Y}$. This would have the disadvantage that $\mathfrak{Y}$ is usually much simpler and that in applications little is known about the totality of functionals. Thus, basing the theory on $\mathfrak{Y}_{enl}$ instead of $\mathfrak{Y}$ would mean a severe restriction. On the other hand, in introducing equivalence classes it makes no formal difference whether we restrict the consideration to elements $y \in \mathfrak{Y}$ or introduce all functionals equivalent to them. However, the two procedures lead one to attribute different norms to the equivalence classes, and here the difference is essential.

Rules for addition and scalar multiplication of equivalence classes are obvious and are formalized by identifying the equivalence classes with the points of the quotient space

$$\mathfrak{Y}' = \mathfrak{Y}_{enl}/\mathfrak{Y}_0 \tag{4.3}$$

where $\mathfrak{Y}_0$ is the subspace spanned by all functionals equivalent to zero.

In the usual way we introduce the norm

$$\| y' \| = \inf_{x^* \in y'} \| x^* \|. \tag{4.4}$$

It is well known[7] that *with this norm the space* $\mathfrak{Y}'$ *of equivalence classes becomes a Banach space.*

It is clear from the definition that $y_1 \sim y_2$ implies $y_1 T_t \sim y_2 T_t$ for each $t > 0$. Thus $\{T_t\}$ *induces a semi-group of transformations on* $\mathfrak{Y}'$. We shall use the symbol $y'T_t$ for the transform of y'. *The norm of the induced semi-group for equivalence-classes will be denoted by* $\| T_t \|^1$. That is

$$\| T_t \|^1 = \sup_{y' \in \mathfrak{Y}'} \frac{\| y' T_t \|}{\| y' \|}. \tag{4.5}$$

We shall use some obvious notations. Thus $y'x$ *is an abbreviation for the class of real functions* $zT_t x$ *where* $z \in y'$. If $y \in \mathfrak{Y}_{enl}$ then

$$\int_a^b (y' T_t)\, ds \tag{4.6}$$

denotes the corresponding equivalence class, etc.

4.2. *Banach limits.* Let $x_h^* \in \mathfrak{X}^*$ be a family of functionals depending on h, where $0 < h \leq h_0$. *We shall say that a functional* x^* *is a Banach limit*[8] *of* x_h^* *as* $h \to 0$, *or*

$$x^* = B\text{-}\lim_{h \to 0} x_h^*, \tag{4.7}$$

---

[7] Cf. [3] p. 30 and 472. Hille refers to (4.4) as the Gelfant-Neumark metric.

[8] Cf. [1] p. 118. Banach introduces the limits for directed sets and calls them transfinite limits.

if for each $x \in \mathfrak{X}$

$$\liminf_{h\to 0} x_h^* x \leqq x^* x \leqq \limsup_{h\to 0} x_h^* x. \tag{4.8}$$

If there exists a constant such that $\| x_h^* \| < M$, then (at least) one Banach limit exists.[9]

Now let $y \in \mathfrak{Y}_{enl}$. Then for fixed $t > 0$ there exists a functional

$$x_t^* = B\text{-}\lim_{h\to 0} \frac{1}{h} \int_0^h (yT_{t+s})\, ds. \tag{4.9}$$

Obviously

$$\| x_t^* \| \leqq \| yT_t \|_E \tag{4.10}$$

(note that on the right side we have the essential norm with respect to the pair $\mathfrak{X}$, $\mathfrak{Y}_{enl}$).

Applying the definition (4.8) to x_t^* using $T_\tau x$ instead of x we get

$$\liminf_{h\to 0} \frac{1}{h} \int_\tau^{\tau+h} (yT_{t+s}x)\, ds \leqq x_t^* T_\tau x \leqq \limsup_{h\to 0} \frac{1}{h} \int_\tau^{\tau+h} (yT_{t+s}x)\, ds. \tag{4.11}$$

Now for each $x \in \mathfrak{X}$ and fixed y, t the two extreme members equal $yT_{t+\tau}x$ for almost all τ. This means that

$$x_t^* \sim yT_t\,. \tag{4.12}$$

We have thus

THEOREM 4.1. *For each* $y \in \mathfrak{Y}_{enl}$ *(and hence for* $y \in \mathfrak{Y}$*) and each* $t > 0$ *the equivalence class* $y'T_t$ *contains all Banach limits of*

$$\frac{1}{h} \int_t^{t+h} (yT_s)\, ds \qquad (h \downarrow 0). \tag{4.13}$$

We express this by writing

$$y'T_t = B\text{-}\lim_{h\to 0} \frac{1}{h} \int_k^{t+h} (y'T_s)\, ds. \tag{4.14}$$

The statement holds also for $t = 0$ *provided* $\| T_t \|_E$ *is bounded near* $t = 0$.

A better continuity property under so general conditions could hardly be expected.

4.3. *Relations for the norms.* One would expect that the norm of the induced semi-group for equivalence classes be the same as $\| T_t \|_E$, at least almost everywhere. Theorem 4.4 will show that this is so for "well-behaved" semi-groups. In general we have only the following result, and the counterexamples below will prove that in (4.15) the inequality can hold for all t.

THEOREM 4.2.

$$\| T_t \|' \leqq \| T_t \|_E\,. \tag{4.15}$$

[9] Banach notes that this is an immediate consequence of the extension theorem applied to the sub-additive functional $p(x) = \limsup x_h^* x$.

PROOF. From (4.10), (4.12), and the definition (4.4) it is clear that

$$\| y'T_t \| \leqq \| y \| \cdot \| T_t \|_E . \tag{4.16}$$

This inequality holds also if y is replaced by any element of its equivalence class. It is therefore permitted to assume that y has been so chosen that

$$\| y \| < (1 + \varepsilon) \cdot \| y' \|. \tag{4.17}$$

The Theorem now follows from (4.5).

THEOREM 4.3. *For each $s > 0$, $t > 0$*

$$\| T_{t+s} \|_E \leqq \| T_t \|^{\text{I}} \cdot \| T_s \|_E . \tag{4.18}$$

This relation is valid also for $s = 0$, provided $\| T_0 \|_E < \infty$.

COROLLARY.

$$\| T_{t+s} \|_E \leqq \| T_t \|_E \cdot \| T_s \|_E \tag{4.19}$$

(that is, $\| T_t \|_E$ behaves like a norm).

PROOF. By definition of $\| T_t \|_E$ we can find a pair $x \in \mathfrak{X}$, $y \in \mathfrak{Y}_{ent}$ such that

$$| yT_{t+\sigma}x | > (1 - \varepsilon) \cdot \| y \| \cdot \| T_{t+s} \|_E \cdot \| x \| \tag{4.20}$$

for all $\sigma \in \Gamma$, where Γ is a set with positive upper density at $\sigma = s$.

Now choose a representative $y_t \in y'T_t$ such that

$$\| y_t \| < (1 + \varepsilon) \cdot \| y'T_t \|. \tag{4.21}$$

For almost all σ

$$yT_{t+\sigma}x = y_t T_\sigma x. \tag{4.22}$$

Accordingly, there exists a set Γ_1 of density 1 at $\sigma = s$ such that for $\sigma \in \Gamma_1$

$$| yT_{t+\sigma}x | < (1 + \varepsilon) \cdot \| y_t \| \cdot \| T_s \|_E \cdot \| x \|. \tag{4.23}$$

Then for $\sigma \in \Gamma_1$

$$\begin{aligned} | yT_{t+\sigma}x | &< (1 + \varepsilon)^2 \cdot \| y'T_t \| \cdot \| T_s \|_E \cdot \| x \| \\ &\leqq (1 + \varepsilon)^2 \cdot \| y \| \cdot \|T_t \|^{\text{I}} \cdot \| T_s \|_E \cdot \| x \|. \end{aligned} \tag{4.24}$$

Since Γ has positive upper density at s, and Γ_1 density 1, the intersection $\Gamma \cap \Gamma_1$ cannot be empty. Thus there exists a value of σ for which both (4.20) and (4.24) hold, and (4.18) follows.

THEOREM 4.4. *If*

$$\| T_0 \|_E = 1 \tag{4.25}$$

then

$$\| T_t \|^{\text{I}} = \| T_t \|_E . \tag{4.26}$$

Note. For (4.25) to hold it is *sufficient* that $\| T_t \|_- \to 1$ as $t \to 0$. The main implication of the Theorem is that (4.25) *guarantees the measurability of $\| T_t \|^{\text{I}}$.*

PROOF. Compare (4.15) with (4.18) for $s = 0$.

4.4 *Relations between* $\| T_t \|_E$ *and* $\| T_t \|_e$. It has now been established that $\| T_t \|_E$ behaves substantially like a norm. The question arises whether the preceding theorems hold also for $\| T_t \|_e$. We shall show that this is not so. In particular, *it is not always true that*

$$(4.27) \qquad \| T_{t+s} \|_e \leqq \| T_t \|_e \cdot \| T_s \|_e ,$$

and the Theorem 4.2 has no analogue if one restricts oneself to $\mathfrak{Y}$. That this misbehavior has no ill effects is due to the

LEMMA.

$$(4.28) \qquad \| T_t \|_e \leqq \| T_t \|_E .$$

This is obvious since when $\mathfrak{Y}$ is increased the sup in (2.2) can not decrease. Counterexamples show that in (4.28) the inequality may hold for all t.

As an application we prove

THEOREM 4.5. *There exists a constant* $p \geqq 0$ *such that*

$$(4.29) \qquad \limsup_{t\to\infty} e^{-pt} \| T_t \|_e \leqq 1.$$

PROOF. From the norm relation (4.19) it follows (cf. [3] p. 135) that $\log \| T_t \|_E / t$ approaches a finite limit or $-\infty$. In view of (4.28) this implies (4.29).

The purpose of passing from $\mathfrak{Y}$ to $\mathfrak{Y}_{enl}$ was to ensure that our Banach limits will be elements of the space. Now this is true of $\mathfrak{Y}$ if it is regularly closed. We have therefore the following theorem which is of some interest, but will not be used in the sequel.

THEOREM 4.6. *If* $\mathfrak{Y}$ *is regularly closed, then* (4.27) *holds.*

4.5. *Counterexamples.* (a) Let $\mathfrak{X}$ and $\mathfrak{Y}$ be reciprocal spaces and let $\mathfrak{Z} \subset \mathfrak{X}$ be a subspace such that $x \in \mathfrak{Z}$, $y \in \mathfrak{Y}$ implies $| yx | < q \cdot \| y \| \cdot \| x \|$, where $q < 1$. Let P be a fixed projection operator from $\mathfrak{X}$ to $\mathfrak{Z}$ with $\| P \| = 1$. Put $T_t = P$ for all t.

Clearly

$$(4.30) \qquad \| T_t \|_e < q < 1,$$

and hence (4.27) cannot be true. Moreover, $\| T_t \|_E = 1$. This shows that in (4.28) the inequality sign can hold for all t.

This example is not convincing, since projections represent a removable structure. (Instances of projections with the required property can be constructed by a slight modification of the following example).

(b) Take the space $\mathfrak{X}$ of the example at the end of Section 1.4 but with the modified norm

$$(4.31) \qquad \| x \| = \sum_{z \text{ irrat.}} | x(z) | + \tfrac{1}{2} \sum_{z \text{ rat.}} | x(z) |,$$

the two sums extending, respectively, over irrational and rational abscissas. For $\mathfrak{Y}$ take the space of bounded measurable functions with the modified norm

$$(4.32) \qquad \| y \| = \max \{ \sup_{z \text{ irrat.}} | y(z) |, \quad 2 \cdot \sup_{z \text{ rat.}} | y(z) | \}.$$

Again $yx = \sum_z y(z)x(z)$ and $T_t x(z) = x(z + t)$.

Here the spaces $\mathfrak{Y}$ and $\mathfrak{Y}_{enl}$ coincide, and therefore $\| T_t \|_e$ and $\| T_t \|_E$ are the same.

Let $x(0) = 1$ and let x vanish elsewhere; let $y(z) = 1$ or 0 according as z is irrational or rational. Then $yT_t x = 1$ for t irrational. One concludes that $\| T_t \|_E = 2$.

On the other hand, the equivalence class y' contains all functions which differ from y only on a set of measure zero. Therefore $\| y' \| =$ ess. sup $| y(z) |$. It is now obvious that $\| T_t \|' = 1$. This shows that in (4.15) the inequality sign can hold.

(c) Let $\mathfrak{X}$ and T_t be as under (b), but choose for $\mathfrak{Y}$ the subspace of functions which differ from a continuous function only on a *denumerable* set. The reader may verify that $\mathfrak{X}$ and $\mathfrak{Y}$ are reciprocal, that $\| T_t \|_e = 1$, $\| T_t \|_E = 2$, and $\| T_t \|' = 1$. This shows the possibility of inequalities both in (4.15) and (4.28).

4.6. A Lemma. *The proof of* (3.11) *depends on the assertion that for each* $y \in \mathfrak{Y}$

$$\| y_{a,b} T_t \|_e \leqq \| T_t \|_e \cdot \| y_{a,b} \|, \tag{4.33}$$

$y_{a,b}$ *being defined in* (3.3)

We now prove (4.33). Let $x \in \mathfrak{X}$ be arbitrary. Since $\mathfrak{X}$ and $\mathfrak{Y}$ are reciprocal, we may for the moment interchange their rôle and apply (4.10) to x. This means that the equivalence class of $T_t x$ contains an element x_t with

$$\| x_t \| \leqq \| T_t x \|_e . \tag{4.34}$$

(In (4.10) we have $\| yT_t \|_E$ instead of $\| yT_t \|_e$, but for elements $y \in \mathfrak{Y}$ the two norms coincide.) Then

$$| y_{a,b} T_t x | = \left| \int_a^b (yT_s x_t)\, ds \right| \leqq \| y_{a,b} \| \cdot \| x_t \| \leqq \| y_{a,b} \| \cdot \| T_t \|_e \cdot \| x \|. \tag{4.35}$$

To prove (4.33) it suffices to note that the essential norm on the left side equals the ordinary norm, since integrals enjoy the strong continuity property.

5. Remetrizations

5.1. A Lemma. *Consider a Banach space* $\mathfrak{Z}$ *and a semi-group of linear operators* $\{T_t\}$ $(0 < t < \infty)$ *from* $\mathfrak{Z}$ *to* $\mathfrak{Z}$ *such that* $\| T_t \|$ *is bounded in every closed subinterval of* $(0, \infty)$. The last condition is certainly satisfied when $\| T_t \|$ is *measurable*. (Cf. [3] p. 133.) The possible unboundedness of $\| T_t \|$ near $t = 0$ creates difficulties, and one is often compelled to introduce the restrictive hypothesis that $\| T_t \|$ is not only bounded near $t = 0$ but bounded by a linear function $1 + at$.

Lemma. *It is possible to remetrize a subset* $\Sigma \subset \mathfrak{Z}$ *which includes the entire range of* $\{T_t\}$ *(i.e. all elements of the form* $T_t z$ *with* $z \in \mathfrak{Z}$, $t > 0$*) so that under the norm:* (a) Σ *becomes a Banach space.* (b) T_t *maps* Σ *into itself.* (c) *The norm* $N(T_t)$ *of* T_t *as an operator on* Σ *satisfies*

$$N(T_t) \leqq e^{pt}, \tag{5.1}$$

where p is a constant.

PROOF. It is well known that there exists a constant such that

$$\lim_{t\to\infty} e^{-pt} \| T_t \| \leqq 1. \tag{5.2}$$

(Cf. [3], p. 135.) Define Σ as the set of all those $z \in \mathfrak{Z}$ for which $e^{-pt} \| T_t z \|$ is bounded over $(0, \infty)$. In view of (5.2) and the boundedness of $\| T_t \|$ in closed subintervals we are sure that Σ contains every element of the form $T_t z$, $t > 0$. It follows in particular that $T_t \Sigma \subset \Sigma$. For $z \in \Sigma$ we put

$$N(z) = \max \{\| z \|, \sup_{t>0} e^{-pt} \cdot \| T_t z \|\}. \tag{5.3}$$

It is obvious that with this norm Σ becomes a Banach space and that

$$N(T_t z) \leqq N(z) e^{pt}. \tag{5.4}$$

We cannot apply this lemma to our case since our semi-group is only weakly ($\mathfrak{Y}$) continuous and $\| T_t \|$ can be unbounded in each interval. It is true that we could remetrize the semi-group, say, on $\mathfrak{Y}$, using the essential norm. However, such a procedure would destroy the relations between $\mathfrak{X}$ and $\mathfrak{Y}$.

The following procedure seems more natural and has the additional advantage to lead to a semi-group of bounded norm on a space which is the *completion* of $\mathfrak{X}$ under a *refined* norm.

5.2. *Equivalence classes as pseudo-functionals.* In the following $\{T_t\}$ is assumed to be weakly ($\mathfrak{Y}$) measurable, but $\mathfrak{X}$ and $\mathfrak{Y}$ need not be reciprocal.

Let y' be an equivalence class. It follows from the definition[10] in Section 2 that $\| yT_t \|_E$ is the same for all $y \in y'$. The common value will be denoted by $\| y'T_t \|_E$.

We know that there exists a number $p \geqq 0$ such that $e^{-pt} \cdot \| T_t \|_E$ is bounded at infinity. (Cf. (4.29) and its proof.) Moreover, $\| T_t \|_E$ is bounded in every finite subinterval of $(0, \infty)$. It follows that for each $y' \in \mathfrak{Y}'$ and each $s > 0$ the function $\| y'T_{t+s} \|_E \cdot e^{-pt}$ is bounded.

Let $\mathfrak{H}$ denote the set of all equivalence classes such that $\| y'T_t \|_E \cdot^{-pt}$ is bounded. We have just shown that $\mathfrak{H}$ includes the entire range of $\{T_t\}$, and hence that $\mathfrak{H} T_t \subset \mathfrak{H}$.

For $y' \in \mathfrak{H}$ we put

$$N(y') = \sup_{t\geqq 0} e^{-pt} \cdot \| y'T_t \|_E. \tag{5.5}$$

Since, by definition, $\| y'T_t \|_E$ is bounded near $t = 0$, we can use (4.10) to conclude that

$$\| y' \| \leqq \| y'T_0 \|_E \tag{5.6}$$

whence

$$N(y') \geqq \| y' \|. \tag{5.7}$$

[10] It will be recalled that $\| yT_t \|_e$ and $\| yT_t \|_E$ are defined in the same way. The capital subscript indicates that y is not necessarily an element of $\mathfrak{Y}$, but $y \in \mathfrak{Y}_{ent}$.

This means that the new topology is coarser than the original one, and hence the norm (5.5) makes $\mathfrak{H}$ into a Banach space. From the very definition

$$N(y'T_t) \leqq e^{pt}N(y'). \tag{5.8}$$

Thus the induced semi-group on $\mathfrak{H}$ has a *true norm* bounded by e^{pt}.

DEFINITION. *For $x \in \mathfrak{X}$, $y' \in \mathfrak{H}$ the inner product $y'x$ is to stand for the class of functions $e^{-pt}\cdot(yT_t x)$ with $y \in y'$. This class is to be considered as element in the familiar Banach space $\mathfrak{B}$ of essentially bounded functions. In this way $y'x$ receives the norm*

$$| y'x | = \text{ess. sup } e^{-pt}\cdot| yT_t x |. \tag{5.9}$$

LEMMA. *The norm of y' as an operator from $\mathfrak{X}$ to $\mathfrak{B}$ coincides with $N(y')$.*

PROOF. First of all we have by (5.7)

$$| y'x | \leqq \| x \|\cdot N(y'). \tag{5.10}$$

It remains to show that to each $y' \in \mathfrak{H}$ there corresponds an $x \in \mathfrak{X}$ such that

$$| y'x | \geqq (1 - \varepsilon)\cdot N(y')\cdot\| x \|. \tag{5.11}$$

First choose $s > 0$ so that for $y \in y'$

$$\| yT_s \|_B > \left(1 - \frac{\varepsilon}{3}\right)\cdot N(y')\cdot e^{ps}. \tag{5.12}$$

There exists then an $x \in \mathfrak{X}$ such that

$$yT_t x \geqq \left(1 - \frac{\varepsilon}{3}\right)^2\cdot N(y')\cdot e^{ps}\cdot\| x \| \tag{5.13}$$

for all t in a set of positive upper density at $t = s$. This proves (5.11).

We are now ready to remetrize $\mathfrak{X}$. Put

$$N(x) = \sup_{y' \in \mathfrak{H}} \frac{| y'x |}{N(y')}. \tag{5.14}$$

In view of (5.10) we have

$$N(x) \leqq \| x \|, \tag{5.15}$$

and it is obvious that

$$N(T_t x) \leqq N(x)\cdot e^{pt}. \tag{5.16}$$

Clearly $N(x_0) = 0$ if and only if for each $y \in \mathfrak{Y}$ we have $yT_t x_0 = 0$ for almost all t.

Let Ξ be the Banach space obtained by introducing in $\mathfrak{X}$ the norm (5.14) *and completing the space under the new norm.*

We shall denote the points of Ξ by primes. Thus $x' \in \Xi$ stands either for an equivalence class in $\mathfrak{X}$ or for the limit of a Cauchy sequence of such classes. If for a sequence $x_n \in \mathfrak{X}$ the corresponding equivalence classes x'_n converge in the new norm to x', then the inner products $y'x_n$ converge in the topology of $\mathfrak{B}$. In this way the inner product $y'x'$ is defined for any pair $x' \in \Xi$, $y' \in \mathfrak{H}$.

From the last lemma and from (5.15) we conclude that

$$N(y') = \sup_{x \in \mathfrak{X}} \frac{|y'x|}{\|x\|} \leqq \sup_{x' \in \Xi} \frac{|y'x|}{N(x')}. \tag{5.17}$$

In view of (5.14) the inequality sign is impossible. We can therefore rewrite (5.17) and (5.14) in the form

$$N(y') = \sup_{x' \in \Xi} \frac{|y'x'|}{N(x')} \qquad N(x') = \sup_{y' \in \mathfrak{H}} \frac{|y'x'|}{N(y')}. \tag{5.18}$$

Summary. We supposed $\{T_t\}$ weakly ($\mathfrak{Y}$) measurable, but $\mathfrak{X}$ and $\mathfrak{Y}$ need not be reciprocal. First of all each of the spaces $\mathfrak{X}$ and $\mathfrak{Y}$ is replaced by a space of equivalence classes (quotient space) by putting into the zero class all x (or y, respectively) for which yT_tx vanishes almost everywhere for each y (or x).

Remetrization then leads to two Banach spaces Ξ and $\mathfrak{H}$ with the following properties.

Ξ is obtained by a refinement of the topology and contains a representative for each $x \in \mathfrak{X}$.

$\mathfrak{H}$ is a subset of the space of equivalence classes $\mathfrak{Y}'$ made into a Banach space by means of a coarser topology. However, $\mathfrak{H}$ contains a representative at least for each y in the range of $\{T_t\}$, i.e. for each element of the form yT_t, $y \in \mathfrak{Y}$.

Ξ and $\mathfrak{H}$ are dual spaces in the sense that for $x' \in \Xi$ and $y' \in \mathfrak{H}$ an inner product $y'x'$ is defined which is a point in the space $\mathfrak{B}$ of essentially bounded functions. Thus a point of one space serves as an operator from the other space to $\mathfrak{B}$ with all essential properties of functionals. The spaces are reciprocal in the sense that (5.18) *holds.*

$\{T_t\}$ induces in each of the spaces a semi-group of transformations. These are adjoints to each other in the sense that $y'(T_tx') \equiv (y'T_t)x'$.

The induced semi-groups have true norms bounded by e^{pt}, that is, (5.8) *and* (5.16) *hold.*

6. The Infinitesimal Generator

We use the notations introduced in Section 3. To simplify formulas we shall now suppose $\mathfrak{Y} = \tilde{\mathfrak{Y}}$ so that integrations can be performed within $\mathfrak{Y}$. This is no restriction since $\mathfrak{Y}$ can always be extended to $\tilde{\mathfrak{Y}}$.

Let again $p \geqq 0$ be chosen so that $e^{-pt} \| T_t \|_B$ is bounded at infinity.

NOTATION. *H shall denote the set of those $y \in \mathfrak{Y}$ for which $\| yT_t \|_B$ is bounded near $t = 0$. This condition is equivalent to the condition that $e^{-pt} \cdot \| yT_t \|_B$ be bounded for all t. The corresponding set of equivalence classes will be denoted by H'.*

(This differs from the set $\mathfrak{H}$ of the preceding section by the norm.)

Every integral of the form (3.3) is in H, but since $\| T_t \|_B$ is not necessarily bounded at the origin, we can *not* conclude that $\mathfrak{Y}_1 \subset H$.

Suppose now that there exists an element $Y \in \mathfrak{Y}_1$ such that there exists a Banach limit (cf. section 4.2)

$$y = B\text{-lim} \frac{YT_h - Y}{h} \qquad (h \downarrow 0) \tag{6.1}$$

and $y \in \mathfrak{H}$. Replacing in the definition x by $T_t x$ we conclude that for each $x \in \mathfrak{X}$ and $t > 0$

$$\begin{aligned}(6.2)\qquad &\liminf h^{-1}\{YT_{t+h}x - YT_t x\} \leqq yT_t x \leqq \\ &\limsup h^{-1}\{YT_{t+h}x - YT_t x\}.\end{aligned}$$

It follows first of all that

$$(6.3)\qquad yT_t = B\text{-lim}\, \frac{YT_{t+h} - YT_t}{h}.$$

Moreover, $yT_t x$ is finitely valued and essentially bounded in each finite interval, while $YT_t x$ is continuous. The upper and lower right derivatives of $YT_t x$ are separated by $yT_t x$. It follows under these conditions (cf. [5] p. 205) that $YT_t x$ is the integral of $yT_t x$, or

$$(6.4)\qquad YT_t - Y = \int_0^t (yT_s)\, ds.$$

From this we conclude that any two functionals satisfying (6.1) belong to the same equivalence class. This justifies the

Definition. *If for some $Y \in \mathfrak{Y}_1$ there exists an equivalence class $y' \in H'$ including a functional*[11] *y for which* (6.1) *holds then we write*

$$(6.5)\qquad y' = \Omega Y.$$

The operator Ω (from $\mathfrak{Y}_1$ to H') will be called the infinitesimal generator of $\{T_t\}$ in $\mathfrak{Y}$.

Note. Various variants of this definition are possible. In particular, it is by no means necessary to restrict the Banach limits to H'. (The theorem leading to (6.4) applies to Perron-integrable functions.) Different definitions will lead to *extensions* of our operator Ω. Our definition has the advantage of avoiding minor convergence proofs.

Lemma. *Necessary and sufficient for Y to belong to the domain $D(\Omega)$ is that it be of form* (6.4). *In particular, $Y \in D(\Omega)$ implies $Y \in H$.*

Ω is a linear operator commuting with T_t.

Note that the linearity follows from (6.4) and not from the definition. The commutativity is expressed in (6.3).

For $y \in H$ we can form the Laplace integral

$$(6.6)\qquad Z = \int_0^\infty e^{-\lambda t}(yT_t)\, dt$$

which will be meaningful at least for $\lambda > p$. Obviously Z is the strong limit of a sequence of elements of the form

$$\sum e^{-\lambda_n t_n} \int_{t_n}^{t_n+1} (yT_t)\, dt,$$

and therefore $Z \in \mathfrak{Y}_1$.

[11] It is important to include all functionals and not only elements of $\mathfrak{Y}$. However the equivalence classes are a priori restricted to H'.

Now

$$\frac{ZT_h - Z}{h} = h^{-1} \int_0^\infty e^{-\lambda t}\{yT_{t+h} - yT_t\}\, dt$$

$$(6.7) \qquad = h^{-1}(e^{\lambda h} - 1)Z - e^{\lambda h} h^{-1} \int_0^h (yT_t)\, dt$$

$$+ e^{\lambda h} h^{-1} \int_0^h (1 - e^{-\lambda t})(yT_t)\, dt.$$

We know from Section 3 that for $h \to 0$ the term in the middle has a Banach limit contained in the equivalence class y'. The last term converges strongly to zero. Accordingly, $Z \in D(\Omega)$ and

$$(6.8) \qquad \lambda Z' - \Omega Z = y'.$$

Conversely, start with an arbitrary $Y \in D(\Omega)$ and *define* y' by

$$(6.9) \qquad \lambda Y' - \Omega Y = y' \qquad (\lambda > p).$$

Then, by definition, $y' \in H'$ and we can define the Laplace integral (6.6) In this case, then, both (6.8) and (6.9) are satisfied. Put $U = Y - Z$. Then $\Omega U = \lambda U'$. By (6.4) we have then for each $x \in \mathfrak{X}$

$$(6.10) \qquad UT_t x - Ux = \lambda \int_0^t (UT_s x)\, ds$$

or

$$(6.11) \qquad UT_t x = (Ux)e^{\lambda t}.$$

However, $e^{-pt} \cdot (UT_t x)$ is bounded, and therefore $Ux = 0$. Since x is arbitrary this implies $U = 0$. This proves the

THEOREM. *Necessary and sufficient that an element $Y \in \mathfrak{Y}_1$ be in the domain of Ω is that for each $\lambda > p$ it be representable as a Laplace integral of form* (6.6) *with $y \in H$. For fixed $y \in H$ the resolvent equation has one and only one solution.*

PRINCETON UNIVERSITY

REFERENCES

[1] S. BANACH: Théorie des Opérations Linéaires. Warszawa 1932.

[2] W. FELLER: *The parabolic differential equations and the associated semi-groups of transformations*. Ann. of Math., vol. 55 (1952), pp. 468–519.

[3] E. HILLE: Functional Analysis and Semi-groups. AMS Colloquium Publ. 31, New York 1948.

[4] R. S. PHILLIPS: *On one-parameter semi-groups of linear transformations*. Proc. Amer. Math. Soc., vol. 2 (1951), pp. 234–237.

[5] S. SAKS: Theory of the Integral. Warszawa-New York 1937.

ANNALS OF MATHEMATICS
Vol. 58, No. 1, July, 1953
Printed in U.S.A.

ON THE GENERATION OF UNBOUNDED SEMI-GROUPS OF BOUNDED LINEAR OPERATORS

BY WILLIAM FELLER[1]

(Received October 2, 1952)

1. Introduction

A fundamental problem in the theory of semi-groups consists in introducing an infinitesimal generator which determines the structure of the semi-group and in characterizing the operators which can serve as infinitesimal generators. This has been done by Hille[2] and Yosida [4] for semi-groups $\{T_t\}$ satisfying two conditions: (1) It is assumed that T_t is strongly continuous not only for $t > 0$ but also for $t = 0$; (2) It is required that for t sufficiently small $\| T_t \| \leq 1 + ct$, where c is a constant.

Strong measurability implies[3] strong continuity for $t > 0$. However, the additional assumption that $T_t x \to x$ as $t \to 0$ is a severe restriction[4] and the same is true of the second condition.

The purpose of this paper is to *generalize the Hille-Yosida theorems to arbitrary strongly measurable semi-groups* (i.e. strong continuity is assumed only for $t > 0$).

The norm of our semi-groups in general will be *unbounded* near $t = 0$. In Section 2 we define the infinitesimal generator in a way which seems most suited for our purposes.

A novel feature is that the "resolvent" $(\lambda I - \Omega)^{-1}$ *will be an unbounded operator* and its domain will *not* coincide with the basic space. In fact, *the right half of the complex λ-plane will be seen to belong to the continuous spectrum of* Ω rather than to the resolvent.

Despite this radical difference the nature of the basic theorems remains essentially the same. The reason lies in the fact that the domain of $(\lambda I - \Omega)^{-1}$ can be remetrized so as to make it a Banach space. With this new norm $(\lambda I - \Omega)^{-1}$ becomes a bounded operator, and the proofs depend on the interplay between the two metrics.

In the last section the theorem is specialized to the case of *semi-groups which are strongly continuous at* $t = 0$ (that is, only condition (2) is dropped). In this case the theorem is a straight forward generalization of the Hille-Yosida theorem.

The inversion formula (4.2) and the idea to base the proof on it are due to Yosida whose elegant proof has served us as a guide. Reduced to his special

[1] Research connected with Office of Ordnance Research Contract No. DA-36-034-ORD-640 at Princeton University.

[2] Cf. [1] Chapter 12. In [2] Hille gives a simplified proof.

[3] This was proved by Phillips [3]. Cf. [1] p. 183.

[4] Cf. Hille's remarks, [1] p. 237 and 242. In the applications of semi-groups to the Cauchy problem for partial differential equations it is important that no restrictive conditions be imposed on the behavior near $t = 0$.

case; the present proof will be seen to give a slightly simplified version of Yosida's proof.

2. Definitions

Let X be a complex Banach space and suppose that for each $t > 0$ there is defined a bounded linear operator T_t from X to X such that $T_{t+s} = T_t T_s$. We shall assume

(a) The set R of elements of the form $T_t x$ with $x \in X$ (the *range*) is dense[5] in X.

(b) $T_{t+h}x \to T_t x$ as $h \downarrow 0$ for each $t > 0$ and $x \in X$.

This implies that $\| T_t \|$ is bounded in every *closed* subinterval of $(0, \infty)$ and that $t^{-1} \log \| T_t \|$ tends to a finite limit or to $-\infty$ as $t \to \infty$ (cf. [1] p. 184 and 135). We can always replace $\{T_t\}$ by the equivalent semi-group $\{e^{-at}T_t\}$ and therefore there is no loss of generality in the assumption

(c) The norm $\| T_t \|$ is bounded at infinity.

This assumption is introduced only to simplify formulas.

Definition 1. *The continuity set Σ of $\{T_t\}$ consists of those elements $x \in X$ for which $T_t x \to x$ as $t \to 0$.*

We note the obvious but important

Lemma. *We have $R \subset \Sigma$ and therefore Σ is dense in X. For each $x \in \Sigma$*

$$N(x) = \sup_{t>0} \| T_t x \| \tag{2.1}$$

is finite and

$$N(x) \geqq \| x \| . \tag{2.2}$$

Definition 2. *If for a fixed $x \in \Sigma$ there exists a $y \in \Sigma$ such that*

$$\frac{T_h x - x}{h} \to y \qquad as \quad h \downarrow 0 \tag{2.3}$$

we shall write

$$y = \Omega x. \tag{2.4}$$

This operator Ω will be called the infinitesimal generator of $\{T_t\}$.

The hypothesis $y \in \Sigma$ is crucial for the sequel. The limit (2.3) can exist without y belonging to Σ but we define Ω as operator from Σ to Σ. Obviously *we have* (2.4) *if and only if*

$$T_t x - x = \int_0^t (T_s y)\, dy, \qquad y \in \Sigma, \tag{2.5}$$

the integral being taken in the sense of Bochner. (Because of the continuity it reduces to the familiar limit of Riemann sums.)

3. The resolvent equation

In this section we put

$$v(t) = \sup_{s>0} \| T_{t+s} \| . \tag{3.1}$$

[5] If this condition is not fulfilled one should replace X by the closure of R.

In consequence of (c) this function is finite for $t > 0$, but it may be unbounded as $t \to 0$.

THEOREM 1. *Let the semi-group* $\{T_t\}$ *satisfy the conditions* (a)–(c) *of Section* 2. *The set* Σ *remetrized by* (2.1) *becomes a Banach space.*

For fixed $x \in \Sigma$ *and* $\lambda > 0$ *the resolvent equation*

$$\lambda\xi - \Omega\xi = x \tag{3.2}$$

has a unique solution

$$\xi = J_\lambda x. \tag{3.3}$$

The (possibly unbounded) operator J_λ *has the following three properties:*

A. *Considered as operator from the Banach space* Σ *to itself,* λJ_λ *is bounded, linear, and has norm*

$$N(\lambda J_\lambda) \leqq 1. \tag{3.4}$$

B.[6] *For* $x \in \Sigma$

$$N(\lambda J_\lambda x - x) \to 0 \qquad \text{as } \lambda \to \infty. \tag{3.5}$$

C. *Let* $x \in \Sigma$ *and* $0 < \delta' < \delta$. *There exist a* $\lambda_0 = \lambda_0(x, \varepsilon)$ *such that if*

$$n \geqq \lambda\delta \tag{3.6}$$

and $\lambda > \lambda_0$, *then*

$$\lambda^n \| J_\lambda^n x \| \leqq \{v(\delta') + \varepsilon\} \cdot \| x \| . \tag{3.7}$$

Furthermore, one has

$$\xi = \int_0^\infty e^{-\lambda s}(T_s x)\, ds \tag{3.8}$$

and

$$N(x) = \sup_{n \geqq 0} \sup_{\lambda > 0} \| \lambda^n J_\lambda^n x \| . \tag{3.9}$$

NOTE. Without introducing Σ in its capacity as Banach space we may restate condition A as follows:

A_1. *For each* $x \in \Sigma$

$$N(\lambda J_\lambda x) \leqq N(x). \tag{3.10}$$

A_2. *If for a sequence* $x_n \in \Sigma$

$$N(x_n - x_m) \to 0 \qquad \text{as } n, m \to \infty \tag{3.11}$$

then $x_n \to x$ *with*

$$x \in \Sigma, \qquad J_\lambda x = \lim_{n\to 0} J_\lambda x_n . \tag{3.12}$$

[6] Condition B requires simply that the domain $D(\Omega)$ be dense in Σ in the sense of the N-norm (cf. the concluding paragraph of the paper).

PROOF. Define ξ by (3.8) where $x \in \Sigma$. An easy calculation (used similarly by Yosida) yields

$$\frac{T_h \xi - \xi}{h} = \frac{e^{\lambda h} - 1}{h} \int_0^\infty e^{-\lambda t}(T_t x)\, dt - e^{\lambda h} \cdot \frac{1}{h} \int_0^h e^{-\lambda t}(T_t x)\, dt. \tag{3.13}$$

It is now clear that as $h \to 0$ the right side tends to $\lambda\xi - x$. This implies that $T_h \xi \to \xi$ so that $\xi \in \Sigma$. Thus $\Omega\xi$ exists and satisfies (3.2).

To prove the uniqueness of this solution assume that $\lambda y - \Omega y = 0$ with $y \in \Sigma$. Then $T_t y$ has a right derivative satisfying

$$\frac{d}{dt} T_t y = \lambda T_t y. \tag{3.14}$$

In view of the assumed continuity this implies

$$T_t y = e^{\lambda t} y. \tag{3.15}$$

But $\| T_t y \|$ is bounded, and therefore $y = 0$.

Next, we have by (2.1)

$$N(\lambda\xi) = \sup \lambda \left\| \int_0^\infty e^{-\lambda s}(T_{t+s})\, ds \right\| \leqq N(x) \tag{3.16}$$

which proves (3.10). Moreover, in view of (2.2) it is clear that (3.11) implies the convergence of x_n to an element x. That $T_h x$ approaches x follows from the triangle inequality and assertion A_2 is now obvious.

For $x \in \Sigma$ we have the uniform continuity of $\| T_t x \|$ in each finite interval $0 \leqq t \leqq a$, and therefore for all t, since $\| T_t \|$ is bounded for $t \geqq a$. To prove property B choose δ so that for all t

$$\| T_{t+s} x - T_t x \| < \varepsilon \qquad \textit{for } s < \delta. \tag{3.17}$$

Then

$$N(\lambda\xi - x) = \sup_{t>0} \lambda \left\| \int_0^\infty e^{-\lambda s}(T_{t+s} x - T_t x)\, ds \right\|. \tag{3.18}$$

The assertion (3.5) follows at once upon splitting the interval of integration into $(0, \delta)$ and (δ, ∞).

To prove property C we note that for $x \in \Sigma$ and $n \geqq 1$

$$J_\lambda^n x = \frac{1}{(n-1)!} \int_0^\infty e^{-\lambda s} s^{n-1}(T_s x)\, ds. \tag{3.19}$$

This can be proved either by induction using the argument which led to (3.13), or from a well-known property of resolvents (cf. (4.5)). Now in view of (3.1)

$$\frac{1}{(n-1)!} \int_{\delta'}^\infty e^{-\lambda t} t^{n-1} \| T_t x \|\, dt \leqq v(\delta') \cdot \| x \| \cdot \lambda^{-n} \tag{3.20}$$

while

$$\frac{1}{(n-1)!} \int_0^{\delta'} e^{-\lambda t} t^{n-1} \| T_t x \|\, dt \leqq \frac{N(x)\lambda^{-n}}{(n-1)!} \int_0^{\delta'\lambda} e^{-s} s^{n-1}\, ds. \tag{3.21}$$

If $n \geqq \lambda\delta$ we get, putting $q = \delta'/\delta$ with $q < 1$,

$$(3.22) \qquad \frac{1}{(n-1)!}\int_0^{\delta'\lambda} e^{-s}s^{n-1}\,ds \leqq e^{-qn}\sum_{r=n}^{\infty}\frac{(qn)^r}{r!} < \frac{q}{(1-q)^2 n},$$

by a known simple estimate (cf., for example, [1] p. 187). Combining (3.19)–(3.22) we get (3.7).

It remains to prove (3.9). Now it is well known that if $f(t)$ is continuous and bounded, and if $\lambda \to \infty$, $n \to \infty$ so that $(n-1)\lambda^{-1} \to t$, then

$$(3.23) \qquad \frac{\lambda^n}{(n-1)!}\int_0^{\infty} e^{-\lambda s}s^{n-1}f(s)\,ds \to f(t).$$

(This is easily verified by the substitution $\lambda s = n - 1 + u(n-1)^{\frac{1}{2}}$.) It follows then from (3.19) that

$$(3.24) \qquad \liminf \lambda^n \| J_\lambda^n x \| \geqq \| T_t x \|.$$

Since t is arbitrary, this means that the right side in (3.9) is greater than or equal to $N(x)$. On the other hand, it cannot exceed $N(x)$ in view of (3.10). Therefore (3.9) holds.

4. The converse

THEOREM 2. *Let Σ be a linear set dense in X and let Ω be a linear operator with domain and range in Σ. Suppose that for $x \in \Sigma$ and $\lambda > 0$ the resolvent equation* (3.2) *has a unique solution $J_\lambda x$, and that the quantity $N(x)$ defined by* (3.9) *is finite. Suppose furthermore that the set Σ remetrized with the norm $N(x)$ becomes a Banach space and that J_λ has the properties* A–C *of Theorem* 1. (*Alternatively, suppose that J_λ has the properties* A_1, A_2, B *and* C.) *Here $v(t)$ is an arbitrary finite non-decreasing function.*

Then there exists a semi-group $\{T_t\}$ of bounded linear operators with domain X and range dense in X which is strongly continuous for $t > 0$ and whose infinitesimal generator coincides with Ω. For it Σ is the set of continuity, (2.1) *holds, and*

$$(4.1) \qquad \| T_{t+s} \| \leqq v(t) \qquad \textit{for } s > 0.$$

For each $x \in X$

$$(4.2) \qquad T_t x = \lim_{\lambda\to\infty} e^{-\lambda t}\sum_{n=0}^{\infty}\frac{(\lambda t)^n}{n!}(\lambda^n J_\lambda^n x).$$

Following Yosida the last expression may be written in the form

$$(4.3) \qquad e^{\lambda t\{\lambda J_\lambda - I\}}x = e^{\lambda t\Omega J_\lambda}x.$$

PROOF. First, notice that in (3.9) zero is admitted as exponent so that $N(x) \geqq \| x \|$. From property B it follows at once that for each n *the domain $D(\Omega^n)$ is dense in Σ in the sense of the norm $N(x)$. Now Σ is dense in X and* $\| x \| \leqq N(x)$. Therefore *$D(\Omega^n)$ is dense in X.*

Although J_λ is not an endomorphism in X, it is an endomorphism in the

Banach space Σ. As such it satisfies the well-known (cf. [1] p. 98) and easily verified relation

$$J_{\lambda+h}x = \sum_{n=0}^{\infty}(-h)^n J_\lambda^{n+1}x, \qquad |h| < \lambda \tag{4.4}$$

and hence

$$\frac{d^n}{d\lambda^n} J_\lambda x = (-1)^n n! J_\lambda^{n+1}x. \tag{4.5}$$

Since $\| x \| \leqq N(x)$ it is clear that *for fixed* $x \in \Sigma$ *the relations* (4.4) *and* (4.5) *remain valid also when interpreted in the original norm of* X.

Following Yosida we now introduce the operator

$$U_\lambda(t)x = e^{-\lambda t} \sum_{n=0}^{\infty} \frac{(\lambda t)^n}{n!} (\lambda^n J_\lambda^n x), \tag{4.6}$$

defined for $x \in \Sigma$. Considered in X the operator $U_\lambda(t)$ is unbounded, but as an operator in the remetrized Banach space Σ it is clear from (3.4) that

$$N(U_\lambda(t)) \leqq 1. \tag{4.7}$$

With this interpretation of $U_\lambda(t)$ we have for $x \in D(\Omega^2)$

$$\begin{aligned} \frac{d}{d\lambda} U_\lambda(t)x &= tU_\lambda(t)\{-x + 2\lambda J_\lambda x - \lambda^2 J_\lambda^2 x\} \\ &= -tU_\lambda(t)J_\lambda^2 \Omega^2 x, \end{aligned} \tag{4.8}$$

whence by (4.7) and (3.4)

$$N\left(\frac{d}{d\lambda} U_\lambda(t)x\right) \leqq \frac{tN(\Omega^2 x)}{\lambda^2}. \tag{4.9}$$

It follows by integration that as $\lambda \to \infty$ the elements $U_\lambda(t)x$ form a Cauchy sequence in the Σ-norm.

Accordingly, *for fixed* $t > 0$ *and* $x \in D(\Omega^2)$ *we define*

$$T_t x = \lim_{\lambda\to\infty} U_\lambda(t)x. \tag{4.10}$$

Then

$$T_t x \in \Sigma, \qquad N(T_t x) \leqq N(x). \tag{4.11}$$

We now require an estimate for the norm $\| T_t x \|$. Choose $0 < q' < q < 1$ arbitrarily and put $N = [\lambda q t]$. In consequence of property C we have for $x \in \Sigma$ and $\lambda > \lambda_0(x, \varepsilon)$

$$\left\| e^{-\lambda t} \sum_{n=N+1}^{\infty} \frac{(t\lambda)^n}{n!} (\lambda^n J_\lambda^n x) \right\| \leqq \{v(q't) + \varepsilon\} \cdot \| x \|. \tag{4.12}$$

On the other hand, we have for all $\lambda > 0$

$$\left\| e^{-\lambda t} \sum_{n=0}^{N} \frac{(\lambda t)^n}{n!} (\lambda^n J_\lambda^n x) \right\| \leqq N(x)e^{-\lambda t} \sum_{n=0}^{N} \frac{(\lambda t)^n}{n!} \leqq \frac{N(x)}{\lambda t(1 - q)^2}, \tag{4.13}$$

by the inequality already used in (3.22). Thus for $x \in \Sigma$ and $\lambda > \lambda_0(x, \varepsilon)$

(4.14) $$\| U_\lambda(t)x \| \leqq \{v(q't) + \varepsilon\}\cdot\| x \| + \frac{N(x)}{\lambda t(1 - q)^2}.$$

The first implication of this is that for fixed $0 < t' < t$ and $x \in D(\Omega^2)$

(4.15) $$\| T_t x \| \leqq v(t')\cdot\| x \|.$$

Thus T_t is a bounded linear transformation defined on $D(\Omega^2)$, which set is dense in X.

By continuity *we now extend the definition of* $T_t x$ *to all* $x \in X$. Then (4.15) continues to hold and this proves (4.1).

We proceed to show that (4.10) *holds for all* $x \in \Sigma$ *if the convergence is interpreted in the original* X*-norm*. To prescribed ε and $t' < t$ find an $x_0 \in D(\Omega^2)$ such that

(4.16) $$v(t')\cdot\| x - x_0 \| < \varepsilon.$$

Then

(4.17) $$\| T_t x - U_\lambda(t)x \| \leqq \| T_t x - T_t x_0 \| + \| U_\lambda(t)(x - x_0) \| + \| T_t x_0 - U_\lambda(t)x_0 \|.$$

The first term on the right is less than ε by (4.16) and (4.1). By (4.14) the second term comes arbitrarily close to ε for λ sufficiently large. The last term tends to zero since $x_0 \in D(\Omega^2)$. This establishes (4.10) for $x \in \Sigma$.

We are now able to *prove the semi-group property* $T_{s+t}x = T_s T_t x$. It suffices to prove it for $x \in D(\Omega^2)$. Now

(4.18) $$U_\lambda(s + t)x = U_\lambda(s)U_\lambda(t)x.$$

Put $U_\lambda(t)x - T_t x = y_\lambda$. Then $y_\lambda \in \Sigma$ and we know from (4.9) that $N(y_\lambda) = OH(\lambda^{-1})$. Hence the right side in (4.18) differs from $U_\lambda(s)T_t x$ by a quantity for whose norm we have the relation

(4.19) $$\| U_\lambda(s)y_\lambda \| \leqq N(U_\lambda(s)y_\lambda) \leqq N(y_\lambda) = O(\lambda^{-1}).$$

It follows that the right side in (4.18) tends to $T_s T_t x$ while the left side approaches $T_{s+t}x$.

Our next assertion is that for $x \in D(\Omega)$

(4.20) $$T_t x - x = \int_0^t (T_s \Omega x)\, ds.$$

In fact, for fixed λ we have

(4.21) $$\frac{d}{ds} U_\lambda(s) = U_\lambda(s)\{\lambda J_\lambda x - x\} = U_\lambda(s)(\lambda J_\lambda \Omega x).$$

Putting $\lambda J_\lambda \Omega x - \Omega x = y_\lambda$ we have $N(y_\lambda) \to 0$ by property B. An estimate of the form (4.19) then shows that as $\lambda \to \infty$

$$U_\lambda(t)x - x = \int_0^t (U_\lambda(s)\Omega x)\, ds + O(1), \tag{4.22}$$

and (4.20) follows by the bounded convergence principle.

From (4.20) we draw several conclusions. First we see that the left side depends continuously on t (which fact follows directly from the equicontinuity of $U_\lambda(t)x$). Now T_t is a bounded operator for each fixed $t > 0$ and, since $D(\Omega)$ is dense in X, it follows from the Banach-Steinhaus theorem ([1] p. 25), or the triangle inequality, that $\{T_t\}$ *is strongly continuous for* $t > 0$.

Second, for $x \in D(\Omega)$ we have $T_t x \to x$ and therefore *the range of* $\{T_t\}$ *is dense* in X. Thus the conditions of Section 2 are satisfied and we may introduce the *true infinitesimal generator* Ω^* *of* $\{T_t\}$. From (4.20) we can conclude that $\Omega x = \Omega^* x$ if $x \in D(\Omega)$ *and if* Ωx belongs to the continuity set of $\{T_t\}$. The latter condition is certainly satisfied if $\Omega x \in D(\Omega)$. Hence if $x \in D(\Omega^2)$ then $\Omega x = \Omega^* x$. In this case $\xi = J_\lambda x$ is given by (3.8) and we conclude from (3.19)

$$N(x) = \sup_n \sup_\lambda \lambda^n \, \| J_\lambda^n x \| \leqq \sup_{t>0} \| T_t x \| . \tag{4.23}$$

On the other hand, (4.11) gives the reverse inequality and therefore (2.1) holds for $x \in D(\Omega^2)$. This together with (4.1) shows that if $x_n \in D(\Omega^2)$ and if $\{x_n\}$ forms a Cauchy sequence in the X-norm, then $\{T_t x_n\}$ is a Cauchy sequence in the Σ-norm. In particular, if $x_n \to x$ then $N(T_t x - T_t x_n) \to 0$. *We conclude that* $T_t x \in \Sigma$ *for each* $x \in X$ *and* $t > 0$. Moreover, if x is of the form $x = T_s y$, then (2.1) holds.

Suppose now that $T_h x \to x$ as $h \to 0$. Using the fact of uniform continuity (cf. (3.17)) we conclude that $N(T_h x - x) \to 0$, whence $x \in \Sigma$. On the other hand, we may now consider the contraction of $\{T_t\}$ to Σ and study this semi-group in the Σ-norm. Then $N(T_t) \leqq 1$ and $T_h x \to x$ for $x \in D(\Omega^2)$ which is a dense set. Hence $T_h x \to x$ for all $x \in \Sigma$ and we have proved that Σ *is the continuity set of* $\{T_t\}$.

Now (4.20) implies that Ω^* *is an extension of* Ω. To prove that it cannot be a proper extension let $x \in D(\Omega^*)$ and put $x - \Omega^* x = y$. By our definition of the infinitesimal generator, y belongs to the continuity set of $\{T_t\}$ and hence $y \in \Sigma$. Accordingly, we can find a sequence $y_n \in D(\Omega)$ such that $N(y - y_n) \to 0$. Then $x_n = J_1 y_n$ converges to $J_1 y$. But since $x_n \in D(\Omega)$, we have $\Omega^* x_n = \Omega x_n = x_n - y_n$ or

$$x_n = \int_0^\infty e^{-t}(T_t y_n)\, dt \to \int_0^\infty e^{-t}(T_t y)\, dy = x = J_1(y) \tag{4.24}$$

whence $x \in D(\Omega)$ as asserted.

5. Continuity at the origin

THEOREM 3. *In order that an operator* Ω *be the infinitesimal generator of a semi-group* $\{T_t\}$ *from* X *to* X *which is strongly continuous for* $t \geqq 0$ *and whose norm* *is bounded at infinity, it is necessary and sufficient that for each $x \in X$ and $\lambda > 0$ the resolvent equation* (3.2) *have a unique solution $J_\lambda x$ and that*

(1) *the range of Jx be dense in X*

(2) *there exists a constant C such that*

$$\lambda^n \| J_\lambda^n x \| \leqq C \cdot \| x \|. \tag{5.1}$$

Proof. The necessity of the condition follows immediately from (3.19). The proof of Theorem 2 goes through under the new conditions with simplifications and omitting all reference to condition C. Alternatively, we may verify that the conditions of Theorem 2 are satisfied. From (3.9) and (5.1) we see that $N(x) \leqq C \| x \|$ so that $\Sigma = X$ and the two norms are equivalent. To prove (3.10) we start from the following simple generalization of (4.4)

$$J_{\lambda+h}^p x = \sum_{n=p-1}^{\infty} \binom{n}{p-1} (-h)^{n-p+1} J_\lambda^{n+1} x. \tag{5.2}$$

For $0 < \mu < \lambda$ we get using (5.1)

$$\| \lambda^q \mu^p J_\lambda^q J_\mu^p x \| \leqq (\eta/\lambda)^p \sum_{n=p-1}^{\infty} \binom{n}{p-1} \left(1 - \frac{n}{\lambda}\right)^{n-p+1} \cdot N(x) = N(x) \tag{5.3}$$

which proves (3.10). Thus condition A holds.

If $x \in D(\Omega)$ then by (5.1)

$$\| \lambda J_\lambda x - x \| = \| J_\lambda \Omega x \| \leqq \frac{C \cdot \| \Omega x \|}{\lambda} \tag{5.4}$$

and hence $\lambda J_\lambda x \to x$ as $\lambda \to \infty$ for $x \in D(\Omega)$. Now the domain of Ω is the range of J_λ and therefore dense in X. Thus $\lambda J_\lambda x \to x$ for all x by the Banach-Steinhaus theorem. Finally, condition C holds with $v(t) = C$.

Princeton University

References

[1] E. Hille. Functional Analysis and Semi-Groups. American Mathematical Society Colloquium Publications, vol. 31. New York 1948.

[2] E. Hille. On the generation of semi-groups and the theory of conjugate functions. Kungl. Fysiografiska Sällskapets i Lund Förhandlingar. Vol. 21, no. 14 (1952).

[3] R. S. Phillips. *On one-parameter semi-groups of linear transformations.* Proc. Amer. Math. Soc. ,Vol. 2 (1951), pp. 234–237.

[4] K. Yosida. On the differentiability and the representation of one-parameter semi-groups of linear operators. J. Math. Soc. of Japan, vol. 1 (1948), pp. 15–21.

Annals of Mathematics
Vol. 60, No. 3, November, 1954
Printed in U.S.A.

THE GENERAL DIFFUSION OPERATOR AND POSITIVITY PRESERVING SEMI-GROUPS IN ONE DIMENSION

By William Feller

(Received July 13, 1954)

1. Introduction

The main purpose of this paper is to give an *abstract definition of a diffusion process* and to derive the *most general form of a diffusion operator* in one dimension. The latter is a generalized differential operator which includes, as a special case, the operator appearing on the right of the classical diffusion equation (1.1) (also when the coefficients vanish or are singular). It is noteworthy that we get our operator essentially from the sole requirement that the resolvent be positivity-preserving. *Thus the diffusion equation (1.1) is now characterized as the only differential equation of the form* $u_t = Au$ *which preserves positivity.*

We begin (Section 2) by defining "processes," that is, semi-groups of transformations of positive measures. It is shown that, and why, only such semi-groups qualify which are adjoints to semi-groups on continuous functions. This fact is of fundamental importance for probability theory (and, in various disguises, for many applications of semi-groups to analysis). Section 4 contains a discussion of this and of the relation of our approach to the more established theory of semi-groups based on the strong topology.

When our transformations are *measure preserving* (that is, when probability measures are transformed into probability measures) then our processes reduce to *stochastic processes of Markovian character*. However, in the general case, probabilities are replaced by *arbitrary measures*. These may be interpreted as variable expectations; the physicist would speak of diffusion with *creation or destruction of masses.*[1]

In Section 3 we define *diffusion processes by the property that their infinitesimal generator is of local character. A Markov process is a diffusion process if, and only if, the path functions are continuous.* This shows that our definition refers to a natural class of processes.[2]

Our problem now reduces to *finding the most general form of an infinitesimal generator A of local character of a semi-group* $\{T_t\}$ *which transforms non-negative continuous functions into non-negative continuous functions.* (No restrictions on the norm $\| T_t \|$ are imposed.)

It is clear that our problem is not affected by topological transformations of the x-axis or, what amounts to the same, by the choice of the scale. However,

---

[1] The restriction of the theory of stochastic processes to strict probability measures appears from many points of view as a serious default. The more general situation is physically equally intuitive and important. Unfortunately an adequate probability interpretation in the sample space of all possible paths does not yet exist.

[2] The equivalence of the two properties in one dimension has been proved by D. Ray. He is engaged in proving the same fact for arbitrary compact spaces.

differential operators are not necessarily transformed into differential operators. Therefore, our infinitesimal generator A need not be a differential operator in a classical sense. It turns out to be a *generalized* differential operator involving derivatives with respect to arbitrary measures.[3]

The *characteristic property* of our operators A is given in Lemma 2 (Section 5). In [6] we have determined the most general operator with this property. In the present paper we have only to find the (rather minute) additional regularity properties due to the fact that A is an infinitesimal generator.

The honest diffusion operator in the case of a pure stochastic process (i.e. when measures are preserved) is of the form of a *mixed derivative* $D_v D_u$ with respect to two monotonic functions u and v, of which u (not necessarily v) is continuous (Section 10, Theorem 8). This is a direct generalization of the classical $A = d^2/dx^2$. If measures are not preserved, we get not only an additional term corresponding to c in (1.1), but have to introduce a regularizing function ϕ. (This is due to the fact that the constant function is not necessarily in the domain of A.) The general form is contained in Theorem 7.

This form applies only when there are no *irregular points*. These are of two kinds: *absorbing points* and *translation points* (Sections 8 and 9). Very roughly, these correspond to the cases $a = b = 0$ and $a = 0, b \neq 0$ in the classical diffusion equation (1.1). In other words, they are points where the differential operator degenerates. It must be borne in mind, however, that whether a in (1.1) does or does not vanish depends largely on the arbitrary choice of the x-scale (cf. the last footnote). Our characterization is, of course, *intrinsic*.

In Section 11 it is shown that *for a diffusion process* (equations (1.1) and (1.2)) *the norm of the solution may tend to infinity as* $t \to 0$ even if $u(t, x) \to f(x)$ and f is smooth. This remark influences the formulation of Cauchy problems in general.

The paper concludes with a study of *isolated translation points*.

Section 6 describes a method of obtaining families of *isomorphic semi-groups*. This device is used for a regularization of the generator.

The traditional diffusion equations are

$$\frac{\partial v}{\partial t} = a(x)\frac{\partial^2 u}{\partial x^2} + b(x)\frac{\partial u}{\partial x} + c(x)u \tag{1.1}$$

and

$$\frac{\partial v}{\partial t} = \frac{\partial^2}{\partial x^2}[av] - \frac{\partial}{\partial x}[bv] + cv \tag{1.2}$$

[3] The case where A is a differential operator in the old sense was treated in [5]. Essentially such an A is a second order operator (the right side of (1.1)) but on a set of measure zero the operator may degenerate into an operator of *higher order in the classical sense*. This is due to the fact that the order of a differential operator depends on the choice of the scale. [Compare differentiation with respect to x and to $\xi = x^2$ at the origin.] This shows the advantage of our *invariant characterization* of operators. [For example, first order operators are invariantly characterized in Definition 7.] The investigation of [5] has been continued by S. Leader [9]. However, the present paper uses no previous results.

of which the first refers to continuous functions, the second to density functions (measures). The probability literature treats only the case $c = 0$.

Equation (1.1) (with $c = 0$) has been derived from probabilistic assumptions by Kolmogorov [8] and, using the condition of local character, in [1]. In [5] we have shown that (1.1) is essentially the *only positivity preserving differential equation* of the parabolic type. This was, up to now, the most general justification of (1.1).

In reality, (1.1) is void of probabilistic, or physical, significance and serves only as an intermediary in the derivation of (1.2) (the so-called Fokker-Planck equation). Years of investigation have shown the futility of the effort to avoid this annoying circuit. Finally, in [2] it turned out that (1.2) is *not the general form* of the Fokker-Planck equation and that other terms must be added. The discussion of Section 4 now explains the theoretical reason for the dependence of (1.2) on (1.1).

As has been stated, using only the condition of preservation of positivity, we derive (1.1) with the right side replaced by a generalized differential operator. These generalized operators enable us to express the adjoint (1.2) even in the case where the coefficients a and b are not differentiable. Another advantage of the new symbolism is the fact that we require no special consideration for cases where a vanishes on a null-set, and the like.

It turns out that probabilistic phenomena, which were known only for end-points (elastic barriers) can also occur at *interior points*. This seems of interest for the behavior of the path-functions.

In [2] we have constructed all the semi-groups (diffusion processes) generated by the right side of (1.1) with $c = 0$. The same analysis carries over to our generalized operators and becomes more natural inasmuch as all quantities are now intrinsically connected with the process and have a probabilistic significance. We shall not discuss that side of the problem in this paper.

Notations

We consider a compact interval $E = [r_1, r_2]$ where $-\infty \leqq r_1 < r_2 \leqq \infty$. We shall denote by $\mathfrak{M}_+$ the aggregate of all non-negative bounded measures on the Borel sets of E, and for $\mu \in \mathfrak{M}_+$ we put $\| \mu \| = \mu(E)$. The aggregate of continuous functions on E will be denoted by C and the aggregate of the non-negative continuous functions by C_+. For $f \in C$ we put $\| f \| = \max | f(x) |$. The unitary function (which equals 1 at all points) will be denoted by $\mathbf{1}$.

2. Processes and semi-groups

We shall be concerned with analytical set-ups which to each $\mu \in \mathfrak{M}_+$ and each $t > 0$ make correspond a new measure $\mu_t \in \mathfrak{M}_+$. [If $\mu_t(E) \equiv 1$ for all t we may speak of the probability distribution of a random variable $X(t)$. The general set-up refers intuitively to variable mass or heat distributions. Probabilistically, $\mu_t(\Gamma)$ can be interpreted as the expected number of points (the mass) contained at time t in Γ.] We assume that the mapping $\mu \to \mu_t$ is given by a linear transformation T_t:

$$\mu_t = \mu T_t. \tag{2.1}$$

Now consider two time points, t and $t + s$, and the corresponding measures μ_t and μ_{t+s} produced by the same initial measure μ. The measure μ_t, taken as new initial distribution, gives rise at time $t + s$ to the measure $(\mu_t)T_s = (\mu T_t)T_s$. We impose the *unicity condition* that this measure be identical with μ_{t+s}. This means simply that our transformations T_t must form a *semi-group*:

$$T_{t+s} = T_t T_s. \tag{2.2}$$

It is now indispensable to introduce a condition which will express that whenever two initial measures μ' and μ'' are, in some sense, close together ("practically indistinguishable"), then the same will be true of the corresponding distributions $\mu' T_t$ and $\mu'' T_t$ for any fixed t. The appropriate definition of "closeness" is implicitly underlying all classical theories. We introduce it here formally and defer to Section 4 the discussion of the condition and of the difference between this and the customary set-up.

DEFINITION 1. *A sequence μ_n of measures is said to converge to μ, in symbols $\mu_n \to \mu$, if for each $f \in C$*

$$\int_E f(x)\mu_n(dx) \to \int_E f(x)\mu(dx). \tag{2.3}$$

[This is sometimes called the "*weak star convergence*." If I_x denotes the closed interval $[r_1, x]$, we may replace μ and μ_n by the corresponding monotonic functions $\mu_n(I_x) = F_n(x)$. Then $\mu_n \to \mu$ means that $F_n(x) \to F(x)$ *at each point x of continuity of F*.]

DEFINITION 2. *By a process we mean a one-parametric family of linear transformations $\{T_t\}$ such that to each $\mu \in \mathfrak{M}_+$ and each $t > 0$ there corresponds a measure $\mu T_t \in \mathfrak{M}_+$ and:* (a) *the semi-group property* (2.2) *holds, and* (b)

$$\mu_n \to \mu \quad \textit{implies} \quad \mu_n T_t \to \mu T_t. \tag{2.4}$$

[The last condition states that T_t is, for fixed t, continuous in the "weak star topology."]

To each $x \in E$ consider the measure μ^x which attributes unit weight to x and zero weight to the complement:

$$\mu^x(\Gamma) = \begin{cases} 1 & \text{if } x \in \Gamma \\ 0 & \text{if } x \notin \Gamma. \end{cases} \tag{2.5}$$

We put

$$P(t, x, \cdot) = \mu^x T_t, \tag{2.6}$$

(so that $P(t, x, \Gamma)$ is the mass of Γ at time t due to the presence, at time 0, of a unit mass at x). For each $f \in C$ and each fixed x, $t > 0$ the number

$$f_t(x) = \int_E P(t, x, dy)f(y) \tag{2.7}$$

is well defined. If $x_n \to x$, then (2.4) shows that $f_t(x_n) \to f_t(x)$. This means that (2.7) is, for fixed t, *continuous* as a function of x. The mapping $f \to f_t$ is f linear transformation which we denote by

$$T_t f = f_t. \tag{2.8}$$

Obviously these transformations have again the semi-group property (2.2). Moreover, if $f \geqq 0$ then also $T_t f \geqq 0$. This implies that when $f \leqq g$ also $T_t f \leqq T_t g$. Since $T_t \mathbf{1}$ is bounded, it follows that for fixed t

$$\| T_t f \| \leqq \| f \| \cdot \| T_t \mathbf{1} \| \tag{2.9}$$

so that T_t (as transformation from C to C) is bounded.

Conversely, suppose that we are given a semi-group $\{T_t\}$ of transformations from C to C with the properties just mentioned. For fixed t, x the value $T_t f(x)$ is a non-negative bounded linear functional on f. This means that to each fixed t, x there exists a measure $P(t, x, \cdot) \in \mathfrak{M}_+$ such that (2.7) holds (with $f_t = T_t f$). Then for each open interval $I \subset E$ we have

$$P(t, x, I) = \sup_f \int_I P(t, x, dy) f(y), \tag{2.10}$$

the sup extending over all $f \in C_+$ which vanish outside I and for which $f \leqq 1$. For each such f the integral in (2.10) is a continuous function of x and therefore $P(t, x, I)$ is, for fixed t, I, a semi-continuous function of x. This implies the measurability of $P(t, x, \Gamma)$ as a function of x for each fixed Borel set Γ and $t > 0$. For any $\mu \in \mathfrak{M}_+$ we can therefore define a new measure $\mu_t = \mu T_t$ by

$$\mu T_t = \int_E \mu(dx) P(t, x, \cdot). \tag{2.11}$$

This is the *adjoint semi-group* to the given semi-group $\{T_t\}$ of transformations from C to C. It is obvious that it has the properties postulated in Definition 2 and we have thus proved

THEOREM 1. *Each process in the sense of Definition 2 is generated by a kernel transformation of the form* (2.11). *Such processes are in a one-to-one correspondence with semi-groups* $\{T_t\}$ *which to each* $f \in C_+$ *make correspond a* $T_t f \in C_+$ *where for fixed* $t > 0$ *each transformation* T_t *is bounded.*

(We shall see in Section 11 that it is possible that $\| T_t \| \to \infty$ as $t \to 0$.)

We shall require the following lemma given by Hille [7], p. 183.

LEMMA 1. *If for fixed* $\mu \in \mathfrak{M}_+$ *and* $f \in C$ *the function*

$$\begin{aligned} S(t) &= \iint_{E \times E} \mu(dx) P(t, x, dy) f(y) \\ &= \int_E \mu(dx) T_t f(x) \end{aligned} \tag{2.12}$$

is measurable, then for each fixed $t > 0$

$$T_{t+h} f \to T_t f \qquad h \downarrow 0 \tag{2.13}$$

uniformly in E.

This is the *strong continuity* property. It does not necessarily hold for $t = 0$.

3. Local character. Diffusion processes

The general problem of the theory consists in describing the most general form of a process[4] in the sense of Definition 2. We propose to give a complete solution for a particularly interesting class of processes.

DEFINITION 3. *A process (in the sense of Definition 2) is called of local character, or a diffusion process, if* (a) *for each* $f \in C$ *which vanishes identically in some neighborhood of any point* $x_0 \in E$ *one has*

$$\frac{T_t f - f}{t}(x_0) \to 0 \qquad \text{as } t \to 0, \tag{3.1}$$

(b) *the strong continuity condition* (2.13) *holds for* $t > 0$ (*not necessarily for* $t = 0$).

NOTE. By Lemma 1 the latter condition is equivalent to requiring that $P(t, x, I)$ be measurable as a function of t for each fixed x and each interval I.

The condition (3.1) can be restated thus: *for each open interval* I *containing* x_0 *one has*

$$t^{-1}P(t, x, E - I) \to 0 \qquad t \to 0. \tag{3.2}$$

From now on $\{T_t\}$ *shall stand for a semi-group from* C *to* C *with the properties stated in Definition* 3.

The Probability Interpretation

Consider the special case

$$P(t, x, E) \equiv 1 \qquad (x \in E,\ t > 0). \tag{3.3}$$

With this restriction, $P(t, x, E)$ is the *transition probability of a Markov process* and, conversely, each Markov process is a "process" in the sense of Definition 2. According to Doob each Markov process leads to a probability measure on the space of all functions $X(t)$ with values in E. (We omit the trivial reference to measurability conditions and "versions.") We have now the following result due to Ray (not yet published):

The functions $X(t)$ *are continuous with probability one if, and only if, the process is of a local character.*

In other words, diffusion processes are the Markov processes with continuous path functions. Of course, our definition is more general, since we do not require (3.3) so that, physically speaking, masses may be created and destroyed.

The condition (3.2) was first introduced in [1] under the name of *Lindeberg condition*. However, in [1] it was supplemented by the additional requirement that

$$P(t, x, I) = 1 + o(t) \qquad (t \to 0) \tag{3.4}$$

[4] The same problem arises in an arbitrary (not necessarily topological) space. The definition of processes of local character is meaningful in locally compact spaces.

It is of great interest that *no condition of type* (3.4) *is required.*

In [1] the Lindeberg condition was introduced to derive the differential equation (1.1) (with $c \equiv 0$). At that time additional hypotheses were introduced. It now turns out that (3.2) alone suffices to derive a generalized differential equation replacing (1.1).

4. Why weak continuity?

Before discussing the condition (2.4) and its relation to the more usual procedure, we want to show by a typical example what is bound to happen without it.

EXAMPLE 1. We consider a semi-group which translates absolutely continuous measures to the right, and singular measures to the left. More precisely, we let $E = [-\infty, +\infty]$. Then, if $\mu \in \mathfrak{M}_+$ and μ is absolutely continuous, we define $\mu T_t \equiv \mu_t$ by $\mu_t(\Gamma) = \mu(\Gamma + t)$; for singular measures we put $\mu_t(\Gamma) = \mu(\Gamma - t)$. Here $\Gamma + t$ is the set of points $x + t$ with $x \in \Gamma$. *This semi-group is not generated by kernel transformations.*[5]

The "pathology" of this semi-group depends on the intuitive feeling that a singular measure can be approximated by absolutely continuous measures and vice versa. This is not true if one uses the familiar Banach metric for measures, but is true with Definition 1. We have here an illustration to Hadamard's principle that a well posed problem must not be sensitive to small changes of initial data.[6]

Our example is typical in the sense that a similar situation necessarily arises in connection with semi-groups of measures, and in particular in connection with the *differential equation* (1.2). In it $v(t, x)$ is to be interpreted as a density, and the corresponding measure is defined by

$$\mu_t(\Gamma) = \int_\Gamma v(t, x)\, dx. \tag{4.1}$$

Classically, of course, one has always assumed that $v(t, x)$ is expressed in terms of the initial density by means of a *kernel transformation*

$$v(t, x) = \int_E v(y) p(t, y, x)\, dy \tag{4.2}$$

---

[5] One may say that it defines, for each individual initial probability distribution μ, a Markov process *without transition probabilities*. The condition (2.4) simply requires that there be an obvious intuitive relation between these processes.

[6] It goes without saying that our definition of "closeness" can be weakened or sharpened, that is, different weak topologies can be introduced. For example, it is possible to require only that T_t *be a kernel transformation*. In this case $T_t f$ may map a continuous f into a discontinuous function. For the relations among these topologies cf. Leader [10]. Alternatively, one may take the semi-group as given and look for the most general weak topology in which it is continuous (cf. [3]). In the present case this leads to an *imbedding* of the given semi-group into a bigger group which operates on the measures on *two lines and translates one line to the right, the other to the left.*

which then automatically defines solutions

$$(4.3) \qquad m_t(x) = \int_E \mu(dy)p(t, y, x), \qquad \mu_t(\Gamma) = \int_\Gamma m_t(x)\, dx$$

for an arbitrary initial measure μ. This is stronger than our weak continuity condition.

The situation is different with the *Hille-Yosida theory* of integration of differential equations by means of semi-groups. To begin with, this theory uses only the Banach metric. In practice this means that one cannot consider semi-groups operating on arbitrary measures but is compelled to restrict the analysis to the subspace of absolutely continuous measures. In the integration of (1.2) the initial distribution must be absolutely continuous. It has been emphasized by Phillips [11] that the extension of the resulting semi-group to all measures can be done in many ways and that the extended semi-groups therefore should not be considered as obtained by the integration process.[7]

Thus our weak continuity condition (2.4) is avoided in the Hille-Yosida theory. Without it there is no equivalence between the "processes"—i.e. transformations of measures—and the adjoint semi-group operating on continuous functions. Without (2.4) therefore *our abstract derivation of the diffusion equation is impossible.*[8]

Instead of (2.4), which is a continuity condition for fixed t, the Hille-Yosida theory introduces the assumption that the strong continuity condition (2.13) (for fixed f) holds also for $t = 0$. If the semi-group is considered on C, then this requirement excludes the familiar ordinary diffusion with an absorbing barrier. To include it, one has to replace C by an appropriate subspace. On the other hand, if one permits the introduction of subspaces, then the integration of (1.1) leads to absurdly many queer semi-groups without resemblance to the original notion of diffusion. For our problem the degree of arbitrariness thus introduced is greater than the admittedly arbitrary choice of an appropriate continuity condition like (2.4).

From our point of view, finally, the behavior of T_t as $t \to 0$ poses a *problem* which can not be settled by a postulate. For a diffusion process it is possible that $\| T_t \| \to \infty$ as $t \to 0$ (Section 11).

5. The local infinitesimal generator

We now consider an arbitrary semi-group $\{T_t\}$ with the properties stated in Definition 3. That is to say, to each $f \in C_+$ there corresponds a $T_t f \in C_+$ and (3.1) and (2.13) hold.

---

[7] This remark shows how the definition of the infinitesimal generator influences the whole point of view. To the classical analyst, and in our weak topology, the passage from (4.2) to (4.3) does not appear so arbitrary.

[8] It should be borne in mind that, even under strong additional assumptions, the Fokker-Planck equation (1.2) (which relates to measures) has been derived only as a consequence of (1.1). This is another indication that our reduction of the problem to semi-groups on continuous functions—i.e. the weak topology—is indispensable.

DEFINITION 4. *Suppose that for a particular $F \in C$ and a particular neighborhood N of a point x_0 one has as $t \to 0$*

$$\frac{T_tF - F}{t}(x) \to f(x), \qquad x \in N \tag{5.1}$$

where f is continuous in N and the convergence is bounded (that is

$$\left|\frac{T_tF - F}{t}\right| < M \qquad \text{for } t < \epsilon). \tag{5.2}$$

Then we write

$$f(x) = AF(x) \qquad x \in N. \tag{5.3}$$

The operator A thus defined is called the local infinitesimal generator of T_t at x_0. If it exists for F we write

$$F \in D(A; x_0). \tag{5.4}$$

When $N = E$ then A will be called the (global) infinitesimal generator.

We have the following simple, but important

LEMMA 2. *Let $F \in D(A, x_0)$. If $F(x_0) = 0$ and F has a local minimum at x_0 then*

$$AF(x_0) \geqq 0. \tag{5.5}$$

PROOF. In view of the local character condition (3.1) any function G which agrees with F in some neighborhood of x_0 belongs to the domain of A at x_0 and $G(x_0) = F(x_0)$. For the lemma it is therefore irrelevant whether and how F is defined outside some neighborhood of x_0. Accordingly, we assume without loss of generality, that $F \geqq 0$ everywhere. Then $T_tF \geqq 0$ everywhere, and as $F(x_0) = 0$ we get (5.5) from (5.1).

The preceding Lemma is our true point of departure: in [6] we have constructed the most general operators of local character with the property expressed in the lemma. It remains only to investigate the further regularity properties imposed by the fact that A is an infinitesimal generator.

We shall require the following lemma which follows easily from Hille's results and is contained in [4].[9]

LEMMA 3. *There exists a number $p \geqq 0$ such that for each $f \in C$, $\lambda > p$, $\varepsilon > 0$*

$$F_\lambda(x) = \int_0^\infty e^{-\lambda t} T_{\varepsilon+t} f(x)\, dt = \int_0^\infty e^{-\lambda t}\, dt \int_E P(t + \varepsilon, x, dy) f(y) \tag{5.6}$$

exists and $F_\lambda \in C$. Moreover F_λ is in the domain of the (global) infinitesimal generator A and

$$\lambda F_\lambda - AF_\lambda = T_\varepsilon f \qquad \lambda > p \tag{5.7}$$

$$\lambda F_\lambda \to T_\varepsilon f \qquad \lambda \to \infty. \tag{5.8}$$

[9] For the purpose of [4] uniform convergence in (5.1) was used instead of bounded convergence. For the infinitesimal generator using weak convergence cf. the last section of [3].

If $\| T_t f \|$ *is bounded near* $t = 0$ *then these relations hold also for* $\varepsilon = 0$ *with* $T_0 f = f$.

6. Null-points

DEFINITION 5. *The point* x_0 *is called null-point if* $T_t f(x_0) = 0$ *for each* $f \in C$ *and all* $t > 0$.

Clearly x_0 is a null-point if and only if $P(t, x_0, E) \equiv 0$ for all $t > 0$. Null-points are of an absurd triviality and it is most natural to exclude them by definition. Absorbing barriers in classical diffusion are null-points, however. Unfortunately, it is in principle possible to have an arbitrary set consisting of null-points.

EXAMPLE 2. Let K be Cantor's ternary set in [0, 1], and denote the intervals of the complement by $I_1, I_2, \cdots$. For an arbitrary $f \in C$ put $T_t f(x) = 0$ whenever $x \in K$. For $x \in I_k$ put $T_t f(x) = u(t, x)$ where $u(t, x)$ is the classical solution of the heat equation $u_t = u_{xx}$ which vanishes on the boundary of I_k and for which $u(0, x) = f(x)$ when $x \in I_k$. It is easily verified that there exists a constant M such that in I_k one has $| u(t, x) | \leqq M \cdot e^{-\pi^2 t/L^2} \cdot \| f \|$ where L is the length of I_k. It follows that $T_t f$ is continuous in [0, 1]. The semi-group property of T_t is obvious.

THEOREM 2. *For* x_0 *to be a null-point it is necessary and sufficient that for each* $F \in D(A, x_0)$ *one has* $AF(x_0) = F(x_0) = 0$.

PROOF. If x_0 is a null-point, then the left side of (5.1) at x_0 reduces to $-t^{-1}F(x_0)$ which can converge only if $F(x_0) = 0$. In this case $f(x_0) = 0$. Conversely, if $F \in D(A)$ implies $F(x_0) = AF(x_0) = 0$, then (5.7) implies that $T_t f = 0$ for each $f \in C$.

LEMMA 4. *If* $f \in C$ *and* f *vanishes outside a closed set* Γ *not containing any null-points, then* $\| T_t f \|$ *is bounded*[10] *near* $t = 0$ *and Lemma* 3 *applies to* f *with* $\varepsilon = 0$. *There exists a* $f \in C_+$ *which is strictly positive outside the set of null-points and for which the same conclusions hold.*

PROOF. If x_0 is not a null-point then there exists a $t_0 > 0$ such that $T_{t_0}\mathbf{1}$ is positive in some interval I around x_0. In view of the continuity property (2.13) we have then $T_t\mathbf{1}(x) > 0$ for $x \in I$ and $t_0 < t < t_0 + h$. Now $T_{t_0}\mathbf{1}$ is bounded by a multiple of $\mathbf{1}$ and therefore $T_t\mathbf{1}(x) > 0$ for $x \in I$ and $t < h$. Now finitely many such intervals cover Γ. Therefore there exists an $\varepsilon > 0$ such that $T_\varepsilon\mathbf{1}$ is positive on Γ. Let m be the minimum of $T_\varepsilon\mathbf{1}$ on Γ. Then, if f vanishes outside Γ,

$$| f(x) | \leqq \| f \| \cdot \mathbf{1} \leqq m^{-1} \cdot \| f \| \cdot T_\varepsilon \mathbf{1}.$$

The boundedness of $\| T_{t+\varepsilon}\mathbf{1} \|$ near 0 implies that of $\| T_t f \|$. Now let N be the set of null-points. Then N is closed. Construct a function $f_n \in C$ such that

[10] It is known (Hille [7] p. 183) that $\| T_t \|$ is bounded in every closed interval excluding 0 and ∞. If for a particular f the norm $\| T_t f \|$ is bounded near 0, it is automatically bounded for $0 < t \leqq a < \infty$.

$0 \leqq f_n \leqq 1$ and $f(x)$ is zero or positive according as the distance from x to N is $\leqq 1/n$ or $>1/n$. Then, as we have just shown,

$$\sup_{0<t<1} \| T_t f_n \| = a_n < \infty .$$

Clearly

$$f = \sum \frac{1}{2^n(1 + a_n)} f_n$$

has the desired properties.

A simple consequence is

THEOREM 3. *If for the semi-group of Definition 3 there exist no null-points, then* $\| T_t \|$ *is bounded in every finite interval* $0 \leqq t \leqq a < \infty$.

PROOF. By the last lemma $\| T_t f \|$ is bounded for $t \leqq a$ for each fixed $f \in C$. The theorem now follows from the principle of uniform boundedness (cf. [7], p. 26).

7. Isomorphisms

Let ϕ be a non-negative continuous function on E and denote by B the set of all functions f such that $f\phi$ is continuous on E; we make B into a Banach space by introducing the norm

$$N(f) = \sup_{x \in E} | f(x)\phi(x) |. \tag{7.1}$$

If ϕ is strictly positive then C and B represent the same set with different, but equivalent, norms. In general B will contain unbounded functions, but the elements of B and C stand in a one-to-one correspondence $f \leftrightarrow f\phi$.

THEOREM 4. *Let* $\{T_t\}$ *be a semi-group from* C *to* C. *For each* $f \in B$ *put*

$$T_t^* f = \frac{1}{\phi} T_t(f\phi). \tag{7.2}$$

Then $\{T_t^*\}$ *is a semi-group from* B *to* B, *and*

$$N(T_t^*) \leqq \| T_t \|. \tag{7.3}$$

If T_t *is positivity preserving (of local character, or strongly continuous), then* T_t^* *has the same property.*

If T_t *is of local character and* $\phi(x_0) \neq 0$, *then the local infinitesimal generator of* T_t^* *at* x_0 *is given by*

$$A^* f \leqq \frac{1}{\phi} A(f\phi) \tag{7.4}$$

and f *belongs to its domain if and only if* $f\phi$ *belongs to the domain of* A.

All these assertions follow directly from the definitions.

The main use of the last theorem is that it affords the possibility of regularizing operators. It is reasonable to ask that the constant function **1** belong to the domain of an operator of local character, but this need not be true of our local infinitesimal generator (cf. the concluding remark of the paper). However, by means of (7.4) it becomes possible to map A onto an operator A^* with this desirable property.

THEOREM 5. *Let* $\{T_t\}$ *be a semi-group of bounded linear operators on* C *such that* $f \in C_+$ *implies* $T_t f \in C_+$, *and suppose that the two conditions* (2.13) *and* (3.1) *of Definition* 3 *hold. Then*

(a) *There exists a function* $\phi \in C_+$ *such that* $\phi(x) > 0$ *for each* x *which is not a null-point, and such that*

$$A\phi \in C, \qquad \lambda_0\phi - A\phi \geqq 0 \tag{7.5}$$

for some $\lambda_0 > 0$.

(b) *If there are no null-points, then the semi-group defined by* (7.2) *has all the properties enumerated above for* $\{T_t\}$, *and*

$$A^*\mathbf{1} = \frac{A\phi}{\phi} \tag{7.6}$$

is bounded.

(c) *If there exist null-points, the statement remains true except that the domain and range of* T_t^* *is (not* C *but) the Banach space of continuous functions with norm* (7.4); *moreover*

$$A^*\mathbf{1} = \frac{A\phi}{\phi} \leqq \lambda_0, \tag{7.7}$$

but $A^*\mathbf{1}$ *need not be bounded below.*

PROOF. Choose $f \in C_+$ in accordance with Lemma 4 and so that $f > 0$ outside the set of null-points. By Lemma 3 the equation

$$\lambda_0\phi - A\phi = f \tag{7.8}$$

will have a solution for each $\lambda_0 > p$. If there are no null-points, then f has a positive minimum m, and in view of (5.6) we have

$$\phi(x) \geqq m \int_0^\infty e^{-\lambda_0 t} P(t, x, E)\, dt. \tag{7.9}$$

Since $P(t, x, E)$ is continuous in t, the vanishing of ϕ at x_0 would imply $P(t, x_0, E)$, or that x_0 is a null-point. Hence ϕ has a positive minimum. The remaining assertions now follow directly from Theorem 4.

8. Absorption points

We begin with a description of two types of singular points which are possible for the semi-groups of Definition 3.

Intuitively speaking, x_0 is an absorption point if no mass can flow out of x_0 into the complement $E - x_0$; mass that existed at x_0 at time 0, or that has arrived later, is subject to an exponential growth or decay ($c > 0$ and $c < 0$, respectively).

DEFINITION 6. *The point x_0 is called absorption point (trap) if one, and therefore each, of the following equivalent conditions is satisfied:*

(i) *For each f in the domain of A at x_0*

$$Af(x_0) = cf(x_0) \tag{8.1}$$

(ii)

$$T_t f(x_0) = e^{ct} f(x_0) \tag{8.2}$$

(iii)

$$P(t, x_0, E - x_0) = 0. \tag{8.3}$$

NOTE. From (8.3) and the semi-group property it follows that

$$P(t + s, x_0, x_0) = P(t, x_0, x_0)P(s, x_0, x_0) \tag{8.4}$$

which is the Hamel equation. As $P \geqq 0$ the only solution is

$$P(t, x_0, x_0) = e^{ct}. \tag{8.5}$$

9. Translation points

This type of singularity generalizes the degeneracy which occurs in the classical case of equation (1.1) when the leading coefficient a vanishes identically.[11] For illustration, cf. Section 11.

DEFINITION 7. *The point x_0 is called a right [left] translation point if*

$$A\{f(x)g(x_0) - f(x_0)g(x)\} > 0 \qquad |x - x_0| < \varepsilon \tag{9.1}$$

implies that in some neighborhood of x_0 the ratio f/g strictly increases [decreases].

It goes without saying that the condition (9.1) refers only to $f, g \in D(A, x_0)$, and that it is supposed that the inequality actually occurs. (If the left side is identically zero for all choices of f, g, then x_0 is an absorption point.)

THEOREM 6. *Let the interior point x_0 be a right translation point.* (i) *It is possible to choose a $\phi \in D(A, x_0)$ such that $\phi > 0$ in a neighborhood of x_0, and to introduce an independent variable $\xi = \xi(x)$ so that*

$$Af = \phi \frac{d^+}{d\xi} \frac{f}{\phi} + \lambda_0 f \tag{9.2}$$

where λ_0 is a constant and $d^+/d\xi$ denotes the right hand derivative of f/ϕ (its existence for $f \in D(A, x_0)$ is part of the assertion).

(ii) *x_0 cannot be point of accumulation of left translation points. If it is point of accumulation of translation points, then* (9.2) *holds for all translation points in a neighborhood with the same choice of ϕ, ξ.*

[11] A translation point x_0 can be *isolated*, as shown in Section 12. In classical form (1.1), the vanishing of a at an isolated point (or a non-dense set) may or may not present a singularity, depending on the choice of the independent variable.

PROOF. Since x_0 is not an absorption point, there exists an open interval I not containing x_0 such that $P(t, x_0, I) > 0$ for some t (cf. (8.3)). This interval can be chosen so that it contains no null-points.[12] Then there exists an $f \in C_+$ which vanishes outside I such that $T_{t_0}f(x_0) > 0$ for some t_0 and therefore for all t of an interval $t_0 \leqq t < t_0 + \delta$. By Lemma 4 the Laplace transform (5.6) exists for this f and $\varepsilon = 0$. It satisfies (5.7) (with the right side equal to f). However f vanishes in a neighborhood of x_0. Accordingly, in this neighborhood we have a strictly positive continuous ϕ with

$$(9.3) \qquad \lambda_0\phi - A\phi = 0 \qquad \lambda_0 > p.$$

Now choose ψ so that

$$(9.4) \qquad A(\psi\phi(x_0) - \phi\psi(x_0)) > 0$$

near x_0. (If the expression vanished at x_0 for each $\psi \in D(A, x_0)$ the point x_0 would be an absorption point.) This ψ can be so chosen that the ratio ψ/ϕ is strictly increasing in some neighborhood of x_0. This assertion will be proved in the next lemma. Put

$$(9.5) \qquad \lambda_0\psi - A\psi = -g.$$

Then g is continuous and positive near x_0. Finally, define the strictly increasing $\xi(x)$ by

$$(9.6) \qquad \xi = \int \frac{g}{\phi}\, d\,\frac{\psi}{\phi}.$$

Now consider the operator

$$(9.7) \qquad \Omega h = \frac{1}{\phi} A(h\phi) - \lambda_0 h.$$

Lemma 2 describes the characteristic property of this operator. It is the same property which was used in [6] to define the operator Ω. Moreover, if for some h one has $\Omega h(x_0) > 0$ then, by (9.3), near x_0

$$(9.8) \qquad \phi(x_0)A(h\phi) - h(x_0)\phi(x_0)A(\phi) > 0.$$

From the definition of a right translation point we see then that h cannot decrease. In the classification of Section 4 of [6] x_0 is therefore an irregular point of type (ii) and by Theorem 5.2 of [6] we have

$$(9.9) \qquad \Omega h(x_0) = \frac{d^+h(x_0)}{du},$$

where u is a continuous monotonic function of x.

---

[12] The set N of null-points is closed. When the assertion is not true, one has

$$P(t, x_0, E - N) = 0$$

for all t. But for $y \in N$ one has $P(t, y, N) = 0$, hence by the semi-group property

$$P(t + s, x_0, N) = \int_E P(t, x_0, dy)P(s, y, N) = 0,$$

so that x_0 is an absorbing point.

Applying (9.9) to $h = \psi/\phi$ we get from (9.7) and (9.5)

$$\Omega h = \frac{g}{\phi}. \tag{9.10}$$

Comparing (9.9) and (9.10) we get, using (9.6),

$$\Omega h(x_0) = \frac{d^+h(x_0)}{d\xi}. \tag{9.11}$$

This proves (9.2). Now with our choice of ϕ, ψ the inequality (9.4) will hold also if x_0 is replaced by any point x sufficiently near x_0. Since ψ/ϕ is strictly increasing, this implies that no left translation points can accumulate to x_0. Clearly the argument applies, with the same choice of ϕ and ψ, to all translation points sufficiently near x_0.

This finishes the proof except that we have postponed the proof that ψ can be chosen so that ψ/ϕ increases. We show this for arbitrary points (not necessarily translation points) and shall thus prove that for our semi-groups the absorption and translation points are the only irregular points.

LEMMA 5. *Let x_0 be an interior point which is not absorbing. Then it is possible to choose f and g so that in a neighborhood of x_0 the inequality* (9.1) *holds and that the ratio f/g is strictly monotonic.*

PROOF. As in the proof of the last theorem, choose $\phi > 0$ so that (9.3) holds. Then there exists a ψ such that (9.4) is true. Let $h = \psi/\phi$. For the operator defined in (9.7) we have near x_0

$$\Omega h = \frac{\phi A\psi - \psi A\phi}{\phi^2} > 0. \tag{9.12}$$

It follows then from the lemma of Section 4 of [6] that there exists an $\eta > 0$ such that h is monotonic both in $x_0 - \eta < x < x_0$ and in $x_0 < x < x_0 + \eta$. Moreover, $\Omega h(x_0) > 0$ implies that h cannot have a maximum at x_0.

We have therefore only to prove that (9.3) cannot possibly imply that ψ/ϕ has a minimum at x_0 for *every* admissible choice of ψ.

Assume the contrary. Choose $f \in C$ so that f vanishes outside an interval I around x_0 containing no null-points. Moreover, let $f(x_0) = 0$, and let f be increasing in some interval around x_0. Without loss of generality we may take this interval to be $(x_0 - \eta, x_0 + \eta)$. To this f Lemmas 3 and 4 apply. For sufficiently large λ therefore

$$\lambda\psi - A\psi = f \tag{9.13}$$

has a solution, and as $\lambda \to \infty$

$$\lambda\psi \to f. \tag{9.14}$$

There are now three possibilities, each of which will lead to a contradiction.

(a) Suppose that for a sequence $\lambda \to \infty$ one has $\psi(x_0) > 0$. Because of $f(x_0) = 0$ the quantity on the left in (9.4) reduces, at x_0, to $(\lambda - \lambda_0)\phi(x_0)\psi(x_0)$, which is positive when $\lambda > \lambda_0$. By the assumption of the indirect proof, ψ/ϕ has therefore a minimum at x_0. However, in the interval $x_0 - \eta < x < x_0$ we have $f(x) < 0$ and therefore for $x_0 - \eta < x_1 < x_0$

$$A\left(\psi - \frac{\psi(x_1)}{\phi(x_1)}\phi\right) > \lambda\psi - \frac{\psi(x_1)}{\phi(x_1)}\lambda_0\phi > \lambda_0\left\{\psi - \frac{\psi(x_1)}{\phi(x_1)}\phi\right\}. \tag{9.15}$$

Therefore, according to the fundamental Lemma 2, the function

$$\psi - \frac{\psi(x_1)}{\phi(x_1)}\phi$$

can have no maximum at x_1. In other words, the ratio ψ/ϕ can have no maximum in $x_0 - \eta < x < x_0$. Since it decreases near x_0, it must decrease in the whole interval. Thus in this interval $\psi > 0$ and $f < 0$, which contradicts (9.14).

(b) If, for a sequence $\lambda \to \infty$ one has $\psi(x_0) < 0$, one arrives at the same contradiction using the interval $x_0 < x < x_0 + \eta$.

(c) The last possibility is that $\psi(x_0) = A\psi(x_0) = 0$ for all λ sufficiently large. We shall show that this can happen only at an absorbing point, and this was excluded.

Let z be an arbitrary continuous function which is constant in some interval around x_0 and vanishes outside a closed set containing no null-points. Then Lemmas 3 and 4 again apply, and there exists therefore a solution of

$$\lambda Z - AZ = z. \tag{9.16}$$

Define ψ again by (9.13), but this time let $f(x_0) \neq 0$. The old argument now applies (instead of to f) to $z(x_0)f - f(x_0)z$, and one concludes that

$$z(x_0)\psi(x_0) - f(x_0)Z(x_0) = A\{z(x_0)\psi - f(x_0)Z\}_{x=x_0} = 0. \tag{9.17}$$

When $\psi(x_0) \neq 0$, this implies

$$\frac{AZ(x_0)}{Z(x_0)} = \frac{A\psi(x_0)}{\psi(x_0)} = c_\lambda \tag{9.18}$$

where the constant c_λ in no ways depends on the choice of z. Consider now two choices z_1 and z_2 for z, with $z_1(x_0) = z_2(x_0)$. In a neighborhood of x_0 the difference $z_1 - z_2$ vanishes identically, and therefore $Z_1 - Z_2 = u$ satisfies $\lambda u - Au = 0$. It follows that $c_\lambda = \lambda$. Then (9.18) contradicts (9.13) since $f(x_0) \neq 0$.

Accordingly, (9.18) is impossible and we must have $\psi(x_0) = 0$. Then, by (9.17), also $Z(x_0) = AZ(x_0) = 0$ which contradicts (9.16). This accomplishes the proof.

COROLLARY. *In the interior of E all irregular points of A in the sense of* [6] *Section* 4, *are absorption or translation points.*

10. The main theorem

The general form of operators with the property indicated in Lemma 2 has been derived in [6]. In the preceding section we have seen that there exist only two types of irregular points. Finally, Theorem 5 gives us a solution ϕ with special properties. Combining these facts we get our main result. To formulate it we need to recall (from [6]) the definition of *the derivative* $D_v z$ *of* z *with respect to an arbitrary monotonic* v. For it to exist it is required that z be continuous except, perhaps, at the points of discontinuity of v, where the one-sided limits $z(x+)$ and $z(x-)$ still exist. Then *at points of discontinuity*

$$D_v z(x) = \frac{z(x+) - z(x-)}{v(x+) - v(x-)} = \frac{\Delta z}{\Delta v}, \tag{10.1}$$

whereas at points of continuity

$$D_v z(x) = \lim_{h \to 0} \frac{z(x + h) - z(x)}{v(x + h) - v(x)}. \tag{10.2}$$

THEOREM 7. *Let* A *be the infinitesimal generator of a semi-group of a diffusion process* (*Definition* 3). *Let* I *be an open interval containing no absorption and no translation points.*

Then there exist in I *a strictly increasing* (*not necessarily continuous or bounded*) *function* v, *a* $\phi \in C$, *and a choice of the independent variable* (*scale*) x *so that in* I

$$Af \equiv \phi \cdot D_v \left(\frac{f}{\phi}\right)' + cf \tag{10.3}$$

in the following sense:

If, in a neighborhood N *of* x, f *belongs to the domain of* A, *then* f/ϕ *has at each point of* N *a right and a left derivative, and the two are identical except, perhaps, at points of discontinuity of* v. *This derivative is differentiable with respect to* v *and* (10.3) *holds. Here*

$$c = \frac{A\phi}{\phi} \tag{10.4}$$

is bounded unless one of the endpoints of I *is a null-point. And* c *has a finite upper bound in any case.*

Conversely, if f *is such that the right side of* (10.4) *exists and is continuous near* x_0, *then* f *belongs locally near* x_0 *to the domain of* A.

THEOREM 8. *If the diffusion process preserves masses, that is, if* $P(t, x, E) \equiv 1$, *then* $\phi = 1$ *and* $c \equiv 0$, so that

$$Af = D_v f'. \tag{10.5}$$

11. Unbounded semi-groups

We show that for the semi-group of a diffusion process the norm $\| T_t \|$ *need not be bounded as* $t \to 0$. For simplicity, we show this by a *generalized translation semi-group* (but it is easy to add a diffusion term).

Let $E = [-\infty, +\infty]$ and define

$$T_t f(x) = f(x + t) \cdot \exp\left\{\int_x^{x+t} c(z)\, dz\right\}. \tag{11.1}$$

This represents a translation combined with local deformation. (Physically, a deterministic motion coupled with changes of mass.) The function $u(t, x) = T_t f(x)$ is the semi-group solution of the degenerate diffusion equation

$$\frac{\partial u}{\partial t} = \frac{\partial u}{\partial x} + cu. \tag{11.2}$$

That is, the right side generates (11.1), but u is a classical solution only if

$$u(0, x) = f(x)$$

is differentiable.

Clearly

$$\begin{aligned} \| T_t \| &= \left\| \exp\left(\int_x^{x+t} c(z)\, dz\right) \right\| \\ &= \sup_{-\infty<x<\infty} \exp\left(\int_x^{x+t} c(z)\, dz\right). \end{aligned} \tag{11.3}$$

Define

$$\begin{aligned} c(z) &= -1 && \text{for} && z < 2 \\ c(z) &= n^2 && \text{for} && 2 \leqq n < z < n + \frac{1}{n} \\ c(z) &= -(n+1)^2 && \text{for} && n + \frac{1}{n} < z < n + 1. \end{aligned} \tag{11.4}$$

Then it is easily seen that for $t = n \geqq 2$ and all x

$$\int_x^{x+t} c(z)\, dz \leqq \int_n^{n+(1/n)} c(z)\, dz = n \tag{11.5}$$

whence

$$\| T_{1/n} \| = e^n \longrightarrow \infty. \tag{11.6}$$

12. Isolated translation points

Several interesting points are illustrated by the semi-group connected with a diffusion process which may be described in physical terms as follows. In $(0, \infty)$ we have ordinary diffusion ($u_t = u_{xx}$) with a reflecting barrier at $x = 0$. In $(-\infty, 0)$ we have ordinary diffusion with an absorbing barrier at $x = 0$. The two intervals are in communication in such a way that a particle reaching the origin from the left passes into the positive half-line.

Here we have a *deterministic motion at a single point*, and this point is therefore an *isolated translation point*.

In analytical terms the semi-group is defined as follows. For $x \geqq 0$

$$T_t f(x) = \frac{1}{2(\pi t)^{1/2}} \int_0^\infty \left\{ \exp \frac{-(x-y)^2}{4t} + \exp \frac{-(x+y)^2}{4t} \right\} f(y)\, dy. \tag{12.1}$$

For $x \leqq 0$ one has (denoting by $*$ convolution in t)

$$\begin{aligned} T_t f(x) &= \frac{1}{2(\pi t)^{1/2}} \int_{-\infty}^0 \left\{ \exp \frac{-(x-y)^2}{4t} - \exp \frac{-(x+y)^2}{4t} \right\} f(y)\, dy \\ &\quad + \frac{1}{\pi t} \int_{-\infty}^x \exp \frac{-s^2}{4t}\, ds * \int_0^\infty \left\{ \exp \frac{-y^2}{4t} \right\} f(y)\, dy. \end{aligned} \tag{12.2}$$

Here

$$A = \frac{d^2}{dx^2} \qquad \text{for} \quad x \neq 0 \tag{12.3a}$$

$$A = \frac{d^+}{d\xi} \qquad \text{for} \quad x = 0 \quad \textit{with} \quad \xi = \frac{x^2}{2}. \tag{12.3b}$$

The existence of $d^+f/d\xi$ at the origin means

$$f'(0) = 0, \qquad f''(0+) = \frac{d^+f}{d\xi}(0). \tag{12.4}$$

Thus 0 is a *translation point and the single condition* (12.3b) *expresses the two "boundary conditions"* (12.4). Classically, our semi-group would be described as the solution of $u_t = u_{xx}$ with (12.4) serving as transition condition between $x < 0$ and $x > 0$.

Note that for our solutions $f'(0-) \neq f'(0+) = 0$. The continuity of Af at $x = 0$ requires that $f''(0-) = f''(0+)$. However the limit of $(T_t f - f)/t$ at $x = 0$ will exist even without this condition and this explains why A involves only conditions on right derivatives.

13. True diffusion plus translation

We consider a process which in physical terms is described as follows. From $x < 0$ a particle moves at unit speed towards the origin. On reaching the origin, it commences a reflecting barrier process in $x > 0$.

For an analytical definition denote the reflecting barrier kernel occurring in (12.1) by $K(t, x, y)$. Then we put

$$T_t f(x) = \int_0^\infty K(t, x, y) f(y)\, dy \qquad (x \geqq 0) \tag{13.1}$$

$$T_t f(x) = \int_0^\infty K(t + x, 0, y) f(y)\, dy \qquad -t \leqq x \leqq 0 \tag{13.2}$$

$$T_t f(x) = f(x + t) \qquad x \leqq -t. \tag{13.3}$$

(The physical significance becomes clear only with the adjoint semi-group operating on measures!)

Here $A = d^2/dx^2$ for $x > 0$ and $A = d/dx$ for $x < 0$. At the origin any f in the domain of A satisfies the conditions

$$f'(0+) = 0, \qquad f'(0-) = f''(0+). \tag{13.4}$$

$$\begin{aligned} \xi(x) &= e^x & x &\leqq 0 \\ \xi(x) &= ch^x & x &\geqq 0. \end{aligned} \tag{13.5}$$

The two conditions (13.4) *together are equivalent to the existence of the derivative* $df/d\xi = A$ *at* $\xi = 1$.

The corresponding example for an elastic barrier process has the additional interesting feature that the constant function belongs to the domain of A locally at each point *except the origin*. Thus the condition $\mathbf{1} \in D(A)$ is not necessarily a regularity condition in the usual sense.

PRINCETON UNIVERSITY

LITERATURE

[1] W. FELLER. *Zur Theorie der stochastischen Prozesse (Existenz und Eindeutigkeitssätze)*. Math. Ann., 113 (1936) 113–160.

[2] W. FELLER. *The parabolic differential equation and the associated semi-groups of transformations*. Ann. of Math., 55 (1952) pp. 468–519.

[3] W. FELLER. *Semi-groups of transformations in general weak topologies*. Ann. of Math., 57 (1953) 287–308.

[4] W. FELLER. *On the generation of unbounded semi-groups of bounded linear operators*. Ann. of Math., 58 (1953) 166–174.

[5] W. FELLER. *On positivity preserving semi-groups of transformations on* $C[r_1, r_2]$. Ann. de la Soc. Polonaise de Math., t. XXV (1952) 85–94.

[6] W. FELLER. *On second order differential operators*. Ann. of Math. (to appear).

[7] E. HILLE. Functional Analysis and Semi-groups. Amer. Math. Soc. Colloquium Publications, vol. 31. New York, 1948.

[8] A. KOLMOGOROFF. *Über die analytischen Methoden in der Wahrscheinlichkeitsrechnung*. Math. Ann., 104 (1931) 415–458.

[9] S. LEADER. *On the infinitesimal generator of a semi-group of positive transformations with local character condition*. Proc. Amer. Math. Soc. 5 (1954), 401–401.

[10] S. LEADER. Convergence of measures and the existence of transition probabilities. To appear in Pacific J. Math.

[11] R. S. PHILLIPS. *The adjoint semi-group*. To appear in Pacific J. Math.

DIFFUSION PROCESSES IN ONE DIMENSION

BY

WILLIAM FELLER[1]

1. **Introduction.** We consider the so-called *backward equation of diffusion theory*

$$(1.1) \qquad u_t(t, x) = \Omega u(t, x), \qquad \Omega = a(x)\frac{d^2}{dx^2} + b(x)\frac{d}{dx}$$

in a finite or infinite interval $r_1 < x < r_2$. Recently [2] the theory of semi-groups of transformations was used to formulate the integration problem and to define the notion of *lateral conditions*. It will now be shown that the semi-groups constructed in [2] permit us to describe the *totality of all stochastic processes* connected with (1.1). To each there corresponds a *forward*, or Fokker-Planck, equation. It reduces only in simplest cases to the familiar "adjoint"

$$(1.2) \qquad v_t = \Omega^* v(t, x), \qquad \Omega^* = \frac{d}{dx}\left\{a(x)\frac{d}{dx} + b(x)\cdot\right\},$$

but the general Fokker-Planck equation is not a differential equation.

The present paper consists of two almost independent parts. Its main purpose is a probabilistic discussion of the general diffusion process in one dimension. An informal description of such processes is given in §2, which may be read independently of this introduction.

First it is necessary to give a strict definition of the notion of a *diffusion process obeying* (1.1). This is done in §3 in a formal way convenient for the application of semi-group theory. To each diffusion process there corresponds a semi-group of transformations. In [2] we investigated the totality of all such semi-groups transforming functions which are continuous in $[r_1, r_2]$ into functions of the same type. In §4 it is shown that *each such semi-group yields a diffusion process*. However, as is shown in §13, *the semi-group associated with a diffusion process may transform a continuous function into a function which is discontinuous at the boundaries*. It is for this reason that the semi-groups constructed in [2] yield the most general diffusion process in the open interval (r_1, r_2), but not in $[r_1, r_2]$. For the description of the latter, a rather obvious generalization of the construction of [2] is required.

In §5 the *probabilistic background of diffusion processes* is discussed. Prob-

Received by the editors May 9, 1953.

[1] Research connected with an Office of Ordnance Research Project at Princeton University, OOR Contract No. DA-36-034-ORD-640.

ably it soon will be possible to define diffusion processes as Markov processes with continuous path functions. Even the present discussion makes it clear why it is unavoidable to start with the backward equation although the forward equation is intuitively and physically more natural.

These preparations are not required for a formal understanding of the following, and main, part of the paper (at least if manipulations with Laplace transforms are accepted at face value).

In the *probabilistic part* we make the (unproved but safe) assumption that the path functions $X(t)$ ("the position of the particle at time t") are continuous except, perhaps, at the boundaries (which may be at infinity(2)). There exists then a well-defined probability that a path, starting at x, will reach a point ρ_2 without first crossing ρ_1.

In §6 these *first passage probabilities* are derived from the familiar renewal principle. For points in the interior they are the same for all processes obeying (1.1). In other words, (1.1) *determines a class of processes which are not distinguishable as long as no boundary is reached.*

Now the interval (r_1, r_2) may be finite or infinite, and the diffusion equation (1.1) may be singular in the sense that the coefficients may be unbounded or may vanish near r_j. Accordingly, the first passage probabilities lead to a classification of *accessible* and *inaccessible boundaries.*

With this notion we can proceed to the main purpose of the paper, namely a *probabilistic derivation and description of the diffusion processes connected with* (1.1). An informal summary is given in the next section.

2. **Qualitative description of the processes.** In order that there exist *one and only one process* satisfying (1.1) it is necessary and sufficient that *both boundaries be inaccessible*; that is, the probability that either boundary be reached within a finite time interval must be zero. A typical example is $u_t = u_{xx}$ in $(-\infty, +\infty)$.

If at least one boundary is *accessible*, then there are infinitely many processes obeying (1.1). A typical example is the case where r_j is finite and $a(x)$ and $b(x)$ are continuous at r_j with $a(r_j) \neq 0$.

The *"absorbing barrier"* process is defined by the rule that the process terminates at the moment when a boundary is reached.

Next simplest is the *instantaneous return process*, which is analogous to a denumerable Markov chain process discussed by Doob [1]. In this process, whenever the boundary r_j is reached, an instantaneous return into the interior is effected and the process starts afresh (independently of the past history) from a point $\mathbf{x}_j$ which is a random variable with distribution function $p_j(x)$. In other words, if $X(s) \to r_j$ as $s \to t-$, then one puts by definition $X(t) = \mathbf{x}_j$, where

(2) This is done only for the purpose of a probabilistic interpretation of the quantities which appear in the formulas derived in [2]. If this interpretation is granted, the main part of the paper (described in §2) becomes independent of the preceding.

$$\text{(2.1)} \qquad \Pr\{\mathbf{x}_j \leqq x\} = p_j(x), \qquad r_1 < x < r_2.$$

The process now starts from scratch.

More generally, there may be a *finite sojourn time* associated with each boundary, that is, one may have $X(s) = r_j$ for a finite time interval $t \leqq s < t + T$. The Markovian character of the process requires that T be a random variable with

$$\text{(2.2)} \qquad \Pr\{T > t\} = \exp(-t/\sigma_j)$$

where σ_j is a positive constant. (The limiting cases $\sigma_j = 0$ and $\sigma_j = \infty$ correspond to the instantaneous return process and the absorbing barrier process, respectively.) Moreover, T must be independent of the past. We define now the following set of rules for the

ELEMENTARY RETURN PROCESS. *If at time t the boundary r_j is reached*(3), *then $X(s) = r_j$ for $t \leqq s < t + T$ where T is a random variable independent of the past and distributed according to* (2.2). *At time $t + T$ a jump occurs to a point $\mathbf{x}_j$ in the interior or to the boundary*(4) *r_k according to the rule*

$$\text{(2.3)} \qquad \begin{aligned} \Pr\{X(t+T) = r_k\} &= p_{jk}, \\ \Pr\{r_1 < X(t+T) \leqq x\} &= \tau_j p_j(x), \end{aligned}$$

and the process starts from scratch. Here

$$\text{(2.4)} \qquad p_{j1} + p_{j2} + \tau_j \leqq 1,$$

$p_{jk} \geqq 0$, $\tau_j \geqq 0$, $\sigma_j \geqq 0$; *and $p_j(x)$ are monotonic functions with $p_j(x) \to 0$ as $x \to r_1$ and $p_j(x) \to 1$ as $x \to r_2$.*

(The difference between the two sides in (2.4) accounts for the possibility of the process terminating or, in physical language, of masses disappearing.)

It should be noticed that the absorbing barrier process and the instantaneous return process take place on the *open* interval (r_1, r_2); whereas the elementary return process involves the two boundaries as well. (The elastic and reflecting barrier processes below are also processes in (r_1, r_2).)

In §9 we construct the Laplace transforms of *the transition probabilities* of such a process from probabilistic considerations. In abstract terminology, they represent the *resolvent* of a semi-group, and they can be identified with a class of resolvents constructed in [2].

While the elementary return process obeys the backward diffusion equation (1.1), the corresponding *Fokker-Planck equation is not* (1.2). It has been derived in [2], and in §10 it is shown that its obvious physical interpretation

(3) I.e. if $X(s) \to r_j$ as $s \to t-$ and $r_1 < X(s) < r_2$ for $t - \epsilon < s < t$.

(4) The possibility of a transition from r_j to itself is introduced in order to arrive formally at the most general boundary condition of [2]. A simple calculation shows that the *same* set of rules can be reformulated with $p_{kk} = 0$ and a modified mean sojourn time σ_k. The corresponding boundary conditions are analytically equivalent to ours.

essentially restates our probabilistic definition of the process.

We come next to a class of processes associated in classical diffusion theory with *elastic* or *reflecting barriers*. A direct description of the properties of the path functions $X(t)$ of such processes and their behavior near the boundaries is very desirable. It represents an unsolved problem even in the simple case of $u_t = u_{xx}$. (The term "reflecting" with its implication of velocities is misleading.)

The elementary return process represents the most general diffusion process in (r_1, r_2) except if at least one boundary is regular in the sense of [2]. In this case we obtain analogues to the classical *elastic and reflecting barrier processes by a direct passage to the limit from an instantaneous return process.* When the boundary r_j is reached, with probability $1-\tau_j$ we let the process terminate, and with probability τ_j we let it recommence from a point ρ_j near r_j. (In other words, $p_j(x)$ has mass one concentrated at ρ_j.) We now let $\rho_j \to r_j$ and at the same time $\tau_j \to 1$ in such a way as to keep certain absorption probabilities constant (cf. §11).

If the boundary r_j is regular, this limiting procedure is shown to lead to the class of processes which, in the terminology of [2], depend on regular Green functions. For them, and only for them, does the true Fokker-Planck equation reduce to (1.2). This limiting procedure sheds some new light on the elastic barrier processes, but does not give a measure-theoretic description of the path functions near the boundaries.

For the elastic barrier process one can again define the probabilities that the process terminates at a boundary r_j, but this event must now be distinguished from the event of reaching the boundary.

The most general diffusion process associated with semi-groups (cf. [2]) is now obtained by starting from an elastic barrier process and superimposing on it the elementary return process in the same way as we have before done with the absorbing barrier process. In other words, when the elastic barrier process in itself would terminate at r_j, we let the process recommence according to the set of rules described above.

We can calculate the Laplace transforms of all these processes directly. In this way we get a *new, and independent, derivation of all the resolvents of the semi-groups constructed in* [2]. However, we require the results of [2] for the derivation of the Fokker-Planck equations and of the boundary conditions. Moreover, only the semi-group methods of [2] give us the assurance that we have obtained *all* diffusion processes of a well-defined class. In particular, the diffusion processes constructed in §12 represent the *most general type of diffusion process in the open interval* (r_1, r_2). For the closed interval we have no such complete result, since the associated semi-group may take a continuous function into a discontinuous one. The nature of the most general process appears clear enough (§13), and explicit formulas for the transition probabilities can easily be calculated by the procedure of §9. However, we shall not go through the formalities of this derivation nor do we give a formal proof that we are actually dealing with the most general process.

3. **Definitions.** We shall be concerned with Markovian processes on an interval E which may be finite or infinite, open, half-open, or closed. The end points of E will be denoted by r_1 and r_2, where $-\infty \leqq r_1 < r_2 \leqq \infty$. Let $X(t)$ represent the path function of the process. We assume that a *transition probability*

$$(3.1) \qquad P(t, x; \tau, \Gamma) = \Pr\{X(\tau) \in \Gamma \mid X(t) = x\}$$

is defined for all $x \in E$, *all Borel sets* $\Gamma \subset E$, *and* $0 < t < \tau < \infty$. *We do not require that* E *has assigned probability one*, that is, we permit that

$$(3.2) \qquad P(t, x; \tau, E) \leqq 1.$$

The difference between the two sides is the probability that the path functions with $X(t) = x$ are defined only in a subinterval of $[t, \tau]$. We assume that the *transition probabilities are stationary*, that is

$$(3.3) \qquad P(t, x; \tau, \Gamma) = P(\tau - t, x, \Gamma),$$

with $P(t, x, \Gamma)$ defined for all $t > 0$.

For simplicity(5) we introduce the *continuity requirement* that for each $x \in E$ and each *neighborhood* Γ of x

$$(3.4) \qquad P(t, x, \Gamma) \to 1, \qquad t \downarrow 0.$$

The Markovian character of the process requires the validity of the *Chapman-Kolmogorov* identity

$$(3.5) \qquad P(t + s, x, \Gamma) = \int_E P(t, x, dy) P(s, y, \Gamma)$$

for $t, s > 0$. Conversely, suppose that $P(t, y, \Gamma)$ is, for fixed Γ, Borel measurable in both t and x and, for fixed (t, x), a probability measure on the Borel set of E, and also that (3.5) holds. Then $P(t, x, \Gamma)$ can serve as the transition probability of a Markovian process in $[r_1, r_2]$. All probability relations in the space of path functions $X(t)$ are then essentially uniquely determined(6) by the knowledge of $P(t, x, \Gamma)$ and the initial probability distribution

$$(3.6) \qquad \mu(\Gamma) = \Pr\{X(0) \in \Gamma\}.$$

In particular

$$(3.7) \qquad \Pr\{X(t) \in \Gamma\} = \int_E \mu(dx) P(t, x, \Gamma) = V(t, \Gamma).$$

(5) For weak continuity properties which always hold cf. [3]. In the present case (3.4) actually follows from the assumed form of the infinitesimal generator.

(6) For the qualification "essentially" cf. Doob, *Stochastic processes*, New York, Wiley, 1953.

Note. If $E=[r_1, r_2]$ the event $X(t)=r_j$ may have a positive probability. Even if $E=(r_1, r_2)$, the range of $X(t)$ may coincide with $[r_1, r_2]$ but the event $X(t)=r_j$ will have probability zero for each fixed t. When we refer to a process in the open interval, we refer to the transition probabilities, and only in this wide sense to $X(t)$.

The terminology is not well established, and we shall use the term "process" also to denote the whole class of individual processes with the same transition probabilities $P(t, x, \Gamma)$ and arbitrary initial distributions.

In diffusion theory one usually starts from the assumption that $P(t, x, \Gamma)$ has a probability density $p(t, x, y)$ which for fixed y satisfies the backward equation (1.1). (Cf. §5.) Now the concept of integrating (1.1) as suggested by the semi-group theory leads to a definition of diffusion processes more adapted to our purposes. It is slightly more general, but the main advantage is a considerable simplification of the analytical apparatus.

In effect, the application of semi-group theory to differential equations amounts to a justification of formal operations with Laplace transforms. Classically, the function

$$u(t, x) = \int_E f(y)P(t, x, dy) \tag{3.8}$$

is supposed to represent the appropriate solution of (1.1) which reduces to $f(x)$ as $t \to 0$. Its Laplace transform

$$F(x) = \int_0^\infty e^{-\lambda t}u(t, x)dt, \qquad \lambda > 0, \tag{3.9}$$

is supposed to be a solution of the ordinary differential equation

$$\lambda F(x) - \Omega F(x) = f(x), \qquad r_1 < x < r_2. \tag{3.10}$$

In order to avoid further explanations and discussions which are irrelevant to the present paper, we shall start with these properties as our basic definition. This procedure is not elegant, but is simplest for our purposes (and becomes more acceptable when translated into semi-group language).

DEFINITION. *A Markovian process on E will be said to obey* (1.1) *if it possesses transition probabilities satisfying* (3.4) *and* (3.5), *and if for every function f which is continuous and bounded on E the Laplace transforms* (3.9) *satisfy* (3.10)(7).

A process will be said to be of the diffusion type if it obeys some equation of the form (1.1).

---

(7) Note that if E is closed, we require f to be continuous at the end points, but (3.10) puts a restriction only in the interior. This makes it possible that $F(x)$ becomes discontinuous at the end points even if f is continuous.

(A more satisfactory definition will be proposed in §5.)

For simplicity *we shall from now on assume that for* $r_1 < x < r_2$ *the functions* $a(x)$, $a'(x)$, $b(x)$ *are defined and continuous*, $a(x) > 0$. Note that $a(x)$ and $b(x)$ need not be bounded, and that it is permissible that $a(x) \to 0$ as $x \to r_j$.

[*Note.* The preceding definition is sufficient for our purposes, but some readers may wish to see its connection with general semi-group theory and with [2]. To explain this we now anticipate the notations of the following section. The continuity requirement (3.4) implies that for $f \in C$, with the notations of §4,

$$T_h f(x) \to f(x) \qquad (h \to 0+) \tag{3.11}$$

for each $x \in (r_1, r_2)$ (not necessarily for the boundaries), the convergence being *bounded*. Unfortunately, in [2] the traditional strong (uniform) continuity was required. Now only the contraction of the semi-group to its range space is strongly continuous, and this circumstance necessitated the clumsy references to the range space in [2]. The above definition would have resulted in greater simplicity and elegance.

With the new continuity requirement it is natural also to modify the usual definition of the infinitesimal generator (for a more general and more systematic approach cf. [3]). If for some $F \in C$ and each $x \in (r_1, r_2)$ one has

$$\frac{T_h F - F}{h}(x) \to \Phi(x) \qquad (\Phi \in C) \tag{3.12}$$

the convergence being *bounded*, then we put $\Phi = \bar{\Omega} F$. The operator $\bar{\Omega}$ thus defined will now be called *infinitesimal generator* of the semi-group. In the Hille-Yosida theory uniform convergence is required, but no difficulties arise. For every $f \in C$ the Laplace transform (3.9) is in the domain of $\bar{\Omega}$ and satisfies

$$\lambda F - \bar{\Omega} F = f; \tag{3.13}$$

for given f this equation has only one solution, and the totality of functions for which $\bar{\Omega} F$ is defined is thus identical with the totality of solutions of (3.13).

Our definition of a diffusion process now amounts to requiring that the infinitesimal generator $\bar{\Omega}$ *be a contraction of the operator* Ω (that is, if $\bar{\Omega} F$ is defined, then so is ΩF and $\bar{\Omega} F = \Omega F$, but it is possible that $\Omega F \in C$ without F being of the form (3.9)).

Of course, not every solution of (3.10) defines a semi-group. The totality of solutions with this property has been constructed in [2]; in fact, the arguments of [2] remain valid with the new definition of infinitesimal generator, and the formulations become simpler. However, we can appeal directly to the theorems in [2] as they stand since every semi-group with our continuity property is strongly continuous in its range space and thus satisfies the conditions imposed in [2].]

4. Diffusion processes as semi-groups. Let B be the Banach space of bounded Borel measurable functions on E with $\|f\| = \sup |f(x)|$. Let $f \in B$ and consider (3.8). For each fixed $t > 0$ we have $u(t, x) \in B$ and $\|u\| \leq \|f\|$. If we put

$$u(t, x) = T_t f(x) \tag{4.1}$$

then T_t is a linear operator from B to B which (1) preserves positivity, (2) has norm $\|T_t\| \leq 1$, and (3) has the semi-group property $T_{t+s} = T_t T_s$. The last assertion is a direct consequence of (3.5). The three properties are expressed in the statement that the transformations $\{T_t\}$ *form a contraction semi-group from B to B.*

Every Markov process on E thus defines a contraction semi-group from B to B. In [2] we have considered such semi-groups subjected to the restriction that they are continuity preserving, that is, that $T_t f$ is continuous whenever f is continuous. In this case it is useless to admit discontinuous functions and it is preferable to consider $\{T_t\}$ as a semi-group from C to C where *C is the subspace of functions f* which are continuous in the closed(8) interval $r_1 \leq x \leq r_2$.

We want now to show that each such semi-group defines a Markov process on $[r_1, r_2]$.

LEMMA. *Let $\{T_t\}$ be a contraction semi-group from C to C such that for each fixed x the function $T_t f(x)$ is continuous in t. Then for each $t > 0$ there exists a kernel $P(t, x, \Gamma)$ which for fixed x is a probability measure on the Borel sets Γ. For fixed Γ this kernel is Borel measurable in t and x. Finally*

$$T_t f(x) = \int_{r_1-}^{r_2+} f(y) P(t, x, dy). \tag{4.2}$$

Note that the semi-group property implies the validity of the Chapman-Kolmogorov equation (3.5).

Proof. Consider the adjoint semi-group to T_t in the following way. If $\mu(\Gamma)$ is any measure on the Borel sets Γ in $[r_1, r_2]$ the transform $T_t^* \mu$ is the uniquely defined measure with which

$$\int_{r_1-}^{r_2+} T_t^* \mu(dy) f(y) = \int_{r_1-}^{r_2+} \mu(dy) T_t f(y) \tag{4.3}$$

for all $f \in C$. Let $\mu_x(\Gamma)$ equal 1 or 0 according as x is, or is not, a point of Γ. Put

(8) Even if E is open we require that $f(x)$ tends to finite limits as $x \to r_j$. This looks like an essential restriction, but it will be shown in §13 that it can be removed and does no harm.

$$P(t, x, \Gamma) = T_t^* \mu_x(\Gamma). \tag{4.4}$$

The fact that T_t^* is positivity preserving and $\|T_t^*\| \leq 1$ implies that $P(t, x, \Gamma)$ is a probability measure for each fixed t and x. When the definition (4.4) is inserted in (4.3) one sees that (4.2) holds for all $f \in C$. Finally, if Γ is an interval and $\chi_\Gamma(x)$ its characteristic function, then there exists a sequence $f_n \in C$ such that $f_n \downarrow \chi_\Gamma$. Then

$$P(t, x, \Gamma) = \lim_{n \to \infty} \int_{r_1-}^{r_2+} P(t, x, dy) f_n(y). \tag{4.5}$$

For each n the right side is continuous in t and in x, and this guarantees the required measurability properties of $P(t, x, \Gamma)$ for all Γ.

Now in [2] the equation (3.10) for the Laplace transforms appears as basic requirement for the resolvents. The strong continuity property of the semi-groups used in [2] obviously implies (3.4) and we have thus

THEOREM 1. *Each semi-group from C to C constructed in* [2] *defines a diffusion process obeying* (1.1).

As will be shown in §13, the converse is true only for diffusion processes in the open interval (r_1, r_2). For a diffusion process in $[r_1, r_2]$ it is possible that $f \in C$ but that $u(t, x)$ is not continuous in $[r_1, r_2]$.

5. **The probabilistic interpretation.** As is well known, the backward equation (1.1) has first been derived from probabilistic considerations by Kolmogorov [7]. He showed also that under strong additional conditions the Fokker-Planck equation in the special form (1.2) must hold. Kolmogorov's conditions were relaxed in [4]. The principal step towards generalization was the introduction of the so-called *Lindeberg condition* which requires that for each fixed x and $\epsilon > 0$

$$\int_{|y-x|>\epsilon} P(t, x, dy) = o(t) \tag{5.1}$$

as $t \to 0$. This, of course, is a strengthening of the continuity condition (3.4).

It is easily seen that a condition of the Lindeberg type is *necessary* for the continuity of the path functions.

On the other hand, it has been shown recently [5] that under very mild differentiability conditions the *Lindeberg condition*(9) *alone suffices to derive*

(9) The abstract formulation is as follows. Let $\{T_t\}$ be a semi-group from C to C. The "Lindeberg condition" requires that if f and g agree in a neighborhood of x_0 then $(T_t(f-g))(x_0) = o(t)$, $t \to 0$. Let Ω_0 be the infinitesimal generator. The differentiability condition mentioned in the text requires that the Laplace transforms (3.9) be twice differentiable whenever $f'' \in C$. The theorem states that under these two conditions and if $\{T_t\}$ *is a contraction semi-group*, Ω_0 *must be a contraction of an operator* Ω *of the form* (1.1) with a term $c(x)I$ added—except for the occurrence of *singular points* where Ω is of a different form. This theorem characterizes the *parabolic differential* equations.

the backward equation (1.1). This is at present the most general derivation of (1.1).

The as yet not completely attained aim of the present method of attack is to *define diffusion processes as those Markovian processes for which the path functions $X(t)$ are continuous with probability one as long as they are in the interior of the basic interval E* (that is, a discontinuity may occur when a boundary is reached). This definition carries over to locally compact topological spaces.

That all our diffusion processes have this property has apparently never been proved rigorously, but may be assumed with impunity. It is also safe to assume that the continuity of $X(t)$ implies that the corresponding transition probabilities define a contraction semi-group which satisfies the "Lindeberg condition" of the preceding footnote.

This condition expresses essentially that the infinitesimal generator Ω_0 is of a *local character*.

Now the local character of Ω_0 and the continuity of $X(t)$ are topological properties, whereas a continuous change of the scale may change differentiable functions into nondifferentiable ones. Hence Ω_0 need not be a differential operator. However, the surmise is that in each case the semi-group induces a scale on the x-axis such that the functions in the domain $D(\Omega_0)$ become twice differentiable and Ω_0 the contraction of a differential operator(10).

Even though the program is as yet unattained, the annoying difference between the forward and backward equations has become understandable. A Markovian process is, essentially, a semi-group in the Banach space of measures. However, *not every contraction semi-group on measures defines a Markovian process, whereas each contraction semi-group from C to C does.* Only the backward equation refers to semi-groups in C and this explains why only the backward equation can be derived from simple probabilistic conditions.

The corresponding forward equations are much more complicated. They will now be derived probabilistically, but only under the assumption that $X(t)$ is continuous with probability one.

6. **First passage times.** We consider a diffusion process obeying (1.1) with the coefficients satisfying the conditions stated at the end of §3. As explained in §5 we shall from now on suppose that the path function $X(t)$ is continuous with probability one as long as $r_1 < X(t) < r_2$ (so far no conditions are imposed for the case that a boundary is reached).

DEFINITION. *Let $r_1 < \rho_1 < \rho_2 < r_2$ be fixed. For any $X(t)$ with $\rho_1 < X(0) = x < \rho_2$ we define the random variable T_X as the moment when for the first time $X(t) = \rho_1$*

(10) Essential parts of the proof have been obtained by S. Leader and the author. When proved, the statement will generalize the result described in footnote 9. It now appears possible to generalize the differential equation (1.1) and to reformulate all the results of [2] in a topologically invariant form.

or $X(t)=\rho_2$. *We put*

$$(6.1)\qquad A(t, x; \rho_2, \bar{\rho}_1) = \Pr\{T_X < t,\ X(T_X) = \rho_2 \mid X(0) = x\},$$

$$(6.2)\qquad A(t, x; \rho_1, \bar{\rho}_2) = \Pr\{T_X < t,\ X(T_X) = \rho_1 \mid X(0) = x\}.$$

The corresponding Laplace transforms will be called $\xi(x; \rho_2, \bar{\rho}_1)$ *and* $\xi(x; \rho_1, \bar{\rho}_2)$. *Thus*

$$(6.3)\qquad \xi(x; \rho_2, \bar{\rho}_1) = \int_0^\infty e^{-\lambda t} A(t, x; \rho_2, \bar{\rho}_1)dt \qquad (\lambda > 0).$$

In words, $A(t, x; \rho_2, \bar{\rho}_1)$ is the probability that a path starting from the position x will within time t reach ρ_2 without previously passing through ρ_1. Such probabilities occur in the theory of *ruin*. In physical language $A(t, x; \rho_2, \bar{\rho}_1)$ is the probability *that a first passage through* ρ_2 *occurs before time* t *when an absorbing barrier is placed at* ρ_1. Even under the present general conditions one has the familiar

THEOREM 2. *Putting* $z(x)=\xi(x; \rho_2, \bar{\rho}_1)$ *one has* $\lambda z - \Omega z = 0$, *and* $z(x)$ *is the only solution of this equation with* $z(\rho_1)=0$ *and* $z(\rho_2)=\lambda^{-1}$. *This solution is monotone.*

Before proving this theorem we shall consider some implications. We interpret "the event that *the boundary* r_j *has been reached* at time t from the interior" (11) as a verbal description of $X(s)\to r_j$ as $s\to t-0$.

As a corollary of Theorem 2 we shall derive the following results.

THEOREM 3. *Let* $X(0)=x$, *with* $r_1<x<r_2$. *The probability that* r_j *will be reached within finite time from the interior is positive if and only if all solutions of* $\lambda z - \Omega z = 0$ *remain bounded as* $z\to r_j$ $(\lambda>0)$.

Thus the probability in question is zero for all x or for no x. This justifies the

DEFINITION. *The boundary* r_j *will be called accessible if and only if there is a positive probability that it will be reached from the interior within a finite time*(12) *(for all* $X(0)=x$*).*

Thus r_j is accessible if and only if all solutions of $\lambda z - \Omega z = 0$ are bounded near r_j. Comparing this with the criteria of [2, p. 488], we see that r_j *is accessible if it is either a regular or an exit boundary.*

THEOREM 4. *Suppose that* r_2 *is accessible. Then* (a) *If* r_1 *is inaccessible, there exists one and only one bounded solution* $z(x)$ *of* $\lambda z - \Omega z = 0$ *with* $z(x)\to\lambda^{-1}$ *as* $x\to r_2$.

This solution is positive and monotonic. Let $A(t, x)$ *be the probability that*

(11) Later on we shall consider instantaneous transitions from r_j to r_k.

(12) Cf. footnote 13.

r_2 will be reached before time t assuming that the starting position is $X(0)=x$. Then

$$z(x) = \int_0^\infty e^{-\lambda t} A(t, x)dt. \tag{6.4}$$

(b) *If r_1 is accessible there exists one and only one solution of $\lambda z - \Omega z = 0$ with $z(x) \to 0$ as $x \to r_1$ and $z(x) \to \lambda^{-1}$ as $x \to r_2$. It is monotonic and for it* (6.4) *holds provided that $A(t, x)$ is interpreted as the probability that r_2 will be reached before time t and before r_1 is reached. (If r_1 is accessible, then there exist many processes obeying* (1.1) *with different probabilities of reaching r_2.)*

From Theorem 4 it will be easy to derive the following analogues of classical results concerning *probabilities of ruin.*

THEOREM 5. *Let r_2 be accessible and let $\Phi(x) = A(\infty, x)$ be the probability of ultimate absorption at r_2. Then $\Phi(x)$ is the smallest positive solution of $a\Phi''(x) + b\Phi' = 0$ with $\Phi(x) \to 1$ as $x \to r_2$.*

More explicitly: If r_1 is an accessible boundary, then $\Phi(x)$ is the strictly increasing solution with $\Phi(r_1) = 0$, $\Phi(r_2) = 1$. The same holds true whenever

$$\exp\left\{-\int_x^{x_0} \frac{b}{a} ds\right\} \tag{6.5}$$

is integrable over (r_1, x_0). Otherwise $\Phi(x) \equiv 1$. This case arises if r_1 is an entrance boundary, and may arise if r_1 is a natural boundary(13).

Note. By the same method one can prove that *in the case of two accessible boundaries the expected duration*

$$\psi(x) = \int_0^\infty t d_t A(t, x) \tag{6.6}$$

is finite and characterized as the unique solution of $a\psi'' + b\psi' = -1$ with $\psi(r_1) = \psi(r_2) = 0$. However, if r_1 is inaccessible, the boundary condition $\psi(r_1) = 0$ can not be imposed, and it can happen that $\psi(x)$ is finite but unbounded, and also that it is infinite for all x.

7. **Proofs.** We start with a few preliminaries. Let

$$\Pi(x, \Gamma) = \int_0^\infty e^{-\lambda t} P(t, x, \Gamma)dt, \qquad \lambda > 0, \tag{7.1}$$

and

(13) [Footnote added September 1953.] Doob has discovered the possibility that for a *natural* boundary r_1 there may be a positive probability that $X(t) \to r_1$ as $t \to \infty$ even though r_1 cannot be reached in finite time. This explains the possibility that $\Phi < 1$ even though r_1 is inaccessible.

$$F(x) = \int_E f(y)\Pi(x, dy). \tag{7.2}$$

This is defined for all bounded measurable f, and agrees with the definition (3.9). Hence, at least when f is continuous, we have by definition

$$\lambda F - \Omega F = f. \tag{7.3}$$

To solve (7.3) for F we require the fact proved by Hille (cf. [2, p. 483]) that there exist two solutions ξ_1 and ξ_2 of

$$\lambda\xi - \Omega\xi = 0 \tag{7.4}$$

satisfying the conditions

$$\xi_1(x) \downarrow, \qquad \xi_2(x) \uparrow, \qquad \xi_j \geqq 0 \qquad \text{for } r_1 < x < r_2. \tag{7.5}$$

In terms of two such solutions we construct the classical *Green* function

$$K(x, y) = \begin{cases} \xi_2(x)\eta_1(y) & \text{for } r_1 < x \leqq y < r_2, \\ \xi_1(x)\eta_2(y) & \text{for } r_1 < y \leqq x < r_2 \end{cases} \tag{7.6}$$

where

$$\eta_j(y) = \frac{\xi_j(y)}{\xi_1(y)\xi_2'(y) - \xi_1'(y)\xi_2(y)}. \tag{7.7}$$

It is known that

$$F(x) = \int_{r_1}^{r_2} K(x, y)f(y)dy \tag{7.8}$$

is a solution of (7.3). *If* it is unique, then a comparison of (7.2) and (7.8) shows that

$$\Pi(x, \Gamma) = \int_\Gamma K(x, y)dy. \tag{7.9}$$

Now the difference of any two admissible solutions of (7.3) is a bounded solution of (7.4). On the other hand, every solution of (7.4) is a linear combination of ξ_1 and ξ_2. As each ξ_j is bounded near one boundary, we conclude that (7.4) has a bounded solution only if either ξ_1 or ξ_2 is bounded. Independently of the choice of ξ_1 and ξ_2 we can therefore state:

(a) If both ξ_1 and ξ_2 are unbounded, then the Laplace transform $\Pi(x, \Gamma)$ of the transition probabilities is uniquely determined by (7.3).

(b) If, say, ξ_2 is bounded but ξ_1 unbounded, then necessarily

$$\Pi(x, \Gamma) = \int_\Gamma K(x, y)dy + \xi_2(x)\Phi_2(\Gamma). \tag{7.10}$$

(c) If both ξ_1 and ξ_2 are bounded, then

$$\Pi(x, \Gamma) = \int_\Gamma K(x, y)dy + \xi_1(x)\Phi_1(\Gamma) + \xi_2(x)\Phi_2(\Gamma). \tag{7.11}$$

Here the $\Phi_j(\Gamma)$ are measures on the Borel sets and depend on λ, but not on x. The most general form of these Φ_j was the main topic of [2], but for the present proof we require no special knowledge about Φ_j.

Proof of Theorem 2. Let $r_1 < x < \rho < r_2$ and let

$$H(t, x; \rho) = \Pr \{\operatorname*{Max}_{s<t} X(s) > \rho \mid X(0) = x\} \tag{7.12}$$

be the probability distribution of the first passage time from x to ρ. Then, if Γ is any set in (ρ, r_2), the event $X(t) \in \Gamma$ can occur only if the first passage has occurred at some time $s<t$. Thus (always assuming the continuity of $X(t)$)

$$P(t, x, \Gamma) = \int_0^t P(t - s, \rho, \Gamma)d_sH(s, x; \rho). \tag{7.13}$$

For the corresponding Laplace transform

$$\xi(x; \rho) = \int_0^\infty e^{-\lambda t}H(t, x; \rho)dt \qquad \lambda > 0, \tag{7.14}$$

this means

$$\Pi(x, \Gamma) = \lambda\Pi(\rho, \Gamma)\xi(x; \rho). \tag{7.15}$$

Now a glance at (7.9)–(7.11) shows that for any fixed set $\Gamma \subset (\rho, r_2)$ and $x \leqq \rho$ the function $\Pi(x, \Gamma)$ reduces to a linear combination of ξ_1 and ξ_2, and hence to a solution of (7.4). Therefore:

$\xi(x; \rho)$ is a solution of (7.4) and $\xi(x; \rho) \to \lambda^{-1}$ as $x \to \rho$.

Note that $\xi(x; \rho)$ depends on the individual process and cannot be determined without knowing the boundary conditions.

Now let $\rho_1 < x < \rho_2$ and define the first passage transform $\xi(x; \rho_1)$ in obvious analogy with (7.14). The first passage from x to ρ_2 can take place either before or after a first passage from x to ρ_1. This consideration leads to a relation which, translated into the language of Laplace transforms, reads

$$\xi(x; \rho_2) = \xi(x; \rho_2, \bar{\rho}_1) + \lambda\xi(x; \rho_1, \bar{\rho}_2)\cdot\xi(\rho_1; \rho_2) \tag{7.16}$$

with $\xi(x; \rho_2, \bar{\rho}_1)$ defined in (6.3). Similarly

$$\xi(x; \rho_1) = \xi(x; \rho_1, \bar{\rho}_2) + \lambda\xi(x; \rho_2, \bar{\rho}_1)\xi(\rho_2; \rho_1). \tag{7.17}$$

We have here two linear equations for the two unknowns $\xi(x; \rho_1, \bar{\rho}_2)$ and $\xi(x; \rho_2, \bar{\rho}_1)$. From the very definition (7.14) we have $0 \leqq \xi(\rho_i; \rho_j) < \lambda^{-1}$ and therefore the determinant of the system does not vanish. Hence

$$\xi(x;\rho_2,\bar\rho_1) = \frac{\xi(x;\rho_2) - \lambda\xi(\rho_1,\rho_2)\xi(x;\rho_1)}{1-\lambda^2\xi(\rho_1;\rho_2)\xi(\rho_2;\rho_1)}. \tag{7.18}$$

Remembering that $\xi(\rho_j;\rho_j)=\lambda^{-1}$ we find

$$\lim_{x\to\rho_1}\xi(x;\rho_2,\bar\rho_1) = 0, \qquad \lim_{x\to\rho_2}\xi(x;\rho_2,\bar\rho_1) = \lambda^{-1}, \tag{7.19}$$

and $\xi(x;\rho_2,\bar\rho_1)$, being a linear combination of two solutions of (7.4), is itself a solution of (7.4). These conditions determine $\xi(x;\rho_2,\bar\rho_1)$ uniquely, and Theorem 2 is proved.

Proof of Theorems 3 and 4. Let $\xi_j(x)$ be two solutions of (7.4) satisfying the boundary conditions (7.5). From Theorem 2 we have

$$\xi(x;\rho_2,\bar\rho_1) = \frac{\xi_2(x)\xi_1(\rho_1) - \xi_1(x)\xi_2(\rho_1)}{\lambda\{\xi_2(\rho_2)\xi_1(\rho_1) - \xi_2(\rho_1)\xi_1(\rho_2)\}} \qquad (\rho_1 < x < \rho_2). \tag{7.20}$$

Suppose first that $\xi_1(x)\uparrow\infty$ as $x\downarrow r_1$. Since $\xi_2(x)$ is positive and nondecreasing, $\xi_2(x)$ remains bounded, and the ratio $\xi_2(x)/\xi_1(x)$ tends to zero. Hence in this case

$$\lim_{\rho_1\to r_1}\xi(x;\rho_2,\bar\rho_1) = \frac{\xi_2(x)}{\lambda\xi_2(\rho_2)}. \tag{7.21}$$

Next, if $\xi_1(x)\uparrow\xi_1(r_1)<\infty$ as $x\downarrow r_1$, then

$$\xi^*(x) = \xi_2(x)\xi_1(r_1) - \xi_1(x)\xi_2(r_1) \tag{7.22}$$

is a solution of (7.4) and

$$\lim_{\rho_1\to r_1}\xi(x;\rho_2,\bar\rho_1) = \frac{\xi^*(x)}{\lambda\xi^*(\rho_2)}. \tag{7.23}$$

Note that $\xi^*(x)$ is obviously nondecreasing, and $\xi^*(x)\to 0$ as $x\to r_1$. Moreover, if $\xi_2(x)\uparrow\xi_2(r_2)<\infty$ as $x\to r_2$, then ξ^* has a finite limit and *all* solutions of (7.4) remain bounded near r_2. Hence, as $\rho_2\to r_2$, either each of the right-hand members in (7.21) and (7.23) tends to zero, or each tends to a monotone solution of (7.4) which tends to λ^{-1} as $x\to r_2$.

Proof of Theorem 5. Put $\xi_\lambda(x)=\lambda z(x)$. Then

$$\xi_\lambda(x) = \int_0^\infty e^{-\lambda t}d_tA(t,x) \tag{7.24}$$

and hence[14] $\xi_\lambda(x)\uparrow\Phi(x)$ as $\lambda\downarrow 0$. Solving the differential equation $\Omega u=f$ and substituting $u=f=\xi_\lambda$ gives an integral equation for ξ_λ which shows that

[14] Alternatively, this can be seen directly from $\lambda(\xi_\lambda-\xi_\nu)-\Omega(\xi_\lambda-\xi_\nu)=(\nu-\lambda)\xi_\nu$ $(\nu>\lambda)$ which shows (since $\xi_\nu>0$) that $\xi_\lambda-\xi_\nu$ can have no negative minimum. If r_1 is accessible, then $\xi_\lambda-\xi_\nu$ vanishes at both ends. If r_1 is inaccessible the result follows by approximation.

$\xi_\lambda \uparrow \Phi$ implies that $\Omega\Phi$ exists and $\Omega\Phi = \lim \Omega\xi_\lambda = \lim \lambda\xi_\lambda = 0$. Clearly $\Phi(r_2) = 1$ since $\xi_\lambda(x) \uparrow 1$ as $x \to r_2$. Suppose now that the integral (6.5) diverges as $x \to r_1$. Then $\Phi \equiv 1$ is the only solution of $\Omega\Phi = 0$ with $\Phi(r_2) = 1$. On the other hand, if the integral (6.5) converges, there exists exactly one solution of $\Omega u = 0$ with $u(r_1) = 0$, $u(r_2) = 1$. From $\Omega(u - \xi_\lambda) = -\lambda\xi_\lambda < 0$ it is seen that the function $u - \xi_\lambda$ can have no minimum. Since it vanishes at $x = r_1$, r_2 it follows that $\xi_\lambda(x) \leqq u(x)$, whence $\Phi(x) \leqq u(x)$. The inequality would require Φ to assume negative values, and this proves the theorem.

8. The absorbing barrier process. Consider a process obeying (1.1) and assume, as before, that the path functions $X(t)$ are continuous.

DEFINITION. *Using the random variable T_X of the definition of §6 we define the absorbing barrier process*(15) *corresponding to our process and (ρ_1, ρ_2) as the process of the path functions $X_{\mathrm{abs}}(t)$ defined by*

$$X_{\mathrm{abs}}(t) = \begin{cases} X(t) & \text{for } t \leqq T_X, \\ X(T_X) & \text{for } t \geqq T_X \end{cases} \tag{8.1}$$

with the obvious probability relations induced by the original process.

We shall refer to T_X as the *absorption time.* With the notation of §6 we have

$$\Pr\{T_X \leqq t\} = A(t, x; \rho_1, \bar{\rho}_2) + A(t, x; \rho_2, \bar{\rho}_1). \tag{8.2}$$

THEOREM 6(16). *Let $P_{\mathrm{abs}}(t, x, \Gamma)$ be the transition probability of the absorbing barrier process and $\Pi_{\mathrm{abs}}(x, \Gamma)$ its Laplace transform* (7.1). *With the notations of §6 we have*(17) *for $r_1 < \rho_1 < x < \rho_2 < r_2$*

$$\begin{aligned} \Pi_{\mathrm{abs}}(x, \rho_1) &= \xi(x; \rho_1, \bar{\rho}_2), \\ \Pi_{\mathrm{abs}}(x, \rho_2) &= \xi(x; \rho_2, \bar{\rho}_1). \end{aligned} \tag{8.3}$$

For any Borel set $\Gamma \subset (\rho_1, \rho_2)$ we have

$$\Pi_{\mathrm{abs}}(x, \Gamma) = \int_\Gamma K(x, y)\,dy \tag{8.4}$$

where $K(x, y)$ is the Green function (cf. (7.6)) of $\lambda u - \Omega u = f$ formed by means of two solutions ξ_1 and ξ_2 of $\lambda\xi - \Omega\xi = 0$ satisfying the boundary conditions

$$\xi_1(\rho_1) = 0, \qquad \xi_2(\rho_2) = 0. \tag{8.5}$$

Proof. Clearly, (8.3) is only a restatement of Theorem 2. From this

(15) By (8.1) the absorbing barrier process is defined in the *closed* interval $[\rho_1, \rho_2]$. If we let $X_{\mathrm{abs}}(t)$ be undefined for $t \geqq T_X$ we get the corresponding process for the *open* interval.

(16) This theorem restates a well known fact. Also the renewal method for the proof is familiar. Both theorem and proof are (under slightly different hypotheses) found in a recent paper by Darling and Siegert [0]. (This paper reached the author after completion of the present manuscript.)

(17) In (8.3) and similar formulas later on the letters ρ or r are used not only to denote a number, but also for the set consisting of the single point ρ or r.

theorem one has also that for $\rho_1 < x < y < \rho_2$

$$\xi(x;\, y,\, \bar{\rho}_1) = \frac{\xi_1(x)}{\lambda\xi_1(y)}\,. \tag{8.6}$$

Again, the renewal argument used in (7.15) shows that for $\Gamma \subset (y, \rho_2)$ one has

$$\begin{aligned} \Pi_{\text{abs}}(x, \Gamma) &= \lambda\xi(x;\, y,\, \bar{\rho}_1)\Pi_{\text{abs}}(y, \Gamma) \\ &= \frac{\xi_1(x)}{\xi_1(y)}\,\Pi_{\text{abs}}(y, \Gamma). \end{aligned} \tag{8.7}$$

Thus

$$\Pi_{\text{abs}}(x, \Gamma) = \xi_1(x)\cdot F(\Gamma) \tag{8.8}$$

where $F(\Gamma)$ is independent of λ. This relation holds only for sets $\Gamma \subset (x, \rho_2)$, but a similar relation holds for sets $\Gamma \subset (\rho_1, x)$. The theorem now follows on comparing this result with the general form of the transition probabilities given in (7.11).

DEFINITION. *We shall say the transition probabilities of a process are defined by a Green function if for* $r_1 < x < r_2$ *and* $\Gamma \subset (r_1, r_2)$ *the corresponding Laplace transform* $\Pi(x, \Gamma)$ *is of the form* (8.4) *where* $K(x, y)$ *is a regular Green function*(18).

Note that no restriction is placed on the transitions $\Pi(x, \rho_j)$ and $\Pi(\rho_j, \Gamma)$.

THEOREM 7. (a) *If both* r_1 *and* r_2 *are inaccessible, then as* $\rho_1 \to r_1$ *and* $\rho_2 \to r_2$ *the transition probabilities of the absorbing barrier process for* (ρ_1, ρ_2) *converge to transition probabilities in* (r_1, r_2) *which are defined by the unique regular Green function for* (r_1, r_2). *In this case there exists one and only one process in* (r_1, r_2) *obeying* (1.1).

(b) *If* r_1 *is inaccessible but* r_2 *accessible, convergence takes place to transition probabilities defined by that regular Green function which is formed by means of a solution with* $\xi(x) \downarrow 0$ *as* $x \to r_2$.

(c) *If* r_1 *and* r_2 *are accessible, then limiting transition probabilities exist and are defined by that regular Green function which is formed by means of solutions* $\xi_j(x)$ *with* $\xi_j(x) \downarrow 0$ *as* $x \to r_j$.

This theorem extends the notion of absorbing barrier processes to intervals (r_1, r_2) with accessible boundaries.

Proof. In the proof of Theorem 3 (§7) it has been shown that as $\rho_1 \to r_1$ and $\rho_2 \to r_2$ the solutions $\xi(x;\, \rho_1,\, \bar{\rho}_2)$ and $\xi(x;\, \rho_2,\, \bar{\rho}_1)$ tend to certain limits. Using these limiting forms the theorem becomes trivial.

It follows in particular that *whenever both* r_1 *and* r_2 *are accessible there exists*

(18) Cf. (7.6). The notion of *regular* Green function was introduced in [2] and refers to Green functions formed by means of solutions satisfying the monotonicity conditions (7.5). The restriction to *regular* Green functions is essential for the validity of some of our statements.

an absorbing barrier process in (r_1, r_2) *obeying* (1.1).

The relevant statements concerning the boundary conditions and the Fokker-Planck equations are contained in Theorem 10, as the limiting case $c_i=0$.

Note on the corresponding densities. It would be pleasing to assert that the transition probabilities of the last theorem are determined by densities $p(t, x, y)$, and that the latter satisfy both the backward equation

$$p_t(t, x, y) = a(x)p_{xx}(t, x, y) + b(x)p_x(t, x, y) \tag{8.9}$$

and the forward equation

$$p_t(t, x, y) = [a(y)p(t, x, y)]_{yy} - [b(y)p(t, x, y)]_y. \tag{8.10}$$

When both boundaries are inaccessible, $p(t, x, y)$ should be uniquely determined by either of these equations; if r_j is accessible, one should expect that for (8.9) the boundary condition $p(t, x, y)\to 0$ as $x\to r_j$ must be imposed; in the case of (8.10) the corresponding boundary condition is

$$a(y) \exp\left[-\int b(y)a^{-1}(y)dy\right] p(t, x, y) \to 0 \tag{8.11}$$

as $y\to r_j$.

These assertions are contained in known results about differential equations provided that both boundaries are regular and that more stringent conditions on $a(x)$ and $b(x)$ are imposed. Unfortunately, the theory of semi-groups leads directly only to a slightly weaker result. We should prove that for fixed x, y

$$K(x, y) = \int_0^\infty e^{-\lambda t}p(t, x, y)dt \tag{8.12}$$

where $p(t, x, y)$ has a bounded derivative with respect to t. Actually it appears that without deeper analysis one can assert only that[19]

$$K(x, y) = \int_0^\infty e^{-\lambda t}V(dt, x, y) \tag{8.13}$$

where V is a nondecreasing function of t.

Now we have seen that for $x<z<y$

$$K(x, y) = \frac{\xi(x)}{\xi(z)} K(z, y). \tag{8.14}$$

Here the second factor on the right is of the form (8.13), while the first

[19] (8.13) is a direct consequence of the Bernstein-Hausdorff-Widder theorem (cf. [8]) and the well known property of resolvents according to which $(-1)^n d^n K(x, y)/d\lambda^n = n! \cdot K_{n+1}(x, y) \geq 0$ where K_n denotes the iterated kernels; cf. [6, p. 99].

factor is the Laplace transform of the first passage time from x through z. But the convolution of two monotonic functions will be continuous, or differentiable, whenever one of the two components has this property. It follows therefore that if $p(t, z, y)$ is, for a fixed z, y, a continuous (differentiable) function of t, then the same is true for any x, z. Moreover, (8.14) shows that for the density $p(t, x, y)$ to satisfy (8.9) it is necessary and sufficient that the first passage time density be a solution of the backward equation (1.1). The duality principle of [2] permits us to translate these assertions into corresponding statements referring to $p(t, x, y)$ as function of t and y, and to (8.10) instead of (8.9).

Accordingly, to prove the opening surmise of this note it would suffice to prove that for fixed $x<z$ the ratio $\xi(x)/\xi(z)$ is the Laplace transform of a differentiable function or, alternatively, that this is true of $K(x, x)$. It seems plausible that this is provable by the methods of the general theory. Unfortunately, the author sees at present only a method of proof which is based on laborious estimates. This does not seem natural, in particular since under slightly stronger conditions the facts are known. Moreover, the assertion would improve only the aesthetical appeal and simplify some formulations, without being essential.

9. **The elementary return process.** We now proceed to calculate the transition probabilities of the process defined in §2. For definiteness we assume for the time being that both boundaries are *accessible*, and $\sigma_1>0$, $\sigma_2>0$. The starting position $X(0)=x$ is assumed to be the interior (r_1, r_2).

(a) The probability that up to time t no transitions from a boundary have occurred and that $X(t)=r_1$ (i.e. the first boundary to be reached is r_1, and the *first* sojourn time has not expired) is

$$\int_0^t d_sA(t-s, x; r_1, \bar{r}_2)\cdot e^{-s/\sigma_1} \tag{9.1}$$

where $A(t, x; r_1, \bar{r}_2)$ is the distribution function of the absorption time defined in (6.1). For the corresponding Laplace transforms (6.3) we introduce the notations

$$\xi_1(x) = \xi(x; r_1, \bar{r}_2), \quad \xi_2(x) = \xi(x; r_2, \bar{r}_1). \tag{9.2}$$

Then (9.1) has the Laplace transform

$$\lambda\xi_1(x)\frac{\sigma_1}{1+\lambda\sigma_1}\cdot \tag{9.3}$$

(b) At the expiration of the sojourn time a "jump" occurs. The probabilities that it takes $X(t)$ to the interior, to r_1, or to r_2, are τ_1, p_{11}, p_{12}, respectively. The probability *density* that this jump occurs at time t is the expression (9.1) multiplied by σ_1^{-1}.

(c) Let $P^{(n)}(t, x, \Gamma)$ be the probability that up to time t exactly n "jumps" occur and that $X(t) \in \Gamma$. Let $\Pi^{(n)}(x, \Gamma)$ be the corresponding Laplace transform. Then

$$\Pi^{(0)}(x, \Gamma) = \Pi_{\text{abs}}(x, \Gamma), \tag{9.4}$$

provided Γ is a set in the interior. If Γ reduces to a boundary point then, as shown under (a),

$$\Pi^{(0)}(x, r_j) = \lambda \xi_j(x) \frac{\sigma_j}{1 + \lambda \sigma_j} \cdot \tag{9.5}$$

(d) The probability *density* that the nth jump occurs at r_j and at time t is $P^{(n-1)}(t, x, r_j) \cdot \sigma_j^{-1}$. It has the Laplace transform

$$\Pi^{(n-1)}(x, r_j)\sigma_j^{-1}. \tag{9.6}$$

(e) If the nth jump occurs at r_j at time s, then for every set $\Gamma \subset (r_1, r_2)$

$$\Pr\{X(s+) \in \Gamma\} = \tau_j \int_\Gamma dp_j(y). \tag{9.7}$$

Therefore the probability that $X(t+s) \in \Gamma$ and that no further absorption has taken place is

$$\tau_j \int_{r_1}^{r_2} P_{\text{abs}}(t, y, \Gamma) \cdot dp_j(y). \tag{9.8}$$

For the corresponding Laplace transform we introduce the abbreviation[20]

$$U_j(\Gamma) = \tau_j \int_{r_1}^{r_2} \Pi_{\text{abs}}(y, \Gamma) dp_j(y). \tag{9.9}$$

Thus for $\Gamma \subset (r_1, r_2)$

$$\Pi^{(n)}(x, \Gamma) = \sum_{j=1}^{2} \Pi^{(n-1)}(x, r_j)\sigma_j^{-1} U_j(\Gamma). \tag{9.10}$$

(f) If at any time a jump takes place at r_i, the probability density that it leads into the interior and then, after a time t, to a first absorption at r_j has the Laplace transform

$$\lambda \tau_{ij} = \lambda \tau_i \int_{r_1}^{r_2} \xi_j(x) dp_i(x). \tag{9.11}$$

(g) From (d) and (f) we conclude

$$\Pi^{(n)}(x, r_j) = \sum_{i=1}^{2} \Pi^{(n-1)}(x, r_i)\sigma_i^{-1}\{p_{ij} + \lambda \tau_{ij}\} \frac{\sigma_j}{1 + \lambda \sigma_j} \cdot \tag{9.12}$$

(20) In comparing our formulas with those of [2], put $U_j(\Gamma) = \int_\Gamma Q_j(x)dx$ where $Q_j(x)$ is the density defined in (21.14).

For convenience of writing we introduce the *matrix notation*

$$(9.13)\qquad \mathbf{p} = (p_{ij}), \qquad \boldsymbol{\tau} = (\tau_{ij}), \qquad \boldsymbol{\sigma} = \begin{pmatrix} \sigma_1 & 0 \\ 0 & \sigma_2 \end{pmatrix}.$$

For row and column matrices we write

$$(9.14)\qquad [x_i] = \begin{matrix} x_1 \\ x_2 \end{matrix}, \qquad [x_i]' = x_1\ x_2.$$

Now (9.12) takes on the form

$$(9.15)\qquad [\Pi^{(n)}(x, r_j)]' = [\Pi^{(n-1)}(x, r_i)]'\boldsymbol{\sigma}^{-1}\{\mathbf{p} + \lambda\boldsymbol{\tau}\}\{\mathbf{I} + \lambda\boldsymbol{\sigma}\}^{-1}\boldsymbol{\sigma}$$

where $\mathbf{I}$ is the identity matrix. By definition we have

$$(9.16)\qquad \Pi(x, \Gamma) = \sum_{n=0}^{\infty} \Pi^{(n)}(x, \Gamma).$$

Thus, using (9.5) and (9.15), we get

$$\begin{aligned} [\Pi(x, r_j)]' &= \lambda[\xi_j]'(\mathbf{I} + \lambda\boldsymbol{\sigma})^{-1} \sum_{n=0}^{\infty} (\{\mathbf{p} + \lambda\boldsymbol{\tau}\}\{\mathbf{I} + \lambda\boldsymbol{\sigma}\}^{-1})^n\boldsymbol{\sigma} \\ (9.17)\qquad &= \lambda[\xi_j]'(\mathbf{I} + \lambda\boldsymbol{\sigma})^{-1}(\mathbf{I} - \{\mathbf{p} + \lambda\boldsymbol{\tau}\}\{\mathbf{I} + \lambda\boldsymbol{\sigma}\}^{-1})^{-1}\boldsymbol{\sigma}. \end{aligned}$$

For abbreviation put[21]

$$(9.18)\qquad \mathbf{R} = (\mathbf{I} + \lambda\boldsymbol{\sigma} - \mathbf{p} - \lambda\boldsymbol{\tau})^{-1}.$$

Then (9.17) reduces to

$$(9.19)\qquad [\Pi(x, r_j)]' = \lambda[\xi_j]'\mathbf{R}\boldsymbol{\sigma}.$$

Having thus $\Pi(x, r_j)$ we get $\Pi(x, \Gamma)$ from (9.16) and (9.10). For $\Gamma \subset (r_1, r_2)$ we get

$$(9.20)\qquad \Pi(x, \Gamma) = \Pi_{\text{abs}}(x, \Gamma) + \lambda[\xi_j]'\mathbf{R}[U_j(\Gamma)].$$

Finally, for completeness we give the corresponding expressions for the case where the process starts at a boundary. The argument which led to (9.19) shows that

$$(9.21)\qquad (\Pi(r_i, r_j)) = \mathbf{R}\boldsymbol{\sigma},$$

and this is also evident from (9.19). Similarly, for $\Gamma \subset (r_1, r_2)$

$$(9.22)\qquad [\Pi(r_i, \Gamma)] = \mathbf{R}[U_i(\Gamma)],$$

as is again clear from (9.20).

---

(21) The proof that $\mathbf{R}$ exists is given in [2, p. 512]. It should have been mentioned there that the proof would formally break down if $\pi_i = \sigma_i = \tau_i = 0$ and $p_{ii} = 1$. However, that case is impossible since it would reduce a lateral condition to $0 = 0$.

In (9.19)–(9.22) we have now *expressions for all the transition probabilities*. The very derivation shows that they have the necessary properties like positivity and that they determine a process obeying (1.1).

We observe that our process is a special case of the one treated in [2, §21]. If one puts there $\pi_1=\pi_2=0$, the formulas agree with ours. The new derivation is more direct, but we still must refer to [2] for the derivation of the lateral condition and the proof that they determine the process.

Summing up we have

THEOREM 8. *If both boundaries are accessible, then there exists exactly one diffusion process in* $[r_1, r_2]$ *obeying* (1.1) *and having the properties of the "elementary return process" described in* §2. *If, and only if,* $\sigma_1=\sigma_2=0$ *the process takes place in the open interval* (r_1, r_2). *The solutions of* (1.1) *corresponding to our process are characterized by the two lateral conditions*

$$(9.23)\quad \begin{aligned} (1-p_{11})u(t, r_1) - p_{12}u(t, r_2) &= \tau_1 \int_{r_1}^{r_2} u(t, y)dp_1(y) - \sigma_1\Omega u(t, r_1), \\ -p_{21}u(t, r_1) + (1-p_{22})u(t, r_2) &= \tau_2 \int_{r_1}^{r_2} u(t, y)dp_2(y) - \sigma_2\Omega u(t, r_2). \end{aligned}$$

Note. For lucidity of exposition we have been assuming that both boundaries are accessible. The starting point was the absorbing barrier process defined in the *open* interval (r_1, r_2). The new process plays in the *closed* interval $[r_1, r_2]$, unless either σ_1 or σ_2 vanishes (in which case the corresponding boundary is excluded).

If both r_1 and r_2 are inaccessible, then no return process is possible. If r_1 is *inaccessible* and r_2 *accessible*, the above construction applies except that no returns can occur from r_1.

10. **The corresponding Fokker-Planck equations.** If $\mu(\Gamma)$ is the probability distribution for the initial point $X(0)$, then the distribution of the position $X(t)$ is given by (3.7). Consider now in particular the transition probabilities of the elementary return process determined by (9.20). The set functions $U_j(\Gamma)$ are defined in (9.8) and (9.9). As explained in §8, the distribution with the Laplace transform $U_j(\Gamma)$ has a density, and the same is then true of our transition probability. We can therefore write

$$(10.1)\quad V(t, \Gamma) = \int_\Gamma v(t, y)dy, \qquad \Gamma \subset (r_1, r_2),$$

$$(10.2)\quad V(t, r_j) = V_j(t).$$

In §§18 and 22 of [2] we have studied, in the Banach space of measures, the semi-group which is the adjoint to the semi-group from C to C corresponding to the process of §9. The following theorem is a restatement of these results in terms of differential equations(22).

THEOREM 9. (a) *In the process of Theorem* 8 *let the mean sojourn times* σ_j *be positive. Then*

$$a(x)v_x(t, x) - b(x)v(t, x) = \Phi_v(t, x) \tag{10.3}$$

exists and is of bounded variation. For every set $\Gamma \subset (r_1, r_2)$ *one has*(23)

$$\frac{\partial}{\partial t}\int_\Gamma v(t, x)dx = \int_\Gamma d_x\left\{\Phi_v(t, x) + \frac{\tau_1}{\sigma_1}V_1(t)p_1(x) + \frac{\tau_2}{\sigma_2}V_2(t)p_2(x)\right\} \tag{10.4}$$

and

$$V_1'(t) = -\frac{1-p_{11}}{\sigma_1}V_1(t) + \frac{p_{21}}{\sigma_2}V_2(t) + \Phi_v(t, r_1), \tag{10.5}$$

$$V_2'(t) = \frac{p_{12}}{\sigma_1}V_1(t) - \frac{1-p_{22}}{\sigma_2}V_2(t) - \Phi_v(t, r_2) \tag{10.6}$$

where

$$\Phi_v(t, r_j) = \lim_{x\to r_j}\Phi_v(t, x). \tag{10.7}$$

(b) *If* $\sigma_1 = \sigma_2 = 0$ *put* $V_1(t) = V_2(t) = 0$ *and* (10.4) *is to be replaced*(24) *by*

$$\frac{\partial}{\partial t}\int_\Gamma v(t, x)dx = \int_\Gamma d_x\{\Phi_v(t, x) + \tau_1\Phi_v(t, r_1)p_1(x) - \tau_2\Phi_v(t, r_2)p_2(x)\}. \tag{10.8}$$

(c) *These equations are the Fokker-Planck (or forward) equations of our process. Depending on the nature of the boundaries*(25) r_j, *they either determine the process uniquely, or one or two boundary conditions of the form* (8.11) *are satisfied and characterize the process.*

Physical interpretation. Let $\Gamma = (\rho_1, \rho_2)$ be an interval contained in (r_1, r_2).

(22) In order not to obscure a simple situation, the formulation of the theorem avoids an explicit reference to the fact (described at the end of §8) that the existence and the necessary regularity properties of the density $v(t, x)$ have been proved in [2] explicitly only for the case when the initial distribution $\mu(\Gamma)$ is sufficiently smooth (that is, belongs to the range of the resolvent). For an arbitrary initial distribution the theorem has been completely proved only under slightly more stringent conditions on the coefficients $a(x)$ and $b(x)$.

(23) Note that in the case $\tau_1 = \tau_2 = 0$ equation (10.4) reduces to the classical differential equation (1.2). If $p_1(x)$ and $p_2(x)$ are differentiable, the integrals in (10.4) may be omitted and we still have a differential equation.

(24) Similarly, if $\sigma_1 = 0$ and $\sigma_2 > 0$ the term $\sigma_1^{-1}V_1(t)$ in (10.4) and (10.6) is to be replaced by $\Phi_v(t, r_1)$.

(25) No lateral conditions are required if r_j are "exit" boundaries. If r_j is a "regular" boundary, a lateral condition is required. In [2, §22] both boundaries are supposed to be regular. The modification for exit boundaries is contained in §16 of [2].

The left side of (10.4) represents the rate of change of the mass in Γ. On the right the term $\Phi_v(t, \rho_2) - \Phi_v(t, \rho_1)$ represents the *net flux* through the boundaries into Γ. The term $\tau_j V_j(t)\{p(\rho_2) - p(\rho_1)\}\sigma_j^{-1}$ expresses that a direct flow takes place into Γ from the boundary r_j. The flow of mass $V_j(t)$ at the boundaries is described as follows. The boundary r_1 loses mass at the rate $(1-p_{11})\sigma_1^{-1}$ times the instantaneous mass (exponential decay). Of this mass the proportion $p_{12}/(1-p_{11})$ flows to r_2, and $\tau_1/(1-p_{11})$ into the interior. On the other hand, r_1 gains mass at a rate $p_{21}\sigma_2^{-1}V_2(t)$ from r_2, and at a rate $\Phi_v(t, r_1)$ through a continuous influx from the interior. For our particular process this gain can be shown to be *positive*. Thus the Fokker-Planck equations give a restatement in physical terms of the original probabilistic description of the process in §2.

The case $\sigma_j = 0$ means that no mass can accumulate at r_j and the return to the interior, if it takes place, is instantaneous. Therefore the flow into Γ is no longer proportional to $V_j(t)$ but to the instantaneous flux at the boundary.

11. The elastic barrier process.

DEFINITION. *The boundary r_j is regular if the functions*

$$(11.1) \qquad W(x) = \exp\left\{-\int^x b(s)a^{-1}(s)ds\right\} \quad \textit{and} \quad \frac{1}{a(x)W(x)}$$

are integrable in some (and therefore every) neighborhood of r_j.

According to [2, p. 516] *every regular boundary is accessible*, but an "exit" boundary is accessible without being regular. For the understanding it is preferable to express the regularity as a condition on the functions ξ_j and η_j of (7.5) and (7.7) by means of which the Green function is formed: *r_j is regular if and only if $\xi_j(x)$ is bounded and $\eta_j(y)$ is integrable in (r_1, r_2).*

Our purpose is now to derive the classical elastic barrier process by a passage to the limit from a diffusion process in (r_1, r_2), obeying (1.1). This process is defined as the following modification of the absorbing barrier process:

(11.2) *When the boundary r_j is reached there is probability $\tau_j \geq 0$ of an instantaneous transfer to the point ρ_j in the interior, and probability $1-\tau_j$ that the process terminates.*

In the return process thus described let γ_j *be the probability that a path starting at ρ_j will be absorbed at r_j.*

For simplicity, we shall assume that both boundaries are regular and shall study the following

Passage to the limit. We let $\rho_j \to r_j$ and at the same time $\tau_j \to 1$ in such a way that the probability $\gamma_j < 1$ remains constant(26). (There is no harm in putting, say, $\tau_1 = 0$ and letting only $\tau_2 \to 1$.)

(26) The *same result* would be obtained if instead of γ_j one keeps constant the expected duration of the time to the ultimate absorption at r_j.

(a) We start by calculating γ_j. Put

$$F(x) = \int_{r_1}^{x} W(s)ds. \tag{11.3}$$

Then $aF''+bF'=0$ and by Theorem 5 the probability, in the *absorbing barrier process*, that a path starting at ρ_2 will terminate at r_2 is $F(\rho_2)/F(r_2)$. Accordingly

$$\gamma_2 = \frac{F(\rho_2)}{F(r_2)}(1-\tau_2)\sum_{n=0}^{\infty}\left\{\tau_2\frac{F(\rho_2)}{F(r_2)}\right\}^n. \tag{11.4}$$

For abbreviation put

$$\theta_2 = \{F(r_2) - F(\rho_2)\}/F(r_2). \tag{11.5}$$

Then from (11.4)

$$1-\tau_2 = \frac{\gamma_2}{1-\gamma_2}\,\frac{F(r_2)}{F(\rho_2)}\cdot\theta_2 \tag{11.6}$$

and thus

$$\lim_{\rho_2\to r_2}\frac{1-\tau_2}{\theta_2} = \frac{\gamma_2}{1-\gamma_2} = c_2. \tag{11.7}$$

A similar calculation applies to r_1.

(b) In the notations of §9 our return process is characterized by $\sigma_j=0$, $p_{ij}=0$, while $p_j(x)$ is a unitary distribution function with the jump at ρ_j. The process is restricted to the interior of (r_1, r_2) and its transition probabilities are given in (9.20). In the present case we have

$$\boldsymbol{R} = (I - \lambda\boldsymbol{\tau})^{-1} \tag{11.8}$$

where the matrix $\boldsymbol{\tau}$ is defined (cf. (9.11)) by

$$\tau_{ij} = \tau_i\xi_j(\rho_i). \tag{11.9}$$

The set function $U_j(\Gamma)$ is defined in (9.9). If $\Gamma\subset(\rho_1, \rho_2)$ it is easily seen that

$$U_1(\Gamma) = \tau_1\xi_2(\rho_1)H_1(\Gamma), \qquad U_2(\Gamma) = \tau_2\xi_1(\rho_2)\,H_2(\Gamma) \tag{11.10}$$

with

$$H_j(\Gamma) = \int_{\Gamma}\eta_j(y)dy. \tag{11.10a}$$

Since $\rho_j\to r_j$, it is clear that it suffices to consider sets $\Gamma\subset(\rho_1, \rho_2)$ so that we may use (11.10). We now proceed to investigate the asymptotic behavior of (11.9) and (11.10).

(c) We recall that $\xi_2(x)$ is a nondecreasing solution of $\lambda\xi-\Omega\xi=0$ with $\xi(r_1)=0$, $\xi(r_2)=\lambda^{-1}$. From [2] we know (formula (20.2) after eliminating the restriction $a(x)\equiv 1$) that for *regular* boundaries the quantities

$$\omega_{ij} = (-1)^j \lim_{x\to r_j} W^{-1}(x)\xi_i'(x) \tag{11.11}$$

exist, and form a symmetric, nonsingular matrix $\boldsymbol{\omega}$.

Now, for example,

$$\begin{aligned} 1-\tau_{22} &= 1-\tau_2\lambda\xi_2(\rho_2) = 1-\tau_2+\tau_2\lambda\int_{\rho_2}^{r_2}\xi_2'(x)dx \\ &= 1-\tau_2+\lambda\tau_2\int_{\rho_2}^{r_2}\{\omega_{22}+o(1)\}W(x)dx. \end{aligned} \tag{11.12}$$

Using (11.5) and (11.7) we get from this

$$\frac{1-\tau_{22}}{\theta_2}\to c_2+\lambda\omega_{22}. \tag{11.13}$$

To simplify the writing put

$$\boldsymbol{\theta}=\begin{pmatrix}\theta_1 & 0\\ 0 & \theta_2\end{pmatrix},\qquad \mathbf{c}=\begin{pmatrix}c_1 & 0\\ 0 & c_2\end{pmatrix}. \tag{11.14}$$

Then (11.13) and its analogues can be combined into the matrix equation

$$\boldsymbol{\theta}^{-1}(I-\boldsymbol{\tau})\to\mathbf{c}+\lambda\boldsymbol{\omega}. \tag{11.15}$$

The same calculation applied to (11.10) yields

$$\theta_j^{-1}U_j(\Gamma)\to -\,\omega_{12}H_j(\Gamma). \tag{11.16}$$

Hence

$$\boldsymbol{R}[U_j] = (I-\boldsymbol{\tau})^{-1}\boldsymbol{\theta}\boldsymbol{\theta}^{-1}[U_j]\to -\,(\mathbf{c}+\lambda\boldsymbol{\omega})^{-1}[H_j(\Gamma)]\omega_{12}. \tag{11.17}$$

Thus

$$\Pi(x,\Gamma)\to\Pi_{\mathrm{abs}}(x,\Gamma)-\omega_{12}[\xi_j]'(\mathbf{c}+\lambda\boldsymbol{\omega})^{-1}[H_j(\Gamma)]. \tag{11.18}$$

Now the right side can be obtained from the resolvent (21.12) of [2] by the specialization $\sigma_i=\sigma_j=0$, $(1-\pi_i)\pi_i^{-1}=c_i$. (It should be noted that our solutions $\eta_j(x)$ in terms of which $H_j(\Gamma)$ is defined in (11.10a) differ from the corresponding solutions $\eta_j(x)$ introduced in [2, formula (20.3)] by a factor $-\omega_{12}^{-1}$ as can be seen by comparing the Green function (20.9) of [2] and (7.6) above.) We therefore need no further verification that the right side of (11.18) really represents transition probabilities.

This proves that part of the following theorem which refers to the backward equation (1.1). The statements concerning the Fokker-Planck equation are a consequence of Theorem 9. We have thus

THEOREM 10. *Our limiting process leads to a diffusion process in the open interval* (r_1, r_2) *which obeys* (1.1). *The corresponding Fokker-Planck equation is given by* (1.2). *The solutions of* (1.1) *are in this case characterized by the lateral conditions*

$$c_i u(t, x) = (-1)^{i-1} \lim_{x \to r_i} W^{-1}(x) u_x(t, x) \tag{11.19}$$

and those of (1.2) *by*

$$c_i \lim_{x \to r_i} W(x) a(x) v(t, x) = (-1)^i \Phi_v(t, x_i) \tag{11.20}$$

(*cf.* (10.3) *and* (11.1)).

For $c_i = 0$ the boundary r_i becomes "reflecting," for $c_i \to \infty$ the process reverts into the absorbing barrier process. (The limiting boundary conditions remain correct.)

It is clear that the resulting formulas do not depend on the assumption that *both* boundaries are regular: one elastic barrier can be combined with any other type at the other end. However, if r_1 is inaccessible, then the probability of absorption at r_2 may be identically one for all ρ, and the mean duration of the absorbing barrier process *may* be infinite for all ρ. In this case, therefore, one needs a new interpretation for γ_i, but one may get it trivially by a new passage to the limit.

12. **The general process induced by semi-groups in $C[r_1, r_2]$.** The elastic barrier process takes place in the *open* interval $E = (r_1, r_2)$. One has $P(t, x, E) \to 0$ as $t \to \infty$, except when both boundaries are reflecting ($c_j = 0$), or when one boundary is inaccessible and the other reflecting. If $P(t, x, E) \to 0$ there is probability one that the process terminates either at r_1 or r_2, and we speak again of an "absorption." Note, however, that *the boundary r_j can be reached without an absorption taking place.* In fact, the process can *start* from a boundary(27).

It is now possible to superimpose the elementary return process described in §2 to any elastic barrier process in the same way as we have superimposed it to the absorbing barrier process. An alternative way, requiring no new calculations, consists in a direct passage to the limit. For that purpose we replace (11.2) by the following conditions.

(27) Let $\xi_j^*(x)$ denote the Laplace transform of the probability that an absorption occurs at r_j before time t, given that $X(0) = x$. If both boundaries are regular, it can be shown without difficulties that the $\xi_j^*(x)$ are solutions of $\lambda\xi - \Omega\xi = 0$ satisfying the four boundary conditions $\xi_1^*(r_1) + \xi_2^*(r_1) = \xi_1^*(r_2) + \xi_2^*(r_2) = \lambda^{-1}$, $c_1\xi_2^*(r_1) - \lim_{x\to r_1} W^{-1}(x)\xi_2^{*\prime}(x) = 0$, $c_2\xi_1^*(r_2) + \lim_{x\to r_2} W^{-1}(x)\xi_1^{*\prime}(x) = 0$. In particular, $\xi_2^*(r_1+)$ is the (positive) probability of an absorption at r_2 given that the process starts at r_1.

(12.1) *Consider an elementary return process in which r_i has a mean sojourn time $\epsilon_i\sigma_i$. When a jump from r_i occurs, it leads with probability $\epsilon_i p_{ij}$ to r_j, with probability $1-\epsilon_i$ to the point ρ_i, and with probability $\epsilon_i\tau_i$ to a random point of (r_1, r_2) with probability distribution $p_i(x)$. Here $p_{i1}+p_{i2}+\tau_i\leqq 1$. We effect the same passage to the limit $\epsilon_i\to 0$, $\rho_i\to r_i$ as described in* §11.

It is not necessary to repeat the calculations since a simple inspection shows that the only change is that the term c_i is replaced by another expression which equally remains unaffected by the passage to the limit. Assuming both boundaries to be regular, the procedure of §11 shows that *the transition probabilities of the limiting process are determined by*

$$\Pi(x, \Gamma) = \Pi_{\text{abs}}(x, \Gamma) - \omega_{12}\lambda[\xi_j(x)]'\boldsymbol{R}\{[c_i^{-1}H_i(\Gamma)] + [U_i(\Gamma)]\} \tag{12.2}$$

with $U_i(\Gamma)$ *defined in* (9.9), $H_i(\Gamma)$ *in* (11.10a), *and*[28]

$$\boldsymbol{R} = \{\boldsymbol{I} - \boldsymbol{p} + \lambda\boldsymbol{\sigma} - \lambda\boldsymbol{\tau} + \lambda\boldsymbol{c}^{-1}\boldsymbol{\omega}\}^{-1}. \tag{12.3}$$

Furthermore

$$[\Pi(x, r_i)]' = -\omega_{i2}\lambda[\xi_j(x)]'\boldsymbol{R}\boldsymbol{\sigma}, \tag{12.4}$$

$$[\Pi(r_i, \Gamma)] = -\omega_{i2}\boldsymbol{R}\{[c^{-1}H_i(\Gamma)] + [U_i(\Gamma)]\}, \tag{12.5}$$

$$\Pi(r_i, r_k) = -\omega_{12}\boldsymbol{R}\boldsymbol{\sigma}. \tag{12.6}$$

If only r_2 is regular, c_1^{-1} must be replaced by 0, and if r_1 is inaccessible, one has to put $p_{11}=p_{12}=\tau_1=\sigma_1=0$.

It has been shown in [2] that in the case $\sigma_j\neq 0$ *the Fokker-Planck equation is still given by* (10.4)–(10.6) *but that the boundary conditions*

$$c_jV_j(t) = \sigma_j \lim_{x\to r_j} \Phi_v(t, x) \tag{12.7}$$

must be added. On the other hand, the lateral conditions (9.23) for the backward equation are changed by the addition of the term $-c_i \lim (-1)^i W^{-1}(x)u_x(t, x)$ to the right-hand member of the ith equation.

13. The most general process.

THEOREM 11. *The process with transition probabilities* (12.2)–(12.6) *represents the most general diffusion process in* $[r_1, r_2]$ *with the property that the associated semi-group* (4.2) *transforms each* $f\in C[r_1, r_2]$ *into a function which is again continuous in* $[r_1, r_2]$. *However, there exist diffusion processes without the last property.*

Proof. The first assertion can be verified by comparing the formulas

[28] In trying to verify that this result agrees with the equivalent for the resolvent given in (22.2) of [2] one should put $c_i=(1-\pi_i)/\pi_i$ and observe the following change in notations: The quantities p_{ij}, τ_i, σ_i of [2] correspond to $p_{ij}(1-\pi_i)$, $\tau_i(1-\pi_i)$, and $\sigma_i(1-\pi_i)$ in the present notations. Furthermore, the solutions $\eta_j(x)$ of [2] are the same as $-\omega_{12}\eta_j(x)$ in the present notation.

(12.3)–(12.6) with the resolvent of the most general contraction semi-group from C to C given in [2, (21.12)]. To substantiate the last assertion consider the

EXAMPLE. Let $P_{abs}(t, x, \Gamma)$ be the transition probability of the process corresponding to $u_t = u_{xx}$ in $E = (0, \infty)$ with an absorbing barrier at $x=0$. We now stipulate that if $x=0$ is reached at time t, then with probability 1/2 we shall have $X(\tau)=0$ for $\tau \geqq t$, while with probability 1/2 the process terminates. In other words, we put

$$P(t, x, \Gamma) = P_{abs}(t, x, \Gamma) \qquad \text{for } \Gamma \subset (0, \infty),$$
$$P(t, 0, \Gamma) = 0,$$
$$P(t, x, 0) = (1/2)\{1 - P_{abs}(t, x, E)\}, \qquad P(t, 0, 0) = 1.$$

Then

$$u(t, x) = \int_{0+}^{\infty} P_{abs}(t, x, dy)f(y) + (1/2)\{1 - P_{abs}(t, x, E)\}f(0)$$
$$\to (1/2)f(0) \qquad \text{as } x \to 0+$$

while $u(t, 0) = f(0)$.

THEOREM 12. *The transition probabilities of the most general diffusion process in the open interval* (r_1, r_2) *are given by* (12.2) *with* $\sigma_1 = \sigma_2 = 0$. *The corresponding Fokker-Planck equation is* (10.8), *subjected to the lateral conditions*

$$[\psi_j]'c(I - p) + [(-1)^j\Phi_v(t, r_j)]' = 0 \tag{13.1}$$

where Φ_v *is defined in* (10.3) *and*

$$\psi_i = \lim_{x \to r_i} W(x)a(x)v(t, x) \tag{13.2}$$

with $W(x)$ *given in* (11.1).

The interesting feature is that although the boundaries r_j are excluded, the lateral conditions (13.1) include the terms p_{ij} which correspond to passages from[29] r_i to r_j.

Proof. We have seen that the Laplace transform F of (4.1) satisfies (7.3). If f is continuous in (r_1, r_2) and if $f(x)$ approaches limits as $x \to r_j$, then the same is true of F. This follows [2, Theorem 13.1] and the fact that the bounded solutions of $\lambda\xi - \Omega\xi = 0$ (if they exist) approach limits as $x \to r_j$. The Hille-Yosida theorem now implies that our semi-group takes C into C. The most general diffusion process in (r_1, r_2) is therefore among those constructed in §12, and it is obvious that $\sigma_1 = \sigma_2 = 0$ is necessary and sufficient for the

(29) They account for events such as $X(t) \to r_1$ as $t \to t_0-$ and $X(t) \to r_2$ as $t \to t_0+$, with $X(t_0)$ undefined.

process to be restricted to the open interval. The lateral condition (13.1) is not explicitly derived in [2], but is obtained by a calculation analogous to that given on p. 514.

To return to diffusion processes in the closed interval, let C^* denote the subspace of B consisting of those $f(x)$ $(r_1 \leqq x \leqq r_2)$ which are continuous for $r_1 < x < r_2$ and approach finite limits as $x \to r_j$ (these limits need not coincide with $f(r_j)$). Using the definition of the infinitesimal generator proposed in [3] it is easily seen that *for a process obeying* (1.1) *the semi-group* (4.1) *transforms* C^* *into itself, that is, that* $f \in C$ *implies* $T_t f \in C$. Using the methods of [2], it is possible to construct the totality of contraction semi-groups from C^* to C^* and then to choose among them those which correspond to a stochastic process (that is, where the adjoint semi-group transforms measures into measures).

It is to be expected that the most general diffusion process in $[r_1, r_2]$ is obtained by superimposing an elementary return process to any process in (r_1, r_2) in which there is a positive probability (and therefore certainty) that the process will terminate after a finite time(30). The existence of such processes can be proved, and their transition probabilities can be obtained, by a straightforward generalization of the arguments of §9. However, the proof that this leads to the *most general* diffusion process seems to require a lengthy argument.

REFERENCES(31)

0. D. A. Darling and A. J. F. Siegert, *The first passage problem for a continuous Markoff process*, Ann. Math. Statist. vol. 24 (1953) pp. 624–639.

1. J. L. Doob, *Markoff chains—Denumerable case*, Trans. Amer. Math. Soc. vol. 58 (1945) pp. 455–473.

2. W. Feller, *The parabolic differential equations and the associated semi-groups of transformations*, Ann. of Math. vol. 55 (1952) pp. 468–519.

3. ———, *Semigroups of transformations in general weak topologies*, Ann. of Math. vol. 57 (1953).

4. ———, *Zur Theorie der stochastischen Prozesse*, Math. Ann. vol. 113 (1936) pp. 113–160.

5. ———, *On positivity preserving semi-groups of transformations on* $C[r_1, r_2]$, Annales Société Mathématique Pologne.

(30) Note that in such a process four things may happen when $X(t) \to r_i$. (1) The process may terminate, (2) $X(t)$ may continue to vary continuously in $r_1 < x < r_2$ as in the elastic barrier process, (3) $X(t)$ may instantaneously change to a random point in the interior, (4) $X(t)$ may for a time remain constant, $X(t) = r_i$, and *then* change to a random point in the interior. Thus there are two types of jumps into the interior. The condition that the associated semi-group T_t changes C into C makes it impossible that both types occur in the same process from the same boundary. The general lateral condition contains four functions $p_i(x)$ of bounded variation.

(31) [September 1953] I am obliged to J. L. Doob for a reference to W. Doeblin, *Sur certains mouvements saléatoires*, C. R. Acad. Sci. Paris vol. 208 (1939) pp. 249–250. In this note our criteria for accessibility are announced. Doeblin's method seems to be based on a direct study of the path functions and the law of the iterated logarithm. It appears that the proof has never been published.

6. E. Hille, *Functional analysis and semi-groups*, Amer. Math. Soc. Colloquium Publications, vol. 31, New York, 1948.

7. A. Kolmogorov, *Ueber die analytischen Methoden in der Wahrscheinlichkeitsrechnung*, Math. Ann. vol. 104 (1931) pp. 415–458.

8. D. V. Widder, *The Laplace transform*, Princeton University Press, 1941.

PRINCETON UNIVERSITY,
PRINCETON, N. J.
STOCKHOLM UNIVERSITY,
STOCKHOLM, SWEDEN.

ANNALS OF MATHEMATICS
Vol. 61, No. 1, January, 1955
Printed in U.S.A.

ON SECOND ORDER DIFFERENTIAL OPERATORS[1]

BY WILLIAM FELLER

(Received May 24, 1954)

1. Introduction

The theory of the operator

$$(1.1)\qquad Af = a(x)\frac{d^2}{dx^2}f + b(x)\frac{d}{dx}f + c(x)f \qquad a > 0$$

is closely related to that of its formal adjoint

$$(1.2)\qquad \frac{d^2}{dx^2}(af) - \frac{d}{dx}(bf) + cf.$$

Nevertheless, (1.2) is meaningful only under differentiability conditions which are unnatural for (1.1). An adjoint to (1.1) exists always, but it cannot be written in terms of derivatives *with respect to* x.

The *characteristic property* of A appears to be (1) that it is of *local character*, (2) that *whenever* f *has a local minimum at* x_0 *and* $f(x_0) = 0$, *then* $Af(x_0) \geqq 0$ (provided Af is meaningful).

The purpose of this paper is to derive the most general operator with this characteristic property. (For a rigorous formulation see section 2.) It will be seen that we are led not only to a useful generalization, but also to an appreciable simplification of the classical theory of second order differential operators.[2]

The impetus to this investigation came from a reconsideration of the now classical problem to derive the most general diffusion operator.[3] It turns out that the two problems are almost identical.

Note that our "characteristic property" remains unaffected by an arbitrary topological transformation of the x-axis. In principle our operators can therefore not be expressed in terms of differentiations with respect to one fixed scale.

---

[1] Sponsored by the Office of Ordnance Research, U. S. Army [Contract DA-36-034-ORD-1296].

[2] This formulation of the problem and its solution are unfortunately tied to one dimension. The actual problem (see the next footnote) is meaningful in an arbitrary locally compact Hausdorff space. In other words, we have a natural definition for what corresponds to *second-order differential operators in arbitrary locally compact Hausdorff spaces*, and also some vague indications as to their nature. The construction of such operators would have far reaching consequences.

[3] Purely analytically this is the following problem: to construct the *most general semi-group of operators which transforms each bounded continuous non-negative function into a function of the same class and whose infinitesimal generator is of local character*. Such an infinitesimal generator is easily seen to have the "characteristic property" of the text. That this problem is equivalent to the general diffusion problem (which is a problem in measures) is shown by means of the theory of adjoint semi-groups. (Cf. these Annals last issue, 60:417.) The problems are meaningful in general spaces, but a complete solution is obtained only in one dimension.

Accordingly, we introduce the derivative D_uf of f *with respect to an arbitrary monotonic* u (cf. Definition 2.3); this u need not be continuous, and a differentiable f will in general jump at points of discontinuity of u.

We consider in particular "*reduced operators*" of the form

$$\Omega f = D_v D_u f \tag{1.3}$$

where u and v are monotonic, u (but not necessarily v) even continuous. We can now loosely describe the principal results as follows.

One has to face the fact that A, at points or at intervals, may degenerate to what corresponds classically to an operator of first or zero order. We call such points *irregular*.[4] Apart from such irregular points *every operator* A *of the type considered can be expressed by means of a reduced operator of type* (1.3) with appropriate u and v. The operator A itself is of the form (1.3) if, and only if, it acts on the function 1 (which is identically equal to unity) and $A1 = 0$.

Some of the advantages of the new approach even in classical problems can now be enumerated.

(1) The formal adjoint[5] of (1.3) is simply $D_u D_v$.

(2) The functions u and v are intrinsically connected with the operator A and all operations become more intuitive when expressed by u and v. For example, if the operator A is taken in the ordinary Hilbert space, it is not self-adjoint, but this Hilbert space depends on the arbitrary choice of the independent variable. Writing the operator in the form (1.3) it is at once apparent that the appropriate Hilbert space is that with the norm

$$\|f\|^2 = \int f^2(s)\, dv(s), \tag{1.4}$$

and in it Ω becomes (formally) self-adjoint.

(3) The problem whether and how a boundary point or a zero of the coefficient $a(x)$ in (1.1) is singular becomes now simple and natural (cf. the next to the last footnote).

(4) With the usual restrictions on (1.1) it is frequently necessary to piece together, in two adjacent intervals, operators of type (1.1) and to construct appropriate compatibility or boundary conditions at the common boundary point. Much pain has been devoted to such piecemeal work which in each case requires generalizations and justifications for eigen-function expansions etc. The intrinsic

[4] It is interesting and important to realize that even in the special case of (1.1) a point x_0 need not be irregular if $a(x_0) = 0$. In fact, whether or not a vanishes at x_0 depends on the choice of the independent variable x, whereas our characterization is invariant. This remark elucidates some of the known difficulties connected with singular points (and boundaries) which can now be eliminated.

[5] For simplicity of notation we do not distinguish between u and v. However, conceptually and in higher spaces, u plays the rôle of a *point function* (or scale), v that of a *measure*. (This explains the possible discontinuities of v.) In diffusion theory, (1.3) as such acts in the Banach space of continuous functions, whereas the formal adjoint $D_u D_v$ is defined for *measures*.

way of writing operators leads to a unique differential equation without boundary conditions, and all expansion theorems apply at once. Two examples may illustrate this important point.

EXAMPLE I. A typical classical problem in heat conductivity is concerned with the equation $w_t = w_{xx}$ for $x < 0$ and, say, $w_t = 2w_{xx}$ for $x > 0$. Here

$$Af = f'' \quad \text{for} \quad x < 0, \qquad Af = 2f'' \quad \text{for} \quad x > 0 \tag{1.5}$$

and one justifies in some way the "boundary condition"

$$2f'(0+) = f'(0-). \tag{1.6}$$

The new derivation of the heat (or diffusion) equation takes no account of the physicist's inclination to speak of two materials joining at 0. The operator A is given for $\infty < x < \infty$ by (1.3) where $v(x) \equiv x$ while $u(x) = x$ for $x \leqq 0$ and $u(x) = \frac{1}{2}x$ for $x \geqq 0$. There is no boundary condition or singularity at $x = 0$: the condition (1.6) expresses that $D_u f(x)$ exists at $x = 0$, which condition holds for all x. One may take $u(x)$ as independent variable.

EXAMPLE II. Let $u(x) = x$ and $v(x) = x$ for $x < 0$ and $v(x) = x + 1$ for $x > 0$. Then (1.3) reduces to

$$\begin{aligned} \Omega f(x) &= f''(x) \qquad\qquad x \neq 0 \\ \Omega f(0) &= f'(0+) - f'(0-). \end{aligned} \tag{1.7}$$

Again *one* equation corresponds to what looks like a system of two equations with a boundary condition.

The connection of a general operator A with the corresponding reduced operator is discussed in §3. Regular and irregular points are introduced in §4, and all representation theorems are listed in §5. The reduction of (1.1) to the form (1.3) is given in the note preceding Theorem 5.2.

The existence proofs, Green function, etc. become actually simpler for the generalized operators in their canonical form than for ordinary differential operators. One is impeded only by the fact that in general $D_u f \cdot D_v g \neq D_v f \cdot D_u g$ whereas for formal manipulations one would like to have equality even in cases when only $D_u f$ and $D_v g$ (but not $D_v f$ and $D_u g$) exist. In §12 we give a substitute formula which renders the same services and permits, in particular, to generalize the classical result that the *Wronskian of two solutions of* $Af + \lambda f = 0$ *is constant.*

2. Definitions and notations

Consider a linear operator A whose domain and range are real functions of a real variable. We shall say that f *belongs to the domain of* A *at* x_0, in symbols

$$f \in \mathfrak{D}(A, x_0), \tag{2.1}$$

if both f and Af are defined and *continuous* in some neighborhood of x_0. This domain, of course, is supposed to be linear, that is

$$f_i \in \mathfrak{D}(A, x_0) \quad \textit{implies} \quad p_1 f_1 + p_2 f_2 \in \mathfrak{D}(A, x_0) \tag{2.2}$$

for any constants p_j.

DEFINITION 2.1. *The operator A is of local character if whenever $f \in \mathfrak{D}(A, x_0)$ and $g(x) = f(x)$ for all x in some neighborhood of x_0, then also $g \in \mathfrak{D}(A, x_0)$ and $Ag(x_0) = Af(x_0)$.*

For a study of Af near x_0 we need not know whether and how f is defined outside an arbitrarily small neighborhood of x_0.

DEFINITION 2.2. *The operator A has property $\mathcal{P}$ if it is of local character and if*

$$(2.3)\qquad f \in \mathfrak{D}(A, x_0), \qquad f(x) \geqq 0, \qquad f(x_0) = 0$$

imply $Af(x_0) \geqq 0$.

DEFINITION 2.3. *Let u be continuous at x_0 and strictly monotonic for $x > x_0$ and for $x < x_0$. We define the right derivative of f with respect to u by*

$$(2.4)\qquad D_u^+ f(x_0) = \lim_{h \to 0+} \frac{f(x_0 + h) - f(x_0)}{u(x_0 + h) - u(x_0)}$$

provided the limit exists. The left derivative $D_u^- f$ and the derivative are defined similarly. If u is strictly increasing and discontinuous at x_0 we put

$$(2.5)\qquad D_u f(x_0) = \frac{f(x_0 +) - f(x_0 -)}{u(x_0 +) - u(x_0 -)}.$$

NOTE. The function u is permitted to have a minimum at x_0, but this is only to simplify the formulation of Theorem 5.2. Everywhere else u will be supposed to be strictly increasing. Conceptually, however, one should distinguish between differentiation with respect to point functions and with respect to measures (which are only for convenience represented by monotonic functions). This distinction becomes important in higher spaces.

In each case the existence of $D_u f$ requires the continuity of f with respect to u. If $D_u f$ is continuous and u increases, then $D_u f > 0$ implies that f increases. On the other hand, if u has a minimum at x_0 and $D_u f$ is continuous and positive, then f too has a minimum at x_0.

3. Reduced operators

If A were of the form (1.1) we would wish to concentrate our attention to the purely differential part $af'' + bf'$, disregarding the effect of $c = A1$. Unfortunately,[6] the constant function 1 need not even be in the domain of A.

DEFINITION. *Suppose that φ and $A\varphi$ are continuous in $r_1 < x < r_2$ and $\varphi > 0$. We put*

$$(3.1)\qquad A^*f = \frac{1}{\varphi} A(f\varphi), \qquad Af = \varphi A^* \frac{f}{\varphi}.$$

[6] If A is of the form (1.1) then, according to Definition (2.1), $1 \in \mathfrak{D}(A)$ only if c is continuous. However, one should beware of the plausible fallacy that the requirement $1 \in \mathfrak{D}(A)$ is necessarily a *regularity* restriction. In diffusion theory one encounters operators A with all desirable regularity properties which nevertheless imply for all $f \in \mathfrak{D}(A)$ that whenever $f(0) > 0$ also $f'(0) > 0$. Such an operator cannot act on 1.

The domain of A^* *consists of all* f *such that* $f\varphi \in \mathfrak{D}(A)$, *and vice versa. Put*

$$A^*1 = \frac{A\varphi}{\varphi} = c \tag{3.2}$$

and

$$\Omega f = A^*f - cf. \tag{3.3}$$

By definition c is continuous in $r_1 < x < r_2$.

THEOREM 3.1. *The operator* A^* *has property* $\mathcal{P}$ *if, and only if,* A *has. The property* $\Omega 1 \equiv 0$ *implies: whenever* f *has a local minimum at* x_0 *and* $f \in \mathfrak{D}(\Omega, x_0)$, *then* $\Omega f(x_0) \geqq 0$.

The first statement is clear since $f \geqq 0, f(x_0) = 0$ implies $\varphi f \geqq 0$ and $\varphi(x_0)f(x_0) = 0$ and vice versa. If $f \in \mathfrak{D}(\Omega, x_0)$ and f has a local minimum at x_0 then

$$\begin{aligned} \Omega f(x_0) &= A^*f(x_0) - c(x_0)f(x_0) \\ &= A^*\{f - f(x_0)1\}(x_0) \end{aligned}$$

and this is non-negative since the function $f - f(x_0)1$ vanishes at x_0 and has there a local minimum.

4. Regular points

These are defined so that at irregular points A reduces to a first order operator (Theorem 5.2).

DEFINITION 4.1. *Let* Ω *be an operator with property* $\mathcal{P}$ *and* $\Omega 1 = 0$. *The point* x_0 *is regular if there exist both increasing and decreasing functions* $f \in \mathfrak{D}(\Omega, x_0)$ *with* $\Omega f(x_0) > 0$. *An interval is regular if all its points are regular.*

The regular points always form an open set. We shall prove (§10) the

CRITERION. *An interval* $\mathcal{J}$ *is regular if, and only if,* $\Omega z = 0$ *has a strictly increasing solution and* Ω *does not vanish*[7] *at any point of* $\mathcal{J}$. *(The latter condition implies that* $\Omega v = 1$ *has a solution and vice versa.)*

LEMMA. *If* $\Omega f(x_0) \neq 0$, *then there exist one-sided neighborhoods* $x_0 \leqq x < x_1$ *and* $x_{-1} < x \leqq x_0$ *in which* f *is monotonic.*

PROOF. If x_0 is a point of accumulation of maxima and minima of f, then Ωf assumes in every neighborhood of x_0 both non-positive and non-negative values. As Ωf is continuous, this implies $\Omega f(x_0) = 0$.

This simple observation shows that there exist only the following

Types of irregular points.

(i) $\Omega f(x_0) > 0$ *implies that* f *is increasing [decreasing].*

(ii) $\Omega f(x_0) > 0$ *implies that* f *either increases [decreases] or has a minimum at* x_0.

(iii) $\Omega f(x_0) > 0$ *implies that* f *has a minimum at* x_0.

(iv) $\Omega f(x_0) = 0$ *for all* $f \in \mathfrak{D}(\Omega, x_0)$.

---

[7] Ω *vanishes* at x_0 *if* $\Omega f(x_0) = 0$ *for each* $f \in \mathfrak{D}(\Omega, x_0)$.

5. Representation theorems

THEOREM 5.1. *If $\mathfrak{g}$ is a regular interval then there exist two strictly increasing functions u and v of which u (but not necessarily v) is continuous and such that:*

For each $f \in \mathfrak{D}(\Omega)$ the one-sided derivatives $D_u^+ f$ and $D_u^- f$ exist at each point of $\mathfrak{g}$, and one has $D_u^+ f = D_u^- f = D_u f$ except, perhaps, at points of discontinuity of v. These derivatives are in turn differentiable with respect to v and one has

$$\Omega f = D_v D_u f. \tag{5.1}$$

In view of the continuity of u one may introduce u as independent variable, that is, without loss of generality one may assume

$$u(x) \equiv x. \tag{5.2}$$

Then (5.1) reduces to

$$\Omega f = D_v f'. \tag{5.3}$$

From the proof it will be seen that all admissable u are of the form $u = px + q$, which means that u and v are essentially *uniquely determined.*

Note that the role of u and v cannot be reversed: the derivative $D_v f$ need not exist. In fact, the operator $D_u D_v$ turns out to be the formal adjoint of Ω.

NOTE. The operator

$$\Omega = a\frac{d^2}{dx^2} + b\frac{d}{dx}, \tag{5.4}$$

can be reduced to the form (5.1) by finding monotonic solutions of

$$\Omega u = 0, \qquad \Omega V = 1, \tag{5.5}$$

and putting

$$v = D_u V = V'/u'. \tag{5.6}$$

If u and V are a pair of solutions, the most general solutions are $pu + q$ and $V + pu + q$. This means replacing v by $1/pv$ + const. Note that we get the formal adjoint to (5.4) even if a and b are discontinuous.

THEOREM 5.2. *Suppose that x_0 is not regular, but that $\Omega f(x_0)$ does not vanish for all $f \in \mathfrak{D}(\Omega, x_0)$. Then there exists a continuous function u such that*

$$\Omega f(x_0) = D_u^+ f(x_0) \quad or \quad \Omega f(x_0) = -D_u^- f(x_0). \tag{5.7}$$

More precisely, in the classification of §4, if x_0 is of type (i), *one may choose the x-scale so that $u(x) = x$ and*

$$\Omega f(x_0) = f'(x_0) \tag{5.8}$$

where f' is a two-sided derivative; if x_0 is of type (ii), *one may choose $u(x) = x$ and then*

$$\Omega f(x_0) = f^+(x_0) \tag{5.9}$$

or $\Omega f(x_0) = -f^-(x_0)$ *according as* $\Omega f > 0$ *is compatible with increasing or decreasing* f; *if* x_0 *is of type* (iii), *the* x-*scale may be so chosen that* $u(x) = |x - x_0|$, *and in this case*

$$\Omega f(x_0) = f^+(x_0) = -f^-(x_0) = D_u^- f(x_0). \tag{5.10}$$

We shall derive Theorem 5.1 from the following one, which permits us to combine regular and irregular points into one formula.

THEOREM 5.3. *Let the interval* $\mathcal{J}$ *be regular. Then there exist in* $\mathcal{J}$ *two strictly increasing functions* u *and* v *such that at each point of continuity of* v

$$\Omega f = aD_vD_uf + bD_uf \tag{5.11}$$

whereas at points of discontinuity

$$\begin{aligned}\Omega f(x) &= a(x+)\frac{D_u^+f(x) - D_u^-f(x)}{v(x+) - v(x-)} + b(x)D_u^+f(x) \\ &= a(x-)\frac{D_u^+f(x) - D_u^-f(x)}{v(x+) - v(x-)} + b(x)D_u^-f(x).\end{aligned} \tag{5.12}$$

Here b *is continuous,* a *is continuous except for jumps where*

$$a(x+) - a(x-) = -b(x)\{v(x+) - v(x-)\}. \tag{5.13}$$

Moreover $a > 0$.

Of course, the x-scale may again be chosen so that $u(x) \equiv x$.

We can use the preceding theorems to express an arbitrary operator A with property $\mathcal{P}$ by means of Ω. However, this representation depends on the arbitrary choice of φ in (3.2); and the question arises as to how the different representations are connected.

THEOREM 5.4. *Let* A *be an operator with property* $\mathcal{P}$ *and suppose that* $A\varphi$ *exists for some* $\varphi > 0$. *Then*

$$Af = \varphi\Omega\frac{f}{\varphi} + cf \tag{5.14}$$

where Ω *is an operator of the type described in Theorems* 5.1 *and* 5.2 *and* $c = A\varphi/\varphi$.

Suppose, in particular, that

$$Af = \varphi D_vD_u\frac{f}{\varphi} + cf. \tag{5.15}$$

If φ *is replaced by* $\varphi_1 > 0$ *with* $c_1 = A\varphi_1/\varphi_1$ *one passes from the representation* (5.15) *to the corresponding representation*

$$Af = \varphi_1 D_{v_1}D_{u_1}\frac{f}{\varphi_1} + c_1f \tag{5.16}$$

by means of the change of variables

$$u_1 = \int \frac{\varphi^2}{\varphi_1^2}\, du \tag{5.17}$$

$$v_1 = \int \frac{\varphi_1^2}{\varphi^2}\, dv. \tag{5.18}$$

Note that one may achieve $c \equiv 0$ if, and only if, there exists a positive solution of $A\varphi = 0$.

6. Preparations

In this section we consider only the case where there exists a strictly increasing u with $\Omega u \geqq 0$. As indicated before, we may introduce this u as new independent variable. In other words, without loss of generality we shall assume that $x \in \mathfrak{D}(\Omega)$ and

$$\Omega x = b(x) \geqq 0, \qquad (\rho_1 < x < \rho_2). \tag{6.1}$$

LEMMA 6.1. *Assume* (6.1) *and that*

$$\Omega z(x) = \xi(x) > 0, \qquad z(x) \downarrow \qquad (\rho_1 < x < \rho_2). \tag{6.2}$$

Then z is convex, and z^+ strictly increasing.

PROOF. Note first that z is strictly decreasing since in any interval of constancy one would have $\Omega z = 0$, against assumption.

Assume that $z(x)$ is not convex. Then there exists a straight line $y = px + q$ which intersects the curve $y = z(x)$ in two points $(x_1, z(x_1))$ and $(x_2, z(x_2))$ so that

$$g(x) = z(x) - px - q \tag{6.3}$$

assumes positive values for some x between x_1 and x_2. Thus $g(x)$ reaches a positive maximum at some point x_0. Since both z and $px + q$ belong to the domain of Ω we conclude that $\Omega g(x_0) \leqq 0$ or

$$\xi(x_0) - pb(x_0) \leqq 0. \tag{6.4}$$

Now this is impossible since obviously $p < 0$ whereas $\xi > 0$, $b \geqq 0$. Hence f is convex, and one-sided derivatives exist everywhere.

LEMMA 6.2. *Assume* (6.1) *and* (6.2). *Each $f \in \mathfrak{D}(\Omega)$ is the difference of two convex functions. In particular, the one-sided derivatives f^+ and f^- exist everywhere, and $f^+ = f^- = f'$ except, perhaps, at denumerably many points.*

PROOF. It suffices to prove the assertion for an arbitrary subinterval $\rho_1 < x < x_0 < \rho_2$. Let $F(x) = f(x) + pz(x)$. For sufficiently large p we shall have $\Omega F > 0$, whence F can have no maximum in the interior of the interval. Again, for sufficiently large p and an arbitrarily chosen $x_1 > x_0$ we shall have $F(x_1) < F(x_0)$ in view of the fact that z decreases strictly. If then F had a minimum between ρ_1 and x_0 it would have a maximum between it and x_1, which is impossible.

Thus F decreases and the preceding lemma shows that F is convex between ρ_1 and x_0. Then $f = F - pz$ proves the assertion.

Lemma 6.3. *Assume* (6.1) *and* $f \in \mathfrak{D}(\Omega, x_0)$, $f(x_0) = f'(x_0) = 0$. *If f changes sign at x_0 then* $\Omega f(x_0) = 0$.

Proof. Without loss of generality we may assume that f assumes positive values for some $x > x_0$ and negative values for some $x < x_0$. For sufficiently small $\varepsilon > 0$ the line $y = \varepsilon(x - x_0)$ will intersect the curve $y = f(x)$ at $x_1 > x_0$, at x_0, and at $x_{-1} < x_0$. The function $f(x) - \varepsilon(x - x_0)$ has therefore a minimum between x_0 and x_1 and a maximum between x_{-1} and x_0. In an arbitrary neighborhood of x_0 there exist thus points at which $\Omega f \geqq \varepsilon b$ and also points at which $\Omega f \leqq \varepsilon b$. Since b and Ωf are continuous, the lemma follows.

Lemma 6.4. *Assume* (6.1). *Let* f, $g \in \mathfrak{D}(\Omega, x_0)$ with

$$\begin{aligned} &f \geqq 0, \qquad f'(x_0) = 0 \\ &g \geqq 0, \qquad g'(x_0) = 0, \qquad \Omega g(x_0) > 0. \end{aligned} \tag{6.5}$$

Then

$$\lim_{x \to x_0} \frac{f(x) - f(x_0)}{g(x) - g(x_0)} = \frac{\Omega f(x_0)}{\Omega g(x_0)}. \tag{6.6}$$

Proof. Without loss of generality we may assume $f(x_0) = g(x_0) = 0$. Let

$$r_* = \underline{\lim}_{x \to x_0} \frac{f(x)}{g(x)}, \qquad r^* \overline{\lim}_{x \to x_0} \frac{f(x)}{g(x)}. \tag{6.7}$$

For $p > r^*$ the function $f - pg$ is non-positive near x_0 and zero at x_0. Thus $\Omega\{f(x_0) - pg(x_0)\} \leqq 0$. This argument shows that

$$r_* \Omega g(x_0) \leqq \Omega f(x_0) \leqq r^* \Omega g(x_0). \tag{6.8}$$

Now if $r_* < p < r^*$ then $F = f - pg$ changes sign in every neighborhood of x_0. The preceding lemma shows that in this case $\Omega F(x_0) = 0$, or $\Omega f(x_0) = p\Omega g(x_0)$. As there can be only one value of p for which this is true we have $r_* = r^*$.

Lemma 6.5. *Let* f, $g \in \mathfrak{D}(\Omega, x_0)$ *and suppose that f and g have one-sided derivatives such that*

$$f^+(x_0) = -f^-(x_0), \qquad g^+(x_0) = -g^-(x_0) > 0. \tag{6.9}$$

Then

$$\Omega f(x_0) = \frac{f^+(x_0)}{g^+(x_0)} \Omega g(x_0). \tag{6.10}$$

Proof. The preceding proof applies with the simplification that we know in advance that $r_* = r^* = f^+/g^+ = f^-/g^-$.

7. Proof of Theorem 5.3

In this section we prove the theorem "in the small," that is, in a neighborhood of a regular point x_0. There is no loss of generality in assuming that (6.1) holds.

By definition, there exists also a decreasing $V(x)$ with

$$\Omega V(x) = \omega(x) > 0. \tag{7.1}$$

According to Lemma 6.1 this $V(x)$ is convex. We put

$$v(x) = V'(x) \tag{7.2}$$

at all points where this derivative exists and

$$a(x) = \omega(x) - b(x)v(x) > 0. \tag{7.3}$$

Both v and a are continuous except for jumps. Since ω and b are continuous, the relation (5.13) is true.

We begin by proving (5.12) where, with our present convention, $D_u^+ f$ and $D_u^- f$ are simply the one-sided derivatives f^+ and f^- with respect to x.

Let x_0 be a point of discontinuity of $v = V'$. If $f \in \mathfrak{D}(\Omega, x_0)$ then the one-sided derivatives f^+ and f^- exist by Lemma 6.2. In order to apply Lemma 6.5 we introduce for the moment

$$\begin{aligned} V_*(x) &= V(x) - (x - x_0)\,\frac{V^+(x_0) + V^-(x_0)}{2} \\ f_*(x) &= f(x) - (x - x_0)\,\frac{f^+(x_0) + f^-(x_0)}{2}. \end{aligned} \tag{7.4}$$

Then

$$\begin{aligned} V_*^+(x_0) &= -V_*^-(x_0) = \frac{V^+(x_0) - V^-(x_0)}{2} \neq 0 \\ f_*^+(x_0) &= -f_*^-(x_0) = \frac{f^+(x_0) - f^-(x_0)}{2}. \end{aligned} \tag{7.5}$$

Clearly the functions (7.4) are in the domain of Ω and

$$\Omega V_*(x_0) = \omega(x_0) - b(x_0)\,\frac{V^+(x_0) + V^-(x_0)}{2}. \tag{7.6}$$

Since $\omega > 0$, $b > 0$, and V is decreasing, we have $\Omega V_*(x_0) > 0$. Thus Lemma 6.5 applies and

$$\frac{\Omega f_*(x_0)}{\Omega V_*(x_0)} = \frac{f^+(x_0) - f^-(x_0)}{V^+(x_0) - V^-(x_0)}. \tag{7.7}$$

Using (7.6) and the analogous formula for f_* we get

$$\begin{aligned} \Omega f(x_0) &= \frac{f^+(x_0) - f^-(x_0)}{V^+(x_0) - V^-(x_0)}\,\omega(x_0) \\ &\quad + \frac{b(x)}{V^+(x_0) - V^-(x_0)}\,\{V^+(x_0)f^-(x_0) - V^-(x_0)f^+(x_0)\}. \end{aligned} \tag{7.8}$$

In view of (7.2) and (7.3) this is equivalent to (5.12).

It remains to prove that (5.11) holds at each point where $V'(x)$ exists (where $v(x)$ is continuous). Let

$$\mathfrak{D}(x) = V(x) - V(x_0) - (x - x_0)\, V'(x_0). \tag{7.9}$$

Then

$$\Omega\mathfrak{D}(x) = \omega(x) - b(x)V'(x_0) > 0. \tag{7.10}$$

By Lemma 6.1 $V(x)$ is convex, and therefore $\mathfrak{D}(x)$ strictly increasing. We show first that $f'(x_0)$ exists. Indeed, if $f^+(x_0) > f^-(x_0)$ consider the function f_* defined in (7.4). For it $f_*^+ = -f_*^- > 0$, and hence for constant $k > 0$ and sufficiently small $|\, x - x_0 \,|$ one would have

$$f_*(x) - f_*(x_0) > k\mathfrak{D}(x) > 0 \tag{7.11}$$

or

$$\Omega f_*(x_0) > k\mathfrak{D}(x_0)$$

which is impossible.

Thus $f'(x_0)$ exists, and we put

$$N(x) = f(x) - f(x_0) - (x - x_0)f'(x_0). \tag{7.12}$$

Now Lemma 6.4 applies to $N(x)$ and $\mathfrak{D}(x)$ and we get

$$\lim_{x \to x_0} \frac{N(x)}{\mathfrak{D}(x)} = \frac{\Omega N(x_0)}{\Omega\mathfrak{D}(x_0)} = \frac{\Omega f(x_0) - b(x_0)f'(x_0)}{\omega(x_0) - b(x_0)v(x_0)} . \tag{7.13}$$

To complete the proof of (5.11) it remains only to show that

$$\lim_{x \to x_0} \frac{N(x)}{\mathfrak{D}(x)} = \lim_{x \to x_0} \frac{f^*(x) - f'(x_0)}{v(x) - v(x_0)} = D_v f'(x_0) \tag{7.14}$$

(where at places of discontinuity f' and v are to be replaced by f^+, $v(x+)$, or by f^-, $v(x-)$). It will be remembered that because of the convexity of $V(x)$ we have $v(x) \neq v(x_0)$ whenever $x \neq x_0$.

We prove (7.14) when x approaches x_0 from the right. For $x_0 \leqq y \leqq x$ put

$$R(y) = N(y)\mathfrak{D}(x) - \mathfrak{D}(y)N(x). \tag{7.15}$$

Then $R(x_0) = R(x) = 0$ and $R(y)$ reaches an extremum between x_0 and x. At a place of a maximum one has

$$R^-(y) \geqq 0, \qquad R^+(y) \leqq 0 \tag{7.16}$$

or

$$\frac{N^+(y)}{\mathfrak{D}^+(y)} \leqq \frac{N(x)}{\mathfrak{D}(x)} \leqq \frac{N^-(y)}{\mathfrak{D}^-(y)} \tag{7.17}$$

whereas at a minimum the reversed inequalities take place. It follows that for *some* sequence $y \to x_0$

$$\lim_{x \to x_0} \frac{N(x)}{\mathfrak{D}(x)} = \lim_{y \to x_0} \frac{f'(y) - f'(x_0)}{V'(y) - V'(x_0)}. \tag{7.18}$$

This is formally the same as the assertion (7.14), but it remains to be proved that (7.18) holds for *every* sequence $y \to x_0$.

Suppose, then, that

$$\underline{\lim}_{y \to x_0+} \frac{N'(y)}{\mathfrak{D}'(y)} < p < \overline{\lim}_{y \to x_0+} \frac{N'(y)}{\mathfrak{D}'(y)}. \tag{7.19}$$

Then the function $N(y) - p\mathfrak{D}(y)$ has infinitely many maxima and minima accumulating at x_0. It follows that

$$\Omega N(x_0) = p\Omega\mathfrak{D}(x_0). \tag{7.20}$$

As shown in (7.10), the factor of p is positive, and there can exist only one value of p for which (7.20) holds. Thus $N'/\mathfrak{D}'$ approaches a unique limit and (7.18) holds as $y \to x_0+$.

8. Solutions of $\Omega Z = 0$

The main purpose of the following Theorem is to show that we can choose $b \equiv 0$.

THEOREM. *If Ω is the operator of Theorem* 5.3 *then there exists a strictly increasing solution of*

$$\Omega Z = 0. \tag{8.1}$$

Define z by

$$z(x) = \frac{b(x)}{a(x)} \tag{8.2}$$

$$z(x) = \frac{b(x)}{a(x+) - a(x-)} \log \frac{a(x+)}{a(x-)} \tag{8.3}$$

at points of continuity and discontinuity of $a(x)$, respectively. If x_0 is a point of continuity, then

$$Z(x) = \int_{x_0}^{x} \exp\left\{-\int_{x_0}^{t} z(s)\, dv(s)\right\} du(t) \tag{8.4}$$

satisfies (8.1) *with*[8] $Z(x_0) = 0$, $D_u Z(x_0) = 1$.

PROOF. We begin by proving that $z(x)$ *is continuous except at the jumps of $a(x)$ and that everywhere*

---

[8] The uniqueness will be obvious from the representation (5.1). When x_0 is a point of discontinuity of $a(x)$ one may still define Z and prescribe *one* one-sided derivative.

$$z(x-) \leqq z(x) \leqq z(x+). \tag{8.5}$$

Suppose (as was the case with our preceding conventions) that $b(x) \geqq 0$. At a point x of discontinuity of a we have from (5.13) that $b(x) > 0$ and

$$\Delta a = a(x+) - a(x-) < 0, \qquad 0 < \frac{-\Delta a}{a(x-)} < 1. \tag{8.6}$$

We may now rewrite (8.3) in either of the forms

$$z(x) = \frac{b(x)}{\Delta a} \log\left\{1 + \frac{\Delta a}{a(x-)}\right\} \tag{8.7}$$

or

$$z(x) = -\frac{b(x)}{\Delta a} \log\left\{1 - \frac{\Delta a}{a(x+)}\right\}. \tag{8.8}$$

The well-known inequalities for $-\log(1 - t)$ and $\log(1 + t)$ then show that

$$\frac{b(x)}{a(x-)} \leqq z(x) \leqq \frac{b(x)}{a(x+)}. \tag{8.9}$$

Thus at each point of discontinuity of a the equation (8.3) yields a value between the right and the left limit of the function (8.2), and (8.5) follows for all points.

To accomplish the proof, note that the right derivative

$$D_u^+ Z(x) = \exp\left\{-\int_{x_0}^{x+} z(s)\, dv(s)\right\} \tag{8.10}$$

exists everywhere and equals the left derivative except at points of discontinuity of v. At such points we have

$$\frac{D_u^+ Z(x)}{D_u^- Z(x)} = \exp\{-z(x)\Delta v\}. \tag{8.11}$$

From (5.13) it is obvious that a point x of discontinuity of v is also a point of discontinuity of a except if $b(x) = 0$. In this case $z(x) = 0$ and the ratio in (8.11) is unity as required. If x is a point of discontinuity of a we get from (8.3) and (5.13)

$$\frac{D_u^+ Z(x)}{D_u^- Z(x)} = \frac{a(x+)}{a(x-)} \tag{8.12}$$

or

$$a(x+)\{D_u^+Z(x) - D_u^-Z(x)\} - \Delta a \cdot D_u^+Z(x) = 0. \tag{8.13}$$

The left side equals Δv times the right side of (5.12) which shows that $\Omega Z(x) = 0$ at each point of discontinuity of a.

If x is a point of continuity of a we have seen that $D_u Z(x)$ exists (either in consequence of $\Delta v = 0$ or $z(x) = 0$). We get

$$\frac{D_u Z(x+h) - D_u Z(x)}{v(x+h) - v(x)} = \frac{D_u Z(x)}{v(x+h) - v(x)} \cdot \left\{\exp\left\{-\int_x^{x+h} z(s)\, dv(s)\right\} - 1\right\} \tag{8.14}$$

provided both x and $x + h$ are points of continuity of v. Since z is continuous at x and $v(x + h) - v(x) \to 0$, the right side of (8.14) tends to

$$-D_u Z(x) \cdot z(x) = -\frac{b(x)}{a(x)} D_u Z(x). \tag{8.15}$$

This proves (8.1) for all points of continuity of a.

9. Proof of Theorem 5.1

Let x_0 be a regular point. We know that in a neighborhood of x_0 the operator Ω can be brought to the form indicated in Theorem 5.3. Here u was an arbitrarily chosen increasing function in the domain of Ω subjected to the condition $\Omega u \geqq 0$. In the preceding section we saw that we may choose u so that $\Omega u = 0$, that is, $b = 0$. There is no loss of generality to introduce u as independent variable. Now $a(x)$ is positive and continuous, in view of (5.13), and $b = 0$. Putting

$$\mu(x) = \int_{x_0}^{x} \frac{dv(s)}{a(s)} \tag{9.1}$$

we get directly

$$\Omega f = D_\mu D_x f. \tag{9.2}$$

This gives us the desired representation (5.1) in the neighborhood of an arbitrary regular x_0. To prove the Theorem 5.1 it remains to show that the scale x and the function μ can be extended to the whole interval $\mathcal{I}$ whenever the latter contains only regular points.

It is seen from (9.2) that $\Omega f = 0$ implies $D_x f = $ const., and hence that the most general solution of $\Omega f = 0$ is $f = px + q$. Given two overlapping intervals $\mathcal{I}_1$ and $\mathcal{I}_2$ in each of which we have a representation of the form (9.2), the scales x_1 and x_2 in the intersection $\mathcal{I}_1 \cap \mathcal{I}_2$ are related by $x_2 = px_1 + q$ with $p > 0$. Changing the scale appropriately we get (9.2) valid in the union $\mathcal{I}_1 \cup \mathcal{I}_2$. This shows that a representation of the form (9.2) holds in every closed subinterval of a regular interval. Thus Theorem 5.1 (and a fortiori Theorem 5.3) is proved "in the large."

10. Proof of the criterion (§4)

The *necessity* of the condition is obvious from the form (5.1) which shows that u is an increasing solution of $\Omega u = 0$. On the other hand, to derive (5.1) we required only an increasing u with $\Omega u \geqq 0$ and a decreasing V with $\Omega V > 0$. To prove the criterion it suffices to show:

If z increases, if $\Omega z = 0$, and if there exists a function f with $\Omega f(x_0) > 0$, then there exists a *decreasing* V with $\Omega V > 0$ in a neighborhood of x_0.

To prove this we may again introduce z as independent variable, that is, suppose $\Omega x = 0$. Without loss of generality suppose $f(x_0) = 0$. Since $\Omega f > 0$ in a neighborhood of x_0, the same is true of $f - p(x - x_0)$ and this shows that $f - p(x - x_0)$ can have no maximum near x_0. Thus, if the line $y = p(x - x_0)$ intersects $y = f(x)$ at x_0 and x_1, then between x_0 and x_1 the graph of f lies below the chord. It follows that f has at x_0 one-sided derivatives with $f^+ < \infty$. Let $V = f - p(x - x_0)$ where $p > f^+(x_0)$. Then $\Omega V = \Omega f > 0$. Moreover, $V^+(x_0) < 0$ so that $V(x) < 0$ for $x - x_0 > 0$ sufficiently small. Since $\Omega V > 0$, there can be no maximum at or near x_0, so that $V(x)$ must decrease near x_0, as asserted.

11. Proof of Theorem 5.2

If there exists an increasing function u with $\Omega u > 0$, we may again introduce it as independent variable, that is, we may suppose that near x_0

$$\Omega x = b(x) > 0, \qquad b(x_0) = 1. \tag{11.1}$$

Let $f \in \mathfrak{D}(\Omega, x_0), f(x_0) = 0, \Omega f(x_0) > 0$. Put

$$F(x) = f(x) - p(x - x_0). \tag{11.2}$$

By assumption of the Theorem, f is either increasing or has a minimum at x_0. If F had infinitely many zeros near x_0, its maxima and minima would accumulate towards x_0 and we would have $\Omega F(x_0) = 0$ or $\Omega f(x_0) = p$. It follows that the one-sided derivatives $f^+(x_0)$ and $f^-(x_0)$ exist. Choosing $p > f^+(x_0)$ (unless $f^+(x_0) = \infty$) we get $F^+(x_0) < 0$, so that F either decreases or has a maximum. In either case $\Omega F(x_0) \leqq 0$, whence $\Omega f(x_0) \leqq p$. Letting $p \to f^+(x_0)$ we find $\Omega f(x_0) \leqq f^+(x_0)$. Choosing $p < f^+(x_0)$ we find in the same way $\Omega f(x_0) \geqq f^+(x_0)$. This proves (5.9).

Suppose now that x_0 is a point of type (i) so that $\Omega f(x_0) > 0$ implies that f is increasing. We have to prove that $f^+(x_0) = f^-(x_0)$. Assume the contrary. Put $p = \frac{1}{2}\{f^+(x_0) + f^-(x_0)\} + \varepsilon$. Then

$$F^+(x_0) = \tfrac{1}{2}\{f^+(x_0) - f^-(x_0)\} - \varepsilon > 0$$

$$F^-(x_0) = \tfrac{1}{2}\{f^+(x_0) - f^-(x_0)\} - \varepsilon > 0$$

for all $|\varepsilon|$ sufficiently small. Hence F has a minimum at x_0 and thus $\Omega F(x_0) = 0$ or $\Omega f(x_0) = p$ for any admissable choice of ε which is impossible. This proves (5.8).

There remains only the case where $\Omega\omega(x_0) > 0$ implies that ω has a minimum at x_0. If ω is such a function with $\omega(x_0) = 0$ we may change the scale so that $\omega(x) = |x - x_0|$. The preceding argument applies without change and shows that $\Omega f = f^+$. This time, however, the positive and negative direction play a symmetric role, and therefore the same argument shows also that $\Omega f = -f^-$. For points of type (iii) we have therefore (5.10).

12. A commutation rule (generalized Wronskians)

Our generalized differential operators D_u and D_v behave like ordinary differentiation. However, if one tries to calculate for our Ω analogs to the classical Green function, etc. the formal analogy seems to break down unless one can conclude that

$$D_v g \cdot D_u f = D_u g \cdot D_v f \tag{12.1}$$

even when f is only known to be in the domain of D_u but not necessarily of D_v. The following lemma gives an integrated version of (12.1) which serves the same purpose.

LEMMA. *Suppose that*

$$D_v D_u f = F, \qquad D_v D_u g = G \tag{12.2}$$

(where $D_v D_u = \Omega$ is the operator of Theorem 5.1*). Then*[9] *$gD_u f - fD_u g$ is in the domain of D_v and*

$$D_v\{gD_u f - fD_u g\} = gF - fG. \tag{12.3}$$

COROLLARY. *If ω_1 and ω_2 are two solutions of*

$$\Omega\omega - c\omega = 0, \tag{12.4}$$

then their Wronskian is constant:

$$\omega_1 D_u \omega_2 - \omega_2 D_u \omega_1 = \text{const.} \tag{12.5}$$

PROOF. One has

$$\int_{x_0}^{x} \{gF - fG\}\, dv = \int_{x_0}^{x} \{gd(D_u f) - fd(D_u g)\}$$
$$= [gD_u f - fD_u g]_{x_0}^{x} - \int_{x_0}^{x} \{dg \cdot D_u f - df \cdot D_u g\}.$$

The last integrand is identically zero, and the lemma is proved.

13. Proof of Theorem 5.4

One has

$$\varphi_1 D_{v_1} D_{u_1} \frac{f}{\varphi_1} = \frac{\varphi^2}{\varphi_1} D_v \left\{\frac{\varphi_1^2}{\varphi^2} D_u \frac{f/\varphi}{\varphi_1/\varphi}\right\} = \frac{\varphi^2}{\varphi_1} D_v \left\{\frac{\varphi_1}{\varphi} D_u \frac{f}{\varphi} - \frac{f}{\varphi} D_u \frac{\varphi_1}{\varphi}\right\} \tag{13.1}$$

according to (5.17), (5.18) and the rule for the differentiation of a quotient. We can now apply formula (12.3) to show that (13.1) reduces to

$$\frac{\varphi^2}{\varphi_1}\left\{\frac{\varphi_1}{\varphi^2}[Af - cf] - \frac{f}{\varphi^2}[A\varphi_1 - c\varphi_1]\right\}.$$

The first part equals $Af - cf$, the second $c_1 f - cf$, so that the whole equals $Af - c_1 f$, which proves (5.17).

PRINCETON UNIVERSITY

[9] Note that we have not supposed f and g to be in the domain of D_v.

Annals of Mathematics **61** (1955) 90–105

COMMUNICATIONS ON PURE AND APPLIED MATHEMATICS, VOL. VIII, 203–216 (1955)

On Differential Operators and Boundary Conditions

By WILLIAM FELLER
Princeton University

1. *Introduction*

We consider the differential operator

$$(1.1) \qquad Af = a \frac{d^2}{dx^2} f + b \frac{d}{dx} f + cf, \qquad a > 0,$$

as an operator from continuous functions of a real variable to continuous functions. It has the two obvious properties:

(i) *It is of local character.*
(ii) *If $f(x_0) = 0$ and f has a local minimum at x_0 then $Af(x_0) \geq 0$.*

We propose to discuss the *most general class of operators with these two characteristic properties.* One would expect that essentially only differential operators are of local character. However, it should be noticed that both our properties (i) and (ii) remain unchanged under the most general continuous transformation $\xi = L(x)$ of scale, whereas such a transformation will in general take differentiable functions into non-differentiable ones and vice versa.

We shall see that the operators with properties (i)–(ii) are a natural generalization of (1.1). They can be represented by means of repeated differentiations with respect (not to one fixed scale x but) to two monotonic functions $u(x)$ and $v(x)$, of which only one need be continuous. The classical operators (1.1) are obtained when u and v are differentiable with respect to one another.

Our generalized differential operators were introduced to solve an outstanding problem of diffusion theory, but we shall not be concerned with this side of the problem. The main purpose of this lecture is to show by examples that our new operators and the new point of view are useful even for perfectly classical problems.

We shall not dwell on any advantage of the generalization as such. The point to be stressed is that we achieve a considerable simplification and unification of the theory. Where before one had to piece together solutions in various intervals and justify the corresponding Sturm-Liouville expansions, we can write the whole as one equation and apply the fundamental existence theorems. Similarly (Example III) new type of boundary conditions which required special theories become part of the familiar Sturm-Liouville theory by the simple expedient of attaching finite masses to the endpoints.

R.L. Schilling et al. (eds.), *Selected Papers II*,

For the reader's convenience we discuss these things restricting ourselves to "reduced" operators—those which correspond to the classical case $c = 0$ and which are characterized by the property that, when operating on a constant, they yield zero. (In general, the constant function need not be in the domain of our operators.) The general operator will be described in Section 7. Since we repeatedly refer to classical and newer type boundary conditions, we shall end by a discussion of the notion of lateral conditions and some less known aspects of it.[1] In Section 8 an application is given to the problem of which parabolic equations preserve positivity.

2. *Generalized Derivatives*

We wish to define a *derivative $D_u f$ of a function f with respect to an arbitrary strictly increasing function u.*

When u is continuous, it may be introduced as independent variable, and $D_u f$ will reduce to the ordinary derivative. More generally, we shall conceive of u as defining a measure (or mass distribution) on the x-axis. Then u will jump at all points carrying a positive mass. The obvious notion *of continuity* of *f with respect to u* requires f to be continuous at each place where u is continuous; but if u jumps at x_0 it is only required that the limits $f(x_0+)$ and $f(x_0-)$ exist.

If f is continuous with respect to u *we define at each point of discontinuity of u*

$$D_u f(x) = \frac{f(x+) - f(x-)}{u(x+) - u(x-)}, \tag{2.1}$$

whereas at points of continuity

$$D_u f(x) = \lim \frac{f(x + h) - f(x)}{u(x + h) - u(x)}, \tag{2.2}$$

provided the limit exists. If h is restricted to positive or negative values only, the corresponding limits are the *one-sided derivatives $D_u^+ f$ and $D_u^- f$*, respectively.

3. *Reduced Operators*

In order to explain the main result in the simplest special case, let us begin by discussing the general operator corresponding to the classical operator (1.1) without constant term, that is, with $c \equiv 0$. Such an operator has the basic properties (i)–(ii) and

[1]The construction of the most general operators with properties (i)–(ii) is contained in [6] to which paper the reader is referred for more precise statements. The method of Example III (Section 6) is of wider applicability and is here described for the first time The ideas of the last section are contained (or hidden) in [5].

(iii) *The function* 1 (*which equals* 1 *for all* x) *is in the domain of* A *and* $A1 \equiv 0$.

We wish to concentrate our attention to operators which are *genuinely of second order*. If the operator is of the form (1.1) one could require that $a \neq 0$. However, this condition depends on the choice of the independent variable and has no intrinsic meaning. The following definition is therefore more natural.

Let Ω *be an operator with properties* (i)–(iii). *We say that* x_0 *is a first order point for* Ω *if either* $\Omega f(x_0) = 0$ *for all* f *in the domain of* Ω, *or if for all* f, g *in the domain of* Ω

$$\frac{\Omega f(x_0)}{\Omega g(x_0)} = \lim_{h \to 0} \frac{f(x+h) - f(x)}{g(x+h) - g(x)} \tag{3.1}$$

provided $\Omega g(x_0) \neq 0$.

It can be shown that in a neighborhood of a first order point the independent variable can be so chosen that Ω *reduces to the right hand derivative.*

The value of this definition depends on the fact that first order points represent the *only type of singularities for our operators.* In fact, we have the following

THEOREM. *Suppose that* Ω *has properties* (i)–(iii) *and that no point of the interval* I *is a first order point. Then there exists in* I *a strictly increasing continuous* u *and a strictly increasing* (*not necessarily continuous*) v *such that*

$$\Omega f \equiv D_v D_u f \tag{3.2}$$

in the following sense: if f *is in the domain of* Ω *then* f *is continuous with respect to* u, *the one sided derivatives* $D_u^+ f$ *and* $D_u^- f$ *exist everywhere* (*are equal except, perhaps, at points of discontinuity of* v), *and these derivatives are in turn differentiable with respect to* v *so that at each point*

$$D_v(D_u^+ f) = D_v(D_u^- f) = \Omega f. \tag{3.3}$$

Of course, u *may be chosen as independent variable*, but v need not be continuous or differentiable with respect to u. It goes without saying that one may choose other scales and measures to obtain for Ω expressions of the form

$$\Omega f = a D_v D_u f + b D_u f. \tag{3.4}$$

However, the normal form (3.2) which reduces Ω to a simple repeated differentiation is obviously most convenient. The only advantage of (3.4) is that it permits to include first order points.

4. *Discussion*

The operator

$$\Omega = a \frac{d^2}{dx^2} + b \frac{d}{dx}, \qquad a > 0, \tag{4.1}$$

can be reduced to the canonical form (3.2) by means of the substitutions

$$(4.2)\qquad \begin{aligned} u(x) &= \int_{x_0}^{x} \exp\left\{-\int_{x_0}^{x} \frac{b(s)}{a(s)}\,ds\right\}\cdot dx, \\ v(x) &= \int_{x_0}^{x} \frac{1}{a(x)}\cdot \exp\left\{\int_{x_0}^{x} \frac{b(s)}{a(s)}\,ds\right\}\cdot dx. \end{aligned}$$

(The most general admissible transformation is obtained on replacing u by $k_1u + k_2$ and v by $k_1^{-1}v + k_3$ with k_i constants.) On the other hand, if v is discontinuous, or not differentiable with respect to u, then (3.2) cannot be brought to the form (4.1). Even if the coefficients a, b are continuous, their form depends on the choice of the independent variable, whereas the scale u and the measure v in (3.2) are intrinsically connected with the operator.

It is well known that in applications (in particular in the theory of heat conduction), we meet constantly situations where the coefficients in (4.1) jump at certain places (cf. Example I below). In each case one has to consider the appropriate transition conditions, etc., and a great literature is devoted to special problems of this kind. It will be seen by our examples that the artifice of writing the operator in the form (3.2) eliminates the necessity of such piecemeal work: the general existence and representation theorems, and eigenfunction expansions, apply in each case directly. The separate transition (or continuity) conditions disappear altogether.

A different field of usefulness for the representation (3.2) is encountered in connection with boundary value problems. In Example III we shall discuss a special boundary value problem connected with the simple operator $\Omega = d^2/dx^2$. The boundary condition is not of the classical type, and as Bauer's investigation [1] shows, the classical methods require special adaptation to yield the existence and expansion theorems. We shall see that with the generalized operator (3.2) all that is required is to put $u = v = x$ in the interior interval, but to attach a *finite mass to the endpoints.* The required expansion theorems then appear as a special case of the canonical theorem (cf. end of Section 5).

These examples illustrate the advantage, also in classical problems, of introducing the generalized differential operator (3.2). Even when (3.2) can be written in the special form (4.1), the fact that u and v are intrinsically connected with the operator renders the theory more intuitive. Thus the classification of boundaries and the investigation of the types of singularities due to the vanishing of the leading coefficient become simpler and more understandable.

The theory of (4.1) is intimately connected with that of its *formal adjoint*

$$(4.3)\qquad \Omega^* f = \frac{d^2}{dx^2}(af) - \frac{d}{dx}(bf).$$

Every operator should have an adjoint, and yet (4.3) is meaningful only under conditions on the coefficients which are unnatural for (4.1). This is particularly disturbing in diffusion theory. Now it turns out that the formal adjoint to (3.2) is simply

$$\Omega^* = D_u D_v \, . \tag{4.4}$$

This is meaningful, for example, if a and b in (4.1) are bounded: the adjoint to any ordinary differential operator is of the form (3.2) but need not be a differential operator in the traditional restricted sense.

The most satisfactory theory of boundary value and expansion problems has been derived for self-adjoint operators (i.e. $b = a'$) using *Hilbert space* methods (cf., e.g., Stone [9]). It turns out that it is most natural to consider the operator (3.2) in a Hilbert space with weight function v, that is, introduce the norm

$$\| f \|^2 = \int f^2(x)\, dv(x). \tag{4.5}$$

With this convention *every operator* (3.2) *becomes formally self-adjoint*, the Green functions become *symmetric*, and the classical theory carries over without difficulties and less formalism. The usual conditions $b = a'$ simply state (cf. (4.2)) that $v \equiv x$, that is, that Ω is self-adjoint with respect to the weight function $v = x$. As Elliott [2] has shown,* the classical results for the Hilbert space (4.5) can then be used to obtain similar expansion theorems for the operators (3.2) and (4.4) considered, respectively, as operators in the Banach spaces of continuous functions and measures. It is in this form that the operators must be considered in the theory of heat condition and diffusion.

5. *Green Functions and Wronskians*

If Ω is the operator (3.2), the existence of solutions of

$$\lambda\xi - \Omega\xi = 0 \tag{5.1}$$

is easily established by the classical method of successive approximations. In fact, if u is introduced as independent variable the construction becomes particularly intuitive. Thus, if $\lambda > 0$ and, say, $\xi(0) = 0$, $\xi'(0) > 0$ one sees trivially that ξ is *convex* and positive for $x > 0$. Similarly, if $\lambda < 0$ the solution will oscillate provided v increases sufficiently (or $-\lambda$ is sufficiently large).

The only small trouble arises in connection with the Wronskian of two solutions ξ_1 and ξ_2 of (5.1). If Ω is of the traditional form (4.1), the Wronskian $\xi_1'\xi_2 - \xi_1\xi_2'$ is easily seen to be constant. When Ω is of the general form (3.2) we may without loss of generality introduce u as independent variable. Then $\xi_1'\xi_2 - \xi_1\xi_2'$ exists (at points of discontinuity of v the derivatives are to be interpreted as right- or left-hand derivatives). However, ξ_i need not be differentiable with respect to v, and thus we cannot use the classical procedure of differentiating

*Added in proof: Cf. also Hille [7a].

$W = \xi_1'\xi_2 - \xi_1\xi_2'$ with respect to v. And even if this formal differentiation were permitted, to see that $D_v W = 0$ we would have to know that

$$D_x\xi_1 \cdot D_v\xi_2 = D_v\xi_1 \cdot D_x\xi_2 , \tag{5.2}$$

which is not correct. Nevertheless, the classical result does carry over, and the *Wronskian*

$$\xi_2 \cdot D_u\xi_1 - \xi_1 \cdot D_u\xi_2 \tag{5.3}$$

of two solutions of (5.1) *is always constant.* This is easily verified using integrations by part instead of differentiations.

Once we have this result, we can choose two solutions ξ_i whose Wronskian is identically 1, and form the *Green funvtion*

$$K(x, y) = \begin{cases} \xi_1(x)\xi_2(y) & \text{for} \quad x \leq y \\ \xi_2(x)\xi_1(y) & \text{for} \quad x \geq y. \end{cases} \tag{5.4}$$

If f is continuous and bounded it is readily seen that

$$F(x) = \int K(x, y)f(y)\, dv(y) \tag{5.5}$$

is a solution of

$$\lambda F - \Omega F = f. \tag{5.6}$$

As was mentioned in the preceding section, *if* (5.5) *is considered as a transformation in the Hilbert space* (4.5), *the kernel is symmetric*[2] and the classical theory applies. For the spaces and norms appropriate to diffusion theory, the analysis of [5] applies even in a simplified form.

So far $v(x)$ was considered only in the interior of the given interval I. However, we may extend the definition of v by attaching finite masses to the endpoints of the interval; formula (5.5) remains meaningful, and defines a solution of (5.6). However, the nature of this solution will depend on the masses carried by the boundaries of the interval. We have here a flexible device of obtaining at no extra cost solutions (and eigen function expansions) for new types of boundary value problems. Example III will illustrate the advantage of this method for an equation of the traditional type.

6. *Illustrative Examples*

Example I. A familiar situation in the theory of heat conduction is typified by the system of the two heat equations

$$\begin{aligned} w_t &= w_{xx} \quad &\text{for} \quad x < 0, \\ w_t &= 4w_{xx} \quad &\text{for} \quad x > 0 \end{aligned} \tag{6.1}$$

[2]With the usual interpretation the kernel is $K(x, y)v'(y)$ which is *unsymmetric.*

together with the "continuity conditions"

$$w(t, 0+) = w(t, 0-), \qquad w_x(t, 0-) = 4w_x(t, 0+). \tag{6.2}$$

These equations are supposed to describe two materials with different conductivities joining at $x = 0$. Now, whatever the nature of heat, it should not be affected by the reference to "two materials," and actually the general derivation of the heat equation includes the point $x = 0$ as a regular point.

As a matter of fact, (6.1) refers to the operator

$$\Omega = \frac{d^2}{dx^2} \quad \text{for} \quad x < 0, \qquad \Omega = 4\frac{d^2}{dx^2} \quad \text{for} \quad x > 0 \tag{6.3}$$

which is reduced to the canonical form $D_v D_u$ by putting

$$\begin{aligned} u(x) &= x \quad \text{for} \quad x < 0, \qquad u(x) = \tfrac{1}{4}x \quad \text{for} \quad x > 0, \\ v(x) &= x. \end{aligned} \tag{6.4}$$

With this notation the equation

$$w_t = D_v D_u w \tag{6.5}$$

holds at *all* points, and no special consideration is required for $x = 0$. *The second equation in* (6.2) *expresses only the fact that* $D_u w$ *exists at* $x = 0$.

If one wishes to solve (6.1) by means of Laplace transforms, one is lead to (5.6). Again, no piecing together of two solutions and no boundary condition at $x = 0$ is required. If the fundamental interval is $-\infty < x < \infty$ the unique solution is given by (5.5) with

$$\begin{aligned} \xi_1(x) &= \begin{cases} \exp\{\lambda^{1/2}x\} & \text{for} \quad x < 0 \\ \exp\{\lambda^{1/2}x/2\} & \text{for} \quad x > 0, \end{cases} \\ \xi_2(x) &= \begin{cases} \exp\{-\lambda^{1/2}x\} & \text{for} \quad x < 0 \\ \exp\{\lambda^{1/2}x/2\} & \text{for} \quad x > 0. \end{cases} \end{aligned} \tag{6.6}$$

Example II. Let us put

$$\begin{aligned} u(x) &= x, & -\infty < x < \infty, \\ v(x) &= x, & -\infty < x < 0, \\ v(x) &= x + 1, & 0 < x < \infty, \end{aligned} \tag{6.7}$$

that is, the origin carries a unit mass. In conventional notation our operator is

$$\begin{aligned} \Omega f(x) &= f''(x), & x \neq 0, \\ \Omega f(0) &= f'(0+) - f'(0-). \end{aligned} \tag{6.8}$$

Equations such as (6.5) or (5.6) could be solved by piecing together appropriate solutions for $x > 0$ and $x < 0$ and using the second equation in (6.8) as "boundary condition." However, the direct method is preferable.

The solutions $\xi_1(x)$ and $\xi_2(x)$ occurring in the Green function (5.4) must be bounded at $-\infty$ and ∞, respectively. For x negative, ξ_1 must therefore be a multiple of $\exp(\lambda^{1/2}x)$ whence we have up to a constant factor

$$\begin{aligned} \xi_1(x) &= \exp\{\lambda^{1/2}x\} && \text{for} \quad x \leq 0, \\ \xi_1(x) &= \exp\{\lambda^{1/2}x\} + \lambda^{1/2}\sinh\{\lambda^{1/2}x\} && \text{for} \quad x \geq 0. \end{aligned} \tag{6.9}$$

Thus the solution $F(x)$ of (5.5) is uniquely determined, and a simple calculation shows that

$$\begin{aligned} F(x) &= \frac{1}{2\lambda^{1/2}} \int_{-\infty}^{+\infty} \exp\{-\lambda^{1/2} \mid x - y \mid\} \cdot f(y)\, dy \\ &\quad - \frac{\exp\{-\lambda^{1/2}x\}}{2(2 + \lambda^{1/2})} \cdot \left\{ \int_{-\infty}^{+\infty} \exp\{-\lambda^{1/2} \mid y \mid\} \cdot f(y)\, dy - \frac{2}{\lambda^{1/2}} f(0) \right\}. \end{aligned} \tag{6.10}$$

The first term on the right represents the classical solution of $\lambda F - F'' = f$. The second term becomes particularly interesting if F is interpreted as the Laplace transform of the solution of the generalized heat equation $w_t = \Omega w$ with $w(0, x) = f(x)$. For, the second term on the right in (6.10) is formally analogous to the term occurring in classical heat conduction in the presence of an elastic barrier.

The behavior of the path-function in a diffusion process near an elastic barrier is a puzzling unsolved problem of probability theory. It is therefore interesting to note that when (6.8) is taken as the diffusion operator we obtain a process in which *the origin behaves as an elastic barrier, even though it is an interior point* and particles can pass through it. This is an interesting phenomenon which may well facilitate the investigation of the behavior of path functions near elastic barriers.

Example III. In [1], W. F. Bauer treats the boundary value problem

$$\lambda F - \{F'' - cF\} = f, \qquad 0 < x < 1, \tag{6.11}$$

$$\begin{aligned} k_1 F(1) - F'(1) - k_2 F''(1) &= 0, \\ F(0) &= 0, \end{aligned} \tag{6.12}$$

which differs from the classical boundary conditions[3] by the appearance of the term $F''(1)$. Bauer has shown that the classical theory can be retraced so as to obtain Sturm-Liouville expansions for this case.

We propose to show that boundary value problems of this type require no special treatment. All that is necessary is to *extend the definition of v so as to include a mass at the endpoints of the interval.* We still get a solution of the form

[3]In [5] the author has constructed the *most general* boundary conditions for second order differential operators, and Bauer's condition is actually a special case. The Green function constructed by Bauer can be obtained by specialization, e.g., from formula (19.5) of [5]. However, the Fourier expansion obtained by Bauer is new.

(5.5) with a symmetric kernel $K(x, y)$, and the required expansions reduce to the standard equipment of Hilbert space theory.[4]

We consider an arbitrary operator Ω of the form (3.2). To simplify the writing we introduce u as independent variable which, as has been explained, is no loss of generality.

We are then concerned with the equation

$$\lambda F - D_v F' = f, \qquad \lambda > 0. \tag{6.13}$$

(In Bauer's case $v = x$, but an additional term cF appears on the left side. Its inclusion in no way changes the following reasoning.)

We suppose that v is *bounded* and that we deal with a *finite* interval[5] $x_1 < x < x_2$. We impose the two *boundary conditions*

$$\begin{aligned} r_1\Omega F(x_1) - q_1 F'(x_1) + p_1 F(x_1) &= 0, \\ r_2\Omega F(x_2) + q_2 F'(x_2) + p_2 F(x_2) &= 0, \end{aligned} \tag{6.14}$$

where the r_i, etc. are non-negative constants.[6]

Now let ξ_1 and ξ_2 be solutions of

$$\lambda\xi - D_v\xi' = 0 \tag{6.15}$$

which satisfy, respectively, the first and the second boundary condition (6.14). The existence of such solutions is easily verified, and just as in [5] one sees that they are bounded, positive, that ξ_1 increases and ξ_2 decreases. These solutions are determined only up to a multiplicative constant, and we norm them so that the Wronskian (which is a positive constant) equals 1.

With these solutions we form the Green function (5.4) and then the solution (5.5). Clearly F is a solution of (6.13). At the right boundary we have

$$\begin{aligned} F(x_2) &= \xi_2(x_2) \int_{x_1}^{x_2} \xi_1(y) f(y)\, dv(y), \\ F'(x_2) &= \xi_2'(x_2) \int_{x_1}^{x_2} \xi_1(y) f(y)\, dv(y), \\ \Omega F(x_2) &= \lambda F(x_2) - f(x_2). \end{aligned} \tag{6.16}$$

---

[4]For the corresponding expansion theorems in the spaces of continuous functions and measures, cf. Elliott [2].

[5]In the classification of [5] these assumptions mean that the boundaries are *regular*. For *exit* boundaries the argument applies equally, but one must put $q_i = 0$ in the boundary conditions. For other types of boundaries no boundary condition can be imposed and (6.13) has a unique solution.

[6]It has been shown in [5] that under these conditions the solution of (6.13) is positivity preserving and that $\lambda \cdot \max F \leq \max f$. In all formulas $F(x_2)$ stands for the $F(x_2-)$, etc. The existence of the limits was proved in [5].

Since ξ_2 satisfies the second boundary condition (6.14) and $\Omega\xi_2(x_2) = \lambda\xi_2(x_2)$ we get from this

$$r_2\Omega F(x_2) + q_2F'(x_2) + p_2F(x_2) = -r_2f(x_2). \tag{6.17}$$

Thus F does not satisfy the required boundary conditions (unless $r_2 = 0$, which represents the classical-type condition).

However, consider the function

$$\begin{aligned} &F(x) + \frac{\xi_1(x)}{(\lambda r_2 + p_2)\xi_1(x_2) + q_2\xi_1'(x_2)} r_2 f(x_2) \\ &\quad + \frac{\xi_2(x)}{(\lambda r_1 + p_1)\xi_2(x_1) - q_1\xi_2'(x_1)} r_1 f(x_1) \equiv \Phi(x) \end{aligned} \tag{6.18}$$

(the denominators cannot vanish since ξ_1 increases, ξ_2 decreases, and both are positive). Since Φ differs from F by a linear combination of solutions of the homogeneous equation, it is also a solution of (6.13) and it is obvious that Φ satisfies the boundary conditions (6.12).

Thus Φ is the required solution. However, we can write

$$\Phi(x) = \int_{r_1-}^{r_2+} K(x, y)f(y)\,dv(y) \tag{6.19}$$

provided we extend v to the boundaries so that x_i carries mass r_i and define

$$\begin{aligned} K(x, x_2) &= \frac{\xi_1(x)}{(\lambda r_2 + p_2)\xi_1(x_2) + q_2\xi_1'(x_2)}, \\ K(x, x_1) &= \frac{\xi_2(x)}{(\lambda r_1 + p_1)\xi_2(x_1) - q_1\xi_2'(x_1)}, \\ K(x_i, x) &= K(x, x_i). \end{aligned} \tag{6.20}$$

We see thus that the solution is again given by a positive symmetric kernel. The existence of a complete set of eigen-functions is contained in the fundamentals of Hilbert space theory, and we may expand either F or the solutions of the corresponding "heat equation" $w_t = \Omega w$.

This illustrates the usefulness of introducing Stieltjes integrals even when dealing with differential operators of the classical type.

7. *The General Operator*

So far we have restricted our attention to operators having the three properties (i)–(iii). Only a simple modification is required to *eliminate the condition* (iii).

Let A be any operator with properties (i)–(ii). Suppose, for the moment, that the constant function **1** is in the domain of A, that is, that

$$A\mathbf{1} = c \tag{7.1}$$

exists. Then the operator Ω defined by

$$\Omega f = Af - cf \tag{7.2}$$

obviously has all three properties (i)–(iii). Since we have found the most general form of an operator Ω with these properties, we have in (7.2) (with an arbitrary c) the *most general form of an operator A with properties* (i)–(ii) and with the regularity condition that $A1$ exists.

For all practical purposes the last condition is no restriction, but it can be removed. Suppose that ϕ is any positive continuous function for which $A\phi$ exists. Then we define a new operator

$$A^*f = (1/\phi)A(f\phi). \tag{7.3}$$

The domain of A^* consists of all f such that $f\phi$ is in the domain of A, and vice versa. Put

$$A^*1 \equiv A\phi/\phi = c. \tag{7.4}$$

Then we may apply A^* to (7.2) and obtain

$$Af = \phi\Omega(f/\phi) + cf, \tag{7.5}$$

where Ω is an operator of the type described in Section 3. In (7.5) we have the desired *general form of operators with properties* (i)–(ii). *Conversely*, $\phi > 0$, c, and Ω may be arbitrarily chosen and A will have properties (i)–(ii).

8. *Connection with Partial Differential Equations and Diffusion*

Consider the parabolic differential equation

$$w_t = Aw, \qquad w = w(t, x), \tag{8.1}$$

where A is the differential operator (1.1). The *Cauchy problem* consists in finding a solution for $t > 0$ such that (in a sense to be made precise)

$$w(t, x) \to f(x) \quad \text{as} \quad t \to 0+ \tag{8.2}$$

where f is an arbitrary continuous function; if the solution is not unique, then appropriate boundary conditions are imposed to insure uniqueness.

We say that the operator A together with the given boundary conditions is *positivity preserving* if $f \geq 0$ implies $w \geq 0$ for $t > 0$.

We now ask the question: *Which operators of local character are positivity preserving*? (In particular, which differential equations of form (8.1) are positivity preserving?)

The *answer* is that A *must have property* (ii). That is, A must be of the form described in the preceding and, more precisely, the coefficient c in (7.5) must have a finite upper bound. Conversely, these conditions are *sufficient*: one may choose u, v, ϕ, c arbitrarily and, with appropriate boundary conditions, the operator (7.5) will be positivity preserving. In other words: we have obtained the *most general positivity preserving operator of local character*; *in particular, no higher order equation of form* (8.1) *is positivity preserving.*[7]

Our problem and its solution is of particular relevance for diffusion. If A is defined by (1.1) then the differential equation (8.1) and its formal adjoint

$$w_t = (aw)_{xx} - (bw)_x + cw \tag{8.3}$$

are the fundamental equations of diffusion theory (the latter known as Fokker-Planck equation). As is well known, Kolmogorov was the first to attempt a derivation of these equations from general probabilistic principles.[8] Despite much work on the problem, a satisfactory solution seemed for a long time out of reach. It was also puzzling and irritating that (8.3) could be derived only as a consequence of (8.1), and yet requires differentiability conditions on a, b which seem unnatural in the derivation of (8.1).

It will be shown in another paper that with our generalized differential operators we have *the most general diffusion equation*. It can be derived from fundamental principles and no regularity conditions need be imposed.

9. *What Are Lateral Conditions?*

The theory of differential operators depends on the notion of boundary conditions. As we have seen, with our generalized differential operators we may sometimes eliminate boundary conditions which were necessary for the traditional operators. At other places reference was made to new type boundary conditions. It is therefore natural to conclude this survey by a few words concerning the notion of lateral conditions in general. The problem can be approached from many angles, but a particularly simple point of departure is afforded by the Cauchy problem for an equation of type (8.1).

For a first orientation suppose that to every f of a certain class $\mathfrak{B}$ there corresponds a *unique* solution w of (8.1) for which (8.2) holds. For any fixed $t > 0$ we may consider w as a transform of f:

$$w(t, x) = T_t f(x). \tag{9.1}$$

As Hadamard has emphasized, the uniqueness of the solution implies that $w(t + s, x)$ may be calculated either from the original data, or by using $w(s, x)$ as given function. Equating the two results we get the semi-group property

$$T_{t+s} = T_t T_s . \tag{9.2}$$

[7]The description of the text is vague in so far as we have not defined the notions of solution, boundary conditions, and convergence in (8.2). In analytical terms the problem solved is: to construct *the most general semi-group of operators which transforms each bounded continuous function* $f \geq 0$ *on an interval* I *into a function of the same class and which has an infinitesimal generator of local character.*

[8]Cf. [8] or, for a more general derivation, [3]. An account of subsequent developments is contained in [4].

If $\mathfrak{B}$ is a Banach space and certain simple conditions are imposed, then the operator A is the infinitesimal generator of the semi-group T_t in the sense of Hille. As long as the Cauchy problem has a unique solution, the problems of integrating (9.1) and of constructing a semi-group with infinitesimal generators A are equivalent.

The classical method of integrating (9.1) consists in introducing the Laplace transform

$$F(x) = \int_0^\infty e^{-\lambda t} w(t, x)\, dt > 0. \tag{9.3}$$

It satisfies the equation

$$\lambda F - AF = f, \tag{9.4}$$

which under the present conditions has a unique solution. Given this solution, F, the classical theory was confronted with the necessity to prove that it actually corresponds to a solution $w(t, x)$ of the original problem.

It is one of the achievements of the Hille-Yosida theory that it dispenses with this necessity. Hille in [7] defines solutions of (8.1) and of (9.4) so that the two integration problems become completely equivalent. At the same time this notion of "solutions" is precise and most natural, and also (8.2) is made precise.

Hille has emphasized that this method is limited to the case where (9.1) has a unique solution. The more intricate problems arise in connection with non-uniqueness, and it has been shown in [5] how the semi-group theory can be adapted to this case.

In the case of non-uniqueness we can no longer prove that for all solutions of (8.1) the semi-group property (9.2) holds true. However, we *postulate* (9.2) as a rigorous definition of what the physicist would call *time-homogeneous* conditions: we decide to consider only solutions with the property that if $w(t_0, x)$ is taken as initial data the new solution at time t will coincide with $w(t_0 + t, x)$. The function of a lateral condition is to ensure the (existence and) uniqueness of such a solution for every f in $\mathfrak{B}$. Given an equation (8.1) our problem is to find the most general possible lateral condition.

Now suppose that there exists a set Σ of functions which is independent of λ and such that for each $f \in \mathfrak{B}$ *there exists exactly one solution* $F \in \Sigma$ of (9.4). The Hille-Yosida arguments can then be used to show that this F is the Laplace transform (9.3) of a solution of (8.1)–(8.2). Moreover, for all "reasonable" initial functions f and each fixed $t > 0$ the solution $w(t, x)$ will, as a function of x, be an element of Σ. Conversely, to each solution of our time-homogeneous Cauchy problems there corresponds such a set Σ. (With an appropriate change of the Hille-Yosida definition of generation of a semi-group by an operator A we may say that we are describing the totality of all semi-groups generated by A.)

This sketchy survey will indicate one justification of the following *definition* (cf. [5]):

A conditioning set Σ *for an operator* A *is a set (in no way depending on* λ*) such that for each* f *of a certain class* $\mathfrak{B}$ *there exists a unique solution* $F \in \Sigma$ *of* (9.4). *A lateral condition is simply the condition that a solution be an element of* Σ.

We have tried to justify this definition by its great usefulness for Cauchy problems and semi-groups. Actually, once stated, the definition appears most natural also from the point of view of Banach spaces. The classical boundary conditions of the Sturm-Liouville problem too can be considered from this point of view. In fact, Stone [9] has given a derivation of these conditions within the framework of pure Hilbert space theory. While his terminology and purpose are formally different, a perusal of the method will show that it boils down to the construction of a conditioning set in our case. Of course, Stone is interested primarily only in self-adjoint differential operators. As we have seen in Example III, even here we can get new boundary conditions and Sturm-Liouville expansions if we modify the weight function so as to include masses at the endpoints of the interval.

The classical differential operators appear in a new light in probability and diffusion theory, where they operate in the Banach space of measures, and where the notion of self-adjointness becomes irrelevant. A complete enumeration of all possible boundary conditions appropriate to diffusion theory has been given in [5]. It should be noticed that such boundary conditions may well be of a global character (depend on integrals) and that the adjoint to a differential operator need not be of local character.

Bibliography

[1] Bauer, W. F., *Modified Sturm-Liouville systems*, Quart. Appl. Math., Vol. 11, 1953, pp. 273–284.

[2] Elliott, J., *Eigen-function expansions associated with singular differential operators*, Trans. Amer. Math. Soc. (to appear).

[3] Feller, W., *Zur Theorie der stochastischen Prozesse (Existenz und Eindeutigkeitssätze)*, Math. Ann., Vol. 113, 1936, pp. 113–160.

[4] Feller, W., *Some recent trends in the mathematical theory of diffusion*, Proc. International Congress Math., Vol. 2, 1950, pp. 322–329.

[5] Feller, W., *The parabolic differential equations and the associated semi-groups of transformations*, Ann. Math., Vol. 55, 1952, pp. 468–519.

[6] Feller, W., *On second order differential operators*, Ann. Math., (to appear).

[7] Hille, E., *Functional analysis and semi-groups*, Amer. Math. Soc. Colloquium Publications, No. 31, New York, 1948.

[7a] Hille, E., *The abstract Cauchy problem and Cauchy's problem for parabolic differential equations*, Jour. d'Analyse Math., Vol. 3, 1953/54, pp. 81–198. [Added in proof.]

[8] Kolmogoroff, A., *Über die analytischen Methoden in der Wahrscheinlichkeitsrechnung*, Math. Ann., Vol. 104, 1931, pp. 415–458.

[9] Stone, M. H., *Linear transformations in Hilbert space*, Amer. Math. Soc. Colloquium Publications, No. 15, New York, 1932.

Received August 13, 1954.

BOUNDARIES INDUCED BY NON-NEGATIVE MATRICES

BY

WILLIAM FELLER[1]

1. Introduction. The object of this paper is to explore certain properties of positivity preserving operators on either functions or measures in a space E. Although our method applies to more general cases (cf. the end of this introduction) we take up explicitly only the case of a *denumerable* E so that the operators reduce to matrices. In this way the basic features of the theory will not be obscured by an unfamiliar formalism or by an a priori imposed topology.

Let E stand for the set of positive integers, and Π for a matrix with elements $\Pi(i, j) \geqq 0$, where $i, j \in E$. Then Π acts as an operator by premultiplication on column-vectors, and by postmultiplication on row-vectors. It will be shown that the relevant properties of Π are intimately connected with the solutions of the infinite system of equations

$$x(i) = \sum_{j \in E} \Pi(i, j)x(j) \quad \textit{or} \quad \Pi\mathbf{x} = \mathbf{x} \tag{1.1}$$

and the dual (or adjoint) system

$$\xi(j) = \sum_{i \in E} \xi(i)\Pi(i, j) \quad \textit{or} \quad \boldsymbol{\xi} = \boldsymbol{\xi}\Pi. \tag{1.2}$$

Here $\mathbf{x}$ and $\boldsymbol{\xi}$ stand for a column- or row-vector, respectively.

The matrix Π is called *stochastic* if all row sums are unity. Now if $\mathbf{x}$ is a strictly positive[2] solution of (1.1), the transformation

$$\Pi'(i, j) = \Pi(i, j)x(j)/x(i)$$

defines a stochastic matrix Π' and it will be shown (§14) that there exists an isomorphism between the solutions of (1.1) and the solutions of the transformed equation $\Pi'\mathbf{x}' = \mathbf{x}'$. For our purposes the matrices Π and Π' are in every respect equivalent, and there is no loss of generality in supposing that Π is stochastic. However, since we shall be dealing also with submatrices of Π, it is most convenient to suppose only that

Received by the editors January 25, 1956.

(1) Research begun in the summer of 1954 at Cornell University with support by the U. S. Air Force under Contract No. AF 18(600)-685 and continued in Princeton, supported by the U. S. Army Ordnance, Office of Ordnance Research, Contract No. DA-36-034-ORD-1296.

(2) Strict positivity is assumed only to simplify formulations.

R.L. Schilling et al. (eds.), *Selected Papers II*,

$$\Pi(i, j) \geqq 0 \quad \sum_{j \in E} \Pi(i, j) \leqq 1 \qquad i, j \in E. \tag{1.3}$$

Such matrices will be called *sub-stochastic*. From now on we assume (1.3).

Probabilistically, a sub-stochastic matrix defines a *random walk*, and we shall explain the random walk interpretations of our boundaries, solutions, etc. However, our development is purely analytic and no probabilistic arguments or results are used.

It will be shown that to the typical solution $\mathbf{x}$ or $\boldsymbol{\xi}$ there corresponds a family of subsets of E contracting to the empty set. It is natural and useful to introduce these sets as neighborhoods of new points or sets. In this way E *will be enlarged by a boundary corresponding to* (1.1), *and by an adjoint boundary corresponding to* (1.2). Frequently these two topologies will be equivalent, but *there exist matrices* Π *such that the two topologies have no connection whatever* (cf. Example III, §17). The two boundaries can have arbitrary topological structures(3). The boundaries will be compact Hausdorff spaces, but the enlarged space need *not* be compact.

In a subsequent paper these boundaries will be applied to the theory of the Kolmogorov differential equations in E. It will be shown that our boundaries play exactly the rôle of ordinary boundaries in potential theory (except that in the latter the two boundaries coincide because of the symmetry of the underlying operators). We shall also find analogues to the classical Green functions and boundary conditions(4) of the theory of harmonic functions and diffusion.

Our development proceeds in *three* stages. We begin by a study of *bounded solutions of* (1.1). Each such solution is the difference of two positive solutions, and we are concerned with the solutions $\mathbf{x}$ such that $0 \leqq x(i) \leqq 1$. They form a convex set $\mathfrak{P}$ and, at the same time, a linear lattice (§4). In §5 we introduce the basic notion of a *sojourn set*. To each such set there corresponds an element $\mathbf{s}_A$ of $\mathfrak{P}$, and these form the lattice $\mathfrak{S}$ of *sojourn solutions*. Analytically they are the *extremals of* $\mathfrak{P}$ in the sense of Krein and Milman (cf., for example, [2, Livre 5, Ch. 2]). This fact is proved in §10, but not used explicitly.

To each sojourn set A there corresponds a sojourn solution $\mathbf{s}_A$, but to different sets there may correspond the same sojourn solution. Two sets are called *equivalent* if $\mathbf{s}_A = \mathbf{s}_B$. It is shown in §8 that (with a trivial exception) *each sojourn set* A *contains a sequence of equivalent sojourn sets* A_ϵ *which contract monotonically to the empty set as* $\epsilon \to 0$. We shall introduce them as

(3) Cf. the absurdly simple example IV, §17, where the meaning of a boundary homeomorphic to the interval (0, 1) is intuitively clear. The construction can be modified so as to obtain almost any type of boundary.

(4) The normal derivatives occurring in the boundary conditions for harmonic functions are, of course, meaningless in a discrete space E. They will be replaced by expressions depending on the given basic operator and which in turn reduce to normal derivatives, etc. in classical cases.

neighborhoods of a set A of the boundary (which will be both open and closed). For the random walk governed by Π and starting at i the value $s_A(i) = s_{A_\epsilon}(i)$ is (for each ϵ) the probability that the path ultimately (after finitely many steps) enters and remains contained in A_ϵ. In other words, $s_A(i)$ is the *probability of an asymptotic approach to the boundary set A*. (See §5 and Theorem 9.4.) The simplest situation is described in §12: here the boundary consists of denumerably many isolated points, each being represented by one sojourn solution. In general, however, no sojourn solutions correspond to the individual *points* of the boundary. We have, in effect, a measure (or capacity) induced on the boundary, and only *sets of positive measure are represented by sojourn solutions*. To introduce *points* of the boundary we use a variation of the well-known method of *maximal ideals* (§13). The enlarged space $E+\mathfrak{B}$ is a Hausdorff space in which *all elements of $\mathfrak{P}$ are continuous and possess continuous boundary values*.

The boundary defined in terms of the bounded solutions of (1.1) sets the stage for a study of the *unbounded positive solutions of* (1.1). They form a vector lattice(5) $\mathfrak{M}$. The transformation mentioned above and studied in §14 defines *a family $\mathfrak{F}$ of stochastic matrices such that there exists an isomorphism between the corresponding vector lattices $\mathfrak{M}$*. To each element $\mathbf{x} \in \mathfrak{M}$ there corresponds a matrix of the family such that the image of $\mathbf{x}$ is the unit vector $\mathbf{1}$. In other words, the vectors dominated by $\mathbf{x}$ are mapped into a set corresponding to the cone $\mathfrak{P}$ of bounded solutions. In this way *each vector $\mathbf{x} \in \mathfrak{M}$ can be made to play the rôle of the unit vector*. These transformations have a simple probabilistic interpretation and also (a less obvious) counterpart in the transformation theory of differential equations of the Sturm-Liouville type.

Each matrix of the family $\mathfrak{F}$ defines a boundary for E, and these boundaries do partly overlap and the topologies agree. It turns out that it is possible *to endow the space E with a boundary $\mathfrak{B}^*$ common to all matrices of the family $\mathfrak{F}$. The boundary $\mathfrak{B}$ defined by the bounded solutions of* (1.1) *is a subset of $\mathfrak{B}^*$ and* $\mathfrak{B}^*$ is the union of the boundaries defined in terms of bounded solutions of (1.1), as Π runs through the members of the family $\mathfrak{F}$. Both $\mathfrak{B}$ and $\mathfrak{B}^*$ have analytic and probabilistic significance. The construction of $\mathfrak{B}^*$ is analogous to the procedure outlined for $\mathfrak{B}$, but in $\mathfrak{M}$ we have no maximal ideals at our disposal. Instead we introduce *lattice ideals maximal relative to a fixed element*. They are characterized in §15.

At last (§16) we consider the *adjoint system* (1.2). Fortunately it requires no special theory since it can be reduced to (1.1) by a simple device which has been used for other purposes by Kolmogorov and Derman. Just as the solutions of (1.1) induce the boundary $\mathfrak{B}^*$ so do the solutions of (1.2) induce an *adjoint boundary*. Example III (§17) shows that *the two topologies are in-*

(5) Unfortunately, $\mathfrak{M}$ is in general not compact, and we have therefore neither extremals nor maximal ideals at our disposal. If each row of Π has only finitely many nonzero elements then $\mathfrak{M}$ is compact and a boundary may be defined in terms of the extremals of $\mathfrak{M}$.

dependent of each other. This phenomenon is new because one usually deals with operators which are symmetric or nearly so. It leads to interesting relations and boundary conditions for stochastic processes in E.

§11 stands somewhat apart from the main part of the paper. In the ergodic theory of stochastic matrices the points and subsets of E are classified as *recurrent* or *transient.* A recurrent subset R in no way contributes to the boundary (and corresponds analytically to a *closed manifold*). For the matrix Π it has only nuisance value, since Π is partitioned in the form (11.3) and the theory effectively reduces to a study of a submatrix corresponding to the set $E-R$. For our purposes it was necessary to correlate these facts with properties of the solutions of (1.1). §11 contains a direct derivation of the basic properties of transient and recurrent sets(6). A reader who is disturbed by this interruption of the theory may omit this section and simply assume that Π is not partitioned (of the form (11.3)).—Examples for the several phenomena are collected in §17.

Harmonic Functions—Aspect

We conclude this introduction by indicating the *relationship of the present theory with classical harmonic functions.* To illustrate the meaning of the new boundaries, let G be a simply connected domain of the Euclidean plane with a complicated boundary Γ (containing prime ends, etc.). A conformal mapping reduces the theory of *bounded harmonic functions* in G to the theory of bounded harmonic functions in the circular disc. The inverse map from the disc into G induces for G an ideal boundary Γ^* which is not topologically equivalent to Γ, and which is more natural for the study of harmonic functions. In several dimensions the "natural" boundary is still less adequate, and a boundary appropriate for the study of (or induced by) the Laplacian operator has been defined in the classical paper [8] by R. S. Martin.The formal analogue for harmonic functions to our boundary is not the *Martin boundary* but a larger one: our topology would make the bounded harmonic functions continuous up to and including the ideal boundary. On the other hand, our boundary is smaller than the *Čech boundary* constructed by means of maximal ideals in the algebra of bounded real functions continuous in C. It constitutes a worth-while program to reduce our boundary by an appropriate identification of points to an analogue of the Martin boundary where points would stand in a one-to-one correspondence to the linearly independent positive solutions of (1.1).

A *direct treatment of harmonic functions by our methods* can proceed in two ways.

(a) As is well known, the theory of harmonic functions can be approached

(6) It is now easy to go a step further and to derive the ergodic theory from our results. Incidentally, the uniqueness Theorem 14.4 properly belongs to §11, but is proved more naturally using the isomorphisms introduced in §14.

using the diffusion equation $u_t = \Delta u$. The latter corresponds to the Kolmogorov differential equations mentioned above, and (using Laplace transforms) can be treated in the same way.

(b) Completely within the framework of the present paper we can proceed as follows. Let Γ be an open circular disc and for each point $P \in \Gamma$ denote by C_P the *greatest* open circular disc with center at P and contained in Γ. Let $|C_P|$ be the area of C_P and define a kernel $K(P, Q)$ by

$$K(P, Q) = \begin{cases} 0 & \text{for } Q \notin C_P \\ \dfrac{1}{|C_P|} & \text{for } Q \in C_P \end{cases} \tag{1.4}$$

where $P, Q \in \Gamma$. Such a kernel plays the rôle of the matrix Π and the equation

$$x(P) = \int_\Gamma K(P, Q)x(Q)dQ \tag{1.5}$$

is analogous to (1.1). Each *harmonic* $x(P)$ is a solution of (1.5).

Probabilistically (1.4) describes a random walk where a step leads from P to a random point Q which has a uniform probability distribution in C_P. If A is a compact set in Γ, the probability that after n steps the moving point will be contained in A tends to zero as $n \to \infty$ (independently of the initial position). In other words, the moving point approaches in probability the (ordinary) boundary.

A bounded harmonic function is representable by means of the Poisson integral in terms of the boundary values $f(\theta)$ on the boundary of Γ. To our sojourn solutions (extremals) there correspond the solutions x where f assumes the value one on an arc α and 0 on the complement. Then $x(P)$ represents the probability that, starting from P, the random walk will asymptotically tend to the arc α. As explained above, a natural topological connection between the boundary and the interior is induced by K. For example, if the radii of C_P are chosen so as to decrease rapidly as P approaches the center of Γ and so that C_P excludes this center, then the center may become a boundary set in the topology induced by the kernel K.

2. **Notations and conventions.** We shall use *bold face* to denote *column vectors*. Thus $\mathbf{x}$ stands for a vector with components $x(1), x(2), \cdots$. Occasionally these components $x(i)$ will be defined only for $i \in C$, where C is a subset of the space E. For column vectors we use the conventional norm

$$\|\mathbf{x}\| = \sup_{i \in E} x(i). \tag{2.1}$$

Row vectors will not occur before §16. By $\mathbf{0}$ and $\mathbf{1}$ we denote the vectors all of whose components equal 0 or 1, respectively. An inequality such as $\mathbf{x} \leqq \mathbf{y}$ has the obvious meaning: $x(i) \leqq y(i)$ for all i. Thus the second inequality in (1.3) may be rewritten as $\Pi\mathbf{1} \leqq \mathbf{1}$.

DEFINITION. *We denote by* $\mathfrak{P}$, $\mathfrak{P}^*$, *and* $\mathfrak{P}_*$ *the aggregates of vectors satisfying*

$$0 \leqq z \leqq 1 \tag{2.2}$$

and, respectively, the conditions

$$\Pi z = z, \tag{2.3}$$

$$\Pi z \geqq z, \tag{2.4}$$

$$\Pi z \leqq z. \tag{2.5}$$

For any set $A \subset E$ we write

$$\Pi(i, A) = \sum_{j \in A} \Pi(i, j). \tag{2.6}$$

With this notation the second inequality in (1.1) reads $\Pi(i, E) \leqq 1$. In the *random walk interpretation* $\Pi(i, A)$ is the transition probability from the point i to the set A.

The restriction Π_A *of* Π *to a set* A *is the matrix defined by*

$$\Pi_A(i, j) = \Pi(i, j) \qquad \text{for } i, j \in A \tag{2.7}$$

and undefined for $i, j \in E - A$. Clearly, Π_A is again a sub-stochastic matrix and every theorem concerning Π applies equally to Π_A. For the probability interpretation cf. §5.

A matrix product $\Pi_A \mathbf{x}$ makes sense only when $\mathbf{x}$ is a vector defined on A. For convenience we shall use the same notation even if $\mathbf{x}$ is defined on the full space E. Thus *we write*

$$y = \Pi_A \mathbf{x} \tag{2.8}$$

to indicate that

$$y(i) = \sum_{j \in A} \Pi(i, j) x(j), \qquad \text{for } i \in A. \tag{2.9}$$

The components $x(i)$ and $y(i)$ for $i \in E - A$ may, but need not, be defined; equation (2.8) contains no statement concerning them.

3. **The basic lemmas.** We begin with the observation that to each $z \in \mathfrak{P}^*$ there exists a *smallest element* $\mathbf{a} \in \mathfrak{P}$ such that $\mathbf{a} \geqq z$. More precisely we have

LEMMA 3.1. *For each* $z \in \mathfrak{P}^*$ *the limit*

$$\mathbf{a} = \lim_{n \to \infty} \Pi^n z \tag{3.1}$$

exists. One has $\mathbf{a} \in \mathfrak{P}$ *and* $\mathbf{a} \geqq z$; *moreover,*

$$z \leqq \mathbf{a} \leqq \mathbf{x}, \tag{3.2}$$

for each $\mathbf{x} \in \mathfrak{P}$ *such that* $\mathbf{x} \geqq z$.

Proof. Put

$$a_0 = z, \qquad a_{n+1} = \Pi a_n = \Pi^n z. \tag{3.3}$$

Then $z \leqq a \leqq 1$ in consequence of (2.4) and (2.2). It follows by induction that $a_n \leqq a_{n+1} \leqq 1$ and therefore by bounded convergence $a_n \to a$; here $a = \Pi a \geqq z$. To conclude the proof suppose that $x \in \mathfrak{P}$ and $x \geqq z = a_0$. Then by induction $x \geqq a_n$ so that (3.2) holds.

The same argument leads to

LEMMA 3.2. *For each* $z \in \mathfrak{P}_*$ *the limit* (3.1) *exists, and* $a \in \mathfrak{P}$. *One has* $a \leqq z$. *Each* $x \in \mathfrak{P}$ *such that* $x \leqq z$ *satisfies*

$$x \leqq a \leqq z. \tag{3.4}$$

The possibility $a = 0$ *is not excluded.*

THEOREM[7] 3.1. *Each bounded solution of* $\Pi x = x$ *is a linear combination of two elements of* $\mathfrak{P}$.

Proof. Without loss of generality we assume $\|x\| \leqq 1$. For each i put $z(i) = |x(i)|$. Then $z \in \mathfrak{P}^*$. With a defined by Lemma (3.1) we have

$$x = \frac{1}{2}(a + x) - \frac{1}{2}(a - x) \tag{3.5}$$

where $(a \pm x)/2 \in \mathfrak{P}$.

4. **Lattice properties of** $\mathfrak{P}$. It is hardly necessary to point out that $\mathfrak{P}$ is a convex set; that is, if $u, v \in \mathfrak{P}$ and $p \geqq 0$, $q \geqq 0$, $p+q \leqq 1$, then $pu + qv \in \mathfrak{P}$. We now prove that $\mathfrak{P}$ is a vector lattice.

THEOREM. *Let* $x, y \in \mathfrak{P}$. *Then* $\mathfrak{P}$ *contains a uniquely determined least upper bound*[8] $x \cup y$ *and a greatest lower bound* $x \cap y$. (*The latter may be zero.*)

Proof. For each i put $z(i) = \max\{x(i), y(i)\}$. Then $z \in \mathfrak{P}^*$, and the vector a defined by (3.1) has the properties required of $x \cup y$. Similarly, putting $z(i) = \min\{x(i), y(i)\}$ the construction of Lemma 3.2 leads to $x \cap y$.

The following two lemmas are known in more general contexts (cf. e.g. [1]).

LEMMA 4.1. *If*

$$x + u = y + v = w, \tag{4.1}$$

then

$$x \cup y + u \cap v = w. \tag{4.2}$$

(7) An *unbounded* solution of $\Pi x = x$ is not necessarily the difference of two non-negative solutions.

(8) This means, $a = x \cup y$ is defined as an element of $\mathfrak{P}$ such that $a \geqq x$ and $a \geqq y$ and with the property that if $u \in \mathfrak{P}$, $u \geqq x$, $u \geqq y$ then $u \geqq a$. A similar definition applies to $x \cap y$.

In particular, one has identically

$$x + y = x \cup y + x \cap y. \tag{4.3}$$

Proof. Clearly $w - u \cap v \geqq w - u = x$ and equally $w - u \cap v \geqq y$. Therefore $w - u \cap v \geqq x \cup y$. On the other hand $w - x \cup y \leqq w - x = u$ and so $w - x \cup y \leqq u \cap v$.

LEMMA 4.2. *One has*

$$(x + y) \cap z \leqq (x \cap z) + (y \cap z). \tag{4.4}$$

If $x \cap y = 0$ *then the distributive law*

$$(x + y) \cap z = (x \cap z) + (y \cap z) \tag{4.5}$$

holds.

Proof. For each i we have

$$\min \{x(i) + y(i), z(i)\} \leqq \min \{x(i), z(i)\} + \min \{y(i), z(i)\}.$$

The assertion (4.4) is an immediate consequence of this and the definition of $x \cap y$ (cf. the construction of Lemma 3.2). Next suppose $x \cap y = 0$. From (4.3)

$$(x + y) \cap z = (x \cup y) \cap z \geqq (x \cap z) \cup (y \cap z) = (x \cap z) + (y \cap z) \tag{4.6}$$

where the last step consists in a repeated application of (4.3). The two inequalities (4.4) and (4.6) together imply (4.5).

Note on unbounded solutions. The operations $x \cup y$ *and* $x \cap y$ *are well defined for any two non-negative solutions of* (1.2). In fact, the construction of $x \cup y$ depends only on the existence of some solution u of (1.2) such that $u \geqq x$ and $u \geqq y$. Now $u = x + y$ is such a solution and may replace 1 in our construction. Note however the footnote to §3 which shows that $x \cup y$ need not exist for arbitrary unbounded solutions. The theory of unbounded *non-negative* solutions is developed in §14.

5. **Sojourn sets.** Since $1 \in \mathfrak{P}_*$ we may apply Lemma 3.2 to $z = 1$. We conclude that

$$s_E(i) = \lim_{n \to \infty} \Pi^n(i, E) \tag{5.1}$$

exists for each i, and $s_E \in \mathfrak{P}$. For each $x \in \mathfrak{P}$ we have $x \leqq 1$, and hence by induction $x \leqq \Pi^n 1 = \Pi^n(i, E)$. This proves

THEOREM 5.1. *The vector* s_E *defined by* (5.1) *is the maximal element of* $\mathfrak{P}$, *that is,* $s_E \in \mathfrak{P}$ *and* $x \leqq s_E$ *for each* $x \in \mathfrak{P}$.

Of course, it may happen that $s_E = 0$, in which case $\mathfrak{P}$ contains 0 as the only element. If Π is strictly stochastic, then $s_E = 1$ (and conversely).

Now let $A \subset E$ be an arbitrary set and apply the above argument to the restriction Π_A of Π to A. To $\mathbf{s}_E$ there corresponds the vector $\boldsymbol{\sigma}_A$ which is defined on A only by

$$(5.2)\qquad \sigma_A(i) = \lim_{n\to\infty} \overset{n}{\Pi}_A(i, A), \qquad i \in A,$$

and which satisfies

$$(5.3)\qquad \boldsymbol{\sigma}_A = \Pi_A \boldsymbol{\sigma}_A.$$

By Theorem 5.1 *this* $\boldsymbol{\sigma}_A$ *may be characterized as the maximal solution of* (5.3) subject to the condition $\mathbf{0} \leqq \boldsymbol{\sigma}_A \leqq \mathbf{1}$.

For convenience we extend the definition of $\boldsymbol{\sigma}_A$ throughout E by putting

$$(5.4)\qquad \sigma_A(i) = 0 \qquad \textit{for } i \in E - A.$$

Clearly $\boldsymbol{\sigma}_A \in \mathfrak{P}^*$, and applying Lemma 3.1 to $\mathbf{z} = \boldsymbol{\sigma}_A$ we see that

$$(5.5)\qquad s_A(i) = \lim_{n\to\infty} \sum_j \Pi^n(i, j)\sigma_A(j)$$

exists for all i. The vector

$$(5.6)\qquad \mathbf{s}_A = \lim_{n\to\infty} \Pi^n \boldsymbol{\sigma}_A$$

is uniquely defined as the smallest vector satisfying

$$(5.7)\qquad \mathbf{s}_A \in \mathfrak{P}, \qquad \mathbf{0} \leqq \boldsymbol{\sigma}_A \leqq \mathbf{s}_A \leqq \mathbf{1}.$$

When $A = E$ the definitions (5.1) and (5.6) agree, and $\boldsymbol{\sigma}_E = \mathbf{s}_E$.

DEFINITION 5.1. *The set A is called sojourn set if $\mathbf{s}_A \neq \mathbf{0}$ (or, what amounts to the same, if $\boldsymbol{\sigma}_A \neq \mathbf{0}$).*

In view of the fact that the limit in (5.2) is attained monotonically, we have the following obvious

CRITERION. *For A to be a sojourn set it is necessary and sufficient that there exist an $i \in A$ and an $\eta < 0$ such that*

$$(5.8)\qquad \overset{n}{\Pi}_A(i, A) > \eta$$

for all n.

PROBABILITY INTERPRETATION. Consider the random walk with stationary transition probabilities $\Pi(i, j)$ and interpret $1 - \Pi(i, E)$ as the probability that the random walk does not continue (terminates). Then $1 - \Pi^n(i, E)$ is the probability that the random walk with initial position i terminates at or before the nth step. From (5.1) we then conclude:

$s_E(i)$ *is the probability that the random walk with initial position i continues indefinitely (= does not terminate after finitely many steps).*

From the definition, $\overset{n}{\Pi}_A(i, A)$ *is the probability that, starting from $i \in A$,* *the random walk will continue for at least n steps without leaving the set A*. Accordingly, for $i \in A$ we see that:

$\sigma_A(i)$ *is the probability that starting from i the random walk will continue indefinitely without ever leaving the set A*.

Finally, in (5.5) the sum on the right equals the probability that from an arbitrary starting point i the nth step in our random walk leads to a position $j \in A$, and that from then on the random walk continues indefinitely without ever leaving A. (During the first n steps the random walk may lead in and out of A.) Therefore:

$s_A(i)$ *is the probability that from the starting point i the random walk will after finitely many steps lead into A and from then on continue indefinitely without ever leading out of A*.

DEFINITION 5.2. *Two sets A and B are equivalent if* $\mathbf{s}_A = \mathbf{s}_B$.

DEFINITION 5.3. *The aggregate of all vectors* $\mathbf{s}_A$ *and* $\mathbf{0}$ *will be denoted by* $\mathfrak{S}$. *The elements of* $\mathfrak{S}$ *will be referred to as sojourn solutions.*

LEMMA 5.1. *If A and B are nonoverlapping, then*

$$\mathbf{s}_A \cap \mathbf{s}_B = \mathbf{0}, \tag{5.9}$$

$$\mathbf{s}_A \cup \mathbf{s}_B = \mathbf{s}_A + \mathbf{s}_B \leqq \mathbf{s}_{A \cup B}. \tag{5.10}$$

Note. It will be shown in §9 that A and B can be replaced by equivalent sets such that the inequality sign in (5.10) becomes a strict equality sign.

Proof. Since, by definition, $\boldsymbol{\sigma}_A + \boldsymbol{\sigma}_B$ equals $\boldsymbol{\sigma}_A$ in A and $\boldsymbol{\sigma}_B$ in B we have

$$\mathbf{s}_A \cup \mathbf{s}_B \geqq \boldsymbol{\sigma}_A + \boldsymbol{\sigma}_B. \tag{5.11}$$

Premultiplying by Π^n and letting $n \to \infty$ we get

$$\mathbf{s}_A \cup \mathbf{s}_B \geqq \mathbf{s}_A + \mathbf{s}_B. \tag{5.12}$$

But the reversed inequality is trivially true, and therefore the equality sign holds. This proves the first half of (5.10). A comparison with (4.1) shows the truth of (5.9). Finally, the inequality $\mathbf{s}_{A \cup B} \geqq \mathbf{s}_A \cup \mathbf{s}_B$ follows directly from the definitions.

6. **Relativization.** Let A be a sojourn set, and $A \subset B \subset E$. In §5 we have considered A as a subset of E and Π_A as a submatrix of Π. Obviously the same considerations apply if we replace E and Π by B and Π_B, respectively. To the sojourn solution $\mathbf{s}_A$ there corresponds the vector $\overset{B}{\mathbf{s}}_A$ defined by

$$\begin{aligned} \overset{B}{s}_A(i) &= \lim_{n \to \infty} \sum_j \overset{n}{\Pi}_B(i, j) \sigma_A(j), && i \in B, \\ \overset{B}{s}_A(i) &= 0, && i \in E - B. \end{aligned} \tag{6.1}$$

With this notation we have $\mathbf{s}_A = \overset{E}{\mathbf{s}}_A$ and $\boldsymbol{\sigma}_A = \overset{A}{\mathbf{s}}_A$. Clearly

$$\boldsymbol{\sigma}_A \leqq \overset{B}{\mathbf{s}}_A \leqq \mathbf{s}_A. \tag{6.2}$$

Premultiply (6.2) by Π^n. The right side remains unchanged, and (5.6) leads to

LEMMA 6.1. *We have*

$$\mathbf{s}_A = \lim_{n\to\infty} \Pi^n \mathbf{s}_A^B. \tag{6.3}$$

We shall repeatedly use the following corollary.

LEMMA 6.2. *Let $A\subset B$ and $\mathbf{A}\subset B$. If $\mathbf{s}_A^B=\mathbf{s}_{\mathbf{A}}^B$, then also $\mathbf{s}_A=\mathbf{s}_{\mathbf{A}}$.*

That is, if A and $\mathbf{A}$ are equivalent relative to B, then they are equivalent.

7. Auxiliary lemmas.

LEMMA 7.1. *Let $z\in\mathfrak{P}_*$ and let A be an arbitrary set. For each n and each $i\in A$ we have*

$$z(i) \geqq \sum_{j\in A} \Pi_A^n(i, j)z(j) + \sum_{\nu=0}^{n-1} \sum_{j\in A} \sum_{k\in E-A} \Pi_A^\nu(i, j)\Pi(j, k)z(k). \tag{7.1}$$

If $z\in\mathfrak{P}$, then the equality sign holds in (7.1).

Proof. For $n=1$ the relation (7.1) reduces to $\boldsymbol{z}\geqq\Pi\boldsymbol{z}$. Assume (7.1) to hold for some n. Using $\boldsymbol{z}\geqq\Pi\boldsymbol{z}$ we get

$$\sum_{j\in A} \Pi_A^n(i, j)z(j) \geqq \sum_{j\in A} \Pi_A^{n+1}(i, j)z(j) + \sum_{j\in A} \sum_{k\in E-A} \Pi_A^n(i, j\Pi)(j, k)z(k). \tag{7.2}$$

Substituting this for the first term on the right in (7.1) we get the assertion (7.1) with n replaced by $n+1$. When $\boldsymbol{z}=\Pi\boldsymbol{z}$ then each of the above inequalities is replaced by an equality, and the lemma is proved.

Choosing in particular $\boldsymbol{z}=\mathbf{1}$ we get the

COROLLARY 1. *For $i\in A$*

$$\Pi_A^n(i, A) + \sum_{\nu=0}^{n-1} \sum_{j\in A} \Pi_A(i, j)\Pi(j, E - A) \leqq 1. \tag{7.3}$$

If Π is strictly stochastic, then the equality sign holds.

Letting $n\to\infty$ we get

COROLLARY 2. *For $i\in A$*

$$\sigma_A(i) + \sum_{\nu=0}^{\infty} \sum_{j\in A} \Pi_A(i, j)\Pi(j, E - A) \leqq 1. \tag{7.4}$$

PROBABILITY INTERPRETATION. The nth term in the outer sum equals the probability that, starting from i, the random walk remains for n steps inside A whereas the step number $n+1$ leads into $E-A$ (i.e. the *first passage* into

$E-A$ occurs at the $(n+1)$st step). The difference between the two sides in (7.4) is the probability that the random walk terminates after finitely many steps.

With each vector $\mathbf{x}$ there are associated its *contour sets*, that is, the sets of those i for which $x(i)$ exceeds a preassigned constant. Concerning these we prove

LEMMA 7.2. *Let* $\mathbf{x}$ *be a bounded solution of* $\Pi\mathbf{x}=\mathbf{x}$, *and* $\mathbf{x}\geqq 0$, $\|\mathbf{x}\|>0$. *For fixed* $0<\eta<\|\mathbf{x}\|$ *put*

$$X_\eta = \{i : x(i) > \|\mathbf{x}\| - \eta\}. \tag{7.5}$$

Then X_η *is a sojourn set. If*

$$\frac{\delta}{\eta} < \epsilon, \qquad \delta > 0, \tag{7.6}$$

then

$$\sigma_{X_\eta}(i) > 1 - \epsilon \qquad \textit{for } i \in X_\delta. \tag{7.7}$$

Proof. Applying Lemma 7.1 with $\mathbf{z}=\mathbf{x}$ and $A=X_\eta$ we get for $i\in X_\delta$

$$\|\mathbf{x}\|-\delta<x(i)\leqq\|\mathbf{x}\|\cdot\Pi^n_{X_\eta}(i, X_\eta)+\{\|\mathbf{x}\|-\eta\}\cdot\sum_{\nu=0}^{n-1}\sum_{j\in X_\eta}\Pi^\nu_{X_\eta}(i, j)\Pi(j, E-X_\eta). \tag{7.8}$$

Using (7.3) with $A=X_\eta$ this leads to

$$\|\mathbf{x}\| - \delta \leqq \|\mathbf{x}\|\cdot\Pi^n_{X_\eta}(i, X_\eta) + \{\|\mathbf{x}\| - \eta\}\cdot\{1 - \Pi^n_{X_\eta}(i, X_\eta), \tag{7.9}$$

or

$$\Pi^n_{X_\eta}(i, X_\eta) \geqq 1 - \frac{\delta}{\eta}\cdot \tag{7.10}$$

The assertion (7.7) now follows letting $n\to\infty$.

LEMMA 7.3. *For each sojourn set* $\|\mathbf{\sigma}_A\|=1$ and $\|\mathbf{s}_A\|=1$.

Proof. Since $\mathbf{s}_E$ is the maximal sojourn solution we conclude from (7.7) directly that $\|\mathbf{s}_E\|=1$. Applying this conclusion to the matrix Π_A instead of Π we see that $\|\mathbf{\sigma}_A\|=1$. Finally, $\mathbf{\sigma}_A\leqq\mathbf{s}_A\leqq 1$ and the lemma is proved.

LEMMA 7.4. *With the notations of Lemma* 7.2 *one has*

$$\mathbf{x} \geqq \{\|\mathbf{x}\| - \eta\}\cdot\mathbf{s}_{X_\eta}. \tag{7.11}$$

Proof. By definition $\mathbf{x}\geqq\{\|\mathbf{x}\|-\eta\}\cdot\mathbf{\sigma}_{X_\eta}$. The assertion follows on pre-multiplying this inequality by Π^n and letting $n\to\infty$.

8. **Nests of equivalent sojourn sets.** The next theorem supplements

Lemma 7.2 for the special case of sojourn solutions and is of prime importance for the sequel.

THEOREM(9) 8. *For any sojourn set A and* $0<\eta<1$ *put*

$$A_\eta = \{i \in A : s_A(i) > 1 - \eta\} \tag{8.1}$$

and

$$\mathbf{A}_\eta = \{i : \sigma_A(i) > 1 - \eta\}. \tag{8.2}$$

Then

$$\mathbf{s}_A = \mathbf{s}_{A_\eta} = \mathbf{s}_{\mathbf{A}_\eta}, \tag{8.3}$$

that is, A, A_η, $\mathbf{A}_\eta$ *are equivalent.*

Proof. When $A=E$ the sets defined by (8.1) and (8.2) are identical and the lemma reduces to the proposition

$$\mathbf{s}_E = \mathbf{s}_{E_\eta}. \tag{8.4}$$

We begin now by showing that (8.3) is more general than (8.4) in appearance only. First note that $\mathbf{A}_\eta \subset A_\eta \subset A$, so that (8.3) really reduces to $\mathbf{s}_A = \mathbf{s}_{\mathbf{A}_\eta}$. Now recall Lemma 6.2 according to which A and $\mathbf{A}_\eta$ are equivalent whenever they are equivalent relative to A (that is, if $\mathbf{s}^A_{\mathbf{A}_\eta} = \boldsymbol{\sigma}_A$). Now this statement differs from (8.4) only notationally, and so it suffices to prove (8.4).

For that purpose put

$$\mathbf{s}_E - \mathbf{s}_{E_\eta} = u \tag{8.5}$$

and suppose that $\|u\| > \alpha > 0$. The set

$$B = \{i : u(i) > \alpha\} \tag{8.6}$$

is a sojourn set, by Lemma 7.2. We now apply that Lemma to $\mathbf{x} = \mathbf{s}_E$. Then $X_\eta = E_\eta$ and by (7.7) we have

$$s_{E_\eta}(i) > 1 - \alpha \qquad \text{for } i \in E_\delta \tag{8.7}$$

provided $\delta < \alpha\eta$. For $i \in E_\delta \cap B$ we have both (8.7) and $u(i) > \alpha$, and therefore $s_E(i) > 1$ which is impossible. Thus $B \subset E - E_\delta$, and hence by the definition of E_δ

$$s_E(i) \leqq 1 - \delta \qquad \text{for } i \in B. \tag{8.8}$$

But $\mathbf{s}_E \geqq \mathbf{s}_B \geqq \boldsymbol{\sigma}_B$ and from (7.7) we see that $\sigma_B(i) > 1 - \delta$ for *some* $i \in B$. This contradicts (8.8), and so $u = 0$ as asserted.

An important implication of the last theorem is that each sojourn set A contains an *equivalent* sojourn set $\mathbf{A}$ such that

---

(9) Compare with Lemma 9.2 at the end of §9.

$$s_A(i) > 1 - \eta \qquad \text{for } i \in A. \tag{8.9}$$

When dealing with a particular sojourn solution $\mathbf{x} = \mathbf{s}_A$ we may therefore always choose the representative set so that (8.9) holds. In fact, it is necessary to do this if set theoretical operations on sojourn sets are to correspond to lattice operations on sojourn solutions. For example, (8.10) is *false* without the assumption that A satisfies (8.9). We therefore introduce the

DEFINITION. *A sojourn set A is called representative if* (8.9) *holds for some* $\eta > 0$.

As was already observed, we have

LEMMA 8.1. *Each sojourn set A contains an equivalent subset* A *which is representative.*

LEMMA 8.2. *Let A be representative, and* $A \subset B$. *Then*

$$\mathbf{s}_B = \mathbf{s}_A + \mathbf{s}_{B-A}. \tag{8.10}$$

Proof. We begin with the special case $B = E$. Put

$$\mathbf{s}_E - \mathbf{s}_A - \mathbf{s}_{E-A} = \boldsymbol{u}. \tag{8.11}$$

We propose to show that $\boldsymbol{u} = 0$. From (5.10) we have $\boldsymbol{u} \geqq 0$. Suppose $\|\boldsymbol{u}\| > 0$ and define three families of sets as follows:

$$\begin{aligned} A_\delta &= \{ i \in A : s_A(i) > 1 - \delta \}, \\ C_\delta &= \{ i \in E - A : s_{E-A}(i) > 1 - \delta \}, \\ U_\delta &= \{ i : u(i) \geqq \|\boldsymbol{u}\| - \delta \} \end{aligned} \tag{8.12}$$

(where C_δ may be empty). Choose $2\delta < \|\boldsymbol{u}\|$. At any point i common to A_δ and C_δ we would have $s_E(i) > 2 - 2\delta > 1$ which is impossible. Therefore A_δ and C_δ are nonoverlapping. The same argument shows that

$$A_\delta \cap C_\delta = 0, \qquad A_\delta \cap U_\delta = 0, \qquad C_\delta \cap U_\delta = 0. \tag{8.13}$$

It follows then from (5.10) that

$$\mathbf{s}_E \geqq \mathbf{s}_{A_\delta} + \mathbf{s}_{C_\delta} + \mathbf{s}_{U_\delta}. \tag{8.14}$$

In view of Theorem 8 this is the same as

$$\mathbf{s}_E \geqq \mathbf{s}_A + \mathbf{s}_{E-A} + \mathbf{s}_{U_\delta}. \tag{8.15}$$

Thus $\boldsymbol{u} \geqq \mathbf{s}_{U_\delta}$ and therefore $\|\boldsymbol{u}\| = 1$. This means that for $i \in A \cap U_\delta$ we have $s_E(i) \geqq 2 - \eta - \delta$ and therefore $A \cap U_\delta = 0$ provided $\delta < 1 - \eta$. With this choice of δ we see that $U_\delta \subset E - A$ and hence

$$\mathbf{s}_E \geqq \mathbf{s}_{E-A} + \boldsymbol{u} \geqq \mathbf{s}_{E-A} + \mathbf{s}_{U_\delta} \geqq 2\mathbf{s}_{U_\delta}. \tag{8.16}$$

But this is absurd since $\|\mathbf{s}_{U_\delta}\| = 1$ by Lemma 7.3. We conclude therefore finally that $\boldsymbol{u} = 0$ or $\mathbf{s}_E = \mathbf{s}_A + \mathbf{s}_{E-A}$.

Now consider the case of an arbitrary set $B \supset A$. Applying the last result to the matrix Π_B (instead of Π) we see that there exists a subset $\mathbf{A} \subset A$ such that

$$(8.17) \qquad \mathfrak{d}_B = \overset{B}{\mathbf{s}}_{\mathbf{A}} + \overset{B}{\mathbf{s}}_{B-\mathbf{A}}, \qquad \overset{B}{\mathbf{s}}_A = \overset{B}{\mathbf{s}}_{\mathbf{A}}$$

(see (6.1) for the notation). Lemma 6.1 shows at once that (8.17) is equivalent to

$$(8.18) \qquad \mathbf{s}_B = \mathbf{s}_A + \mathbf{s}_{B-\mathbf{A}}, \qquad \mathbf{s}_A = \mathbf{s}_{\mathbf{A}}.$$

Now $B-\mathbf{A} \supset B-A$, and the difference is contained in A. But for $i \in A$ we have $s_{B-\mathbf{A}}(i) \leqq 1 - s_{\mathbf{A}}(i) \leqq 1-\eta$ and hence $B-\mathbf{A}$ is equivalent to $B-A$.

9. Properties of sojourn solutions.

THEOREM 9.1 (CRITERION). *For an element $\mathbf{x} \in \mathfrak{P}$ to be a sojourn solution it is necessary and sufficient that*

$$(9.1) \qquad \mathbf{x} \cap (\mathbf{s}_E - \mathbf{x}) = \mathbf{0}.$$

Equivalently it is necessary and sufficient for any sojourn set C

$$(9.2) \qquad \mathbf{x} > t\mathbf{s}_C, \qquad t > 0 \quad \textit{implies} \quad \mathbf{x} \geqq \mathbf{s}_C.$$

Proof. (1) Let $\mathbf{x} = \mathbf{s}_A$. Then A may be supposed to be representative. By Lemma 8.2 we have in this case $\mathbf{s}_E - \mathbf{x} = \mathbf{s}_{E-A}$, and (9.1) is contained in (5.9). Thus (9.1) is necessary.

(2) Proof that (9.1) implies (9.2). In consequence of (9.1) we may apply (4.3) to obtain

$$(9.3) \qquad \mathbf{s}_C = \mathbf{s}_C \cap \{\mathbf{x} + (\mathbf{s}_E - \mathbf{x})\} = \mathbf{s}_C \cap \mathbf{x} + \mathbf{s}_C \cap (\mathbf{s}_E - \mathbf{x}).$$

Suppose now that $t\mathbf{s}_C \leqq \mathbf{x}$ where $t > 0$. By (9.1) the last term in (9.3) vanishes, and thus (9.3) reduces to $\mathbf{s}_C = \mathbf{s}_C \cap \mathbf{x}$ or $\mathbf{s}_C \leqq \mathbf{x}$.

(3) Suppose now that (9.2) holds. We have to show that $\mathbf{x} = \mathbf{s}_X$ for some sojourn set X.

Let $\mathbf{x} \in \mathfrak{P}$ and consider the set X_η defined in (7.5). Then (7.11) holds, and therefore $\mathbf{x} \geqq \mathbf{s}_{X_\eta}$ by (9.2). Thus

$$(9.4) \qquad \mathbf{x} - \mathbf{s}_{X_\eta} = \mathbf{u}, \qquad \mathbf{u} \in \mathfrak{P}.$$

Applying the preceding remark to $\mathbf{u}$ instead of $\mathbf{x}$ we conclude that if $\mathbf{u} \neq 0$ then there exists a sojourn set A such that $\mathbf{u} \geqq t\mathbf{s}_A$ with $t > 0$. It follows then from (9.4) and (9.2) that $\mathbf{x} \geqq \mathbf{s}_A$.

Let A_η be defined as in (8.1). For $i \in A_\eta$ we have $x(i) \geqq s_A(i) \geqq 1-\eta$ and thus $A_\eta \subset X_\eta$. But then from (9.4)

$$(9.5) \qquad \mathbf{x} \geqq \mathbf{s}_{A_\eta} + \mathbf{u} \geqq (1 + t)\mathbf{s}_{A_\eta}$$

or $\|\mathbf{x}\| \geqq 1 + t$ against assumption. Therefore $\mathbf{u} = 0$ and $\mathbf{x} = \mathbf{s}_{X_\eta}$.

THEOREM 9.2. *If A and B are sojourn sets, then*

$$(9.6) \qquad \mathbf{s}_{A\cap B} = \mathbf{s}_A \cap \mathbf{s}_B.$$

Warning. The formal analogue for $A \cup B$ is false.

Proof. Since obviously $\mathbf{s}_{A\cap B} \leqq \mathbf{s}_A \cap \mathbf{s}_B$ it suffices to prove that there exists a set $C \subset A \cap B$ such that $\mathbf{s}_C = \mathbf{s}_A \cap \mathbf{s}_B$. We are therefore permitted to replace A and B by equivalent subsets and to prove the relation (9.6) for them. This means that there is no loss of generality in supposing that both A and B are representative. Accordingly, we shall assume that

$$(9.7) \qquad s_A(i) > 3/4 \quad \textit{for} \quad i \in A, \qquad\qquad s_B(i) > 3/4 \quad \textit{for} \quad i \in B.$$

From the probability interpretation(10) of sojourn probabilities one sees that

$$(9.8) \qquad \mathbf{s}_{A\cup B} \leqq \mathbf{s}_{A\cap B} + \{\mathbf{s}_E - \mathbf{s}_A\} + \{\mathbf{s}_E - \mathbf{s}_B\}.$$

We see from this and (9.7) that $s_{A\cap B}(i) > 1/4$ when $i \in A \cap B$. Thus $A \cap B$ satisfies an inequality of the form (8.9) and we may apply Lemma 8.2 to obtain

$$(9.9) \qquad \mathbf{s}_A = \mathbf{s}_{A\cap B} + \mathbf{s}_{A-A\cap B}, \qquad \mathbf{s}_B = \mathbf{s}_{A\cap B} + \mathbf{s}_{B-A\cap B}.$$

From this we get

$$(9.10) \qquad \mathbf{s}_A \cap \mathbf{s}_B = \mathbf{s}_{A\cap B} + \mathbf{s}_{A-A\cap B} \cap \mathbf{s}_{A\cap B} + \mathbf{s}_{A\cap B} \cap \mathbf{s}_{B-A\cap B} + \mathbf{s}_{A-A\cap B} \cap \mathbf{s}_{B-A\cap B}$$

(the distributive law (4.5) is applicable in consequence of (5.9)). The last three terms in (9.10) vanish since the two sets involved are in each case non-overlapping (cf. (5.9)). Thus (9.10) reduces to (9.6).

THEOREM 9.3. *Let* $\mathbf{x}$, $\mathbf{y}$, *and* $\mathbf{x}_n$ *be sojourn solutions* (*elements of* $\mathfrak{S}$). *Then*

(a) $\mathbf{x} \cap \mathbf{y} \in \mathfrak{S}$,

(b) $\mathbf{y} - \mathbf{x} \in \mathfrak{S}$ *provided* $\mathbf{x} \leqq \mathbf{y}$.

(c) $\mathbf{x} \cup \mathbf{y} \in \mathfrak{S}$,

(d) $\mathbf{x} + \mathbf{y} \in \mathfrak{S}$ *provided* $\mathbf{x} \cap \mathbf{y} = 0$.

(e) *If either* $\mathbf{x}_n \downarrow u$ *or* $\mathbf{x}_n \uparrow u$ *then* $u \in \mathfrak{S}$.

Proof. Theorem 9.2 contains a statement which is stronger (and deeper) than (a). Also, (a) is an immediate consequence of the criterion (9.2). Similarly, (b) is contained in Lemma 8.2. To prove (c) note that by (4.2) we have

$$(9.11) \qquad \mathbf{x} \cup \mathbf{y} = \mathbf{s}_E - (\mathbf{s}_E - \mathbf{x}) \cap (\mathbf{s}_E - \mathbf{y}),$$

and the right side is a sojourn solution in consequence of (b) and (a). Next,

(10) For the reader unacquainted with (or distrusting) probability arguments we give a direct proof of (9.8). For the purpose of this proof we set $\Pi_A(i, j) = 0$ when $i \notin A$. A simple induction shows that $\Pi^n_{A\cup B}(i, A\cup B) \leqq \Pi^n_{A\cap B}(i, A\cap B) + \{1 - \Pi^n_A(i, A)\} + \{1 - \Pi^n_B(i, B)\}$ (note that the right side is $\geqq 1$ when $i \notin A \cap B$). Letting $n \to \infty$ we get $\sigma_{A\cup B} \leqq \sigma_{A\cap B} + (1 - \sigma_A) + (1 - \sigma_B)$. Premultiplying by Π^n and letting $n \to \infty$ we obtain (9.8).

(d) is but a special case of (a). Finally, if $\mathbf{x}_n \downarrow u$ then the criterion (9.2) shows trivially that $u \in \mathfrak{S}$. For the case $\mathbf{x}_n \uparrow u$ the proof goes by complementation as under (c).

The following theorem is listed for its intrinsic interest and will not be used in the sequel.

THEOREM 9.4. *Let A be representative. Then*

$$s_A(i) = \lim_{n\to\infty} \Pi^n(i, A). \tag{9.12}$$

Warning. For an arbitrary sojourn set (9.12) need not be true.

For the *random walk* with initial position i the relation (9.12) states that *the probability of finding the moving point at time n in the set A approaches $s_A(i)$ as $n\to\infty$.*

Proof. From the definition (5.5) we have

$$s_A(i) \leqq \liminf_{n\to\infty} \Pi^n(i, A), \tag{9.13}$$

and equally

$$s_{E-A}(i) \leqq \liminf_{n\to\infty} \Pi^n(i, E - A). \tag{9.14}$$

Now (8.10) applies, and so the left sides add to $s_E(i)$. Thus

$$s_A(i) + s_{E-A}(i) = \lim_{n\to\infty} \left\{\Pi^n(i, A) + \Pi^n(i, E - A)\right\}, \tag{9.15}$$

and the theorem follows trivially.

We conclude with a lemma which supplements Theorem 8 inasmuch as it associates with each sojourn solution $\mathbf{s}_A$ a *sequence of sets which is independent of the representative set A* and thus intrinsically connected with $\mathbf{s}_A$. However, this lemma will not be used in the sequel.

LEMMA 9.1. *Let A be a sojourn set and*

$$S_\eta = \left\{i : s_A(i) > 1 - \eta\right\}, \qquad 0 < \eta < 1. \tag{9.16}$$

Then S_η is equivalent to A.

Proof. Once more we are permitted to replace A by any equivalent set, in particular by the set A_η defined in (8.1). Now $A_\eta \subset S_\eta$ and Lemma 8.2 applies. Thus

$$\mathbf{s}_{S_\eta} = \mathbf{s}_{A_\eta} + \mathbf{s}_{S_\eta - A_\eta}. \tag{9.17}$$

However, for $i \in S_\eta$ we have

$$s_{S_\eta - A_\eta}(i) \leqq s_{E-A_\eta}(i) = s_E(i) - s_{A_\eta}(i) \leqq 1 - (1 - \eta) \tag{9.18}$$

and therefore $\|\boldsymbol{\sigma}_{S_\eta - A_\eta}\| \leqq \eta$. Accordingly $\mathbf{s}_{S_\eta - A_\eta} = \mathbf{0}$ in consequence of Lemma 7.3.

10. Sojourn solutions as extremals of $\mathfrak{P}$.

THEOREM 10. *In order that an element $\mathbf{x} \in \mathfrak{P}$ be a sojourn solution it is necessary and sufficient that the relations*

$$\mathbf{x} = t\mathbf{u} + (1-t)\mathbf{v}, \qquad 0 < t < 1, \quad \mathbf{u}, \mathbf{v} \in \mathfrak{P}, \tag{10.1}$$

imply $\mathbf{u} = \mathbf{v} = \mathbf{x}$.

This amounts to saying that the sojourn solutions coincide with the extremals of $\mathfrak{P}$ (cf. [2, Livre V, Ch. 2]).

Proof. (1) Suppose that $\mathbf{x} \in \mathfrak{S}$ and that (10.1) holds. We have

$$\mathbf{s}_E - \mathbf{x} = t \cdot (\mathbf{s}_E - \mathbf{u}) + (1-t) \cdot (\mathbf{s}_E - \mathbf{v}). \tag{10.2}$$

Since $t\mathbf{u} \leqq \mathbf{x}$ and $t(\mathbf{s}_E - \mathbf{u}) \leqq \mathbf{s}_E - \mathbf{x}$ we conclude from (9.1) that $\mathbf{u} \cap (\mathbf{s}_E - \mathbf{u}) = \mathbf{0}$ and so $\mathbf{u} \in \mathfrak{S}$ by Theorem 9.1. For the same reason $\mathbf{v} \in \mathfrak{S}$. Accordingly, (9.2) assures us that $\mathbf{x} \geqq \mathbf{u}$ and also $\mathbf{x} \geqq \mathbf{v}$. Therefore

$$(1-t)\mathbf{v} = \mathbf{x} - t\mathbf{u} \geqq (1-t)\mathbf{u}.$$

In this way $\mathbf{v} \geqq \mathbf{u}$ and, by the same argument, $\mathbf{u} \geqq \mathbf{v}$ and so $\mathbf{u} = \mathbf{v}$.

(2) If $\mathbf{x}$ is not a sojourn probability then by Theorem 9.1

$$\mathbf{x} \cap (\mathbf{s}_E - \mathbf{x}) = \mathbf{z} \qquad \mathbf{z} \neq \mathbf{0}. \tag{10.3}$$

We show that

$$\mathbf{x} = \frac{1}{2}(\mathbf{x} + \mathbf{z}) + \frac{1}{2}(\mathbf{x} - \mathbf{z}) \tag{10.4}$$

is a decomposition of the form (10.1) with $t = 1/2$. Clearly $\mathbf{0} \leqq \mathbf{x} - \mathbf{z} \leqq \mathbf{1}$, and so $\mathbf{x} - \mathbf{z} \in \mathfrak{P}$. Also, for each i we have $z(i) \leqq s_E(i) - x(i) \leqq 1 - x(i)$, so that $\mathbf{0} \leqq \mathbf{x} + \mathbf{z} \leqq \mathbf{1}$ and $\mathbf{x} + \mathbf{z} \in \mathfrak{P}$.

Note. The Krein-Milman theorem leads to an alternative proof of the *uniform approximation Theorem* 13.4.

11. Recurrent and transient sets. For the further development we require a few elementary facts about *partitioned matrices*. As was stated in the introduction, part of the results could be obtained from the ergodic theory of stochastic matrices. It is simpler and more natural to derive all required facts in one sweep purely analytically. The present method opens a new access to the ergodic theory. The following terminology will, perhaps, appear artificial, but it comes closest to established usage in probability theory.

DEFINITION 11. *A set $R \subset E$ is called indecomposable-recurrent if it is a sojourn set, but no proper subset of R is a sojourn set. A set is called recurrent if it is the union of indecomposable-recurrent sets. A point i is called recurrent if there exists a recurrent set R such that $i \in R$. A point which is not recurrent is transient.*

Warning. The set C_i which consists of the single point i obviously is recurrent if, and only if, $\Pi(i, i)=1$. Thus i can be recurrent without C_i being recurrent. It is hoped that this will cause less confusion than would the introduction of two new terms.

THEOREM 11.1. *In order that r be recurrent, it is necessary and sufficient that*[11]

$$s_E(r) = 1, \qquad s_{E-r}(r) = 0. \tag{11.1}$$

PROBABILITY INTERPRETATION. For a random walk starting at r the relation $s_E(r)=1$ attributes probability zero to the event that the random walk terminates after finitely many steps. On the other hand, $s_{E-r}(r)=0$ means that the probability of only finitely many returns to r is zero. Thus, according to Theorem 11.1, *the point r is recurrent if and only if there is probability one that, starting from r, the random walk leads infinitely often back to r*. Our definition therefore agrees with the definition used in probability theory.

Proof. (1) **Necessity.** Consider first the special case where E is indecomposable-recurrent. By Theorem 8 the set where $s_E(i)>1-\eta$ is a sojourn set, and since no proper subset of E is a sojourn set we have $s_E(i)=1$ for all i. Furthermore, $E-i$ is not a sojourn set, and therefore $\mathbf{s}_{E-i}=\mathbf{0}$ for each i. Thus (11.1) holds for all points. Moreover, since $\mathbf{s}_E=\mathbf{1}$ we see that Π *is strictly stochastic*.

Turning to the general case, let r be a fixed recurrent point. By definition there exists an *indecomposable-recurrent* set R such that $r\in R$. Applying our last conclusion to Π_R instead of Π we see first of all that Π_R *is strictly stochastic*. Therefore

$$\Pi_R(i, R) = 1, \qquad \Pi_R(i, E - R) = 0 \quad \textit{for} \quad i \in R. \tag{11.2}$$

This means that Π is of the form of a *partitioned* matrix

$$\Pi = \begin{pmatrix} \Pi_R & 0 \\ M & \Pi_{E-R} \end{pmatrix}. \tag{11.3}$$

When the whole space is indecomposable-recurrent we have shown that $s_E(j)=1$, $s_{E-i}(j)=0$ for all i, j. Using the notation of §6 we can write the corresponding equations for our subspace R in the form

$$\sigma_R(i) = 1, \qquad \overset{R}{s}_{R-i}(j) = 0, \qquad i \in R, j \in E. \tag{11.4}$$

Applying (6.3) we get at once

$$s_R(i) = 1, \qquad s_{R-i}(j) = 0, \qquad i \in R, \quad j \in E. \tag{11.5}$$

Thus $s_E(i)=1$. For $r\in R$ we have from (11.3) clearly $s_{E-R}(r)=0$, and thus

[11] We denote the complement of r by $E-r$ rather than by the more correct $E-\{r\}$.

$s_{E-i}(r) \leqq s_{R-i}(r) + s_{E-R}(r) = 0$ for each $i \in R$. This statement includes (11.1).

(2) **Sufficiency.** To each $r \in E$ define a set $R \subset E$ as follows: $k \in R$ if and only if

$$\Pi^n(r, k) > 0 \qquad \textit{for some } n \geqq 0. \tag{11.6}$$

For $k \in R$ and $j \in E - R$ we have obviously $\Pi(k, j) = 0$ and so $\Pi(k, E - R) = 0$ for each $k \in R$. This means that Π is again a partitioned matrix of the form (11.3), but in general Π_R will not be strictly stochastic.

We begin with the following simple remark (which will be used also in the proof of Theorem 11.3). *If $s_E(r) = 1$ for one $r \in R$, then Π_R is strictly stochastic* (so that (11.2) holds). In fact, for each n we have

$$1 = s_E(r) = \sum_{j \in R} \Pi^n(r, j) s_E(j); \tag{11.7}$$

the right side can equal unity only if each positive $\Pi^n(r, j)$ is multiplied by unity. From the definition (11.6) then $s_E(k) = 1$ for each $k \in R$ and thus $\Pi_R 1 = 1$.

Now to the proof that (11.1) is sufficient. If (11.1) holds for some r, then Π_R is strictly stochastic and so R is a sojourn set. We have to prove: *if $A \subset R$ is a sojourn set, then $A = R$.* Assume the contrary and choose $k \in R - A$. Then $R - k \supset A$ is a sojourn set. Choose n so that (11.6) holds. Then

$$\sigma_{R-k}(r) = \sum_j \Pi^n_{R-k}(r, j) \sigma_{R-k}(j) \leqq \Pi^n_{R-k}(r, R - k) \leqq 1 - \Pi^n(r, k) < 1. \tag{11.8}$$

By Theorem 8 the set B of all j such that $\sigma_{R-k}(j) > \sigma_{R-k}(r)$ is a sojourn set, and r is not in B. Therefore $R - r \supset B$ is a sojourn set, and hence there exists an $i \in R - r$ such that $s_{R-r}(i) > 0$. Now

$$s_{R-r}(r) = \sum_j \Pi^m(r, j) s_{R-r}(j) \geqq \Pi^m(r, i) s_{R-r}(i). \tag{11.9}$$

By an appropriate choice of m the last term can be made positive. Thus $s_{R-r}(r) > 0$ against assumption, and the proof is completed.

THEOREM 11.2. *The space E is partitioned into mutually nonoverlapping sets $T, R_1, R_2, \cdots$ as follows: T is the set of all transient points; each R_n is indecomposable-recurrent. Each submatrix R_n satisfies* (11.2). *One has*

$$s_{R_n}(i) = \delta_{n,m} \qquad \textit{for } i \in R_m \tag{11.10}$$

where $\delta_{n,m}$ is the Kronecker symbol. For each sojourn set A

$$\mathbf{s}_A = \mathbf{s}_{A \cap T} + \mathbf{s}_{A \cap R} \tag{11.11}$$

where $R = \cup R_n$ is the set of recurrent points. Furthermore

$$s_{A \cap T}(i) = 0 \qquad \textit{for } i \in R; \tag{11.12}$$

$$\mathbf{s}_{A \cap T} = \Pi_T \mathbf{s}_{A \cap T}, \qquad \mathbf{s}_{A \cap R} = \sum_{R_n \subset A} \mathbf{s}_{R_n}. \tag{11.13}$$

Proof. In the sufficiency proof to Theorem 11.1 we have constructed the set R which is obviously the *unique* indecomposable-recurrent set containing the recurrent point r. It follows that no two indecomposable-recurrent sets can overlap. It is then clear that Π is again of the form (11.3), where $E-R=T$ and Π_R is completely partitioned

$$\Pi_R = \begin{pmatrix} R_1 & 0 & \cdots \\ 0 & R_2 & \cdots \\ \cdot & \cdot & \cdot\ \cdot\ \cdot \end{pmatrix}. \tag{11.14}$$

The remaining assertions are immediate consequences of this and the definitions.

The main implication of Theorem 11.2 is that in the study of a sojourn set A we need only worry about the transient part $A\cap T$. Since $\mathbf{s}_{A\cap T}$ vanishes on R we are in substance dealing with the submatrix Π_T only. In other words, *we have reduced our problem to the case where all points are transient.* (Each indecomposable-recurrent class contributes one single element to $\mathfrak{P}$.)

We proceed to the most important

THEOREM 11.3. *Let the sojourn set A be transient. Then there exists a sequence of sets A_n equivalent to A such that*

$$\mathbf{s}_{A_n} = \mathbf{s}_A, \qquad A \supset A_1 \supset A_2 \cdots, \qquad A_n \to 0. \tag{11.15}$$

COROLLARY. *The complement of any finite set in A is equivalent to A.*

Proof. It suffices to prove that A and $A-r$ are equivalent for each r. If $\sigma_A(r)<1$ then this statement is amply contained in Theorem 8. Accordingly, assume $\sigma_A(r)=1$. Define the set R as above by the property (11.6). As has been remarked in the sufficiency proof of Theorem 11.1, the relation $\sigma_A(r)=1$ implies that Π_R is *strictly stochastic*, so that Π is of the form (11.3). Also $A\supset R$ and we have only to prove that $R-r$ is equivalent to R.

Since Π_R is strictly stochastic, R is a sojourn set. It contains a sojourn set B as a proper subset, for otherwise r would be recurrent. By Lemma 8.1 we can choose B so that $\mathbf{s}_R=\mathbf{s}_B+\mathbf{s}_{R-B}$. If $r\in R-B$ then $R-r\supset B$ is a sojourn set and the statement is proved. If $r\in B$ then $\sigma_B(r)<1$. In fact, choose $k\in R-B$ and choose n so that (11.6) holds. Then $\sigma_B(r)\leqq 1-\Pi^n(r,k)<1$. By Theorem 8 the set B' of all $i\in B$ such that $\sigma_B(i)>\sigma_B(r)$ is a sojourn set equivalent to B. Therefore $B-r$ is squivalent to B. But then

$$\mathbf{s}_R = \mathbf{s}_{B-r} + \mathbf{s}_{R-B} \leqq \mathbf{s}_{R-r} \leqq \mathbf{s}_R \tag{11.16}$$

and so R and $R-r$ are equivalent.

12. The discrete part of the boundary.

DEFINITION 12.1. *A sojourn solution $\mathbf{s}_A\neq 0$ corresponding to a transient*[12]

[12] The restriction to transient sets is introduced only to avoid clumsy formulations and trivialities.

set A is called minimal if $\mathbf{s}_B \leqq \mathbf{s}_A$ implies that $\mathbf{s}_A = \mathbf{s}_B$ or $\mathbf{s}_B = \mathbf{0}$. A sojourn solution $\mathbf{s}_A$ is continuous if there exists no minimal sojourn solution $\mathbf{s}_B \leqq \mathbf{s}_A$.

Clearly if $\mathbf{s}_A$ is minimal and $\mathbf{x} \leqq \mathbf{s}_A$ (where $\mathbf{x} \in \mathfrak{P}$), then $\mathbf{x} = t\mathbf{s}_A$.

LEMMA 12.1. *There exist at most denumerably many minimal sojourn solutions $\mathbf{s}^{(1)}, \mathbf{s}^{(2)}, \cdots$.*

Proof. For two distinct minimal solutions one has $\mathbf{s}^{(i)} \cap \mathbf{s}^{(k)} = \mathbf{0}$, by the very definition. Theorem 8 shows that we may choose sojourn sets S_n so that

$$\mathbf{s}_{S_n} = \mathbf{s}^{(n)} \tag{12.1}$$

and that the S_n are mutually nonoverlapping, in fact so that $s_{S_n}(i) > 1 - \epsilon_i$ for $i \in S_n$.

DEFINITION 12.2. *We enlarge the space E by adding for each $\mathbf{s}^{(n)}$ a new point γ_n. In the set $E + \Gamma$ thus obtained we introduce a topology as follows. The set Ω is open if for each $\gamma_n \in \Omega$ there exists a sojourn set $S_n \subset \Omega$ such that* (12.1) *holds.*

According to this definition each subset of E is open.

THEOREM 12.1. *With this topology $E + \Gamma$ is a Hausdorff space. Each of the subsets E and Γ in itself has the discrete topology*(13).

Proof. If $\mathbf{s}_A = \mathbf{s}_B = \mathbf{s}^{(n)}$ then also $\mathbf{s}_{A \cap B} = \mathbf{s}^{(n)}$ by Theorem 9.2. Thus the union of open sets and the intersection of finitely many open sets are again open. That γ_n and γ_m $(n \neq m)$ have nonoverlapping neighborhoods was shown above in the proof to Lemma 12.1. Clearly each point $i \in E$ represents an open set, and each γ is relatively open in Γ.

THEOREM 12.2. *In order that*

$$\mathbf{x} = \sum \alpha_n \mathbf{s}^{(n)} \tag{12.2}$$

be an element of $\mathfrak{P}$ it is necessary and sufficient that $0 \leqq \alpha_n \leqq 1$. Then

$$x(i) \to \alpha_n \quad \text{as} \quad i \to \gamma_n, \qquad i \in E. \tag{12.3}$$

If $\mathfrak{P}$ contains no continuous sojourn solutions(14) *and all points of E are transient then each $\mathbf{x} \in \mathfrak{P}$ is of the form* (12.2).

Proof. If $0 \leqq \alpha_n \leqq 1$ then $\mathbf{x} \in \mathfrak{P}$. When all the α_n are bounded, (12.2) represents a bounded solution of $\Pi \mathbf{x} = \mathbf{x}$. Relation (12.3) follows then from the fact that there exists a neighborhood of γ_n in which $s^{(n)}(i) \geqq 1 - \epsilon$, and hence

(13) It will be noticed that the recurrent part of E in no way influences the boundary. For a better understanding consider the case where E is indecomposable-recurrent, that is, where E is the only sojourn set. Then $\mathfrak{S}$ contains the unique element $\mathbf{1}$. If one added a boundary point γ, it would have no neighborhood except the whole space, and thus we would not have a Hausdorff space.

(14) By Theorem 11.2 each indecomposable-recurrent set in E contributes exactly one trivial sojourn solution, and these should be added to (12.2) in the general case.

$s_E(i) - s^{(n)}(i) \leqq \epsilon$. The necessity of the condition $0 \leqq \alpha_n \leqq 1$ is a trivial consequence of this. Finally, given an arbitrary $\mathbf{x} \in \mathfrak{P}$, we have

$$\alpha_n \mathbf{s}^{(n)} = \mathbf{s}^{(n)} \cap \mathbf{x}. \tag{12.4}$$

With this definition of α_n the right side of (12.2) is $\leqq \mathbf{x}$, and clearly the difference is a continuous solution.

Once the boundary is defined we may say that $s^{(n)}(i)$ *is the probability that, starting from i, the random walk will asymptotically approach the boundary point* γ_n.

13. **The maximal ideal space.** We wish to define the boundary so that the sojourn solutions $\mathbf{s} \in \mathfrak{S}$ will stand in one-to-one correspondence with certain sets of boundary points. In anticipation of the final outcome these sets may be described as sets of positive measure or capacity([15]). For each *continuous* sojourn solution (Definition 12.1) $\mathbf{s} \in \mathfrak{S}$ there exist null-sequences such that $\mathbf{s} \geqq \mathbf{s}_1 \geqq \cdots$, $\mathbf{s}_n \to 0$, and such a sequence corresponds to a monotone sequence of sets of the boundary. The intersection may be an arbitrary set of measure zero, and not necessarily a point. It seems therefore hopeless to define points of the boundary directly in terms of sequences of sojourn solutions. Instead, we adapt to our purposes ideas widely used in representation theory.

We recall (cf. [1]) that a set J of sojourn solutions is a *lattice ideal*([16]) in $\mathfrak{S}$ if the following is true: whenever $\boldsymbol{x} \in J$ and $\boldsymbol{y} \in J$ then also $\boldsymbol{x} \cup \boldsymbol{y} \in J$ and $\boldsymbol{z} \in J$ where $\boldsymbol{z} \in \mathfrak{S}$ and $\boldsymbol{z} \leqq \mathbf{x}$. An ideal is *maximal* if $\mathfrak{S}$ is the only ideal containing J as a proper subset, and $J \neq \mathfrak{S}$. There exist maximal ideals whenever $\mathfrak{S}$ contains more than one element.

DEFINITION 13. *Suppose that all points of E are transient*([17]). *Let $\mathfrak{B}$ be the set of maximal ideals. In the space $E + \mathfrak{B}$ we introduce a topology as follows. A set $\Omega \subset E + \mathfrak{B}$ is open if to each maximal ideal $\omega \in \Omega$ there corresponds an element $\mathbf{s}_A \in \mathfrak{S}$ such that $\mathbf{s}_A \notin \omega$ and $A \subset \Omega$, and moreover Ω contains each maximal ideal β such that $\mathbf{s}_{E-A}$.*

Before proceeding it is necessary to show that the discrete boundary introduced in §12 is really part of the boundary $\mathfrak{B}$ as defined in the last definition and that there is no contradiction between the two definitions of open sets.

([15]) Formula (12.2) expresses the value $x(i)$ as an integral over the boundary values α_n with respect to the measure which attaches weight $s^{(n)}(i)$ to the point γ_n. This is a representation of the Green-function type. An abstract representation of this type is possible also in the most general case, using the Krein-Milman theorem. For this purpose it is not necessary to introduce a boundary, but only a measure defined on $\mathfrak{S}$. See, for example, the excellent representation in Choquet [3, Chapter 7].

([16]) For our purposes it would be more natural to use dual ideals instead of ideals.

([17]) Alternatively, replace in the sequel $\mathfrak{S}$ by the sublattice of sojourn solutions corresponding to transient sets, that is, of sojourn solutions such that $\mathbf{s}_A = \mathbf{s}_{A_n}$ where $A_n \downarrow 0$. Cf. the footnote to Theorem 12.1.

LEMMA 13.1. *The two definitions* 13 *and* 12.2 *are mutually consistent.*

Proof. Let s_A be a *minimal* sojourn solution and define J as the set of all elements $x \in \mathfrak{S}$ such that $x \cap s_A = 0$. It is obvious that $s_A \notin J$, that J is a maximal ideal, and that J but no other maximal ideal contains $s_{E-A} = s_E - s_A$. It follows then that the set $A + J$ is open according to either of the definitions.

For the proof that $E + \mathfrak{B}$ is a Hausdorff space we require two lemmas.

LEMMA 13.2. *For any maximal ideal either* $s_A \in J$ *or* $s_{E-A} \in J$.

Proof. Suppose that $s_A \notin J$ and define I as the set of all elements of the form $x \cup y$, where $x \in J$ and $y \leqq s_A$. It is clear that I is an ideal and that $J \subset I$. As J is maximal we have $I = \mathfrak{S}$, and therefore $s_{E-A} \in I$. This means that $s_{E-A} = x \cup y$ with $x \in J$ and $y \leqq s_A$. Now $y \cap s_{E-A} = (x \cup y) \cap y = y$ or $y \leqq s_{E-A}$. Therefore $y \leqq s_A \cap s_{E-A} = 0$ or $s_{E-A} = x \in J$ as asserted.

LEMMA 13.3. *Let J be a maximal ideal and suppose that* $s_A \notin J$ *and* $s_B \notin J$. *Then* $s_{A \cap B} \notin J$.

Proof. In accordance with Lemma 8.2 we may assume without loss of generality that the sets A and B have been chosen so that

$$s_E = s_A + s_{E-A} = s_B + s_{E-B}.$$

Then, by Lemma 4.1

$$s_A \cap s_B = s_E - s_{E-A} \cup s_{E-B}. \tag{13.1}$$

Now in consequence of the preceding lemma both s_{E-A} and s_{E-B} are in J, and therefore $u = s_{E-A} \cup s_{E-B} \in J$. Again, $s_E = u \cup (s_E - u)$, and therefore $u \in J$ implies that $s_E - u \notin J$. A recall of Theorem 9.2 now completes the proof.

THEOREM 13.1. *The space* $E + \mathfrak{B}$ *of Definition* 13 *is a Hausdorff space.*

Proof. The union of open sets is obviously open. We prove that the intersection of two open sets Ω_1 and Ω_2 is open. Every subset of E is open and therefore there is nothing to be proved in case $\Omega_1 \cap \Omega_2 \subset E$. Suppose then that there exists a maximal ideal $\omega \in \Omega_1 \cap \Omega_2$, and let $A_i \subset \Omega_i$ be a set as described in the Definition 13. Then $s_{A_1} \notin \omega$ and $s_{A_2} \notin \omega$, and so $s_{A_1 \cap A_2} \notin \omega$ by Lemma 13.3. Therefore $s_{A_1 \cap A_2} \neq 0$, and $A_1 \cap A_2$ is a sojourn set contained in $\Omega_1 \cap \Omega_2$. Finally, if β is a maximal ideal such that $s_{E-A_1 \cap A_2} \in \beta$, then a fortiori $s_{E-A_1} \in \beta$ and $s_{E-A_2} \in \beta$, so that $\beta \in \Omega_1 \cap \Omega_2$. This proves that $\Omega_1 \cap \Omega_2$ is open.

To verify the separation postulate, let α and β be two distinct maximal ideals. Then there exists a sojourn solution $s_A \in \alpha$ such that $s_A \notin \beta$. By Lemma 13.2 we have in this case $s_{E-A} \in \beta$. Again, the set A is determined only up to an equivalence class and may be chosen so that $s_E = s_A + s_{E-A} = s_A \cup s_{E-A}$ (Lemma 8.2). Then no maximal ideal can contain both s_A and s_{E-A}. Thus if we put $\Omega_1 = (E - A) +$all maximal ideals containing s_A, and similarly $\Omega_2 = A$ $+$all maximal ideals containing s_{E-A}, then Ω_1 and Ω_2 are nonoverlapping open sets such that $\alpha \in \Omega_1$ and $\beta \in \Omega_2$. If α is maximal ideal and $i \in E$, then i is its own open neighborhood and α has a neighborhood excluding i by virtue of Theorem 11.3([18]).

THEOREM 13.2. *Let* $\mathbf{s}_A \in \mathfrak{S}$, *and let* ω *be a boundary point (maximal ideal). Then, as* $i \to \omega$ *(where* $i \in E$*)*

$$s_A(i) \to \begin{matrix} 1 & \text{if } \mathbf{s}_A \notin \omega \\ 0 & \text{if } \mathbf{s}_A \in \omega. \end{matrix} \tag{13.2}$$

Proof. Among the equivalent sets defining $\mathbf{s}_A$ choose A so that (8.9) holds. If $\mathbf{s}_A \notin \omega$ choose as a neighborhood of ω the set Ω_2 defined in the preceding proof. Then $s_A(i) > 1 - \eta$ for $i \in \Omega_2 \cap E$, and this proves the first relation in (13.2). The second follows on replacing A by $E - A$.

We see thus that each sojourn solution $\mathbf{s}_A$ can be extended to a *continuous function* on the whole space $E + \mathfrak{B}$. On the boundary such a function assumes only the values 0 and 1. Finite linear combinations of sojourn solutions induce on the boundary $\mathfrak{B}$ continuous functions assuming only finitely many values. We call such functions *step-functions*. The last theorem is now supplemented by

THEOREM 13.3. *Each* $\mathbf{x} \in \mathfrak{P}$ *has continuous boundary values.*

This is an immediate consequence of:

THEOREM 13.4. *Each* $\mathbf{x} \in \mathfrak{P}$ *can be approximated uniformly in E by finite linear combinations of sojourn solutions.*

In other words: in order that a function $f(\beta)$ *defined on* $\mathfrak{B}$ *represent boundary values of some* $\mathbf{x} \in \mathfrak{P}$ *it is necessary and sufficient that* $0 \leqq f \leqq 1$ *and that f be the uniform limit of step functions.*

The proof of Theorem 13.4 will be based on the following lemma which will be used again for the proof of Theorem 15.1. With a view to this latter application the lemma is formulated so as to cover also the case of *unbounded solutions.*

LEMMA 13.4. *Let* $\mathbf{x} \geqq \mathbf{0}$ *be a solution of* $\Pi \mathbf{x} = \mathbf{x}$ *and put* $X = \{i: x(i) > \eta\}$ *where* η *is a positive constant. Then*

$$\mathbf{x} \geqq \eta \mathbf{s}_X. \tag{13.3}$$

Moreover, letting $\mathbf{s}_A = \mathbf{s}_E - \mathbf{s}_X$, *one has*

$$\mathbf{x} \cap \mathbf{s}_A \leqq \eta \mathbf{s}_A. \tag{13.4}$$

NOTE. If $\mathbf{x}$ is bounded and $\eta < \|\mathbf{x}\|$, then X is a sojourn set and $\mathbf{s}_X \neq \mathbf{0}$. For an unbounded $\mathbf{x}$ it may happen that $\mathbf{s}_X = \mathbf{0}$ for all η,

([18]) Note that sojourn solutions corresponding to recurrent sets have been excluded only to establish this point.

Proof. Clearly $\mathbf{x} \geqq \eta \delta_X$. Premultiplying by Π^n and letting $n \to \infty$ we obtain (13.3).

Next put $\mathbf{x} \cap \mathbf{s}_A = \mathbf{y}$ and suppose for the moment that $\|\mathbf{y}\| > \eta$. By Lemma 7.2 the set $Y = \{i: y(i) > \eta\}$ is a sojourn set and applying (13.3) we see that $\mathbf{y} \geqq \eta \mathbf{s}_Y$. Now $\mathbf{y} \leqq \mathbf{s}_A$ and therefore $\mathbf{s}_Y \leqq \mathbf{s}_A$. This, however, is impossible, for $\mathbf{y} \leqq \mathbf{x}$ implies that $Y \subset X$, whence $\mathbf{s}_Y \leqq \mathbf{s}_X$ and therefore $\mathbf{s}_Y \leqq \mathbf{s}_A \cap \mathbf{s}_X = \mathbf{0}$. It follows that $\|\mathbf{y}\| \leqq \eta$ and hence $\mathbf{y} \leqq \eta \cdot \mathbf{1}$. Premultiplying by Π^n and letting $n \to \infty$ we get $\mathbf{y} \leqq \eta \mathbf{s}_E$, that is, $\mathbf{x} \cap \mathbf{s}_A \leqq \eta \cdot \mathbf{s}_E$. This is the same as (13.4), and the lemma is proved.

Proof of Theorem 13.4. Without loss of generality assume that $\|\mathbf{x}\| = 1$. Choose $\eta = 1/2$ and define the sets X and A as in the proof of Lemma 13.4. By this lemma

$$\frac{1}{2}\mathbf{s}_X \leqq \mathbf{x} \leqq \mathbf{s}_X + \frac{1}{2}\mathbf{s}_A \tag{13.5}$$

or

$$\mathbf{0} \leqq \mathbf{x} - \frac{1}{2}\mathbf{s}_X \leqq \frac{1}{2}\mathbf{s}_E. \tag{13.6}$$

Thus $\mathbf{x} - 2^{-1}\mathbf{s}_X \in \mathfrak{P}$ and $\|\mathbf{x} - 2^{-1}\mathbf{s}_X\| \leqq 1/2$. Applying the same procedure to $\mathbf{x} - 2^{-1}\mathbf{s}_X$ we get a linear combination $L = 2^{-1}\mathbf{s}_X + \alpha \mathbf{s}_Y$ such that $\mathbf{x} - L \in \mathfrak{P}$, and $\|\mathbf{x} - L\| \leqq 1/4$, etc.

14. **Unbounded solutions. Isomorphisms.** It has been remarked at the end of §4 that *non-negative* unbounded solutions of $\Pi \mathbf{x} = \mathbf{x}$ enjoy lattice properties similar to those of bounded solutions. We shall now outline a new approach to the theory of unbounded solutions which has analytical advantages and important probabilistic implications. It has a close analogue in a familiar transformation of the Sturm-Liouville differential equations, although this analogy is hidden by the altogether different formalism. For simplicity of formulations we shall consider only strictly stochastic matrices, that is, we assume $\Pi \mathbf{1} = \mathbf{1}$.

Let $\mathbf{z}$ be a strictly positive[19] solution of $\Pi \mathbf{z} = \mathbf{z}$ and define a matrix Π' by

$$\Pi'(i, j) = \Pi(i, j)\frac{z(j)}{z(i)} \cdot \tag{14.1}$$

Clearly Π' is again strictly stochastic and for its powers one has

$$\Pi'^n(i, j) = \Pi^n(i, j)\frac{z(j)}{z(i)} \cdot \tag{14.2}$$

(19) The restriction to strictly positive solutions is introduced only to simplify formulations and represents no serious loss of generality. For, if $\mathbf{z} \geqq 0$ is a solution which is not strictly positive, let R be the set of all i such that $z(i) = 0$. It is obvious that Π is *partitioned* in the form (11.3). All the matrices similar to Π in the sense of Definition 14 will be of the same partitioned form and there will be a one-to-one correspondence between the solutions of $\Pi \mathbf{x} = \mathbf{x}$ and $\Pi' \mathbf{x}' = \mathbf{x}'$ which vanish on R.

If $\mathbf{x}$ is any non-negative solution of $\Pi\mathbf{x}=\mathbf{x}$, then the vector $\mathbf{x}$ defined by

$$x'(i) = \frac{x(i)}{z(i)} \tag{14.3}$$

is a non-negative solution of $\Pi'\mathbf{x}'=\mathbf{x}'$.

NOTATION. *We denote by* $\mathfrak{M}$ *the aggregate of all strictly positive solutions of* $\Pi\mathbf{x}=\mathbf{x}$, *and by* $\mathfrak{M}'$ *the corresponding set for* Π'.

DEFINITION 14. *Two strictly stochastic matrices* Π *and* Π' *will be called similar if they stand in the relationship* (14.1), *where* $\mathbf{z}\in\mathfrak{M}$. *The vector* $\mathbf{x}'$ *defined in* (14.3) *will be called the (canonical) image of* $\mathbf{x}$.

We have now the obvious but important

THEOREM 14.1. *The similarity is a symmetric and transitive relationship. The transformation* (14.3) *is a one-to-one mapping of* $\mathfrak{M}$ *onto* $\mathfrak{M}'$ *which is a lattice isomorphism, that is, for* $\mathbf{x}, \mathbf{y}\in\mathfrak{M}$

$$(\mathbf{x}\cap\mathbf{y})' = \mathbf{x}'\cap\mathbf{y}', \qquad (\mathbf{x}\cup\mathbf{y})' = \mathbf{x}'\cup\mathbf{y}'. \tag{14.4}$$

In particular, the vector $\mathbf{z}$ itself is mapped into $\mathbf{z}'=\mathbf{1}$. In this way an arbitrary positive solution of $\Pi\mathbf{z}=\mathbf{z}$ can be made to play the rôle of the unit solution. It is possible to introduce notions of relative boundedness, relative sojourn solutions, etc. However, we can proceed in a more direct manner.

THEOREM 14.2. *Let both* $\mathbf{z}$ *and* $\mathbf{x}$ *be sojourn solutions. Then the image* $\mathbf{x}'$ *is a sojourn solution if* $\mathbf{x}\leqq\mathbf{z}$, *and is unbounded otherwise.*

Proof. To begin with, let $\mathbf{x}\leqq\mathbf{z}$. From (9.1) we have $\mathbf{x}\cap(\mathbf{z}-\mathbf{x})=\mathbf{0}$, and therefore $\mathbf{x}'\cap(\mathbf{1}-\mathbf{x}')=\mathbf{0}$. By Theorem 9.1 this is equivalent to the assertion that $\mathbf{x}$ is a sojourn solution. Next, suppose that $\mathbf{x}\leqq\mathbf{z}$ does not hold. Then there exists a nonzero sojourn $\mathbf{y}\leqq\mathbf{1}-\mathbf{z}$ such that $\mathbf{x}\geqq\mathbf{y}$. By Theorem 8 there exists a sojourn set Y such that $y(i)>1-\epsilon$ for $i\in Y$. Now $x'(i)>\epsilon^{-1}(1-\epsilon)$ for $i\in Y$ and $\mathbf{x}'$ is therefore unbounded.

We see thus that when $\mathbf{z}$ is a sojourn solution, *the boundary of E defined by the matrix* Π' *consists of a part of the boundary defined by* Π. On the other hand, if $\mathbf{z}$ is not a sojourn solution then the sojourn solutions of $\Pi'\mathbf{x}'=\mathbf{x}'$ need not be images of the sojourn solutions of $\Pi\mathbf{x}=\mathbf{x}$. In the next section it will be shown that nevertheless there exists a simple connection between the boundaries of all similar stochastic matrices.

Probability interpretation. For simplicity let us begin with the case where $\mathbf{z}$ is a *minimal* sojourn solution (Definition 12.1). To $\mathbf{z}$ there corresponds an isolated boundary point γ and, for the random walk defined by Π, we know that $z(i)$ represents the probability of an ultimate asymptotic approach to γ if the initial position is i. It is then clear that $\Pi'(i, j)$, as defined in (14.1) represents *the conditional one-step transition probability from i to j evaluated* *on the hypothesis of an ultimate asymptotic approach to* γ. In statistical terminology: out of a sample of mutually independent random walks obeying Π we consider the subpopulation of those which ultimately land at γ. *The process defined in this way is the same as a random walk with transition probabilities* Π'. It is now clear that in this new process the boundary point γ will be approached with probability one, so that γ must be the *unique boundary point for* Π'.

Next consider the case where $z = p_1 z_1 + p_2 z_2$, where the z_i are minimal sojourn solutions and $p_i > 0$. If $p_1 = p_2 = 1$, then z is again a sojourn solution, and the above probabilistic interpretation requires only a slight rephrasing. In particular, the boundary of Π' will now consist of the two boundary points γ_1 and γ_2 corresponding to z_1 and z_2. If $p_1 \neq p_2$ different weights are attached to the two boundary points: if a particle ultimately approaches γ_i it has probability $p_i/(p_1+p_2)$ to belong to our subpopulation defining Π'.

Obviously a similar probabilistic interpretation can be given in the most general case. Using the notions of boundary and real valued functions on this boundary our description requires only a trivial rephrasing.

We conclude this section by two theorems which, though of considerable interest will not be used in the sequel. The second illustrates the power of the present method.

THEOREM 14.3. *If the point* i *is transient [recurrent] for the matrix* Π, *then it is transient [recurrent] for each similar matrix* Π'.

Proof. It is known (see [6, Chapter 15, §5]), that i is transient if, and only if, $\sum_n \Pi^n(i, i) < \infty$. A glance at (14.2) completes the proof.

THEOREM 14.4. *Suppose that* E *is indecomposable-recurrent*(20). *Then* $x(i) = \text{const.}$ *is the only non-negative solution of* $\Pi \mathbf{x} = \mathbf{x}$.

Proof. By assumption no proper subset of E is a sojourn set, and therefore by Lemma 7.2 there can exist no nonconstant *bounded* solution $\mathbf{x} \geqq \mathbf{0}$. (That $\mathbf{x} = \mathbf{1}$ is a solution follows from Theorem 11.1.) Suppose now that $\mathbf{x} > \mathbf{0}$ is an unbounded solution of $\Pi \mathbf{x} = \mathbf{x}$, and consider the matrix Π' of (14.1) with $\mathbf{z} = \mathbf{1} + \mathbf{x}$. Let $\mathbf{u}_1'$ and $\mathbf{u}_2'$ be the vectors with components

$$u_1'(i) = \frac{1}{1 + x(i)}, \qquad u_2'(i) = \frac{x(i)}{1 + x(i)} \cdot$$

Then $\mathbf{u}_1'$ and $\mathbf{u}_2'$ are independent bounded solutions of $\Pi' \mathbf{u}' = \mathbf{u}'$. But this is impossible, since E is indecomposable-recurrent not only relative to Π, but relative to Π' as well.

15. **Relatively maximal ideals. The total boundary.** The preceding considerations show that, probabilistically and analytically, similar matrices

(20) Cf. Definition 11. The character of E remains unchanged when Π is replaced by a similar matrix.

are closely related. Theorem 14.2 establishes in some cases an obvious connection between parts of the boundaries $\mathfrak{B}$ and $\mathfrak{B}'$ defined by two similar matrices Π and Π'. It is easily seen by the same method (and will be shown in a different way) that in general the boundaries Π and Π' have a common part (defined by solutions such that both $\mathbf{x}$ and $\mathbf{x}'$ are bounded) but that each boundary contains points which do not have an image on the other.

It will now be shown that it is possible, and natural, using the totality $\mathfrak{M}$ of *all* positive solutions of $\Pi\mathbf{x}=\mathbf{x}$, *to introduce a boundary* $\mathfrak{B}^*$ *which is the same for all matrices similar to* Π; *the boundary* $\mathfrak{B}$ *as introduced in* §13 *by means of bounded solutions will be a subset of* $\mathfrak{B}^*$, *and* $E+\mathfrak{B}$ *embedded in* $E+\mathfrak{B}^*$.

The notion of sojourn *set* can be formulated invariantly for the whole family of similar matrices (Definition 15.2), but the notion of sojourn *solution* has no intrinsic meaning. The procedure of §13 requires therefore two modifications. Instead of ideals in $\mathfrak{S}$ we have to use ideals in $\mathfrak{M}$, but here we change the definition so as not to distinguish between $\mathbf{x}$ and the scalar multiples $p\mathbf{x}$. Moreover, in $\mathfrak{S}$ we had maximal ideals because $\mathfrak{S}$ has a maximal element (lattice unit) $\mathbf{s}_E$. Maximal ideals in $\mathfrak{M}$ need not exist, and are in any case not usable for our purposes. We use a *relative maximality*.

DEFINITION 15.1. *A subset* $I\subset\mathfrak{M}$ *will be called an ideal in* $\mathfrak{M}$ *if*

(1) $\mathbf{x}\cup\mathbf{y}\in I$, *whenever* $\mathbf{x}, \mathbf{y}\in I$;

(2) $\mathbf{z}\in I$, *whenever* $\mathbf{z}\leqq\mathbf{x}$, $\mathbf{x}\in I$, $\mathbf{z}\in\mathfrak{M}$;

(3) $p\mathbf{x}\in I$, *whenever* $p\geqq 0$, $\mathbf{x}\in I$.

The ideal I will be called maximal relative to $\mathbf{u}\in\mathfrak{M}$ *if* $\mathbf{u}\notin I$ *but* $\mathbf{u}\in I_1$ *for each ideal* I_1 *properly containing I.*

It will be noticed that *the transformation* (14.3) *takes an ideal which is maximal relative to* $\mathbf{u}$ *into an ideal maximal relative to* $\mathbf{u}'$. In this sense the notion of relatively maximal ideals refers to the *family of similar matrices* rather than to an individual matrix. Note that if *I is maximal relative to both* $\mathbf{u}$ *and* $\mathbf{v}$ *then it is maximal also relative to* $\mathbf{u}\cup\mathbf{v}$ *and to* $\mathbf{u}\cap\mathbf{v}$.

We proceed to prove the existence of relatively maximal ideals and to describe them. If Π is strictly stochastic, we may simplify the language by taking $\mathbf{u}=\mathbf{1}$. In fact, the transformation (14.3) permits us to reduce the general case to this apparently special case.

THEOREM 15.1. *Let* Π *be strictly stochastic, and I an ideal in* $\mathfrak{M}$ *maximal relative to* $\mathbf{1}$. *Let J be the set of all sojourn solutions in I, that is,* $J=I\cap\mathfrak{S}$. *Then*: (a) *J is a maximal ideal in* $\mathfrak{S}$. (b) *in order that an element* $\mathbf{x}\in\mathfrak{M}$ *belong to I it is necessary and sufficient that to each* $\epsilon>0$ *there exist a sojourn solution* $\mathbf{s}$ *such that*

$$\mathbf{x}\cap\mathbf{s}<\epsilon\mathbf{s} \qquad \mathbf{s}\notin J. \tag{15.1}$$

(c) *Conversely, if J is an arbitrary maximal ideal in* $\mathfrak{S}$ *and I is defined as the set of all* $\mathbf{x} \in \mathfrak{M}$ *which satisfy condition* (b), *then* I *is an ideal in* $\mathfrak{M}$, *maximal relative to* $\mathbf{1}$.

Proof. (1) Obviously J is an ideal in $\mathfrak{S}$ and we have to prove that it is maximal. Let J_1 be any ideal in $\mathfrak{S}$ such that $J \subset J_1$. If J is a proper subset of J_1 there exists an element $\mathbf{s}$ such that $\mathbf{s} \in J_1$ but $\mathbf{s} \notin J$. Denote by I_1 the set of all elements of $\mathfrak{M}$ of the form $\mathbf{x} \cup \mathbf{y}$, where $\mathbf{x} \in I$ and $\mathbf{y} \leqq p\mathbf{s}$. Clearly I_1 is an ideal, and $I \subset I_1$. Now I is maximal relative to $\mathbf{1}$, and therefore $\mathbf{x}$ and $\mathbf{y}$ can be chosen so that $\mathbf{x} \cup \mathbf{y} = \mathbf{1}$. We have then([21])

$$(15.2) \quad \mathbf{1} - \mathbf{s} = (x \cup y) \cap (1 - \mathbf{s}) = \{(1 - \mathbf{s}) \cap \mathbf{x}\} \cup \{(1 - \mathbf{s}) \cap \mathbf{y}\}.$$

But $\mathbf{y} \leqq p\mathbf{s}$ and so $(\mathbf{1}-\mathbf{s}) \cap \mathbf{y} = \mathbf{0}$ by Theorem 9.2. Accordingly, (15.2) reduces to $\mathbf{1}-\mathbf{s} = (\mathbf{1}-\mathbf{s}) \cap \mathbf{x}$ or $\mathbf{1}-\mathbf{s} \leqq \mathbf{x}$. Now $\mathbf{x} \in I$ and since I is an ideal, we have $\mathbf{1}-\mathbf{s} \in I$ and therefore $\mathbf{1}-\mathbf{s} \in J$. Again, $J_1 \subset J$, so that $\mathbf{1}-\mathbf{s} \in J_1$. It is seen that J_1 contains both $\mathbf{s}$ and $\mathbf{1}-\mathbf{s}$, and therefore also $\mathbf{s} \cup (\mathbf{1}-\mathbf{s}) = \mathbf{1}$. Thus $J_1 = \mathfrak{S}$, and J is maximal as asserted.

(2) Next let J be an *arbitrary* maximal ideal in $\mathfrak{S}$ and define I by the property (b). We show that *I is an ideal*. Requirements (2) and (3) of Definition 15.1 are trivially fulfilled. To verify that (1) holds, choose $\mathbf{x}, \mathbf{y} \in I$. By (15.1) there exist two sojourn solutions $\mathbf{s}_A$ and $\mathbf{s}_B$ such that

$$(15.3) \qquad \mathbf{x} \cap \mathbf{s}_A \leqq \epsilon \cdot \mathbf{s}_A, \qquad \mathbf{y} \cap \mathbf{s}_B \leqq \epsilon \cdot \mathbf{s}_B$$

and $\mathbf{s}_A \notin J$, $\mathbf{s}_B \notin J$. By Lemma 13.3 this implies that also $\mathbf{s}_{A \cap B} \notin J$. On the other hand, using Theorem 9.2 we get from (15.3)

$$(15.4) \qquad \mathbf{x} \cap \mathbf{s}_{A \cap B} \leqq \epsilon \cdot \mathbf{s}_{A \cap B}, \qquad \mathbf{y} \cap \mathbf{s}_{A \cap B} \leqq \epsilon \cdot \mathbf{s}_{A \cap B}$$

and therefore (using the distributive law as in (15.2))

$$(15.5) \qquad (\mathbf{x} \cup \mathbf{y}) \cap \mathbf{s}_{A \cap B} = (\mathbf{x} \cap \mathbf{s}_{A \cap B}) \cup (\mathbf{y} \cap \mathbf{s}_{A \cap B}) \leqq \epsilon \cdot \mathbf{s}_{A \cap B}.$$

Thus $\mathbf{x} \cup \mathbf{y} \in I$ and so I is an ideal in $\mathfrak{M}$.

(3) We show that *I is maximal relative to* $\mathbf{1}$. Let $\mathbf{x} \in \mathfrak{M}$ be an arbitrary element such that $\mathbf{x} \notin I$. By definition this means that there exists an $\eta > 0$ such that for each sojourn solution $\mathbf{s}_A$

$$(15.6) \qquad \textit{if } \mathbf{x} \cap \mathbf{s}_A \leqq \eta \cdot \mathbf{s}_A \quad \textit{then} \quad \mathbf{s}_A \in J.$$

Now define $\mathbf{s}_X$ and $\mathbf{s}_A$ as in Lemma 13.4. A comparison of (15.6) and (13.4) shows that $\mathbf{s}_A \in J$ (possibly $\mathbf{s}_A = \mathbf{0}$). On the other hand, $\mathbf{1} = \mathbf{s}_A + \mathbf{s}_X = \mathbf{s}_A \cup \mathbf{s}_X$, and therefore $\mathbf{s}_X \notin J$. Moreover, $\mathbf{s}_X$ satisfies (13.3). In other words, if $\mathbf{x} \notin I$, *then there exist sojourn solutions* $\mathbf{s}_A$ *and* $\mathbf{s}_X$ *such that*

$$(15.7) \qquad \mathbf{x} \geqq \eta \mathbf{s}_X, \qquad \mathbf{s}_X \notin I, \qquad \mathbf{s}_A \notin I, \qquad \mathbf{s}_X + \mathbf{s}_A = \mathbf{1}.$$

([21]) The distributive law used in (15.2) holds in every vector-lattice; see [1, Theorem 7.6] or [2].

(This result will be used in the proof of Theorem 15.2.)

To complete the proof, consider an arbitrary ideal I_1 in $\mathfrak{M}$ such that $I_1 \supset J$ and $\mathbf{1} \notin I_1$. Then $\mathbf{s}_A \in I_1$ and consequently $\mathbf{s}_X \notin I_1$. It follows now from the first inequality in (15.7) that $\mathbf{x} \notin I_1$. We see thus that $\mathbf{x} \notin I$ implies $\mathbf{x} \in I_1$, and hence $I_1 \subset I$. This proves the asserted maximality of I relative to $\mathbf{1}$, and also that if I is maximal relative to $\mathbf{1}$ it is necessarily of the form described in the theorem.

We propose to define a boundary $\mathfrak{B}^*$ by a procedure analogous to that of §13, except that ideals in $\mathfrak{M}$ are to be used instead of ideals in $\mathfrak{S}$. For that purpose it is necessary to define an intrinsic analogue to sojourn sets.

DEFINITION 15.2. *Let* Π *be strictly stochastic and let* $\mathbf{z}$ *be an arbitrary element of* $\mathfrak{M}$. *A set* $A \subset E$ *will be called carrier of* $\mathbf{z}$ *if for the matrix* Π' *of* (14.1) *the sets* A *and* E *are equivalent.*

Spelled out in detail and without reference to the transformation (14.1) our definition amounts to the following. As in §3 it is seen that

$$(15.8)\qquad \begin{aligned} \zeta(i) &= \lim_{n\to\infty} \Pi_A^n(i, j) z(j), && i \in A; \\ \zeta(i) &= 0, && i \notin A, \end{aligned}$$

always defines a vector $\boldsymbol{\zeta} \leqq \mathbf{z}$ (the limit being attained monotonically). Similarly

$$(15.9)\qquad \mathbf{z}^* = \lim_{n\to\infty} \Pi^n \boldsymbol{\zeta}$$

exists, and $\mathbf{z}^* \leqq \mathbf{z}$. *The set* A *is carrier of* $\mathbf{z}$ *if and only if* $\mathbf{z}^* = \mathbf{z}$.

It will be noticed that *if* A *is a carrier for* $\mathbf{z}$, *then it is also carrier of the image* $\mathbf{z}'$ *of* $\mathbf{z}$ *for any similar matrix* Π'. Moreover, if E is transient, then by Theorem 11.3 there exists a sequence $A_1 \supset A_2 \supset \cdots, \bigcap A_n = 0$, of carriers of $\mathbf{z}$. *Each sojourn set* A *is, of course, carrier of* $\mathbf{s}_A$.

We let $\mathfrak{B}^*$ *stand for the set of all ideals in* $\mathfrak{M}$ *which are maximal relative to some* $\mathbf{x} \in \mathfrak{M}$.

DEFINITION 15.3. *For a strictly stochastic* Π *and a transient*(22) E *we introduce in* $E + \mathfrak{B}^*$ *a topology as follows.*

A set $\Omega \subset E + \mathfrak{B}^*$ *is open if to each point* $\omega \in \Omega \cap \mathfrak{B}^*$ *there corresponds a vector* $\mathbf{x} \in \mathfrak{M}$ *such that*

(a) *The ideal* ω *is maximal relative to* $\mathbf{x}$.

(b) *The set* Ω *contains a subset of* E *which is carrier of* $\mathbf{x}$.

(c) *The set* $\Omega \cap \mathfrak{B}^*$ *contains each* $\beta \in \mathfrak{B}^*$ *which is maximal relative to some* $\mathbf{u} \leqq \mathbf{x}$.

THEOREM 15.2. *The space* $E + \mathfrak{B}^*$ *of Definition* 15.3 *is a Hausdorff space.*

(22) As in Definition 13, the assumption that all points be transient is introduced for convenience of formulations only. Note that E is transient with respect to any matrix similar to Π (Theorem 14.3).

Proof. The union of open sets is trivially open. Also, every subset of E is open. To prove that the intersection of two open sets Ω_1 and Ω_2 is open, consider an element $\omega \in \mathfrak{B}^*$ common to Ω_1 and Ω_2. Let x and y be the corresponding elements described in the definition. In view of the isomorphisms described in Theorem 14.1 there is no loss of generality (but only change of notation) in supposing that $x \cup y = 1$. Then ω is maximal relative to 1. We know from (15.7) that there exist two sojourn sets X and Y such that

$$(15.10) \qquad x \geqq \eta s_X, \qquad y \geqq \eta s_Y, \qquad s_X \notin \omega, \ s_Y \notin \omega.$$

By the definition there exist two sets $A \subset \Omega_1$ and $B \subset \Omega_2$ which are carriers of x and y, respectively. Now $\|x\| \leqq 1$ and it is seen from the definition of a carrier set (see (15.8)–(15.9)) that $x \leqq s_A$. Hence $s_X \leqq s_A$ by (15.10). Furthermore, the set X is defined only up to an equivalence, and in view of Theorem 9.2 we may choose $X \subset A$. A similar argument holds for Y. We have therefore

$$(15.11) \qquad X \subset \Omega_1, \qquad Y \subset \Omega_2,$$

and obviously X and Y are carriers of s_X and s_Y, respectively. As before, let $J = \omega \cap \mathfrak{S}$ be the set of all sojourn solutions contained in ω. Theorem 15.1 shows that J is a maximal ideal. Then $s_X \notin J$ and $s_Y \notin J$ by (15.10), and hence $s_{X \cap Y} \notin J$ by Lemma 13.3. Now $J \subset \omega$, and thus $s_{X \cap Y} \notin J$. The set $X \cap Y$ is a carrier of $s_{X \cap Y}$, and is contained in the intersection of Ω_1 and Ω_2 (see (15.11)). Thus $\Omega_1 \cap \Omega_2$ satisfies the conditions (a) and (b) of Definition 15.3 with $s_{X \cap Y}$ playing the rôle of x. Condition (c) is trivially satisfied since if β is maximal relative to some $u \leqq s_{X \cap Y}$ then by definition $\beta \in \Omega_1$ and $\beta \in \Omega_2$ and so $\beta \in \Omega_1 \cap \Omega_2$.

It remains to prove that to any two points of $E + \mathfrak{B}^*$ one may find two nonoverlapping neighborhoods. By Theorem 11.3 this is trivial unless both points are in $\mathfrak{B}^*$. Let $\omega_1 \neq \omega_2$ be two points of $\mathfrak{B}^*$. If ω_i is maximal relative to x_i, then automatically ω_i is maximal relative to $x_1 \cup x_2$. As before there is no loss of generality in assuming that $x_1 \cup x_2 = 1$. In other words, it suffices to verify the separation property for two elements ω_i both of which are maximal with respect to 1.

According to Theorem 15.1 the intersections $J_i = \omega_i \cap \mathfrak{S}$ are two distinct maximal ideals in $\mathfrak{S}$. As in the proof of Theorem 13.1 there exist two sojourn sets A_1 and $A_2 = E - A$, such that $s_{A_i} \in \omega_i$, but $s_{A_1} \notin \omega_2$ and $s_{A_2} \notin \omega_1$. Moreover, $s_{A_1} + s_{A_2} = 1$. Let $\Omega_i = A_i +$all elements of $\mathfrak{B}^*$ which are maximal relative to some $x \leqq s_{A_i}$. Then Ω_i is an open set containing ω_i. If $\omega \in \mathfrak{B}^*$ is maximal with respect to some $x \leqq s_{A_1}$, then there exists an element $s_A \leqq s_{A_1}$ such that ω is maximal relative to s_A (see (15.7)). Now the ideal $J = \omega \cap \mathfrak{S}$ does not contain s_A and hence it contains s_{A_2} by Lemma 13.2. Accordingly ω can not be maximal relative to any element $y \leqq s_{A_2}$, and so $\omega \in \Omega_1$ implies $\omega \notin \Omega_2$. Then Ω_1 and Ω_2 are nonoverlapping, and the theorem is proved.

The points of $\mathfrak{B}^*$ are ideals in $\mathfrak{M}$ and therefore $\mathfrak{B}^*$ formally depends on the matrix Π. However, if Π is replaced by a similar matrix Π' then the image I' of an ideal I is again an ideal and it is clear that the boundary as such remains unchanged. Therefore we have

THEOREM 15.3. *Each matrix of the family* $\mathfrak{F}$ *induces the same boundary* $\mathfrak{B}^*$ *and the same topology.*

The space $E+\mathfrak{B}$ *of Definition* 13 *is an open set in* $E+\mathfrak{B}^*$.

In other words, $\mathfrak{B}^*$ may be considered as the union of the boundaries, constructed by means of bounded solutions, for all matrices similar to the given matrix Π.

16. **The adjoint boundary.** Instead of considering Π as an operator on column vectors, we now consider the dual operator. That is, if $\boldsymbol{\alpha}=\{\alpha(1), \alpha(2), \cdots\}$ is a *row vector*, we consider the new row vector $\boldsymbol{\alpha}\Pi$, provided it is meaningful. This is certainly the case whenever $\sum|\alpha(j)|<\infty$.

We are interested in particular in the *eigenvectors* $\boldsymbol{\alpha}=\boldsymbol{\alpha}\Pi$, that is, the solutions of (1.3). To the set $\mathfrak{P}$ there correspond the eigenvectors satisfying the conditions

$$\alpha(j) \geqq 0, \qquad \sum \alpha(j) \leqq 1, \tag{16.1}$$

to $\mathfrak{M}$ all eigenvectors such that $\alpha(j)\geqq 0$. It is well known (see, for example [6, p. 329]) that in the case of a matrix Π which is not partitioned (i.e., not of the form (11.3)) there exists at most one eigenvector $\boldsymbol{\alpha}$ satisfying (16.1). However, Derman [5] has shown that there may exist many unbounded positive eigenvectors.

If $\boldsymbol{\alpha}$ is any strictly positive([23]) solution of $\boldsymbol{\alpha}=\boldsymbol{\alpha}\Pi$ we define a new matrix

$$\Pi^*(i, j) = \frac{\alpha(j)}{\alpha(i)}\Pi(j, i) \tag{16.2}$$

which is strictly stochastic. In the case where $\boldsymbol{\alpha}$ is a probability vector, $\Pi^*(i, j)$ is simply the transition probability of the *time reversed random walk*, or the inverse probability to Π. This notion has been introduced by Kolmogorov [7] (see also [6, p. 321]). To Derman [5] is due the simple, but important, observation that the transformation (16.2) remains probabilistically meaningful and useful even if $\sum\alpha(j)=\infty$.

Suppose now that there exists a second strictly positive solution $\boldsymbol{\beta}$ of $\boldsymbol{\beta}=\boldsymbol{\beta}\Pi$. Put

$$x^*(i) = \frac{\beta(i)}{\alpha(i)}. \tag{16.3}$$

Then

$$\Pi^*\mathbf{x}^* = \mathbf{x}^*, \qquad \mathbf{x}^* > 0. \tag{16.4}$$

([23]) Cf. the footnote to §14.

Conversely, if (16.4) has a solution $\mathbf{x}^*>0$, then the vector $\boldsymbol{\beta}$ defined by (16.3) is a solution of $\boldsymbol{\beta}=\boldsymbol{\beta}\Pi$. Also

$$\frac{\beta(j)}{\beta(i)}\Pi(j,i) = \Pi^*(i,j)\frac{x^*(j)}{x^*(i)}. \tag{16.5}$$

We have thus

THEOREM 16. *If* $\boldsymbol{\alpha}$ *runs through the set of all positive solutions of* $\boldsymbol{\alpha}=\boldsymbol{\alpha}\Pi$, *then the matrix* Π^* *of* (16.2) *runs through a family of similar matrices. Conversely, each matrix similar to* Π^* *is of the form* (16.5) *where* $\boldsymbol{\beta}$ *is a positive solution of* $\boldsymbol{\beta}=\boldsymbol{\beta}\Pi$.

It is thus seen that the study of positive solutions of $\boldsymbol{\alpha}=\boldsymbol{\alpha}\Pi$ is reduced to that of solutions $\mathbf{x}=\Pi\mathbf{x}$.

DEFINITION 16. *The boundary of E as defined in* §15 *by the similar matrices* Π^* *will be called the adjoint boundary defined by* Π.

The probabilistic meaning is given inasmuch as (16.2) corresponds to a time-reversal for the random walk.

17. Examples.

I. *Unbounded solutions*: the unsymmetric one-dimensional random walk. Changing the notation slightly, we let i run through all integers, and put

$$\Pi(i, i+1) = p, \qquad \Pi(i, i-1) = q, \qquad -\infty < i < \infty \tag{17.1}$$

where $p>q>0$, $p+q=1$. Clearly $\Pi\mathbf{x}=\mathbf{x}$ has the unique bounded solution $\mathbf{x}=1$, and the unique unbounded solution defined by $x(i)=(q/p)^i$. Each set $i\geqq a$ is a carrier set for the former, and $i\leqq a$ for the latter. The boundary $\mathfrak{B}$ consists of a single point, $\mathfrak{B}^*$ of two. The equation $\boldsymbol{\alpha}=\boldsymbol{\alpha}\Pi$ has the two unbounded solutions $\alpha(i)\equiv 1$, and $\alpha(i)=(p/q)^i$. *The two topologies coincide in this case.*

II. *Symmetric random walk in $N\geqq 3$ dimensions.* It has been proved by Murdoch [9] that in this case $\mathbf{x}=1$ is the only non-negative solution of $\Pi\mathbf{x}=\mathbf{x}$. Moreover, $\boldsymbol{\alpha}=\boldsymbol{\alpha}\Pi$ has the solution $\alpha(i)=1$, and with this solution the matrices Π and Π^* are identical. Both the boundary and the adjoint boundary reduce to a single point each and again the two topologies are identical.

III. We show that *the boundary and the adjoint boundary need have no relationship to each other.*

Relabeling the points, let E consist of the points A_n, C_n $(n\geqq 0)$, and B_n $(n\geqq 1)$. The reader is asked to draw a diagram with the points ordered on three parallel lines A, B, C. We define

$$\begin{aligned} \Pi(A_n, A_{n+1}) &= \Pi(C_n, C_{n+1}) = 1-\epsilon_n, && n\geqq 0,\\ \Pi(A_n, B_n) &= \Pi(C_n, B_n) = \epsilon_n, && n\geqq 1,\\ \Pi(B_n, A_{n-1}) &= \Pi(B_n, C_{n-1}) = 1/2, && n\geqq 1 \end{aligned} \tag{17.2}$$

where

$$\epsilon_0 = 0, \qquad \epsilon_i > 0 \ (i \geqq 1), \qquad \sum \epsilon_i < \infty. \tag{17.3}$$

The equations $\Pi\mathbf{x} = \mathbf{x}$ admit of the two linearly independent solutions

$$\begin{aligned} x_{1,2}(A_n) &= \frac{1}{2} \pm \frac{1}{2} \sum_{k=n}^{\infty} (1 - \epsilon_k), \qquad x_{1,2}(B_n) = \frac{1}{2}, \\ x_{1,2}(C_n) &= \frac{1}{2} \mp \frac{1}{2} \sum_{k=n}^{\infty} (1 - \epsilon_k). \end{aligned} \tag{17.4}$$

There exists no solution, bounded or unbounded, of $\Pi\mathbf{x} = \mathbf{x}$ linearly independent of $\mathbf{x}_1$ and $\mathbf{x}_2$.

The boundary $\mathfrak{B} = \mathfrak{B}^*$ *consists of two points* α *and* γ; *a set is a neighborhood of* α *if it contains* α *and all points* A_n *with* $n \geqq N$. Intuitively α and γ represent the "points at infinity" of the lines A and C, respectively, and $\mathbf{x}_1$ gives the probability that the random walk will be ultimately constrained to the line A. *The set* $E - B_1 - B_2 - \cdots$ *is equivalent to* E.

On writing down the *adjoint equations* $\alpha = \alpha\Pi$ one notices immediately that $\alpha(C_n) = \alpha(A_n)$ and hence that $\alpha = \alpha\Pi$ has *a solution which is unique* up to a scalar multiplier. This solution is

$$\begin{aligned} &\alpha(A_0) = \alpha(C_0) = 1, \\ &\alpha(A_n) = \alpha(C_n) = \frac{1}{2\epsilon_n}\, \alpha(B_n) = \frac{(1 - \epsilon_1) \cdots (1 - \epsilon_{n-1})}{\epsilon_1 \cdots \epsilon_n}. \end{aligned} \tag{17.5}$$

Now the inverse matrix is

$$\begin{aligned} \Pi^*(A_n, B_{n+1}) &= \Pi^*(C_n, B_{n+1}) = 1 - \epsilon_n, \\ \Pi^*(A_n, A_{n-1}) &= \Pi^*(C_n, C_{n-1}) = \epsilon_n, \\ \Pi^*(B_n, A_n) &= \Pi^*(B_n, C_n) = 1/2. \end{aligned} \tag{17.6}$$

The uniqueness of the solution of $\boldsymbol{\alpha} = \boldsymbol{\alpha}\Pi$ shows that the *adjoint boundary consists of a single point.* From (17.6) it is clear that with probability one the random walk will pass infinitely often through each of the lines A, B, C. Thus *the adjoint boundary consists of a unique point*: *to each neighborhood of it there exists an integer* N *such that the neighborhood contains all points* A_n, B_n, C_n *with* $n \geqq N$. That is to say, each neighborhood contains the complete neighborhood of two boundary points of $\mathfrak{B}$ plus infinitely many other points.

IV. *Continuous solutions.* Imagine the points of E ordered according to a dyadic branching scheme and labelled accordingly as 0, 1, 00, 01, 10, $\cdots$. Generally, if $\boldsymbol{\delta}$ stands for any finite sequence of zeros and ones, then $\boldsymbol{\delta}$ represents a point of E and we define

$$\Pi(\delta, \delta 0) = \Pi(\delta, \delta 1) = 1/2. \tag{17.7}$$

Obviously the set of all points starting with $\boldsymbol{\delta}$ is a sojourn set, and the corresponding sojourn solution is easily written down. For example, if $\delta = 0110$ then $\mathbf{s}$ equals 1 at all points δ, $\delta 0$, $\delta 1$, etc. It equals 1/2, 1/4, 1/8 respectively at the points 011, 01, and 0 and $\mathbf{s}$ equals 0 at all other points.

Clearly there are no minimal sojourn solutions, so *all sojourn solutions are continuous*. We have seen that to each dyadic interval of (0, 1) there corresponds a sojourn solution. Taking linear combinations it is readily seen that *the sojourn solutions are in one-to-one correspondence with sets of positive Lebesgue measure in* (0, 1) so that to sets on (0, 1) which are equivalent in the sense of Lebesgue there corresponds the same element $\mathbf{s} \in \mathfrak{S}$. Likewise, *the elements* $\mathbf{x} \in \mathfrak{P}$ *are in one-to-one correspondence with the measurable functions on* (0, 1) *which are non-negative and whose essential upper bound does not exceed unity*.

In this particular case it is natural to define a boundary isomorphic to the interval (0, 1). The boundary which we have defined in terms of maximal ideals is larger: on it the boundary values of $\mathbf{x} \in \mathfrak{P}$ are continuous instead of merely being bounded.

References

1. G. Birkhoff, *Lattice theory*, Amer. Math. Soc. Colloquium Publications, vol. 25, New York; 1940.

2. N. Bourbaki, *Éléments de mathématique*. Première partie. *Les structures fondamentales de l'analyse*, Paris, Hermann, 1939–1955.

3. G. Choquet, *Theory of capacities*, Annales de l'Institut Fourier vol. 5 (1955) pp. 131–295.

4. C. Derman, *A solution to a set of fundamental equations in Markov chains*, Proc. Amer. Math. Soc. vol. 5 (1954) pp. 332–334.

5. ———, *Some contributions to the theory of Markov chains*, Thesis, and Technical Report No. 7, Columbia University, Department of Mathematical Statistics, 1955.

6. W. Feller, *An introduction to probability theory and its applications*, New York, Wiley, 1950.

7. A. Kolmogoroff, *Zur Theorie der Markoffschen Ketten*, Math. Ann. vol. 112 (1935) pp. 155–160.

8. R. S. Martin, *Minimal positive harmonic functions*, Trans. Amer. Math. Soc. vol. 49 (1941) pp. 137–172.

9. B. H. Murdoch, *Preharmonic functions*, Princeton thesis, 1954.

Added in proof.

10. D. Blackwell, *On transient Markov processes with a countable number of states and stationary transition probabilities*, Ann. Math. Statist. vol. 26 (1955) pp. 654–658.

Princeton University,
Princeton, N. J.

A DIFFUSION EQUIVALENT TO A COUNTABLE MARKOV CHAIN*

BY WILLIAM FELLER AND HENRY P. MCKEAN, JR.

PRINCETON UNIVERSITY

Communicated by S. Bochner, April 5, 1956

1. *Introduction.*—Let $K = [0, 1]$, let $C(K)$ be the space of continuous functions on K to the real numbers, let $m(dx)$ be a strictly positive Borel measure on K, and let B be the second-order differential operator $D_m D_x$. Subject to suitable boundary conditions, the problem

$$(\mu - B)w = v \qquad \mu > 0,\ v \in C(K), \tag{1}$$

has precisely one solution:

$$w = (G_\mu v)(x) = \int_K G(\mu, x, s)v(s)m(ds), \tag{2}$$

the kernel $G(\mu, \cdot, \cdot)$ being the appropriate Green function. One has

$$w = \int_0^{+\infty} e^{-\mu t}\, u(t, x)\, dt, \tag{3}$$

where $u(t, x)$ is the solution to the corresponding parabolic equation,

$$u_t(t, x) = Bu(t, x) \qquad u(0+, x) = v(x), \tag{4}$$

which satisfies the same boundary conditions. Also, this solution $u(t, x)$ has the representation,

$$u(t, x) = (T_t v)(x) = \int_K P(t, x, ds)v(s), \tag{5}$$

in which the operators $(T_t: t > 0)$ constitute a semigroup mapping $C(K)$ into $C(K)$ and the kernel $P(\cdot, \cdot, \cdot)$ satisfies the Chapman-Kolmogorov identity

$$P(t_1 + t_2, x, M) = \int_K P(t_1, x, ds)P(t_2, s, M), \tag{6}$$

each $P(t, x, \cdot)$ being a Borel probability measure on K, and one can construct a Markov process (diffusion) with sample functions x_t and transition probabilities,[1]

$$PR(x_{t+s} \in M \mid x_s = x) = P(t, x, M). \tag{7}$$

Here we shall suppose that $m(dx)$ *is concentrated on the rational numbers* $R = (r_i: i > 0)$ *in* $(0, 1)$, the mass $m_i = m(r_i)$ being >0,[2] *so that equation* (2) *becomes*

R.L. Schilling et al. (eds.), *Selected Papers II*,

$$(G_\mu v)(r_i) = \sum_{k>0} G(\mu, r_i, r_k)v(r_k)m_k. \tag{8}$$

This makes one suspect that equation (5) *is about the same as*

$$(T_t v)(r_i) = \sum_{k>0} P(t, r_i, r_k)v(r_k) \tag{9}$$

and that the Markov process, advertised above, is (roughly speaking) nothing but a countable Markov chain on R.

To support these contentions, we shall show that

$$P(t, x, K - R) = PR(x_t \notin R | x_0 = x) = 0 \qquad t > 0, x \in K, \tag{10}$$

which makes it clear that relation (9) is true and that the matrix $p(t)$, whose entries are $p_{ik}(t) = P(t, r_i, r_k)$, has row sums 1 and satisfies the Chapman-Kolmogorov identity

$$p(t_1 + t_2) = p(t_1)p(t_2) \qquad t_1, t_2 > 0. \tag{11}$$

This shows that the irrational numbers have no importance and that one can lump them into a single state, setting

$$\begin{aligned} \hat{x}_t &= i & x_t &= r_i & i > 0 \\ &= +\infty & x_t &\in K - R, \end{aligned} \tag{12}$$

this being a Markov chain on R with stationary transition probabilities,

$$PR(\hat{x}_{t+s} = k | \hat{x}_s = i) = p_{ik}(t). \tag{13}$$

Also, making $t \downarrow 0$, we shall show that

$$p(t) \to 1, \tag{14}$$

$$t^{-1}p_{ik}(t) \to 0 \qquad |i - k| > 0, \tag{15}$$

and

$$t^{-1}(1 - p_{ii}(t)) \to +\infty, \tag{16}$$

so that, in the terminology of P. Lévy,[3] *the states i are instantaneous,* $+\infty$ *is fictitious, and no state is stable.*[4]

The counterparts to the Kolmogorov backward and forward differential equations are

$$p_{ij}'(t) = m(r_i)^{-1}\left[\lim_{r_k \downarrow r_i} \frac{p_{kj} - p_{ij}}{r_k - r_i} - \lim_{r_k \uparrow r_i} \frac{p_{ij} - p_{kj}}{r_i - r_k}\right] \tag{17}$$

and

$$p_{ij}'(t) = m(r_j)^{-1}\left[\lim_{r_k \downarrow r_j} \frac{p_{ik} - p_{ij}}{r_k - r_j} - \lim_{r_k \uparrow r_j} \frac{p_{ij} - p_{ik}}{r_j - r_k}\right], \tag{18}$$

which shows how hopeless it is to try to write down backward and forward equations for an arbitrary Markov chain.

2. *Construction.*—Given $v \in C(K)$, write $v \in D(B)$, provided that v satisfies the classical boundary conditions

$$v^+(0) = 0 = v^-(1) \tag{19}$$

and Bv belongs to $C(K)$. This domain $D(B)$ is dense in $C(K)$; the operator B contracted to $D(B)$ generates a semigroup $(T_t\colon\ t > 0)$ mapping $C(K)$ into $C(K)$; the connection between B and the semigroup is

$$T_t v = v + tBv + c(t) \qquad (t \downarrow 0),\ v \in D(B) \qquad (20)$$

(these statements being meant in the customary strong topology); and, setting

$$G_\mu = \int_0^{+\infty} e^{-\mu t}\, T_t\, dt, \qquad \mu > 0, \qquad (21)$$

we have the situation described in Section 1. Also, and this will be important below, the Green function $G(\mu, .\ , .)$ in equation (2) is >0 on $K \times K$ and

$$\int_0^{+\infty} e^{-\mu t}\, P(t, x, M)\, dt = \sum_{r_i \in M} G(\mu, x, r_i) m_i \qquad \mu > 0.^5 \qquad (22)$$

Our next step is to prove relation (10). Given $x \in K$, equation (22) makes it clear that $P(t, x, R) = 1$, provided that t is not contained in some exceptional set, N, of measure 0. Choose $t \in N$, and put $M = (t_0\colon\ P(t_0, x, r_i) < 1$, some $i > 0)$. We have

$$1 > P(t, x, R) \geq \int_R P(t - t_0, x, ds) P(t_0, s, R) = P(t - t_0, x, R) \qquad t_0 \notin M,$$

and, since M has measure 0, this is a contradiction, N is empty, and relation (10) is true.

This brings us to relations (14), (15), and (16). Choose positive $v \in D(B)$, vanishing near $x \in K$ and such that $v = 1$ outside some neighborhood J containing x. Keeping equation (20) in mind, we have

$$P(t, x, K - J) \leq (T_t v)(x) = o(t) \qquad (t \downarrow 0), \qquad (23)$$

and this proves relation (15). To prove relation (14), we have merely to show that $p_{ii}(t) \to 1\ (t \downarrow 0)$. Remembering relations (11) and (15), we have

$$\begin{aligned} o(1) + G(\mu, r_i, r_i) m_i &= \int_0^{+\infty} e^{-\mu t}\, p_{ii}(t + \epsilon)\, dt \\ &= \sum_{k > 0} G(\mu, r_i, r_k) m_k\, p_{ki}(\epsilon) \\ &= o(1) + G(\mu, r_i, r_i)\, m_i\, p_{ii}(\epsilon) \quad (\epsilon \downarrow 0), \end{aligned}$$

and, since $G(\mu, r_i, r_i) > 0$, we see that relation (14)[6] is true.

Coming to relation (16), take $i > 0$ and a positive $v \in D(B)$ such that $v(r_i) = 1$. We have

$$t^{-1}(1 - p_{ii}(t)) \geq t^{-1}(1 - (T_t v)(r_i)) = -(Bv)(r_i) + o(1) \qquad (t \downarrow 0),$$

and, since $-(Bv)(r_i)$ can be taken as large as we please, relation (16) is true.

Remark: According to a theorem of D. Ray,[7] relation (23) implies that the sample functions x_t can be taken continuous with probability 1. Consequently, the sets

$$S_i = (t\colon\ \hat{x}_t = i) = (t\colon\ x_t = r_i) \qquad i > 0, \qquad (24)$$

are closed and, by a theorem of P. Lévy,[8] nowhere dense.

* Work supported by the Office of Ordnance Research, United States Army, Contract No. DA-36-034-ORD-1296.

1 W. Feller, "The General Diffusion Operator and Positivity-preserving Semi-groups in One Dimension," *Ann. Math.*, **60**, 417–436, 1954; "The Parabolic Differential Equations and the Associated Semi-groups of Transformations," *ibid.*, **55**, 468–519, 1952; J. L. Doob, *Stochastic Processes* (New York: John Wiley & Sons, 1953).

2 Given such a measure $m(dx)$, we make the following definition: $Bv = w$ means that the one-sided derivatives v^+ and v^- exist on (0, 1), that $v^+ = v^-$ on $K - R$, and that $v^+(b) - v^-(a) = \Sigma\, w(r_i)m_i\ (a < b)$, the summation being taken over $(i: r_i \in [a, b])$.

3 P. Lévy, "Systémes markoviens et stationnaires: Cas dénombrable," *Ann. sci. École norm. supér.*, **68**, 327–381, 1951.

4 We note that this contradicts P. Lévy's statement (*ibid.*, p. 375) that a chain must have stable states.

5 Compare W. Feller, "The Parabolic Differential Equations and the Associated Semi-groups of Transformations," *Ann. Math.*, **55**, 468–519, 1952.

6 This is a special case of a theorem due to J. Doob, "Topics in the Theory of Markov Chains," *Trans. Am. Math. Soc.*, **52**, 37–64, 1942.

7 D. Ray, "Stationary Markov Processes with Continuous Paths," to appear in *Trans. Am. Math. Soc.* (1956).

8 P. Lévy, *op. cit.*, p. 373. His argument contains a slip, but it is easy to correct.

STOCHASTIC PROCESSES CONNECTED WITH HARMONIC FUNCTIONS

BY

JOANNE ELLIOTT[1] AND WILLIAM FELLER

0. **Introduction.** The family of transformations

$$T_t f(x) = \pi^{-1} \int_{-\infty}^{+\infty} \frac{t}{t^2 + (x-y)^2} f(y)dy, \qquad t > 0 \tag{0.1}$$

define a semi-group of bounded linear transformations from $C[-\infty, +\infty]$ to itself. The salient feature of this semi-group is that the function

$$u(t, x) = T_t f(x)$$

is harmonic in the half-plane $t>0$. The infinitesimal generator of this semi-group is of the form

$$\Omega F(x) = \pi^{-1} P. \int_{-\infty}^{+\infty} \frac{F'(y)}{y-x} dy, \tag{0.2}$$

the integral being taken in the sense of a Cauchy principal value. The precise definition of this generator is found in Theorem 3.3 as a byproduct of the present investigation[2].

In [2] the first of the present authors studied a class of semi-groups from $C[-a, a]$ to itself with $0<a<\infty$, whose infinitesimal generator Ω_a is obtained from (0.2) by truncation:

$$\Omega_a F(x) = \pi^{-1} P. \int_{-a}^{+a} \frac{F'(y)}{y-x} dy; \tag{0.3}$$

the exact definition of this operator is found in Definition 1.1 of the present paper. These semi-groups share with (0.1) the property of being positivity preserving and norm-not-increasing. Among them there is a minimal semi-group $T_t^{(a)}$ characterized by the boundary conditions:

$$T_t^{(a)} f(\pm a) = 0. \qquad f \in C[-a, +a]. \tag{0.4}$$

Although the present paper in no way depends on the consideration of

Received by the editors November 21, 1955.

[1] Research sponsored by the Office of Ordnance Research under Contract Number DA-19-059-ORD-1611 at Mount Holyoke College.

[2] Hille [6] has investigated (0.1) as a semi-group from $L_p(-\infty, +\infty)$ to itself when $p>1$ and found the infinitesimal generator. These semi-groups on $C[-\infty, +\infty]$ and $L(-\infty, +\infty)$ apparently have not been investigated previously.

function spaces, it is now desirable to give the *probabilistic background* of our semi-groups.

To (0.1) there corresponds probabilistically the *Cauchy process*, a separable Markovian process with stationary transition probabilities,

$$(0.5) \qquad P(t, x, S) = \pi^{-1} \int_S \frac{t}{t^2 + (x - y)^2} dy$$

where S stands for a Borel set. It is easily seen that the path functions $X(t)$ of this process are discontinuous. In fact, if S is an interval not containing x then (0.5) implies that as $\Delta t \downarrow 0$,

$$(0.6) \quad (\Delta t)^{-1} \Pr \{X(t + \Delta t) \in S \mid X(t) = x\} = \pi^{-1} \int_S \frac{dy}{(y - x)^2} + 0(1).$$

This relation determines the probability distribution of the saltuses of $X(t)$ and characterizes the Cauchy process.

In like manner the semi-group determined by (0.3)–(0.4) depends on an integral transformation, whose kernel(³) $P_a(t, x, S)$ defines the transition probabilities of a new stochastic process with path functions confined to the interval $(-a, a)$. One of the purposes of the present paper is to investigate the connection between this process and the Cauchy process.

M. Kac [7] was the first to study the *absorbing barrier process connected with the Cauchy process*: this is the process confined to $(-a, a)$ whose path functions coincide with those of a Cauchy process as long as $|X(t)| < a$, but stop being defined from the first moment when $|X(t)| \geqq a$. Kac showed how to calculate *the probability $F_a(t, x)$ that the lifetime of this process will exceed t when the starting point is x*. For the Cauchy process itself $1 - F_a(t, x)$ is the probability distribution of the *first passage time* from x into the set $|s| > a$.

The investigation in [2] was inspired partly by Kac's result. The fact that Kac was led to the operator (0.3) combined with the minimality of the semi-group determined by the boundary condition (0.4) suggests that $P_a(t, x, S)$, represents the *transition probabilities of the absorbing barrier process*. If this is true one should have

$$(0.7) \qquad F_a(t, x) = P_a(t, x, E), \qquad E = (-a, a).$$

Purely formally this equation can be written in the form

$$F_a = \overset{a}{T_t} 1$$

where 1 is the unit function. Since Ω_a is the infinitesimal generator of $T_t^{(a)}$, this means that at least in some operational sense, $F_a(t, X)$ is the solution of the functional equation

(³) The Laplace transform (1.6) of $P_a(t, x, S)$ is described in §1.

(0.8) $$\frac{\partial}{\partial t} F_a(t, x) - \Omega_a F_a(t, x) = 0$$ (4)

with the lateral conditions

(0.9) $$F_a(t, \pm a) = 0, \qquad F_a(0, x) = 1.$$

The Laplace transforms of these equations are valid and are given at the end of §1.

We now come to the contents of the present paper. If $P_a(t, x, S)$ is indeed the transition probability of the absorbing barrier process we can draw several conclusions.

(a) *Concerning first passage times.* The quantity $F_a(t, x) - F_a(t+h, x)$ represents the probability in the Cauchy process of a path starting at $x \in (-a, a)$ to stay within $(-a, a)$ for a time t and then with the time interval $(t, t+h)$ to leave $(-a, a)$. The position at time t has, by assumption, the probability distribution $P_a(t, x, S)$, and the probability of a saltus leading out of $(-a, a)$ is governed by (0.6). Combining these results and letting $h \to 0$ one is led to expect that(5)

(0.10) $$\begin{aligned} -\frac{\partial}{\partial t} F_a(t, x) &= \pi^{-1} \int_{-a}^{a} P_a(t, x, dy) \int_{|z|>a} \frac{dz}{(z-y)^2} \\ &= 2a\pi^{-1} \int_{-a}^{a} P_a(t, x, dy)(a^2 - y^2)^{-1}. \end{aligned}$$

The Laplace transform version of this equation is contained in (1.15).

Now this equation is curious and contains an apparent contradiction. For, it states that $-\partial F_a(t, x)/\partial t$ is the transform, under the semi-group $T_t^{(a)}$, of the *unbounded* function $\phi(y) = 2\pi^{-1}(a^2 - y^2)^{-1}$. From the general behavior of the semi-group for bounded functions one should accordingly expect that

(0.11) $$-\frac{\partial}{\partial t} F_a(t, x) \to 2\pi^{-1}(a^2 - x^2)^{-1}, \qquad t \to 0.$$

This, however, leads to immediate contradictions. *It has been proved in* [3] *that actually* (0.10) *holds (at least in the Laplace transform sense) with*

(4) Kac's fundamental equation (8.12) may be interpreted as the statement that F_a is a "weak" solution of (0.8) in an appropriate Hilbert space. The uniqueness theorem proved by Kac shows that his boundary conditions are essentially equivalent to (0.9). It in no way contradicts the existence of the other semi-groups and processes generated by Ω_a. It was the study of Kac's equation (8.12) that led to the analysis of Ω_a in [2].

(5) Equation (0.10) presupposes that with probability one the jump occurs from a point in the interior of $(-a, a)$ to a point z with $|z| > a$. That this is so follows indirectly from the validity of (0.10), but has been proved previously by McKean [9]. A similar statement is true for all stable processes with index $\alpha \leqq 1$. For $\alpha > 1$ the equation corresponding to (0.10) contains an additional term which accounts for the possibility that $z = \pm a$ is the *first* point outside $(-a, a)$ actually to be reached.

$$(0.12)\qquad -\frac{\partial}{\partial t} F_a(t, x) \to 0, \qquad t \to 0,$$

and not (0.11).

This result shows that the extension of the semi-group from bounded to unbounded functions is by no means obvious. The corresponding extension of the resolvent and of Ω_a was studied in [3] and reveals surprising analytical features. The definition of the extended operator corresponding to Ω_a is found in Definition 1.3.

(b) *Relations between transition probabilities.* The kernel $P_a(t, x, S)$ defines, for each $x \in (-a, a)$ and each $t > 0$, a measure on the Borel subsets of $(-a, a)$. We find it convenient to extend P_a to a measure on $(-\infty, \infty)$ by stipulating that $P_a(t, x, \cdot)$ ascribes measure 0 to Borel sets contained in the complement of $(-a, a)$.

For the Cauchy process, a passage from a point $x \in (-a, a)$ at time t to a set $S \subset (-a, a)$ can occur in two ways: either during the entire time interval $(0, t)$, the path is constrained to $(-a, a)$, or a first passage into the exterior of $(-a, a)$ occurs at some time $s < t$. Using the argument which led to (0.11), we conclude that we should have

$$(0.13)\quad P(t, x, S) = P_a(t, x, S) + \int_0^t ds \int_{-a}^a P_a(s, x, dy) \int_{|z|>a} \frac{P(t-s, z, S)}{\pi(z-y)^2} dz$$

provided $x \in (-a, a)$ and $S \subset (-\infty, \infty)$.

By the same argument and under the same conditions, we should have for $a < b$, $|x| < a$, and $S \subset (-\infty, \infty)$,

$$(0.14)\quad \begin{aligned} P_b(t, x, S) &= P_a(t, x, S) \\ &+ \int_0^t ds \int_{-a}^a P_a(s, x, dy) \int_{a<|z|<b} \frac{P_b(t-s, z, S)}{\pi(z-y)^2} dz. \end{aligned}$$

We shall not attempt to derive equations (0.13) and (0.14) directly for the absorbing barrier processes. Instead, we *start* from the kernel P_a, using the fact that it describes a Markov process in $(-a, a)$. In Theorems 3.1 and 2.1 we prove the relations corresponding to (0.13) and (0.14) for the Laplace transforms of the densities of P_a, P, and P_b, and in Corollaries 3.1 and 2.1 we show that (0.13) and (0.14) actually hold for our kernels.

With these relations at hand, we can now check that the kernel P_a gives indeed the transition probabilities for the absorbing barrier process of the Cauchy process. For arbitrary x, denote by n_x the integer such that $(2n_x - 1)a \leq x < (2n_x + 1)a$, and by I_x the interval $((2n_x - 1)a, (2n_x + 1)a)$. Then put

$$(0.15)\qquad P_{per}(t, x, S) = P_a(t, x - 2n_x a, S - 2n_x a).$$

Probabilistically, P_{per} describes, for fixed I_x, a replica of the P_a process restricted to the interval I_x. In the sequel, we shall refer to this as the P_a process in I_x.

Since $P(t, x, S) = P(t, x-2na, S-2na)$, it is easily seen that if (0.13) holds, then

$$\begin{aligned} P(t, x, S) &= P_{per}(t, x, S) \\ &\quad + \int_0^t ds \int_{-\infty}^{\infty} P_{per}(s, x, dy) \int_{z \notin I_x} \frac{P(t-s, z, S)}{\pi(z-y)^2} dz, \end{aligned} \tag{0.16}$$

for $x \in I_x$.

A standard iteration procedure applied to (0.16) gives

$$P(t, x, S) = \sum_{k=0}^{n} P_{per}^{(k)}(t, x, S) + r_n(t, x, S), \tag{0.17}$$

where the terms in the sum are defined recursively by

$$\begin{aligned} P_{per}^{(0)}(t, x, S) &= P_{per}(t, x, S), \\ P_{per}^{(k+1)}(t, x, S) &= \int_0^t ds \int_{-\infty}^{\infty} P_{per}^{(k)}(s, x, dy) \int_{z \notin I_x} \frac{P_{per}(t-s, z, S)}{\pi(z-y)^2} dz. \end{aligned} \tag{0.18}$$

Now define,

$$P_{per}^{*}(t, x, S) = \sum_{k=0}^{\infty} P_{per}^{(k)}(t, x, S). \tag{0.19}$$

The series converges since all the terms in (0.17) are positive and $P(t, x, S) \leqq 1$.

We next show that P coincides with P^* for almost all x. For each t, x, and S the sequence $\{r_n\}$ is a decreasing, positive sequence. From Theorem 4.1, (7.1)–(7.4), and Theorem 7.1, it follows that if f is continuous in $[-a, a]$ and periodic 2a, then for $x \in I_x$

$$\int_{-\infty}^{\infty} P(t, x, dy) f(y) = \int_{-\infty}^{\infty} P_{per}^{*}(t, x, dy) f(y). \tag{0.20}$$

In particular if $f(y) \equiv 1$, we must have

$$\lim_{n \to \infty} \int_{-\infty}^{\infty} r_n(t, x, dy) = 0. \tag{0.21}$$

Hence $\lim_{n\to\infty} r_n(t, x, S) = 0$ and $P(t, x, S) = P_{per}^{*}(t, x, S)$ for all $x \in (-\infty, \infty)$, except the odd multiples of $\pm a$.

Probabilistically, P_{per}^{*} gives the transition probabilities of a Markov process consisting of an infinite succession of P_a processes, such that whenever one terminates at y a new initial position z is chosen according to (0.6), and the process starts afresh in the interval I_z. It is clear that P_a gives the transition probabilities for the absorbing barrier process in $(-a, a)$ for the P^*_{per} process, and hence also for the Cauchy process in view of the preceding paragraph.

(c) *Return processes.* In [2] there was constructed a whole family of semi-groups generated by Ω_a, and each yields the transition probabilities of a stochastic process. Among these, the simplest to describe are the *return processes*(6). They are formally and probabilistically analogous to corresponding processes in diffusion theory. Now, in diffusion the path functions are continuous, and the infinitesimal generator is a differential operator. It is therefore a priori to be expected that all processes with the same local behavior should be governed by the same infinitesimal generator. No such reasoning applies in the case of the Cauchy process. And in fact, we shall show that in the latter case, there exist more general return processes than those considered in [5]; and the infinitesimal generator does not necessarily coincide with Ω_a.

The general return process considered here may be described as follows. Consider a particle moving in the interval $(-a, a)$ according to the transition probabilities $P(t, x, S)$ of the Cauchy process as long as the particle remains within $(-a, a)$. When it for the first time leaves the interval, it jumps to a point z with $|z| > a$. We now associate with each point $|z| > a$ a probability $\tau(z)$ $(0 \leqq \tau(z) \leqq 1)$, that the particle is returned to a random point $Y \in (-a, a)$ with a given probability distribution $\Pr\{Y \in S\} = p(z, S)$. The process then continues as before from the point Y to which it has been returned.

The special case where

$$p(z,S) = \begin{cases} p_1(S) & \text{for } z > a, \\ p_2(S) & \text{for } z < a \end{cases} \tag{0.22}$$

has been treated in [2].

The argument which led to (0.13) now leads to an integral equation for the transition probabilities $\mathfrak{P}_a$ of this return process. The Laplace transform of this integral equation is (4.2). It is shown in §4 that to arbitrarily prescribed p and τ the integral equation has a solution which, in a certain sense, is minimal. This solution determines a semi-group from $C[-a, a]$ to itself and, accordingly, a stochastic process. These are discussed in §5. An example in §4 shows that *our integral equation may have several probabilistically admissible solutions*. Also, *even when $\tau(z) \equiv 1$ one may have a norm-decreasing semi-*

(6) All other processes can be obtained by a passage to the limit analogous to that described in [5]. We have thus a one-one correspondence between the processes in [2] and the diffusion processes constructed in [4]. The interpretation given in [5] applies in both cases. However, a probabilistic investigation of the behavior of the path functions of the processes formally analogous to the elastic and reflecting barrier processes of diffusion theory remains an open problem.

group, with $\mathfrak{P}_a(t, x, E) < 1$(7). Equation (4.4) gives a sufficient condition for this not to occur. It will be seen that the infinitesimal generators of our return processes coincide with Ω_a only if (0.22) holds.

(d) *Connection with harmonic functions.* Consider the semi-group (0.1) applied only to continuous functions which are *odd* and *periodic with period* 2a. Obviously such a function f can be prescribed arbitrarily in $(0, a)$ and we are dealing therefore in effect with a semi-group from $C[0, a]$ to itself. Again $T_t f$ is harmonic, odd and periodic, and therefore admits of a Fourier sine series. In fact if $f \sim \sum_{n=1}^{\infty} a_n \sin (n\pi/a)x$ is the formal Fourier series of f, then

$$(0.23) \qquad T_t f = \sum_{n=1}^{\infty} e^{-(n\pi/a)t} \sin \frac{n\pi}{a} x.$$

Here $T_t f$ is a function harmonic in $0 < x < a$, $t > 0$, uniquely determined by the boundary conditions $T_t f(0) = T_t f(a) = 0$.

Now these are the boundary conditions of the absorbing barrier process on $(0, a)$ and one might suspect that (0.23) represents the semi-group corresponding to the absorbing barrier process. This conjecture may appear strengthened by the fact that in the case of the homogeneous diffusion (governed by the parabolic equation $u_t = u_{xx}$) the above construction does indeed lead to the absorbing barrier process, and that the latter admits of a Fourier representation of the form (0.23) (with the exponent replaced by $n^2\pi^2 t/a^2$).

Actually (0.23) *is a transition semi-group and determines a Markovian process, but it is not the absorbing barrier process.* Its infinitesimal generator is of the form (7.28) with tan replaced by cot. It is not clear whether and how the process is related to the Cauchy process.

More generally, the class of functions *u harmonic in the semi-strip* $\alpha_1 \leqq x \leqq \alpha_2$, $t > 0$, *and subject to the classical boundary conditions*

$$(0.24) \qquad p_i u(\alpha_i, t) + (-1)^i q_i \frac{\partial}{\partial x} u(\alpha_i, t) = 0$$

defines a transition semi-group from $C[\alpha_1, \alpha_2]$ *to itself,* and therefore a Markovian process in $[\alpha_1, \alpha_2]$, provided $p_i \geqq 0$, $q_i \geqq 0$. The semi-group (0.23) corresponds to $q_1 = q_2 = 0$. The corresponding cosine series,

$$(0.25) \qquad T_t f = \sum_{n=0}^{\infty} a_n e^{-(n\pi/a)t} \cos \frac{n\pi}{a} x$$

(7) The probabilistic explanation of this curious phenomenon is as follows: The probability distribution $p(z, S)$ may make it probable that each jump out of the interval E will bring the path closer to a boundary point $\pm a$; if T_n are the time distances between successive jumps out of E, then it may occur that the expectation of $\sum T_n$ is finite, so that infinitely many jumps out of E may occur in a finite time. When this occurs the path function of our return process is, of course, no longer determined. This phenomenon is well-known in the theory of Markov processes in denumerable spaces and has been discussed by Doob [1].

where $f \sim \sum_{n=0}^{\infty} a_n \cos (n\pi/a)x$, represents a transition semi-group from $C[0, a]$ to itself such that $u(x, t) = T_t f(x)$ is a harmonic function satisfying the boundary conditions (0.24) with $p_1 = p_2 = 0$. The formal analogue for the heat equation is the reflecting barrier solution; it satisfies the same boundary conditions and has a representation of the form (0.25) (with $n\pi/a$ in the exponent replaced by $n^2\pi^2/a^2$). An alternative representation for the semi-group (0.25) is obtained by extending the definition of $f(x)$ from $(0, a)$ to all x in such a way that f becomes even and periodic with period 2a. With this convention (0.1) is equivalent to (0.25). The semi-groups associated with the other boundary conditions admit of representations analogous to (0.25) and (0.23) by means of trigonometric series.

The boundary conditions (0.24) are not the most general ones assuring positivity. For example, the boundary condition $u(a, t) = u(-a, t)$ for the interval $(-a, a)$ leads to the semi-group

$$(0.26) \qquad T_t f(x) = \sum_{n=0}^{\infty} e^{-(n\pi/a)t} \left\{ a_n \cos \frac{n\pi}{a} x + b_n \sin \frac{n\pi}{a} x \right\},$$

which is a transition semi-group.

It is not clear that any of these semi-groups should have a more intimate connection with the Cauchy process. However, in §7, we show that *both* (0.25) *and* (0.26) *can be interpreted as return processes*. The corresponding return probabilities are $r(z) \equiv 1$ and $p(z, S)$ given by (7.1) and (7.23), respectively.

1. **Preliminaries.** In this section we collect some results to which we shall refer frequently in the sequel.

DEFINITION 1.1. *If* (i) $F \in C[-a, +a]$ *is absolutely continuous on* $(-a, +a)$, (ii) $[F'(x)]^2(a^2 - x^2)^{1/2} \in L(-a, +a)$, *and* (iii) *for some* $h \in C[-a, +a]$,

$$(1.1) \qquad \pi^{-1} P. \int_{-a}^{a} \frac{F'(t)}{t - x} dt = h(x)$$

almost everywhere on $[-a, +a]$, *then we define* $\Omega_a F = h$ *as a linear transformation from* $C[-a, +a]$ *to itself.*

THEOREM 1.1. *For each* $\lambda > 0$, *the equation*

$$(1.2) \qquad \lambda F_a - \Omega_a F_a = f$$

with $f \in C[-a, +a]$ *has a unique solution* $F_a \in C[-a, +a]$ *under the boundary condition*

$$(1.3) \qquad F_a(-a) = F_a(a) = 0.$$

This solution is of the form

$$(1.4) \qquad F_a(x) = \int_{-a}^{a} \Gamma_a(x, y; \lambda) f(y) dy$$

where $\Gamma_a(x, y; \lambda) \geq 0$ *and is symmetric in* x *and* y. *The function* F_a *has a continuous derivative in* $(-a, +a)$.

Proof. The proof of these statements may be found in [2]. Note that $F_a(x)$ depends on λ. We shall find it convenient to omit this dependence in writing, however. The kernel $\Gamma_a(x, y; \lambda)$ has the following connection with the absorbing barrier transition probabilities described in the first part of the Introduction. We define

$$\gamma_a(x, S; \lambda) = \int_S \Gamma_a(x, y; \lambda)dy \tag{1.5}$$

with S a Borel subset of $[-a, +a]$. Then if $P_a(t, x, S)$ denotes, as before, the transition probability for the absorbing barrier process, we have

$$\gamma_a(x, S; \lambda) = \int_0^\infty e^{-\lambda t} P_a(t, x, S)dt. \tag{1.6}$$

As we pointed out in the Introduction, it will be necessary to consider Γ_a operating on certain unbounded functions. This class of functions is described as follows:

DEFINITION 1.2. *Let* $U[-a, +a]$ *denote the Banach space of functions* f *of the form*

$$f(x) = \phi(x)(a^2 - x^2)^{-1} \tag{1.7}$$

where $\phi \in C[-a, +a]$ *and* $\|f\|_U = \|\phi\|_C$.

In [3, §3], we showed that the transformation $f \rightarrow F_a$ defined by (1.4) is a bounded linear transformation from $U[-a, +a]$ to itself or to $C[-a, +a]$.

If $f \in U[-a, +a]$, then F_a as defined by (1.4) does not necessarily satisfy (1.2). In [3, Theorem 4.1] we derived the functional equation analogous to (1.2) which is satisfied in the more general situation. These results are summed up in the following:

DEFINITION 1.3. *We define*

$$(a^2 - x^2)\tilde{\Omega}_a F(x) = \pi^{-1} P. \int_{-a}^{a} \frac{F'(t)(a^2 - t^2)}{t - x} dt - \pi^{-1} \int_{-a}^{a} F(t)dt \tag{1.8}$$

for those $F \in C[-a, +a]$ *which are absolutely continuous on any proper subinterval of* $(-a, +a)$ *with* $[F'(x)]^2(a^2 - x^2)^{3/2} \in L(-a, +a)$ *and for which the right-hand side of* (1.8) *is almost everywhere equal to an element of* $C[-a, +a]$.

We then have

THEOREM 1.2. *If* F_a *is defined by* (1.4) *with* $f \in U[-a, +a]$ *for some* $\lambda > 0$, *then* F_a *is the unique solution in* $C[-a, +a]$ *of*

$$\lambda F_a - \tilde{\Omega}_a F_a = f. \tag{1.9}$$

Proof. See [3, Theorem 4.1].

We now go on to the Laplace transform version of the discussion in section (a) of the Introduction.

THEOREM 1.3. *The equation*

$$\lambda u - \Omega_a u = 0, \qquad \lambda > 0, \tag{1.10}$$

has exactly two linearly independent solutions in $C[-a, +a]$ *given by*

$$\xi_a^{\pm}(x) = \pi^{-1}\int_{-a}^{a}\frac{\Gamma_a(x, y; \lambda)}{a \pm y}\,dy \tag{1.11}$$

which satisfy

$$\xi_a^{+}(a) = \xi_a^{-}(-a) = 0, \ \xi_a^{+}(-a) = \xi_a^{-}(a) = 1. \tag{1.12}$$

Proof. See [3, Theorem 3.1].

In the language of §(a) *of the Introduction,* ξ_a^+ *and* ξ_a^- *are the Laplace transforms of the right and left escape probability densities, respectively.*

We shall usually make use of $\xi_a^+ + \xi_a^- = \xi_a$. Interpreted according to (0.10),

$$\begin{aligned}\xi_a(x) &= -\int_0^{\infty} e^{-\lambda t}\frac{\partial}{\partial t}F_a(t, x)dt \\ &= 2a\pi^{-1}\int_{-a}^{a}\frac{\Gamma_a(x, y; \lambda)}{a^2 - y^2}\,dy.\end{aligned} \tag{1.13}$$

On the other hand, if we integrate by parts

$$\xi_a(x) = F_a(0, x) - \lambda\int_0^{\infty} e^{-\lambda t}F_a(t, x)dt = 1 - \lambda\int_{-a}^{a}\Gamma_a(x, y; \lambda)dy, \tag{1.14}$$

using (0.7) and (0.9). We should thus have the relation

$$\xi_a(x) = 1 - \lambda\int_{-a}^{a}\Gamma_a(x, y; \lambda)dy = 2a\pi^{-1}\int_{-a}^{a}\frac{\Gamma_a(x, y; \lambda)}{a^2 - y^2}\,dy. \tag{1.15}$$

Formula (1.15) is proved analytically in [3, Theorem 3.1].

2. **The identity between kernels.** This section contains the Laplace transform version of (0.13) and (0.14) namely,

THEOREM 2.1. *If* Γ_a *and* Γ_b *are defined as in Theorem* 1.1 *with* $a<b$, *then*

$$\begin{aligned}\Gamma_a(x, y; \lambda) &= \Gamma_b(x, y; \lambda) \\ &\quad - \pi^{-1}\int_{-a}^{a}\Gamma_a(x, u; \lambda)\left\{\int_{a<|z|<b}\frac{\Gamma_b(z, y; \lambda)}{(z - u)^2}\,dz\right\}du\end{aligned} \tag{2.1}$$

when $|x| < a$, $|y| < a$ *and*

$$\Gamma_b(x, y; \lambda) = \pi^{-1} \int_{-a}^{a} \Gamma_a(x, u; \lambda) \left\{ \int_{a<|z|<b} \frac{\Gamma_b(z, y; \lambda)}{(z-u)^2} dz \right\} du \tag{2.2}$$

when $|x| < a$, $a < |y| < b$.

Proof. We can reduce (2.1) and (2.2) to a single relation by setting $\Gamma_a(x, y; \lambda) = 0$ for $|x| > a$, $|y| > a$. We define F_a and F_b as in (1.4). To prove (2.1) and (2.2), it is sufficient to prove that

$$F_a(x) = F_b(x) - \pi^{-1} \int_{-a}^{a} \Gamma_a(x, u; \lambda) \left\{ \int_{a<|z|<b} \frac{F_b(z)}{(z-u)^2} dz \right\} du \tag{2.3}$$

for each $f \in C[-b, +b]$ and $|x| < a$.

Now, defining Ω_b as in Definition 1.1, we have for each $\lambda > 0$ and $x \in (-a, a)$,

$$\begin{aligned} \lambda F_b(x) - \Omega_b F_b(x) = f(x) = \lambda F_b(x) - \pi^{-1} P.\int_{-a}^{a} \frac{F_b'(t)}{t-x} dt \\ - \pi^{-1} \int_{a<|z|<b} \frac{F_b'(t)}{t-x} dt, \end{aligned} \tag{2.4}$$

the second equality holding almost everywhere in $[-b, +b]$.

For convenience we introduce the notations,

$$H(u) = \int_{a<|z|<b} \frac{F_b(z)}{(z-u)^2} dz \tag{2.5}$$

and

$$F^*(x) = \pi^{-1} \int_{-a}^{a} \Gamma_a(x, u; \lambda) H(u) du. \tag{2.6}$$

Since $H \in U[-a, +a]$ (see Definition 1.2), we have from Theorem 1.2

$$\lambda F^*(x) - \tilde{\Omega}_a F^*(x) = H(x), \tag{2.7}$$

where $\tilde{\Omega}_a$ is the operator of Definition 1.3.

We next show that (2.7) can be rewritten as

$$\begin{aligned} \lambda F^*(x) - \pi^{-1} P.\int_{-a}^{a} \frac{d/dt F^*(t)}{t-x} dt \\ - F^*(a)(a-x)^{-1} - F^*(-a)(a+x)^{-1} = H(x). \end{aligned} \tag{2.8}$$

First we prove that

$$[d/dx F^*(x)]^2 (a^2 - x^2)^{1/2} \in L(-a, +a); \tag{2.9}$$

this assures that $[F^*(x)]' \in L(-a, +a)$ and an application of [3, Theorem 4.2] accomplishes the proof of (2.8).

To prove (2.9), we note that, by the last statement in Theorem 1.1, $H(x)$ is of the form,

$$H(x) = F_b(-a)(a + x)^{-1} + F_b(a)(a - x)^{-1} + H_1(x) \tag{2.10}$$

where $H_1 \in U[-a, +a]$ and $H_1 \in L_p(-a, +a)$ for every $p > 1$. Hence

$$F^*(x) = F_b(-a)\xi_a^+(x) + F_b(a)\xi_a^-(x) + \int_{-a}^{+a} \Gamma_a(x, y; \lambda) H_1(y) dy. \tag{2.11}$$

Now $\xi_a^\pm$ by Theorem 1.3 are in the domain of Ω_a and hence satisfy condition (2.9) (with F^* replaced by $\xi_a^\pm$). That the integral in (2.11) satisfies this condition follows easily from [3, (2.9), Theorem 1.2 and Lemma 3.1].

We have now proved (2.8). An integration by parts in the integral on the right-hand side of (2.8) gives

$$\lambda F^*(x) - \pi^{-1} P. \int_{-a}^{a} \frac{d/dt F^*(t)}{t - x} dt = \pi^{-1} \int_{a<|z|<b} \frac{F_b'(z)}{z - x} dz. \tag{2.12}$$

Next put

$$\phi(x) = F_b(x) - F_a(x) - F^*(x). \tag{2.13}$$

It follows from the definition of F_b and F_a and (2.9) that $[\phi'(x)]^2(a^2 - x^2)^{1/2} \in L[-a, +a]$ and that

$$\lambda\phi - \Omega_a\phi = 0. \tag{2.14}$$

Finally from [3, Corollary 3.1] we get

$$\lim_{x \to \pm a} F^*(\pm a) = F_b(\pm a). \tag{2.15}$$

Since (1.3) holds, $\phi(\pm a) = 0$. But we showed in [2] that the only solution of (2.14) under these boundary conditions is $\phi(x) \equiv 0$, which proves (2.3).

COROLLARY 2.1. *If P_a and P_b are defined as in* §0, *then* (0.14) *holds for $b > a$, $x \in (-a, a)$ and S a Borel subset of* $(-\infty, \infty)$.

Proof. Denote the difference between the right and left sides of (0.14) by $Q(t, x, S)$. Let f be an element of $C[-\infty, \infty]$ for which $f(\pm a) = f(\pm b) = 0$, and form the function

$$F(t, x) = \int_{-\infty}^{\infty} Q(t, x, dy) f(y). \tag{2.16}$$

We have shown in Theorem 2.1 that the Laplace transform of F with respect to t is identically 0. Since for each $x \in (-a, a)$, the function F is continuous for $t \geqq 0$, we conclude that $F(t, x) = 0$ for $x \in (-a, a)$ and $t \geqq 0$. From this it follows that $Q(t, x, S) = 0$, except perhaps when S contains the points $\pm a$, $\pm b$. However, since for any $c > 0$, $P_c(t, x, S) = 0$ if S is contained in the complement of $(-c, c)$, it can be checked directly that (0.14) holds in this case also.

3. The relation between the Cauchy process and the absorbing barrier processes. Define

$$(3.1)\qquad \Gamma(x, y; \lambda) = \int_0^\infty e^{-\lambda t} p(t, x, y)dt$$

for $\lambda > 0$ where

$$(3.2)\qquad p(t, x, y) = \frac{t}{(x-y)^2 + t^2}.$$

The main object of this section is to prove the following theorems:

THEOREM 3.1. *If $\Gamma_a(x, y; \lambda)$ is defined as in Theorem 1.1 and $\Gamma(x, y; \lambda)$ as in (3.1), then for each $a > 0$*

$$(3.3)\quad \Gamma(x, y; \lambda) = \Gamma_a(x, y; \lambda) + \pi^{-1}\int_{-a}^{a} \Gamma_a(x, u; \lambda)\left\{\int_{|z|>a} \frac{\Gamma(z, y; \lambda)}{(z-u)^2}dz\right\}du$$

when $|x| < a$, $|y| < a$ and

$$(3.4)\qquad \Gamma(x, y; \lambda) = \pi^{-1}\int_{-a}^{a} \Gamma_a(x, u; \lambda)\left\{\int_{|z|>a} \frac{\Gamma(z, y; \lambda)}{(z-u)^2}dz\right\}du$$

when $|x| < a$, $|y| > a$.

This theorem expresses relation (0.13) in Laplace transforms.

Our second main theorem is:

THEOREM 3.2. *If*

$$(3.5)\qquad F(x) = \int_{-\infty}^{\infty} \Gamma(x, y; \lambda)f(y)dy$$

for some $\lambda > 0$ and some $f \in C[-\infty, +\infty]$, then

$$(3.6)\qquad F(x) = \lim_{b\to\infty} F_b(x)$$

where F_b is given in (1.4). The convergence is uniform in any finite interval.

We first take up the

Proof of Theorem 3.1. Defining $F(x)$ as in (3.5) it follows from [6] that for fixed λ, there exists a set of f dense in $C[-\infty, +\infty]$ for which

$$(3.7)\qquad \lambda F(x) - \pi^{-1}P.\int_{-\infty}^{\infty} \frac{F'(t)}{t-x}dt = f(x)$$

almost everywhere. (This can also be shown easily by choosing $f(x)$ of the form $u(\alpha, x)$, harmonic in α and x with $\int_{-\infty}^{\infty} [u(\alpha, x)]^2 dx < \infty$.) Now $F \in C[-\infty, +\infty]$, and we can rewrite (3.7) as

$$\lambda F(x) - \tilde{\Omega}_a F(x) = \pi^{-1} \int_{|z|>a} \frac{F(z)}{(z-x)^2} dz + f(x) \tag{3.8}$$

for any $a > 0$ and $\tilde{\Omega}_a$ as in Definition 1.8.

The function on the right of (3.8) is an element of $U[-a, +a]$. Thus by Theorem 1.2,

$$F(x) = \int_{-a}^{a} \Gamma_a(x, y; \lambda) f(y) dy + \pi^{-1} \int_{-a}^{a} \Gamma_a(x, u; \lambda) \left\{ \int_{|z|>a} \frac{F(z)}{(z-u)^2} dz \right\} du \tag{3.9}$$

for every $a > 0$, whenever f belongs to our dense set for which (3.8) holds, and $|x| < a$.

Each side of (3.9) defines a bounded linear transformation from $C[-\infty, +\infty]$ to $C[-a, +a]$, and thus the identity holds for all $f \in C[-\infty, +\infty]$. This in turn implies that (3.3) and (3.4) hold.

Corollary 3.1. *If P and P_a are defined as in §0, then* (0.13) *holds.*

Proof. The proof is similar to that of Corollary 2.1.

To prove Theorem 3.2, we require a lemma:

Lemma 3.1. *If ξ_a is given by* (1.15) *then*

$$\lim_{a \to \infty} \xi_a(x) = 0 \tag{3.10}$$

and the convergence is uniform in any finite interval.

Proof. By [2, formula (2.9) and Lemma 5.1], it follows that

$$\int_{-a}^{a} \Gamma_a(x, y; \lambda) dy + \lambda \int_{-a}^{a} \Gamma_a(x, y; \lambda)(a^2 - y^2)^{1/2} dy = (a^2 - x^2)^{1/2}. \tag{3.11}$$

Therefore

$$\int_{-a}^{a} \Gamma_a(x, y; \lambda) dy = \lambda^{-1}[1 - \xi_a(x)] \geqq (a^2 - x^2)^{1/2}(1 + a\lambda)^{-1}. \tag{3.12}$$

It was shown in [2] that

$$0 \leqq \xi_a(x) \leqq 1. \tag{3.13}$$

Combining (3.12) and (3.13), we obtain

$$1 \geqq 1 - \xi_a(x) \geqq \lambda(a^2 - x^2)^{1/2}(1 + a\lambda)^{-1}, \tag{3.14}$$

which proves the lemma.

We now proceed to:

Proof of Theorem 3.2. We proved in [2, Theorem 4.2], that

(3.15) $$\lambda \mid F_b(x) \mid \leqq \max_{x\in[-b,+b]} \mid f(x) \mid .$$

Therefore by (2.3) and (1.15), it follows that for $b>a$ and $|x|<a$,

(3.16) $$\mid F_b(x) - F_a(x) \mid \leqq \lambda^{-1}\|f\|\xi_a(x).$$

Thus by Lemma 3.1 the sequence $\{F_b(x)\}$ converges uniformly in any finite interval to a limit function $\Phi(x)$. We have now only to show that Φ coincides with the function F defined by (3.5). Since the convergence of F_b to Φ is bounded, we conclude from (2.3) that (3.9) holds with F replaced by Φ. Hence $F-\Phi=\Psi$ is a bounded solution of

(3.17) $$\Psi(x) = \pi^{-1}\int_{-a}^{a}\Gamma_a(x, u; \lambda)\left\{\int_{|z|>a}\frac{\Psi(z)}{(z-u)^2}dz\right\}du,$$

for each $a>0$. But this implies that

$$\mid \Psi(x) \mid \leqq M\xi_a(x) \text{ for } \mid x \mid < a,$$

and by Lemma 3.1 we must have $\Psi(x)\equiv 0$. This completes the proof of the theorem.

As a by-product of the investigations of this section, we determine the infinitesimal generator of the semi-group $\{T_t\}$ from $C[-\infty, +\infty]$ to itself defined by (0.1).

DEFINITION 3.1. *We define* $\Omega F=G$ *if and only if* (i) $F\in C[-\infty, +\infty]$ *and* F *is absolutely continuous over any finite interval*; (ii) F' *is square integrable over any finite interval* (iii) $G\in C[-\infty, +\infty]$ *and*

(3.18) $$G(x) = \lim_{\alpha\to-\infty,\beta\to\infty} \pi^{-1}P.\int_{\alpha}^{\beta}\frac{F'(t)}{t-x}dt$$

almost everywhere.

THEOREM 3.3. *The operator* Ω *defined in Definition* 3.1 *is the infinitesimal generator of the semi-group* $\{T_t\}$ *from* $C[-\infty, +\infty]$ *to itself defined by* (0.1).

Proof. The transformation $f\to F$ defined by (3.5) is the resolvent of this semi-group. To prove our theorem we need only show that (3.5) holds if and only if $\lambda F-\Omega F=f$, where $f\in C[-\infty, +\infty]$ and $\lambda>0$. First, suppose that F is defined by (3.5). Then from Theorem 3.1 we have (3.9) for every $a>0$. Thus, $F(x)$ is of the form (1.4) with $f\in U[-a, +a]$ (the f in (1.4) is here to be replaced by $f+\pi^{-1}\int_{|z|>a}F(z)(z-x)^{-2}dz$ in the notation of this theorem); by Theorem 1.2, it follows that F also satisfies (3.8) and conditions (i) and (ii) of Definition 3.1. Since F is absolutely continuous over any finite interval, it follows from [3, Theorem 4.2] that we may rewrite (3.8) as

$$\begin{aligned}\lambda F(x) - \pi^{-1}P.\int_{-a}^{a}\frac{F'(t)}{t-x}dt + \pi^{-1}[F(a)(a-x)^{-1} + F(-a)(a+x)^{-1}]\\ = f(x) + \pi^{-1}\int_{|z|>a}\frac{F(z)}{(z-x)^2}dz,\end{aligned} \tag{3.19}$$

almost everywhere. An integration by parts on the right of (3.19) over $(\alpha, -a)$ and (a, β) gives $\lambda F - \Omega F = f$.

Conversely if $F \in$ domain Ω, then define $f = \lambda F - \Omega F$. Clearly $F \in$ domain $\tilde{\Omega}_a$ for each $a > 0$, and we may retrace our steps to obtain (3.19), (3.8) and finally (3.9). Since F satisfies (3.9) for every $a > 0$, it must coincide with (3.5), since we proved in the proof of Theorem 3.2 that (3.17) has no bounded solutions.

4. **The integral equation of return processes.** We now turn to the processes described in §(c) of the Introduction. We thus consider a fixed finite interval $(-a, a)$ with return probabilities $\tau(z)p(z, S)$ defined for $|z| > a$.

Here for fixed z (where $|z| > a$) the function $p(z, S)$ is a probability measure on the Borel sets of the open interval $(-a, a)$. For fixed S it is Borel-measurable in z. Finally $\tau(z)$ is a Borel-measurable function, $0 \leq \tau(z) \leq 1$. We complete the definition of p by putting $p(z, \{a\}) = p(z, \{-a\}) = 0$.

Let $\mathfrak{P}_a(t, x, S)$ denote the transition probabilities associated with this process, and put

$$\Pi_a(x, S; \lambda) = \int_0^\infty e^{-\lambda t}\mathfrak{P}_a(t, x, S)dt. \tag{4.1}$$

An argument similar to that used in §(b) of the Introduction to obtain (0.13) and (0.14) leads to the following

Integral equation for Π_a.

$$\begin{aligned}\Pi_a(x, S; \lambda) = \gamma_a(x, S; \lambda) + \pi^{-1}\int_{-a}^{a}\Gamma_a(x, u; \lambda)\\ \cdot\left\{\int_{|z|>a}\frac{\tau(z)}{(z-u)^2}\left[\int_{-a}^{a}p(z, dv)\Pi_a(v, S; \lambda)\right]dz\right\}du.\end{aligned} \tag{4.2}$$

Here γ_a is defined in (1.5); the interval of integration is open, but may be equally taken as closed provided one puts

$$p(z, \{a\}) = p(z, \{-a\}) = 0. \tag{4.3}$$

In this section we show that there exists a *minimal solution* to (4.2) which enjoys all properties required of Laplace transforms of transition probabilities. However, there may be additional admissible solutions of (4.2) and to prove *uniqueness* we introduce the following

Condition A. *We say that condition A is satisfied if there exist numbers*

$q<1$ and $0<\epsilon<a$ such that

$$\tau(z)p(z, T_\epsilon) < q < 1 \tag{4.4}$$

where

$$T_\epsilon = (-a, -a+\epsilon) \cup (a-\epsilon, a). \tag{4.5}$$

We now prove

THEOREM 4.1. *Define the iterated kernels K_n inductively as follows:*

$$K_0(x, S; \lambda) = \pi^{-1} \int_{-a}^{a} \Gamma_a(x, u; \lambda) \left\{ \int_{|z|>a} \frac{\tau(z)}{(z-u)^2} p(z, S) dz \right\} du,$$

$$\begin{aligned} K_{n+1}(x, S; \lambda) &= \int_{-a}^{a} K_0(x, dv; \lambda) K_n(v, S; \lambda) \\ &= \int_{-a}^{a} K_n(x, dv; \lambda) K_0(v, S; \lambda), \end{aligned} \tag{4.6}$$

$$K_n(x, \{a\}; \lambda) = K_n(x, \{-a\}; \lambda) = 0.$$

Then

$$\Pi_a(x, S; \lambda) = \gamma_a(x, S; \lambda) + \sum_{n=0}^{\infty} \int_{-a}^{a} K_n(x, dv; \lambda) \gamma_a(v, S; \lambda) \tag{4.7}$$

is a minimal solution of the integral equation (4.2) *in the following sense.*

For each $x \in E$ this Π_a is a completely additive function on the Borel sets S of E, and

$$0 \leqq \Pi_a(x, S; \lambda) \leqq \lambda^{-1}. \tag{4.8}$$

Every other non-negative solution Π_a' of (4.2) *satisfies the inequality*

$$\Pi_a' \geqq \Pi_a. \tag{4.9}$$

If the condition A *is satisfied, then* (4.7) *is the only solution of* (4.2) *which is bounded in x and S for each fixed $\lambda > 0$. Furthermore, in this case the series in* (4.7) *converges uniformly in x and S.*

Finally, if condition A *is satisfied and $\tau(z) = 1$ for each $|z| > a$, then*

$$\Pi_a(x, E; \lambda) = \lambda^{-1}. \tag{4.10}$$

Note. By definition

$$\Pi_a(x, \{a\}; \lambda) = \Pi_a(x, \{-a\}; \lambda) = 0. \tag{4.11}$$

That Π_a actually defines transition probabilities is an easy consequence of the Hille-Yosida theory and will be shown in §5 (at least under a slight additional condition).

Proof. Let

$$(4.12)\qquad \Pi_{N,a}(x, S; \lambda) = \gamma_a(x, S; \lambda) + \sum_{n=0}^{N} \int_{-a}^{a} K_n(x, dv; \lambda)\gamma_a(v, S; \lambda).$$

Clearly, for each $v \in E$ and S

$$(4.13)\qquad 0 \leqq \Pi_{N,a}(v, S; \lambda) \leqq \Pi_{N+1,a}(v, S; \lambda),$$

$$(4.14)\qquad \Pi_{N+1,a}(x, S; \lambda) = \gamma_a(x, S; \lambda) + \int_{-a}^{a} K_0(x, dv; \lambda)\Pi_{N,a}(v, S; \lambda).$$

To prove that (4.8) holds we first note that by (1.5) and (1.15)

$$(4.15)\qquad \begin{aligned} 0 &\leqq \Pi_{N,a}(x, S; \lambda) \leqq \Pi_{N,a}(x, E; \lambda) \\ &= \lambda^{-1}\{1 - \xi_a(x)\} + \lambda^{-1} \sum_{n=0}^{N} \int_{-a}^{a} K_n(x, dv; \lambda)\{1 - \xi_a(v)\}. \end{aligned}$$

Now, since $0 \leqq \tau(z)p(z, S) \leqq 1$, we have

$$(4.16)\qquad 0 \leqq K_0(x, S; \lambda) \leqq \xi_a(x) \leqq 1$$

by (1.15) and (3.13). Therefore

$$(4.17)\qquad \sum_{n=0}^{N} \int_{-a}^{a} K_n(x, dv; \lambda)\{1 - \xi_a(v)\} \leqq \sum_{n=0}^{N} \{K_n(x, E; \lambda) - K_{n+1}(x, E; \lambda)\}.$$

Combining the last three relations we have

$$(4.18)\qquad 0 \leqq \Pi_{N,a}(x, S; \lambda) \leqq \lambda^{-1}\{1 - \xi_a(x) + K_0(x, E; \lambda) - K_{N+1}(x, E; \lambda)\} \leqq \lambda^{-1}.$$

We conclude from (4.13) and (4.18) that for each fixed x and S the sequence $\Pi_{N,a}(x, S; \lambda)$ converges; it follows from monotonic convergence that the limit function (4.7) is a solution of (4.2) and satisfies (4.8). This proves the first part of the theorem.

For any non-negative solution of (4.2) we have obviously

$$(4.19)\qquad \begin{aligned} \Pi_a(x, S; \lambda) &= \gamma_a(x, S; \lambda) + \sum_{n=0}^{N} \int_{-a}^{a} K_n(x, dv; \lambda)\gamma_a(v, S; \lambda) \\ &\quad + \int_{-a}^{a} K_{N+1}(x, dv; \lambda)\Pi_a(v, S; \lambda). \end{aligned}$$

The minimality assertion (4.9) is a direct consequence of this. Also, it follows from (4.19) that Π_a as defined by (4.7) is the *unique* bounded solution whenever

$$(4.20)\qquad \lim_{N \to \infty} K_N(x, E; \lambda) = 0 \text{ for each } x.$$

Suppose now that condition A is satisfied. From (4.6) and (4.16) we get

(4.21) $$K_N(x, E; \lambda) \leqq \int_{-a}^{a} K_{N-2}(x, dw; \lambda)\left\{\int_{-a}^{a} K_0(w, dv; \lambda)\xi_a(v)\right\}.$$

Choose ϵ so that (4.4) holds, and put for this fixed ϵ

(4.22) $$\eta = \sup_{x\in E-T_\epsilon} \xi_a(x).$$

It follows from (3.12) that $0<\eta<1$. Clearly

(4.23) $$\int_{-a}^{a} K_0(w, dv; \lambda)\xi_a(v) \leqq (1-\eta)K_0(w, T_\epsilon; \lambda) + \eta K_0(w, E; \lambda).$$

From the first equation in (4.6) we get using condition A and (1.15)

(4.24) $$K_0(w, T_\epsilon; \lambda) \leqq q\cdot\xi_a(w) \leqq q,$$

and so by (4.23)

(4.25) $$\int_{-a}^{a} K_0(w, dv; \lambda)\xi_a(v) \leqq q(1-\eta) + \eta = \alpha < 1.$$

From (4.21) we see that

(4.26) $$K_N(x, E; \lambda) \leqq \alpha K_{N-2}(x, E; \lambda)$$

which proves that (4.20) holds and that the series in (4.7) converges uniformly.

Finally, if $\tau(z)=1$ for each $|z|>a$, then the equality sign holds in (4.17) and this together with (4.20) implies (4.10). This completes the proof.

We conclude this section by an example which shows that *in the absence of condition* A *the solution* Π_a *of* (4.7) *need not be the only bounded solution of the integral equation* (4.2).

If this example is modified by defining $\tau(z)\equiv 1$ for $|z|\geqq a$, then the new solution shows that *even if* $\tau(z)=1$ *for each* $|z|>a$ *the minimal solution* Π_a *does not necessarily satisfy* (4.10). The probabilistic implication of this is described in the Introduction.

EXAMPLE. We first choose two sequences $\{\epsilon_k\}$ and $\{\eta_k\}$ of positive numbers for which

(4.27) $$\sum_{k=0}^{\infty} \epsilon_k < \infty, \qquad \epsilon_k \downarrow 0, \qquad \eta_k \downarrow 0.$$

Let

(4.28) $$A_0 = \{x\in[-a, a]/\xi_a(x) > 1-\epsilon_0\},$$
$$A_k = \left\{x\in[-a, a]/\pi^{-1}\int_{-a}^{a} \Gamma_a(x, u; \lambda)\left[\int_a^{a+\eta_k} \frac{dz}{(z-u)^2}\right]du > 1-\epsilon_k\right\}.$$

Each A_k contains an interval $[a-\delta, a]$ since, for each $\eta>0$,

$$\lim_{x\to\pm a}\left\{\pi^{-1}\int_{-a}^{a}\Gamma_a(x,u;\lambda)\left[\int_a^{a+\eta}\frac{dz}{(z-u)^2}\right]du\right\} = 1 \tag{4.29}$$

by [3, Corollary 3.1]. It is also clear that $A_{k+1}\subset A_k$.

We now choose $\tau(z)$ and $p(z, S)$ to satisfy the following conditions:

$$\begin{aligned} \tau(z) &\equiv 0, && z \leqq -a,\\ \tau(z) &\equiv 1, && z \geqq a,\\ p(z, A_{k+1}) &= 1, && a < z < a+\eta_k. \end{aligned} \tag{4.30}$$

The sets A_k contract as $k\to\infty$ to the end points $\pm a$, and so condition A is violated.

A bounded solution to the equation

$$F(x, S) = \int_{-a}^{a} F(v, S)K_0(x, dv; \lambda) \tag{4.31}$$

may be constructed by successive approximation as follows: Put

$$\begin{aligned} F_0(x, S) &\equiv 1,\\ F_{n+1}(x, S) &= \int_{-a}^{a} K_0(x, dv; \lambda)F_n(v, S). \end{aligned} \tag{4.32}$$

By (4.16)

$$0 \leqq F_1(x, S) \leqq 1 = F_0(x, S), \tag{4.33}$$

and by induction

$$0 \leqq F_{n+1}(x, S) \leqq F_n(x, S) \leqq 1. \tag{4.34}$$

Thus

$$F(x, S) = \lim_{n\to\infty} F_n(x, S) \tag{4.35}$$

exists for each fixed x and S and satisfies (4.31).

We have only to show that $F(x, S)\not\equiv 0$. First,

$$\begin{aligned} F_n(x, S) &\geqq \int_{-a}^{a}\Gamma_a(x,u;\lambda)\left\{\int_a^{a+\eta_k}(z-u)^{-2}\int_{A_{k+1}} p(z, dv)F_{n-1}(v, S)\right\} du\\ &\geqq \left\{\min_{x\in A_{k+1}} F_{n-1}(x, S)\right\}\left\{\int_{-a}^{+a}\Gamma_a(x,u;\lambda)\left[\int_a^{a+\eta_k}\frac{dz}{(z-u)^2}\right]du\right\}. \end{aligned} \tag{4.36}$$

Therefore, for $x\in A_k$,

$$F_n(x, S) \geqq (1-\epsilon_k)\min_{x\in A_{k+1}} F_{n-1}(x, S). \tag{4.37}$$

Now for $x\in A_0$,

$$
\begin{aligned}
F_n(x, S) &\geqq (1 - \epsilon_0) \min_{x \in A_1} F_{n-1}(x, S) \\
&\geqq \prod_{k=0}^{n-1}(1 - \epsilon_k) \min_{x \in A_n} F_0(x, S) \\
&= \prod_{k=0}^{n-1} (1 - \epsilon_k),
\end{aligned} \tag{4.38}
$$

by repeated application of (4.37). Hence, for $x \in A_0$,

$$F(x, S) \geqq \prod_{k=0}^{\infty} (1 - \epsilon_k) > 0. \tag{4.39}$$

But A_0 is a set of positive measure, and thus F is a nontrivial bounded solution of (4.31).

5. **The associated semi-group.** Let $\Pi_a(x, S; \lambda)$ be the solution (4.7) of the integral equation (4.2). For $f \in C[-a, a]$ put

$$R_\lambda f(x) = \int_{-a}^{a} \Pi_a(x, dv; \lambda) f(v), \qquad x \in E. \tag{5.1}$$

We introduce a condition on the function p which assures that (5.1) defines a bounded linear transformation from $C[-a, a]$ to itself.

CONDITION B. *For each* $f \in C[-a, a]$

$$\operatorname*{essential\ limit}_{z \to \pm a} \tau(z) \int_{-a}^{a} p(z, dv) f(v) \tag{5.2}$$

exists.

The two main theorems of this section are:

THEOREM 5.1. *Let* Π_a *be defined by* (4.7); *under conditions* A *and* B *the transformation* R_λ *defined by* (5.1) *is a bounded linear transformation from* $C[-a, +a]$ *to itself and*

$$\|R_\lambda\| \leqq \lambda^{-1}. \tag{5.3}$$

If $\tau(z) \equiv 1$, *then equality holds in* (5.3). (*See* §4 *for definition of* τ.)

THEOREM 5.2. *Under conditions* A *and* B, *if* $F = R_\lambda f$ *for some* $f \in C[-a, +a]$ *and some* $\lambda < 0$, *then* F *is the unique solution in* $C[-a, +a]$ *of*

$$\lambda F(x) - \tilde{\Omega}_a F(x) - \int_{|z|>a} \frac{\tau(z) \int_{-a}^{a} p(z, dv) F(v)}{(z - x)^2} = f(x) \tag{5.4}$$

where p *and* τ *are defined in* §4 *and* $\tilde{\Omega}_a$ *is the operator of Definition* 1.3.

Define

$$AF(x) = \tilde{\Omega}_a F(x) + \int_{|z|>a} \frac{\tau(z)\int_{-a}^{a} p(z, dv)F(v)}{(z-x)^2}\, dz \tag{5.5}$$

and let $\mathfrak{D}(A)$ denote the set of $F \in C[-a, +a]$ for which the right side of (5.5) is in $C[-a, +a]$. Thus A defines a transformation on $C[-a, +a]$ to itself with domain $\mathfrak{D}(A)$. Theorem 5.2 shows that for each $\lambda > 0$, the range of R_λ coincides with $\mathfrak{D}(A)$. From the Hille-Yosida theorem and Theorems 5.1 and 5.2 we conclude that A is the infinitesimal generator of a semi-group $\{U_t\}$ from $\overline{\mathfrak{D}(A)}$ to itself, which is strongly continuous for $t \geqq 0$. This semi-group can be written in the form

$$U_t f(x) = \int_{[-a,+a]} \mathfrak{P}_a(t, x, dv) f(v). \tag{5.6}$$

Since U_t is positivity preserving and $\|U_t\| \leqq 1$ (again by the Hille-Yosida theorem), we see that for each x, $\mathfrak{P}_a(t, x, S)$ defines a probability measure on the Borel subsets of $[-a, +a]$. The function $\mathfrak{P}$ gives the transition probabilities of the return process of §(c) in the Introduction.

The backward equation for the return process is thus given by

$$\frac{\partial}{\partial t} U_t f(x) = A U_t f(x) \tag{5.7}$$

and if $f \in \overline{\mathfrak{D}(A)}$, then $\lim_{t \to 0} U_t f(x) = f(x)$ uniformly in x.

We shall also show, under our conditions A and B, that $\overline{\mathfrak{D}(A)}$ is actually a proper subspace of $C[-a, +a]$:

THEOREM 5.3. *Under conditions* A *and* B *the range of* R_λ *is contained in the subspace of* $C[-a, +a]$ *of functions* F *satisfying*

$$F(\pm a) = \text{ess.} \lim_{z \to \pm a} \tau(z) \int_E p(z, dv) F(v). \tag{5.8}$$

The set defined by (5.8) is a closed linear manifold in $C[-a, +a]$. Clearly, under condition A, this manifold cannot coincide with $C_1[-a, a]$.

We now proceed to prove the theorems of this section.

Proof of Theorem 5.1. From (4.7), we have for each $f \in C[-a, +a]$,

$$\begin{aligned} R_\lambda f(x) &= \int_{-a}^{a} \Gamma_a(x, u; \lambda) f(u)\, du \\ &\quad + \sum_{n=0}^{\infty} \int_{-a}^{a} f(u) \left\{ \int_{-a}^{a} \Gamma_a(v, u; \lambda) K_n(x, dv; \lambda) \right\} du, \end{aligned} \tag{5.9}$$

where the series on the right converges uniformly in x. Let us consider the first term of the series in (5.9),

$$S_0(x) = \int_{-a}^{a} K_0(x, dv; \lambda) \left\{ \int_{-a}^{a} \Gamma_a(v, u; \lambda) f(u) du \right\} \tag{5.10}$$
$$= \int_{-a}^{a} K_0(x, dv; \lambda) F_a(v).$$

Now $F_a(v) \in C[-a, +a]$, and it is easily seen by the definition of K_0 and condition B that S_0 is of the form (1.4) with $f \in U[-a, +a]$. Hence $S_0 \in C[-a\ +a]$. Letting S_n represent the nth term of the series we have

$$S_{n+1}(x) = \int_{-a}^{a} K_0(x, dv; \lambda) S_n(v) \tag{5.11}$$

and it follows by induction that $S_n \in C[-a, +a]$ for every n. Thus, since the series $\sum_{n=0}^{\infty} S_n(x)$ converges uniformly in x, we conclude that $R_\lambda f \in C[-a, +a]$. Finally (5.3) follows from (4.8).

Proof of Theorem 5.2. It is now very simple to prove Theorem 5.2. If $F = R_\lambda f$ for some $f \in C[-a, +a]$, then from (4.2) we have

$$F(x) = F_a(x)$$
$$+ \pi^{-1} \int_{-a}^{a} \Gamma_a(x, u; \lambda) \left\{ \int\int_{|z|>a} \frac{\tau(z)}{(z-u)^2} \left[\int_{-a}^{a} p(z, dv) F(v) \right] dz \right\} du \tag{5.12}$$

with F_a given by (1.4). Since $F \in C[-1, +1]$, it follows from condition B that the function in braces in (5.12) is an element of $U[-a, +a]$. Hence (5.4) follows from Theorem 1.2. The uniqueness follows from (4.20).

Proof of Theorem 5.3. We showed in [3, Corollary 3.1], that if F_a is given by (1.4) with $f \in U[-a, +a]$, then

$$\lim_{x \to \pm a} F(x) = (2a)^{-1} \cdot \pi \lim_{x \to \pm a} f(x)(a^2 - x^2).$$

From this it is easily seen that (5.8) holds.

6. The adjoint semi-group. In this section we shall consider the transformation

$$R_\lambda^* \mu(S) = \int_{-a}^{+a} \Pi_a(x, S; \lambda) \mu(dx) \tag{6.1}$$

where μ is a measure on the *open* interval $(-a, +a)$ and Π_a is defined by (4.7). We shall prove the following two theorems:

THEOREM 6.1. *Under conditions* A *and* B, *the transformation* R_λ^* *given by* (6.1) *is a bounded linear transformation from the space* $\mathfrak{M}$ *of totally finite measures on the open interval* $(-a, +a)$ *to itself, for each* $\lambda > 0$, *and*

$$\|R_\lambda^* \mu\| \leqq \lambda^{-1} \|\mu\|. \tag{6.2}$$

Furthermore, $R_\lambda^* \mu$ is an absolutely continuous measure for which

$$\int_{-a}^{+a} \frac{R_\lambda^* \mu(dx)}{a^2 - x^2} < \infty \tag{6.3}$$

and

THEOREM 6.2. *If $M = R_\lambda^* \mu$ for some $\lambda > 0$ and some $\mu \in \mathfrak{M}$, then under conditions* A *and* B *we have for almost all intervals $I \subset (-a, +a)$,*

$$\lambda M(I) - A^* M(I) = \mu(I) \tag{6.4}$$

where

$$A^* M(I) = \pi^{-1} \int_I \frac{d}{dx} P.\int_{-a}^{a} \frac{M(dy)}{y - x} + \int_{-a}^{a} M(du) \int_{|z|>a} \frac{\tau(z) p(z, I)}{(z - u)^2} dz. \tag{6.5}$$

These two theorems show that R_λ^* is the resolvent of a semi-group $\{U_t^*\}$ from $\mathfrak{M}$ to itself, whose infinitesimal generator is of the form (6.5). Thus the "forward equation" for our return processes can be written as

$$\frac{\partial}{\partial t} U_t \mu(I) = A^* U_t \mu(I).$$

Proof of Theorem 6.1. Consider the function

$$\Psi(S) = \sum_{n=0}^{\infty} \int_{-a}^{a} K_n(x, S; \lambda) \mu(dx) \tag{6.6}$$

for a fixed measure μ on $(-a, +a)$. Here Ψ is defined on Borel subsets S of $(-a, +a)$. From (4.26) we conclude that Ψ is a non-negative, totally finite measure on $(-a, +a)$. Moreover,

$$M(S) = R_\lambda^* \mu(S) = \int_{-a}^{a} \Psi(dv) \gamma_a(v, S; \lambda) + \int_{-a}^{a} \mu(dx) \gamma_a(x, S; \lambda). \tag{6.7}$$

By Lemma 11.1 in [2] it follows that M is an absolutely continuous measure. To obtain (6.3) we note that

$$\begin{aligned} \int_{-a}^{a} \frac{M(dx)}{a^2 - x^2} &= \int_{-a}^{a} \{\Psi(dv) + \mu(dv)\} \int_{-a}^{a} \frac{\Gamma_a(v, x; \lambda)}{(a^2 - x^2)} dx \\ &\leqq \{\|\Psi\| + \|\mu\|\} \cdot \pi (2a)^{-1} \end{aligned} \tag{6.8}$$

by (1.15) and (3.13). This completes the proof.

Proof of Theorem 6.2. Using (6.7) and Lemma 11.1 in [2], we have

$$(6.9)\qquad M(I) - \pi^{-1}\int_I \left\{\frac{d}{dx}\, P.\int_{-a}^{a} \frac{M(dy)}{y-x}\right\} dx = \Psi(I) + \mu(I),$$

for almost all intervals I. To obtain (6.4), we shall show that

$$(6.10)\qquad \Psi(I) = \int_{-a}^{a} M(du)\left\{\int_{|z|>a} \frac{\tau(z)p(z, I)}{(z-u)^2}\, dz\right\}.$$

Note that the integral on the right of (6.10) exists by (6.3). It is easily verified that

$$(6.11)\qquad \Psi(I) = \int_{-a}^{a} \Psi(dv)K_0(v, I; \lambda) + \int_{-a}^{a} \mu(dv)K_0(v, I; \lambda).$$

On the other hand, letting

$$(6.12)\qquad L(u, S) = \int_{|z|>a} \frac{\tau(z)p(z, S)}{(z-u)^2}\, dz,$$

we have

$$(6.13)\qquad \begin{aligned} \int_{-a}^{a} \Psi(dv)K_0(v, I; \lambda) &= \int_{-a}^{a} \Psi(dv)\int_{-a}^{a} \gamma_a(v, du; \lambda)L(u, I) \\ &= \int_{-a}^{a} L(u, I)\int_{-a}^{a} \Psi(dv)\gamma_a(v, du; \lambda) \\ &= \int_{-a}^{a} L(u, I)\left\{M(du) - \int_{-a}^{a} \gamma_a(x, du; \lambda)\mu(dx)\right\} \end{aligned}$$

by (6.7). This completes the proof.

7. **Return processes leading to harmonic functions.** In this section we deal with return processes of the type described in §(d) of the Introduction.

We first take up a return process which leads to the semi-group (0.26). For simplicity we shall consider the process on $(-1, +1)$. In this case $\tau(z) \equiv 1$ and

$$(7.1)\qquad \begin{aligned} p(z, S) &= 1 \quad z - 2n \in S \\ &= 0 \quad z - 2n \notin S \end{aligned}\Bigg\} z \in (2n-1, 2n+1)$$

where S is a Borel subset of $(-1, 1)$. *Note that condition* A *is not satisfied for this choice of* τ *and* p. Using the periodicity of p, we may write the integral equation (4.2) as

$$(7.2)\qquad \begin{aligned} \Pi_1(x, S; \lambda) &= \gamma_1(x, S; \lambda) \\ &\quad + \pi^{-1}\int_{-1}^{1} \Gamma_1(x, u; \lambda)\left\{\int_{-1}^{1} \Pi_1(z, S; \lambda)R(z, u)dz\right\} du, \end{aligned}$$

where

$$(7.3)\quad R(z, u) = \sum_{n\neq 0} \frac{1}{(z-2n-u)^2} = \frac{\pi^2}{4}\left[\sin\frac{\pi}{2}(z-u)\right]^{-2} - (z-u)^{-2}.$$

It is easily seen from (3.3) and (3.4) that a solution of (7.2) is given by

$$(7.4)\qquad \Pi_1(x, S; \lambda) = \int_S \sum_{k=-\infty}^{+\infty} \Gamma(x, y+2k; \lambda)$$

for $\lambda > 0$. However, since condition A is violated we cannot use the argument of Theorem 4.1 to conclude that (7.4) is the only admissible solution. The next theorem establishes this uniqueness (up to sets of measure zero in x).

THEOREM 7.1. *For each* $\lambda > 0$, *the function* Π_1 *defined by* (7.4) *is the only solution of* (7.2) *for which* $0 \leqq \Pi_1(x, S; \lambda) \leqq M < \infty$ *with* M *independent of* x *and* S.

Proof. Define a linear transformation T by

$$(7.5)\qquad T\phi(x) = \pi^{-1}\int_{-1}^{1} \Gamma_1(x, u; \lambda)\left\{\int_{-1}^{1} \phi(z)R(z, u)dz\right\} du.$$

If there existed two bounded solutions to (7.2), then their difference would be, for each fixed S, a bounded solution of

$$(7.6)\qquad T\phi(x) = \phi(x).$$

We may assume $|\phi(x)| \leqq 1$; since T is a positivity preserving transformation $|\phi(x)| = |T^n\phi(x)| \leqq T^n 1$, where 1 denotes the function identically equal to 1 in $[-1, +1]$. To show that (7.6) implies $\phi(x) \equiv 0$, it is therefore sufficient to show that $T^n 1 \to 0$ as $n \to \infty$ for each $x \in (-1, +1)$. Now

$$(7.7)\qquad 0 \leqq T1 = \xi_1 \leqq 1$$

where ξ_1 is defined by (1.15). Hence,

$$(7.8)\qquad 0 \leqq T^{n+1}1 \leqq T^n 1 \leqq 1, \qquad n \geqq 1.$$

We conclude from this that $T^n 1(x)$ converges to a limit function $\psi \geqq 0$.

Moreover,

$$(7.9)\qquad \psi(x) = \pi^{-1}\int_{-1}^{1} \Gamma_1(x, u; \lambda)\left\{\int_{-1}^{1} \psi(z)R(z, u)dz\right\} du.$$

Using (1.15), we have

$$(7.10)\qquad 2\int_{-1}^{1} \psi(x)(1-x^2)^{-1}dx = \int_{-1}^{1} \xi_1(u)\left\{\int_{-1}^{1} \psi(z)R(z, u)dz\right\} du.$$

But

$$\int_{-1}^{1} R(x, u)\, du = 2(1 - x^2)^{-1}, \tag{7.11}$$

and (7.10) becomes

$$\int_{-1}^{1} \psi(x) \left\{ \int_{-1}^{1} R(x, u) du \right\} dx = \int_{-1}^{1} \xi_1(u) \left\{ \int_{-1}^{1} \psi(z) R(z, u) dz \right\} du, \tag{7.12}$$

and thus

$$0 = \int_{-1}^{1} \psi(z) \left\{ \int_{-1}^{1} R(z, u) [1 - \xi_1(u)] du \right\} dz. \tag{7.13}$$

But the function in braces is a positive function and $\psi(z) \geqq 0$. Hence, $\psi(z) = 0$ almost everywhere.

THEOREM 7.2. *If, for some* $\lambda > 0$ *and some* $f \in C[-1, +1]$,

$$F(x) = \int_{-1}^{1} \Pi_1(x, dy; \lambda) f(y), \tag{7.14}$$

where Π_1 *is given by* (7.4), *then*

$$\lambda F(x) - \frac{1}{2} P. \int_{-1}^{1} F'(t) \cot \frac{\pi}{2} (t - x) dt = f(x) \tag{7.15}$$

almost everywhere on $[-1, +1]$.

Proof. By (7.2), F satisfies

$$\begin{aligned} F(x) &= \int_{-1}^{1} \Gamma_1(x, u; \lambda) f(u) du \\ &+ \pi^{-1} \int_{-1}^{1} \Gamma_1(x, u; \lambda) \left\{ \int_{-1}^{1} F(z) R(z, u) dz \right\} du, \end{aligned} \tag{7.16}$$

and therefore is of the form (1.4) for some $f \in U[-a, +a]$. By Theorem 1.2,

$$\lambda F(x) - \tilde{\Omega}_1 F(x) = f(x) + \int_{-1}^{1} R(z, u) F(u) du. \tag{7.17}$$

By [3, Theorem 4.2], we can simplify this expression if F is absolutely continuous on $(-1, +1)$. That this is so can be easily verified; let f have the Fourier expansion

$$f(x) \sim \sum_{n=0}^{\infty} a_k \cos k\pi x + b_k \sin k\pi x \tag{7.18}$$

then

$$F(x) = \sum_{n=0}^{\infty} (\lambda + n\pi)^{-1}\{a_n \cos k\pi x + b_k \sin k\pi x\}, \tag{7.19}$$

the series converging uniformly. From (7.19) we see that F is absolutely continuous on $(-1, +1)$. Hence [3, Theorem 4.2] we can rewrite (7.17) in the form

$$\begin{aligned} F(x) - \pi^{-1}P.\int_{-1}^{1} \frac{F'(t)}{t-x}\,dt \\ &= f(x) + \pi^{-1}\int_{-1}^{1} F(z) \left\{\frac{\pi^2}{4}\left[\sin\frac{\pi}{2}(z-u)\right]^{-2} - (z-u)^{-2}\right\} dz \\ &\quad - F(1)(1-x)^{-1} - F(-1)(1+x)^{-1} \\ &= f(x) + \pi^{-1}\int_{-1}^{1} F'(z)\left\{\frac{\pi}{2}\cot\frac{\pi}{2}(z-u) - (z-u)^{-1}\right\} dz \end{aligned} \tag{7.20}$$

which, upon rearrangement gives (7.15).

It follows from (7.18) and (7.19) that *the semi-group associated with this process has the representation*

$$T_t f = \sum_{n=0}^{\infty} e^{-n\pi t}\{a_n \cos n\pi x + b_n \sin n\pi x\}. \tag{7.21}$$

The infinitesimal generator is

$$\Omega F = \frac{1}{2}\, P.\int_{-1}^{1} F'(t) \cot\frac{\pi}{2}(t-x)dt, \tag{7.22}$$

where the domain of Ω consists of those F which are absolutely continuous on $[-1, 1]$, and for which the right side of (7.22) is equivalent to an element of $C[-1, +1]$.

A return process similar to the one described above is obtained by choosing $\tau(z) \equiv 1$ and

$$\left.\begin{aligned} p(z, S) &= 1 \qquad (-1)^n(z-2n) \in S \\ &= 0 \qquad (-1)^n(z-2n) \notin S \end{aligned}\right\} z \in (2n-1, 2n+1). \tag{7.23}$$

In this case the resolvent kernel satisfies,

$$\begin{aligned} \Pi_1(x, S; \lambda) &= \gamma_1(x, S; \lambda) \\ &+ \pi^{-1}\int_{-1}^{1} \Gamma_1(x, u; \lambda)du \int_{-1}^{1} \Pi_1(z, S; \lambda) \left\{\sum_{n\neq 0} [(-1)^n z - u + 2n]^{-2}\right\} dz. \end{aligned} \tag{7.24}$$

Now

$$(7.25)\quad \sum_{n\neq 0}[(-1)^n z-u+2n]^{-2}=\frac{\pi^2}{16}\left[\left\{\sin\frac{\pi}{4}(z-u)\right\}^{-2}+\left\{\cos\frac{\pi}{4}(z+u)\right\}^{-2}\right]-(z-u)^{-2}$$

and if F is defined by (7.14) we obtain as before

$$(7.26)\quad \lambda F(x)-\frac{1}{4}P.\int_{-1}^{1}F'(t)\left\{\cot\frac{\pi}{4}(t-x)+\tan\frac{\pi}{4}(t+x)\right\}dt=f.$$

The semi-group on $C[-1,+1]$ associated with this process is

$$(7.27)\quad T_t f(x)=\sum_{n=0}^{\infty}e^{-n\pi t/2}a_n\cos\frac{n\pi}{2}(x-1)$$

where f has the formal expansion (7.27) with $t=0$. The infinitesimal generator is given by

$$(7.28)\quad \Omega F=\frac{1}{4}P.\int_{-1}^{1}F'(t)\left\{\cot\frac{\pi}{4}(t-x)+\tan\frac{\pi}{4}(t+x)\right\}dt,$$

the domain of Ω consisting of those $F\in C[-1,+1]$ which are absolutely continuous on $(-1,+1)$ and for which the right side of (7.28) is equal almost everywhere to an element of $C[-1,+1]$.

References

1. J. L. Doob, *Markoff chains—denumerable case*, Trans. Amer. Math. Soc. vol. 58 (1945) pp. 455–473.

2. J. Elliott, *The boundary value problems and semi-groups associated with certain integro-differential operators*, Trans. Amer. Math. Soc. vol. 76 no. 2 (1954) pp. 300–331.

3. ———, *On an integro-differential operator of the Cauchy type*, Proc. Amer. Math. Soc. vol. 7 (1956) pp. 616–626.

4. W. Feller, *The parabolic differential equations and the associated semi-groups of transformations*, Ann. of Math. vol. 55 (1952) pp. 468–519.

5. ———, *Diffusion processes in one dimension*, Trans. Amer. Math. Soc. vol. 77 no. 1 (1954) pp. 1–31.

6. E. Hille, *On the generation of semi-groups and the theory of conjugate functions*, Kungl. Fysiografiska Sällskapets i Lund Förhandlingar vol. 21 (1952).

7. M. Kac, *On some connections between probability theory and differential and integral equations*, Proceedings of the 2nd Symposium on Probability and Statistics, Berkeley, 1950, pp. 189–215.

8. M. Kac and H. Pollard, *The distribution of the maximum of partial sums of independent random variables*, Canadian Journal of Mathematics vol. 2 (1950) pp. 375–384.

9. H. P. McKean, Jr., *Sample functions of stable processes*, Ann. of Math. vol. 61 (1955) pp. 564–579.

Columbia University,
New York, N. Y.
Princeton University,
Princeton, N. J.

GENERALIZED SECOND ORDER DIFFERENTIAL OPERATORS AND THEIR LATERAL CONDITIONS

BY WILLIAM FELLER[1]

1. Introduction

The differential operator

$$\mathfrak{A} = aD_s^2 + bD_s, \qquad a > 0, \quad D_s = \frac{d}{ds}, \tag{1.1}$$

defined on an interval $s_1 < s < s_2$ may be reduced to the form

$$\mathfrak{A} = \alpha D_x^2, \qquad \alpha > 0 \tag{1.2}$$

by introducing a new variable x, namely, a solution of $\mathfrak{A}x = 0$. All solutions of this differential equation are monotonic, and x is determined up to a linear transformation. Putting

$$m(x) = \int_{x_0}^{x} \alpha^{-1}(t)\, dt, \tag{1.3}$$

we can further reduce $\mathfrak{A}$ to a succession of two differentiations

$$\mathfrak{A} = D_m D_x. \tag{1.4}$$

It turns out that the canonical form (1.4) has considerable formal advantages over (1.1).

It will be observed that (1.4) is much more general than (1.1). In fact, *the operator* $\mathfrak{A} = D_m D_x$ *is meaningful for an arbitrary strictly increasing function* m, not necessarily continuous or bounded. (See Section 2 for the definitions.) This generalization introduces a remarkable simplification and unification of the theory.[2] It is usual to transform (1.1) in several arbitrary ways, but the

Received March 28, 1957.

[1] This research was done while the author was on sabbatical leave of absence with the support of the United States Air Force through the Air Force Office of Scientific Research of the Air Research and Development Command.

[2] One has frequently to deal with functions defined, in two contiguous intervals, by two differential equations joined by so-called "transition conditions". In the new notation such a "system" reduces to *one* differential equation, and no transition conditions occur. For a simple example define x and m by

$$x(s) = \begin{cases} s \text{ for } s < 0 \\ 2s \text{ for } s > 0, \end{cases} \qquad m(s) = s. \tag{*}$$

Then (1.4) reduces to D_s^2 when $s < 0$ and $\frac{1}{2}D_s^2$ when $s > 0$, with the "transition condition" $D_s^- f(0) = \frac{1}{2}D_s^+ f(0)$. Interchanging x and m in (*) leads to the *same* differential system with the transition condition $D_s^- f(0) = D_s^+ f(0)$. A yearly crop of papers devoted to series expansions under conditions of this kind now becomes redundant, since the same general eigenfunction expansion is valid for *all* our operators. Moreover, due to the intrinsic characterization of the operators (1.4), it is possible to write down all possible transition conditions and render them more plausible.

R.L. Schilling et al. (eds.), *Selected Papers II*,

scale x and the "canonical measure" m are essentially unique and intrinsically related to the operator (1.4). The latter plays an important role in the theory of the vibrating string, m representing the distribution of the actual mass. With the new notation, strings with a positive mass concentrated at individual points require no special treatment. As a matter of curiosity it is possible to treat the motion of a continuous string with the entire mass concentrated at the rational points.[3]

It is easy to understand why a treatment of differential operators in the special form (1.1) is never satisfactory. The *formal adjoint* of (1.1) is given by $\mathfrak{A}^*g = D_s^2(ag) - D_s(bg)$ provided the coefficients a and b are sufficiently regular. Without this highly undesirable proviso no formal adjoint of the form (1.1) exists. By contrast, no difficulties arise in connection with (1.4). Considered as operator on functions continuous in an interval $\beta_1 \leqq x \leqq \beta_2$, it admits of the formal adjoint $\mathfrak{A}^* = D_x D_m$, which is an operator on measures. (The true adjoint will be discussed in Section 13.)

A perhaps more familiar point of view consists in considering $\mathfrak{A}$ as an operator in the traditional Hilbert space $\mathbf{H}_m$ defined by the norm

$$\| f \|_m^2 = \int_{\beta_1+}^{\beta_2-} f^2(x)\, dm(x). \tag{1.5}$$

The operator $\mathfrak{A}$ then becomes formally self-adjoint. In Section 10 we shall discuss how to make $\mathfrak{A}$ a true self-adjoint operator. In Section 11 we shall take up the same problem for the more desirable (though less familiar) Hilbert space $\overline{\mathbf{H}}_m$ over the *closed* interval $\beta_1 \leqq x \leqq \beta_2$ when m is extended to the boundaries by attributing a positive weight m_i to β_i; the norm in this space is defined by

$$\overline{\| f \|_m^2} = \| f \|_m^2 + m_1 f^2(\beta_1) + m_2 f^2(\beta_2). \tag{1.6}$$

The operators of the form (1.4) are still of a special form, and one is automatically led to the study of the more general operator defined by $\mathfrak{A}f + cf$, where c is a function. However, the class of operators of this type does not admit of an elegant theory free of artificial restrictions. A more natural class may be obtained as follows. Let ψ be a solution of the differential equation

$$\mathfrak{A}\psi + c\psi = 0. \tag{1.7}$$

In any interval not containing a zero of ψ we may define a new operator $\mathfrak{A}_\psi$ by

$$\mathfrak{A}_\psi f = \frac{1}{\psi} D_m \psi^2 D_x \frac{f}{\psi} \tag{1.8}$$

[3] A *continuous* diffusion process in which the total mass is at any time t concentrated in the set of rationals was discussed by Feller and McKean [4]. It is noteworthy that in this case the parabolic *partial differential equation* $D_t u = D_m D_x u$ *is equivalent to an infinite system of ordinary differential equations.*

(precise definitions are postponed to Section 2). It turns out (see Section 5) that

(1.9) $$\mathfrak{A}_\psi f = \mathfrak{A}f + cf = D_m D_x f + cf.$$

Thus $\mathfrak{A} + c$ may be reduced to the form $\mathfrak{A}_\psi$, but operators of this form are more general. An *arbitrary continuous* $\psi > 0$ *defines an operator* $\mathfrak{A}_\psi$ (depending on m). *If* ψ has a derivative $D_x \psi = \psi'$, then

(1.10) $$\mathfrak{A}_\psi f = \frac{1}{\psi} D_m\{\psi f' - \psi' f\},$$

implying that each f in the domain of $\mathfrak{A}_\psi$ possesses a derivative $D_x f = f'$. *If* ψ' is continuously differentiable with respect to m, then (1.10) reduces further to (1.9) with

(1.11) $$c = -\psi^{-1}\mathfrak{A}\psi = -\psi^{-1} D_m D_x \psi.$$

If ψ' is of bounded variation (without $D_m \psi'$ existing), the coefficient c can no longer be expressed as a function, but an *integrated version* of (1.9) remains valid: for each open subinterval

(1.12) $$\int_{x_1}^{x_2} (\mathfrak{A}_\psi f)\, dm = D_x f \Big|_{x_1}^{x_2} - \int_{x_1}^{x_2} \psi^{-1} f \cdot d\psi'.$$

It is thus seen that operators of the form (1.8) give a meaning to the notion of $\mathfrak{A}f + cf$ for the case where c is, say, a *Dirac function*.[4] They permit us to treat, for example, a vibrating string with an elastic force acting at a *single point* just as if the coefficients were continuous. However, the operators $\mathfrak{A}_\psi$ are more general than the integrated version (1.12) of $D_m D_x + c$, and more easy to treat than $D_m D_x + c$.

After these explanatory remarks we may describe the *main contents* of this paper as follows. Suppose we are given a (not necessarily finite) interval $I: (\beta_1, \beta_2)$ and in it a strictly increasing function m (i.e. a positive measure), not necessarily bounded at the endpoints β_i. Furthermore, we suppose given in I a continuous function $\psi > 0$. With these data we define an operator

[4] Such is the case for ψ defined by $2\psi(x) = 1 + |x|$. As an *illustrative example* consider the operator $\mathfrak{A}_\psi$ with this ψ and $m(x) = x$ $(-\infty < x < \infty)$. A trivial calculation shows that $\mathfrak{A}_\psi f(x) = D_x^2 f(x)$ for each $x \neq 0$. At the origin both f and $D_x(f\psi^{-1})$ must be continuous, i.e. for each f in the domain of $\mathfrak{A}_\psi$:

(**) $$\mathfrak{A}_\psi f(x) = f''(x), \qquad x \neq 0$$
$$f^+(0) - f^-(0) = f(0),$$

while the value of $\mathfrak{A}_\psi f$ at the origin is defined by continuity. Now (**) is equivalent to (1.12) with the *Dirac function* at the origin: the integral on the left equals $f'(x_2) - f'(x_1)$ if (x_1, x_2) does not include the origin, but equals $f'(x_2) - f'(x_1) - f(0)$ if $x_1 < 0 < x_2$. Observe that $\mathfrak{A}_\psi$ in no way involves discontinuous functions, and that its theory is trivially reducible to special operators of the form $D_m D_x$.

$\mathfrak{A}_\psi$ as in (1.8). Observe that ψ is a solution of the differential equation $\mathfrak{A}_\psi \phi = 0$. The most general solution of this equation is defined by

$$\phi(x) = k_1 \psi(x) \int_{x_0}^{x} \psi^{-2}(s)\, ds + k_2 \psi(x), \qquad x_0 \in I \tag{1.13}$$

where k_1 and k_2 are constants. Such a ϕ can have at most one zero, and in any interval not containing it, we have

$$\mathfrak{A}_\phi = \mathfrak{A}_\psi . \tag{1.14}$$

In other words, our operator is completely determined by the interval I, the measure m, and the two-parametric family of functions annihilated by it. The representation in the form (1.8) is therefore essentially unique.

Operators $\mathfrak{A}_\psi$ of this sort are a natural generalization[5] of the familiar second order differential operators. It is the purpose of this paper to show that the classical Sturm-Liouville and other theories carry over in a natural way leading to a theory which is simpler and more uniform.

Before reviewing the main results of the paper we remark that the general operators $\mathfrak{A}_\psi$ are by no means more complicated than the operators $\mathfrak{A}$ of the special form (1.4). In fact (see Section 4), each operator $\mathfrak{A}_\psi$ *can be mapped onto a "related" operator* $D_\mu D_\xi$ in such a way that no special theory is required. For all intents and purposes it suffices therefore to study operators of the special form (1.4). Their analysis simplifies greatly in consequence of their pleasing property that $\mathfrak{A}f > 0$ *implies the convexity* of f.

In *Section* 3 we introduce a *boundary classification.* The boundary points which we call *active* correspond to the Grenzkreis type in H. Weyl's now classical disjunction (see [7]). However, we subdivide Weyl's Grenzpunkt type by introducing the notion of *semiactive* boundaries. This subclass is introduced for the discussion of *bounded* solutions and has no bearing or meaning when the Hilbert space norm (1.5) is adopted (which explains why

[5] The operator $\mathfrak{L} = a\, D_x^2 + b\, D_x + c$, $a > 0$, has the two obvious properties: (1) it is of local character, (2) the minimum property: if f and $\mathfrak{L}f$ are continuous in a neighborhood of x_0 and $f \geqq 0$, $f(x_0) = 0$, then $\mathfrak{L}f(x_0) \geqq 0$. It has been shown that the most general operator with these properties has the form of $\mathfrak{A}_\psi$ except for the possible presence of singular points at which $\mathfrak{L}$ reduces to a *first* order operator. If condition (2) is replaced by the stronger requirement that at a place of a local minimum x_0 always $\mathfrak{L}f(x_0) \geqq 0$ (i.e. if the hypothesis $f(x_0)$ be deleted), then $\mathfrak{A}_\psi$ is of the special form $\mathfrak{A}$ of (1.4). (Cf. [1] and [2].)

A point x_0 is nonsingular if the equation $\mathfrak{L}\phi = 0$ admits near x_0 of *two* independent solutions. If ψ is such a solution and $\psi(x_0) = 0$, the right side in (1.13) defines $\phi(x)$ near and at x_0 by continuity. If ψ has two or more zeros, so does each solution ϕ of $\mathfrak{L}\phi = 0$, but the connection (1.13) remains valid.

The condition that $\psi > 0$ introduced in the present paper is a *restriction introduced to avoid going into the complex domain.* The most general operator $\mathfrak{L}$ without singular points is obtained by permitting ψ to have zeros, but requiring that the right side in (1.13) remain meaningful. Near a zero of ψ one has then to use the representation $\mathfrak{L} = \mathfrak{A}_\phi$, or use a complex solution $\phi + i\psi$ which is automatically free of zeros.

it does not occur in the Weyl-Stone theory). It goes without saying that the type of a boundary point is preserved under the above mentioned mapping of $\mathfrak{A}_\psi$ onto $D_\mu D_\xi$.

To continue, consider first the *Hilbert space* theory with norm (1.5). For each fixed λ the simple mapping $f \to f\psi^{-1}$ carries a pair f^*, F^* satisfying

$$\lambda F^* - \mathfrak{A}_\psi F^* = f^* \tag{1.15}$$

into a pair f, F satisfying

$$\lambda F - D_\mu D_\xi F = f, \tag{1.16}$$

and the mapping is isometric: $\| f^* \|_m = \| f \|_\mu$. Accordingly, as far as Hilbert space theory is concerned, it suffices to study the equation (1.16).

In *Section* 8 it is shown that for each $\lambda > 0$ there exists a positivity preserving linear transformation $\mathfrak{G}_\lambda$ of norm $\leqq 1$ carrying each $f \in \mathbf{H}_m$ into a $\mathfrak{G}_\lambda f = F \in \mathbf{H}_m$ satisfying (1.16). The solution of (1.16) is unique except when at least one boundary is active. In each case the range R of $\mathfrak{G}_\lambda$ is independent of λ; the operators $\mathfrak{G}_\lambda$ and $\mathfrak{A}/R$ (the restriction of $\mathfrak{A}$ to R) are *self-adjoint*, $\mathfrak{G}_\lambda = (\lambda - \mathfrak{A}/R)^{-1}$. (Here $\mathfrak{A} = D_\mu D_\xi$.)

In *Section* 10 we study arbitrary symmetric bounded linear transformations $\mathfrak{G}_\lambda$ such that $F = \mathfrak{G}_\lambda f$ satisfies (1.16) for each $f \in \mathbf{H}_m$. For a fixed λ we describe *the most general possible range of such a transformation* by appropriate "*boundary conditions*." Conversely, given a possible range R (i.e. admissible boundary conditions), we show that *to each* $\lambda > \lambda_0$ *there corresponds a* $\mathfrak{G}_\lambda$ *whose range coincides with* R. Also we find conditions for $\mathfrak{G}_\lambda$ to be positive or to have norm $\leqq \lambda^{-1}$. Both $\mathfrak{G}_\lambda$ and $\mathfrak{A}/R$ are self-adjoint operators. This analysis corresponds to, and amplifies, the well-known discussion in Chapter X of Stone's treatise [6], but uses simpler methods. (A treatment of unsymmetric operators could proceed following the methods in Section 12.)

New light is shed on the classical boundary conditions in *Section* 11, where we treat the same problem in the enlarged Hilbert space over the closed interval $[\beta_1, \beta_2]$ with norm (1.6). The preceding analysis applies and analogous results hold except that we get a *new type of boundary conditions*.[6] On inspection it is seen that they in reality amount to *a definition of the operator* $\mathfrak{A}$ at the boundary, and that no boundary conditions in the ordinary sense occur as long as the points of the boundary carry a positive weight. The classical boundary conditions are obtained as a degenerate limiting case of this definition of $\mathfrak{A}$ when $m_i \to 0$.

[The Hilbert space theory is contained in Theorem 8.1 and Sections 10, 11. Sections 9, 12–13 have no relation to it, and in the introductory sections all

[6] In a disguised form these new boundary conditions occur in the literature as "time dependent" or "λ-dependent" boundary conditions. They now require no special treatment. The same boundary conditions for continuous and for measure-solutions were discussed in [3].

parts referring to boundedness may be omitted. For the spectral analysis see McKean [5].[7])

As usual in analysis, the Hilbert space approach leads to the most elegant results. However, our differential operators play an important role in probability theory, the heat conduction, etc., and it is therefore unavoidable to consider them as operators on set functions and on continuous functions. A satisfactory theory can be developed, but new problems arise.

For most purposes it is natural to use the space **C** of functions continuous in the closed interval $\bar{I}$ with the traditional norm

$$\| f \| = \sup_{x \in I} | f(x) |. \tag{1.17}$$

This norm lends itself splendidly for the study of $\mathfrak{A}$, but not equally well for $\mathfrak{A}_\psi$. As in the case of the Hilbert space $\mathbf{H}_m$, there exists (see Section 8) for each $\lambda > 0$ a positive transformation $\mathfrak{G}_\lambda$ from **C** to **C** such that for each $f \in \mathbf{C}$ the transform $F = \mathfrak{G}_\lambda f$ has a norm $\| F \| \leqq \lambda^{-1} \| f \|$. However, the mapping which connects (1.15) and (1.16) is no longer norm-preserving. It takes the space **C** into the Banach space $\mathbf{C}_\psi$ of functions such that $f\psi^{-1}$ is continuous in the closed interval, $\mathbf{C}_\psi$ being metrized by the norm

$$\| f \|_\psi = \| f\psi^{-1} \| = \sup | f(x)\psi(x) | . \tag{1.18}$$

Thus all theorems concerning (1.16) and the classical norm $\| f \|$ translate into equivalent theorems concerning (1.15) and the norm[8] (1.18). The following theorem (proved in *Section* 9) is therefore of interest:

In order that the equation (1.15) *admit for each* $f^* \in \mathbf{C}$ *a solution* $F^* \in \mathbf{C}$ *such that* $\| F^* \| \leqq \lambda \| f^* \|$, *it is necessary and sufficient that* ψ *be convex upward.*[9]

By analogy with Sections 10–11 we treat in *Section* 12 the *most general boundary conditions* for transformations from **C** to **C** inverting $\lambda - \mathfrak{A}$. However, while in the case of the Hilbert space norm (1.5) it was natural to restrict the theory to symmetric operators, the requirement of symmetry is meaningless in **C**. Accordingly, we study arbitrary transformations, and we

[7] McKean considers only Green function transformations and symmetric boundary conditions introduced in [3] and guaranteeing the boundedness of the solutions of (1.16). The mappings of Section 4 automatically extend his results to *all* transformations from $\mathbf{H}_m$ to $\mathbf{H}_m$ provided each boundary condition involves only one boundary point (see Note 3 to Theorem 8.1). The newer type boundary conditions applying to the enlarged space $\bar{\mathbf{H}}_m$ require only a redefinition of the Green function on the boundary, and thus McKean's analysis could be extended to them by obvious modifications.

[8] Actually several norms are equally applicable to each of the equations (1.15) and (1.16). For example, for $\mathfrak{A} = D_m D_x$ in the interval $0 < x < \infty$ the norm $\| f \|_x = \sup | f(x)x^{-1} |$ is fully as good as (1.17).

[9] I owe this observation to H. P. McKean, Jr. If ψ is convex upward, so is each positive ϕ satisfying $\mathfrak{A}_\psi \phi = 0$. In the special case of operators $\mathfrak{A}_\psi = \mathfrak{A} + c$ the convexity of ψ is equivalent to $c \leqq 0$.

encounter *new "boundary conditions" depending on global functionals*, and not only on the boundary.

Section 13 deals with the transformations on *measures* which are *adjoint* to the transformations constructed in Section 12. We discover here that in general the adjoint $\mathfrak{A}^*$ to $\mathfrak{A}$ need not be a differential operator.[10] For $\mathfrak{A}^*$ to be of local character it is necessary and sufficient that the boundary conditions for $\mathfrak{A}$ do not involve global functionals, that is, they have to be of the type of the boundary conditions discussed in Section 11 for $\overline{\mathbf{H}}_m$. [Sections 12–13 are independent of Sections 10–11.]

It may be observed that the results of Sections 8–13 can be rephrased in the terminology of *semigroups*. In fact, saying that the transformations $\mathfrak{G}_\lambda$ have, for $\lambda > \lambda_0$, a common range R is equivalent to saying that the restriction $\mathfrak{A}/R$ generates a semigroup.[11]

2. Definitions and preparations

In a fixed open interval $I:(\beta_1, \beta_2)$ (where $-\infty \leqq \beta_1 < \beta_2 \leqq \infty$) we suppose given a *strictly increasing* function m, which is continuous to the right: $m(x) = m(x+)$. One should conceive of $m(x)$ as the *measure* of the half-open interval $[\beta_1, x]$, and the notation using measures and derivatives of measures is really preferable for our purposes, but it seems less popular. We call m the *canonical measure* for our operators. Needless to say that $m(x)$ is finite for each x, but m need not be bounded near the endpoints β_i. (When $\beta_2 < \infty$, $m(\beta_2 -) < \infty$, it is possible, and often desirable, to enlarge the interval I by adding to it the point β_2 carrying a positive weight m_2.)

If m happens to be continuous, the interval I can be referred to m, instead of x, as a scale parameter. In this case the right derivative of f with respect to m is defined, as usual, by

$$D_m^+ f(x) = \lim_{h \downarrow 0} \frac{f(x+h) - f(x)}{m(x+h) - m(x)}. \tag{2.1}$$

In the general case we keep this definition at each point of continuity of m. (It is clear that at such points $m(x+h)$ in the denominator may be replaced by $m(x+h-)$ without affecting the limit.) At points of discontinuity we

[10] The contents of Sections 12 and 13 are a generalization, but also a great simplification, of the developments in [3] concerning operators of the special type (1.1). Unfortunately, a notational confusion has introduced an *error* into [3], which is explained below in footnote 20.

Added in proof. While the present paper was in press the author has received a recent paper [8] by *A. D. Ventcelj* in which the error mentioned above is pointed out and corrected for operators of the special form $D_m D_x$. In the text we treat lateral conditions for more general operators and also for the case when the Hilbert space norm is used and/or the requirement of positivity is dropped. All results of [3] are superseded by the present paper except that here we refrain from using the terminology of semigroups.

[11] On the generation of semigroups by restrictions of operators and the abstract definition of boundary conditions see [3].

replace the denominator by the increment of m over the *closed* interval $[x, x + h]$. Then

$$D_m f(x) = \frac{f(x+) - f(x-)}{m(x+) - m(x-)} \tag{2.2}$$

is not only the right, but also the left, derivative at x. For a derivative $D_m f$ to exist it is necessary that f be *continuous with respect to* m, jumps being permitted at points of discontinuity of m.

As usual D_m exists if $D_m^+ = D_m^-$. *The derivative* D_m *is defined by* (2.1) *and* (2.2), respectively, at points of continuity and discontinuity of m.

When no confusion can arise, we shall occasionally simplify the notation by inconsistently using the symbol D_m in the meaning of: *either* D_m^+ *or* D_m^- even if $D_m^+ \neq D_m^-$ or if only one is defined. (Thus it is clear that when $x = \beta_2$ is substituted in $D_m f$ we should read $D_m^- f(\beta_2)$.)

We agree that in some interval I_0 the function f *can belong to the domain of the operator* $D_m D_x$ *only if the one-sided derivatives* $f^+ = D_x^+ f$ *and* $f^- = D_x^- f$ *exist everywhere and are continuous with respect to* m. The value of $D_m D_x f$ is then defined at each point where the derivative exists. The definition of $\mathfrak{A} = D_m D_x$ will be completed by *specifying the domain*: in connection with the space $\mathbf{C}$ of continuous functions we shall require that $D_m D_x f \in \mathbf{C}$; when the Hilbert space norm (1.5) is used, we shall naturally require that $F = D_x f$ be absolutely continuous with respect to m and $\| D_m F \|_m < \infty$. Under any circumstances

$$\int_{\alpha+}^{\beta-} \mathfrak{A} f \cdot dm = f^-(\beta) - f^+(\alpha). \tag{2.3}$$

Note. It is best to conceive of f', which is a function of bounded variation, as inducing a signed measure. The operator D_x then changes point functions into set functions, and D_m (a Radon-Nikodym derivative) changes the measure back into a point function. The formal adjoint $D_x D_m$ operates on set functions, and D_m maps a set function into a point function.

Coming to *notations* it will be noticed that, as usual, 0 and x denote not only numbers, but also the obvious functions. However, to avoid confusion, *we shall use the symbol* $\mathbf{1}$ *to denote the function which equals* 1 *everywhere.*

Existence of solutions. Let c be a real function such that

$$\int_{I_0} | c | \, dm < \infty$$

for each subinterval I_0 of I. For arbitrarily prescribed constants k_1, k_2 the classical method of successive approximations shows that there exists a unique function u defined throughout I and satisfying almost everywhere (m) the differential equation

$$D_m D_x u = cu \tag{2.4}$$

with the initial conditions

$$u(\beta) = k_1, \qquad u^+(\beta) = k_2. \tag{2.5}$$

More precisely,

$$u(x) = k_1 + k_2 x + \int_\beta^x ds \int_\beta^s cu\, dm. \tag{2.6}$$

In fact, putting $u^{(0)}(x) = k_1 + k_2(x - \beta)$ and

$$u^{(n+1)}(x) = k_1 + k_2(x - \beta) + \int_\beta^x ds \int_\beta^s cu^{(n)}\, dm \tag{2.7}$$

it is seen by induction that

$$\begin{aligned} | u^{(n+1)}(x) - u^{(n)}(x) | &\leqq \int_\beta^x ds \int_\beta^s | c | \cdot | u^{(n)} - u^{(n-1)} | \cdot dm \\ &\leqq K \cdot q^n \end{aligned} \tag{2.8}$$

provided $| x - \beta | < \delta$ where δ is chosen so that

$$2\delta \int_{\beta-\delta}^{\beta+\delta} | c |\, dm < q < 1. \tag{2.9}$$

It follows that $u^{(n)}$ converges in the interval $(\beta - \delta, \beta + \delta)$ uniformly to a solution of (2.6), and the argument shows at the same time that $u = 0$ is the *only* solution when $k_1 = k_2 = 0$. This solution u can be continued by the same procedure to an arbitrary subinterval of I.

3. Minimal solutions. Classification of boundaries

Consider the operator $\mathfrak{A}_\psi$ defined by (1.8) where $\psi > 0$ is continuous in $I{:}(\beta_1, \beta_2)$. A function ϕ satisfies $\mathfrak{A}_\psi \phi = 0$ if, and only if,

$$\psi^2(\phi/\psi)' = \text{const.} \tag{3.1}$$

Note that this relationship is really symmetric in ϕ and ψ since in each interval not containing a zero of ϕ

$$\psi^2(\phi/\psi)' = -\phi^2(\psi/\phi)'. \tag{3.2}$$

The ratio ϕ/ψ is necessarily monotonic, and hence ϕ can have at most one zero.

LEMMA 3.1. *Among the functions ϕ satisfying* (3.1) (*or, what amounts to the same,* (1.13)) *there exists one, ϕ_*, such that*

$$\int_\beta^{\beta_2} \phi_*^{-2}(s)\, ds = \infty \tag{3.3}$$

for each $\beta \in I$. Each solution ϕ of (3.1) *which is not a multiple of ϕ_* satisfies*

$$\int_\beta^{\beta_2} \phi^{-2}(s)\, ds < \infty, \qquad \frac{\phi_*(x)}{\phi(x)} \to 0 \quad as \quad x \to \beta_2 \tag{3.4}$$

provided $\phi(x) \neq 0$ for $x > \beta$.

We shall say that ϕ_* is *minimal* at β_2. The solution which is minimal at β_1 is defined in like manner.

Proof. Let ϕ be a solution of (3.1) which is independent of ψ. It is seen from (1.13) that as $x \to \beta_2$

$$(3.5) \qquad |\phi(x)\psi^{-1}(x)| \to \infty \qquad \textit{if, and only if,} \qquad \int_\beta^{\beta_2} \psi^{-2}(s)\,ds = \infty$$

for some (and therefore for each) $\beta \in I$. Because of symmetry this statement holds also when ϕ and ψ are interchanged. Since it is impossible that both $\phi\psi^{-1}$ and $\psi\phi^{-1}$ tend to $\pm\infty$, it follows that the minimal solution, if it exists, is determined up to a constant multiplier. On the other hand, if the integral in (3.5) converges, put

$$(3.6) \qquad \phi_*(x) = \psi(x) \int_x^{\beta_2} \psi^{-2}(s)\,ds.$$

Then $\psi(x)\phi_*^{-1}(x) \to \infty$ as $x \to \beta_2$, and therefore (3.3) holds. Thus either ψ or the solution defined by (3.6) is minimal. The lemma is proved.

(Note that ψ may be minimal both at β_1 *and* β_2. In this case for each solution ϕ of (3.1) which is independent of ψ, the ratio ϕ/ψ is monotonic and goes from $-\infty$ to ∞. On the other hand, if the solutions ϕ_{*1} and ϕ_{*2} which are minimal at β_1 and β_2, respectively, are independent, both may be taken as positive, and there exist then infinitely many positive solutions of (3.1).)

Definition 3.1. *Let $\psi > 0$ be continuous in I. A solution ϕ_* of $\mathfrak{A}_\psi \phi_* = 0$ will be called minimal at β_2 if* (3.3) *holds.*

Definition 3.2. *Let ϕ_* be minimal for $\mathfrak{A}_\psi$ at β_2, and let ϕ be any solution of $\mathfrak{A}_\psi \phi = 0$ which is independent of ϕ_*.*

The endpoint β_2 will be called active for $\mathfrak{A}_\psi$ if for some (and consequently for each) $\beta \in I$

$$(3.7) \qquad \int_\beta^{\beta_2} \phi^2\,dm < \infty.$$

A nonactive endpoint β_2 will be called semiactive if

$$(3.8) \qquad \int_\beta^{\beta_2} |\phi\phi_*|\,dm < \infty.$$

Each solution being a linear combination of ϕ and ϕ_*, it is clear that this definition does not depend on the choice of ϕ. The definition applies to the left boundary with the integrals extended over (β_1, β).

In the case of the simple operator $\mathfrak{A} = D_m D_x$ we can put

$$(3.9) \qquad \begin{array}{lll} \phi_*(x) = \beta_2 - x, & \phi(x) = 1 & \textit{if} \quad \beta_2 < \infty \\ \phi_*(x) = 1, & \phi(x) = x & \textit{if} \quad \beta_2 = \infty. \end{array}$$

The endpoint β_2 is active for $\mathfrak{A} = D_m D_x$ if, and only if,

(3.10) $$\int_{\beta}^{\beta_2} x^2\, dm(x) < \infty .$$

It is semiactive if the integral in (3.10) *diverges but*

(3.11) $$\int_{\beta}^{\beta_2} (\beta_2 - x)\, dm(x) < \infty \qquad \textit{in case} \quad \beta_2 < \infty$$

(3.12) $$\int_{\beta}^{\beta_2} x\, dm(x) < \infty \qquad \textit{in case} \quad \beta_2 = \infty .$$

Obviously the first inequality in (3.11) is equivalent to[12]

(3.13) $$\int_{\beta}^{\beta_2} m(y)\, dy < \infty \qquad \beta_2 < \infty .$$

4. Related operators

Let $\psi > 0$ be continuous in $I:(\beta_1, \beta_2)$. For a fixed $\beta \in I$ define a function ξ by

(4.1) $$\xi(x) = \int_{\beta}^{x} \psi^{-2}(s)\, ds.$$

This maps I onto an interval $I^*:(\beta_1^*, \beta_2^*)$, and in it we define a monotone right-continuous function $\mu(\xi)$ by

(4.2) $$\mu(\xi) = \int_{\beta}^{x} \psi^2(s)\, dm(s)$$

where x stands for the point mapped on ξ. In I^* we define the operator

(4.3) $$\mathfrak{A}^* = D_\mu D_\xi .$$

We shall say that $(\mathfrak{A}, I)$ and $(\mathfrak{A}^*, I^*)$ are related by the transformation (4.1) – (4.2). Needless to say that this term will be employed also if ξ is replaced by a linear function $k_1 \xi + k_2$, say, $-\xi$ (cf. (4.9)).

---

[12] In [3] a classification of boundaries was introduced for operators of the special type $aD_x^2 + bD_x$. It was intended for the study of boundedness and/or integrability of solutions, and did not pay attention to the Hilbert space norm (1.5). This classification may be carried over to our more general operators $\mathfrak{A} = D_m D_x$. It is based only on the criteria (3.11) and (3.12), and according to it β_2 is for $\mathfrak{A}$

regular if $\beta_2 < \infty$, $m(\beta_2-) < \infty$;
exit if not regular, $\beta_2 < \infty$, *and* (3.11) *holds*;
entrance if not regular, $\beta_2 = \infty$, *and* (3.12) *holds*;
natural otherwise.

Thus:

$$\textit{an active } \beta_2 \textit{ is} \begin{cases} \textit{regular} & \text{if } \beta_2 < \infty,\ m(\beta_2-) < \infty \\ \textit{exit} & \text{if } \beta_2 < \infty,\ m(\beta_2-) = \infty \\ \textit{entrance} & \text{if } \beta_2 = \infty \end{cases}$$

$$\textit{a semiactive } \beta_2 \textit{ is} \begin{cases} \textit{exit} & \text{if } \beta_2 < \infty \\ \textit{entrance} & \text{if } \beta_2 = \infty . \end{cases}$$

A β_2 which is neither active nor semiactive is *natural*, and conversely.

To each function f continuous on I we let correspond a function f^* defined on I^* by means of the inverse mapping to (4.1) as $f^* = f\psi^{-1}$, i.e. by

$$f^*(\xi) = f(x(\xi))\cdot\psi^{-1}(x(\xi)). \tag{4.4}$$

Clearly

$$\mathfrak{A}_\psi = \psi\cdot\mathfrak{A}^* \tag{4.5}$$

in the obvious sense. The following theorem is now trivially verified:

THEOREM 4.1. *For each* f

$$\int_{I^*} |f^*|^2\, d\mu = \int_I f^2\, dm. \tag{4.6}$$

If $\lambda F - \mathfrak{A}_\psi F = f$, *then* $\lambda F^* - \mathfrak{A}^* F^* = f^*$. *With the norms introduced by* (1.17), (1.18) *we have*

$$\| f^* \| = \| f \|_\psi\,. \tag{4.7}$$

In particular, $\psi^* = \mathbf{1}$ and $\mathfrak{A}_\psi\,\phi = 0$ implies $\mathfrak{A}^*\phi^* = 0$. Recalling that ϕ is minimal for $\mathfrak{A}_\psi$ at β_2 if, and only if, for each solution ϕ_1 of $\mathfrak{A}_\psi\,\phi_1 = 0$ which is independent of ϕ the ratio $\phi(x)\phi_1^{-1}(x) \to 0$ as $x \to \beta_2$, we get

LEMMA 4.1. *The mapping* $f \to f^*$ *carries solutions of* $\mathfrak{A}_\psi\phi = 0$ *into solutions of* $\mathfrak{A}^*\phi^* = 0$, *and the property of minimality is preserved.*

The type of the boundary β_2 for $\mathfrak{A}_\psi$ depends on the integral (3.7), i.e. on the behavior of m near the endpoint β_2. To investigate the relationship between the types of β_2 and β_2^*, we must therefore replace the interval I by a subinterval (β, β_2) and apply the mapping to it. The convergence of all integrals occurring in the criterion then depends only on a vicinity of the right endpoint, and (4.6) together with Lemma 4.1 show directly the truth of

THEOREM 4.2. *The endpoint* β_i *is for* $\mathfrak{A}_\psi$ *of the same type (active, semiactive, or neither) as the endpoint* β_i^* *is for* $\mathfrak{A}^*$.

Note that $\mathfrak{A}_\psi\phi = 0$ may admit of several positive solutions in which case $\mathfrak{A}_\psi = \mathfrak{A}_\phi$ (see Theorem 5.1). Thus $\mathfrak{A}_\psi$ may be mapped in several ways on operators of the simpler type $D_\mu D_\xi$. (Under any circumstances different such mappings are available in each proper subinterval of I, and in particular, for neighborhoods of the endpoints.) For example, in $(0, \infty)$

$$D_m D_x f = x^{-1} D_m\, x^2 D_x(f x^{-1}). \tag{4.8}$$

Thus *the operator* $\mathfrak{A} = D_m D_x$ *in* $(0, \infty)$ *is related to the operator of the same type* $\mathfrak{A}^* = D_\mu D_\xi$ *by means of*

$$\xi = x^{-1}, \qquad \mu(\xi) = -\int_1^{\xi^{-1}} s^2\, dm(s). \tag{4.9}$$

Here $f^*(\xi) = f(\xi^{-1})\xi$, and the endpoint ∞ corresponds to 0. This transformation enables us always to *reduce the study of the behavior of solutions near a boundary* $\beta_2 = \infty$ *to the neighborhood*[13] *of* 0.

5. A basic identity and its corollaries

We suppose that in the interval I the monotonic function m and the continuous $\psi > 0$ are given; we consider the operator $\mathfrak{A}_\psi$ defined by (1.8) and other operators of the same type obtained by varying ψ while keeping m fixed.

LEMMA 5.1. *Let* $U > 0$ *and* V *be functions such that* $\mathfrak{A}_\psi U$ *and* $\mathfrak{A}_\psi V$ *exist and are continuous in* I. *Then*

$$U \cdot \mathfrak{A}_\psi V - V \cdot \mathfrak{A}_\psi U = U \cdot \mathfrak{A}_U V. \tag{5.1}$$

Proof. For each subinterval (α, β) of I we have

$$\int_\alpha^\beta U \cdot \mathfrak{A}_\psi V \cdot dm = \int_\alpha^\beta U\psi^{-1} \cdot d\{\psi^2(V/\psi)'\}, \tag{5.2}$$

where the prime indicates differentiation with respect to the variable of integration. A simple integration by parts leads hence to

$$\begin{aligned} \int_\alpha^\beta \{U \cdot \mathfrak{A}_\psi V - V \cdot \mathfrak{A}_\psi U\} \cdot dm &= \psi^2 \cdot \left\{ \frac{U}{\psi} \left(\frac{V}{\psi}\right)' - \frac{V}{\psi}\left(\frac{U}{\psi}\right)' \right\} \Bigg|_\alpha^\beta \\ &= U^2 \cdot \left(\frac{V/\psi}{U/\psi}\right)' \Bigg|_\alpha^\beta = U^2 \cdot \left(\frac{V}{U}\right)' \Bigg|_\alpha^\beta . \end{aligned} \tag{5.3}$$

This is the integrated version of (5.1), and the lemma is proved.

THEOREM 5.1. *If* ϕ *is a positive solution of* $\mathfrak{A}_\psi \phi = 0$, *then*

$$\mathfrak{A}_\psi = \mathfrak{A}_\phi . \tag{5.4}$$

(That is to say, the representation (1.8) has an intrinsic character, independent of the choice of the null-solution ψ.)

Proof. Set $U = \phi$ in (5.1).

LEMMA 5.2. *We have*

$$\mathfrak{A}_\psi V - cV = \mathfrak{A}_U V \qquad \textit{where} \quad c = U^{-1} \cdot \mathfrak{A}_\psi U. \tag{5.5}$$

In particular: given an arbitrary integrable function c, *in the neighborhood of any point it is possible to reduce the operator* $\mathfrak{A}_\psi - c$ *to the form* $\mathfrak{A}_U$ *involving the same canonical measure* m.

[13] In the terminology of [3] explained in the preceding footnote this means that an *exit boundary may be transformed into an entrance boundary, and vice versa.* A theorem such as "at an exit boundary a solution of $\lambda u - \mathfrak{A}u = 0$ approaches a finite limit" is equivalent to "at an entrance boundary the ratio $u(x)/x$ approaches a finite limit" etc. This may serve to explain the formal advantages and simplicity of the present approach.

Proof. Divide (5.1) by U to obtain (5.5). We are free to choose U as a solution of $\mathfrak{A}_\psi U - cU = 0$ (see Section 2 concerning its existence).

LEMMA 5.3. (Wronski identity). *If* $\lambda u_i - \mathfrak{A}_\psi u_i = 0$, *where* $i = 1, 2$ *and* λ *is a constant, then*

$$u_1^2(u_2/u_1)' = \text{const.} \tag{5.6}$$

In particular, $\lambda u_i - D_m D_x u_i = 0$ *implies*

$$u_1 u_2' - u_2 u_1' = \text{const.} \tag{5.7}$$

Proof. Clearly (5.6) is a special case of (5.3). When u_1 and u_2 are differentiable with respect to x, then (5.6) reduces to (5.7). [Note that (5.7) holds both for the right and the left derivative.]

6. The homogeneous equation

For simplicity of exposition we first consider the reduced operator $\mathfrak{A}$ of (1.4) in an interval $I: -\infty \leqq \beta_1 < x < \beta_2 \leqq \infty$. We are interested in solutions u of the equation

$$\lambda u - D_m u' = 0, \qquad \lambda > 0 \tag{6.1}$$

defined throughout I; here λ is a constant.

For an arbitrary point $\beta \in I$ there exists (see Section 2) a solution u satisfying the initial conditions $u(\beta) = 1, u^+(\beta) = 0$. Remembering that $D_m u' > 0$ implies that u' increases, i.e. that u is *convex*, we see that $u \geqq 1$. It follows from Lemma 5.2 that the operator $\lambda - D_m D_x$ is identical with the reduced operator $\mathfrak{A}_u$, and we can apply Lemma 3.1 to conclude that there exist solutions u_{*1} and u_{*2} of (6.1) which are minimal, respectively, at β_1 and β_2. For each $\beta \in I$ we have then

$$\int_{\beta_1}^{\beta} u_{*1}^{-2}(x)\, dx = \infty, \qquad \int_{\beta}^{\beta_2} u_{*2}^{-2}(x)\, dx = \infty. \tag{6.2}$$

Near β_2 the solution u_{*2} is positive, and hence *convex*. If it were increasing, there should exist an $\varepsilon > 0$ such that $u_{*2}(x) > \varepsilon(x - \alpha)$ for $x > \alpha$, which contradicts the condition that the second integral in (6.2) diverges for each β. Thus u_{*2} is positive and decreasing near β_2; this implies that u_{*2} is convex and decreasing throughout I. We have thus

THEOREM 6.1. *For each* $\lambda > 0$ *the equation* $\lambda u - \mathfrak{A}u = 0$ *admits of two positive convex solutions* u_{*1} *and* u_{*2} *such that* $u_{*1} \uparrow$ *while* $u_{*2} \downarrow$ *throughout the interval* I, *and that* (6.2) *holds.*

(This condition determines the solutions up to a multiplicative constant. We shall say that u_{*i} *is minimal* at β_i.)

The remaining part of this section is devoted to a study of the solutions near β_2, and the interval (β_1, β_2) may be replaced by an arbitrary subinterval

(β, β_2). Needless to say that the left boundary is subject to the same analysis.

Integrate (6.1) to get

$$u^+(x) = u^+(\beta) + \int_{\beta+}^{x+} \lambda u \, dm, \qquad \beta < x. \tag{6.3}$$

For a positive, decreasing u the left side is < 0 while the integrand is > 0. Accordingly $-u^+(\beta)$ is an upper bound for the integral for all $x > \beta$. We have thus

LEMMA 6.1. *For each positive decreasing solution u of* (6.1) *and each $\beta \in I$*

$$\int_{\beta}^{\beta_2} u \, dm < \infty. \tag{6.4}$$

This lemma applies, in particular, to the solution u_* minimal at β_2 and will be used prominently in the next section.

For the reader's convenience we shall first take up the case $\beta_2 < \infty$ and reduce the case $\beta_2 = \infty$ to it. However, this separation is by no means necessary, and the case $\beta_2 = \infty$ can be treated directly by means of the simple estimates here used for $\beta_2 < \infty$.

Consider again the solution u of (6.1) satisfying the initial conditions $u(\beta) = 1$, $u^+(\beta) = 0$. It is obvious from (6.3) that $u^+(x) > \lambda\{m(x) - m(\beta)\}$ for each $x > \beta$. Therefore u^+ can be integrable over (β, β_2) only if

$$\infty > \int_{\beta}^{\beta_2} \{m(x) - m(\beta)\} \, dx = \int_{\beta}^{\beta_2} (\beta_2 - y) \, dm(y). \tag{6.5}$$

On the other hand, (6.3) shows that

$$u^+(x) < \lambda u(x)\{m(x) - m(\beta)\}, \qquad x > \beta, \tag{6.6}$$

and therefore (6.5) implies that log u has an integrable derivative. In other words, u remains bounded at β_2 if, and only if, (6.5) holds. Remembering that every solution of (6.1) is a linear combination of u and u_{*2} we see that in this case all solutions of (6.1) remain bounded at β_2.

Suppose now that m remains bounded at β_2, that is, suppose $\beta_2 < \infty$ and $m(\beta_2-) < \infty$. Then (6.5) holds, and each solution u remains bounded. Moreover, in view of (6.3) the derivative $u^+(x)$ approaches a finite limit $u^-(\beta_2)$ as $x \to \beta_2$. In this case the boundary point β_2 may be treated just as an interior point, and the classical method of successive approximations (see Section 2) shows that to arbitrary constants A, B there exists exactly one solution of (6.1) such that

$$u(\beta_2) = A, \qquad u^-(\beta_2) = B. \tag{6.7}$$

In particular, $u_{*2}(\beta_2) = 0$ and $u_{*2}^-(\beta_2) < 0$.

Next suppose that $m(\beta_2-) = \infty$ but that (6.5) holds. We have seen that in this case all solutions u remain bounded at β_2. From (6.3) we see that $u^+(x) \to \infty$ for each increasing u, and therefore no solution u independent of

u_{*2} has a bounded derivative; in particular, there does not exist a positive decreasing solution independent of u_{*2}. It is clear from (6.4) that $u_{*2}(\beta_2) = 0$. For the sake of completeness we show that $u^-_{*2}(\beta_2) < 0$. Because of the convexity of u_{*2} we have $u_{*2}(s)(\beta_2 - s)^{-1} < -u^+_{*2}(s) < -u^+_{*2}(\beta)$ for each $s > \beta$, and hence we conclude from (6.3) that

$$-u^+_{*2}(x) > -u^+_{*2}(\beta)\left\{1 - \lambda \int_\beta^x (\beta_2 - s)\, dm(s)\right\}. \tag{6.8}$$

We may choose β so close to β_2 that the integral in (6.5) is smaller than $(2\lambda)^{-1}$. Then $-u^+_{*2}(x) > -\frac{1}{2}u^+_{*2}(\beta)$ for each $x > \beta$, and hence

$$-u^-_{*2}(\beta_2) \geqq -\tfrac{1}{2}u^+_{*2}(\beta) > 0,$$

as asserted.

Finally, when (6.5) does *not* hold, each solution u independent of u_{*2} is unbounded at β_2, while (6.4) implies that $u_{*2}(\beta_2) = u^-_{*2}(\beta_2) = 0$.

Recalling Definition 3.2 we can summarize these findings in the following theorem (only part (i) has a bearing on the Hilbert space approach).

THEOREM 6.2. *Let* $\beta_2 < \infty$.

(i) *If* β_2 *is active (that is, if* $m(\beta_2-) < \infty$*), each solution* u *of* (6.1) *and its derivative* u^+ *approach finite limits at* β_2. *Exactly one* u *satisfies the terminal condition* (6.7), *and hence there exist independent positive decreasing solutions. For the minimal* u_* *we have* $u_*(\beta_2) = 0$ *and* $u^-_*(\beta_2) \neq 0$.

(ii) *If* β_2 *is semiactive (that is,* $m(\beta_2-) = \infty$ *but* (6.5) *holds), each* u *is bounded at* β_2, *but* $|u'(x)| \to \infty$ *as* $x \to \beta_2$ *except if* u *is a multiple of* u_*. *The latter is the only positive decreasing solution and* $u_*(\beta_2) = 0$, $0 > u^-_*(\beta_2) > -\infty$.

(iii) *If* β_2 *is neither active nor semiactive, then* $u_*(\beta_2) = u^-_*(\beta_2) = 0$. *Each* u *independent of* u_* *is unbounded.*

It has been pointed out in Section 4 that the transformation (4.9) permits us to reduce the case $\beta_2 = \infty$ to the case $\beta_2 < \infty$. Without loss of generality we may suppose $\beta_1 < 0 < \beta_2 = \infty$. Then (4.9) carries the boundary point ∞ for $D_m D_x$ into the left endpoint 0 for $D_\mu D_\xi$, and the type of the boundary does not change. There exists in $(0, \infty)$ a one-to-one correspondence between the solutions u of $\lambda u - D_m D_x u = 0$ and the solutions u^* of $\lambda u^* - D_\mu D_\xi u^* = 0$ established by the relation $u^*(\xi) = \xi u(\xi^{-1})$. Applying the last theorem to u^* and reading it in terms of u we have

THEOREM 6.3. *Let* $\beta_2 = \infty$.

(i) *If* ∞ *is an active boundary (that is, if* $\int^\infty x^2\, dm < \infty$*), then for each solution* u *of* (6.1)

$$u(x) = Ax + B + o(1), \qquad x \to \infty \tag{6.9}$$

where A *and* B *are constants. To prescribed* A, B *there exists exactly one solution* u *satisfying the terminal condition* (6.9). *The minimal solution* $u_* \downarrow B > 0$.

(ii) *If* ∞ *is semiactive (i.e.* $\int^\infty x^2\, dm = \infty$ *but* $\int^\infty x\, dm < \infty$*), then for each solution* u *which is near* ∞ *increasing*

$$(6.10)\qquad x^{-1}u(x) \to A < \infty, \qquad but \qquad Ax - u(x) \to \infty \text{ as } x \to \infty.$$

The minimal solution $u_* \downarrow B > 0$, *and no solution independent of* u_* *is bounded at* ∞.

(iii) *If* $\int^\infty x\, dm = \infty$, *then* $u_* \downarrow 0$. *For each solution* u *independent of* u_* *we have* $x^{-1}u(x) \to \pm\infty$.

In reality the last two theorems can be formulated jointly using the notation (3.9). With this notation, however, the formulation applies at once to the more general operator $\mathfrak{A}_\psi$ of (1.8). A glance at Theorem 4.1 and Lemma 4.1 will show that the last two theorems are a special case of the following (which, for brevity, we state explicitly only for active boundaries).

THEOREM 6.4. *The equation*

$$(6.11)\qquad \lambda u - \psi^{-1}D_m(\psi^2 D_x(u\psi^{-1})) = 0, \qquad \lambda > 0$$

admits of two independent positive solutions u_{*i} *such that* (6.2) *holds.*

Let ϕ *be a solution of* $\mathfrak{A}_\psi \phi = 0$ *which is at* β_2 *positive but not minimal, and let* ϕ_* *be the solution of* $\mathfrak{A}_\psi\, \phi_* = 0$ *which is minimal at* β_2. *Then:*

If β_2 *is active, for each solution* u *of* (6.11)

$$(6.12)\qquad \frac{u(x)}{\phi(x)} \to A, \qquad \frac{A\phi(x) - u(x)}{\phi_*(x)} \to B \qquad as\ x \to \beta_2$$

where A *and* B *are constants. To arbitrarily prescribed* A, B *there exists exactly one* u *satisfying* (6.12). *In particular* $u_{*2}\, \phi^{-1} \to 0$.

(The parts of Theorems 6.2–6.3 relating to nonactive boundaries can be translated in like manner.)

It follows in particular that when β_2 is active

$$(6.13)\qquad \int_\beta^{\beta_2} u^2\, dm < \infty$$

for each solution u of (6.11). In other words, if β_2 is active for $\mathfrak{A}_\psi$, it is active also for $\lambda - \mathfrak{A}_\psi$. The corresponding statement for classical differential operators was formulated by H. Weyl [7].

7. Green functions

As before let $\psi > 0$ be continuous in the open interval $I: (\beta_1, \beta_2)$. We recall from Lemma 5.3 that for any two solutions of $\lambda u - \mathfrak{A}_\psi u = 0$ we have

$$u_2^2(u_1/u_2)' = \text{const.}$$

and this constant is zero only if u_1 and u_2 are dependent.

DEFINITION 7.1. *For fixed* λ *let* u_1 *and* u_2 *be two independent solutions of* $\lambda u - \mathfrak{A}_\psi u = 0$ *normed by the condition*

$$(7.1)\qquad u_2^2(u_1/u_2)' = \mathbf{1}.$$

The Green function G_λ for $(\mathfrak{A}_\psi, I)$ corresponding to the ordered pair u_1, u_2 is the function of two variables defined by

$$G_\lambda(x, y) = \begin{cases} u_1(x)u_2(y) & \text{for } \beta_1 < x < y < \beta_2 \\ u_2(x)u_1(y) & \text{for } \beta_1 < y < x < \beta_2 . \end{cases} \tag{7.2}$$

We denote by $\mathfrak{G}_\lambda$ the transformation which takes a function defined in I into the function $F = \mathfrak{G}_\lambda f$ defined for $x \in I$ by

$$F(x) = \int_{\beta_1}^{\beta_2} G_\lambda(x, y)f(y)\, dm(y), \tag{7.3}$$

the domain of $\mathfrak{G}_\lambda$ consisting of those f for which the integral exists in the sense of absolute convergence.

The Green function G_λ, and the transformation $\mathfrak{G}_\lambda$, will be called minimal if u_i is minimal at the boundary β_i, $i = 1, 2$. (Cf. Theorem 6.1.)

The basic mapping described in Section 4 takes the operator $\mathfrak{A}_\psi$ into the operator $\mathfrak{A}^* = D_\mu D_\xi$ defined in an interval $I^*: (\beta_1^*, \beta_2^*)$. The variable x becomes a function of ξ, and the solutions u_i are mapped into solutions u_i^* of

$$\lambda u_i^* - D_\mu D_\xi u_i^* = 0, \tag{7.4}$$

where $u_i^*(\xi) = u_i(x)/\psi_i(x)$.

The norming condition (7.1) can be rewritten as

$$\psi^2 \left\{ \left(\frac{u_1}{\psi}\right)' \frac{u_2}{\psi} - \frac{u_1}{\psi} \left(\frac{u_2}{\psi}\right)' \right\} = 1 \tag{7.5}$$

or

$$u_2^* \cdot D_\xi u_1^* - u_1^* \cdot D_\xi u_2^* = 1. \tag{7.6}$$

Thus the ordered pair u_1^*, u_2^* satisfies the norming condition (7.1) relative to the operator $\mathfrak{A}^*$, and we can form the Green function G_λ^* corresponding to $(\mathfrak{A}^*, I^*)$, namely

$$G_\lambda^*(\xi, \eta) = \begin{cases} u_1^*(\xi)u_2^*(\eta) & \text{for } \beta_1^* < \xi < \eta < \beta_2^* \\ u_2^*(\xi)u_2^*(\eta) & \text{for } \beta_1^* < \eta < \xi < \beta_2^*. \end{cases} \tag{7.7}$$

It induces the transformation $\mathfrak{G}_\lambda^*$ taking f^* into the function F^* defined by

$$F^*(\xi) = \int_{\beta_1^*}^{\beta_2^*} G_\lambda^*(\xi, \eta)f^*(\eta)\, d\mu(\eta). \tag{7.8}$$

A simple inspection shows now that the function F^* defined in I^* by (7.8) is identical with the image F/ψ of $F = \mathfrak{G}_\lambda f$ under the mapping (4.4). We have thus the simple but useful

THEOREM 7.1. *The mapping of Section 4 taking the operator $\mathfrak{A}_\psi$ into the related operator $\mathfrak{A}^* = D_\mu D_\xi$ induces an isomorphism between the corresponding* *transformations* $\mathfrak{G}_\lambda$ *and* $\mathfrak{G}_\lambda^*$. *If* $\mathfrak{G}_\lambda$ *is minimal for* $\mathfrak{A}_\psi$, *then* $\mathfrak{G}_\lambda^*$ *is minimal for* $\mathfrak{A}^*$.

Consequently, we may restrict further consideration to operators of the simple form $\mathfrak{A} = D_m D_x$ and reformulate all results for the general $\mathfrak{A}_\psi$. The advantage of considering only the operator $\mathfrak{A}$ is merely formal: the formulas become more pleasing to the eye, and the argument is more intuitive due to the fact that the positive solutions of $\lambda u - \mathfrak{A}u = 0$ are convex.

THEOREM 7.2. *Let* G_λ *be a Green function for* $\mathfrak{A} = D_m D_x$ *and* $I:(\beta_1, \beta_2)$, *formed by means of two solutions* u_i *such that* u_1 *is increasing and* u_2 *decreasing. Then*

(i) *Each bounded continuous* f *is mapped into a bounded continuous* F *such that*

$$\lambda F - D_m F' = f. \tag{7.9}$$

Furthermore,

$$0 \leqq f \leqq \mathbf{1} \quad \textit{implies} \quad 0 \leqq \lambda F \leqq \mathbf{1}. \tag{7.10}$$

(ii) *If* f *is square integrable with respect to* m, *then* F *is bounded and absolutely continuous with respect to* m; *the equation* (7.9) *holds* m*-almost everywhere, and*

$$\lambda^2 \int_{\beta_1}^{\beta_2} F^2 \, dm \leqq \int_{\beta_1}^{\beta_2} f^2 \, dm. \tag{7.11}$$

Proof. We have

$$F(x) = u_1(x) \int_{x+}^{\beta_2} f(y)u_2(y) \, dm(y) + u_2(x) \int_{\beta_1}^{x+} f(y)u_1(y) \, dm(y). \tag{7.12}$$

The solutions u_i being monotonic, it follows directly from Lemma 6.1 that the integrals converge absolutely for each bounded f. For the special choice $f = \mathbf{1}$ we get, remembering that $u_1^+ u_2 - u_1 u_2^+ = \mathbf{1}$ and the monotonicity of u_i,

$$\begin{aligned} \lambda F(x) &= u_1(x) \int_{x+}^{\beta_2} du_2'(y) + u_2(x) \int_{\beta_1}^{x+} du_1'(y) \\ &= 1 + u_1(x)u_2^-(\beta_2) - u_2(x)u_1^+(\beta_1) \leqq 1. \end{aligned} \tag{7.13}$$

This proves (7.10).

If f is continuous and m has a continuous derivative with respect to x, then (7.12) may be differentiated formally to obtain

$$F^+(x) = u_1^+(x) \int_{x+}^{\beta_2} f(y)u_2(y) \, dm(y) + u_2^+(x) \int_{\beta_1}^{x+} f(y)u_1(y) \, dm(y). \tag{7.14}$$

To verify the general validity of (7.14) denote the right hand member by $\Phi(x)$. A simple integration by parts shows that

$$\int_\alpha^\beta \Phi(x) \, dx = F(\beta) - F(\alpha), \tag{7.15}$$

showing that F^+ equals almost everywhere a derivative of F. The right continuity of Φ implies that $\Phi = F^+$ everywhere. Now each term in (7.14) is continuously differentiable with respect to m; remembering that $D_m u_i^+ = \lambda u_i$ and $u_1^+ u_2 - u_2^+ u_1 = \mathbf{1}$, we see that differentiation of (7.14) leads to (7.9). This proves the first part of the theorem.

Suppose that f is square integrable. Lemma 6.1 implies that u_2 is square integrable over each interval (x, β_2), and therefore the first integral in (7.12) converges absolutely. A similar argument applies to the second integral, and a trivial modification of the above reasoning shows the validity of (7.13) and the assertion relative to (7.9). It remains to prove (7.11). Now by Schwarz' inequality and (7.10)

$$\begin{aligned} \lambda^2 F^2(x) &\leqq \lambda^2 \int_{\beta_1}^{\beta_2} G_\lambda(x, y) f^2(y)\, dm(y) \cdot \int_{\beta_1}^{\beta_2} G_\lambda(x, y)\, dm(y) \\ &\leqq \lambda \int_{\beta_1}^{\beta_2} G_\lambda(x, y) f^2(y)\, dm(y). \end{aligned} \tag{7.16}$$

Integrate this inequality with respect to m using Fubini's theorem. Since G_λ is symmetric, a renewed application of (7.10) leads to (7.11). The theorem is proved.

8. The minimal transformation

In this section λ is an arbitrary *positive* number.

THEOREM 8.1. *Let $\mathbf{H}_m$ be the Hilbert space defined by* (1.5). *Let G_λ be the minimal Green function for the operator $\mathfrak{A} = D_m D_x$ and the interval*

$$I: (\beta_1 < x < \beta_2).$$

The corresponding transformation[14] *$\mathfrak{G}_\lambda$ of* (7.3) *is a symmetric, self-adjoint transformation with norm $\| \mathfrak{G}_\lambda \|_m \leqq \lambda^{-1}$. Its domain is the entire Hilbert space $\mathbf{H}_m$, and its range R is defined as follows:*

(i) *If neither β_1 nor β_2 is active, the range R coincides with the set Δ of all functions F such that $\mathfrak{A}F$ exists m-almost everywhere and both F and $\mathfrak{A}F$ are square integrable* (m).

(ii) *If β_i is active, the range R is the subset of Δ consisting of those functions for which*

$$\begin{aligned} &F(x) \to 0 && \text{as } x \to \beta_i \quad \text{if } |\beta_i| < \infty \\ &F(x) = o(|x|) && \text{as } x \to \beta_i \quad \text{if } |\beta_i| = \infty. \end{aligned} \tag{8.1}$$

In other words, for arbitrary $f \in \mathbf{H}_m$ there exists exactly one solution $F \in R$ of $\lambda F - \mathfrak{A}F = f$. The restriction of $\mathfrak{A}$ to R is a self-adjoint operator.

Note. It is clear from Theorem 7.1 that the present theorem applies

[14] Strictly speaking we are dealing with the restriction of $\mathfrak{G}_\lambda$ to $\mathbf{H}_m$ (and, similarly, in the next theorem with the restriction to $\mathbf{C}$). It seems foolish, however, to introduce a new letter in each case.

equally to arbitrary operators $\mathfrak{A}_\psi$, the general formulation of the *boundary condition* (8.1) being: *if β_i is active, then*

$$(8.2) \qquad F(x) = o(\phi(x)) \quad as \quad x \to \beta_i$$

where ϕ is an arbitrary solution of $\mathfrak{A}_\psi \phi = 0$ which is positive but not minimal at β_i. The range R is independent of λ, and thus $\mathfrak{G}_\lambda$ is the *resolvent* of $\mathfrak{A}$ if no boundary is active, and the *resolvent of the restriction* $\mathfrak{A}/R$ otherwise. The range R is obviously *dense* in the Hilbert space $\mathbf{H}_m$.

Proof. By Theorem 7.2 the transformation $\mathfrak{G}_\lambda$ is positive, symmetric, and has $\mathbf{H}_m$ for domain. It is therefore self-adjoint. The inequality (7.11) implies $\| \mathfrak{G}_\lambda \|_m \leqq \lambda^{-1}$. We next show that the range of $\mathfrak{G}_\lambda$ is contained in R. This statement is tautological if neither boundary is active. Suppose, then, that β_2 is active. Formula (7.12) applies with u_i standing for the solution u_{*i} which is minimal at β_i. By Theorems 6.2 and 6.3 the the solution u_1 is square integrable over (β_1, β_2), and therefore the first integral in (7.12) approaches 0 as $x \to \beta_2$ while the second remains bounded. If $\beta_2 < \infty$, then u_1 is bounded and $u_2 \to 0$ as $x \to \beta_2$. If $\beta_2 = \infty$, then $u_1 = O(x)$ and $u_2 = O(\mathbf{1})$. In the first case $F(x) \to 0$, in the second $F(x) = O(\mathbf{1})$, as asserted. It remains to prove that, conversely, R is contained in the range of $\mathfrak{G}_\lambda$. Choose an arbitrary $F_0 \in R$, and set $f = \lambda F_0 - \mathfrak{A}F_0$. Then f is square integrable, $F_0 \in R$, and $F = \mathfrak{G}_\lambda f \in R$. It follows that $u = F_0 - F$ is a solution of $\lambda u - \mathfrak{A}u = 0$ and $u \in R$. However, by Theorems 6.2 and 6.3, $u = 0$ is the only solution with these properties, and hence $F = F_0$ as asserted.

THEOREM 8.2. *Let $\mathbf{C}$ be the Banach space of functions continuous in the closed interval $[\beta_1, \beta_2]$ with the norm* (1.17). *Let G_λ be the minimal Green function for $\mathfrak{A} = D_m D_x$ and $I: (\beta_1, \beta_2)$.*

The transformation $\mathfrak{G}_\lambda$ is a positive linear transformation from $\mathbf{C}$ to $\mathbf{C}$ with norm $\| \mathfrak{G}_\lambda \| \leqq \lambda^{-1}$. Its domain is $\mathbf{C}$, and the range R is defined as follows: Let Δ be the set of all functions $F \in \mathbf{C}$ such that $\mathfrak{A}F \in \mathbf{C}$.

(i) *$R = \Delta$ if, and only if, the boundaries are at infinity or, in case of $|\beta_i| < \infty$, this β_i is neither active nor semiactive.*[15]

(ii) *If $|\beta_i| < \infty$ and β_i is either active or semiactive, then*

$$(8.3) \qquad F(x) \to 0 \quad as \quad x \to \beta_i$$

for each $F \in R$, and R consists of all $F \in \Delta$ satisfying (8.3).

In other words, for each $f \in \mathbf{C}$ there exists exactly one $F \in R$ satisfying the equation $\lambda F - \mathfrak{A}F = f$. The operator $\mathfrak{G}_\lambda$ is the resolvent of $\mathfrak{A}$ or of the restriction $\mathfrak{A}/R$.

Proof. Referring again to Theorem 7.2, we first prove that $F \in \mathbf{C}$, that is, that $F(x)$ approaches limits as $x \to \beta_i$. Consider two special cases. (i) Suppose that f vanishes identically in some neighborhood of β_2. In this neigh-

[15] In the nomenclature of footnote 12 it is required that neither boundary be regular or exit.

borhood F reduces to a constant multiple of u_2, and therefore a finite limit $F(\beta_2)$ exists. In particular, $F(\beta_2) = 0$ when $\beta_2 < \infty$. (ii) Let $f = \mathbf{1}$. By (7.13) in this case $\lambda F = \mathbf{1} - p_1 u_1 - p_2 u_2$ (where $p_1 = 0$ if u_1 is unbounded at β_2). Again it is clear that F approaches a limit $F(\beta_2)$ and $0 \leqq F(\beta_2) \leqq \lambda^{-1}$. We are now in a position to consider an arbitrary $f \in \mathbf{C}$. It can be split into a sum $f = f(\beta_2)\mathbf{1} + f_1 + f_2$ of three functions, of which the first is constant, the second vanishes identically in some neighborhood of β_2, and the third satisfies the inequality $|f_2| < \varepsilon$. The transform of $f(\beta_2)\mathbf{1} + f_1$ has been shown to approach a limit as $x \to \beta_2$. By (7.10) the transform $F_2 = \mathfrak{G}_\lambda f_2$ satisfies the inequality $\lambda \| F_2 \| < \varepsilon$. Accordingly, the oscillation of $F = \mathfrak{G}_\lambda f$ near β_2 is smaller than $\varepsilon\lambda^{-1}$, and it follows that the limit $F(\beta_2)$ exists.[16]

From (7.10) we see that $\| \mathfrak{G}_\lambda \| \leqq \lambda^{-1}$. To show that each $F = \mathfrak{G}_\lambda f$ satisfies (8.3) suppose that $\beta_2 < \infty$ and that β_2 is semiactive. Then $u_2 \downarrow 0$, while u_1 remains bounded at β_2. It follows that the first term on the right in (7.12) approaches 0 as $x \to \beta_2$, and it suffices to prove that $u_2(x)m(x) \to 0$. Now for $x > \beta$

$$u_2(x)m(x) < u_2(x)m(\beta) + \int_\beta^x u_2(y)\, dm(y) < u_2(x)m(\beta) + \varepsilon \tag{8.4}$$

provided β is chosen close enough to β_2. It follows that lim sup $u_2(x)m(x) < \varepsilon$ or $u_2 m \to 0$ as asserted. When β_2 is finite and active, the fact that $F \to 0$ is trivial since m remains bounded. We have thus proved that the range of $\mathfrak{G}_\lambda$ is contained in R. The last step of the proof of Theorem 8.1 applies here without change to show that R is contained in the range of $\mathfrak{G}_\lambda$, and this completes the proof.

The contrast between the last two theorems is striking. An infinite active boundary gives rise to nonuniqueness (i.e. a boundary condition) in the Hilbert space $\mathbf{H}_m$, but not in $\mathbf{C}$; a finite semiactive boundary gives a boundary condition in $\mathbf{C}$ but not in $\mathbf{H}_m$. Actually a finite semiactive boundary may be transformed into an infinite semiactive boundary (Section 4), and it is clear therefore that in the latter case the absence of a boundary condition is imposed only by the extraneous norm condition.

To be specific, the last theorem asserts that

$$0 \leqq \lambda \mathfrak{G}_\lambda \mathbf{1} \leqq \mathbf{1} \tag{8.5}$$

and the second equality sign holds only in the case of uniqueness (i). Exactly the same argument can be applied not only to the function $\mathbf{1}$, but to any positive solution ϕ of $\mathfrak{A}\phi = 0$. For example, if the basic interval I is $(0, \infty)$ we get

$$0 \leqq \lambda \mathfrak{G}_\lambda x \leqq x, \tag{8.6}$$

[16] For an application in Section 9 we note that we have proved $\lambda F(\beta_2) \leqq f(\beta_2)$ if $f > 0$ and $\beta_2 < \infty$. [Incidentally, this is not true if $\beta_2 = \infty$ and ∞ is semiactive.]

with the second equality sign holding if infinity is neither an active nor a semiactive boundary, in which case $\lambda \mathfrak{G}_\lambda x = o(x)$ as $x \to \infty$. Theorem 8.2 is contained in the following which does not refer to any particular norm and shows how $\mathfrak{G}_\lambda$ acts on *unbounded* functions.

THEOREM 8.3. *Suppose* $-\infty \leqq \beta_1 < 0 < \beta_2 \leqq \infty$. *Let* $\mathbf{L}$ *be the set of all functions* f *which are continuous in the open interval* $I:(\beta_1, \beta_2)$ *and for which* $f(x)x^{-1}$ *tends to a finite limit as* $x \to \beta_i$. *Let* Δ *be the subset of those* $F \in \mathbf{L}$ *for which* $\mathfrak{A}F \in \mathbf{L}$ *(where* $\mathfrak{A} = D_m D_x$*). Let* $\mathfrak{G}_\lambda$ *be the transformation* (7.3) *associated with the minimal Green function for* $(\mathfrak{A}, I)$.

Then $\mathfrak{G}_\lambda$ *is a closed linear transformation from* $\mathbf{L}$ *to* $\mathbf{L}$. *Its domain is* $\mathbf{L}$, *and its range* R *coincides with* Δ *if, and only if, neither boundary is active or semiactive. If* β_i *is active or semiactive, then* R *coincides with the subset of those* $F \in \Delta$ *for which*

$$F(x)x^{-1} \to 0 \qquad x \to \beta_i. \tag{8.7}$$

The proof is a mere repetition of the preceding one; the assertion follows also from Theorem 8.2 by the standard mapping procedure introduced at the end of Section 4.

COROLLARY TO THEOREM 8.2. *Let* $\mathbf{C}_\psi$ *be the Banach space of functions continuous in* I *metrized by the norm* (1.18). *The minimal Green function for* $\mathfrak{A}_\psi$ *induces a bounded transformation* $\mathfrak{G}_\lambda$ *from* $\mathbf{C}_\psi$ *to* $\mathbf{C}_\psi$ *with norm* λ^{-1}.

It is clear that in this statement ψ may be replaced by an arbitrary positive solution of $\mathfrak{A}_\psi \phi = 0$, and that the boundary conditions are inferred from (8.3).

Note on the term "minimal". It is readily seen from Theorems 8.1 and 8.2 that for any *positive* bounded transformation $\bar{\mathfrak{G}}_\lambda$ from $\mathbf{C}$ to $\mathbf{C}$ (or from $\mathbf{H}_m$ to $\mathbf{H}_m$) which transforms each f into a solution F of $\lambda F - \mathfrak{A}F = f$ one has $\bar{\mathfrak{G}}_\lambda f \geqq \mathfrak{G}_\lambda f$ for each $f \geqq 0$. This observation will be amplified in Sections 10 and 12.

9. The role of convexity

We have just seen that for the most general operator $\mathfrak{A}_\psi$, defined by an arbitrary continuous $\psi > 0$, there exists a minimal Green function for each $\lambda > 0$. It leads to an inversion $\mathfrak{G}_\lambda$ of $\lambda - \mathfrak{A}_\psi$, and $\mathfrak{G}_\lambda$ is a positive operator such that $\lambda \mathfrak{G}_\lambda \psi \leqq \psi$. This means that with the norm (1.18) $\lambda \mathfrak{G}_\lambda$ becomes norm nondecreasing. Unfortunately we are not always free to accept this norm, and very often we are tied to the traditional Banach norm (1.17). The question then arises (in particular in probability theory and heat conduction) whether the operator $\lambda \mathfrak{G}_\lambda$ corresponding to $\mathfrak{A}_\psi$ is norm preserving or decreasing ("a transition operator") when the familiar norm (1.17) is used. In other words, we ask: what conditions must ψ satisfy in order that

$$\lambda \mathfrak{G}_\lambda \mathbf{1} \leqq \mathbf{1}. \tag{9.1}$$

The answer is found in

THEOREM 9.1. *Let $\mathfrak{G}_\lambda$ be the transformation* (7.3) *associated with the minimal Green function for* $(\mathfrak{A}_\psi, I)$ *where* $\psi > 0$ *in* I. *In order that* (9.1) *hold for all* $\lambda > \lambda_0$, *it is necessary and sufficient that* ψ *be convex downward* (*the graph of* ψ *lying below the chords*).

In this case each positive solution ϕ *of* $\mathfrak{A}_\psi \phi = 0$ *is convex downward, and* (9.1) *holds for each* $\lambda > 0$.

Proof. In the special case $\psi = \mathbf{1}$ the operator $\mathfrak{A}_\psi$ reduces to $\mathfrak{A} = D_m D_x$: the relation (9.1) is then contained in (7.10) and all solutions of $\mathfrak{A}_\psi \phi = 0$ are linear functions, hence convex. The assertion is therefore true in the special case $\psi =$ const., and we shall from now on assume that ψ *is not constant.* It is convenient to break up the remainder of the proof into minor lemmas.

LEMMA 9.1. *If* ψ *is convex downward, so is each positive solution* ϕ *of* $\mathfrak{A}_\psi \phi = 0$.

Proof. Since ψ is convex downward and not constant, it either decreases near β_1 or increases near β_2. Assume the latter. Being convex, ψ increases at least linearly, and hence the integral over $\psi^{-2}(x)\,dx$ converges at β_2. It follows that

$$\phi(x) = \psi(x) \int_x^{\beta_2} \psi^{-2}(s)\,ds \tag{9.2}$$

defines that solution of $\mathfrak{A}_\psi \phi = 0$ which is minimal at β_2. We prove that this ϕ is convex downward.

In view of the assumed convexity, ψ possesses a monotonically increasing right derivative ψ^+. From (9.2) we see that also the right derivative

$$\phi^+(x) = \psi^+(x) \int_x^{\beta_2} \psi^{-2}(s)\,ds - \psi^{-1}(x) \tag{9.3}$$

exists for each $x \in I$. For $x < x + h < \beta_2$ we have

$$\begin{aligned} \phi^+(x+h) - \phi^+(x) &= \{\psi^+(x+h) - \psi^+(x)\} \int_{x+h}^{\beta_2} \psi^{-2}(s)\,ds \\ &\quad - \psi^+(x) \int_x^{x+h} \psi^{-2}(s)\,ds - \psi^{-1}(x+h) + \psi^{-1}(x) \qquad (9.4) \\ &\geqq -\int_x^{x+h} \psi^+(s)\cdot\psi^{-2}(s)\,ds - \psi^{-1}(x+h) + \psi^{-1}(x) = 0. \end{aligned}$$

Hence ϕ^+ increases and ϕ is convex. If ψ is not minimal at β_1, then the same argument applies at β_1, and we see that the two minimal solutions are linearly independent and convex. Every positive solution of $\mathfrak{A}_\psi \phi = 0$ being a positive linear combination of the minimal solutions, the lemma is proved.

LEMMA 9.2. *Let* $F_\lambda = \mathfrak{G}_\lambda f$ *be the transform defined by* (7.3) *when* G_λ *is the minimal Green function for* $\mathfrak{A} = D_m D_x$. *If* f *is continuous in* I *and* $f(x) \to 0$ *as* $x \to \beta_i$ $(i = 1, 2)$, *then* $\lambda F_\lambda \to f$ *as* $\lambda \to \infty$ *uniformly in* I.

Proof. (This lemma is known and proved for completeness only.) Suppose first that $f \in R$, where R is the range described in Theorem 8.2. Then $\mathfrak{A}f$ exists, and we conclude from

$$\lambda F_\lambda - \mathfrak{A}F_\lambda = f \tag{9.5}$$

that

$$\lambda \mathfrak{A}F_\lambda - \mathfrak{A}^2 F_\lambda = \mathfrak{A}f, \tag{9.6}$$

and that $\mathfrak{A}F_\lambda \in R$. It follows that $\mathfrak{A}F_\lambda = \mathfrak{G}_\lambda(\mathfrak{A}f)$, and Theorem 8.2 asserts that $\lambda \| \mathfrak{A}F_\lambda \| \leqq \| \mathfrak{A}f \|$. Thus from (9.5)

$$\| \lambda F_\lambda - f \| \leqq \lambda^{-1} \| \mathfrak{A}f \| \rightarrow 0, \qquad \lambda \rightarrow \infty. \tag{9.7}$$

The assertion is therefore proved for $f \in R$. However, each f which vanishes at the boundaries is in the *closure* of R, and a routine application of the triangle inequality shows that $\| \lambda F_\lambda - f \| \rightarrow 0$ for each f in this closure.

LEMMA 9.3. *Adhering to the notations of the preceding lemma, suppose that there exists a function $f > 0$ such that*

$$\lambda F_\lambda \leqq f \tag{9.8}$$

for each $\lambda > \lambda_0$. Then $f > 0$ is convex upward.

Conversely, suppose that f is convex upward and that the interval I is finite $(-\infty < \beta_1 < \beta_2 < \infty)$. Then (9.8) holds for all $\lambda > 0$.

Proof. A comparison of (9.8) and (9.5) shows that (9.8) is equivalent to $\mathfrak{A}F_\lambda \leqq 0$. In this case $D_x F_\lambda$ decreases, and F_λ is convex upward. Let $I^{(0)}$ be a closed subinterval of I, and construct a function $f^{(0)}$ such that $f^{(0)}(x) = f(x)$ for $x \in I^{(0)}$ and $0 \leqq f^{(0)}(x) \leqq f(x)$ for $x \in I$ and $f^{(0)}(x) \rightarrow 0$ as $x \rightarrow \beta_i$. Put $F_\lambda^{(0)} = \mathfrak{G}_\lambda f^{(0)}$. By Lemma 9.2

$$\lambda F_\lambda(x) \geqq \lambda F_\lambda^{(0)}(x) \rightarrow f^{(0)}(x) = f(x), \qquad x \in I^{(0)}. \tag{9.9}$$

It follows that $\lambda F_\lambda(x) \rightarrow f(x)$ for each $x \in I$ and hence f is convex upward. This proves the first part of the lemma.

Conversely, suppose that f is convex upward and $| \beta_i | < \infty$. We recall the footnote 15 to Theorem 8.2 according to which

$$\lambda F_\lambda(\beta_i) \leqq f(\beta_i), \qquad i = 1, 2. \tag{9.10}$$

Suppose now that (9.8) is false, and let $I' = (x_1, x_2)$ be a *maximal* subinterval of I such that $\lambda F_\lambda(x) > f(x)$ for $x \in I'$. In view of (9.10) the endpoints of I' must be interior points, i.e.

$$\lambda F_\lambda(x_i) - f(x_i) = 0, \qquad i = 1, 2. \tag{9.11}$$

But inside I' we have $\mathfrak{A}F_\lambda > 0$. Hence F_λ is convex downward, and the same is true of the difference $\lambda F_\lambda - f$. Remembering (9.11) we conclude that $\lambda F_\lambda(x) < f(x)$ for $x \in I'$, which contradicts the definition of I'. The lemma is proved.

Completion of the proof. Suppose that (9.1) holds for all $\lambda > \lambda_0$, and put $\mathfrak{G}_\lambda \mathbf{1} = H_\lambda$. Using Theorem 7.1 we wish to reformulate (9.1) in terms of the related operator $\mathfrak{A}^* = D_\mu D_\xi$. The variable x becomes a function of ξ, and H_λ and $\mathbf{1}$ are mapped, respectively, into the functions defined by $F_\lambda(\xi) = H_\lambda(x)\psi^{-1}(x)$ and $f(\xi) = \psi^{-1}(x)$. Here $F_\lambda = \mathfrak{G}_\lambda^* f$, where $\mathfrak{G}_\lambda^*$ corresponds to the minimal Green function for $\mathfrak{A}^*$. Now (9.1) states that $\lambda F_\lambda \leqq f$ for $\lambda > \lambda_0$, and the preceding lemma shows that f is convex upward, i.e. that $D_\xi^+ f$ is decreasing. But $D_\xi^+ f = \psi^2 D_x^+(\psi^{-1}) = -D_x^+\psi$, and therefore ψ is convex downward.

Conversely, suppose that ψ is convex downward. By Lemma 9.1 all positive solutions of $\mathfrak{A}_\psi \phi = 0$ are convex downward, and since $\mathfrak{A}_\psi = \mathfrak{A}_\phi$, there is no loss of generality in assuming that ψ is not minimal either at β_1 or β_2. The transformation (4.1) then maps the interval $I = (\beta_1, \beta_2)$ on a finite interval $I^* = (\beta_1^*, \beta_2^*)$, and f becomes a function of ξ which is convex upward. The last lemma then asserts that $\lambda F_\lambda \leqq f$ for all $\lambda > 0$, and the theorem is proved.

10. Self-adjoint operators

We consider the Hilbert space $\mathbf{H}_m$ defined by (1.5) and propose to investigate the most general closed symmetric linear operator $\mathfrak{G}_\lambda$ which, for a fixed $\lambda > 0$, associates with each $f \in \mathbf{H}_m$ an element $F \in \mathbf{H}_m$ such that

$$\lambda F - \mathfrak{A}F = f, \qquad \mathfrak{A} = D_m D_x \tag{10.1}$$

for almost all $x \in I = (\beta_1, \beta_2)$. Since the domain of $\mathfrak{G}_\lambda$ is the entire space, $\mathfrak{G}_\lambda$ is a *bounded self-adjoint* operator.

Let $\mathfrak{G}_\lambda^{\min}$ be the minimal Green function for $(\mathfrak{A}, I)$, and $\mathfrak{G}_\lambda^{\min}$ the induced transformation. Both $F = \mathfrak{G}_\lambda f$ and $F^{\min} = \mathfrak{G}_\lambda^{\min} f$ satisfy (10.1), and therefore $w = F - F^{\min}$ must be a solution of

$$\lambda w - \mathfrak{A}w = 0. \tag{10.2}$$

Let w_1 and w_2 be two independent solutions of (10.2). It is then clear that

$$\mathfrak{G}_\lambda f = \mathfrak{G}_\lambda^{\min} f + w_1 \cdot \Phi_f^{(1)} + w_2 \cdot \Phi_f^{(2)}, \tag{10.3}$$

where $\Phi_f^{(i)}$ stands for a bounded linear functional, i.e. an element of $\mathbf{H}_m$. It follows that $\mathfrak{G}_\lambda$ is a kernel transformation with $F = \mathfrak{G}_\lambda f$ being defined by

$$F(x) = \int_{\beta_1}^{\beta_2} G_\lambda(x, y) f(y)\, dm(y), \qquad x \in I \tag{10.4}$$

where

$$G_\lambda(x, y) = G_\lambda^{\min}(x, y) + w_1(x)\Phi^{(1)}(y) + w_2(x)\Phi^{(2)}(y). \tag{10.5}$$

The further condition that $\mathfrak{G}_\lambda$ be symmetric reduces (10.5) to the form

$$G_\lambda(x, y) = G_\lambda^{\min}(x, y) + \sum_{i,j=1}^{2} p_{ij}(\lambda) w_i(x) w_j(y), \tag{10.6}$$

where the $p_{ij}(\lambda)$ are constants, $p_{12}(\lambda) = p_{21}(\lambda)$.

If w_1 and w_2 (i.e. if all solutions of (10.2)) are square integrable (m), then the symmetric matrix $P(\lambda) = (p_{ij}(\lambda))$ may be chosen arbitrarily, and (10.6) defines a transformation of the required form. Such is the case when both boundaries are *active*. If only β_2 is active, three coefficients $p_{ij}(\lambda)$ must vanish; if neither endpoint is active, the minimal $\mathfrak{G}_\lambda^{\min}$ is the only transformation of the desired type.

When for each $\lambda > 0$ the coefficients $p_{ij}(\lambda)$ are chosen in an arbitrary manner, no natural connection between the corresponding $\mathfrak{G}_\lambda$ must be expected. In particular, even though $(\lambda - \mathfrak{A})\mathfrak{G}_\lambda$ reduces to the identity, $\mathfrak{G}_\lambda$ will not represent the resolvent of any fixed operator. *Only if the range R of $\mathfrak{G}_\lambda$ is independent of λ does $\mathfrak{G}_\lambda$ represent the resolvent*[17] *of an operator, namely* $\mathfrak{G}_\lambda = \{\lambda - \mathfrak{A}/R\}^{-1}$ where $\mathfrak{A}/R$ is the restriction of $\mathfrak{A}$ to R. This $\mathfrak{A}/R$ as well as $\mathfrak{G}_\lambda$ is then *self-adjoint*.

We are thus confronted with the following questions:

(1) Given a self-adjoint $\mathfrak{G}_\lambda$, determine its range R.

(2) Describe the most general possible range R.

(3) Given R, determine the coefficients $p_{ij}(\lambda)$ in such a way that $\mathfrak{G}_\lambda$ will have the prescribed range R.

(4) Find the conditions under which the transformation $\mathfrak{G}_\lambda$ will be positive, or norm-decreasing, etc.

The case of one active boundary is simpler than the case of two active boundaries, and can be derived from it in trivial fashion by equating certain coefficients to zero. We shall therefore suppose that both β_1 and β_2 are active. First we take up the case where $|\beta_i| < \infty$. It will be observed that the *same* calculations apply to infinite intervals (and, indeed, to the general operator $\mathfrak{A}_\psi$). The only advantage (and only purpose) in a separate treatment of finite intervals is the appearance of the classical boundary conditions under their traditional form, unobscured by a new symbolism.

THEOREM 10.1. *Let β_1 and β_2 be active boundaries, $|\beta_i| < \infty$. For some fixed $\lambda > 0$ let $\mathfrak{G}_\lambda$ be a bounded self-adjoint transformation taking each $f \in \mathbf{H}_m$ into $\mathfrak{G}_\lambda f = F \in \mathbf{H}_m$ such that* (10.1) *holds almost everywhere* (m). *Then:*

(a) *There exist two boundary conditions which are either of the form*[18]

$$-F^+(\beta_1) + \pi_{11} F(\beta_1) + \pi_{12} F(\beta_2) = 0, \tag{10.7}$$
$$F^-(\beta_2) + \pi_{21} F(\beta_1) + \pi_{22} F(\beta_2) = 0,$$

where

$$\pi_{12} = \pi_{21}, \tag{10.7a}$$

[17] The so-called resolvent equation $\mathfrak{G}_\lambda - \mathfrak{G}_\nu = (\nu - \lambda)\mathfrak{G}_\nu\mathfrak{G}_\lambda$ is an obvious consequence of (10.1): substracting (10.1) for two values of the parameter we get $\nu\{\mathfrak{G}_\nu - \mathfrak{G}_\lambda\} - A\{\mathfrak{G}_\nu - \mathfrak{G}_\lambda\} = (\lambda - \nu)\mathfrak{G}_\lambda$. This is equivalent to the resolvent equation *if, and only if, the ranges of $\mathfrak{G}_\lambda$ and $\mathfrak{G}_\nu$ coincide.*

[18] In (10.7) it is understood that F is continuous in the closed interval and $F^-(\beta_2) = \lim F^+(x) = \lim F^-(x)$. In (10.8) the case $\pi_1 = 0$ is included with the obvious interpretation that it implies $F(\beta_1) = 0$. If $\pi_1 = \pi_2 = 0$, the boundary conditions (10.8) are $F(\beta_1) = F(\beta_2) = 0$, i.e. $\mathfrak{G}_\lambda = \mathfrak{G}_\lambda^{\min}$.

or of the form

$$(10.8) \qquad \alpha\pi_1^{-1}F(\beta_1) = \alpha\pi_2^{-1}F(\beta_2) = \pi_1 F^+(\beta_1) - \pi_2 F^-(\beta_2).$$

The range R of $\mathfrak{G}_\lambda$ consists of those $F \in \mathbf{H}_m$ which satisfy the boundary conditions and the trivial requirement $\mathfrak{A}F \in \mathbf{H}_m$.

(b) *Conversely: to arbitrarily prescribed coefficients π_{ij} (or π_i, α) there corresponds a λ_0 such that for each $\lambda > \lambda_0$ there exists exactly one transformation $\mathfrak{G}_\lambda$ of the described sort whose range R is characterized by the prescribed boundary condition.*

Note 1. Given the coefficients in the boundary conditions, the explicit form of the coefficients $p_{ij}(\lambda)$ is found in (10.22) for the boundary conditions (10.7), and in (10.23)–(10.25) for the conditions (10.8).

Note 2. If $\pi_{12} = \pi_{21} = 0$ in (10.7), the transformation $\mathfrak{G}_\lambda$ is induced by the *Green function* G_λ formed by means of two solutions satisfying one boundary condition each.

Note 3. The condition $\pi_{12} = \pi_{21}$ is imposed merely by the requirement of symmetry. The arguments go through also when $\pi_{12} \neq \pi_{21}$. See Section 12 for more general unsymmetric transformations.

THEOREM 10.2. (a) *In order that the transformation $\mathfrak{G}_\lambda$ of Theorem* 10.1b *be positive for all λ sufficiently large, it is necessary and sufficient that $\pi_{12} \leqq 0$ (in the case of the boundary condition* (10.7)) *or $\pi_1\pi_2 \geqq 0$ (in the case* (10.8)).

(b) *In order that $\mathfrak{G}_\lambda$ be positive and*

$$(10.9) \qquad \lambda\mathfrak{G}_\lambda \mathbf{1} \leqq \mathbf{1}$$

for all λ sufficiently large, it is necessary and sufficient that

$$(10.10) \qquad \pi_{12} \leqq 0, \qquad \pi_{i1} + \pi_{i2} \geqq 0, \qquad i = 1, 2$$

or

$$(10.11) \qquad \pi_1 \geqq 0, \qquad \pi_2 \geqq 0, \qquad \alpha > 0,$$

respectively. In this case $\mathfrak{G}_\lambda$ is positive, and (10.9) *holds for all $\lambda > 0$.*

For $\lambda\mathfrak{G}_\lambda \mathbf{1} = \mathbf{1}$ it is necessary and sufficient that $\pi_{ii} + \pi_{12} = 0$ or $\pi_1 = \pi_2 = \frac{1}{2}$, $\alpha = 0$, respectively.

Proof. The two solutions w_i in (10.6) may be chosen at will, but it is most convenient to let w_i be the solution of (10.2) determined by the conditions

$$(10.12) \qquad w_1(\beta_1) = w_2(\beta_2) = 1, \qquad w_1(\beta_2) = w_2(\beta_1) = 0.$$

(In other words, $w_1(x) = u_{*2}(x)/u_{*2}(\beta_1)$ and $w_2(x) = u_{*1}(x)/u_{*1}(\beta_2)$ where u_{*i} is minimal at β_i.) We shall require the derivatives

$$(10.13) \qquad \begin{aligned} \omega_{11} &= -w_1^+(\beta_1), & \omega_{12} &= w_1^-(\beta_2), \\ \omega_{21} &= -w_2^+(\beta_1), & \omega_{22} &= w_2^-(\beta_2). \end{aligned}$$

It is clear that all these quantities depend on λ.

LEMMA 10.1. *The matrix Ω of the elements ω_{ij} satisfies the conditions*

$$\omega_{11} > 0, \qquad \omega_{22} > 0, \qquad \omega_{12} = \omega_{21} < 0. \tag{10.14}$$

As $\lambda \to \infty$

$$\omega_{ii} \to \infty, \qquad \omega_{12} \to 0. \tag{10.15}$$

Proof. By (5.7) the Wronskian $w_1 w_2' - w_1' w_2$ is constant. At the left boundary it reduces to $-\omega_{21}$, at the right to $-\omega_{12}$. Hence $\omega_{12} = \omega_{21}$. The signs in (10.14) are determined by the fact that w_1 decreases, while w_2 increases. Finally, if $\beta_1 < x < x + h < \beta_2$, then (10.2) and the convexity of w_2 imply

$$1 \geqq w_2(x+h) \geqq \lambda \int_{x+}^{x+h-} w(s)\, dm(s) \geqq \lambda w^+(x) \int_{x+}^{x+h-} (s-x)\, dm(s), \tag{10.16}$$

and it follows that $w^+(x) \to 0$ as $\lambda \to \infty$ for each fixed $x < \beta_2$. This implies that $\omega_{12} \to 0$ and also that $\omega_{22} \to \infty$. The latter can be concluded equally from the fact that the Wronskian is constant.

Proof of Theorem 10.1(a). For each $f \in \mathbf{H}_m$ we have from (10.6)

$$F = F^{\min} + \sum_{i,j} p_{ij}(\lambda) w_i \cdot (w_j, f), \tag{10.17}$$

where (g, f) is an abbreviation for the inner product

$$(g, f) = \int_{\beta_1}^{\beta_2} gf\, dm. \tag{10.18}$$

By Theorem 8.1 the function $F^{\min}$ has the boundary values 0, and hence

$$F(\beta_i) = \sum_{j=1}^{2} p_{ij}(\lambda)(w_j, \lambda F - \mathfrak{A}F). \tag{10.19}$$

Repeated integration by parts leads to

$$\lambda(w_j, F) = \int_{\beta_1}^{\beta_2} F\, dw_j' = \omega_{j1} F(\beta_1) + \omega_{j2} F(\beta_2) - (-1)^j F'(\beta_j) + (w_j, \mathfrak{A}F), \tag{10.20}$$

and (10.19) takes on the form

$$F(\beta_i) = \sum_{j,k=1}^{2} p_{ij}(\lambda)\omega_{jk} F(\beta_k) - \sum_{j=1}^{2} (-1)^j p_{ij}(\lambda) F'(\beta_j). \tag{10.21}$$

We have now to distinguish two cases.

Case I. The matrix $P(\lambda)$ with elements $p_{ij}(\lambda)$ is not singular. Premultiplying (10.21) by $P^{-1}(\lambda)$, we are led to the boundary conditions (10.7) with the matrix $\Pi = (\pi_{ij})$ given by

$$\Pi = P^{-1}(\lambda) - \Omega. \tag{10.22}$$

Thus each F in the range satisfies (10.7). Conversely, suppose that $F \in \mathbf{H}_m$, $\mathfrak{A}F \in \mathbf{H}_m$, and that (10.7) is true. Set $f = \lambda F - \mathfrak{A}F$ and $\bar{F} = \mathfrak{G}_\lambda f$. Both

F and $\bar{F}$ satisfy (10.1), and hence $F - \bar{F}$ is a solution of (10.2), i.e. $F - \bar{F} = \rho_1 w_1 + \rho_2 w_2$. Now both F and $\bar{F}$ satisfy (10.7), and hence $p_{i1}(\lambda)\rho_1 + p_{i2}(\lambda)\rho_2 = 0$. Since P is nonsingular, this implies $\rho_1 = \rho_2 = 0$, and it follows that (10.7) is a necessary and sufficient condition on the range.

Case II. The matrix $P(\lambda)$ is singular. Since it is symmetric, it is necessarily of the form

$$(10.23) \qquad P(\lambda) = T\begin{pmatrix} \pi_1^2 & \pi_1 \pi_2 \\ \pi_1 \pi_2 & \pi_2^2 \end{pmatrix}.$$

Multiply the equations in (10.21) by π_2 and $-\pi_1$, respectively, to obtain

$$(10.24) \qquad \pi_2 F(\beta_1) = \pi_1 F(\beta_2).$$

If $\pi_1 = \pi_2 = 0$ we have $F(\beta_1) = F(\beta_2) = 0$. Otherwise each equation in (10.21) reduces to (10.8) with

$$(10.25) \qquad T^{-1} = \alpha + \sum_{i,j=1}^{2} \omega_{ij}\, \pi_i\, \pi_j\,.$$

This completes the proof of the first part.

Proof of Theorem 10.1(b). Prescribe an arbitrary symmetric matrix $\Pi = (\pi_{ij})$. In view of (10.15) the diagonal elements in the matrix $\Pi + \Omega$ will be dominant for all λ sufficiently large. It follows that for such λ a matrix $P(\lambda)$ satisfying (10.22) exists, and the kernel (10.6) defines the transformation $\mathfrak{G}_\lambda$ of the desired kind. This argument applies equally to (10.8) and (10.23).

Proof of Theorem 10.2(a). It is clear from (10.17) and (10.19) that $\mathfrak{G}_\lambda$ is positive if, and only if, $p_{ij}(\lambda) \geqq 0$. Referring again to (10.15) we see that, for sufficiently large λ, the diagonal elements of $\Pi + \Omega$ are positive, while the corner elements approach π_{12}. For $P(\lambda) = (\Pi + \Omega)^{-1}$ to be positive, it is then necessary and sufficient that $\pi_{12} \leqq 0$. Similarly, in the case (10.23) the condition $\pi_1 \pi_2 \geqq 0$ is necessary and sufficient.

Proof of Theorem 10.2(b). We have

$$(10.26) \qquad \lambda \int_{\beta_1}^{\beta_2} w_i\, dm = \int_{\beta_1}^{\beta_2} dw_i' = \omega_{i1} + \omega_{i2}\,.$$

Substituting $f = \mathbf{1}$ into (10.17) we find therefore

$$(10.27) \qquad \lambda F(x) = 1 - w_1(x) - w_2(x) + \sum_{i,j,k} w_i(x) p_{ij}(\lambda)\omega_{jk}\,.$$

This shows that (10.9) holds if, and only if

$$(10.28) \qquad \sum_{j,k} p_{ij}(\lambda)\omega_{jk} \leqq 1, \qquad i = 1, 2.$$

In other words, in order that $\mathfrak{G}_\lambda$ be positive and (10.9) hold, it is necessary and sufficient that the row sums of $P(\lambda)\Omega$ be $\leqq 1$ and $P(\lambda) \geqq 0$. If (10.10) holds, then $P(\lambda) \geqq 0$ for all $\lambda \geqq 0$, and the row sums of $P(\lambda)\Pi$ are positive. However $P(\lambda)\Pi = 1 - P(\lambda)\Omega$, and therefore (10.10) is a *sufficient* condition for each $\lambda > 0$. On the other hand, in the case of the boundary condition (10.7) the matrix $P(\lambda)$ converges to a diagonal matrix, and hence $P(\lambda)\Omega = 1 - P(\lambda)\Pi$ can have row sums $\leqq 1$ only if Π (which is independent of λ) has nonnegative row sums. Thus (10.10) is necessary and sufficient. The same argument applies, in a simpler form, to the case of the matrix (10.23). The two theorems are proved.

If $\beta_2 = \infty$ the symbols $F(\beta_2)$ and $F^-(\beta_2)$ are senseless, and the same is true of formulas (10.12)–(10.13). However, the theorems and the proofs remain valid with a slight reinterpretation of the symbols. Instead of the basic Lemma 10.1 we use

LEMMA 10.2. *If $\beta_2 = \infty$, let w_2 be that solution of* (10.2) *for which*

$$x^{-1}w_2(x) \to 1 \quad \text{as} \quad x \to \infty, \tag{10.29}$$

and define w_1 similarly. Put

$$\omega_{i2} = \lim_{x\to\infty} \{xw_i'(x) - w_i(x)\}, \qquad i = 1, 2, \tag{10.30}$$

and define ω_{i1} similarly in case $\beta_1 = -\infty$ (and by (10.13) *if $\beta_1 > -\infty$). If both boundaries are active, these quantities are finite and satisfy* (10.14)–(10.15).

Proof. By Theorem 6.3 the limit of $x^{-1}w(x)$ exists for each positive solution of (10.2). Now

$$\lambda \int_\beta^x sw_2 \, dm(s) = \int_\beta^x s \, dw_2'(s) = \{sw_2^-(s) - w_2(s)\} \Big|_\beta^x . \tag{10.31}$$

From (6.9) it follows that sw_2 is square integrable over (β_1, β_2) and therefore $xw_2^-(x) - w_2(x)$ approaches a limit monotonically as $x \to \infty$. Rewriting the Wronskian in the form

$$\begin{aligned} &w_2'(x)w_1(x) - w_1'(x)w_2(x) \\ &\quad = \{xw_2'(x) - w_2(x)\}\cdot x^{-1}w_1(x) - \{xw_1'(x) - w_1(x)\}\cdot x^{-1}w_2(x), \end{aligned} \tag{10.32}$$

we see the validity of the arguments in the proof of Lemma 10.1.

Consider now (10.17), and remember that the derivative of $F^{\min}$ satisfies (7.14). It is seen at once that in the case of an active boundary at ∞ the limits

$$\lim_{x\to\infty} x^{-1}F(x) \quad \text{and} \quad \lim_{x\to\infty} \{xF'(x) - F(x)\} \tag{10.33}$$

exist and are finite. With these quantities the proof of Theorem 10.1 applies without change and we have

THEOREM 10.3. *Theorem* 10.1 *applies also to the case $\beta_2 = \infty$ provided the symbols $F(\beta_2)$ and $F^-(\beta_2)$ appearing in the boundary conditions are replaced by the limits appearing in* (10.33).

In the case of an operator $\mathfrak{A}_\psi$, *Theorem* 10.1 *still applies with* $F(\beta_2)$ *and* $F^-(\beta_2)$ *to be replaced, respectively, by*

$$(10.34)\qquad \lim_{x\to\beta_2}\frac{F(x)}{\phi(x)} \quad \text{and} \quad \lim_{x\to\beta_2}\phi^2(x)\cdot D_x\frac{F(x)}{\phi(x)},$$

where ϕ *is a solution of* $\mathfrak{A}_\psi\,\phi = 0$ *which is at* β_2 *positive but not minimal.*

The last formulation combines Theorems 10.1 and 10.2: in the case $\beta_2 < \infty$ we set $\phi = 1$; in the case $\beta_2 = \infty$, however, $\phi = x$.

11. Self-adjoint operators in closed intervals

Consider a *finite* interval $I:(\beta_1, \beta_2)$ in which m remains *bounded*: in other words, we suppose that both boundaries are finite and active. It is then natural to extend the measure m to the closed interval $\bar{I}:[\beta_1, \beta_2]$ by attributing the weights m_1 and m_2 to the two points of the boundary. This leads to introducing the Hilbert space $\bar{\mathbf{H}}_m$ over $\bar{I}$ defined by the inner product

$$(11.1)\qquad (f, g) = \int_{\beta_1-}^{\beta_2+} fg\,dm = \int_{\beta_1+}^{\beta_2-} fg\,dm + \sum_{i=1}^{2} m_i f(\beta_i)g(\beta_i).$$

For sake of definiteness we shall suppose $m_1 > 0$ and $m_2 > 0$, although the argument applies equally when only one boundary point carries a positive weight.

So far the operator $\mathfrak{A} = D_m D_x$ is defined only in the interior of $\bar{I}$. We shall presently see that, essentially, there exists only one way of defining $\mathfrak{A}$ at the boundaries, provided we require that all functions in the domain of $\mathfrak{A}$ be continuous in $\bar{I}$.

A linear transformation $\bar{\mathfrak{G}}_\lambda$ mapping each $f \in \bar{\mathbf{H}}_m$ into a solution F of $\lambda F - \mathfrak{A}F = f$ is necessarily of the form (10.5) where $\Phi^{(i)}$ is a functional in the enlarged Hilbert space $\bar{\mathbf{H}}_m$. If such $\bar{\mathfrak{G}}_\lambda$ is bounded and symmetric, it must be induced by a kernel $\bar{G}_\lambda$ obtained by extending the kernel G_λ of (10.6) to the boundaries in accordance with the requirement that F be continuous. Supposing that the solutions w_i are specified by (10.12) we shall have

$$(11.2)\qquad \begin{aligned} \bar{G}_\lambda(x, y) &= G_\lambda(x, y) && \beta_1 < x, y < \beta_2\\ \bar{G}_\lambda(\beta_i, y) &= p_{i1}(\lambda)w_1(y) + p_{i2}(\lambda)w_2(y), && \beta_1 < y < \beta_2\\ \bar{G}_\lambda(x, \beta_j) &= p_{1j}(\lambda)w_1(x) + p_{2j}(\lambda)w_2(x), && \beta_1 < x < \beta_2\\ \bar{G}_\lambda(\beta_i, \beta_j) &= p_{ij}(\lambda), && i, j = 1, 2. \end{aligned}$$

(For example, the function f defined by $f(x) = 0$ for $x < \beta_2$ and $f(\beta_2) = 1$ is mapped into $p_{11}(\lambda)w_1 + p_{21}(\lambda)w_2$.)

The matrix $P(\lambda)$ of the coefficients $p_{ij}(\lambda)$ completely determines the transformation $\bar{\mathfrak{G}}_\lambda$. Given $\bar{\mathfrak{G}}_\lambda$, the operator $\mathfrak{A}$ at β_i is defined for each F in the range of $\bar{\mathfrak{G}}_\lambda$ by $\mathfrak{A}F(\beta_i) = \lambda F(\beta_i) - f(\beta_i)$. Formally this definition depends on λ. However, repeating the argument of Section 10 we are able to find all possible ranges, that is, all possible definitions of $\mathfrak{A}$. Given a possible range R (i.e. a possible definition), we can construct the corresponding $\bar{\mathfrak{G}}_\lambda$.

THEOREM 11.1. *The Theorems* 10.1 *and* 10.2 *remain valid for the enlarged Hilbert space with inner product* (11.1) *provided that the boundary conditions* (10.7) *are replaced by*

$$(11.3)\qquad \begin{aligned} -F^+(\beta_1) + \pi_{11} F(\beta_1) + \pi_{12} F(\beta_2) + m_1 \mathfrak{A}F(\beta_1) &= 0 \\ F^-(\beta_2) + \pi_{21} F(\beta_1) + \pi_{22} F(\beta_2) + m_2 \mathfrak{A}F(\beta_2) &= 0, \end{aligned}$$

and (10.8) *by*

$$(11.4)\qquad \begin{aligned} \alpha\pi_1^{-1}F(\beta_1) = \alpha\pi_2^{-1}F(\beta_2) = \pi_1 F^+(\beta_1) - \pi_2 F^-(\beta_2) \\ + \pi_1 \mathfrak{A}F(\beta_1) + \pi_2 \mathfrak{A}F(\beta_2). \end{aligned}$$

It should be observed that when $m_1 \neq 0$, $m_2 \neq 0$, equations (11.3) represent *a definition of* $\mathfrak{A}$ *at the boundary* rather than *"boundary conditions"*. The classical boundary conditions appear as a *degenerate limiting case.* The theorem asserts that, for sufficiently large λ, an operator $\mathfrak{G}_\lambda$ inverting $\lambda - \mathfrak{A}$ and acting from $\mathbf{H}_m$ to *continuous* functions is possible only if $\mathfrak{A}$ is defined as indicated. In the case $m_2 = 0$ the boundary point β_2 has measure zero, and no definition of $\mathfrak{A}$ is required. However, nonuniqueness persists, and the range R must be specified by means of a lateral condition.

The *proofs* of Section 10 carry over to the present situation almost without change. In the construction of $\mathfrak{G}_\lambda$ to given boundary conditions the relation (10.22) defining the matrix $P(\lambda)$ in the nonsingular case now takes on the form

$$(11.5)\qquad P^{-1}(\lambda) = \Pi + \Omega + M, \qquad M = \begin{pmatrix} m_1 & 0 \\ 0 & m_2 \end{pmatrix}.$$

In the singular case of a matrix of the form (10.23) the coefficient T is now given by

$$(11.6)\qquad T^{-1} = \alpha + \sum_{i,j=1}^{2} \omega_{ij}\,\pi_i\,\pi_j + \lambda \sum_{i=1}^{2} m_i\,\pi_i^2,$$

replacing (10.25).

12. Positive operators in C

We turn to the problem of finding the general form of the solutions of $\lambda F - \mathfrak{A}F = f$ when f, F are bounded continuous functions, $\mathfrak{A} = D_m D_x$, $\lambda > 0$. Theorem 7.1 enables us to translate the results into statements concerning the general operator $\mathfrak{A}_\psi$.

The minimal Green function of Section 8 provides a solution $F^{\min}$. This solution is unique except if at least one boundary is finite *and* either active or semiactive. In this case new solutions F are formed by adding to $F^{\min}$ a bounded solution w of $\lambda w - \mathfrak{A}w = 0$. Such a w has finite limits at both boundaries, and therefore F and $\mathfrak{A}F$ have finite limits at β_i whenever $f(x)$

tends to a limit as $x \to \beta_i$. Thus we are dealing with linear transformations from **C** to **C** (see (1.17)): the operator $\mathfrak{A}$ *is defined at the boundary points by the requirement of continuity.*

The object of our study is linear transformations $\mathfrak{G}_\lambda$ $(\lambda > 0)$ mapping each $f \in \mathbf{C}$ into a solution $\mathfrak{G}_\lambda f = F$ of $\lambda F - \mathfrak{A}F = f$ such that

$$0 \leqq f \leqq \mathbf{1} \quad \textit{implies} \quad 0 \leqq \lambda F \leqq \mathbf{1}. \tag{12.1}$$

Our problem is to construct *the most general family* $\{\mathfrak{G}_\lambda\}$ *of such transformations having a common range* R. The restriction $\mathfrak{A}/R$ of $\mathfrak{A}$ to R is then an operator independent of λ, and

$$\mathfrak{G}_\lambda = (\lambda - \mathfrak{A}/R)^{-1} \tag{12.2}$$

coincides with the resolvent of $\mathfrak{A}/R$. In the language of semigroup theory, $\mathfrak{A}/R$ generates a transition semigroup, and we are constructing the most general transition semigroup. This has important applications in the theories of diffusion and Markov chains.

We proceed to describe the general form of the possible ranges R and then to construct the corresponding $\mathfrak{G}_\lambda$.

DEFINITION 12.1. *By boundary conditions of the general type we mean two independent*[19] *equations*

$$\begin{aligned} \rho_1 F(\beta_1) - \sigma_1 F^+(\beta_1) + \tau_1 \mathfrak{A}F(\beta_1) + \int_{\beta_1}^{\beta_2} F'(x)\zeta_1(x)\,dx &= 0, \\ \rho_2 F(\beta_2) + \sigma_2 F^-(\beta_2) + \tau_2 \mathfrak{A}F(\beta_2) + \int_{\beta_1}^{\beta_2} F'(x)\zeta_2(x)\,dx &= 0, \end{aligned} \tag{12.3}$$

where

$$\rho_i \geqq 0, \qquad \sigma_i \geqq 0, \qquad \tau_i \geqq 0, \tag{12.4}$$

$$\zeta_1 \downarrow, \qquad \zeta_2 \uparrow, \qquad \zeta_i \geqq 0, \qquad \int_{\beta_1}^{\beta_2} \zeta_i(x)\,dx < \infty, \tag{12.5}$$

and, of course, $\rho_i + \sigma_i + \tau_i + \zeta_i \neq 0$.

Boundary conditions of the special type are two equations

$$F(\beta_1) = F(\beta_2), \tag{12.6a}$$

$$\begin{aligned} \rho F(\beta_i) &- \sigma_1 F^+(\beta_1) + \sigma_2 F^-(\beta_2) \\ &+ \tau_1 \mathfrak{A}F(\beta_1) + \tau_2 \mathfrak{A}F(\beta_2) + \int_{\beta_1}^{\beta_2} F'(x)\zeta(x)\,dx = 0, \end{aligned} \tag{12.6b}$$

where $\rho > 0$, $\sigma_i > 0$, $\tau_i > 0$,

$$\zeta \uparrow, \qquad \int_{\beta_1}^{\beta_2} |\zeta(x)|\,dx < \infty, \tag{12.7}$$

and $\rho + \sigma_1 + \sigma_2 + \tau_1 + \tau_2 + \zeta \neq 0$.

---

[19] Since ζ_1 decreases and ζ_2 increases, the two equations (12.3) are necessarily independent unless $\sigma_i = \tau_i = 0$ and $\zeta_i =$ const. In this case (12.6a) holds.

A range R of the general [special] type consists of all $F \in \mathbf{C}$ such that $\mathfrak{A}F \in \mathbf{C}$ and F satisfies boundary conditions of the general [special] type.

THEOREM 12.1. *Suppose that $|\beta_i| < \infty$ and that both boundaries are active.*

(a) *Let $\{\mathfrak{G}_\lambda\}$ $(\lambda > \lambda_0)$ be a family of operators from $\mathbf{C}$ to $\mathbf{C}$ with a common range R such that* (12.1)–(12.2) *hold. Then R is necessarily*[20] *of the form described in Definition* 12.1.

(b) *Conversely, to each R of Definition* 12.1 *and each $\lambda > 0$, there corresponds a $\mathfrak{G}_\lambda$ whose range is R and such that* (12.1)–(12.2) *hold.*

(c) *In order that $\lambda\mathfrak{G}_\lambda \mathbf{1} = \mathbf{1}$, it is necessary and sufficient that $\rho_1 = \rho_2 = 0$ (or $\rho = 0$ in the special case* (12.6)).

Proof. (a) *Set-up.* As in the proof of Theorem 10.1, let w_i be the solutions of $\lambda w - \mathfrak{A}w = 0$ satisfying the boundary conditions $w_i(\beta_i) = 1$, $w_1(\beta_2) = w_2(\beta_1) = 0$. We are again concerned with transformations of the form (10.3) where $\mathfrak{G}_\lambda^{\min}$ corresponds to the minimal Green function and $\Phi_f^{(i)}$ is a bounded linear functional. Each such functional is represented by a function of bounded variation (signed measure) in the *closed* interval $\bar{I}:[\beta_1, \beta_2]$. To simplify the writing we introduce the following *notation for the inner product*:

$$(f, d\zeta) = \int_{\beta_1-}^{\beta_2+} f\, d\zeta \tag{12.8}$$

where ζ is of bounded variation (we always suppose ζ continuous on the right). Note that the integral contains contributions from the two boundaries.

Putting

$$F_\lambda = \mathfrak{G}_\lambda f \qquad F_\lambda^{\min} = \mathfrak{G}_\lambda^{\min} f, \tag{12.9}$$

we can rewrite (10.3) in the form

$$F_\lambda = F_\lambda^{\min} + w_i\cdot(f, d\Phi_1^{(\lambda)}) + w_2\cdot(f, d\Phi_2^{(\lambda)}) \tag{12.10}$$

where (for fixed λ, i) the functions $\Phi_i^{(\lambda)}$ are of bounded variation. Since $F_\lambda^{\min}(\beta_i) = 0$, we have

$$F_\lambda(\beta_i) = (f, d\Phi_i^{(\lambda)}). \tag{12.11}$$

Thus $\mathfrak{G}_\lambda$ is a positive transformation if, and only if, the $\Phi_i^{(\lambda)}$ are nondecreasing. Furthermore

$$\lambda\mathfrak{G}_\lambda^{\min}\mathbf{1} = \mathbf{1} - w_1 - w_2, \tag{12.12}$$

[20] Theorem 12.1 generalizes the main result of [3] concerning operators of the special form $aD_x^2 + bD_x$. Unfortunately the formulation in [3] contains an *error*: Due to a notational confusion it was erroneously claimed that the functions ζ_i are necessarily bounded whereas, in reality, only their integrals need be finite. The old proof is more complicated and is completely superseded by the present one.

which shows that $\lambda\mathfrak{G}_\lambda \mathbf{1} \leqq \mathbf{1}$ if, and only if, the variation of $\lambda\Phi_i^{(\lambda)}$ is $\leqq 1$, i.e.

$$\lambda\{\Phi_i^{(\lambda)}(\beta_2 +) - \Phi_i^{(\lambda)}(\beta_1 -)\} \leqq 1. \tag{12.13}$$

Each pair of nondecreasing $\Phi_i^{(\lambda)}$ satisfying this condition defines a positive transformation $\mathfrak{G}_\lambda$ such that $\lambda\mathfrak{G}_\lambda \mathbf{1} \leqq \mathbf{1}$, and all transformations of the desired kind are obtained in this way. A function $F \in \mathbf{C}$ *is in the range of this* $\mathfrak{G}_\lambda$ *if, and only if,* $\mathfrak{A}F \in \mathbf{C}$ *and*

$$F(\beta_i) = (\lambda F - \mathfrak{A}F, d\Phi_i^{(\lambda)}), \qquad i = 1, 2. \tag{12.14}$$

In fact, the necessity of this condition is obvious from (12.11). Conversely, define $f = \lambda F - \mathfrak{A}F$ and $F_\lambda = \mathfrak{G}_\lambda f$. Then $F - F_\lambda = w$ is a solution of $\lambda w - \mathfrak{A}w = 0$ vanishing at both boundaries, and hence $F = F_\lambda$.

(b) *Construction of the ranges.* Suppose that for each $\lambda > \lambda_0$ we are given two nondecreasing $\Phi_i^{(\lambda)}$ of variation $\leqq \lambda^{-1}$ such that (12.10) defines a transformation $\mathfrak{G}_\lambda$ with range R independent of λ. We want to show that R is defined by boundary conditions as described in Definition 12.1.

Choose a fixed $F \in R$. For each $\lambda > \lambda_0$ equation (12.14) holds. Letting $\lambda \to \infty$ through an appropriate sequence, we may suppose that $\lambda\Phi_2^{(\lambda)} \to \zeta_2$ where ζ_2 is nondecreasing and of variation $\leqq 1$. (The convergence is to be interpreted in the usual way as convergence at all points of continuity of ζ_2, i.e. as "weak star" convergence.) Then F satisfies the boundary condition $F(\beta_2) = (F, d\zeta_2)$ which is obviously a special case of (12.3) *except* if it reduces to the tautology $F(\beta_2) = F(\beta_2)$, i.e. except when $\lambda\Phi_2^{(\lambda)}(x) \to 0$ for $x < \beta_2$ and $\lambda\Phi_2^{(\lambda)}(\beta_2 +) \to 1$. In this case the limiting function ζ_2 attributes unit mass to β_2 and mass zero to the interval $[\beta_1, \beta_2]$.

Let this be true. Then

$$\lambda(\mathfrak{A}F, d\Phi_2^{(\lambda)}) \to \mathfrak{A}F(\beta_2). \tag{12.15}$$

An integration by parts permits us to rewrite (12.14) in the form

$$a(\lambda)F(\beta_2) + \int_{\beta_1}^{\beta_2} F'(x)\cdot\lambda\Phi_2^{(\lambda)}(x)\,dx + (\mathfrak{A}F, d\Phi_2^{(\lambda)}) = 0, \qquad 0 \leqq a(\lambda) \leqq 1, \tag{12.16}$$

(where we disposed of the arbitrary additive constant by setting $\Phi_2^{(\lambda)}(\beta_1 -) = 0$). Finally put

$$A(\lambda) = \lambda \int_{\beta_1}^{\beta_2} \Phi_2^{(\lambda)}(x)\,dx. \tag{12.17}$$

Letting $\lambda \to \infty$ through an appropriate sequence we may suppose that

$$A^{-1}(\lambda)\lambda\Phi_2^{(\lambda)} \to \zeta \tag{12.18}$$

where $\zeta \geqq 0$ is an increasing function and

$$\int_{\beta_1}^{\beta_2} \zeta(x)\,dx = 1 - \sigma \leqq 1. \tag{12.19}$$

Clearly

$$A^{-1}(\lambda)\cdot \int_{\beta_1}^{\beta_2} F'(x)\cdot\lambda\Phi_2^{(\lambda)}(x)\,dx \to \int_{\beta_1}^{\beta_2} F'(x)\zeta(x)\,dx + \sigma F^-(\beta_2). \tag{12.20}$$

Divide (12.16) by $\sigma(\lambda) = a(\lambda) + A(\lambda) + \lambda^{-1}$, and let $\lambda \to \infty$ in such a way that (12.18) holds. Taking, if necessary, a further subsequence $\lambda \to \infty$ such that

$$a(\lambda)\sigma^{-1}(\lambda) \to \rho_2, \qquad A(\lambda)\sigma^{-1}(\lambda) \to \delta, \qquad \lambda^{-1}\sigma^{-1}(\lambda) \to \tau_2, \tag{12.21}$$

we see that the second equation in (12.3) holds with $\zeta_2 = \delta\zeta$ and $\sigma_2 = \delta\sigma$. Since $\rho_2 + \tau_2 + \delta = 1$, we have $\rho_2 + \tau_2 + \sigma_2 + \zeta_2 \neq 0$, and we have thus verified the validity of the two equations (12.3).

However, we have to consider the possibility that these two equations are linearly dependent, i.e. that both reduce to the statement $F(\beta_1) = F(\beta_2)$. This can occur only if $\Phi_1^{(\lambda)}$ and $\Phi_2^{(\lambda)}$ are the same and (12.10) takes on the special form

$$F_\lambda = F_\lambda^{\min} + (w_1 + w_2)\cdot(f,\ d\Phi^{(\lambda)}). \tag{12.22}$$

The above argument can now be repeated with trivial changes. The first passage to the limit $\lambda\Phi^{(\lambda)} \to \zeta$ will reduce to a tautology whenever ζ is constant in the interior of (β_1, β_2) but attributes unit mass to the set consisting of the two boundary points. In this case the right side in (12.15) is to be replaced by $p_1\,\mathfrak{A}F(\beta_1) + p_2\,\mathfrak{A}F(\beta_2)$, where $p_i \geqq 0$, $p_1 + p_2 = 1$. In the integration by parts leading to (12.16) we replace[21] the norming $\Phi^{(\lambda)}(\beta_1 -) = 0$ by $\Phi^{(\lambda)}(\beta) = 0$ where β is an arbitrary fixed point, $\beta_1 < \beta < \beta_2$. The limiting function ζ is capable of negative values and may be unbounded at both ends.

(c) *Uniqueness.* We have shown that our ranges satisfy the described boundary conditions, but it remains to show that each F satisfying the boundary conditions is an element of the range (assuming, of course, that $\mathfrak{A}F \in \mathbf{C}$). This is equivalent to saying that no solution $w \neq 0$ of $\lambda w - \mathfrak{A}w = 0$ satisfies the boundary conditions. A direct verification of this assertion is easy, but not necessary for our purposes. In fact, we shall prove that for each $\lambda > 0$ there exists *exactly* one transformation $\mathfrak{G}_\lambda$ whose range satisfies the prescribed boundary conditions. On the other hand, $\mathfrak{G}_\lambda$ is determined only up to a multiple $\Phi_f\cdot w$ of a bounded solution w of $\lambda w - \mathfrak{A}w = 0$ which satisfies the boundary conditions. It follows that no such solution $\neq 0$ can exist.

Before proceeding to the existence proof, we digress to give

(d) *An alternative form*[22] *of the boundary conditions. The boundary con-*

[21] Otherwise (12.18) would reduce to $A^{-1}(\lambda)\Phi^{(\lambda)}(x) \to p_1$ for each $x \in (\beta_1, \beta_2)$, and, instead of deriving a new boundary condition, we would once more verify that $F(\beta_1) = F(\beta_2)$.

[22] It is equally possible to proceed with a direct calculation using the original form of the boundary conditions and the quantities ω_{ij} of (10.13). The new form of the boundary conditions has the advantage of leading to the general form (12.44) when no positivity is required.

ditions (12.3) *are equivalent to*

$$(12.23)\qquad \sum_{j=1}^{2} \alpha_{ij} F(\beta_j) + \tau_i \mathfrak{A}F(\beta_i) + \int_{\beta_1}^{\beta_2} \mathfrak{A}F(x)\cdot z_i(x)\cdot dm(x) = 0,$$

where

$$(12.24)\qquad \begin{pmatrix} \alpha_{11} & \alpha_{12} \\ \alpha_{21} & \alpha_{22} \end{pmatrix} = \begin{pmatrix} \rho_1 + c_1 & -c_1 \\ -c_2 & \rho_2 + c_2 \end{pmatrix}$$

and the z_i are convex upward and such that

$$(12.25)\qquad \begin{aligned} z_i(\beta_i) &= \sigma_i, & z_1(\beta_2) &= z_2(\beta_1) = 0, \\ -z_1^-(\beta_2) &\leqq c_1, & z_2^+(\beta_1) &\leqq c_2. \end{aligned}$$

To pass from (12.3) to (12.23), define the constants c_i by

$$(12.26)\qquad \int_{\beta_1}^{\beta_2} (c_i - \zeta_i)\, dx = \sigma_i,$$

and put

$$(12.27)\qquad z_1(x) = \int_x^{\beta_2} (c_1 - \zeta_1(x))\, dx, \qquad z_2(x) = \int_{\beta_1}^{x} (c_2 - \zeta_2(x))\, dx.$$

A simple integration by parts now transforms (12.3) into (12.23). Conversely, it is clear that the coefficients α_{ij} may be prescribed subject to the conditions $\alpha_{i1} + \alpha_{i2} \geqq 0$, $\alpha_{12} \leqq 0$, $\alpha_{21} \leqq 0$; and the convex z_i may be chosen arbitrarily subject only to (12.25). Then an integration by parts leads backward from (12.23) to (12.3).

Similarly, *the boundary condition* (12.6b) *may be rewritten in the form*

$$(12.28)\qquad \rho F(\beta_i) + \tau_1 \mathfrak{A}F(\beta_1) + \tau_2 \mathfrak{A}F(\beta_2) + \int_{\beta_1}^{\beta_2} \mathfrak{A}F(x)\cdot z(x)\cdot dm(x) = 0$$

where

$$(12.29)\qquad z(x) = \sigma_1 + \int_{\beta_1}^{x} (c - \zeta(x))\, dx \qquad\qquad z(\beta_2) = \sigma_2.$$

(e) *The construction.* For a given $\lambda > 0$ we propose to determine the $\Phi_i^{(\lambda)}$ in (12.10) so that the ensuing $\mathfrak{G}_\lambda$ will have a range satisfying the boundary conditions (12.23). For this purpose we introduce the notation

$$(12.30)\qquad Z_i(x) = \int_{\beta_1}^{\beta_2} G_\lambda^{\min}(x, y) z_i(y)\, dm(y)$$

and

$$(12.31)\qquad q_{ij} = \lambda \int_{\beta_1}^{\beta_2} z_i w_j\, dm$$

(remember that m is defined in the open interval and hence that no contributions from the boundaries appear in (12.31)).

Multiply (12.10) by z_i and integrate to obtain the two equations

$$\lambda \int_{\beta_1}^{\beta_2} F_\lambda\, z_i\, dm = \lambda \int_{\beta_1}^{\beta_2} fZ_i\, dm + \sum_{j=1}^{2} q_{ij} \cdot (f, d\Phi_j^{(\lambda)}). \tag{12.32}$$

By using the relation $\mathfrak{A}F_\lambda = \lambda F_\lambda - f$ it becomes evident that F_λ *satisfies the boundary conditions* (12.23) *if, and only if,*

$$\sum_{j=1}^{2} M_{ij}(f, d\Phi_j^{(\lambda)}) = \int_{\beta_1}^{\beta_2} f\cdot\{z_i - \lambda Z_i\}\, dm + \tau_i f(\beta_i) \tag{12.33}$$

where the matrix M is defined by

$$\begin{aligned} M &= \begin{pmatrix} \alpha_{11} + q_{11} + \lambda\tau_1 & \alpha_{12} + q_{12} \\ \alpha_{21} + q_{21} & \alpha_{22} + q_{22} + \lambda\tau_2 \end{pmatrix} \\ &= \begin{pmatrix} \rho_1 + \lambda\tau_1 + c_1 + q_{11} & -c_1 + q_{12} \\ -c_2 + q_{21} & \rho_2 + \lambda\tau_2 + c_2 + q_{22} \end{pmatrix}. \end{aligned} \tag{12.34}$$

We next prove that for each $\lambda > 0$ *this matrix possesses an inverse* M^{-1} *with elements* $M_{ij}^{-1} \geqq 0$.

Clearly $M_{ii} > 0$. Remembering that $w_1(\beta_2) = 0$ and $z_2(\beta_1) = 0$ we get

$$M_{21} = -c_2 + \int_{\beta_1}^{\beta_2} z_2\, dw_1' = -c_2 - \int_{\beta_1}^{\beta_2} w_1' \cdot z_2' \cdot dx. \tag{12.35}$$

Now $w_1' < 0$, while z_2' decreases from $z_2^{+}(\beta_1) \leqq c_2$. It follows that

$$M_{21} < -c_2 + c_2 \int_{\beta_1}^{\beta_2} (-w_1')\, dx = 0, \tag{12.36}$$

and similarly $M_{12} < 0$. Hence $M_{ij}^{-1} \geqq 0$.

Premultiplication of (12.33) by M^{-1} leads to an explicit form for $\Phi_i^{(\lambda)}$ and we have the following result:

In order that the range of the transformation (12.10) *satisfy the boundary condition* (12.3) [*or the equivalent form* (12.23)] *it is necessary and sufficient that* $\Phi_i^{(\lambda)}$ *be absolutely continuous* (*m*) *within the open interval* (β_1, β_2) *with a density*

$$\frac{d\Phi_i^{(\lambda)}}{dm} = \sum_{j=1}^{2} M_{ij}^{-1}\, \{z_j - \lambda Z_j\} \tag{12.37}$$

and that $\Phi_i^{(\lambda)}$ *attaches weight* $M_{ij}^{-1}\tau_j$ *to the boundary* β_j.

In the case of the special boundary condition (12.6) [or (12.28)] *and the transformation* (12.22), *condition* (12.37) *is replaced by*

$$\frac{d\Phi^{(\lambda)}}{dm} = M^{-1}(z - \lambda Z) \tag{12.38}$$

and the weights are $M^{-1}\tau_i$ *where*

$$M = \rho + \lambda\tau_1 + \lambda\tau_2 + \lambda \int_{\beta_1}^{\beta_2} (w_1 + w_2) z\, dm. \tag{12.39}$$

Note. The above result may be formulated more elegantly in the case $\sigma_i > 0$, by extending the measure m to the boundary in such a way that (12.37) holds also on the boundary. We know that Z_i vanishes at both boundary points. In view of (12.25) the function $z_i - \lambda Z_i$ equals σ_i at β_i and 0 at the other boundary. It follows that *if m is extended to the closed interval $[\beta_1, \beta_2]$ by attributing weight $m_i = \tau_i/\sigma_i$ to β_i, then $\Phi_i^{(\lambda)}$ is defined by the density* (12.37) [*or* (12.38)] *throughout the closed interval,* and no special reference to β_i is required.

(f) *Verification of the norm conditions.* Lemma 9.3 guarantees that $z_j - \lambda Z_j \geqq 0$. Since $M_{ij}^{-1} \geqq 0$, it follows that (12.37) defines nondecreasing $\Phi_i^{(\lambda)}$. Multiplying (12.12) by z_i and integrating, we get, remembering (12.30) and (12.31),

$$\lambda \int_{\beta_1}^{\beta_2} \{z_i - \lambda Z_i\}\, dm = q_{i1} + q_{i2}\,. \tag{12.40}$$

For the special choice $f = \mathbf{1}$ the right side in (12.33) reduces therefore to

$$0 \leqq \{q_{i1} + q_{i2} + \tau_i\}\lambda^{-1} = \{M_{i1} + M_{i2} - \rho_i\}\lambda^{-1} \leqq \{M_{i1} + M_{i2}\}\lambda^{-1} \tag{12.41}$$

with the equality sign holding only if $\rho_i = 0$. Premultiplying the two equations (12.41) by the positive matrix M^{-1}, we find that the variations of $\Phi_i^{(\lambda)}$ are $\leqq 1$, and equal to 1 only if $\rho_1 = \rho_2 = 0$.

The same argument applies to (12.38) and completes the proof.

THEOREM 12.2. *In the case where β_2 is semiactive, Theorem* 12.1 *remains true except that in the boundary conditions $\sigma_2 = 0$.*

Proof. The novel feature is that with a semiactive boundary $m(\beta_2) = \infty$, and therefore the boundedness of $\mathfrak{A}F$ no longer implies that F' is bounded. The above proof requires no change up to (12.19), but (12.20) makes no sense. Now

$$F^-(x) - F^+(\beta) = \int_{\beta+}^{x-} \mathfrak{A}F \cdot dm, \tag{12.42}$$

and hence

$$F'(x)m^{-1}(x) \to \mathfrak{A}F(\beta_2) \qquad x \to \beta_2\,. \tag{12.43}$$

Thus the behavior of F' near β_2 is determined by $\mathfrak{A}F(\beta_2)$ instead of by $F^-(\beta_2)$. It is possible that for each F in the range $\mathfrak{A}F(\beta_2) = 0$. This then constitutes a special boundary condition of the form (12.3), and no further argument is required. Otherwise choose F so that $\mathfrak{A}F(\beta_2) > 0$. It is then obvious that (12.20) holds with $\sigma F^-(\beta_2)$ replaced by $\sigma'\mathfrak{A}F(\beta_2)$. Thus the term containing $F^-(\beta_2)$ drops out. The actual construction of $\mathfrak{G}_\lambda$ requires no change (and simplifies somewhat due to the absence of a term in the boundary condition).

Note. From (12.14) it is easy to deduce the *general form* of the boundary conditions when $\mathfrak{G}_\lambda$ is not positivity preserving or norm-decreasing. Integrating twice by parts and adjusting the two arbitrary constants so as to eliminate the terms $F(\beta_i)$ we see easily that in general the range of $\mathfrak{G}_\lambda$ will be determined by two equations of the form

$$F(\beta_i) = \int_{\beta_1-}^{\beta_2+} (\mathfrak{A}F) \cdot dZ_i. \tag{12.44}$$

The boundary conditions (12.22) are a special case and show the particular form required for $\mathfrak{G}_\lambda$ to be positive.

13. The adjoint transformations

Let **M** be the Banach space of functions of bounded variation (signed measures) on the closed interval $\bar{I}:[\beta_1, \beta_2]$, the total variation serving as norm. Considered as transformation from **C** to **C**, each $\mathfrak{G}_\lambda$ has an adjoint $\mathfrak{G}_\lambda^*$ mapping each $\pi \in \mathbf{M}$ into another element $\Pi = \mathfrak{G}_\lambda^* \pi \in \mathbf{M}$. By using the notation (12.8) for inner products, the adjoint $\mathfrak{G}_\lambda^*$ is defined by $(\mathfrak{G}_\lambda f, d\pi) = (f, d\pi)$. If $\mathfrak{G}_\lambda$ is defined by a kernel

$$\mathfrak{G}_\lambda f = \int_{\beta_1}^{\beta_2} G_\lambda(\cdot, y) f(y)\, dm(y), \tag{13.1}$$

then the equation

$$P(y) = \int_{\beta_1}^{\beta_2} G_\lambda(x, y)\, d\pi(x) \tag{13.2}$$

defines a continuous m-density, and

$$\Pi(x_2-) - \Pi(x_1+) = \int_{x_1}^{x_2} P(y)\, dm(y) \tag{13.3}$$

defines the transform $\Pi = \mathfrak{G}_\lambda^* \pi$.

Consider first the case of a *symmetric* kernel G_λ, for example, the minimal Green function $G_\lambda^{\min}$ of Section 8. If π has a continuous density $p = d\pi/dm$, then $P = \mathfrak{G}_\lambda p$, and hence

$$\lambda P - \mathfrak{A}P = p. \tag{13.4}$$

Integrating with respect to m we find

$$\lambda \Pi - D_x P = \pi + \text{const.} \tag{13.5}$$

or

$$\lambda \Pi - D_x D_m \Pi = \pi + \text{const.} \tag{13.6}$$

In other words, $\mathfrak{A}^* = D_x D_m$ *appears as the formal adjoint of the differential operator* $\mathfrak{A} = D_m D_x$. It is easily verified by integration by parts that (13.6) holds almost everywhere when G is an arbitrary Green function and π an arbitrary function of bounded variation.

The minimal Green function $G_\lambda^{\min}$ *provides for each* $\lambda > 0$ *a solution of* (13.6) *for given* $\pi \in \mathbf{M}$. This solution is unique except if there exists a v of bounded variation such that almost everywhere (m)

$$\lambda v - D_x D_m v = 0. \tag{13.7}$$

In this case it is readily seen that $u = D_m v$ is a solution of

$$\lambda u - D_m D_x u = 0, \qquad \int_{\beta_1}^{\beta_2} | u | \, dm < \infty \tag{13.8}$$

and, conversely, each u satisfying (13.8) gives rise to a solution of (13.7) which is of bounded variation. Referring to Section 6 we see that *the case of non-uniqueness arises if, and only if, a boundary β_i is either active or infinite and semiactive.*

Thus *an active boundary causes nonuniqueness for the inversion both of $\lambda - \mathfrak{A}$ in* **C** *and for $\lambda - \mathfrak{A}^*$ in* **M**; *a semiactive boundary β_i causes nonuniqueness in the first case when $| \beta_i | < \infty$, and in the second case when $| \beta_i | = \infty$.*

Whenever nonuniqueness occurs, the range R^* of $\mathfrak{G}_\lambda^*$ is determined by appropriate boundary conditions. With *symmetric* kernels these boundary conditions are the same in the spaces **C** and **M** so that no further discussion is required. However, an altogether new situation confronts us when $\mathfrak{G}_\lambda$ has an *unsymmetric* kernel G.

For sake of definiteness let us consider the transformation $\mathfrak{G}_\lambda$ of Theorem 12.1 determined by the boundary conditions (12.3). The formulas are simpler and more pleasing in the nondegenerate case $\sigma_1 \neq 0$, $\sigma_2 \neq 0$ when formal elegance can be achieved by *extending the measure m to the closed interval $\bar{I}$: $[\beta_1, \beta_2]$; by definition, the points β_i carry the weights*

$$m_1 = \tau_1/\sigma_1, \qquad m_2 = \tau_2/\sigma_2, \tag{13.9}$$

respectively. (The case $\sigma_i = 0$ may be treated directly by the same method, or by a passage to the limit.)

THEOREM 13.1. *Let m be defined in the closed interval $\bar{I}$ in accordance with* (13.9). *Let $\mathfrak{G}_\lambda$ be that inverse to $\lambda - \mathfrak{A}$, $\lambda > 0$, whose range R is determined by the boundary conditions* (12.3) *supposing* $0 < m_1, m_2 < \infty$ ($\mathfrak{G}_\lambda$ *is described in Theorem* 12.1). *Then:*

(i) *The adjoint $\mathfrak{G}_\lambda^*$ maps each function of bounded variation (signed measure) π on $\bar{I}$ into $\Pi = \mathfrak{G}_\lambda^* \pi$ such that Π is absolutely continuous with respect to m and its density $P = d\Pi/dm$ is continuous in $\bar{I}$. (It is given by* (13.19)–(13.20).)

(ii) *There exists an operator $\mathfrak{A}^*$, independent of λ, such that*

$$\mathfrak{G}_\lambda^* = (\lambda - \mathfrak{A}^*)^{-1} \tag{13.10}$$

(i.e. *$\mathfrak{G}_\lambda^*$ is the resolvent of $\mathfrak{A}^*$ and $\mathfrak{A}^*$ a formal adjoint to $\mathfrak{A} = D_m D_x$*).

In the case when the monotone functions ζ_1 and ζ_2 occurring in the boundary conditions (12.3) *as well as π have densities*

$$\eta_1 = -\frac{d\zeta_1}{dm}, \qquad \eta_2 = \frac{d\zeta_2}{dm}, \qquad p = \frac{d\pi}{dm}, \tag{13.11}$$

the identity $\lambda\Pi - \mathfrak{A}^*\Pi = \pi$ *may be expressed in terms of the density*[23] $P = d\Pi/dm$ *as follows:*

$$\lambda P(x) - D_m D_x P(x) - P(\beta_1)\sigma_1^{-1}\eta_1(x) - P(\beta_2)\sigma_2^{-1}\eta_2(x) = p(x) \tag{13.12}$$

for $\beta_1 < x < \beta_2$, *and*

$$\lambda P(\beta_k) - P(\beta_k)\rho_k\tau_k^{-1} - m_k^{-1}\Lambda_k P = p(\beta_k), \qquad k = 1, 2 \tag{13.13}$$

where the operator Λ_i *is defined by*[24]

$$\Lambda_k P = (-1)^k \lim_{x\to\beta_k} \{P'(x) - P(\beta_1)\sigma_1^{-1}\zeta_1(x) + P(\beta_2)\sigma_2^{-1}\zeta_2(x)\}. \tag{13.14}$$

Note 1. Observe that in the interior I the operator $\mathfrak{A}^*$ reduces to the differential operator $D_x D_m$ if, and only if, $\eta_i = 0$, that is, if $\zeta_i =$ const. in the open interval I. In this case the boundary conditions (12.3) take on the form (11.3) known from the Hilbert space theory (except that the symmetry condition $\pi_{12} = \pi_{21}$ is no longer required). In other words, *the classical boundary conditions with the added terms* $\tau_i \mathfrak{A}F(\beta_i)$ *are the only boundary conditions making the adjoint* $\mathfrak{A}^*$ *again a differential operator.*

Note 2. Integrating (13.12) and adding to (13.13) we find the identity

$$\lambda \int_{\beta_1-}^{\beta_2+} P(y)\, dm(y) + \sum_{k=1}^{2} P(\beta_k)\sigma_k^{-1}\rho_k = \int_{\beta_1-}^{\beta_2+} p(y)\, dm(y). \tag{13.15}$$

Since $\mathfrak{G}_\lambda^*$ is positive, (13.15) shows that $\lambda \| \mathfrak{G}_\lambda^* \| = 1$ if, and only if, $\rho_1 = \rho_2 = 0$. This, of course, follows also from Theorem 12.1, (c).

Proof. The transformation $\mathfrak{G}_\lambda$ is induced by the kernel

$$G_\lambda(x, y) = G_\lambda^{\min}(x, y) + \sum_{i,j=1}^{2} w_i(x)M_{ij}^{-1}\{z_j(y) - \lambda Z_j(y)\} \tag{13.16}$$

where $\beta_1 \leqq x, y \leqq \beta_2$ and $G_\lambda^{\min}$ is the minimal Green function for A and the interval I (see (12.37) and the note following (12.39)). In other words, the transform $\Pi = \mathfrak{G}_\lambda^* \pi$ has the density $P = d\Pi/dm$ defined by

$$P(y) = P^{\min}(y) + \sum_{i,j=1}^{2} (w_i, d\pi)M_{ij}^{-1}\{z_j(y) - \lambda Z_j(y)\} \tag{13.17}$$

where $\beta_1 \leqq y \leqq \beta_2$ and

$$P^{\min}(y) = \int_{\beta_1}^{\beta_2} G_\lambda^{\min}(x, y)\, d\pi(y); \tag{13.18}$$

the notation $(f, d\pi)$ for inner products was introduced in (12.8).

[23] The explicit form of $\mathfrak{A}^*$ in the general case without densities is, of course, obtained by integrating (13.12) and introducing the original set functions Π, ζ_i, π instead of their densities. Observe that this integration changes the term $D_m D_x P$ into $D_x D_m \Pi$.

[24] A special case of Theorem 13.1 is treated in [3]. However, due to the error mentioned in footnote 20 only the case of *bounded* ζ_i was treated, and therefore the analogue to (13.13) in [3], Section 22, has a more restricted form.

For $y = \beta_j$ we get from (13.17) (see again the note following (12.39))

$$P(\beta_j) = \sum_{i=1}^{2} (w_i, d\pi) M_{ij}^{-1} \sigma_j, \tag{13.19}$$

and (13.17) may more conveniently be rewritten in the form

$$P(y) = P^{\min}(y) + \sum_{j=1}^{2} P(\beta_j)\sigma_j^{-1} \{z_j(y) - \lambda Z_j(y)\}. \tag{13.20}$$

The continuity of P is obvious. To prove (13.12) we recall the definition (12.27) of z_j according to which

$$D_x z_1 = -c_1 + \zeta_1, \qquad D_x z_2 = c_2 - \zeta_2. \tag{13.21}$$

Thus $\eta_j = D_m D_x z_j$, by the definition (13.11). Since $\lambda P^{\min} - \mathfrak{A}P^{\min} = p$ and $\lambda Z_j - \mathfrak{A}Z_j = z_j$, the assertion (13.12) follows on operating with $\mathfrak{A}$ on (13.20).

The relations (13.13) are less obvious. The minimal Green function is formed by means of the two solutions w_2, w_1, which are minimal at β_1 and β_2, respectively. They satisfy the boundary conditions $w_1(\beta_2) = w_2(\beta_1) = 0$, $w_i(\beta_i) = 1$, and their Wronskian equals $w_2^+(\beta_1) = -w_1^-(\beta_2)$. It follows (cf. (7.14)) that

$$D_x^+ P^{\min}(x) = -\frac{w_1^+(x)}{w_1^-(\beta_2)} \int_{\beta_1+}^{x+} w_2(y)\, d\pi(y) + \frac{w_2^+(x)}{w_2^+(\beta_1)} \int_{x+}^{\beta_2-} w_1(y)\, d\pi(y). \tag{13.22}$$

Hence

$$(-1)^j \lim_{x \to \beta_j} D_x P^{\min}(x) = -\int_{\beta_1+}^{\beta_2-} w_j(y)\, d\pi(y). \tag{13.23}$$

Remembering that π attaches weight $m_i p(\beta_i)$ to β_i we can rewrite (13.23) as

$$\begin{aligned} (-1)^j \lim_{x \to \beta_j} D_x P^{\min}(x) &= m_j p(\beta_j) - (w_j, d\pi) \\ &= m_j p(\beta_j) - \sum_{k=1}^{2} P(\beta_k)\sigma_k^{-1} M_{kj} \end{aligned} \tag{13.24}$$

(cf. (13.19)).

Next, $Z_i = \mathfrak{G}_\lambda^{\min} z_i$, and hence we have by analogy with (13.23)

$$(-1)^j \lim_{x \to \beta_j} D_x Z_i(x) = -\int_{\beta_1+}^{\beta_2-} w_j(y) z_i(y)\, dm(y) = -\lambda^{-1} q_{ij} \tag{13.25}$$

(using the notation (12.31)).

We now differentiate (13.20) with respect to y using (13.24), (13.21), and (13.25). In this way we find for the operators defined in (13.14)

$$\Lambda_k P = m_k p(\beta_k) + \sum_{i=1}^{2} P(\beta_i)\sigma_i^{-1}\{-M_{ik} + (-1)^{i+k} c_k + q_{ik}\}. \tag{13.26}$$

Finally, by using the definition (12.34) of the matrix M it is seen that within the braces we have the elements of a diagonal matrix, $\rho_i + \lambda\tau_i$, and thus (13.26) reduces to

$$\Lambda_k P = m_k p(\beta_k) - \lambda P(\beta_k) m_k - P(\beta_k)\rho_k \sigma_k^{-1}. \tag{13.27}$$

Dividing by m_k we get (13.13).

To complete the proof of the theorem it remains to verify (13.10), namely, that (13.12)–(13.13) admit of only *one* solution. Now consider the $P(\beta_j)$ for the moment as two parameters. If these are given, (13.12) admits of exactly one solution such that $P(x) \to P(\beta_j)$ as $x \to \beta_j$. Equations (13.13) then represent two algebraic linear equations for the two unknowns $P(\beta_1)$ and $P(\beta_2)$. We know that to arbitrarily prescribed $p(\beta_j)$ there exists at least one solution, namely the one furnished by $\mathfrak{G}_\lambda^*$. It follows that the determinant of the system can not vanish, and hence the solution is unique. The theorem is proved.

Note. We have shown that the operator $\mathfrak{A}^*$ is completely determined by (13.12)–(13.13) without any boundary conditions. In the limiting case $m_2 = 0$ the boundary β_2 carries no mass, and it is senseless to talk about a density $P = d\Pi/dm$ at the individual point β_2. However, $P(\beta_2)$ remains defined by the requirement of continuity $P(\beta_2) = \lim P(x)$ as $x \to \beta_2$. Multiply the second equation in (13.13) by m_k and pass to the limit $m_k \to 0$. We get formally

$$\rho_2 P(\beta_2) + \sigma_2 \Lambda_2 P = 0. \tag{13.28}$$

In this case it is senseless to talk about $\mathfrak{A}^*$ at β_2. However, the operator $\mathfrak{A}^*$ *defined by* (13.12)–(13.13) *is now to be restricted by the boundary condition* (13.28). Thus, exactly as in the case of the Hilbert space (see Section 11) *the boundary conditions appear as a limiting case of the definition of operators on the boundary.* Incidentally, when the functions ζ_i appearing in the boundary conditions (12.3) are bounded, then $\Lambda_k P$ reduces to a linear combination of $P'(\beta_k)$, $P(\beta_1)$, $P(\beta_2)$, and (13.28) reduces to a boundary condition of the classical type.

References

1. W. Feller, *On second order differential operators*, Ann. of Math. (2), vol. 61 (1955), pp. 90–105.
2. ———, *On the intrinsic form for second order differential operators*, to appear in Illinois J. Math., vol. 2 (1958).
3. ———, *The parabolic differential equations and the associated semi-groups of transformations*, Ann. of Math. (2), vol. 55 (1952), pp. 468–519.
4. W. Feller and H. P. McKean, Jr., *A diffusion equivalent to a countable Markov chain*, Proc. Nat. Acad. Sci. U. S. A., vol. 42 (1956), pp. 351–354.
5. H. P. McKean, Jr., *Elementary solutions for certain parabolic partial differential equations*, Trans. Amer. Math. Soc., vol. 82 (1956), pp. 519–548.
6. M. H. Stone, *Linear transformations in Hilbert space*, Amer. Math. Soc. Colloquium Publications, vol. 15 (1932).

7. H. Weyl, *Über gewöhnliche Differentialgleichungen mit Singularitäten und die zugehörige Entwicklungen willkürlicher Funktionen*, Math. Ann., vol. 68 (1910), pp. 220–269.

Added in proof. Cf. footnote 10.

8. A. D. Ventcelj, *Semigroups of operators generated by a generalized differential operator of second order* [in Russian], Doklady Akad. Nauk SSSR, vol. 111 (1956), pp. 269–272.

Princeton University
Princeton, New Jersey

Annals of Mathematics
Vol. 65, No. 3, May, 1957
Printed in U.S.A.

ON BOUNDARIES AND LATERAL CONDITIONS FOR THE KOLMOGOROV DIFFERENTIAL EQUATIONS*

By William Feller

(Received June 15, 1956)

0. Introduction

It has been shown in [4] that a positive matrix Π defined over the set of integers $E = \{1, 2, \cdots\}$ induces a *boundary* for E. In the present paper this notion will be applied to families of matrices $P(t)$, with elements $P(t; i, k)$, depending on the parameter $t > 0$ and having the semi-group property

$$(0.1) \qquad P(t + s) = P(t)P(s).$$

Kolmogorov [13] has shown that under appropriate regularity conditions such a $P(t)$ will satisfy a pair of matrix differential equations (adjoint to each other). Since then a great many investigations have been concerned with problems of non-uniqueness, with solutions which fail to satisfy the adjoint equation, and with other "pathological" situations.

It is the purpose of the present paper to show that the unbounded matrix operator Ω on which the Kolmogorov equations depend shares the essential properties of second order elliptic differential operators, such as the Sturm-Liouville operator in one dimension or the Laplacian Δ in two dimensions. The Kolmogorov equations are then the exact counterpart of the heat equation $u_t = \Delta u$, except that due to the essential lack of symmetry of Ω the adjoint equation bears little resemblance to the original.

Our solutions will depend on boundary conditions in which one easily recognizes the classical boundary conditions of, say, the theory of harmonic functions. Of course, the latter depend on normal derivatives and no analogue to partial derivatives exists in our case. However, our boundary conditions are expressed in terms of functionals which remain meaningful under the most general circumstances, and can be applied to harmonic functions. There they reduce to the normal derivatives whenever the boundary is sufficiently smooth, but they give the proper expression for an *arbitrary* boundary.

The perfect analogy of the Kolmogorov equations with diffusion equations explains the many phenomena and problems connected with the non-uniqueness of the solutions, and leads to new insights. Consider, for simplicity, an isolated point ω of the boundary. It is possible to extend Ω (that is, the Kolmogorov equations) and the matrix $P(t)$ to the countable set $E + \omega$. This extension is not uniquely determined, but it involves very little arbitrariness. It adds a new row and a new column to Ω, and these bear no similarity to the remaining rows and columns of Ω. This leads in the most natural manner to generalizations of the

* Sponsored by the Office of Ordnance Research U. S. Army Contract DA-36-034-ORD-2001, OOR Project No. 264.

Kolmogorov equations, and the reader will find exact counterparts to various examples which have recently been constructed to show that the Kolmogorov equations are not necessary.

Again we gain new insights into classical situations. When Ω and P are extended to $E + \omega$, the new differential equations no longer involve any boundary conditions. In fact, the old boundary condition now appears simply as the residual, or limiting form, left of the row in Ω corresponding to ω when the point ω is eliminated and P restricted to the original E. Exactly the same phenomenon can be discerned in connection with the classical boundary value problems for the Sturm-Liouville equation.

We encounter a novel feature in that the transpose of the matrix $P(t)$ is entirely different from $P(t)$ and thus induces an entirely different boundary. The Green functions etc. depend on the intimate interplay between the two boundaries, a fact obscured in classical theories where one usually deals only with operators which are self-adjoint or nearly symmetric. This shows the advantage of going back to a countable E where one is not prejudiced by an a priori given topology and where one, accordingly, encounters operators of the most general nature. Note, however, that the two boundaries are not necessarily countable.

An outline of the procedure and summary of the results will be found in Section 2.

1. The general set-up

We shall say that the matrix $P(t)$ is Markovian if (0.1) *holds and moreover*

$$P(t; i, k) \geqq 0, \qquad \textstyle\sum_k P(t; i, k) \leqq 1, \tag{1.1}$$

$$P(t) \to I, \qquad t \downarrow 0 \tag{1.2}$$

where I is the identity matrix. Clearly (1.2) and (0.1) together imply the continuity of $P(\ )$. Doob has shown ([2], cf. also [3], p. 258) that P possesses a derivative at the origin, in the sense that

$$\begin{aligned} \lim_{t\downarrow 0} t^{-1}\{1 - P(t; i, i)\} &= q_i \leqq \infty, \\ \lim_{t\downarrow 0} t^{-1}\cdot P(t; i, k) &= q_{ik} < \infty \qquad (i \neq k) \end{aligned} \tag{1.3}$$

always exist and $\sum_j q_{ij} \leqq q_i$. When $0 < q_i < \infty$ for all i it is more convenient to write

$$q_{ij} = q_i \Pi(i, j) \tag{1.4}$$

where Π is a substochastic matrix, that is,

$$\Pi(i, j) \geqq 0 \qquad \textstyle\sum_j \Pi(i, j) \leqq 1. \tag{1.5}$$

A *formal* differentiation of (0.1) with respect to t leads to the infinite system of differential equations

(1.6) $$P'(t; i, k) = -q_i P(t; i, k) + q_i \sum_j \Pi(i, j) P(t; j, k)$$

(known as the *Kolmogorov backward* equation). Similarly, differentiation with respect to s leads *formally* to the so-called *Kolmogorov forward* equations

(1.7) $$P'(t; i, k) = -P(t; i, k) q_k + \sum_j P(t; i, j) q_j \Pi(j, k).$$

It is not stated that these equations are really satisfied, but we shall take them as our *starting* point.

We now introduce a matrix Π *with elements* $\Pi(i, j)$; *a diagonal matrix* $\mathfrak{q}$ *with elements* q_i ; *and finally the matrix*

(1.8) $$\Omega = -\mathfrak{q} + \mathfrak{q}\Pi$$

with elements $\Omega(i, k) = -q_i + q_i \Pi(i, k)$. Note that Ω represents an *unbounded* operator and that the associative law of matrix multiplication does not necessarily apply.

The Kolmogorov differential equations (1.6) *and* (1.7) *can be written compactly in the form*

(1.9) $$P' = \Omega P, \qquad P' = P\Omega,$$

respectively.

We shall embark on the study of these two equations, with the matrices $\mathfrak{q}$ and Π considered given.[1] For convenience of notation[2] *we shall in the sequel assume that*

(1.10) $$q_i > 0 \qquad \sum_j \Pi(i, j) = 1.$$

Notational conventions. As far as possible we shall use bold face *latin* letters for *column* vectors, and *greek* ones for *row* vectors. It is understood that the vector $\mathbf{c}$ has components $c(1)$, $c(2)$, $\cdots$. The symbols $\mathbf{0}$ and $\mathbf{1}$ will be used indiscriminately to denote either a row or a column vector all of whose components equal 0 or 1, respectively. As is customary, we introduce for column and for row vectors the two norms

(1.11) $$\| \mathbf{c} \| = \sup | c(i) | \qquad \| \boldsymbol{\rho} \| = \sum | \rho(i) |.$$

Accordingly one should interpret *column vectors as representing functions on* E, *and row vectors as signed measures*. The word bounded is always to be interpreted in the sense of (1.11). The *Banach space of bounded column vectors will be denoted by* $\mathfrak{C}$, *that of row vectors by* $\mathfrak{R}$.

A Markovian $P(t)$ induces the transformations $\mathbf{c} \to \mathbf{C}(t) = P(t)\mathbf{c}$ on $\mathfrak{C}$, and $\boldsymbol{\rho} \to \mathbf{R}(t) = \boldsymbol{\rho} P(t)$ on $\mathfrak{R}$ and we have $\| \mathbf{C}(t) \| \leqq \| \mathbf{c} \|$ and $\| \mathbf{R}(t) \| \leqq \| \boldsymbol{\rho} \|$. At least for sufficiently smooth $\mathbf{c}$ and $\boldsymbol{\rho}$ the Kolmogorov equations (1.9) imply[3]

[1] For a Markovian $P(t)$ generated by a *diffusion* operator rather than equations resembling the Kolmogorov equations cf. [7].

[2] All results require only an obvious rephrasing in the general case.

[3] The two equations (1.12) are the analogues, respectively, of the backward diffusion

(1.12) $$\mathbf{C}'(t) = \Omega\mathbf{C}(t) \qquad \mathbf{R}'(t) = \mathbf{R}(t)\Omega$$

with $\mathbf{C}(0) = \mathbf{c}$ and $\mathbf{R}(0) = \boldsymbol{\rho}$.

Note that in the sense of matrix algebra $\mathbf{c}\boldsymbol{\rho}$ represents the infinite matrix with elements $c(i)\rho(k)$, whereas $\boldsymbol{\rho}\mathbf{c}$ is the inner product $\sum \rho(i)c(i)$. *For this inner product we shall occasionally use the alternative notation*

(1.13) $$[\boldsymbol{\rho}, \mathbf{c}] = \boldsymbol{\rho}\mathbf{c} = \sum \rho(i)c(i).$$

As usual, δ_{ik} stands for the *Kronecker symbol*, so that (δ_{ik}) is the identity matrix I.

2. Outline and summary

In Section 3 the familiar Laplace transform is introduced, and the integration of the Kolmogorov differential equations is reduced to the study of the operator $\lambda I - \Omega$ and its inverses. Note that this operator differs only notationally from $I - \Pi_\lambda$ where Π_λ is the matrix

(2.1) $$\Pi_\lambda = \mathfrak{q}(\lambda + \mathfrak{q})^{-1}\Pi = (\lambda + \mathfrak{q})^{-1}\mathfrak{q}\Pi$$

with elements

(2.2) $$\Pi_\lambda(i, k) = \frac{q_i}{\lambda + q_i}\,\Pi(i, k).$$

This is a substochastic matrix which will play a fundamental role in the sequel.

Section 4 contains the basic theorem that there exists a uniquely determined Markovian $P(t)$ satisfying both Kolmogorov differential equations and which is the *minimal solution* for each of them.[4] This minimal solution is denoted by $P^{\min}(t)$. It represents the *unique* solution for $P' = \Omega P$ if, and only if, it is strictly stochastic, and this in turn is the case if, and only if, the vector equation $\Pi_\lambda\mathbf{u} = \mathbf{u}$ admits of no bounded solution.

The further study therefore hinges on the properties of the solutions of $\Pi_\lambda\mathbf{u} = \mathbf{u}$. We denote by $\mathfrak{P}$ and $\mathfrak{P}_\lambda$, respectively, the set of all column vectors satisfying

(2.3) $$\begin{aligned} \mathfrak{P}&: \quad \Pi\mathbf{u} = \mathbf{u} \\ \mathfrak{P}_\lambda&: \quad \Pi_\lambda\mathbf{u} = \mathbf{u} \end{aligned} \qquad \mathbf{0} \leqq \mathbf{u} \leqq \mathbf{1},$$

---

equation $u_t = au_{xx} + bu_x$ for continuous functions and of the so-called Fokker-Planck equation $v_t = (av)_{xx} - (bv)_x$ for probability densities. *If* $P(t)$ satisfies *both* Kolmogorov equations then $P(t)$ is the analogue of the common Green function for these parabolic differential equations.

[4] This is a very simple special case of a Theorem in [5] where arbitrary (non-countable) spaces are treated and the coefficients are permitted to depend on t. The present proof rephrases the old one in terms of Laplace transforms. It appears somewhat simpler than the version given by Hille [8], and considerably simpler than the proof by Kato [10]. The difficulties of the latter stem from the use of the Hille-Yosida theorem instead of direct arguments. The basic condition of the Hille-Yosida theorem, namely the denseness of the range space, is here not fulfilled, and the description of the range space (i.e. the appropriate boundary condition) presupposes that the exit and entrance boundaries have been introduced.

where $\lambda > 0$. It is shown in Section 5 that there exists a one-to-one correspondence (lattice isomorphism) between the elements of $\mathfrak{P}_\lambda(\lambda > 0)$, and equally between the elements of $\mathfrak{P}_\lambda$ and a certain subset $\bar{\mathfrak{P}} \subset \mathfrak{P}$. Moreover, $\mathfrak{P}$ admits, in a unique manner, a representation as a direct sum[5] $\mathfrak{P} = \bar{\mathfrak{P}} + \mathfrak{P}^{(0)}$. It goes without saying that these sets may be empty.

It has been shown in [4] that the set $\mathfrak{P}$ induces a boundary $\mathfrak{B}$ for the space E, and a topology in $E + \mathfrak{B}$. The same is, of course, true of each $\mathfrak{P}_\lambda$, and also of $\bar{\mathfrak{P}}$. It is shown in Section 6 that *each* $\mathfrak{P}_\lambda$ *as well as* $\bar{\mathfrak{B}}$ *induce the same boundary* $\bar{\mathfrak{B}}$, and the same topology in $E + \bar{\mathfrak{B}}$. We call $\bar{\mathfrak{B}}$ *the exit boundary*, and the set $\mathfrak{B} - \bar{\mathfrak{B}} = \mathfrak{B}^{(0)}$ *the passive boundary* of E. The probabilistic interpretation of these notions in terms of the stochastic process governed by $P^{\min}(t)$ is described in Section 7.

Clearly these notions reflect the action of the matrices Π, Π_λ, and $P(t)$ on column vectors (functions). To the premultiplication of a column vector by Π_λ there corresponds dually the postmultiplication of a row vector by Π_λ, and we require also properties of solutions of $\varrho = \varrho\Pi_\lambda$. This equation would be equivalent to $\Pi_\lambda \mathbf{u} = \mathbf{u}$ were it not for the fact that the column vector $\mathbf{1}$ always satisfies $\Pi_\lambda \mathbf{1} \leqq \mathbf{1}$ and provides a natural upper bound in the sets $\mathfrak{P}_\lambda$. No analogue to such a vector exists for $\varrho = \varrho\Pi_\lambda$. Nevertheless Section 10 contains analogues to the main results of Section 5 for the solutions of $\varrho = \varrho\Pi_\lambda$. These results are used analytically in Sections 11–14, but the dual of the exit boundary, namely the *entrance boundary* $\mathfrak{B}^*$ is introduced only in Section 15. There also a general *duality principle* is described which shows the essential symmetry between the pair of equations (1.12); in fact, it enables us (at least in all cases of non-uniqueness) to translate every statement and each admissible boundary condition from $\mathbf{C}'(t) = \Omega\mathbf{C}(t)$ to $\mathbf{R}'(t) = \mathbf{R}(t)\Omega$ and vice versa.

We now turn to the description of the various Markovian[6] $P(t)$ consistent with one or both of the Kolmogorov equations. The *minimal solution* $P^{\min}(t)$ *is characterized by the fact that each* $\mathbf{C}$ *in the range of* $P^{\min}(t)$ *vanishes on the exit boundary*; that is, if $\mathbf{C} = P^{\min}(t)\mathbf{c}$ then $C(i) \to 0$ as i approaches the exit boundary. The same statement applies to the range of the resolvent. This theorem can be translated into one (Theorem 15.1) describing the behavior of $\varrho P^{\min}(t)$ at the *entrance boundary*. It follows that $P' = \Omega P$ *has a unique admissible solution if, and only if, the exit boundary is empty.*[7]

The exit boundary can have a very complicated structure and need not be countable. For the reader's convenience, however, we formulate the following results for the case where the exit boundary consists of finitely many points only. To each point ω there corresponds (for each $\lambda > 0$) a solution $\mathbf{u}_\lambda$ of $\mathbf{u}_\lambda = \Pi_\lambda \mathbf{u}_\lambda$. This is

---

[5] $\mathfrak{P}^{(0)}$ contains exactly those $\mathbf{u} \in \mathfrak{P}$ which satisfy $P^{\min}(t)\,\mathbf{u} = \mathbf{u}$ for some, and therefore all t.

[6] For non-stochastic solutions see the footnote to Section 15.

[7] If the exit boundary is empty, $P^{\min}(t)$ will be strictly stochastic and, because of the minimality, every other solution of $P' = \Omega P$ with $P(0) = I$ will be *norm-increasing*, that is, $\sum_k P(t; i, k) > 1$ for some i. Such solutions may exist and can be constructed by the present methods as indicated in the footnote to Section 15.

the Laplace transform of the solution $\mathbf{U}(t)$ (a column vector with components $\mathbf{u}(t; i)$) of the matrix differential equation $\mathbf{U}'(t) = \Omega\mathbf{U}(t)$ with initial values $\mathbf{U}(0) = \mathbf{0}$. For a probabilistic interpretation cf. Section 7). Given the solution $\mathbf{C}(t) = P^{\min}(t)\mathbf{c}$ of the first initial value problem in (1.12), we obtain infinitely many others of the form $\mathbf{C}(t) + \mathbf{U}(t)\Phi_{\mathbf{c}}$, where $\Phi_{\mathbf{c}}$ is a functional of $\mathbf{c}$. The problem is to determine this $\Phi_{\mathbf{c}}$ so that the new solution will be of the form $P(t)\mathbf{c}$, with $P(t)$ Markovian. It is shown in Section 8 that this is always possible.

Theorem 8.1 shows the existence of such a Markovian $P(t)$ *satisfying the Kolmogorov equation* $P' = \Omega P$ *but not* $P' = P\Omega$. (The analogue to $P' = P\Omega$ is given in Theorem 8.2).

Considering that we are dealing with the countable space $E = \{1, 2, \cdots\}$ and the additional point ω, the question arises inevitably as to why ω should be treated differently from the other points. In Section 9 we consider the *enlarged space* $E + \omega$, and show that the Kolmogorov differential equations (and the operator Ω) can be extended to cover ω. We obtain an enlarged *Markovian* $P(t)$ *with elements defined for* $i, k = 1, 2, \cdots, \omega$. For $i = 1, 2, \cdots$ the Kolmogorov equations $P' = \Omega P$ are satisfied, but for $i = \omega$ a new line is to be added. Now the solution depends on an arbitrary row-vector $\boldsymbol{\varrho}$. *If* $\boldsymbol{\varrho}$ *is chosen bounded, the new row of* Ω *looks exactly like the others, and we get a replica of the Kolmogorov equations* $P' = \Omega P$ *for the enlarged space.*[8] *However, if* $\boldsymbol{\varrho}$ *is unbounded, then in the notation* (1.3)

$$q_\omega = \infty \qquad \textstyle\sum_k q_{\omega k} = \infty. \tag{2.4}$$

In its essential features the process thus obtained resembles one which has been constructed by Kolmogorov[9] [14] by probabilistic methods and analyzed by D. G. Kendall and Reuther [12].

We next turn to the construction of Markovian $P(t)$ satisfying *both* Kolmogorov differential equations. This is possible in terms of solutions of the equations $\mathbf{U}'(t) = \Omega\mathbf{U}(t)$ and $\mathbf{A}'(t) = \mathbf{A}(t)\Omega$, with initial values $\mathbf{0}$. For clarity, we begin (Section 12) with solutions $P(t)$ involving just one $\mathbf{U}(t)$ and one $\mathbf{A}(t)$ or, as one might say, *solutions involving one exit and one entrance boundary point*. As before, we may or may not adjoin this point to E. If we do, the row in (1.3) corresponding to this new point takes on the form

$$q_\omega \leqq \infty \qquad q_{\omega i} = 0 \qquad\qquad i \neq \omega, \tag{2.5}$$

and the form for Ω corresponding to ω is also given. The refusal to consider ω as a point again leads to a *lateral condition*[10] as a degeneration of Ω.

[8] That is, the matrices $\mathfrak{q}$ and Π extend to ω. Note however that $\Pi(i, \omega) = 0$ for each $i \neq \omega$, and that this implies $P^{\min}(t; i, \omega) = 0$.

[9] It should be noted that all the various phenomena occurring in Markov chains have been discussed in measure theoretic terms in a well known paper by P. Lévy [6]. This investigation has been continued by Chung [1]. In P. Lévy's terminology the boundary point ω represents an *instantaneous* state in the example of Section 9, a *fictitious* state in Section 8.

[10] The lateral conditions depend on unbounded functionals generalizing the *normal derivatives* of the classical boundary conditions. They are introduced in Section 11.

The occurrence of (2.5) has been described probabilistically by P. Lévy [16] and Kolmogorov [14], and the Kolmogorov example has been subjected to a thorough analytical investigation by Kendall and Reuther [12], and Kendall [11].[11]

In Sections 13, 14 and 16 we treat the case where the *exit boundary consists of m points and the entrance boundary of n points.*[12] We give the most general form of Markovian $P(t)$ satisfying both Kolmogorov equations. It raises interesting, but unanswered, questions concerning the analytical and topological relationship between the two boundaries.

In the last section the reader will find the simplest example of the Kolmogorov equations where each boundary consists of exactly one point, and they have non-overlapping neighborhoods. A more interesting example where the two boundaries (topologies) coincide is provided by the birth and death process,[13] that is the process where $\Pi(i, i+1) + \Pi(i, i-1) = 1$ (transitions are possible only to neighbors). Here there are four possibilities: (1) There exists no boundary, (2) the point at ∞ is an exit boundary, (3) it is an entrance boundary, (4) it is both an exit and an entrance boundary point. Karlin and McGregor [9] have given necessary and sufficient conditions for the uniqueness of the solutions of the Kolmogorov equations, and these conditions are criteria for the four cases. It is noteworthy that these criteria are an exact counterpart to those encountered in diffusion theory, and that in the terminology of diffusion theory the four cases correspond to: (1) natural, (2) exit, (3) entrance, (4) regular boundary. (One could also introduce a Hilbert space and obtain the analogue of the Weyl classification.)

3. Reformulation in terms of Laplace transforms

For a Markovian matrix $P(t)$ we introduce the matrix P_λ whose elements are the Laplace transforms

$$P_\lambda(i, j) = \int_0^\infty e^{-\lambda t} P(t; i, j)\, dt, \qquad \lambda > 0. \tag{3.1}$$

[11] We have at the same time obtained a simple answer to a question raised by Kendall [11], namely whether there can exist two distinct *strictly stochastic* $P(t)$ satisfying *both* Kolmogorov equations. The situation is obviously as follows: If the *exit* boundary is empty, $P^{\min}(t)$ is strictly stochastic. If an *exit* boundary exists but the entrance boundary is *empty*, then there exist infinitely many solutions of $P' = \Omega P$, but $P^{\min}$ is the unique bounded solution of $P' = P\Omega$. If *neither of the boundaries is empty, there exist infinitely many Markovian $P(t)$ satisfying both* Kolmogorov equations. Among them there are infinitely many strictly *stochastic* ones, *except if each of the two boundaries reduces to a single point*: then there exists exactly one strictly stochastic $P(t)$ satisfying *both* equations.

[12] In the general case of non-countable boundaries our sums should be replaced by integrals, but we do not investigate the details.

[13] Ledermann and Reuter [15], Karlin and McGregor [9]. The latter note is to be expanded in a paper to appear in the Transactions of the American Mathematical Society. There exists also a completed but unpublished thorough investigation of this process by B. O. Koopman.

Clearly

$$P_\lambda(i, j) \geqq 0, \qquad \lambda \sum_j P_\lambda(i, j) \leqq 1. \tag{3.2}$$

The Kolmogorov differential equations (1.5)–(1.7) take on the form

$$(\lambda + \mathfrak{q})P_\lambda = \mathfrak{q}\Pi P_\lambda + I \tag{3.3}$$

$$P_\lambda(\lambda + \mathfrak{q}) = P_\lambda \mathfrak{q}\Pi + I. \tag{3.4}$$

Finally, the semi-group property (1.1) is equivalent to the so-called *resolvent* equation[14]

$$P_\lambda P_\nu = \frac{P_\nu - P_\lambda}{\lambda - \nu} \qquad \lambda, \nu > 0. \tag{3.5}$$

Conversely, let to each $\lambda > 0$ correspond a matrix P_λ such that (3.2), (3.3) and (3.5) hold. In view of (3.5) there exists a derivative of P_λ with respect to λ, and by induction we derive the familiar relations for resolvents

$$\frac{d^n}{d\lambda^n} P_\lambda = n!(-1)^n P_\lambda^{n+1}, \qquad n = 1, 2, \cdots. \tag{3.6}$$

Therefore

$$(-1)^n \frac{d^n}{d\lambda^n} P_\lambda(i, j) \geqq 0, \qquad \lambda^{n+1} \left| \frac{d^n}{d\lambda^n} \sum_j P_\lambda(i, j) \right| \leqq n!. \tag{3.7}$$

This implies that $P_\lambda(i, j)$ is the Laplace transform of a function $P(t; i, j)$ such that (1.2) holds (cf. Widder [17], pp. 315 and 310). Now (3.3) is equivalent to

$$P(t; i, k) - \delta_{ik} = q_i \int_0^t \{-P(s; i, k) + \textstyle\sum_j \pi(i, j)P(s; j, k)\}\, ds. \tag{3.8}$$

Hence $P(t; i, k)$ depends continuously on t (and is therefore continuously differentiable). Clearly (1.6) holds and $P(t; i, k) \to \delta_{ik}$ as $t \to 0$. The argument leading from (1.1) to (3.5) is now reversible and we have therefore:

THEOREM 3.1. *In order that a family of matrices P_λ defined for $\lambda > 0$ represent the Laplace transform of a Markovian matrix $P(t)$ which satisfies the first Kolmogorov differential equation $P' = \Omega P$, it is necessary and sufficient that* (3.2), (3.3) *and* (3.5) *hold.*

Needless to say that a corresponding statement holds true for the second equation $P' = P\Omega$ and (3.4), but we shall not require this fact explicitly.

---

[14] In fact, we have the obvious identity

$$\int_0^\infty e^{-\nu s} P(t + s)\, ds = e^{\nu t} P_\nu - \int_0^t e^{\nu(t-s)} P(s)\, ds. \tag{$*$}$$

By virtue of (1.1) the Laplace transform of the left side of $(*)$ equals $P_\lambda P_\nu$. For $\lambda > \nu$ the Laplace transform of $e^{\nu t}$ equals $(\lambda - \nu)^{-1}$, and by the convolution theorem the transform of the right side in $(*)$ equals the right hand member in (3.5).

For brevity we shall say that a matrix P_λ satisfying (3.2) and (3.3) represents an *admissible solution* of (3.3). Obviously it induces the bounded transformation

$$\mathbf{c} \to \mathbf{C}: \quad \mathbf{C} = P_\lambda \mathbf{c}, \qquad \mathbf{c} \in \mathfrak{C}. \tag{3.9}$$

Clearly $\mathbf{C}$ is *one* solution of

$$\lambda \mathbf{C} - \Omega \mathbf{C} = \mathbf{c}. \tag{3.10}$$

THEOREM 3.2. *For each $\lambda > 0$ let P_λ satisfy* (3.2) *and* (3.3). *For the resolvent equation* (3.5) *to hold it is necessary and sufficient that the transformations* (3.9) *have a common range $\mathfrak{L}$ (independent of λ); that is to say, for each $\lambda > 0$ and each $\mathbf{c} \in \mathfrak{C}$ there exists exactly one solution $\mathbf{C} \in \mathfrak{L}$ of* (3.10).

PROOF. Put for the moment $\bar{\mathbf{c}} = \mathbf{c} + (\nu - \lambda)P_\lambda \mathbf{c}$. If (3.5) holds, then $P_\nu \bar{\mathbf{c}} = P_\lambda \mathbf{c}$ and hence the range of P_ν contains the range of λ. Since λ and ν are interchangeable this implies that the two ranges coincide. Thus the condition is necessary. Conversely, with $\mathbf{C}$ defined by (3.9) we have $\nu\mathbf{C} - \Omega\mathbf{C} = \bar{\mathbf{c}}$. If the range of P_ν coincides with the range of P_λ, then $\mathbf{C}$ is in the range of P_ν and hence $P_\nu \bar{\mathbf{c}} = \mathbf{C}$. Since $\mathbf{c}$ is arbitrary this implies (3.5).

To avoid repetitions we introduce the

DEFINITION 3. *A set $\mathfrak{L}$ (independent of λ) will be called conditioning set for* (3.10) *if for each $\lambda > 0$ and each $\mathbf{c} \in \mathfrak{C}$ there exists exactly one solution $\mathbf{C} = P_\lambda \mathbf{c} \in \mathfrak{L}$ of* (3.10), *and the matrix P_λ defining it satisfies* (3.2). *The condition $\mathbf{C} \in \mathfrak{L}$ will be called a lateral condition for* (3.10) *(or* (3.3)).

In other words, *the equation* (3.10) *together with a lateral condition determines the Laplace transform of a Markovian $P(t)$ and conversely.* This $P(t)$ satisfies (3.3). Needless to say that similar considerations apply to the second equation (3.4) and to operators different from Ω.

A criterion for $P(t)$ to be *strictly stochastic* is contained in

THEOREM 3.3. *Let $P(t)$ be Markovian and let P_λ stand for the corresponding Laplace transform. Then either*

$$P(t)\mathbf{1} = \mathbf{1} \qquad \lambda P_\lambda \mathbf{1} = \mathbf{1} \tag{3.11}$$

for all $t > 0$ and $\lambda > 0$, or else both equations are false for all $t > 0$, $\lambda > 0$.

PROOF. In view of (1.1) and (1.2) the vector $P(t)\mathbf{1}$ is non-increasing, and $P(t_0)\mathbf{1} = \mathbf{1}$ implies $P(2t_0)\mathbf{1} = \mathbf{1}$. In this case (3.11) holds for all $t > 0$, $\lambda > 0$. If $\lambda P_\lambda \mathbf{1} = \mathbf{1}$ then also $\nu P_\nu \mathbf{1} = \mathbf{1}$ in consequence of the resolvent equation (3.5), and again (3.11) holds identically.

4. The minimal solution

With the matrix Π_λ of (2.1) we form

$$S_\lambda^{(n)} = \sum_{p=0}^{n} \Pi_\lambda^p \qquad S_\lambda = \sum_{p=0}^{\infty} \Pi_\lambda^p \tag{4.1}$$

$$P_\lambda^{(n)} = S_\lambda^{(n)} \{\lambda + \mathfrak{q}\}^{-1}. \tag{4.2}$$

THEOREM 4.1. *For each* $\lambda > 0$ *the matrix*

$$P_\lambda^{\min} = \lim_{n\to\infty} P_\lambda^{(n)} = S_\lambda\{\lambda + \mathfrak{q}\}^{-1} \tag{4.3}$$

satisfies (3.2), (3.3) *and* (3.4). *If* P_λ *is a non-negative matrix satisfying either* (3.3) *or* (3.4), *then*

$$P_\lambda \geqq P_\lambda^{\min}. \tag{4.4}$$

PROOF. For $n \geqq 1$ we have obviously

$$S_\lambda^{(n)} = \{\lambda + \mathfrak{q}\}^{-1}\mathfrak{q}\Pi S_\lambda^{(n-1)} + I \tag{4.5}$$

$$S_\lambda^{(n)} = S_\lambda^{(n-1)}\{\lambda + \mathfrak{q}\}^{-1}\mathfrak{q}\Pi + I \tag{4.6}$$

or

$$\{\lambda + \mathfrak{q}\}P_\lambda^{(n)} = \mathfrak{q}\Pi P_\lambda^{(n-1)} + I \tag{4.7}$$

$$P_\lambda^{(n)}\{\lambda + \mathfrak{q}\} = P_\lambda^{(n-1)}\mathfrak{q}\Pi + I. \tag{4.8}$$

To calculate the row sums of $P_\lambda^{(n)}$ postmultiply (4.8) by the column vector $\mathbf{1}$. Since $\Pi\mathbf{1} = \mathbf{1}$ we get

$$\lambda P_\lambda^{(n)}\mathbf{1} + \Pi_\lambda^n\{\lambda + \mathfrak{q}\}^{-1}\mathfrak{q}\mathbf{1} = \mathbf{1}. \tag{4.9}$$

Thus each row sum in $\lambda P_\lambda^{(n)}$ is at most 1. It follows that the limit in (4.3) exists and $\lambda P_\lambda^{\min}$ is substochastic. The convergence is monotone and so a passage to the limit in (4.7) and (4.8) is permitted. It leads, respectively, to (3.3) and (3.4). Finally, if $P_\lambda \geqq 0$ satisfies (3.3) then $P_\lambda \geqq P_\lambda^{(0)}$ and by induction $P_\lambda \geqq P_\lambda^{(n)}$. The Theorem is proved.

THEOREM 4.2. *The matrix* $P_\lambda^{\min}$ *satisfies the resolvent equation* (3.5).

PROOF. We propose to prove that

$$\Pi_\lambda^n\{\lambda + \mathfrak{q}\}^{-1} - \Pi_\nu^n\{\nu + \mathfrak{q}\}^{-1} = (\nu - \lambda)\sum_{a=0}^n \Pi_\lambda^a\{\lambda + \mathfrak{q}\}^{-1}\Pi_\nu^{n-a}\{\nu + \mathfrak{q}\}^{-1}. \tag{4.10}$$

Summation over n leads to (3.5), and hence (4.10) represents a stronger assertion. To prove it, call the right hand member A_n. Then

$$\Pi_\lambda A_n = A_{n+1} - (\nu - \lambda)\{\lambda + \mathfrak{q}\}^{-1}\Pi_\nu^{n+1}\{\nu + \mathfrak{q}\}^{-1}. \tag{4.11}$$

Now assume (4.10) to be true for some n (it is clearly true for $n = 0$). Operating on the left side of (4.10) we find

$$\begin{aligned} \Pi_\lambda A_n &= \Pi_\lambda^{n+1}\{\lambda + \mathfrak{q}\}^{-1} - \Pi_\lambda\Pi_\nu^n\{\nu + \mathfrak{q}\}^{-1} \\ &= \Pi_\lambda^{(n+1)}\{\lambda + \mathfrak{q}\}^{-1} - \Pi_\nu^{n+1}\{\nu + \mathfrak{q}\}^{-1} \\ &\quad - (\nu - \lambda)\{\lambda + \mathfrak{q}\}^{-1}\Pi_\nu^{n+1}\{\nu + \mathfrak{q}\}^{-1}. \end{aligned} \tag{4.12}$$

Subtracting (4.12) from (4.11) we get (4.10) with n replaced by $n + 1$.

LEMMA. *We have*

$$\Pi_\lambda^n\mathbf{1} \downarrow \bar{\mathbf{x}}_\lambda \tag{4.13}$$

where $\bar{\mathbf{x}}_\lambda$ *is the maximal element of* $\mathfrak{P}_\lambda$. *That is, if* $\mathbf{x} \in \mathfrak{P}_\lambda$ *then* $\mathbf{x} \leqq \bar{\mathbf{x}}_\lambda$. (For the definition cf. (2.3).)

PROOF. Put $\mathbf{v}_n = \Pi_\lambda^n \mathbf{1}$. Then $\mathbf{v}_1 \leqq \mathbf{v}_0 = \mathbf{1}$, and so $\mathbf{v}_n \downarrow \bar{\mathbf{x}}_\lambda \in \mathfrak{P}_\lambda$. Conversely, if $\mathbf{x} \in \mathfrak{P}_\lambda$ then by definition $\mathbf{x} \leqq \mathbf{1} = \mathbf{v}_0$, and by induction $\mathbf{x} = \Pi_\lambda \mathbf{x} \leqq \Pi_\lambda \mathbf{v}_n = \mathbf{v}_{n+1}$. This proves the lemma.

Notice that $\mathbf{v}_1 = \{\lambda + \mathfrak{q}\}^{-1}\mathfrak{q}\mathbf{1}$ and that the second term on the left in (4.9) reduces to $\Pi_\lambda^n \mathbf{v}_1 = \Pi_\lambda^{n+1}\mathbf{1}$. This proves the truth of

THEOREM 4.3. *One has*[15]

$$\lambda P_\lambda^{\min}\mathbf{1} = \mathbf{1} - \bar{\mathbf{x}}_\lambda \tag{4.14}$$

where $\bar{\mathbf{x}}_\lambda$ *is the maximal element of* $\mathfrak{P}_\lambda$. *In particular*, $\lambda P_\lambda^{\min}\mathbf{1} = \mathbf{1}$ *if, and only if, the equation* $\Pi_\lambda \mathbf{x} = \mathbf{x}$ *admits of no non-trivial solution*[16] $\mathbf{0} \leqq \mathbf{x} \leqq \mathbf{1}$.

COROLLARY. (*Uniqueness criterion*). *If for some, and therefore for each,* λ *the vector* $\mathbf{0}$ *is the only element of* $\mathfrak{P}_\lambda$, *then* $P_\lambda^{\min}$ *represents the only Markovian* $P(t)$ *satisfying either of the Kolmogorov differential equations* $P' = \Omega P$ *or* $P' = P\Omega$.

We shall see (Section 8) that the existence of a solution $\neq \mathbf{0}$ of (4.14) implies that there are infinitely many Markovian $P(t)$ satisfying $P' = \Omega P$. However, this statement is false for $P' = P\Omega$. Moreover, even in our uniqueness case $P' = P\Omega$ may admit of solutions $P(t)$ with *increasing* row-sums.

5. Solutions of $\mathbf{v} = \Pi_\lambda \mathbf{v}$

We proceed to investigate the relationship between the sets $\mathfrak{P}$ and $\mathfrak{P}_\lambda$ introduced in (2.3). Consider a fixed $\lambda > 0$ and $\mathbf{v} \in \mathfrak{P}_\lambda$. Then $\mathbf{v} = \Pi_\lambda \mathbf{v} \leqq \Pi \mathbf{v} \leqq \mathbf{1}$ and therefore $\Pi^{n-1}\mathbf{v} \leqq \Pi^n \mathbf{v} \leqq \mathbf{1}$. Accordingly

$$\Pi^n \mathbf{v} \uparrow \mathbf{v}_0 \tag{5.1}$$

where $\mathbf{v}_0 = \Pi \mathbf{v}_0$. Moreover, $\mathbf{v}_0$ is *minimal* among the elements $\mathbf{u} \in \mathfrak{P}$ such that $\mathbf{u} \geqq \mathbf{v}$. For $\mathbf{u} \geqq \mathbf{v}$ implies $\mathbf{u} = \Pi^n \mathbf{u} \geqq \Pi^n \mathbf{v} \uparrow \mathbf{v}_0$, and so $\mathbf{u} \geqq \mathbf{v}_0$.

DEFINITION 5.1. *The canonical map of* $\mathfrak{P}_\lambda$ *into* $\mathfrak{P}$ *is the map which sends the element* $\mathbf{v} \in \mathfrak{P}_\lambda$ *into the element* $\mathbf{v}_0 \in \mathfrak{P}$ *defined by* (5.1); *that is,* $\mathbf{v}$ *is mapped into the minimal element* $\mathbf{v}_0$ *of the set of vectors* $\mathbf{u} \in \mathfrak{P}$ *such that* $\mathbf{u} \geqq \mathbf{v}$.

Now the equation $\mathbf{v} = \Pi_\lambda \mathbf{v}$ may be rewritten in the form

$$\mathbf{v} + \lambda \mathfrak{q}^{-1}\mathbf{v} = \Pi \mathbf{v}. \tag{5.2}$$

Premultiplying by Π^a and adding we get

$$\mathbf{v} + \lambda \textstyle\sum_{a=0}^{n-1} \Pi^a \mathfrak{q}^{-1}\mathbf{v} = \Pi^n \mathbf{v}. \tag{5.3}$$

Letting $n \to \infty$ and observing (5.1) we are led to

LEMMA 5.1. *The canonical image* $\mathbf{v}_0$ *of the element* $\mathbf{v} \in \mathfrak{P}_\lambda$ *is given by*

$$\mathbf{v}_0 = \mathbf{v} + \lambda \Gamma \mathbf{v} \tag{5.4}$$

[15] For the probabilistic interpretation cf. end of Section 7.

[16] This is equivalent to the requirement that there exist no bounded solution $\mathbf{x} \neq \mathbf{0}$ of $\Pi_\lambda \mathbf{x} = \mathbf{x}$; equivalently, that there exist no nontrivial solution $\mathbf{x} \geqq \mathbf{0}$ of $\Pi_\lambda \mathbf{x} \geqq \mathbf{x}$.

where the matrix Γ *is defined by*[17]

(5.5) $$\Gamma(i, j) = \sum_{a=0}^{\infty} \Pi^a(i, j) \frac{1}{q_j}.$$

For an analysis of the inverse mapping we observe that the proof of Theorem 4.3 is purely algebraic and therefore applies not only to the vector **1** but to each $\mathbf{u} \geqq \mathbf{0}$ which satisfies $\mathbf{u} = \Pi\mathbf{u}$. We have thus[18]

LEMMA 5.2. *If* $\mathbf{u} \in \mathfrak{P}$ *then for each* $\lambda > 0$

(5.6) $$\lambda P_\lambda^{\min}\mathbf{u} = \mathbf{u} - \mathbf{u}_\lambda$$

where $\mathbf{u}_\lambda \in \mathfrak{P}_\lambda$ *and* $\mathbf{u}_\lambda$ *is the maximal element of the set of vectors* $\mathbf{v} \in \mathfrak{P}_\lambda$ *such that* $\mathbf{v} \leqq \mathbf{u}$.

The resolvent equation (3.5) now leads immediately to

LEMMA 5.3. *The vectors* $\mathbf{u}_\lambda$ *defined by* (5.6) *satisfy for* $\lambda, \nu > 0$

(5.7) $$\mathbf{u}_\lambda - \mathbf{u}_\nu = (\nu - \lambda)P_\lambda^{\min}\mathbf{u}_\nu = (\nu - \lambda)P_\nu^{\min}\mathbf{u}_\lambda .$$

In particular, $\mathbf{u}_\lambda = \mathbf{0}$ implies $\mathbf{u}_\nu = \mathbf{0}$ and we introduce the

DEFINITION 5.2. *A vector* $\mathbf{u} \in \mathfrak{P}$ *will be called passive if for some, and consequently for each,* $\lambda > 0$

(5.8) $$\mathbf{u} = \lambda P_\lambda^{\min}\mathbf{u};$$

this amounts to saying that $\mathbf{v} \in \mathfrak{P}_\lambda$ and $\mathbf{v} \leqq \mathbf{u}$ imply $\mathbf{v} = \mathbf{0}$.

Now fix for the moment λ and $\mathbf{v} \in \mathfrak{P}_\lambda$. Let $\mathbf{v}_0 \in \mathfrak{P}$ be the canonical image of $\mathbf{v}$ and let $\mathbf{v}' \in \mathfrak{P}_\lambda$ be the maximal element of the set $\mathbf{w} \in \mathfrak{P}_\lambda$, $\mathbf{w} \leqq \mathbf{v}_0$. By definition $\mathbf{v}' \geqq \mathbf{v}$. On the other hand, $\mathbf{v}_0 \geqq \mathbf{v}'$ and therefore $\mathbf{v}_0$ is the canonical image not only of $\mathbf{v}$ but also of $\mathbf{v}'$. Hence the canonical mapping takes $\mathbf{v}' - \mathbf{v}$ into zero, and so $\mathbf{v}' = \mathbf{v}$. We formulate this in the form of

LEMMA 5.4. *Let* $\mathbf{v} \in \mathfrak{P}_\lambda$ *and let* $\mathbf{v}_0 \in \mathfrak{P}$ *be the canonical image of* $\mathbf{v}$ (*i.e.* $\mathbf{v}$ *is given by* (5.5).) *Then*

(5.9) $$\lambda P_\lambda^{\min}\mathbf{v}_0 = \mathbf{v}_0 - \mathbf{v}.$$

Choose $\mathbf{u}' \in \mathfrak{P}$ arbitrary and define $\mathbf{v}$ by

(5.10) $$\lambda P_\lambda^{\min}\mathbf{u}' = \mathbf{u}' - \mathbf{v}.$$

Let $\mathbf{v}_0$ be the canonical image of $\mathbf{v}$. In view of the minimal character of $\mathbf{v}_0$ we have $\mathbf{u}' \geqq \mathbf{v}_0$. Put $\mathbf{u}' - \mathbf{v}_0 = \mathbf{u}$. For our $\mathbf{v}$ both (5.9) and (5.10) hold, and subtracting (5.9) from (5.10) we get (5.8). Accordingly when $\mathbf{u}'$ is not the canonical image of $\mathbf{v}$, then there exists a passive non-null vector $\mathbf{u} \leqq \mathbf{u}'$. Conversely, if $\mathbf{u}$

[17] If $\Gamma(i, j) = \infty$ then automatically $v(j) = 0$, and, by definition, $\Gamma(i, j)v(j) = 0$.
For a fixed j the column $\Gamma(i, j)$ is finite if, and only if, j is *transient* (cf. e.g. [6] p. 321). For the present paper non-transient points have only a nuisance value, and the reader is advised to assume from the start that *all* j are transient.

[18] We formulate and use Lemma 5.2 only for *bounded* $\mathbf{u}$. However, (5.6) applies equally to *unbounded* solutions of $\mathbf{u} = \Pi\,\mathbf{u}$ and shows that the domain of $P_\lambda^{\min}$ may exceed $\mathfrak{C}$.

is such a vector, subtract (5.8) from (5.10) to see that $\mathbf{u}' - \mathbf{u} \geqq \mathbf{v}$ and that $\mathbf{u}'$ is therefore not the canonical image of $\mathbf{v}$. This proves

LEMMA 5.5. *For a* $\mathbf{u}' \in \mathfrak{P}$ *to be the canonical image of some* $\mathbf{v} \in \mathfrak{P}_\lambda$ *it is necessary and sufficient that there exist no passive non-null* $\mathbf{u} \in \mathfrak{P}$ *such that* $\mathbf{u} \leqq \mathbf{u}'$.

COROLLARY. *The range of the canonical mapping from* $\mathfrak{P}_\lambda$ *to* $\mathfrak{P}$ *is independent of* λ.

Now remember that each $\mathfrak{P}_\lambda$ has a uniquely determined maximal element $\bar{\mathbf{x}}_\lambda$, as given by (4.13). The canonical image $\bar{\mathbf{x}}$ of $\bar{\mathbf{x}}_\lambda$ is obviously the maximal element of the range of the canonical mapping $\mathfrak{P}_\lambda \to \mathfrak{P}$, and therefore $\bar{\mathbf{x}}$ is independent of λ. By Lemma 5.4

$$\lambda P_\lambda^{\min} \bar{\mathbf{x}} = \bar{\mathbf{x}} - \bar{\mathbf{x}}_\lambda . \tag{5.11}$$

Subtracting (5.11) from (4.14) we see that

$$\mathbf{x}^{(0)} = \mathbf{1} - \bar{\mathbf{x}} \tag{5.12}$$

is passive.

To conclude, we remark that (as is easily verified and was shown in [4]) $\mathfrak{P}$ (and similarly $\mathfrak{P}_\lambda$) is a *lattice* in the sense that to any $\mathbf{x}, \mathbf{y} \in \mathfrak{P}$ there corresponds uniquely a minimal element $\mathbf{x} \cup \mathbf{y}$ of the set of all vectors $\mathbf{u} \in \mathfrak{P}$ such that $\mathbf{u} \geqq \mathbf{x}$ and $\mathbf{u} \geqq \mathbf{y}$, and a maximal element $\mathbf{x} \cap \mathbf{y}$ of the set $\mathbf{u} \in \mathfrak{P}$ such that $\mathbf{u} \leqq \mathbf{x}$ and $\mathbf{u} \leqq \mathbf{y}$. If $\mathbf{u}$ is passive then by Lemma (5.5) $\mathbf{u} \cap \bar{\mathbf{x}} = \mathbf{0}$, and so $\mathbf{u} \leqq \mathbf{x}^{(0)}$. We are now in a position to summarize our results in

THEOREM 5.1. *There exist two elements* $\mathbf{x}^{(0)}, \bar{\mathbf{x}} \in \mathfrak{P}$ *with the following properties:*

(i) *A vector* $\mathbf{u} \in \mathfrak{P}$ *is passive (i.e.* (5.8) *holds) if and only if* $\mathbf{u} \leqq \mathbf{x}^{(0)}$.

(ii) *For each* $\lambda > 0$ *the range of the canonical mapping* $\mathfrak{P}_\lambda \to \mathfrak{P}$ *coincides with the set* $\bar{\mathfrak{P}}$ *of elements* $\mathbf{u} \in \mathfrak{P}$ *such that* $\mathbf{u} \leqq \bar{\mathbf{x}}$.

(iii) *The correspondence between* $\mathfrak{P}_\lambda$ *and* $\mathfrak{P}$ *is one-to-one and a lattice isomorphism.*

(iv) *For an arbitrary* $\mathbf{v}_0 \in \bar{\mathfrak{P}}$ *the relation* (5.9) *implies* (5.4).

In order to simplify the further exposition we conclude this section by showing that there is no serious loss of generality in supposing that $\bar{\mathbf{x}}$ and $\bar{\mathbf{x}}_\lambda$ are strictly positive. Let $\mathfrak{E}$ be the set of all points i such that $\bar{x}_\lambda(i) = 0$ for some fixed λ, and denote the complementary set by $\mathfrak{E}'$. The equation $\bar{\mathbf{x}}_\lambda = \Pi_\lambda \bar{\mathbf{x}}_\lambda$ implies that the matrix Π_λ decomposes in such a way that $\Pi_\lambda(i, j) = 0$ when $i \in \mathfrak{E}$, $j \in \mathfrak{E}'$. Clearly the same is then true of Π and of $P_\lambda^{\min}$. It follows then from (4.15) that

$$\lambda \textstyle\sum_{j \in \mathfrak{E}} P_\lambda^{\min}(i, j) = 1, \qquad i \in \mathfrak{E} \tag{5.13}$$

and hence that $\bar{x}(i) = 0$ for $i \in \mathfrak{E}$. In short, we have the rather trite

THEOREM 5.2. *In order that* $\bar{x}_\lambda(i) = 0$ *for some* λ *it is necessary and sufficient that* $\bar{x}(i) = 0$ *(in which case* $\bar{x}_\lambda(i) = 0$ *for each* λ*). If* $\mathfrak{E}$ *is the set of all* i *such that* $\bar{x}(i) = 0$, *and* $\mathfrak{E}'$ *its complement then each of the Kolmogorov equations* (3.3) *and* (3.4) *decomposes into two equations of the same type for the submatrices of* P_λ *with elements* $i, j \in \mathfrak{E}$ *and* $i, j \in \mathfrak{E}'$, *respectively. For the former we have* (5.13), *that is, each admissible solution* P_λ *coincides on* $\mathfrak{E}$ *with* $P_\lambda^{\min}$.

In constructing P_λ the set $\mathfrak{E}$ plays no role and we are compelled to consider the restriction of P_λ to $\mathfrak{E}'$. In order to avoid trivialities,[18a] *we shall therefore from now on suppose that* $\bar{x}(i) > 0$ *for each* i. *This convention excludes all recurrent* i (cf. footnote to Lemma 5.1).

6. The exit boundary

We now require the notion of the boundary induced by the matrix Π, or more precisely, by the vectors $\mathbf{u} \in \mathfrak{P}$. Fortunately only few and easily explainable facts are needed for our present purposes. We assume that $\bar{\mathbf{x}}$ *is strictly positive*, but this is no restriction inasmuch as the set on which $\bar{\mathbf{x}}$ vanishes in no way affects the exit boundary.

An element $\mathbf{u} \in \mathfrak{P}$ *is a sojourn solution if, and only if,*[19] $\mathbf{u} \cap (\mathbf{1} - \mathbf{u}) = \mathbf{0}$, that is, if $\mathbf{0} \leqq \mathbf{v} \leqq \mathbf{u}$ and $\mathbf{v} \leqq \mathbf{1} - \mathbf{u}$ implies $\mathbf{v} = \mathbf{0}$. In particular, $\mathbf{x}^{(0)}$ and $\bar{\mathbf{x}}$ are sojourn solutions. Each sojourn solution $\mathbf{u}$ induces a set or point ω on the boundary; ω is a *point* if $\mathbf{u}$ is minimal, that is if $\mathbf{0} \leqq \mathbf{v} \leqq \mathbf{u}$ and $\mathbf{v} \in \mathfrak{P}$ implies $\mathbf{v} = \mathbf{0}$ or $\mathbf{v} = \mathbf{u}$. If $\mathbf{u}^{(1)}$ and $\mathbf{u}^{(2)}$ are two sojourn solutions, the corresponding boundary sets are non-overlapping if and only if $\mathbf{u}^{(1)} \cap \mathbf{u}^{(2)} = \mathbf{0}$.

To each sojourn solution $\mathbf{u}$ there corresponds a family of sets of points i called *representative*[20] *sets* for $\mathbf{u}$. Each such set serves as a *typical neighborhood for the boundary set* ω. If $\mathbf{u}^{(1)} \cap \mathbf{u}^{(2)} = \mathbf{0}$ then there exist disjoint neighborhoods for $\omega^{(1)}$ and $\omega^{(2)}$. In the present paper we shall make explicit use of representative sets only in the proof of Theorem 6.1, and no advantage is gained by explaining their properties. The basic fact on which the probability interpretation of Section 7 is based, is as follows. *If* $\mathfrak{U}$ *is a set which is representative for* $\mathbf{u}$, *then*[21] *for each* i

$$u(i) = \lim_{n\to\infty} \textstyle\sum_{j\in\mathfrak{U}} \Pi^n(i, j). \tag{6.1}$$

In the sequel $\mathbf{x}^{(0)}$ and $\bar{\mathbf{x}}$ will always denote the vectors defined in Theorem 5.1 and we shall use the following

NOTATION. *E stands for the set of positive integers (each constituting its own neighborhood). The boundary of E induced by $\mathfrak{P}$ will be denoted by $\mathfrak{B}$, and the subsets of $\mathfrak{B}$ corresponding to $\mathbf{x}^{(0)}$ and $\bar{\mathbf{x}}$ by $\mathfrak{B}^{(0)}$ and $\bar{\mathfrak{B}}$, respectively.* ($\mathfrak{B} = \mathfrak{B}^{(0)} \cup \bar{\mathfrak{B}}$).

DEFINITION. *$\bar{\mathfrak{B}}$ will be called the exit boundary, $\mathfrak{B}^{(0)}$ the passive boundary.*

Note that $\bar{\mathfrak{B}}$ (but not $\mathfrak{B}$) depends on $\mathfrak{q}$.

In the preceding considerations one may replace Π and $\mathfrak{P}$ by Π_λ and $\mathfrak{P}_\lambda$, thus introducing a boundary for each $\lambda > 0$. Fortunately we have the basic

---

[18a] Note however that the restriction of Π to $\mathfrak{E}'$ need not be strictly stochastic. We have assumed Π to be strictly stochastic and thus we loose (formally at least) some generality in favor of simplicity of notations.

[19] This is not the definition, but the criterion of [4], Theorem 9.1. Each $\mathbf{x} \in \mathfrak{P}$ is in the linear span of sojourn solutions, that is, can be approximated by linear combinations of sojourn solutions.—In case of doubt the reader is asked to consider only the typical case where $\mathfrak{P}$ contains only finitely many linearly independent elements. Then each $\mathbf{x} \in \mathfrak{P}$ is a linear combination of finitely many sojourn solutions, and the boundary contains finitely many points only.

[20] Cf. [4], Definition in Section 8.

[21] [4], Theorem 9.4.

THEOREM 6.1. *For each* $\lambda > 0$ *the boundary induced by* $\mathfrak{P}_\lambda$ *is identical with the exit boundary* $\bar{\mathfrak{B}}$, *and the topologies are equivalent.*

PROOF. By virtue of Theorem 5.1 there exists a one-to-one correspondence between the sojourn solutions in $\mathfrak{P}_\lambda$ and $\bar{\mathfrak{P}}$. Accordingly, we have only to prove the following two assertions: (a) Let $\mathfrak{U}^{(\lambda)}$ be a set representative for the sojourn solution $\mathbf{u}_\lambda \in \mathfrak{P}_\lambda$, and let $\mathbf{u}_0$ be the canonical image of $\mathbf{u}_\lambda$. Then $\mathfrak{U}^{(\lambda)}$ contains a set which is representative for $\mathbf{u}_0$. Conversely: (b) If $\mathfrak{U}$ is representative for $\mathbf{u}_0$ then it contains a set $\mathfrak{U}^{(\lambda)}$ representative for $\mathbf{u}_\lambda$.

To prove (a) observe that $\Pi_\lambda < \Pi$ implies that $\mathfrak{U}^{(\lambda)}$ is representative for *some* sojourn solution $\mathbf{u} \in \mathfrak{P}$ (not necessarily in $\bar{\mathfrak{P}}$), and that $\mathbf{u} \geqq \mathbf{u}_\lambda$. Because of the minimal character of $\mathbf{u}_0$ we have $\mathbf{u} \geqq \mathbf{u}_0$. Let $\mathfrak{U}^{(0)}$ be any set which is representative of $\mathbf{u}_0$. Then also the intersection $\mathfrak{U}^{(0)} \cap \mathfrak{U}^{(\lambda)}$ is representative of $\mathbf{u}_0$ (cf. [4], Theorem 9.2). This proves (a).

To prove (b)[22] let $\mathbf{v}_0 = \bar{\mathbf{x}} - \mathbf{u}_0$. Then $\mathbf{v}_0$ is a sojourn solution and we can choose a set $\mathfrak{V}$ representative of $\mathbf{v}_0$ so that $\mathfrak{U}$ and $\mathfrak{V}$ are non-overlapping. Now let $\mathbf{s}_{\mathfrak{U}}^{(\lambda)}$ and $\mathbf{s}_{\mathfrak{V}}^{(\lambda)}$ be the sojourn solutions (possibly equal to $\mathbf{0}$) as introduced in [4], Section 5, except that Π is replaced by Π_λ. Then $\mathbf{s}_{\mathfrak{U}}^{(\lambda)} \leqq \mathbf{u}_0$ and $\mathbf{s}_{\mathfrak{V}}^{(\lambda)} \leqq \mathbf{v}_0$ as a trivial consequence of $\Pi_\lambda \leqq \Pi$. If $\mathbf{u}_0$ and $\mathbf{v}_0$ are the canonical images, respectively, of $\mathbf{u}_\lambda \in \mathfrak{P}_\lambda$ and $\mathbf{v}_\lambda \in \mathfrak{P}_\lambda$, then (because of the maximality property) $\mathbf{s}_{\mathfrak{U}}^{(\lambda)} \leqq \mathbf{u}_\lambda$ and $\mathbf{s}_{\mathfrak{V}}^{(\lambda)} \leqq \mathbf{v}_\lambda$. But in consequence of Theorem 5.1 we have $\mathbf{u}_\lambda + \mathbf{v}_\lambda = \bar{\mathbf{x}}_\lambda$, whereas $\mathbf{s}_{\mathfrak{U}}^{(\lambda)} + \mathbf{s}_{\mathfrak{V}}^{(\lambda)} \leqq \bar{\mathbf{x}}_\lambda$ (since $\mathfrak{U}$ and $\mathfrak{V}$ are non-overlapping, cf. [4], Lemma 5.1). It follows that $\mathbf{s}_{\mathfrak{U}}^{(\lambda)} = \mathbf{u}_\lambda$, and therefore $\mathfrak{U}$ contains a set representative of the sojourn solution $\mathbf{u}_\lambda$ (cf. [4], Lemma 8.1). This accomplishes the proof.

Now let $\mathbf{u}_0 \in \bar{\mathfrak{P}}$ be a sojourn solution and $\mathbf{u}_\lambda$ its canonical image in $\mathfrak{P}_\lambda$. Then $\mathbf{u}_\lambda$ is a sojourn solution for Π_λ and by [4], Theorem 13.2, we have the following

BASIC LEMMA. *Both* $\mathbf{u}_0$ *and* $\mathbf{u}_\lambda$ *can be extended continuously to* $E + \bar{\mathfrak{B}}$. *More precisely: if* ω *is the boundary set corresponding to* $\mathbf{u}_0$ *and* $i \in E$ *approaches a boundary point* β *then for* $\lambda > 0$

$$u_\lambda(i) \to \begin{matrix} 1 & \text{if } \beta \in \omega \\ 0 & \text{if } \beta \notin \omega. \end{matrix} \tag{6.2}$$

Consider the mapping

$$\mathbf{c} \to \mathbf{C}_\lambda^{\min}, \qquad \mathbf{C}_\lambda^{\min} = P_\lambda^{\min}\mathbf{c}, \qquad \mathbf{c} \in \mathfrak{C}. \tag{6.3}$$

If $\| \mathbf{c} \| \leqq 1$ then $\| \mathbf{C}_\lambda^{\min} \| < \| P_\lambda^{\min}\mathbf{1} \|$ and hence we deduce from (4.15) and (6.2) that $C_\lambda^{\min}(i) \to 0$ as i approaches the exit boundary. Next take the more general mapping (3.9) where P_λ is an arbitrary admissible solution of the Kolmogorov equation $\lambda P_\lambda = \Omega P_\lambda + I$. Then $\mathbf{C}$ and $\mathbf{C}_\lambda^{\min}$ differ by a bounded solution of $\lambda \mathbf{v} = \Omega \mathbf{v}$, that is, by a linear combination of elements of $\mathfrak{P}_\lambda$. However, there does not exist such an element vanishing on $\bar{\mathfrak{B}}$ (cf. [4] Theorem 13.3) and we have thus

THEOREM 6.2. *Among the admissible solutions of* $\lambda P_\lambda = \Omega P_\lambda + I$ *the matrix* $P_\lambda^{\min}$ *is completely characterized by the following boundary condition: for the vector*

[22] In the present paper only the assertion (a) will be used, and part (b) is proved only for completeness and to justify a more elegant formulation of the theorem.

$\mathbf{C}_\lambda^{\min}$ *defined in* (6.3) *we have* $C_\lambda^{\min}(i) \to 0$ *as* i *approaches a point of the exit boundary.*

Finally let $P^{\min}(t)$ be the Markovian matrix whose Laplace transform is $P_\lambda^{\min}$. Theorem 3.3 obviously applies not only to the vector $\mathbf{1}$ but to any $\mathbf{u} \in \mathfrak{P}$. Combining this with Theorem 5.1 (i) we see that for a vector $\mathbf{u}$ such that $\mathbf{0} \leqq \mathbf{u} \leqq \mathbf{1}$ to satisfy $P^{\min}(t)\mathbf{u} = \mathbf{u}$ for some (and consequently for each) t, it is necessary and sufficient that $\mathbf{u} \in \mathfrak{P}^{(0)}$. We have thus[23]

THEOREM 6.3. *For each* t *the boundary induced by* $P^{\min}(t)$ *is identical with the passive boundary* $\mathfrak{B}^{(0)}$.

7. Probability interpretation

Every stochastic process associated with the Kolmogorov differential equations may be viewed as follows. The matrix Π describes a *road system* connecting the points $i \in E$ and determines the possibility and probability of a movement from i to j *if and when* the "particle" moves from i. The coefficients q_i determine the *speed* of the actual motion. The probability that the sojourn time at i exceeds t is $e^{-q_i t}$. For a particle which at time 0 is at i the probability density for its leaving i at time t is $q_i e^{-q_i t}$ and its Laplace transform is $q_i(\lambda + q_i)^{-1}$. Thus[24] $\Pi_\lambda(i, j)$ is the Laplace transform of the probability density that the particle moves for the first time at t and moves to j. From the convolution theorem we now see that:

$\Pi_\lambda^n(k, j)$ *is the Laplace transform of the probability density of the following compound event: a "particle" which at time* 0 *is assumed to be at* k, *moves within* $(0, t)$ *exactly* $n - 1$ *times, the* n^{th} *jump occurs at time* t *and leads to* j.

Obviously, then $\Pi_\lambda^n(i, j)(\lambda + q_j)^{-1}$ is the Laplace transform of the probability that at time t the particle is at j and that exactly n jumps have occurred. Thus the quantity $P_\lambda^{(n)}(i, j)$ defined in (4.2) describes the probability of a *passage from* i *at time* 0 *to* j *at time* t *such that the number of jumps does not exceed* n. Finally, $P^{\min}(t; i, j)$ *is the probability of a passage from* i *to* j *in finitely many jumps.*[25]

This explains the minimal character of $P_\lambda^{\min}$. In fact, it will be seen presently that if infinitely many jumps occur within a finite time, the particle reaches the exit boundary. The several solutions of the Kolmogorov differential equations

[23] Remember that only non-recurrent i contribute to the boundary. The powers of $P^{\min}(t)$ are given by $P^{\min}(nt)$, and i is recurrent if, and only if, $\sum_n P^{\min}(nt; i, i) = \infty$. In view of the ergodic properties the sum may be replaced by the corresponding integral, which is formally $P_0^{\min}(i, i)$. Thus *i is recurrent for $P^{\min}(t)$ if, and only if, it is recurrent for Π.*

[24] An attractive *alternative interpretation* of Π_λ is as follows. Modify the process so as to attribute to the "particle" a constant probability density of *dying*. That is, the process has finite duration and the probability that it exceeds t is, by definition, $e^{-\lambda t}$. In this modified process, $\Pi_\lambda(i, j)$ is *the probability of the compound event that the particle (starting from i) ever moves from i* (instead of dying at i) *and the first move leads to j*. In this way Π_λ associates with $P(t)$ the transition probabilities of a *timeless* random walk. [This interpretation has been used by physicists for a long time. I do not know whose the priority is, but I have learned the model from the late G. Placzek.]

[25] Formula (4.10) translated back to probabilities reads: a path leading from i at time 0 to k at time $t + s$, and this in n jumps, must lead in $m \leqq n$ jumps to j at time t and thence to k in $n - m$ jumps. This is formula (52) of [5].

describe various modes of return from the boundary to E, and therefore presuppose infinitely many jumps within a finite time.

For the (discrete) random walk associated with the matrix Π formula (6.1) provides a probability interpretation for the sojourn solution $\mathbf{u}$: let ω be the boundary set corresponding to $\mathbf{u}$ and $\mathfrak{U}$ a neighborhood of ω. Then, according to (6.1) the probability of a particle starting from i to be at the nth step in $\mathfrak{U}$ tends to $u(i)$ as $n \to \infty$. A much stronger statement has been proved in [4], namely that $u(i)$ as well is the probability of finding our particle in $\mathfrak{U}$ at *all* times $n \geqq N(i)$. If $\mathbf{u} \leqq \mathbf{x}^{(0)}$ then $\mathbf{u}$ is a sojourn solution for the matrix $P^{\min}(t)$ for each t, and we have the following

Interpretation of the passive boundary $\mathfrak{B}^{(0)}$. *Let* $\mathbf{u} \leqq \mathbf{x}^{(0)}$ *be a sojourn solution, and* ω *the corresponding set in* $\mathfrak{B}^{(0)}$. *Then (in the process associated with* $P^{\min}(t)$) *the probability of an asymptotic approach to* ω *is given by* $\mathbf{u}$ *in the sense that to each neighborhood* $\mathfrak{U}$ *of* ω *there corresponds a* $T(i, \varepsilon) > 0$ *such that*[26]

$$(7.1) \qquad \sum\nolimits_{j \in \mathfrak{U}} P^{\min}(t; i, j) > 1 - \varepsilon \qquad \textit{for } t > T(i, \varepsilon).$$

We have here an asymptotic approach, but the boundary itself is never reached. Quite different is the situation for the *exit boundary*. Let $\mathbf{u} \in \mathfrak{P}$ be an arbitrary sojourn solution for Π. For the corresponding sojourn solution $\mathbf{u}_\lambda$ for Π_λ, as defined in (5.6), we have

$$(7.2) \qquad u_\lambda(i) = \lim_{n\to\infty} \sum\nolimits_j \Pi_\lambda^n(i, j) u(j).$$

Now $\lambda^{-1}\Pi_\lambda^n(i, j)$ is the Laplace transform of the probability that the nth jump occurs before time t and takes the particle to j. In (7.2) this quantity is multiplied by the probability $u(j)$ of an asymptotic approach to the boundary set ω which corresponds to $\mathbf{u}$. Thus for an arbitrary neighborhood $\mathfrak{U}$ of ω we see that $\lambda^{-1}u_\lambda(i)$ gives the probability of the compound event that (starting from i) an infinite number of jumps have occurred before time t and that, after finitely many steps, the particle has reached $\mathfrak{U}$ and stayed in $\mathfrak{U}$. In other words, the particle *has reached the boundary set* ω. Now if $\mathbf{u} \leqq \mathbf{x}^{(0)}$ then $\mathbf{u}_\lambda = \mathbf{0}$. We have thus:

For any sojourn solution $\mathbf{u} \in \mathfrak{P}$ *the solution* $\mathbf{u}_\lambda$ *defined in* (7.2) *represents the Laplace-Stieltjes transform of the probability* $F(t, i)$ *that, starting from* i, *the boundary set* ω *corresponding to* $\mathbf{u}$ *will be reached before time* t. *The passive boundary can never be reached. The probability of reaching the exit boundary before time* t *is* $\bar{\mathbf{x}}_\lambda(i)$, *in obvious agreement with* (4.13).

Needless to emphasize that the column vector $\mathbf{F}(t)$ satisfies the homogeneous differential equation $\mathbf{F}' = \Omega\mathbf{F}$ with the initial condition $\mathbf{F}(0) = \mathbf{0}$.

8. The instantaneous return process

Let P_λ be a matrix such that each column of P_λ differs from the corresponding column of $P_\lambda^{\min}$ by a solution of $\lambda\mathbf{u} - \Omega\mathbf{u} = \mathbf{0}$. Then P_λ obviously satisfies the

[26] In full rigor we have only proved that (7.1) will hold for all $t = n\tau$ for $n \geqq N(i, \varepsilon)$ where τ is *arbitrary, but fixed*. We omit the continuity argument required for a full justification.

Kolmogorov equation $\lambda P_\lambda - \Omega P_\lambda = I$ (though not necessarily the adjoint equation $\lambda P_\lambda - P_\lambda \Omega = I$). We now consider for each fixed λ a matrix with elements of the form

$$P_\lambda(i, k) = P_\lambda^{\min}(i, k) + \mathbf{x}_\lambda(i) \cdot m_\lambda \cdot \mathbf{A}_\lambda(k) \tag{8.1}$$

and inquire into the conditions on $\mathbf{x}_\lambda$, m_λ, $\mathbf{A}_k$ under which (8.1) will represent the Laplace transforms of a Markovian $P(t)$, that is, satisfy (3.2) and (3.5). We shall consider the following situation without at present attempting to prove that no other set-up is possible.

(i). Let ω be an arbitrary point or set of the exit boundary $\mathfrak{B}$ and $\mathbf{x}_\lambda$ the corresponding sojourn solution in $\mathfrak{P}_\lambda$. In other words, $\mathbf{x}_\lambda \in \mathfrak{P}_\lambda$ is the solution with boundary values 1 on ω and 0 on $\mathfrak{B} - \omega$.

(ii). Let $\mathbf{A}_\lambda = \boldsymbol{\alpha} P_\lambda^{\min}$ be the *row-vector* with elements

$$\mathbf{A}_\lambda(k) = \textstyle\sum_i \alpha(i) P_\lambda^{\min}(i, k). \tag{8.2}$$

Here the row-vector $\boldsymbol{\alpha}$ is assumed independent of λ and we suppose that for each $\lambda > 0$

$$\alpha(i) \geqq 0 \qquad \bar{a}(\lambda) \equiv \textstyle\sum \alpha(i)\{1 - \bar{x}_\lambda(i)\} < \infty. \tag{8.3}$$

A glance at (4.15) shows that under these conditions

$$\mathbf{A}_\lambda \geqq \mathbf{0}, \qquad \textstyle\sum_k A_\lambda(k) = \bar{a}(\lambda)\lambda^{-1}. \tag{8.4}$$

It is, of course, *sufficient* that $\sum \alpha(i) < \infty$, but this condition is *not necessary* (cf. the example (17.2) where $\mathbf{x}_\lambda = \bar{\mathbf{x}}_\lambda$).

For an arbitrary *bounded* column vector the transformation $\mathbf{c} \to \mathbf{C}_\lambda = P_\lambda \mathbf{c}$ introduced in (3.9) takes on the form

$$C_\lambda(i) = \textstyle\sum_j P_\lambda^{\min}(i, j)c(j) + x_\lambda(i) \cdot m_\lambda \cdot (\sum_j A_\lambda(j)c(j)). \tag{8.5}$$

Now $x_\lambda(i) \to 1$ as i approaches the boundary set ω and by Theorem 6.2 there exists a limit $C_\lambda(\omega)$ of $C_\lambda(i)$. Rewriting (8.5) in the form

$$\mathbf{C}_\lambda = \mathbf{C}_\lambda^{\min} + \mathbf{x}_\lambda \cdot C_\lambda(\omega) \tag{8.6}$$

(with $\mathbf{C}_\lambda^{\min}$ as defined in (6.3)) we get

$$C_\lambda(\omega) - C_\lambda(i) = -C_\lambda^{\min}(i) + \{1 - x_\lambda(i)\} \cdot C_\lambda(\omega). \tag{8.7}$$

Multiply by $\alpha(i)$ and add to obtain

$$\begin{aligned} \textstyle\sum \alpha(i)\{C_\lambda(\omega) - C_\lambda(i)\} &= -\textstyle\sum A_\lambda(i)c(i) + a(\lambda) \cdot C_\lambda(\omega) \\ &= \left\{-\frac{1}{m_\lambda} + a(\lambda)\right\} \cdot C_\lambda(\omega) \end{aligned} \tag{8.4}$$

where we put for abbreviation

$$a(\lambda) = \textstyle\sum \alpha(i)\{1 - x_\lambda(i)\}. \tag{8.9}$$

We have thus shown that each $P_\lambda \mathbf{c}$ obeys (8.8). Of course, on the part $\mathfrak{B} - \omega$ the boundary values of $\mathbf{C}_\lambda$ are 0. Conversely, suppose that a vector $\mathbf{C}$ in the domain of Ω satisfies the boundary condition (8.8) and has the boundary values 0 on $\bar{\mathfrak{B}} - \omega$. Put $\mathbf{c} = \lambda\mathbf{C} - \Omega\mathbf{C}$ and define $\mathbf{C}_\lambda$ as above. Then $\lambda\mathbf{C} - \Omega\mathbf{C} = \mathbf{c} = \lambda\mathbf{C}_\lambda - \Omega\mathbf{C}_\lambda$ and thus $\mathbf{u} = \mathbf{C} - \mathbf{C}_\lambda$ is a solution of $\lambda\mathbf{u} - \Omega\mathbf{u} = \mathbf{0}$. Since it vanishes on $\bar{\mathfrak{B}} - \omega$ we have $\mathbf{u} \leqq \mathbf{x}_\lambda$. But $\mathbf{u}$ satisfies (8.8); this requires that the boundary values of $\mathbf{u}$ along ω be constant, and so $\mathbf{u}$ is a multiple of $\mathbf{x}_\lambda$. However $\mathbf{x}_\lambda$ does not satisfy (8.8) for any value of $m_\lambda < \infty$, and therefore (8.8) is *characteristic of the range of the transformation* $\mathbf{c} \rightarrow \mathbf{C}_\lambda = P_\lambda\mathbf{c}$. By Theorem 3.2 the resolvent equation (3.5) will hold if, and only if, this *range is independent of* λ, that is, if

$$(8.10)\qquad m_\lambda = \frac{1}{m + a(\lambda)} = \frac{1}{m + \sum \alpha(i)\{1 - x_\lambda(i)\}}$$

where m does not depend on λ.

If $m \geqq 0$ the matrix P_λ will be non-negative. As for the norm-condition, we get from (8.1), (8.4), and the basic equality (4.15)

$$(8.11)\qquad \begin{aligned} \lambda \textstyle\sum_k P_\lambda(i, k) &= 1 - \bar{x}_\lambda(i) + x_\lambda(i) \cdot \frac{a(\lambda)}{m + a(\lambda)} \\ &= 1 - \frac{m}{m + a(\lambda)}\, x_\lambda(i) - \{\bar{x}_\lambda(i) - x_\lambda(i)\}. \end{aligned}$$

We have thus:

THEOREM 8.1. *For each constant* $m \geqq 0$ *the matrix* P_λ *defined by* (8.1) *and* (8.8) *is an admissible solution of the first Kolmogorov equation* $\lambda P_\lambda - \Omega P_\lambda = I$.

The purely stochastic case $\lambda P_\lambda\mathbf{1} = \mathbf{1}$ *arises if, and only if,* $m = 0$ *and* $\omega = \bar{\mathfrak{B}}$ *is the entire exit boundary* (i.e. $\mathbf{x}_\lambda = \bar{\mathbf{x}}_\lambda$). *The range of the transformation* $\mathbf{c} \rightarrow \mathbf{C}_\lambda = P_\lambda\mathbf{c}$ *is characterized by the lateral condition*

$$(8.12)\qquad m\cdot C(\omega) + \textstyle\sum_i \alpha(i)\{C(\omega) - C(i)\} = 0$$

together with the requirement that $C(i) \rightarrow 0$ *as* i *approaches a point of* $\bar{\mathfrak{B}} - \omega$.

Note. Instead of restricting the construction to one single $\mathbf{x}_\lambda$ we could have taken finitely many non-overlapping sets $\omega^{(1)}, \cdots, \omega^{(m)}$ on the exit boundary, and replaced $\mathbf{x}_\lambda$ by $\sum p^{(i)}\mathbf{x}_\lambda^{(i)}$, where the $p^{(i)}$ are arbitrary weights and $\mathbf{x}_\lambda^{(i)}$ the sojourn solution determining $\omega^{(i)}$. More generally, we could have used an arbitrary weight-distribution along the boundary.

It is clear that P_λ does not satisfy the second Kolmogorov equation. To find the analogue to (3.4) we first evaluate the inner product of the ith row of $\lambda P_\lambda - I$ with $\mathbf{x}$, namely

$$(8.13)\qquad \begin{aligned} &\lambda \textstyle\sum_k P_\lambda(i, k)x(k) - x(i) \\ &\quad = \lambda \textstyle\sum_k P_\lambda^{\min}(i, k)x(k) - x(i) + x_\lambda(i)m_\lambda\cdot \sum_{j,k} \alpha(j)P_\lambda^{\min}(i, k)x(k) \\ &\quad = -x_\lambda(i) + x_\lambda(i)m_\lambda\cdot a(\lambda) \\ &\quad = -m\cdot m_\lambda\cdot x_\lambda(i). \end{aligned}$$

Now consider P_λ as an operation on row vectors, that is, as resolvent of *adjoint semi-group*. Let $\varrho \in \mathfrak{R}$ be a row vector and $\mathbf{R}_\lambda = \varrho P_\lambda$ its transform. ̄ range of this transformation is dense in $\mathfrak{R}$, and the operator $\bar{\Omega}$ defined by

$$\mathbf{R}_\lambda \bar{\Omega} = \lambda \mathbf{R}_\lambda - \varrho \tag{8.14}$$

is independent of λ, namely the infinitesimal generator. It is clear from (8 and (8.13) that (using the notation (1.13))

$$\lambda \mathbf{R}_\lambda - \varrho = \mathbf{R}_\lambda \Omega - m \cdot [\lambda \mathbf{R}_\lambda - \varrho, \mathbf{x}]\boldsymbol{\alpha} \tag{8.15}$$

or

$$\mathbf{R}_\lambda \bar{\Omega} = \mathbf{R}_\lambda \Omega - m[\mathbf{R}_\lambda \bar{\Omega}, \mathbf{x}]\boldsymbol{\alpha}. \tag{8.16}$$

Applying this, in particular, to the ith row of P_λ we get

$$\begin{aligned}(P_\lambda \bar{\Omega})(i, k) = &-P_\lambda(i, k)q_k + \textstyle\sum_j P_\lambda(i, j)q_j\Pi(j, k) \\ &- m \cdot \alpha(k) \cdot \textstyle\sum_j P_\lambda\bar{\Omega}(i, j)x(j).\end{aligned} \tag{8.17}$$

Clearly (8.16) defines $\bar{\Omega}$ uniquely, and we have

THEOREM 8.2. *The analogue of the second Kolmogorov equation* (3.4) *for matrix P_λ of Theorem* 8.1 *is given by* (8.17).

8a. Probabilistic interpretation and direct construction

We give a simple interpretation of the process of the last Theorem, but o for the case $\sum \alpha(i) = 1$. Consider a Markovian process defined as follows:

A "particle" starts from i and moves in accordance with the $P^{\min}(t)$ process and when the boundary is reached at a point ω_0 the process stops whenever $\omega_0 \in \mathfrak{B} -$ If $\omega_0 \in \omega$ the process stops with probability $m/(m + 1)$ and continues with pı ability $p = 1/(m + 1)$. In the latter case a new starting point k is chosen sub to the probability distribution $\{\alpha(k)\}$, and the process starts from scratch.

Consider the minimal process starting at i, and denote by $p_i(t)$ the probabi density of the event that the boundary set ω is reached at time t. The Lapl transform of $p_i(t)$ is $x_\lambda(i)$. If the initial position is chosen randomly accord to the probability distribution $\boldsymbol{\alpha}$, the corresponding probability density $\sum \alpha(i)p_i(t)$, and its Laplace transform $\sum \alpha(i)x_\lambda(i)$. Now our compound proc coincides with the minimal process until the boundary is reached and when t happens it may start from scratch. The probability that before t the bound is reached exactly n times and at time t the actual position is k can be calcula by an $(n + 1)$-fold convolution, and has the Laplace transform

$$x_\lambda(i) \cdot (p \textstyle\sum_j \alpha(j)x_\lambda(j))^{n-1} \cdot A_\lambda(k).$$

Summing over $n = 1, 2, \cdots$ we see that the probability of the position k time t after a finite number of crossings of the boundary has the Laplace tra form

$$P_\lambda^{\min}(i, k) + x_\lambda(i) \cdot \frac{x_\lambda(i)A_\lambda(k)}{1 - p \sum \alpha(j)x_\lambda(j)}$$

which is identical with the expression (8.1) for $P_\lambda(i, k)$.

9. Extension to the boundary

The arbitrariness of the procedure outlined in Section 8 is, perhaps, best seen if we write the transformation (8.5) in the form

$$\mathbf{C}_\lambda = \mathbf{C}_\lambda^{\min} + \mathbf{x}_\lambda \cdot \Phi_\lambda(\mathbf{c}) \tag{9.1}$$

where Φ_λ is a certain functional. Even with an *arbitrary* choice of this functional $\mathbf{C}_\lambda$ will be a solution of $\lambda\mathbf{C}_\lambda - \Omega\mathbf{C}_\lambda = \mathbf{c}$, and we should find the most general functional compatible with our requirements.

Whatever the functional Φ_λ, *the range of the transformation* (9.1) *contains only vectors with boundary values* 0 *along* $\bar{\mathfrak{B}} - \omega$ *and constant along* ω. *In the present section, therefore, we shall restrict the consideration to the Banach space* $\mathfrak{C}_\omega$ of bounded vectors with these properties, and suppose that $\mathbf{c} \in \mathfrak{C}_\omega$. It appears then that the only eligible functional is of the form $\sum A(i)c(i) + Ac(\omega)$ where $c(\omega)$ is the value of c assumed on ω. Accordingly, we propose to generalize the procedure of Section 8 to transformations of the form (9.1) where Φ_λ is a functional such that

$$\frac{1}{m_\lambda} \Phi_\lambda(c) = \textstyle\sum_k A_\lambda(k)c(k) + A_\lambda(\omega)c(\omega), \tag{9.2}$$

the components $A_\lambda(k)$ being defined[27] by (8.2)–(8.3).

It is now to be noticed that in order to represent our transformation by means of a matrix P_λ *we must add a row and a column for* ω.

Accordingly, we consider instead of (8.1) the enlarged matrix

$$\begin{aligned} P_\lambda(i, k) &= P_\lambda^{\min}(i, k) + x_\lambda(i)\cdot m_\lambda \cdot A_\lambda(k) \\ P_\lambda(i, \omega) &= x_\lambda(i)\cdot m_\lambda\cdot A_\lambda(\omega) \\ P_\lambda(\omega, k) &= m_\lambda \cdot A_\lambda(k), \qquad P_\lambda(\omega, \omega) = m_\lambda A_\lambda(\omega) \end{aligned} \tag{9.3}$$

where $i, k = 1, 2, \cdots$. The transformation (9.1)–(9.2) can now again be written in the form $\mathbf{C}_\lambda = P_\lambda \mathbf{c}$ and we are once more able to proceed as in Section 8.

Equations (8.6) and (8.7) remain valid without change, and so does the first part in (8.8). However, this time

$$A_\lambda(i)c(i) = -A_\lambda(\omega)c(\omega) + \frac{1}{m_\lambda} C_\lambda(\omega) \tag{9.4}$$

and so (8.8) reads

$$\sum \alpha(i)\{C_\lambda(\omega) - C_\lambda(i)\} = \left\{-\frac{1}{m_\lambda} + a(\lambda)\right\} \cdot C_\lambda(\omega) + A_\lambda(\omega)c(\omega). \tag{9.5}$$

[27] It should be proved that no other form for A_λ is possible. Here this fact is accepted by analogy with one-dimensional diffusion where a rigorous proof exists.

We are not yet acquainted with the value of $\Omega\mathbf{C}_\lambda$ at ω but all vectors are continuous at ω and so

$$\lambda C_\lambda(\omega) - \Omega C_\lambda(\omega) = c(\omega). \tag{9.6}$$

If this is substituted back into (9.5) we get an equation which *determines the range of our transformation* and plays the role of (8.8). As in Section 8 we conclude that this range must be independent of λ, that is, the coefficients must stand in a constant proportion. We must therefore have

$$A_\lambda(\omega) = p \qquad m_\lambda = \frac{1}{m + \lambda p + a(\lambda)}, \tag{9.7}$$

and the lateral condition now assumes the form

$$p\Omega C(\omega) + m\cdot C(\omega) + \textstyle\sum_{i=1}^{\infty} \alpha(i)\{C(\omega) - C(i)\} = 0. \tag{9.8}$$

Exactly as in Section 8 we verify that (9.8) really guarantees the resolvent equation and the norm condition, and we have thus

THEOREM 9.1. *For arbitrary constants $m \geqq 0$, $p \geqq 0$ the matrix P_λ defined by* (9.3) *and* (9.7) *represents the Laplace transform of a Markovian $P(t)$ defined on the set of states* $1, 2, \cdots, \omega$. *The purely stochastic case* $\lambda \sum_k P_\lambda(i, k) + \lambda P_\lambda(i, \omega) = 1$ (*where* $i = 1, 2, \cdots, \omega$) *arises if and only if* $m = 0$ *and* $\omega = \mathfrak{B}$.

It will be noticed that this time no mention was made of lateral conditions. The reason is that (9.8) *does not represent a lateral condition except when* $p = 0$ (which is the case covered by Theorem 8.1). In fact, from the very construction of $\mathbf{C}_\lambda$ we know that for $i = 1, 2, \cdots$ we have $\lambda\mathbf{C}_\lambda - \Omega\mathbf{C}_\lambda = \mathbf{c}$ where the operator Ω is defined as before (that is, does not involve the value $C_\lambda(\omega)$). But for $i = \omega$ equation (9.8) simply defines the meaning of Ω. We state this in the form

THEOREM 9.2. *The matrix P_λ of the preceding theorem satisfies the following equations* (*extending the Kolmogorov equations* (3.3) *to* ω)

$$\begin{aligned} \lambda P_\lambda(i, k) &= -q_i P_\lambda(i, k) + q_i \textstyle\sum_{j=1}^{\infty} \Pi(i, j) P_\lambda(j, k) + \delta_{ik} \\ \lambda P_\lambda(\omega, k) &= -\frac{m}{p} P_\lambda(\omega, k) - \frac{1}{p} \textstyle\sum_{j=1}^{\infty} \alpha(j)[P_\lambda(\omega, k) - P_\lambda(j, k)] + \delta_{\omega,k} \end{aligned} \tag{9.9}$$

where $k = 1, 2, \cdots, \omega$.

In particular, if $\sum \alpha(j) = 1$ *the last equation may be rewritten in the form*

$$\lambda P_\lambda(\omega, k) = -\frac{m+1}{p} P_\lambda(\omega, k) + \frac{1}{p} \textstyle\sum_{j=1}^{\infty} \alpha(j) P_\lambda(j, k) + \delta_{\omega,k} \tag{9.10}$$

which is of the canonical form with

$$\frac{m+1}{p} = q_\omega \qquad \frac{1}{m+1}\,\alpha(j) = \Pi(\omega, j). \tag{9.11}$$

It remains to calculate the derivatives $q_{i\omega}$, $q_{\omega k}$, $q_{\omega\omega}$ defined in (1.1) and corresponding to the Markovian $P(t)$ determined by our P_λ. There is no doubt[28]

[28] The continuity can be proved as in Section 4 but is evident from (9.3): we know that $x_\lambda(i)$ and $A_\lambda(k)$ are Laplace transforms of positive continuous functions. Now m_λ is clearly

about the continuity of the elements of $P(t)$ and therefore we can determine, in the familiar manner, the behavior of $P(t)$ near 0 from the behavior of P_λ as $\lambda \to \infty$. From the definition (8.9) we have

$$\lim_{\lambda\to\infty} a(\lambda) = \sum \alpha(i) \leqq \infty \tag{9.12}$$

and using (5.7) to calculate the derivative of $\mathbf{x}_\lambda$ we find also

$$\frac{d}{d\lambda} a(\lambda) = \boldsymbol{\alpha} P_\lambda^{\min} \mathbf{x}_\lambda = \sum_j A_\lambda(j) x_\lambda(j). \tag{9.13}$$

Here each term decreases monotonically to 0 and so $a'(\lambda) \to 0$ and hence $a(\lambda) = \mathbf{o}(\lambda)$. Substituting this into the defining relations (9.3), (9.7) we get

$$\lim_{\lambda\to\infty} \lambda P_\lambda(\omega, \omega) = \lim_{\lambda\to\infty} \frac{\lambda p}{m + \lambda p + a(\lambda)} = 1 \tag{9.14}$$

$$\lim_{\lambda\to\infty} \lambda\{1 - \lambda P_\lambda(\omega,\omega)\} = \lim_{\lambda\to\infty} \frac{\lambda\cdot\{m + a(\lambda)\}}{m + \lambda p + a(\lambda)} = \frac{m}{p} + \frac{1}{p} \sum \alpha(i) \leqq \infty \tag{9.15}$$

$$\lim_{\lambda\to\infty} \lambda^2 P_\lambda(\omega, k) = \lim_{\lambda\to\infty} \lambda A_\lambda(k) \cdot \frac{\lambda}{m + \lambda p + a(\lambda)} = \frac{\alpha(k)}{p}. \tag{9.16}$$

We have thus proved

THEOREM 9.3. *For the Markovian $P(t)$ defined by P_λ we have*

$$\lim_{t\to 0} P(t; \omega, \omega) = 1$$

$$q_\omega = \lim_{t\to 0} t^{-1}\{1 - P(t; \omega, \omega)\} = mp^{-1} + p^{-1} \sum \alpha(i) \leqq \infty \tag{9.17}$$

$$q\omega_k = \lim_{t\to 0} t^{-1} P(t; \omega, k) = p^{-1}\alpha(k).$$

To sum up: if $\sum \alpha(i) < \infty$ the row corresponding to ω looks just like any other row in the scheme (1.3) or (1.6). However, even in this case ω enjoys the peculiarity that $\Pi(i, \omega) = 0$ for all i, that is, *ω can not be reached in finitely many steps from anywhere.*

In general we have $q_\omega = \infty$, $\sum q_{\omega k} = \infty$. Then the state ω is *instantaneous in the terminology of P. Lévy* [16]: *the time which the "particle" spends at ω is a set of positive measure but nowhere dense.* (Once arrived, the particle leaves ω at infinite speed, only to return at an equally fast rate.)

The analogue to the second Kolmogorov equation (3.4) is obtained just as in Theorem 8.2, but takes on a more intelligible form.

THEOREM 9.4. *The matrix P_λ of* (9.3) *satisfies the following analogue to the second Kolmogorov equation* (3.4): *For $k = 1, 2, \cdots$ and $i = 1, 2, \cdots, \omega$*

$$\begin{aligned} \lambda P_\lambda(i, k) - \delta_{ik} &= -P_\lambda(i, k) q_k + \sum_j P_\lambda(i, j) q_j \Pi(j, k) + \frac{\alpha(k)}{p} P_\lambda(i, \omega) \\ \lambda P_\lambda(i, \omega) - \delta_{i\omega} &= -\frac{m}{p} P_\lambda(i, \omega) - \sum_j (P_\lambda \Omega)(i, j) x(j). \end{aligned} \tag{9.18}$$

completely monotonic and therefore the Laplace-Stieltjes transform of a monotone function. The convolution of the three functions is continuous.

Clearly for $k \neq \omega$ the new term is equivalent to writing the equations in the old form by defining $q_\omega \Pi(\omega, k) = (\alpha(k))/p$. The last equation, as we have seen, is reducible to the classical form if α is bounded.

The next point in our program consists in replacing $\mathbf{A}_\lambda$ by a solution $\boldsymbol{\xi}$ of the homogeneous equation $\lambda\boldsymbol{\xi} - \boldsymbol{\xi}\Omega = 0$. Then P_λ will satisfy not only $\lambda P_\lambda - \Omega P_\lambda = I$ but also $\lambda P_\lambda - P_\lambda \Omega = I$. We shall have $q_{\omega k} = 0$ $(k \neq \omega)$ so that ω will lack direct communications both from and to all other points.

10. Solutions of $\lambda\mathfrak{n} - \mathfrak{n}\Omega = 0$ (with $\mathfrak{n}$ a row vector)

We turn to the study of the linear equations adjoint to those of Section 5. The principal difference is due to the circumstance that before the vector $\mathbf{1}$ provided a natural upper bound and also a solution of $\Pi\mathbf{u} = \mathbf{u}$; no standard vector of this sort is available for our present purposes.

DEFINITION[29] 10.1. *$\mathfrak{P}^*$ will stand for the set of all row vectors such that*

$$\mathfrak{n} \geqq \mathbf{0} \qquad \mathfrak{n}\mathfrak{q} = \mathfrak{n}\mathfrak{q}\Pi. \tag{10.1}$$

Similarly, $\mathfrak{P}^*_\lambda$ stands for the set of vectors satisfying the conditions

$$\mathfrak{n} \geqq \mathbf{0} \qquad \textstyle\sum_k \eta(k) < \infty, \qquad \mathfrak{n}\{\lambda + \mathfrak{q}\} = \mathfrak{n}\mathfrak{q}\Pi.) \tag{10.2}$$

Note that $\mathfrak{P}^*$ includes unbounded vectors, $\mathfrak{P}^*_\lambda$ is restricted to bounded vectors. This difference is important.

LEMMA 10.1. *If $\mathfrak{n} \in \mathfrak{P}^*$ then for each $\lambda > 0$*

$$\lambda\mathfrak{n}P_\lambda^{\min} = \mathfrak{n} - \mathfrak{n}_\lambda \tag{10.3}$$

where $\mathfrak{n}_\lambda \geqq \mathbf{0}$ is a solution of $\mathfrak{n}_\lambda\{\lambda + \mathfrak{q}\} = \mathfrak{n}_\lambda\mathfrak{q}\Pi$ (but not necessarily bounded). Moreover

$$\mathfrak{n}_\lambda - \mathfrak{n}_\nu = (\nu - \lambda)\mathfrak{n}_\nu P_\lambda^{\min}. \tag{10.4}$$

PROOF. Following the pattern of the proof of (5.6) let us premultiply (4.7) by $\mathfrak{n}$ to obtain

$$\lambda\mathfrak{n}P_\lambda^{(n)} + \mathfrak{n}\mathfrak{q}\Pi_\lambda^n(\lambda + \mathfrak{q})^{-1} = \mathfrak{n}. \tag{10.5}$$

Thus $\mathfrak{n}\mathfrak{q}\Pi_\lambda^n \downarrow \boldsymbol{\xi}$ where $\boldsymbol{\xi} = \boldsymbol{\xi}\Pi_\lambda$. Put $\boldsymbol{\xi} = \mathfrak{n}_\lambda\{\lambda + \mathfrak{q}\}$ and (10.3) follows. The resolvent equation (3.5) leads directly to (10.4).

We turn to the analogue of Lemma 5.1. It is not a priori clear that the series $\sum \eta_\lambda(i)\Gamma(i, k)$ converges for each $\mathfrak{n}_\lambda \in \mathfrak{P}^*_\lambda$ and each k. However, Lemma 5.1 applied to $\bar{\mathbf{x}}$ shows that

$$\bar{\mathbf{x}}_\lambda + \lambda\Gamma\bar{\mathbf{x}}_\lambda = \bar{\mathbf{x}} \tag{10.6}$$

and we are assuming that $\bar{\mathbf{x}}$ is strictly positive (cf. end of Section 5). The same is then true of $\bar{\mathbf{x}}_\lambda$. Since $\bar{\mathbf{x}} \leqq \mathbf{1}$ the inner product $\mathfrak{n}_\lambda\bar{\mathbf{x}} = \sum \eta_\lambda(i)\bar{x}(i) < \infty$ and we may premultiply (10.6) by $\mathfrak{n}_\lambda$ to see that $\mathfrak{n}_\lambda\Gamma$ *is a finite* (not necessarily bounded) row-vector.

[29] The interpretation in terms of an *entrance boundary* is postponed to Section 15.

THEOREM 10.1. *Let* $\mathfrak{n}_\lambda \in \mathfrak{P}_\lambda^*$ *for some fixed* $\lambda > 0$. *Then (always assuming that* $\bar{\mathbf{x}}$ *is strictly positive)*

$$\mathfrak{n}_\lambda + \lambda \mathfrak{n}_\lambda \Gamma = \mathfrak{n} \qquad \mathfrak{n} \in \mathfrak{P}^*. \tag{10.7}$$

Moreover

$$\lambda \mathfrak{n} P_\lambda^{\min} = \mathfrak{n} - \mathfrak{n}_\lambda . \tag{10.8}$$

If, keeping this $\mathfrak{n}$ *fixed, one lets* λ *vary in* (10.8) *then each* $\mathfrak{n}_\lambda$ *so defined will satisfy* (10.7) *and* $\mathfrak{n}_\lambda \in \mathfrak{P}_\lambda^*$.

PROOF. From the definition

$$\mathfrak{n}_\lambda + \lambda \mathfrak{n}_\lambda \mathfrak{q}^{-1} = \mathfrak{n}_\lambda \mathfrak{q} \Pi \mathfrak{q}^{-1}. \tag{10.9}$$

Postmultiply by $\mathfrak{q}\Pi^a\mathfrak{q}^{-1}$ to get on summing

$$\mathfrak{n}_\lambda + \lambda \mathfrak{n}_\lambda \textstyle\sum_{a=0}^{n-1} \Pi^a \mathfrak{q}^{-1} = \mathfrak{n}_\lambda \mathfrak{q} \Pi^n \mathfrak{q}^{-1}. \tag{10.10}$$

We know that the left side converges monotonically to that of (10.7). Consequently $\mathfrak{n}_\lambda \mathfrak{q}\Pi^n\mathfrak{q}^{-1} \uparrow \mathfrak{n}$ where $\mathfrak{n}\mathfrak{q}\Pi\mathfrak{q}^{-1} = \mathfrak{n}$, that is, $\mathfrak{n} \in \mathfrak{P}^*$. Next define $\mathfrak{n}_\lambda$ by (10.8). One sees from (10.4) directly that $\sum \eta_\lambda(k) < \infty$ for all λ (since we have assumed it for one particular value of λ). Finally, (10.7) follows by the simple argument which led to Lemma 5.4.

DEFINITION 10.2. *The vector* $\mathfrak{n}$ *defined in* (10.7) *is the canonical image of* $\mathfrak{n}_\lambda$. *The set of the canonical images of all vectors in* $\mathfrak{P}_\lambda^*$ *is independent of* λ *and will be denoted by* $\bar{\mathfrak{P}}^*$.

As in the case of the $\mathfrak{P}_\lambda$ we have a one-to-one correspondence (lattice isomorphism) between the elements of $\mathfrak{P}_\lambda^*$ for different λ simply by pairing the vectors with the same canonical image.

THEOREM 10.2. *For each* λ *let* $\bar{\mathbf{x}}_\lambda \in \mathfrak{P}_\lambda$ *and* $\mathfrak{n}_\lambda \in \mathfrak{P}_\lambda^*$ *be elements with fixed canonical images* $\mathbf{x} \leqq \bar{\mathbf{x}}$ *and* $\mathfrak{n} \in \mathfrak{P}^*$, *respectively. Then the several inner products satisfy the following conditions*[30]

$$\mathfrak{n}\mathbf{x}_\lambda = \mathfrak{n}_\lambda \mathbf{x} \tag{10.11}$$

$$\lambda \mathfrak{n}\mathbf{x}_\lambda \uparrow L \leqq \infty \qquad \lambda \mathfrak{n}_\lambda \mathbf{x}^{(0)} = c = \text{const.} \tag{10.12}$$

$$\lambda \textstyle\sum_k \eta_\lambda(k) = c + \lambda \mathfrak{n}\bar{\mathbf{x}}_\lambda . \tag{10.13}$$

PROOF. Postmultiplying (10.7) by $\mathbf{x}_\lambda$ and remembering that $\mathbf{x}_\lambda + \lambda\Gamma\mathbf{x}_\lambda = \mathbf{x}$ we get (10.11) (implying in particular, that $\mathfrak{n}\bar{\mathbf{x}}_\lambda < \infty$). Next we see, postmultiplying (10.4) by $\mathbf{x}^{(0)}$, that $\lambda(\mathfrak{n}_\lambda\mathbf{x}^{(0)} - \mathfrak{n}_\nu\mathbf{x}^{(0)}) = (\nu - \lambda)\mathfrak{n}_\nu \lambda P_\lambda^{\min}\mathbf{x}^{(0)} = (\nu - \lambda)\mathfrak{n}_\nu\mathbf{x}^{(0)}$ or $\lambda\mathfrak{n}_\lambda\mathbf{x}^{(0)} = \nu\mathfrak{n}_\nu\mathbf{x}^{(0)}$ as asserted in (10.12). The same calculations with $\mathbf{x}^{(0)}$ re-

[30] It will be seen in Section 13 that a pair $\mathfrak{n}$, $\mathbf{x}$ such that $L = \infty$ plays an entirely different role than does a pair for which $L < \infty$. As $\mathbf{x}$ determines a set ω of the *exit* boundary, so does $\mathfrak{n}$ induce a set γ on the *entrance* boundary (Section 15). It would be *interesting to investigate the implication of* $L = \infty$ *for the relationship between the neighborhoods of* ω *and* γ. (It appears probable that $L < \infty$ implies that there exists non-overlapping neighborhoods, i.e. that ω and γ are essentially independent. See also the note to Theorem 14.2.)

placed by $\mathbf{x}$ shows that $\lambda \mathfrak{n}_\lambda \mathbf{x} \leqq \nu \mathfrak{n}_\nu \mathbf{x}$ for $\nu \geqq \lambda$. Finally, (10.13) follows from the preceding since $\sum \eta_\lambda(k) = \mathfrak{n}_\lambda \mathbf{1} = \mathfrak{n}_\lambda(\mathbf{x}^{(0)} + \bar{\mathbf{x}})$.

11. The "normal derivative" analogue

Let $\mathfrak{n} \in \bar{\mathfrak{P}}^*$, that is, let (10.3) hold with $\mathfrak{n}_\lambda \neq \mathbf{0}$ and $\mathfrak{n}_\lambda \in \mathfrak{P}_\lambda^*$. Let $\mathbf{C}$ be a column vector such that $\Omega\mathbf{C}$ exists and is bounded. Since $\mathfrak{n}$ need not be bounded, the formal inner product

$$[\mathfrak{n}, \Omega\mathbf{C}] = \sum_i \eta(i)\Omega C(i) \tag{11.1}$$

seems to be meaningless. However, if $\mathbf{C}$ happens to be of the form $\mathbf{C} = P_\lambda^{\min}\mathbf{c}$ where $\mathbf{c}$ has a compact support (i.e. only finitely many $c(i)$ are non-zero), then the following manipulations are legitimate:

$$[\mathfrak{n}, \Omega\mathbf{C}] = [\mathfrak{n}, \lambda P_\lambda^{\min}\mathbf{c} - \mathbf{c}] = [\mathfrak{n} - \mathfrak{n}_\lambda, \mathbf{c}] - [\mathfrak{n}, \mathbf{c}] = -[\mathfrak{n}_\lambda, \mathbf{c}]. \tag{11.2}$$

By the definition (10.2) the vector $\mathfrak{n}_\lambda$ is bounded, and therefore $[\mathfrak{n}_\lambda, \mathbf{c}]$ makes sense for all $\mathbf{c}$. Thus it seems possible to give a meaning to (11.1) by an appropriate passage to the limit. However, it is more convenient to use the relations (11.2) to *define* $[\mathfrak{n}, \Omega\mathbf{C}]$.

LEMMA 11.1. *Let* $\mathbf{C}$ *be in the range of* $P_\lambda^{\min}$ *(for some and therefore all* λ*) and put*

$$\lambda\mathbf{C} - \Omega\mathbf{C} = \mathbf{c}_\lambda . \tag{11.3}$$

Then $[\mathfrak{n}_\lambda, \mathbf{c}_\lambda]$ *is independent of* λ.

PROOF. From (10.4) we get at once

$$\begin{aligned} [\mathfrak{n}_\lambda, \mathbf{c}_\lambda] &= [\mathfrak{n}_\nu + (\nu - \lambda)\mathfrak{n}_\nu P_\lambda^{\min}, \mathbf{c}_\lambda] \\ &= [\mathfrak{n}_\nu, \mathbf{c}_\lambda + (\nu - \lambda)P_\lambda^{\min}\mathbf{c}_\lambda] \\ &= [\mathfrak{n}_\nu, \mathbf{c}_\lambda + (\nu - \lambda)\mathbf{c}] = [\mathfrak{n}_\nu, \mathbf{c}_\nu] \end{aligned} \tag{11.4}$$

as asserted.

We next show that the general case can be reduced to the preceding one.

LEMMA 11.2. *Let* $\mathbf{0} \leqq \mathbf{C} \leqq \mathbf{1}$ *and let* $\Omega\mathbf{C}$ *be bounded. There exists a solution* $\mathbf{u} \in \bar{\mathfrak{P}}$ *such that*

$$\mathbf{C} - \mathbf{u} = P_\lambda^{\min}(\mathbf{c}_\lambda - \lambda\mathbf{u}) \tag{11.5}$$

where $\mathbf{c}_\lambda$ *is defined by* (11.3). *Thus*

$$\mathbf{C} = P_\lambda^{\min}\mathbf{c}_\lambda + \mathbf{u}_\lambda \tag{11.6}$$

where $\mathbf{u}_\lambda \in \mathfrak{P}_\lambda$ *is the element with canonical image* $\mathbf{u}$.

Since $\mathbf{u}$ is automatically continuous along the exit boundary (and $\mathbf{u}_\lambda$ has the same boundary values) (11.5) expresses simply that *each* $\mathbf{C}$ *in the domain of* Ω *has continuous boundary values along the exit boundary* $\mathfrak{B}$.

PROOF. It is clear that (11.6) holds with some $\mathbf{u}_\lambda \geqq \mathbf{0}$. Define $\mathbf{u}$ as the canonical image of $\mathbf{u}_\lambda$, and (11.5) follows from (11.6).

We can now apply Lemma 11.1 to $\mathbf{C} - \mathbf{u}$ observing that $\Omega(\mathbf{C} - \mathbf{u}) = \Omega\mathbf{C}$. This leads us to

THEOREM (AND DEFINITION) 11.1. *Let* $\Omega\mathbf{C}$ *be bounded and* $\mathfrak{n} \in \bar{\mathfrak{P}}^*$. *We set*

$$[\mathfrak{n}, \Omega\mathbf{C}] = -[\mathfrak{n}_\lambda, \mathbf{c}_\lambda] + \lambda[\mathfrak{n}_\lambda, \mathbf{u}] \tag{11.7}$$

where $\mathbf{c}_\lambda = \lambda\mathbf{C} - \Omega\mathbf{C}$ *and* $\mathbf{u}$ *is the solution of* $\Omega\mathbf{u} = \mathbf{u}$ *which agrees with* $\mathbf{C}$ *along the exit boundary. The right side of* (11.7) *is independent of* λ.

In view of Theorem 10.2 the last equation may be rewritten as

$$[\mathfrak{n}, \Omega\mathbf{C}] = -[\mathfrak{n}_\lambda, \mathbf{c}_\lambda] + \lambda[\mathfrak{n}, \mathbf{u}_\lambda] = -\mathfrak{n}_\lambda\mathbf{c}_\lambda + \lambda\mathfrak{n}\mathbf{u}_\lambda . \tag{11.8}$$

Note that the right sides in (11.7) and (11.8) involve only absolutely convergent inner products.

Note. Perhaps the simplest counterpart to the two Kolmogorov equations (3.3) and (3.4) is the Sturm-Liouville equation

$$\lambda F - F'' = f \qquad \left(\Omega = \frac{d^2}{dx^2}\right) \tag{11.9}$$

on the interval (0, 1). (Because of the symmetric nature of Ω the two adjoint equations here become identical.) In the present example $\bar{\mathfrak{P}}^*$ becomes the set of non-negative solutions of $\eta'' = 0$, that is, $\bar{\mathfrak{P}}^*$ is the set of positive linear combinations of x and $1 - x$. Accordingly, we have at disposal only the two independent functionals $[x, F'']$ and $[1 - x, F'']$. Now

$$[x, F''] = \int_0^1 xF''(x)\, dx = F'(1) - \{F(1) - F(0)\}, \tag{11.10}$$

and we see how the introduction of $[x, F'']$ is equivalent to the use of the derivatives occurring in the classical boundary conditions of the Sturm-Liouville theory. Here the two methods of writing are still equivalent. In the theory of *harmonic functions*, however, the normal derivatives occurring in the boundary conditions exist only when the boundary is sufficiently smooth. They can *always* be replaced by operators of the form, $[\eta, \Delta F]$, and the latter lend themselves also to a formulation of more general boundary conditions.

12. Regular solutions constant along the exit boundary

Let $\bar{\mathbf{x}}$ and $\bar{\mathbf{x}}_\lambda$ as in Section 6 be the maximal elements of $\bar{\mathfrak{P}}$ and $\mathfrak{P}_\lambda$, respectively. Furthermore, let $\mathfrak{n} \in \bar{\mathfrak{P}}^*$ be fixed and let $\mathfrak{n}_\lambda$ be the element of $\mathfrak{P}_\lambda^*$ corresponding to $\mathfrak{n}$.

In this section we shall construct admissible matrices of the form

$$P_\lambda^{\min} + m_\lambda \cdot \bar{\mathbf{x}}_\lambda \mathfrak{n}_\lambda ,$$

that is, matrices with elements

$$P_\lambda(i, k) = P_\lambda^{\min}(i, k) + \bar{x}_\lambda(i) \cdot m_\lambda \cdot \eta_\lambda(k) \tag{12.1}$$

where m_λ is a scalar. The transformation $\mathbf{c} \to \mathbf{C}_\lambda = P_\lambda\mathbf{c}$ takes on the form

$$\mathbf{C}_\lambda = \mathbf{C}_\lambda^{\min} + \bar{\mathbf{x}}_\lambda \cdot m_\lambda[\mathfrak{n}_\lambda, \mathbf{c}], \tag{12.2}$$

and it is seen that *as i tends to the exit boundary, $c_\lambda(i)$ converges to a limit*[31] *which we denote by*

$$C_\lambda(\omega) = m_\lambda \cdot [\mathfrak{n}_\lambda, \mathbf{c}]. \tag{12.3}$$

Applying Theorem 11.1 in the form (11.8) we have

$$[\mathfrak{n}, \Omega \mathbf{C}_\lambda] = -[\mathfrak{n}_\lambda, \mathbf{c}] + \lambda[\mathfrak{n}, \bar{\mathbf{x}}_\lambda] \cdot C_\lambda(\omega). \tag{12.4}$$

Substituting from (12.4) into (12.3) one gets

$$\{1 - m_\lambda \cdot \lambda[\mathfrak{n}, \bar{\mathbf{x}}_\lambda]\} \cdot C_\lambda(\omega) + m_\lambda \cdot [\mathfrak{n}, \Omega \mathbf{C}_\lambda] = 0. \tag{12.5}$$

We see thus that each $\mathbf{C}_\lambda$ in the range of P_λ satisfies (12.5). On the other hand, if $\Omega \mathbf{C}$ is bounded, put $\mathbf{c} = \lambda \mathbf{C} - \Omega \mathbf{C}$ and define $\mathbf{C}_\lambda$ by (12.2). Supposing that both $\mathbf{C}$ and $\mathbf{C}_\lambda$ satisfy (12.5), the difference $\mathbf{C} - \mathbf{C}_\lambda$ has to be a multiple of $\bar{\mathbf{x}}_\lambda$ satisfying the same condition, and so $\mathbf{C}_\lambda = \mathbf{C}$. In other words, the range of P_λ consists exactly of all $\mathbf{C}$ such that $\Omega \mathbf{C}$ is bounded and that (12.5) holds for $\mathbf{C} = \mathbf{C}_\lambda$. Therefore the range is independent of λ if, and only if, the coefficients in (12.5) are in a constant ratio, that is, if

$$m_\lambda = \frac{1}{m + \lambda[\mathfrak{n}, \bar{\mathbf{x}}_\lambda]} \tag{12.6}$$

(or if $m_\lambda = 0$ in which case $P_\lambda = P_\lambda^{\min}$). If (12.6) holds the range of P_λ consists of all $\mathbf{C}$ in the domain of Ω satisfying the lateral condition

$$[\mathfrak{n}, \Omega \mathbf{C}] + mC(\omega) = 0. \tag{12.7}$$

It remains to verify the norm-conditions (3.2). Now from (12.1) and Theorem 10.2

$$\begin{aligned} \lambda \textstyle\sum_k P_\lambda(i, k) &= 1 - \bar{x}_\lambda(i) + \bar{x}_\lambda(i) \cdot m_\lambda \cdot \lambda[\mathfrak{n}_\lambda, \mathbf{1}] \\ &= 1 - \bar{x}_\lambda(i) + \bar{x}_\lambda(i) \cdot m_\lambda(\lambda[\mathfrak{n}_\lambda, \bar{\mathbf{x}}] + \lambda[\mathfrak{n}_\lambda, \mathbf{x}^{(0)}]) \\ &= 1 - \bar{x}_\lambda(i) \frac{m - \tau}{m + \lambda[\mathfrak{n}, \bar{\mathbf{x}}_\lambda]}. \end{aligned} \tag{12.8}$$

It follows that the norm conditions require $m \geqq c$, where c stands for the constant $\lambda[\mathfrak{n}_\lambda, \mathbf{x}^{(0)}]$.

THEOREM[32] 12.1. *Choose $m \geqq \lambda[\mathfrak{n}_\lambda, \mathbf{x}^{(0)}] =$ const. The matrix P_λ defined by*

[31] It follows from Lemma 11.2 that if each $\mathbf{C}_\lambda$ has constant boundary values, then each column of P_λ differs from the corresponding column of $P_\lambda^{\min}$ by a multiple of $\bar{\mathbf{x}}_\lambda$. If P_λ is to be a solution of both Kolmogorov equations (3.3) and (3.4) then it must be of the form (12.1) with $\mathfrak{n}_\lambda$ *some* bounded solution of $\lambda \mathfrak{n}_\lambda - \mathfrak{n}_\lambda \Omega = 0$. That $\mathfrak{n}_\lambda$ must be chosen so as to have a fixed canonical image $\mathfrak{n} \in \bar{\mathfrak{P}}^*$ is easily verified from the resolvent equation (3.5). However, this fact will become clearer as a consequence of the duality principle. At any rate, in this section we start with the ansatz (12.1) with $\mathfrak{n}_\lambda$ having a fixed image $\mathfrak{n}$.

[32] Since $\mathfrak{n}$ can be chosen arbitrarily we have here a variety of examples where the two Kolmogorov differential equations can have many common solutions, all of them strictly stochastic. D. G. Kendall [11] constructed an example to prove the possibility of such occurrence.

(12.1) *and* (12.6) *satisfies both Kolmogorov equations* (2.3) *and* (2.4) *and represents the Laplace transform of a Markovian* $P(t)$. *The strictly stochastic case* $\lambda P_\lambda \mathbf{1} = \mathbf{1}$ *arises if, and only if,* $m = \lambda[\mathbf{n}_\lambda, \mathbf{x}^{(0)}]$. *The range of the transformation* $\mathbf{c} \rightarrow P_\lambda \mathbf{C}$ *is characterized by* (12.7), *where* $C(\omega)$ *denotes the value of* $\mathbf{C}$ *along the exit boundary.*

(The range of the adjoint transformation will be described in Section 15.)

[*Note.* For the stochastic process depending on our $P(t)$ the boundary plays the role of a *fictitious state*, in the terminology of P. Lévy: the particle spends a time of measure 0 at ω, but this exceptional set is indispensable to a proper definition of the process.]

The meaning of the lateral condition (12.7) will appear clearer if (as in Section 9) we enlarge the space E by an element ω and *consider* $C(\omega)$ *not only as a limit, but as a value assumed by* $\mathbf{C}$ *on the new element* ω. It is then natural to restrict the domain $\mathfrak{C}$ to such vectors $\mathbf{c}$ which are *continuous* on $E + \omega$, that is, for which $c(\omega) = \lim c(i)$ as i approaches the exit boundary. Then one is led to replace (12.2) by the more general transformation[33]

$$C_\lambda(i) = C_\lambda^{\min}(i) + \bar{x}_\lambda(i) \cdot m_\lambda([\mathbf{n}_\lambda, \mathbf{c}] + pc(\omega)) \tag{12.9}$$

which will still represent a solution of $\lambda \mathbf{C}_\lambda - \Omega \mathbf{C}_\lambda = \mathbf{c}$. For $i = \omega$ the vector $\mathbf{C}_\lambda$ is defined by continuity. Now (12.9) amounts to extending the definition of our vectors to ω as follows

$$\bar{x}(\omega) = \bar{x}_\lambda(\omega) = 1 \qquad \eta(\omega) = \eta_\lambda(\omega) = p \geqq 0 \tag{12.10}$$

and to extend the definition (12.1) of $P_\lambda(i, k)$ to $i, k = 1, 2, \cdots, \omega$ where, of course,

$$P_\lambda^{\min}(i, \omega) = P_\lambda^{\min}(\omega, k) = 0. \tag{12.11}$$

All row- and column-vectors being extended to ω, the *inner product in the enlarged space is defined by*

$$[\boldsymbol{\alpha}, \mathbf{c}]_e = [\boldsymbol{\alpha}, \mathbf{c}] + \alpha(\omega)c(\omega) = \textstyle\sum_{i=1}^{\infty} \alpha(i)c(i) + \alpha(\omega)c(\omega) \tag{12.12}$$

(so that (12.9) reads $\mathbf{C}_\lambda = \mathbf{C}_\lambda^{\min} + \bar{\mathbf{x}}_\lambda \cdot m_\lambda [\mathbf{n}_\lambda, \mathbf{c}]_e$).

Now the value of $\Omega \mathbf{C}_\lambda$ at ω is well defined by continuity so that

$$\lambda C_\lambda(\omega) - \Omega C_\lambda(\omega) = c(\omega), \tag{12.13}$$

but we shall presently derive an independent definition of Ω at ω.

The argument which led to the last Theorem now requires hardly any change. Multiply (12.13) by p and subtract from (12.4) to obtain

$$[\mathbf{n}, \Omega \mathbf{C}_\lambda]_e = -[\mathbf{n}_\lambda, \mathbf{c}]_e + \lambda[\mathbf{n}, \bar{\mathbf{x}}_\lambda]_e \cdot C_\lambda(\omega). \tag{12.14}$$

This is the same as (12.4) with all old inner products $[\quad]$ replaced by $[\quad]_e$. The same is true of the remaining equations. Instead of (12.6) we get

[33] In principle, p could be made dependent on λ. However, just as in Section 9 it would turn out that the resolvent equation requires p to be constant. The deeper reason for this will appear from the duality principle.

$$(12.15)\qquad m_\lambda = \frac{1}{m + \lambda[\mathfrak{n}, \bar{\mathbf{x}}_\lambda]_e} = \frac{1}{m + \lambda p + \lambda[\mathfrak{n}, \bar{\mathbf{x}}_\lambda]}$$

and instead of (12.8) we get $mC(\omega) + [\mathfrak{n}, \Omega\mathbf{C}]_e = 0$ which is the same as

$$(12.16)\qquad p\cdot\Omega C(\omega) = -mC(\omega) - [\mathfrak{n}, \Omega\mathbf{C}].$$

The point of interest is that *when* $p > 0$ *we no longer deal with a lateral condition but, instead,* (12.16) *simply defines the value of* $\Omega\mathbf{C}$ *at* ω, *that is,* (12.16) *now represents a new row in the Kolmogorov equations* (3.3). (The corresponding column in (3.4) will be determined in Section 15.)

To summarize, we have

THEOREM 12.1. *The matrix* P_λ *with elements* $P_\lambda(i, k)$, $i, k = 1, 2, \cdots, \omega$ *defined by* (12.1), (12.10), (12.15) *represents the Laplace transform of a Markovian matrix* $P(t)$ *defined on* $E + \omega$, *provided* $m \geqq \tau \equiv \lambda\sum_1^\infty \eta(j)x_\lambda^{(0)}(j)$. *The purely stochastic case* $\lambda\sum_{k=1}^\infty P_\lambda(i, k) + \lambda P_\lambda(i, \omega) = 1$ *arises if, and only if,* $m = \tau$. *For* $i, k \neq \omega$ *the elements* $P_\lambda(i, k)$ *satisfy both Kolmogorov equations* (3.3) *and* (3.4). *The row corresponding to* $i = \omega$ *in* (3.3) *is*

$$(12.17)\qquad (\lambda + mp^{-1})P_\lambda(\omega, k) + p^{-1}[\mathfrak{n}, (-\mathfrak{q} + \mathfrak{q}\Pi)P_\lambda(\cdot, k)] = \delta_{\omega,k}$$

where $\delta_{\omega,k}$ *denotes the Kronecker symbol.*

The equation (12.17) does not show clearly the behavior of $P(t; \omega, k)$ as $t \to 0$. We prove therefore:

THEOREM 12.3. *As* $t \to 0$ *one has*

$$(12.18)\qquad \frac{1 - P(t; \omega, \omega)}{t} \to \frac{m + L}{p} \leqq \infty \qquad\qquad \frac{P(t; \omega, k)}{t} \to 0$$

where $L = \lim_{\lambda\to\infty} \lambda[\mathfrak{n}, \bar{\mathbf{x}}_\lambda] \leqq \infty$ *and* $k \neq \omega$.

PROOF. As in the proof of Theorem 9.3 one sees that (12.18) is equivalent to

$$(12.19)\qquad \lambda\{1 - \lambda P_\lambda(\omega, \omega)\} = \lambda\cdot\frac{m + \lambda[\mathfrak{n}, \bar{\mathbf{x}}_\lambda]}{m + \lambda p + \lambda[\mathfrak{n}, \bar{\mathbf{x}}_\lambda]} \to \frac{m + L}{p}$$

$$(12.20)\qquad \lambda^2 P_\lambda(\omega, k) = m_\lambda\cdot\lambda^2\eta_\lambda(k) \to 0$$

as $\lambda \to \infty$. Now we know from Theorem 10.2 that $\lambda[\mathfrak{n}, \bar{\mathbf{x}}_\lambda] \uparrow L \leqq \infty$. On the other hand, $[\mathfrak{n}, \bar{\mathbf{x}}_\lambda] = [\mathfrak{n}_\lambda, \bar{\mathbf{x}}]$. Since $\mathfrak{n}_\lambda \downarrow \mathbf{0}$ (cf. (10.4)), we have $\lambda[\mathfrak{n}, \bar{\mathbf{x}}_\lambda] = o(\lambda)$, and the relations (12.19) and (12.20) follow at once.

In the terminology of P. Lévy the state ω will be *instantaneous* if $L = \infty$. Under any circumstances $q_{\omega,k} = 0$ for $k \neq \omega$.

13. General solutions: preparations

If $\mathbf{x}_\lambda^{(1)}, \cdots, \mathbf{x}_\lambda^{(m)}$ are solutions of $\lambda\mathbf{x}_\lambda - \Omega\mathbf{x}_\lambda = \mathbf{0}$ and $\mathfrak{n}_\lambda^{(1)}, \cdots, \mathfrak{n}_\lambda^{(n)}$ solutions of $\lambda\mathfrak{n}_\lambda - \mathfrak{n}_\lambda\Omega = \mathbf{0}$ then the matrix P_λ with elements

$$(13.1)\qquad P_\lambda(i, k) = P_\lambda^{\min}(i, k) + \sum_{a=1}^m \sum_{b=1}^n x_\lambda^{(a)}(i)M_\lambda^{(a,b)}\eta_\lambda^{(b)}(k)$$

is a common solution of the two Kolmogorov equations (3.3) and (3.4). Our problem is to find the conditions on the $\mathbf{x}_\lambda^{(a)}$, $\mathbf{n}_\lambda^{(b)}$ and $M_\lambda^{(a,b)}$ which will ensure that (13.1) is an *admissible* solution, that is, conditions under which the norm relations (3.1) and the resolvent equation (3.5) will be valid.

Clearly (13.1) represents the *most general* common solution of the two Kolmogorov equations in the special case where the equations $\lambda\mathbf{x}_\lambda - \Omega\mathbf{x}_\lambda = \mathbf{0}$ and $\lambda\mathbf{n}_\lambda - \mathbf{n}_\lambda\Omega = \mathbf{0}$ possess only a finite number of linearly independent solutions (in other words, if both exit and entrance boundaries are *finite*). In the general case the summation in (13.1) should be replaced by an integral, and the matrix $M_\lambda^{(a,b)}$ by a kernel depending on two arguments, the first ranging over the exit boundary, the second over the entrance boundary (Section 15). However, to avoid introducing a new symbolism and to discuss the interesting phenomena in the simplest possible manner, we shall limit ourselves to solutions of the form (13.1) involving only finitely many free parameters.

We consider again P_λ primarily as an operator on column vectors $\mathbf{c}$ taking $\mathbf{c}$ into $\mathbf{C}_\lambda = P_\lambda\mathbf{c}$ where

$$C_\lambda(i) = C_\lambda^{\min}(i) + \sum_{a=1}^{m} \sum_{b=1}^{n} x_\lambda^{(a)}(i) M_\lambda^{(a,b)}(\mathbf{n}_\lambda^{(b)}\mathbf{c}) \tag{13.2}$$

where $\mathbf{n}_\lambda^{(b)}\mathbf{c} = [\mathbf{n}_\lambda^{(b)}, \mathbf{c}]$ is the inner product of $\mathbf{n}_\lambda^{(b)}$ and $\mathbf{c}$, and $\mathbf{C}_\lambda^{\min} = P_\lambda^{\min}\mathbf{c}$. To simplify the clumsy appearance of (13.1) and (13.2) we introduce the

Notational convention. For arbitrary symbols $u^{(1)}, \cdots, u^{(n)}$ (numbers or vectors) we shall denote by $\{u\}_1^n$ the $n \times 1$ matrix (column) with elements $u^{(a)}$. Similarly $\{u\}_1'^n$ will be the corresponding transpose, that is, the $1 \times n$ matrix with elements $u^{(a)}$. A matrix with elements $M^{(a,b)}$ etc. *will be denoted by the corresponding German capital.*

With this convention, formulas (13.1) take on the compact form[34]

$$P_\lambda = P_\lambda^{\min} + \{\mathbf{x}_\lambda\}_1'^m \mathfrak{M}_\lambda \{\mathbf{n}_\lambda\}_1^n \tag{13.3}$$

and

$$\mathbf{C}_\lambda = \mathbf{C}_\lambda^{\min} + \{\mathbf{x}_\lambda\}_1'^m \mathfrak{M}_\lambda \{\mathbf{n}_\lambda \mathbf{c}\}_1^n \tag{13.4}$$

$\mathfrak{M}_\lambda$ being an $m \times n$ matrix.

We know that $\mathbf{C}_\lambda^{\min}$ vanishes at the exit boundary and that each $\mathbf{x}_\lambda^{(a)}$ has continuous boundary values. For simplicity of expressions let us assume that for some particular value of λ the $\mathbf{x}_\lambda^{(a)}$ are *sojourn* solutions, that is, that *there exist sets $\omega^{(1)}, \cdots, \omega^{(a)}$ such that*[35]

[34] Note that $\mathbf{x}_\lambda^{(a)}\mathbf{n}_\lambda^{(b)}$ represents the matrix over E with elements $x_\lambda^{(a)}(i)\eta_\lambda^{(b)}(k)$. It is clear that $\{\mathbf{n}_\lambda\mathbf{c}\}_1^n$ stands for the column with elements $\mathbf{n}_\lambda^{(a)}\mathbf{c}$.

[35] To an *arbitrary* $\mathbf{x}_\lambda^{(a)} \geq \mathbf{0}$ there corresponds the open set $\omega^{(a)}$ of the exit boundary $\mathfrak{B}$ on which the boundary values of $\mathbf{x}_\lambda^{(a)}$ are positive. Again the $\mathbf{x}_\lambda^{(a)}$ can be chosen so that the sets $\omega^{(a)}$ will be non-overlapping (that is, $\mathbf{x}_\lambda^{(a)} \cap \mathbf{x}_\lambda^{(b)} = \mathbf{0}$ for $a \neq b$). Then (13.6) is to be replaced by

$$C_\lambda(\omega)^{(a)} = \lim \frac{C_\lambda(i)}{x_\lambda^{(a)}(i)}.$$

$$(13.5)\qquad x_\lambda^{(a)}(i) \to 1 \text{ as } i \text{ approaches } \omega^{(a)}$$
$$x_\lambda^{(a)}(i) \to 0 \text{ as } i \text{ approaches } \bar{\mathfrak{B}} - \omega^{(a)}.$$

There is no loss in assuming that the sets $\omega^{(a)}$ are non-overlapping. It follows then that the limits as i approaches $\omega^{(a)}$

$$(13.6)\qquad C_\lambda(\omega^{(a)}) = \lim C_\lambda(i)$$

exist for $a = 1, \cdots, m$. Suppose then that P_λ satisfies the resolvent equation (3.5). In this case the range of P_λ is independent of λ, and hence $P_\lambda \mathbf{c}$ must assume constant boundary values along $\omega^{(1)}, \cdots, \omega^{(a)}$. In other words, for each λ the solutions $\mathbf{x}_\lambda^{(1)}, \cdots, \mathbf{x}_\lambda^{(m)}$ used in (13.1) or (13.3) must be sojourn solutions corresponding to the same boundary sets $\omega^{(1)}, \cdots, \omega^{(m)}$, that is, $\mathbf{x}_\lambda^{(a)}$ must have a fixed canonical image $\mathbf{x}^{(a)} \in \bar{\mathfrak{P}}$, namely the sojourn solution in $\mathfrak{P}$ which induces $\omega^{(a)}$.

We shall see in Section 15 that there is a complete symmetry between the action of P_λ as operator on column vectors or row-vectors, except that in the latter case the *entrance boundary* takes over the role of the exit boundary. We can then repeat the argument to show that the solutions $\mathbf{n}_\lambda^{(b)}$ used in (13.1) or (13.3) must have fixed canonical images $\mathbf{n}^{(b)} \in \bar{\mathfrak{P}}^*$. Again there is no loss in supposing that for $a \neq b$ the solutions $\mathbf{n}^{(a)}$ and $\mathbf{n}^{(b)}$ are non-overlapping in the sense that $\mathbf{n}^{(a)} \cap \mathbf{n}^{(b)} = \mathbf{0}$ (meaning that the corresponding open sets on the entrance boundary are non-overlapping). These considerations lead us to introduce the following

Set-up. We shall be concerned with matrices P_λ *of the form* (13.3) *under the following conditions: The exit boundary* $\bar{\mathfrak{B}}$ *is supposed to be partitioned*[36] *into* m *non-overlapping sets* $\omega^{(1)}, \cdots \omega^{(m)}$ *where* $\omega^{(a)}$ *is induced by the sojourn solution* $\mathbf{x}^{(a)} \in \bar{\mathfrak{P}}$. *The vector* $\mathbf{x}_\lambda^{(a)}$ *appearing in* (13.3) *is the element of* $\mathfrak{P}_\lambda$ *with canonical image* $\mathbf{x}^{(a)}$. *Then* (cf. [5])

$$(13.7)\qquad \bar{\mathbf{x}} = \mathbf{x}^{(1)} + \cdots + \mathbf{x}^{(m)}, \qquad \bar{\mathbf{x}}_\lambda = \mathbf{x}_\lambda^{(1)} + \cdots + \mathbf{x}_\lambda^{(m)}$$

are the maximal elements of $\bar{\mathfrak{P}}$ *and* $\mathfrak{P}_\lambda$, *and* $\mathbf{x}^{(0)} = \mathbf{1} - \bar{\mathbf{x}}$ *is the maximal passive element (including the passive boundary).* Similarly, $\mathbf{n}_\lambda^{(1)}, \cdots, \mathbf{n}_\lambda^{(n)}$ *will be* n *non-overlapping elements of* $\mathfrak{P}_\lambda^*$ *with fixed canonical images* $\mathbf{n}^{(1)}, \cdots, \mathbf{n}^{(n)} \in \bar{\mathfrak{P}}^*$.

Note. We shall be interested only in families P_λ satisfying the resolvent equation (3.5). The range of the transformations $\mathbf{c} \to P_\lambda \mathbf{c}$ is then independent of λ, and the sets $\omega^{(a)}$ are the boundary sets along which each $\mathbf{C}$ in this range is constant. Now if one had $C(\omega^{(1)}) = C(\omega^{(2)})$ for each such $\mathbf{C}$, then the union $\omega^{(1)} \cup \omega^{(2)}$ would be a set of constancy. For each λ the first two rows of $\mathfrak{M}_\lambda$ would agree, and it appears that the notation in such a case is redundant: one should simply replace the m sets $\omega^{(1)}, \cdots, \omega^{(m)}$ by $\omega^{(1)} \cup \omega^{(2)}, \omega^{(3)}, \cdots, \omega^{(m)}$, or $\mathbf{x}^{(1)}, \cdots, \mathbf{x}^{(m)}$ by $\mathbf{x}^{(1)} + \mathbf{x}^{(2)}, \mathbf{x}^{(3)}, \cdots, \mathbf{x}^{(m)}$. In other words there is no loss of generality in supposing that for $a \neq b$ there exists a $\mathbf{C}$ in the range of P_λ such that

[36] The condition that $\omega^{(1)}, \cdots, \omega^{(m)}$ *cover* $\mathfrak{B}$ is no restriction: if the first row of $\mathfrak{M}_\lambda$ vanishes then $\mathbf{x}_\lambda^{(1)}$ will not really appear in (13.3), and $\mathbf{C}_\lambda$ will *vanish* along $\omega^{(1)}$.

$$C(\omega^{(a)}) \neq C(\omega^{(b)}). \tag{13.8}$$

A similar remark applies to the $\mathbf{n}^{(b)}$. If two columns of $\mathfrak{M}_\lambda$ stand in a fixed proportion, for example

$$M_\lambda^{(\nu,a)} : M_\lambda^{(\nu,b)} = \rho \qquad a \neq b, \tag{13.9}$$

where $\nu = 1, \cdots, m$ and ρ is independent of λ, then one can introduce the vector $\rho\mathbf{n}^{(a)} + \mathbf{n}^{(b)}$ instead of the pair $\mathbf{n}^{(a)}, \mathbf{n}^{(b)}$, thus reducing the n vectors $\mathbf{n}^{(1)}, \cdots, \mathbf{n}^{(n)}$ to $n - 1$ vectors doing the same services.[37] We shall say that the $\mathbf{x}^{(a)}$ and $\mathbf{n}^{(b)}$ *have been chosen without redundancy if it does not happen that two columns of* $\mathfrak{M}_\lambda$ *stand in a fixed ratio* (i.e. (13.9)) *or that two rows are identical.*

In the sequel we shall consistently use the following

Notations for the inner products. Remembering Theorem 10.2[38] *we set*

$$U_\lambda^{(a,b)} = \lambda \mathbf{n}_\lambda^{(a)} \mathbf{x}^{(b)} = \lambda \mathbf{n}^{(a)} \mathbf{x}_\lambda^{(b)} \qquad \begin{array}{l} a = 1, \cdots, n \\ b = 1, \cdots, m \end{array} \tag{13.10}$$

$$\tau^{(a)} = \lambda \mathbf{n}_\lambda^{(a)} \mathbf{x}^{(c)} \tag{13.11}$$

(this is independent of λ*). The* $n \times m$ *matrix* $\mathfrak{U}_\lambda$ *with elements* $U_\lambda^{(a,b)}$ *increases monotonically, and we put*

$$\mathfrak{U}_\lambda \uparrow \mathfrak{U}, \tag{13.12}$$

where the elements $U^{(a,b)}$ *are possibly infinite.*

LEMMA 13.1. *In order that the matrix* P_λ *be (for some fixed* λ*) non-negative and satisfy the norm condition* $\lambda P_\lambda \mathbf{1} \leqq \mathbf{1}$ *it is necessary and sufficient that* $\mathfrak{M}_\lambda$ *be non-negative and that*[39]

$$\mathfrak{M}_\lambda \mathfrak{U}_\lambda \{1\}_1^m + \mathfrak{M}_\lambda\{\tau\}_1^m \leqq \{1\}_1^m. \tag{13.13}$$

Moreover, $\lambda P_\lambda \mathbf{1} = \mathbf{1}$ *if, and only if, the equality sign holds in* (13.13).

COROLLARY. *If* P_λ *is an admissible solution, then the* $m \times m$ *matrix*

$$\mathfrak{S}_\lambda = \mathfrak{M}_\lambda \mathfrak{U}_\lambda \tag{13.14}$$

is substochastic.

PROOF. It is known ([4], Section 16) that a point i may be chosen so that the ratios $\eta^{(1)}(i)/\eta^{(b)}(i)$ for $b = 2, \cdots, n$ will be as large as we please. (Indeed, these ratios approach infinity as i approaches the set of entrance boundary induced by $\mathbf{n}^{(1)}$). Choosing $c(i) = 1$ and $c(j) = 0$ for $j \neq i$ we see at once from (13.4) that $M_\lambda^{(a,1)}$ must be non-negative if $C_\lambda(\omega^{(a)})$ is to be non-negative. Since

[37] There is a slight lack of symmetry in this formulation inasmuch as we eliminated a *row* of $\mathfrak{M}_\lambda$ only if two rows are identical, but eliminate a *column* if two columns are in a constant proportion.

[38] Remember that Theorem 10.2 presupposes that $\bar{\mathbf{x}}$ be strictly positive, and this is assumed in the sequel (cf. end of Section 5).

[39] It is hardly necessary to explain that $\{1\}_1^m$ stands for a column with m ones, so that $\mathfrak{M}_\lambda \mathfrak{U}_\lambda\{1\}_1^m$ stands for the m row sums of $\mathfrak{M}_\lambda \mathfrak{U}_\lambda$ arranged as a column vector.

this argument applies equally to all columns of $\mathfrak{M}_\lambda$ it is seen that $\mathfrak{M}_\lambda$ must be non-negative.

To test the norm condition, apply (13.4) to $\mathbf{c} = \mathbf{1}$. By (4.15) we have in this case $\lambda \mathbf{C}_\lambda^{\min} = \mathbf{1} - \bar{\mathbf{x}}_\lambda$. Furthermore from (13.7), (13.10), and (13.11) we have for $b = 1, \cdots, n$

$$\lambda \mathbf{n}_\lambda^{(b)} \mathbf{1} = \sum_{t=1}^{m} U_\lambda^{(b,t)} + \tau^{(b)}. \tag{13.15}$$

On the right we recognize the row sums of $\mathfrak{U}_\lambda$, and using our notational convention we may condense (13.15) into

$$\lambda \{\mathbf{n}_\lambda \mathbf{1}\}_1^n = \mathfrak{U}_\lambda \{1\}_1^m + \{\tau\}_1^n. \tag{13.16}$$

Thus

$$\lambda P_\lambda \mathbf{1} = \mathbf{1} - \bar{\mathbf{x}}_\lambda + \{\mathbf{x}_\lambda\}_1^{\prime m} \mathfrak{M}_\lambda (\mathfrak{U}_\lambda \{1\}_1^m + \{\tau\}_1^n). \tag{13.17}$$

Remembering that $\bar{\mathbf{x}}_\lambda = \sum \mathbf{x}_\lambda = \{\mathbf{x}_\lambda\}_1^{\prime m} \{1\}_1^m$ we see that the right side of (13.17) is dominated by $\mathbf{1}$ if, and only if, (13.13) holds. The Lemma is proved.

14. General solution: construction

It is understood that in this section we adhere to the set-up and to the notations as described in the last section. Unexpected, but interesting, complications arise if the matrix $\mathfrak{U}$ of (13.12) is permitted to contain infinite elements. It is therefore desirable to treat the simpler finite case separately.

THEOREM 14.1. *Suppose that $\mathfrak{U}$ is finite. Choose a non-negative $m \times n$ matrix $\mathfrak{M}$ and define the $m \times m$ matrix $\mathfrak{S} = \mathfrak{M}\mathfrak{U}$. Suppose that*[40]

$$\mathfrak{S}\{1\}_1^m + \mathfrak{M}\{\tau\}_1^n \leqq \{1\}_1^m. \tag{14.1}$$

Then the matrix P_λ of (13.3) *with*

$$\mathfrak{M}_\lambda = (I - \mathfrak{S} + \mathfrak{M}\mathfrak{U}_\lambda)^{-1} \mathfrak{M} \tag{14.2}$$

is an admissible solution of both Kolmogorov equations (3.3) *and* (3.4). *The strictly stochastic case $\lambda P_\lambda \mathbf{1} = \mathbf{1}$ arises if, and only if, the equality sign holds in* (14.1).

The range of the transformation $\mathbf{c} \to P_\lambda \mathbf{c}$ is characterized by the lateral condition

$$\{C(\omega)\}_1^m - \mathfrak{S}\{C(\omega)\}_1^m + \mathfrak{M}\{[\mathbf{n}, \Omega \mathbf{C}]\}_1^n = 0. \tag{14.3}$$

This P_λ represents the most general solution compatible with our set-up.

(The lateral condition for P_λ as operator on row vectors is described in Theorem 16.3).

PROOF. To begin with, assume that P_λ of (13.3) is an admissible solution, in particular that (13.13) holds. From (13.4) we get for the boundary values of $\mathbf{C}_\lambda$

$$\{C_\lambda(\omega)\}_1^m = \mathfrak{M}_\lambda \{\mathbf{n}_\lambda \mathbf{c}\}_1^n \tag{14.4}$$

[40] In explicit form (14.1) reads

$$\sum_{t=1}^{m} S^{(a,t)} + \sum_{t=1}^{n} M^{(a,t)} \tau^{(t)} \leqq 1 \qquad a = 1, \cdots, m.$$

and introducing this into (13.3) we get

$$(14.5)\qquad \begin{aligned} \mathbf{C}_\lambda &= \mathbf{C}_\lambda^{\min} + \{\mathbf{x}_\lambda\}_1'^m \{C_\lambda(\omega)\}_1^m \\ &= \mathbf{C}_\lambda^{\min} + \textstyle\sum_{a=1}^m \mathbf{x}_\lambda^{(a)} C_\lambda(\omega^{(a)}). \end{aligned}$$

Applying Theorem 11.1 in the form (11.8) to $\mathbf{C} = \mathbf{C}_\lambda$, $\mathbf{c}_\lambda = \mathbf{c}$, and $\mathfrak{n} = \mathfrak{n}^{(b)}$ to obtain

$$(14.6)\qquad [\mathfrak{n}^{(b)}, \Omega\mathbf{C}_\lambda] = -\mathfrak{n}_\lambda^{(b)}\mathbf{c} + \textstyle\sum_{a=1}^m U_\lambda^{(b,a)} C_\lambda(\omega^{(a)})$$

which may be rewritten in the compact form

$$(14.7)\qquad \{[\mathfrak{n}, \Omega\mathbf{C}_\lambda]\}_1^n = -\{\mathfrak{n}_\lambda\mathbf{c}\}_1^n + \mathfrak{U}_\lambda\{C_\lambda(\omega)\}_1^n.$$

Substituting back from (14.7) into (14.4) we get

$$(14.8)\qquad \{C_\lambda(\omega)\}_1^m - \mathfrak{S}_\lambda\{C_\lambda(\omega)\}_1^m + \mathfrak{M}_\lambda\{[\mathfrak{n}, \Omega\mathbf{C}_\lambda]\}_1^n = 0$$

where we used the abbreviation $\mathfrak{S}_\lambda = \mathfrak{M}_\lambda\mathfrak{U}_\lambda$ introduced in (13.14).

We have thus proved that *each* $\mathbf{C}_\lambda$ *in the range of* P_λ *satisfies* (14.8). We have supposed that (13.13) holds, and so $\mathfrak{S}_\lambda$ is substochastic. Remembering that $\mathfrak{U}_\lambda$ is strictly positive and increases monotonically to $\mathfrak{U}$ we see at once that for an appropriate subsequence $\lambda \to \infty$

$$(14.9)\qquad \mathfrak{S}_\lambda \to \mathfrak{S} = \mathfrak{M}\mathfrak{U} \qquad \mathfrak{M}_\lambda \to \mathfrak{M}$$

where $\mathfrak{M}$ and $\mathfrak{S}$ satisfy conditions (14.1). The lateral condition (14.8) goes over into (14.3) and we have thus proved the first part of the Theorem, namely: *If* P_λ *is an admissible solution, then each* $\mathbf{C}$ *satisfies the lateral condition* (14.3) *where* $\mathfrak{M}$, $\mathfrak{S}$ *are defined by* (14.9) *and satisfy* (14.1).

We next prove: *the lateral condition* (14.1) *determines the range uniquely*. More precisely: suppose P_λ is such that its range satisfies (14.3) and suppose that some $\mathbf{C}$ satisfies (14.3) and $\Omega\mathbf{C}$ is bounded. Then $\mathbf{C}$ is in the range of P_λ. In fact, put $\mathbf{c}_\lambda = \lambda\mathbf{C} - \Omega\mathbf{C}$ and $\mathbf{C}_\lambda = P_\lambda\mathbf{c}_\lambda$. Then $\mathbf{x}_\lambda = \mathbf{C} - \mathbf{C}_\lambda$ is a solution of $\lambda\mathbf{x}_\lambda - \Omega\mathbf{x}_\lambda = \mathbf{0}$ which satisfies (14.3). We show that $\mathbf{x}_\lambda = \mathbf{0}$. Assume contrarywise that $\mathbf{x}_\lambda = \sum p^{(a)}\mathbf{x}_\lambda^{(a)}$. Then

$$(14.10)\qquad [\mathfrak{n}^{(b)}, \Omega\mathbf{x}_\lambda] = \lambda[\mathfrak{n}^{(b)}, \textstyle\sum_{a=1}^m p^{(a)}\mathbf{x}_\lambda^{(a)}] = \sum_{a=1}^m U_\lambda^{(b,a)} p^{(a)}$$

and so (14.3) applied to $\mathbf{x}_\lambda$ gives

$$(14.11)\qquad \{p\}_1^m - \mathfrak{S}\{p\}_1^m + \mathfrak{M}\mathfrak{U}_\lambda\{p\}_1^m = \mathbf{0}.$$

Here $\mathfrak{S}$ is substochastic. Now the matrix $\mathfrak{U}_\lambda$ is strictly positive, and so a row of $\mathfrak{M}\mathfrak{U}_\lambda$ vanishes only if the same row of $\mathfrak{M}$ vanishes. But since $\mathfrak{S} = \mathfrak{M}\mathfrak{U}$ in this case that row of $\mathfrak{S}$ vanishes also. Therefore the $m \times m$ matrix $\mathfrak{S} - \mathfrak{M}\mathfrak{U}_\lambda$ is non-negative and has row sums strictly less than 1. Under such circumstances (14.11) implies $p^{(a)} = 0$ for $a = 1, \cdots, m$ and our assertion is proved.

Now prescribe $\mathfrak{M}$ arbitrarily so that (14.1) holds. We have just shown that $\mathfrak{S} - \mathfrak{M}\mathfrak{U}_\lambda = \mathfrak{M}\mathfrak{U} - \mathfrak{M}\mathfrak{U}_\lambda$ is non-negative and has row sums strictly less than

1. The inverse matrix occurring in (14.2) exists therefore and equals

$$\sum_n(\mathfrak{S} - \mathfrak{M}\mathfrak{U}_\lambda)^n.$$

Accordingly (14.2) defines a non-negative matrix. With this $\mathfrak{M}_\lambda$ define P_λ as in (13.3). Then each $\mathbf{C}_\lambda$ in the range of P_λ satisfies (14.8). On the left there appears the matrix

$$\begin{aligned} I - \mathfrak{S}_\lambda &= I - \mathfrak{M}_\lambda\mathfrak{U}_\lambda \\ &= I - (I - \mathfrak{M}\mathfrak{U} + \mathfrak{M}\mathfrak{U}_\lambda)^{-1}\mathfrak{M}\mathfrak{U}_\lambda \\ &= (I - \mathfrak{M}\mathfrak{U} + \mathfrak{M}\mathfrak{U}_\lambda)^{-1}(I - \mathfrak{M}\mathfrak{U}). \end{aligned} \tag{14.12}$$

Thus premultiplication of (14.8) by $I - \mathfrak{M}\mathfrak{U} + \mathfrak{M}\mathfrak{U}_\lambda$ shows that *each* $\mathbf{C}$ *in the range of* P_λ *satisfies the prescribed lateral condition* (14.3).

It remains to show that P_λ satisfies the norm-condition $\lambda P_\lambda \mathbf{1} \leqq \mathbf{1}$, or in other words, that (13.13) holds. Rewrite (14.1) as

$$(I - \mathfrak{M}\mathfrak{U})\{1\}_1^m - \mathfrak{M}\{\tau\}_1^m \geqq 0. \tag{14.13}$$

Then

$$(I - \mathfrak{M}\mathfrak{U} + \mathfrak{M}\mathfrak{U}_\lambda)\{1\}_1^m \geqq \mathfrak{M}\mathfrak{U}_\lambda\{1\}_1^m + \mathfrak{M}\{\tau\}_1^n. \tag{14.14}$$

Premultiplying (14.14) by the inverse of the matrix appearing on the left we obtain (13.13), the inverse being a positive matrix. The Theorem is proved.

The situation becomes more interesting if one has $U^{(a,b)} = \infty$ for one or more pairs $\mathbf{x}^{(a)}$, $\mathbf{n}^{(b)}$. In this connection we achieve a great simplification by requiring that the basic vectors $\mathbf{x}^{(a)}$ and $\mathbf{n}^{(b)}$ be chosen *without redundancy* (cf. the Note to the *set-up* described in Section 13; the preceding Theorem holds also if the $\mathbf{x}^{(a)}$ and $\mathbf{n}^{(b)}$ are chosen in a redundant way).

THEOREM 14.2. *Suppose that the matrix* P_λ *of* (13.3) *is an admissible solution of the Kolmogorov equations. If the basic vectors* $\mathbf{x}^{(a)}$ *and* $\mathbf{n}^{(b)}$ *have been chosen without redundancy and are numbered in an appropriate order, then the matrix* $\mathfrak{U}$ *can contain infinite elements only in the main diagonal:*

$$U^{(a,a)} = \infty \quad \textit{for} \quad a \leqq r; \qquad U^{(a,b)} < \infty \quad \textit{for} \quad a \neq b, \textit{ and } a > r. \tag{14.15}$$

Theorem 14.1 *now remains true provided the matrices* $\mathfrak{S}$ *and* $\mathfrak{M}$ *in* (14.1), (14.2), *and* (14.3) *are replaced by non-negative matrices* $\bar{\mathfrak{S}}$ *and* $\bar{\mathfrak{M}}$ *with the following properties*

$$\bar{S}^{(a,a)} = 0, \tag{14.16}$$

$$\bar{M}^{(a,b)} = 0 \quad \textit{for} \quad b \leqq r, \qquad b \neq r \tag{14.17}$$

and[41]

[41] In (14.18) a factor $U^{(t,b)}$ is infinite only if $t = b \leqq r$, in which case $\bar{M}^{(a,t)} = 0$ and the *product* $\bar{M}^{(a,t)}\, U^{(t,b)}$ is to be interpreted as 0. The equations (14.16)–(14.18) state that *off the main diagonal* $\bar{\mathfrak{S}}$ dominates the formal matrix product $\bar{\mathfrak{M}}\mathfrak{U}$, while all the entries of $\bar{\mathfrak{S}}$ on the main diagonal are zero.

(14.18) $$\sum_{t=1}^{n} \bar{M}^{(a,t)} U^{(t,b)} \leqq \bar{S}^{(a,b)} \qquad \textit{if } b \neq a.$$

Note. To understand the curious implications of this Theorem consider the case $m = 3$, $n = 1$, and write $\mathbf{n}$ instead of $\mathbf{n}^{(1)}$. The Theorem implies that *if*

(14.19) $$\lambda \mathbf{n} \mathbf{x}_\lambda^{(1)} \uparrow \infty \qquad \lambda \mathbf{n} \mathbf{x}_\lambda^{(2)} \uparrow \infty$$

then one of the boundary conditions is automatically[42]

(14.20) $$C(\omega^{(1)}) = C(\omega^{(2)}).$$

In this case the matrix P_λ involves only the sum $\mathbf{x}_\lambda^{(1)} + \mathbf{x}_\lambda^{(2)}$, and we have lost one *degree of freedom* by comparison with the case where the limits (14.19) are finite. It is seen that (14.19) expresses a sort of mutual interdependence between $\mathbf{n}$, $\mathbf{x}^{(1)}$ and $\mathbf{x}^{(2)}$, which presumably is an expression for the overlapping of the neighborhoods of both $\omega^{(1)}$ and $\omega^{(2)}$ with each neighborhood of the *entrance* boundary set corresponding to $\mathbf{n}$. Cf. the concluding note to Section 10.

Proof of Theorem 14.2

The first part of the proof of Theorem 14.1 remains valid, and we see that the boundary conditions (14.8) apply without change. Even the passage to the limit (14.9) can be effected. The trouble is that, the matrix $\mathfrak{U}$ being infinite, it is possible that $\mathfrak{M}_\lambda \to 0$ and $\mathfrak{M}_\lambda \mathfrak{U}_\lambda \to I$. In this case the limiting form of (14.8) reduces to the identity $0 = 0$ which is of no use. To avoid degeneracies of this nature we shall multiply each row of (14.8) by a factor such as to render the coefficients of the diagonal terms unity.

The diagonal terms of the matrix $I - \mathfrak{S}_\lambda$ in (14.8) are

(14.21) $$\delta_\lambda^{(a)} = 1 - S_\lambda^{(a,a)} = 1 - \sum_{t=1}^{n} M_\lambda^{(a,t)} U_\lambda^{(t,a)}.$$

We begin by proving that $\delta_\lambda^{(a)} > 0$ for $a = 1, \cdots, m$. By Lemma 13.1 the matrix $\mathfrak{S}_\lambda = \mathfrak{M}_\lambda \mathfrak{U}_\lambda$ is substochastic, and therefore $S_\lambda^{(a,a)} \leqq 1$. If $S_\lambda^{(a,a)} = 1$ for some a then $S_\lambda^{(a,b)} = 0$ for $b \neq a$. But $\mathfrak{U}_\lambda$ being strictly positive, $S_\lambda^{(a,b)} = 0$ implies that the ath row of $\mathfrak{M}_\lambda$ vanishes, and in this case $S_\lambda^{(a,a)} = 0$ also. Thus $S_\lambda^{(a,a)} = 1$ is impossible.

It is therefore legitimate to divide the ath equation in (14.8) by $\delta_\lambda^{(a)}$. We write the resulting system in the form

(14.22) $$(I - \bar{\mathfrak{S}}_\lambda)\{C_\lambda(\omega)\}_1^m + \bar{\mathfrak{M}}_\lambda\{[\mathbf{n}, \Omega \mathbf{C}_\lambda]\}_1^n = 0$$

[42] The formulation of Theorem 14.2 then requires us to replace the three vectors $\mathbf{x}^{(1)}$, $\mathbf{x}^{(2)}$, $\mathbf{x}^{(3)}$ by the equivalent system $\mathbf{x}^{(1)} + \mathbf{x}^{(2)}$, $\mathbf{x}^{(3)}$. Denoting these again by $\mathbf{x}^{(1)}$, $\mathbf{x}^{(2)}$ we have the case $m = 2$, $n = 1$ with the matrix $\mathfrak{U}$ a 1×2 row with elements (∞, U). If $U < \infty$, the matrix $\mathfrak{M}$ is a 2×1 column with elements $\bar{M}$, 0, and the boundary conditions (14.3) take on the form

$$C(\omega^{(1)}) - pC(\omega^{(2)}) + \bar{M}[\mathbf{n}, \Omega \mathbf{C}] = 0$$
$$-qC(\omega^{(1)}) + C(\omega^{(2)}) = 0$$

where $p \geqq \bar{M}U$, and $q \geqq 0$. The norm conditions (14.1) require $p + \bar{M}\tau \leqq 1$ and $q \leqq 1$.

where $\mathfrak{M}_\lambda$ and $\mathfrak{S}_\lambda$ have the elements

$$\bar{M}_\lambda^{(a,b)} = M_\lambda^{(a,b)}/\delta_\lambda^{(a)} \tag{14.23}$$

and

$$\bar{S}_\lambda^{(a,a)} = 0 \qquad \bar{S}_\lambda^{(a,b)} = \sum_{t=1}^m \bar{M}_\lambda^{(a,t)} U_\lambda^{(t,b)} \qquad b \neq a. \tag{14.24}$$

In other words, off the main diagonal $\bar{\mathfrak{S}}_\lambda$ coincides with the formal matrix product $\bar{\mathfrak{M}}_\lambda \mathfrak{U}_\lambda$. By its definition $\bar{\mathfrak{S}}_\lambda$ is again a substochastic matrix, and $\mathfrak{U}_\lambda$ is strictly positive. As in (14.9) we can let $\lambda \to \infty$ in such a manner that $\bar{\mathfrak{S}}_\lambda \to \bar{\mathfrak{S}}$ and $\bar{\mathfrak{M}}_\lambda \to \mathfrak{M}$, and the boundary conditions (14.8) go over into (14.3) with $\mathfrak{M}$, $\mathfrak{S}$ replaced by $\bar{\mathfrak{M}}$ and $\bar{\mathfrak{S}}$.

It is clear that $\bar{\mathfrak{S}}$ is substochastic, and in fact, that the norm conditions (14.1) hold for $\bar{\mathfrak{M}}$ and $\bar{\mathfrak{S}}$. Moreover, (14.24) shows that

$$U^{(\nu,b)} = \infty \quad \textit{implies} \quad \bar{M}^{(a,\nu)} = 0 \qquad \textit{for each } a \neq b, \tag{14.25}$$

and

$$\bar{S}^{(a,b)} \geqq \sum_{t=1}^n \bar{M}^{(a,t)} U^{(t,b)} \qquad \textit{for } a \neq b \tag{14.26}$$

(where on the right terms of the form $0 \cdot \infty$ are set $=0$, which accounts for the inequality sign).

We next show that the boundary conditions (14.1) thus obtained *completely describe the range*[43] *of* P_λ. As in the proof of Theorem 14.1 this assertion is equivalent to saying that *the equations* (14.12) (with $\bar{\mathfrak{M}}$ and $\bar{\mathfrak{S}}$ replacing $\mathfrak{M}$ and $\mathfrak{S}$) admit of no non-trivial solution $p^{(a)}$. Assume that $\{p^{(a)}\}$ is a solution. To simplify notations suppose for the moment that the $\mathbf{x}^{(a)}$ are numbered so that the $p^{(a)}$ appear in decreasing order, say $p^{(1)} = \cdots = p^{(r)} > p^{(r+1)} \geqq \cdots \geqq p^{(m)}$ and $p^{(1)} > 0$. For $a \leqq r$ we have

$$\begin{aligned} p^{(a)} &\leqq p^{(a)} + \sum_{s=1}^n \sum_{t=1}^m \bar{M}^{(a,s)} U_\lambda^{(s,t)} p^{(t)} \\ &= \sum_{t=1}^m \bar{S}^{(a,t)} p^{(t)} \leqq p^{(a)} \sum_{t=1}^r \bar{S}^{(a,t)} + p^{(r+1)} \sum_{t=r+1}^m \bar{S}^{(a,t)} \\ &\leqq p^{(a)}. \end{aligned} \tag{14.27}$$

Here the extreme members can be equal only if

$$\bar{M}^{(a,s)} = 0, \qquad \sum_{t=1}^r \bar{S}^{(a,t)} = 1, \qquad \sum_{t=r+1}^m \bar{S}^{(a,t)} = 0, \quad a = 1, \cdots, r. \tag{14.28}$$

Under such circumstances the first r equations of the boundary conditions (14.3) reduce to

$$C(\omega^{(a)}) = \sum_{t=1}^r \bar{S}^{(a,t)} C(\omega^{(t)}), \qquad a = 1, \cdots, r \tag{14.29}$$

the $r \times r$ matrix on the right being strictly stochastic. There is no loss of generality in supposing that it contains no strictly stochastic submatrix.[44] But then

[43] It is interesting that this need not be true if redundancies have not been removed.

[44] If there exists a strictly stochastic submatrix of order $r' < r$ we are free to renumber the $\mathbf{x}^{(a)}$ so that this matrix appears in the left upper corner, and we get the system (14.29) with r replaced by r'.

(14.29) implies $C(\omega^{(i)}) = \cdots = C(\omega^{(i)})$. Here $r > 1$ since the diagonal terms $\bar{S}^{(a,a)} = 0$. Thus the choice of $\mathbf{x}^{(1)}, \cdots, \mathbf{x}^{(m)}$ is *redundant*, and we have supposed that all redundancies are removed. It follows that our boundary conditions are complete.

We proceed to show that *each row and each column of* $\mathfrak{U}$ *contains at most one infinite element.* Suppose, for example, that $U^{(1,1)} = U^{(1,2)} = \infty$. From (14.25) we see that in this case the first column of $\mathfrak{M}$ vanishes identically. Then the vector $\mathfrak{n}^{(1)}$ in no way affects the boundary conditions and is redundant. Next, assume $U^{(1,1)} = U^{(2,1)} = \infty$. Then $\bar{M}^{(a,1)} = \bar{M}^{(a,2)} = 0$ for $a > 1$. The first two columns of $\mathfrak{M}$ are linearly dependent and the system $\mathfrak{n}^{(1)}, \cdots, \mathfrak{n}^{(n)}$ is again redundant: the pair of vectors $\mathfrak{n}^{(1)}, \mathfrak{n}^{(2)}$ should be replaced by the single vector $\bar{M}^{(1,1)}\mathfrak{n}^{(1)} + \bar{M}^{(1,2)}\mathfrak{n}^{(2)}$.

We have thus proved that the conditions enumerated in Theorem 14.2 are *necessary*. The proof of the converse proceeds exactly as in Theorem 14.1. The matrix $\mathfrak{M}_\lambda$ of (14.2) now takes on the form $\mathfrak{M}_\lambda = (I - \bar{\mathfrak{S}} + \mathfrak{M}\mathfrak{U}_\lambda)^{-1}\mathfrak{M}$. In $I - \bar{\mathfrak{S}} + \mathfrak{M}\mathfrak{U}_\lambda$ the diagonal terms are $\geqq 1$. Since $\mathfrak{U}_\lambda \leqq \mathfrak{U}$, the terms off the main diagonal are non-negative and have sums strictly less than unity. The matrix therefore possesses a non-negative inverse.

15. Duality. The entrance boundary

Let π be a strictly positive row-vector such that $\pi\mathfrak{q}\Pi = \pi\mathfrak{q}$ (that is, $\pi \in \mathfrak{P}^*$ in the sense of Definition 10.1).

Definition 14.1. *We denote by* T_π *the transformation under which*

(15.1) *the column vector* $\mathbf{x}$ *maps into the row vector* $\mathbf{x}^*$: $x^*(i) = x(i)\pi(i)$
the row vector $\mathfrak{n}$ *maps into the column vector* $\mathfrak{n}^*$: $\eta^*(i) = \eta(i)\pi(i)$.

Put

$$\Pi^*(i, k) = \frac{\pi(k)q_k}{\pi(i)q_i}\,\Pi(k, i). \tag{15.2}$$

It is clear that

$$\Pi\mathbf{x} = \mathbf{x} \quad \textit{implies} \quad \mathbf{x}^*\mathfrak{q}\Pi^* = \mathbf{x}^*\mathfrak{q} \tag{15.3}$$

$$\mathfrak{n}\mathfrak{q}\Pi = \mathfrak{n}\mathfrak{q} \quad \textit{implies} \quad \Pi^*\mathfrak{n}^* = \mathfrak{n}^*. \tag{15.4}$$

Thus Π^* is a kind of transpose of Π which enables us to translate conclusions concerning transformations $\mathbf{c} \to P_\lambda\mathbf{c}$ into theorems on the dual transformation $\varrho \to \varrho P_\lambda$.

The transformation T_π applied either to the columns or to the rows of P_λ leads us to introduce the matrix P_λ^* defined by

$$P_\lambda^*(i, k) = \frac{\pi(k)}{\pi(i)}\,P_\lambda(k, i). \tag{15.5}$$

Also, we introduce the operator Ω^* which is the dual to our Ω by

(15.6) $$\Omega^*(i, k) = -q_i + q_i\Pi^*(i, k).$$

We note the following facts, which may be verified by trivial arithmetic.

THEOREM 15.1.

(15.7) $$\lambda\mathbf{C} - \Omega\mathbf{C} = \mathbf{c} \quad \textit{implies} \quad \lambda\mathbf{C}^* - \mathbf{C}^*\Omega^* = \mathbf{c}^*$$

(15.8) $$\lambda\mathbf{H} - \mathbf{H}\Omega = \mathfrak{n} \quad \textit{implies} \quad \lambda\mathbf{H}^* - \Omega^*\mathbf{H}^* = \mathfrak{n}^*.$$

In particular, the first Kolmogorov equation $(\lambda + \mathfrak{q})P_\lambda = \mathfrak{q}\Pi P_\lambda + I$ *implies the analogue of the second, namely* $P_\lambda^*(\lambda + \mathfrak{q}) = P_\lambda^*\mathfrak{q}\Pi^* + I$, *and similarly*

$$P_\lambda(\lambda + \mathfrak{q}) = P_\lambda\mathfrak{q}\Pi + I$$

goes over into $(\lambda + \mathfrak{q})P_\lambda^* = \mathfrak{q}\Pi^*P_\lambda^* + I$.

Obviously minimal solutions are mapped into minimal solutions.

In Section 5 we have introduced the *boundary* $\mathfrak{B}$ *of our countable space* E *as induced by* $\mathfrak{P}$, that is, by the *bounded* positive solutions of $\Pi\mathbf{x} = \mathbf{x}$. It was shown that a subset $\bar{\mathfrak{B}}$ of $\mathfrak{B}$ coincides with the boundary induced by each Π_λ taken individually. While $\mathfrak{B}$ depends on Π only, the exit boundary $\bar{\mathfrak{B}}$ depends both on Π and $\mathfrak{q}$. All these notions are tailored to the unit vector $\mathbf{1}$ on which the very definition of a substochastic Π depends. For the dual problem, when our matrices operate on row vectors, there exists no counterpart to $\mathbf{1}$. In fact, the *bounded* vectors $\mathfrak{n}_\lambda \in \mathfrak{P}_\lambda^*$ used in the construction of our matrices P_λ depend on *unbounded* solutions $\mathfrak{n} \in \bar{\mathfrak{P}}^*$.

Now it has been shown in [4] that the restriction to bounded $\mathbf{x}$ is not necessary. The boundary $\mathfrak{B}$ defined in terms of the bounded vectors in $\mathfrak{P}$ is a subset of a larger set, the *total boundary* of E defined by Π: each positive solution $\mathbf{x}$ of $\Pi\mathbf{x} = \mathbf{x}$ (bounded or unbounded) induces a subset of the total boundary and a class of neighborhoods, and these definitions are mutually consistent.[45]

Under the mapping T_π the vector π goes into $\mathbf{1}$, and each $\varrho \in \mathfrak{P}^*$ is mapped into a positive solution ϱ^* of $\Pi^*\varrho^* = \varrho^*$. The *adjoint boundary* $\mathfrak{B}^*$ defined by Π has been introduced in [4] as synonymous with the total boundary defined by Π^*; this is independent of the choice of π and also of $\mathfrak{q}$. We are, of course, not

[45] The mapping $\mathbf{y} \to \tilde{\mathbf{y}}$ where $\tilde{y}(i) = y(i)/x(i)$ sends each solution of $\Pi\mathbf{y} = \mathbf{y}$ into a solution of $\tilde{\Pi}\tilde{\mathbf{y}} = \tilde{\mathbf{y}}$ where $\tilde{\Pi}(i, k) = \Pi(i, k)x(k)/x(i)$. In particular, $\mathbf{x}$ itself is mapped into $\mathbf{1}$. The ordinary boundary (as described in Section 6) defined by $\tilde{\Pi}$ becomes the subset of the total boundary induced by $\mathbf{x}$.

The same mapping applied to the columns of P_λ sends each solution of

$$(\lambda + \mathfrak{q})P_\lambda = \mathfrak{q}\Pi P + I$$

into a matrix $\tilde{P}_\lambda$ with elements $\tilde{P}_\lambda(i, k) = P_\lambda(i, k)x(k)/x(i)$ which is a solution of the Kolmogorov equation $(\lambda + \mathfrak{q})\tilde{P}_\lambda = \mathfrak{q}\tilde{\Pi}\tilde{P}_\lambda + I$. In this way *each solution* $\mathbf{x}$ *of* $\Pi\mathbf{x} = \mathbf{x}$ *can be made to play the role of* $\mathbf{1}$, *and this relativization permits one to generalize all results of this paper and to construct semi-groups generated by the Kolmogorov equations where the matrices* $P(t)$ *are not necessarily substochastic.*

interested in the whole adjoint boundary, but only in that part of it which is induced by the vectors in $\bar{\mathfrak{P}}^*$, the canonical images of vectors in $\mathfrak{P}_\lambda^*$. (That part naturally depends on $\mathfrak{q}$). Now if $\mathbf{n} \in \bar{\mathfrak{P}}^*$ is the canonical image of $\mathbf{n}_\lambda \in \mathfrak{P}_\lambda^*$ then $\mathbf{n}_\lambda$ is the *maximal* element of $\mathfrak{P}_\lambda^*$ dominated by $\mathbf{n}$. The mapping $T_\mathbf{n}$ sends $\mathbf{n}$ into $\mathbf{1}$ and, accordingly, $\mathbf{n}_\lambda$ into the maximal solution of $\lambda \mathbf{n}_\lambda^* - \Omega^* \mathbf{n}_\lambda^* = \mathbf{0}$ dominated by $\mathbf{1}$. This is the element (denoted in Section 5 by $\bar{\mathbf{x}}_\lambda$) defining the *exit boundary* for Π^* (with $\boldsymbol{\pi} = \mathbf{n}$). We now introduce the

DEFINITION 15.1 *The entrance boundary defined by the matrices* Π *and* $\mathfrak{q}$ *is the union of the exit boundaries* $\bar{\mathfrak{B}}$ *defined by* $\mathfrak{q}$ *and the matrices* Π^* *of* (15.2) *as* $\boldsymbol{\pi}$ *runs through* $\bar{\mathfrak{P}}^*$. *The exit boundary for a particular* Π^* *is called the subset of the entrance boundary induced by* $\boldsymbol{\pi} \in \bar{\mathfrak{P}}^*$. *The definition of neighborhoods is taken over without change.*

These notions enable us to describe the action of our matrices P_λ as operators on row vectors or measures.

16. Lateral conditions for the second Kolmogorov equation

$$\varrho \rightarrow \mathbf{R}_\lambda = \varrho P_\lambda \tag{16.1}$$

where P_λ is any of the matrices constructed above. We can use the duality described in Section 15 to determine the range of (16.1) in all cases where P_λ satisfies the Kolmogorov equation (3.4).

Let us begin with $P_\lambda^{\min}$. If $\mathbf{n} \in \bar{\mathfrak{P}}^*$ (Definition 10.2), then $\lambda \mathbf{n} P_\lambda^{\min} = \mathbf{n} - \mathbf{n}_\lambda < \mathbf{n}$. Accordingly, the transformation of Section 15 applied using $\boldsymbol{\pi} = \mathbf{n}$ will transform $P_\lambda^{\min}$ into a matrix P_λ^* such that $\lambda P_\lambda^* \mathbf{1} \leqq \mathbf{1}$. We can then translate Theorem 6.2 as follows.

THEOREM 16.1. *Let* $\mathbf{R}_\lambda^{\min} = \varrho P_\lambda^{\min}$. *If* γ *is a set of the entrance boundary and* $\mathbf{n} \in \bar{\mathfrak{P}}^*$ *the element which induces it, then as* i *approaches* γ

$$\lim \frac{R_\lambda^{\min}(i)}{\eta(i)} = 0. \tag{16.2}$$

Conversely, if $\mathbf{R}$ *is row-vector such that* $\mathbf{R}\Omega$ *is bounded, and if* $R(i)/\eta(i)$ *approaches* 0 *as* i *approaches* γ *(where* γ *is arbitrary), then* $\mathbf{R}$ *is in the range of* $P_\lambda^{\min}$.

As in Section 11 we see that whenever $\mathbf{R}\Omega$ is bounded there exists exactly one $\mathbf{n} \in \bar{\mathfrak{P}}^*$ such that $\mathbf{R} - \mathbf{n}$ is in the range of $P_\lambda^{\min}$. The inner product $[\mathbf{R}\Omega, \mathbf{x}]$ where $\mathbf{x} \in \bar{\mathfrak{P}}$ converges since both $\mathbf{R}\Omega$ and $\mathbf{x}$ are bounded, and therefore we do not require the special definition of Section 11. Keeping the notations of Section 11, put $\varrho_\lambda = \lambda \mathbf{R} - \mathbf{R}\Omega$. Then $\mathbf{R} = \varrho_\lambda P_\lambda^{\min} + \mathbf{n}_\lambda$ and hence for $\mathbf{x} \in \bar{\mathfrak{P}}$

$$\begin{aligned}[\mathbf{R}\Omega, \mathbf{x}] &= -[\varrho_\lambda, \mathbf{x}] + [\varrho_\lambda, \lambda P_\lambda^{\min} \mathbf{x}] + [\mathbf{n}_\lambda, \mathbf{x}] \\ &= -[\varrho_\lambda, \mathbf{x}_\lambda] + [\mathbf{n}_\lambda, \mathbf{x}]\end{aligned} \tag{16.3}$$

just as in (11.7).

It is now easy to verify:

THEOREM 16.2. *If* P_λ *is the matrix of Theorem* 12.1 *then the range of the transformation* (16.1) *is characterized by the lateral condition*

$$(16.4)\qquad [\mathbf{R}\Omega, \bar{\mathbf{x}}] + m\cdot\lim \frac{R(i)}{\eta(i)} = 0$$

the limit being taken as i approaches the set of the entrance boundary induced by $\mathfrak{n}$.
(This is the analogue to (12.7)).

For the general case of Section 14 let us denote by $\gamma^{(b)}$ *the set of the entrance boundary induced by* $\mathfrak{n}^{(b)}(b = 1, 2, \cdots, n)$ *and put*

$$(16.5)\qquad \mathbf{R}^{(b)} = \lim \frac{R(i)}{\eta^{(b)}(i)}$$

where i approaches $\gamma^{(b)}$. Then $\{\mathbf{R}\}_1'^n$ represents the $1 \times n$ matrix whose elements are the row-vectors $\mathbf{R}^{(b)}$, and we have

THEOREM 16.3. *For P_λ of Theorem* 13.1 *the range of* (16.1) *is determined by the lateral condition (derived from* (13.2))

$$(16.6)\qquad \{\mathbf{R}\}_1'^n - \{\mathbf{R}\}_1'^n\mathfrak{U}\mathfrak{M} + \{[\mathbf{R}\Omega, \mathbf{x}]\}'\mathfrak{M} = 0.$$

Note that in (13.2) there occurred the $m \times m$ matrix $\mathfrak{S} = \mathfrak{M}\mathfrak{U}$ which was substochastic. In (15.6) the matrix $\mathfrak{U}\mathfrak{M}$ is an $n \times n$ matrix. The m equations (13.2) go over into n equations.

The same rephrasing applies to Theorem 14.2. It is clear how a combination of Theorems 12.2 and 13.2 permits to enlarge the space E by a finite number of elements and to construct processes on the space thus enlarged.

17. The bilateral birth process

The simplest illustrative example is obtained with a slight change of notation by taking for E the space of *all* integers. Define for $i, k = 0, \pm 1, \pm 2$ constants q_k and a matrix Π so that

$$(17.1)\qquad \sum_{k=-\infty}^{+\infty} \frac{1}{q_k} < \infty \qquad \Pi(i, i+1) = 1$$

(whence $\Pi(i, j) = 0$ for $j \neq i + 1$). The only solution of $\Pi\mathbf{x} = \mathbf{x}$ is $\mathbf{x} = \mathbf{1}$, while the only solution of $\mathfrak{n}q\Pi = \mathfrak{n}q$ is given by $\eta(k) = q_k^{-1}$. It is easily seen that

$$(17.2)\qquad x_\lambda(i) = \prod_{j=i}^{\infty} \frac{q_j}{\lambda + q_j} \qquad \eta_\lambda(k) = \frac{1}{\lambda + q_k} \prod_{j=-\infty}^{k-1} \frac{q_j}{\lambda + q_j}$$

$$(17.3)\qquad P_\lambda^{\min}(i, k) = \begin{cases} 0 \quad \text{if} \quad k < i \\ \dfrac{q_i}{\lambda + q_1} \cdots \dfrac{q_{k-1}}{\lambda + q_{k-1}} \cdot \dfrac{1}{\lambda + q_k} & \text{if } k \geqq i. \end{cases}$$

Note that $P_\lambda^{\min}(i, k)q_k$ represents the Laplace transform of the flux out of k and so

$$(17.4)\qquad x_\lambda(i) = \lim_{k\to\infty} P_\lambda^{\min}(i, k)$$

neatly expresses $x_\lambda(i)$ as the Laplace transform of the "time of arrival at $-\infty$". Similarly

$$\eta_\lambda(k) = \lim_{i\to-\infty} P_\lambda^{\min}(i, k) \tag{17.5}$$

shows that $\eta_\lambda(k)$ is indeed the Laplace transform of the probability of being at k if the starting point is at $-\infty$ (= the entrance boundary).

In order to discuss the matrix P_λ of Theorem 12.1 note that in our case

$$\lambda[\mathbf{n}, \mathbf{x}_\lambda] = \lambda \sum_{-\infty}^{+\infty} \frac{x_\lambda(i)}{q_i} \sum_{-\infty}^{+\infty} \{x_\lambda(i+1) - x_\lambda(i)\} = (1 - \delta) \tag{17.6}$$

where we have put for abbreviation

$$\delta = \prod_{j=-\infty}^{+\infty} \frac{q_j}{\lambda + q_j} = x_\lambda(-\infty). \tag{17.7}$$

Accordingly, the matrix of Theorem 12.1 becomes

$$P_\lambda(i, k) = P_\lambda^{\min}(i, k) + \frac{x_\lambda(i)\eta_\lambda(k)}{m + 1 + \delta}, \qquad m \geqq 0. \tag{17.8}$$

The boundary condition (12.7) reduces to

$$\begin{aligned} 0 = mC(\infty) + [\mathbf{n}, \Omega\mathbf{C}] &= mC(-\infty) + \sum_{i=-\infty}^{+\infty} \frac{1}{q_i} \Omega C(i) \\ &= (m+1)C(\infty) - C(-\infty) \end{aligned} \tag{17.9}$$

and the adjoint boundary condition (16.4) reduces similarly to

$$(m + 1) \lim_{i\to-\infty} \frac{R(i)}{q_i} = \lim_{i\to+\infty} \frac{R(i)}{q_i}. \tag{17.10}$$

Now $R(i)/q_i$ is probabilistically the Laplace transform of the flux at i (mass times q^{-1}) and so (17.10) expresses the lateral conditions: "Flux at $-\infty$ equals $1/(m + 1)$ flux at $+\infty$." Probabilistically this leads to the following interpretation of the process: starting from i "the particle" moves in simple steps to the right until it reaches ∞. At that moment it has probability $m/(m + 1)$ to disappear (the process terminating), and probability $1/(m + 1)$ to recommence the movement starting from $-\infty$. The same is expressed by (17.8). In fact, the Laplace transform of being at time t at k after having passed through ∞ exactly n times is obviously $x_\lambda(i)\eta_\lambda(k)\cdot((1 - \delta)/(m + 1))^n$ and summing over n leads to (17.8). It is now evident that we can modify the process by an identification of the exit boundary ∞ and the entrance boundary $-\infty$, and permitting a finite sojourn time at this new point. Theorem 12.2 describes the corresponding P_λ. Similarly the Theorems of Sections 8, 9 describe the possibility of a direct return from ∞ to k without passing through $-\infty$.

In the present example the situation is particularly simple and intuitive due to the fact that $\mathbf{C}$, $\mathbf{x}$, etc. are *continuous not only at the exit boundary, but also on the entrance boundary* $-\infty$. This leads, for example, to the pleasing direct interpretation of the functional $\lambda[\mathbf{n}, \mathbf{x}_\lambda]$ in (17.6) by $x_\lambda(-\infty)$ which is the Laplace transform of the distribution function of the time spent on a complete trip from $-\infty$ to

$-\infty$. Whether similar relations can be established in the general case, remains an open question.

PRINCETON UNIVERSITY

REFERENCES

[1] K. L. CHUNG, *Foundations of the theory of continuous parameter Markov chains*, to appear in Proceedings Third Berkeley Symposium on Probability and Statistics.

[2] J. L. DOOB, *Markoff chains—denumerable case*, Trans. Amer. Math. Soc., 58 (1945), 455–473.

[3] ———, Stochastic processes, J. Wiley and Sons, New York 1953.

[4] W. FELLER, *Boundaries induced by positive matrices*, Trans. Amer. Math. Soc., 83 (1956), 19–54.

[5] ———, *On the integro-differential equations of purely discontinuous Markoff processes*, Trans. Amer. Math. Soc., 48 (1940), 488–515 and Errata 58 (1945), 474.

[6] ———, An introduction to probability theory and its applications, Vol. 1, J. Wiley and Sons, New York, 1950.

[7] W. FELLER and H. P. MCKEAN JR., *A diffusion equivalent to a countable Markov chain*, Proc. Nat. Acad Sci. U. S. A., 42 (1956), 351-354.

[8] E. HILLE, *Perturbation methods in the study of Kolmogoroff's equations*, Proceedings International Congress of Mathematicians, 3 (1956), 365–376.

[9] S. KARLIN and J. MCGREGOR, *Representation of a class of stochastic processes*, Proc. Nat. Acad. Sci. U.S.A. (6), 4 (1955), 387-391.

[10] T. KATO, *On the semi-groups generated by Kolmogoroff's differential equations*, J. Math. Soc. Jap., 6 (1954), 1-15.

[11] D. G. KENDALL, *Some further pathological examples in the theory of denumerable Markov processes*, Quart. J. Math. Oxford Ser., 7 (1956), 39–56.

[12] D. G. KENDALL and G. E. H. REUTER, *Some pathological Markov processes with a denumerable infinity of states and the associated semi-groups of operators on l*, Proceedings International Congress of Mathematicians, Amsterdam, 1954.

[13] A. N. KOLMOGOROFF, *Über die analytischen Methoden in der Wahrscheinlichkeitsrechnung*, Math. Ann., 104 (1931), 415–458.

[14] A. N. KOLMOGOROV, *On some problems concerning the differentiability of the transition probabilities in a temporally homogeneous Markov process having a denumerable set of states*, Učenye Zapiski Moskovskogo Gos. Univ. (4), 148 (1951), 53–59.

[15] W. LEDERMANN and G. E. REUTER, *Spectral theory for the differential equations of simple birth and death processes*, Philos. Trans. Roy. Soc. London, Ser. A, 246 (1954), 321-369.

[16] P. LÉVY, *Systèmes Markoviens et stationnaires; cas dénombrable*, Ann. École Norm. (3), 68 (1951), 327-381.

[17] D. V. WIDDER, The Laplace transform, Princeton University Press, Princeton, 1946.

ANNALS OF MATHEMATICS
Vol. 68, No. 3, November, 1958
Printed in Japan

NOTES TO MY PAPER

"On boundaries and lateral conditions for the Kolmogorov differential equations"

These Annals, 65 (1957), 527–70

BY WILLIAM FELLER

(Received June 25, 1958)

I am indebted to Professors K. L. Chung and G. E. H. Reuter for critical remarks which led to the following elucidations and amendments.

Footnote 11, p. 533, contains a rephrasing of a correct theorem of the text, but the italicized condition "if each of the two boundaries reduces to a point" should read: "if the entrance boundary reduces to a single point". The paper makes no further reference to this incidental remark.

On p. 541, Theorem 6.1 (b), the proof of the equivalence of the several topologies is based on the inequality $\mathbf{s}_{\mathfrak{U}}^{(\lambda)} + \mathbf{s}_{\mathfrak{B}}^{(\lambda)} \leqq \bar{\mathbf{x}}_\lambda$ which is stated as obvious. Even the equality follows by a straightforward lattice theoretic argument whose probabilistic version may be stated in words as follows.

Consider the two Markov chains governed, respectively, by Π_λ and by Π. Since $\Pi_\lambda \leqq \Pi$, the probability of any event in the common sample space is smaller under the Π_λ chain. It follows that in this chain the probability of the system's passing infinitely often in and out of the set $\mathfrak{U}$ is zero: if the process continues forever it must ultimately stay either in $\mathfrak{U}$ or in $\mathfrak{B}$. The probabilities of these three events are given by the vectors $\bar{\mathbf{x}}_\lambda$, $\mathbf{s}_{\mathfrak{U}}^{(\lambda)}$, and $\mathbf{s}_{\mathfrak{B}}^{(\lambda)}$ and hence $\bar{\mathbf{x}}_\lambda = \mathbf{s}_{\mathfrak{U}}^{(\lambda)} + \mathbf{s}_{\mathfrak{B}}^{(\lambda)}$.

p. 543, formula (7.1). Replace 1 by $u(i)$.

pp. 545–546. A regrettable notational confusion led to the following misstatements which fortunately have no further effect: in (8.11) the numerator $a(\lambda)$ should be replaced by $\bar{a}(\lambda)$, in (8.13) the factor m should be $\overline{m} = m + [\alpha, \mathbf{1} - \mathbf{x}]$, and in (8.16) and (8.17) the factor m should be replaced by $\overline{m}^{-1}$.

p. 548, footnote 28. The complete monotonicity of $a(\lambda)$ follows on clearing (9.7) of the denominator and differentiating using Leibniz' rule.

p. 549, last line of formula (9.18). The constant m should again be replaced by $\overline{m}$. More serious is that the operator Ω refers to the minimal semigroup and is therefore senseless in the present context. Actually Ω should be replaced by Ω^*, the generator of the P_λ-semigroup, and the formula simply restates its definition: the matrix Ω^* is defined by the identity $P_\lambda \Omega^* \mathbf{x} = (\lambda P_\lambda - I)\mathbf{x}$, where x is an arbitrary column vector.

Sections 13 and 14 are based on the assumption that the matrices $\mathfrak{U}_\lambda$ are strictly positive. As explained in Section 5 this assumption means no loss of generality beyond the elimination of trivial cases of decomposable matrices which require more words than thoughts.

In Theorem 14.2 the case $m = n = 1$ (dealt with previously) is excluded : the reasoning following (14.21) does not apply to it. To cover this case the boundary condition (14.13) should be rewritten in the homogeneous form to include the case of " infinite " constants.

In Sections 8 and 9 the construction of semi-groups uses the minimal solutions as point of departure. Perhaps I should have emphasized more explicitly that the *same construction can be superimposed to the more general solutions of Section* 14. When the exit boundary consists of $m < \infty$ points we obtain in this way *the most general admissible solution*[1]. The paper concentrates on solutions satisfying both systems of Kolmogorov's equations because its main purpose was to discover an abstract formulation of the classical boundary conditions of potential theory (e.g., to find the general nature of "normal derivatives "); the particular point of interest was to discover the relations of adjointness when the operators are strongly unsymmetric and the two induced topologies are essentially different.

PRINCETON UNIVERSITY

[1] I am informed by G. E. H. Reuter that he found an alternative construction of the most general solution in the case $m = 1$. In this case the admissible semigroups are in one-to-one correspondence with those for harmonic functions and those generated by Sturm-Liouville operators: both the construction and the uniqueness proof may be carried over from my paper: *Generalized second order differential operators and their lateral conditions*, Illinois Journal of Mathematics vol. 1 (1957) pp. 459–504. (This paper completely supersedes my previous work in this direction). In a paper to appear in the Journal des Mathématiques dedicated to M. Fréchet certain discrete systems are treated by the methods of partial differential equations showing that the analogies mentioned are not purely formal.

ON BOUNDARIES DEFINED BY STOCHASTIC MATRICES[1]

BY

WILLIAM FELLER

An operator A defined on a space E may in a natural way define a boundary for E. If A is unsymmetric, the adjoint operator may define a different boundary. The case of a denumerable space E is particularly instructive because of the absence of a topology defined independently of A. In this abstract we consider only the case of a denumerable E. The two boundaries need not be denumerable, and their topological structure may be complicated. These boundaries are useful for the formulation of lateral conditions and the construction of semigroups. These show surprisingly close analogies with harmonic functions, diffusion, etc.

1. The cone $\mathfrak{P}$. Let E be the set of points $1, 2, 3, \cdots$. Boldface small letters will denote column vectors, that is, $\mathbf{x}$ stands for a vector with components $x(1), x(2), \cdots$. We put $\|\mathbf{x}\| = \sup |x(i)|$. In particular $\mathbf{1}$ stands for the vector all of whose components equal 1.

Consider now a matrix Π with elements $\Pi(i,j)$ such that

$$\Pi(i,j) \geqq 0, \qquad \sum_j \Pi(i,j) \leqq 1. \tag{1}$$

An equation of the form

$$\mathbf{y} = \Pi\mathbf{x}, \qquad \text{that is,} \qquad y(i) = \sum_j \Pi(i,j)x(j), \tag{2}$$

states that the value of $y(i)$ is an average of the values $x(j)$. Such an equation therefore has many properties in common with the relation

$$v = \Delta u, \tag{3}$$

where Δ is the Laplacian. In particular, the solutions of

$$\mathbf{x} = \Pi\mathbf{x} \tag{4}$$

are close analogues of harmonic functions.

We denote by $\mathfrak{P}$ *the totality of all nonnegative solutions* $\mathbf{x}$ *of* (4) *such that* $\|\mathbf{x}\| \leqq 1$.

Each bounded solution of (4) is a linear combination of solutions in $\mathfrak{P}$. The set $\mathfrak{P}$ is a *positive cone* in the sense that if $\mathbf{x}_i \in \mathfrak{P}$, then also $p_1\mathbf{x}_1 + p_2\mathbf{x}_2 \in \mathfrak{P}$ whenever $p_i \geqq 0$ and $p_1 + p_2 \leqq 1$.

It is well known that such a cone is completely determined by its *extremal elements* in the sense of Krein and Mil'man. We shall now describe these extremal elements and their probabilistic interpretation. It will be seen that

[1] Abstract.

R.L. Schilling et al. (eds.), *Selected Papers II*,

the extremal elements determine a *boundary* for E which can be of a very complicated structure.

2. Extremal elements and sojourn probabilities. For an arbitrary set $\Gamma \subset E$ write

$$\Pi(i,\Gamma) = \sum_{j\varepsilon\Gamma} \Pi(i,j). \tag{5}$$

It is easily seen that the quantity $\Pi^n(i,E)$ decreases monotonically as $n \to \infty$, and we put

$$s_E(i) = \lim_{n\to\infty} \Pi^n(i,E).\text{[2]} \tag{6}$$

Clearly $0 \leqq \mathbf{s}_E \leqq \mathbf{1}$, and $\mathbf{s}_E = \mathbf{1}$ if, and only if, $\Pi(i,E) = 1$, that is, if Π is stochastic. If $\mathbf{s}_E = 0$, then $\mathfrak{P}$ is empty, and (4) has no bounded solutions. Under any circumstances $\mathbf{s}_E$ is the *maximal element in* $\mathfrak{P}$, that is,

$$\Pi\mathbf{s}_E = \mathbf{s}_E, \quad \text{and} \quad \mathbf{x} \leqq \mathbf{s}_E \qquad (\text{for each } \mathbf{x} \,\varepsilon\, \mathfrak{P}). \tag{7}$$

Probabilistically the matrix Π defines a random walk (or Markov chain), and $\mathbf{s}_E(i)$ is the probability that the processes starting at i will not terminate after finitely many steps.

We now define the *restriction of* Π *to the set* Γ by

$$\Pi_\Gamma(i,j) = \begin{cases} \Pi(i,j), & (\text{if } j \,\varepsilon\, \Gamma), \\ 0, & (\text{if } j \not\varepsilon\, \Gamma). \end{cases} \tag{8}$$

By analogy with (6) we put

$$\sigma_\Gamma(i) = \begin{cases} \lim_{n\to\infty} \Pi_\Gamma^n(i,\Gamma), & (\text{if } i \,\varepsilon\, \Gamma), \\ 0, & (\text{if } i \not\varepsilon\, \Gamma) \end{cases} \tag{9}$$

and

$$s_\Gamma(i) = \lim_{n\to\infty} \sum_j \Pi^n(i,j)\sigma_\Gamma(j), \qquad (i \,\varepsilon\, E). \tag{10}$$

Of course, for most sets Γ we have $\boldsymbol{\sigma}_\Gamma = \mathbf{s}_\Gamma = 0$. *If* $\mathbf{s}_\Gamma$ *(and therefore* $\boldsymbol{\sigma}_\Gamma$*) does not vanish identically, then* Γ *is called a sojourn set and* $\mathbf{s}_\Gamma$ *the sojourn probability corresponding to* Γ.

For our random walk, $\sigma_\Gamma(i)$ is the probability that, starting from $i \,\varepsilon\, \Gamma$, the system will never leave Γ. And $s_\Gamma(i)$ is the probability that, starting from an arbitrary i, the system will, after a finite number of steps, pass into Γ and stay there forever. If the chain is recurrent and irreducible, then E is the only sojourn set. If all states are transient, then the complement of any finite set is a sojourn set.

The sojourn probabilities coincide with the extremals of $\mathfrak{P}$ *in the sense of Krein and Mil'man.* Thus every $x \,\varepsilon\, \mathfrak{P}$ is essentially a positive linear combination of sojourn probabilities.

[2] Here and below, Π^n is the nth (matrix) power of Π.

3. Boundary points. If all states i are recurrent, then $\mathfrak{P}$ contains only scalar multiples of the solution **1**, and E has no boundary.[3] In order not to burden our statements by obvious exceptions, we shall now assume that *all* states are transient. Analytically, this condition is equivalent to

$$\sum_n \Pi^n(i,i) < \infty, \qquad (i = 1, 2, \cdots). \tag{11}$$

Suppose then that Γ is a sojourn set. There exists a monotonically decreasing sequence of sets $\Gamma_n \subset \Gamma$ contracting to the empty set such that for each i

$$s_{\Gamma_n}(i) = s_\Gamma(i). \tag{12}$$

Probabilistically this means: the event "the system ultimately goes into Γ and stays forever in Γ" is equivalent to each of the events "the system ultimately stays in Γ_n." In other words, if this event takes place, the system passes successively into $\Gamma_n, \Gamma_{n+1}, \cdots$. Now $\Gamma_n \downarrow 0$, and therefore the net $\{\Gamma_n\}$ may be described as an *escape route to infinity*.

To describe the simplest situation, consider the case where there exist exactly N linearly independent bounded solutions of (4). Then there exist exactly N *nonoverlapping sojourn sets* $\Gamma^{(1)}, \cdots, \Gamma^{(N)}$, *and*

$$\mathbf{s}_E = \mathbf{s}_{\Gamma^{(1)}} + \cdots + \mathbf{s}_{\Gamma^{(N)}}. \tag{13}$$

Starting from an arbitrary point i, the process either stops after finitely many steps [with probability $1 - s_E(i)$], or it passes into one of the sets $\Gamma^{(k)}$ and ultimately remains restricted to this set. In the latter case the system moves along the escape route corresponding to $\Gamma^{(k)}$. In other words, there are N mutually exclusive escape routes to infinity, and the system is bound to move along one of them.

In this case we speak of a *boundary consisting of N isolated points.* If $\Gamma_1 \supset \Gamma_2 \supset \cdots$ is an escape route, each Γ_k is a neighborhood of the corresponding boundary point.

The simplicity of the case just considered stems from the fact that with an *isolated* boundary point there is associated a solution $\mathbf{x} \in \mathfrak{P}$, that is, a *positive* probability of the system converging to it. Unfortunately, $\mathfrak{P}$ may contain a continuum of independent solutions, and the boundary may have a very complicated topological structure. To give a precise definition of the boundary in terms of $\mathfrak{P}$ it is necessary to consider $\mathfrak{P}$ as a lattice and define the boundary as the *maximal ideal space*. In this abstract we shall describe only the analogy with harmonic functions.

Let $\mathfrak{P}'$ be the cone of bounded positive functions harmonic, say, in the circular disk $r < 1$. Each $u \in \mathfrak{P}'$ can be represented by means of the Poisson integral in terms of its boundary values. A harmonic function u is an *extremal*

[3] The set E corresponds to a closed manifold if the chain is irreducible and to a union of closed manifolds if the chain is reducible.

of $\mathfrak{P}'$ if its boundary values are 1 on a set Λ of the unit circle and 0 almost everywhere on the complement of Λ. Again such an extremal admits of a probabilistic interpretation: in the Wiener process (or ordinary diffusion) $u(P)$ represents the probability that a particle starting at P will ultimately be absorbed in Λ (rather than in the complement).

Thus the elements $u \in \mathfrak{P}'$ are in one-to-one correspondence with the sets of positive measure on the circle. To each *point* Q of the circle there corresponds an *unbounded* harmonic function, *viz.*, the radial derivative of the Green function $G(P,Q)$. In the trite case of a circular domain there is, of course, no point in defining the boundary in terms of $\mathfrak{P}'$. However, for more complicated domains, *e.g.*, when the boundary contains prime ends, it has been well known since the pioneer work of Martin that it is convenient to disregard the "ordinary" topology of the boundary and replace it by one more intimately connected with properties of the Δ-operator. Finally, in our case there is no a priori defined boundary, and we are compelled to introduce one in terms of Π.

4. Equivalent operators; unbounded solutions. Up to now we have considered only bounded solutions of (4). There exist also unbounded ones, but their study and their probabilistic meaning can be reduced to the bounded case. Let $\mathbf{z}$ be an arbitrary positive[4] solution of $\Pi\mathbf{z} = \mathbf{z}$, and put

$$\Pi'(i,j) = \Pi(i,j)\,\frac{z(j)}{z(i)}. \tag{14}$$

Then Π' is a strictly stochastic matrix, that is, Π' satisfies the conditions (1) with strict equality at the second place. To each solution $\mathbf{x}$ of (4) the transformation

$$\mathbf{x} \to \mathbf{x}', \qquad x'(i) = \frac{x(i)}{z(i)} \tag{15}$$

makes correspond a solution $\mathbf{x}'$ of $\mathbf{x}' = \Pi'\mathbf{x}'$. Clearly $\mathbf{x}'$ is bounded if, and only if, $\mathbf{x}$ is dominated by a scalar multiple of $\mathbf{z}$. If Π is strictly stochastic, then the relation (15) with $\mathbf{x} = \mathbf{1}$ shows that (14) is reversible and the relation between Π and Π' symmetric.

We see thus that (14) establishes an equivalence relation between strictly stochastic matrices and that there is a one-to-one correspondence between the solutions (4) for equivalent matrices. If in (14) the solution $\mathbf{z}$ is unbounded, the boundary of Π as defined in Sec. 3 can be enlarged by adding to it the boundary of Π'. In short, a bigger and better boundary may be defined by considering the *total* boundary of E relative to *all* solutions of $\Pi x = x$ or, what amounts to the same, relative to the set of all equivalent matrices Π'.

In this formulation the procedure is bound to appear artificial. Its use in the formulation of the appropriate lateral conditions for the Kolmogoroff differential equations shows that notions are natural and applicable. More-

[4] Only obvious conventions are required if $z(i) = 0$ for certain i.

over, (14) is a close analogue of certain transformations frequently used for Sturm-Liouville operators and in diffusion theory. Probabilistically, Π' is obtained from Π by a conditioning process. The reader is asked to interpret (14) in the simple case where $\mathbf{z}$ is a sojourn probability.

5. The adjoint boundary. Up to now Π was considered as an operator by premultiplication on column vectors. The adjoint operation consists in post-multiplying row vectors by Π. To (4) there corresponds the equation

$$\mathbf{a} = \mathbf{a}\Pi, \qquad \text{that is,} \qquad a(j) = \sum_i a(i)\Pi(i,j). \tag{16}$$

For the case of a probability vector, that is, when $a(i) \geqq 0$, $\Sigma a(i) = 1$, Kolmogoroff has introduced the matrix

$$\Pi^*(i,j) = \Pi(j,i)\,\frac{a(j)}{a(i)} \tag{17}$$

as the time reversion of Π. Clearly Π^* is a strictly stochastic matrix. Recent investigations by Derman, Harris, and Robbins show that it is useful and meaningful to consider (17) also when $\Sigma a(i) = \infty$.

We can now define the boundary of Π^*, and in this way *endow E with two boundaries*. Note that if $\mathbf{a}$ in (17) is replaced by another solution of (16), then Π^* is replaced by an *equivalent* stochastic matrix so that the new boundary of E is independent of the choice of $\mathbf{a}$.

In Sec. 3 we have seen that the random walk defined by Π moves *toward* the boundary as there defined. Probabilistically, therefore, this boundary is an *exit boundary*. For the matrix Π^* the direction of motion is reversed, and therefore the exit boundary for Π^* is an *entrance boundary*. In principle, the two boundaries occur in ordinary diffusion theory, say, in an interval $a \leqq x \leqq b$. In the classification introduced by the author, b may play the role of an exit boundary, an entrance boundary, of both (then it was called "regular"), or of neither ("natural"). In the spirit of the present theory a natural boundary is no boundary at all.

However, there is an essential and interesting difference between our case and all classical cases: usually the underlying space has a natural topology, and a point of the "ordinary" boundary may or may not belong to either of the two boundaries defined by an operator. Our space E is not endowed by any natural topology, and *there is absolutely no connection between the two boundaries*. Three points of the exit boundary together may form two points of the entrance boundary, etc. The more remarkable is the fact that in the case of differential equations our boundaries lead to boundary conditions in every respect analogous to classical boundary conditions.

6. The Kolmogoroff differential equations. Up to now we have considered only iterations of the operator Π. Instead of the integral exponents we now introduce a continuous time parameter, *i.e.*, we deal with matrices $P(t;i,j)$

which satisfy the conditions (1) and have the semigroup property

$$P(t+s) = P(t)P(s), \qquad (t,s > 0). \tag{18}$$

As Π defined a random walk, so P defines a stochastic process in E.

Kolmogoroff has pointed out that an interesting class of matrices P satisfies the two infinite systems of differential equations,

$$P'(t) = -pP(t) + p\Pi P(t), \tag{19}$$

and

$$P'(t) = -P(t)p + P(t)p\Pi, \tag{20}$$

together with the initial condition $P(0) =$ identity matrix. Here Π stands for the matrix (1), and p is a diagonal matrix with elements $p_1, p_2, \cdots$. For our stochastic process the transition probabilities $\Pi(i,j)$ represent a road system which determines where the system can move to from i. The p_i regulate the speed of the movement, inasmuch as $1/p_i$ is the expected duration of an uninterrupted stay of the system at i.

The author showed in 1940 that there always exists a matrix P satisfying both (19) and (20) but there can exist infinitely many admissible common solutions to these differential equations. This phenomenon is neatly explained by the notion of the two boundaries. In fact, in connection with (19) and (20) we encounter all the phenomena known for parabolic partial differential equations. Thus we encounter the classical boundary conditions of the Sturm-Liouville theory: absorbing, reflecting, elastic barriers, etc. The semigroups and processes connected with (19) and (20) are the same as those of diffusion theory, except that the boundaries can be more complicated, so that we find a much greater variety. Of course, there are no normal or partial derivatives in a denumerable space, and the boundary conditions must be written in a new analytical form. However, this form carries over to partial differential equations, and we are able to view many known phenomena from a more comprehensive point of view.

PRINCETON UNIVERSITY,
PRINCETON, N.J.

ON THE INTRINSIC FORM FOR SECOND ORDER DIFFERENTIAL OPERATORS

Dedicated to Paul Lévy
on the occasion of his seventieth birthday

BY

WILLIAM FELLER[1]

1. Introduction

A function f will be said to be *in the local domain of the linear operator* $\mathfrak{A}$ at the point s, (in symbols: $f \in D(\mathfrak{A}, s)$), if f *and* $\mathfrak{A}f$ are defined and *continuous* in some neighborhood of s. Similarly, the domain $D(\mathfrak{A}, I)$ of $\mathfrak{A}$ for the interval I consists of all f such that f and $\mathfrak{A}f$ are continuous in I.

The differential operator

$$\mathfrak{A} = aD_s^2 + bD_s + c, \qquad a > 0, \quad D_s = \frac{d}{ds} \tag{1.1}$$

enjoys the following obvious properties:

(1) *Local character*: If $f(s) = 0$ for all s in a neighborhood of the point s_0, then $f \in D(\mathfrak{A}, s_0)$ and $\mathfrak{A}f(s_0) = 0$.

(2) $\mathfrak{A}$ *is nontrivial*: To each point s of the interval of definition there exists an $f \in D(\mathfrak{A}, s)$ such that $f(a) \neq 0$ and $\mathfrak{A}f(s) \neq 0$.

(3) *Weak minimum property*: Let $f \in D(\mathfrak{A}, s)$ be nonnegative in a neighborhood of s and $f(s) = 0$. Then $\mathfrak{A}f(s) \geqq 0$.

In other words, if the point s is both a zero and a local minimum for f, then $\mathfrak{A}f(s) \geqq 0$. For the *pure* differential operator $\mathfrak{A} = aD_s^2 + bD_s$ (where $a > 0$) the property (3) may be sharpened to

(3′) *Strong minimum property*: If $f \in D(\mathfrak{A}, s)$ has a local minimum at s, then $\mathfrak{A}f(s) \geqq 0$.

Various problems have led the author to derive the general form of linear operators in one dimension having these properties [1]. The class of such operators forms a natural generalization of the classical second order differential operators. It has been shown elsewhere, [1], [2], [3], that their use has considerable advantages. The use of the new canonical form renders the theory more satisfactory and at the same time simpler; it achieves an unexpected unification and is more adapted for many applied problems.

Received June 26, 1957.

[1] This research was done while the author was on sabbatical leave of absence with the support of the United States Air Force through the Air Force Office of Scientific Research of the Air Research and Development Command.

It is the purpose of this paper to present an improved derivation of a canonical form of operators with the properties (1), (2), (3) and to provide additional information on various points.

The problem may be viewed under two different aspects. It is easily shown that the equation $\mathfrak{A}u = 0$ admits of *at most* two independent solutions (provided independence is defined as in footnote 3). For certain purposes it is natural to *postulate* the existence (at least locally) of two independent solutions, and the general form of our operators can then be derived with ease. If $\mathfrak{A}$ has the *strong minimum* property (3′), one of the two solutions may be taken as the constant function, and the other is automatically strictly monotonic. It may be used to reparametrize the interval of definition in such a way that the general solution of $\mathfrak{A}u = 0$ is a *linear function*.

The crucial Section 3 contains the derivation of a canonical form $D_m D_x$ of our operators with this additional property; Section 4 reduces the general case of operators with the *strong minimum property and two independent solutions* to the simple case of Section 3. The analogous problems for operators with the *weak* minimum property are treated in Sections 7 and 8 leading to the canonical representations (8.1) and (*). These four sections are independent of the remainder of the paper.

Unfortunately, when only the properties (1), (2), (3) (or (3′)) are postulated, the equation $\mathfrak{A}u = 0$ need not have any solutions. The nonexistence may be due to two conditions:

(a) The original operator $\mathfrak{A}$ may be defined on too small a domain.

(b) The operator $\mathfrak{A}$ behaves essentially as a first order operator (the case of a vanishing coefficient a in (1.1)).

In case (a) the definition of $\mathfrak{A}$ may be extended, without violating the postulated properties (1), (2), (3), in such a way that the preceding theory applies to the extended operator. This is the less elegant part of the derivation and is contained in Sections 5 and 9, respectively, for operators with the strong and weak minimum property.

The *singular* (or first order) points at which the $\mathfrak{A}$ degenerates into a first order operator are described intrinsically in Definition 2.2, and are treated in Section 6. They may, but need not, be isolated points, but the regular points always form an open set.

Section 10 contains three illustrative examples.

2. The operator $D_m D_x$

In a fixed interval $I\colon -\infty \leqq x_- < x < x_+ \leqq \infty$ let *m be a strictly increasing function* continuous on the right:

$$(2.1) \qquad m \uparrow \qquad m(x) = \lim_{h \downarrow 0} m(x + h);$$

it is not required that m be bounded at either end. The difference $m(y) - m(x)$ (where $x < y$) should be thought of as a measure of the half-open interval $(x, y]$.

If m is continuous, the interval I may be parametrized by m instead of x, and differentiation with respect to m is in this case defined in the familiar fashion. For a discontinuous m we adopt the same definition with the agreement to consider increments only over *closed* intervals. Thus, if g is a function defined in I such that the right and left limits $g(x+)$ and $g(x-)$ exist for each $x \in I$, we define *the right derivative of g with respect to m by*

$$D_m^+ g(x) = \lim_{y \downarrow x} \frac{g(y+) - g(x-)}{m(y+) - m(x-)}, \tag{2.2}$$

provided, of course, the limit exists. The left derivative $D_m^- g$ is defined symmetrically. Clearly at *a point of discontinuity of* m

$$D_m^+ g(x) = D_m^- g(x) = \frac{g(x+) - g(x-)}{m(x+) - m(x-)}, \tag{2.3}$$

and a differentiable g is continuous at each point of continuity of m.

We now define an operator which will prove to be the natural generalization of $aD_s^2 + bD_s$.

DEFINITION 2.1. *We say that a function f is in the domain $D(\mathfrak{A}, I)$ of the operator $\mathfrak{A} = D_m D_x$ in the interval I if:*

(i) *The right and left derivatives $f^+ = D_x^+ f$ and $f^- = D_x^- f$ exist and*

$$\begin{aligned} f^+(x) &= \lim_{y \downarrow x} f^+(y) = \lim_{y \downarrow x} f^-(y) \\ f^-(x) &= \lim_{y \uparrow x} f^+(y) = \lim_{y \uparrow x} f^-(y) \end{aligned} \tag{2.4}$$

for each $x \in I$.

(ii) *$D_m D_x^+ f$ exists and is continuous in I.*

The definition (2.2) implies that $D_m D_x^+ f$ is identical with $D_m D_x^- f$ and we shall write

$$\mathfrak{A}f = D_m D_x f = D_m D_x^- f = D_m D_x^+ f. \tag{2.5}$$

Clearly our definition applies equally to any subinterval of I.

LEMMA 2.1. *The operator $\mathfrak{A} = D_m D_x$ has the properties* (1), (2), (3′).

Proof. Only (3′) requires comment. Suppose that $f \in D(\mathfrak{A}, x)$ has a local minimum at the point x. Then $f^-(x) \leqq 0$ and $f^+(x) \geqq 0$. It follows now from (2.4) that f^+ cannot strictly decrease in any neighborhood N of x as would be the case if $\mathfrak{A}f < 0$ in N.

The *converse* of this lemma is true in the sense that the operators of the form $\mathfrak{A} = D_m D_x$ are essentially the only ones[2] with the properties (1), (2), (3′). An exception occurs only in connection with *singular points*, where

[2] To reduce $aD_s^2 + bD_s$ to the canonical form $D_m D_x$, choose for x an increasing solution of $aD_s^2 x + bD_s x = 0$. Then $aD_s^2 + bD_s = \alpha D_x^2$, where $\alpha > 0$ if $a > 0$. Now put $m = \int \alpha^{-1}\, dx$.

$\mathfrak{A}$ has the character of a first order differential operator. To give an intrinsic characterization of such points note that a function f in the domain of $\mathfrak{A}$ has strong regularity properties and, in particular, cannot oscillate widely. More precisely, we have

LEMMA 2.2. *Let $\mathfrak{A}$ have properties* (1), (2), (3′). *Suppose that $f \in D(\mathfrak{A}, x_0)$ and $\mathfrak{A}f(x_0) > 0$. Then there exist one-sided neighborhoods $x_0 \leqq x < x_0 + \delta$ and $x_0 - \delta < x \leqq x_0$ in each of which f is strictly monotonic.*

Proof. Assume the assertion to be false, that is, suppose that there exists a sequence of points $x_n \to x_0$ such that f has a local maximum at x_{2n} and a local minimum at x_{2n+1}. By the strong minimum property (3′) we have $\mathfrak{A}f(x_{2n}) \leqq 0$ and $\mathfrak{A}f(x_{2n+1}) \geqq 0$, and the assumed continuity of $\mathfrak{A}f$ requires therefore that $\mathfrak{A}f(x_0) = 0$, against hypothesis.

According to this lemma the relation $\mathfrak{A}f(x_0) > 0$ implies that either f is strictly monotonic in some neighborhood of x_0, or f has a local minimum at x_0. In the special case of a first order differential operator $\mathfrak{A} = bD_x$ with $b > 0$, the relation $\mathfrak{A}f(x_0) > 0$ implies that f is increasing, but for an honest second order operator $\mathfrak{A}f(x_0) > 0$ is possible both for increasing and for decreasing f. These remarks will explain the

DEFINITION 2.2. *Let $\mathfrak{A}$ have the properties* (1), (2), (3′). *The point x_0 is regular for $\mathfrak{A}$ if $D(\mathfrak{A}, x_0)$ contains both an increasing and a decreasing f such that $\mathfrak{A}f(x_0) > 0$.*

A point which is not regular is called a first order point.

Theorem 6.1 shows that at first order points $\mathfrak{A}$ reduces to a first derivative. In the absence of first order points $\mathfrak{A}$ is of the nature of a second order differential operator. This is seen from the following two basic theorems concerning operators with the strong minimum property (3′).

THEOREM 2.1. *Suppose that in an interval I: $s_- < s < s_+$ the operator $\mathfrak{A}$ has the properties* (1), (2), (3′) *and that all points of I are regular.*

There exist in I a strictly increasing continuous function x and a strictly increasing right-continuous function m such that $\mathfrak{A}f = D_m D_x f$ for each f in a local domain of $\mathfrak{A}$.

The variable x may be used as scale parameter for I (so that I: $x_- < x < x_+$) and is determined uniquely up to a trivial linear transformation $x \to px +$ const. For given x the "canonical measure" m is determined up to an irrelevant additive constant.

Note that $\mathfrak{A}$ may be defined on a small domain whence $D_m D_x$ may be a proper extension of $\mathfrak{A}$.

A different aspect of the situation is covered by

THEOREM 2.2. *The assertions of the preceding theorem remain valid if the condition that all points be regular is replaced by the following:*

To each $s \in I$ there corresponds a neighborhood N_s of s such that in N_s there exist two linearly independent[3] *solutions u_i of $\mathfrak{A}u_i = 0$.*

3. A special case

We shall prove the most interesting special case of Theorem 2.1 directly and then reduce the general case to it.

THEOREM 3.1. *Suppose that $\mathfrak{A}$ is defined in the interval $I\colon -\infty \leqq x_- < x < x_+ \leqq \infty$ and has properties* (1), (2), (3′). *Suppose furthermore that*[4]

$$\mathfrak{A}\mathbf{1} = 0, \qquad \mathfrak{A}\mathbf{x} = 0. \tag{3.1}$$

Then there exists in I a strictly increasing right-continuous m such that $\mathfrak{A}f = D_m D_x f$ for each f in a domain of $\mathfrak{A}$. This m is determined up to an irrelevant additive constant.

The *proof* will be based on the following

LEMMA 3.1. *Adhering to the conditions of the preceding theorem, let $\mathfrak{A}f$ be continuous and strictly positive. Then f is convex downward.*

Proof. Let p and q be constants, and define u by $u(x) = f(x) - px - q$. Then $\mathfrak{A}u = \mathfrak{A}f > 0$ so that u can have no local maximum in I. If $x_1 < x_2$ are two zeros of u, then u must be strictly negative in the interval $x_1 < x < x_2$ and hence the graph of f lies below each of its chords.

Proof of Theorem 3.1. By property (2) there exists a function g such that $\mathfrak{A}g > 0$ in some neighborhood of an arbitrarily prescribed point x_0. In this neighborhood g is convex. An arbitrary $f \in D(\mathfrak{A}, x_0)$ may be represented as a difference, $f = (f + tg) - tg = f_1 - f_2$ of two functions f_i, and choosing t sufficiently large we shall have $\mathfrak{A}f_i > 0$ in some neighborhood N of x_0. It follows that f_i is convex in N, and hence the one-sided derivatives f^+ and f^- exist everywhere in N and satisfy the continuity requirement (2.4). Since x_0 is arbitrary, it follows that f^+ and f^- exist and satisfy (2.4) for each $x \in I$ and $f \in D(\mathfrak{A}, I)$.

Suppose now that $\mathfrak{A}f = \phi$ and $\mathfrak{A}g = \gamma$ are continuous and positive in the interval $I_0 \subset I$. Then

$$\mu(x) = g^+(x) \tag{3.2}$$

defines a strictly increasing right-continuous function μ defined in I_0.

Let t be a real parameter and consider the function

$$u_t = f - tg. \tag{3.3}$$

[3] Two functions u_i are called *independent* in an interval I_0 if a nontrivial relationship $p_1u_1 + p_2u_2 = 0$ does not exist *in any subinterval* of I_0. See example (c) of Section 10 concerning this convention.

[4] We denote by bold face $\mathbf{x}$ the function whose value equals x for each x, and by $\mathbf{1}$ the function which equals 1 everywhere. No new symbol was thought necessary for the zero function.

If

$$t < \min_{x \in I_0} \frac{\phi(x)}{\gamma(x)}, \tag{3.4}$$

then $\mathfrak{A}u_t$ is strictly positive in I_0, and the convexity of u_t implies that its right derivative *increases*. In other words, (3.4) implies that

$$f^+(\xi_2) - f^+(\xi_1) > t\{\mu(\xi_2) - \mu(\xi_1 +)\}, \qquad \xi_1 < \xi_2 . \tag{3.5}$$

If $t > \max \phi(x)/\gamma(x)$, the inequality in (3.5) is reversed. Thus the increment of f^+ is an interval function with the mean value property

$$\begin{aligned} \min_{\xi_1 \leqq x \leqq \xi_2} \frac{\phi(x)}{\gamma(x)} \cdot \{\mu(\xi_2) - \mu(\xi_1+)\} &\leqq f^+(\xi_2) - f^+(\xi_1+) \\ &\leqq \max_{\xi_1 \leqq x \leqq \xi_2} \frac{\phi(x)}{\gamma(x)} \cdot \{\mu(\xi_2) - \mu(\xi_1+)\}. \end{aligned} \tag{3.6}$$

This being the property on which the construction of the Riemann-Stieltjes integral is based, we conclude that

$$f^+(\xi_2) - f^+(\xi_1+) = \int_{\xi_1+}^{\xi_2+} \frac{\phi(x)}{\gamma(x)}\, d\mu(x). \tag{3.7}$$

Finally, putting

$$m = \int \frac{d\mu}{\gamma}, \tag{3.8}$$

(3.7) takes on the form

$$f^+(\xi_2) - f^+(\xi_1+) = \int_{\xi_1+}^{\xi_2+} \mathfrak{A}f \cdot dm. \tag{3.9}$$

This is the integrated version of $\mathfrak{A}f = D_m D_x f$, and we have proved Theorem 3.1 *locally*: to each $x \in I$ there exists a neighborhood N_x such that in it $\mathfrak{A}f = D_m D_x f$ for each f in the domain of $\mathfrak{A}$. A glance at (3.9) shows that the weight function m is determined up to an irrelevant additive constant.

To prove the theorem in the large, choose among these N_x denumerably many, say I_k, $k = 0, \pm 1, \pm 2, \cdots$, which cover the interval I and such that I_k has an overlap with I_{k-1} and I_{k+1}, but no other interval I_k. It is then clear that, starting from I_0, the free additive constants in $I_1, I_{-1}, I_2, I_{-2}, \cdots$ can be adjusted so that the definitions of m agree in the common parts $I_k \cap I_{k+1}$, and Theorem 3.1 is thus proved.

4. Proof of Theorem 2.2

In this section $\mathfrak{A}$ *is assumed to have properties* (1), (2), (3′) *and to be defined in the interval* I: $s_- < s < s_+$. We shall reduce Theorem 2.2 to Theorem 3.1.

LEMMA 4.1. *To each* $s \in I$ *there corresponds a neighborhood* N_s *with the following property: If* u *is a function such that* $\mathfrak{A}u = 0$ *in* N_s *and* $u(s_1) = u(s_2)$ *for two points* $s_1 < s_2$ *in* N_s, *then* $u =$ const. *in the interval* (s_1, s_2).

Proof. We take for N_s any interval around s for which there exists an f

such that $\mathfrak{A}f > 0$ in N_s. Property (2) guarantees the existence of such a neighborhood. Suppose the assertion false, e.g. that $u(s_1) = u(s_2) < u(s_0) - \varepsilon$ for some s_0 such that $s_1 < s_0 < s_2$ and $\varepsilon > 0$. Put for abbreviation $F = f + \nu\varepsilon^{-1}u$, where ν is a constant exceeding the oscillation of f in N_s. Then $F(s_0) > F(s_i)$ for $i = 1, 2$, and hence the maximum of F in (s_1, s_2) is attained at an interior point although $\mathfrak{A}F = \mathfrak{A}f > 0$. This contradicts the property (3′).

LEMMA 4.2. *If* $\mathbf{1}$ *is in the domain of* $\mathfrak{A}$, *then* $\mathfrak{A}\mathbf{1} = 0$. *Otherwise* $\mathfrak{A}$ *may be extended by defining* $\mathfrak{A}\mathbf{1} = 0$, *and the thus extended operator has again the properties* (1), (2), (3′).

Proof. The functions $\mathbf{1}$ and $-\mathbf{1}$ have a local minimum at each point, and the minimum property (3′) therefore requires that $\mathfrak{A}\mathbf{1} = 0$. If $\mathbf{1}$ is not in the domain $D(\mathfrak{A}, s)$, let D' be the set of all functions of the form $f + c\mathbf{1}$ where $f \in D(\mathfrak{A}, s)$ and c is a constant. The representation $f + c\mathbf{1}$ is obviously unique, and $f + c\mathbf{1}$ has extrema at the same places as f. It is therefore possible to put $\mathfrak{A}(f + c\mathbf{1}) = \mathfrak{A}f$ without violating any of the three requirements (1), (2), (3′).

LEMMA 4.3. *Suppose that to each* $s \in I$ *there corresponds a neighborhood* N_s' *of* s *such that in* N_s' *there exist two independent solutions* u_i *of* $\mathfrak{A}u_i = 0$. *Then there exists in* I *a strictly increasing continuous* x *such that* $\mathfrak{A}x = 0$. *Furthermore* $\mathfrak{A}\mathbf{1} = 0$. *If in some interval* $I' \subset I$ *the function* u *satisfies* $\mathfrak{A}u = 0$, *then* $u = px + q\mathbf{1}$ *where* p *and* q *are constant.*

Proof. In the intersection $N_s \cap N_s'$ the functions u_i are monotonic in the weak sense by virtue of Lemma 4.1. If need be, let us extend $\mathfrak{A}$ in accordance with the preceding lemma by setting $\mathfrak{A}\mathbf{1} = 0$. Let $s_1 < s_2$ be two arbitrary points of $N_s \cap N_s'$, and put $v_i = u_i - u_i(s_1)\mathbf{1}$. Then (using the letter $\mathfrak{A}$ for the extended operator) $\mathfrak{A}v_i = 0$, and hence $v = v_2(s_2)v_1 - v_1(s_2)v_2$ is a solution of $\mathfrak{A}v = 0$ such that $v(s_1) = v(s_2) = 0$. By Lemma 4.1 then $v = 0$ in the interval (s_1, s_2), and, since s_1 and s_2 are arbitrary, $v = 0$ in $N_s \cap N_s'$. Thus there exists a linear relationship between u_1, u_2, and $\mathbf{1}$. It follows that $\mathbf{1}$ was in the domain of the original (unextended) operator $\mathfrak{A}$. Furthermore, u_1 and u_2 being independent, there exists a linear combination $x = p_1u_1 + p_2u_2$ which is independent of $\mathbf{1}$ and therefore *strictly* monotonic, say increasing.

To prove the lemma "in the small", suppose that the function u satisfies $\mathfrak{A}u = 0$ in some subinterval $I' \subset N_s \cap N_s'$. In this interval also $v = u - px - q\mathbf{1}$ satisfies $\mathfrak{A}v = 0$, and the constants p, q may be chosen so that $v(s_1) = v(s_2)$. It follows then from Lemma 4.1 that v vanishes in (s_1, s_2), and thus $u = px + q\mathbf{1}$.

It remains to prove the lemma "in the large." For that purpose choose among the intervals $N_s \cap N_s'$ a doubly infinite sequence, say $\cdots, I_{-1}, I_0, I_1, \cdots$, whose union covers I and such that I_k overlaps I_{k-1} and I_{k+1}, but no other interval I_j. We have shown that in each I_k there exists a strictly increasing function x_k satisfying $\mathfrak{A}x_k = 0$. In I_0 we put $x = x_0$. In the intersection $I_0 \cap I_1$ we have $x = p_1 x_1 + q_1 \mathbf{1}$, and we use this equation as a definition of x throughout I_1. Then $\mathfrak{A}x = 0$ in the union $I_0 \cup I_1$, and x is strictly increasing. Proceeding in this way by induction we construct a strictly increasing solution x of $\mathfrak{A}x = 0$ defined in the union $\bigcup_{i=-k}^{k} I_i$, and hence in I. Finally, if $\mathfrak{A}u = 0$ in some interval $I' \subset I$, then in each $I_k \cap I'$ the function u is of the form $u = p_k x + q_k \mathbf{1}$. Since I_k and I_{k+1} overlap, we have $p_k = p_{k+1}$, $q_k = q_{k+1}$, and the lemma is proved.

To complete the proof of Theorem 2.2 we have merely to observe that the conditions (3.1) of Theorem 3.1 hold in consequence of the preceding lemma.

5. Proof of Theorem 2.1

We shall again reduce this theorem to Theorem 3.1. As in the preceding section we suppose that $\mathfrak{A}$ *is defined in the interval* I: $s_- < s < s_+$ *and has the properties* (1), (2), (3′). We denote by $\mathbf{s}$ the function which at each point s equals s.

LEMMA 5.1. *Suppose that in an interval* $I_0 \subset I$ *one has* $\mathfrak{A}\mathbf{s} = \beta \geqq 0$. *If* f *decreases and* $\mathfrak{A}f > 0$ *in* I_0, *then the graph of* f, *as a function of* s, *is convex downward.*

Proof. We refer to the proof of Lemma 3.1 replacing the scale parameter x by s. We have then $\mathfrak{A}u = \mathfrak{A}f - p\beta$. All chords are inclined downward, whence $p < 0$ and $\mathfrak{A}u > 0$. Thus the argument of the original proof applies in the present case.

LEMMA 5.2. *Let* s_0 *be a regular point. A neighborhood* N *of* s_0 *can be parametrized by a parameter* x *(a continuous strictly increasing function of* s*) in such a way that whenever* $\mathfrak{A}f$ *is continuous in some subinterval of* N, *the one-sided derivatives* $D_x^+ f$ *and* $D_x^- f$ *exist and satisfy the continuity conditions* (2.4).

Proof. By Definition 2.2 of regular points we can find a neighborhood N of s_0 in which there exist an increasing function x and a decreasing function v such that $\mathfrak{A}x > 0$ and $\mathfrak{A}v > 0$. These functions are *strictly* monotonic (see Lemma 2.2). We choose x as a local parameter in N. In terms of it N becomes the interval $x_1 < x < x_2$, and in N we have

$$\mathfrak{A}\mathbf{x} = \beta \geqq 0, \qquad \mathfrak{A}v > 0, \qquad v\downarrow. \tag{5.1}$$

To prove the lemma it suffices to show that (5.1) implies the following: *if* $\mathfrak{A}f$ *is continuous in the open interval* $N' \subset N$, *then in any closed subinterval* $N'' \subset N'$ *the function* f *admits a representation as the difference of two convex functions of* x.

Let $N': x_1' < x < x_2'$ and $N'': x_1'' < x < x_2''$. Since $\mathfrak{A}v$ is strictly positive and v strictly decreasing in N, we may choose the real parameter t so large that for $F = f + tv$ we shall have $\mathfrak{A}F > 0$ in N' and $F(x_2') < F(x_2'')$. It follows then that F *decreases* in N''. In fact, if $x_1' < x < x_2'' < x_2'$, then the maximum of F in the interval (x, x_2') cannot be attained at an interior point since there $\mathfrak{A}F > 0$. This maximum is not attained at the right endpoint x_2' since $F(x_2') < F(x_2'')$. Hence the maximum is attained at the left endpoint x, and thus F decreases in (x_1', x_2''). By Lemma 5.1 both F and v are, as functions of x, convex, and since $f = F - tv$, the assertion is proved.

Completion of the proof. Using the notation of the last lemma define in N a new operator $\mathfrak{A}^*$ by

$$\mathfrak{A}^*f = \mathfrak{A}f - \beta \cdot D_x^- f. \tag{5.2}$$

This operator $\mathfrak{A}^*$ *has again the characteristic properties* (1), (2), (3′). The local character is obvious. That $\mathfrak{A}^*$ is nontrivial is seen from the fact that v being decreasing we have $D_x^- v \leqq 0$ and hence $\mathfrak{A}^* v \geqq \mathfrak{A}v > 0$. Finally, at a place of a local minimum of f we have $D_x^- f \leqq 0$ and therefore $\mathfrak{A}^* f \geqq \mathfrak{A}f \geqq 0$ for each f in the local domain of $\mathfrak{A}^*$.

It is clear from the definition that $\mathfrak{A}^*\mathbf{x} = 0$, and Lemma 4.2 shows that either $\mathfrak{A}^*\mathbf{1} = 0$ or we are free to supplement the definition of $\mathfrak{A}^*$ by setting $\mathfrak{A}^*\mathbf{1} = 0$. The conditions (3.1) of Theorem 3.1 are then satisfied, and we can apply this theorem to describe the structure of $\mathfrak{A}^*$. It is true that $\mathfrak{A}^*f$ need not be continuous since $D_x^- f$ may have jumps. However, in consequence of the continuity conditions (2.4) and the continuity of β, it is clear that for each f in its domain $\mathfrak{A}^*f$ is continuous on the left, and that even the right limits exist at each point. The proof of Theorem 3.1 applies under these circumstances without change and shows that there exists a function μ such that $\mathfrak{A}^*f = D_\mu D_x f$. If then in N we define the operator

$$\bar{\mathfrak{A}} = D_\mu D_x + \beta D_x^-, \tag{5.3}$$

we see that $\bar{\mathfrak{A}}$ *is, in* N, *an extension of* $\mathfrak{A}$.

We wish to apply Theorem 3.1 to reduce $\bar{\mathfrak{A}}$ to the canonical form

$$\bar{\mathfrak{A}} = D_m D_u. \tag{5.4}$$

For that purpose we require in N a strictly increasing function u satisfying $\bar{\mathfrak{A}}u = 0$. Introducing u as local parameter in N we get (5.4) as an immediate consequence of Theorem 3.1. To construct the desired function u, we use the familiar method of successive approximations to show that the differential equation

$$D_\mu^- v + \beta v = 0 \tag{5.5}$$

admits, in N, of a positive solution v. (This v is continuous except, perhaps, at points of discontinuity of μ.) Clearly $u = \int v\, dx$ is a strictly increasing solution of $\bar{\mathfrak{A}}u = 0$ (defined up to a linear transformation $u \to pu + q\mathbf{1}$).

We have thus proved Theorem 2.1 "in the small", namely that to each regular point $s_0 \in I$ there corresponds a neighborhood N in which $\mathfrak{A}f = D_m D_u f$ for each f in the domain of $\mathfrak{A}$. To show that a similar representation holds "in the large" throughout I we have merely to repeat the concluding argument of the last section.

6. Singular points

We proceed to show that at irregular points $\mathfrak{A}$ degenerates into a *first order* differential operator.

THEOREM 6.1. *Suppose that in the interval* $I: s_- < s < s_+$ *the operator* $\mathfrak{A}$ *has the properties* (1), (2), (3′).

Then there exists in I *a strictly increasing continuous function* x *such that for each nonregular point* $s \in I$ *either*

$$\mathfrak{A}f(s) = D_x^+ f(s) \quad \text{for all} \quad f \in D(\mathfrak{A}, s) \tag{6.1}$$

or

$$\mathfrak{A}f(s) = -D_x^+ f(s) \quad \text{for all} \quad f \in D(\mathfrak{A}, s). \tag{6.2}$$

This theorem was formulated so as to reveal the salient features in the simplest manner. Actually the proof will automatically provide further information embodied in the following

AMPLIFICATION. *If* s *is an irregular point, then exactly one of the following five mutually exclusive situations prevails*:

For $f \in D(\mathfrak{A}, s)$ *the relation* $\mathfrak{A}f(s) > 0$ *implies that*:

(i) *f is strictly increasing near s*;

(ii) *f is either increasing or has a strict local minimum at s, each possibility actually occurring for some f*;

(iii) *f has a strict local minimum at s*;

(iv)–(v) *same as* (i), (ii) *with "increasing" replaced by "decreasing."*

In the cases (i) *and* (v) *we can replace* (6.1)–(6.2) *by*

$$\mathfrak{A}f(s) = \pm D_x f(s), \tag{6.3}$$

where D_x *is a two-sided derivative. In the case* (iii)

$$\mathfrak{A}f(s) = D_x^+ f(s) = - D_x^- f(s). \tag{6.4}$$

Proof "in the small." We prove that to each singular point $s_0 \in I$ there corresponds a neighborhood N such that Theorem 6.1 and its amplification are true when I is replaced by N.

The fact that there exist only the five types of singular points enumerated above is an immediate consequence of Lemma 2.2.

Let s be a singular point. For reasons of symmetry it suffices to consider the case where $\mathfrak{A}f(s_0) > 0$ *implies that f is not strictly decreasing* (types (i)–(iii) above). In view of property (2) we may choose a neighborhood of s_0 in which there exists a function v such that

$$\mathfrak{A}v = \beta > 0. \tag{6.5}$$

Then v *is strictly* increasing in a right neighborhood $s_0 \leqq s < s_0 + \varepsilon$ and strictly monotone in the left neighborhood $s_0 - \varepsilon < s \leqq s_0$. We choose ε so small that (6.5) holds in the neighborhood N: $s_0 - \varepsilon < s < s_0 + \varepsilon$.

We next prove that

$$\mathfrak{A}f(s_0) = \beta(s_0)D_v^+f(s_0) \tag{6.6}$$

for each $f \in D(\mathfrak{A}, s_0)$. For this purpose we introduce the function defined in N by

$$F_t = \beta(s_0)f - tv, \tag{6.7}$$

depending on the real parameter t. Clearly $F_t \in D(\mathfrak{A}, s_0)$. If $t < \mathfrak{A}f(s_0)$, we have $\mathfrak{A}F_t(s_0) > 0$ implying that F_t increases in a right neighborhood of s_0, i.e. that $F_t(s) > F_t(s_0)$ for $s_0 < s < s_0 + \delta$. Then

$$\beta(s_0)\{f(s) - f(s_0)\} > t\{v(s) - v(s_0)\}, \tag{6.8}$$

or

$$\beta(s_0) \liminf_{s \downarrow s_0} \frac{f(s) - f(s_0)}{v(s) - v(s_0)} \geqq \mathfrak{A}f(s_0). \tag{6.9}$$

For $t > \mathfrak{A}f(s_0)$ the same argument yields the reversed inequality for the limit superior, and thus (6.6) holds.

In the interval N we now introduce the *strictly increasing function* x defined by

$$x(s) = \int_{s_0}^{s} \frac{|dv|}{\beta} + \text{const.} \tag{6.10}$$

Then $D_x^+ f(s_0) = \beta(s_0)D_v^+ f(s_0)$, and (6.1) is proved at the point s_0.

More precisely, if s_0 is of the type (i), then v is strictly increasing, and $\mathfrak{A}F_t > 0$ implies that also F_t increases near s_0. In this case the situation is perfectly symmetric for right and left neighborhoods, and we have not only (6.6) but also $\mathfrak{A}f(s_0) = \beta(s_0)D_0^-f(s_0)$ and hence (6.3) with the positive sign. Similarly, if s_0 is of type (iii), v has a minimum at s_0, and $D_x^- f(s_0) = -\beta(s_0)D_v^-f(s_0)$. In this case (6.4) holds. Finally, when s_0 is of type (ii), then v may be chosen strictly increasing, but it is impossible to strengthen (6.1) by making a generally valid statement about the left derivative (which, however, exists in any case; see example (c), Section 10).

Finally, note that when $s \in N$ is any other singular point, the argument applies *with the same* v *and* x, and hence the theorem is proved for the neighborhood N.

Proof "in the large." (This part of the theorem is of minor interest and is proved for completeness only.) We proceed in two steps. First consider the special case where all points of I are singular. We can cover I by denumerably many open intervals $\cdots, N_{-1}, N_0, N_1, \cdots$ such that in N_k there exists a strictly increasing function x_k in terms of which $\bar{\mathfrak{A}}f = \pm D_{x_k}^{\pm} f$ as indicated in the theorem. Removing, if necessary, superfluous intervals, we suppose that N_k overlaps N_{k-1} and N_{k+1}, but no other N_j. Clearly x_k is determined uniquely up to an additive constant, and we can choose these additive constants in $N_0, N_1, N_{-1}, N_2, N_{-2}, \cdots$ successively in such a way that x_k and x_{k+1} coincide in the overlap $N_k \cap N_{k+1}$. Then each x_k is the restriction of an increasing function x defined throughout I, and for it Theorem 6.1 holds true.

Next, consider the general case where I contains both regular and irregular points. If an interval $I_0 \subset I$ contains only irregular points, then we have shown that there exists an interval $N \supset I_0$ in which the theorem is true. In other words, to each irregular $s \in I$ we can find a neighborhood N such that the theorem is valid for N and *the endpoints of* N *are regular points* (or endpoints of the basic interval I). By definition the regular points form an open set, and hence the set S of singular points is closed in I. We can therefore cover S by denumerably many intervals N_k such that in N_k the theorem is true for a certain function x_k and the endpoints of N_k are regular. We may suppose the N_k *nonoverlapping and ordered*, i.e. N_k lies entirely to the left of N_{k+1}. Then $I - \bigcup N_k$ consists of denumerably many closed intervals (containing only regular points), and Theorem 6.1 has no bearing on them: In them x may be defined in a perfectly arbitrary manner. Thus we may in N_k define $x = x_k + c_k$ and dispose of the constants $c_1, c_{-1}, c_2, \cdots$ successively in such a manner that x will be increasing in $\bigcup N_k$. In the complementary set we define x by linear interpolation and the theorem holds for the function x thus defined.

7. The natural scale

We have seen that for operators $\mathfrak{A}$ with the strong minimum property (3′) an interval I not containing first order points may be referred to a parameter x in terms of which $\mathfrak{A}$ assumes a particularly simple form. This result will now be extended to operators enjoying only the weak minimum property (3). It will be shown that the equation $\mathfrak{A}u = 0$ can have *at most* two independent solutions (provided independence is defined as in footnote 3; cf. example (c) of Section 10). At present we shall assume that, at least locally, there actually exist two such solutions. Later on we shall remove this restriction.

THEOREM 7.1. *Suppose that in the interval* I: $s_- < s < s_+$ *the operator* $\mathfrak{A}$ *has the properties* (1), (2), (3), *and assume that to each* $s \in I$ *there corresponds a neighborhood* N *in which the equation* $\mathfrak{A}u = 0$ *possesses two independent positive solutions.*

Then there exist two independent functions ϕ_i *defined in* I *and satisfying throughout* I *the equations* $\mathfrak{A}\phi_i = 0$. *If* $\mathfrak{A}u = 0$ *in some interval* $I_0 \subset I$, *then*

$$u = p_1\phi_1 + p_2\phi_2,$$

where the p_i *are constants.*

The interval I *may be parametrized by a variable* x *(a strictly increasing continuous function of* s*) in such a way that for two arbitrary functions* ϕ *and* ψ *satisfying the equations* $\mathfrak{A}\phi = 0$ *and* $\mathfrak{A}\psi = 0$ *the ratio* ψ/ϕ *is differentiable with respect to* x *in any interval* I_0 *not containing a zero of* ϕ. *Furthermore*

$$\phi^2(\psi/\phi)' = \text{const.}, \tag{7.1}$$

or, equivalently,

$$\psi = k_1 \phi \int \frac{dx}{\phi^2} + k_2 \phi, \tag{7.2}$$

where k_i = const. *The parameter* x *is defined uniquely up to a trivial linear transformation* $x \to px + q\mathbf{1}$.

The proof will be based on the following

LEMMA 7.1. *Let* $\mathfrak{A}$ *have the properties* (1), (2), (3), *and suppose that in the interval* I *we have* $\phi > 0$ *and* $\mathfrak{A}\phi = 0$.

Define an operator $\mathfrak{A}^*$ *by*

$$\mathfrak{A}^* f = \mathfrak{A}(f\phi). \tag{7.3}$$

Then $\mathfrak{A}^*$ *has the properties* (1), (2), (3′).

Proof. Only the strong minimum property (3′) requires comment. The domain of $\mathfrak{A}^*$ contains all f such that $f\phi$ is in the domain of $\mathfrak{A}$. Let f be such a function with a local minimum at the point s_0. The function F defined by

$$F(s) = \{f(s) - f(s_0)\}\phi(s) \tag{7.4}$$

vanishes at s_0 and is nonnegative in a neighborhood of s_0. Clearly $F \in D(\mathfrak{A}, s_0)$, and by the property (3) we have $\mathfrak{A}F(s_0) \geqq 0$. Now

$$\mathfrak{A}^* f = \mathfrak{A}F + f(s_0)\mathfrak{A}\phi = \mathfrak{A}F, \tag{7.5}$$

and therefore $\mathfrak{A}^*$ has the strong minimum property (3′).

Proof of Theorem 7.1. Let N be a neighborhood of $s_0 \in I$ such that in N we have $u_i > 0$ and $\mathfrak{A}u_i = 0$, where u_1 and u_2 are linearly independent. In N we define $\mathfrak{A}^*$ as in (7.3) putting $\phi = u_1$. Then $\mathfrak{A}^*\mathbf{1} = 0$, and $\mathfrak{A}^*(u_2/u_1) = 0$. Thus $\mathfrak{A}^* z = 0$ has two linearly independent solutions, and Theorem 2.2 applies to $\mathfrak{A}^*$. It implies that the ratio u_2/u_1 is strictly monotonic, in fact

$$u_2 = k_1 \xi u_1 + k_2 u_1, \tag{7.6}$$

where ξ is the natural scale for $\mathfrak{A}^*$ in the interval N and k_i = const.

Unfortunately this ξ depends on the arbitrary choice of the two solutions u_i, and we require a parameter x essentially independent of this choice. A very elementary calculation will show that the function x defined by

$$x = \int_{s_0}^{s} u_1^2 \, d\frac{u_2}{u_1} \tag{7.7}$$

has this property. More precisely: if $v_i > 0$ and $\mathfrak{A}v_i = 0$, then the functions $w_i = v_i/\phi$ satisfy $\mathfrak{A}^* w_i = 0$ and are in consequence of Theorem 2.2 linear combinations of ξ and $\mathbf{1}$. It follows that

$$v_1 = \alpha_1 u_1 + \alpha_2 u_2, \qquad v_2 = \beta_1 u_1 + \beta_2 u_2, \tag{7.8}$$

where the α_i and β_i are constants. Now clearly

$$\int_{s_0}^{s} v_1^2 \, d\frac{v_2}{v_1} = (\beta_2\alpha_1 - \alpha_2\beta_1)x. \tag{7.9}$$

We have thus proved the theorem "in the small," namely when I is replaced by the interval N. A simple repetition of the argument used at the end of Section 4 shows then the validity of the theorem "in the large."

8. The first representation theorem

THEOREM 8.1. *Under the assumptions of Theorem* 7.1 *we have, whenever* $\mathfrak{A}f$ *exists*,[5]

$$\mathfrak{A}f = \frac{1}{\phi} D_m\left(\phi^2 D_x \frac{f}{\phi}\right) = \frac{1}{\psi} D_m\left(\psi^2 D_x \frac{f}{\psi}\right). \tag{8.1}$$

Here m is a strictly increasing right-continuous function defined in I. For a fixed choice of the scale parameter x the "canonical measure" m is determined uniquely up to an irrelevant additive constant.

Proof. It suffices to prove the assertion for the fixed neighborhood N introduced in the proof of Theorem 7.1. In consequence of Theorem 2.2 we have in N the representation

$$\mathfrak{A}^*f = D_\mu D_\xi f \tag{8.2}$$

with an appropriate monotone μ. The scale parameter ξ depends on an arbitrary multiplicative constant, and we may dispose of it in such a way that in (7.6) we have $k_1 = 1$. A comparison of (7.6) and (7.7) then shows that

$$D_\xi = u_1^2 D_x, \tag{8.3}$$

and hence (putting as before $u_1 = \phi$)

$$\mathfrak{A}f = \mathfrak{A}^* \frac{f}{\phi} D_\mu\left(\phi^2 D_x \frac{f}{\phi}\right). \tag{8.4}$$

[5] At a zero of ψ the last term in (8.1) is senseless, and the corresponding statement in (8.1) should be interpreted as empty. Since ϕ and ψ cannot vanish at the same point, we have in (8.1) always a valid representation for $\mathfrak{A}$. The formal definition of the right side in (8.1) is in all respects analogous to Definition 2.1.

Added in proof. The operator Ω defined by $\Omega f = D_m D_x f + cf$ is always reducible to the canonical form (8.1), but the converse is true only if ϕ is sufficiently regular. Thus if $\phi' = n$ exists and is of bounded variation (not necessarily continuous), then the integrated version of (8.1) reduces to

$$\int_{x_1+}^{x_2-} \mathfrak{A}f \cdot dm = f^-(x_2) - f^+(x_1) - \int_{x_1+}^{x_2-} \frac{1}{\phi} f \, dn. \tag{*}$$

If n is absolutely continuous with respect to m, then (*) is equivalent to $\mathfrak{A}f = D_m D_x f + cf$ with $c = \phi^{-1}dn/dm$. The case $c \leq 0$ corresponds in the most general case to a convex ϕ. For details see [2].

Here μ still depends on the choice of the solution ϕ; to prove (8.1) we have to show that

$$m = \int \frac{d\mu}{\phi} \tag{8.5}$$

is independent of this choice, and thus intrinsically connected with $\mathfrak{A}$. For a particular choice of $\phi > 0$ let us define m by (8.5). Then the first formula in (8.1) holds, and we have to prove that for arbitrary $s_1 < s_2 \in N$ we have

$$\int_{s_1}^{s_2} \frac{\psi}{\phi}\, d\left(\phi^2 D_x \frac{f}{\phi}\right) = \int_{s_1}^{s_2} d\left(\psi^2 D_x \frac{f}{\psi}\right) = \psi^2 D_x \frac{f}{\psi}\Big|_{s_1}^{s_2}. \tag{8.6}$$

The truth of (8.6) follows simply by using integration by parts remembering that ϕ and ψ are related by (7.1).

9. The general case

We turn to an investigation of operators $\mathfrak{A}$ with the properties (1), (2), (3) without assuming the existence of solutions of $\mathfrak{A}u = 0$. We shall show that a representation of the form (8.1) holds except when the interval contains singular points. These can be characterized intrinsically by analogy with Definition 2.2, but for simplicity of formulations we shall be satisfied with the following purely formal

DEFINITION 9.1. *Let $\mathfrak{A}$ be defined in the interval I: $s_- < s < s_+$ and have properties* (1), (2), (3). *A point $s_0 \in I$ is called of zero order if*

$$\mathfrak{A}f(s_0) = kf(s_0) \quad \textit{for each} \quad f \in D(\mathfrak{A}, s_0). \tag{9.1}$$

A point $s_0 \in I$ is of first order if there exist in a neighborhood N of s_0 a strictly increasing continuous function y and a continuous function λ such that

$$\mathfrak{A}f(s_0) = \pm D_y^{\pm}(f\lambda)(s_0) \quad \textit{for each} \quad f \in D(\mathfrak{A}, s_0). \tag{9.2}$$

A point is regular if it is neither of zero nor of first order.

With this convention we have

THEOREM 9.1. *Suppose that $\mathfrak{A}$ is defined in the interval I and has the properties* (1), (2), (3). *Suppose furthermore that all points of I are regular.*

Then $\mathfrak{A}$ can be extended to an operator as described in Theorem 8.1. (*In other words, a scale parameter x and a canonical measure m can be introduced so that for each f in the domain of $\mathfrak{A}$ formula* (8.1) *holds, and ϕ, ψ stand in the relation* (7.1).)

Proof. It obviously suffices to show that the assertion is true for some neighborhood of an arbitrary point $s_0 \in I$. Let N be an interval around s_0 in which there exists a function w such that $w > 0$ and $\mathfrak{A}w \neq 0$. The existence of such an interval is guaranteed by property (2). In N we define (by analogy with (7.3)) an operator $\mathfrak{A}^*$ by

$$\mathfrak{A}^*f = \mathfrak{A}(fw) - f\cdot\mathfrak{A}w. \tag{9.3}$$

Obviously $\mathfrak{A}^*$ is of local character and has the property (3). Now $\mathfrak{A}^*\mathbf{1} = 0$, and therefore $\mathfrak{A}^*$ *enjoys even the strong minimum property* (3′). We have now to distinguish *three cases*:

(a) At some point $s \in N$ the operator $\mathfrak{A}^*$ is *trivial*, that is, $\mathfrak{A}^*$ does not have the property (2). Since $\mathfrak{A}^*\mathbf{1} = 0$, this is equivalent to saying that $\mathfrak{A}^*f(s) = 0$ for each $f \in D(\mathfrak{A}^*, s)$, or that

$$\mathfrak{A}f(s) = kf(s) \quad \text{where} \quad k = w^{-1}(s)\cdot\mathfrak{A}w(s) \tag{9.4}$$

for each $f \in D(\mathfrak{A}, s)$. In this case *s is a point of zero order.*

In the absence of zero order points $\mathfrak{A}^*$ has the properties (1), (2), (3′), and the theorems of Sections 2 and 6 apply. The next possibility is then:

(b) The operator $\mathfrak{A}^*$ has properties (1), (2), (3′), but some point $s \in N$ is a first order point. In this case there exists a strictly increasing continuous function ξ such that either

$$\mathfrak{A}^*f(s) = D_\xi^+ f(s) \quad \text{for each} \quad f \in D(\mathfrak{A}^*, s), \tag{9.5}$$

or an analogous equation with D_ξ^+ replaced by $-D_\xi^+$ or $\pm D_\xi^-$ holds. Now (9.5) is equivalent to saying that for each $f \in D(\mathfrak{A}, s)$

$$\mathfrak{A}f = D_\xi^+ \frac{f}{w} + \frac{f}{w}\cdot\mathfrak{A}w \tag{9.6}$$

holds at the particular point s. Put

$$W = \exp\left\{\int (\mathfrak{A}w)\cdot d\xi\right\} \tag{9.7}$$

and

$$y = \int W\, d\xi. \tag{9.8}$$

Then (9.6) reduces to

$$\mathfrak{A}f = D_y^+\left(\frac{f}{w}\, W\right), \tag{9.9}$$

and thus s is *a first order point.*

(c) If N contains neither zero- nor first-order points, then $\mathfrak{A}^*$ has the properties (1), (2), (3′), and all points are regular. By Theorem 2.1 in this case for each $f \in D(\mathfrak{A}^*, N)$

$$\mathfrak{A}^*f = D_\mu D_\xi f \tag{9.10}$$

for some monotone functions ξ and μ. Formula (9.3) now shows that in this case *the operator $\bar{\mathfrak{A}}$ defined in N by*

$$\bar{\mathfrak{A}}f = D_\mu D_\xi \frac{f}{w} - \frac{f}{w}\, c, \qquad c = w^{-1}\cdot\mathfrak{A}w \tag{9.11}$$

is an extension of $\mathfrak{A}$ (i.e. $\mathfrak{A}f = \bar{\mathfrak{A}}f$ whenever the left side is continuous).

The argument of Lemma 2.1 shows directly that $\bar{\mathfrak{A}}$ has the minimum property (3), and it is obvious that $\mathfrak{A}$ has the properties (1), (2). The classical method of successive approximations proves the existence of a solution of the differential equation $\bar{\mathfrak{A}}u = 0$ such that $u(s_0) = \alpha$ and $D_\xi^+ u(s_0) = \beta$, where α and β are arbitrary constants. Therefore, in a neighborhood N' of s_0 there exist two positive independent solutions u_i of $\bar{\mathfrak{A}}u = 0$, and in N' Theorem 8.1 applies to $\bar{\mathfrak{A}}$. Thus to each point $s_0 \in I$ there corresponds a neighborhood N' in which $\mathfrak{A}$ has an extension $\bar{\mathfrak{A}}$ of the form described in Theorem 8.1, and Theorem 9.1 is proved.

10. Examples

(a) *Removable singularities.* Let $\mathfrak{A}$ be defined only on the set of polynomials in x^2 (even polynomials), and for each such polynomial $\mathfrak{A}p = p''$. Here *the origin is a first order point of type* (iii) in the classification of Section 6: if $p''(0) > 0$, then p has a strict local minimum at the origin.

Obviously the second derivative D_x^2 is an extension of $\mathfrak{A}$ for which all points are regular. In this sense the singularity of the origin for $\mathfrak{A}$ is due to the fact that $\mathfrak{A}$ is defined on too small a domain. However, one should not jump to the hasty conclusion that D_x^2 is the "natural" or unique extension of our $\mathfrak{A}$. The next example shows a *different extension of* $\mathfrak{A}$ which appears in diffusion theory and for which the origin is a first order point of type (ii).

(b) *An isolated first order point.* Let $\mathfrak{A}$ be defined in $(-\infty, \infty)$ as follows: for $x \neq 0$ we put $\mathfrak{A}f(x) = f''(x)$. For f to be in the domain of $\mathfrak{A}$ *at the origin we require that* f *be continuous,*

$$(10.1) \qquad f^+(0) = 0, \qquad \lim_{x \to 0} f''(x) = f^{++}(0) = \mathfrak{A}f(0).$$

The operator $\mathfrak{A}$ so defined has applications in diffusion theory[6] and shares many essential properties of the ordinary second derivative. For example, it is easily shown that for a bounded continuous f the equation

$$(10.2) \qquad \lambda F - \mathfrak{A}F = f \qquad \lambda > 0$$

has exactly one bounded solution F.

Despite their strange appearance equations (10.1) state simply that *the origin is a first order point.* If we define a strictly monotonic function by $\xi(x) = \frac{1}{2}x^2$ for $x \geq 0$ (and arbitrarily for $x < 0$), then (10.1) *is equivalent to* $\mathfrak{A}f(0) = D_\xi^+ f(0)$. Thus the origin is of type (ii). Note that this $\mathfrak{A}$ extends the operator of example (a).

[6] Equation (10.2) is the Laplace transform of the diffusion equation for the following situation. In the half interval $x < 0$ we have an ordinary homogeneous diffusion with an *absorbing* barrier at the origin; in $x > 0$ we have an ordinary diffusion with a *reflecting* barrier at the origin. If, starting from $x < 0$, the particle reaches the origin, it immediately *penetrates* into the right half-line. Ordinarily (10.1) would be interpreted as a "transition condition" for two separate diffusion processes.

(c) *Linear independence.* Let $\mathfrak{A}$ be the operator of example (b), and put $\mathfrak{A}^* = \mathfrak{A} - \lambda^2$ where λ is a constant. Clearly $\mathfrak{A}^*$ has the properties (1), (2), (3), and the origin is a first order point for $\mathfrak{A}^*$. Define u_i by

$$u_1(x) = \begin{cases} \exp(\lambda x) & \text{for} \quad x \leqq 0 \\ \cosh(\lambda x) & \text{for} \quad x \geqq 0 \end{cases} \tag{10.3}$$

and

$$u_2(x) = \begin{cases} \sinh(\lambda x) & \text{for} \quad x \leqq 0 \\ 0 & \text{for} \quad x \geqq 0. \end{cases} \tag{10.4}$$

Then $\mathfrak{A}^* u_i = 0$. There exists no linear combination $p_1 u_1 + p_2 u_2$ which vanishes identically in a neighborhood of 0, but nevertheless u_1 *and* u_2 *are not independent in the sense of our definition of* footnote 3.

References

1. W. Feller, *On second order differential operators*, Ann. of Math. (2), vol. 61 (1955), pp. 90–105.
2. ———, *Generalized second order differential operators and their lateral conditions*, Illinois J. Math., vol. 1 (1957), pp. 459–504.
3. H. P. McKean, Jr., *Elementary solutions for certain parabolic partial differential equations*, Trans. Amer. Math. Soc., vol. 82 (1956), pp. 519–548.
4. *Added in proof.* For applications to the theory of the vibrating string see a forthcoming note in the Lévy testimonial volume of the Publications de l'Institut Statistique de l'Université de Paris.

Princeton University
Princeton, New Jersey

The birth and death processes as diffusion processes;

By William FELLER [1].

(Princeton University.)

1. Introduction. — A family of matrices $P(t)$ with elements $P_{ik}(t)$ defined for all real $t \geq 0$ will be called Markovian if

$$(1.1) \qquad P_{ik}(t) \geq 0, \qquad \sum_{k=0}^{\infty} P_{ik}(t) \leq 1 \qquad (i = 0, 1, 2, \ldots)$$

and the semi-group relation

$$(1.2) \qquad P(t+s) = P(t)P(s) \qquad (t, s) \geq 0)$$

holds where

$$(1.3 \qquad \lim_{t \to 0} P(t) = P(0) = I$$

is the identity matrix.

In analytic [2] terms the theory of the birth and death process reduces to *the construction of Markovian matrices* $P(t)$ *such that each column satisfies the infinite system of differential equations*

$$(1.4) \qquad F'_i(t) = -(\delta_i + \beta_i)F_i(t) + \delta_i F_{i-1}(t) + \beta_i F_{i+1}(t)$$

[1] This paper was written while the author was on a sabbatical leave of absence with the support of the United States Air Force and the Office of Ordnance Research.

[2] The main emphasis of this paper lies on *analytical methods* and we therefore postpone *probabilistic interpretations and applications* to the end [*cf.* Section 16 (a)].

and each row satisfies the adjoint system

$$G'_k(t) = -(\delta_k + \beta_k) G_k(t) + \beta_{k-1} G_{k-1}(t) + \delta_{k+1} G_{k+1}(t), \tag{1.5}$$

here the coefficients (³) are constants such that

$$\delta_0 \geqq 0, \quad \beta_{-1} = 0, \quad \beta_0 > 0, \quad \beta_i > 0, \quad \delta_i > 0, \quad (i = 1, 2, \ldots). \tag{1.6}$$

In terms of the infinite matrix

$$A = \begin{bmatrix} -(\delta_0 + \beta_0) & \beta_0 & 0 & 0 & \ldots \\ \delta_1 & -(\delta_1 + \beta_1) & \beta_1 & 0 & \ldots \\ 0 & \delta_2 & -(\delta_2 + \beta_2) & \beta_2 & \ldots \\ . & .. & .... & .. & \ldots \end{bmatrix} \tag{1.7}$$

the differential equations (1.5) and (1.6) for P may be rewritten in the form

$$P'(t) = AP(t), \qquad P'(t) = P(t)A, \tag{1.8}$$

respectively.

It has been shown in [4] that there *always* exists a Markovian $P(t)$ satisfying *both* differential equations (1.8) but that its row-sums need not add up to unity. The problem of uniqueness and of criteria which will ensure that $\sum_k P_{ik}(t) = 1$ were left open. In recent years these problems have been attacked with remarkable analytical skill by a variety of powerful tools (⁴).

It is the purpose of the present paper to treat, and extend, the problem using exceedingly elementary methods and starting from scratch without relying on any known results. The surprisingly simple new approach is made possible by exploiting fully the analogy of (1.8) with a pair of adjoint *diffusion equations* and copying sla-

(³) The letters were chosen in the hope that β will be suggestive of birth, δ of death.

(⁴) W. Ledermann and G. E. H. Reuter [15] used spectral analysis and approximating finite matrices. In addition to various critera they derived the two cases $q = 0$ and $q' = 0$ of the solutions given below in Section 11.

S. Karlin and J. McGregor [9] derived necessary and sufficient conditions using orthogonal polynomials and the moment problem.

B. Koopman [14] treated the problem by continued fraction expansions.

vishly the methods developed for second order differential operators [8].

The basic feature of the present approach consists in introducing a *metric* in the space E of states 0, 1, 2, ... and to use the distance as the *natural scale*. We introduce furthermore (Section 3) a canonical measure in the space E. With our new notations *the two systems of differential equations* (1.4) *and* (1.5) *reduce to the same form*. The common right side has the appearance of the *canonical form of the generalized second order differential operators* in one dimension except that in the denumerable space E derivatives are naturally replaced by difference ratios. The uniqueness criteria revolve about the question whether or not the space E and the associated canonical measure are *bounded*.

The case of *non-uniqueness* (which used to be regarded as a curious anomaly) now becomes most natural : it is, both analytically and probabilistically, the twin brother of the diffusion in a *finite* interval. This parallelism goes so far that, in Section 11, we find the *classical boundary condition* $qz+q'z'=0$ known from diffusion and the Weyl-Kodaira theory. Here z' is replaced by a true derivative at the *boundary of* E (namely a point $x_\infty < \infty$ added by completing E in its natural metric). The processes governed by (1.4) and (1.5) stand in a one-to-one correspondence with the processes determined by a diffusion operator. This is true also of the processes (Section 13) which satisfy (1.4) but not (1.5) or *vice versa*.

If the space E is *bounded* it is natural to complete it by adding the *boundary point* x_∞ *and to treat the latter just as all other points*. This procedure requires us (Sections 12, 14, 15) to add a row indexed ∞ to each of the systems (1.4) and (1.5); however, the new lines have no formal resemblance to the others. The resulting processes are natural from an analytic point of view, but they exhibit the types of « pathology » which were described and studied in special examples constructed by Kolmogorov, Kendall, and Kendall and Reuter.

In theory, the processes of Sections 10-15 could be obtained by specialization from [7], where the general Kolmogorov differential equations are treated. However, the going would be rough and the analytical apparatus would lack the simplicity made possible by the simple ordering and the natural metric in E. On the other hand, the present treatment of the birth and death processes may serve as a simple and intuitive illustration of (or introduction to) the more complicated topologies and tools used for denumerable systems in [7]. It illustrates also the treatment of second order differential operators and the diffusion equations in [8] and [5]. It seemed therefore reasonable to make the exposition self-contained and elementary, *without presupposing any known probabilistic results.*

2. THE NATURAL SCALE. — We define a sequence of points on the x-axis by

$$(2.1)\quad \begin{cases} x_0 = \dfrac{1}{\delta_0} \quad \text{if } \delta_0 > 0, \qquad x_0 = 0 \quad \text{if } \delta_0 = 0, \\ x_1 = x_0 + \dfrac{1}{\beta_0}, \\ \ldots\ldots\ldots\ldots \\ x_n = x_0 + \dfrac{1}{\beta_0} + \ldots + \dfrac{\delta_1\delta_2\ldots\delta_{n-1}}{\beta_0\beta_1\ldots\beta_{n-1}} \qquad (n = 1, 2, \ldots), \\ x_\infty = \lim\limits_{n\to\infty} x_n. \end{cases}$$

The set $\{0, 1, 2, \ldots\}$ of states (or indices) will be mapped onto the space E whose points are $x_0, x_1, x_2, \ldots$. *The state n is identified with x_n : if f is a function assuming the value f_n at the state n we put*

$$(2.2)\qquad f(x_n) = f_n$$

and consider f as a function on E. *The right difference ratio f^+ of f is a function assuming at x_n the value*

$$(2.3)\qquad f^+(x_n) = \frac{f(x_{n+1}) - f(x_n)}{x_{n+1} - x_n} \qquad (n = 0, 1, 2, \ldots).$$

To simplify formulas *we extend this definition by the convention*

$$(2.4)\qquad f^+(x_{-1}) = \delta_0 f(x_0).$$

In the case $x_\infty < \infty$ we shall consider also the *closed space* $\bar{\mathrm{E}}$ containing the points $x_0, x_1, \ldots, x_\infty$ with the obvious definition of convergence. Functions defined on E and their difference ratios will be extended to $\overline{E}$ by

(2.5) $$f(x_\infty) = \lim_{i \to \infty} f(x_i),$$

(2.6) $$f^-(x_\infty) = \lim_{i \to \infty} f^+(x_i),$$

provided that the limits exist. When $f(x_\infty)$ is finite we have obviously

(2.7) $$f^-(x_\infty) = \lim_{i \to \infty} \frac{f(x_\infty) - f(x_i)}{x_\infty - x_i},$$

which shows that $f^-(x_\infty)$ *represents the ordinary left derivative of* f *at* x_∞.

By *the graph of a function* f is meant the polygon joining the points $(x_n, f(x_n))$ of the Cartesian plane.

For the probabilistic interpretation, *see* Section 16 (*b*).

3. The canonical measure. Densities. — In the space E we introduce a *measure* μ *by attributing to the point* x_n *the weight*

(3.1) $$\mu_n = \frac{\beta_0 \beta_1 \ldots \beta_{n-1}}{\delta_1 \delta_2 \ldots \delta_n} \qquad (\mu_0 = 1).$$

If m is a bounded measure on E attributing to x_n the weight m_n, the *density of* m *is the function* f *defined by*

(3.2) $$f(x_n) = \frac{m_n}{\mu_n} \qquad (n = 0, 1, \ldots).$$

A function f is a density if, and only if, it is *summable*, that is, if

(3.3) $$\sum |f(x_n)| m_n < \infty.$$

In the sequel *all measures will be replaced by their densities*.

For an arbitrary function g on E we define *the difference ratio* (5) $D_\mu g$ *of* g *with respect to* μ *as the function which at* x_n *assumes the value*

(3.4) $$D_\mu g(x_n) = \frac{g(x_n) - g(x_{n-1})}{\mu_n} \qquad (n = 1, 2, \ldots).$$

(5) Note that (3.4) represents the ordinary *blackward* difference ratio of g with respect to the measure of the closed intervals $[x_0, x_n]$. In (2.3) we used forward difference ratios with respect to x_n.

With these notations we have the basic theorem whose vérification is trivial.

THEOREM 3.1. — *Let Ω be the operator mapping f into the function Ωf which at x_n assumes the value*

$$\Omega f(x_n) = -(\delta_n + \beta_n) f(x_n) + \delta_n f(x_{n-1}) + \beta_n f(x_{n+1}) \qquad (n = 0, 1, \ldots). \tag{3.5}$$

Then Ω is an iterated difference ratio

$$\Omega f(x_n) = D_\mu f^+(x_n) \qquad (n = 0, 1, \ldots). \tag{3.6}$$

[At the particular point x_0 the right hand member is defined in accordance with (2.4) as

$$\begin{aligned} D_\mu f^+(x_0) &= \mu_0^{-1} [f^+(x_0) - f^+(x_{-1})] = f^+(x_0) - \delta_0 f(x_0) \\ &= -(\delta_0 + \beta_0) f(x_0) + \beta_0 f(x_1) \end{aligned} \tag{3.7}$$

in agreement with the definition of Ω.]

4. THE SYMMETRIC DIFFERENTIAL EQUATIONS FOR MARKOV DENSITIES. — It is natural to consider each *row* of a Markovian matrix $P(t)$ as defining a measure (probability distribution) on the space E, whereas the *columns* are real functions on E. In accordance with our program we shall therefore *replace each Markovian matrix* $P(t)$ *by the matrix* $p(t)$ *of corresponding densities; for fixed* t *the elements of* $p(t)$ *are defined by*

$$p(x_i, x_k; t) = P_{ik}(t) \mu_k^{-1}. \tag{4.1}$$

The basic conditions (1.1)-(1.3) may be restated as

$$p(x_i, x_k; t) \geq 0, \qquad \sum_{k=0}^{\infty} p(x_i, x_k; t) \mu_k \leq 1, \tag{4.2}$$

$$p(x_i, x_k; t+s) = \sum_{j=0}^{\infty} p(x_i, x_j; t) p(x_j, x_k; s) \mu_j, \tag{4.3}$$

$$p(x_i, x_k; 0) = \delta_{ik} \mu_k^{-1}, \tag{4.4}$$

where δ_{ik} *is the Kronecker* symbol. It is, of course, possible to rewrite (4.3) in matrix notation, but it is preferable to recognize the summation sign as an integration with respect to μ and the right hand side as representing the familiar *composition of two kernels*.

We proceed to show that, when transcribed from P to p, *the two systems of differential equations* (1.4) *and* (1.5) *reduce to the same form.* For this reason we shall have to deal almost exclusively with *symmetric* kernels.

For fixed k and t the sequences of numbers $p(x_i, x_k; t)$ and $P_{ik}(t)$ are proportional. Therefore, if each *column* of P satisfies (1.4), the same relations hold for p; in other words, the function f defined for fixed k by $f(x_i, t) = p(x_i, x_k; t)$ satisfies the system of differential equations

$$(4.5) \qquad \frac{d}{dt} f(x_i, t) = -(\delta_i + \beta_i) f(x_i, t) + \delta_i f(x_{i-1}, t) + \beta_i f(x_{i+1}, t) \\ = \Omega F(x_i, t).$$

However, if the *rows* of P satisfy the adjoint equations (1.5) the corresponding rows of p defined by $g(x_k, t) = p(x_i, x_k; t) = P_{ik}(t)\,\mu_k^{-1}$ (where i is fixed) satisfy

$$(4.6) \qquad \frac{d}{dt} g(x_k; t) = -(\delta_k + \beta_k) g(x_k, t) \\ + \beta_{k-1} \frac{\mu_{k-1}}{\mu_k} g(x_{k-1}, t) + \delta_{k+1} \frac{\mu_{k+1}}{\mu_k} g(x_{k+1}, t).$$

It is now seen that (4.6) reduces to

$$(4.7) \qquad \frac{d}{dt} g = \Omega_g$$

which is the same [6] as (4.5). We have thus proved :

Teorem 4.1. — *Let* $P(t)$ *denote a family of Markovian matrices. If the columns of* P *satisfy* (1.4) [*the rows satisfy* (1.5)] *then the*

([6]) In self-explanatory matrix notation the matrix p satisfies the two equations $p' = AP$ and $p' = p\,MAM^{-1}$ corresponding to (1.8); here M is the diagonal matrix with elements μ_k. It is asserted that *the transpose* of MAM^{-1} equals A or, in other words, that MA is symmetric.

columns [rows] of the corresponding density kernel p satisfy

$$\frac{d}{dt}f = \mathrm{D}_\mu f^+ = \Omega f. \tag{4.8}$$

From now on we shall have to deal with (4.8) only, replacing the two systems of differential equations (1.4) and (1.5). We may look upon (4.8) in the traditionnal manner as an infinite system of ordinary differential equations, but it has considerable advantages to treat (4.8) as the analogue of a diffusion equation $\mathrm{D}_t u = \mathrm{D}_\mu \mathrm{D}_x u$ in one dimension (here D_x stands for ordinary differentiation with respect to x and D_μ for differentiation with respect to an arbitrary positive measure, not necessarily continuous). In (4.8) the iterated derivative of the diffusion equation is replaced by an iterated difference ratio. This change is so unessential, that we can use the methods developed in [8] with the simplications due to the present special conditions.

Note that the symmetry between the rows and columns of the kernel p is not complete : the rows have to be *summable* solutions of (4.8), whereas the columns must be *bounded*.

5. The resolvent (Laplace transforms). — We introduce the kernel of *Laplace transforms* that is, the function π_λ defined on $\mathrm{E}\times\mathrm{E}$ by

$$\pi_\lambda(x_i, x_k) = \int_0^\infty e^{-\lambda t} p(x_i, x_k; t)\, dt = \mu_k^{-1}\int_0^\infty e^{-\lambda t}\, \mathrm{P}_{ik}(t)\, dt, \tag{5.1}$$

where $\lambda > 0$ is arbitrary, but fixed. Equations (4.2) imply

$$\pi_\lambda(x_i, x_k) \geq 0, \qquad \lambda\sum_{k=0}^{\infty} \pi_\lambda(x_i, x_k)\,\mu_k \leq 1. \tag{5.2}$$

On multiplying (4.3) by $e^{-\lambda t - \nu s}$ and integrating we get the wellknown *resolvent equation*

$$\frac{\pi_\lambda - \pi_\nu}{\lambda - \nu} = -\,\pi_\lambda \pi_\nu \qquad (\lambda,\ \nu > 0), \tag{5.3}$$

where $\pi_\lambda \pi_\nu$ stands for the *composition of the two kernels* defined by

$$(\pi_\lambda \pi_\nu)(x_i, x_k) = \sum_{j=0}^{\infty} \pi_\lambda(x_i, x_j)\, \pi_\nu(x_j, x_k)\, \mu_j. \tag{5.4}$$

THEOREM 5.1 (7). — *The transformation defined in* (5.1) *establishes a one-to-one correspondence between Markovian matrices* $P(t)$ *whose columns satisfy* (1.4) [*row satisfy* (1.5)] *and kernels* π_λ *with the properties* (5.2), (5.3), *whose k-th column* [*k-th row*] *satisfies the relations*

$$\lambda F(x_i) - D_\mu F^+(x_i) = \delta_{ik}\mu_k^{-1} \qquad (i = 0, 1, 2, \ldots), \tag{5.5}$$

where δ_{ik} *is the Kronecker symbol.*

Proof. — Clearly (5.5) is the Laplace transform version of (4.8) together with the initial condition (4.4). It remains to prove that each kernel π_λ with the stated properties corresponds to a $P(t)$ as asserted. Now (5.3) implies that π_λ may be differentiated with respect to λ and we get by induction the well-known relation

$$\frac{d^n}{d\lambda^n}\pi_\lambda = (-1)^n n!\,\pi_\lambda^{(n+1)} \qquad (n = 1, 2, \ldots), \tag{5.6}$$

where $\pi_\lambda^{(n)}$ stands for the n-fold composition of the kernel π_λ with itself [defined as in (5.4)]. Accordingly for $n = 0, 1, 2, \ldots$

$$(-1)^n \frac{d^n}{d\lambda^n}\pi_\lambda(x_i, x_k) \geq 0, \qquad \frac{1}{n!}\left|\frac{d^n}{d\lambda^n}\sum_{k=0}^{\infty}\pi_\lambda(x_i, x_k)\mu_k\right| \leq \lambda^{-(n+1)}. \tag{5.7}$$

Thus $\pi_\lambda(x_i, x_k)$ is a completely monotonic function of λ and therefore the Laplace transform of a function $p(x_i, x_k; .)$ such that (4.2) holds (*cf.* Widder [18], p. 315 and 310). Equation (5.5) is equivalent to

$$\begin{aligned} p(x_i, x_k; t) - \delta_{ik}\mu_k^{-1} \\ = \int_0^t \{ -(\delta_i + \beta_i)\,p(x_i, x_k; s) + \delta_i\,p(x_{i-1}, x_k; s) \\ + \beta_i\,p(x_{i+1}, x_k; s)\}\,ds. \end{aligned} \tag{5.8}$$

It follows that $p(x_i, x_k; t)$ depends coutinuously on t; differentiation of (5.8) with respect to t then shows that the columns [rows] of p satisfy (4.8) with the initial conditions (4.4). Finally, as we have

(7) It is hardly necessary to say that this is a standard theorem. We give the proof only because of its extreme simplicity in the present case.

seen, (5.3) impies that the double Laplace transforms of the two sides in (4.3) are equal. It follows that (4.3) holds almost everywhere and hence everywhere, p depending continuously on t.

6. Boundary classification. — *We shall say that the boundary x_∞ is :*

(6.1) *Regular if* $x_\infty < \infty$, $\sum \mu_k < \infty$;

(6.2) *Exit if it is not regular but* $x_\infty < \infty$, $\sum (x_\infty - x_k)\mu_k < \infty$;

(6.3) *Entrance if it is not regular but* $\sum x_k \mu_k < \infty$;

Natural otherwise.

These terms correspond to those introduced in [5] for the diffusion equations. For the probabilistic interpretation, *see* Section 16 (*e*).

Note that in the case of an entrance boundary $\sum \mu_k < \infty$; hence if both (6.2) and (6.3) are satisfied, the boundary is regular. Thus the *four types are mutually exclusive*. Rearranging the terms, we have

$$(6.4) \qquad \sum_{k=0}^{\infty} (x_\infty - x_k)\mu_k = \sum_{i=0}^{\infty} (x_{i+1} - x_i) \sum_{k=0}^{i} \mu_k.$$

This shows the symmetry of the two sets of conditions (6.2) and (6.3): For an *entrance* boundary the total measure is finite and the scale x_k is summable with respect to the measure; for an *exit* boundary the scale variable is bounded and the measure of the segments (x_0, x_i) is integrable with respect to the scale.

7. The increasing solution of $\lambda u - \Omega u = 0$. — In this Section we use the conventions (2.5) and (2.6) to define the value of u and of its derivative at x_∞. These limits, as well as x_∞, may be finite or infinite. In order not to impair the readability of the formulas we refrain from indicating explicity the dependance of u on λ.

Theorem 7.1. — *For each fixed $\lambda > 0$ there exists exactly one function u defined on* E *such that*

$$(7.1) \qquad \lambda u - D_\mu u^+ = 0, \qquad u(x_0) = 1.$$

Both u and u^+ are strictly increasing (that is, the graph of u is convex). Moreover

$$u(x_\infty) < \infty \tag{7.2}$$

if, and only if, x_∞ is regular or exit; similarly,

$$\sum u(x_k)\mu_k < \infty \tag{7.3}$$

if, and only if, x_∞ is regular or entrance. If x_∞ is regular,

$$u^-(x_\infty) = \lim u^+(x_k) < \infty. \tag{7.4}$$

(Thus u is bounded and summable if, and only if, x_∞ is regular. Note that *the behavior of u is independent of λ.*)

Proof. — From the definition (2.4) we have $u^+(x_{-1}) = \delta_0 \geq 0$ and hence from (7.1) by induction

$$u^+(x_n) - \delta_0 = \sum_{k=0}^{n} D_\mu u^+(x_k)\mu_k = \lambda \sum_{k=0}^{n} u(x_k)\mu_k. \tag{7.5}$$

Also, by the definition of u^+,

$$u(x_n) - 1 = \sum_{i=0}^{n-1} u^+(x_i)(x_{i+1} - x_i). \tag{7.6}$$

From this recursive system we can calculate successively $u^+(x_0)$, $u(x_1)$, $u^+(x_1)$, $u(x_2)$, ... It is then obvious that $u(x_{n+1}) > u(x_n)$ and $u^+(x_{n+1}) > u^+(x_n)$ which proves the first part of the theorem.

An explicit formula for $u(x_n)$ is obtained by substituting from (7.5) into (7.6) and reversing the order of summation. The result is

$$u(x_n) = 1 + \delta_0(x_n - x_0) + \lambda \sum_{k=0}^{n-1} u(x_k)(x_n - x_k)\mu_k. \tag{7.7}$$

This equation makes it clear that the conditions (6.2) are *necessary* for $u(x_\infty) < \infty$. Conversely, suppose that (6.2) holds. From (7.5) we have

$$u(x_{n+1}) - u(x_n) \leq \delta_0(x_{n+1} - x_n) + \lambda u(x_n)(x_{n+1} - x_n) \sum_{k=0}^{n} \mu_k. \tag{7.8}$$

Divide by $u(x_n)$ to obtain

$$\frac{u(x_{n+1})}{u(x_n)} - 1 < \delta_0(x_{n+1} - x_n) + \lambda(x_{n+1} - x_n) \sum_{k=0}^{n} \mu_k. \tag{7.9}$$

Remembering (6.4) we see that the right side represents the terms of a convergent series. It follows that $\sum \log\left\{\frac{u(x_{n+1})}{u(x_n)}\right\}$ converges and therefore $u(x_\infty) < \infty$. Thus (6.2) and (7.2) are equivalent.

Next we observe that in consequence of (7.5), the relations (7.3) and (7.4) imply each other. The graph of u being convex we have

$$u^+(x_0) \leq \frac{u(x_n) - 1}{x_n - x_0} \leq u^+(x_{n-1}) \qquad (n = 1, 2, \ldots). \tag{7.10}$$

Thus $u(x_n) > cx_n + c'$ where c and c' are positive constants. A glance at (7.5) now shows that the condition (6.3) is *necessary* for $u^+(x_n)$ to be bounded, that is, for (7.4) and hence for (7.3). Conversely, assume (6.3) to hold. Using the second inequality ni (7.10) we have

$$u^+(x_n) - u^+(x_{n-1}) = \lambda u(x_n)\mu_n \leq \lambda \mu_n[1 + u^+(x_{n-1})x_n] \tag{7.11}$$

and thus

$$\sum \left\{\frac{u^+(x_n)}{u^+(x_{n-1})} - 1\right\} < \infty. \tag{7.12}$$

It follows that $\sum \log\left\{\frac{u^+(x_n)}{u^+(x_{n-1})}\right\} < \infty$ and therefore (7.4), and hence also (7.3), hold.

8. The decreasing solution of $\lambda v - \Omega v = 0$. — Up to a multiplicative constant u is obviously the only function satisfying $\lambda u - \Omega u = 0$ at *all* points of E. However, x_0 plays the role of a boundary point and there exists a two-parametric family of functions satisfying the equation at all other points. We single out one among them for closer examination [8]. For its probabilistic significance, *see* Section 16 (*d*).

---

[8] The following theorem is not of an accidental character but suggested by the analogy with the differential equation $\lambda w - w'' = 0$. For any two solutions w_i of this equation the Wronskian $w_1 w'_2 - w_2 w'_1$ is constant. Similarly, if w_i are

THEOREM 8.1. — *Let u be as in Theorem 7.1 and $\lambda > 0$. The equations*

$$v(x_n) - u(x_n) \sum_{j=n}^{\infty} \frac{x_{j+1} + x_j}{u(x_j)\,u(x_{j+1})} \qquad (n = 0, 1, \ldots) \tag{8.1}$$

define a strictly decreasing function v on E. For it

$$\begin{cases} \lambda v(x_n) - D_\mu v^+(x_n) = 0 & \text{if } n \geq 1, \\ \lambda v(x_0) - D_\mu v^+(x_0) = 1. \end{cases} \tag{8.2}$$

Moreover

$$v(x_n)\,u^+(x_n) - u(x_n)\,v^+(x_n) = 1 \qquad (n \geq 0) \tag{8.3}$$

and

$$v^-(x_\infty) = \lim_{n \to \infty} v^+(x_n) = -\frac{1}{u(x_\infty)} \tag{8.4}$$

[*here $v^-(x_\infty) = 0$ if $u(x_\infty) = \infty$*].

Proof. — Since u^+ is increasing we have

$$\begin{aligned} v(x_n) &= u^+(x_n) \sum_{j=n}^{\infty} \frac{x_{j+1} - x_j}{u(x_{j+1}) - u(x_j)} \left\{ \frac{1}{u(x_j)} - \frac{1}{u(x_{j+1})} \right\} \\ &\leq \frac{1}{u^+(x_n)}\, u(x_n) \sum_{j=n}^{\infty} \left\{ \frac{1}{u(x_j)} - \frac{1}{u(x_{j+1})} \right\} \leq \frac{1}{u^+(x_n)} \end{aligned} \tag{8.5}$$

which shows the convergence of the series.

Taking differences it is seen at once that

$$v^+(x_n) = u^+(x_n) \sum_{j=n}^{\infty} \frac{x_{j+1} - x_j}{u(x_j)\,u(x_{j+1})} - \frac{1}{u(x_n)}. \tag{8.6}$$

---

two functions on E satisfying $\lambda w_i - \Omega w_i = 0$ we have for their « Wronskian » $w_1 w_2^+ - w_2 w_1^+ = C$. Choose $w_2 = u$ and set the $C = 1$. Integration of the ensuing difference equation leads to solutions of $\lambda w - \Omega w = 0$ and (8.1) is obtained by an appropriate choice of the arbitrary constant. The most general solution is of the form $c_1 u + c_2 w$.

It follows

$$(8.7)\quad -\frac{1}{u(x_n)} \leq v^+(x_n) \leq \sum_{j=n}^{\infty} u^+(x_j)\frac{x_{j+1}-x_j}{u(x_j)\,u(x_{j+1})} - \frac{1}{u(x_n)}$$

$$= \sum_{j=n}^{\infty}\left\{\frac{1}{u(x_j)} - \frac{1}{u(x_{j+1})}\right\} - \frac{1}{u(x_n)} = -\frac{1}{u(x_\infty)}.$$

Thus v decreases, and (8.4) is true. On multiplying (8.6) by $u(x_n)$ we get (8.3).

It remains to verify (8.2). From (8.3) we have for $n = 1, 2, \ldots$

$$(8.8)\quad v(x_n)\,u^+(x_n) - u(x_n)\,v^+(x_n) = v(x_{n-1})\,u^+(x_{n-1}) - u(x_{n-1})\,v^+(x_{n-1}).$$

This is equivalent to

$$(8.9)\quad v(x_n)\{u^+(x_n) - u^+(x_{n-1})\} - u(x_n)\{v^+(x_n) - v^+(x_{n-1})\}$$
$$= v^+(x_{n-1})\{u(x_n) - u(x_{n-1})\} - u^+(x_{n-1})\{v(x_n) - v(x_{n-1})\}.$$

The right side is obviously zero and therefore (8.9) reduces to

$$(8.10)\quad 0 = \mu_n\{v(x_n)\,D_\mu u^+(x_n) - u(x_n)\,D_\mu v^+(x_n)\}$$
$$= \mu_n u(x_n)\{\lambda v(x_n) - D_\mu v^+(x_n)\}\qquad (n \geq 1)$$

which proves the first line in (8.2). To complete the proof we have to evaluate

$$(8.11)\quad \lambda v(x_0) - \Omega v(x_0) = (\lambda + \delta_0 + \beta_0)\,v(x_0) - \beta_0 v(x_1).$$

Now from the definition (8.1)

$$(8.12)\quad v(x_1) = v(x_0)\,u(x_1) - \frac{x_1 - x_0}{u(x_0)} = v(x_0)\,u(x_1) - \beta_0^{-1}.$$

But by the definition of u

$$(8.13)\quad \beta_0 u(x_1) = (\lambda + \delta_0 + \beta_0)\,u(x_0) = \lambda + \delta_0 + \beta_0.$$

Substituting back into (8.12) and (8.11) we get the last line in (8.2).

9. The Green function. — Let u and v be the functions defined in the preceding two sections, $\lambda > 0$ being fixed.

Definition. — *The minimal Green function is the symmetric kernel defined by*

$$(9.1)\qquad \pi_\lambda(x_i, x_k) = \begin{cases} u(x_i)\,v(x_k) & \text{for } k \geq i, \\ v(x_i)\,u(x_k) & \text{for } k \leq i. \end{cases}$$

With it we associate the transformation $\mathcal{G}_\lambda$ *wich maps the function* f *on* E *into the function* F_λ *defined by*

$$(9.2)\qquad \begin{aligned} F_\lambda(x_i) &= \sum_{k=0}^{\infty} \pi_\lambda(x_i, x_k) f(x_k)\mu_k \\ &= v(x_i)\sum_{k=0}^{i} u(x_k) f(x_k)\mu_k + u(x_i)\sum_{k=i+1}^{\infty} v(x_k) f(x_k)\mu_k. \end{aligned}$$

(The transformation is defined if the series converges absolutely.)

THEOREM 9.1. — *If* f *is bounded* [*summable*] *so is* F_λ, *and*

$$(9.3)\qquad \lambda F_\lambda - D_\mu F_\lambda^+ = f.$$

In particular, the transform of the constant function 1 *satisfies*

$$(9.4)\qquad \lambda\sum_{k=0}^{\infty} \pi(x_i, x_k)\mu_k = 1 - \frac{u(x_i)}{u(x_\infty)} - \delta_0 v(x_i).$$

Note. — If $u(x_\infty) = \infty$ the second term on the right is to be interpreted as zero. It will be observed that (9.4) *is stronger than* (5.2). Note that $u(x_\infty) < \infty$ if, and only if, x_∞ is regular or exit.

Proof. — *a.* We start by proving (9.4). The right side in (9.2) is meaningful (finite or infinite) whenever $f \geq 0$. We can rewrite the terms using $\lambda u(x_k)\mu_k = \mu_k D_\mu u^+(x_k) = u^+(x_k) - u^+(x_{k-1})$ and the corresponding expression for v. For $f = 1$ we get [remembering the convention (2.4)]

$$(9.5)\qquad \begin{aligned} \lambda\sum_{k=0}^{\infty} \pi_\lambda(x_i, x_k)\mu_k &= v(x_i)\sum_{k=0}^{i}\{u^+(x_k) - u^+(x_{k-1})\} + u(x_i)\sum_{k=i+1}^{\infty}\{v^+(x_k) - v^+(x_{k-1})\} \\ &= v(x_i)u^+(x_i) - u(x_i)v^+(x_i) - \delta_0 v(x_i) + u(x_i)v^+(x_\infty). \end{aligned}$$

In view of (8.3) and (8.4) this is the same as (9.4).

b. It follows from (9.4) that the transformation $\mathcal{G}_\lambda$ carries a bounded f into a bounded F_λ. Remembering that π_λ is symmetric and that the right side in (9.4) is ≤ 1 we get

$$(9.6)\quad \lambda\sum_{i=0}^{\infty}|F_\lambda(x_i)|\mu_i \leq \sum_{k=0}^{\infty}|f(x_k)|\mu_k\lambda\sum_{i=0}^{\infty}\pi(x_i, x_k)\mu_i \leq \sum_{k=0}^{\infty}|f(x_k)|\mu_k.$$

Thus a summable f is carried into a summable F_λ.

c. *Proof of* (9.3). — Taking differences we get from (9.2)

$$(9.7)\quad F_\lambda^+(x_i) = v^+(x_i)\sum_{k=0}^{i}u(x_k)f(x_k)\mu_k + u^+(x_i)\sum_{k=i+1}^{\infty}v(x_k)f(x_k)\mu_k$$

and hence for $i \geq 1$

$$(9.8)\quad F_\lambda^+(x_i) - F_\lambda^+(x_{i-1}) = \{v^+(x_i) - v^+(x_{i-1})\}\sum_{k=0}^{i}u(x_k)f(x_k)\mu_k$$
$$+\{u^+(x_i) - u^+(x_{i-1})\}\sum_{k=i+1}^{\infty}v(x_k)f(x_k)\mu_k$$
$$-\{v(x_i)u^+(x_{i-1}) - u(x_i)v^+(x_{i-1})\}f(x_i)\mu_i.$$

A simple manipulation shows that

$$(9.9)\quad v(x_i)u^+(x_{i-1}) - u(x_i)v^+(x_{i-1}) = v(x_i)u^+(x_i) - u(x_i)v^+(x_i) = 1,$$

the last equality being due to the Wronski identity (8.3). This permits us to rewrite (9.8) in the form

$$(9.10)\quad D_\mu F_\lambda^+(x_i) = D_\mu v^+(x_i)\sum_{k=0}^{i}u(x_k)f(x_k)\mu_k$$
$$+ D_\mu u^+(x_i)\sum_{k=i+1}^{\infty}v(x_k)f(x_k)\mu_k - f(x_i)$$

which proves (9.3) at x_i for $i \geq 1$. For $i = 0$ the relation (9.8) holds provided the symbols $F_\lambda^+(x_{-1})$, ... are defined in accordance with (2.4). The last braces on the right in (9.8) vanish for $i = 0$ and therefore

$$(9.11)\quad D_\mu F_\lambda^+(x_0) = f(x_0)D_\mu v^+(x_0) + D_\mu u^+(x_0)\sum_{k=1}^{\infty}v(x_k)f(x_k)\mu_k.$$

In view of (8.2) this proves the validity of (9.3) at the point x_0.

LEMMA 9.1. — *If x_∞ is regular or exit, then*

$$\lim_{i\to\infty} F_\lambda(x_i) = 0 \tag{9.12}$$

for each bounded and for each summable f.

Proof. — Since $x_\infty < \infty$ and $u(x_\infty) < \infty$ it follows from the definition (8.1) at once that $v(x_n) \to 0$. The last series in (9.2) converges for $f = 1$ and hence it converges absolutely whenever f is bounded or summable. Thus (9.12) follows directly from the definition (9.2).

LEMMA 9.2. — *If $\lambda G - D_\mu G^+ = f$, then $F_\lambda - G$ is a multiple of u.*

Proof. — This is obvious since u is, up to a multiplicative constant, the only solution of the homogeneous equation.

THEOREM 9.2. — *The kernel π_λ satisfies the resolvent equation* (5.3).

Proof. — Choose f bounded and summable. From (9.3) we have

$$\lambda\{F_\lambda - F_\nu\} - D_\mu\{F_\lambda - F_\nu\}^+ = (\nu - \lambda) F_\nu \tag{9.13}$$

and hence by the last lemma

$$F_\lambda(x_i) - F_\nu(x_i) - (\nu - \lambda)\sum_{k=0}^{\infty} \pi_\lambda(x_i, x_k) F_\nu(x_k)\mu_k = pu(x_i), \tag{9.14}$$

where p does not depend on i. Here the left side is bounded and summable while u has this property only if x_∞ is regular. However, for a regular x_∞ the left side tends to 0 as $i \to \infty$ while $u(x_\infty) > 0$. Therefore $p = 0$. For the particular choice $f(x_k) = \delta_{kj}\mu_j^{-1}$ (with j fixed) we have $F_\nu(x_i) = \pi_\nu(x_i, x_j)$ and (9.14) reduces to the resolvent equation.

10. *The minimal solution.* — We proceed to collect and reformulate the information contained in the last section. *Cf.* Section 16 (*e*).

THEOREM 10.1 ([9]). — *The Green function of Theorem* 9.1 *represents*

([9]) The criteria of this theorem were first obtained by Karlin and McGregor [9].

the Laplace transform of a Markovian family of matrices $P(t)$ *whose columns satisfy the differential equations* (1.4) *and whose rows satisfy* (1.5).

a. This is the unique common solution except if x_∞ *is regular.*
b. In order that [10]

$$\sum_{k=0}^{\infty} P_{ik}(t) = 1 \qquad (i = 0, 1, \ldots) \tag{10.1}$$

it is necessary and sufficient that $\delta_0 = 0$ *and that* x_∞ *be neither regular nor exit.*

(In the case of a regular x_∞ a one parametric family of common solutions will be constructed in the next section.)

Proof. — The first part is a restatement of Theorem 5.1. To prove *a* let $P^\star$ be a common Markovian solution of (1.4) and (1.5) and let $\pi_\lambda^\star(x_i, x_k)$ be the Laplace transform of the density $P_{ik}^\star(t)\,\mu_k^{-1}$. By Lemma 9.2 the rows and the colums of $\pi_\lambda^\star$ differ from those of π_λ by a multiple of u. But unless x_∞ is regular the solution u is either unbounded or not summable, and hence $\pi_\lambda^\star = \pi_\lambda$.

The assertion *b* is an immediate consequence of (9.4).

THEOREM 10.2 (*Minimal character*). — *Let* $P^\star(t)$ *represent a Markovian family of matrices whose columns satisfy* (1.4) *or whose rows satisfy* (1.5) (*it is not required that both differential equations hold*). *Then*

$$P^\star(t) \geqq P(t). \tag{10.2}$$

Proof. — Define $\pi_\lambda^\star$ as in the last proof. Suppose first that the *columns* of $P^\star$ satisfy (1.4). Then $\pi_\lambda^\star(., x_k) - \pi_\lambda(., x_k) = pu$ for each fixed k, where p is a constant depending on λ and k. When x_∞ is entrance or natural, then u is unbounded and therefore $p = 0$.

---

(10) The semi-group property (1.2) implies that the left side is a monotonic function of t, and one concludes that (10.1) holds either for each $t > 0$ or for no $t > 0$. In the special case of the birth and death process it is easily seen that (10.1) holds for each $i \geqq 0$ or for no $i \geqq 0$.

When $x_\infty < \infty$ then $\pi_\lambda(x_i, x_k) \to 0$ as $i \to \infty$ and therefore $p \geq 0$ as asserted.

Consider next the case where the *rows* of $P^\star$ satisfy (1.5). Then $\pi_\lambda^\star(x_i, .) = \pi_\lambda(x_i, .) = pu$ for each fixed i. When x_∞ is exit or natural, u is not summable and therefore $p = 0$. When x_∞ is regular or entrance we have

$$(10.3) \qquad \lim_{k \to \infty} \frac{u(x_k)}{v(x_k)} = \infty, \qquad \lim_{k \to \infty} \frac{\pi(x_i, x_k)}{v(x_k)} = u(x_i)$$

and therefore the positivity of $\pi^\star$ requires that $p \geq 0$.

This argument not only proves the theorem but furnishes additional information concerning the uniqueness problem for each of the two systems of differential equations; it enables us also to *drop the condition that* P *be Markovian.* To summarize the situation let us agree for the moment to say that a family of matrices $P(t)$ represents *a solution* of (1.4) if the semi-group relation (1.2) holds and if each *column* of $P(t)$ is *bounded* and satisfies (1.4); similarly for (1.5) we require that the *rows* represent *summable* solutions. We have then four cases to distinguish.

a. Natural x_∞. — Neither (1.4) nor (1.5) admit of any solution except the common solution of Theorem 10.1. Thus each system is a consequence of the other and implies the Markovian character of the solution. The equality (10.1) holds if $\delta_0 = 0$.

b. Exit x_∞. — The matrix P of Theorem 10.1 is the *unique solution* of (1.5), *but* (1.4) *admits of infinitely many Markovian solutions* (to be constructed in Section 13). Thus (1.5) implies (1.4), but not *vice versa.* The equality (10.1) does not hold for the common solution.

c. Entrance x_∞. — The matrix P of Theorem 10.1 is the *unique solution of* (1.4), *but* (1.5) *admits of infinitely many positive solutions* (which can be constructed by the method of Section 13). Now the differential equations (1.4) imply (1.5) and the Markovian character of the solution. The strictly stochastic case (10.1) arises if, and only if, $\delta_0 = 0$. In this case Theorem 10.1 provides the only *Markovian* solution of (10.5).

d. Regular x_∞. — Here complete non-uniqueness prevails, and we proceed to examine this case in detail.

11. THE « ELASTIC BARRIER » SOLUTIONS. — In this Section we suppose that x_∞ is *regular* and propose to construct *the most general family of Markovian matrices* $P^\star(t)$ *whose columns satisfy* (1.4) *and whose rows satisfy* (1.5). Adhering to our notations we introduce the kernel $p^\star$ of densities and their Laplace transforms by

$$(11.1)\qquad p^\star(x_i, x_k; t) = P^\star_{ik}(t)\,\mu_k^{-1}, \qquad \pi^\star_\lambda(x_i, x_k) = \int_0^\infty e^{-\lambda t} p^\star(x_i, x_k; t)\,dt.$$

The letters P, p, π_λ will always refer to the *minimal solution* of Theorem 10.1.

For each fixed λ the rows and columns of $\pi^\star_\lambda$ differ from those of π_λ only by a multiple of u, and therefore

$$(11.2)\qquad \pi^\star_\lambda(x_i, x_k) = \pi_\lambda(x_i, x_k) + u(x_i)\,\rho_\lambda\, u(x_k),$$

where ρ_λ does not depend on i and k. We are to determine ρ_λ in its dependence on λ in such a way that the conditions enumerated in Section 5 will be fulfilled.

For convenience we introduce *the notation*

$$(11.3)\qquad [f, g] = \sum_{k=0}^{\infty} f(x_k)\,g(x_k)\,\mu_k$$

for the inner product of two functions.

The kernel $\pi^\star_\lambda$ induces the transformation $\mathcal{G}^\star_\lambda$ mapping f into $F^\star_\lambda = \mathcal{G}^\star_\lambda f$, where

$$(11.4)\qquad F^\star_\lambda(x_i) = \sum_{k=0}^{\infty} \pi^\star_\lambda(x_i, x_k)\,f(x_k)\,\mu_k.$$

Thus

$$(11.5)\qquad F^\star_\lambda = F_\lambda + \rho_\lambda [f, u]\,u$$

with F_λ defined in (9.2). Since u is both bounded and summable each bounded [summable] f is mapped into a bounded [summable] $F^\star_\lambda$. Clearly $\lambda F^\star - \Omega F^\star = f$ and hence the rows and columns of $\pi^\star_\lambda$

satisfy (5.5). It remains to discuss the norm-conditions (5.2) and the resolvent equation (5.3)

LEMMA 11.1. — *In order that*

(11.6) $$0 \leq \lambda \sum_{k=0}^{\infty} \pi_\lambda^*(x_i, x_k)\,\mu_k \leq 1$$

it is necessary and sufficient that

(11.7) $$0 \leq \rho_\lambda \leq \frac{1}{u(x_\infty)\{u^-(x_\infty) - \delta_0\}},$$

where $u^-(x_\infty) = \lim u^+(x_n)$ *as in* (7.4).

Proof. — It follows from (7.5) that

(11.8) $$\lambda[u, 1] = u^-(x_\infty) - \delta_0$$

(the quantity on the right is thus seen to be positive). Remembering (9.4) we have therefore

(11.9) $$\lambda \sum_{k=0}^{\infty} \pi_\lambda^*(x_i, x_k)\,\mu_k = 1 - \delta_0 v(x_i) - u(x_i)\left\{\frac{1}{u(x_\infty)} - \rho_\lambda(u^-(x_\infty) - \delta_0)\right\}.$$

Since $\frac{v(x_i)}{u(x_i)} \to 0$ as $i \to \infty$ the second inequality in (11.6) necessitates that the expression within braces be non-negative and the first requires that $\rho_\lambda \geq 0$. On the other hand, if $\rho_\lambda \geq 0$ then $\pi_\lambda^* > 0$. The lemma is proved.

LEMMA 11.2. — *For bounded as well as for summable* f *the transform* F_λ^* *and its difference ratios* F_λ^{*+} *have finite limits at* x_∞ :

(11.10) $$F_\lambda^*(x_\infty) = \lim_{i \to \infty} F_\lambda^*(x_i), \qquad F_\lambda^{*-}(x_\infty) = \lim_{i \to \infty} F_\lambda^{*+}(x_i).$$

When $\rho_\lambda > 0$ *these limits stand in the relationship*

(11.11) $$F_\lambda^{*-}(x_\infty) + \alpha_\lambda F_\lambda^*(x_\infty) = 0 \qquad (\alpha_\lambda \geq 0),$$

where

(11.12) $$\rho_\lambda = \frac{1}{u(x_\infty)\{\alpha_\lambda u(x_\infty) + u^-(x_\infty)\}}.$$

Proof. — By Lemma 9.1 and (11.5) we have

$$F_\lambda^*(x_\infty) = \rho_\lambda u(x_\infty)[u, f], \tag{11.13}$$

On the other hand, x_∞ being regular, the derivative $u^-(x_\infty)$ is finite and $v^-(x_\infty)u(x_\infty) = 1$ by (8.4). It follows therefore from (9.7) at once that

$$F_\lambda^-(x_\infty) = \lim_{i\to\infty} F_\lambda^+(x_i) = \frac{-[u, f]}{u\ (x_\infty)}. \tag{11.14}$$

Accordingly,

$$F_\lambda^{*-}(x_\infty) = \left\{-\frac{1}{u(x_\infty)} + \rho_\lambda u^-(x_\infty)\right\}[u, f]$$

and a comparison of this formula with (11.13) accomplishes the proof. (As $\rho_\lambda \to 0$ the « *boundary condition* » (11.11) approaches the relation $F_\lambda^*(x_\infty) = 0$ which characterizes the *minimal* solution of Section 10.)

LEMMA 11.3. — *The range of the transformation $\mathcal{G}_\lambda^*$ is characterized by the « boundary condition »*

$$\Phi^-(x_\infty) + \alpha_\lambda \Phi(x_\infty) = 0 \tag{11.15}$$

in the following sense ; In order that a bounded [summable] Φ be of the form $\Phi = \mathcal{G}_\lambda^* f$ with f bounded [summable] it is necessary and sufficient that both Φ and $D_\mu\Phi^+$ be bounded [summable] and (11.15) hold.

Proof. — The last lemma expresses the necessity of the condition. Conversely, for a Φ which satisfies the condition define f by $f = \lambda\Phi - D_\mu\Phi^+$, and F_λ^* as above. Then $\Phi - F_\lambda^* = pu$ by Lemma 9.2. However $u^-(x_\infty) + \alpha_\lambda u(x_\infty) > 0$ and therefore $p = 0$.

LEMNA 11.4 ([11]). — *For the kernels π_λ^* to satisfy the resolvent equation it is necessary and sufficient that the range of the transformation $\mathcal{G}_\lambda^*$ be independent of λ [that is, $\alpha_\lambda = \alpha =$ Cte].*

([11]) This lemma is of general character and will be used repeatedly. It contains in substance the abstract definition of boundary conditions for semi-groups introduced in [5].

Proof. — The resolvent equation is equivalent to

$$(11.16) \qquad F_\lambda^\star - F_\nu^\star + (\lambda - \nu)\,\mathcal{G}_\lambda^\star F_\nu^\star = 0 \qquad (\lambda, \nu > 0).$$

In fact, (11.16) is a direct consequence of (5.3), and conversely (5.3) is contained in (11.16) as a special case, namely when $f(x_i) = \delta_{ik}\mu_k^{-1}$, where δ_{ik} is the Kronecker symbol.

If (11.16) holds then $F_\nu^\star$ is a linear combination of two functions in the range of $\mathcal{G}_\lambda^\star$ and therefore the range of $\mathcal{G}_\nu^\star$ is contained in the range of $\mathcal{G}_\lambda^\star$. But λ and μ are interchangeable and hence the two ranges are identical. On the other hand, the left side of (11.16) is a solution of the homogeneous equation $\lambda z - \Omega z = 0$ and therefore zero if it is an element of the range of $\mathcal{G}_\lambda^\star$.

Summing up, we have shown that we get all admissible kernels from (11.2) if we choose $\alpha_\lambda = \alpha$ and define ρ_λ by (11.12). The norm condition (11.7) requires that $\alpha u(x_\infty) + \delta_0 \geq 0$. Now u depends on λ and obviously $u(x_\infty) \to \infty$ as $\lambda \to \infty$. It follows that (11.7) holds for all $\lambda > 0$ if, and only if, $\alpha \geq 0$. The minimal solution of te last section corresponds formally to $\alpha = \infty$.

Equation (11.1) associates with $\pi_\lambda^\star$ the kernel of densities $p^\star$ whose columns satisfy the differential equations (5.5). Each column of $\pi_\lambda^\star$ satisfies the boundary condition (11.15), and the integrated version (5.8) of the differential equations shows that the same is true of the columns of $p^\star$. This proves

THEOREM 11.1. — *Let x_∞ be regular. The totality of all Markovian families of matrices $P^\star(t)$ whose columns satisfy the differential equations (1.4) and whose rows satisfy (1.5) stand in a one-to-one correspondence with the different boudary conditions of the form*

$$(11.17) \qquad q\Phi(x_\infty) + q'\Phi^-(x_\infty) = 0 \qquad (q, q' > 0)$$

for the rows and columns of the corresponding kernel of densities $p^\star$. More precisely, for each $t > 0$ the kernel $p^\star$ is symmetric and

$$(11.18) \qquad \lim_{i \to \infty} \left\{ q.p^\star(x_i, x_k; t) + q' \frac{p^\star(x_{i+1}, x_k; t) - p^\star(x_i, x_k; t)}{x_{i+1} - x_i} \right\} = 0.$$

In order that $\sum_k P_{ik}^\star(t) = 1$ it is necessary and sufficient that

$$(11.19) \qquad \delta_0 = 0, \qquad q = 0.$$

We recall that the boundary condition (11.18) may be rewritten by expressing the derivative $\Phi^-(x_\infty)$ in accordance with (2.7) rather than (2.6).

12. Extension to the closed space $\overline{\mathrm{E}}$. — It is both natural and simple to extend by continuity the kernel π_λ^* and the transformation $\mathcal{G}_\lambda^*$ of the preceding section to the space $\overline{\mathrm{E}}$ containing the points $x_0, x_1, \ldots x_\infty$ (*see* Section 2). This amounts to adding to the matrices $\mathrm{P}^*(t)$ a row and a column corresponding to x_∞ and extending the differential equations (1.4) and (1.5) accordingly. We shall see that significant new results are abtained in this way. We use the following :

Set-up. We suppose x_∞ regular, that is, $x_\infty < \infty$, $\sum \mu_k < \infty$: The measure μ is extended to $\overline{\mathrm{E}}$ by attributing to x_∞ the weight $\mu_\infty > 0$. Inner products in the extended space will be denoted by $[\quad]_e$, thus

$$(12.1) \qquad [u, f]_e = [u, f] + u(x_\infty) f(x_\infty) \mu_\infty = \sum_{k=0}^{\infty} u(x_k) f(x_k) \mu_k + u(x_\infty) f(x_\infty) \mu_\infty.$$

As before, $u(x_\infty) \lim u(x_i)$ and the kernels π_λ^* of (11.2) and π_λ of (9.1) are extended to x_∞ by continuity.

By Lemma 9.12 we have $\pi_\lambda(x_i, x_\infty) = \pi_\lambda(x_\infty, x_i) = 0$ for $i = 0, 1, \ldots, \infty$. The transformation $\mathcal{G}_\lambda^*$ induced by π_λ^* maps f into

$$(12.2) \qquad \mathrm{F}_\lambda^* = \mathrm{F}_\lambda + \rho_\lambda u[u, f]_e,$$

where $\mathrm{F}_\lambda = \mathcal{G}_\lambda f$ is the transform (9.2) of f under the minimal Green function π_λ.

As in the last section we wish to define ρ_λ for all $\lambda > 0$ in such a manner that the kernels π_λ^* will satisfy the conditions enumerated in Section 5 for the Laplace transforms of Markovian families.

At each point $x_i \neq x_\infty$ we have as before

$$(12.3) \qquad \lambda \mathrm{F}_\lambda^*(x_i) - \Omega \mathrm{F}_\lambda^*(x_i) = f(x_i),$$

where $\Omega \mathrm{F} = \mathrm{D}_\mu \mathrm{F}^+$. This is true even if f is not continuous at x_∞, but (12.3) is meaningless for x_∞. For a fixed λ we can extend the definition of Ω by putting

$$(12.4) \qquad \lambda F_\lambda^*(x_\infty) - \Omega F_\lambda^*(x_\infty) = f(x_\infty),$$

but this definition depends on λ.

LEMMA 12.1. — *The kernels π_λ^* satisfy the resolvent equation if, and only if, Ω can be defined at x_∞ in such a way that for each $\lambda > 0$ and each bounded [summable] f the equation*

$$(12.5) \qquad \lambda\Phi - \Omega\Phi = f$$

possesses a unique solution $\Phi = F_\lambda^* = \mathcal{G}_\lambda^* f$.

Proof. — Consider $\mathcal{G}_\lambda^*$ as a transformation, say, on bounded functions. We know from Lemma 11.4 that the resolvent equation holds if, and only if, the range of $\mathcal{G}_\lambda^*$ is independent of λ. If (12.5) has a unique solution then the range of $\mathcal{G}_\lambda^*$ consists of all Φ such that Φ and $\Omega\Phi$ are bounded; this range is independent of λ. Conversely, suppose that the resolvent equation holds and consider it in the form (11.16). For each $\lambda > 0$ define Ω_λ by $\Omega_\lambda F_\lambda^* = \lambda F_\lambda^* - f$. The domain of Ω_λ is the range of $\mathcal{G}_\lambda^*$ and hence independent of λ. From (11.16) we get at once

$$(12.6) \qquad \Omega_\lambda F_\nu^* = \Omega_\lambda \{ F_\lambda^* + (\lambda - \nu)\mathcal{G}_\lambda^* F_\nu^* \} = \nu F_\nu^* - f = \Omega_\nu F_\nu^*$$

and we see that Ω_λ is independent of λ. We put $\Omega = \Omega_\lambda$, this notation being consistent with the old definition of $\Omega F(x_j) = D_\mu F^+(x_j)$ for $x_j \neq x_\infty$. Then (12.5) has a unique solution and the lemma is proved.

The argument of Section 11 may now be repeated with trite modifications. Equation (11.13) takes on the form

$$(12.7) \qquad F_\lambda^*(x_\infty) = \rho_\lambda u(x_\infty)[u, f]_e$$

while (11.14) goes over into

$$(12.8) \qquad F_\lambda^{*-}(x_\infty) = -\frac{1}{u(x_\infty)}[u, f] + \rho_\lambda u^-(x_\infty)[u, f]_e$$
$$= \left\{ -\frac{1}{u(x_\infty)} + \rho_\lambda u^-(x_\infty) \right\} [u, f]_e - f(x_\infty)\mu_\infty.$$

Keeping for the moment λ fixed and using the definition (12.4) we conclude

(12.9) $$\mu_\infty \Omega F_\lambda^*(x_\infty) + \alpha_\lambda F_\lambda^*(x_\infty) + F_\lambda^{*-}(x_\infty) = 0,$$

where α_λ anx ρ_λ are related by

(12.10) $$\rho_\lambda = \frac{1}{u(x_\infty)\{\alpha_\lambda u(x_\infty) + u^-(x_\infty) + \lambda\mu_\infty u(x_\infty)\}}.$$

Equation (12.9) replaces (11.10) but instead of a " boundary condition " it represents a *definition* of $\Omega F_\lambda^*(x_\infty)$. In order that the definition be independent of λ it is necessary and sufficient that $\alpha_\lambda = \alpha = C$. As in Lemma 11.1 we conclude then that with $\alpha \geqq 0$ we have $\pi_\lambda^* > 0$ and for $i = 0, 1, \ldots,$

(12.11) $$\lambda \sum_{k=0}^{\infty} \pi_\lambda^*(x_i, x_k)\mu_k + \lambda\pi_\lambda^*(x_i, x_\infty)\mu_\infty$$
$$= 1 - \delta_0 v(x_i) - \rho_\lambda\{\alpha u(x_\infty) + \delta_0\} u(x_i) \leqq 1$$

and, because of continuity, the same inequality holds for $i = \infty$. It is clear from (12.11) that the condition $\alpha \geqq 0$ is not only sufficient, but also necessary for the norm condition to hold.

Definition. — *A function Φ defined on the closed space $\overline{E}$ belongs to the domain of the extented operator Ω if it is continuous* [*i. e. if* $\Phi(x_\infty) = \lim \Phi(x_i)$] *and if*

(12.12) $$\Phi^-(x_\infty) = \lim_{i \to \infty} \Phi^+(x_i)$$

exists (is finite). Then

(12.13) $$\Omega\Phi(x_\infty) = -\frac{\alpha}{\mu_\infty}\Phi(x_\infty) - \frac{1}{\mu_\infty}\Phi^-(x_\infty),$$

where $\alpha > 0$ is a constant; for $i = 0, 1, 2, \ldots,$ we put as before $\Omega\Phi(x_i) = D_\mu \Phi^+(x_i)$.

Theorem 12.1. — *With the above definition of Ω there exists exactly one Markovian family of matrices $P^*(t)$ with rows and columns defined for $i, j = 0, 1, \ldots, \infty$ such that rows and columns of the corresponding* *density kernel $p^\star$ satisfy for each fixed j*

$$\frac{d}{dt}p^\star(\,.\,, x_j;\, t) = \Omega p^\star(\,.\,, x_j;\, t), \tag{12.14}$$

$$\frac{d}{dt}p^\star(x_j, .\,;\, t) = \Omega p^\star(x_j, .\,;\, t). \tag{12.15}$$

Note 1. — Equation (12.13) extends the two systems of differential equations (1.4) and (1.5) to x_∞. As $\mu_\infty \to 0$ this equation approaches the boundary condition (11.18) and our $P^\star$ approaches, in an obvious sense, the matrix $P^\star$ of Section 11 defined over the original space E. Thus the *boundary conditions appear as a degenerate limiting case of the definition of Ω in the closed space*, and we see that the extension of Ω to $\overline{E}$ leaves very little freedom of choice.

Note 2. — The question arises whether we have obtained *the most general Markovian family of matrices with rows numbered* 0, 1, ..., ∞ *all of whose columns and rows satisfy the differential equations* (1.4) *and* (1.5), *respectively, for all finite values of the subscript.* The answer is negative since it is not necessary to assume the symmetry $\pi_\lambda^\star(x_i, x_\infty) = \pi_\lambda^\star(x_\infty, x_i)$ nor the continuity $\pi_\lambda^\star(x_i, x_\infty) = \lim \pi_\lambda^\star(x_i, x_k)$ which we have assumed without further motivation. However, it is easily seen that these two assumptions represent the only arbitrary elements of our derivation and hence we have obtained the most general solution such that the kernel $p^\star$ is symmetric and its columns are continuous on $\overline{E}$.

The last theorem is closely connected with a problem which has recently attracted wide attention. It has been proved by Doob [2] (*cf.* also Chung [1]) that for each Markovian family the limits

$$q_i = \lim_{t\to 0}\{1 - P_{ii}(t)\}t^{-1}, \qquad q_{ij} = \lim t^{-1}P_{ij}(t) \qquad (i \neq j) \tag{12.16}$$

exist, but they may be infinite. « Pathological » cases where $q_i = \infty$ for some indices i have recently been studied by Chung [1], Kendall [11], Kendall and Reuter [12], Kolmogorov [13], P. Lévy [16], and probably others. In particular, Theorem 12.1 contains an analogue to an example studied by Kendall. For completeness we prove

THEOREM 12.2. — *For the matrix of the last theorem*

$$\lim_{t\to 0} t^{-1}\{1-P^*_{\infty\infty}(t)\}=\infty, \tag{12.17}$$

$$\lim_{t\to 0} t^{-1}P^*_{i\infty}(t)=\lim_{t\to 0} t^{-1}P^*_{\infty i}(t)=0 \qquad (i=0, 1, \ldots). \tag{12.18}$$

(In the terminology of P. Levy this implies that ∞ is an *instantaneous* state. A different type of instantaneous state will be encountered in Section 15.)

Proof. — Theorem 12.1 asserts, in particular, that $P^*_{\infty\infty}(t)\to 1$ as $t\to 0$. Let us, as a preliminary step, verify this relation by the routine method of Laplace transforms. It is obvious that

$$\lim_{t\to 0} P^*_{\infty\infty}(t)=\lim_{\lambda\to\infty}\lambda\pi^*_\lambda(x_\infty, x_\infty)\mu_\infty \tag{12.19}$$

provided that the limit on the left exists. Now

$$\lambda\pi^*_\lambda(x_\infty, x_\infty)\mu_\infty=\frac{\lambda\mu_\infty u(x_\infty)}{\alpha u(x_\infty)+u^-(x_\infty)+\lambda\mu_\infty u(x_\infty)}. \tag{12.20}$$

In view of the convexity of the graph of u we have

$$\lim_{\lambda\to\infty}\frac{u(x_i)}{u(x_\infty)}=0, \qquad \lim_{\lambda\to\infty} u(x_\infty)=\infty \qquad (i=0, 1, 2, \ldots) \tag{12.21}$$

and hence, in consequence of (7.5),

$$u^-(x_\infty)=\delta_0+\lambda\sum_{k=0}^{\infty}u(x_k)\mu_k=o(\lambda u(x_\infty)) \tag{12.22}$$

as $\lambda\to\infty$. Therefore the right side of (12.20) tends to 1 as asserted.

Since the limits in (12.16) are known to exist, we have similarly

$$\begin{aligned}\lim t^{-1}\{1-P^*_{\infty\infty}(t)\}&=\lim_{\lambda\to\infty}\lambda\{1-\lambda\pi^*_\lambda(x_\infty, x_\infty)\mu_\infty\}\\ &=\lim_{\lambda\to\infty}\lambda\frac{\alpha u(x_\infty)+u^-(x_\infty)}{(\alpha+\lambda\mu_\infty)u(x_\infty)+u^-(x_\infty)}.\end{aligned} \tag{12.23}$$

Again the convexity of the graph of u shows that for each finite i

$$\frac{u(x_\infty)-u(x_i)}{x_\infty-x_i}<u^-(x_\infty) \tag{12.24}$$

and hence

$$\lim_{\lambda \to \infty} \frac{u^-(x_\infty)}{u(x_\infty)} = \infty. \tag{12.25}$$

It follows that the limit in (12.24) is infinite, as asserted. Finally, for $i < \infty$,

$$\lim_{t \to 0} t^{-1} P_{i\infty}(t) = \lim_{\lambda \to \infty} \lambda^2 \pi_\lambda^\star(x_i, x_\infty) \mu_\infty. \tag{12.26}$$

13. THE SIMPLEST UNSYMMETRIC SOLUTION. — We turn to the problem of constructing Markovian families of matrices $P^\star(t)$ whose columns satisfy the differential equations (1.4) but whose rows do *not* satisfy (1.5). For that purpose we must suppose that x_∞ *is either regular or exit*, for otherwise the solution of (1.4) is unique. We begin with a special case which will be generalized in the next section. This separation is convenient because the family of this section is analytically simpler and illustrates a particular phenomenon of probabilistic interest.

We introduce again the kernel $\pi_\lambda^\star$ of Laplace transforms defined by (11.1). It is required that each column of $\pi_\lambda^\star$ differs from the corresponding column of the minimal Green function π_λ by a multiple of u, and hence

$$\pi_\lambda^\star(x_i, x_k) = \pi_\lambda(x_i, x_k) + u(x_i) C_\lambda(x_k) \qquad (i, k = 0, 1, 2, \ldots). \tag{13.1}$$

This kernel induces a transformation $\mathcal{G}_\lambda^\star$ on *bounded* functions defined by (11.4). Using again the notation (11.3) for inner products, the transform $F_\lambda^\star = \mathcal{G}_\lambda^\star f$ is given by

$$F_\lambda^\star = F_\lambda + [C_\lambda, f] u \tag{13.2}$$

with $F_\lambda = \mathcal{G}_\lambda f$ defined by (9.2).

Our problem consists in determining for each $\lambda > 0$ the coefficients $C_\lambda(x_k)$ in such a way that the kernell $\pi_\lambda^\star$ satisfies the norm conditions (5.2) and the resolvent equation (5.3).

Since we suppose that x_∞ is exit or regular, we have $F_\lambda(x_i) \to 0$ as $i \to \infty$ [*cf.* (9.12)] and it is clear from (13.2) that $\pi_\lambda^\star \geq 0$ if, and only if, $C_\lambda \geqq 0$. Using (9.4) we have

$$(13.3)\qquad \lambda \sum_{k=0}^{\infty} \pi_\lambda^\star(x_i, x_k)\mu_k = 1 - \delta_0 v(x_i) - u(x_i)\left\{\frac{1}{u(x_\infty)} - \lambda[C_\lambda, 1]\right\}$$

and hence *the norm conditions* (5.2) *are satisfied by* $\pi_\lambda^\star$ *if, and only if,*

$$(13.4)\qquad C_\lambda \geqq 0, \qquad \lambda u(x_\infty)\sum_{k=0}^{\infty} C_\lambda(x_k)\mu_k \leqq 1.$$

By Lemma 11.4 the resolvent equation holds if, and only if, the range of the transformations $\mathcal{G}_\lambda^\star$ is independent of λ (their domain being the set of bounded f). It is clear from (13.2) that

$$(13.5)\qquad F_\lambda^\star(x_\infty) = \lim_{i\to\infty} F_\lambda^\star(x_i) = u(x_\infty)[C_\lambda, f],$$

and hence *the range of* $\mathcal{G}_\lambda^\star$ *consists of all functions* Φ *such that* Φ *and* $\Omega\Phi$ *are bounded and* $\Phi(x_\infty) = \lim \Phi(x_i)$ *exists and satisfies*

$$(13.6)\qquad \Phi(x_\infty) = u(x_\infty)[C_\lambda, \lambda\Phi - \Omega\Phi].$$

Suppose now that (13.4) holds and that the range of $\mathcal{G}_\lambda^\star$ is independent of λ. Let Φ satisfy (13.6) for some, and therefore all λ. In view of (13.4) we may let $\lambda \to \infty$ through an appropriate sequence of values in such a way that

$$(13.7)\qquad \lambda u(x_\infty) C_\lambda(x_k) \to a(x_k) \qquad (k = 0, 1, 2, \ldots).$$

To derive the simplest possible boundary condition ([12]) let us assume that the passage to the limit is uniform. Then

$$(13.8)\qquad a(x_k) \geqq 0, \qquad \sum_{k=0}^{\infty} a(x_k)\mu_k \leqq 1$$

and from (13.6) we get

$$(13.9)\qquad \Phi(x_\infty) = \sum_{k=0}^{\infty} a(x_k)\Phi(x_k)\mu_k$$

or $\Phi(x_\infty) = [a, \Phi]$.

([12]) Another possibility will be discussed in Section 15, *cf.* the analogous development in [8].

Although (13.9) has been derived only heuristically, we are able to show that it can serve as a boundary condition : we shall determine C_λ in such a way that each F^*_λ satifies the boundary condition (13.9). For that purpose put

$$(13.10)\qquad A_\lambda = \mathcal{G}_\lambda a = \sum_{k=0}^{\infty} \pi_\lambda(.,\, x_k)\, a(x_k)\, \mu_k.$$

Then

$$(13.11)\qquad [a,\, F^*_\lambda] = [A_\lambda,\, f] + [C_\lambda,\, f][a,\, u]$$

and hence F^*_λ will satisfy the boundary condition (13.9) if, and only if, the right side in (13.11) equals (13.5), that is, if

$$(13.12)\qquad C_\lambda(x_i) = \frac{A_\lambda(x_i)}{u(x_\infty) - [a,\, u]} \qquad (i = 0,\, 1,\, 2,\, \ldots).$$

Here the denominator is positive since, u being strictly increasing,

$$(13.13)\qquad [a,\, u] = \sum_{k=0}^{\infty} a(x_k)\, u(x_k)\, \mu_k < u(x_\infty) \sum_{k=0}^{\infty} a(x_k)\, \mu_k \leqq u(x_\infty).$$

Using once more (9.4) we see that

$$(13.14)\qquad \lambda[A_\lambda,\, 1] \leqq 1 - \frac{1}{u(x_\infty)}[a,\, u]$$

and hence that (13.4) is satisfied.

It has been shown that each F^*_λ satisfies (13.9). Conversely suppose that a function Φ satisfies (13.9) and that Φ and $\Omega\Phi$ are bounded. Put $\lambda\Phi - \Omega\Phi = f$. Since also $\lambda F^*_\lambda - \Omega F^*_\lambda = f$ we see that $F^*_\lambda - \Phi = qu$ where q is a constant. Moreover qu must satisfy (13.9). But we have shown in (13.13) that u does *not* satisfy (13.9) and we conclude that $q = 0$ or $\Phi = F^*_\lambda$. In other words, the common range of all $\mathcal{G}^*_\lambda$ consists of the functions Φ satisfying (13.9) and the trivial conditions that Φ and $\Omega\Phi$ be bounded. We have thus proved :

THEOREM 13.1. — *Suppose that x_∞ is regular or exit and choose $a(x_k)$ satisfying* (13.8). *There exists exactly one Markovian family of matrices $P^*(t)$ such that each column satisfies the differential equations*

(1.4) *together with the boundary condition* (13.9), *namely*,

$$(13.15)\qquad \lim_{i\to\infty} P^{\star}_{ij}(t) = \sum_{k=0}^{\infty} a(x_k)\, P^{\star}_{kj}(t)\, \mu_k \qquad (j = 0, 1, 2, \ldots).$$

In order that $\sum_j P^{\star}_{ij}(t) = 1$ *it is necessary and sufficient that*

$$(13.16)\qquad \delta_0 = 0, \qquad \sum a(x_k)\, \mu_k = 1.$$

For a probabilistic interpretation, *see*, Section 16 (*f*).

Since the rows of $P^{\star}$ do not satisfy the differential equations (1.5) we must find the functional equations replacing (1.5).

THEOREM 13.2. — *Let*

$$(13.17)\qquad \Phi(x_k;\, t) = P^{\star}_{\cdot k}(t)\, \mu_k^{-1}$$

represent a fixed row of the density matrix of the last theorem. Then

$$(13.18)\qquad \frac{d}{dt}\Phi(x_k;\, t) = \Omega\Phi(x_k;\, t) - a(x_k) \lim_{j\to\infty} \frac{\Phi(x_{j+1};\, t) - \Phi(x_j;\, t)}{x_{j+1} - x_j}.$$

[These equations replace (1.5) but are *not* differential equations in the ordinary sense.]

Proof. — The adjoint to the transformation $\mathcal{G}^{\star}_{\lambda}$ maps a *summable* function h into the summable function $H^{\star}_{\lambda}$ defined by

$$(13.19)\qquad H^{\star}_{\lambda}(x_k = \sum_{i=0}^{\infty} h(x_i)\, \pi^{\star}_{\lambda}(x_i, x_k)\, \mu_i = H_{\lambda}(x_k) + [h, u]\, G_{\lambda}(x_k),$$

where $H_{\lambda} = \mathcal{G}_{\lambda} h$ is the transform of h under the minimal Green function (which is a symmetric kernel). We shall prove that

$$(13.20)\qquad \lambda H^{\star}_{\lambda} - h = \Omega H^{\star}_{\lambda} - a \lim_{j\to\infty} H^{\star +}_{\lambda}(x_j).$$

If we choose, in particular, $h(x_k) = \delta_{ik}\, \mu_i^{-1}$, where δ_{ik} is the Kronecker symbol, then (13.20) reduces to the Laplace transform version of the assertion (13.18). Accordingly, (13.20) is equivalent to an integrated version of (13.18). But we know from the last theorem that

$\Phi(x_k; t)$ is a continously differentiable function of t, and hence (13.20) implies (13.18).

The proof of (13.20) is simple. We know that $\lambda H_\lambda - h = \Omega H_\lambda$ and $\gamma A_\lambda - a = \Omega A_\lambda$ and hence we have from (13.19)

$$\lambda H_\lambda^* - h = \Omega H_\lambda^* + \frac{[h, u]}{u(x_\infty) - [a, u]} a. \tag{13.21}$$

In order to express $[h, u]$ in terms of H_λ^* we use (11.14) first with $f = h$ then with $f = a$. In this way we find

$$\begin{aligned} \lim_{j \to \infty} H_\lambda^{*-}(x_j) &= -\frac{[h, u]}{u(x_\infty)} \left\{ 1 + \frac{[a, u]}{u(x_\infty) - [a, u]} \right\} \\ &= -\frac{[h, u]}{u(x_\infty) - [a, u]}. \end{aligned} \tag{13.22}$$

The last two equations together yield (13.20).

14. Extension to $\overline{E}$. — It is exceedingly simple to modify the example of the last section so as to obtain a kernel π_λ^* defined over the closed space $\overline{E}$ and corresponding to a Markovian family of matrices $P^*(t)$ with a row and a column indexed ∞. The procedure is the same as in Section 12 and the following lemma may be verified by trite changes in the proof of Theorem 13.1.

Lemma 14.1. — *Let x_∞ be regular and attach a weight $\mu_\infty > 0$ to it. Let $a(x_k)$ satisfy the norm canditions* (13.8).

Define the kernel π_λ^* by (13.1) *for $i, k = 0, 1, 2, \ldots, \infty$, where*

$$C_\lambda(x_i) = \frac{A_\lambda(x_i)}{u(x_\infty)\{1 + \lambda\mu_\infty\} - [a, u]} \qquad (i = 0, 1, \ldots), \tag{14.1}$$

$$C_\lambda(x_\infty) = \frac{1}{u(x_\infty)\{1 + \lambda\mu_\infty\} - [a, u]} \tag{14.2}$$

[*here A_λ is defined in* (13.10) *and* $\pi_\lambda(x_i, x_\infty) = \pi_\lambda(x_\infty, x_k) = 0$].

This π_λ^* corresponds to a Markovian family of matrices $P^*(t)$. It induces a transformation $\mathcal{G}_\lambda^*$ mapping a bounded f into $F_\lambda^* = \mathcal{G}_\lambda^* f$, where

$$F_\lambda^*(x_i) = \sum_{k=0}^{\infty} \pi_\lambda^*(x_i, x_k) f(x_k) \mu_k + \pi_\lambda^*(x_i, x_\infty) f(x_\infty) \mu_\infty. \tag{14.3}$$

This tranform (and, in particular, each column of π^*_λ) satisfies the equations $\lambda F^*_\lambda - \Omega F^*_\lambda = f$, where the operator Ω is defined for continuous Φ by

$$\Omega\Phi(x_i) = D_\mu \Phi^+(x_i) \qquad (i = 0, 1, \ldots), \tag{14.4}$$

$$\Omega\Phi(x_\infty) = \frac{1}{\mu_\infty}\left\{ -\Phi(x_\infty) + \sum_{k=0}^{\infty} a(x_k)\Phi(x_k)\mu_k \right\}. \tag{14.5}$$

The kernel π^*_λ is uniquely determined by this property.

In order that $\lambda \mathcal{G}^*_\lambda 1 = 1$ it is necessary and sufficient that $\delta_0 = 0$ and $\sum a(x_k)\mu_k = 1$.

This lemma is equivalent to

THEOREM 14.1. — *Under the assumptions of Lemma 14.1 there exists exactly one Markovian family of matrices $P^*(t)$ (with rows and columns numbered $0, 1, 2, \ldots, \infty$) such that if $F_i(t) = P^*_{ij}(t)$ represents the elements of a fixed column ($j = 0, 1, \ldots, \infty$) the F_i satisfy the differential equations (1.4) for $i < \infty$ and*

$$F'_\infty(t) = -\frac{1}{\mu_\infty} F_\infty(t) + \frac{1}{\mu_\infty}\sum_{k=0}^{\infty} a(x_k) F_k(t)\mu_k, \tag{14.6}$$

where

$$F_\infty(t) = \lim_{i\to\infty} F_i(t).$$

For the probabilistic interpretation and the theory of the Kolmogorov differential equations it is of interest to know the asymptotic behavior of $P^*_{i\infty}(t)$ and $P^*_{\infty j}(t)$ as $t \to 0$.

THEOREM 14.2. — As $t \to 0$

$$t^{-1}\{1 - P^*_{\infty\infty}(t)\} \to \mu_\infty^{-1}, \tag{14.8}$$

$$t^{-1} P^*_{\infty k}(t) \to \mu_k a(x_k)\mu_\infty^{-1} \qquad (k \neq \infty), \tag{14.9}$$

$$t^{-1} P^*_{i\infty}(t) \to 0 \qquad (i \neq \infty). \tag{14.10}$$

Proof. — As in the proof of Theorem 12.2 the limits are known to exist and may be evaluated by the routine method of Laplace transforms. Thus (12.19) together with (14.2) shows trivially that

$P^{*}_{\infty\infty}(t) \to 1$, and the first equality in (12.23) leads without further calculations to (14.8). Similarly, since $P^{*}_{\infty k}(t) \to 0$ for $k < \infty$ we have

(14.11) $$\lim_{t \to 0} t^{-1} P^{*}_{\infty k}(t) = \lim_{\lambda \to \infty} \lambda^2 \pi^{*}_{\lambda}(x_\infty, x_k)\, \mu_k$$
$$= \lim_{\lambda \to \infty} \frac{\lambda A_\lambda(x_k)\, \mu_k}{\lambda^{-1} + \mu_\infty - \lambda^{-1} u^{-1}(x_\infty)\,[u, u]}.$$

To show the truth of (14.9) it suffices now to prove that for each $k < \infty$

(14.12) $$\lambda A_\lambda(x_k) \to a(x_k) \qquad (\lambda \to \infty).$$

Since the minimal solution $P(t)$ of Section 10 tends to the identity matrix as $t \to 0$, we have for the minimal Green function

(14.13) $$\lim_{\lambda \to \infty} \lambda \pi_\lambda(x_i, x_k)\, \mu_k = \delta_{ik}.$$

The relation (14.12) is a direct consequence of this and the definition (13.10). The proof of (14.10) proceeds in the same way in a routine manner.

It remains to find the equations replacing the differential equations (1.5). We proceed as in the proof of Theorem 13.2 and introduce the transform H^{*}_{λ}; however, instead of the inner product $[h, u]$ we have now to use the inner product $[h, u]_e$ introduced in (12.1).

LEMMA. 14.2. — *Let h be summable and*

(14.14) $$H^{*}_{\lambda} = H_\lambda + [h, u]_e G_\lambda$$

(H^{*}_{λ} *is the image of h under the transformation adjoint to* $\mathcal{G}^{*}_{\lambda}$). *Then*

(14.15) $$\lambda H^{*}_{\lambda}(x_k) - h(x_k) = \Omega H^{*}_{\lambda}(x_k) + a(x_k)\, H^{*}_{\lambda}(x_\infty) \qquad (k < \infty),$$

(14.16) $$\lambda H^{*}_{\lambda}(x_\infty) - h(x_\infty) = -\frac{1}{\mu_\infty} H^{*}_{\lambda}(x_\infty) - \frac{1}{\mu_\infty} H^{*-}_{\lambda}(x_\infty),$$

where, as before, $H^{*-}_{\lambda}(x_\infty) = \lim H^{*+}(x_i)$.

Proof. — These equations replace (13.20) and are proved by the same argument.

As in Theorem 13.2 we can translate the equations (14.15)-(14.16) into equivalent equations for the densities $p^{*}(x_i, x_k; t)$. For the convenience of readers used to the familiar notations we formulate the result directly for the matrices $P^\star(t)$.

THEOREM 14.3. — *The i-th column* ($i \leqq \infty$) *of the matrix* $P^\star(t)$ *of Theorem* 14.1 *satisfies the equations*

$$(14.17)\quad \frac{d}{dt}P^\star_{ik}(t) = -(\delta_k+\beta_k)P^\star_{ik}(t)+\beta_{k-1}P^\star_{ik-1}(t) + \delta_{k+1}P^\star_{ik+1}(t)+\frac{\mu_k}{\mu_\infty}a(x_k)P^\star_{i\infty}(t)\qquad (k<\infty)$$

$$(14.18)\quad \frac{d}{dt}P^\star_{i\infty}(t) = -\frac{1}{\mu_\infty}P^\star_{i\infty}(t)-\lim_{j\to\infty}\frac{P^\star_{ij+1}(t)\,\mu^{-1}_{j+1}-P^\star_{ij}(t)\,\mu^{-1}_{j}}{x_{j+1}-x_j}.$$

In (14.17) we recognize the differential equations with an added term, but (14.18) is a functional equation of a novel type. For a probabilistic interpretation, *see* Section 16 (g).

15. A " PATHOLOGICAL " EXTENSION. — It is both surprising and interesting that the conditions (13.8) that a be summable is unnecessarily restrictive : the examples and constructions remain meaningful even for certain unbounded a. In considering this possibility we are led to a " pathological " Markovian family related to one of the examples due to Kolmogorov [13] and studied analytically by Kendall and Reuter [12].

For simplicity we shall assume x_∞ regular, although the method can be extended to cover the case of an exit x_∞.

LEMMA 15.1. — *Let* x_∞ *be regular and*

$$(15.1)\qquad a(x_k)\geqq 0,\qquad \sum_{k=0}^{\infty}a(x_k)\{x_\infty - x_k\}<\infty.$$

Then for each $\lambda>0$

$$(15.2)\qquad a^\star_\lambda=\sum_{k=0}^{\infty}a(x_k)\{u(x_\infty)-u(x_k)\}<\infty.$$

The transform $A_\lambda=\mathcal{G}_\lambda a$ *of* (13.10) *satisfies*

$$(15.3)\qquad \lambda\sum_{k=0}^{\infty}A_\lambda(x_k)=\sum_{k=0}^{\infty}a(x_k)\left\{1-\frac{u(x_k)}{u(x_\infty)}-\delta_0 v(x_k)\right\}<\infty$$

Proof. — Since the graph of u is convex we have

$$(15.4)\qquad a_N^* < u\ (x_\infty)\sum_{k=0}^{\infty} a(x_k)\{x_\infty - x_k\} < \infty.$$

It follows from (15.2) that the series on the right side in (15.3) converges, and (15.3) is a direct consequence of (9.4) and the definition (13.10) of A_λ.

LEMMA 15.2. — *Let x_∞ be regular, let the $a(x_k)$ satisfy* (15.1), *and let $q \geq 0$ be an arbitrary constant.*

Define π_λ^* by (13.1) *for $i, k = 0, 1, \ldots, \infty$, where*

$$(15.5)\qquad C_\lambda(x_i) = \frac{A_\lambda(x_i)}{u(x_\infty)\{q + \lambda\mu_\infty\} + a_\lambda^*} \qquad (i = 0, 1, 2, \ldots),$$

$$(15.6)\qquad C_\lambda(x_\infty) = \frac{1}{u(x_\infty)\{q + \lambda\mu_\infty\} + a_\lambda^*}.$$

Then the conclusions of Lemma 14.1 *remain valid except that in the definition of $\Phi\Omega(x_\infty)$ the equation* (14.5) *is to be replaced by*

$$(15.7)\qquad \Omega\Phi(x_\infty) = -\frac{1}{\mu_\infty}\left\{q\Phi(x_\infty) + \sum_{k=0}^{\infty} a(x_k)\{\Phi(x_\infty) - \Phi(x_k)\}\right\}\mu_k$$

and that the necessary and sufficient condition for $\lambda\mathcal{G}_\lambda^* 1 = 1$ now reads $\delta_0 = q = 0$.

Proof. — Equation (14.3) may be rewritten in the form

$$(15.8)\qquad F_\lambda^* = F_\lambda + u[C_\lambda, f]_e.$$

As in the proof of Theorem 13.1 we have to show that for an arbitrary bounded f the transform F_λ^* satisfies (15.7), that is

$$(15.9)\quad \lambda F_\lambda^*(x_\infty) - f(x_\infty) = -\frac{1}{\mu_\infty}\left\{qF_\lambda^*(x_\infty) + \sum a(x_k)\{F_\lambda^*(x_\infty) - F_\lambda^*(x_k)\}\right\}\mu_k.$$

Now $F_\lambda(x_\infty) = 0$ and hence

$$(15.10)\qquad F_\lambda^*(x_\infty) = u(x_\infty)\, C_\lambda, f]_e$$

and

$$(15.11)\qquad \sum_{k=0}^{\infty} a(x_k)\{F_\lambda^\star(x_\infty)-F_\lambda^\star(x_k)\}\mu_k = -[a, F_\lambda]+a_\lambda^\star[C_\lambda, f]_e$$
$$= -[A_\lambda, f]+a_\lambda^\star[C_\lambda, f]_e.$$

If, in this equation, we express the $A_\lambda(x_k)$ by $C_\lambda(x_k)$ in accordance with (15.5) we are led to (15.9).

Finally, from (15.3) and (9.4) we get

$$(15.12)\qquad \lambda \mathcal{G}_\lambda^\star 1 = 1-\delta_0 v - u\frac{q+\delta_0[a, v]}{u(x_\infty)\{q+\lambda\mu_\infty\}+a_\lambda^\star}$$

which shows that the right side equals 1 if, and only if, $\delta_0 = q = 0$.

This lemma is equivalent to

THEOREM 15.1. — *Under the assumptions of the last lemma, there exists exactly one Markovian family of matrices* $P^\star(t)$ (*with rows and columns numbered* 0, 1, ..., ∞) *such that if* $F_i(t) = P_{ij}^\star(t)$ *represents the elements of a fixed column* ($j = 0, 1, \ldots, \infty$) *the* F_i *satisfy the differential equations* (1.4) *for* $i<\infty$ *and*

$$(15.13)\qquad F'_\infty(t) = -qF_\infty(t)+\sum_{k=0}^{\infty} a(x_k)\{F_\infty(t)-F_k(t)\}\mu_k,$$

where

$$(15.14)\qquad F_\infty(t) = \lim_{i\to\infty} F_i(t).$$

THEOREM 15.2. — *If* $\sum a_k = \infty$, *then as* $t\to 0$

$$(15.15)\qquad t^{-1}\{1-P_{\infty\infty}^\star(t)\}\to\infty$$

while (14.9) *and* (14.10) *remain true without change.*

Proof. — We proceed as in the proofs of Theorems 12.2 and 14.2. The left side of (15.15) approaches the same limit as

$$(15.16)\qquad \lambda\{1-\lambda\pi_\lambda^\star(x_\infty, x_\infty)\mu_\infty\} = \lambda\frac{u(x_\infty)q+a_\lambda^\star}{u(x_\infty)\{q+\lambda\mu_\infty\}+a_\lambda^\star}.$$

However, in consequence of (12.21), (12.22), and (15.4),

$$a_\lambda^\star u(x_\infty)^{-1}\to\sum a_k = \infty,\qquad a_\lambda^\star = o(\lambda u(x_\infty))\quad \text{as}\quad \lambda\to\infty,$$

and therefore the right side of (15.16) approaches ∞ as asserted.

To prove that (14.9) holds we have to evaluate the limit of

$$\lambda^2 \pi_\lambda^*(x_\infty, x_k)\,\mu_k = \lambda A_\lambda(x_k)\,\frac{\lambda u(x_\infty)\,\mu_k}{u(x_\infty)\{q+\lambda\mu_k\}+a_\lambda^*}. \tag{15.17}$$

The right side of (15.17) has the same limit as $\lambda\,A_\lambda(x_k)$. We have therefore to prove that (14.12) holds, but this does not follow trivially since $\sum a(x_k)\,\mu_k=\infty$. However, since v^+ is a (negative) increasing function we have

$$v(x_j)=v(x_j)-v(x_\infty)<-v^+(x_j)\{x_\infty-x_j\} \tag{15.18}$$

and therefore for $N>k$

$$\lambda\sum_{j=N}^{\infty} a(x_j)\,\pi_\lambda(x_j, x_k)\,\mu_j<-\lambda\,u(x_k)\,v^-(x_N)\sum_{j=N}^{\infty} a(x_j)\{x_\infty-x_j\}\,\mu_j < \text{Cte}\,\frac{v^+(x_N)}{v(x_N)}\,\lambda\,\pi_\lambda(x_k, x_N)\to 0 \tag{15.19}$$

by (14.13). It follows then from (13.10) and (14.13) that

$$\lim_{\lambda\to\infty}\lambda\,A_\lambda(x_k)=\lim_{\lambda\to\infty}\lambda\sum_{i=0}^{N-1} a(x_j)\,\pi_\lambda(x_j, x_k)\,\mu_j\to a(x_k) \tag{15.20}$$

as asserted.

Note. — Equations (14.16) and (14.18) are meaningless under the present circumstances since $H^{\star+}(x_i)$ does not necessarily approach a finite limit. The correct form of these relations may be found in [7], Theorem 8.2.

16. Probabilistic interpretation. — *a. The basic differential equations* (1.4) and (1.5) are obtained starting from the postulate that as $t\to 0$

$$\begin{cases} P_{ii}(t)=1-(\delta_i+\beta_i)\,t+o(t),\\ P_{ii+1}(t)=\beta_i t+o(t), \qquad P_{ii-1}(t)=\delta_i t+o(t). \end{cases} \tag{16.1}$$

These relations explain the use of the words birth (β_i) and death (δ_i). The conditions (16.1) enable us to differentiate (1.2) with respect to t and thus to derive the differential equations (1.4) *rigorously*. Formal differentiation with respect to s leads to the equations (1.5), but this operation can not be justified without an additional assumption of uniformity in (16.1). Accordingly, the conditions (16.1) *imply* (1.4) *but not* (1.5), and the example of Section 13 shows that (1.5) need not be satisfied.

On the other hand, (1.4) as such is probabilistically meaningless, while (1.5) is the so-called *Fokker-Planck* equation, or the *continuity* equation in physical terminology. The conditions (16.1) may be interpreted to the effect that the probability mass present in state i flows out at a constant relative rate $\delta_i + \beta_i$ and is distributed to $i+1$ and $i-1$ in the ratio $\beta_i : \delta_i$. Equation (1.5) expresses the rate of change of the probability mass at k as the sum of the three flows present.

b. The meaning of the natural scale. — Consider a Markovian family of matrices $P(t)$ satisfying (16.1). With it we associate, in the usual manner, the ordinary denumerable *Markov chain with a matrix* T *of transition probabilities* whose elements t_{ik} are defined by

$$t_{ii+1} = \frac{\beta_i}{\beta_i + \delta_i}, \qquad t_{ii-1} = \frac{\delta_i}{\beta_i + \delta_i} \tag{16.2}$$

(implying $t_{ik} = 0$ for all other combinations of indices). If $X(t)$ is the random variable of a Markovian process on E regulated by $P(t)$, then T^n determines the position of $X(t)$ after n jumps irrespective of the times when these jumps occur. The T-process takes a "particle" from an initial position i to other positions, except that at the state 0 a *destruction* takes place with probability $\delta_0(\beta_0 + \delta_0)^{-1}$. In other words, the Markov chain (T) either continues forever or ends after finitely many steps at the state 0.

If the elements of T^n are denoted by $t_{ik}^{(n)}$, then the probability $z_i^{(n)}$ that, starting from i, the chain continues for at least n steps equals

$$z_i^{(n)} = \sum_{k=0}^{\infty} t_{ik}^{(n)}. \tag{16.3}$$

Using bold face for column vectors we rewrite (16.3) in the form of a matrix equation $\mathbf{z}^{(n)} = T^n \mathbf{1}$. Thus $\mathbf{z}^{(n+1)} = T\,\mathbf{z}^{(n)}$ with $\mathbf{z}^{(0)} = \mathbf{1}$ and hence $\mathbf{z}^{(n)} \downarrow \mathbf{z}$ where $T\,\mathbf{z} = \mathbf{z}$.

For $\delta_0 = 0$ we have $z = 1$. If $\delta_0 > 0$, then $Tx = x$ and, up to a multiplicative constant, x is the only solution of this equation. Therefore : $x_\infty = \infty$ implies $z = 0$; if $x_\infty < \infty$ then z is a multiple of x. Now by induction $x \leq z^{(n)} x_\infty \leq 1 . x_\infty$ and hence when $\delta_0 > 0$

$$(16.4) \qquad z = \frac{x}{x_\infty}.$$

In words : $\frac{x_i}{x_\infty}$ is the probability that, starting from i, the process continues forever (no destruction takes place at 0). A trivial translation of the subscripts allows us to reformulate this statement as follows : *For each integer $r > 0$ the probability that, starting from $r + i$, the chain never leads to $r - 1$ equals* $\frac{x_{r+i} - x_r}{x_\infty}$.

In other words : *the probability that the "particle" eventually drifts out to infinity without returning to 0 is positive if, and only if, $x_\infty < \infty$.*

c. *The meaning of u.* — Let $r > 0$ be a fixed integer. In the process regulated by $P(t)$ let $F_i(t)$ for $i < r$ denote the probability that, starting from i at time 0, the state r will be reached before time t. [Abstractly : $F_i(t) = P\left\{\sup_{\tau \leq t} X(\tau) > r \mid X(0) = i\right\}$.] Put $F_r(t) = 1$. For $i < r$ the F_i satisfy the differential equations (1.4) with the initial conditions $F_i(0) = 0$. The Laplace transforms

$$(16.5) \qquad \Phi_\lambda(x_i) = \int_0^\infty e^{-\lambda t}\, dF_i(t)$$

therefore satisfy $\lambda \Phi_\lambda(x_i) - \Omega \Phi_\lambda(x_i) = 0$ for $i < r$ and $\Phi_\lambda(x_r) = 1$. It follows that $\Phi_\lambda(x_i) = \frac{u(x_i)}{u(x_r)}$. Thus :

For $i < r$ the tatio $\frac{u(x_i)}{u(x_r)}$ is the Laplace transform of the first passage time from i to r.

Letting $r \to \infty$ we may interpret the result in following way :

If $u(x_\infty) < \infty$ there exists a finite prabability that x_∞ will be reached within a finite time. (This is true for each initial state.)

If $u(x_\infty) = \infty$, the boundary will not be reached, that is, with unit *probability* $X(t)$ *will have only finitely many jumps within any finite time invertal. However, if* $x_\infty < \infty$, *the boundary point* x_∞ *is approached asymptotically as* $t \to \infty$ *unless the process terminates at a finite time.*

d. The meaning of v. — Let $0 < i < r$ and denote by $F_i(t)$ the probability that, starting from i at time 0, the state 0 will be reached before time t and before r has been reached. Put $F_0(t) = 1$ and $F_r(t) = 0$. Then the differential equations (1.4) are again satisfied for $0 < i < r$, and $F_i(0) = 0$. The Laplace transforms (16.5) therefore satisfy the equations $\lambda \Phi_\lambda(x_i) - \Omega \Phi_\lambda(x_i) = 0$ with $\Phi_\lambda(x_0) = 1$ and $\Phi_\lambda(x_r) = 0$. It follows that Φ_λ is a linear combination of u and v, namely

$$\Phi_\lambda(x_i) = \frac{u(x_r)\,v(x_i) - v(x_r)\,u(x_i)}{u(x_r)\,v(x_0) - v(x_r)\,u(x_0)} \qquad (0 \leq i \leq r). \tag{16.6}$$

Letting $r \to \infty$ we get $\Phi_\lambda(x_i) \to \frac{v(x_i)}{v(x_0)}$. We describe this result as follows :

The ratio $\frac{v(x_i)}{v(x_0)}$ *represents the Laplace transform of the probability that, starting from i at time* 0, *the state* 0 *will be reached before time t and before* x_∞ *has been reached.*

e. The minimal solution. Types of boundary. — Consider any process $X(t)$ whose transition probabilities satisfy (16.1) and, therefore, the differential equations (1.4). Define a new random variable $\overline{X}(t)$ by setting $\overline{X}(t) = X(t)$ if X has only finitely many jumps within the time interval $(0, t)$ and $\overline{X}(t) = x_\infty$ otherwise. This is a formal way of saying that *the process is stopped at the moment when* x_∞ *is reached* for the first time. In the physicist's language $\overline{X}$ represents the *absorbing barrier process.*

It is clear intuitively that $\overline{X}$ is again a Markovian process and that its transition probabilities are given by the minimal solution of Section 10. *All processes compatible with the conditions* (16.1) *agree up to the moment when* x_∞ *is reached for the first time; the minimal solution represents this common part of the processes :* $P_{ik}(t)$ *is the probability of reaching k from the starting point i in finitely many steps.*

Formula (9.4) is now seen to express the following relation : the probability that only finitely many jumps have occurred within the time interval $(0, t)$ equals 1 minus the probability that either x_∞ has been reached or the process has terminated [13] at the state 0.

We have seen above under c that x_∞ *can be reached only if it is regular or exit.* This explains the uniqueness feature of the differential equations (1.4) in the other cases. The probabilistic difference between an *entrance boundary* and a *natural boundary* is contained in the fact that $v(x_\infty) > 0$ in the first case, $v(x_\infty) = 0$ in the second: the probability that 0 will be reached before a fixed time t has a positive lower bound in the case of an entrance boundary and in this sense the process may start from an entrance x_∞.

f. The example of Section 13. — We shall give a probabilistic interpretation of this example and, at the same time, a *new derivation* of all pertinent formulas.

We define a process as follows. The starting state i is fixed. When x_∞ is reached for the first time, the process recommences immediately from a new state j, where j is a random variable with distribution $\{a(x_k)\mu_k\}$. The defect $1-\sum a(x_k)\mu_k$ is the probability that the process stops when x_∞ is reached. Needless to say that the same rule applies when x_∞ is reached for the second, third, ... time.

The probability of a passage from i to k in finitely many steps is given by the minimal solution P or — in terms of Laplace transforms — by the minimal Green function π_λ. The probability that x_∞ is reached before time t has the Laplace transform $\frac{u(x_i)}{u(x_\infty)}$. The probability that t time units later the process is at state k without x_∞ having been reached for a second time is $\sum a(x_j)P_{jk}(t)\mu_j$ and the corresponding Laplace transform is $A_\lambda(x_k)\mu_k$. By the convolution theorem the probability that, starting from i, at time t the process is at k having passed through x_∞ exactly once has the Laplace transform $u(x_i)n^{-1}(x_\infty)A_\lambda(x_k)\mu_k$. The times elapsed between two consecutive

---

(13) This interpretation of the term $\delta_0 v(x_i)$ may be derived also from the meaning of $\frac{v(x_i)}{v(x_0)}$ described above under d, or directly by the same procedure.

passages through x_∞ are independent random variables with the Laplace transform

$$(16.7)\qquad \gamma_\lambda = \frac{1}{u(x_\infty)} \sum a(x_j)\, u(x_j)\, \mu_j = \frac{1}{u(x_\infty)} [a, u].$$

The probability of finding at time t the state k after 1, 2, 3,... passages through x_∞ has therefore the Laplace transform

$$(16.8)\qquad \frac{u(x_i)}{u(x_\infty)} \frac{1}{1-\gamma} A_\lambda(x_k)\, \mu_k,$$

and this is precisely the content of formulas (13.1) and (13.12).

g. The other examples. — The example of Section 14 is essentially the same except that we now permit that the process stays at x_∞ (i. e. $X = x_\infty$) during a time which is a random variable with an exponential distribution. This makes x_∞ a regular state whereas in the preceding example it was " fictitious ".

The example of Section 15 is more perplexing. From a purely analytical point of view it does not differ much from the example of Section 14, but it does not admit of a direct probabilistic interpretation due to the fact that $\sum a(x_k)\, \mu_k = \infty$.

As a consequence, x_∞ becomes an " instantaneous " state : the set of points t of the time axis where $X(t) = x_\infty$ is of positive measure, but contains no intervals. The probability that the process *stays* at x_∞ during an uninterrupted time interval is zero.

The absorbing barrier process of Section 11 is a limiting case : to obtain it *analytically* it suffices to choose a sequence of distributions $\{a^{(n)}\}$ such that

$$\sum_k a^{(n)}(x_k)(x_\infty - x_k)\, \mu_k = C \qquad \text{and} \qquad a^{(n)}(x_k) \to 0$$

for each fixed k as $n \to \infty$. (*See* [6] for the corresponding diffusion processes.)

REFERENCES.

[1] K. L. Chung, *Foundations of the theory of continuous parameter Markov chains* (*Proceedings of the Third Berkeley Symposium on Mathematical Statistics and Probability*, vol. 2, 1956, p. 29-40).

[2] J. L. Doob, *Topics in the theory of Markov chains* (*Trans. Amer. Math. Soc.*, vol. 52, 1942, p. 37-64).

[3] J. L. Doob, *Stochastic processes*, J. Wiley and Sons, 1953.

[4] W. Feller, *On the integro-differential equations of purely discontinuous Markoff processes* (*Trans. Amer. Math. Soc.*, vol. 48, 1940, p. 488-815 and Errata, vol. 58, 1945, p. 474).

[5] W. Feller, *The parabolic differential equations and the associated semi-groups of transformations* (*Annals of Math.*, vol. 55, 1952, p. 468-519).

[6] W. Feller, *Diffusion processes in one dimension* (*Trans. Amer. Math. Soc.*, vol. 77, 1954, p. 1-31).

[7] *On boundaries and lateral conditions for the Kolmogorov differential equations* (*Annals of Math.*, vol. 00, 1957, p. 000).

[8] *Generalized second order differential operators and their lateral conditions* (To appear in the *Illinois Journal of Math.*).

[9] S. Karlin and J. McGregor, *Representation of a class of stochastic processes* (*Proc. Nat. Acad. Sc. U. S. A.*, vol. 41, 1955, p. 387-391).

[10] S. Karlin and J. McGregor, *The differential equations of birth and death processes and the Stieltjes moment problem* (To appear in *Trans. Amer. Math. Soc.*).

[11] D. G. Kendall, *Some further pathological examples in the theory of denumerable Markov processes* (*Quart. J. Math. Oxford Ser.*, vol. 7, 1956, p. 39-56).

[12] D. G. Kendall and G. E. H. Reuter, *Some pathological Markov processes with a denumerable infinity of states and the associated semi-groups of operators on l* (*Proc. International Congress of Mathematicians*, Amsterdam 1954).

[13] A. N. Kolmogorov, *On some problems concerning the differentiability of the transition probabilities in a temporally homogeneous Markov process having a denumerable set of states* (in Russian) (*Učenye Zapiski Moskovskogo Gos. Univ.*, (4), t. 148, 1951, p. 53-59).

[14] B. O. Koopman, to appear.

[15] W. Ledermann and G. E. Reuter, *Spectral theory for the differential equations of simple birth and death processes* (*Phil. Trans. Roy. Soc. London*, Ser. A, vol. 246, 1954, p. 321-369).

[16] P. Lévy, *Systèmes markoviens et stationnaires. Cas dénombrable* (*Ann. Sc. Éc. Norm. Sup.*, (3), vol. 68, 1951, p. 327-381).

[17] P. Lévy, *Complément à l'étude des processus de Markoff* (*ibid.*, vol. 69, 1952, p. 203-212).

[18] D. V. Widder, *The Laplace transform*, Princeton University Press, Princeton 1946.

(Manuscrit reçu le 26 juin 1957.)

NON-MARKOVIAN PROCESSES WITH THE SEMIGROUP PROPERTY

By William Feller[1]

Princeton University

1. Introduction. Every $N \times N$ stochastic matrix P defines the transition probabilities of a Markovian process with positive discrete time parameter. Its n-step transition probabilities satisfy the Chapman-Kolmogorov, or semigroup, relation $P^{n+m} = P^n P^m$. We shall show that *for $N \geqq 3$ there exist non-Markovian processes with N states whose transition probabilities satisfy the same equation.*[2] All elements of P will equal N^{-1}. The process may be chosen *strictly stationary.* A simple modification leads to non-Markovian processes with *continuous time* parameter and the semigroup property with N states or a continuum of states.

The triviality of the following example should not obscure the interest of the problem concerning the existence of non-Markovian processes satisfying the Chapman-Kolmogorov equation. As so many other basic problems in probability theory, it has been formulated by P. Lévy who with his usual ingenuity gave the first counter-example to the obvious conjecture.

2. Let $\mathfrak{P}$ be the sample space whose points $(x^{(1)}, \cdots, x^{(N)})$ are the random permutations of $(1, 2, \cdots, N)$ each carrying probability $1/N!$ Let $\mathfrak{R}$ be the set of the N points $(x^{(1)}, \cdots, x^{(N)})$ such that $x^{(i)} = \nu$ for all $1 \leqq i \leqq N$ where ν is a fixed integer $1 \leqq \nu \leqq N$; each point of $\mathfrak{R}$ carries probability $1/N$. Finally, Let $\mathfrak{S}$ be the mixture of $\mathfrak{P}$ and $\mathfrak{R}$ with $\mathfrak{P}$ carrying weight $1 - N^{-1}$ and $\mathfrak{R}$ weight N^{-1}.

More formally, $\mathfrak{S}$ contains the $N! + N$ arrangements $(x^{(1)}, x^{(2)}, \cdots, x^{(N)})$ which represent either a premutation of $(1, 2, \cdots, N)$ or the N-fold repetition of an integer ν, $1 \leqq \nu \leqq N$. To each point of the first class we attribute probability $(1 - N^{-1})(N!)^{-1}$, to each point of the second class probability N^{-2}.

Then clearly

$$(1) \qquad P\{x^{(i)} = \nu\} = N^{-1}, \qquad P\{x^{(i)} = \nu, \quad x^{(j)} = \mu\} = N^{-2}$$

for all $i \neq j$. Thus all transition probabilities are equal:

$$(2) \qquad P\{x^{(i)} = \nu \mid x^{(j)} = \mu\} = N^{-1}.$$

Given, say, that $x^{(1)} = 1$, $x^{(2)} = 1$ the probability that $x^{(3)} \neq 1$ is zero, and hence the process is non-Markovian.

3. To extend the process to all integral values of the time parameter consider, in the usual manner, a double infinity of independent repetitions of the described

Received February 4, 1959.

[1] Research supported by the Office of Ordnance Research U. S. Army Contract DA-36-034-ORD-2001.

[2] [Added in proof.] D. Blackwell has pointed out to me that the variables of our process represent a sequence of *random variables which are pairwise independent without being mutually independent.*

sample space. In other words, we consider the product space $\cdots \mathfrak{S} \times \mathfrak{S} \times \mathfrak{S} \cdots$ with product measure; its points are the doubly infinite sequences $x = \{x^{(i)}\}$ such that for each integer r the N-dimensional block $(x^{(rN+1)}, x^{(rN+2)}, \cdots, x^{(r+1)N})$ represents the projection of x onto a coordinate space $\mathfrak{S}$. This represents a non-Markovian process with the stationary transition probabilities (2). However, the process itself is not stationary in view of the periodicity modulo N.

To obtain a *stationary* process of the same type it suffices to introduce N replicas of our process with time shifts 0, 1, 2, $\cdots$, $N - 1$ and define a new process as their mixture with equal weights.

4. To construct a process of a *similar character defined for all* $t \geqq 0$ consider the above discrete process and a Poisson process $\{N(t)\}$ with mean t independent of it. Define a new process by

$$x(t) = x^{(N(t))}. \tag{3}$$

Its absolute probabilities are given by

$$P\{x(t) = \nu\} = \sum_{i=0}^{\infty} P\{N(t) = i\} \cdot P\{x^{(i)} = \nu\} = N^{-1}. \tag{4}$$

The joint probabilities for $0 \leqq s < t$ are calculated in like manner, but the possibility that $N(t) = N(s)$ makes it necessary to consider separately the cases $\nu = \mu$ and $\nu \neq \mu$. Clearly

$$\begin{aligned} P_{\nu\mu}(t) &= P\{x(s + t) = \mu \mid x(s) = \nu\} \\ &= N^{-1}(1 - e^{-t}), \qquad \text{if } \nu \neq \mu, \\ P_{\nu\nu}(t) &= e^{-t} + N^{-1}(1 - e^{-t}) \end{aligned} \tag{5}$$

and a simple calculation shows that

$$P_{\nu\mu}(s + t) = \sum_{\lambda=1}^{N} P_{\nu\lambda}(s) P_{\lambda\mu}(t) \tag{6}$$

even though the process is clearly non-Markovian.

Finally, one might replace the N states by N intervals and an appropriate motion within them.

REFERENCE

[1] P. Lévy, "Exemples de processus pseudo-Markoviens," *Comptes Rendus Académie Sciences*, Paris, Vol. 228, (1949), pp. 2004–2006.

DIFFERENTIAL OPERATORS WITH THE POSITIVE MAXIMUM PROPERTY

BY

WILLIAM FELLER[1]

1. Introduction

Let I be a fixed open interval, and consider a linear functional operator whose domain and range consist of real functions continuous in a subinterval of I. We say that f is in the local domain of Ω at the point $s \in I$, in symbols $f \in D(\Omega, s)$, if both f and Ωf are continuous in a neighborhood of s. Similarly, the global domain $D(\Omega, J)$ for an interval $J \subset I$ consists of the functions f such that both f and Ωf are continuous in J. We shall suppose that Ω is of *local character* in the following sense: if f vanishes identically in some neighborhood of s, then $f \in D(\Omega, s)$ and $\Omega f(s) = 0$. For such an operator the restriction of Ωf to an interval J depends only on the behavior of f in J.

We say that Ω has the *positive maximum property if one has* $\Omega f(s) \leqq 0$ *for each* $s \in I$ *and each* $f \in D(\Omega, s)$ *which attains a positive local maximum at* s. The classical differential operator defined by $af'' + bf' + cf$ with $a > 0$ has the positive maximum property if $c \leqq 0$. In view of the well-known role of such operators in many theories, we propose in this note to find a canonical form for the general operators of local character with the positive maximum property.

Let x be a continuous strictly increasing function in I, and m a strictly increasing right continuous function (not necessarily bounded). We shall view x as a scale, the increments of m as a measure on the Borel sets of I. The symbol $D_x f$ will be used indiscriminately for right and left derivatives provided they exist at each point and are continuous except for jumps. Differentiation with respect to m has the obvious meaning provided we agree to consider increments only for *closed* intervals. The operator $\Omega_0 = D_m D_x$ has been discussed in [1] as a natural generalization of the classical operator $aD_s^2 + bD_s$; it is characterized by the *strong* maximum property that $\Omega_0 f(x) \leqq 0$ for each x such that f attains a local maximum at x.

Operators of the form $\Omega_0 + c$ with $c \leqq 0$ have the positive maximum property, and we shall show that the most general operator with this property is of a similar, though slightly more intricate, form.

We parametrize I by x and consider an arbitrary *convex* continuous function ω of x; that is, $\omega > 0$ and $D_x \omega \uparrow$ throughout I. The operator A defined by

$$(1.1) \qquad Af = \frac{1}{\omega} D_m \left\{ \omega^2 D_x \frac{f}{\omega} \right\}$$

Received September 10, 1958.

[1] Research sponsored by the Office of Ordnance Research.

is a special case of the operators studied in [1]. We recall in particular that the representation (1.1) is *unique* in the sense that if

$$\frac{1}{\omega} D_m \left(\omega^2 D_x \frac{f}{\omega}\right) = \frac{1}{\varphi} D_\mu \left(\varphi^2 D_\xi \frac{f}{\varphi}\right) \tag{1.2}$$

for all $f \in D(A)$, then $\xi = px + q$, $\mu = p^{-1}m +$ const., and $A\varphi = 0$. Conversely, if $\varphi > 0$ is an arbitrary solution of $A\varphi = 0$, then

$$Af = \frac{1}{\varphi} D_m \left\{\varphi^2 D_x \frac{f}{\varphi}\right\}. \tag{1.3}$$

In other words, the canonical scale x and measure m are determined up to a trivial linear transformation, and the right side of (1.1) remains unchanged if ω is replaced by an arbitrary positive annihilator of A. The following lemma shows that operators of the form (1.1) represent a simple generalization of the operators $\Omega_0 +$ const.

LEMMA. *If $\varphi > 0$ and $A\varphi = 0$, then φ is convex, and*

$$d\omega'/\omega = d\varphi'/\varphi \qquad (\omega' = D_x\, \omega). \tag{1.4}$$

Thus the measure γ defined by $d\gamma = d\omega'/\omega$ is independent of the choice of the annihilator ω. The operator A may be redefined by

$$Af\cdot dm = df' - f\, d\gamma \tag{1.5}$$

in the sense that the integrals of the two sides are equal for each f in the domain of A. The derivative $D_x f = f'$ is continuous except for jumps, and these can occur only at the points of discontinuity of either m or γ.

Conversely, if γ is an arbitrary nonnegative measure on I, then there exists in I a two-parameter family of positive convex annihilators ω of (1.5), *and with each of them* (1.5) *is equivalent to* (1.1).

If ω is in the domain of $\Omega_0 = D_m Dx$, then (1.5) reduces to the simpler form $Af = D_m D_x - cf$ where $c \geqq 0$.

The main result of the present paper is contained in

THEOREM 1. *Let Ω have local character and the positive maximum property. Suppose that Ω nowhere degenerates into a first order operator.*[2] *Then there exists a uniquely determined operator A of the form* (1.1) *or* (1.5) *which is an extension of Ω, that is, $\Omega f = Af$ wherever Ωf is continuous.*

Every operator of the form (1.1) *or* (1.5) *has the positive maximum property.*

In [3] it is shown that operators of the form (1.5) with discontinuous measures m and γ occur naturally in the theory of the vibrating string. Examples will be found in that paper. The next theorem shows the role of

[2] A differential operator A is of first order if there exists a scale parameter y such that $Af = bD_y f + cf$ for each $f \in D(A)$. For an intrinsic characterization see [2].

our operators for the diffusion with possible *destruction of masses*. Let **C** be the usual Banach space of continuous functions in I.

THEOREM 2.[3] *Let* $\{T_t\}$ *be a strongly continuous positivity-preserving semi-group of operators from* **C** *to* **C** *such that* $T_t \mathbf{1} \leqq \mathbf{1}$ *for* $t > 0$, *and suppose that its generator* Ω *defined by*

$$\Omega f(x) = \lim_{t \to 0} \frac{T_t f - f}{t}(x) \tag{1.6}$$

is of local character. Then Ω *has the positive maximum property.*

In other words, the most general diffusion process without creation of masses is generated by an operator of the form (1.5).

2. Proof of the lemma

Let A be defined by (1.1). If Af is continuous in J, the one-sided derivatives $D_x(f/\omega) = (f/\omega)'$ exist everywhere, and

$$Af = \frac{1}{\omega} D_m\{f'\omega - f\omega'\}. \tag{2.1}$$

On integration by parts

$$\int_\alpha^\beta Af\, dm = f' - f\frac{\omega'}{\omega}\Big|_\alpha^\beta + \int_\alpha^\beta \omega' \cdot d\frac{f}{\omega}, \tag{2.2}$$

and another integration by parts shows that (1.5) holds with $d\gamma = d\omega'/\omega$.

It follows that if $\varphi > 0$ and $A\varphi > 0$, then $D_m\varphi' \geqq 0$, so that φ' is non-decreasing, and hence φ is convex. On the other hand if

$$\varphi = \omega \int \frac{dx}{\omega^2} + \text{const.}\, \omega, \tag{2.3}$$

then $A\varphi = 0$, and conversely. Clearly the two arbitrary constants in (2.3) may be chosen so that at a prescribed point, φ will be positive and have a local minimum. Such φ can have no positive maximum, and therefore will be positive and convex throughout I. That with such φ the relations (1.3) and (1.1) are equivalent has been shown in [1] and follows easily on using integrations by parts. The remaining assertions of the lemma are now obvious.

3. Proof of Theorem 1

We assume that Ω is of local character, that it has the positive maximum property, and that it nowhere degenerates into a first order differential operator. Let the interval I be parametrized arbitrarily by s.

We prove first: suppose that in a subinterval $J \subset I$ we have $\Omega\phi = 0$,

[3] By means of this theorem the arguments of [1], Section 9, may be simplified considerably.

$\phi > 0$. *Then ϕ can have no strict maximum in the interior* (that is, the maximum of ϕ in J is assumed at a boundary point). Assume that ϕ has a strict maximum at the interior point s. Since Ω does not degenerate, we can find an $f \epsilon D(\Omega, s)$ such that $\Omega f(s) > 0$, and therefore $\Omega f > 0$ in a neighborhood $N \subset J$ of s. It is possible to choose λ so large that the value of $f + \lambda\phi$ at s exceeds the value at either boundary point, and hence $f + \lambda\phi$ attains a local maximum at some interior point $s' \epsilon N$. However, this contradicts the positive maximum property since $\Omega(f + \lambda\phi)(s') = \Omega f(s') > 0$. This proves the assertion.

The same argument shows that Ω has the weak maximum property: if $g \leqq 0$ in a neighborhood of s and $g(s) = 0$, then $\Omega g(s) \leqq 0$ for each $g \epsilon D(\Omega, s)$. We know therefore from [2] that the domain of definition of Ω may be enlarged to include two functions ϕ, ψ such that $\Omega\omega = 0$ if and only if ω is a linear combination of ϕ and ψ. The functions ϕ and ψ are linearly independent in each subinterval of I. Furthermore, in every subinterval in which $\omega > 0$ the operator Ω may be represented in the form (1.1), x and m having the properties described in Section 1.

To two arbitrary points $\alpha < \beta$ choose a linear combination $\omega = p\phi + q\psi$ such that $\omega(\alpha) = \omega(\beta) = A > 0$. Then ω cannot have a strict positive maximum or a strict negative minimum at an interior point of (α, β), and hence we conclude that $0 \leqq \omega(s) \leqq A$ for $\alpha < s < \beta$. Furthermore, ω cannot decrease in the interval $s > \beta$, and cannot increase in the interval $s < \alpha$. Therefore $\omega \geqq 0$ everywhere, and a zero of ω is possible only in the interior of (α, β). The sum of two such functions will be strictly positive, and we conclude that *it is possible to choose two independent functions ϕ and ψ strictly positive throughout I such that $\Omega\phi = 0$ and $\Omega\psi = 0$.*

If for fixed λ the function $\omega = \psi - \lambda\phi$ had two zeros, it would somewhere in between attain a positive maximum or a negative minimum, which is impossible since $\Omega\omega = 0$. Therefore the ratio ψ/ϕ is strictly monotone, and without loss of generality we may assume that

$$\xi = \psi/\phi \tag{3.1}$$

is strictly increasing.

After these preliminaries we come to the main point of the proof. We introduce ξ as a new variable, and show that

As functions of ξ the reciprocals ϕ^{-1} and ψ^{-1} are concave.

In other words, it is asserted that for arbitrary points $\alpha < \beta$ and $\alpha < s < \beta$

$$\{\xi(s) - \xi(\alpha)\}\phi^{-1}(\beta) + \{\xi(\beta) - \xi(s)\}\phi^{-1}(\alpha) \leqq \{\xi(\beta) - \xi(\alpha)\}\phi^{-1}(s), \tag{3.2}$$

and similarly for ψ. To prove (3.2) define ω by

$$\begin{aligned} &\{\xi(\beta) - \xi(\alpha)\}\omega(s) \\ &\quad = \phi(s)\{\xi(s) - \xi(\alpha)\}\phi^{-1}(\beta) + \phi(s)\{\xi(\beta) - \xi(s)\}\phi^{-1}(\alpha). \end{aligned} \tag{3.3}$$

Then ω is a linear combination of ϕ and $\psi = \phi\xi$ such that $\omega(\alpha) = \omega(\beta) = 1$. As we have seen, this implies $\omega(s) \leqq 1$ for $\alpha < s < \beta$, which proves (3.2).

It has been shown in [2] that our Ω is of the form given in (1.1) where ω is an arbitrary positive annihilator of Ω. We choose $\omega = \phi$. The relation $\Omega\psi = 0$ is then equivalent to $\phi^2 D_x \xi = \lambda$ where $\lambda > 0$ is a constant. The concavity of ϕ^{-1} implies that a one-sided derivative $D_\xi \phi$ exists at all points, and we have

$$(3.4) \qquad D_x \phi = D_\xi \phi \cdot D_x \xi = \lambda\phi^{-2} D_\xi \phi = -\lambda D_\xi \phi^{-1}.$$

Since the last term is increasing, we have proved that ϕ as a function of x is *convex*.

This proves the first part of Theorem 1. Given an arbitrary operator (1.5) and an f in its domain, $Af < 0$ together with $f > 0$ implies $D_m f' < 0$, so that f can have no positive maximum in an interval where $Af < 0$. Thus A has the positive maximum property.

4. Proof of Theorem 2

Let $f \in D(\Omega)$, and suppose that f attains a local positive maximum at a point s where $f(s) > 0$. Put

$$(4.1) \qquad F = f - f(s)\mathbf{1}.$$

Then $F(0) = 0$ and $F \leqq \mathbf{0}$ in a neighborhood of s. Therefore $G = F \cup \mathbf{0}$ is identically zero in a neighborhood of s. In consequence of the local character of Ω we have therefore

$$(4.2) \qquad 0 = \Omega G(s) = \lim_{t\to 0} t^{-1}\{T_t G - G\}(s) = \lim_{t \to 0} t^{-1} T_t G(s).$$

On the other hand

$$(4.3) \qquad T_t F \geqq T_t f - f(s)\mathbf{1},$$

and therefore

$$\liminf_{t\to 0} t^{-1} T_t F(s) \geqq \lim_{t\to 0} t^{-1}\{T_t f - f\}(s) = \Omega f(s),$$

which proves the assertion.

References

1. W. Feller, *Generalized second order differential operators and their lateral conditions*, Illinois J. Math., vol. 1 (1957), pp. 459–504.
2. ———, *On the intrinsic form for second order differential operators*, Illinois J. Math., vol. 2 (1958), pp. 1–18.
3. ———, *On the equation of the vibrating string*, J. Math. Mech., to appear in vol. 8 (1959).

Princeton University
Princeton, New Jersey

SOME NEW CONNECTIONS BETWEEN PROBABILITY AND CLASSICAL ANALYSIS

By WILLIAM FELLER†

Recent research has revealed the intimate relationship between potential theory and Markov processes, and has supplied new examples of the fertility of a probabilistic approach to problems of classical analysis. Choquet's work on capacities, the Beurling–Deny theory of general potentials, Doob's probabilistic approach to the Dirichlet problem, and Hunt's basic results concerning potentials and Markov processes are closely related, despite the diversity of formal appearances and methods. I had hoped in this address to discuss the interconnections between these theories, but the task proved too overwhelming for the limited time and my own limitations. I am therefore compelled unashamedly to restrict this talk to some related aspects of my own work. I propose to describe the two boundaries and topologies induced by the annihilators of certain operators; to discuss their justification, their use for an abstract theory of so-called boundary value problems, and their connection with an invariant theory of operators of local character (which generalize the ill-defined notion of differential operators).

I shall not endeavor to develop a theory or even to state results in a precise form. Rather I shall try to explain the background and the purpose of the theory by means of a few simple examples, preferably using classical harmonic functions. Although everything will be interpreted probabilistically, the main emphasis is purely analytic.

In the sequel D will always denote a topological space and $\mathbf{C}$ the familiar Banach space of continuous functions in D with

$$\|f\| = \sup_{p\in D} |f(p)|.$$

Unless otherwise stated all operators will act on continuous functions.

1. Boundaries induced by positive operators

1.1. Harmonic functions. It is convenient to start with a probabilistic interpretation of harmonic functions by means of an *ad hoc* constructed random walk. In § 2 we shall approach the Laplacian more directly and more naturally.

† Research in part supported by the Office of Ordnance Research, U.S. Army.

R.L. Schilling et al. (eds.), *Selected Papers II*,

Let D be the open unit disc in the plane and for each point $p \in D$ let $D_p \subset D$ be the greatest open disc centered at p and contained in D. We shall study the operator T from $\mathbf{C}$ to $\mathbf{C}$ for which $Tf(p)$ *equals the arithmetic mean* of f over D_p. Thus

$$Tf = \int_D K(.,q) f(q)\, dq, \tag{1}$$

where $K(p,q)$ equals $|D_p|^{-1}$ or 0 according as $q \in D_p$ or $q \bar{\in} D_p$. Similarly, T^n is induced by a kernel $K^{(n)}$.

This T determines a discrete *random walk* with arbitrary initial position $p \in D$ in which $K^{(n)}(p, .)$ is the density of the random position $Q_n \in D$ after n steps. We obtain a well-defined measure in the space of all sequences $\{Q_n\}$ $(Q_0 = p, Q_n \in D)$ and it can be shown that almost all sequences are convergent to a point of the boundary circle. For our purposes a less refined purely analytical statement will suffice. For any set $A \subset D$ the probability that $Q_n \in A$ equals $T^n\chi(p)$, where χ is the characteristic function of A. It is easily verified that $T^n f \to 0$ for each $f \in \mathbf{C}$ vanishing at the boundary. Therefore $T^n\chi \to 0$ for each compact A, and this is equivalent to the statement that Q_n approaches the boundary in probability.

This result can be rendered more precise as follows. Let $\Gamma \subset B$ be a set of the boundary circle B, and u_Γ the harmonic function determined, in the classical sense, by the boundary values 1 on Γ and 0 on $B - \Gamma$. Then $u_\Gamma(p)$ *is the probability that* Q_n *approaches* Γ as $n \to \infty$.

Harmonic functions appear in this context because they are eigenfunctions satisfying $\phi = T\phi$. The set $\mathfrak{P}$ of all solutions of this equation such that $0 \leqslant \phi \leqslant 1$ is a convex set and the harmonic functions u_Γ coincide with its *extremals*. The sets

$$\Gamma_\epsilon = \{q \in D \mid u_\Gamma(q) > 1 - \epsilon\} \tag{2}$$

are a system of deleted neighborhoods of Γ. If χ is the characteristic function of Γ_ϵ then $T^n\chi \to u_\Gamma$, and hence

$$u_\Gamma(p) = \lim_{n\to\infty} \operatorname{prob}\{Q_n \in \Gamma_\epsilon\}. \tag{3}$$

From this it follows easily that $u_\Gamma(p)$ is the probability of an actual asymptotic approach to Γ, but for our purposes the weaker statement (3) fully suffices. [An alternative proof is given in the next section.]

The point to be emphasized is that this set-up is not of an analytic nature but can be carried out abstractly for a large class of operators in an arbitrary topological space D. No boundary need be defined *a priori*, and it is natural to *define* a boundary in such a way that the sets Γ_ϵ become neighborhoods of the corresponding boundary sets. Again, D may possess a boundary which does not admit of the interpretation of Γ_ϵ as neighborhoods, and it may be necessary to introduce a new boundary appropriate to the study of the transformation T. The simplest example is obtained by mapping the unit disc D conformally onto a domain $\tilde{D}$ whose boundary $\tilde{B}$ is of a complicated structure with prime ends, etc. Under this conformal map T and its random walk are carried over to $\tilde{D}$, the new eigenfunctions are again harmonic, but obviously the convergence properties of the random walk remain true only if we replace the 'natural' boundary of $\tilde{D}$ by the boundary and topology induced by the conformal map. This, of course, is a probabilistic version of the now familiar observation due to Martin[17] that the study of harmonic functions in complicated domains requires the introduction of an appropriate boundary.

We pass to a more interesting example of a different kind.

1.2. Relativization and isomorphisms. For an arbitrary (not necessarily bounded) $\psi > 0$ harmonic in the disc D we define a new transformation by

$$T_\psi f = \psi^{-1} T(f\psi). \tag{4}$$

Its kernel is given by

$$K_\psi(p,q) = \psi^{-1}(p)\, K(p,q)\, \psi(q). \tag{5}$$

A function $v > 0$ satisfies $v = T_\psi v$ if and only if $\phi = v\psi$ satisfies $\phi = T\phi$ (is harmonic). We have thus a 1-1 correspondence between the positive eigenfunctions of T and T_ψ with 1 and ψ corresponding to ψ^{-1} and 1, respectively.

Operators of this form will be called *similar* to T. Clearly similarity is transitive, symmetric and reflexive. A closely related transformation has been used by Brelot[1] for harmonic functions. We shall see that the notion of similarity is exceedingly useful, and has its counterpart in similar semigroups and differential equations. Here we use it to illustrate the notion of the boundary induced by T_ψ and to derive a new proof for our interpretation of the harmonic function u_Γ.

Denote by $\mathfrak{P}$ and $\mathfrak{P}_\psi$, respectively, the sets of positive eigenfunctions of T and T_ψ bounded by 1. For simplicity of exposition consider the case $\psi = u_\Gamma$ where ψ is an extremal of $\mathfrak{P}$. The mapping $u\psi \leftrightarrow \phi$ establishes a 1-1 correspondence between $\mathfrak{P}_\psi$ and the subset of $\mathfrak{P}$ of elements such that $\phi \leqslant u_\Gamma$; this correspondence preserves extremals. Now the structure of the set $\mathfrak{P}$ of harmonic functions is best described in terms of the 'natural' boundary of D. Since relations between T, $\mathfrak{P}$, and $D \cup B$ with the natural topology will carry over to T_ψ, $\mathfrak{P}_\psi$ and $D \cup \Gamma$, the set Γ plays for T_ψ the role that B plays for T and may be considered the *boundary induced by* T_ψ.

The same conclusion may be reached probabilistically. From $\psi^{-1} = T_\psi \psi^{-1}$ it follows that $T_\psi^n f \to 0$ for each $f \in \mathbf{C}$ vanishing along Γ. Hence if χ is the characteristic function of a neighborhood U of Γ we have $T_\psi^n \chi \to 1$: thus in the random walk associated with T_ψ the paths converge in probability to the boundary set Γ.

This remark leads to a new proof of the interpretation of u_Γ given in the preceding section. Using (5) it is seen that the relation $T_\psi^n \chi \to 1$ may be rewritten as

$$\int_U K^{(n)}(p,q)\, \psi(q)\, dq \to \psi(p). \tag{6}$$

Now the neighborhood U of Γ may be chosen so small that in it $\psi = u_\Gamma$ is arbitrarily close to 1 and we conclude that in the T-random walk $\mathbf{P}\{Q_n \in U\} \to u_\Gamma(p)$ for each neighborhood of Γ. This is a slightly weakened version of the interpretation of $u_\Gamma(p)$ as *probability of an asymptotic approach to* Γ.

Given this interpretation of u_Γ we see that in the T-random walk K_ψ represents the *conditional* transition probability density given the event that the paths converge to Γ. Probabilistically, then, *the* T_ψ*-random walk is obtained from the* T*-random walk by conditioning or relativization*: in the T-walk we pay attention only to paths converging to Γ. More precisely, let $\mathfrak{S}$ be the set of all sequences $\{Q_n\}$, $(Q_0 = p, Q_n \in D)$, with the measure $\mathbf{P}$ induced by T, and let $\mathfrak{S}_\Gamma$ be the subset of sequences converging to Γ. Then the T_ψ-walk assigns zero probability to $\mathfrak{S} - \mathfrak{S}_\Gamma$ and probability $\mathbf{P}\{\mathfrak{A}\} \div \mathbf{P}\{\mathfrak{S}\} = \mathbf{P}\{\mathfrak{A}\}\, \psi^{-1}(p)$ to the subsets $\mathfrak{A} \subset \mathfrak{S}$.

1.3. Abstract construction; restricted and total boundaries. We pass to the extreme case where D is the set of integers $1, 2, \ldots$, with the discrete topology. This has the advantage that no preconceived intuitive notion of boundary obscures the view. The boundaries may be of a complicated topological structure and the present case will clearly reveal the features and problems of the most general set-up.

$\mathbf{C}$ is now the space of bounded functions and we write $\mathbf{f} \in \mathbf{C}$ as a column matrix with elements $f(i)$. We consider an operator T defined by a matrix Π with elements $\Pi(i,j)$ so that in matrix notation $T\mathbf{f} = \Pi\mathbf{f}$. The matrix Π is supposed to be substochastic, that is, its elements are $\geqslant 0$, its row sums $\leqslant 1$. We denote by $\mathfrak{P}$ the set of all eigenvectors $\boldsymbol{\phi}$ such that $\boldsymbol{\phi} = \Pi\boldsymbol{\phi}$

and $\mathbf{0} \leqslant \boldsymbol{\phi} \leqslant \mathbf{1}$, and by $\mathfrak{P}^{\infty}$ the set of all (possibly unbounded) solutions $\boldsymbol{\phi} > \mathbf{0}$ of $\boldsymbol{\phi} = \Pi\boldsymbol{\phi}$. To avoid trivialities we shall assume: (i) each $\boldsymbol{\phi}$ is strictly positive, (ii) $\mathfrak{P}$ contains at least two independent vectors. [Condition (i) eliminates the nuisance of partitioned matrices requiring words rather than thoughts; (ii) eliminates empty and single-point boundaries.]

Again Π may be interpreted as the matrix of transition probabilities in a *random walk* (Markov chain), the row defects $1-\Sigma\Pi(.,j)$ accounting for the possibility of a termination of the process. For an arbitrary initial position $i \in D$ we have a probability measure on the set $\mathfrak{S}$ of all terminating or infinite sequences of integers $\{Q_n\}$. The subset $\mathfrak{S}^{(n)}$ of sequences of length $\geqslant n$ has probability $\Sigma_j \Pi^n(i,j)$, and hence the probability of the set $\mathfrak{S}^{(\infty)}$ of infinite sequences is given by the ith element of $\bar{\boldsymbol{\phi}} = \lim \Pi^n \mathbf{1}$. Note that $\bar{\boldsymbol{\phi}}$ is the *maximal* element of $\mathfrak{P}$.

We proceed to introduce a *restricted* or $\mathfrak{P}$-*boundary* B, and a *total* or $\mathfrak{P}^{\infty}$-*boundary* $B^{\infty} \supset B$. We begin with the extremely simple special case where $\boldsymbol{\phi} = \Pi\boldsymbol{\phi}$ has only *finitely* many independent solutions.

(*a*) *The* $\mathfrak{P}$-*boundary*. Let $\mathfrak{P}$ be spanned by N non-zero vectors $\boldsymbol{\phi}^{(1)}, \ldots, \boldsymbol{\phi}^{(N)}$. These can be chosen as extremal elements of $\mathfrak{P}$, which amounts to saying that $\|\boldsymbol{\phi}^{(k)}\| = 1$ and $\bar{\boldsymbol{\phi}} = \boldsymbol{\phi}^{(1)} + \ldots + \boldsymbol{\phi}^{(N)}$. For fixed k and $\epsilon > 0$ put $\Gamma_{\epsilon}^{(k)} = \{i \mid \phi^{(k)}(i) > 1-\epsilon\}$. As $\epsilon \to 0$ we get a nest of non-empty sets with empty intersection; from $\bar{\boldsymbol{\phi}} \leqslant \mathbf{1}$ we conclude that for fixed $\epsilon > \frac{1}{2}$ the sets $\Gamma_{\epsilon}^{(j)}$ and $\Gamma_{\epsilon}^{(k)}$ are non-overlapping ($j \neq k$).

The *restricted boundary* B consists of N points $\beta^{(1)}, \ldots, \beta^{(N)}$ such that $\Gamma_{\epsilon}^{(k)}$ is a deleted neighborhood of $\beta^{(k)}$. We can extend the definition of each $\boldsymbol{\phi} \in \mathfrak{P}$ to $D \cup B$ by putting $\phi^{(j)}(\beta^{(k)}) = 1$ or 0 according as $j = k$ or $j \neq k$. Then each $\boldsymbol{\phi}$ is continuous in $D \cup B$, and the 'Dirichlet problem' is soluble: to prescribed boundary values there corresponds exactly one $\boldsymbol{\phi} \in \mathfrak{P}$.

Finally with an obvious notation, $\Pi^n(i, \Gamma_{\epsilon}^{(k)}) \to \phi^{(k)}(i)$, as $n \to \infty$ for each fixed ϵ. From this one deduces that in the random walk starting at i the sample sequences $\{Q_n\}$ converge with probability $\phi^{(k)}(i)$ to $\beta^{(k)}$; with probability $1-\bar{\phi}(i)$ they terminate; and the probability of no convergence is 0.

(*b*) *The* $\mathfrak{P}^{\infty}$-*boundary* can be introduced directly, but it is more convenient to use the similarity transformation introduced in § 1.2. A matrix Π_{ψ} is *similar* to Π if either $\Pi_{\psi} = \Pi$ or

$$\Pi_{\psi}(i,j) = \psi^{-1}(i)\,\Pi(i,j)\,\psi(j), \tag{7}$$

where $\boldsymbol{\psi} \in \mathfrak{P}^{\infty}$. We recall that similarity is transitive, reflexive, and symmetric and that the mapping $\boldsymbol{\phi} \leftrightarrow \mathbf{v}$ where $v(i) = \phi(i)\psi^{-1}(i)$ establishes a 1-1 correspondence between $\mathfrak{P}^\infty$ and $\mathfrak{P}_\psi^\infty$.

To see the relation between B and the corresponding restricted boundary B_ψ induced by Π_ψ consider the typical case $\boldsymbol{\psi} = \boldsymbol{\phi}^{(1)} + \boldsymbol{\phi}^{(2)}$. To $\boldsymbol{\phi}^{(3)}, \ldots, \boldsymbol{\phi}^{(N)}$ and to each unbounded $\boldsymbol{\phi} \in \mathfrak{P}^\infty$ there correspond unbounded vectors in $\mathfrak{P}_\psi^\infty$ and the boundary B_ψ reduces to two points whose neighborhoods coincide with the neighborhoods of $\beta^{(1)}$ and $\beta^{(2)}$. For the interpretation of this Π_ψ in terms of *conditional probabilities* see § 1.2.

In general, if B_ψ is the restricted boundary induced by Π_ψ we shall identify points of B and B_ψ with coinciding systems of neighborhoods. With this identification we define the *total boundary* B^∞ *induced by* Π *as the union of the boundaries* B_ψ as $\boldsymbol{\psi}$ ranges over $\mathfrak{P}^\infty$. All *similar matrices induce the same total boundary*, and if $\mathfrak{P}^\infty$ is spanned by M independent vectors, then B^∞ contains exactly M points.

Note that $D \cup B^\infty$ need not be compact. (No compactification seems natural or desirable for our purposes.)

(c) *The maximal ideal boundaries*. When $\mathfrak{P}$ is not spanned by denumerably many elements the extremal elements of $\mathfrak{P}$ do not correspond to points of the prospective boundary, but rather to sets of positive capacity. No satisfactory definition of points and neighborhoods is known. Now both $\mathfrak{P}$ and $\mathfrak{P}^\infty$ have a lattice structure similar to that of harmonic functions, and the correspondence between $\mathfrak{P}^\infty$ and $\mathfrak{P}_\psi^\infty$ is a lattice isomorphism. This makes it possible to define points of B and B^∞ by maximal ideals in $\mathfrak{P}$ and $\mathfrak{P}^\infty$, respectively (see[5]). Unfortunately these boundaries are absurdly large. For example, sample sequences converge to sets rather than to points; each $\boldsymbol{\phi} \in \mathfrak{P}$ has continuous boundary values which is at variance with the desirable model of harmonic function in a disc D with the natural topology. That the lattices $\mathfrak{P}$ and $\mathfrak{P}^\infty$ are isomorphic to lattices of continuous functions on some Hausdorff spaces is, of course, well known (see, for example, Kadison[15]). To us the main point is that this Hausdorff space appears as a *boundary* of D and is useful as such.

Our maximal ideal boundaries serve well for an orientation and as a guide, but the introduction of a less clumsy and more appropriate boundary is an open, and promising, problem.

[*Postscript*. A partial solution has now been obtained by J. L. Doob[19] who uses Martin's original construction. The Martin boundary is sufficiently small for sample sequences to converge toward points. In the finite case, however, this boundary may be bigger than ours; its neighborhoods are larger and this could lead to difficulties in connection with boundary value problems and non-minimal semigroups.]

2. Semigroups and differential equations

2.1. Orientation. We turn to the more interesting study of a family of positive operators $\{T(t)\}$, $t \geqslant 0$, from $\mathbf{C}$ to $\mathbf{C}$ with $\|T(t)\| \leqslant 1$, and with the semigroup property $T(t+s) = T(t)\,T(s)$. The probabilistic counterpart to a fixed $T(t)$ is a (possibly terminating) random walk with jumps taking place at times $t, 2t, \ldots$. To the whole semigroup there corresponds a random motion (Markov process) in D with continuous time: the sample space $\mathfrak{S}$ is the space of functions Q defined in an interval $0 \leqslant t < \tau \leqslant \infty$ such that $Q(t) \in D$, and $Q(0) = p \in D$ is a given point. The semigroup induces a $\mathbf{P}$-measure in $\mathfrak{S}$ such that for each Borel set $A \subset D$ we have $\mathbf{P}\{Q(t) \in A\} = T(t)\,\chi(p)$ where χ is the characteristic function of A. Needless to say with the $\mathbf{P}$-measure almost all paths are reasonably regular. In particular, they are continuous when the semigroup is generated by a differential operator ('diffusion processes').

If a path Q is defined only within a finite interval $0 \leqslant t < \tau$ (that is, if the process terminates at time τ) then, except on a null set of paths, as $t \uparrow \tau$ either $Q(t) \to q \in D$ or $Q(t)$ has no point of accumulation in D. In the first case we say that the process terminates at q, in the second that it *terminates 'at the boundary'*. However, it remains to justify this expression.

For this purpose we shall have to introduce a boundary B induced by the *generator* Ω of the semigroup. It will be seen that all operators $T(t)$, $t > 0$, *induce the same boundary* $\pi \subset B$, and that the process *either terminates at* $B - \pi$ *or approaches* π *asymptotically* without reaching it.

We say that the semigroup is *generated by* Ω if for a dense set of 'smooth' $f \in \mathbf{C}$

$$\lim_{t \to 0} t^{-1}\{T(t)\,f - f\} = \Omega f \in \mathbf{C} \tag{8}$$

in the sense of pointwise convergence. According to this definition, introduced in[11], Ω may generate many semigroups; the infinitesimal generator in the sense of Hille[13] is a contraction $\Omega|\Sigma$ of Ω obtained by imposing *lateral conditions* (see § 4.3).

Putting $T(t)\,f = u(t, .)$ the function u will satisfy the functional equation $u_t = \Omega u$ with the 'initial condition' $u(0, .) = f$. (This is literally true for smooth f, and in an operational sense for all f.) In classical terms we are concerned with 'solving' this equation. We consider first a family of 'similar' generators and then the relation between generators with the same annihilators.

2.2. The Laplacian. Isomorphisms. We return to harmonic functions and take as example the familiar *heat equation* $u_t = \Delta u$ in the unit disc D. Among the positive semigroups generated by Δ there exists a *minimal semigroup*. In classical terms $u(t,.) = T(t)f$ is the solution of $u_t = \Delta u$ with initial condition $u(0,.) = f$ and *boundary condition* that u vanishes at the circle B. Associated with it is the *Wiener process* in D terminating at B. Physically the process represents a homogeneous diffusion in D with 'absorbing boundaries', or heat conduction with zero temperature at the boundary.

For this minimal semigroup $\|T(t)\| < 1$ and therefore the boundary π induced by $T(t)$ is empty, but we are concerned with the boundary induced by harmonic functions, that is, the *annihilators* of the generator Δ. In this connection, of course, we adhere to the natural topology of the closed disc. As in § 1.1 let u_Γ be the harmonic function in D determined by the boundary values 1 on the set $\Gamma \subset B$ and 0 on $B-\Gamma$. For the process starting at the point $p \in D$ we have the following analogue to the results of § 1.1.

$u_\Gamma(p)$ is the probability that the process terminates at the boundary set Γ. The probability that this happens before time t is $v(t,p)$ where v is the solution of $v_t = \Delta v$ with zero initial values and boundary values 1 on Γ and 0 on $B-\Gamma$.

This assertion becomes plausible on observing that for bounded harmonic ψ obviously

$$T(t)\psi = \psi - v(t,.), \tag{9}$$

where $v_t = \Delta v$ and v has zero initial values and boundary values coinciding (almost everywhere) with those of ψ. Now $T(t)\,1(p) = \mathbf{P}\{Q(t) \in D\}$ is the probability that the process does not terminate before it. Together with (9) this implies the assertion for the particular case $u_\Gamma = 1$, or $\Gamma = B$, at least if we accept as fact that the process does not terminate in the interior of D. We shall only indicate how the general assertion may be reduced to this particular case by a generalization of the method of *similarity transformations* or isomorphisms, introduced in § 1.2.

For positive harmonic ψ we define a new semigroup of positive operators from $\mathbf{C}$ to $\mathbf{C}$ by

$$T_\psi(t)f = \psi^{-1}T(t)(f\psi). \tag{10}$$

If $0 \leqslant f \leqslant 1$ then $0 \leqslant T_\psi(t)f \leqslant \psi^{-1} \leqslant T(t)\psi \leqslant 1$ and hence $\|T_\psi(t)\| \leqslant 1$. A glance at (8) shows that the semigroup $\{T_\psi\}$ is generated by the operator Δ_ψ defined by

$$\Delta_\psi f = \psi^{-1}\Delta(f\psi). \tag{11}$$

For simplicity of exposition we now restrict the consideration to $\psi = u_\Gamma$. First note that ψ^{-1} is unbounded near $B-\Gamma$ and close to 1 near Γ, and that $T_\psi(t)^{-1} \leqslant \psi^{-1}$. From this it is easy to conclude that *the random process corresponding to* $\{T_\psi\}$ *terminates at* Γ. Our interpretation of u_Γ is obtained from this by adapting the argument following (6). As at the end of § 1.2 we remark that the kernel of the T_ψ-semigroup represents the conditional transition probability densities of the T-process given that the path terminates at Γ. In the sense explained above the T_ψ-*process is therefore simply the restriction of the* T-*process to the paths terminating at* Γ.

Returning to the analytical relationship between the T- and the T_ψ-semigroup note that $\Delta_\psi \phi = 0$ if and only if $\phi\psi$ is harmonic. The positive (possible unbounded) annihilators of Δ and Δ_ψ stand in a 1-1 correspondence (which is a lattice isomorphism). The bounded annihilators of Δ_ψ correspond to the harmonic functions dominated by $\psi = u_\Gamma$, and thus the set Γ is the appropriate boundary for $\{T_\psi\}$ just as the circle B is for $\{T\}$. In short we find the same situation as in § 1.3.

The circle B, *induced by the harmonic functions, represents the 'total' boundary for the family of all similar semigroups* $\{T_\psi(t)\}$ [*or all differential equations* $u_t = \Delta_\psi u$]. *The boundary induced by the bounded annihilators of* Δ_ψ *corresponds to the subset* $\Gamma \subset B$.

That Γ is the appropriate boundary for the T_ψ-semigroup reflects the fact that the common *range of the transformations* $T_\psi(t)$ *is characterized by the side condition that* $T_\psi(t)f$ *vanishes along* Γ just as $T(t)f$ vanishes along B.

We have here the simplest example of a 'boundary condition' and see that it relates to our boundary rather than the 'natural' one. In the terminology of classical differential equations, Δ_ψ is a differential operator with coefficients singular along $B - \Gamma$, and boundary conditions can be imposed only along Γ. How vague and unsatisfactory such descriptions can be is known from the simple Sturm–Liouville theory in one dimension.

2.3. The active boundary. Singular operators. A new phenomenon may be described in connection with the operator $\Omega = \omega\Delta$ in the disc D where $\omega > 0$ is continuous in D but may tend to zero or infinity near the circle B. This operator has the same annihilators as the Laplacian Δ and it is interesting to compare the semigroups generated by $\omega\Delta$ with those generated by Δ. In classical terms we are concerned with the integration of the parabolic differential equation $u_t = \omega\Delta u$ which may be singular owing to the behavior of ω near B. We describe here the main features of the theory carried out in[6] for denumerable spaces (the Kolmogoroff differential equations) by methods of much wider applicability.

There exists a uniquely determined *minimal* positive semigroup $\{T(t)\}$ generated by $\omega\Delta$ and for it $\|T(t)\| \leqslant 1$. However, for an appropriate choice of ω it may happen that $T(t)\,1 = 1$: the minimal semigroup is in this case *unique*; by contrast to the heat equation no boundary conditions need or can be imposed in this case, and the induced random process (diffusion) does not terminate at a finite time.

With an arbitrary $\omega > 0$ let $\{T(t)\}$ be the minimal semigroup generated by $\omega\Delta$ in the unit disc D. Then the following is true.

(*a*) *The passive boundary* π. There exists a set $\pi \subset B$ of the unit circle (determined up to a null set) such that for each $t > 0$ the eigenfunctions ϕ of $\phi = T(t)\,\phi$ such that $0 \leqslant \phi \leqslant 1$ coincide with the positive harmonic functions (the annihilators of $\omega\Delta$) dominated by u_π, the harmonic function with boundary values 1 on π and 0 on $B-\pi$. In this sense *the boundary induced by each* $T(t)$ *coincides with* π. In the case $\omega = 1$ (the heat equation) π is empty. For any set $\Gamma \subset \pi$ the value $u_\Gamma(p)$ is the probability in the $T(t)$-process starting at the point $p \in D$ that Γ is *asymptotically approached as* $t \to \infty$; the probability of reaching Γ at a finite time is zero.

(*b*) *The active boundary* $A = B-\pi$. For each set $\Gamma \subset A$ the value $u_\Gamma(p)$ *is the probability that the process will terminate at* Γ. The probability that this occurs before time t equals $v(t,p)$ where v is a solution of $v_t = \omega\Delta v$ with zero initial values and boundary values 1 on Γ and 0 on $A-\Gamma$.

No boundary conditions can be imposed along the passive boundary. This summary explains the relations between minimal *semigroups generated by operators with the same annihilators.*

In analytic terms we may characterize the active and passive boundaries as follows. For $\lambda \geqslant 0$ the bounded positive solutions of

$$\lambda\phi - \omega\Delta\phi = 0 \tag{12}$$

form a convex set $\mathfrak{P}_\lambda$ endowed by a lattice structure similar to that of harmonic functions. Now there exists a 1-1 correspondence (lattice isomorphism) between the $\mathfrak{P}_\lambda$ for $\lambda > 0$ on one hand, and the harmonic functions dominated by u_A on the other hand. Thus the *active boundary is induced by each* $\mathfrak{P}_\lambda$ *for* $\lambda > 0$. An alternative interpretation may be given in terms of the resolvent $(\lambda-\omega\Delta)^{-1}$ and a discrete random walk associated with it.

3. The adjoint boundary

3.1. Duality. So far we have restricted consideration to operators acting on functions. Actually the study of an operator T from **C** to **C** cannot be separated from the study of the adjoint operator T^* which takes *measures* into measures. In probability T^* is the primary notion, although we are compelled to take T as the basic tool of the theory of Markov processes. The reasons are discussed in[7]; see also Dynkin[3].

To avoid new notations let D be an open domain of the plane and suppose that T is of the form (1) with an arbitrary *positive* kernel. The adjoint transformation acts on all measures, but it is convenient to consider only absolutely continuous measures and treat T^* as an operator on densities. Let then **L** be the Banach space of integrable functions in D with the usual norm

$$N(u) = \int_D |u|\, dq. \tag{13}$$

Then T^* as an operator from **L** to **L** carries $u \in \mathbf{L}$ into

$$T^*(u) = \int_D u(p)\, K(p, .)\, dp. \tag{14}$$

It is positive and $N(T^*) \leqslant 1$. The transformation (14) remains meaningful for all $u > 0$, although the integral may diverge.

In principle the construction of a boundary induced by T^* should follow the method used for T, but fortunately *an extremely simple trick will save us the trouble of a repeated construction.*

We start from the set $\mathfrak{P}^*$ of all finite eigenfunctions $u > 0$ of $u = T^*u$. They need not be integrable, but for simplicity we shall suppose that each $u \in \mathfrak{P}^*$ is strictly positive and continuous in D.

For an arbitrary $u \in \mathfrak{P}^*$ define a kernel $\tilde{K}$ by

$$\tilde{K}(p, q) = u(q)\, K(q, p)\, u^{-1}(p), \tag{15}$$

where $p \in D$, $q \in D$. Clearly

$$\int_D \tilde{K}(p, q)\, dq = 1 \tag{16}$$

for each p, and thus $\tilde{K}$ represents the kernel of a new transformation $\tilde{T}$ from **C** to **C**.

If $v \in \mathfrak{P}^*$ then $\phi = vu^{-1}$ is a continuous eigenfunction of $\tilde{T}\phi = \phi$, not necessarily bounded. Conversely, to each positive eigenfunction of $\tilde{T}\phi = \phi$ there corresponds the element $v = \phi u \in \mathfrak{P}^*$. This establishes a 1-1 correspondence between $\mathfrak{P}^*$ and the set $\tilde{\mathfrak{P}}^\infty$ of positive eigenfunctions of $\tilde{T}$. (These sets are endowed with lattice structures and the correspondence is a lattice isomorphism, but we omit these details.) In § 1.3 we have seen that $\tilde{T}$ induces a *total boundary* relative to the set $\tilde{\mathfrak{P}}^\infty$ of all positive eigenfunctions of $\tilde{T}\phi = \phi$. If in (15) we replace u by another element of $\mathfrak{P}^*$ then $\tilde{T}$ is replaced by a *similar* operator (as defined in § 1.3) and the total boundary remains unchanged. This justifies the

Definition. The adjoint boundary B^* induced by T is the total boundary induced by the operator $\tilde{T}$ acting from **C** to **C**. It is independent of the choice of $u \in \mathfrak{P}^*$.

Probabilistic interpretation. Consider first the case where $N(u) = 1$ and interpret u as the stationary probability density of the position (at any time) in a Markov chain with transition probability densities K. This process is defined for all integral values of the time parameter from $-\infty$ to ∞. In[16] Kolmogoroff pointed out that in this process $\tilde{K}(p,q)$ represents the conditional probability density of the position q at time n given that at time $n+1$ the position is p. Moreover, the same relationship exists between the higher order transition probability densities of the K-chain and the $\tilde{K}$-chain. In other words, for the K-chain, $\tilde{K}$ represents the transition probability densities in the negative time direction: the $\tilde{K}$-chain is obtained from the K-chain by *reversing the time direction.*

Roughly speaking, then, the boundary induced by T represents the directions towards which the sample sequences can converge, and the adjoint boundary the directions from which the process can originate. This description applies to, and becomes more concrete in connection with, continuous time parameter processes and leads to an interpretation of the boundary conditions for differential equations.

Analytically there is no change in the situation when the integral (13) diverges, and it is therefore annoying that Kolmogoroff's intuitive interpretation of $\tilde{K}$ breaks down. However, as Derman[2] pointed out, it may be salvaged for non-integrable u by considering a whole family of chains.

3.2. Relations between the two topologies. It is natural to ask whether and how the topology of $D \cup B$ induced by T is related to the topology of $D \cup B^*$ induced by T^*. A first answer is that every imaginable situation can arise. For the familiar symmetric operators of analysis the two spaces are identical. Still simpler is the other extreme, where B and B^* are disjoint and have disjoint neighborhoods. Most vexing are the intermediate cases. Among the examples given in[5] there appears the following configuration:

The boundary B consists of m points $\beta^{(1)}, \ldots, \beta^{(m)}$ and B^* consists of n points $\gamma^{(1)}, \ldots, \gamma^{(n)}$. Each deleted neighborhood of B is a deleted neighborhood of B^* and vice versa. However, each neighborhood of $\gamma^{(1)}$ contains a neighborhood of $\beta^{(1)} \cup \beta^{(2)}$ and each neighborhood of $\beta^{(1)}$ or $\beta^{(2)}$ is contained in a neighborhood of $\gamma^{(1)}$. Roughly speaking, the point $\gamma^{(1)}$ is the same as the set $\beta^{(1)} \cup \beta^{(2)}$. Similarly, two points of B may be equivalent to the union of the three points of B^*, etc.

These phenomena lead to many new problems connecting topological and analytical problems and are interesting in connection with boundary conditions; see the end of § 4.3.

4. Background and program

4.1. The problems. Given a topological space D an important problem of probability theory is to find the most general Markov process on D. (Hunt's beautiful results in [14] permit us to reformulate this in terms of *potentials*.) Space does not permit us here to analyze why and how this problem is reduced to that of finding semigroups of operators from **C** to **C**. Anyhow, the following slightly more general problem is of obvious interest in itself.

Find all operators Ω generating [*in the sense of* (8)] *positive semigroups* $\{T(t)\}$ *from* **C** *to* **C**. We have omitted the restriction $\|T(t)\| \leqslant 1$, which becomes more and more untenable even for probability theory and excludes semigroups of interest in potential theory, in diffusion (with creation of masses), and heat conduction. [We note in passing that our method of isomorphisms cannot be fully exploited as long as one adheres to the conventional Banach norm. It would be interesting and highly desirable to reformulate the whole theory free from restriction to normed spaces utilizing the *Köthe–Mackey* concept of dual spaces.]

Given a generator Ω we face the problem of *finding all positive semigroups generated by it and to discover the analytic and probabilistic relations among them.* The first part leads to a strict formulation of the vague and unsatisfactory notion '*boundary conditions*' for differential equations. The problem is to construct all possible lateral conditions; in this form it is analogous to the construction of self-adjoint contractions in Hilbert space, but it leads to new angles.

Finally, it is important to *find the generator of the adjoint semigroup.* (In special cases this amounts to finding the physicist's Fokker–Planck, or continuity, equation.)

This point of view links together operators which classically seemed worlds apart. For example, when D is the set of integers we are led to infinite systems of ordinary differential equations that can be treated precisely as partial differential equations; in fact, we obtain the boundary conditions with an analogue of '*normal derivatives*' at the boundary in a form which is applicable also, say, to harmonic functions in a domain with non-differentiable boundary. Again, when D is the real line, the semigroup for which $u(t, .) = T(t)f$ is harmonic in the half plane $t > 0$ is generated by a Riesz potential, and Elliott[4] has shown that its restriction to finite intervals is closely related to a second-order differential operator.

It is possible now to see, in rough outlines, the way to a rather general solution of the problems stated. Space does not permit us to go into details, and we proceed instead to indicate how the first problem is connected with an intrinsic theory of differential operators (or their analogue) in arbitrary spaces.

4.2. Operators of local character. The concept of a differential operator is defined only in special spaces and depends on a co-ordinate system. We replace it by the more meaningful and more general notion of an operator of local character. The positive semigroups generated by an operator of local character present an obvious analytic interest. (The corresponding Markov processes are the only ones whose path functions are continuous with probability one; see Ray[18].) They are a natural generalization of the second-order elliptic differential operators in Euclidean spaces: on one hand, such operators share the main properties of the Laplacian and generate positive semigroups. On the other hand, the derivation of the diffusion equation based on the local character condition (which I regret having introduced in [12] under the misleading name of Lindeberg condition) shows that no differential operator of higher order shares this property.

For an operator Ω of local character to generate a positive semigroup such that $\|T(t)\| \leqslant 1$ it is necessary, and very likely also sufficient, that it have the following *positive maximum property*: for each f in the (local) domain of Ω such that f attains a positive local maximum at the point p one has $\Omega f(p) \leqslant 0$. Dropping the norm condition $\|T(t)\| \leqslant 1$ leads to the *weak maximum condition* which requires $\Omega f(p) \leqslant 0$ only at points p such that $f(p) = 0$ and $f \leqslant 0$ in a neighborhood of p.

Operators of local character with the maximum property promise to be a fertile analogue in topological spaces of the Laplacian and general elliptic operators. It is a challenging problem (now within reach) to find a canonical expression for the operators of this class. A complete answer exists only in one dimension[9], but unpublished results of McKean point toward a solution in Euclidean spaces.

The notion of *induced boundaries* plays an important role in this connection. To explain it, consider an operator Ω of local character with the maximum property in a one-dimensional interval I. It is easily seen that Ω can have at most two independent annihilators. We may suppose, if necessary, that the domain of Ω has been extended so that Ω possesses the maximal number of annihilators compatible with its definition. We say then that the point $p \in I$ is regular if in a neighborhood N of p there exist two independent annihilators, that is, if the local topology induced by Ω agrees with the given topology. At a point p near which there exists only one annihilator, Ω induces a topology in which each interval has but one boundary point (is half open). This is reflected both in the analytic and the probabilistic properties of the corresponding processes. For example, all paths starting at a deficient point go in one direction (see Dynkin[3]). Similarly, in an arbitrary topological space D it will be desirable to avoid singularities and pathologies by requiring that in the interior of D the *local topology* induced by Ω agrees with the given topology.

An indication of the general character of our operators is obtained from the *canonical form* of an operator Ω of local character with the positive maximum property in an open interval I without singular points. It is given by the following theorem[9]. The interval I can be parametrized by a 'canonical scale' x in such a way that each f in the domain of Ω has one-sided derivatives with respect to x and they are continuous except for jumps. (We denote them indiscriminately by f'.) Moreover, there exist two Borel measures m and γ in I, the m-measure of each interval being positive, such that for each f in the domain of Ω,

$$\Omega f . dm = df' - f d\gamma \tag{17}$$

in the sense that the integrals are equal. Of course, if m and γ are absolutely continuous this canonical form reduces to $\Omega f = af'' - cf$. However, this traditional form requires artificial restrictions on the coefficients, whereas (17) is of an intrinsic nature and its theory is simpler and more flexible. Thus, in the equation of the vibrating string our m represents the mechanical mass and γ an attractive elastic force[10]. The form (17) now covers cases where either the mass or the force are concentrated at individual points, which are usually treated by artificial passages to the limit, whereas the generalized form of Ω makes available once and for all the basic existence and expansion theorem.

6-2

4.3. Lateral conditions. A *conditioning set* Σ for Ω is a set such that for $f \in \mathbf{C}$ and each $\lambda > \lambda_0$ there exists exactly one solution $F \in \Sigma$ of

$$\lambda F - \Omega F = f. \tag{18}$$

Using the Hille–Yosida theory it can be shown that each semigroup generated by Ω corresponds to a conditioning set. For example, the minimal semigroups generated by $\Omega = \omega\Delta$ (§ 2.3) correspond to restricting the domain of Ω to functions vanishing at the active boundary. The problem 'to find a semigroup generated by the differential operator Ω restricted by the conditioning set Σ' is the rigorous formulation of the vague and not always soluble problem of 'integrating the differential equation $u_t = \Omega u$ with the boundary condition $u \in \Sigma$'.

For simplicity consider an operator Ω such that for $\lambda > 0$ the equation $\lambda\xi - \Omega\xi = 0$ admits exactly m independent solutions $\xi \in \mathbf{C}$. Then the *active boundary* A induced by Ω consists of m points $\beta^{(1)}, \ldots, \beta^{(m)}$. For example, if Ω is a second-order differential operator in one dimension, say of the form (17), $m \leqslant 2$. All F in the domain of Ω will be continuous in $D \cup A$. The solution $F^{\min}$ of (18) corresponding to the *minimal* semigroup generated by Ω satisfies the boundary condition $F^{\min}(\beta^{(j)}) = 0$. Every other solution is of the form

$$F = F^{\min} + \sum_{j=1}^{m} \Phi^{(j)}\xi^{(j)}, \tag{19}$$

where $\xi^{(j)}$ is the solution of $\lambda\xi - \Omega\xi = 0$ such that $\xi^{(j)}(\beta^{(k)}) = 0$ or 1 according as $j \neq k$ or $j = k$, and where $\Phi^{(j)}$ is a certain functional of f.

Our problem now consists in determining these arbitrary functionals in their dependence on λ in such a way that *the range* Σ *of the transformation* $f \to F$ *is independent of* λ, and to describe all possible such ranges (conditioning sets). In [6] and [8] this problem is solved by a method of wide applicability. It is interesting that even in the case of differential equations the lateral condition $F \in \Sigma$ need not be of local character; hence the classical notion of boundary condition is too restrictive.

The *adjoint* semigroup is generated by an operator Ω^* acting on measures. It is noteworthy that even for a differential operator Ω the adjoint Ω^* need not be of local character. It appears, however, that for the minimal semigroup Ω^* shares the local character of Ω.

It is customary to consider (in Euclidean spaces) only differential operators Ω which are symmetric or so nearly symmetric that Ω and Ω^* induce the same boundary. New and interesting phenomena occur for truly unsymmetric operators. An abstract formulation of many classical problems is to find semigroups generated by Ω whose adjoint is generated by a given formal adjoint Ω^* of Ω.

Suppose that the minimal semigroup generated by Ω has an adjoint generated by Ω^*. Let the *adjoint boundary* consist of n points $\gamma^{(1)}, \ldots, \gamma^{(n)}$ corresponding to n independent solutions $\eta^{(k)}$ of $\lambda\eta - \Omega^*\eta = 0$. For the semigroup corresponding to (19) to be generated by Ω^* it is necessary and sufficient that each functional $\Phi^{(j)}$ is a linear combination of the $\eta^{(k)}$, so that the general solution of our problem involves mn free parameters to be determined in such a way that the range of the transformation (19) be independent of λ.

The solution of this problem given in [6] shows clearly the abstract generalization of the *normal derivatives* appearing in the mixed boundary value problem for harmonic functions. (One advantage of the abstract approach is to derive the boundary conditions in a form that is always meaningful, whereas the classical normal derivatives impose a regularity condition on the boundary.)

The general solution seems also to indicate that the topological question whether $\beta^{(j)}$ and $\gamma^{(k)}$ have disjoint neighborhoods is related to the behavior of the inner product of $\xi^{(j)}$ with $\eta^{(k)}$ as $\lambda \to \infty$ (see § 3.2). However, the precise way in which the relations between the two topologies induced by Ω are reflected in the analytical properties of Ω is an open and promising problem.

REFERENCES

[1] Brelot, M. Le problème de Dirichlet. Axiomatique et frontière de Martin. *J. Math. pures appl.* (9), 35, 297–335 (1956).

[2] Derman, C. A solution to a set of fundamental equations in Markov chains. *Proc. Amer. Math. Soc.* 5, 332–334 (1954).

[3] Dynkin, E. B. The infinitesimal operators of Markov processes. [In Russian, English summary.] *Teoria Verojatnostei i le Primenenija*, 1, 38–60 (1956).

[4] Elliott, J. The boundary value problems and semi-groups associated with certain integro-differential operators. *Trans. Amer. Math. Soc.* 75, 300–331 (1954).

[5] Feller, W. Boundaries induced by non-negative matrices. *Trans. Amer. Math. Soc.* 83, 19–54 (1956).

[6] Feller, W. On boundaries and lateral conditions for the Kolmogorov differential equations. *Ann. Math.* (2), 65, 527–570 (1957).

[7] Feller, W. The general diffusion operator and positivity preserving semigroups in one dimension. *Ann. Math.* (2), 60, 417–436 (1954).

[8] Feller, W. Generalized second order differential operators and their lateral conditions. *Ill. J. Math.* 1, 459–504 (1957).

[9] Feller, W. On the intrinsic form for second order differential operators. *Ill. J. Math.* 2, 1–18 (1958).

[10] Feller, W. On the equation of the vibrating string. *J. Math. Mech.* 8, 339–348 (1959).

[11] Feller, W. The parabolic differential equation and the associated semi-groups of transformations. *Ann. Math.* (2), 55, 468–519 (1952).

[12] Feller, W. Zur Theorie der stochastischen Prozesse. *Math. Ann.* 113, 113–160 (1936).

[13] Hille, E. and Phillips, R. S. *Functional Analysis and Semi-groups.* Revised edition. Amer. Math. Soc. Colloquium Publ. no. 31 (1957).

[14] Hunt, G. A. Markoff processes and potentials. I–III. *Ill. J. Math.* 1, 44–93 (1957); 2, 151–213 (1958).

[15] Kadison, R. V. A representation theory for commutative topological algebra. *Mem. Amer. Math. Soc.* no. 7 (1951).

[16] Kolmogoroff, A. Zur Theorie der Markoffschen Ketten. *Math. Ann.* 112, 155–160 (1935).

[17] Martin, R. S. Minimal positive harmonic functions. *Trans. Amer. Math. Soc.* 49, 137–172 (1941).

[18] Ray, D. Stationary Markov processes with continuous paths. *Trans. Amer. Math. Soc.* 82, 452–493 (1956).

[*Added in proof*

[19] Doob, J. L. Discrete potential theory and boundaries. To appear in the *J. Math. Mech.*]

A Renewal Theorem

WILLIAM FELLER & S. OREY*

In this note we present a rather simple proof of a renewal theorem which contains known theorems[1] and gives new information in case the random variables do not possess expectations. The proof is based on a Fourier representation generalizing that obtained by D. G. KENDALL [2] for denumerable Markov chains.

Let $\{X_n\}$, $n = 1, 2, \cdots$, be a sequence of mutually independent random variables with a common probability distribution F (considered as measure on R_1). We exclude the case of lattice variables, that is, we assume that F does not concentrate the entire mass on the multiples of a real number. In the lattice case the theorem requires an obvious modification, but the proof is simpler.

Let $X_0 = 0$ and $S_n = \sum_0^n X_k$ for $n = 0, 1, \cdots$. The probability distribution of S_n is given by the n-fold convolution F^{n*}. If A is a Borel set on the real line and x is a real number, the sets $-A$ and $A + x$ are defined in the obvious way by symmetry and by translation. For any interval I we put formally

$$U(I) = \sum_0^\infty F^{n*}(I) \tag{1}$$

with the understanding that the series may diverge.

Theorem. *There exist only two possibilities. Either $U(I) = \infty$ for every bounded interval $I = (-h, h)$, $h > 0$, or else there exists a number $0 \leqq \alpha < \infty$ such that for every bounded interval I*

$$\lim_{x \to \infty} U(x + I) = \alpha \mid I \mid \tag{2}$$

(where $\mid I \mid$ denotes the length of I). In particular, if

$$m = \int_{-\infty}^{+\infty} xF(dx) > 0, \tag{3}$$

then $\alpha = m^{-1}$; in all other cases $\alpha = 0$.

---

* Research connected with projects sponsored by the OOR at Princeton University and by the Office of Scientific Research at the University of Minnesota.

[1] For an account of previous work and references see [3].

Journal of Mathematics and Mechanics, Vol. 10, No. 4 (1961).

If $U(I) = \infty$ the sequence $\{S_n\}$ will be called *interval recurrent*; in this case with probability one S_n enters every open interval infinitely often (see [1]). Needless to say, for reasons of symmetry a relation analogous to (2) holds as $x \to -\infty$, with $\alpha = |m|^{-1}$ if $m < 0$.

We begin by proving an analogous theorem for the symmetric measure V defined by

$$V(I) = \tfrac{1}{2}\{U(I) + U(-I)\}. \tag{4}$$

Lemma 1. *Either* $V(I) = \infty$ *for every bounded interval* $I = (-h, h)$, $h > 0$, *or there exists a* β *with* $0 \leqq \beta < \infty$ *such that as* $x \to \infty$

$$V(I + x) \to \beta \, |I|. \tag{5}$$

Proof. For $0 < s < 1$ put

$$V_s(I) = \tfrac{1}{2} \sum_0^\infty s^n[F^{n*}(I) + F^{n*}(-I)]. \tag{6}$$

Then V_s represents a finite symmetric measure on the real line. If φ denotes the Fourier transform of F, then the Fourier transform of V_s is given by

$$\begin{aligned}\int_{-\infty}^{+\infty} e^{izx} V_s(dx) &= \tfrac{1}{2} \sum_0^\infty s^n[\varphi^n(z) + \varphi^n(-z)] \\ &= \Re \frac{1}{1 - s\varphi} = \Re \frac{1 - s\varphi}{|1 - s\varphi|^2} \geqq 0.\end{aligned} \tag{7}$$

Let $g \in L_1(-\infty, +\infty)$ be an even real function such that its transform

$$\gamma(z) = \int_{-\infty}^{+\infty} e^{izx} g(x)\, dx \tag{8}$$

is also in $L_1(-\infty, +\infty)$. We have then the Parseval identity

$$\int_{-\infty}^{+\infty} g(t - x) V_s(dx) = \frac{1}{2\pi} \int_{-\infty}^{+\infty} e^{-itz} \gamma(z) \Re \frac{1}{1 - s\varphi(z)}\, dz. \tag{9}$$

Choosing, in particular,

$$\begin{aligned} g(x) &= \frac{3}{2h}\left(1 - \frac{|x|}{h}\right)^2 \quad \text{for} \quad |x| \leq h \\ &= 0 \qquad\qquad\qquad\qquad \text{for} \quad |x| \geqq h \end{aligned} \tag{10}$$

we have

$$\gamma(z) = \frac{6}{h^2 z^2}\left(1 - \frac{\sin zh}{zh}\right), \tag{11}$$

and we get from (9) for $t = 0$

$$\frac{3}{2h} \int_{-h}^{h} \left(1 - \frac{|x|}{h}\right)^2 V_s(dx) = \frac{3}{\pi h^2} \int_{-\infty}^{+\infty} \left(1 - \frac{\sin zh}{zh}\right) z^{-2}\, \Re \frac{1}{1 - s\varphi(z)}\, dz. \tag{12}$$

From this we see that

$$\int_{|z|>2h^{-1}} \frac{1}{1+z^2} \Re \frac{1}{1-s\varphi(z)} dz \leq \pi h \int_{-h}^{h} \left(1 - \frac{|x|}{h}\right)^2 V_s(dx) \tag{13}$$

$$\leq \pi h V_s((-h, h)).$$

Suppose now that there exists an interval $I_0 = (-h_0, h_0)$ such that

$$V(I_0) = \lim_{s \to 1} V_s(I_0) < \infty. \tag{14}$$

In this case the left side in (13) tends to 0 as $h \to 0$. Moreover, since F is not of the lattice type, $(1 - \varphi)^{-1}$ remains bounded in every compact interval excluding the origin. We conclude thus from (13), letting $s \to 1$:

In order that (14) *be true for some* $h_0 > 0$ *it is necessary and sufficient that the measure whose density is*

$$\frac{1}{1+z^2} \Re \frac{1}{1-s\varphi(z)}$$

converges (weakly and in variation) to a bounded measure as $s \to 1$. *In every interval excluding the origin the limit measure is automatically absolutely continuous with density given by*

$$\frac{1}{1+z^2} \Re \frac{1}{1-\varphi(z)};$$

the weight assigned by the limit measure to the origin will be denoted by ν $(0 \leq \nu < \infty)$.

If this convergence takes place, then the left side of (12) remains bounded for *every* $h > 0$, and hence (14) implies that $V(I) < \infty$ for every bounded interval. Thus (14) *is necessary and sufficient for interval recurrence*, and we have found a *criterion for interval recurrence.*[2]

If (14) holds for some $h > 0$, we have from (9) the following *Fourier representation*[3] *for* V: *For each even* g *with two continuous derivatives and vanishing for* $|x| > h$

$$\int_{-\infty}^{+\infty} g(t-x) V(dx) = \frac{1}{2\pi} \int_{-\infty}^{+\infty} \cos tx \cdot \gamma(x) \cdot \Re \frac{1}{1-\varphi(x)} dx + \frac{\nu}{2\pi} \gamma(0), \tag{15}$$

[2] In [1] a criterion is given equivalent to the convergence of the measure with density $\Re(1 - s\varphi)^{-1}$ to a finite measure in some neighborhood of the origin. We see that this implies such convergence for the density

$$\frac{1}{1+z^2} \Re \frac{1}{1-s\varphi(z)}$$

on the entire line.

[3] In [2] KENDALL gives an integral representation for the transition probabilities $P_{ij}(t)$ of a denumerable Markov chain. In the case $i = j$, j stable, his result follows from formula (15). To see this let F be the probability distribution of the time between the first and second departure from j, A the (exponential) distribution of the time elapsed from 0 to the first change away from j, and B the distribution of the negative of the (exponentially distributed) time one must wait between t and the next departure from j. Then $P_{jj} = A * V * B$ and (15) applies with $g = A * B$.

with

$$\gamma(0) = \int_{-h}^{h} g(x)\,dx. \tag{16}$$

From (15) the lemma follows easily. By the Riemann-Lebesgue lemma

$$\lim_{t\to\infty} \int_{-h}^{h} g(x)\,V(dx + t) = \frac{\nu}{2\pi} \int_{-h}^{h} g(x)\,dx. \tag{17}$$

Since the twice differentiable functions g are dense among continuous functions, (17) holds for every even continuous function g with $g(h) = 0$. It follows that the measure on $(-h, h)$ defined for fixed t by $V(t + x)$ converges, as $t \to \infty$, to the measure with density $\nu(2\pi)^{-1}$, which proves the lemma with $\beta = \nu(2\pi)^{-1}$.

Incidentally, the proof shows that if, and only if, $\{S_n\}$ is interval recurrent,

$$\lim_{\epsilon\to 0} \lim_{s\to 1} \int_{-\epsilon}^{\epsilon} \Re \frac{1}{1 - s\varphi(z)}\,dz = \infty. \tag{18}$$

Straightforward calculation gives[4]

$$\nu = \pi \mid m \mid^{-1} \quad \text{if} \quad 0 < \mid m \mid < \infty, \tag{19}$$

and $\nu = 0$ if the X_i are non-negative and have an infinite expectation. It will follow from the proof of the main theorem that $\nu = 0$ in all other cases.

Proof of the theorem. We suppose that $\{S_n\}$ is not interval recurrent, and that the X_i are neither positive nor negative variables, for otherwise nothing is to be proved. In other words, if $\Delta = [0, \delta)$ we assume that for some Δ we have $0 < U(\Delta) < \infty$ and also $0 < U(-\Delta) < \infty$.

We recall that for any real Borel set $U(B) = \sum_0^\infty \mathrm{P}\{S_n \in B\}$ is the expected number of times S_n visits B. We shall also require the probability

$$W(B) = \mathrm{P}\left\{\bigcup_{n=0}^{\infty} \{S_n \in B\}\right\} \tag{20}$$

that $S_n \in B$ for some $n \geqq 0$. Finally, for any interval I and a Borel set $B \subset I$ let $W_I(B)$ be the probability that the first entry into I occurs at a point $x \in B$. Then W_I is a measure in I and $W_I(I) = W(I)$. For a fixed $\delta > 0$ let $\Delta = [0, \delta]$ and $\Delta^* = \Delta \cup (-\Delta) = (-\delta, \delta)$. If I is any interval of length δ, we have

$$U(I) = \int_I W_I(dx) U(I - x) \leqq U(\Delta^*)\cdot W(I), \tag{21}$$

showing that for all intervals of length δ

$$W(I) \leqq U(I) \leqq \omega\cdot W(I), \qquad \omega = U(\Delta^*). \tag{22}$$

The lemma asserts $V(\Delta + t_n) \to \beta\delta$ for each sequence $t_m \uparrow \infty$. If

$$U(\Delta + t_n) \to \lambda\,\delta, \tag{23}$$

[4] The hypothesis that the mean is m is used only to infer $\varphi'(0) = m$. We see that if $\varphi'(0) = m$ and $\{S_n\}$ is not recurrent, the first moment of F must exist and equal m.

then clearly $\lambda \leqq 2\beta$. If $\lambda = 2\beta$ for every choice of $t_m \uparrow \infty$, then the theorem holds with $\alpha = 2\beta$. If $\lambda = 0$ for every choice, the theorem holds with $\alpha = 0$. Accordingly, we assume that there exists a $t_n \uparrow \infty$ such that (23) holds with $0 < \lambda < 2\beta$.

Now from (22)

$$\liminf_{n\to\infty} W(\Delta + t_n) \geqq \omega^{-1}\cdot\lambda\,\delta > 0. \tag{24}$$

Also, since

$$2V(\Delta + t_n) = U(\Delta + t_n) + U(-\Delta - t_n) \to 2\beta\,\delta,$$

we have

$$U(-\Delta - t_n) \to (2\beta - \lambda)\,\delta, \tag{25}$$

and hence

$$\liminf_{n\to\infty} W(-\Delta - t_n) \geqq \omega^{-1}\cdot(2\beta - \lambda)\,\delta > 0. \tag{26}$$

Now let

$$B_n = \bigcup_{m=0}^{\infty}\bigcup_{j=0}^{\infty}\{S_m \geqq t_n\ ,\ S_{m+j}\ \varepsilon\ \Delta^*\}$$

be the event that $S_m \geqq t_n$ for some m and that $-\delta < S_k < \delta$ for a later index k. Clearly $B_m \supset B_{m+1}$ and

$$\mathrm{P}\{B_n\} \geqq \int_{\Delta+t_n} W_{\Delta+t_n}(dx)W(\Delta^* - x) \geqq W(-\Delta - t_n)W(\Delta + t_n). \tag{27}$$

Thus (24), (26), and (27) together show that $\mathrm{P}\{\cap B_n\} > 0$. This, however, implies that $\{S_n\}$ is interval recurrent, contradicting our assumption. Thus whenever (23) holds $\lambda = 0$ or $\lambda = 2\beta$. If there existed a choice of $t_m \uparrow \infty$ such that (23) held with $\lambda = 0$ and another choice of $t_m \uparrow \infty$ satisfying (23) with $\lambda = 2\beta$, then it is not hard to see that there could be interpolated a third $t_m \uparrow \infty$ making (23) true for some $0 < \lambda < 2\beta$. This shows that *as* $t \uparrow \infty$ *either*

$$U(\Delta + t) \to 2\beta\,\delta \quad \textit{and} \quad U(-\Delta - t) \to 0 \tag{28}$$

or

$$U(\Delta + t) \to 0 \quad \textit{and} \quad U(-\Delta - t) \to 2\beta\,\delta. \tag{29}$$

It is well known that if a finite expectation m exists, $\{S_n\}$ is interval recurrent in case $m = 0$; according as $m > 0$ or $m < 0$ either (28) or (29) holds with $2\beta = |m|^{-1}$. It is easily verified that $\beta = 0$ if $m = \pm\infty$ in the sense that the integral converges at one limit, diverges at the other. To conclude the argument we shall show that $\beta = 0$ *also in the case where the expectation fails to exist* in the sense that the integral diverges at both limits.

For reasons of symmetry we may suppose that (28) holds. The assertion $\beta = 0$ is then trivial unless $\mathrm{P}\{\limsup S_n = \infty\} = 1$. In this case let $N_0 = 0$

and let N_{n+1} be the least integer k such that $S_k > S_{N_n}$. Then with probability one the N_n are defined for all n, and $S_{N_n} = Z_n$ defines a sequence of random variables. The random variables $Z_{n+1} - Z_n$ are mutually independent, positive, and identically distributed. Since $P\{Z_1 > x\} \geqq P\{X_1 > x\}$, the expectation of $Z_{n+1} - Z_n$ is infinite. Hence the theorem applies to $Z_{n+1} - Z_n$ with $\alpha = 0$ and shows that as $t \to \infty$

$$P\left\{\bigcup_{n=0}^{\infty} \{Z_n \varepsilon \Delta + t\}\right\} \to 0. \tag{30}$$

Let $w(t)$ be the least integer k such that $S_k > t$. Then (30) implies that

$$P\{S_{w(t)} > t + \delta\} \to 1 \tag{31}$$

for each $\delta > 0$. Now

$$W(\Delta + t) = P\left\{\bigcup_{n=0}^{\infty} \{S_n \varepsilon \Delta + t \mid S_{w(t)} \leqq t + \delta\}\right\} \cdot P\{S_{w(t)} \leqq t + \delta\}$$
$$+ P\left\{\bigcup_{n=0}^{\infty} \{S_{n+w(t)} \varepsilon \Delta + t \mid S_{w(t)} > t + \delta\}\right\} \cdot P\{S_{w(t)} > t + \delta\}. \tag{32}$$

As a consequence of (31) the first term on the right tends to zero as $t \to \infty$. The second term on the right is bounded by

$$\sup_{h>\delta} P\left\{\bigcup_{n=0}^{\infty} \{S_n \varepsilon \Delta - h\}\right\} \leqq \sup_{h>\delta} U(\Delta - h). \tag{33}$$

In view of (28) this can be made arbitrarily small by choosing δ large. Thus $\beta = 0$, and we have proved the

Corollary. *In the non-recurrent case we have either* (28) *or* (29), *and the necessary and sufficient condition for* $\beta > 0$ *is the existence of a finite expectation with* $|m| > 0$.

References

[1] K. L. Chung & W. H. J. Fuchs, *On the distribution of values of sums of random variables*, Amer. Math. Soc. Memoir **6** (1951) pp. 1–12.

[2] D. G. Kendall, *Integral representation for Markov transition probabilities*, A. M. S. Bull. **42** (1958) pp. 358–362.

[3] W. L. Smith, *Renewal theory and its ramifications*, J. of the Royal Stat. Soc., Ser. B, 20 (1958), pp. 243–302.

Princeton University
and
University of Minnesota

COMMUNICATIONS ON PURE AND APPLIED MATHEMATICS, VOL. XIV, 285–293 (1961)

K. O. Friedrichs anniversary issue

A Simple Proof for Renewal Theorems*

WILLIAM FELLER

1. Introduction

Let F be a one-dimensional distribution function, that is, a right-continuous non-decreasing function defined on the extended real line with $F(-\infty)=0$, $F(\infty)=1$. We shall generally use the same letter for a distribution function and the associated measure, but when the argument is a set it will be put into braces. Thus for $-\infty \leqq a < b \leqq \infty$ if $\Delta = (a, b]$ we write $F\{\Delta\} = F(b)-F(a)$.

We say that F is *arithmetic* if all points of increase occur at multiples of a fixed number $d > 0$. The distribution F is called *transient* if

$$U\{\Delta\} = \sum_{n=0}^{\infty} F^{n*}\{\Delta\} \tag{1.1}$$

is finite for every bounded interval[1]; as usual, F^{n*} denotes the n-fold convolution of F with itself. For a continuous g with compact support we may form the convolution

$$\varphi = g * U \tag{1.2}$$

to obtain a continuous solution of the integral equation

$$\varphi = g + \varphi * F. \tag{1.3}$$

When $g > 0$, successive approximations show that (1.2) represents the *minimal* solution of (1.3) and that a solution exists only if F is transient. By Lemma 2.1 the solution φ is unique up to an additive constant, except when F is arithmetic (in which case an arbitrary function of period d may be added to φ). A distribution F supported by the positive half-line (that is, for which $F(0) = 0$) is always transient. If also $g(x) = 0$ for $x < 0$, then (1.3) reduces to the Volterra equation

*The research for this paper was supported in part by the U. S. Army Ordnance, Office of Ordnance Research, Contract DA-36-034-ORD-2001. Reproduction in whole or in part permitted for any purpose of the United States Government.

[1]According to an often used theorem (Chung-Fuchs [5]) it suffices that (1.1) converges for *one* open interval containing the origin. This fact follows also from (4.6).

$$\varphi(x) = g(x) + \int_0^x \varphi(x-y)F\{dy\}. \tag{1.4}$$

Various Tauberian theorems have a bearing on the asymptotic behavior of the solution φ at infinity. The sharpest result is given by the *renewal theorem for distributions on the positive half-line*[2]: *φ is bounded. In the non-arithmetic case as $x \to \infty$*

$$\varphi(x) \to \mu^{-1}E(g); \tag{1.5}$$

in the arithmetic case as $n \to \infty$

$$\varphi(nd) \to \mu^{-1}dE(g). \tag{1.6}$$

Here

$$E(g) = \int g(x)dx, \qquad \mu = \int_0^\infty xF\{dx\}, \tag{1.7}$$

and μ^{-1} is to be interpreted as 0 if $\mu = \infty$. In terms of the measure U we may restate (1.5) as follows: *for each interval Δ of finite length $|\Delta|$ as $t \to \infty$*

$$U\{\Delta + t\} \to \mu^{-1} \cdot |\Delta|. \tag{1.8}$$

(As usual, $\Delta+t$ is the interval obtained by translating Δ through t.) In the arithmetic case U attributes weights u_n to the multiples of d and $u_n \to d\mu^{-1}$.

We shall give a surprisingly simple proof of this theorem having the advantage that it does not distinguish between arithmetic and non-arithmetic distributions. The basic feature applies also to the more complicated case of distributions with two tails, and to several dimensions. It seems to be of wider applicability [12].

The bilateral case has been treated by several authors [3] by different methods. In its full generality the following theorem has been proved only recently by Feller and Orey [9] using a combination of Fourier analytic and probabilistic methods.

THEOREM. *For an arbitrary transient F we have $U\{\Delta+t\} \to 0$ as $t \to \pm\infty$ for each bounded interval Δ except if the moment*

$$\mu = \int_{-\infty}^{+\infty} xF\{dx\} \tag{1.9}$$

---

[2] Proved in 1948 by Erdös-Feller-Pollard [8] for arithmetic distributions and by Blackwell [1] for non-arithmetic ones. The arithmetic case can also be derived from Kolmogorov's results [10] on Markov chains. A huge older literature proves the theorem under additional assumptions on F.

[3] For arithmetic distributions with one-sided finite moment by Chung and Wolfowitz [7]; for non-arithmetic distributions with finite moment and an absolutely continuous component Chung and Pollard [6]; for arbitrary F with finite one-sided moment Blackwell [2] and Harris [unpublished].

exists. In this case necessarily $\mu \neq 0$ *and for non-arithmetic distributions*

$$U\{\varDelta+t\} \to \begin{cases} |\mu^{-1}| \cdot |\varDelta| & \text{as} \quad \mu t \to \infty, \\ 0 & \text{as} \quad \mu t \to -\infty. \end{cases} \tag{1.10}$$

(The obvious amendment holds for arithmetic distributions.)

Our simple proof requires no essential change if μ is finite, and applies even in the case where at least one tail in (1.9) converges (see the end of Section 5). Unfortunately a more intricate argument is required in the case where both tails diverge, and it seemed most natural to formulate it probabilistically in terms of the associated random walk. It will be seen that most of Section 5 is devoted to this argument.

With the present arrangement of the proof we obtain as an immediate consequence the following generalization of a theorem proved by Chung [4] in the case of finite moments.

THEOREM. *Let* F *be an* r*-dimensional distribution function,* $r \geqq 2$*, with transient marginal distributions. Then for each fixed vector* $\mathbf{c}$ *and each bounded interval* $\varDelta$ *we have*

$$U\{\varDelta+t\mathbf{c}_n\} \to 0 \quad \text{as} \quad t \to \infty. \tag{1.11}$$

2. Basic Lemmas

The following lemma is easily verified and known, but the only reference seems to be to Choquet and Deny [3] who prove a generalization to arbitrary groups.

LEMMA 2.1. *If* z *is a bounded solution of* $z = z * F$*, then* $z(x) =$ const. *in the non-arithmetic case, and* $z(nd) =$ const. *in the arithmetic case.*

As usual we put $||g|| = \sup |g(x)|$.

LEMMA 2.2. *Let* $\varDelta = (0, h)$ *be fixed and suppose that there exists a constant* c *such that for each continuous* g *supported by* $\varDelta$ *the solution* $\varphi = g * U$ *of* (1.3) *satisfies*

$$||\varphi|| \leqq c \cdot ||g||. \tag{2.1}$$

Suppose that g *has a continuous derivative* g'*. Then for each fixed* x

$$\varphi(t+x) - \varphi(t) \to 0 \tag{2.2}$$

as $t \to \pm\infty$*, except that in the case of an arithmetic distribution* t *must be restricted to multiples of* $\pm d$.

Proof: Define a family of functions z_t by

$$z_t(x) = \varphi(t+x). \tag{2.3}$$

Since differentiation commutes with convolution we have $\|\varphi'\| \leqq c \cdot \|g'\|$, and so the family $\{z_t\}$ is equicontinuous. Out of any denumerable subset we may therefore choose a sequence $\{z_{t_n}\}$ such that

$$z_{t_n}(x) \to z(x) \tag{2.4}$$

uniformly in every bounded interval. From (1.3) we see that $z = z * F$, and the assertion now follows from Lemma 1.

COROLLARY. *Suppose that for a particular sequence* $\{t_n\}$ *tending to* $\pm\infty$ *and a fixed interval* Δ

$$U\{\Delta+t_n\} \to \gamma \cdot |\Delta|. \tag{2.5}$$

Then $U\{\Gamma+t_n\} \to \gamma \cdot |\Gamma|$ *for every finite interval* Γ.

Proof: By (2.1) we may consider $\varphi(t)$ as a bounded linear functional on g (defined in Δ). It equals the expectation of g with respect to the restriction of the measure U to the interval $\Delta+t$. Any limiting functional must commute with translations, and so any convergent sequence $\{\varphi(t_n)\}$ necessarily converges to a multiple of the ordinary integral of g. Now (2.5) implies that the assertion is true of every interval $\Gamma = \Delta+x$. In view of the additivity of both sides we conclude that the assertion holds for every interval whose length is a rational multiple of $|\Delta|$, and hence for all Γ.

3. Proof for the Half-Line

We suppose $F(0) = 0$ and so φ satisfies (1.4). Since convolutions commute with translations we may suppose that g is supported by the interval $(0, h)$. For $x > h$ the value $\varphi(x)$ is then a (defective) average of the values assumed by φ in Δ, and so the maximum[4] of φ is assumed in Δ. Thus Lemma 2.2 applies. Integrate (1.4) from 0 to $t > h$. After an integration by parts we get

$$\int_0^t [1-F(y)]\varphi(t-y)dy = E(g). \tag{3.1}$$

Letting $t \to \infty$ through a sequence such that $\varphi(t-y) \to \xi$ for some (and therefore all) y, we find

$$\xi \int_0^\infty [1-F(y)]dy = E(g). \tag{3.2}$$

The integral on the left equals μ, and (2.5) is proved.

[4] It suffices to consider the case where $F(h) < 1$. Then $\|\varphi\| \leqq \|g\| \cdot (1-F(h))^{-1}$.

4. Hitting Probabilities in Random Walks

It is often difficult to decide whether a distribution is transient; and even if (1.3) is known to have a solution, the boundedness condition of Lemma 2.2 is by no means obvious. We must therefore depend on deeper tools. Formulated purely analytically they would appear exceedingly intricate, but they become intuitively obvious when formulated probabilistically.

Consider a sequence $X_1, X_2, \cdots$ of mutually independent random variables with the common distribution F. (Analytically the X_j are the coordinate variables in an infinitely dimensional space with a product measure induced by F.) We put

$$S_0 = 0, \qquad S_n = X_1 + \cdots + X_n, \tag{4.1}$$

and refer to S_n as the position at time n of a "particle" in a random walk. Then $F^{n*}\{\Delta\} = P\{S_n \in \Delta\}$, and $U\{\Delta\}$ is the expected number of integers n such that $S_n \in \Delta$.

For an arbitrary interval Δ we define

$$\rho(\Delta) = P\{S_n \in \Delta \text{ for some } n\} \tag{4.2}$$

as the probability that the random walk ever enters Δ. We shall also use the measure $\rho(\Delta, dx)$, supported by Δ, giving the probability that the *first entry* [5] into Δ occurs at x:

$$\rho\{\Delta, \Gamma\} = \sum_{n=1}^{\infty} P\{S_n \in \Delta \cap \Gamma, \qquad S_j \notin \Delta \quad \text{for} \quad j = 1, \cdots, n-1\}. \tag{4.3}$$

Then $\rho(\Delta) = \rho\{\Delta, \Delta\}$.

For any two intervals $\Delta^{(1)}$ and $\Delta^{(2)}$ denote by $\Delta^{(1)} \dot{+} \Delta^{(2)}$ the interval containing all points of the form $x_1 + x_2$ with $x_k \in \Delta^{(k)}$. Since

$$P\{S_{n+m} \in \Delta^{(1)} \dot{+} \Delta^{(2)}\} \geqq P\{S_n \in \Delta^{(1)}\} \cdot P\{S_m \in \Delta^{(2)}\}, \tag{4.4}$$

we have

$$\rho(\Delta^{(1)} \dot{+} \Delta^{(2)}) \geqq \rho(\Delta^{(1)})\rho(\Delta^{(2)}). \tag{4.5}$$

Finally we recall the frequently used formula

$$U\{\Delta\} = \int_{\Delta} \rho\{\Delta, dx\} U\{\Delta - x\}, \tag{4.6}$$

obtained by considering the random walk from the moment when it enters Δ and taking the point of entry as new origin.

---

[5] Used also in [9].

5. Proof in the General Case

It is convenient to introduce special symbols for three types of intervals as follows. For $h > 0$ we put

$$(5.1) \qquad \Delta_h = (-h, h), \qquad \Delta_h^+ = (0, h), \qquad \Delta_h^- = (-h, 0).$$

In this section we suppose that *F is transient.*

LEMMA 5.1. *The condition of Lemma 2.2 is satisfied for all Δ.*

Proof: We apply (4.6) to $\Delta = t+\Delta_h$. For each x in the domain of integration $\Delta - x \subset \Delta_{2h}$ and therefore,

$$(5.2) \qquad U\{t+\Delta_h\} \leqq \rho(t+\Delta_h)\, U\{\Delta_{2h}\} \leqq U\{\Delta_{2h}\}.$$

This provides a uniform bound for $U\{\Delta\}$ for all intervals of a fixed length, and so (2.1) holds.

It follows from Lemma 2.2 that

$$(5.3) \qquad \limsup_{t\to\pm\infty} \frac{U\{t+\Delta\}}{|\Delta|} = \lambda$$

is *independent of Δ* and finite. If $\lambda = 0$, there is nothing to be proved, and so we assume $\lambda > 0$.

LEMMA 5.2.[5] *For each $h > 0$*

$$(5.4) \qquad \lim_{t\to\infty} \rho(t+\Delta_h)\cdot\rho(-t+\Delta_h) = 0.$$

Proof: Choose t so large that the probabilities of entering the intervals $\pm t+\Delta_h$ before the n-th step are smaller than $\frac{1}{2}\rho(\pm t+\Delta_h)$. By (4.5) the probability of re-entering Δ_{2h} after time n is then $\geqq \frac{1}{4}\rho(t+\Delta_h)\rho(-t+\Delta_h)$. This probability is smaller than the n-th remainder in the series $U\{\Delta_{2h}\} = \sum F^{n*}\{\Delta_{2h}\}$ and tends to 0 as $n \to \infty$.

In view of (4.6) we have also

$$(5.5) \qquad U\{t+\Delta_h\}\cdot U\{-t+\Delta_h\} \to 0,$$

and so for large t one of the terms is bound to be small. We have to prove that the leading term does not switch sides.

LEMMA 5.3. *As $s \to \infty$ either*

$$(5.6) \qquad s^{-1}U\{\Delta_s^+\} \to \lambda \quad \textit{and} \quad s^{-1}U\{\Delta_s^-\} \to 0,$$

or the corresponding relations hold with Δ_s^+ and Δ_s^- interchanged.

Proof: From (5.5)

$$\limsup_{s\to\infty} \frac{U\{\Delta_s\}}{s} = \limsup_{s\to\infty} \frac{U\{\Delta_s^+\}+U\{\Delta_s^-\}}{s} \leqq \lambda. \tag{5.7}$$

In conjunction with (5.2) this implies that to each fixed $\lambda_1 > \lambda$ there exists a constant a so large that for *every* interval Γ

$$U\{\Gamma\} \leqq a+\lambda_1|\Gamma|. \tag{5.8}$$

We now apply (4.6) to a large interval $\Delta = t+\Delta_s$, where t and s will be chosen appropriately. The domain of integration is $t-s < x < t+s$. For $x < t$ we have $\Delta_s^+ \subset \Delta - x$ and therefore,

$$U\{\Delta-x\} \leqq U\{\Delta_s^+\}+a+\lambda_1 s; \tag{5.9}$$

for $x > t$ the analogous inequality holds with Δ_s^+ replaced by Δ_s^-. *Per absurdum* suppose that we may pick arbitrarily large s such that

$$s^{-1}U\{\Delta_s^+\} < \lambda q \quad \textit{and} \quad s^{-1}U\{\Delta_s^-\} < \lambda q \tag{5.10}$$

with some fixed $q < 1$. From (4.6)

$$U\{t+\Delta_s\} \leqq [a+(\lambda q+\lambda_1)s]\rho(t+\Delta_s). \tag{5.11}$$

Choose λ_1 so that $\lambda q+\lambda_1 < 2\lambda$ and then s so large that $a+(\lambda q+\lambda_1)s < 2\lambda s$. Letting $t \to \infty$ through values such that the left side tends to $2\lambda s$, we are led to a contradiction. Thus for all s sufficiently large one of the quantities in (5.10) will be close to λ, the other close to 0. Since they depend continuously on s the lemma is proved.

We have proved more: the inequality (5.11) holds with $q = \lambda_1$, and we have thus the

COROLLARY.[6] *To each $\varepsilon > 0$ there exists a σ such that for $s > \sigma$*

$$U\{\Delta\} > (1-\varepsilon)\lambda|\Delta| \quad \textit{implies} \quad \rho(\Delta) > 1-2\varepsilon. \tag{5.12}$$

LEMMA 5.4. *If (5.6) holds, then for each fixed interval Δ as $t \to \infty$*

$$U\{t+\Delta\} \to \lambda\cdot|\Delta|, \qquad U\{-t+\Delta\} \to 0. \tag{5.13}$$

Proof: Choose two numbers s_1 and s_2 large enough to apply (5.12). Since $t^{-1}U\{\Delta_t^+\} \to \lambda$ we can, for t sufficiently large, find two points x and $t-x$ such that the condition of (5.12) holds for each of the intervals

[6]To aid the intuition we remark that we have proved much more. With (5.6) the integrand in (4.6) will be small for $x > t$ and so the probability tends to 1 that the first entry occurs in the lower half of the interval $t+\Delta_s$. An obvious sharpening of the estimate shows that for large t and s the first entry is most likely to occur near the left endpoint $t-s$. Intuitively speaking, the random walk drifts upward.

$\Gamma_1 = x+\Delta_{s_1}$ and $\Gamma_2 = t-x+\Delta_{s_2}$. Put $\Gamma = \Delta_{s_1}\dot{+}\Delta_{s_2}$. Then, as in (4.6),

$$(5.14)\qquad U\{t+\Gamma\} = U\{\Gamma_1\dot{+}\Gamma_2\} \geqq \rho(\Gamma_1)U\{\Gamma_2\} \leqq (1-2\varepsilon)^2\cdot\lambda\cdot 2s_2.$$

Thus

$$(5.15)\qquad \liminf U\{t+\Gamma\} \geqq (1-2\varepsilon)^2\frac{s_2}{s_1+s_2}\cdot\lambda|\Gamma|.$$

By Lemma 2.2 the same inequality holds for *every* fixed interval, and since $s_2(s_1+s_2)^{-1}$ may be chosen arbitrarily close to unity this implies the first relation in (5.13). By (5.5) the second is a consequence of the first.

The last lemma is equivalent to our theorem except that we have to identify λ with the first moment of F. It may be worthwhile to reformulate our result so far probabilistically as follows.

THEOREM. *If $\lambda > 0$, then to each $\varepsilon > 0$ there exists an L such that*

$$(5.16)\qquad P\{t < S_n < t+L \text{ for some } n\} > 1-\varepsilon$$

either for all $t > t_0$ or all $t < -t_0$.

LEMMA 5.5. *If* (5.13) *holds, then*

$$(5.17)\qquad \int_{-\infty}^{0}|x|F\{dx\} < \infty.$$

Proof: We know that (5.16) holds for all t sufficiently large. Given that $(\nu-1)L \leqq S_n < \nu L$ the probability that $S_{n+1} < 0$ is $\geqq F(-\nu L)$. For ν sufficiently large the event that the random walk enters the interval $((\nu-1)L, \nu L)$ and immediately following the first entry jumps to a negative value has therefore a probability $> (1-\varepsilon)F(-\nu L)$. When the random walk starts from a negative value it is practically sure to pass through $(0, L)$. If the series $\sum F(-\nu L)$ diverged, the expected number of such passages would be infinite. Therefore the series converges and (5.17) is true.

CONCLUSION OF THE PROOF

Once we know (5.17) we can proceed as in Section 3 (and do not even require the existence of the limits in (5.13)). Integrating (1.3) and applying an integration by parts leads to

$$(5.18)\qquad \int_0^t g(s)ds = \int_{-\infty}^{+\infty}[H(x)-F(x)]\cdot[\varphi(t-x)-\varphi(-x)]dx,$$

where $H(x) = 1$ and 0 for $x > 0$ and $x < 0$. If (5.17) holds, the left tail in (5.18) remains bounded and $\varphi(-x)\to 0$ as $x\to\infty$. Thus if $\varphi(t)$ tends to a positive limit as $t\to\infty$ the integral of $H-F$ must converge and so F has a first moment. Then

$$[\varphi(\infty)-\varphi(-\infty)]\mu = \int_{-\infty}^{+\infty} g(s)ds \tag{5.19}$$

and the theorem is proved.

6. The Two-Dimensional Case

Let Δ_s be the rectangle of points x, y such that $0 < x < 1$, $0 < y < s$. The renewal theorem for the marginal distribution shows that $U\{\Delta_s+tc\}$ remains bounded as $t \to \infty$, whereas from Lemma 2.2 we conclude that the asymptotic behaviour of the ratio $s^{-1}U\{\Delta_s+tc_n\}$ as $t \to \infty$ does not depend on s. This implies (1.11).

Bibliography

[1] Blackwell, D., *A renewal theorem*, Duke Math. J., Vol. **15**, **1948**, pp. **145–150**.
[2] Blackwell, D., *Extension of a renewal theorem*, Pacific J. Math., Vol. **3**, **1953**, pp. **315–320**.
[3] Choquet, G., and Deny, J., *Sur l'équation de convolution* $\mu = \mu * \sigma$, C. R. Acad. Sci. Paris, Vol. 250, 1960, pp. 799–801.
[4] Chung, K. L., *On the renewal theorem in higher dimensions*, Skand. Aktuarietidskr., Vol. 35, 1952, pp. 188–194.
[5] Chung, K. L., and Fuchs, W. H. J., *On the distribution of values of sums of random variables*, Mem. Amer. Math. Soc., No. **6**, **1951**, pp. **1–12**.
[6] Chung, K. L., and Pollard, H., *An extension of renewal theory*, Proc. Amer. Math. Soc., Vol. **3**, **1952**, pp. **303–309**.
[7] Chung, K. L., and Wolfowitz, J., *On a limit theorem in renewal theory*, Ann. of Math., Vol. 55, **1952**, pp. 1–16.
[8] Erdös, P., Feller, W., and Pollard, H., *A theorem on power series*, Bull. Amer. Math., Soc., Vol. 55, **1949**, pp. 201–204.
[9] Feller, W., and Orey, S., *A renewal theorem*, *J.* Math. Mech., to appear.
[10] Kolmogoroff, A., *Anfangsgründe der Theorie der Markoffschen Ketten mit unendlich vielen möglichen Zuständen*, Mat. Sbornik N. S., Vol. **1**, **1936**, pp. 607–610.
[11] Smith, W. L., *Renewal theory and its ramifications*, J. Roy. Statist Soc., Ser. B, Vol. **20**, 1958, pp. **243–302**.
[12] Spitzer, F., *On hitting probabilities*, J. Math. Mech., to appear.

Received February, 1961.

On the Classical Tauberian Theorems*)

By

William Feller

1. Introduction. This note contains an elementary proof of the now classical Tauberian theorem in the refined form due to Karamata. As a further application of the method we derive (section 5) a *ratio* limit theorem of a form recently introduced by Doob [1] (in the "Abelian" version) for a variety of transforms. To elucidate the nature of the Tauberian relationship we derive in section 4 a slightly improved version of the classical theorem for arbitrary functions of bounded variation.

The emphasis of the paper is not on new results, but on the extreme simplicity of the proof. It has the advantage of symmetry, proving in one sweep the so-called "Tauberian" theorems and their "Abelian" converses. More important is that the argument of section 3 leads in a natural and compelling fashion to the peculiar form of Karamata's theorem; it is found to be, in a certain sense, the best possible.

It is surprising that a proof known for its depth and ingenuity may be reduced to so simple terms, in particular since one basic feature of the proof is not new[1]).

We shall formulate the theorems in terms of Laplace transforms. Their equivalent for power series $\sum u_n z^n$ is obtained by the substitution $z = e^{-\lambda}$ and letting U represent a step function with jumps u_n at n $(n = 0, 1, \ldots)$.

2. Slowly varying functions. To avoid later interruptions we collect here a few known facts. A *positive function L defined on* $(0, \infty)$ *is called slowly varying at* ∞ [*or* 0] *if for each* $x > 0$

$$\frac{L(tx)}{L(t)} \to 1 \qquad as \qquad t \to \infty\,[or\ 0]. \tag{2.1}$$

(We shall abbreviate the relation expressed by (2.1) to $L(tx) \sim L(t)$.) If L is slowly varying at ∞ the function defined by $L(t^{-1})$ is slowly varying at 0. In particular, any L tending to a positive limit is slowly varying.

The importance of slowly varying functions is due to the following well-known[2])

---

*) This paper was prepared for a Symposium at Yale University, June 4–6, 1962, in honor of Einar Hille on the occasion of his formal retirement. Research supported by U.S. Army Research Office (Durham).

[1]) See H. König [3] with references to similar arguments in R. Schmidt (1925) and A. Beurling (1945).

[2]) The notion of slowly varying functions is due to Karamata. In [2] he uses an equivalent definition which leads to lemma 1 only by intricate calculations. In the meantime the lemma and

R.L. Schilling et al. (eds.), *Selected Papers II*,

Lemma 1. *Let v be a positive monotone function on $(0, \infty)$. Suppose that for a dense set of points x as $t \to \infty$ [or $t \to 0$]*

$$\lim \frac{v(tx)}{v(t)} = \zeta(x) < \infty \tag{2.2}$$

exists and is not identically 0. *Then*

$$v(x) = x^p L(x), \tag{2.3}$$

where p is a constant and L varies slowly at ∞ [or 0], *and*

$$\frac{v(tx)}{v(t)} \to x^p \tag{2.4}$$

for all positive x. Conversely (2.3) *implies* (2.4).

Proofs. Let Σ be the set for which (2.2) is true. For $x, y \in \Sigma$ we have

$$\frac{v(txy)}{v(t)} = \frac{v(txy)}{v(ty)} \cdot \frac{v(ty)}{v(x)} \tag{2.4}$$

and conclude that $xy \in \Sigma$ and

$$\zeta(xy) = \zeta(x)\zeta(y). \tag{2.5}$$

Hence $\zeta(1) = 1$ and it is seen by induction that $\zeta(e^s) = (\zeta(e))^s$ for all rational s. Since ζ is monotone, this relation holds for all $s > 0$ with $p = \log \zeta(e)$. The remaining assertions are obvious.

Although we shall not use it explicitly it may be interesting to note Karamata's theorem[3]) according to which (2.3) at ∞ implies $v(x) = o(x^{p+\varepsilon})$ and $x^{p-\varepsilon} = o(v(x))$.

3. Positive measures. We consider a measure carried by $[0, \infty)$ and finite on bounded sets and denote by $U(x)$ its value on $(-\infty, x]$. In other words U is a right-continuous non-decreasing function defined on $(-\infty, \infty)$ with $U(x) = 0$ for $x < 0$, and $U(x) > 0$ for some x. Its formal Laplace transform is defined by

$$\omega(\lambda) = \int_0^\infty e^{-\lambda x} U(dx) = \lambda \int_0^\infty e^{-\lambda x} U(x)\, dx. \tag{3.1}$$

It is assumed that it converges for all $\lambda > 0$.

---

its simple proof were noticed independently by several authors. Needless to say that (2.5) is equivalent to the Hamel equation and that therefore the assumed monotonicity of ζ is not necessary: it suffices to assume ζ measurable or bounded on compacts.

3) The theorem is easily proved for non-decreasing v. For, if to a given $\varepsilon > 0$,

$$\frac{v(te)}{v(t)} < e^{\alpha+\varepsilon} \qquad \text{for} \quad t > t_0$$

then for all $y > 1$ and $t > t_0$ putting $e^{n-1} < y \leq e^n$

$$\frac{v(ty)}{v(t)} \leq \frac{v(te^n)}{v(t)} \leq e^{n(\alpha+\varepsilon)} \leq e^{\alpha+\varepsilon} \cdot y^{\alpha+\varepsilon}.$$

For fixed $t > 0$ define functions U_t and ω_t by a simple change of scales:

$$U_t(x) = \frac{U(tx)}{\omega(t^{-1})}, \quad \omega_t(\lambda) = \frac{\omega(t^{-1}\lambda)}{\omega(t^{-1})}. \tag{3.2}$$

Obviously ω_t is the Laplace transform of U_t.

The functions ω_t are non-increasing and $\omega_t(1) = 1$. By the usual compactness argument we may therefore select a sequence $\{t_k\}$ such that $t_k \to \infty$ [or $t_k \to 0$] and such that the sequence $\{\omega_{t_k}\}$ is convergent on the interval $\lambda > 1$ in the weak sense:

$$\omega_{t_k}(\lambda) \to \Phi(\lambda) \tag{3.3}$$

at least at each point of continuity[4]) of Φ. The same compactness argument applied to $\{U_{t_k}\}$ shows[5]) in view of the uniqueness of Laplace transform that

$$U_{t_k} \to F \tag{3.4}$$

and that for $\lambda > 1$ the Laplace transform of F exists and equals $\Phi(\lambda)$. In other words, for almost all x

$$\frac{U(t_k x)}{\omega(t_k^{-1})} \to F(x). \tag{3.5}$$

We shall resist the temptation to restate (3.5) in the form of an "all inclusive Tauberian theorem" and remark merely that (3.5) relates the asymptotic behavior of $U(t)$ to that of $\omega(t^{-1})$. A serious interest attaches to (3.5) only if the limit is independent of the particular choice of the sequence $\{t_k\}$. Now lemma 1 shows that (3.3) holds for all sequences $t_k \to \infty$ [or $\to 0$] if, and only if,

$$\omega(\lambda) = \lambda^{-\alpha} L(\lambda^{-1}) \tag{3.6}$$

with L slowly varying at ∞ [or 0]. (Here $\alpha \geqq 0$ since ω is decreasing.) With (3.6)

$$\frac{\omega(t^{-1}\lambda)}{\omega(t^{-1})} \to \lambda^{-\alpha} \tag{3.7}$$

which corresponds to $F(x) = x^\alpha/\Gamma(\alpha + 1)$. By (3.5) in this case

$$U(t) \sim \frac{1}{\Gamma(\alpha + 1)} t^\alpha L(t) \quad \text{at } \infty \text{ [or 0]}. \tag{3.8}$$

The fact that (3.6) implies (3.8) constitutes KARAMATA's Tauberian theorem. Before formulating it we prove its "Abelian" converse, that

$$U(t) = \frac{1}{\Gamma(\alpha + 1)} t^\alpha L(t) \tag{3.9}$$

with L slowly varying at ∞ [or 0] implies

$$\omega(\lambda) \sim \lambda^{-\alpha} L(\lambda^{-1}) \quad \text{at 0 [or } \infty]. \tag{3.10}$$

---

[4]) It will be seen that Φ is a Laplace transform and therefore continuous.

[5]) The unboundedness of the U_{t_k} plays no role. The measure U_t^* defined by $U^*(dx) = e^{-x} U_t(dx)$ has the Laplace transform ω_t^* given by $\omega_t^*(\lambda) = \omega_t(\lambda + 1)$ $(\lambda > 0)$, and by construction the U_t^* are bounded by 1.

For that purpose it suffices to prove that

$$\frac{\omega(t^{-1}\lambda)}{\omega(t^{-1})} \to \lambda^{-\alpha}. \tag{3.11}$$

For in this case $\omega(\lambda) = \lambda^{-\alpha} L_1(\lambda^{-1})$ by lemma 1, and this together with (3.9) implies $L \sim L_1$.

We prove (3.11) first when $t \to \infty$. If U is bounded, then $\alpha = 0$ [cf. footnote 4], and the assertion (3.11) is obvious. For each fixed $\varepsilon > 0$ and $\lambda > 0$ we can find an A so large that for $x > A$

$$(1-\varepsilon) L(x) \leqq L\left(\frac{x}{\lambda}\right) \leqq (1+\varepsilon) L(x). \tag{3.12}$$

If ω is unbounded at the origin we have then as $t \to \infty$

$$\omega(t^{-1}\lambda) = o(1) + t^{-1}\lambda \int_{A/\lambda}^{\infty} e^{-t^{-1}\lambda x} x^{\alpha} L(x)\,dx = o(1) + \lambda^{-\alpha} t^{-1} \int_{A}^{\infty} e^{-t^{-1}} y^{\alpha} L\left(\frac{y}{\lambda}\right) dy, \tag{3.13}$$

and hence by (3.12)

$$(1-\varepsilon)\lambda^{-\alpha}\omega(t^{-1}) + o(1) \leqq \omega(t^{-1}\lambda) \leqq (1+\varepsilon)\lambda^{-\alpha}\omega(t^{-1}) + o(1) \tag{3.14}$$

as was to be proved. The same argument applies as $t \to 0$.

We have thus the combined Tauberian and Abelian

Theorem 1. *If L is slowly varying either at ∞ or 0, then each of the relationships* (3.8) *and* (3.10) *implies the other.*

In particular, $\omega(\lambda) \sim \lambda^{-\alpha}$ if, and only if, $U(t) \sim t^{\alpha}/\Gamma(\alpha+1)$.

Our argument shows at the same time that we have found the most general case in which a relationship of the given form exists between a distribution and its transform.

4. Signed measures. Our method remains applicable when U represents a *signed* measure, but the Tauberian estimates may assume a different form. To illustrate the point we prove [6])

Theorem 2. *Suppose that U is of bounded variation in every finite interval and that*

$$V(x) = \int_0^x y\,U(dy) = o(x), \quad x \to \infty. \tag{4.1}$$

If the Laplace transform ω is slowly varying at 0, then

$$U(t) - \omega(t^{-1}) = o(1). \tag{4.2}$$

In particular $\omega(\lambda) \to 1$ as $\lambda \to 0$ implies $U(t) \to 1$ as $t \to \infty$.

---

[6]) This is a slightly improved version of theorem 3b of Chapter V in Widder [4]. By an argument developed by Littlewood and Landau the condition (4.1) (when $\omega(\lambda) \to 1$) may be replaced by the requirement that in the canonical decomposition $V = V^+ - V^-$ of V into positive and negative parts $V^-(b) - V^-(a) \leqq \text{Const} \cdot (b-a)$ for each interval (a, b).

Proof. We have to show that the integral $\int_0^\infty e^{-y}[U(ty) - U(t)]\,dy = \omega(t^{-1}) - U(t)$ tends to 0 as $t \to \infty$. Now

$$U(ty) - U(t) = \int_t^{ty} \frac{1}{x} V(dx) = \frac{V(ty)}{ty} - \frac{V(t)}{t} + \int_t^{ty} \frac{V(x)}{x^2}\,dx. \tag{4.3}$$

It follows from (4.1) that for each fixed y this expression tends to 0. Moreover, there exists a constant C such that $|V(x)| < C|x|$, and the right side in (4.3) is dominated by $2C + C|\log y|$. Therefore (4.2) holds by dominated convergence.

Note that (4.2) is *stronger than the assertion*

$$\frac{U(t)}{\omega(t^{-1})} \to 1 \tag{4.4}$$

provided $\liminf \omega(t^{-1}) > 0$. If $\omega(t^{-1}) \to 0$ then (4.2) is *weaker*. Obviously a variation of the condition (4.1) would lead to related theorems with different estimates.

5. A ratio limit theorem. We revert to the limiting procedure of section 3, but shall require that the limit $\Phi(\lambda)$ of (3.3) is positive for some $\lambda > 1$. This amounts to requiring that for some (and therefore all) $\lambda > 1$

$$\liminf_{t \to 0} \frac{\omega(t\lambda)}{\omega(t)} > 0. \tag{5.1}$$

A measure U satisfying this condition will be called *tame at* ∞; a similar definition applies to 0. For a tame measure the limit Φ of (3.3) is positive, and therefore the limit F of (3.5) finite. Thus when U is tame at ∞

$$\liminf_{t \to \infty} \frac{U(t)}{\omega(t^{-1})} > 0 \tag{5.2}$$

and it is easily seen that conversely (5.2) implies (5.1). Obviously tameness imposes a restriction on the rate of growth. Typical measures that are not tame at ∞ and 0, respectively, are defined by the densities $x^{1/2}\,\mathrm{ch}\,x^{1/2}$ and $x^{-(1/2)}\exp x^{-1}$. Their transforms are given by $\pi^{1/2}\lambda^{-1/2}\exp(4\lambda)^{-1}$ and $\pi^{1/2}\exp(-2\lambda^{1/2})$.

Theorem 3. *Let* $U^{(1)}$ *and* $U^{(2)}$ *represent two measures that are tame at* ∞ [*or* 0] *and whose Laplace transforms* $\omega^{(1)}(\lambda)$ *and* $\omega^{(2)}(\lambda)$ *exist for all* $\lambda > 0$. *For any slowly oscillating* L

$$\frac{U^{(1)}(t)}{U^{(2)}(t)} \sim L(t) \tag{5.3}$$

at ∞ [*or* 0] *if, and only if,*

$$\frac{\omega^{(1)}(t^{-1})}{\omega^{(2)}(t^{-1})} \sim L(t). \tag{5.4}$$

In particular, the relations,

$$\frac{U^{(1)}(t)}{U^{(2)}(t)} \to 1 \quad \textit{and} \quad \frac{\omega^{(1)}(t^{-1})}{\omega^{(2)}(t^{-1})} \to 1 \tag{5.5}$$

are equivalent.

Proof. We use the notation introduced in (3.2). Choose a sequence $\{t_k\}$, say $t_k \to \infty$, such that for both $j = 1$ and 2 and all $\lambda \geqq 1$

$$\omega_{t_k}^{(j)}(\lambda) \to \Phi^{(j)}(\lambda). \tag{5.6}$$

Then

$$U_{t_k}^{(j)} \to F^{(j)} \tag{5.7}$$

(pointwise at all points of continuity). On account of the tameness $\Phi^{(j)} \neq 0$ and therefore $F^{(j)} \neq 0$.

Now assume (5.4). Then $\Phi^{(1)} = \Phi^{(2)}$ and hence $F^{(1)} = F^{(2)}$, and (5.7) reduces to

$$\frac{U^{(1)}(t_k)}{U^{(2)}(t_k)} \div L(t_k) \to 1. \tag{5.8}$$

Now *every* sequence of points $t_k' \to \infty$ contains a subsequence $\{t_k\}$ for which (5.8) holds, and hence (5.3) is true. The same argument leads from (5.3) to (5.4).

Note. The case where (5.4) is replaced by

$$\frac{\omega^{(1)}(\lambda)}{\omega^{(2)}(\lambda)} \sim \lambda^{-\alpha} L(\lambda^{-1}) \qquad \lambda \to 0 \tag{5.9}$$

is only formally more general, since $\omega^{(2)}(\lambda)\lambda^{-\alpha}$ is the Laplace transform of the convolution of $U^{(2)}$ with $x^\alpha/\Gamma(\alpha+1)$. However, it is *not* true that

$$\frac{\omega^{(1)}(\lambda)}{\omega^{(2)}(\lambda)} \sim \lambda^{-\alpha} \qquad \lambda \to 0 \tag{5.10}$$

implies that $U^{(1)}/U^{(2)}$ behaves as $x^\alpha/\Gamma(\alpha+1)$. For example with $\omega^{(1)}(\lambda) = \lambda^{-\alpha-p}$, $\omega^{(2)}(\lambda) = \lambda^{-p}$ the relation (5.10) holds, but

$$\frac{U^{(1)}(x)}{U^{(2)}(x)} = \frac{\Gamma(p)}{\Gamma(p+\alpha+1)} x^\alpha, \tag{5.11}$$

which depends on p.

Literature

[1] J. L. DOOB, Relative limit theorems in analysis. J. Anal. Math. 8, 289–306 (1960/61).
[2] J. KARAMATA, Sur un mode de croissance régulière des fonctions. Mathematica (Cluj) 4, 38–53 (1930).
[3] H. KÖNIG, Neuer Beweis eines klassischen Tauber-Satzes. Arch. Math. 11, 278–279 (1960).
[4] D. V. WIDDER, The Laplace transform. Princeton 1946.

Eingegangen am 19. 7. 1962

Anschrift des Autors:
William Feller
Department of Mathematics
Princeton University
Princeton (N.J.), USA

ON SEMI-MARKOV PROCESSES*

BY WILLIAM FELLER

PRINCETON UNIVERSITY

Communicated February 10, 1964

1. *Introduction.*—This paper generalizes the theory of the Kolmogorov equations to a class of semi-Markov processes introduced by P. Lévy[1] and W. L. Smith.[2] (The general theory of semi-Markov processes has been developed by R. Pyke.[3]) The restriction to the state space of integers and to stationarity in time is introduced for notational convenience only. The theory goes through for the semi-Markov analogue of the completely discontinuous processes in arbitrary probability spaces.[4] The starting point is the system of infinitely many convolution equations (2.2) representing the obvious generalization of the *Kolmogorov backward* differential equations. As in the Markovian case there exists a substochastic *minimal* solution. If it is strictly stochastic, there exist no other solutions, but otherwise additional solutions may be constructed. Necessary and sufficient conditions for this situation are given; they are connected with the boundary theory for an associated discrete Markov chain, and it becomes apparent that the author's previous constructions of abstract exit boundaries and superimposed return processes for the classical Kolmogorov equations carry over.[5]

The minimal solution satisfies an adjoint system of convolution equations playing the role of Kolmogorov's *forward equations.* Their structure and probabilistic meaning is far from obvious, however. The last theorem describes the asymptotic behavior as $t \to \infty$.

2. *The Backward Equations.*—We consider a stochastic process on the state space of integers 1, 2, ... with right continuous paths $\chi(\cdot)$. The paths will be piecewise constant, and if jumps accumulate to a finite T, then $\chi(t_n) \to \infty$ for an appropriate sequence $t_n \uparrow T$. We take T as a stopping time and shall decide whether or not $T < \infty$.

Just as in true Markov processes, the jumps are supposed to be regeneration points obliterating the influence of the past. The successive sojourn times at i are mutually independent and—because of the assumed stationarity—they have a common distribution S_i. The novel feature is that not only can S_i be chosen arbitrarily, but the duration of a sojourn at i depends also on the terminal state.

The mechanism of the process is now determined by an infinite matrix of defective probability distributions F_{ik} (with $F_{ik}(0) = 0$, $F_{ik}(\infty) \leq 1$), *where $F_{ik}(t)$ is the probability that a sojourn time at i has a duration $\leq t$ and ends by a jump to k.* We put

$$q_{ik} = F_{ik}(\infty), \qquad S_i(t) = \sum_k F_{ik}(t). \tag{2.1}$$

Then S_i is the absolute distribution of the sojourn times at i and q_{ik} the probability that a jump to i is followed by a jump to k. We assume $S_i(\infty) = 1$ for all i, and so the matrix $Q = (q_{ik})$ is strictly stochastic.

Denote by $P_{ik}(t)$ the (conditional) probability of the event $\chi(t + \tau) = k$ *given that a jump to i occurs at τ.* Considering the first jump following τ it is seen that

$$\begin{aligned} P_{ik}(t) &= \sum_j \int_0^t F_{ij}\{dx\} P_{jk}(t - x), \qquad (k \neq i) \\ P_{ii}(t) &= 1 - S_i(t) + \sum_j \int_0^t F_{ij}\{dx\} P_{ji}(t - x). \end{aligned} \tag{2.2}$$

In the Markovian case $F_{ik}(t) = q_{ik}(1 - e^{-c_i t})$, and (2.2) is equivalent to Kolmogorov's system of backward differential equations.

3. *The Minimal Solution.*—We pass to Laplace transforms and matrix notation. For $\lambda > 0$ put

$$\varphi_{ij} = \int_0^\infty e^{-\lambda x} F_{ij}\{dx\}, \qquad \sigma_i = \sum_j \varphi_{ij} \tag{3.1}$$

$$\Pi_{ij} = \int_0^\infty e^{-\lambda x} P_{ij}(x)dx. \tag{3.2}$$

(The dependence on λ of these and other quantities denoted by Greek letters should be borne in mind.) For the corresponding infinite matrices we put $\Pi = (\Pi_{ij})$ and $\Phi = (\Phi_{ij})$, whereas Σ denotes the *diagonal* matrix with elements σ_i. In terms of Laplace transforms the backward equations (2.2) take on the form

$$\Pi = \lambda^{-1}[I - \Sigma] + \Phi\Pi, \tag{3.3}$$

where I is the identity matrix.

To solve this equation put

$$\Pi^{(n)} = (I + \Phi + \ldots + \Phi^n)(I - \Sigma)\lambda^{-1}. \tag{3.4}$$

The diagonal matrix $(I - \Sigma)\lambda^{-1}$ is positive, and so $\Pi^{(0)} \leq \Pi^{(1)} \leq \ldots$ .Obviously every nonnegative solution of (3.3) is $\geq$ than the formal limit

$$\Pi = (I + \Phi + \Phi^2 + \ldots)(I - \Sigma)\lambda^{-1} \tag{3.5}$$

of (3.4), and if this matrix is finite and has bounded row sums, then it satisfies (3.3). We refer to (3.5) as the *minimal* solution.

Denote by **1** the column vector with unit elements. Then $\Phi\mathbf{1}$ represents the row sums of Φ, and by definition $\Sigma\mathbf{1} = \Phi\mathbf{1}$. The row sums of $\lambda\Pi^{(n)}$ therefore satisfy

$$\lambda\Pi^{(n)}\mathbf{1} = \mathbf{1} - \Phi^{n+1}\mathbf{1}. \tag{3.6}$$

But $\Phi\mathbf{1} \leq \mathbf{1}$, and hence the sequence of vectors $\Phi^n\mathbf{1}$ is monotone and approaches a vector ξ (depending on λ) such that

$$\Phi\xi = \xi \qquad 0 \leq \xi \leq 1. \tag{3.7}$$

For *any* such eigenvector ξ we have by induction $\xi \leq \Phi^n\mathbf{1}$, and hence ξ is the *maximal* column vector satisfying (3.7). Thus,

$$\lambda\Pi\mathbf{1} = \mathbf{1} - \xi. \tag{3.8}$$

It follows in particular that the matrix $\Sigma\Phi^n$ is finite. Being limits of completely monotone functions, its elements are completely monotone. Since $(1 - \sigma_i)\lambda^{-1}$ is the Laplace transform (with respect to Lebesgue measure) of the point function $1 - S_i$, it follows that Π_{ik} is the Laplace transform of a point function P_{ik}. Now (3.8) expresses that the row sums $\Sigma_k P_{ik}(t)$ are ≤ 1 and identically equal to 1 only if $\xi = 0$. (Because of the completely monotone character of $\xi_i(\lambda)$, either $\xi = 0$ for all λ or for no λ.) We have thus

THEOREM 1. *The backward equations* (2.2) *possess a minimal solution, and its Laplace transform is given by* (3.5). *The matrix* $(P_{ik}(t))$ *is stochastic or substochastic. It is stochastic for all* t *if and only if for some* $\lambda > 0$ (3.7) *implies* $\xi = 0$.

Because of its minimal character the solution is (at least probabilistically) unique whenever it is strictly stochastic. In all other cases it is possible to construct additional acceptable solutions leading to return processes; the method is the same as in the Markovian case.[5]

COROLLARY. *The solution is unique if, and only if, it is strictly stochastic.*

THEOREM 2. *(Forward equation) The minimal solution satisfies the system of convolution equations with Laplace transform*

$$\Pi = \lambda^{-1}\{I - \Sigma\} + \Pi\,\{I - \Sigma\}^{-1}\Phi\{I - \Sigma\}. \tag{3.9}$$

The proof is immediate upon substituting from (3.5) into the right side of (3.9). This equation is quite unexpected. The diagonal matrix $\{I - \Sigma\}^{-1}$ contains the Laplace transforms of the renewal functions ΣS_i^{n*}. In the Markovian case the forward equations are obtained formally by considering the *last* jump preceding t (although they may hold even when no last jump exists). In appearance (3.9) is similar, but it is difficult to see why the corresponding terms in (3.9) represent the probabilities of the last jump. Needless to say, the numerical solution does *not* have the semigroup property $P(t + s) = P(t)P(s)$.

4. *Criteria.*—The matrix Φ is substochastic and induces a transient discrete Markov chain, which can terminate. The ith row sum of Φ^n is the probability that, starting from i, the chain survives n steps, and so ξ_i is the probability that

the chain never dies. As $\lambda \downarrow 0$ we have $\Phi \uparrow Q$, and it is possible to relate ξ to the boundary induced by Q. For that purpose write

$$\varphi_{ik} = q_{ik} - \lambda\psi_{ik} \tag{4.1}$$

where

$$\Psi_{ik} = \lambda^{-1} \int_0^\infty [1 - e^{-\lambda x}] F_{ik}\{dx\}. \tag{4.2}$$

The equation $\xi = \Phi\xi$ now takes on the form

$$Q\xi = \xi + \lambda\Psi\xi \tag{4.3}$$

whence by induction

$$Q^{n+1}\xi = \xi + \lambda \sum_{r=0}^{n} Q^r\Psi\xi. \tag{4.4}$$

It is well known that an element a_{ik} of the formal matrix ΣQ^r is finite if, and only if, in the Q-chain, k is transient. From (4.4) we have therefore

THEOREM 3. *The minimal solution is strictly stochastic (and hence unique) if all states of the Markov chain induced by Q are persistent.*

By the usual decomposition this theorem applies to indecomposable subclasses, and in studying nonuniqueness we may therefore assume that all states are transient. Clearly $Q\xi \geq \xi$, and hence the sequence of vectors $Q^n\xi$ is monotone, and so $Q^n\xi \uparrow x$ where

$$Qx = x, \qquad \xi \leq x \leq 1. \tag{4.5}$$

Obviously x is the smallest vector satisfying (4.5,) and from (4.4) we get

$$x = \xi + \lambda\Omega\xi, \quad \text{where } \Omega = \sum_{r=0}^{\infty} Q^r\Psi. \tag{4.6}$$

The column vectors x and ξ determine each other since $Q^n\xi \uparrow x$ and $\Phi^n x \downarrow \xi$. (It can be shown that x is independent of λ and that the theory of canonical mappings[5] carries over.)

To develop criteria for deficiency we consider the associated discrete Markov chains with transition probabilities Q. Its *restriction* to a set A of integers is the chain obtained by stopping the Q-chain as soon as it leaves A. This restriction is governed by the substochastic $_AQ$ matrix obtained on replacing by Q all elements q_{ij} with $i \notin A$ or $j \notin A$. The row sums of $_AQ^n$ represent the probabilities of surviving n steps, and so

$$_AQ^n 1 \rightarrow y \qquad n \rightarrow \infty \tag{4.7}$$

where y_i is the probability that starting from i the restricted process will never terminate. A set A is a *sojourn set*[6] if $y \neq 0$. Such sojourn sets are typical neighborhoods of sets on the Q-exit boundary. A truly substochastic minimal solution is possible only if the boundary can be reached within a finite time, and this presupposes that the sojourn times at states near the boundary are likely to be short. As an appropriate measure we introduce the truncated moments

$$m_{ij} = \int_0^\infty \frac{x}{1+x} F_{ij}\{dx\}. \tag{4.8}$$

(For sufficient conditions it will suffice to consider the moments (4.17), but these can be large due to negligible masses near infinity.) Put $M = (m_{ij})$ and

$$U = \sum_{n=0}^{\infty} Q^n M. \tag{4.9}$$

The matrix $U = (u_{ij})$ is finite if all states are transient.

Theorem 4.—*The minimal solution is deficient (not stochastic) if, and only if, there exists a sojourn set A such that*

$$\sum_{j \in A} u_{ij} < \infty \tag{4.10}$$

for each $i \in A$.

Proof: (a) *Necessity:* If the minimal solution is defective, there exists a nontrivial vector ξ satisfying (4.6) and $\xi = \lim \Phi^n \mathbf{1}$. Thus, ξ_i can be interpreted as the probability that, starting from i, the discrete chain governed by the substochastic matrix Φ never terminates. It follows that for every $\epsilon > 0$ the set $A = \{i | \xi_i > 1 - \epsilon\}$ is a sojourn set for the Φ-chain. But $\varphi_{ik}(\lambda) \uparrow q_{ik}$ as $\lambda \downarrow 0$, and so A is *a fortiori* a sojourn set for the Q-chain. For fixed λ there exists a constant γ such that $\Psi_{ik} \geq \gamma\, m_{ik}$ because the ratio of the integrands is bounded away from 0. Thus, $\Omega \geq \gamma U$, and from (4.6) se get for *all* i

$$x_i \geq \lambda(1 - \epsilon)\gamma \sum_{k \in A} u_{ik}. \tag{4.11}$$

(b) *Sufficiency:* Consider a Markov chain governed by Φ and starting at i. The elements of the ith row of $\Sigma\Phi^r$ represent the expected number of visits to 1, 2, ... In the familiar manner we consider separately the visits prior to the first visit outside A and those following such a visit. For this purpose let τ_{ik} be the probability that the first visit to the complement A' occurs at k (thus, $\tau_{ii} = 1$ if $i \in A'$). From (3.5) we get then

$$\Pi = {}_A\Pi + T\,\Pi \tag{4.12}$$

where $T = (\tau_{ij})$ and ${}_A\Pi$ is the matrix obtained from (3.5) on replacing Φ by the restriction ${}_A\Phi$.

For $i \in A$ the row sum $\sum_k \tau_{ik}$ equals the probability of ever leaving A. Since $\varphi_{ik} \leq q_{ik}$, this probability is less than the corresponding probability for the Q-chain, namely, $1 - y_i$ [see (4.7)]. In view of (3.8) we get therefore from (4.12)

$$\lambda\Pi\mathbf{1} \leq \mathbf{1} - y + ({}_AI + {}_AQ + {}_AQ^2 + \ldots)(I - \Sigma)\mathbf{1}. \tag{4.13}$$

Now

$$(I - \Sigma)\mathbf{1} = \lambda\Psi\mathbf{1} = \lambda\Psi\mathbf{1}_A + \lambda\Psi\mathbf{1}_B \tag{4.14}$$

where $\mathbf{1}_A$ and $\mathbf{1}_B$ are the indicators of A and $B = A'$, respectively. $(\mathbf{1}_A + \mathbf{1}_B = \mathbf{1})$. Now $\lambda\Psi_{ik} \leq q_{ik}$, and it is therefore apparent that the contribution of $\mathbf{1}_B$ to (4.13) is a column vector whose ith component in A is less than the probability that the Q-random walk will ever leave A, that is $\mathbf{1} - y_i$. Thus,

$$\lambda\Pi\mathbf{1} \leq 2(1 - y) + ({}_AI + {}_AQ + {}_AQ^2 + \ldots)\,\lambda\Psi\mathbf{1}_A. \tag{4.15}$$

For the final estimate we decompose the intervals of integration in (4.2) and (4.8) into $0 < x \leq t$ and $x > t$. For the induced decomposition of the matrices write $\Psi = \Psi' + \Psi''$ and $M = M' + M''$. Since $(1 - e^{-\lambda x}) \leq \lambda x$, we have $\Psi' \leq (1 + t)M'$. For $x > t$ we use $1 - e^{-\lambda x} < 1$ to obtain $\lambda\Psi' \leq (1 + t^{-1})M''$. These estimates are valid uniformly for all $\lambda > 0$. With obvious notations we have now from (4.15) and (4.9)

$$\lambda\Pi 1 \leq 2(1 - y) + \lambda(1 + t)U'1_A + (1 + t^{-1})U''1_A. \tag{4.16}$$

Now choose $i\epsilon A$ such that $1 - y_i < \epsilon$. In view of the convergence conditions (4.10) it is possible to choose $t > 1$ so large that the ith component of $U''1_A$ is $< \epsilon$. Since $U' \leq U$ and the ith component of $U\,1_A$ is finite, we see that as $\lambda \to 0$ the ith component of $\lambda\Pi 1$ is $<4\epsilon + O(\lambda)$, and this accomplishes the proof.

Sufficient conditions can be derived more easily. For example, let the sojourn times have finite expectations

$$m_i = \int_0^\infty x\, S_i\{dx\}, \qquad m_i > 0 \tag{4.17}$$

and denote by m the column vector with components m_i. Without the tedious estimates the last proof yields easily the

CRITERION. *If there exists a sojourn set A such that the components of the vector $\sum_r {}_AQ^r m$ are finite in A, then the minimal solution is not strictly stochastic.*

5. *Asmptotic Behavior.*—If Q is indecomposable and persistent, there exists, as Derman[7] has shown, a unique strictly positive row vector $r = (r_1, r_2, \ldots)$ such that

$$rQ = r, \qquad r_1 = 1. \tag{5.1}$$

THEOREM 5. *If Q is indecomposable and persistent, we have for the (unique) solution of the backward equations as $t \to \infty$*

$$P_{ik}(t) \to a_k \tag{5.2}$$

where $a = (a_1, a_2, \ldots)$,

$$a_k = Cr_k m_k, \qquad \Sigma a_i = 1 \tag{5.3}$$

if $\Sigma r_i m_i < \infty$, and $a_k = 0$ otherwise.

It is noteworthy that positive ergodic limits a_k can exist even when the Q-chain has null states.

Proof: Consider the transient chain with matrix Φ and initial state i. For $k > 1$ denote by ϵ_k the expected number of passages through k prior to the first return to 1, and put $\epsilon_1 = 1$. Let τ be the probability of such a return. The expected number of visits to k before the second return to 1 is $(1 + \tau)\epsilon_k$ and so on. The first row of $\lambda\Pi$ has therefore the components

$$\frac{\lambda}{1 - \tau} \cdot \epsilon_k \cdot \frac{1 - \sigma_i}{\lambda}. \tag{5.4}$$

From the Derman construction it is clear that $\epsilon_k \uparrow r_k$ as $\lambda \downarrow 0$. From the definition it is seen that $\lambda^{-1}(1 - \sigma_i) \uparrow m_i$. The vector with components (5.4) being a probability vector, it follows that if $\Sigma r_k m_k = \infty$, the first factor in (5.4) tends to 0, and that otherwise the first row of $\lambda\Pi$ approaches the probability vector occurring in

(5.3). The same argument applies to the other rows. The basic Tauberian theorem now shows that the integral averages of $P_{ik}(t)$ behave as they should. The convergence of $P_{ik}(t)$ itself follows from a renewal equation for P_{ik} involving the first passage time to k. (This renewal equation can be obtained purely analytically by considering the absorbing barrier process obtained formally from our process changing $F_{kj}(t)$ to 0 for $j \neq k$ and to 1 for $j = k$. The kth column of the new solution being proportional to that of Π_λ, we get the Laplace transform version of the renewal equations and expressions for the first passage time.)

* This paper is connected with research sponsored by the U.S. Army Research Office (Durham) at Princeton University.

[1] Lévy, P., *Proc. Intern. Congr. Math.* (Amsterdam) **3,** 416 (1954).

[2] Smith, W. L., *Proc. Roy. Soc.* (*London*), **A232,** 6 (1955).

[3] Pyke, R., *Ann. Math. Statist.*, **32,** 1231 and 1243 (1961). The second paper treats finite state spaces.

[4] Feller, W., *Trans Am. Math. Soc.*, **48,** 488 (1940).

[5] *Ibid.*, *Ann. Math.*, **65,** 527 (1957). The behavior of the sample functions is analyzed in Chung, K. L., *Acta Math.*, **110,** 19 (1963).

[6] Feller, W., *Trans. Am. Math. Soc.*, **83,** 19 (1956).

[7] Derman, C., *Proc. Am. Math. Soc.*, **5,** 332 (1954).

On the Fourier Representation for Markov Chains and the Strong Ratio Theorem

WILLIAM FELLER*

1. Introduction. For unity of exposition we consider first only discrete parameter denumerable chains; for the continuous parameter analogue see section 7. As usual, the n step transition probabilities are denoted by $p_{jk}^{(n)}$, the first passage probabilities by $f_{jk}^{(n)}$ (here $n = 0, 1, \cdots$ and $f_{jk}^{(0)} = 0$, for all j, k). Following Chung [1] we introduce the corresponding quantities with a taboo state r and denote them by ${}_rp_{jk}^{(n)}$ and ${}_rf_{jk}^{(n)}$; in terms of the sample paths

$$(1.1) \qquad {}_rp_{jk}^{(n)} = P\{X_n = k, X_1 \neq r, \cdots, X_{n-1} \neq r \mid X_0 = j\}$$

and similarly for ${}_rf_{jk}^{(n)}$. For the generating functions we use capital letters. Thus

$$(1.2) \qquad P_{jk}(z) = \sum_{n=0}^{\infty} p_{jk}^{(n)} z^n$$

and so on. (Following Chung we should write $\hat{p}_{jk}$ instead of P_{jk} . The change is introduced for typographical convenience since we shall use the complex conjugate $P_{jk}^*(z)$.)

From the theory of recurrent events one has the familiar relation

$$(1.3) \qquad P_{kk}(z) = \frac{1}{1 - F_{kk}(z)}, \qquad |z| < 1.$$

For the sequel it is crucial that there exists the alternative representation

$$(1.4) \qquad P_{kk}(z) = \frac{{}_jP_{kk}(z)}{1 - F_{jk}(z)F_{kj}(z)}, \qquad |z| < 1.$$

[For an analytic proof see formula (3) in I, §10 of [1]. A direct probabilistic proof is as follows. Consider the recurrent event "the system returns to k having passed through $j \neq k$". Its interarrival times have the generating function $F_{kj}F_{jk}$. Now the system can be in k after 0, 1, 2, $\cdots$ such occurrences, whence

$$P_{kk} = {}_jP_{kk} \sum_{n=0}^{\infty} (F_{kj}F_{jk})^n,$$

* This research was supported by the Army Research Office in Probability at Princeton University.

Journal of Mathematics and Mechanics, Vol. 15, No. 2 (1966).

R.L. Schilling et al. (eds.), *Selected Papers II*,

as asserted.] Indeed, perhaps the main result of this paper is the fact that with an appropriate technique (1.4) may lead to simple proofs of relations which otherwise require lengthy calculations. We shall deal with two only vaguely related topics, treated by real and complex variable methods, respectively.

To avoid trivialities we assume the chain irreducible, that is, transitions are possible between any pair of states r, k. Then ${}_r p_{kk}^{(n)}$ is a transition probability in a transient chain and hence we may define numbers ${}_r\mu_k < \infty$ by

$$(1.5) \qquad {}_r\mu_k = {}_rP_{kk}(1) = \sum_{n=0}^{\infty} {}_r p_{kk}^{(n)}.$$

An extremely simple argument will yield lemma 4 of section 3 which trivially implies

Theorem 1. *If the chain is irreducible and persistent then for all combinations* j, k, r

$$(1.6) \qquad -{}_j\mu_k \leqq {}_k\mu_j \sum_{n=0}^{N} p_{kk}^{(n)} - {}_j\mu_k \sum_{n=0}^{N} p_{jj}^{(n)} \leqq {}_k\mu_j$$

$$(1.7) \qquad 0 \leq \sum_{n=0}^{N} (p_{kk}^{(n)} - p_{jk}^{(n)}) \leqq {}_j\mu_k$$

$$(1.8) \qquad 0 \leqq \frac{1}{{}_j\mu_k} \sum_{n=0}^{N} p_{kk}^{(n)} - \frac{1}{{}_k\mu_j} \sum_{n=0}^{N} p_{kj}^{(n)} \leqq 1.$$

(Except for the precise bounds, (1.6) and (1.8) are due to S. Orey, and (1.7) to W. Hoeffding. See [9] for proofs and further references.)

An immediate consequence of (1.6) is the well-known ratio limit theorem in the form

$$(1.9) \qquad \lim_{N\to\infty} \frac{\sum_{n=0}^{N} p_{kk}^{(n)}}{\sum_{n=0}^{N} p_{jj}^{(n)}} = \frac{{}_j\mu_k}{{}_k\mu_j}$$

which shows that there exist numbers λ_k such that

$$(1.10) \qquad \frac{{}_j\mu_k}{{}_k\mu_j} = \frac{\lambda_k}{\lambda_j}.$$

These relations can be found in [1] and are mentioned only because they appear naturally in the present context and throw some light on the Fourier representation to which we now turn. First we recall that a sequence of positive numbers λ_k is said to represent a *subinvariant measure* if

$$(1.11) \qquad \sum_r \lambda_r p_{rk} \leqq \lambda_k .$$

The measure is invariant if in (1.11) the equality sign holds for all k. As was first observed by C. Derman, and as is easily verified, the numbers λ_k of (1.10) form an invariant measure if the chain is persistent.

A theorem of D. G. Kendall [6] asserts that to each subinvariant measure of an aperiodic irreducible chain there corresponds a unique Hermitian system of complex valued integrable functions w_{jk} such that

$$w_{jk} = w_{kj}^* \tag{1.12}$$

(the star denoting the complex conjugate) and

$$\left(\frac{\lambda_j}{\lambda_k}\right)^{1/2} (p_{jk}^{(n)} - p_{jk}^{(\infty)}) = \int_{-\pi}^{\pi} e^{in\vartheta} w_{jk}(\vartheta)\, d\vartheta. \tag{1.13}$$

where

$$p_{jk}^{(\infty)} = \lim_{n\to\infty} p_{jk}^{(n)}.$$

In particular, $w_{kk} \geqq 0$.

Kendall's proof depends on a skillful use of a deep theorem of Sz-Nagy concerning the unsymmetric operator induced by the matrix (p_{jk}) on the Hilbert space defined on the integers by the measure $\{\lambda_k\}$; it is asserted that this Hilbert space can be imbedded in a larger one, in such a way that our operator becomes the projection of a unitary operator in the larger space. Despite their formal elegance, Hilbert space methods are not natural for the problem, and, in fact, the Fourier representation (1.13) has nothing to do with subinvariant measures. In fact, we shall see that *for transient chains, Kendall's result* (1.13) *holds for arbitrary positive numbers* λ_k . *On the other hand, a persistent chain admits of no truly subinvariant measure and* (1.10) *defines the unique invariant measure.* [As Kingman [8, p. 342] observed, this follows easily from the construction of the invariant measure given in Chung's treatise [1]. Alternatively, the assertion becomes obvious on passing from the chain determined by the stochastic matrix (p_{jk}) to the "reversed" chain with matrix ${}^{\#}p$ defined in (3.7). If $\{\lambda_k\}$ is a subinvariant measure $({}^{\#}p_{jk})$ is a substochastic matrix and to each subinvariant measure $\{\zeta_k\}$ there corresponds a column vector x with components $x_k = \zeta_k/\lambda_k$ such that ${}^{\#}px \leqq x$. As is well-known from boundary theory [3] a non-constant positive solution of ${}^{\#}px \leqq x$ can exist only if ${}^{\#}p$ is transient, and since

$$\sum_n {}^{\#}p_{jk}^{(n)} = \frac{\lambda_j}{\lambda_k} \sum_n p_{jk}^{(n)} < \infty$$

this is impossible]

We shall see that, for the non-diagonal terms of a transient chain the assertion follows very easily from the convergence of the generating function $P_{jk}(z)$ along the circle $|z| = 1$. For the diagonal terms of any chain Kendall's Fourier representation turns out to be a special case of a representation for arbitrary renewal intensities, which is easily derived by a method introduced by Chung and Fuchs [section 4]. More delicate is the situation for the non-diagonal terms of a persistent chain; for this case we shall obtain a slight sharpening of Kendall's result using only elementary estimates connected with the relation

(1.4) (see section 6). For the analogue for chains depending on continuous time see section 7.

2. Preliminaries. Transposed to the chain with transition probabilities ${}_rp_{jk}^{(n)}$, the basic relation (1.3) reads

$$ {}_rP_{kk}(z) = \frac{1}{1 - {}_rF_{kk}(z)}. \tag{2.1} $$

A trite calculation now shows that with the numbers ${}_r\mu_k$ defined in (1.5), the relation (1.4) may be rewritten in the form

$$ \frac{{}_j\mu_k}{1 - F_{jk}(z)F_{kj}(z)} - P_{kk}(z) = \Phi_{jk}(z), \tag{2.2} $$

where

$$ \Phi_{jk}(z) = \frac{{}_jF_{kk}(1) - {}_jF_{kk}(z)}{1 - F_{kk}(z)}. \tag{2.3} $$

Finally we recall the obvious identity

$$ P_{jk}(z) = F_{jk}(z)P_{kk}(z). \tag{2.4} $$

The following section will not be used in the sequel.

3. Concerning theorem 1. The proof depends on the following simple lemma, in which we adhere to the convention to denote generating functions by the obvious capital letters.

Lemma 1. *For* $n = 0, 1, \cdots$ *let* $u_n \geqq 0$, $v_n \geqq 0$ *and*

$$ r_n = u_n + u_{n+1} + \cdots, \quad s_n = v_n + v_{n+1} + \cdots. \tag{3.1} $$

Put

$$ A(z) = \frac{1}{1 - z} \cdot \frac{U(1) - U(z)}{V(1) - V(z)} = \frac{R(z)}{V(1) - V(z)}. \tag{3.2} $$

If $s_0 < \infty$ *and* $r_n \leqq s_n$ *for all* n, *then the coefficients* a_n *of* $A(z)$ *satisfy* $0 \leqq a_n \leqq 1$.

Proof. From the second representation in (3.2) it is obvious that $a_n \geqq 0$ and that increasing the coefficients r_n leads to increased values for the a_n. It follows that the a_n are dominated by the coefficients in the power series

$$ \frac{S(z)}{V(1) - V(z)} = \frac{1}{1 - z}, \tag{3.3} $$

and this proves the lemma.

Lemma 2. *Put*

$$ C_{jk}(z) = \frac{1}{1 - F_{jk}(z)F_{kj}(z)} \tag{3.4} $$

and denote the coefficients by $c_{jk}^{(n)}$. *If the chain* (p_{jk}) *is irreducible and persistent, then*

$$(3.5) \qquad 0 \leq {}_j\mu_k \sum_{n=0}^{N} c_{jk}^{(n)} - \sum_{n=0}^{N} p_{kk}^{(n)} \leq 1.$$

Proof. The middle term in (3.5) is the Nth coefficient of $(1-z)\Phi_{jk}(z)/(1-z)$ and so the assertion represents a special case of the preceding lemma.

Proof of theorem 1. (a) The inequalities (1.6) follow trivially on comparing (3.5) with the relation obtained by interchanging j and k, and remembering that $c_{jk} = c_{kj}$.

(b) From (2.4) and (1.4) it is clear that the middle term in (1.7) is the Nth coefficient of

$$(3.6) \qquad {}_jP_{kk}(z)\cdot\frac{1}{1-z}\,\frac{1-F_{jk}(z)}{1-F_{jk}(z)F_{kj}(z)}.$$

It is obvious that the last factor satisfies the conditions of lemma 1, and so the coefficients of (3.6) are bounded by ${}_jP_{kk}(1)$.

(c) Consider the "reversed" chain with transition probabilities

$$(3.7) \qquad {}^{\#}p_{jk}^{(n)} = \frac{\lambda_k}{\lambda_j}\, p_{kj}^{(n)},$$

where $\{\lambda_k\}$ is the invariant measure defined in (1.10). The new chain is again irreducible and transient, and, from the very definition (1.5), it is obvious that ${}^{\#}_j\mu_k = {}_j\mu_k$. Evidently, therefore, (1.8) coincides with (1.7) applied to the reversed chain, and this concluded the proof.

Note *on the convergence of*

$$(3.8) \qquad \sum \{{}_j\mu_k c_{jk}^{(n)} - p_{kk}^{(n)}\}.$$

It is clear that the series (3.7) converges [converges absolutely] if the formal series for $\Phi_{jk}(1)$ converges [converges absolutely]. Now the convergence of the latter series depends on the behavior of ${}_jF_{kk}$, and it is not difficult to prove that the series for $\Phi_{jk}(1)$ may converge while that for $\Phi_{kj}(1)$ diverges. In Kendall's suggestive terminology the convergence of the series (3.7) is *not* subject to a solidarity theorem, and the same is true of the series arising from (1.6).

Suppose, however, that the recurrence distribution $\{f_{kk}^{(n)}\}$ has a finite expectation $F_{kk}(1)$. Then the series for $\{1-F_{kk}(z)\}^{-1}$ converges absolutely at $z=1$ by a theorem due to Wiener and P. Lévy (see [3]), and, hence, *the series* (3.11) *converges absolutely for all k if the chain is persistent and ergodic.* For the series arising from (1.6) this was proved differently by Orey [9].

4. Representation of renewal intensities. Let $\{f^{(n)}\}$ be a probability distribution on the positive integers with true span 1, that is, suppose that 1 is the greatest common divisor of the indices n for which $f^{(n)} > 0$. As is well known, the generating function

$$F(z) = \sum_{1}^{\infty} f^{(n)} z^n$$

satisfies the inequality $|F(z)| \leqq 1$ for all $|z| \leqq 1$, and $|F(z)| = 1$ for $z = 1$. The generating function

$$P(z) = \sum p^{(n)} z^n$$

of the renewal intensities $p^{(n)}$ is given by

$$P(z) = \frac{1}{1 - F(z)}. \tag{4.1}$$

Except at $z = 1$, it is continuous in the closed unit disk $|z| \leqq 1$. A Fourier representation for $p^{(n)}$ may be obtained by a method devised by Chung and Fuchs [2], and used also by Kendall to show that his measures have densities.

Letting $F = u + iv$ one has for $|z| < 1$

$$2 \operatorname{Re} P(z) = \frac{2(1 - u)}{(1 - u)^2 + v^2} \geqq 1 + \frac{1 - u^2 - v^2}{(1 - u)^2 + v^2} > 1. \tag{4.2}$$

Writing $z = re^{i\vartheta}$ it follows for fixed $r < 1$ and $n \geqq 0$,

$$r^n p^{(n)} = \frac{1}{2\pi} \int_{-\pi}^{\pi} e^{-in\vartheta} P(re^{i\vartheta}) \, d\vartheta \tag{4.3}$$

whereas for $n < 0$ the integral equals 0. Adding these relations for $\pm n$ one gets

$$r^n p^{(n)} = \frac{1}{2\pi} \int_{-\pi}^{\pi} e^{in\vartheta} [2 \operatorname{Re} P(re^{i\vartheta}) - 1] \, d\vartheta \tag{4.4}$$

for $n = 0, 1, \cdots$. By virtue of (4.2) the expression within brackets is positive, and for $n = 0$ one sees that $r^n p^{(n)}$ is the Fourier coefficient of a probability distribution W_r on the unit circle. The convergence of these Fourier coefficients implies the existence of a probability distribution W_1 such that $W_r \to W_1$ as $r \to 1$. Outside a neighborhood of $\vartheta = 0$ the densities of W_r converge boundedly to a limit, which is therefore the density of W_1 except for a possible atom of weight a at the origin. Thus

$$p^{(n)} = \alpha + \frac{1}{2\pi} \int_{-\pi}^{\pi} e^{in\vartheta} \frac{1 - u^2(e^{i\vartheta}) - v^2(e^{i\vartheta})}{[1 - u(e^{i\vartheta})]^2 + v^2(e^{i\vartheta})} \cdot d\vartheta. \tag{4.5}$$

By the Riemann–Lebesgue lemma $p^{(n)} \to \alpha$ as $n \to \infty$, and the familiar renewal theorem asserts that $\alpha = 0$ except if the distribution $\{f^{(n)}\}$ has a finite expectation $\alpha^{-1} = F'(1)$. We have thus the desired *representation of arbitrary renewal intensities* $p^{(n)}$ *as Fourier coefficients of a probability distribution* W_1.

The same argument applies when $\{f^{(n)}\}$ is defective (that is, if $F(1) < 1$). Then (4.5) holds with $\alpha = 0$.

This result contains as a special case the Kendall representation (1.13) for the diagonal terms.

5. Transient chains.

Theorem 2. *Let the chain be transient and choose* $\lambda_k > 0$ *arbitrary. There exist bounded continuous functions* w_{jk} *such that* $w_{kk}(\vartheta) \geqq 0$ *and*

$$(5.1)\qquad \int_{-\pi}^{\pi} e^{in\vartheta} w_{jk}(\vartheta)\, d\vartheta = \begin{cases} \left(\dfrac{\lambda_j}{\lambda_k}\right)^{1/2} p_{jk}^{(n)} & \text{for } n \geqq 0 \\ \left(\dfrac{\lambda_k}{\lambda_j}\right)^{1/2} p_{kj}^{(n)} & n \leqq 0 \end{cases}$$

(implying $w_{jk}(-\vartheta) = w_{kj}(\vartheta) = w_{jk}^*(\vartheta)$*).*

Proof. Since $\sum p_{jk}^{(n)} < \infty$ the power series for $P_{jk}(z)$ converges uniformly for $|z| = 1$, and hence, when $j \neq k$,

$$(5.2)\qquad \frac{1}{2\pi}\int_{-\pi}^{\pi} e^{-in\vartheta} P_{jk}(e^{i\vartheta})\, d\vartheta = \begin{cases} p_{jk}^{(n)} & \text{for } n \geqq 0 \\ 0 & n \leqq 0. \end{cases}$$

An analogous relation holds with j and k interchanged, and so (5.1) holds with

$$(5.3)\qquad w_{jk}(\vartheta) = \frac{1}{2\pi}\left[\left(\frac{\lambda_j}{\lambda_k}\right)^{1/2} P_{jk}(e^{-i\vartheta}) + \left(\frac{\lambda_k}{\lambda_j}\right)^{1/2} P_{kj}(e^{i\vartheta})\right].$$

The case $j = k$ is covered by the result of the last section.

6. Persistent chains. From section 4 we know that if k is an aperiodic persistent state $p_{kk}^{(n)}$ is the nth Fourier coefficient of a probability measure ($n = 0, 1, 2, \cdots$). The following theorem provides the slight additional information concerning (6.1). It will be obvious how this can be improved under additional regularity conditions (such as the existence of moments for the recurrence times), but we shall not dwell on this point.

Theorem 3. *Let the chain be irreducible, aperiodic, and persistent. Let* $\{\lambda_k\}$ *stand for the (essentially unique) invariant measure. There exists a unique system of integrable functions* w_{jk} *for which* (1.12) *and* (1.13) *hold, and* $w_{kk} \geqq 0$*. These functions are continuous outside every neighborhood of* $\vartheta = 0$ *and*

$$(6.1)\qquad \frac{1}{\lambda_j} w_{jj} - \frac{1}{\lambda_k} w_{kk}$$

is square integrable.

Proof. Consider first the diagonal terms. It was proved in section 4 that (1.13) holds with

$$(6.2)\qquad 2\pi w_{kk}(\vartheta) = 2\,\mathrm{Re}\, P_{kk}(e^{i\vartheta}) - 1 \geqq 0.$$

Referring to (2.2), it is now obvious that to prove the square integrability of (6.1) it suffices to prove the proposition that *the boundary values* $\Phi_{jk}(e^{i}\vartheta)$ *are square integrable over* $(-\pi, \pi)$. Now by Schwarz's inequality

$$(6.3)\qquad |{}_jF_{kk}(1) - {}_jF_{kk}(z)|^2 \leqq {}_jF_{kk}(1)\cdot \sum_{n=0}^{\infty} {}_jf_{kk}^{(n)}\ |1 - r^n e^{in\vartheta}|^2$$

$$\leqq 2 \sum_{n=0}^{\infty} {}_jf_{kk}^{(n)}[1 - r^n \cos n\vartheta] \leqq 2\ \text{Re}\ [1 - F_{kk}(z)].$$

Thus

$$|\Phi_{jk}(z)|^2 \leqq 2\,\frac{\text{Re}\ [1 - F_{kk}(z)]}{[1 - F_{kk}(z)]^2} = 2\ \text{Re}\,\frac{1}{1 - F_{kk}(z)}\ ,$$

and we know, from section 4, that the right side is a non-negative function whose integral over $-\pi, \pi$ does not exceed 2.

Turning to the non-diagonal terms put

$$(6.5)\qquad 2\pi u_{j,k}(z) = \left(\frac{\lambda_j}{\lambda_k}\right)^{1/2} P_{jk}^*(z) + \left(\frac{\lambda_k}{\lambda_j}\right)^{1/2} P_{kj}(z).$$

For fixed $r < 1$ it is clear that the nth Fourier coefficient of $u_{jk}(re^{i}\vartheta)$ equals $p_{jk}^{(n)}$ when $n > 0$, and $p_{kj}^{(n)}$ when $n \leqq 0$. Let

$$w_{jk}(\vartheta) = u_{jk}(e^{i\vartheta}).$$

Then $w_{kj}(\vartheta) = w_{jk}^*(\vartheta)$ and, to prove the representation (1.13), it suffices to prove the proposition that u_{jk} *is integrable over* $|z| = 1$.

Put for abbreviation

$$(6.6)\qquad C(z) = \frac{1}{1 - F_{jk}(z)F_{kj}(z)}.$$

We refer again to (2.2) and recall that $\Phi_{jk}(z)$ has integrable boundary values. Using (2.4) and (1.10) it is clear that it suffices to prove that *the function*

$$(6.7)\qquad F_{kj}C + F_{jk}^*C^* = (F_{kj} - F_{jk}^*)C + 2F_{jk}^*\cdot \text{Re}\ C$$

is integrable over $|z| = 1$. As in (6.3) and (6.4) we get, for $|z| < 1$,

$$(6.8)\qquad |(F_{kj} - F_{jk}^*)C|^2 \leqq 2[1 - \text{Re}\ F_{jk}F_{kj}]\cdot |C|^2 = 2\ \text{Re}\ C.$$

But C is the generating function of renewal intensities and therefore the boundary values of Re $C(z)$, for $|z| = 1$, are integrable and non-negative. It follows that (6.7) is integrable, and this concludes the proof.

7. Continuous parameter chains. In [7] Kendall showed that his theorem is applicable also to denumerable chains depending on a continuous time parameter. For this purpose he used a combination of direct Hilbert space methods with a passage to the limit from discrete parameter chains. We show that also in the new situation subinvariant measures play no role and that our elementary approach can be used.

Consider a chain with *continuous* transition probabilities $p_{jk}(t)$ (where $t \geqq 0$ and $p_{kk}(0) = 1$). The role of the generating function is now played by the Fourier transform

$$P_{jk}(\zeta) = \int_0^\infty e^{i\zeta t} p_{jk}(t)\, dt \tag{7.1}$$

where $\zeta = x + iy$ is a complex variable; the integral converges at least in the open half-plane $y > 0$. It is convenient to treat the diagonal terms separately. [It is of some interest that the renewal argument remains applicable if k is stable, that is, if a finite derivative $p'_{kk}(0) = q$ exists. In this case the sojourn time at k has an exponential distribution with expectation q^{-1}. The time to the first return to k has the characteristic function

$$\psi(\zeta) = \frac{q}{q - i\zeta}\, \Phi(\zeta)$$

where $\Phi(\zeta)$ is the characteristic function of the time between the first exit and the first return to k. By the familiar renewal argument the transform (7.1) of $p_{kk}(t)$ equals

$$\frac{1}{1 - \psi(\zeta)} \cdot \frac{1}{q - i\zeta}.$$

It is easily verified that the real part of this function is non-negative and integrable at infinity, and so the argument of section 4 applies without change.]

Theorem 4. *For each diagonal term $p_{kk}(|t|)$ is the characteristic function of a probability distribution on the line which is absolutely continuous except for a possible atom at the origin.*

In other words, it is asserted that there exists an even integrable function $w_{kk}(\alpha) > 0$ such that for $t > 0$

$$p_{kk}(t) - p_{kk}(\infty) = \int_{-\infty}^{+\infty} \cos t\alpha \cdot w_{kk}(\alpha)\, d\alpha. \tag{7.2}$$

Proof. For fixed $h > 0$ and $y > 0$ the Riemann sums

$$\sum_{n=0}^{\infty} e^{inhx} e^{-nhy} p_{kk}(nh) h \tag{7.3}$$

represent the generating function (with $r = e^{-hy}$) of a discrete parameter chain with transition probabilities $e^{-hy} p_{jk}(h)$. Passing to the limit $h \to 0$ it follows that $\operatorname{Re} P_{kk}(x + iy) \geqq 0$. Denoting this function of x by $\frac{1}{2}\gamma_{kk}^{(y)}$ we have

$$\gamma_{kk}^{(y)}(x) = \int_{-\infty}^{+\infty} \cos tx \cdot e^{-y|t|} p_{kk}(|t|)\, dt. \tag{7.4}$$

Now a non-negative characteristic function is integrable if the corresponding probability distribution has a bounded density. It follows, therefore, from (7.4) that $\gamma_{kk}^{(y)}$ is integrable and hence by the Fourier inversion formula

$$e^{-y|t|}p_{kk}(|t|) = \frac{1}{2\pi}\int_{-\infty}^{+\infty} \cos tx \cdot \gamma_{kk}^{(y)}(x)\, dx. \tag{7.5}$$

For fixed $y > 0$ the left side is thus seen to be the characteristic function of a probability distribution and letting $y \to 0$ it follows that $p_{kk}(|t|)$ is the characteristic function of a probability distribution Ω. But we know also that $p_{kk}(|n|h)$ is the nth Fourier coefficient of a probability distribution Ω_h on the unit circle. It is obvious that $\Omega_h\{dx\}$ is obtained in the familiar manner from $\Omega\{h^{-1}\, dx\}$ by rolling up the line on the unit circle. Now, apart from a possible atom at the origin, Ω_h has a density, and this implies the absolute continuity of Ω, except for a possible atom at the origin. (It seems possible in this way to prove that outside a neighborhood of the origin Ω has a continuous density.)

Theorem 5. *Suppose that the chain with continuous transition probabilities $p_{jk}(t)$ is irreducible. If it is transient choose $\lambda_k > 0$ arbitrary, otherwise let $\{\lambda_k\}$ stand for the unique invariant measure. There exists then a unique Hermitian system of integrable functions w_{jk} such that*

$$w_{jk}(-\alpha) = w_{kj}(\alpha) = w_{jk}^*(\alpha)$$

and

$$\left(\frac{\lambda_j}{\lambda_k}\right)^{1/2}[p_{jk}(t) - p_{jk}(\infty)] = \int_{-\infty}^{+\infty} e^{it\alpha}w_{jk}(\alpha)\, d\alpha. \tag{7.6}$$

Proof. Only minor changes are required in the set-up for discrete parameter chains. For $j \neq k$ the first passage time distributions f_{jk} are well defined, and we denote their characteristic functions by F_{jk}. The recurrence relation (1.3) is now inapplicable, but the alternative representation (1.4) remains valid. (See formula (6) §II, 12 of Chung's book. Also the probabilistic derivation outlined after (1.4) remains valid if the simplest regularity properties of the sample paths are taken for granted.) Accordingly the functions $\Phi_{j,k}$ defined in (2.2) can no longer be written in the form (2.3), but their square integrability follows easily from (6.4) by the passage to the limit used in the proof of theorem 4.

The only point in the old argument that requires an amendment is the proof of the integrability of (6.5). In this proof we relied on the integrability of Re $C(\alpha)$ which followed from a renewal argument. In the present situation we can prove only that Re $C(\alpha)$ *is integrable over every finite interval.* Since the same is true of Φ_{jk} the proof of section 6 leads to the local integrability of w_{jk} and we must verify separately that *the boundary values $P_{jk}(x)$ are integrable at* $\pm\infty$. It remains to verify the italicized assertions.

The first italicized assertion is amply contained in the following lemma implicit in [2] and the proof is given only because of its simplicity

Lemma. *Let χ be the characteristic function of a probability distribution H concentrated on* $(0, \infty)$. *Then for each* $\lambda > 0$

$$(7.7) \qquad \frac{1}{\pi}\int_{-\infty}^{+\infty} \mathrm{Re}\,\frac{1}{1-\chi(x)}\cdot\frac{\lambda}{\lambda^2+x^2}\,dx < \infty$$

the integrand being non-negative.

Proof. For fixed $y \geqq 0$ it is clear that $1/(1 - \chi(x + iy))$ is the Fourier–Stieltjes transform of the renewal measure

$$U = \sum H^{n*}.$$

The real part is, therefore, the transform of the symmetric measure U_0 attributing unit mass to the origin and mass $U_0(I) = \frac{1}{2}[U(I) + U(-I)]$ to any interval I excluding the origin. It follows that

$$(7.8) \qquad \frac{1}{\pi}\int_{-\infty}^{+\infty} \mathrm{Re}\,\frac{1}{1-\chi(x+iy)}\cdot\frac{\lambda}{\lambda^2+x^2}\,dx = \int_{-\infty}^{+\infty} e^{-(\lambda+y)|t|}U_0\{dt\}.$$

The positivity of the integrand is easily verified, and letting $y \to 0$ one gets the assertion (7.7).

To prove the second italicized assertion we use the fact that the transition probabilities have bounded densities (see [7], p. 266). By the Plancherel theorem this implies the square integrability of the chatacteristic functions $F_{ik}(x)$ over $-\infty < x < \infty$ and by the Riemann–Lebesgue lemma $F_{ik}(x) \to 0$ as $x \to \pm\infty$. In (1.4), therefore, the denominator tends to 1 while the numerator ${}_iP_{kk}(x)$ is the transform (7.1) of the bounded integrable function ${}_ip_{kk}(t)$. It follows that $P_{kk}(x)$ is square integrable over a neighborhood of $\pm\infty$, and the same is therefore true of $P_{ik} = F_{ik}P_{kk}$.

REFERENCES

[1] K. L. CHUNG, *Markov Chains with Stationary Transition Probabilities*, Springer Verlag, 1960.

[2] K. L. CHUNG - W. H. J. FUCHS, On the distribution of values of sums of independent random variables, *Mem. Amer. Math. Soc.*, **6** (1950) 1–12.

[3] P. ERDÖS, H. POLLARD, & W. FELLER, A property of power series with positive coefficients, *Bull. Amer. Soc.*, **55** (1949) 201–204.

[4] W. FELLER, Boundaries induced by stochastic matrices, *Trans. Amer. Math. Soc.*, **83** (1956) 19–54.

[5] G. H. HARDY & W. W. ROGOSINSKI, *Fourier Series*, Cambridge Tracts no. 38, (1944).

[6] D. G. KENDALL, Unitary dilations of Markov transition operators, and the corresponding integral representations for transition-probability matrices, *Surveys in Probability and Statistics dedicated to H. Cramér*, Stockholm, 1959.

[7] ———, Unitary dilations of one-parameter semigroups of Markov transition operators, and the corresponding integral representations for Markov processes with a countable infinity of states, *Proc. London Math. Soc.* (*3rd series*), **9** (1959) 417–431.

[8] J. F. C. KINGMAN, The exponential decay of Markov transition probabilities, *Proc. London Math. Soc.* (*3rd series*), **13** (1963) 337–358.

[9] S. OREY, Sums arising in the theory of Markov chains, *Proc. Amer. Math. Soc.*, **12** (1961) 847–856.

Princeton University

ON THE INFLUENCE OF NATURAL SELECTION ON POPULATION SIZE

By William Feller

THE ROCKEFELLER UNIVERSITY AND PRINCETON UNIVERSITY

Communicated February 2, 1966

1. *Introduction.*—We are concerned with a problem first considered by J. B. S. Haldane[1] and by M. Kimura.[2] These authors introduced a measure for the loss in population size caused by a slow natural selection; this loss is called selective death by Haldane, and substitutional (or evolutional) load by Kimura. Their calculations lead to the conclusion that the presumed loss may become so enormous as to preclude a simultaneous selection at several loci at a reasonable rate. My attention was drawn to these calculations by Th. Dobzhansky who explained to me the grave consequences of such a result. Other eminent biologists were worried by Haldane's conclusions.[3]

Fortunately for modern evolution theory, the calculations turn out to contain an error which invalidates the result. As has been pointed out many times, the application of the familiar formulas for gene frequencies to natural populations presupposes that the total population size remains constant. A changing population size introduces an inaccuracy which is in practice negligible over reasonably short periods, but the error is cumulative and cannot be neglected in calculations involving long periods of time. This point is of considerable importance because it is customary in evolution theory to think exclusively in terms of relative frequencies and relative fitnesses with a complete neglect of population sizes and absolute rates of change. The dangers of this habit are discussed in a paper[4] which contains also a detailed analysis of Haldane's calculations. The object of the present paper is to present the solution of Haldane's problems in an interesting special case.

2. We consider the selection at one autosomal locus in a sexually reproducing diploid without mutations under random mating. We associate with the three genotypes AA, Aa, and aa *absolute* fitnesses w_1, w_2, and w_3. In other words, w_1 is the expected number of offspring (counted at an appropriate stage of development) resulting from the pairing of two A-alleles. Then

R.L. Schilling et al. (eds.), *Selected Papers II*,

$$w(q) = w_1p^2 + 2w_2pq + w_3q^2 \tag{2.1}$$

is the (absolute) fitness of a population containing the two alleles in the proportion $p:q$ (where $p + q = 1$). To simplify the notations we put, as usual,[5] $w_1 = 1$. We are interested in the development of the population size when the a-allele is at a disadvantage, and hence we put

$$w_1 = 1, \qquad w_2 = 1 - \beta \leq 1, \qquad w_3 = 1 - \gamma \leq 1. \tag{2.2}$$

Assume that we know the size N_0 of the parental population and the relative frequencies p_0, q_0 of the A- and a-alleles in it. The corresponding frequencies in the $(n + 1)$st generation are given by

$$p_{n+1} = p_n\frac{p_n + w_2q_n}{w(q_n)}, \qquad q_{n+1} = q_n\frac{w_2p_n + w_3q_n}{w(q_n)}, \tag{2.3}$$

and the expected size of the $(n + 1)$st generation is determined from

$$N_{n+1} = N_nw(q_n). \tag{2.4}$$

It follows that

$$N_{n+1} = N_0w(q_0)w(q_1)\ \ldots\ w(q_n). \tag{2.5}$$

The assumption (2.2) implies that the fitness $w(q_n)$ never exceeds one, and so the population size N_n *decreases monotonically to a limit* N_∞ *which represents the ultimate population size.* Our problem is to calculate this limit N_∞ under a variety of conditions.

3. *The So-Called Unstable Equilibrium.*—This case deserves special attention because it involves widespread misconceptions, and because it represents the most unfavorable situation. To fix ideas we show first that *if* $w_1 = w_3 = 1$ *but* $w_2 < 1$, *then*

$$N_\infty = N_0(p_0 - q_0), \tag{3.1}$$

provided $q_0 \leq {}^1/_2$. (Otherwise, of course, $N_\infty = N_0(q_0 - p_0)$.)

It is generally assumed that $({}^1/_2, {}^1/_2)$ represents a stationary gene distribution because with $p_0 = q_0 = {}^1/_2$ the formulas (2.3) imply under the present conditions that $p_n = q_n = {}^1/_2$ for all n. This argument neglects the inevitable chance fluctuations in population size (the sampling variance in the terminology of S. Wright) which will disturb the balance already in the first generation. The disturbance may be ever so small, but (3.1) shows that the smaller the disturbance, the smaller the ultimate population size. The practical conclusion is that when $w_3 = w_1 = 1$, *a population starting from the so-called unstable equilibrium* $p_0 = q_0 = {}^1/_2$ *is bound to die out.*

In other words, *an absolutely lethal allele* (a) *would be more favorable for the population.* This is true if $w_2 < 1$, no matter how small the difference $\beta = 1 - w_2$ is. The apparent paradox becomes understandable when one notes that the selection acts only on the heterozygotes; with a small β the loss in any one generation is small, but effective selection takes place over a long period. On the other hand, if the a-allele were absolutely lethal, the population size would remain constant at the level $N_0p_0^2$.

This qualitative argument cannot establish the exact relation (3.1). Before proving it, we remark that the results of section 5 imply that a similar relation holds for an arbitrary "unstable equilibrium," namely, when $w_2 < w_3$. The critical point is then represented by the frequencies

$$p = \frac{1 - w_2}{1 + w_3 - 2w_2} = \frac{\beta}{2\beta - \gamma}, \qquad q = \frac{w_3 - w_2}{1 + w_3 - 2w_2} = \frac{\beta - \gamma}{2\beta - \gamma}. \tag{3.2}$$

Here $q_n \to 0$ if $q_0 < q$, whereas $q_n \to 1$ if $q_0 > q$. The analogue to (3.1) is given by (5.2) for $s = -1$.

Proof of formula (3.1): Suppose $q_0 < 1/2$. When $w_3 = 1$, it follows from (2.3) that

$$w(q_n) = \frac{p_n^2 - q_n^2}{p_{n+1} - q_{n+1}} = \frac{p_n - q_n}{p_{n+1} - q_{n+1}}. \tag{3.3}$$

When the factors in (2.5) are expressed in this way, most terms cancel, and one obtains

$$N_{n+1} = N_0 \frac{p_0 - q_0}{p_{n+1} - q_{n+1}}. \tag{3.4}$$

But $p_{n+1} \to 1$ and so (3.1) is true.

4. *Comparison with Lethals.*—We now consider the general situation where $q_n \to 0$ and show that

$$\textit{if } w_3 \geq w_2^2, \textit{ then } N_\infty \leq N_0 p_0^2 \textit{ while } w_3 \leq w_2^2 \textit{ implies } N_\infty \geq N_0 p_0^2. \tag{4.1}$$

As previously pointed out, if $w_2 = w_3 = 0$ (that is, if a is lethal), the expected population size remains fixed at the level $N_0 p_0^2$ representing the number of homozygotes in the parental generation. We see thus that $w_3 = w_2^2$ represents a critical value in the following sense: *when $w_3 > w_2^2$, the ultimate population size is smaller than it would be if (a) were totally lethal, whereas with $w_3 < w_2^2$ the situation is more favorable than with lethals.* Thus, *an increase in fitness does not normally produce an increase of the ultimate population size.*

To prove the first half of (4.1), we use the first relation in (2.3) which we rewrite in the form

$$w^2(q_n) = \frac{p_n^2}{p_{n+1}^2} (p_n + w_2 q_n)^2 < \frac{p_n^2}{p_{n+1}^2} w(q_n). \tag{4.2}$$

Thus, $w(q_n) < p_n^2/p_{n+1}^2$. Substituting into (2.5), we get because of the obvious cancellation of terms $N_n < N_0 p_0^2/p_{n+1}^2$, and since by assumption $p_{n+1} \to 1$, this proves the assertion. The second half of (4.1) follows by a similar argument.

5. *The General Case.*—Referring to the notation (2.2), we now put

$$\gamma = (s + 1)\beta. \tag{5.1}$$

Then $w_3 \leq w_2$ if $s \geq 0$, and otherwise there exists a state of "unstable equilibrium" given by (3.2). Under any circumstances $s \geq -1$. Following Haldane we are now considering the case of a mild selection, that is, we suppose that β is small. We shall prove that with an error of the order of magnitude β, if $q_n \to 0$, then *one has approximately*[6]

$$N_\infty \approx N_0 p_0 \left(\frac{p_0}{p_0 + sq_0}\right)^{1/s}, \tag{5.2}$$

except if $s = 0$, in which case

$$N_\infty \approx N_0 p_0 e^{-q_0/p_0}. \tag{5.3}$$

If $w_3 = 1$, one has $s = -1$, and in this case the right side in (5.2) reduces to $N_0(p_0 - q_0)$. The condition $q_n \to 0$ is satisfied if (and only if) $q_0 < p_0$; and we saw in section 4 that under these circumstances (5.2) provides an exact formula for all possible values of β. Again $N_\infty = 0$ if q_0 is chosen at the "unstable equilibrium" (3.2).

For $s = 1$ the right side of (5.2) reduces to $N_0 p_0^2$. As we saw in the preceding section, the relation (5.2) reduces to an exact formula when $w_3 = w_2^2$, that is, when $\gamma = 2\beta - \beta^2$, even if β is not small.

It can be shown that the right side in (5.2) is a monotone function of s, that is, *the ultimate population size is (for fixed initial gene frequencies p_0, q_0) the smaller, the greater the fitness of the homozygotes aa.* For s exceedingly large, (5.2) reduces to the approximation $N_\infty \approx N_0 p_0$. The a-alleles are eliminated, and the ultimate population size equals approximately the number of a-alleles in the parental population.

Proof of (5.2): Put for abbreviation $r_n = q_n/p_n$. From (2.3) one sees easily that

$$r_{n+1} = r_n \frac{w_2 p_n + w_3 q_n}{p_n + w_2 q_n} = r_n \frac{w_2 + w_3 r_n}{1 + w_2 r_n}. \tag{5.4}$$

Using the notations (2.2) and (5.1), one concludes easily that

$$r_n - r_{n+1} = \beta r_n \cdot \frac{1 + sr_n}{1 + (1 - \beta) r_n}. \tag{5.5}$$

Writing

$$\lambda_n = \frac{\beta r_n}{1 + (1 - \beta) r_n} = \frac{\beta q_n}{1 - \beta q_n}, \tag{5.6}$$

we may rewrite (5.5) in the form

$$\lambda_n = \frac{r_n - r_{n+1}}{1 + sr_n}. \tag{5.7}$$

By the mean value theorem one has approximately

$$\lambda_n \approx \frac{1}{s} \log \frac{1 + sr_n}{1 + sr_{n+1}} \tag{5.8}$$

with an error term of the order of $(r_n - r_{n+1})^2$, that is, β^2. Summing over n we get, because of the obvious cancellation of terms and because $r_n \to 0$,

$$\sum_{n=0}^{\infty} \lambda_n \approx \frac{1}{s} \log (1 + sr_0) = \frac{1}{s} \log \frac{p_0 + sq_0}{p_0}, \tag{5.9}$$

provided $s \neq 0$.

From the first relation in (2.3) we get

$$w(q_n) = \frac{p_n}{p_{n+1}}(1 - \beta q_n) \approx \frac{p_n}{p_{n+1}} \cdot \frac{1}{1 + \lambda_n} \approx \frac{p_n}{p_{n+1}} e^{-\lambda_n} \tag{5.10}$$

with a relative error of the order of magnitude λ_n^2. Introducing this into (2.5), we get

$$N_{n+1} \approx N_0 \cdot \frac{p_0}{p_{n+1}} e^{-(\lambda_0 + \lambda_1 + \ldots + \lambda_n)}. \tag{5.11}$$

In view of (5.9) the assertion (5.2) follows letting $n \to \infty$ in (5.11). The special case (5.3) is obtained in like manner, or else by a passage to the limit $s \to 1$ in (5.2).

6. *The Error Term.*—The relative error in the approximation formulas (5.2) and (5.3) is at most of the order of magnitude of β and is of no particular interest for the theory of evolution. We shall therefore be satisfied with a brief indication how the next term of the approximation may be obtained, reducing the relative error to the order of magnitude β^2.

In (5.10) it is clear that $e^{-\lambda_n}$ must be replaced by $\exp\{-\lambda_n + {}^1/_2\lambda_n^2\}$. Similarly, the use of a two-term Taylor approximation replaces (5.8) by the more accurate formula

$$\frac{1}{s}\log\frac{1 + sr_n}{1 + sr_{n+1}} \approx \lambda_n + {}^1/_2\frac{(r_n - r_{n+1})^2}{(1 + sr_n)^2} = \lambda_n + {}^1/_2\lambda_n^2. \tag{6.1}$$

To obtain a better approximation, we have to evaluate $\lambda_0^2 + \lambda^2 + \ldots$ up to an error of magnitude $O(\beta^2)$. From (5.6) and (5.7) we obtain

$$\lambda_n^2 = \frac{\beta r_n}{1 + (1 - \beta)r_n} \cdot \frac{r_n - r_{n+1}}{1 + sr_n}, \tag{6.2}$$

and hence

$$\sum_{n=0} \lambda_n^2 \approx \int_0^{r_0} \frac{\beta r dr}{(1 + (1 - \beta)r)\cdot(1 + sr)}. \tag{6.3}$$

The integral can be evaluated using integration by parts. The final result is that when $s \neq 1, 2$,

$$N_\infty \approx N_0 p_0 \left(\frac{p_0}{p_0 + sq_0}\right)^{\frac{1}{s} + \beta\frac{s+1}{s(s-1)}} \cdot q_0^{{}^1/_2\beta(s+1)/(s-1)}, \tag{6.4}$$

with a relative error of $O(\beta^2)$.

The author is greatly indebted to Th. Dobzhansky and Wyatt W. Anderson for stimulating discussions.

[1] Haldane, J. B. S., "The cost of natural selection," *J. Genet.*, **55**, 511–524 (1957). More precise expressions for the cost of natural selection in *J. Genet.*, **57**, 351–360 (1960).

[2] Kimura, M., "Optimum mutation rate and the degree of dominance as determined by the principle of minimum of genetic load," *J. Genet.*, **57**, 21–34 (1960); "Natural selection as the process of accumulating genetic information in adaptive evolution," *Genet. Res. Cambridge*, 127–140 (1961).

[3] See, e.g., Brues, A. M., "The cost of evolution versus the cost of non-evolving," *Evolution*, **18**, 379–383 (1964); Mayr, E., *Animal Species and Evolution* (Harvard Univ. Press, 1963).

[4] Feller, W., "On fitness and the cost of natural selection," to appear in *Genet. Res.*

[5] The general case $w_1 \neq 1$ is covered by our formulas if one replaces N_n by $N_n w_1^{-n}$. The true population size is for large n given by $N_\infty w_1^n$, approximately, and N_∞ / N_0 represents the per cent loss in populatioh size caused by the initial gene frequencies (p_0, q_0) compared to the case when the entire population consists of AA-homozygotes. For the importance of considering the fitness, $w_1 > 1$, see the paper cited in ref. 4.

[6] When $s < 0$, there exists an "unstable equilibrium" (3.2) and $q_n \rightarrow 0$ only if $p_0 + sq_0 > 0$.

J. SIAM Appl. Math.
Vol. 14, No. 4, July, 1966
Printed in U.S.A.

INFINITELY DIVISIBLE DISTRIBUTIONS AND BESSEL FUNCTIONS ASSOCIATED WITH RANDOM WALKS*

WILLIAM FELLER†

1. Introduction. The reflections of this paper are elementary, but they may present some amusement value in that we get simple proofs for useful results previously obtained only by hard methods. We shall discuss two topics that are only vaguely connected. Despite the importance of infinitely divisible distributions surprisingly few such distributions are known explicitly, and we shall add to this collection by showing that certain first passage times and hitting points in ordinary random walks are infinitely divisible. This result may be of some interest to workers in difference equations and Monte Carlo methods because they imply that in various approximation- and simulation-schemes ordinary random walks may be replaced (theoretically at least) by stochastic processes of an entirely different nature and depending on a continuous time parameter. In this connection it will be observed that the probabilist may view boundary value problems of the Dirichlet type in a manner departing radically from the orthodox analytic point of view.

In §7 we turn to another method of replacing random walks by stochastic processes depending on continuous time. The new process is obtained by randomization of the times at which the jumps of the random walk occur. To probabilities in the original random walk there correspond probabilities in the new process which can be written down explicitly in terms of the former. Many such probabilities involve Bessel functions, and we obtain in this way probabilistic interpretations of some basic relations for Bessel functions. It will be shown in the appendix that the method leads to a simple proof of a formula to which access is otherwise difficult, although it is of constant use in differential equations and queuing theory. §§7 and 8 depend only on the results of §2. No acquaintance with infinitely divisble distributions is assumed.

2. First passage times. We begin with the familiar random walk in one dimension with unit steps. For notational simplicity we assume that the

* Received by the editors July 7, 1965. Presented by invitation at the Courant Symposium, held at the Courant Institute of Mathematical Sciences, New York, June 8–9, 1965.

† Department of Mathematics, Princeton University, Princeton, New Jersey. This research was supported in part by the United States Army Research Office, Project on probability and differential equations at Princeton University.

process is symmetric.[1] As usual, we pretend that the nth step occurs at time n, and let $\mathbf{S}_n$ stand for the position at time n; in particular, $\mathbf{S}_0 = 0$.

For $r = 1, 2, \cdots$, we denote by $\mathbf{T}^{(r)}$ the (chance dependent) *time of the first visit to the point* r, and by $\{f_n^{(r)}\}$ its probability distribution. More formally, $f_0^{(r)} = 0$ for all r whereas for $n > 0$,

$$(2.1)\quad f_n^{(r)} = P\{\mathbf{T}^{(r)} = n\} = P\{\mathbf{S}_1 < r, \cdots, \mathbf{S}_{n-1} < r, \mathbf{S}_n = r\}.$$

For abbreviation we put $f_n^{(1)} = f_n$ and introduce the generating function

$$(2.2)\qquad F(\zeta) = \sum_{n=1}^{\infty} f_n \zeta^n.$$

To derive the distribution $\{f_n^{(r)}\}$ of $\mathbf{T}^{(r)}$ we note that a visit to a point $r + s > 0$ is necessarily preceded by a visit to r. Taking the point r as new origin it is seen that the time $\mathbf{T}^{(r+s)} - \mathbf{T}^{(r)}$ between the first visits to r and $r + s$ is stochastically independent of $\mathbf{T}^{(r)}$ and has the same distribution as $\mathbf{T}^{(s)}$. Analytically this fact is reflected in the convolution equation

$$(2.3)\qquad f_n^{(r+s)} = \sum_{k=0}^{n} f_k^{(r)} f_{n-k}^{(s)}$$

(which is a special case of the Chapman-Kolmogorov equation). It implies an obvious multiplicative property for the generating function $\sum f_n^{(r)} \zeta^n$ of $\mathbf{T}^{(r)}$ and we conclude that *the generating function of* $\mathbf{T}^{(r)}$ *is given by* F^r.

It used to be considered hopeless to obtain explicit expressions for the probabilities $f_n^{(r)}$ themselves, and rather intricate papers were devoted to the derivation of approximations and asymptotic properties. It turns out, however, that elementary combinatorial arguments (see [2, Chap. III, §4]) (independent of generating functions) lead to the explicit formula

$$(2.4)\qquad f_{2k+r}^{(r)} = \frac{r}{2k+r} \binom{2k+r}{r} \frac{1}{2^{2k+r}}$$

for $k = 0, 1, \cdots$ (obviously $f_{n+r}^{(r)} = 0$ if n is odd). We give here a new proof depending on generating functions; it is more in the spirit of the present paper and explains the generalizations to higher dimensions.

We begin by calculating the generating function F. If the first step in the random walk is positive we have $\mathbf{T}^{(1)} = 1$ while the argument preceding (2.3) shows that in the contrary case $\mathbf{T}^{(1)} = 1 + \mathbf{Y}$, where $\mathbf{Y}$ is a random variable with the same distribution as $\mathbf{T}^{(2)}$. The generating function of this

[1] With trite changes of parameters our considerations apply equally to random walks in which the steps to the right and left have probabilities p and q, respectively. In particular, the generating function $F(\zeta)$ in (2.6) is replaced by $\sqrt{p/q}F(2\sqrt{pq}\zeta)$, and in (2.8) the factor 2^{-r-2k} is replaced by $p^{r+k} q^k$.

variable being F^2 we see that F satisfies the quadratic equation

$$F(\zeta) = \tfrac{1}{2}\zeta + \tfrac{1}{2}\zeta F^2(\zeta). \tag{2.5}$$

Since $0 < F(\zeta) < 1$ for $0 < \zeta < 1$ this implies that

$$F(\zeta) = \frac{1 - \sqrt{1 - \zeta^2}}{\zeta}, \tag{2.6}$$

and thus we are in possession of an explicit expression for the generating function F^r. To find its coefficients we introduce the probabilities and generating functions

$$u_n^{(r)} = P\{\mathbf{S}_n = r\}, \qquad U^{(r)}(\zeta) = \sum_{n=0}^{\infty} u_n^{(r)}\zeta^n. \tag{2.7}$$

The event $\{\mathbf{S}_n = r\}$ occurs if and only if among the first n steps exactly $(n + r)/2$ are positive, and hence

$$u_{r+2k}^{(r)} = \binom{r + 2k}{k} 2^{-r-2k}, \tag{2.8}$$

while $u_n^{(r)} = 0$ if $n - r$ is odd. The generating functions $U^{(r)}$ are therefore known, and we have in particular[2]

$$U^{(0)}(\zeta) = \frac{1}{\sqrt{1 - \zeta^2}} = \frac{1}{1 - \zeta F(\zeta)}. \tag{2.9}$$

Now $\mathbf{S}_n = r$ if the first visit to r occurred at some time $\nu \leqq n$ and then $\mathbf{S}_n - \mathbf{S}_\nu = 0$. Thus $u_n^{(r)}$ satisfies a convolution formula equivalent to the identity

$$U^{(r)} = U^{(0)}F^r. \tag{2.10}$$

It follows that

$$F^r(\zeta) = U^{(r)}(\zeta) - \zeta U^{(r+1)}(\zeta), \tag{2.11}$$

or

$$f_n^{(r)} = u_n^{(r)} - u_{n-1}^{(r+1)}, \tag{2.12}$$

which is the same as (2.4).

3. Interpolation for nonintegral *r*. The relations established in the preceding section are known, but they appear in a new light when one observes

[2] A direct probabilistic proof of (2.9) is as follows. It is easily seen that the probability of a *first return to* 0 at time $n + 1$ equals f_n. The recurrence times to the origin have therefore the generating function $\zeta F(\zeta)$. Now $\mathbf{S}_n = 0$ if and only if a return to 0 occurs at time n, and hence $U^{(0)}(\zeta) = \sum_{n=0}^{\infty} [\zeta F(\zeta)]^n$, which is the same as (2.9).

that they remain meaningful also for nonintegral r. We show that for every positive r the function F^r may be interpreted as the generating function of a random variable $\mathbf{T}^{(r)}$. We begin by a formal argument postponing a probabilistic discussion to the next section.

The random variable $\mathbf{T}^{(r)}$ assumes the values $r, r + 2, r + 4, \cdots$, and the interpolated random variables are therefore not integral-valued. This is of no serious consequence, but the resulting notational inconvenience can be avoided by an appropriate centering, namely by considering the random variables $\mathbf{T}^{(r)} - r$ instead of $\mathbf{T}^{(r)}$. The generating function of $\mathbf{T}^{(r)} - r$ is g^r, where g stands for the generating function of $\mathbf{T}^{(1)} - 1$, namely

$$g(\zeta) = \zeta^{-1}F(\zeta) = \frac{1 - \sqrt{1 - \zeta^2}}{\zeta^2}. \tag{3.1}$$

With this change of notation we prove the following:

THEOREM. *For fixed* $t > 0$ *the function* g^t *represents the generating function of the distribution of a random variable* $\mathbf{T}(t)$ *assuming the values* $0, 2, 4, \cdots$ *with probabilities*

$$P\{\mathbf{T}(t) = 2k\} = \frac{t}{2k+t}\binom{2k+t}{k}\frac{1}{2^{2k+t}}. \tag{3.2}$$

Proof. It is seen that

$$\frac{d}{d\zeta}\log g(\zeta) = \frac{1}{\zeta}\left[-1 + \frac{1}{\sqrt{1-\zeta^2}}\right] = \sum_{n=1}^{\infty}\binom{2n}{n}\frac{1}{2^{2n}}\zeta^{2n-1}, \tag{3.3}$$

and hence

$$\log\frac{g(\zeta)}{g(0)} = \sum_{n=1}^{\infty}\binom{2n}{n}\cdot\frac{1}{2n}\left(\frac{\zeta}{2}\right)^{2n}. \tag{3.4}$$

But $g(0) = \frac{1}{2}$ and so for $0 < \zeta < 1$,

$$g^t(\zeta) = \exp\left([t\log 2]\cdot[-1 + h(\zeta)]\right), \tag{3.5}$$

where h is the probability generating function defined by

$$h(\zeta) = \frac{1}{\log 2}\sum_{n=1}^{\infty}\binom{2n}{n}\frac{1}{2n}\left(\frac{\zeta}{2}\right)^{2n}. \tag{3.6}$$

Now (3.5) represents the canonical form for the generating function of an infinitely divisible integral-valued random variable, but for our purposes no appeal to the general theory is necessary. It suffices to rewrite (3.5) in the form

$$g^t(\zeta) = e^{-t\log 2}\sum_{n=0}^{\infty}\frac{(t\log 2)^n}{n!}h^n(\zeta), \tag{3.7}$$

which exhibits g^t as a weighted average of the generating functions $h^0, h^1, h^2, \cdots$. It follows that g^t corresponds to a mixture of the convolutions H^{n*} of the distribution determined by h. In other words, g^t generates a compound Poisson distribution.

It remains to show that the coefficients of g^t are given by the right side in (3.2). A direct analytic verification would be hopeless, but we know that the assertion is true for $t = 0, 1, 2, \cdots$, and prove it in general by an interpolation argument.

Recall that a function $\phi(t)$ defined for $t > 0$ is absolutely monotone if it has derivatives of all orders and $\phi^{(n)}(t) > 0$ for all positive n and t. We use the well-known fact that such a function is uniquely determined by the values $\phi(0), \phi(1), \phi(2), \cdots$.

For fixed ζ with $0 < \zeta < 1$ we have

$$2g(\zeta) = \frac{2}{1 + \sqrt{1 - \zeta^2}} > 1,$$

and hence $(2g(\zeta))^t$ is an absolutely monotone function of t. Keeping ζ fixed, put

$$a(t) = \sum_{k=0}^{\infty} \frac{t}{2k + t} \binom{2k + t}{k} \frac{1}{2^{2k}} \zeta^{2k}. \tag{3.8}$$

Each individual term of this series is a polynomial in t with nonnegative coefficients, and hence a is an absolutely monotone function of t. But comparing (3.8) with (2.4) it is seen that $a(r) = 2^r g^r(\zeta)$ for all *integral* $r > 0$. By the uniqueness property of absolutely monotone functions this implies that $a(t) = 2^t g^t(\zeta)$ for all $t > 0$, and this completes the proof.

At the same time we have proved:

COROLLARY. *g^t is of the form* (3.7), *where h is the probability generating function defined in* (3.6).

4. $\{\mathbf{T}(t)\}$ process. The last result may be restated probabilistically as follows.

There exists a stochastic process $\{\mathbf{T}(t), t \geqq 0\}$ with independent increments proceeding by discrete jumps so that the number of jumps between 0 *and t is the variable of a Poisson process with intensity* $\log 2$. *The successive jumps are mutually independent random variables assuming positive even values. The probability that a jump has magnitude $2n$ equals*

$$h_{2n} = \frac{1}{\log 2} \binom{2n}{n} \frac{1}{2n} \cdot 2^{-2n}. \tag{4.1}$$

This distribution of $\mathbf{T}(t)$ is given in (3.2). In other words: whatever the value $\mathbf{T}(t)$ and the history of the process up to time t, the probability that a jump occurs within the short time interval $(t, t + \tau)$ is $\tau \log 2 + o(\tau)$; if a jump occurs it equals $2n$ with probability h_{2n}, and no other values are possible. Here $n = 1, 2, \cdots$.

It should be borne in mind that for integral $t = r$ the distribution of $\mathbf{T}(t)$ coincides with the distribution of the first passage time variable $\mathbf{T}^{(r)} - r$. Now all probabilities relating to the random walk can be expressed in terms of the first passage probabilities $f_n^{(r)}$, and therefore in terms of the distributions of the new process. Furthermore, the new process $\{\mathbf{T}(t)\}$ can be simulated by Monte Carlo techniques quite independently of the random walk. It follows that whenever the random walk is used for approximations to differential equations, for asymptotic solutions, and similar purposes, it is in principle possible to replace the random walk by the new process $\{\mathbf{T}(t)\}$. Whether this method is practical remains an open question. The theoretical possibilities and implications will appear more clearly in connection with random walks in higher dimensions.

Before passing to them one side remark is in order. We have added an example to the small collection of explicitly known divisible distributions. From it, it is easy to obtain many new ones. In fact, if ϕ is any probability generating function it is easily verified that $g^t(\phi(\zeta))$ is again the generating function of an infinitely divisible variable. Examples for this procedure occur naturally in the next section.

5. Higher dimensions. The random walk in a plane regular lattice may serve as a typical example for random walks in higher dimensions. The state at time n is given by the vector $\mathbf{S}_n = (\mathbf{X}_n, \mathbf{Y}_n)$ with integral components, and for notational simplicity we again assume symmetry: a step to each of the 4 neighboring points has probability $\frac{1}{4}$. Instead of visits to the point $r > 0$ we consider visits to the line $x = r$, but to the random variable $\mathbf{T}^{(r)}$ there correspond two new random variables:

(a) *The hitting time* $\mathcal{A}^{(r)}$ is the time of the first visit to the line $x = r$.

(b) *The hitting point* $\mathcal{B}^{(r)}$ is the ordinate of the point at which the first visit to the line $x = r$ occurs.

More formally, for $r = 1, 2, \cdots$, the pair $(\mathcal{A}^{(r)}, \mathcal{B}^{(r)})$ is defined by the condition that

$$(5.1) \quad \{\mathcal{A}^{(r)} = m, \mathcal{B}^{(r)} = n\} = \{X_1 < r, \cdots, \mathbf{X}_{m-1} < r, \mathbf{X}_m = r, \mathbf{Y}_m = n\}.$$

We begin with the variables $\mathcal{A}^{(r)}$ because they are positive and therefore require no change in notations. It will be shown that they are in every way similar to the first passage times $\mathbf{T}^{(r)}$. This result can be derived directly or else from the results of §2.

First, it is clear that the argument of §2 applies without essential change. In particular, $\mathcal{A}^{(r)}$ is again the sum of r independent random variables distributed like $\mathfrak{A}^{(1)}$. Accordingly, if $A(\zeta)$ stands for the generating function of $\mathfrak{A}^{(1)}$, then $\mathfrak{A}^{(r)}$ has the generating function A^r. Furthermore, the argument that led to the quadratic equation (2.5) shows that

$$A(\zeta) = \tfrac{1}{4}\zeta + \tfrac{1}{2}\zeta A(\zeta) + \tfrac{1}{4}\zeta A^2(\zeta). \tag{5.2}$$

It follows that

$$A(\zeta) = \frac{2 - \zeta - 2\sqrt{1 - \zeta}}{\zeta}, \tag{5.3}$$

which is the same as

$$A(\zeta) = F\left(\frac{\zeta}{2 - \zeta}\right). \tag{5.4}$$

This result can be derived also using the concluding remark of the preceding section. Indeed, let $\mathbf{N}$ be the number of steps in the x-direction up to (and including) the first visit to the line $x = 1$. Obviously $\mathbf{N}$ is a random variable with the same distribution as the first passage time $\boldsymbol{T}^{(1)}$, and hence $\mathbf{N}$ has the generating function F. Between the $(n - 1)$th and the nth step in the x-direction there occur $\mathbf{Z}_n$ steps in the y-direction. Thus, adding the one step in the x-direction, the nth step in the x-direction increases the total number of steps by $\mathbf{Z}_n + 1$. The random variables $\mathbf{Z}_n + 1$ being independent of each other and of $\mathbf{N}$, they have a common geometric distribution with generating function $\zeta(2 - \zeta)^{-1}$. The number $\mathfrak{A}^{(1)}$ of steps up to the first visit to the line $x = 1$ is the sum $\mathbf{Z}_1 + \cdots + \mathbf{Z}_N + \mathbf{N}$ with a random number of elements, and its generating function is therefore given by (5.4).

As was remarked at the end of the preceding section, it follows from (5.4) that *the variables* $\mathfrak{A}^{(r)}$ *are again infinitely divisible* and may be imbedded in a compound Poisson process.

Substantially the same considerations apply to the variables $\mathfrak{B}^{(r)}$, but, since these assume also negative values, the method of generating functions is not applicable and we must use the characteristic function of $\mathfrak{B}^{(1)}$, namely

$$B(\zeta) = \sum_{k=-\infty}^{+\infty} P\{\mathfrak{B}^{(1)} = k\}e^{ik\zeta}. \tag{5.5}$$

The argument used for F and A shows that $\mathfrak{B}^{(r)}$ has the characteristic function $B^r(\zeta)$ and that B satisfies the quadratic equation

$$B(\zeta) = \tfrac{1}{4} + \tfrac{1}{2}(\cos\zeta)B(\zeta) + \tfrac{1}{4}B^2(\zeta). \tag{5.6}$$

Accordingly,

$$B(\zeta) = 2 - \cos\zeta - \sqrt{(2 - \cos\zeta)^2 - 1} = F\left(\frac{1}{2 - \cos\zeta}\right). \tag{5.7}$$

The latter representation may be derived also by the argument used above for $A(\zeta)$. It follows again that the variables $\mathcal{B}^{(r)}$ *are infinitely divisible* and can be imbedded in a continuous time parameter process. In the present case, however, the variables $\mathcal{B}^{(r)}$ are of independent interest in connection with harmonic functions, and we turn to a discussion of this because it sheds some new light on all our processes.

6. On boundary value problems for harmonic functions. The Dirichlet problem for harmonic functions in the halfplane $x > 0$ seeks a harmonic function u defined for $x > 0$ and all y, and such that $u(x, y) \to f(y)$ as $x \to 0+$, where f is a prescribed function. Under appropriate conditions on f the solution is given by

$$u(x, y) = \frac{1}{\pi} \int_{-\infty}^{+\infty} f(\rho) \frac{x}{x^2 + (y - \rho)^2} \, d\rho. \tag{6.1}$$

In analysis

$$\pi^{-1} \frac{x}{x^2 + y^2} \tag{6.2}$$

is known as the Poisson kernel, but to probabilists (6.2) represents the Cauchy density. The Cauchy process is a process in one dimension with independent increments and transition probabilities given by (6.2) with x serving as time parameter. The point of interest in the present context is that the (one-dimensional) Cauchy process is related to the two-dimensional Brownian motion (Wiener process) just as our variables $\mathcal{B}^{(r)}$ are to the symmetric two-dimensional random walk: *In a two-dimensional Brownian motion starting at the origin the ordinate of the point at which the path for the first time meets the line with abscissa $x > 0$ has the Cauchy density* (6.2). (This assertion can be proved directly by dimensional analysis, or indirectly by the standard passage to the limit from random walks to Brownian motion.)

A basic property of the Cauchy density (6.2) is that it is *harmonic* in the halfplane $x > 0$. Similarly, the probabilities $b_n^{(r)} = P\{\mathcal{B}^{(r)} = n\}$ are *preharmonic*, that is, they satisfy the difference equation

$$b_n^{(r+1)} + b_n^{(r-1)} + b_{n+1}^{(r)} + b_{n-1}^{(r)} = 4b_n^{(r)} \tag{6.3}$$

(which states substantially the same as the quadratic equation (5.6)). Accordingly, if $\{f_k\}$ is any bounded doubly infinite sequence then

$$u(r, m) = \sum_{k=-\infty}^{\infty} b_{m-k}^{(r)} f_k \tag{6.4}$$

defines a solution of the preharmonic equation

$$u(r, m) = \tfrac{1}{4}[u(r + 1, m) + u(r, m + 1) + u(r, m - 1) + u(r - 1, m)]. \tag{6.5}$$

For this reason the distributions of $\mathfrak{B}^{(r)}$ may play the same role for the preharmonic difference equation (6.5) as the Cauchy density (6.2) does in the theory of harmonic functions.[3]

It has become standard practice to relate harmonic functions to preharmonic functions, and the latter to the symmetric random walk with unit steps. We see now that there exists an entirely different approach which is more natural from the probabilistic point of view. In interpreting the classical formula (6.1) it is not necessary (or desirable) to insist on the fact that a Cauchy process can be derived from a higher dimensional Brownian motion in the manner described. The natural probabilistic interpretation of (6.1) is related to a Cauchy process in *one* dimension with sample paths $\mathbf{C}(t)$; (6.1) states that $u(x, y)$ is the expectation of the random variable defined for fixed x, y by $f(y + \mathbf{C}(x))$. The discrete analogue is given by (6.4) and refers to *a random walk in which the individual steps assume the value k with probability $b_k^{(1)}$* (here $k = 0, \pm 1, \pm 2, \cdots$). In other words, instead of the two-dimensional random walk with unit steps, Monte Carlo methods for our Dirichlet problem could as well use the random walk $\mathfrak{B}^{(1)}, \mathfrak{B}^{(2)}, \cdots$. It is not clear how practical such a procedure is, but the theoretical stability and approximation properties are better than with the old approach.

There exists another possibility, namely to *discretize only the space variable* y while keeping the time parameter x continuous. This corresponds to using the interpolated process $\mathfrak{B}(t)$ which, as we have seen, is a compound process defined as follows. The number of jumps in a time interval of duration t is a Poisson variable with expectation $t \log 2$. The successive jumps are integral-valued random variables with the characteristic function $h(1/[2 - \cos \zeta])$. (This procedure is related to the random walks with randomized time discussed in the next section.)

Similar considerations apply to more general boundary value problems. Analysts are used to seeing an essential difference between parabolic initial value problems and elliptic problems such as the Dirichlet problem for the unit disk. From a probabilistic point of view the difference is rather superficial, and the latter may be related to a one-dimensional stochastic process (of the Cauchy type) with $\log (1/r)$ as time parameter. The given boundary values $f(\theta)$ determine the values of a function $u(r, \theta)$ for every fixed $r > 0$ which is easily seen to be harmonic in the punctured disk. The circle being compact, the ergodic theorem for Markov processes guarantees the continuity of u in the entire open disk.

7. Randomized time and Bessel functions. We return to the one-dimensional symmetric random walk with unit steps but drop the assumption that

[3] An equivalent result in purely analytical terms has been obtained by Allen and Murdoch [1]. Their argument depends on intricate techniques and it is interesting to note how the probabilistic approach simplifies and elucidates the theory.

the steps occur regularly in time. Instead we assume that the time intervals between successive steps are random variables with a common probability distribution L and independent of each other as well as of the random walk itself. We denote by $v_k(t)$ the probability of the state k at time t; thus $k = 0, \pm 1, \pm 2, \cdots$, and $v_0(0) = 1$.

With an arbitrary distribution L the future development depends not only on the present state, but also on the time elapsed since the present state was entered. This means that the process is not Markovian, and the Chapman-Kolmogorov equation

$$(7.1) \qquad v_n(s + t) = \sum_{k=-\infty}^{+\infty} v_k(s)v_{n-k}(t)$$

does not necessarily hold. Due to the well-known properties of the exponential distribution, however, the process is Markovian and (7.1) holds when L is exponential. From now on we assume therefore that the times between successive jumps are exponentially distributed with unit expectation. Under these circumstances (7.1) is true and the $v_k(t)$ serve also as transition probabilities in our process.

The number of jumps up to time $t > 0$ is a Poisson variable with expectation t and so

$$(7.2) \qquad v_k(t) = e^{-t} \sum_{n=0}^{\infty} \frac{t^n}{n!} u_n^{(k)},$$

where $u_n^{(k)}$ is the probability that the nth step of a random walk leads to k. This probability is given in (2.8) and direct substitution into (7.2) shows that

$$(7.3) \qquad v_k(t) = e^{-t} I_k(t),$$

where I_ρ stands for the so-called modified Bessel function of order $\rho > 0$ defined by

$$(7.4) \qquad I_\rho(t) = \sum_{\nu=0}^{\infty} \frac{1}{\nu!\,\Gamma(\nu + \rho + 1)} \left(\frac{t}{2}\right)^{2\nu+\rho}.$$

It will be noticed that the identity (7.1) reduces to

$$(7.5)^* \qquad I_n(t + s) = \sum_{k=-\infty}^{\infty} I_k(s)I_{n-k}(t),$$

which formula is well-known in the theory of Bessel functions where it is known as K. Neumann's identity. Many other important relations for Bessel functions appear in like manner as a simple byproduct of proba-

* *Added in proof.* J. Keilson [4] obtained an analog to (7.5) for generalized Bessel functions starting from a Chapman-Kolmogorov equation appearing in queuing theory.

bilistic considerations. For example, if we drop the assumption of symmetry and assign to the steps to the right and left probabilities p and q, respectively, we get

$$(7.6) \qquad v_k(t) = \left(\frac{p}{q}\right)^{\frac{1}{2}k} e^{-t} I_k(2\sqrt{pqt}).$$

With a slight change of notations the obvious identity $\sum v_k(t) = 1$ takes on the form

$$(7.7) \qquad e^{\frac{1}{2}\tau(x+x^{-1})} = \sum_{-\infty}^{+\infty} x^k I_k(\tau).$$

This is the famous formula of Schlömilch.

To every formula for the original random walk there corresponds in like manner a formula referring to our process with randomized time. As an illustration consider the first passage time formula (2.4). The probability distribution of the time of the nth jump is the n-fold convolution of the exponential distribution with itself and has density $e^{-t}t^{n-1}/(n-1)!$. Accordingly, in our time dependent process *the time of the first visit to* $r > 0$ *is a random variable with density*

$$(7.8) \qquad w_r(t) = e^{-t} \sum_{n=1}^{\infty} \frac{t^{n-1}}{(n-1)!} f_n^{(r)},$$

which is the same as

$$(7.9) \qquad w_r(t) = e^{-t} \frac{r}{t} I_r(t).$$

We have thus obtained the analytically nontrivial result that for integral $r > 0$ the right side in (7.9) represents a probability density in $(0, \infty)$ satisfying the convolution equation

$$(7.10) \qquad w_{r+s}(t) = \int_0^t w_r(t-x) v_s(x)\, dx.$$

The interpolation procedure of §3 implies that this statement remains true for nonintegral r and s. In other words, *the densities* (7.9) *are infinitely divisible.* We shall return to this point in the appendix.

Many readers will notice that our formulas involving Bessel functions occur in queuing theory. It is easy to explain the reason for this. Consider a single server queue in which interarrival and service times are exponentially distributed with expectations α and σ, respectively. The state of the system is represented by the number of people in the queue including the person being served. Under these conditions it is clear that the process is a random walk with (exponentially) randomized time, except that the walk is not symmetric. The probabilities of positive and negative steps are in the ratio $p/q = \sigma/\alpha$. With a trite adjustment of the time scale the probability of the state k at time t is given by (7.6). On the other hand, if there are r people in the queue the ensuing busy period is the length of the time interval up to the first time the state 0 is reached, and so busy periods are covered by our first passage densities (7.9).

8. Appendix. On a Laplace transform. The Laplace transform of (7.9) is given by

$$\psi_r(\lambda) = \int_0^\infty e^{-(\lambda+1)t} \frac{r}{t} I_r(t)\, dt = [\lambda + 1 - \sqrt{(\lambda + 1)^2 - 1}]^r. \tag{8.1}$$

This formula can be found in many books, but despite its very frequent use no simple proof seems to exist.[4] On the other hand, from the defining power series (7.4) one gets by termwise integration trivially

$$\begin{aligned} \psi_r(\lambda) &= r \sum_{\nu=0}^{\infty} \frac{\Gamma(2\nu + r)}{\nu!\Gamma(\nu + r + 1)} \cdot [2(\lambda + 1)]^{-2\nu-r} \\ &= \sum_{\nu=0}^{\infty} f_{2\nu+r}^{(r)}(\lambda + 1)^{-2\nu-r}, \end{aligned} \tag{8.2}$$

with $f_{2\nu+r}^{(r)}$ defined in (2.4). It follows that the assertion (8.1) is equivalent to the assertion that for fixed $r > 0$ the numbers $f_{2\nu+r}^{(r)}$ have the generating function F^r defined in (2.6). This fact was established in §3, and so we have a simplified proof for the important relation (8.1).

We show how this proof can be further simplified and made more direct. As stated in §2, for integral r the form (2.4) for $f_{2k+r}^{(r)}$ can be derived by simple arguments either combinatorially or from the generating function. In other words, (8.1) is quite elementary when r is a positive integer. From this the validity of (8.1) for all $r > 0$ may be derived as follows. Define $\psi_r(\lambda)$ by the first power series in (8.2). Then $2^r\psi_r(\lambda)$ is, for fixed $\lambda > 0$, a linear combination of polynomials in r with positive coefficients, and therefore an absolutely monotone function of r (see §3). An elementary calculation shows that also $2^r(\lambda + 1)^r[\lambda + 1 - \sqrt{(\lambda + 1)^2 - 1}]^r$ is absolutely monotone. These two absolutely monotone functions agree for $r = 1, 2, \cdots$, and are therefore identical.

REFERENCES

[1] A. C. Allen and B. H. Murdoch, *A note on preharmonic functions*, Proc. Amer. Math. Soc., 4(1953), pp. 842–852.

[2] W. Feller, *An Introduction to Probability Theory and Its Applications*, vol. 1, 2nd ed., John Wiley, New York, 1957.

[3] G. N. Watson, *A Treatise on the Theory of Bessel Functions*, Cambridge University Press, Cambridge, 1922.

[4] J. Keilson, *Some comments on single-server queuing methods and some new results*, Proc. Cambridge Philos. Sec., 60(1964), pp. 237–251.

[4] The standard reference is to [3, p. 386] where, however, only references to older papers can be found.

ON REGULAR VARIATION AND LOCAL LIMIT THEOREMS

WILLIAM FELLER
PRINCETON UNIVERSITY

1. Introduction

Recent work on limit theorems in probability is marked by two tendencies. The old limit theorems are being supplemented and sharpened by a variety of so-called local limit theorems (which sometimes take the form of asymptotic expansions). Even more striking is the increasing role played by functions of regular variation. They made their debut in W. Doblin's pioneer work of 1940 where he gave a complete description of the domains of attraction of the nonnormal stable distributions. A long series of investigations started by E. B. Dynkin and continued by J. Lamperti, S. Port, and others have shown that essential limit theorems connected with renewal theory depend on regular variation. The same is true of the asymptotic behavior of the maximal term of a sequence of independent random variables and of D. A. Darling's theorems concerning the ratio of this term to the corresponding partial sum.

It seems that each of these problems still stands under the influence of its own history and that, therefore, a great variety of methods is used. Actually a considerable unification and simplification of the whole theory could be achieved by a systematic exploitation of two powerful tools: J. Karamata's beautiful theory of regular variation and the method of estimation introduced by A. C. Berry in his well-known investigation of the error term in the central limit theorem.

[It seems that proofs of Karamata's theorems can be found only in his paper of 1930 in the Rumanian journal *Mathematica* (Vol. 4), which is not easily accessible. For purposes of probability theory, one requires a generalization from Lebesgue to Stieltjes integrals. A streamlined version is contained in the forthcoming second volume of my *Introduction to Probability Theory*, but this book does not contain the inequalities derived in the sequel.]

Berry's method is of wide applicability and not limited to the normal distribution. It leads to an estimate for the discrepancy between distributions in terms of the discrepancy between the corresponding characteristic functions. In the case of the normal distribution, the latter discrepancy can be estimated in terms of the moments, and the theory of regular variation leads readily to similar

Research supported by Army Research Office (Durham) Project in Probability at Princeton University.

R.L. Schilling et al. (eds.), *Selected Papers II*,

estimates, in the case of convergence to other stable distributions. For example, the asymptotic expansions connected with the central limit theorem have their natural analogues in this general setting.

[See H. Cramér, "On the approximation to a stable probability distribution," pp. 70–76 in *Studies in Mathematical Analysis and Related Topics,* Stanford University Press, 1962, and "On asymptotic expansions for sums of independent random variables with a limiting stable distribution," *Sankhyā,* Vol. 25 (1963), pp. 13–24. The basic relations for regularly varying functions make it possible to avoid Cramér's severe restrictions simplifying at the same time the calculations.] Indeed, we shall see that expansions and error estimates of this sort are more general than the basic limit theorems in the sense that they may apply even when the leading terms do not converge. A typical example is treated in section 9.

We proceed to a brief sketch of the basic properties of regular variation and of its connection with limit theorems. Part of the material of sections 2–7 will be contained in the second volume of my book on probability, but encumbered by details and spread over many places. For a better understanding of the whole theory we shall in this address generalize the notion of regular variation by considering inequalities instead of equalities (section 7). In probabilistic terms, we shall replace the condition that a sequence of distributions F_n converges to a limit by the requirement that it be compact in the sense of the following.

DEFINITION 1. *A family $\{F_n\}$ of probability distributions is stochastically compact if every sequence $\{F_{n_k}\}$ contains a further subsequence converging to a probability distribution not concentrated at one point.*

It is essential that the limit be nondegenerate. We shall see that compactness is related to our one-sided regular variation much in the same way as convergence is to regular variation. Furthermore, the typical local limit theorems and error estimates depend on compactness rather than actual convergence, and they can be formulated in this more general setting.

2. The basic property of regular variation

A positive finite valued function U defined on $(0, \infty)$ is said to *vary regularly at* ∞ if for each x

$$\frac{U(tx)}{U(t)} \to x^p, \qquad t \to \infty \tag{2.1}$$

where p is a constant. If this exponent is 0, one says that U *varies slowly at* ∞. In other words, for a slowly varying function one has

$$\frac{\mathcal{L}(tx)}{\mathcal{L}(t)} \to 1, \qquad t \to \infty, \tag{2.2}$$

and U varies regularly iff it is of the form

$$U(x) = x^p \mathcal{L}(x). \tag{2.3}$$

Regular variation at 0 is defined in like manner, that is, U varies regularly at 0 iff $U(1/x)$ varies regularly at ∞. [Since regular variation at ∞ is not affected by the behavior of U in a finite interval, it suffices to assume that U is defined and positive in some interval (a, ∞). By the same token, regular variation at 0 is a local property.]

At first sight the condition (2.1) appears rather artificial, but it may be replaced by the more natural condition that

$$\frac{U(tx)}{U(t)} \to \psi(x). \tag{2.4}$$

Indeed, if such a limit exists and does not vanish identically, then it is either nonmeasurable or of the form $\psi(x) = x^p$. We prove this assertion together with a variant streamlined for applications to probabilistic limit theorems. It refers to convergence of monotone functions and, as usual, it is understood that convergence need take place only at point of continuity of the limit.

LEMMA 1. (a) *Let U be positive and monotone, and suppose that there exists a sequence of numbers $a_n \to \infty$ such that*

$$nU(a_nx) \to \psi(x) > 0. \tag{2.5}$$

Then $\psi(x) = Cx^p$ and U varies regularly at ∞.

(b) *The same conclusion holds if U and ψ are assumed continuous. (It suffices that ψ is finite valued and positive in some interval. The coefficients n may be replaced by arbitrary $\lambda_n > 0$ such that $\lambda_{n+1}/\lambda_n \to 1$.)*

PROOF. If ψ is monotone there is no loss of generality in assuming that 1 is a point of continuity and $\psi(1) = 1$. For fixed t determine n as the *last* index such that $a_n \leq t$. Then $U(t)$ lies between $U(a_n)$ and $U(a_{n+1})$. Since $nU(a_n) \to 1$, it follows easily that (2.4) holds. But then the relation

$$\frac{U(txy)}{U(t)} = \frac{U(txy)}{U(ty)}\frac{U(ty)}{U(t)} \tag{2.6}$$

implies that $\psi(xy) = \psi(x)\psi(y)$. This equation differs only notationally from the famous Hamel equation, and its unique measurable solution is given by $\psi(x) = x^\alpha$. Part (b) is proved in like manner.

The most usual applications in probability theory refer to the truncated second moment

$$U(x) = \int_{-x}^{x} y^2F\{dy\}, \qquad x > 0 \tag{2.7}$$

of a probability distribution F, or to the tail sum

$$T(x) = 1 - F(x) + F(-x-). \tag{2.8}$$

Generally speaking, a distribution with a regularly varying tail sum T is well-behaved, except if T is slowly varying. This exceptional role is illustrated by the following example which shows that distributions with slowly varying tails can exhibit severe pathologies.

EXAMPLE. Let F be a distribution concentrated on $(0, \infty)$ (that is, let

$F(0) = 0$) and denote its characteristic function by $\varphi = u + iv$. If F has a finite expectation μ then $v'(0) = \mu$, and hence $v(\zeta) > 0$ for all sufficiently small positive values of ζ. This is not so if $\mu = \infty$. Indeed, we shall exhibit *an arithmetic distribution F with slowly varying tail $1 - F$ such that φ has infinitely many zeros accumulating to* 0. *Furthermore*,

$$\limsup_{\zeta \to 0+} \frac{v(\zeta)}{\zeta} = \infty, \qquad \liminf_{\zeta \to 0-} \frac{v(\zeta)}{\zeta} = -\infty. \tag{2.9}$$

In other words, the values of φ oscillate wildly, and the curious nature of these oscillations becomes clear if one reflects that the integral of v over any positive interval $(0, a)$ is strictly positive. The set at which $v(\zeta)/\zeta$ is strongly negative is therefore rather sparse.

To obtain the desired example choose an integer of the form $a = 4\nu + 1$, and let F attribute weight $1/(n(n+1))$ to the point a^n, $(n = 1, 2, \cdots)$. Then

$$\varphi(\zeta) = \sum_{n=1}^{\infty} \frac{1}{n(n+1)} \exp(ia^n\zeta). \tag{2.10}$$

If $\zeta = 3\pi/2$, then $a^n\zeta$ is congruent $(3/2)\pi$ modulo 2π, and hence for every positive integer r,

$$\varphi\left(a^{-r}\tfrac{3}{2}\pi\right) = \sum_{n=1}^{r-1} \frac{1}{n(n+1)} \exp\left(ia^{n-r}\tfrac{3}{2}\pi\right) - \frac{1}{r}. \tag{2.11}$$

Since $|\sin t| \leq |t|$, it is obvious that as $r \to \infty$,

$$v\left(a^{-r}\tfrac{3}{2}\pi\right) = -\frac{1}{r} + O\left(\frac{1}{r^2}\right), \tag{2.12}$$

which proves the second relation in (2.9). The first one is even easier to verify.

3. Applications

(a) *Distribution of maxima.* Let $X_1, X_2, \cdots$ be independent random variables with a common distribution function F such that $F(x) < 1$ for all x. Let $M_n = \max[X_1, \cdots, X_n]$. We inquire whether there exist constants $a_n \to \infty$ such that the random variables M_n/a_n have a nondegenerate limit distribution $\mathcal{G}$, that is,

$$F^n(a_n x) \to \mathcal{G}(x) \tag{3.1}$$

at points of continuity. By assumption there exist values $x > 0$ for which $0 < \mathcal{G}(x) < 1$, and at such points (3.1) is equivalent to

$$n[1 - F(a_n x)] \to -\log \mathcal{G}(x). \tag{3.2}$$

By lemma 1, therefore, *the possible limits are of the form*

$$\begin{aligned} \mathcal{G}(x) &= e^{-Cx^{-\alpha}}, &\quad x > 0, \\ &= 0, &\quad x < 0. \end{aligned} \tag{3.3}$$

A limit distribution exists iff $1 - F$ varies regularly. Indeed, the lemma asserts the necessity of regular variation, and the sufficiency is obvious on choosing a_n such that $n[1 - F(a_n)] \to 1$. What first appeared to be a relatively deep theorem thus becomes a simple corollary to a simple lemma.

(b) *Stable distributions and their domains of attraction.* Let $X_1, X_2, \cdots$ be independent random variables, with the common distribution F and characteristic function $\varphi = u + iv$. Put $S_n = X_1 + \cdots + X_n$. The variables X_j are said to belong to the domain of attraction of a nondegenerate distribution $\mathcal{G}$ iff there exist real constants $a_n > 0$ and b_n such that the distributions of $a_n^{-1}S_n - b_n$ tend to $\mathcal{G}$. If F is symmetric one can put $b_n = 0$. Then φ is real, and convergence takes place iff

$$\varphi^n\left(\frac{\zeta}{a_n}\right) \to \gamma(\zeta), \tag{3.4}$$

where γ is a continuous function. It is obvious that $a_n \to \infty$, and so (3.4) refers to the behavior of φ near the origin. Otherwise there is no essential difference between (3.4) and (3.1), and we see again that for $\zeta > 0$, the limit γ is necessarily of the form $\gamma(\zeta) = \exp(-C\zeta^\alpha)$ where C and α are positive constants. This solves the problem as far as symmetric distributions are concerned: *only stable characteristic functions have a domain of attraction, and for real φ a relation* (3.4) *holds iff* $1 - \varphi$ *varies regularly at the origin.*

In the case of asymmetric distributions, the same argument still applies to the real parts of φ and $\log \gamma$. Now it is well known that for $\zeta > 0$ a stable characteristic function is not necessarily of the form $\log \gamma(\zeta) = -(a + ib)\zeta^\alpha$, but may be also of the form

$$\log \gamma(\zeta) = -a\zeta + i(b\zeta + c\zeta \log \zeta). \tag{3.5}$$

The appearance of the logarithm on the right would seem to preclude the application of lemma 1, but this is not so. We proceed to show that a minor modification of lemma 1 explains the general form of stable characteristic functions. If the random variables $a_n^{-1}S_n - b_n$ have a limit distribution $\mathcal{G}$ with characteristic function γ, then

$$nv\left(\frac{\zeta}{a_n}\right) - ib_n\zeta \to \psi(\zeta) \tag{3.6}$$

where ψ is the imaginary part of $\log \gamma$. (It is easily seen that γ can have no zeros). We now prove the following lemma.

LEMMA 2. *Suppose that* (3.6) *holds for two real continuous functions v and ψ, and that $a_n \to \infty$ and $a_{n+1}/a_n \to 1$. Then for $\zeta > 0$,*

$$\psi(\zeta) = c\zeta^\alpha \quad \textit{or} \quad \psi(\zeta) = c_1\zeta + c_2\zeta \log \zeta. \tag{3.7}$$

PROOF. Choose $\lambda > 0$ arbitrarily and put

$$w(\zeta) = \frac{v(\lambda\zeta)}{\lambda\zeta} - \frac{v(\zeta)}{\zeta}, \qquad \zeta > 0. \tag{3.8}$$

Then

$$\frac{n}{a_n} w\left(\frac{\zeta}{a_n}\right) \to \frac{\psi(\lambda\zeta)}{\lambda\zeta} - \frac{\psi(\zeta)}{\zeta}. \tag{3.9}$$

It follows easily that w *varies regularly at the origin,* and hence

$$\frac{\psi(\lambda\zeta)}{\lambda\zeta} - \frac{\psi(\zeta)}{\zeta} = C\zeta^{p}, \qquad \zeta > 0 \tag{3.10}$$

where C and p are constants. In principle, p could depend on λ, but from the fact that the norming constants a_n are independent of λ, one concludes easily that p is an absolute constant. Put $f(\zeta) = (\psi(\zeta)/\zeta) - 1$. Without loss of generality, one may suppose that $\psi(1) = 1$. Then $C = f(\lambda)$ and (3.10) takes on the form $f(\lambda\zeta) - f(\zeta) = f(\lambda)\zeta^{p}$. Interchanging the roles of λ and ζ, we conclude that

$$f(\zeta)[1 - \lambda^{p}] = f(\lambda)[1 - \zeta^{p}]. \tag{3.11}$$

Thus $f(\zeta) = C[1 - \zeta^{p}]$ unless $p = 1$, in which case (3.10) reduces to $f(\lambda\zeta) = f(\lambda) + f(\zeta)$. Since $f(1) = 0$, the only continuous solution of the last equation is given by $f(\zeta) = \log \zeta$, and this concludes the proof of the lemma.

The proof is admittedly not as simple as the proof of lemma 1, but it is nevertheless remarkable that so elementary an argument leads to the general form of the stable characteristic functions and gives at the same time *the precise conditions under which a given characteristic function φ belongs to a domain of attraction.*

4. Karamata's relations

In this section we denote by U an arbitrary nondecreasing function with $U(0) = 0$. We are interested only in the asymptotic behavior of U at infinity, and so there is no loss of generality in assuming that U vanishes identically in some neighborhood of 0. Together with U we consider the one-parametric family of truncated moments

$$U_p(x) = \int_0^{x+} y^{p} U(dy), \qquad x > 0, \tag{4.1}$$

for all values of p for which the integral diverges (at infinity). When considering U_p it is always understood that $U_p(\infty) = \infty$. For other values we change the notation and consider the tails

$$V_q(x) = \int_{x-}^{\infty} y^{-q} U(dy), \qquad x > 0. \tag{4.2}$$

In probabilistic applications U will be identified with the truncated second moment (2.7) of a probability distribution F. Then V_2 coincides with the tail sum T defined in (2.8), and U_p is the truncated moment of order $2 + p$, provided it diverges. The slightly greater generality will contribute to the understanding of the various phenomena of attraction. The following propositions generalize Karamata's basic relations from Lebesgue to Stieltjes integrals. (Only the limiting cases $\rho = 0$, $\rho = -p$, and $q = \rho$ present new features.) We shall later replace the asymptotic relations by asymptotic inequalities, and the new proof applies also to the following propositions.

The first proposition states that if U is of regular variation, namely

$$U(x) \sim x^{\xi}\mathfrak{L}(x), \qquad x \to \infty, \tag{4.3}$$

then the integrals U_p and V_q are related to U as would be the case if $\mathfrak{L}$ were a constant. (The sign $\sim$ indicates that the ratio of the two sides tends to 1.) Note that $\zeta \geq 0$, for otherwise U could not increase.

PROPOSITION 1. *If U is of the form* (4.3) *with $\mathfrak{L}$ slowly varying, then*

$$U_p(x) \sim \frac{\rho}{p+\rho}\, x^{p+\rho}\mathfrak{L}(x), \qquad p > -\rho, \tag{4.4}$$

$$V_q(x) \sim \frac{\rho}{q-\rho}\, x^{\rho-q}\mathfrak{L}(x), \qquad q > \rho. \tag{4.5}$$

The trouble with these relations is that they break down in the interesting limit cases $p = -\rho$ and $q = \rho$. However, they may be rewritten in the form

$$\frac{U_p(x)}{x^p U(x)} \to \frac{\rho}{p+\rho}, \tag{4.6}$$

$$\frac{x^q V_q(x)}{U(x)} \to \frac{\rho}{q-\rho}, \tag{4.7}$$

and in this form they remain valid for all admissible combinations of the parameters p, q, ρ (with the obvious interpretation when a denominator vanishes. Under any circumstances $p \geq -\rho$, $q \geq \rho$, and $\rho \geq 0$).

PROPOSITION 2. *The relation* (4.3) *implies* (4.6) *and* (4.7).

The most interesting point is that the conditions (4.6) and (4.7) are not only necessary, but also sufficient for the regular variation of U, except in the limiting cases $p = -\rho$ and $q = \rho$ if they arise.

PROPOSITION 3. *If either* (4.6) *or* (4.7) *hold with a nonzero denominator, then U varies regularly.*

If (4.6) holds with $p = -\rho \neq 0$, we may interchange the role of U and U_p to conclude that U_p varies regularly. In other words, if *either U or U_p varies regularly, then*

$$\frac{U_p(x)}{x^p U(x)} \to \lambda \leq \infty \tag{4.8}$$

where $\lambda = \rho/(p+\rho)$. Conversely, if (4.8) *holds with $0 < \lambda < \infty$ then both U and U_p vary regularly. Finally,* (4.8) *with $\lambda = 0$ or ∞ implies regular variation of U and U_p, respectively.* A similar remark applies to the V_q.

5. Compactness and convergence criteria for triangular arrays

In order to explain the application of the preceding propositions and to motivate the proposed generalization of the notion of regular variation, we recall a basic fact concerning triangular arrays of random variables. For each n we consider n mutually independent random variables $X_{1,n}, \cdots, X_{n,n}$ with a common distribution F_n. As usual, we put $S_n = X_{1,n} + \cdots + X_{n,n}$. The familiar convergence theorems for triangular arrays imply the following.

CRITERION. (i) *In order that every sequence $\{S_{n_k}\}$ contains a subsequence whose* *distributions converge to a probability distribution, it is necessary and sufficient that for each $t > 0$,*

$$\limsup_{n\to\infty} n \int_{-t}^{t} y^2 F_n(dy) < \infty, \tag{5.1}$$

$$\limsup_{n\to\infty} n \left| \int_{-t}^{t} y F_n(dy) \right| < \infty, \tag{5.2}$$

and that for given $\epsilon > 0$ there exists a $\tau > 0$ such that

$$n[1 - F_n(\tau) + F_n(-\tau)] < \epsilon \tag{5.3}$$

for all n.

(ii) *If none of the limit distributions is concentrated at a single point, then*

$$\liminf_{n\to\infty} n \int_{-\tau}^{\tau} y^2 F_n(dy) > 0 \tag{5.4}$$

for some $\tau > 0$.

(iii) *The distributions of S_n converge to a probability distribution iff*

$$\lim_{n\to\infty} n \int_{-s}^{t} y^2 F_n\{dy\} = \psi(s, t) < \infty \tag{5.5}$$

exists for almost all positive s and t, and

$$\lim_{n\to\infty} n \int_{-t}^{t} y F_n(dy) \tag{5.6}$$

exists for some (and therefore almost all) $t > 0$.

6. Domains of attraction

We shall now show that Karamata's relations enable us to derive from the preceding criterion not only Doblin's original characterization of the nonnormal stable domains of attraction, but also the analogue for the normal distribution as well as certain variants which were derived by various authors at the expense of cumbersome calculations.

We return to a sequence $\{X_n\}$ of independent random variables with a common distribution F and partial sums S_n. We seek conditions for the existence of a limit distribution of S_n/a_n with appropriately chosen $a_n > 0$. For that purpose we apply the criterion to the triangular array defined by $X_{k,n} = X_k/a_n$ with distribution $F_n(x) = F(a_n x)$. If U denotes the truncated second moment (2.7), then condition (5.5) specialized to $s = t$ requires the existence of a limit of $na_n^{-2}U(a_n t)$ for almost all t. This implies regular variation of U, and hence we can write

$$U(t) = t^{2-\alpha}\mathcal{L}(t) \tag{6.1}$$

with $\mathcal{L}$ slowly varying at ∞.

[The same consideration applies to the variables $a_n^{-1}S_n - b_n$. The variables of the triangular array are then $a_n^{-1}(X_k - \beta_n)$ where $b_n = n\beta_n/a_n$, and since obviously $\beta_n = o(a_n)$ it is easily seen that $n \int_{-t}^{t} y^2 F(a_n\, dy + \beta_n)$ behaves essentially as the integral in (6.2).]

From the definition of U it is clear that $0 \leq \alpha \leq 2$. Now the function V_2 introduced in (4.2) coincides with the tail sum T (defined in (2.8)), and for it the relation (4.7) holds with $\rho = 2 - \alpha$ and $q = 2$. When $\alpha = 0$ it follows that either $nT(a_n\tau) \to \infty$ or else $na_n^{-2}U(a_n\tau) \to 0$, which excludes convergence.

We have thus found that condition (6.1) with $0 < \alpha \leq 2$ is *necessary*. Assume now that it is satisfied. Since U is right continuous we can choose a_n such that $na_n^{-2}U(a_n) = 1$, in which case

$$n \int_{-t}^{t} y^2 F_n(dy) = \frac{n}{a_n^2} U(a_n t) \to t^{2-\alpha}. \tag{6.2}$$

The condition (5.3) is automatically satisfied, since in consequence of (4.7),

$$T(x) \sim \frac{2-\alpha}{\alpha} x^{-\alpha} \mathfrak{L}(x) \qquad \text{if} \quad \alpha < 2 \tag{6.3a}$$

$$T(x) = o(\mathfrak{L}(x)) \qquad \text{if} \quad \alpha = 2. \tag{6.3b}$$

When $\alpha > 1$ the distribution F has an expectation μ, and we may suppose $\mu = 0$. Using (4.7) (with $\rho = 2 - \alpha$ and $q = 1$) when $\alpha > 1$ and (4.6) (with $\rho = 2 - \alpha$ and $p = 1$) when $\alpha < 1$, one sees that the condition (5.6) is an immediate consequence of (6.2). We skip over the case $\alpha = 1$ in which (5.6) is not necessarily satisfied and centering constants may be essential to achieve convergence.

To assure the proper convergence of our distributions it remains to establish the convergence in (5.5) when $s \neq t$. Now when $\alpha = 2$ the left side in (6.2) tends to the constant 1, and this trivially implies that $\psi(s, t) = 1$ for all s, t. Thus (6.1) with $\alpha = 2$ represents the necessary and sufficient condition for convergence to the normal distribution.

When $\alpha < 2$ the existence of the limit $\psi(s, t)$ is equivalent to the existence of the limits

$$\lim_{x \to \infty} \frac{x^2[1 - F(x)]}{U(x)}, \qquad \lim_{x \to -\infty} \frac{x^2 F(-x)}{U(x)}, \tag{6.4}$$

and this requires a certain balance between the right and left tails. Under any circumstances F *belongs to a domain of attraction iff* (6.1) *holds with* $0 < \alpha \leq 2$ *and the limits in* (6.4) *exist.* They are always finite, and they vanish when $\alpha = 2$. In this formulation the only difference between the normal and other stable distributions is that the subsidiary condition relating to (6.4) is automatically satisfied when $\alpha = 2$.

When $\alpha < 2$ the tail sum T varies regularly at ∞, but this is not necessarily true when $\alpha = 2$. However, regularly varying tails play a noticeable role even if F possesses a variance or finite higher order moments. This is not visible in the usual formulation of the central limit theorem because (as we have seen) the norming constants a_n are such as to emphasize the central part of F and to obliterate the extreme tails. An entirely different picture is presented if one introduces norming constants which emphasize the tails. In fact, suppose that the tail sum T varies regularly, say

$$T(x) = x^{-p}\Lambda(x), \qquad x > 0 \tag{6.5}$$

where Λ varies slowly at infinity. Then F has finite absolute moments of order $<p$, but none of order $>p$. Whatever $p > 0$, *the tail sum for* S_n *varies regularly and is* $\sim nx^{-p}\Lambda(x)$. With the standard norming constants a_n, this is noticeable only when $p < 2$, that is, when F does not belong to the normal domain of attraction, but the assertion remains true when the central limit theorem applies. To see the probabilistic consequences, note that our assertion implies that as $t \to \infty$,

$$P\{|X_1| > t \mid |S_2| > t\} \to \tfrac{1}{2}, \tag{6.6}$$

whenever (6.5) *holds*. (This observation is due to B. Mandelbrot.) Because of the symmetry between X_1 and X_2 this may be expressed thus: if the tail sum varies regularly then *a large observed value of* $|X_1 + X_2|$ *is likely to be due entirely to one of the two components*. By contrast, if F is the exponential distribution with density e^{-x}, the left side in (6.6) equals $(1 + t)^{-1}$ and tends to 0. It is thus apparent that as far as the extreme tails are concerned, regular variation plays the same role within the domain of attraction of the normal distribution as it does for other stable distributions.

In conclusion let us remark that the Karamata relations enable us to reformulate the conditions for domains of attraction in terms of truncated moments of arbitrary orders.

7. Local limit theorems

We now show that the regular variation of U enables us to use uniformly for all domains of attraction the methods originally developed for distributions with a variance (or higher order moments). We shall be satisfied to give a typical example which, however, is of special interest.

Denote by $I_{x,h}$ the open interval of length $2h$ centered at the point x. We assume that

$$P\{S_n - b_n \leq a_n t\} \to \mathcal{G}(t) \tag{7.1}$$

where $\mathcal{G}$ is a (necessarily stable) distribution with density g and characteristic function γ. To avoid trivialities we assume F to be nonarithmetic. (A systematic reduction of arithmetic distributions to nonarithmetic ones will be described in chapter XVI of my second volume.) We are interested in the probability

$$p_n(I_{x,h}) = P\{S_n - b_n \in I_{x,h}\} = F^{n*}(x + h + b_n) - F^{n*}(x - h + b_n) \tag{7.2}$$

(at points of continuity). The following theorem states that in the limit, p_n becomes independent of x; it could be sharpened by various estimates.

THEOREM. *As* $n \to \infty$,

$$a_n p_n(I_{x,h}) \to g(0)\cdot 2h. \tag{7.3}$$

PROOF. Let δ_τ denote the density defined by

$$\delta_\tau(x) = \frac{1}{\pi}\frac{1 - \cos \tau x}{\tau x^2} \tag{7.4}$$

whose characteristic function vanishes for $|\zeta| \geq \tau$ and is given by $1 - |\zeta|/\tau$ when $|\zeta| \leq \tau$.

The starting observation is that $(1/2h)p_n(I_{x,h})$ is the value at the point $x + b_n$ of the convolution of F^{n*} and the uniform distribution with characteristic function $\sin h\zeta/h\zeta$. If φ^n were in $\mathfrak{L}_2$, we could apply the Fourier inversion formula directly, but to cover the most general case we take a further convolution with δ_τ. By the Fourier inversion formula,

$$\frac{a_n}{2h}\int_{-\infty}^{+\infty} p_n(I_{x-y,h})\delta_\tau(y)\,dy = \frac{a_n}{2\pi}\int_{-\tau}^{\tau} e^{-i\zeta(x+b_n)}\varphi^n(\zeta)\left(1 - \frac{|\zeta|}{\tau}\right)d\zeta. \tag{7.5}$$

A relation of this form was the starting point of Berry's investigation, and the same technique is of much wider applicability than is generally realized. We proceed to estimate the two sides in (7.5).

(a) *Proof that the right side tends to* $g(0)$. With the obvious change of variables the right side becomes

$$\frac{1}{2\pi}\int_{-a_n\tau}^{a_n\tau} e^{-i\zeta(x+b_n)/a_n}\varphi^n\left(\frac{\zeta}{a_n}\right)\frac{\sin h\zeta/a_n}{h\zeta/a_n}\left(1 - \frac{|\zeta|}{a_n\tau}\right)d\zeta. \tag{7.6}$$

From the assumption (7.1) it follows that for each fixed ζ the integrand tends to $\gamma(\zeta)$. By the Fourier inversion formula the formal limit of (7.6) equals $g(0)$, and to prove the assertion it suffices to show that the contribution of the intervals $|\zeta| > A$ is negligible when A is sufficiently large. More precisely, we show that given ϵ there exists an A such that

$$\int_{A<\zeta<\tau a_n}\left|\varphi^n\left(\frac{\zeta}{a_n}\right)\right|d\zeta < \epsilon. \tag{7.7}$$

Since F is nonarithmetic, $\varphi(\zeta)$ is bounded away from 0 in every closed interval excluding the origin. There exists, therefore, a number $q < 1$ such that the contribution of $\eta a_n < |\zeta| < \tau a_n$ to (7.7) is bounded by $\tau a_n q^n$, which tends to 0. The only difficulty consists in proving that *there exist numbers A and η such that*

$$\int_{A<\zeta<\eta a_n}\left|\varphi^n\left(\frac{\zeta}{a_n}\right)\right|d\zeta < \epsilon. \tag{7.8}$$

Since we can pass from φ to the characteristic function $|\varphi|^2$, there is no loss in generality in assuming φ to be real and positive. Using the inequality $1 - t < e^{-t}$ it is seen that it suffices to prove that for n sufficiently large and η sufficiently small,

$$n\left[1 - \varphi\left(\frac{\zeta}{a_n}\right)\right] > c\zeta^\alpha \qquad \text{for } 0 < \zeta < \eta a_n \tag{7.9}$$

where α is the characteristic exponent of the stable distribution $\mathcal{G}$ and c a positive constant independent of η. Obviously

$$n\left[1 - \varphi\left(\frac{\zeta}{a_n}\right)\right] \geq n\int_{-a_n/\zeta}^{a_n/\zeta}\left(1 - \cos\frac{\zeta x}{a_n}\right)F(dx) \geq \tfrac{1}{3}\,n\,\frac{a_n^2}{\zeta^2}\,U\left(\frac{a_n}{\zeta}\right). \tag{7.10}$$

We know that

$$\lim \frac{n}{a_n} U(a_n) = c_0 \tag{7.11}$$

exists, and so the first inequality in (7.9) will hold with $c = c_0/6$ if

$$\frac{U(a_n)}{U\left(\frac{a_n}{\zeta}\right)} < 2\zeta^{2-\alpha}. \tag{7.12}$$

But U varies regularly and so (7.12) will hold for all n sufficiently large provided only that $\zeta > 1$ and a_n/ζ is sufficiently large.

(b) *The left side in* (7.5). We now describe Berry's method of estimation, which is by no means restricted to our special problem. Put $\tau = 2\epsilon^{-2}$. The density δ_τ attributes mass $<\epsilon$ to $|y| \geq \epsilon$. For $|y| < \epsilon$ the interval $I_{x-y,h}$ contains the interval $I_{x,h-\epsilon}$, and so the integral on the left side is $\geq (1 - \epsilon)p_n(I_{x,h-\epsilon})$. Replacing h by $h + \epsilon$ we have thus obtained an upper estimate of the form

$$\frac{a_n}{2h} p_n(I_{x,h}) \leq (1 + \epsilon)\gamma(0) + \epsilon \tag{7.13}$$

for all n sufficiently large. Using this we get an upper bound for the contribution of $|y| \geq \epsilon$ to the integral on the left in (7.5). For $|y| < \epsilon$ we have $p_n(I_{x-y,h}) \leq p_n(I_{x,h+\epsilon})$, and we get thus a lower bound for $p_n(I_{x,h})$ similar to (7.13).

[For distributions with variance (and therefore belonging to the domain of attraction of the normal distribution), the theorem was proved by L. A. Shepp using different methods: "A local limit theorem," *Ann. Math. Statist.*, Vol. 35 (1964), pp. 419–423. After presenting this address, I noticed that our version of the theorem is contained in more general results recently obtained by Charles Stone in "A local limit theorem for non-lattice multi-dimensional distributed functions," *Ann. Math. Statist.*, Vol. 36 (1965), pp. 546–551. (See also section 9.)]

8. Dominated variation

We proceed to investigate how much of the theory of regular variation remains if the requirement that a unique limit exists is replaced by a compactness condition. For definiteness we focus our attention on measures.

DEFINITION. *A positive monotone function on* $(0, \infty)$ *varies dominatedly at* ∞ *if every sequence* $\{t_k'\}$ *converging to* ∞ *contains a subsequence* $\{t_k\}$ *such that*

$$\frac{U(t_k x)}{U(t_n)} \to \tau(x) < \infty \tag{8.1}$$

for almost all x.

An equivalent requirement is the existence of numbers $a_n \to \infty$ such that every subsequence of $\{nU(a_n x)\}$ contains a convergent subsequence.

We now show the close relationship between dominated and regular variation.

THEOREM 1. *A nondecreasing function* U *varies dominatedly at* ∞ *iff there exist positive constants such that*

$$\frac{U(tx)}{U(t)} < Cx^{\rho} \qquad \text{for} \quad t > \tau, \quad x > 1. \tag{8.2}$$

PROOF. The condition is sufficient by virtue of Helly's selection theorem. If U varies dominatedly, it is possible to choose $\rho > 0$ such that

$$\frac{U(2t)}{U(t)} < 2^{\rho} \qquad \text{for} \quad t > \tau. \tag{8.3}$$

Then

$$\frac{U(2^n t)}{U(t)} < 2^{n\rho}, \tag{8.4}$$

and if $2^{n-1} < x \leq 2^n$, this implies (8.2) with $C = 2^{\rho}$. This condition is therefore necessary.

We now preserve the notations and conventions introduced in (4.1) and (4.2), and proceed to prove the counterpart to the basic relation (4.7). Also (4.6) has a similar counterpart, but the proof is slightly more delicate.

THEOREM 2. *If $q > \rho$, then* (8.2) *implies*

$$\limsup_{n \to \infty} \frac{x^q V_q(x)}{U(x)} \leq \gamma \tag{8.5}$$

with

$$\gamma = -1 + C \frac{q}{q - \rho}. \tag{8.6}$$

Conversely, (8.5) *implies* (8.2) *with*

$$C = 1 + \gamma, \qquad \rho = \frac{\gamma}{\gamma + 1} q. \tag{8.7}$$

PROOF. (i) Assume (8.2) and choose $\lambda > 1$. Then

$$V_q(t) = \sum_{n=1}^{\infty} [V_q(\lambda^{n-1}t) - V_q(\lambda^n t)] \leq \sum_{n=1}^{\infty} (\lambda^{n-1}t)^{-q}[U(\lambda^n t) - U(\lambda^{n-1}t)] \tag{8.8}$$

$$= -t^{-q}U(t) + (\lambda^q - 1)t^{-q} \sum_{n=1}^{\infty} \lambda^{-nq} U(\lambda^n t)$$

therefore, for $t > \tau$ and arbitrary $\lambda > 1$

$$\frac{t^q V_q(t)}{U(t)} \leq -1 + C \frac{\lambda^q - 1}{\lambda^{q-\rho} - 1}. \tag{8.9}$$

Letting $\lambda \to 1$, one gets (8.6).

(ii) Assume that for $y > \tau$,

$$\frac{y^q V_q(y)}{U(y)} \leq \gamma. \tag{8.10}$$

An integration by parts shows that

$$U(y) = -y^q V_q(y) + q \int_0^y s^{q-1} V_q(s)\, ds. \tag{8.11}$$

Putting for abbreviation

$$\int_0^y s^{q-1}V_q(s)\,ds = W_q(y), \tag{8.12}$$

we see, therefore, from (8.10) that for $y > \tau$,

$$\frac{y^{q-1}V_q(y)}{W_q(y)} \leq \frac{\gamma}{\gamma+1}\, q \cdot \frac{1}{y} = \rho\,\frac{1}{y}. \tag{8.13}$$

Integrating between t and tx one gets for $x > 1$ and $t > \tau$

$$\frac{W_q(tx)}{W_q(t)} \leq x^{\rho}. \tag{8.14}$$

Referring again to (8.11) we have, therefore,

$$U(tx) \leq qW_q(tx) \leq qx^{\rho}W_q(x) = x^{\rho}[U(x) + x^qV_q(x)] \leq x^{\rho}U(x)[1+\gamma], \tag{8.15}$$

and so (8.2) holds with the constants, given in (8.7).

Note. The occurrence of the factor C in (8.2) and (8.5) introduces a lack of reciprocity between the constants occurring in (8.2) and (8.5). In fact, starting from the relations (8.5)–(8.6), one does not get (8.2) with the original exponent ρ, but the new exponent is

$$\rho' = \frac{C-1}{C}\, q + \frac{1}{C}\,\rho. \tag{8.16}$$

Thus $\rho' = \rho$ only if $C = 1$. In the theory of regularly varying functions one could choose C arbitrarily close to 1, and this establishes a complete symmetry between (8.2) and (8.5). It is therefore natural to ask whether our inequalities can be improved to obtain more symmetric relations. The following examples show that our inequalities are, in a sense, the best. In both examples F is a probability distribution and U its second truncated moment (2.7). We take $q = 2$ so that V_q coincides with the tail sum T.

EXAMPLES. (a) Let $a > 1$ be fixed, and let F attribute mass $(a-1)a^{-n}$ to the point a^n (here $n = 1, 2, \cdots$). For $a^n < t < a^{n+1}$ one has $U(t) = a^{n+1} - a$ and $V_2(t) = T(t) = a^{-n}$. The left side in (8.5) therefore equals a. Now for every $x > 1$,

$$\limsup \frac{U(tx)}{U(t)} \geq a. \tag{8.17}$$

Whatever the exponent ρ, the constant C in (8.2) is therefore at least a, whereas (8.7) leads to the estimate $(a+1)$. Since a can be chosen arbitrarily large, the estimate (8.7) is essentially the best.

(b) Let F attribute mass $e^{-1}/n!$ to the point $\lambda_n = (2^n n!)^{1/2}$. For $\lambda_n < t < \lambda_{n+1}$, clearly $U(t) = e^{-1}(2^n - 1)$ and $V_2(t) = T(t) \backsim e/(n+1)!$. For every ϵ and t sufficiently large the inequality (8.2) holds with $C = 2 + \epsilon$ and $\rho = \epsilon$, while (8.5) is true with $\gamma = 1$.

9. Stochastic compactness

If a sequence of probability distributions is stochastically compact in the sense of the definition in section 1, the same is true of the sequence of distributions obtained by symmetrization. For our purposes it suffices, therefore, to consider symmetric distributions. This is not essential, but it simplifies the exposition.

THEOREM. *Let F be symmetric. In order that there exist constants $a_n > 0$ such that the family of distributions $F^{n*}(a_n x)$ is stochastically compact, each of the following conditions is necessary and sufficient.*

(a) *The following condition holds:*

$$\limsup_{x\to\infty} \frac{x^2T(x)}{U(x)} < \infty. \tag{9.1}$$

(b) *There exist constants $\alpha > 0$, C, τ such that*

$$\frac{U(tx)}{U(t)} < Cx^{2-\alpha} \qquad \text{for} \quad x > 1, \quad t > \tau. \tag{9.2}$$

One admissible choice of a_n is such that

$$\frac{n}{a_n^2} U(a_n) = 1. \tag{9.3}$$

PROOF. Assume that $\{F^{n*}(a_n x)\}$ is stochastically compact. As stated in section 5, the sequence of numbers $na_n^{-2}U(a_n x)$ is bounded for each $x > 0$, and there exist some x for which it is bounded away from 0. Since a scale factor is inessential, we may suppose that this is the case for $x = 1$. Then

$$A^{-1} < \frac{n}{a_n^2} U(a_n) < A \tag{9.4}$$

for some constant A. Because of the monotonicity of U this implies

$$A^{-1} < \frac{a_{n+1}}{a_n} < A. \tag{9.5}$$

Again, because of the right continuity of U there exist numbers α_n such that $n\alpha_n^{-2}U(\alpha_n) = 1$, and obviously the ratios α_n/a_n remain between A^{-1} and A. It follows that we can replace a_n by α_n without affecting the stochastic compactness. This justifies (9.3).

Assume now that (9.1) is false. In consequence of (9.5) there exists then a sequence $n_1, n_2, \cdots$ such that as n runs through it

$$\frac{a_n^2 F(a_n)}{U(a_n)} \to \infty, \tag{9.6}$$

and hence

$$nF(a_n) \to \infty. \tag{9.7}$$

But by the theory of triangular arrays, $nF(a_n x)$ remains bounded for every fixed $x > 0$, and so the condition (9.1) *is necessary*. Using theorem 2 of section 8 with $q = 2$, it is seen that the conditions (9.1) and (9.2) imply each other. It remains to show that (9.2) is sufficient.

Choose a_n so as to satisfy (9.3). Then $na_n^{-2}U(a_n x) < Cx^{2-\alpha}$ for all $x > 1$. Furthermore, we see from (8.5) that

$$\limsup nT(a_nx) \leq C\gamma\cdot x^{-\alpha}, \tag{9.8}$$

and by the criteria of section 5 these relations suffice to guarantee the stochastic boundedness.

By way of application note that the error estimate in section 7 depended only on (9.2) but not on the regular variation of U. To be sure, if F does not belong to a domain of attraction, then the integral on the right in (7.5) need not converge, but stochastic compactness of $\{F^{n*}(a_nx + b_n)\}$ guarantees that it remains bounded away from 0 and ∞. The argument of section 7 then applies to each convergent subsequence, and the theorem may be replaced by the following more general theorem.

THEOREM 2. *Assume that F is nonarithmetic and that there exist constants a_n, h_n such that the sequence of distributions $F^{n*}(a_nx + b_n)$ is stochastically compact. There exist norming factors α_n such that*

$$\alpha_n p_n(I_{x,h}) \to 2h, \tag{9.9}$$

and $A^{-1} < \alpha_n/a_n < A$.

Genet. Res., Camb. (1967), **9**, *pp.* 1–15
Printed in Great Britain

On fitness and the cost of natural selection

By WILLIAM FELLER
The Rockefeller University, New York, and Princeton University, U.S.A.

(*Received 4 February 1966*)

1. INTRODUCTION

In genetic experiments the population size is normally not a biologically important variable but is determined by extraneous conditions or laboratory facilities; the same is true in sampling natural populations as well as in breeding experiments. The geneticist is therefore compelled to think in terms of relative frequencies of gene variants rather than in absolute numbers of their carriers. Accordingly, the mathematical formulas used by geneticists usually refer only to relative frequencies and relative fitnesses. They describe the structural changes in a population satisfactorily as long as the total population size remains constant. This condition is usually inconsequential because in practice population sizes remain nearly constant over moderate periods of observation. A novel situation confronts us in evolutionary theory when one considers developments over extremely long periods. The familiar formulas then introduce an error which is negligible over moderate time-intervals, but whose cumulative effect may be so great that the whole picture changes when absolute population sizes are taken into account. This seems to be true in particular for the calculation of the cost of natural selection, for problems of survival of genes and for the notion of an unstable equilibrium.

This paper is written by a mathematician, and accordingly no new biological models or hypotheses are advanced. The models we study are the familiar ones obtained by applying the Mendelian laws to a single isolated locus. Obviously such a model is too crude to reflect the deeper problems currently exciting geneticists. In fact, a one-locus Mendelian population is a mathematical fiction in the sense that it cannot be observed in nature whereas it could be simulated on a computer. Despite all this the model continues to be used for purposes of orientation, and certain purely mathematical deductions from it (such as the notion of equilibrium and the tendency to it) continue to play a fruitful role in genetical thinking. It may be useful therefore to present the theoretical conclusions from this model when proper account is taken of the fluctuations in the population size. The point is that (save in certain exceptional configurations) a Mendelian population cannot have a constant rate of increase, that is, it cannot be Malthusian. Given the various fitnesses, it is possible to calculate the size N_n of the nth generation; if only the relative fitnesses are known, the size N_n can be calculated up to a factor μ^n where μ is the maximal absolute fitness. When this is done there seems to be more room and flexibility for evolutionary processes than is sometimes supposed.

For clarity we begin with a review of some fairly obvious facts concerning

R.L. Schilling et al. (eds.), *Selected Papers II*,

Malthusian populations and the notion of environment. Next we turn to the main topic of this paper, namely, an analysis of the calculation of the cost of natural selection made by J. B. S. Haldane and M. Kimura (Haldane, 1957, 1960; Kimura, 1960, 1961). The problem is roughly to estimate the effect on the ultimate population size of a depression in fitness at one particular locus due to an assumed change of the prevailing conditions. The loss in population size is called selective death by Haldane, and substitutional (or evolutional) load by Kimura. This loss was found to be practically independent of the amount of depression (as long as the latter is small) and so enormous as to lead to the conclusion that simultaneous selection at several loci cannot proceed at a reasonable rate. The present scrutiny of these calculations was undertaken at the suggestion of Th. Dobzhansky who commented on the paradoxical nature of Haldane's result and its grave consequences. Other evolutionists were worried by Haldane's conclusions and it was hoped that a revision of the observational data may lead to a more acceptable result (Brues, 1964; Mayr, 1963).*

Haldane calculated the cost of selection for various types of organism, but since the basic principle remains the same it will suffice to analyze in detail only the case of haploid organisms, which is mathematically by far the simplest. Indeed, in this special case no problem arises when one considers the population size. In sections 4–5 we analyze Haldane's definition and the possibility of alternative models assuming constant population size.

Next we consider the selection at an autosomal locus in a sexually reproducing diploid without mutations. The (rather unexciting) result of the formal calculation of the cost of natural selection is presented in a companion paper (Feller, 1966) but here we discuss some noteworthy features requiring only a minimum of mathematics.

In section 6 we indicate how the true population size can be calculated. In section 7 it is shown that the so-called state of *unstable* equilibrium is easily misinterpreted: under the usual assumptions the population is destined to die out rather than to be maintained at an equilibrium.

In section 8 we turn to the problem of survival of a gene, which cannot be treated without reference to absolute population size because the fact that the frequency q_n of the a-gene tends to zero does not necessarily imply the disappearance of this gene. Indeed, when $q_n \to 0$ the absolute number of a-genes may increase from generation to generation at a geometric rate. Even if the total population is kept at a fixed level, the process of elimination of the a-genes may be so slow that these genes are not in practice eliminated from the evolutionary process. The examples of sections 9 and 10 will clarify this point.

2. MALTHUSIAN POPULATIONS AND CONSTANT ENVIRONMENT

Consider a uniform population with non-overlapping generations in which each individual gives rise to a random number of direct descendants. If μ is the expected

* After completion of this paper and the publication of the companion paper (Feller, 1966) I became the grateful recipient of letters with references to other pertinent papers.

number of such descendants and the parental population consists of N_0 individuals, the expected size of the next generation is $N_1 = N_0\mu$. The actual size is a random variable but for large populations the true size is likely to be close to the expected size and it is therefore usual to treat N_1 as if it were the actual size. Assuming unchanged conditions we get for the expected size of the next generation $N_2 = N_1\mu = N_0\mu^2$, and generally $N_n = N_0\mu^n$. This is the well-known Malthusian model for population growth.

When $\mu = 1$ the expected population size remains constant. A large population may be expected to remain constant over a number of generations, but the cumulative effect of chance fluctuations will grow more pronounced and ultimately *a Malthusian population with fitness* $\mu = 1$ *is bound to die out.** A similar remark applies to Mendelian populations and shows that caution is indicated in dealing with very large time intervals and with assumed states of equilibrium.

Obviously evolution could not occur without some species having fertilities exceeding 1, and then the Malthusian model breaks down for a different reason. With increasing density the population is bound to affect the environment, and hence the assumption of constant external conditions becomes biologically untenable. To take this into account one assumes that the fertility μ depends on the population density in such a manner that the expected population size will tend to a maximum value N_∞. Over reasonably short observational periods the fertility will not change appreciably, and over such periods the population will appear as practically Malthusian. However, when the model of density-dependent fertilities is accepted, the observed fertilities at any time cannot be used to predict the ultimate population size: of two populations the one with lower fertility may tend to a higher plateau.

Another limitation of the long-run significance of the Malthusian parameter (and of fitness in general) is inherent in the relativity of the notion of environment. To explain this by a simple example, consider a population which in two separated geographic regions R_1 and R_2 has different fertilities μ_1 and μ_2. We suppose positive migration probabilities s_1 and s_2 with the effect that an offspring of an individual in R_1 may serve as a parent in either R_1 or R_2, with corresponding probabilities $1-s_1$ and s_1. Initially the population may be distributed in an arbitrary manner, for example, be concentrated in region R_1. Assuming constant conditions the two regional subpopulations in R_1 and R_2 tend to become practically Malthusian with the *same fertility rate* μ which is a complicated average† of μ_1 and μ_2. Thus the effective long-run fertility μ depends on the migration probabilities between the two regions, however small these probabilities may be. To an observer in R_1 one

* This is a basic result of the Galton–Fisher theory of branching processes which applies to Malthusian populations. Readers puzzled by the phenomenon that a population with fixed expected size can be condemned to extinction may consider a hypothetical population as follows: the nth generation consists with probability $1/n$ of n individuals, and with probability $1-1/n$ of 0 individuals. The expected size of the nth generation equals 1, but the probability that the population survives n generations is only $1/n$ and tends to zero.

† Letting $a_j = \mu_j(1-s_j)$ one gets (using the theory of Markov chains)

$$2\mu = a_1 + a_2 + \sqrt{[(a_1-a_2)^2 + 4\mu_1\mu_2 s_1 s_2]}$$

population may appear less fit than another, and yet it will ultimately grow faster. This is so because the whole biological environment contains both R_1 and R_2 while any given observer sees only one region and the corresponding fertility. In the long run the hidden parameters turn out to be crucial. With obvious verbal changes this model applies also to temporal chance fluctuations within the same region.

3. HAPLOID ORGANISMS AND MIXED MALTHUSIAN POPULATIONS

The theory of selection in haploid, clonal, or self-fertilizing organisms is concerned with a gene pool of two kinds of genes A and A'. Assuming constant external conditions and disregarding possible mutations one is led to a model in which there is no interaction between the genes. Mathematically the nature of the genes is then irrelevant and in effect we are dealing with a composite (mixed) population consisting of two Malthusian components. To fix ideas, denote fertilities in the two components (or the fitnesses of the A- and A'-genes) by μ and μ', so that the ratio

$$\frac{\mu'}{\mu} = 1-k \tag{3.1}$$

represents the relative fitness of A'. The sizes N_n and N_n' of the two components in the nth generation are given by the geometric progressions

$$N_n = N_0\mu^n, \qquad N_n' = N_0'\mu'^n \tag{3.2}$$

and the total population size equals N_n+N_n'. As usual we denote the relative frequencies of A and A' by p_n and q_n, so that

$$p_n = \frac{N_n}{N_n+N_n'}, \qquad q_n = \frac{N_n'}{N_n+N_n'} \tag{3.3}$$

Since A and A' are representative in the proportion $p_n:q_n$ the average rate of increase* μ_n of the nth generation is given by

$$\mu_n = p_n\mu+q_n\mu' = \mu(1-kq_n) \tag{3.4}$$

This rate of increase varies from generation to generation except in the trite case $k=0$ when A and A' have the same fertility. Accordingly, a mixture of two Malthusian populations is *not* Malthusian, and the expected population size remains constant only in the trite case $\mu=\mu'=1$. On the other hand, if k is positive, the frequency q_n tends to zero, and from (3·4) it is seen that the effective fertility μ_n tends to μ. This means that after a sufficiently long time the population will be practically Malthusian with a rate of increase μ. It should be noticed, however, that it may take an exceedingly long time to complete this process.

Example. In his calculations of the cost of natural selection Haldane assumes that prior to a sudden change in the environment the A genes had been at a disadvantage, and hence their frequency p_0 in the parental generation is exceedingly

* That is, the ratio of the expected sizes of two successive generations.

small. The following calculations are based on the value of $p_0 = 10^{-4}$ which Haldane quotes repeatedly as relatively large. It is assumed that k is positive, but small. The A'-genes are then at a disadvantage and their relative frequency q_n decreases steadily from the initial value $q_0 = 0{\cdot}9999$ to 0. Table 1 shows how many generations it takes to reduce the frequency q_n to 0·99, 0·9, and so on. The values are calculated from (3.5) and do not depend on the absolute fitness μ. When $k = 0{\cdot}01$ it takes some 460 generations to reduce the frequency q_n to 0·99 and some 1000 generations to reduce it to 0·3. The average fitness of the whole population equals initially 0·99001 μ; eventually the population is homozygous with fitness μ. Natural selection increases the fitness by 1%, but after 460 generations it is still only 0·9901 μ and after 1000 generations only 0·997 μ. In other words, 1000 generations cover only seven-tenths of the whole selection process. It is true that for $k = 0{\cdot}1$ the process is slightly more than 10 times faster, but for smaller k it is incredibly slow. In the extreme example, $k = 10^{-3}$, 1000 generations will reduce q_n only to 0·99973 and the total fitness is increased only by 0·0000003 μ.

Table 1. *Number* n *of generations required to reduce the initial frequency* $q_0 = 0{\cdot}9999$ *of the* A'*-gene to the given value* q

$q =$	0·99	0·9	0·7	0·5	0·3	0·1
Assuming:						
$k = 0{\cdot}01$	459	698	832	917	1000	1135
$k = 0{\cdot}1$	44	67	79	87	95	109

Returning to the general situation note that the sizes of the A- and A'-components are given by (3.2) whence

$$q_n = \frac{q_0(1-k)^n}{p_0 + q_0(1-k)^n} \tag{3.5}$$

When $k > 0$ one has $q_n \to 0$. This means that the *relative* size of the A'-component tends to zero, but this is not necessarily true of the absolute size. Several situations are possible.

Case (i). If $\mu > \mu' > 1$ each of the two components increases at a geometric rate. (Biologically this model ultimately breaks down for the reasons discussed in the preceding section.)

Case (ii). If $\mu > 1 > \mu'$ the expected size of the A'-component decreases geometrically, but the total population size exhibits in the long run a geometric increase.

Case (iii). If $\mu = 1 > \mu'$ the A'-component dies out geometrically. The *expected* size of the A-component remains constant (namely N_0), but because of chance fluctuations even the A-component is bound to die out ultimately (see footnote, page 3). Within the deterministic approximation which equates actual population size with its expected value one would say that the ultimate population size equals N_0: the A-component remains constant while the A'-component dies out.

4. HALDANE'S SELECTIVE DEATHS

Haldane calculated the cost of natural selection for various Mendelian populations, but the simplest case appears under the heading 'selection in haploid, clonal, or self-fertilizing organisms' (Haldane, 1957, 1960). The other cases present greater biological interest, but require deeper mathematical techniques. The basic principle being the same in all cases, we shall discuss in detail only the simplest case. As Haldane remarks, low mutation rates do not affect the general picture and are therefore disregarded. We are then concerned with the simple model of the preceding section: there are two types of gene, A and A', with fertilities (absolute fitnesses) μ and $\mu' = 1-k$. The genes do not interact in any way. In the nth generation their expected numbers are N_n and N'_n, respectively, and they appear in the proportion $p_n : q_n$ [see (3.2) and (3.3)]. To comply with the general usage (and Haldane's notation) we put* $\mu = 1$ and $\mu' = 1-k$. The expected size of the A-component then remains for all times equal to N_0, while the number of A'-genes decreases geometrically. The loss in the nth generation equals $N'_n - N'_{n+1}$, and these losses add up to

$$(N'_0 - N'_1) + (N'_1 - N'_2) + \ldots = N'_0 \tag{4.1}$$

This is intuitively obvious seeing that the total population size decreases from $N_0 + N'_0$ to N_0.

It may now be best to analyze Haldane's notion of selective deaths formally before considering its biological background and motivation. Using the size $N_0 + N'_0$ of the parental generation as unit of measurement Haldane defines as selective deaths in the nth generation the quantity $d_n = kq_n$; the sum $D = d_0 + d_1 + \ldots$ then represents the total of selective deaths.† The q_n are given by formula (3.5), and from it Haldane obtains an approximation to the sum D using a subtle and original approach.

It will be noticed that (using the original population size $N_0 + N'_0$ as unit) one has $N'_0 = q_0$, and so $kq_0 = (1-\mu')N'_0 = N'_0 - N'_1$. For the parental population therefore the selective deaths d_0 are the same as the decrement in population size $N'_0 - N'_1$, and this interpretation agrees with Haldane's general explanations. But since every generation can be taken as parental it follows that the number of selective deaths in the nth generation should equal $N'_n - N'_{n+1}$ which is *not* the same as Haldane's $d_n = kq_n$ because the population size has decreased from $N_0 + N'_0 = 1$ to $N_n + N'_n < 1$. Now Haldane was interested only in approximations, and since k is small, the effective population size does remain approximately constant over rather long periods of time. It seems therefore obvious that conceptually Haldane's selective deaths are identical with the true decrements $N'_n - N'_{n+1}$, but that in his

* There is no serious restriction in this convention. With an arbitrary μ it suffices to measure the size of the nth generation in units μ^n to reduce the general model to $\mu = 1$. It must be noted, however, that μ is the fertility of the A-component and not of the whole population. As was remarked above, the fertility of the latter varies from generation to generation.

† Kimura's substitutional (or evolutional) load differs from Haldane's d_n by an adjustment factor a^n to which the remark of the preceding footnote applies.

calculation he used the standard mathematical approximation familiar to geneticists and applicable as long as the population size remains nearly unchanged.*

In other words, Haldane actually defines the selective deaths in a given generation as the loss $N'_n - N'_{n+1}$ in population size. The total selective deaths then amount to the number of A'-genes originally present, in accordance with the fact that the A'-component is ultimately lost. The formula $d_n = kq_n$ expresses the number of selective deaths in units of the population size of the nth generation, and the calculations neglect the effect of the changes in this size. The cumulative effect of this approximation is disastrous. In fact, Haldane concluded that (as long as k is small) the total of selective deaths is practically independent of k and given by

$$D = \log \frac{1}{p_0} \tag{4.2}$$

(in units of the size of the parental population). For reasons discussed in the next section we are interested only in situations where p_0 is small, and hence D large. Indeed, Haldane considers $p_0 = 10^{-4}$ as rather large, and with it $D = 9{\cdot}2$. Haldane's general conclusion is that the selective deaths involve 'a number of deaths equal to about 10 or 20 times the number in a generation, always exceeding this number, and perhaps rarely being 100 times this number. To allow for occasional high values I take 30 as a mean'.

To appreciate the magnitude of this estimate it should be remembered that the deaths are due solely to the disability of the A'-genes and that the total of A'-genes ever to be born equals

$$N'_0 + N'_1 + N'_2 + \ldots = N'_0(1 + (1-k) + (1-k)^2 + \ldots) = N'_0/k$$

In extreme cases the number D of presumed selective deaths may exceed the number of A'-genes ever to be born.

5. DISCUSSION OF THE MODEL

As we have seen, assuming strictly constant conditions, the expected total population size cannot remain constant except when $\mu = \mu' = 1$. Given a fixed relative fitness $(\mu'/\mu) = 1 - k < 1$ it is possible to obtain a constant population size either by postulating that the fertility μ is an appropriate function of the population size, or else by assuming the existence of external forces ensuring the desired stability. One may suspect that Haldane had some such model in mind, in particular since the quoted sentence refers to the number 'in a generation' rather than in the parental generation. It seems to me that Haldane's discussion leaves no doubt that he was thinking in terms of our model with constant environment and constant fertilities, and that notions of steady state and constant populations enter

* The reader may wonder why Haldane should have resorted to subtle approximations when the exact result can be written down in the simple form (4.1). The answer is that Haldane was interested in more general situations which admit of no simple solution. In other words, the general problem did require a new approach, and no general method is universally the best in *every* special case.

the discussion merely as an unintended byproduct of the routine mathematical formulas to which geneticists are used. It is my impression that similar unwarranted extrapolations in evolutionary theory are accepted as unavoidable mathematical truths and render biological thinking more difficult. It may therefore be useful to discuss the alternative models in some detail. However, the following discussion has no bearing on the subsequent sections. Let us consider the following extreme possibilities:

(*a*) The absolute fitnesses are constant, but the environmental conditions are such as to maintain a constant population size (say by immigration or emigration). It suffices to consider the simple case $\mu = 1$, $\mu' = 1 - k < 1$. The loss caused by the disadvantage of the A'-genes must be replaced by immigration or some other method. The cost of natural selection is not borne by the population as such, but by the assumed mechanism of replacement. This cost depends on the method of replacement. The most economical method consists in replacing each A'-gene that is lost by a superior A-gene. The replacements may occur at various times, but each A'-gene in the parental population will ultimately cost exactly one replacement, and the total number of replacements equals the number N_0' of A'-genes originally present. The least economical way consists in replacing each A'-gene lost by another A'-gene. With this system the composition of the population remains unchanged, and each generation requires the same number of replacements (namely $q_0 N_0'$). Among the infinity of intermediate procedures there is one whose total cost is given by Haldane's expression $D = k(q_0 + q_1 + \ldots)$. It consists in adding to the nth generation A- and A'-genes in the proportion $p_n : q_n$ so as not to upset the proportions observed. Under no conditions does the population itself bear any load, and the number of replacements is a measure of the efficiency of the mechanism of replacements rather than of the cost of selection.

(*b*) At the other extreme one may assume a constant environment and a fertility μ depending on the actual population size so as to ensure a steady state. In other words, it is assumed that the population size will change from the original size, say unity, to a steady-state level $N_\infty \leqslant 1$. If $N_\infty = 1$ there is no drop in population size, and hence no cost of natural selection. Otherwise the variable size of the initial generations must be taken into account, and our argument is not affected by the new model. Furthermore, the cost of selection now depends on the manner in which μ depends on the population size; this presents a biological rather than a mathematical problem.

The biologist may think of a variety of conditions producing a nearly constant population size, but mathematically it will lead to a combination of our two extreme situations.* The question of the cost of natural selection appears in a new light in each case.

Finally a word concerning the biological background of the problem which may

* This conclusion may conflict with biological facts and intuition. If it is untenable, so is the basic assumption of the model that the gene pair (A, a) may be treated independently of others. (It must be borne in mind that two strongly interacting species cannot be treated separately as Malthusian populations.)

be described very roughly as follows. It is supposed that originally the population was in a near equilibrium with fertility rates α and α'. At a time when the population size is N_0 and the two genes stand in the proportion $p_0:q_0$ the external conditions are changed with the effect that from now on the fertilities are μ and μ'. It is assumed that the change has a negative effect on the population size and the problem is to find the anticipated new level of equilibrium. Before the change the A'-genes had been at an advantage, and hence the A-genes were maintained at a low level by mutation. This explains the assumed low values for p_0. Note that the development of our population is independent of the fertilities α and α' before the change to the new fertility rates μ and μ', but without keeping the original situation in mind one would obviously never speak of selective *deaths*. Haldane had, of course, good reasons to *assume* an actual depression in the fertility of the A' type. This does not change the fact that his arguments in no way depend on this assumption; and the results should apply for all past fitnesses α and α', and hence *both* fertilities have *increased*. To give an extreme example, the change from $\alpha=\alpha'=\frac{1}{2}$ to $\mu=1$ and $\mu'=0{\cdot}99$ would be called a 'depression' merely because a *relative* fitness is decreased. But how can a doubling of fertilities in Malthusian populations lead to deaths? The whole problem is affected by the habitual reliance on relative fitnesses as if they had an objective meaning.

6. DIPLOID POPULATIONS

Haldane's theory applies to arbitrary Mendelian populations, and although the calculations are more intricate, the principle is the same. We shall therefore be satisfied with an indication of the solution of his problem when absolute population sizes and fitnesses are taken into account.

Consider then a random-mating population with non-overlapping generations and an autosomal pair of alleles A and a not subject to mutations. We wish to study the development of the absolute population size and associate with the three genotypes AA, Aa, aa fixed *absolute* fitnesses w_1, w_2, w_3. In other words, w_1 is the expected number of offspring (counted at an appropriate stage of development) resulting from a pairing of two A-genes. As usual we denote the two gene frequencies in the nth generation by p_n and q_n. We have then the familiar relations

$$p_{n+1} = \frac{p_n(w_1 p_n + w_2 q_n)}{w(p_n)}, \qquad q_{n+1} = \frac{q_n(w_2 p_n + w_3 q_n)}{w(p_n)} \tag{6.1}$$

where

$$w(p_n) = w_1 p_n^2 + 2w_2 p_n q_n + w_3 q_n^2 \tag{6.2}$$

is the average fitness of the nth generation. If N_0 is the size of the parental population then the expected sizes N_n of the successive generations can be calculated recursively from the relation

$$N_{n+1} = w(p_n).N_n \tag{6.3}$$

Our problem is to find out how the choice of the initial frequencies (p_0, q_0) affects the ultimate population size. The principle of the calculations may be illustrated by an example.

Example. Let $w_1 = 0{\cdot}36$, $w_2 = 1{\cdot}16$, and $w_3 = 0{\cdot}96$. For convenience we take the initial population size as unit, that is, we put $N_0 = 1$. For the initial gene frequencies we take $p_0 = 0{\cdot}8$ and $q_0 = 0{\cdot}2$. From (6.2) one gets $w(p_0) = 0{\cdot}64$. The frequencies p_1 and q_1 in the first generation can now be calculated from (6.1), and the population size N_1 is obtained from (6.3). Proceeding in this manner we get the development summarized in Table 2.

Table 2

Generation	Frequency p_n	Fitness $w(p_n)$	Population size
0	0·80	0·64	1
1	0·65	0·80	0·64
2	0·52	0·90	0·51
3	0·43	0·95	0·46
4	0·37	0·97	0·44
5			0·42

The fitness $w(p_n)$ of the nth generation tends to 1, and the absolute population size N_n tends to a limit N_∞ which equals about 0·4. The true fitness varies from generation to generation, and our population is not Malthusian. Nevertheless, after a sufficient number of generations the fitness and the expected population size remain practically constant. The gene frequencies p_n and q_n converge to the stable equilibrium $(\bar{p}, \bar{q})$ given by $\bar{p} = 0{\cdot}2$, $\bar{q} = 0{\cdot}8$. If the population had started at this state of equilibrium all the gene frequencies p_n would have remained unchanged, and the population would have been maintained at the original level. The loss of some 60% of the original population is therefore attributable to the rather unfavourable initial gene frequencies $p_0 = 0{\cdot}8$ and $q_0 = 0{\cdot}2$. In this sense it represents the cost of natural selection (although Haldane preferred to exclude the case of a stable equilibrium).

The general case of a *stable* equilibrium can be treated in like manner, and the calculations will not be spelled out because the result is not exciting. Note, however, that the use of absolute fitness permits us to formulate various conclusions. For example, mutations do not affect the fitnesses w_1, w_2, w_3 but do change gene frequencies $\bar{p}$, $\bar{q}$ in the state of equilibrium. Now the limit value $\bar{p}$ is such that $w(\bar{p}) > w(p)$ for all p near $\bar{p}$. It follows unequivocally that small *mutation* rates always depress the ultimate fitness, and hence also the ultimate population size. As in Table 2 the fitnesses $w(p_n)$ and the population sizes N_n converge much faster than do the gene frequencies p_n.

7. THE SO-CALLED UNSTABLE EQUILIBRIUM

A more interesting situation is encountered when $w_1 > w_2 < w_3$, that is, when the heterozygotes are least fit. In this case there exists a critical point defined by the frequencies

$$\bar{p} = \frac{w_1 - w_2}{w_1 + w_3 - 2w_2}, \qquad \bar{q} = \frac{w_3 - w_2}{w_1 + w_3 - 2w_2} \tag{7.1}$$

As is well known (and easily verified from (6.1)), the development now depends essentially on the initial gene frequencies p_0 and q_0 in the following way:

$$\text{If} \quad \begin{array}{l} q_0 < \bar{q} \\ q_0 > \bar{q} \end{array} \quad \text{then} \quad \begin{array}{l} q_n \to 0 \\ q_n \to 1 \end{array} \tag{7.2}$$

If $q_0 = \bar{q}$ then, theoretically, $q_0 = q_1 = q_2 = \ldots$ and for this reason (7.1) is called a point of unstable equilibrium, but this term is apt to give an erroneous impression. Indeed, the chance fluctuations inherent in the Mendelian model are bound to bring the system off balance in the first generation with the effect that q_1 will not exactly equal $\bar{q}$. This means that an unstable equilibrium cannot be maintained but that always either $q_n \to 0$ or $q_n \to 1$. This result appears in a new light if the population size is taken into account. Only two essentially different cases are possible.

(*a*) Assume $w_1 = w_3 = 1$, but $w_2 < 1$. The so-called unstable equilibrium is represented by $\bar{p} = \bar{q} = \frac{1}{2}$. With any other initial frequencies (p_0, q_0) one has a tendency to homozygosity, but before this state is reached the total population size may have dwindled to a negligible fraction of its original size. Indeed, it can be shown that the expected size of the nth generation is given by*

$$N_n = \frac{p_0 - q_0}{p_n - q_n} \tag{7.3}$$

provided, of course, that $p_0 > q_0$ and the original population size N_0 is taken as unit. But then $p_n \to 1$ and $q_n \to 0$ so that the expected ultimate population size in the homozygous state equals $N_\infty = p_0 - q_0$. Thus if the initial gene distribution (p_0, q_0) is close to the so-called equilibrium $(\frac{1}{2}, \frac{1}{2})$ *the expected population size dwindles to the small fraction* $p_0 - q_0$. This is not quite what one expects from an equilibrium. In this special case the cost of natural selection is indeed exorbitant.

(*b*) Assume $w_1 > 1$ but $w_2 < w_3 < 1$. In the homozygous state *aa* the population decreases at the geometric rate w_3, and hence it is bound to die out whenever $q_n \to 1$, that is, whenever the initial frequency q_0 exceeds the critical value $\bar{q}$. It is interesting that the elimination of the total population could have been avoided by making the homozygotes *aa* less fit than the heterozygote *Aa*, for in this case one would have $p_n \to 1$ and $q_n \to 0$. For our population with the initial gene distribution p_0, q_0 *it would be preferable to have* $w_3 = 0$ *than* $w_3 > w_2$: with a lethal in double dose the population would survive and, indeed, increase at the geometric rate w_1.

* See Feller, 1966. The assertion can be proved by checking that

$$p_{n+1} - q_{n+1} = \frac{p_n - q_n}{w(p_n)}$$

and hence

$$N_{n+1} = \frac{p_n - q_n}{p_{n+1} - q_{n+1}} N_n$$

(see (6.1) and (6.3)). Use this relation with n replaced by $n-1, n-2, \ldots, 0$ and multiply the resulting n equations. Most terms cancel, and one obtains (7.3).

8. SURVIVAL

So far we have considered diploid populations (without mutations) with fitnesses w_1, w_2, w_3 such that there exists a state of stable or unstable equilibrium. In all other cases the population will be ultimately homozygous, and we may suppose that the final state is AA. Then $q_n' \to 0$, but this does *not* imply that the a-genes die out. Indeed, the total number N_n' of a-genes depends ultimately only on w_2, but not on w_1 or w_3. When $w_2 > 1$ the numbers N_n will ultimately increase at a geometric rate even though $q_n \to 0$.

To see this in detail we use the basic relations of section 6. Since we assume that $p_n \to 1$ and $q_n \to 0$ the (absolute) fitness $w(p_n)$ of the nth generation tends to the fitness w_1 of the AA type (see (6.2)). It follows that after a sufficiently large number of generations the population fitness is practically fixed at w_1, which means that, as far as size is concerned, the population *ultimately* behaves like a Malthusian population with fertility w_1. As we saw, the population is bound to extinction unless $w_1 > 1$.

The expected number N_n' of a-genes in the nth generation is given by $N_n' = N_n q_n$, where N_n is the expected population size. Thus

$$\frac{N_{n+1}'}{N_n'} = \frac{N_{n+1}\, q_{n+1}}{N_n\, q_n} = w(p_n)\frac{q_{n+1}}{q_n} \tag{8.1}$$

by the very definition of the fitness $w(p_n)$ (or by (6.3)). From the second relation in (6.1) it is seen that the right side in (8.1) equals $w_2 p_n + w_3 q_n$, which quantity tends to w_2 because $p_n \to 1$ and $q_n \to 0$. This means that after a sufficiently large number of generations the ratio N_{n+1}'/N_n' will be practically indistinguishable from w_2. If $w_2 < 1$ the a-genes are doomed for extinction, but if $w_2 > 1$ this number will ultimately increase at a geometric rate. Of course, the total population size N_n increases at an even faster rate.

We have thus reached the interesting conclusion that *when the absolute fitness w_2 of the heterozygotes Aa exceeds* 1, *the number N_n' of* a*-genes will increase approximately as a geometric sequence with ratio w_2 even though $q_n \to 0$.*

It is surprising that the ultimate rate of increase of N_n' depends only on w_2, *but not on the fitnesses of the homozygotes.** In particular, even when a is lethal in double dose the ultimate number of a-genes is the same as when the relative fitnesses of Aa and aa are the same.

It follows that a gene whose frequency q_n tends to 0 is not necessarily lost for the evolutionary process. We can go a step further. If $w_2 < 1$ the a-genes will be ultimately eliminated, but the process may be so slow that the genes survive for a period long enough to encounter changes of the environment which may modify the fitnesses. To judge the effect of the selection process we must solve two problems:

(a) Estimate the speed of the selection process, (b) estimate its cost, that is, the ratio of the ultimate population size in terms of what this size would be if the

* Under the present conditions, of course, $w_1 > w_2$ and $w_3 \leqslant w_2$.

population had been homozygous from the beginning. Such estimates are given in Feller, 1966, and we shall be satisfied here to give two instructive examples.

9. FIRST EXAMPLE

Assume $w_1 = w_2 > w_3$, that is, only the homozygotes *aa* have a disadvantage. As is well known, under these circumstances $p_n \to 1$ and $q_n \to 0$. It can be shown* that the expected size N_n of the *n*th generation tends to $N_0 p_0$. In other words, the ultimate population size is the same as if all *a*-genes had been removed from the parental population. The selection process in this case causes a loss of a fraction q_0 of the initial population as could be expected from the circumstance that in this example only *a*-genes are being eliminated.

This elimination process is exceedingly slow. Indeed, if $w_1 = w_2 = 1$ and $w_3 = 1 - k$ the gene frequency q_n is given approximately† by

$$q_n = \frac{q_0}{1 + kq_0 n} \tag{9.1}$$

at least when *n* is large. (Under all circumstances the right side underestimates the true value of q_n). The expected number N'_n of *a*-genes is therefore approximately

$$N'_n = \frac{N'_0 p_0}{1 + kq_0 n}$$

It seems that $k = 0{\cdot}1$ would be generally considered a serious disadvantage, and yet it would take 1000 generations to reduce the initial frequency $q_0 = 0{\cdot}01$ to half its value, that is, to eliminate one-half the *a*-genes. A 1000 generations should suffice for many changes in the environment but even if the assumption of constant conditions and fixed fitnesses is taken at face value it must be remembered that a population is spread over a great number of biological niches and that during a very slow process chance fluctuations are likely to produce a variety of results in different niches,

Anyhow, this discussion over-emphasizes one side of the picture in that we attributed fitness $w_1 = 1$ to the homozygotes *AA*, even though we saw in section 2 that a finite population with fitness 1 is bound for extinction. It is therefore necessary to assume that actually $w_1 > 1$, though the difference $w_1 - 1$ may be small. But then the number of *a*-genes will *increase* from generation to generation irrespective of the degree of disability of the *aa*-homozygotes. Assuming a rate of increase of 1% per generation, the size of the *n*th generation will be, approximately, $N_0 p_0 \mu^n$, where $\mu = 1{\cdot}01$. The actual number of *a*-genes in the *n*th genera-

* From (6.1) one sees that $w(p_n) = p_n / p_{n+1}$, and hence $N_{n+1} = N_n p_n / p_{n+1}$. It follows by induction that $N_n = N_0 p_0 / p_n$, and since $p_n \to 1$ the ultimate population size equals $N_0 p_0$.

† From the second relation in (6.1) we get

$$q_{n+1} = q_n \frac{1 - kq_n}{1 - kq_n^2} > q_n \frac{1}{1 + kq_n}$$

If the inequality > were replaced by an equality sign formula (9.1) would represent the exact solution of the recurrence relation.

tion increases at a rate which is practically indistinguishable from an increase at the same rate μ. Even in the extreme case where a is lethal in double dose the actual number of a-genes in the nth generation increases *steadily* (and exceeds $N_0 p_0 q_0 \mu^n/(1+q_0 n)$).

10. SECOND EXAMPLE

The preceding example is extreme in that the heterozygotes Aa have no disadvantage. As a fairly representative example we take $w_1 = 1$, $w_2 = \beta$, $w_3 = \beta^2$ where $\beta < 1$. For it exact formulas can be supplied.

The expected population size now tends* to $N_0 p_0^2$, which means that a fraction $N_0 q_0(1+p_0)$ of the original population is eliminated. In other words, each a-gene lost causes a loss of $p_0 q_0$ genes of the A-type, and so the process of selection is costlier than in the preceding example. The process itself proceeds faster, but for small $1-\beta$ it is still exceedingly slow; the *exact*† expression for q_n is

$$q_n = \frac{q_0}{q_0 + p_0 \beta^{-n}}$$

SUMMARY

A Mendelian population without artificial external constraints does in general not increase at a constant rate. Formulas neglecting the changes in population size introduce an error which is negligible under ordinary circumstances but whose cumulative effect over long periods may be disastrous. Questions relating to the cost of natural selection, the nature of an unstable equilibrium, the survival of genes, etc. cannot be treated without regard to absolute population sizes. The limitation of the notion of relative fitnesses is illustrated by the fact that in some typical situations the survival of the a-gene depends only on the absolute fitness of the Aa-heterozygote, but not on the fitnesses of the homozygotes. Furthermore, a *decrease* of the (absolute or relative) fitness of one genotype may actually increase the viability of the population and its ultimate size.

Even when the relative frequency q_n of the a-gene tends to zero the absolute number of such genes may increase from generation to generation at a geometric rate. Therefore the circumstance that $q_n \to 0$ may be insignificant as compared to the fact that the earth cannot sustain an infinitely increasing population. Ultimately the population size is bound to influence the environment and so the fitnesses will change. Thus we must consider density-dependent fitnesses and then observed fitnesses cannot be used to predict the ultimate fate of a population. It is now known (Dobzhansky, 1965) that relative fitnesses are sometimes very sensitive

* The exact formula is now $N_n = N_0 p_0^2/p_n^2$ as can be seen by induction from the fact that

$$w(p_n) = (p_n + \beta q_n)^2 = p_n^2/p_{n+1}^2$$

† This can be verified from (6.1), but it is difficult to describe the method used to obtain the result.

to small changes in environment and that the same species may occupy a great variety of environmental niches. It is therefore quite likely that at least part of a population will find itself in a modified environment before too many generations have passed. For the evolution of a species and the development of new forms it is then not important that under fixed conditions the relative frequency q_n of the a-gene would tend to zero. The problem is whether the actual number of such genes will increase for a period sufficiently long to encounter changed conditions or to establish itself in new combinations. This question is significant because the convergence of the frequencies q_n to zero may be extremely slow. Thus even in a population of fixed size a disappearing gene could exist long enough to contribute to evolutionary processes.

Speaking generally, the thinking in terms of an assumed steady state and relative fitnesses seems to aggravate the problem of applying the wonderful results of modern genetics to the theory of evolution. For example, various mechanisms which are often considered as eliminating genetic variability may sometimes produce the opposite effect. The theory of evolution should distinguish between what the physicist would call macroscopic and microscopic equilibrium. Even if the world as we see it were in a perfect equilibrium this would not imply an approximate steady state for individual species, not to speak of genes. It is clear that an evolution to higher forms depends on a frequent decrease in fertility rates. If one considers slow changes rather than an unattainable steady state then a loss of fitness may be beneficial in the long run and contribute to genetic variety.

I am indebted to Th. Dobzhansky for drawing my attention to Haldane's theory as well as for many stimulating discussions. Thanks are due also to other members of his laboratory for hospitality and instruction.

REFERENCES

Brues, A. M. (1964). The cost of evolution versus the cost of not evolving. *Evolution, Lancaster, Pa.*, **18**, 379–383.

Dobzhansky, Th. (1965). Genetic diversity and diversity of environment. To appear in *Proc. Fifth Berkeley Symp. Probab. Statist.*

Feller, W. (1966). On the influence of natural selection on population size. *Proc. natn. Acad. Sci. U.S.A.* **55**, 733–738.

Ford, E. B. (1964). *Ecological Genetics.* London: Methuen; New York: Wiley.

Haldane, J. B. S. (1957). The cost of natural selection. *Genetics*, **55**, 511–524.

Haldane, J. B. S. (1960). More precise expressions for the cost of natural selection. *Genetics*, **57**, 351–360.

Kimura, M. (1960). Optimum mutation rate and the degree of dominance as determined by the principle of minimum genetic load. *Genetics*, **57**, 21–34.

Kimura, M. (1961). Natural selection as the process of accumulating genetic information in adaptive evolution. *Genet. Res.* **2,** 127–140.

Mayr, E. (1963). *Animal Species and Evolution.* Harvard Univ. Press.

Z. Wahrscheinlichkeitstheorie verw. Geb. 10, 261–268 (1968)

On the Berry-Esseen Theorem *

WILLIAM FELLER

Received February 1, 1968

Summary. The paper presents a simple derivation of a generalized Berry-Esseen theorem not requiring moments.

1. Notations and Results

We consider a fixed number, n, of mutually independent random variables $X_1, \ldots, X_n$ with distributions $F_1, \ldots, F_n$. Let F denote the distribution of their sum $S = X_1 + \cdots + X_n$. We are interested in estimates for the discrepancy

$$\Delta = \sup |F(xs) - \mathfrak{N}(x)|; \tag{1.1}$$

here $\mathfrak{N}$ stands for the normal distribution with zero expectation and unit variance, while s is an appropriate positive norming factor. Upper bounds for Δ will be obtained in terms of truncated moments. As the derivation is independent of the mode of truncation, no restriction is placed on the latter. Thus, to each X_k we let correspond an interval $-\infty \leqq -\tau_k < 0 < \tau_k' \leqq \infty$ and put

$$\bar{X}_k = \begin{cases} X_k & \text{if} \quad -\tau_k < X_k < \tau_k' \\ 0 & \text{otherwise} \end{cases} \tag{1.2}$$

$$X_k = \bar{X}_k + X_k'. \tag{1.3}$$

Then, throughout this paper,

$$\alpha_k = E(\bar{X}_k), \quad \beta_k = E(\bar{X}_k^2), \quad \gamma_k = E(|\bar{X}_k|^3), \tag{1.4}$$

$$a = \alpha_1 + \cdots + \alpha_n, \quad b = \beta_1 + \cdots + \beta_n, \quad c = \gamma_1 + \cdots + \gamma_n. \tag{1.5}$$

We begin with a strengthening of the classical Berry-Esseen theorem. It is designed to cover variables without third moment, but the estimate (1.8) is sharper than the classical one even when third moments do in fact exist.

Theorem 1. *Suppose that* $E(X_k) = 0$ *and* $E(X_k^2) = \sigma_k^2 < \infty$ *for* $k = 1, \ldots, n$. *Put*

$$s^2 = \sigma_1^2 + \cdots + \sigma_n^2 \tag{1.6}$$

and

$$\beta_k' = E(X_k'^2), \quad b' = \beta_1' + \cdots + \beta_n'. \tag{1.7}$$

Then

$$\Delta \leqq 6\left(\frac{c}{s^3} + \frac{b'}{s^2}\right). \tag{1.8}$$

Note. No attempt at sharp estimates was made. Instead, all calculations are based on simple rational approximations so as to render trivial the verification

* Research connected partly with a Project for research in probability at Princeton University, supported by the Army Research Office.

R.L. Schilling et al. (eds.), *Selected Papers II*,

of each step. No impressive improvement of the constant 6 can be obtained by arguments based on the crude estimate of (2.2). For an improvement of this inequality see ZOLOTAREV [9]. (The direct calculations in [1] for identically distributed variables lead to the constant 33/4 instead of 6.)

From Theorem 1 it is easy to derive an analogous result for arbitrary random variables. To formulate it we introduce the quantities

$$\pi_k = P\{X_k' \neq 0\}, \qquad p = \pi_1 + \cdots + \pi_n \tag{1.9}$$

and

$$\lambda_k = \alpha_k^2/\pi_k. \tag{1.10}$$

If $\pi_k = 0$ we define $\lambda_k = 0$ when $\alpha_k = 0$, and $\lambda_k = \infty$ otherwise. (It will be seen that λ_k is a measure for the appropriateness of the given centering of X_k.)

Theorem 2. *If*

$$s^2 \geqq b + \sum \lambda_k \tag{1.11}$$

then

$$\Delta \leqq 6\left(\frac{c}{s^3} + \frac{s^2 - b}{s^2}\right) + p. \tag{1.12}$$

(In the limit as $s \to \infty$ the right side is $\geqq 6$, and so the theorem is true when $\lambda_k = \infty$ for some k.) Under the conditions of Theorem 1 one has $\alpha_k = -E(X_k')$ and hence by CHEBYSHEV's inequality

$$\lambda_k \leqq E(X_k'^2) = \sigma_k^2 - \beta_k. \tag{1.13}$$

Thus (1.11) is satisfied and $s^2 - b = b'$, so that (1.12) represents a variant of (1.8). The condition (1.11) is not essential because the norming constant s in (1.1) can always be replaced by $s' < s$ at the expense of increasing the bound Δ for the error by

$$\sup |\mathfrak{N}(x) - \mathfrak{N}(s'x/s)| \leq \sqrt{2\pi e}\,\frac{s - s'}{s}.$$

It follows that Theorem 2 is generally applicable, but the unwieldy constants λ_k make it unattractive. It is therefore preferable to replace Theorem 2 by a weakened version in which the constants λ_k are eliminated. To achieve this we introduce condition (1.14) which assures that the centering of the variables X_k is not too absurd. In fact, condition (1.14) is trivially satisfied except when $E(X_k)$ is finite or equals $\pm\infty$ (in the obvious sense that only one tail diverges). In these exceptional cases the validity of (1.14) can be achieved by adding an appropriate centering constant to X_k. In theorem 1 the conditions (1.14) and (1.15) are satisfied with $t_k = t_k' = \infty$. As remarked before, (1.15) imposes no serious restriction.

Theorem 3. *Suppose that*

$$\int_{-t_k}^{\tau'_k} x\,F_k(dx) \leqq 0 \quad \textit{and} \quad \int_{-\tau_k}^{t'_k} x\,F_k(dx) \geqq 0 \tag{1.14}$$

for some $-\infty \leqq -t_k \leqq -\tau_k$ *and* $\tau_k' \leqq t_k' \leqq \infty$. *If*

$$s^2 \geqq \sum_{k=1}^{n} \int_{-t_k}^{t'_k} x^2 F_k(dx), \tag{1.15}$$

then (1.12) *holds.*

2. Discussion

It is clear in principle that the standard method of truncation enables one to apply the Berry-Esseen theorem to arbitrary variables, but the truncated variables are not properly centered and one requires error estimates in terms of quantities relating to the original variables. The case of identically distributed variables with finite variances and the classical norming was first considered by M. L. KATZ [3] who gave a bound for the error in terms of $E(X^2 g(X))$, where g is an even unbounded monotone function. This result was generalized to variable distributions with variances by V. V. PETROV [6]. Now the main application of the Berry-Esseen theorem is to triangular arrays subject to the central limit theorem, and it is therefore desirable to estimate the error by quantities related to the Lindeberg conditions. This point of view is implicit in YU. P. STUDNEV [7] who had previously in [8] derived an asymptotic form of the Berry-Esseen theorem depending on the integral of the function

$$L_n(x) = \frac{1}{s^2} \sum_{k=1}^{n} \int_{|t|>xs} t^2 F_k(dt) \tag{2.1}$$

occurring in the Lindeberg condition. STUDNEV's claim that his theorem implies the Katz-Petrov result is unfortunately based on an oversight since in [7] he has inadvertedly ommitted the term $o(s^{-1})$ which is essential in the asymptotic estimate of [8].

Lindeberg's condition requires that $L_n(x) \to 0$ as $n \to \infty$ for every fixed $x > 0$. In the notations of theorem 1 with truncation at $\pm s_n$ this would imply that $b' = o(s^2)$ and $c = o(s^3)$. Conversely, these two conditions together imply that $L_n(x) \to 0$, but obviously neither of these conditions alone suffices to ensure that $L_n(x) \to 0$. It follows trivially that both terms on the right side in (1.8) are indispensable. This simple argument answers a problem raised by IBRAGIMOV and OSIPOV [2] who constructed a complicated example to show that an estimate of the form $\Delta < C \cdot L_n(x)$ (for a fixed x) is impossible even if the existence of moments of some order $2 + \alpha > 2$ is assumed. A similar remark applies to the estimate (1.12).

Several variants and generalizations of the Katz-Petrov theorem were derived in OSIPOV [4], and OSIPOV and PETROV [5], by increasingly complex method.

Our theorem 1 (with an unspecified constant on the right side) is actually contained in [5]. Theorems 2 and 3 formulate a general version of the Berry-Esseen theorem involving only quantities naturally tied to the central limit theorem.

It is shown in the present paper that the apparent great generality of these theorems is deceptive inasmuch as they can be derived as simple corollaries of theorem 1. This is done by a method of truncation leading to zero expectations and prescribed variances (Section 4). Another purpose of this paper is to show that theorem 1 can be derived in a simple manner using a new technique for bounding products of characteristic functions in the absence of third moments. It is hoped that this method may find wider applications.

The proof of theorem 1 is based on the well-known lemma (due essentially to Berry) that if ϕ denotes the characteristic function of F then for any $T > 0$

$$\pi\Delta \leqq \int_{-T}^{T} \left|\phi\left(\frac{\xi}{s}\right) - \exp\left(-\frac{1}{2}\xi^2\right)\right| \cdot |\xi^{-1}|\, d\xi + \frac{24}{T\sqrt{2\pi}}. \tag{2.2}$$

(See, for example, [1], p. 512.) The integrand will be estimated using the obvious identity

$$u_1 \cdots u_n - v_1 \cdots v_n = \sum_{\nu=1}^{n} u_1 \cdots u_{\nu-1}(u_\nu - v_\nu)\, v_{\nu+1} \cdots v_n \tag{2.3}$$

with

$$u_k = \phi_k(\xi/s), \qquad v_k = \exp(-\tfrac{1}{2}\sigma_k^2 \xi^2/s^2), \tag{2.4}$$

where ϕ_k is the characteristic function of F_k.

3. Proof of Theorem 1

Denote the integral occurring in (2.2) by I. Our first aim is to show that

$$I \leqq \int_{-\infty}^{+\infty} e^{-\xi^2/8} \cdot \sum \left|\phi_k\left(\frac{\xi}{s}\right) - \exp\left(-\frac{\sigma_k^2\xi^2}{2s^2}\right)\right| \cdot |\xi|^{-1}\, d\xi. \tag{3.1}$$

From the familiar inequalities for the partial sums of the Taylor series for e^{ix} we get trivially

$$\left|\phi_k\left(\frac{\xi}{s}\right) - 1 + \frac{\sigma_k^2\xi^2}{2s^2}\right| \leqq \frac{\gamma_k|\xi|^3}{6s^3} + \frac{\beta_k'\xi^2}{s^2}. \tag{3.2}$$

Remembering that $0 \leqq e^{-x} - 1 + x < \frac{1}{2}x^2$ for all $x > 0$ we conclude that

$$\left|\phi_k\left(\frac{\xi}{s}\right) - \exp\left(\frac{-\sigma_k^2\xi^2}{2s^2}\right)\right| \leqq \frac{\gamma_k|\xi|^3}{6s^3} + \frac{\beta_k'\xi^2}{s^2} + \frac{\xi^4}{8s^4}\sum \sigma_k^4. \tag{3.3}$$

Assuming (3.1) to be true this implies

$$I \leqq \int_{-\infty}^{+\infty} e^{-\xi^2/8}\left[\frac{c\,\xi^2}{6s^3} + \frac{b'|\xi|}{s^2} + \frac{|\xi|^3}{8s^4}\sum\sigma_k^4\right] d\xi \leqq \frac{32}{9} + 8\frac{b'}{s^2} + 8\sum\left(\frac{\sigma_k}{s}\right)^4. \tag{3.4}$$

(In the first term we used the inequality $\sqrt{2\pi} < 8/3$.)

To justify (3.1) from (2.3) we need an upper bound for the u_k, and here the difficulty arises that these factors are not necessarily uniformly small in any appropriate interval of integration $(-T, T)$. It is therefore necessary to partition the subscripts into two classes. Leaving the parameter T for the time being indeterminate we say that *a subscript k belongs to the class A iff*

$$\frac{\beta_k^{1/2}T}{s} \leqq \frac{4}{3}. \tag{3.5}$$

(The right side may be replaced by any number that is $\leqq \sqrt{2}$ but close to $\sqrt{2}$.)

Clearly

$$\begin{aligned} \phi_k(\xi/s) - 1 + \frac{\beta_k\xi^2}{2s^2} &= E\left(e^{i\xi\bar{X}_k/s} - 1 - \frac{i\xi\bar{X}_k}{s} + \frac{\xi^2\bar{X}_k^2}{2s^2}\right) \\ &\quad + E\left(e^{i\xi X'_k/s} - 1 - \frac{i\xi X'_k}{s}\right) \end{aligned} \tag{3.6}$$

and hence

$$|\phi_k(\xi/s)| \leq \left|1 - \frac{\beta_k \xi^2}{2s^2}\right| + \frac{\gamma_k |\xi|^3}{6s^3} + \frac{\beta'_k \xi^2}{2s^2}. \tag{3.7}$$

When $k \in A$ and $|\xi| < T$ the first quantity within absolute values on the right is positive, and so

$$|\phi_k(\xi/s)| \leq \exp\left(-\frac{1}{2}\xi^2\left[\frac{\beta_k}{s^2} - \frac{\gamma_k T}{3s^3} - \frac{\beta'_k}{s^2}\right]\right). \tag{3.8}$$

Taking the product over all $k \in A$ we get (with obvious notations) the inequality

$$\prod_{k \in A} |\phi_k(\xi/s)| \leq \exp\left(-\frac{1}{2}\xi^2\left[\frac{b_A}{s^2} - \frac{c_A T}{3s^3} - \frac{b'_A}{s^2}\right]\right) \tag{3.9}$$

valid for $|\xi| < T$. We proceed to find a lower bound for the expression D within brackets.

From the moment inequality we conclude that for any subscript in the complement of A

$$\gamma_k \geq \beta_k^{3/2} \geq \frac{4s}{3T}\beta_k. \tag{3.10}$$

Summing over $k \notin A$ we get

$$c - c_A \geq \frac{4s}{3T}(b - b_A). \tag{3.11}$$

Consider for the moment b_A and c_A as arbitrary non-negative parameters. A routine computation shows that among all such combinations satisfying (3.11) the one minimizing the quantity D within brackets in (3.9) corresponds to the smallest possible b_A, and so the minimum is achieved when

$$b_A = b - \frac{3T}{4s}c \quad \text{and} \quad c_A = 0.$$

Remembering that $b + b' = s^2$ we conclude that

$$\prod_{k \in A} |\phi_k(\xi/s)| \leq \exp\left(-\frac{1}{2}\xi^2\left[1 - \frac{3Tc}{4s^3} - \frac{2b'}{s^2}\right]\right). \tag{3.12}$$

We now return to the identity (2.3). For a subscript $k \in A$ the bound obtained in (3.8) for $|\phi_k(\xi/s)|$ exceeds $\exp\left(-\frac{\sigma_k^2 \xi^2}{2s^2}\right)$ and applies therefore to both u_k and v_k. For subscripts outside A we use the trite inequalities $|u_k| \leq 1$ and $|v_k| \leq 1$. When $\nu \notin A$ it follows that the factor of $u_\nu - v_\nu$ in (2.3) is in norm less than (3.12). When $\nu \in A$ the same is true provided the contribution of $|\phi_\nu(\xi/s)|$ is omitted. But a glance at (3.9) and (3.5) makes it clear that the omission of the ν-th term will increase the exponent by less than

$$\frac{\beta_\nu \xi^2}{2s^2} \leq \frac{8}{9} \cdot \frac{\xi^2}{T^2}, \tag{3.13}$$

and so the right side of (2.3) is in absolute value less than

$$\sum |u_k - v_k| \cdot \exp\left(-\frac{1}{2}\xi^2\left[1 - \frac{3Tc}{4s^3} - \frac{2b'}{s^2} - \frac{16}{9T^2}\right]\right). \tag{3.14}$$

This proves the assertion (3.1) *provided only that T is chosen so that*

$$T \cdot \frac{c}{s^3} + \frac{8}{3}\frac{b'}{s^2} + \frac{64}{27T^2} \leq 1. \tag{3.15}$$

With this proviso we have now the inequality (3.4) and we proceed to express the last sum on the right in terms of the quantities occurring in the theorem. From now on we suppose that

$$\frac{c}{s^3} + \frac{b'}{s^2} \leqq \frac{1}{6}, \tag{3.16}$$

for otherwise there is nothing to prove. By the moment inequality $\beta_k^3 \leqq \gamma_k^2$ and hence

$$\sigma_k^4 \leqq (\gamma_k^{2/3} + \beta_k')^2 \leqq c^{1/3}\gamma_k + (2c^{2/3} + b')\beta_k'. \tag{3.17}$$

Among all linear combinations of this form with coefficients subject to (3.16) the maximum is assumed for $c = s^3/6$ and $b' = 0$. Since $6^{1/3} > 9/5$ we conclude that

$$\sum \left(\frac{\sigma_k}{s}\right)^4 \leqq \frac{5}{9}\frac{c}{s^3} + \frac{50}{81}\frac{b'}{s^2}, \tag{3.18}$$

and hence finally from (3.4)

$$I < 8\frac{c}{s^3} + 13\frac{b'}{s^2}. \tag{3.19}$$

We introduce this estimate into the basic inequality (2.2) using the fact that $(2\pi)^{-1/2} < 2/5$. To prove theorem 1 it suffices to choose T such that

$$8\frac{c}{s^3} + 13\frac{b'}{s^2} + \frac{48}{5T} \leqq 6\pi\left(\frac{c}{s^3} + \frac{b'}{s^2}\right). \tag{3.20}$$

But $6\pi > 94/5$, and hence (3.20) holds with

$$\frac{1}{T} = \frac{9}{8}\frac{c}{s^3} + \frac{5}{8}\frac{b'}{s^2}. \tag{3.21}$$

With this definition the left side in (3.15) reduces to

$$\frac{8}{9} - \left(\frac{5T}{9} - \frac{8}{3}\right)\frac{b'}{s^2} + \frac{64}{27T^2}. \tag{3.22}$$

Now (3.16) implies that $T^{-1} < 3/16$; thus the coefficient of b' is positive and the last term in (3.22) is $< 1/12$. The quantity (3.22) is therefore < 1 as required.

4. Proof of Theorem 2

The theorem is trivially true when $\pi_k = 0$ but $\alpha_k \neq 0$ for some k. Otherwise we define for each k a point $x_k = -\alpha_k/\pi_k$ with the understanding that $x_k = 0$ when $\alpha_k = \pi_k = 0$. With λ_k defined in (1.10) we now define a number $0 \leqq \eta \leqq 1$ by

$$\sum \lambda_k = \eta(s^2 - b). \tag{4.1}$$

For definiteness assume first that $\eta \neq 0$.

Let $U_1, \ldots, U_n$ be random variables independent of each other and of the given variables X_k, and such that each U_k assumes the values 1 and 0 with probabilities η and $1 - \eta$, respectively. Put

$$*X_k = \begin{matrix} \bar{X}_k & \text{if} & X_k = \bar{X}_k \\ U_k x_k/\eta & \text{if} & X_k = X_k'. \end{matrix} \tag{4.2}$$

The distribution of $*X_k$ differs from the distribution of $\bar{X}_k$ in that it attributes a smaller weight to the origin and an extra weight of mass $\eta\pi_k$ to the point

x_k/η. Clearly

$$E(^*X_k) = E(\bar{X}_k) + x_k \pi_k = 0$$
(4.3)
$$E(^*X_k^2) = E(\bar{X}_k^2) + \lambda_k/\eta = \beta_k + \lambda_k/\eta$$

so that the sum $^*S = {}^*X_1 + \cdots + {}^*X_n$ has variance s^2. This sum differs from S only if at least one among the variables $X_1', \ldots, X_n'$ is different from zero, and the probability for this event is $\leqq p$. Thus

$$\Delta \leqq {}^*\Delta + p \tag{4.4}$$

where $^*\Delta$ is the maximum discrepancy between the distribution of $^*S$ and the normal distribution. But theorem 1 applies to the variables $^*X_k$, and it remains to calculate the quantities $^*\gamma_k$ and $^*\beta_k'$ corresponding to the truncated moments γ_k and β_k'. For definiteness suppose $x_k > 0$. If $s \geqq \tau_k'$ we use truncation at τ_k' and see that $^*\beta_k' = \lambda_k/\eta$ or $^*\beta_k' = 0$ according as $x_k > \tau_k'\eta$ or $x_k < \tau_k'\eta$. In the first case $^*\gamma_k = \gamma_k$, in the second

$$^*\gamma_k = \gamma_k + x_k^3 \pi_k/\eta^2 \leqq \gamma_k + s\,\lambda_k/\eta\,.$$

If, on the other hand, $s \leqq \tau_k'$ we use truncation at s and conclude trivially that under any circumstances

$$\frac{^*\gamma_k}{s^3} + \frac{^*\beta_k'}{s^2} \leqq \frac{\gamma_k}{s^3} + \frac{\lambda_k}{\eta\, s^2}\,. \tag{4.5}$$

Summation over k now shows that $^*\Delta$ is indeed bounded by the first term on the right side in (1.12), and this concludes the proof when $\eta > 0$.

The contingency $\eta = 0$ arises if, and only if, $\lambda_k = 0$ for all k. This means that all $\bar{X}_k$ have zero expectation. The preceding argument applies, except that the U_k now assume the values ± 1 each with probability $\pi_k/2$, and $x_k^2 = (s^2 - b)/p$ for all k. The sum $^*S$ has again zero expectation and variance s^2.

5. Proof of Theorem 3

This is an immediate consequence of theorem 2. Suppose $\alpha_k < 0$. By the second condition in (1.14) and Schwarz' inequality

$$\alpha_k^2 \leqq \left(\int_{\tau'_k}^{t'_k} x\, F_k(dx)\right)^2 \leqq \pi_k \int_{\tau_k}^{t'_k} x^2\, F_k(dx)\,. \tag{5.1}$$

If $\alpha_k > 0$ the integration extends between $-t_k'$ and $-\tau_k'$. With the notation of the preceding section this implies that

$$\sum \lambda_k + b \leqq \sum \int_{-t_k}^{t'_k} x^2\, F_k(dx) \leqq s^2 \tag{5.2}$$

and so the condition of theorem 2 is fulfilled.

References

1. Feller, W.: An introduction to probability theory and its applications, Vol. II. New York: J. Wiley 1966.
2. Ibragimov, I. A., and L. V. Osipov: On an estimate of the remainder in Lindeberg's theorem. Theor. Probab. Appl. 11, 125—128 (1966).

3. Katz, M. L.: Note on the Berry-Esseen theorem. Ann. math. Statistics **34**, 1107–1108 (1963).
4. Osipov, L. V.: Refinement of Lindeberg's Theorem. Theor. Probab. Appl. **11**, 299–302 (1966).
5. — and V. V. Petrov: On an estimate of the remainder in the central limit theorem. Theor. Probab. Appl. **12**, 281–286 (1967).
6. Petrov, V. V.: An estimate of the deviation of the distribution of a sum of independent random variables from the normal law. Doklady Akad. Nauk SSSR n. Ser. **160**, 1013 to 1015 (1965).
7. Studnev, Yu. P.: A remark on the Katz-Petrov theorem. Theor. Probab. Appl. **10**, 682–684 (1965).
8. — On the role of Lindeberg's conditions. Dopovidi Akad. Nauk. Ukrain. RSR **3**, 239–242 (1958). Select. Transl. math. Statist. Probab. **4**, 303–306 (1963).
9. Zolotarev, V. M.: An absolute estimate of the remainder in the central limit theorem. Theor. Probab. Appl. **11**, 95–105 (1966).

Prof. W. Feller
Department of Mathematics
Princeton University, Princeton, N. J.
and
Rockefeller University, New York, N.Y.,
USA

*An Extention of the Law of the Iterated Logarithm to Variables without Variance**

WILLIAM FELLER

Dedicated to SALOMON BOCHNER

1. Introduction. Throughout this paper $X, X_1, X_2, \cdots$ will denote a sequence of mutually independent random variables with a common distribution F. In order not to obscure the main point we shall assume F to be *symmetric*, that is $F(-x) = 1 - F(x)$ at points of continuity. The results carry over to the general situation by the standard symmetrization procedure. As usual, we put $S_n = X_1 + \cdots + X_n$.

When $E(X^2) = 1$ the central limit theorem asserts that the distribution of $S_n/\sqrt{n}$ tends to the standard normal distribution, whereas by the law of the iterated logarithm

$$\limsup \frac{S_n}{(2n \log \log n)^{1/2}} = 1 \tag{1.1}$$

with probability one. From the simple lemma 1 below it is obvious that in the absence of a finite variance this lim sup is infinite. [This observation was made by V. Strassen and proved by deeper tools]. However, the central limit theorem holds also when $E(X^2) = \infty$ provided only that

$$H(x) = \int_{|t| \le x} t^2 F(dt) \tag{1.2}$$

varies slowly as $x \to \infty$. If ρ_n is defined as the largest root of

$$\rho_n^2 = nH(\rho_n), \tag{1.3}$$

then the distribution of S_n/ρ_n tends to the standard normal. In this situation it is natural to ask whether there exist any norming constants a_n such that

$$\limsup S_n/a_n = 1 \tag{1.4}$$

with probability one and, in particular, whether this is true with

$$a_n = \rho_n(2 \log \log \rho_n)^{1/2}. \tag{1.5}$$

* Research connected with a Project for research in probability at Princeton University, supported by the Army Research Office.

Journal of Mathematics and Mechanics, Vol. 18, No. 4 (1968).

R.L. Schilling et al. (eds.), *Selected Papers II*,

This problem has an unexpected solution because it turns out that in the absence of variances the applicability of *the central limit theorem is neither necessary nor sufficient for the validity of the relations* (1.4)–(1.5).

Our main results are stated in section 7. Neglecting freak phenomena that can occur when the distribution F and the sequence $\{a_n\}$ exhibit extremely wild fluctuations the typical situation may be described roughly as follows.

If there exists any sequence $\{a_n\}$ *for which* (1.4) *holds, then* (1.5) *will do. Necessary and sufficient for this is that*

$$\int_0^\infty \frac{H(dx)}{H(x) \log \log x} < \infty. \tag{1.6}$$

This condition is rather stringent, but it does not imply the slow variation of H. When it fails, the lim sup in (1.4) necessarily equals either zero or infinity for any reasonably regular sequence $\{a_n\}$. In fact, if the integral (1.6) diverges the two events

$$\{|S_n| > a_n \text{ i.o.}\} \quad \textit{and} \quad \{|X_n| > a_n \text{ i.o.}\} \tag{1.7}$$

have the same probability (zero and one being the only possible values by virtue of the Hewitt–Savage theorem). This equivalence may be taken as an indication that dominating values of S_n are as a rule due to the influence of one single term X_k. To make this observation more precise, put

$$X_n^M = \max [X_1, \cdots, X_n] \quad \text{and} \quad {}^{(1)}S_n = S_n - X_n^M. \tag{1.8}$$

Very roughly speaking, it will be shown that when the integral (1.6) diverges and the constants a_n are chosen so that the series $\sum P|X_n| > a_n\}$ barely diverges, then with probability one

$$\limsup S_n/a_n = \infty \quad \text{but} \quad \limsup {}^{(1)}S_n/a_n = 0. \tag{1.9}$$

Theorem 2 shows in particular that even within the domain of attraction of the normal distribution the generalized *law of the iterated logarithm as expressed by* (1.4)–(1.5) *may fail due to the influence one single component* X_k. *With the norming constants* (1.5) *the* lim sup *in* (1.4) *is then infinite, but the relation* (1.4) *holds if* S_n *is replaced by* ${}^{(1)}S_n$.

[The equivalence of the events in (1.7) was discovered in Feller (1946) assuming that F has moments of all orders $< \alpha < 2$ but none of order $> \alpha$. Feller had to impose stronger regularity conditions on $\{a_n\}$. The same phenomenon was discussed by P. Lévy (1952 and 1953) with relaxed conditions on the a_n but additional restrictions on F. The present paper generalizes these results, but its main purpose is to explore the analogue to the law of the iterated logarithm and the influence of the large components when the distribution F is within the domain of attraction of the normal distribution or close to it.]

2. The basic assumptions. The sequence $\{a_n\}$ occurring in the limit relation (1.4) is not uniquely determined since the relation remains valid when the values a_{n_k} are changed for a sufficiently sparse subsequence $\{n_k\}$. A natural class of norming constants is defined by the conditions that

$$a_n/\sqrt{n} \uparrow \infty \tag{2.1}$$

and there exist a constant M such that

$$\frac{a_{2n}}{a_n} < M. \tag{2.2}$$

Condition (2.1) serves merely to avoid trivialities. Condition (2.2) is used mainly in section 6 to prove that if (1.4) is true for some sequence $\{a_n\}$ then also for the special sequence defined by (1.5). Without (2.2) this assertion is false and $\{a_n\}$ must be replaced by the less pleasing sequence defined in (6.1).

Trite complications can arise when the distribution F varies extremely erratically. (Example *b* shows what can happen if the growth of F is concentrated on a very rapidly increasing sequence $\{x_n\}$). In order not to obscure the main point by side issues we shall assume that there exists a constant M such that for large x

$$\frac{H(2x)}{H(x)} < M. \tag{2.3}$$

This is obviously equivalent to

$$\frac{H(tx)}{H(x)} < Mt^{\mu} \tag{2.4}$$

for all x and t sufficiently large. As is well known, for F to belong to some domain of attraction it is necessary that as $x \to \infty$ the left side in (2.4) actually tends to t^{μ} with $\mu < 2$. In (2.4) values $\mu > 2$ are permitted, and so this condition is mild indeed.

It is not necessary to impose any condition on the distribution F if one assumes instead that the behaviour of $\{a_n\}$ is tuned in a natural way to that of H. For example, it is natural to assume that the ratios a_n/ρ_n increase, and this condition is sufficient. As a matter of fact, a perusal of the arguments shows that (2.3) *can be replaced by the condition that there exists a power σ such that the ratios $a_n/H^{\sigma}(a_n)$ increase*. All naturally defined norming constants satisfy this condition with $\sigma = 1$ and so in practice our results apply to arbitrary distributions. The condition (2.3) serves to simplify formulations.

3. Preparatory lemmas. We shall say that the sequence $\{X_n'\}$ *is obtained from* $\{X_n\}$ *by (symmetric) truncation at* τ_n *and* t_n *if*

$$X_n' = \begin{cases} X_n & \textit{when} \quad \tau_n < |X_n| \leqq t_n \\ 0 & \textit{otherwise}. \end{cases} \tag{3.1}$$

The partial sums will be denoted by S_n'.

Lemma 3.1. For arbitrary (symmetric) truncation and arbitrary constants a_n, *if*

$$P\{S_n' > a_n \text{ i.o.}\} = 1 \tag{3.2}$$

then also

$$P\{S_n > a_n \text{ i.o.}\} = 1. \tag{3.3}$$

Proof. Because of symmetry the two events

$$\{S_n' > a_n, S_n - S_n' \geqq 0 \text{ i.o.}\} \quad \text{and} \quad \{S_n' > a_n, S_n - S_n' \leqq 0 \text{ i.o.}\} \tag{3.4}$$

have the same probability, which equals either zero or one by the Hewitt–Savage theorem. Now the event in (3.2) is the union of the two events in (3.4), whereas the event in (3.3) contains the first event in (3.4), and so (3.2) implies (3.3).

An immediate consequence of this lemma is that *the certainty of the second relation in* (1.7) *is necessary for the first.*

Lemma 3.2. If (2.1) *holds, the series*

$$\sum P\{X_n > \lambda a_n\} \tag{3.5}$$

either diverges for all $\lambda > 0$ *or else it converges for all* $\lambda > 0$.

Proof. Assume divergence when $\lambda = \lambda_0$. The divergence for $\lambda < \lambda_0$ follows from the obvious monotonicity. If $\lambda > \lambda_0$ choose an integer $r > \lambda^2/\lambda_0^2$. Then $\lambda_0 a_{nr} > \lambda a_n$ and hence

$$\sum P\{X_n > \lambda a_n\} \geqq \sum P\{X_{nr} \geqq \lambda_0 a_{nr}\} \geqq r^{-1} \sum P\{X_n > \lambda_0 a_n\},$$

and so the series on the left diverges also for $\lambda > \lambda_0$.

For any sequence $\{a_n\}$ we define a function a_t for all $t > 0$ by monotone interpolation between the integers, and put

$$a_t^2 = tA(t). \tag{3.6}$$

Thus (2.1) implies that $A \geqq 0$ and that A *is monotone.*

Lemma 3.3. The series (3.5) *converges if, and only if,*

$$I = \int_0^\infty \frac{x^2 F(dx)}{A(x)} < \infty. \tag{3.7}$$

Proof. When $\lambda = 1$ the series (3.5) converges obviously if, and only if,

$$\sum n \int_{a_{n-1}}^{a_n} F(dx) < \infty. \tag{3.8}$$

Between a_{n-1} and a_n the integrand in (3.7) varies monotonically from $n - 1$ to n, and so (3.7) and (3.8) are equivalent.

4. The first truncation. We saw that the convergence of the integral (3.7) implies the convergence of the series (3.5). This in turn implies that with probability one the relation $|X_n| > a_n$ will take place only for finitely many subscripts n, and thus we can limit our attention to the variables obtained by truncation between 0 and a_n. Such a variable will be decomposed into a sum of three variables obtained by further truncation. In this section we dispose of the largest values. For this purpose we introduce the abbreviations

$$v(t) = A(t)H^{-1}(t) \tag{4.1}$$

$$b_n = a_n v^{-\eta}(a_n) \tag{4.2}$$

where $\eta > 0$ is a constant. In the sequel we denote by X_n' *the variables obtained by truncation at* b_n *and* ϵa_n. The partial sums will be denoted by S_n' while ${}^{(1)}S_n'$ denotes (as in (1.8)) the random variable obtained by substracting from S_n' the largest among the variables $X_1', \cdots, X_n'$.

We shall be interested only in situations where

$$v(a_n) \to \infty. \tag{4.3}$$

Lemma 4.1. *Assume the convergence of the integral* (3.7) *and that the conditions* (2.1), (2.3), *and* (4.3) *hold. Then there exists a constant* η_0 *such that for* $\eta < \eta_0$ *with probability one*

$$S_n' = o(a_n) \quad \textit{and} \quad {}^{(1)}S_n' = o(b_n). \tag{4.4}$$

Proof. Clearly

$$b_n^2 = nH^{2\eta}(a_n)A^{1-2\eta}(a_n); \tag{4.5}$$

since both H and A are monotone it follows for $\eta \leqq \frac{1}{2}$ that $b_n \uparrow \infty$. We can therefore define recursively an increasing sequence of integers n_k such that

$$b_{n_k} < a_{n_{k-1}} \leqq b_{n_k+1}. \tag{4.6}$$

To avoid double subscripts we put

$$a_{n_k} = \alpha_k, \; b_{n_k} = \beta_k. \tag{4.7}$$

Let

$$\tau_k = \sum_{n_{k-1} < n \leq n_k} P\{X_n' \neq 0\}. \tag{4.8}$$

By Chebyshev's inequality

$$\begin{aligned} \tau_k &< \frac{n_k}{\beta_{k-1}^2}[H(\alpha_k) - H(\beta_k)] \\ &= \frac{\alpha_k^2 v^{2\eta}(\alpha_{k-1})}{\alpha_{k-1}^2} \cdot \frac{H(\alpha_k) - H(\beta_k)}{A(\alpha_k)}. \end{aligned} \tag{4.9}$$

Now v need not be monotone, but from (2.4) we get

$$\frac{v(\alpha_{k-1})}{v(\alpha_k)} \leqq \frac{H(\alpha_k)}{H(\alpha_{k-1})} \leqq M \cdot \left(\frac{\alpha_k}{\alpha_{k-1}}\right)^{\mu}. \tag{4.10}$$

Furthermore, by (4.6)

(4.11) $$\alpha_{k-1} \geqq \beta_k = \alpha_k v^{-\eta}(\alpha_k).$$

Putting for abbreviation $C = M^{2\eta}$ and $\gamma = 2\eta(1 + \mu\eta)$ we conclude from (4.9) that

(4.12) $$\tau_k \leqq C \cdot v^{\gamma}(\alpha_k) I(\beta_{k-1}, \alpha_k],$$

where $I($ $]$ represents the measure attributed by the integral I to the given interval.

We need also the rougher estimate

(4.13) $$\tau_k \leqq C v^{\gamma - 1}(\alpha_k)$$

obtained from (4.9) by neglecting $H(\beta_k)$. In view of (4.3) these two estimates imply that when $\gamma < \frac{1}{2}$ and k is sufficiently large

(4.14) $$\tau_k^2 < I(\alpha_{k-2}, \alpha_k].$$

Since the integral I converges this means that

(4.15) $$\sum \tau_k^2 < \infty \quad \textit{and hence} \quad \sum \tau_k \tau_{k-1} < \infty.$$

It follows now from the Borel–Cantelli lemma that with probability one ultimately each block $n_{k-2} < n \leqq n_k$ contains at most one component $X_n' \neq 0$, and this is by definition $< \epsilon a_n$. For given n let k be the integer such that $n_{k-1} < n \leqq n_k$. Omitting, if necessary, the largest variable with a subscript $> n_{k-2}$ we find that with probability one for n sufficiently large

(4.16) $$^{(1)}S_n' = S_{n_{k-2}}' < \epsilon(\alpha_{k-2} + \alpha_{k-4} + \cdots).$$

But by (4.6)

$$\alpha_{k+1} = v(\alpha_{k+1})\beta_{k+1} \geqq v(\alpha_{k+1})\alpha_{k-1}$$

and hence ultimately the right side in (4.16) is dominated by a geometric series with arbitrarily large ratio. The sum is therefore $O(\alpha_{k-2}) = o(b_n)$. Since $X_n' < \epsilon a_n$ and ϵ is arbitrary this implies (4.4).

Remark. If one assumes (as is natural) that v is monotone the condition (2.3) used in the estimate (4.10) becomes unnecessary. In this case the lemma is valid with $\eta_0 = \frac{1}{4}$.

5. The second truncation. In this section we investigate the contribution of the variables X_n'' obtained by truncation at $c_n = a_n v^{-p}(a_n)$ and $b_n = a_n v^{-\eta}(a_n)$; here we preserved the notation of the preceding section and $p > \eta$ is an arbitrary constant.

Lemma 5.1. *Suppose $I < \infty$ and that the conditions* (2.1)–(2.3) *and* (4.3) *are satisfied. Then with probability one*

(5.1) $$S_n'' = o(a_n v^{-\delta}(a_n))$$

for all $\delta > 0$ sufficiently small. If v is montone $\delta < \eta/2$ suffices.

Proof. In view of (2.2) we can choose a sequence of integers n_k such that

$$M^{2k-1} < a_{n_k}^2 \leqq M^{2k}. \tag{5.2}$$

We adhere to the abbreviations (4.7) and put similarly $\gamma_k = c_{n_k}$. Put

$$U_k = \sum_{n_{k-1} < n \leq n_k} X_n'' \tag{5.3}$$

and denote by U_k^* the maximum of the partial sums on the right. Because of symmetry

$$P\{U_k > t\} \geqq \tfrac{1}{2}P\{U_k^* > t\}. \tag{5.4}$$

On the other hand, the monotonicity of A and (2.3) imply that for $n < m$

$$\frac{v(a_n)}{v(a_m)} \leqq \frac{H(a_m)}{H(a_n)} \leqq M\left(\frac{a_m}{a_n}\right)^\mu \tag{5.5}$$

and hence for $n_{k-1} < n \leqq n_k$ and $\mu\delta < 1$

$$\frac{a_n v^{-\delta}(a_n)}{\alpha_k v^{-\delta}(\alpha_k)} \geqq M^{-\delta}\left(\frac{\alpha_{k-1}}{\alpha_k}\right)^{1-\mu\delta} \tag{5.6}$$

The right side remains bounded away from 0 and ∞, and to prove the assertion (5.1) it suffices therefore to show that for any $\delta > 0$

$$\sum P\{U_k > \alpha_k v^{-\delta}(\alpha_k)\} < \infty. \tag{5.7}$$

This, in turn, is certainly true if

$$\sum \alpha_k^{-4} v^{4\delta}(\alpha_k) E(U_k^4) < \infty. \tag{5.8}$$

Put for abbreviation

$$V = \operatorname{Var}(U_k) = \sum_{n_{k-1} < n \leq n_k} E(X_n''^2). \tag{5.9}$$

The variables occurring in the sum being symmetric and bounded by $\beta_k = \alpha_k v^{-\eta}(\alpha_k)$ one gets easily

$$E(U_k^4) \leqq [\alpha_k^2 v^{-2\eta}(\alpha_k) + 3V]V. \tag{5.10}$$

Now trivially

$$V \leqq n_k H(\alpha_k) = \alpha_k^2 v^{-1}(\alpha_k). \tag{5.11}$$

Using the relation (5.6) with δ replaced by p we get the alternative estimate

$$V \leqq n_k A(\alpha_k) I(\epsilon\gamma_k, \beta_k] = \alpha_k^2 I(\epsilon\gamma_k, \beta_k], \tag{5.12}$$

valid when $\epsilon > 0$ is sufficiently small. Combining the two estimates we find that

$$E(U_k^4) \leqq 4\alpha_k^4 v^{-2\eta}(\alpha_k) I(\epsilon\alpha_k v^{-p}(\alpha_k), \alpha_k]. \tag{5.13}$$

Since $v(\alpha_k) \to \infty$ the factor ϵ can be dropped if it is compensated by increasing the exponent p, which is arbitrary. The convergence of the series (5.8) for $\delta < \eta/2$ is therefore guaranteed by the following lemma, which will conclude the proof.

Remark. The condition (2.3) was used in (5.5), and a sharper estimate is obtained by supposing instead that v is monotone.

Lemma 5.2. *For arbitrary $p > 0$ and $\sigma > 0$*

$$\sum v^{-\sigma}(\alpha_k) I(\alpha_k v^{-p}(\alpha_k), \alpha_k] < \infty. \tag{5.14}$$

Proof. The series may be rewritten in the form

$$\sum_k v^{-\sigma}(\alpha_k) \sum_j I(\alpha_{j-1}, \alpha_j] \tag{5.15}$$

where (for fixed k) the inner sum ranges over all j such that

$$\alpha_k v^{-p}(\alpha_k) \leqq \alpha_j \leqq \alpha_k . \tag{5.16}$$

But by (5.2) the ratio of two consecutive α_γ lies between M and M^3, and hence for any j satisfying (5.16)

$$0 \leqq k - j < C \log v(\alpha_k) \tag{5.17}$$

where $C = p/\log M$. Now

$$\frac{v(\alpha_j)}{v(\alpha_k)} \geqq \frac{A(\alpha_j)}{A(\alpha_k)} \geqq \frac{\alpha_j^2}{\alpha_k^2} \geqq v^{-2p}(\alpha_k) \tag{5.18}$$

whereas (5.5) shows that

$$\frac{v(\alpha_j)}{v(\alpha_k)} \leqq M \cdot \left(\frac{\alpha_k}{\alpha_j}\right)^\mu \leqq M \cdot v^{p\mu}(\alpha_k). \tag{5.19}$$

We now change the order of summation in (5.15) to obtain

$$\sum_j I(\alpha_{j-1}, \alpha_j] \sum v^{-\sigma}(\alpha_k) \tag{5.20}$$

the inner summation extending over all k for which (5.16) holds. The preceding inequalities imply that the number of such terms does not exceed a fixed multiple of $\log v(\alpha_j)$ and that each term is dominated by a fixed negative power of $v(\alpha_j)$. The inner sum therefore tends to 0 as $j \to \infty$, and hence the series is dominated by a multiple of the integral I.

6. The central part. We turn our attention to the variables X_n^T obtained by truncation at 0 and $c_n = a_n v^{-p}(a_n)$. The condition (2.2) guarantees that for fixed $\delta > 0$ the ratios $a_n/a_{\delta n}$ remain bounded. Omitting the first δn variables it follows easily that

$$s_n^2 = \operatorname{Var}(S_n^T) \geqq (1 - \delta) n H(a_n v^{-q}(a_n)) \tag{6.1}$$

for any $q > p$ and n sufficiently large. On the other hand

(6.2) $$s_n^2 \leqq nH(a_n v^{-p}(a_n)).$$

We proceed to show that with the particular choice

(6.3) $$v(x) = \log\log x$$

the fluctuations of S_n^T are subject to the classical law of the iterated logarithm. This asserts that if

(6.4) $$c_n = o(s_n(\log\log s_n)^{-1/2})$$

then with probability one

(6.5) $$\limsup \frac{S_n^T}{s_n(\log\log s_n)^{1/2}} = \sqrt{2}.$$

It is not difficult to see that the condition (6.4) is implied by (6.1) if H is of slow variation (that is, if F belongs to the domain of attraction of the normal distribution). Under our present more general conditions (6.4) need not be true. The relation (6.5) is therefore not assured, and even when it is true it is desirable to replace the clumsy norming constants s_n by simpler ones. This is achieved by

Lemma 6.1. *Assume that the integral* (1.6) *converges and the conditions* (2.1)–(2.2) *hold. Then with probability one*

(6.6) $$\limsup S_n^T/a_n = \sqrt{2}.$$

Proof. Assume for the moment that we could sharpen (6.1)–(6.2) to

(6.7) $$s_n^2 \sim nH(a_n).$$

The right side is the same as $a_n^2 v^{-1}(a_n)$ or $c_n^2 v^{2p-1}(a_n)$, and so (6.4) would be equivalent to (6.6). Unfortunately (6.7) need not be true, but we shall show that this relation becomes true if $n \to \infty$ omitting a set of integers that is so sparse as not to affect the conclusion (6.5). For this purpose we use again the sequence $\{n_k\}$ defined by (5.2) and put again $a_{n_k} = \alpha_k$. From Lemma 5.2 we see that

(6.8) $$\sum \frac{1}{\log^{1+\sigma} k} \cdot \frac{H(\alpha_k) - H(\alpha_k v^{-q}(\alpha_k))}{H(\alpha_k)} < \infty.$$

For arbitrary, but fixed, $\delta > 0$ denote by K the set of all integers k such that

(6.9) $$H(\alpha_k v^{-q}(\alpha_k)) < (1 - \delta)H(\alpha_k).$$

It is clear from (6.8) that

(6.10) $$\sum_{k \in K} \log^{-1-\sigma} k < \infty$$

where $\sigma > 0$ is arbitrary. Now the central limit theorem applies to the truncated variables X_n^T, and (6.10) implies trivially that

$$\sum_{k \epsilon K} P\{S_{n_k}^T > \epsilon \alpha_k\} < \infty \tag{6.11}$$

for every $\epsilon > 0$. Hence

$$\limsup_{k \epsilon K'} S_{n_k}^T / \alpha_k = 0. \tag{6.12}$$

The standard proof of the iterated logarithm applies when k is restricted to the complementary set K', and so

$$\limsup_{k \epsilon K} \frac{S_{n_k}^T}{s_{n_k}(\log \log s_{n_k})^{1/2}} = \sqrt{2}. \tag{6.13}$$

But on K' we have

$$s_{n_k}^2 > (1 - \delta)^2 n_k H(\alpha_k) = (1 - \delta)^2 \alpha_k \log \log \alpha_k \tag{6.14}$$

and the inequality is reversed when the factor $(1 - \delta)^2$ is omitted. Since ϵ is arbitrary we conclude that

$$\limsup S_{n_k}^T / \alpha_k = \sqrt{2}. \tag{6.15}$$

Thus (6.6) holds for arbitrary geometric subsequences, and from this (6.6) follows by the standard argument (used in section 5.).

7. Summary. In the preceding section a_n was defined as the largest root of

$$a_n^2 = nH(a_n) \log \log a_n . \tag{7.1}$$

With this choice the condition (2.1) is satisfied, and the two conditions (2.2) and (2.3) are equivalent.

In the final formulation it is preferable to use the canonical norming constants ρ_n defined in (1.3). If the tail $1 - F$ varies slowly then $a_n^2 \sim \rho_n^2 \log \log \rho_n$, but this relation may be false even for distributions belonging to the domain of attraction of the normal distribution. It is therefore remarkable that we have nevertheless

Theorem 1. *Assume* (2.1) *and* (2.2). *If the integral* (1.6) *converges then with probability one*

$$\limsup \frac{S_n}{a_n} = \limsup \frac{S_n}{\rho_n(\log \log \rho_n)^{1/2}} = \sqrt{2}. \tag{7.2}$$

In case of divergence with probability one

$$\limsup \frac{S_n}{a_n} = \limsup \frac{S_n}{\rho_n(\log \log \rho_n)^{1/2}} = \infty . \tag{7.3}$$

Proof. We apply the preceding results with $v(x) = \log \log x$. This function being monotone the assumption (2.3) is not necessary for lemmas 4.1 and 5.1.

Together with Lemma 3.3 these show immediately that in the case of convergence

$$S_n = S_n^T + o(a_n). \tag{7.4}$$

The first relation in (7.2) now follows from (6.6).

Comparing the definitions (7.1) and (1.3) one sees immediately that $a_n \geqq \rho_n$, and hence

$$a_n^2 \geqq \rho_n^2 \log \log \rho_n . \tag{7.5}$$

As already remarked, the two sides are not necessarily asymptotically equivalent, but we have seen that the integers n for which

$$H(a_n/\log \log a_n) < (1 - \delta)H(a_n) \tag{7.6}$$

are negligible in the sense that if $n \to \infty$ through values for which (7.6) is true then $S_n^T = o(a_n)$, while on the complementary set the limit relations (6.5) and (6.6) apply. But when (7.6) is false the quantity $x_n^2 = (1 - \delta)a_n^2/\log \log a_n$ satisfies $x_n^2 < nH(x_n)$ and is therefore $\leqq \rho_n^2$. Thus $a_n^2 \sim \rho_n^2 \log \rho_n$ except on a negligible set, and so (7.2) is true in case of convergence.

In case of divergence the lemmas of section 3 imply that for arbitrary $\lambda > 0$ with probability one the relation $S_n > \lambda a_n$ takes place for infinitely many n. In view of (7.5) this implies (7.3).

The following theorem illustrates the role of the maximal term in the case of divergence. For the notation see (1.8).

Theorem 2. *Suppose that* (2.3) *holds and*

$$\int_0^\infty \frac{x^2 F(dx)}{H(x) \log \log^{1+\gamma} x} < \infty \tag{7.7}$$

for some $\gamma < \frac{1}{4}$, *but that the integral diverges when* $\gamma = 0$. *Then* (7.3) *holds but*

$$\limsup \frac{{}^{(1)}S_n}{\rho_n(\log \log \rho_n)^{1/2}} = \sqrt{2}. \tag{7.8}$$

Proof. As remarked at the end of section 4, in view of the monotone character of log log Lemma 4.1 applies with $\eta_0 = \frac{1}{4}$, and Lemma 5.1 with $\delta < \eta/2$. The proof of the preceding theorem now applies with the sole change that the iterated logrithm in (7.1) is raised to the power $1 + \gamma$.

We saw that a limit relation of the form (7.2) can hold simultaneously for distinct sequences $\{a_n\}$, but it is easily seen that two such sequences can be essentially different only on subsequences that are rather sparse. Example (b) of the concluding section shows that even in the case of a divergent integral (1.6) a relation analogous to (7.2) may hold for a sequence of norming constants obtained by redefining $\{a_n\}$ on a certain sparse set. In order to avoid such freaks and trivialities we now limit our attention to sequences $\{a_n\}$ such that

$$\frac{a_n}{\rho_n(\log \log \rho_n)^{1/2}} \to \infty . \tag{7.9}$$

Although this is not really essential we simplify the proof by assuming that (2.3) holds for some value $M < 4$; in this case

$$\frac{H(tx)}{H(x)} < t^{\gamma}, \qquad \gamma < 2 \tag{7.10}$$

for all t and x sufficiently large.

Theorem 3. *Assume the conditions* (2.1)–(2.2) *and* (7.10). *For a sequence satisfying* (7.9) *we have* $\limsup S_n a_n^{-1} = 0$ *or* ∞ *according as* $\sum P\{X_n > a_n\}$ *converges or diverges.*

Proof. With v defined by (4.1) we have

$$v(a_n) = \frac{a_n^2}{nH(a_n)} = \frac{a_n^2}{\rho_n^2}\cdot\frac{H(\rho_n)}{H(a_n)} \geqq \left(\frac{a_n}{\rho_n}\right)^{2-\gamma} \geqq (\log\log \rho_n)^{1-\gamma/2} \tag{7.11}$$

for n sufficiently large. The proof of Theorem 1 now applies in a simplified form, since we have only to prove that $S_n^T = o(a_n)$.

Theorem 4. *Assume* (2.3). *Whenever the integral* (1.6) *diverges there exists a sequence* a_n' *such that*

$$\limsup S_n/a_n' = \infty \quad \textit{but} \quad {}^{(1)}S_n = o(a_n'). \tag{7.12}$$

Proof. It is possible to find a monotone function w such that the integral

$$\int_0^{\infty} \frac{H(dx)}{H(x)w^{1+\gamma}(x)} \tag{7.13}$$

diverges for $\gamma = 0$ but converges for all $\gamma > 0$. Since (1.6) diverges we may also suppose $w(x) > \log\log x$. If we then define a_n and a_n' respectively as the largest roots of

$$a_n^2 = nH(a_n)w^{1+\gamma}(a_n), \qquad a_n'^2 = nH(a_n')w(a_n')$$

the first relation in (7.11) follows from Lemma 3.3 and the divergence of (7.12) when $\gamma = 0$. The second relation is true because in the notations of sections 4–6 each of the partial sums S_n^T, ${}^{(1)}S_n'$, and S_n'' is $o(a_n')$.

8. Examples. (a) Let the distribution F be concentrated on $|x| > 1$ with density $|x|^{-3}$. Then $H(x) = \log x$ for $x > 1$, and $2\rho_n^2 \sim n \log n$. The integral (7.7) converges only for $\gamma > 0$, and by Theorem 2 therefore

$$\limsup \frac{S_n}{(n\log n \log\log n)^{1/2}} = \infty, \qquad \limsup \frac{{}^{(1)}S_n}{(n\log n\log\log n)^{1/2}} = 1.$$

In this case the analogue to the classical law of the iterated logarithm has been secured by the expedient of removing the offending maximal term. The same result may be achieved by a different method. With r standing for an arbitrary integer define

$$a_n^2 = \begin{cases} n \log n \log \log n & \text{if } r^2 < \log \log n < r^2 + 1 \\ n \log n\, (\log \log n)^2 & \text{otherwise.} \end{cases}$$

It is easily verified that the series (3.5) converges, and the argument used in Theorem 1 shows directly that

$$\text{(8.1)} \qquad \limsup S_n/a_n = \limsup S_n^T/a_n = 1.$$

To assure that the first part is true we had to modify the sequence $\{a_n\}$, but the second part is true only because it holds on the subsequence on which our new a_n agree with those used in Theorem 1.

(b) As a last example let F be concentrated on the points $\pm x_k$ of a very rapidly increasing sequence, say $x_k = (2k!)^k$. Let $P\{X = x_k\} \sim 1/(k+1)!$. Put

$$a_n = x_k \quad \textit{when} \quad (k-1)! < n \leqq k!.$$

In order that $S_n > (1 + \epsilon)a_n$ it is necessary that at least two components X_j exceed x_k. The probability for this is the order of magnitude k^{-2}, and so with probability one this event occurs only finitely often. But with probability one it will happen infinitely often that one component equals x_k, and one concludes easily that $\limsup S_n/a_n = 1$. Such a relation is therefore possible with norming constants in no way related to those of Theorem 1. This however is due only to the erratic behavior of F.

References

[1] W. Feller, A limit theorem for random variables with infinite moments, *American J. Math.*, **68** (1946) 257–262.

[2] P. Lévy, Loi faible et loi forte des grands nombres, *C. R. Acad. Sci. Paris*, **235** (1952) 1186–1188.

[3] ———, Loi faible et loi forte des grands nombres, *Bull. Sci. Math.*, **77** (1953) 9–40.

[4] V. Strassen, A converse to the law of the iterated logarithm, *Z. Wahrscheinlichkeits theorie*, **4** (1966) 265–268.

Princeton University
and
Rockefeller University
Date communicated: January 20, 1968.

ONE-SIDED ANALOGUES OF KARAMATA's REGULAR VARIATION *)

William Feller

To the memory of J. Karamata

1. Introduction

The monotone functions studied in this paper are assumed to be defined on $(0, \infty)$, and to be non-negative and right continuous. Point functions are introduced for notational convenience only, but we are really concerned with the associated measure which attributes value $U(x)$ to the interval $[0, x]$ when U increases, and to (x, ∞) when U decreases to zero.

Karamata's original theory of regularly varying functions has been generalized in chapter VIII of [1] to measures. A monotone function U *is said to vary regularly at infinity with exponent* α *if*

$$\lim_{t \to \infty} \frac{U(tx)}{U(t)} = x^{\alpha} \tag{1.1}$$

for some α and each $x > 0$. At first glance this definition appears to be artificial, but it is motivated by the fact that if the limit on the left exists, it is necessarily of the form x^{α}. Accordingly *U varies regularly at infinity if, and only if, as $t \to \infty$ the measures $U(t\mathrm{d}x)/U(t)$ converge to a finite measure* in every finite interval. (Here, and in the following, convergence of measures is taken in the usual weak sense.) The function U varies regularly at the origin if $U(t^{-1})$ varies regularly at infinity. From now on we omit the qualification " at infinity ", and it will be tacitly understood that in our passages to the limit the variable tends to ∞.

With an arbitrary measure U on $(0, \infty)$ we may associate the truncated moment functions

$$U_p(x) = \int_x^{\infty} y^{-p}\, U(dy) \tag{1.2}$$

* Work connected with a Project for research in probability theory at Princeton University, supported by the Army Research Office.

R.L. Schilling et al. (eds.), *Selected Papers II*,

and

$$(1.3) \qquad Z_q(x) = \int_0^x y^q \, U(dy)$$

defined whenever the integrals converge. These functions define new measures and they satisfy obvious identities such as

$$(1.4) \qquad U_p(x) = \int_x^\infty y^{-p-q} Z_q(dy) .$$

For convenience of notation and exposition we shall from now on use $U = Z_0$ as representative of the whole family $\{Z_q\}$ and formulate all theorems in terms of U and U_p. Various relations between the diverse Z_q will be implicit in our theorems. As a last piece of notation we introduce the frequently occurring function

$$(1.5) \qquad R_U(t) = \frac{t^p U_p(t)}{U(t)} .$$

The notion of regular variation was introduced, and achieved its greatest success, in connection with Tauberian theorems. In recent years more attention was paid to hitherto little known relations derived by Karamata in [3] and connecting the asymptotic behavior of the various functions Z_q and U_p. The basic theorems may be summarized as follows.

(i) *Let U vary regularly with exponent $\alpha \geqslant 0$. Then U_p exists for $p > \alpha$ (but for no $p < \alpha$). Furthermore*

$$(1.6) \qquad \lim_{t \to \infty} R_U(t) = r$$

exists, and $r = \dfrac{\alpha}{p - \alpha} < \infty$. If $r > 0$ then U_p varies regularly with exponent $\alpha - p$.

(ii) *Let U be such that U_p exists for some fixed $p > 0$ and varies regularly with some exponent $q \leqslant 0$. Then the limit (1.6) exists, $0 < r \leqslant \infty$. If $r < \infty$ then U varies regularly with exponent $p + q$.*

(iii) *Let the limit r exist. If $0 < r < \infty$ then both U and U_p vary regularly with exponents*

$$(1.7) \qquad \alpha = p \frac{r}{r+1} \text{ and } \alpha - p = -p \frac{1}{r+1}$$

respectively ; hence

$$U_p(t) \sim rt^{-p}\, U(t)\,. \tag{1.8}$$

If $r = 0$ *then* U *varies slowly* ($\alpha = 0$) *and* $U_p(t) = o(t^{-p}\, U(t))$. *If* $r = \infty$ *then* U_p *varies slowly and* $U(\mathrm{t}) = o(t^p\, U_p(t))$.

Thus except when either U or U_p varies slowly regular variation is tied to a relation of the form (1.8) and the existence of the limit (1.6) characterizes regular variation.

Karamata considered only measures defined by densities. Simplified proofs and extensions of his results can be found in [1]. In this book it was shown that the measure theoretic version of Karamata's relations introduces coherence and unity in the theory of domains of attraction, and that it leads to a substantial simplification of this theory. (In such connections U is usually the truncated second moment of a probability distribution, and U_2 is then the tail sum of this distribution.)

In [2] it turned out that various compactness arguments and local limit theorems in probability do not depend on the full strength of regular variation, but only on a one sided version of it.

We now proceed to describe this generalization and to show that Karamata's relations carry over to a surprising extent. In section 4 we discuss inequalities going in the opposite direction.

In section 5 we turn to *ratio limit theorems*. Roughly speaking, we show that if two monotone functions U and V stand in the relation $V = UL$ with L slowly varying, then also $V_p \sim U_p\, L$. For regularly varying function this is implicit in Karamata's relation (1.8), but it is surprising that dominated variation should suffice for the conclusion. Furthermore, we obtain a necessary and sufficient condition for the ratio V/U to be slowly varying. In section 6 these results are reformulated in the form of a *Tauberian ratio limit theorem*.

To illustrate the way in which dominated variation occurs naturally in probabilistic contexts we discuss in section 7 the asymptotic behavior of the *tails of infinitely divisible distributions*. This section is independent of the Karamata relations and may be read directly after section 2.

2. Dominated variation

We start from the following

Definition. *A monotone function U varies dominatedly if*

(2.1)
$$\limsup \frac{U(2t)}{U(t)} < \infty \quad \textit{in case} \quad U\uparrow$$
$$\limsup \frac{U(t/2)}{U(t)} < \infty \quad \textit{in case} \quad U\downarrow .$$

This leads immediately to the

Criterion. *A non-decreasing U varies dominatedly if there exist constants* C, γ, *and* t_0 *such that*

$$\frac{U(tx)}{U(t)} < Cx^{\gamma} \qquad x > 1 , \ t > t_0 . \tag{2.2}$$

For non-increasing U the same criterion applies with $x>1$ *replaced by* $x<1$.

Proof. The sufficiency is obvious. Assume (2.1) and choose t_0 and C such that

$$\frac{U(2t)}{U(t)} < C \qquad \text{for} \qquad t > t_0 . \tag{2.3}$$

Put $\gamma = \text{Log}_2 C$. For $x>1$ define n by $2^{n-1}<x\leqslant 2^n$. A repeated application of (2.3) then shows that the left side in (2.2) is $\leqslant C^n \leqslant Cx^{\gamma}$.

Dominated variation of U may be described by saying that *the measures associated with* $U(t\cdot)/U(t)$ *form a sequentially compact family* in the sense that every sequence contains a subsequence converging on finite intervals to a finite measure. As in the case of slowly varying functions, the occurrence of limit measures that vanish identically on $(0, \infty)$ introduces some lack of symmetry. The supplementary condition (4.1) is designed to avoid this anomaly.

3. One-sided version of the Karamata relations

From now on U will stand for a non-decreasing function and $p>0$ will be a fixed number such that the integral U_p converges. Only the case $U(\infty) = \infty$ is of practical interest. We adhere to the notation (1.5) for R_U and put

$$\underline{r} = \liminf R_U(t), \qquad \bar{r} = \limsup R_U(t). \tag{3.1}$$

We shall also use the notation

$$\mathscr{I}_U(t) = \int_t^\infty y^{-p-1} U(y)\, dy. \tag{3.2}$$

Theorem 1. *For U to vary dominatedly it is necessary and sufficient that $\bar{r}<\infty$. Similarly U_p varies dominatedly iff $\underline{r}>0$.*

More precisely: The relation (2.2) with $\gamma<p$ entails

$$R_U(t) \leqslant A \qquad t > t_0 \tag{3.3}$$

with

$$A = \frac{Cp}{p-\gamma} - 1. \tag{3.4}$$

Conversely, (3.3) implies (2.2) with

$$C = A + 1, \qquad \gamma = \frac{A}{A+1} p. \tag{3.5}$$

In like manner, if

$$R_U(t) \geqslant \eta > 0, \qquad t > t_0 \tag{3.6}$$

then

$$\frac{U_p(tx)}{U_p(t)} \geqslant K x^{-q}, \qquad x > 1, \; t > t_0 \tag{3.7}$$

with

$$K = \frac{\eta}{\eta+1}, \qquad q = \frac{p}{\eta+1}. \tag{3.8}$$

Conversely, if (3.7) holds with $q<p$ then

$$r \geqslant \frac{K(p-q)}{Kq+(1-K)p}. \tag{3.9}$$

(Note that necessarily $K \leqslant 1$ as can be seen letting $x \to 1$ in (3.7). On replacing t by tx^{-1} it is seen that (3.7) not only asserts dominated variation of U_p, but implies uniformity away from the origin.)

PROOF. (i) Using integration by parts and the notation (3.2) it is seen that the definition (1.2) of U_p leads to the identity

$$p\,\mathscr{I}_U(t) = U_p(t) + t^{-p}\,U(t) \tag{3.10}$$

valid at all points of continuity. If (2.2) holds with $\gamma<p$ we conclude for $t>t_0$

$$U_p(t) + t^{-p}\,U(t) \leqslant Cp \cdot U(t) \int_t^\infty y^{-p-1}(y/t)^\gamma\,dy = \tag{3.11}$$

$$= C\frac{p}{p-\gamma}\,t^{-p}\,U(t)$$

and so (3.3) holds with A defined in (3.4).

(ii) Assume (3.3). Then by (3.10)

$$pt^p\,\mathscr{I}_U(t) \leqslant (A+1)\,U(t) \tag{3.12}$$

or

$$\frac{s^{-p-1}\,U(s)}{\mathscr{I}_U(s)} \geqslant \frac{p}{A+1}\cdot\frac{1}{s} \qquad s>t_0. \tag{3.13}$$

Integrating between t and $tx>t$ we get

$$\log\frac{\mathscr{I}_U(t)}{\mathscr{I}_U(tx)} \geqslant \frac{p}{A+1}\log x, \qquad t>t_0. \tag{3.14}$$

Thus from (3.12)

$$(A+1)\,t^{-p}\,U(t) \geqslant p\mathscr{I}_U(t) \geqslant p\mathscr{I}_U(tx)\cdot x^{p/(A+1)} \tag{3.15}$$

and by the definition (3.2)

$$p\mathscr{I}_U(tx) \geqslant U(tx)\cdot(tx)^{-p}. \tag{3.16}$$

Accordingly, (2.2) holds with C and γ given in (3.5). (This part of the theorem was proved slightly differently in [2].)

(iii) Assume (3.6). As in the last part we conclude

$$\log \frac{\mathcal{I}_U(t)}{\mathcal{I}_U(tx)} \leqslant \frac{p}{\eta+1} \log x\,, \qquad x > 1\,,\ t > t_0\,. \tag{3.17}$$

A repeated use of (3.10) now shows that

$$\begin{aligned} U_p(t) &\leqslant p\mathcal{I}_U(t) \leqslant p\mathcal{I}_U(tx) \cdot x^{p/(\eta+1)} = \\ &= x^{p/(\eta+1)} \left[U_p(tx) + (tx)^{-p}\, U(tx)\right]. \end{aligned} \tag{3.18}$$

From (3.6) with t replaced by tx it is seen that the expression within brackets is $<(1+\eta^{-1})\, U_p(tx)$, and so the assertion concerning (3.7) is true.

(iv) Assume (3.7) with $q<p$. From the definition (1.2) of U_p we get by Fubini's theorem

$$p \int_0^t y^{p-1}\, U_p(y)\, dy = U(t) + t^p\, U_p(t) \tag{3.19}$$

which proves that the integral on the left converges for all $t>0$. Let B stand for the value of the left side when $t = t_0$. For $y>t_0$ we can apply (3.7) to conclude

$$\begin{aligned} U(t) + t^p\, U_p(t) &\leqslant B + pK^{-1}\, U_p(t) \int_{t_0}^t y^{p-1} (t/y)^q\, dy < \\ &< B + \frac{p}{p-q} K^{-1} t^p\, U_p(t)\,. \end{aligned} \tag{3.20}$$

Divide this inequality by $U(t)$ and let $t \to \infty$. If $U(t) \to \infty$ we get the assertion (3.8). If $U(t)$ remains bounded there is nothing to be proved because (3.7) implies that $t^p\, U_p(t)$ increases at least as fast as t^{p-q}, and hence $\underline{r} = \infty$ whenever U is bounded.

NOTE. Our result lack the perfect symmetry of the original Karamata relations. Starting from (2.2) we get (3.3)-(3.4). However, when we apply the converse with these given values we get (2.2) in the weaker form with γ replaced by a constant $\gamma' > \gamma$. Examples given in [2] show that, in an obvious sense, this result is the best possible.

4. Other conditions

By theorem 1 both U and U_p vary dominatedly whenever $\underline{r}>0$ and $\bar{r}<\infty$. The next theorem gives even simpler criteria for dominated variation that remain applicable in the limiting situations $\underline{r}=0$ and $\bar{r}=\infty$.

Theorem 3. *If there exists a number $\xi>1$ such that*

$$\liminf \frac{U(\xi t)}{U(t)} > 1 \tag{4.1}$$

then U_p varies dominatedly. Similarly the relation

$$\liminf \frac{U_p(t/\xi)}{U_p(t)} > 1 \tag{4.2}$$

implies the dominated variation of U.

Proof. Clearly

$$\frac{t^p U_p(t)}{U(t)} \geqslant \frac{t^p\left[U_p(t)-U_p(t\xi)\right]}{U(t)} \geqslant \xi^{-p}\frac{U(t\xi)-U(t)}{U(t)}. \tag{4.3}$$

When the right side is bounded away from 0 this implies $\underline{r}>0$, and so U_p varies dominatedly by theorem 1.

We can go a step further. If, besides (4.1), it is known that U varies dominatedly with exponent $\gamma<p$, then $t^{-p}\,U(t)/U_p(t)$ is bounded away from 0, and hence the second inequality in (4.3) implies that

$$\liminf \frac{U_p(t)-U_p(t\xi)}{U_p(t)} > 0. \tag{4.4}$$

This is equivalent to (4.2).

Similarly

$$\frac{U_p(t)-U_p(t\xi)}{U_p(t\xi)} \leqslant t^{-p}\frac{U(t\xi)-U(t)}{U_p(t\xi)} = \tag{4.5}$$

$$= \frac{\xi^p}{R_U(t\xi)}\,\frac{U(t\xi)-U(t)}{U(t\xi)}.$$

The second fraction on the right does not exceed 1, and so (4.2) ensures that $R_U(t\xi)$ remains bounded, and hence that U is of dominated variation.

Again, if it is known that R_U is bounded away from 0 then (4.5) shows that (4.2) implies (4.1).

We have thus proved the

COROLLARY. *If U is of dominated variation with exponent $\gamma<p$ then (4.1) implies (4.2). Similarly, if U_p is of dominated variation with exponent $-q$ where $q<p$, then (4.2) entails (4.1).* (In each case both functions are of dominated variation.)

5. Ratio limit theorems

Let U and V be non-decreasing unbounded functions, and suppose that L is slowly varying (= regularly varying with exponent 0).

DEFINITION. *We shall say that U and V are L-equivalent and write*

$$V \leftrightarrow UL \tag{5.1}$$

if the ratio UL/V tends to 1 at all points of continuity.

More precisely, it is required that for each $\varepsilon>0$ and fixed $\lambda>1$

$$(1-\varepsilon)\, L(t)\, U(t/\lambda) \leqslant V(t) \leqslant (1+\varepsilon)\, L(t)\, U(t\lambda) \tag{5.2}$$

for all t sufficiently large.

THEOREM 4. *Let U be of dominated variation. In order that there exist a slowly varying function* L *such that (5.1) holds it is necessary and sufficient that*

$$R_U(t) - R_V(t) \to 0 \qquad \textit{boundedly}. \tag{5.3}$$

Needless to say, R_V and $\mathcal{J}_V$ are defined by analogy with R_U in (1.5) and $\mathcal{J}_U$ in (3.2).

PROOF. (a) *Necessity.* Assume (5.1) and suppose that U satisfies the basic inequality (2.2). Obviously the slow variation of L implies that for t sufficiently large and all $x>1$

$$\frac{V(tx)}{V(t)} < C' x^{\gamma'} \tag{5.4}$$

or any pair of constants $C'>C$ and $\gamma'>\gamma$. Thus V is of dominated variation, and since $p>\gamma$ the function V_p exists.

Let $t_n \to \infty$ in such a way that the measures associated with $U(t_n \cdot)/U(t_n)$ tend (in finite intervals) to a limit measure m. The relation (5.1) implies obviously that the measures associated with $V(t_n \cdot)/V(t_n)$ tend to the same limit m. Thus when t runs through $\{t_n\}$ we have for fixed $x > 1$

$$(5.5) \qquad \frac{U_p(t) - U_p(tx)}{U(t)\, t^{-p}} = \int_1^x y^{-p} \frac{U(tdy)}{U(t)} \to \int_1^x y^{-p}\, m(dy),$$

and the same relation holds with U replaced by V. But (5.4) implies that this passage to the limit is uniform as $x \to \infty$; it remains valid also for $x = \infty$ with the right side being finite. We have thus shown that $R_U(t_n) - R_V(t_n) \to 0$. But the t_n may be picked as elements of an arbitrarily prescribed sequence, and so the limit relation in (5.3) holds pointwise for an arbitrary approach $t \to \infty$. Now we know that the dominated variation of U and V implies the boundedness of both R_U and R_V, and the condition (5.3) holds true.

(b) *Sufficiency*. The variation of U being dominated, R_U remains bounded and so (5.3) implies the boundedness of R_V and hence the dominated variation of V. The calculation of part (ii) in section 3 show that

$$(5.6) \qquad \frac{s^{-p-1}\, U(s)}{\mathscr{I}_U(s)} - \frac{s^{-p-1}\, V(s)}{\mathscr{I}_V(s)} = \frac{p}{t}\left[\frac{1}{1 + R_U(s)} - \frac{1}{1 + R_V(s)}\right].$$

The expression within brackets is in absolute value bounded by $|R_U(s) - R_V(s)|$, and therefore tends to 0 boundedly. Integrating between t and $tx > t$ we conclude therefore that

$$(5.7) \qquad \log \frac{\mathscr{I}_U(t)}{\mathscr{I}_U(tx)} \cdot \frac{\mathscr{I}_V(tx)}{\mathscr{I}_U(t)} \to 0.$$

In other words, the ratio $\mathscr{I}_U/\mathscr{I}_V$ varies slowly, and therefore we can put

$$(5.8) \qquad \mathscr{I}_V(t) = L(t)\, \mathscr{I}_U(t)$$

where L varies slowly.

We now recall the inequality (3.14) which implies that to each $\lambda > 1$ there exists an $\eta < 1$ such that

$$(5.9) \qquad \mathscr{I}_U(\lambda t) < \eta\, \mathscr{I}_U(t)$$

for all t sufficiently large. From (5.8) we conclude therefore that

$$(5.10) \qquad \lim \frac{\mathcal{I}_V(t) - \mathcal{I}_V(\lambda t)}{[\mathcal{I}_U(t) - \mathcal{I}_U(\lambda t)]\, L(t)} =$$

$$= \lim \frac{L(t)\,\mathcal{I}_U(t) - L(\lambda t)\,\mathcal{I}_U(\lambda t)}{L(t)\,\mathcal{I}_U(t) - L(t)\,\mathcal{I}_U(\lambda t)} = 1\,.$$

But the fraction on the left lies between

$$\frac{V(\lambda t)}{U(t)\,L(t)} \qquad \text{and} \qquad \frac{V(t)}{U(\lambda t)\,L(t)}$$

and so (5.1) is true.

6. Application to Tauberian theorems

If the measure U varies regularly at infinity, then its Laplace transform ω varies regularly at the origin. More precisely, Karamata's now classical Tauberian theorem states that for any $\alpha \geqslant 0$ and slowly varying function L the two relations

$$(6.1) \qquad U(x) \sim x^{\alpha} L(x) \qquad\qquad \omega(\lambda) \sim \Gamma(\alpha+1)\,\lambda^{-\alpha} L(\lambda^{-1})$$

imply each other; here $x \to \infty$ but $\lambda \to 0$. [The sign $\sim$ indicates that the ratio of the two sides tends to 1.] For an example of a probabilistic application suppose that

$$(6.2) \qquad U(x) = \int_0^x y^p\, F(dy)$$

is the truncated p^{th} moment of a probability distribution F on the positive half axis. For simplicity let p stand for a positive integer. Then $U_p(x) = 1 - F(x)$ and $\omega = (-1)^p \phi^{(p)}$ where ϕ is the Laplace-Stieltjes transform of F. If ω varies regularly in accordance with (6.1) then Karamata's relation (1.8) implies that

$$(6.3a) \qquad 1 - F(x) \sim \frac{\alpha}{p - \alpha}\, x^{\alpha - p} L(x) \qquad \textit{when} \qquad \alpha < p$$

$$(6.3b) \qquad 1 - F(x) = o(x^{\alpha} L(x)) \qquad \textit{when} \qquad \alpha = p\,.$$

(Note that necessarily $0 \leqslant \alpha \leqslant p$ because the measure F is finite.) In other words, the behavior at the origin of the derivatives of the Laplace transform determines the behavior of the tail $1 - F(x)$, and vice versa.

From the continuity theorem for Laplace transforms one concludes without difficulties that *dominated variation of U at* ∞ *is equivalent to dominated variation of* ω *at* 0. Even in the case of mere dominated variation the behavior of $\omega = (-1)^p \phi^{(p)}$ therefore permits inferences concerning the behavior of the tail $1-F$, but naturally the conclusions will lack the pleasing precision of (6.3). It is therefore remarkable that precise asymptotic equivalence relations can be obtained when comparing two probability distributions F and G with Laplace transforms ϕ and γ.

A typical *Tauberian ratio limit theorem would state that the two relations*

$$\gamma^{(p)}(\lambda) \sim \phi^{(p)}(\lambda)\, L\left(\frac{1}{\lambda}\right) \qquad \lambda \to 0 \tag{6.4}$$

and

$$1 - G(x) \sim [1 - F(x)]\, L(x) \qquad x \to \infty \tag{6.5}$$

imply each other. This is not true in full generality; indeed, (6.3*b*) points to exceptional situations even when the transforms in (6.4) vary regularly. However, our results yield a variety of fairly general sufficient conditions for the validity of the conclusion. Suppose, for example, that for some constants A and $\alpha < p$

$$(-1)^p \phi^{(p)}(\lambda) < A\, \lambda^{-\alpha} \tag{6.6}$$

for λ sufficiently small. It is easily seen in this case that U varies dominatedly with exponent α and (6.4) is equivalent to

$$V \leftrightarrow UL \tag{6.7}$$

in the sense of (5.1). (Here V stands for the truncated moment function of G defined as in (6.2).) Theorem 4 then asserts that (5.3) holds, and this implies

$$V_p \leftrightarrow U_p L \tag{6.8}$$

whenever U is bounded away from 0. Now (6.5) differs from (6.8) only notationally, and we know that the condition (4.1) guarantees that R_U is bounded away from 0 and that $U_p = 1 - F$ varies dominatedly. Again, (4.1) holds if, and only if, each limit of a convergent sequence of measures $U(t_n\, dx)/U(t_n)$ attributes a positive measure to $(0, \infty)$. This requirement is satisfied if, and only if,

$$\liminf_{\varepsilon \to 0} \frac{\phi^{(p)}(\varepsilon \lambda_0)}{\phi^{(p)}(\varepsilon)} < 1 \tag{6.9}$$

for some $\lambda_0 > 1$. Accordingly, *if the conditions (6.6) and (6.9) hold then (6.4) implies (6.6.) as well as the dominated variation of* $1-F$ *and* $1-G$.

Our results permit various paraphrases of the sufficient conditions, and also of the ratio limit theorem itself. That (6.6) by itself is not sufficient is shown by (6.3*b*); without (6.9) certain subsequences may exhibit the pattern of slow variation, and the conclusion (6.5) must be replaced by a weaker conclusion of the form (6.3*b*).

7. On the tails of infinitely divisible distributions

To illustrate the usefulness of the notion of dominated variation in probabilistic contexts we prove the following

Proposition. *Let H stand for an infinitely divisible probability distribution with Lévy measure* $M\{dx\}$. *If M varies dominatedly at* $+\infty$ *then*

$$1 - H(x) \sim M\{(x, \infty)\}, \qquad x \to +\infty \tag{7.1}$$

in the sense that the ratio of the two sides tends to unity at all points of continuity. (A very special case involving regular variation is mentioned in [1], p. 540.)

Proof. We shall show that the general proposition follows easily from the special case where M is supported by the positive half axis and has a finite mass μ. In this case

$$M\{(x, \infty)\} = \mu[1 - F(x)] \qquad x > 0 \tag{7.2}$$

where F is a probability distribution on $(0, x)$, and H reduces to the compound Poisson distribution given by

$$H(x) = e^{-\mu} \sum_{n=0}^{\infty} \frac{\mu^n}{n!} F^{n*}(x) \qquad x > 0\,. \tag{7.3}$$

We proceed to prove the assertion (7.1) for distributions of this form assuming that $1-F$ varies dominatedly. Note that F^{n*} is the distribution of the sum $S_n = X_1 + \dots + X_n$ of n mutually independent random variables with the common distribution F. Since these variables are positive, the event $\{S_n > x\}$ occurs whenever at least one among the n variables exceeds x, and so

$$1 - F^{n*}(x) \geqslant n[1 - F(x)] - \binom{n}{2}[1 - F(x)]^2 \tag{7.4}$$

by an easily verified inequality named after Bonferoni. Substituting into (7.3) it follows that

$$(7.5)\qquad 1 - H(x) \geqslant \mu\left[1 - F(x)\right] - \frac{1}{2}\mu^2\left[1 - F(x)\right]^2 =$$

$$= M\{(x, \infty)\}\left[1 + o(1)\right], \qquad x \to \infty .$$

To obtain an appraisal in the opposite direction choose $0 < \varepsilon < \frac{1}{2}$ and note that the event $\{S_n > x\}$ cannot occur unless either at least one among the variables $X_1, \ldots, X_n$ exceeds $(1-\varepsilon)\,x$, or at least two among them exceed $\varepsilon x/n$. Thus

$$(7.6)\qquad 1 - F^{n*}(x) \leqslant n\left[1 - F((1-\varepsilon)x)\right] + \binom{n}{2}\left[1 - F(\varepsilon x/n)\right]^2 .$$

To apply the argument used in (7.5) we would have to know that the ratio of the two brackets on the right tends to 0 as $x \to \infty$. Because of the assumed dominated variation of $1 - F$ this is true for every fixed n, but to make the ratio $< \delta$ we must have $\varepsilon x/n$ sufficiently large, that is, $n \leqslant ax$, where a is an appropriate constant. On the other hand, if r is the smallest integer exceeding ax and if $ax > 2\mu$ we have trivially

$$(7.7)\qquad \sum_{n=r}^{\infty} \frac{\mu^n}{n!} < 2\left(\frac{\mu}{r}\right)^r < 2\left(\frac{\mu}{ax}\right)^{ax}$$

and the right side tends to zero faster than any power of x^{-1}. In view of the dominated variation of $1 - F$ this implies that the quantity (7.7) is $o(1 - F(x))$, and this together with (7.6) shows as in (7.5) that

$$(7.8)\qquad 1 - H(x) \leqslant \mu\left[1 - F(x)\right]\left(1 + o(1)\right) .$$

This proves the assertion for distributions of the form (7.3).

For the general case we represent the Lévy measure M as a sum of three measures supported by the intervals $(1, \infty)$, $[-1, 1)$, and $(-\infty, 1]$, respectively. This puts H in the form of a triple convolution, and so we may conceive of H as of the distribution of a sum $X + Y + Z +$const. of three infinitely divisible mutually independent random variables such that $X \geqslant 0$, $Z \leqslant 0$, and Y has a Lévy measure supported by $[-1, 1]$. It follows that Y has moments of all orders, and hence for arbitrary $\varepsilon > 0$ and n

$$(7.9)\qquad P\{\,|\,Y\,| > \varepsilon x\,\} = o\,(x^n) \qquad x \to \infty .$$

Because of the assumed dominated variation $M\{(x, \infty)\}$ decreases more slowly than a certain power $x^{-\alpha}$, and hence the quantity (7.9) is $o(M\{(ax,\infty)\}$ for any fixed $a>0$. Since $Z\leqslant 0$ and X has a distribution of the form (7.3) we conclude that

$$(7.10)\qquad P\{X+Y+Z>x\}\leqslant P\{X>(1-\varepsilon)x\}+P\{Y>\varepsilon x\}\sim \\ \sim M\{((1-\varepsilon)x,\infty)\}.$$

On the other hand,

$$(7.11)\qquad P\{X+Y+Z>x\}\geqslant P\{X>(1+\varepsilon)x\}\cdot P\{Y+Z>-\varepsilon x\},$$

and the last factor tends to 1 as $x\to\infty$. The probabilities on the left are therefore $\sim M\{(x,\infty)\}$, as asserted.

REFERENCES

[1] FELLER, W., *An introduction to probability theory and its applications*, vol. II. New York, 1966

[2] —— On regular variation and local limit theorems. *Proc. of the Fifth Berkeley Symposium on Mathematical Statistics and Probability*, 1966, vol. II, part 1, pp. 373-388.

[3] KARAMATA, J., Sur un mode de croissance régulière. *Mathematica* (Cluj), vol. 4 (1930), pp. 38-53.

(Reçu le 28 Mai 1968)

William Feller
Princeton University and
Rockfeller University.

Z. Wahrscheinlichkeitstheorie verw. Geb. 14, 21 – 26 (1969)

General Analogues to the Law of the Iterated Logarithm*

WILLIAM FELLER

Summary. Let $X_1, X_2, \ldots$ be independent and $S_n = X_1 + \cdots + X_n$. Suppose $E(X_k) = 0$ and $s_n^2 = E(S_n^2) < \infty$ and put $\psi_n(h) = \log E(e^{hS_n/s_n})$ for $h > 0$. To avoid trivialities we suppose that $s_{n+1} < s_n (\log s_n)^p$ for some $p > 0$. We assume that there exists a number $q < 1$ such that if

$$q \log \log s_n < \zeta_n < q^{-1} \log \log s_n$$

there exists a (necessarily unique) number $x_n(\zeta_n)$ determined by the parametric equations

$$\zeta_n = h\,\psi_n'(h) - \psi_n(h), \qquad x_n(\zeta_n) = \psi_n'(h).$$

It was shown in the preceding paper that under mild restrictions on the X_k

$$P\{S_n > s_n x_n(\zeta_n)\} = \exp[-\log \log s_n + o(\log \log s_n)].$$

It is now shown that under these conditions with probability one

$$\limsup \frac{S_n}{s_n x_n(\zeta_n)} = 1.$$

It is easy to give examples where $x_n(\zeta_n) \sim C(\log \log x_n)^{\frac{1}{2}}$ with an arbitrary $C > 0$. Other examples, however, exhibit an entirely different behavior of the sequence $\{s_n x_n(\zeta_n)\}$.

1. Introduction

Let $X_1, X_2 \ldots$ be mutually independent random variables with $E(X_n) = 0$ and finite variances. Put $S_n = X_1 + \cdots + X_n$ and $s_n^2 = E(S_n^2)$. When possible, define ψ_n by

$$\psi_n(h) = \log E(\exp h\,S_n/s_n). \tag{1.1}$$

For $h > 0$ the functions ψ_n' and

$$\lambda_n(h) = h\,\psi_n'(h) - \psi_n(h) \tag{1.2}$$

are increasing and so there exists a unique function $x_n(\zeta_n)$ defined by the parametric equations

$$\zeta_n = \lambda_n(h), \qquad x_n = \psi_n'(h). \tag{1.3}$$

A recent paper [1] was concerned with limit theorems of the form

$$P\{S_n \geqq s_n\, x_n(\zeta_n)\} = \exp[-\zeta_n + o(\zeta_n)] \tag{1.4}$$

---

* Work connected with a Project for Research in Probability Theory at Princeton University sponsored by the U.S. Army Research Office.

R.L. Schilling et al. (eds.), *Selected Papers II*,

where $\zeta_n \to \infty$ and $x_n(\zeta_n) \to \infty$. Relations of this sort are valid under mild conditions on the distributions of X_n and the sequence $\{\zeta_n\}$. In the present paper we are interested in situations where (1.4) is valid when ζ_n remains close to

$$\rho_n = \log \log s_n. \tag{1.5}$$

More precisely, we start from

Assumption A. *There exists a number q such that $x_n(\zeta_n)$ is defined for all ζ_n with*

$$q\,\rho_n \leqq \zeta_n \leqq \frac{1}{q}\,\rho_n, \tag{1.6}$$

and (1.4) *is true for any sequence $\{\zeta_n\}$ satisfying* (1.6).

As was shown in [1] this assumption puts only surprisingly mild restrictions on $\{X_n\}$. The same is true of the following conditions which could be weakened still further.

Assumption B. *There exists a number p such that*

$$\frac{s_{n+1}}{s_n} < (\log s_n)^p. \tag{1.7}$$

For convenience we put

$$a_n = s_n\, x_n(\rho_n). \tag{1.8}$$

The purpose of this paper is to prove the

Theorem. *If the Assumptions* A *and* B *hold, then with probability one*

$$\limsup \frac{S_n}{a_n} = 1. \tag{1.9}$$

The examples given in [1] show that it is usually not too difficult to find reasonable estimates for $\{a_n\}$. The relation (1.9) is of the form of the classical law of the iterated logarithm, but the asymptotic properties of $\{x_n(\zeta_n)\}$ will in general bear no resemblance to the classical $(2\log\log s_n)^{\frac{1}{2}}$.

The point in the present approach to (1.9) is the possibility of good estimates for $\{a_n\}$. It is possible to derive a general theorem of the form (1.9) without any assumptions on $\{X_n\}$ whatever; it appears as a simple corollary to criteria to decide whether $P\{S_n > a_n \text{ i.o.}\}$ equals 0 or 1 (see [2]). Such criteria are of theoretical interest, but give little information concerning the probable fluctuations of $\{S_n\}$.

2. A Lemma

If $1 \leqq t < 1/q$ then for every $\tau < t$ ultimately

$$P\{S_n > t\,a_n\} < \frac{1}{(\log s_n)^\tau}. \tag{2.1}$$

Similarly, if $q < t \leqq 1$ and $\tau > t$, then ultimately

$$P\{S_n > t\,a_n\} > \frac{1}{(\log s_n)^\tau}. \tag{2.2}$$

Proof. It follows from the parametric equations (1.3) and the definition (1.2) of λ_n that

$$\frac{dx_n}{d\zeta_n}=\frac{1}{h}, \quad \text{and} \quad \frac{\zeta_n}{x_n}<h. \tag{2.3}$$

Thus

$$\frac{dx_n}{x_n}<\frac{d\zeta_n}{\zeta_n}. \tag{2.4}$$

Suppose first $t>1$. Integrating (2.4) between ρ_n and $t\rho_n$ we get

$$\frac{x_n(t\rho_n)}{x_n(\rho_n)}<t. \tag{2.5}$$

Thus $t a_n \geqq s_n x_n(t\rho_n)$, and hence

$$\begin{aligned} P\{S_n>t a_n\} &\leqq P\{S_n>s_n x_n(t\rho_n)\} \\ &=\exp[-t\rho_n+o(\rho_n)]<\frac{1}{(\log s_n)^{t-\varepsilon}} \end{aligned} \tag{2.6}$$

for every $\varepsilon>0$ provided n is sufficiently large. This proves (2.1) and the case $t<1$ can be treated in like manner.

3. On the Sequence $\{a_n\}$

Consider the inequality (2.2) with $t=1$. Applying Chebyshev's inequality to the probability on the left one sees immediately that as $n\to\infty$

$$x_n(\rho_n)=o(\log s_n)^{\frac{1}{2}}. \tag{3.1}$$

In view of (1.8) this implies that

$$\log s_n \sim \log a_n \tag{3.2}$$

the sign $\sim$ indicating that the ratio of the two sides tends to unity. An important consequence is that *the lemma remains valid with s_n on the right sides replaced by a_n.*

Next we show that *the sequence $\{a_n\}$ is nearly monotone* in the sense that for any fixed $\delta>0$

$$a_n>(1-\delta)\,a_m \tag{3.3}$$

for all $n>m$ provided only m is sufficiently large. Indeed, let $S_{m,n}=S_n-S_m$ and $t>1$. From the lemma we get for ε sufficiently small and m large

$$\begin{aligned} P\{S_m>a_m/t\} &\geqq \frac{1}{(\log s_m)^{1-\varepsilon}} \geqq \frac{1}{(\log s_n)^{1-\varepsilon}} \\ &\geqq P\{S_n>a_n\} \geqq P\{S_m>a_n t\}\cdot P\{S_{m,n}>-(t-1)\,a_n\} \\ &\geqq \left(1-\frac{1}{[(t-1)\,x_n(\rho_n)]^2}\right) P\{S_m>a_n t\}. \end{aligned} \tag{3.4}$$

The factor in parentheses tends to 1, and the string of inequalities therefore implies that $t a_n \geqq a_m/t$. As $t\geqq 1$ is arbitrary, this is equivalent to (3.3).

4. The Sequences $\{n_k\}$ and $\{v_k\}$

To avoid illegible double subscripts we put generally $v_{k+1}=n_k+1$. We define $\{n_k\}$ by induction. Choose $\lambda>1$ arbitrary, but fixed. Let $v_1=1$ and choose n_1 arbitrarily. Given n_{k-1} we have $v_k=n_{k-1}+1$ and we define n_k as the *largest* integer such that

$$a_{n_k}\leqq\lambda\, a_{v_k}. \tag{4.1}$$

Using (3.3) we get for arbitrary $\delta>0$ and sufficiently large k

$$a_{n_{k+1}}>\lambda(1-\delta)\, a_{n_{k-1}}. \tag{4.2}$$

Thus the sequence $\{a_{n_k}\}$ increases ultimately at least as fast as a geometric sequence, and hence there exists an $\eta>0$ such that $\log a_{n_k}>\eta\, k$. As we saw, (2.1) remains valid also with s_n on the right replaced by a_n, and so we see that for $t>1$

$$\sum P\{S_{n_k}>t\, a_{n_k}\}<\infty. \tag{4.3}$$

Next we show that for $t<1$

$$\sum P\{S_{v_k}>t\, a_{v_k}\}=\infty. \tag{4.4}$$

Using (1.7) and (3.1) it is seen that at least for large n

$$\frac{a_{n+1}}{a_n}<\frac{s_{n+1}}{s_n}\cdot x_{n+1}(\rho_{n+1})<(\log a_n)^{p+1}. \tag{4.5}$$

Applying this with n replaced by n_k we get remembering (4.1) for large k

$$a_{v_{k+1}}\leqq a_{n_k}(\log a_{n_k})^{p+1}\leqq a_{v_k}(\log a_{v_k})^{p+2}. \tag{4.6}$$

One concludes easily that ultimately

$$\log a_{v_k}<2(p+2)\, k\log k, \tag{4.7}$$

and the truth of (4.4) is now a direct consequence of (2.2).

5. Proof that $\limsup S_n\, a_n^{-1}<1+\varepsilon$

Let $t>1$. We use again the notation $S_{m,n}=S_n-S_m$. The event $\{S_{n_k}>t\, a_{n_k}\}$ occurs certainly if for some j simultaneously $S_j>t^2\, a_{n_k}$ and $S_{j,n_k}>-(t^2-t)\, a_{n_k}$, and the last two events are independent. From Kolmogorov's inequality it follows that as $k\to\infty$

$$P\{S_{j,n_k}>-(t^2-t)\, a_{n_k}\ \textit{all}\ j\leqq n_k\}\to 1. \tag{5.1}$$

When k is large enough to make this probability $>\frac{1}{2}$ we get therefore

$$P\{S_j>t^2\, a_{n_k}\ \ \textit{some}\ \ v_k\leqq j\leqq n_k\}\leqq 2P\{S_{n_k}>t\, a_{n_k}\}. \tag{5.2}$$

Now (3.3) shows that for large enough k one has $a_{v_k}<t\, a_j$ for all $j>v_k$, and in this case it follows from the definition (4.1) that

$$a_{n_k}\leqq\lambda t\, a_j,\qquad v_k\leqq j\leqq n_k. \tag{5.3}$$

Comparing with (5.2) we infer

$$P\{S_j \geqq \lambda t^3 a_j \ \ some \ \ \nu_k \leqq j \leqq n_k\} \leqq 2P\{S_{n_k} > t\, a_{n_k}\}. \tag{5.4}$$

We saw in (4.3) that the right side represents the term of a convergent series. The integers being partitioned into the non-overlapping blocks $\nu_k \leqq j \leqq n_k$, it follows now from the first Borel-Cantelli lemma that

$$P\{S_j \geqq \lambda t^3 a_j \text{ i.o.}\} = 0. \tag{5.5}$$

But $\lambda > 1$ and $t > 1$ are arbitrary, and hence (5.5) implies lim sup $S_j/a_j \leqq 1$.

6. Proof that lim sup $S_n a_n^{-1} > 1 - \varepsilon$

Let $t < 1$ and put for abbreviation

$$\Gamma_j = \{t\, a_{\nu_j} < S_{\nu_j} < a_{\nu_j}\}. \tag{6.1}$$

Integrating (2.4) over $t\, a_n < x_n < a_n$ it is seen that for arbitrary $\tau < t^{-1}$ and n sufficiently large

$$P\{S_n > a_n\} \leqq (P\{S_n > t\, a_n\})^{\tau}.$$

Hence

$$\sum P\{\Gamma_j\} = \infty. \tag{6.2}$$

Let $\alpha < \beta$ be two arbitrary integers. Then

$$P\left\{\bigcup_{j=\alpha}^{\beta} \Gamma_j\right\} = \sum_{j=\alpha}^{\beta} P\left\{\Gamma_j - \Gamma_j \bigcup_{k=j+1}^{\beta} \Gamma_k\right\}, \tag{6.3}$$

the set on the right representing those points of Γ_j that do not appear in any later Γ_k. To estimate the probability on the right note that for $k > j$

$$\Gamma_j \Gamma_k \subset \Gamma_j \{S_{\nu_j, \nu_k} > t\, a_{\nu_k} - a_{\nu_j}\}, \tag{6.4}$$

and the events on the right are independent. Now $a_{\nu_{j+1}} > \lambda a_{\nu_j}$, and because of the near-monotonicity of $\{a_n\}$ it follows that ultimately

$$(t - t^2)\, a_{\nu_k} > a_{\nu_j}, \qquad k > j \tag{6.5}$$

provided only that λ is chosen so large that $t - \lambda^{-1} > t^2$. In this case (6.3) and (6.4) together yield

$$P\left\{\bigcup_{j=\alpha}^{\beta} \Gamma_j\right\} \geqq \sum_{j=\alpha}^{\beta} P\{\Gamma_j\}[1 - P\{S_{\nu_j, \nu_k} > t^2 a_{\nu_k} \ some \ k > j\}]. \tag{6.6}$$

Again, if (6.5) holds, then

$$\{S_{\nu_k} > t^3 a_{\nu_k} \ some \ k > j\} \supset \{S_{\nu_j, \nu_k} > t^2 a_{\nu_k} \ some \ k > j\}\ \{S_{\nu_j} > -t\, a_{\nu_j}\}. \tag{6.7}$$

The two events on the right are independent, and by Chebyshev's inequality the probability of the latter tends to 1 as $j \to \infty$.

We have thus shown that for $\varepsilon > 0$ and all $\beta > \alpha$

$$P\left\{\bigcup_{j=\alpha}^{\beta} \Gamma_j\right\} \geqq [1 - (1+\varepsilon)\, P\{S_{\nu_k} > t^3 a_{\nu_k} \ some \ k > \alpha\}] \cdot \sum_{j=\alpha}^{\beta} P\{\Gamma_j\}, \tag{6.8}$$

provided only α is sufficiently large. Because of the assumed divergence the sum on the right can be made as large as we please by choosing β sufficiently large. The left side remains bounded, and the probability within the brackets decreases as $\alpha \to \infty$. Thus (6.8) contains a contradiction unless

$$P\{S_{\nu_k} > t^3 a_{\nu_k} \text{ i.o.}\} = 1. \tag{6.9}$$

Since $t<1$ is arbitrary, this implies that lim sup $s_n/a_n \geqq 1$, and this completes the proof.

References

1. Feller, W.: Limit theorems for probabilities of large deviations. Z. Wahrscheinlichkeitsrechnung verw. Geb. **14**, 1 – 20 (1969).
2. – On fluctuations of sums of independent random variables. To appear Proc. Nat. Acad. Sci., USA **62** (1969).

Professor Dr. William Feller
Fine Hall
Princeton University
Princeton, N.J. 08540, USA

(Received April 25, 1969)

Z. Wahrscheinlichkeitstheorie verw. Geb. 14, 1–20 (1969)

Limit Theorems for Probabilities of Large Deviations*

WILLIAM FELLER

Summary. Let $\{X_{k,n};\ k=1,\dots,n\}$ be a triangular array of independent variables with row sums S_n. Suppose $E(X_{k,n})=0$ and $E(S_n^2)=1$ and that $\psi_n(h)=\sum_{k=1}^{n}\log E(e^{hX_{k,n}})$ exists for $0\leqq h\leqq \varepsilon_n$. Under mild conditions we show that

$$P\{S_n>z_n\}=\exp[-r_n+o(r_n)], \qquad r_n\to\infty \tag{1}$$

where the quantities z_n and r_n are related by the parametric equations

$$z_n=\psi_n'(h_n), \qquad r_n=h_n\psi_n'(h_n)-\psi_n(h_n). \tag{2}$$

If the distributions of the $X_{k,n}$ behave reasonably well it is usually not difficult to obtain satisfactory asymptotic estimates for z_n in terms of r_n and vice versa.

The principal application is to sequences $\{X_k\}$. Then $X_{k,n}=X_k/s_n$ and $S_n=(X_1+\cdots+X_n)/s_n$. A familiar special case of (1) is given by

$$P\{X_1+\cdots+X_n>s_n z_n\}\sim[1-\mathfrak{N}(z_n)]\exp[-P_n(z_n)]$$

where $\mathfrak{N}$ is the standard normal distribution and P_n a certain power series. In this case $r_n=z_n^2/2$, but (2) may lead to radically different relationships between r_n and z_n.

1. Preliminaries

Let X be a random variable with distribution F such that $\varphi(h)=E(e^{hX})$ exists for some $h>0$. A new probability distribution $\bar{F}$ may then be defined by

$$\bar{F}(dy)=e^{hy}F(dy)/E(e^{hX}), \tag{1.1}$$

and with this distribution we may associate a random variable $\bar{X}$ on a new probability space. The corresponding characteristic functions need not be analytic around the origin, but they may be written in the form $\varphi(it)$ and $\varphi(h+it)/\varphi(h)$, respectively, and this shows that the transformation $F\leftrightarrow\bar{F}$ commutes with convolutions. This transformation is familiar from queueing theory and random walks (Feller [5]), but we shall be concerned only with its applications to limit theorems.

Consider then a triangular array $\{X_{k,n};\ k=1,\dots,n\}$ in which the variables within each row are mutually independent and such that expectations

$$\varphi_{k,n}(h)=\int_{-\infty}^{+\infty}e^{hy}F_{k,n}(dy) \tag{1.2}$$

exist for some $h>0$. Denote the row sums by S_n and suppose that

$$E(X_{k,n})=0, \qquad E(S_n^2)=1. \tag{1.3}$$

* Work connected with a Project for Research in Probability at Princeton University sponsored by the U.S. Army Research Office.

R.L. Schilling et al. (eds.), *Selected Papers II*,

Within its interval of existence $\varphi_{k,n}$ is a convex function. At the origin it has the value 1 and a vanishing right derivative. Hence

$$\varphi_{k,n}(h) \geqq 1, \tag{1.4}$$

and

$$\psi_{k,n}(h) = \log \varphi_{k,n}(h) \geqq 0 \tag{1.5}$$

is well defined. Clearly

$$E(e^{hS_n}) = e^{\psi_n(h)} \quad \text{where} \quad \psi_n(h) = \sum_{k=1}^{n} \psi_{k,n}(h). \tag{1.6}$$

We now introduce the associated triangular array of variables $\bar{X}_{k,n}$ with distributions $\bar{F}_{k,n}$ defined by the transformation (1.1). Their expectations and variances are given by $\psi'_{k,n}(h)$ and $\psi''_{k,n}(h)$, respectively. (We use the same notation for one-sided derivatives.) For the row sums $\bar{S}_n$ we have therefore

$$E(\bar{S}_n) = \psi'_n(h), \qquad \operatorname{Var}(\bar{S}_n) = \psi''_n(h). \tag{1.7}$$

The distributions G_n and $\bar{G}_n$ of S_n and $\bar{S}_n$ are again connected by the transformation (1.1), and inverting it we get

$$P\{S_n > z\} = \exp[\psi_n(h) - h\psi'_n(h)] \int_{z-\psi'_n(h)}^{\infty} e^{-hy} \bar{G}_n(dy + \psi'_n(h)). \tag{1.8}$$

The integration is with respect to a probability distribution centered at its expectation. Roughly speaking this implies that for large values of h and for z close to $\psi'_n(h)$ the integral depends essentially only on the behavior of $\bar{G}_n$ near its expectation. In particular, if the distribution $\bar{G}_n$ were normal and $z = \psi'_n(h)$ the integral would reduce to

$$\exp[\tfrac{1}{2} h^2 \psi''(h)] [1 - \mathfrak{N}(h\sqrt{\psi''_n(h)})] \approx (2\pi)^{-\frac{1}{2}} (h\sqrt{\psi''_n(h)})^{-1}. \tag{1.9}$$

Here $\mathfrak{N}$ stands for the standard normal distribution, and the indicated approximation is reasonable when $h^2\psi''_n(h)$ is large. It is therefore plausible that appropriate conditions on the array $\{X_{k,n}\}$ will achieve that $\bar{G}_n$ is close enough to the normal distribution to make the approximation (1.9) usable.

This idea was first used by Esscher [2] in risk theory, and was exploited by Cramér [1] to obtain a limit theorem for sums of identically distributed random variables. Cramér's method was generalized by Feller [3] to sequences with variable distributions, that is, to triangular arrays of the form $X_{k,n} = X_k/s_n$ and $F_{k,n}(y) = F_k(y s_n)$. (Here S_n stands for the normalized sum $(X_1 + \cdots + X_n)/s_n$ with $s_n^2 = \operatorname{Var}(S_n)$.) It was shown that if for a monotone sequence $z_n \to \infty$

$$|X_n| < \frac{1}{5} \cdot \frac{s_n}{z_n} \tag{1.10}$$

then as $n \to \infty$

$$\begin{aligned} P\{S_n > z_n\} &\sim (1 - \mathfrak{N}(z_n)) \exp[-P_n(z_n)] \\ &\sim (2\pi)^{-\frac{1}{2}} z_n^{-1} \exp[-\tfrac{1}{2} z_n^2 - P_n(z_n)], \end{aligned} \tag{1.11}$$

the sign $\sim$ indicating that the ratio of the two sides tends to unity. Here

$$\tfrac{1}{2} z^2 + P_n(z) = h\psi'_n(h) - \psi_n(h) \tag{1.12}$$

where h and z are related by the auxiliary equation $z=\psi_n'(h)$. The condition (1.10) ensures that ψ_n is an analytic function and $P_n(z)$ reduces to a power series starting with z^3. If (1.10) is sharpened to $X_n=o(s_n/z_n)$ then $P_n(z_n)\to 0$, but in general it can only be asserted that $|P_n(z_n)|$ is dominated by $\frac{1}{2}z_n^2$. Rather strong inequalities for the coefficients of P_n were derived[1].

2. Outline

The asymptotic relation (1.11) was derived as a complement to the central limit theorem, and it was treated as such in the subsequent literature. In retrospect the form (1.11) appears somewhat deceiving and the reliance on the central limit theorem seems due primarily to the temptation to use the available Berry-Esseen estimate[2]. As an indication that the central limit theorem is not too closely connected with the problem under discussion it may be mentioned that the sequence $\{X_k\}$ may belong to the domain of attraction of the normal distribution without $\{X_k\}$ doing so; more surprisingly, the converse situation is also possible (see Section 11).

The original purpose of the limit theorem (1.11) was to elucidate the role of the condition $|X_n|=o(s_n(\log\log s_n)^{-\frac{1}{2}})$ in Kolmogorov's law of the iterated logarithm (Feller [4]). It was possible to derive criteria for a numerical sequence $\{a_n\}$ to belong to the upper or lower class, and thus to obtain a sharpening of the law of the iterated logarithm as well as to study its gradual breakdown when Kolmogorov's bound is replaced by bounds of the form $|X_n|=O(s_n(\log\log s_n)^{-\eta})$ with $0<\eta\leqq\frac{1}{2}$. However, the very form of the law of the iterated logarithm is closely related to the central limit theorem, and no essential generalization can be expected by means of limit theorems involving the normal distribution.

In this note we shall derive a novel type of limit theorems of the form

$$P\{S_n>z_n\}=\exp[-r_n+o(r_n)],\qquad r_n\to\infty. \tag{2.1}$$

This is cruder than (1.11), but the refined form is necessary only for delicate studies connected with events such as $\{S_n>a_n \text{ i.o.}\}$ whereas the form (2.1) suffices for cruder results of the type of the original law of the iterated logarithm. In fact, for sequences $\{X_k\}$ with *normalized* partial sums $S_n=(X_1+\cdots+X_n)/s_n$ it will be shown in another paper that, under widely applicable conditions, *if* (2.1) *holds with* $r_n=\log\log s_n$ *then with probability one*

$$\limsup\frac{S_n}{z_n}=1. \tag{2.2}$$

1 Petrov [9] derived a result similar to (1.11) valid for $z_n=o(\sqrt{n}/\log n)$. He replaces (1.10) by a weaker condition but obtains less precise information concerning the coefficients of P_n. The appearance of n (rather than s_n) in the bound for z_n is due to two artificial uniformity conditions which with our notations read: $\liminf s_n^2/n>0$ and $U(-\infty,-C/s_n^2)<\varepsilon n s_n^{-2}$ for some fixed C and all n and $\varepsilon>0$. Richter [11–12] extended these results to densities as well as to several dimensions. A new approach (not based on the analyticity of the characteristic functions) was devised by Linnik [7] and extended by Petrov [10]. Here the condition $E(\exp a|X_j|)<C$ is replaced by $E(\exp a|X_j|^\gamma)<C$ for an appropriate $\gamma<1$. All these limit theorems are essentially of the form (1.11).

2 A systematic direct approach would start from the characteristic function $E(\exp(iz\bar{S}_n))$ considered as analytic function in $z=x-ih$. For fixed $h>0$ this function can be expanded about the point $x(h)=E(\exp h\bar{S}_n)$ and limit theorems may be obtained by the standard inversion procedures. The results obtainable in this way seem sharper than ours, but they involve more intricate methods.

1*

The appearance of $r_n = \log\log s_n$ in this formulation is apt to give an erroneous impression. Kolmogorov's law bears its name because $z_n = (2\log\log s_n)^{\frac{1}{2}}$ whereas the sequence $\{r_n\}$ is auxiliary in that it appears in proofs but not in the formulation of the law. When the distribution of $\bar{S}_n$ is sufficiently close to the normal the relation between z_n and r_n is given by $z_n \sim (2r_n)^{\frac{1}{2}}$, but we shall obtain relations of an entirely different form leading to theorems of the form (2.2) with no resemblance to the iterated logarithm law.

The aim of the present paper is to derive limit theorems of the form (2.1). The intended applications are to sequences of random variables and exponents of the form $r_n = (1 \pm \varepsilon)\log\log s_n$, but as this plays no role in the proofs we shall treat arbitrary triangular arrays and unspecified exponents r_n. In practice the r_n may be considered given and it is required to find as much information concerning the z_n as possible. However, it is also possible to consider the z_n as given.

The whole paper is based on Theorem 1. It depends essentially only on the condition (3.4) which is so mild that the theoretical applicability of the theorem is extremely wide. Unfortunately its practical use is extremely narrow due to the intricate nature of the functions involved. The main idea of the present paper consists in introducing three Conditions (A – C) which permit a linearization procedure and lead to Theorem 2 which is much more manageable than the more general Theorem 1. Both theorems involve annoying epsilons and deltas. These can be eliminated if the respective conditions for the theorem are assumed to hold uniformly in a certain small interval. Thus each theorem has a more elegant variant, Theorems 3 and 4 corresponding to Theorems 2 and 1, respectively. The proof for the reduction is the same in both cases.

The Conditions A – C are stated in their general form in order to explain their true nature. Simpler sufficient conditions show that B and C are mild indeed. Thus B is automatically satisfied if $X_{k,n} = o(s_n r_n)$ uniformly in $k \leqq n$. (This contrasts with the condition $X_{k,n} = o(s_n r_n^{-1})$ in the classical iterated logarithm.) Condition C is automatically satisfied if the $X_{k,n}$ are not centered absurdly far to the left of the origin. Crucial is the Condition A which is a softened version of the familiar requirement that the $X_{k,n}$ be asymptotically negligible.

For practical uses Theorem 2, and its variant in Theorem 3, depend on the solvability of the parametric equations (6.7). These look formidable, but they are more tractable than the corresponding implicit equations for the coefficients of the power series P_n. Indeed, mild and natural regularity conditions suffice to derive with ease sufficient asymptotic estimates. Typical examples are found in the last two sections. The first leads to limit laws differing from the iterated logarithm only by a change of the constant $\sqrt{2}$. The second leads to limit laws of a radically different nature.

When the condition of individual negligibility fails, a combination of Theorems 3 and 4 still leads to non-trivial bounds for the probabilities in question. This is illustrated by the example of Section 10, but the method is widely applicable. In this connection attention may be drawn to the one-sided estimate at the end of Section 9. It explains and puts in a general framework a curious phenomenon discovered by Marcinkiewicz and Zygmund [8] in a particular case.

3. The Underlying Inequalities

We return to the identity (1.8) and put for abbreviation

$$\lambda_n(h) = h\,\psi_n'(h) - \psi_n(h), \qquad h > 0. \tag{3.1}$$

When $z = \psi_n'(h)$ the integration variable y runs from 0 to ∞ and so under any circumstances

$$P\{S_n > \psi_n'(h)\} \leqq \exp[-\lambda_n(h)]. \tag{3.2}$$

On the other hand, $\bar{G}_n$ has variance $\sigma_n^2(h) = \psi_n''(h)$, and by Chebyshev's inequality it attributes to the interval $\psi_n'(h) \pm 2\sigma_n(h)$ a probability $\geqq \frac{3}{4}$. But

$$P\{S_n > \psi_n'(h) - 2\sigma_n(h)\} \geqq \exp[-\lambda_n(h)] \int_{-2\sigma_n(h)}^{2\sigma_n(h)} e^{-hy}\,\bar{G}_n(dy + \psi_n'(h))$$

and so

$$P\{S_n > \psi_n'(h) - 2\sigma_n(h)\} \geqq \tfrac{3}{4} \exp[-\lambda_n(h) - 2h\,\sigma_n(h)]. \tag{3.3}$$

Combining these inequalities and recalling that $\lambda_n(h) \leqq h\,\psi_n'(h)$ we get

Theorem 1. *If $\{h_n\}$ is a sequence such that $\lambda_n(h_n) \to \infty$ and*

$$h_n\,\sigma_n(h_n) = o(\lambda_n(h_n)), \tag{3.4}$$

then

$$P\{S_n > \psi_n'(h)\} \leqq \exp[-\lambda_n(h_n)] \tag{3.5}$$

but for every fixed $\varepsilon > 0$ and $\delta > 0$

$$P\{S_n > \psi_n'(h_n)(1-\varepsilon)\} \geqq \exp[-\lambda_n(h_n)(1+\delta)] \tag{3.6}$$

for all n sufficiently large.

The condition (3.4) turns out to be surprisingly mild and, from a purely formal point of view, the theorem is of great generality. Its applicability is limited by the complex nature of the relationship between $\psi_n'(h)$ and $\lambda_n(h)$. For fixed n the function λ_n is continuous and monotone, but it may be bounded. It is therefore difficult to discern the nature of sequences $\{r_n\}$ that can be put into the form $r_n = \lambda_n(h_n)$ and to determine the asymptotic behavior of the corresponding sequences $\{\psi_n'(h_n)\}$. We shall return to Theorem 1 in Section 9, but we shall consider it mainly as a stepping stone to limit theorems of more restricted validity but wider applicability. They are based on approximating $\psi_{k,n}(h) = \log \varphi_{k,n}(h)$ by $\varphi_{k,n}(h) - 1$. Before describing the procedure in Section 5 we list the notations to be used.

4. Notations

The truncated second moment function of S_n induces a measure

$$U_n(dy) = \sum_{k=1}^{n} y^2\,F_{k,n}(dy). \tag{4.1}$$

With our norming *U_n is a probability measure:* It plays an important role in various limit theorems. In particular, the distribution G_n of S_n tends to the normal distribution with unit variance if, and only if, U_n tends to the unit atomic measure concentrated at the origin.

When the quantities in question are meaningful we define

$$\tau_n(h) = \sum_{k=1}^{n} [\varphi_{k,n}(h) - 1], \tag{4.2}$$

$$\zeta_n(h) = \sum_{k=1}^{n} \varphi'_{k,n}(h), \tag{4.3}$$

$$\rho_n(h) = h\,\zeta_n(h) - \tau_n(h). \tag{4.4}$$

It is readily verified that these functions can be written in the form

$$\tau_n(h) = \int_{-\infty}^{+\infty} \frac{e^{hy} - 1 - hy}{y^2}\, U_n(dy), \tag{4.5}$$

$$\zeta_n(h) = \tau'_n(h) = \int_{-\infty}^{+\infty} \frac{e^{hy} - 1}{y}\, U_n(dy), \tag{4.6}$$

$$\rho_n(h) = \int_{-\infty}^{+\infty} \frac{e^{hy}(hy - 1) + 1}{y^2}\, U_n(dy), \tag{4.7}$$

the integrands being defined as continuous at the origin.

For convenience we indicate here the most important properties of these functions.

Lemma 1. (i) *For fixed $h > 0$ the integrands in* (4.5)–(4.7) *are monotone functions going from* 0 *to* ∞.

(ii) *The ratio of the second to the first integrand goes monotonically from h^{-1} to* ∞.

(iii) *The ratio of the third integrand to the second goes monotonically from* 0 *to* h.

Proof. The monotonicity of the integrand in (4.5) is verified by routine differentiation; for the other integrands the assertion follows them from (ii) and (iii). To verify (ii) it suffices to show that

$$\frac{e^t - 1 - t}{(e^t - 1)\,t} = \frac{1}{t} - \frac{1}{e^t - 1}$$

has a negative derivative. A routine calculation reduces this assertion to the inequality $t < e^{t/2} - e^{-t/2}$ valid for $t > 0$. In view of (4.4) the assertion (iii) is implied by (ii).

For notational convenience we introduce the following

Abbreviation. If $f(h)$ stands for any of the integrals (4.5)–(4.7) we denote by $f(h; -\infty, a)$ and $f(h; a, \infty)$ its restrictions to the intervals $(-\infty, a]$ and (a, ∞), respectively. For example

$$\zeta_n(h; -\infty, a) = \int_{-\infty}^{a+} \frac{e^{hy} - 1}{y}\, U_n(dy). \tag{4.8}$$

The measure attributed by U_n to an interval $(a, b]$ will be denoted by $U_n(a, b]$.

5. Linearization

Our first aim is to obtain approximations to the function λ_n defined in (3.1). A simple comparison of the derivatives shows that

$$\frac{x}{1+x} < \log(1+x) < x, \qquad x > 0. \tag{5.1}$$

Applying this to $\psi_{k,n}(h) = \log \varphi_{k,n}(h)$ we get

$$\lambda_n(h) \leqq \sum_{k=1}^{n} \frac{h\,\varphi'_{k,n}(h) - \varphi_{k,n}(h) + 1}{\varphi_{k,n}(h)}. \tag{5.2}$$

The numerator vanishes for $h=0$ and has the positive derivative $h\,\varphi''_{k,n}(h)$; as was pointed out already, because of its convexity the denominator is $\geqq 1$. Each fraction is therefore dominated by its numerator, and so

$$\lambda_n(h) \leqq \rho_n(h) \tag{5.3}$$

for all values $h>0$ for which $\rho_n(h)$ makes sense.

In the reverse direction we have

$$\begin{aligned} \lambda_n(h) &\geqq \sum_{k=1}^{n} \frac{h\,\varphi'_{k,n}(h)}{\varphi_{k,n}(h)} - \varphi_{k,n}(h) + 1 \\ &= \rho_n(h) - h\,\gamma_n(h) \end{aligned} \tag{5.4}$$

where

$$\begin{aligned} \gamma_n(h) &= \sum_{k=1}^{n} \frac{\varphi'_{k,n}(h)\,[\varphi_{k,n}(h) - 1]}{\varphi_{k,n}(h)} \\ &\leqq \sum_{k=1}^{n} \varphi'_{k,n}(h)\,[\varphi_{k,n}(h) - 1]. \end{aligned} \tag{5.5}$$

The two estimates (5.3)–(5.4) suffice for our purposes[3].

The quantity $\gamma_n(h)$ will occur in the sequel also for another reason. Comparing the definitions (1.6) and (4.3) one sees that

$$\zeta_n(h) - \psi'_n(h) = \gamma_n(h). \tag{5.6}$$

6. The Main Theorem

We proceed to combine the inequalities of Sections 3 and 5. For this purpose we introduce three assumptions. The first is crucial, but the other two will turn out to be of an essentially formal nature and surprisingly mild.

3 It is not difficult to derive better estimates. For $x>0$ the difference between $(1+x)^{-1}$ and $1-x+\cdots+(-1)^m x^m$ is alternately positive or negative, and the same is true for the difference of the corresponding integrals. One concludes that for all integers $N \geqq 1$

$$\lambda_n(h) - \rho_n(h) \leqq \sum_{k=1}^{n} \sum_{m=1}^{2N} \left[h\,\varphi'_{k,n}(h) - \frac{\varphi_{k,n}(h)-1}{m+1}\right] [\varphi_{k,n}(h) - 1]^m (-1)^{m-1}$$

and the inequality is reversed when $2N$ is replaced by $2N-1$. By means of these inequalities one may derive sharpened forms of our theorem reminiscent of (1.11). Even in cases where (1.11) is applicable the new development may be simpler, and more efficient. We shall, however, not pursue this line of thought.

In the sequel $\{h_n\}$ denotes a sequence of positive numbers for which the $\varphi_{k,n}(h_n)$ $(k=1, \ldots, n)$ are defined.

Condition A. *We shall say that Condition* A *is satisfied if*

$$\gamma_n(h_n)=\sum_{k=1}^{n} \varphi'_{k,n}(h_n) \frac{\varphi_{k,n}(h_n)-1}{\varphi_{k,n}(h_n)}=o\big(\zeta_n(h_n)\big). \tag{6.1}$$

Seeing that the factors $\varphi'_{k,n}(h_n)$ are positive and add up to $\zeta_n(h_n)$ it is obvious that *Condition* A *is certainly satisfied if*

$$\sup_{k \leqq n} [\varphi_{k,n}(h_n)-1] \to 0. \tag{6.2}$$

However, (6.2) is not necessary: (6.1) requires only that the total contribution of those $\varphi'_{k,n}(h_n)$ for which $\varphi_{k,n}(h_n)>1+\eta$ be *relatively* small as compared to the entire sum.

When applied to sequences (that is, $X_{k,n}=X_k/s_n$), the strengthened version (6.2) is reminiscent of the condition $|X_n|=o(s_n/z_n)$ in the law of the iterated logarithm, but it is actually much milder. This is so not only because of its one-sided nature, but because (as the next corollary shows) $\{h_n\}$ is practically dominated by a multiple of $\{z_n\}$ and in most cases $h_n=o(z_n)$.

Condition B. *We shall say that Condition* B *is satisfied if for every* $\varepsilon>0$

$$\zeta_n\big(h_n; \varepsilon\, \zeta_n(h_n), \infty\big)=o\big(\zeta_n(h_n)\big), \tag{6.3}$$

that is, in the integral (4.6) defining $\zeta_n(h_n)$ the contribution of the interval $(\varepsilon \zeta_n(h_n), \infty)$ is asymptotically negligible. (For the notation for truncated integrals see Section 4.) The sequence $\{\zeta_n(h_n)\}$ may remain bounded, but Theorems 2 and 3 will require that $h_n \zeta_n(h_n) \to \infty$. A glance at (4.6) then shows that (6.3) *is equivalent with*

$$\sum_{k=1}^{n} \int_{\varepsilon \zeta_n(h_n)}^{\infty} e^{h_n y}\, y\, F_{k,n}(dy)=o\big(\zeta_n(h_n)\big). \tag{6.4}$$

Condition (6.4) is certainly satisfied if with probability 1 ultimately

$$X_{k,n}<\varepsilon\, \zeta_n(h_n), \qquad (k=1, 2, \ldots, n) \tag{6.5}$$

and hence the impact of the condition is negligible.

Condition C can be formulated in several ways. We postpone the discussion of its nature to Section 7 and give here a version which is perhaps the least plausible, but serves best for the purposes of the following proof.

We shall say that Condition C *holds if there exists a constant* $c>0$ *such that*

$$c\, h_n\, \zeta_n(h_n) \leqq \rho_n(h_n) \leqq h_n\, \zeta_n(h_n). \tag{6.6}$$

(It will be recalled that the right-hand inequality is automatically satisfied, and the condition refers only to the lower bound.)

The parametric equations (6.7) look forbidding, but in fact they are easily handled whenever the measure U is not too irregular.

Theorem 2. *Let* $\{z_n\}$, $\{r_n\}$, *and* $\{h_n\}$ *be related by the parametric equations*

$$z_n=\zeta_n(h_n), \qquad r_n=\rho_n(h_n) \tag{6.7}$$

with ζ_n *and* ρ_n *defined by* (4.6) *and* (4.7). *Assume*[4] *that Conditions* A – C *hold. If* $r_n \to \infty$, *then for arbitrary* $\varepsilon > 0$, $\delta > 0$ *and n sufficiently large.*

$$P\{S_n > z_n\} < \exp[-r_n(1-\delta)], \tag{6.8}$$

$$P\{S_n > (1-\varepsilon)\, z_n\} > \exp[-r_n(1+\delta)]. \tag{6.9}$$

Proof. Because of (6.6) the Condition (6.1) implies

$$h_n\, \gamma_n(h_n) = o(\rho_n(h_n)) = o(r_n) \tag{6.10}$$

and from (5.4) we see therefore that $\lambda_n(h_n) \geqq r_n - o(r_n)$. Since $\zeta_n(h) \geqq \psi_n'(h)$ (by (5.6)) the assertion (6.8) is an immediate consequence of (3.2).

For the inequality in the reverse direction it is convenient to start from a variant of (3.3) which avoids the use of variances. We use the argument that led to (3.3), but replacing the interval $\psi_n'(h) \pm 2\sigma_n(h)$ by $\psi_n'(h) \pm \eta\, \zeta_n(h)$. This yields

$$P\{S_n > \psi_n'(h_n) - \eta\, \zeta_n(h_n)\} \geqq (1-a_n) \exp[-\rho_n(h_n) - \eta\, h_n\, \zeta_n(h_n)] \tag{6.11}$$

where

$$a_n = P\{|\bar{S}_n - \psi_n'(h_n)| > \eta\, \zeta_n(h_n)\}. \tag{6.12}$$

Choose η smaller than $\varepsilon/2$ and $c\,\delta/2$. In view of (5.6), (6.1) and (6.6) it is clear that (6.11) implies

$$P\{S_n > (1-\varepsilon)\, \zeta_n(h_n)\} \geqq (1-a_n) \exp[-\rho_n(h_n)(1+\delta/2)] \tag{6.13}$$

for all n large enough to make $\gamma_n(h_n) < (\varepsilon - \eta)\, \zeta_n(h_n)$. To prove (6.9) it suffices therefore to show that a_n remains bounded away from one.

We truncate the variables $\bar{X}_{k,n}$ at $\alpha\, \zeta_n(h_n)$, that is, we put $\bar{X}_{k,n} = \bar{X}_{k,n}^{(1)} + \bar{X}_{k,n}^{(2)}$ where $\bar{X}_{n,n}^{(1)} = \bar{X}_{k,n}$ if the latter variable is $< \alpha\, \zeta_n(h_n)$ and $\bar{X}_{k,n}^{(1)} = 0$ otherwise. With obvious notations then

$$|\bar{S}_n - \psi_n'(h_n)| \leqq |\bar{S}_n^{(1)} - E(\bar{S}_n^{(1)})| + |\bar{S}_n^{(2)}| + E(\bar{S}_n^{(2)}). \tag{6.14}$$

Now by (6.3)

$$E(\bar{S}_n^{(2)}) \leqq \zeta_n(h_n;\, \alpha\, \zeta_n(h_n),\, \infty) = o(\zeta_n(h_n)) \tag{6.15}$$

and this entails

$$P\{\bar{S}_n^{(2)} \neq 0\} = o(1). \tag{6.16}$$

Finally

$$\operatorname{Var}(S_n^{(1)}) \leqq \int_{-\infty}^{\alpha \zeta_n(h_n)} e^{h_n y}\, U_n(dy) \leqq \frac{\alpha\, \zeta_n(h_n)}{1 - e^{-\alpha h_n \zeta_n(h_n)}} \int_{-\infty}^{\alpha \zeta_n(h_n)} \frac{e^{h_n y} - 1}{y}\, U_n(dy). \tag{6.17}$$

The integral is $\leqq \zeta_n(h_n)$. The assumption $\rho_n(h_n) \to \infty$ implies that $h_n\, \zeta_n(h_n) \to \infty$ and so by Chebyshev's inequality

$$P\{|S_n^{(1)} - E(S_n^{(1)})| \geqq \tfrac{1}{2}\eta\, \zeta_n(h_n)\} \leqq \frac{8\alpha}{\eta^2} \tag{6.18}$$

for n sufficiently large. Since α is arbitrary it follows from (6.14) – (6.16) and (6.18) that $a_n \to 0$.

Anticipating Corollary 1 of the next section we may restate Theorem 2 in the following weakened, but more intelligible, form.

4 Condition C may be omitted if (6.1) is replaced by the slightly stronger Condition (6.10).

Corollary. *Assume that* $|X_{k,n}|=o(z_n)$ *for* $k=1,\ldots,n$ *and that the probability mass attributed by* U_n *to* $[0,\infty)$ *remains bounded away from* 0. *The conclusions* (6.8)–(6.9) *of the theorem then hold if* $r_n\to\infty$ *and Condition* A *is satisfied.*

In this case z_n/h_n *and* r_n/h_n^2 *remain bounded away from* 0.

7. On the Condition C

The left side inequality in (6.6) was introduced in a purely formal manner and we proceed now to explain its nature. It is, of course, equivalent to

$$\tau_n(h_n)\leqq(1-c)\,h_n\,\zeta_n(h_n),\qquad c>0. \tag{7.1}$$

This relation becomes plausible if one notices that for fixed n the derivative $\tau_n'(h)=\zeta_n(h)$ is a monotone function of $h>0$. This implies $\tau_n(h)\leqq h\,\zeta_n(h)$ and leads one to expect that in most concrete situations the left side should be substantially smaller than the right. As a matter of fact, it is difficult to construct examples where $\tau_n(h_n)\sim h_n\,\zeta_n(h_n)$, and these are extremely artificial. (By contrast, the relation $\tau_n(h_n)=o(\zeta_n(h_n))$ is commonplace.)

We now show that Condition C may be regarded as counterpart to Condition B in that it requires that the integral defining $z_n=\zeta_n(h_n)$ should not be determined essentially by the restriction to an interval of the form $(-\infty,-a\,h_n^{-1})$. More precisely we prove

Lemma 2. *The Condition* C *is satisfied if, and only if* [5], *there exist constants* $a>0$ *and* $p>0$ *such that*

$$\zeta_n(h_n;\,-a\,h_n^{-1},\infty)>p\,\zeta_n(h_n). \tag{7.2}$$

Proof. (a) Assume (6.6). From part (iii) of Lemma 1 it is clear that

$$\begin{aligned}\rho_n(h)&\leqq q_a\,h\,\zeta_n(h;\,-\infty,-a\,h^{-1})+h\,\zeta_n(h;\,-a\,h^{-1},\infty)\\&=q_a\,h\,\zeta_n(h)+(1-q_a)\,h\,\zeta_n(h;\,-a\,h^{-1},\infty)\end{aligned} \tag{7.3}$$

where $q_a=[e^a-1-a]/(e^a-1)\,a$. Clearly $q_a\to0$ as $a\to\infty$, and hence it is possible to choose a so large that $q_a<c$. Letting $h=h_n$ we get then (7.2) with $p=(c-q_a)/(1-q_a)$.

(b) Assume (7.2). Using the same Lemma 1 we get

$$\begin{aligned}\rho_n(h_n)&\geqq\rho(h_n;\,-a\,h_n^{-1},\infty)\geqq q_a\,h_n\,\zeta_n(h_n;\,-a\,h_n^{-1},\infty)\\&\geqq q_a\,p\,h_n\,\zeta_n(h_n)\end{aligned} \tag{7.4}$$

and so (6.6) holds with $c=q_a\,p$.

---

5 We remark in passing that (7.2) implies

$$\rho_n(h_n;\,-a\,h_n^{-1},\infty)>p\,\rho_n(h_n). \tag{$*$}$$

Indeed, by part (iii) of Lemma 1 we have

$$\frac{\rho_n(h_n;\,-a\,h_n^{-1},\infty)}{\zeta_n(h_n;\,-a\,h_n^{-1},\infty)}\geqq\frac{\rho_n(h_n;\,-\infty,-a\,h_n^{-1})}{\zeta_n(h_n;\,-\infty,-a\,h_n^{-1})}$$

or

$$\frac{\rho_n(h_n;\,-a\,h_n^{-1},\infty)}{\rho_n(h_n;\,-\infty,-a\,h_n^{-1})}\geqq\frac{\zeta_n(h_n;\,a\,h_n^{-1},\infty)}{\zeta_n(h_n;\,-\infty,-a\,h_n^{-1})}\geqq\frac{p}{1-p}$$

which is equivalent to ($*$).

Lemma 3. (a) *Condition C is satisfied if there exists a constant $b>0$ such that*

$$\zeta_n(h_n) \geqq b\,h_n. \tag{7.5}$$

(b) *This, in turn, is certainly the case if there exist two constants $a>0$ and $q>0$ such that*

$$U_n(-a\,h_n^{-1}, \infty) > q. \tag{7.6}$$

Corollary 1. *Condition C and (7.5) are satisfied if the mass attributed to $[0, \infty)$ by U_n remains bounded away from 0.*

The latter condition, though far from necessary, covers probably most cases of more than curiosity value.

Proof. Obviously (7.6) implies

$$\zeta_n(h_n) \geqq \zeta_n(h_n;\, -a\,h_n^{-1}, \infty) \geqq q\,\frac{1-e^{-a}}{a}\,h \tag{7.7}$$

and so (7.5) is true. On the other hand, when (7.5) holds we have trivially

$$\zeta_n(h_n;\, -\infty, -a\,h_n^{-1}) \leqq a^{-1} h_n \leqq (a\,b)^{-1}\,\zeta_n(h_n). \tag{7.8}$$

If $a\,b>1$ this is equivalent to (7.2) with $p=1-(a\,b)^{-1}$.

The following criterion may illuminate the nature of Condition C, but it will not be used explicitly in the sequel. Let

$$\omega_n(h) = \int_{-\infty}^{+\infty} e^{hy}\,U_n(dy). \tag{7.9}$$

Criterion. *Suppose that $\{\zeta_n(h_n)\}$ is bounded away from 0. For Condition C to hold it is necessary that $\omega_n(h_n)>\eta$ for some fixed $\eta>0$, and sufficient that $\omega_n(h)>\eta$ for $(1-\delta)\,h_n<h\leqq h_n$.*

Proof. A glance at the integrands shows that if $\omega_n(h_n)\to 0$ then also the left side in (7.2) tends to 0 for any $a>0$. This argument applies also to subsequences and entails the necessity assertion. On the other hand, $\zeta_n'(h)=\omega_n(h)$, and hence

$$\zeta_n(h_n) \geqq \int_{(1-\delta)\,h_n}^{h_n} \omega_n(t)\,dt;$$

when the integrand exceeds η then (7.5) holds with $b=\delta\,\eta$.

Corollary 2. *Condition C is satisfied if for the first moment M_n of U_n the inequality $M_n > -(1-\eta)\,h_n^{-1}$ holds (in particular, if $M_n \geqq 0$).*

This is an immediate consequence of our criterion and the identity

$$\omega_n(h) - 1 - M_n\,h = \int_{-\infty}^{+\infty} [e^{hy} - 1 - h\,y]\,U_n(dy).$$

8. Variants

The outer form of Theorem 2 is probably satisfactory for most applications, but it lacks elegance due to the appearance of the auxiliary argument $(1-\varepsilon)\,z_n$. This can be eliminated at the cost of requiring that the three conditions hold not only for $\{h_n\}$ but also for all sequences in a certain neighborhood. This requirement is so natural that it requires serious effort to show by examples that it is not automatically satisfied. Surprising is that the required neighborhood is so small[6].

In the following we adhere to the notations of Section 6; in particular, $z_n = \zeta_n(h_n)$ and $r_n = \rho_n(h_n)$.

Theorem 3. *Let $\{H_n\}$ be a sequence such that for some $\eta > 0$.*

$$\zeta_n(H_n) > (1+\eta)\, z_n \tag{8.1}$$

or, equivalently,

$$\rho_n(H_n) > (1+\eta')\, r_n. \tag{8.2}$$

Assume that Conditions A – C *are satisfied by all triples $\{t_n\}$, $\{\zeta_n(t_n)\}$, $\{\rho_n(t_n)\}$ with $h_n \leqq t_n \leqq H_n$. Then, if $r_n \to \infty$,*

$$P\{S_n > r_n\} = \exp\{-r_n + o(r_n)]. \tag{8.3}$$

Proof. (a) We begin by verifying that (8.1) and (8.2) are interchangeable. (This is a consequence of assumption C, and is not true in general.) In consequence of Condition C

$$c\, t\, \zeta_n(t) < \rho_n(t) < t\, \zeta_n(t), \qquad h_n \leqq t \leqq H_n. \tag{8.4}$$

But $\rho_n'(t) = t\, \zeta_n'(t)$, and so

$$\frac{\zeta_n'(t)}{\zeta_n(t)} \leqq \frac{\rho_n'(t)}{\rho_n(t)} \leqq \frac{1}{c}\, \frac{\zeta_n'(t)}{\zeta_n(t)}. \tag{8.5}$$

Integrating between h_n and H_n one gets

$$\frac{\zeta_n(H_n)}{z_n} \leqq \frac{\rho_n(H_n)}{r_n} \leqq \left(\frac{\zeta_n(H_n)}{z_n}\right)^{1/c}$$

which proves the equivalence of (8.1) and (8.2).

(b) From Theorem 2 we know that (6.8) will be satisfied for all n sufficiently large, and so we have only to consider the inequality in the reverse direction.

(c) Starting from (8.1) choose $0 < \varepsilon < \eta/(1+\eta)$ and define t_n by

$$z_n = (1-\varepsilon)\, \zeta_n(t_n). \tag{8.6}$$

Then $h_n < t_n < H_n$ and so by (6.8)

$$P\{S_n > z_n\} > \exp[-\rho_n(t_n)(1+\delta)] \tag{8.7}$$

for arbitrary $\delta > 0$ provided n is sufficiently large. But from (8.5)

$$\log \frac{\rho_n(t_n)}{r_n} < \frac{1}{c} \log \frac{\zeta_n(t_n)}{z_n} = \frac{1}{c} \log \frac{1}{1-\varepsilon}. \tag{8.8}$$

Since ε is arbitrary it follows that the right side in (8.7) is $\geqq \exp[-r_n(1+2\delta)]$, and this finishes the proof.

6 The bounds given in Theorem 3 are not the best possible. It suffices that $\zeta_n(H_n) > z_n + 2\sigma_n(h_n)$ where $\sigma_n(h)$ is the variance of $\bar{S}_n$ (see Section 3); we know that $\sigma_n(h_n) = o(z_n)$.

The following weakened version of Theorem 3 is probably preferable in most applications.

Corollary. *Suppose*[7] *that*

$$\sup_{k\leqq n} \frac{1}{z_n} X_{k,n} \to 0 \tag{8.9}$$

and that U_n-measure of $[0, \infty)$ remains bounded away from 0.

Let $\{H_n\}$ be a sequence satisfying at least one of the Conditions (8.1), (8.2), *or* $H_n > (1+\delta)\, h_n$. *If*

$$\sup_{k\leqq n} [\varphi_{k,n}(H_n) - 1] \to 0 \tag{8.10}$$

and $r_n \to \infty$, then the conclusion (8.3) *holds.*

Proof. If $\{H_n\}$ satisfies (8.1) the conditions of the corollary are more stringent than those of the theorem. Indeed, since the $\varphi_{k,n}$ are monotone, (8.10) guarantees that Condition A is satisfied for all sequences $t_n \leqq H_n$. The assumption concerning U_n ensures similarly that Condition C holds (see Corollary 1, Section 7). Finally, (8.9) is stronger than Condition B for $h_n \leqq t_n \leqq H_n$ (see (6.5)). Accordingly, it suffices to prove that under the present circumstances the condition $H_n > (1+\delta)\, h_n$ is stronger than (8.1). We prove this in slightly greater generality so as to make the following lemma applicable also to the original version of the theorem.

Lemma 4. *Suppose that* (8.4) *holds with $H_n > (1+\delta)\, h_n$. Then* (8.1) *is true for some η depending on c.*

Proof. By Lemma 2 the relations (8.4) imply the existence of constants $a > 0$ and $p > 0$ such that

$$\zeta_n(t;\, -a\, t^{-1}, \infty) > p\, \zeta_n(t), \qquad h_n \leqq t \leqq H_n. \tag{8.11}$$

But the left side equals

$$\int_{-a t^{-1}}^{\infty} \frac{1-e^{-ty}}{y}\, e^{ty}\, U_n(dy) < \frac{e^a - 1}{a}\, t \int_{-\infty}^{+\infty} e^{ty}\, U_n(dy) = \frac{e^a - 1}{a}\, t\, \zeta_n'(t). \tag{8.12}$$

Thus

$$\frac{\zeta_n'(t)}{\zeta_n(t)} > \frac{b}{t}, \qquad b = \frac{a}{p(e^a - 1)}. \tag{8.13}$$

Integrating between h_n and H_n we get (8.1) with $1+\eta = (1+\delta)^b$.

9. The General Form

The following theorem is a variant of Theorem 1 and stands to it in the same relation as Theorem 3 does to Theorem 2. As mentioned before, Theorem 4 is theoretically sharper than Theorem 3, but its applicability is limited by its unwieldy nature. Nevertheless, some interesting information can be obtained from

7 The condition $r_n \to \infty$ implies that $h_n z_n \to \infty$ and by (7.5) this is impossible unless $z_n \to \infty$. *Thus uniform boundedness of the $X_{k,n}$ suffices.* The assumption concerning U_n may, of course, be weakened to (7.6).

a comparison of the two theorems. To facilitate such comparisons we now write z_n^* and h_n^* instead of z_n and h_n. For the notations see Section 4.

Theorem 4. *Let z_n^* and r_n be related by*

$$z_n^* = \psi_n'(h_n^*), \qquad r_n = \lambda_n(h_n^*). \tag{9.1}$$

Let $\{H_n\}$ be a sequence such that for some fixed $\eta > 0$

$$\psi_n'(H_n) > (1+\eta)\, z_n^* \tag{9.2}$$

or, equivalently,

$$\lambda_n(H_n) > (1+\eta')\, r_n. \tag{9.3}$$

Assume that

$$\psi_n''(H_n) = o(z_n^2) \tag{9.4}$$

and

$$c\, t\, \psi_n'(t) < \lambda_n(t) \leqq t\, \psi_n'(t), \qquad h_n^* \leqq t \leqq H_n. \tag{9.5}$$

If $r_n \to \infty$ then

$$P\{S_n > z_n^*\} = \exp[-r_n + o(r_n)]. \tag{9.6}$$

The *Proof* does not differ essentially from the proof of Theorem 3. Condition C is replaced by (9.5), and since $\lambda_n'(t) = t\, \psi_n''(t)$ we get from (9.5)

$$\frac{\psi_n''(t)}{\psi_n'(t)} \leqq \frac{\lambda_n'(t)}{\lambda_n(t)} \leqq \frac{1}{c} \frac{\psi_n''(t)}{\psi_n'(t)}$$

which is the analogue to (8.5).

As a rule, a direct analysis of the parametric equations (9.1) will be usually impracticable whereas (as the examples of the following sections show) mild regularity conditions suffice to make the corresponding equations (6.7) of the linearized theory amenable to a surprisingly simple treatment. It is therefore fortunate that it is often possible to use the linearized model of Theorem 3 for purposes of approximations even when the main Condition A is not satisfied. (See Proposition 2 in Section 10.)

As a general result in this direction we shall discuss a curious phenomenon discovered by Marcinkiewicz and Zygmund [8] in a special case. These authors studied sequences $\{X_k\}$ such that X_k assumes the values $\pm a_k$ each with probability $\frac{1}{2}$. The law of the iterated logarithm applies if $a_n = o(s_n(\log\log s_n)^{-\frac{1}{2}})$ where $s_n^2 = a_1^2 + \cdots + a_n^2$. Marcinkiewicz and Zygmund showed that when this condition fails

$$\limsup \frac{X_1 + \cdots + X_n}{\sqrt{2 s_n \log\log s_n}} \leqq 1. \tag{9.7}$$

This shows, roughly speaking, that the presence of large components *decreases* the upper limit, contrary to what might be expected intuitively. The next lemma puts this discovery in a more general setting. It shows that the probabilities of large deviations in Theorem 4 are always *smaller* than those obtained by using formally the algorithm of Theorem 3. Thus the presence of non-negligible components tends to decrease the probabilities of large deviations (and hence also the upper limits for the sums of random variables).

Lemma 5. *Let z_n and r_n be related by $z_n = \zeta_n(h_n)$ and $r_n = \rho_n(h_n)$, while z_n^* and r_n are given by the parametric equations $z_n^* = \psi_n'(h_n^*)$ and $r_n = \lambda_n(h_n^*)$.*

Then $z_n^* \leqq z_n$.

Proof. From (5.3) we know that $\lambda_n(h_n) \leqq \rho_n(h_n)$ and hence $h_n^* \geqq h_n$. Since ψ_n' is monotone it follows that

$$\psi_n(h_n^*) - \psi_n(h_n) \leqq (h_n^* - h_n)\, \psi_n'(h_n^*). \tag{9.8}$$

Thus

$$r_n = \lambda_n(h_n^*) = h_n^*\, \psi_n'(h_n^*) - \psi(h_n^*) \geqq h_n\, \psi_n'(h_n^*) - \psi_n(h_n). \tag{9.9}$$

Now $\psi_n(h_n) \leqq \tau_n(h_n)$ since $\log x < x - 1$ for $x > 1$. Thus

$$r_n \geqq h_n\, \psi'(h_n^*) - \tau_n(h_n) = h_n[\psi'(h_n^*) - \zeta_n(h_n)] + \rho_n(h_n) = h_n[z_n^* - z_n] + r_n, \tag{9.10}$$

and this proves the assertion.

10. Examples Related to the Law of the Iterated Logarithm

Let $\{X_k\}$ be a sequence of independent random variables with $E(X_k) = 0$ and $s_n^2 = E((X_1 + \cdots + X_n)^2)$. Given $0 < r_n \to \infty$ we suppose that[8]

$$X_k < C \frac{s_n}{\sqrt{r_n}} \qquad k = 1, \ldots, n \tag{10.1}$$

where C is a positive constant. The measure U_n of (4.1) is then supported by the interval $(-\infty, C/\sqrt{r_n})$ and it is convenient to introduce a new scale to reduce this interval to $(-\infty, C)$. Accordingly we define a measure V_n by

$$V_n(dx) = U_n\left(\frac{dx}{\sqrt{r_n}}\right) = \frac{1}{r_n} \sum_{k=1}^{n} x^2\, F_k(dx\, s_n/\sqrt{r_n}). \tag{10.2}$$

In Kolmogorov's law of the iterated logarithm (10.1) is replaced by $|X_k| = o(s_n/\sqrt{r_n})$ with $r_n = \log\log s_n$, and the measures V_n then tend to the atomic measure concentrated at the origin. The refined form (Feller [4]) supposes that $|X_n| < \frac{1}{5} s_n/\sqrt{r_n}$. Analogues to the results of that paper cannot be obtained by the present methods because they depend on sharper estimates of the form (1.11). On the other hand, these theoretically sharper estimates provide no concrete information concerning the constants z_n in (10.6). This shows that our new approximations may be useful even in situations to which the refined theory applies.

As mentioned before, analogues to the law of the iterated logarithm depends on the possibility of finding a sequence $\{z_n\}$ such that

$$P\{(X_1 + \cdots + X_n)/s_n > z_n\} \sim e^{-r_n + o(r_n)} \tag{10.3}$$

with $r_n = \log\log s_n$, but we leave the sequence $\{r_n\}$ unspecified.

For simplicity we introduce the unnecessarily restrictive condition that the V_n-mass of the closed interval $[0, C]$ remains bounded away from zero; in symbols

$$V_n([0, C]) > \gamma > 0. \tag{10.4}$$

This condition requires merely that the variables X_k are not absurdly excentric. In view of Corollary 1 (Section 7) this implies that Condition C is satisfied. To

8 Usually r_n is defined as an function of s_n such that $s_n/\sqrt{r_n}$ increases. It suffices then to require (10.1) and (10.5) for the particular value $k = n$.

satisfy Condition A we shall assume that the individual components are relatively small in the sense that for fixed $\varepsilon>0$

$$P\{|X_k|>\varepsilon\, s_n/\sqrt{r_n}\}\to 0 \tag{10.5}$$

uniformly in $k=1,\ldots,n$. (This condition will be dropped later on.)

Proposition 1. *Assume* (10.1), (10.4) *and* (10.5). *Then* (10.3) *holds with*

$$z_n \sim a_n \sqrt{r_n} \tag{10.6}$$

where

$$a_n=\int_{-\infty}^{C} \frac{e^{t_n x}-1}{x}\, V_n(dx) \tag{10.7}$$

and t_n is the unique positive root of the equation

$$\int_{-\infty}^{C} \frac{e^{t_n x}(t_n x-1)+1}{x^2}\, V_n(dx)=1. \tag{10.8}$$

The sequence $\{a_n\}$ remains bounded away from 0 *and* ∞.

In particular, if V_n tends to a limit distribution V, then $z_n\sim a\sqrt{r_n}$ where a is obtained by the obvious passage to the limit in (10.7) and (10.8). Under the conditions for the law of the iterated logarithm the limit distribution is concentrated at the origin. In the limit (10.8) reduces to the equation $\frac{1}{2}t_n^2=1$, and we get the classical relation $z_n\sim\sqrt{2r_n}$. We see now that we can obtain laws of the iterated logarithm with the constant $\sqrt{2}$ replaced by an arbitrary constant $a>0$.

Proof. We begin by recalling that the integrand in (10.8) is a continuous monotone function of x assuming at the origin the value $t_n^2/2$. In view of (10.4) the integral is therefore $\geqq \gamma\, t_n^2/2$, and it is obviously $\leqq e^{t_n C}\, t_n/C$. The roots t_n are therefore uniquely determined and remain bounded away from 0 and ∞. A similar argument now shows that the same is true of $\{a_n\}$.

Now put $h_n=t_n\sqrt{r_n}$. Then (10.8) and (10.7) differ only notationally from the equations $\rho_n(h_n)=r_n$ and $\zeta_n(h_n)=a_n\sqrt{r_n}$ with ζ_n and ρ_n defined in (4.6) and (4.7). Thus (10.6) states merely that z_n and r_n are connected as required by the basic parametric equations (6.7).

The proposition now follows trivially from Theorem 3. We saw already that (10.4) implies that Condition C is satisfied. Condition B is trivially satisfied because $z_n=\zeta_n(h_n)\to\infty$ while the variables $X_{k,n}=X_k/s_n$ remain bounded. This statement remains true also with h_n replaced by $H_n=(1+\eta)\,h_n$. Finally, (10.5) assures that Condition A is satisfied in the strong form (6.7), namely

$$\sup_{k\leqq n} \int_{-\infty}^{C/\sqrt{r_n}} (e^{t_n\sqrt{r_n}\,x}-1)\, F_k(dx\, s_n)\to 0. \tag{10.9}$$

Indeed, in consequence of the boundedness of $\{t_n\}$ the integrand remains bounded, and for $|x|<\varepsilon/\sqrt{r_n}$ it is arbitrarily close to 0 if ε is chosen sufficiently small. This concludes the proof.

We turn to an investigation of what happens when the individual negligibility condition (10.5) is dropped. The main purpose of the following considerations is to show how the forbidding conditions of Theorem 4 disappear when reasonable regularity conditions are imposed. The following method can be adapted to most practical examples.

Denote by $K_n(\varepsilon)$ the class of subscripts $k \leqq n$ such that

$$P\{|X_k| > \varepsilon s_n/\sqrt{r_n}\} > \varepsilon. \tag{10.10}$$

The influence of a class K_n may be measured by its contribution to the probability measure U_n, that is by the measure U_n^* defined by

$$U_n^*(dx) = \sum_{k \in K} x^2 F_k(dx\, s_n). \tag{10.11}$$

Proposition 1 remains valid if the assumption (10.5) is replaced by the milder requirement that $U_n^* \to 0$, since this implies that Condition A holds. We now permit the non-negligible components to play an essential part, but suppose, roughly speaking, that their role is not absolutely dominant. More precisely, we suppose that there exists a constant $\gamma > 0$ such that the measure U_n^* defined by (10.11) with $K_n = K_n(\varepsilon)$ satisfies

$$U_n^*[0, \infty] > \gamma > 0 \tag{10.12}$$

for every $\varepsilon > 0$. This condition may be redundant. In any case, it is so mild that I know of no example where it is not automatically satisfied.

If (10.12) holds for every $\varepsilon > 0$ it is possible to find a diagonal sequence $\varepsilon_n \to 0$ such that (10.12) remains valid for $K_n = K_n(\varepsilon_n)$ (with $>$ replaced by $\geqq$). We now consider the triangular array $\{X_{k,n}, k \in K_n\}$. The variances of its row sums lie between γ and 1, but except for this lack of the proper norming the array satisfies all conditions of Theorem 3, and Proposition 1 applies to it with obvious adjustments of the norming constants.

The following refers to the situation treated in Proposition 1 except that Condition (10.5) is dropped.

Proposition 2. *Suppose that* (10.12) *holds. Then there exists h_n^* such that $\lambda_n(h_n^*) = r_n$. With $z_n^* = \psi_n'(h_n^*)$ one has*

$$P\{(X_1 + \cdots + X_n)/s_n > z_n^*\} = e^{-r_n + o(r_n)}. \tag{10.13}$$

There exist two constants such that

$$0 < a < \frac{z_n^*}{\sqrt{r_n}} < A. \tag{10.14}$$

Proof. The existence of the root h_n^* is not obvious since the individual functions λ_n may be bounded. Fortunately it is easy to find reasonable bounds on λ_n. From (5.3) we know that $\lambda_n(h) \leqq \rho_n(h)$ for all $h > 0$. On the other hand, on restricting all summations to the subscripts $k \in K_n$ we get with self-explanatory notations $\lambda_n(h) \geqq \lambda_n^\#(h)$. But the proof of Proposition 1 applies to the array $\{K_{k,n}, k \in K_n\}$ and shows that there exist constants $h_n^\#$ such that

$$r_n = \rho_n^\#(h_n^\#) \sim \lambda_n^\#(h_n^\#). \tag{10.15}$$

Thus $\lambda_n(h_n^\#) \geqq r_n$ whereas if $h_n = t_n\sqrt{r_n}$ is the root of $\rho_n(h_n) = r_n$ given by (10.8) we have $\lambda_n(h_n) \leqq r_n$. Thus $\lambda_n(h_n^*) = r_n$ for some h_n^* with $h_n \leqq h_n^* \leqq h_n^\#$.

To prove the proposition it remains to show that Theorem 4 is applicable and that (10.14) is true. We know that the ratios $h_n/\sqrt{r_n}$ remain bounded away from 0 and ∞. Since Condition C is satisfied when the summation is restricted to

K_n, we conclude that with appropriate constants γ_j

$$\lambda_n(h_n^*) = \lambda_n^\#(h_n^\#) \geqq \gamma_1 h_n^\# \zeta_n^\#(h_n^\#) \geqq \gamma_2 h_n^2 \geqq \gamma_3 h_n z_n \tag{10.16}$$

with z_n given by (10.6). But Lemma 5 asserts that $z_n \geqq z_n^*$, and so $\lambda_n(h_n^*) \geqq h_n^* \psi_n'(h_n^*)$. The same argument applies with r_n replaced by $(1+\eta) r_n$, and so the condition (9.5) of Theorem 4 is satisfied. Condition (9.4) is trivially satisfied since

$$\psi_n'' = \sum \frac{\varphi_{k,n}'' \varphi_{k,n} - \varphi_{k,n}'^2}{\varphi_{k,n}} \leqq \sum \varphi_{k,n}''$$

and

$$\sum \varphi_{k,n}''(a h_n) = \int_{-\infty}^{\infty} e^{a t_n \sqrt{r_n} x} U_n(dx)$$

remains bounded by virtue of (10.1)

11. A Typical General Example

We adhere to the setup of the preceding section and suppose that there exists a sequence of positive constants c_n such that

$$X_k < \frac{s_n}{c_n}, \qquad k = 1, \ldots, n. \tag{11.1}$$

We do not exclude the case that $c_n \to \infty$, so that (11.1) is far from the usual boundedness conditions. (It is introduced in order to avoid tiresome discussions of the behavior of the tail.) The measure U_n is supported by the interval $(-\infty, 1/c_n)$ and it is again convenient to introduce a new scale. Accordingly we introduce the measure $V_n(dx) = U_n(c_n\, dx)$ which is supported by $(-\infty, 1)$. The case that $r_n = O(c_n^2)$ is covered by the propositions of the preceding section, and hence we may assume that (11.2)

$$\frac{r_n}{c_n^2} \to \infty. \tag{11.2}$$

We propose to show that under very natural conditions *our limit theorems depend only on the behavior of the reduced probability measure V_n near the boundary.* For concreteness we assume[9] that within a fixed interval $(1-\delta, 1)$

$$V_n(1-x, 1) > A_n (1-x)^{a_n} \tag{11.3}$$

where $0 \leqq a_n < a^*$ and $0 < A_* < A_n < A^* < \infty$ for all n.

We shall see that with the present setup the conditions for the validity of Theorem 3 are simple indeed, and the only problem consists in solving the parametric equations $\rho_n(h_n) = r_n$ and $\zeta_n(h_n) = z_n$. We proceed to show that, for given

9 The reader may verify that the argument goes through under the much less restrictive hypothesis that

$$V_n(1-\delta, 1) > \left(\frac{r_n}{c_n^2}\right)^{-\eta}$$

for some fixed $\eta > 0$. The only change is that the precise asymptotic relation (11.4) is replaced by

$$(1-\varepsilon) \log \frac{r_n}{c_n^2} < \frac{h_n}{c_n} < \frac{1+\eta}{1-\delta} \log \frac{r_n}{c_n^2}$$

for n sufficiently large.

$r_n \to \infty$, *these equations possess a unique solution, and for it*

$$\frac{h_n}{c_n} \sim \log \frac{r_n}{c_n^2} \tag{11.4}$$

$$z_n \sim \frac{r_n}{h_n} \sim \frac{r_n}{c_n \log(r_n/c_n^2)}. \tag{11.5}$$

(The sign $\sim$ indicates that the ratio of the two sides tends to 1.)

The equation $\rho_n(h_n) = r_n$ may be written in the form

$$\int_{-\infty}^{1} \frac{e^{x h_n/c_n}(x\, h_n/c_n - 1) + 1}{x^2} V_n(dx) = \frac{r_n}{c_n^2}. \tag{11.6}$$

As a function of the free parameter h_n the left side increases from 0 to ∞, and so there is no question about the existence of a root h_n. Because of the monotonicity of the integrand we have

$$\frac{h_n}{c_n} e^{h_n/c_n} \geqq \frac{r_n}{c_n^2}, \tag{11.7}$$

and in view of (11.2) this requires that $\frac{h_n}{c_n} \to \infty$. An inequality in the opposite direction is obtained by restricting the interval of integration to $(1-\delta, 1)$. An integration by parts then shows that

$$\begin{aligned} \frac{r_n}{c_n^2} &\geqq \left(\frac{h_n}{c_n}\right)^2 \int_{-\infty}^{1} e^{x h_n/c_n} V_n(1-x, 1)\, dx \\ &\geqq A_n \left(\frac{h_n}{c_n}\right)^2 e^{h_n/c_n} \int_{1-\delta}^{1} e^{-(1-x) h_n/c_n} (1-x)^{a_n}\, dx. \end{aligned} \tag{11.8}$$

Since $h_n/c_n \to \infty$ the right side is asymptotically equivalent to

$$A_n \Gamma(1+a_n)\, e^{h_n/c_n} \left(\frac{h_n}{c_n}\right)^{1-a_n}.$$

A comparison with (11.7) shows that for n sufficiently large

$$e^{(1-\varepsilon) h_n/c_n} < \frac{r_n}{h_n^2} < e^{(1+\varepsilon) h_n/c_n} \tag{11.9}$$

and this proves the assertion (11.4).

We now recall from (4.4) and (4.5) that $\lambda_n(h_n) = h_n \zeta_n(h_n) - \tau_n(h_n)$ where with the present notations

$$\tau_n(h_n) = c_n^2 \int_{-\infty}^{1} \frac{e^{x h_n/c_n} - 1 - x\, h_n/c_n}{x^2} V_n(dx). \tag{11.10}$$

As in the case of the integral in (11.5) it is seen that the contribution of the interval $(1-\delta, 1)$ plays a dominant role. In it the ratio of the two integrands is close to h_n/c_n, and hence we conclude that $\tau_n(h_n) = o(\lambda_n(h_n))$. Thus $\lambda_n(h_n) \sim h_n \zeta_n(h_n)$ which is the same as the assertion (11.5).

We have thus solved the parametric equations occurring in Theorem 3. We now use this theorem in the form given in the corollary (Section 8). Our assumption (11.3) guarantees that the measures $U_n([0, \infty]) = V_n([0, 1])$ are bounded away from 0. To ensure that the boundedness condition (8.7) be satisfied we must

2*

require that

$$\log \frac{r_n}{c_n^2} = o(r_n). \tag{11.10}$$

This excludes only the possibility of certain subsequences $\{c_{n_k}\}$ approaching 0 extremely fast. The only serious restriction to the applicability of Theorem 3 is therefore represented by the condition (8.10) requiring that the individual components be asymptotically negligible. In view of (11.4) the sequence $\{h_n\}$ must be expected to increase exceedingly slowly, and so condition (8.10) is mild indeed. It may be further relaxed by the method used for Proposition 2 in the preceding section.

Note on the Central Limit Theorem

For simplicity let us strengthen the condition (11.1) to $|X_k| < s_n/c_n$. The transformed variables $\bar{X}_k$ introduced in Section 1 are subject to the same bounds. We consider the transforms $\bar{X}_{k,n}$ corresponding to $h = h_n$. It is clear that *if* $c_n \to \infty$ *both sequences* $\{X_k\}$ *and* $\{\bar{X}_k\}$ *belong to the domain of attraction of the normal distribution* $\mathfrak{N}$. If $\{c_n\}$ remains bounded the measures U_n do not converge to the unitary measure, and so $\{X_n\}$ does not belong to the domain of attraction of $\mathfrak{N}$. But the $\bar{X}_k$ remain bounded while the variances of the partial sums tend to ∞. Thus we have an example where *the transformed sequence belongs to the normal domain of attraction, but the given sequence* $\{X_k\}$ *does not.* The converse possibility is more commonplace.

References

1. Cramér, H.: Sur un nouveau théorème-limite de la théorie des probabilitiés. Actualités Scientifiques, No. 736 (Colloque Consacré à la Théorie des Probabilitiés, vol. III.), pp. 5–23. Paris, 1938.
2. Esscher, F.: On the probability function in the collective theory of risk. Skand. Aktuarietidskr. **15**, 175–195 (1932).
3. Feller, W.: Generalization of a probability limit theorem of Cramér. Trans. Amer. math. Soc. **54**, 361–372 (1943).
4. – The general form of the so-called law of the iterated logarithm. Trans. Amer. math. Soc. **54**, 373–402 (1943).
5. – An introduction to probability theory and its applications, vol. 2. New York: J. Wiley 1966.
6. – On probabilities of large deviations. Proc. Nat. Acad. Sci. USA **61**, 1224–1227 (1968).
7. Linnik, Yu V.: On the probability of large deviations for the sums of independent variables. Proc. Fourth Berkeley Sympos. Math. Statist. Probab. II, 289–306 (1961).
8. Marcinkiewicz, J., Zygmund, A.: Remarque sur la loi du logarithme itéré. Fundamenta Math. **29**, 215–222 (1937).
9. Petrov, V. V.: Generalization of Cramér's limit theorem to independent variables with variable distributions. Vestnik Leningrad. Univ. No. **8**, 13–25 (1953). [In Russian.]
10. – Limit theorems for large deviations when Cramér's condition fails, I–II. Ibid. No. **19**, 49–68 (1963); and N. **1**, 58–75 (1964). [In Russian.]
11. Richter, W.: Local limit theorems for large deviations. Theor. Probab. Appl. (Translations) **2**, 206–220 (1957).
12. – Multi-dimensional local limit theorems for large deviations. Ibid. **3**, 100–106 (1958).

Professor Dr. William Feller
Fine Hall
Princeton University
Princeton, N.J. 08540, USA

(Received April 2, 1969)

On the oscillations of sums of independent random variables

By WILLIAM FELLER*

1. Summary

Let $X_1, X_2, \cdots$ be mutually independent random variables with $E(X_j) = 0$ and $E(X_j^2) = \sigma_j^2 > 0$. As usual we put $S_n = X_1 + \cdots + X_n$ and $s_n^2 = E(S_n^2)$. Given a numerical sequence $\{\alpha_n\}$ an old probabilistic problem consists in finding the probability of the event $\{S_n > s_n\alpha_n \ i.o.\}$ (namely that there be infinitely many n such that $S_n > s_n\alpha_n$). Under the assumption that $X_n = o(s_n(\log\log s_n)^{-1/2})$ Kolmogorov's law of the iterated logarithm answers the question for the particular sequences $\alpha_n^2 = (1 \pm \varepsilon)\log\log s_n$ when $\varepsilon \neq 0$. A more refined criterion was found in [1], but only under the artificial assumption $|X_n| < s_n/(10\alpha_n)$. It depends on sharp estimates for $P\{S_n > s_n\alpha_n\}$ for large α_n, and such estimates are possible only under severe conditions on the distributions of the X_j.

In the present paper we propose to re-examine the problem from a general point of view. It turns out that a criterion can be derived without imposing any restriction whatever on the underlying distributions (Theorem 1), and an extremely mild regularity condition makes this criterion simpler and more informative (Theorem 2).

We shall be concerned with the class of events of the form

$$\{\alpha_n + L_1\beta_n < S_n/s_n < \alpha_n + L_2\beta_n \ i.o.\} \tag{1.1}$$

where $\{\alpha_n\}$ and $\{\beta_n\}$ are two given numerical sequences, and $I = (L_1, L_2)$ is a fixed interval with $-\infty < L_1 < L_2 \leqq \infty$. It is convenient to introduce the following *notation*:

$$a_n = s_n\alpha_n\,, \qquad b_n = s_n\beta_n\,, \qquad S_n^* = (S_n - a_n)/b_n\,. \tag{1.2}$$

Then

$$p(I) = P\{S_n^* \in I \ i.o.\} \tag{1.3}$$

stands for the probability of the event (1.1). Under our assumptions[1] it turns out that $p(I)$ assumes only the values 0 or 1.

---

* Work connected with a Project for Research in Probability at Princeton University supported by the U.S. Army Research Office.

[1] That $P\{S_n > a_n \ i.o.\}$ is either 0 or 1 was shown by P. Lévy [6] assuming, of course, that $s_n \to \infty$.

R.L. Schilling et al. (eds.), *Selected Papers II*,

It is obviously necessary to impose some conditions on $\{\alpha_n\}$ and $\{\beta_n\}$. It would seem most natural to consider only monotone $\{\alpha_n\}$ but this does not significantly simplify the proofs. It is therefore preferable and more instructive to state the milder property actually used.

A. *Condition on* $\{\alpha_n\}$. *There exist constants* $\lambda > 1$ *and* C *such that*

$$\text{(1.4)} \qquad \textit{if } 1 \leqq a_k/a_i \leqq \lambda, \quad \textit{then} \quad s_k - s_i \leqq C(a_k - a_i)/\alpha_i .$$

Furthermore $\alpha_n \to \infty$, *and* $\{a_n\}$ *is monotone.*

If $\{\alpha_n\}$ is monotone, (1.4) is satisfied with $C = 1$ for *all* λ. More generally, if $\alpha_n = L(s_n)$ is defined by an arbitrary slowly varying function[2] L then (1.4) holds for *all* λ (with C depending on λ). Finally, (1.4) holds whenever there exists a constant q such that $1 \leqq a_k/a_i \leqq \lambda$ implies $\alpha_k/\alpha_i \geqq (s_k/s_i)^{-q}$. Note that pairs with $1 < a_k/a_i < \lambda$ do not necessarily exist, and so the requirement (1.4) may be empty.

The sequence $\{\beta_n\}$ is more or less auxiliary to $\{\alpha_n\}$. The choice $\beta_n = \varepsilon\alpha_n$ occurs when one seeks the upper limit of S_n/a_n, but in more refined questions $\beta_n = o(\alpha_n)$, and so the condition $\beta_n \leqq \alpha_n$ is introduced for convenience only. The condition $\beta_n > \omega\alpha_n^{-1}$ is more enigmatic, but essential for the proofs.[3] Condition (ii) is self-explanatory and essential. As to (1.7), it holds automatically if $\{b_n\}$ is monotone, which is probably the only case of interest. Note that (1.7) imposes no restrictions on fluctuations of $\{\beta_n\}$ upward, but requires merely that $\{b_n\}$ cannot vary too violently in the opposite direction from $\{a_n\}$.

B. *Conditions on* $\{\beta_n\}$. (i) *There exist a constant* $\omega > 0$ *such that*

$$\text{(1.5)} \qquad \alpha_n \geqq \beta_n > \omega\alpha_n^{-1} .$$

(ii) *To each* $\varepsilon > 0$ *there corresponds a* $\delta > 0$ *such that*

$$\text{(1.6)} \qquad \textit{if } 1 \leqq a_k/a_i \leqq 1 + \delta, \quad \textit{then} \quad 1 - \varepsilon < b_k/b_i < 1 + \varepsilon .$$

(iii) *To each* $t > 1$ *there exists a* $\tau_t > 0$ *such that*

$$\text{(1.7)} \qquad \textit{if } 1 \leqq a_k/a_i \leqq 1 + t, \quad \textit{then} \quad b_k/b_i > \tau_t .$$

From now on it will be understood that conditions A and B are satisfied.

Our criteria will depend on the behavior of the series

$$\text{(1.8)} \qquad \Sigma(I) = \sum_{n=1}^{\infty} \min\left[1, \frac{a_{n+1} - a_n}{b_n}\right] P\{S_n^* \in I\} .$$

[2] For a general orientation it may be useful to remark that under any circumstances with probability one $S_n = o(s_n \log^{\eta} s_n)$ for every $\eta > 1/2$. (Indeed, by Kolmogorov's criterion the series $\sum X_n/(s_n \log^{\eta} s_n)$ converges almost surely, and the assertion now follows from Kronecker's lemma.) Only sequences with $\alpha_n < \log^{\eta} s_n$ are therefore of genuine interest, and slowly varying sequences are the natural object of the study.

[3] The criteria of [1] are based on the choice $\beta_n = \alpha_n^{-1}$, which corresponds to the choice (1.14) below. Recently derived limit theorems [2] show that with this choice the condition $\beta_n > \omega\alpha_n^{-1}$ will be satisfied in practically all imaginable regular situations.

Under mild conditions on $\{\alpha_n\}$, the series (1.8) converges or diverges at the same time as

$$\sum_{n=1}^{\infty} \min\left[1, \frac{\sigma_{n+1}^2}{s_n s_{n+1}} \cdot \frac{\alpha_n}{\beta_n}\right] P\{S_n^* \in I\} . \tag{1.9}$$

For convenience we introduce the following *notation*: if $I = (L_1, L_2)$ is an arbitrary interval, then

$$I_\eta = (L_1 - \eta, L_2 + \eta)$$

with the understanding that any interval (u, v) with $u \geqq v$ is considered empty.

THEOREM 1. (a) *If* $\Sigma(I) < \infty$ *for some interval* I, *then* $p(I_{-\varepsilon}) = 0$ *for every* $\varepsilon > 0$.

(b) *If* $I = (L_1, L_2)$ *with* $0 \leqq L_1 < L_2 < \infty$ *and* $\Sigma(I) = \infty$, *then* $p(I_\varepsilon) = 1$ *for every* $\varepsilon > 0$.

The restriction in part (b) to bounded intervals may seem surprising, but in the present generality the theorem is false for unbounded I. (See § 7.) The condition $L_1 \geqq 0$ is not required when $\beta_n = o(\alpha_n)$. (It serves mainly to keep away from events such as $\{S_n < s_n\}$.)

The special choice $\beta_n = \varepsilon\alpha_n$ satisfies conditions B, and leads directly[4] to the following:

COROLLARY.[5] *If there exists a number* $\rho > 0$ *such that*

$$\sum \min\left[1, \frac{a_{n+1} - a_n}{a_n}\right] P\{S_n > x a_n\} \tag{1.10}$$

converges for $x > \rho$ *and diverges for* $x < \rho$, *then with probability one*

$$\limsup \frac{1}{a_n} S_n = \rho . \tag{1.11}$$

If the series converges for all $x > 0$ *then* (1.11) *holds with* $\rho = 0$.

So far nothing can be concluded in the case of divergence for all x. However, if the series diverges even with $P\{S_n > x a_n\}$ replaced by $P\{x a_n < S_n < y a_n\}$ for some $y(x)$ then such divergence implies (1.11) with $\rho = \infty$.

Theorem 1 is marred by the appearance of the intervals I_ε and by the restriction to bounded intervals in part (b). Both defects can be removed if it is possible to choose $\{\beta_n\}$ satisfying the

---

[4] The restriction in part (b) to bounded intervals plays no role since $\Sigma(I)$ will diverge for bounded intervals containing the point ρ.

[5] The stress in this paper is on general criteria, and therefore no precise information concerning ρ in (1.11) is obtained. An alternative approach depends on deriving expressions for the probabilities $P\{S_n^* \in I\}$ and leads to a theorem of different appearance. The details are described in [2].

Condition C. *For fixed $u < v$ the ratios*

$$\frac{P\{S_n > a_n + b_n u\}}{P\{S_n > a_n + b_n v\}} \tag{1.12}$$

remain bounded away from 1 *and* ∞.

Sequences $\{\beta_n\}$ for which condition C is satisfied exist under mild regularity conditions on the distributions of the X_n. Indeed, let

$$P\{S_n > s_n x\} = \exp\left(-f_n(x)\right) . \tag{1.13}$$

If f_n is differentiable and

$$\beta_n = 1/f_n'(\alpha_n) , \tag{1.14}$$

then $f_n(\alpha_n + x\beta_n) = f_n(\alpha_n) + x + \cdots$, and it is plausible that under ordinary circumstances the neglected terms will remain bounded as $n \to \infty$ (with x fixed). Then (1.12) tends to e^{-u+v}. The limit theorems derived recently [2] lend credibility to this conjecture. Under any circumstances, given $\{a_n\}$, if a sequence $\{b_n\}$ satisfying (1.12) exists at all, it is determined uniquely up to obvious asymptotic equivalences or changes in scale.

If condition C holds, then for any two intervals I and J off $-\infty$ the ratios $P\{S_n^* \in I\}$: $P\{S_n^* \in J\}$ remain bounded away from 0 and ∞, and hence Theorem 1 reduces to the more pleasing

THEOREM 2. *If condition* C *holds then* $\Sigma(I)$ *either converges for all* $I = (L_1, L_2)$ *with* $L_1 > -\infty$, *or else for none.*

In case of convergence $p(I) = 0$ *for every* $I = (L_1, L_2)$ *with* $L_1 > -\infty$.

In case of divergence $p(I) = 1$ *for every* $I = (L_1, L_2)$ *with* $0 \leqq L_1 < L_2 \leqq \infty$.

It is possible to reformulate the theorems in a way that sheds some light on the role of the series $\Sigma(I)$. For that purpose we require a definition which serves also to simplify formulations later on.

Definition. An increasing sequence $\{n_k\}$ *of integers will be called tight if there exists an* h *such that*

$$a_{n_{k+1}} - a_{n_k+1} < h b_{n_k+1} . \tag{1.15}$$

It will be called separated if

$$a_{n_{k+1}} - a_{n_k} > h b_{n_k} \tag{1.16}$$

for some $h > 0$. *We shall refer to* h *as a mesh of* $\{n_k\}$.

Note that, in contrast to (1.16), the definition of tightness involves only non-overlapping blocks of integers. A sequence $\{n_k\}$ may be both tight and separated.

Theorem 1 will be reduced to the following Theorem 3 (which is essentially equivalent to it). The main difficulty of the present paper consists in proving part (b) of Theorem 3 and the calculations of §§ 4–5 are devoted to it. They are admittedly intricate, but they are actually simpler than those required for the very special case in [1]; due to their general character, they are also more meaningful.

THEOREM 3. (a) *Suppose that there exist tight sequences* $\{n_k\}$ *with arbitrarily small mesh for which the series*

$$\sum P\{S^*_{n_k} \in I\} \tag{1.17}$$

converges. Then $p(I_{-\varepsilon}) = 0$ *for every* $\varepsilon > 0$. *Otherwise* $p(I_\varepsilon) = 1$, *provided* I *is bounded.*

(b) *Let* $I = (L_1, L_2)$ *with* $0 \leq L_1 < L_2 < \infty$. *If the series* (1.17) *diverges for some separated sequence* $\{n_k\}$, *then not only* $p(I_\varepsilon) = 1$, *but the stronger statement*

$$P\{S^*_{n_k} \in I_\varepsilon \ i.o.\} = 1 \tag{1.18}$$

is true. Otherwise $p(I_{-\varepsilon}) = 0$.

This is the form in which the theorem will be proved, and it is given here to avoid repetition. When condition C applies, the convergence or divergence of the series (1.17) does not depend on I. Using our test for $I_{\pm\varepsilon}$ leads then to the following more appealing form of Theorem 3. It is understood that $I = (L_1, L_2)$ with $L_1 > -\infty$.

THEOREM 4. *Assume that condition* C *holds.*

(a) *In order that* $p(I) = 0$ *it is necessary and sufficient that there exist tight sequences* $\{n_k\}$ *with arbitrarily small mesh for which the series* (1.17) *converges.*

(b) *In order that* $p(I) = 1$ *it is necessary and sufficient that there exist a separated sequence* $\{n_k\}$ *for which the series diverges. In this case* $P\{S^*_{n_k} \in I \ i.o.\} = 1$.

If the condition A on $\{\alpha_n\}$ holds for *all* $\lambda > 1$ (with $C = C_\lambda$ depending on λ) then part (a) may be put into the following sharper form:[6]

$p(I) = 0$ *if and only if* (1.17) *diverges for some tight sequence.*

To illustrate these results we derive in § 6 a general analogue to the law of the iterated logarithm for *normal variables* (Theorem 5 in § 6).

2. On the series $\Sigma(I)$

In this section the interval I is arbitrary, and our main purpose is to construct sequences $\{n_k\}$ such that the series $\Sigma(I)$ and (1.17) converge or

[6] The proof follows the pattern of the proof of Lemma 1 and will not be spelled out in detail.

diverge at the same time. The first lemma serves to compare probabilities appearing in the series $\Sigma(I)$ not too far apart. From now on we use the abbreviation

$$S_{j,k} = S_k - S_j \qquad j < k. \tag{2.1}$$

LEMMA 1. *To given* $\varepsilon > 0$ *there exists a* $\delta > 0$ *such that*

$$P\{S_j^* \in I \quad \text{some} \quad i \leqq j \leqq k\} \leqq 2P\{S_k^* \in I_\varepsilon\} \tag{2.2}$$

for all pairs i, k *satisfying*

$$a_k - a_i < \delta b_i . \tag{2.3}$$

PROOF. Since $b_i \leqq a_i$ it follows from (2.3) that all the ratios a_k/a_j with $i \leqq j \leqq k$ remain between 1 and $1 + \delta$. In view of (1.6) it is possible to choose δ so small that the corresponding ratios b_j/b_k remain constricted to an arbitrary neighborhood of 1. Accordingly, we choose δ so small that (2.3) implies

$$a_j + b_j I \subset a_k + b_k I_{\varepsilon/2} , \qquad i \leqq j \leqq k . \tag{2.4}$$

The event $S_k^* \in I_\varepsilon$ then takes place whenever for some j simultaneously $S_j^* \in I$ and $| S_{j,k} | \leqq b_k\varepsilon/2$. The last event is independent of $X_1, \cdots, X_j$, and considering the smallest j for which $S_j^* \in I$ we conclude that

$$\begin{aligned} &P\{S_k^* \in I_\varepsilon\} \\ &\geqq P\{S_j^* \in I \quad \text{some} \quad i \leqq j \leqq k\} \cdot P\{| S_{j,k} | \leqq b_k\varepsilon/2 \quad \text{all} \quad i \leqq j \leqq k\} . \end{aligned} \tag{2.5}$$

$\geqq$ To prove the lemma it suffices to show that the last probability can be made $>1/2$. Now if $1 + \delta \leqq \lambda$ we can use the condition (1.4) on $\{\alpha_n\}$ and conclude from Kolmogorov's inequality that

$$\begin{aligned} P\Big\{| S_{j,k} | > \frac{\varepsilon b_k}{2} \quad \text{some} \quad i \leqq j \leqq k\Big\} &\leqq \frac{4(s_k^2 - s_i^2)}{\varepsilon^2 b_k^2} \\ &\leqq C\frac{8s_k(a_k - a_i)}{\varepsilon^2 b_k^2 \alpha_i} \leqq \frac{8C\delta}{\varepsilon^2 \omega} \cdot \Big(\frac{b_i}{b_k}\Big)^2 \frac{s_k}{s_i} . \end{aligned} \tag{2.6}$$

From (1.4) it is clear that

$$\frac{s_k}{s_i} < 1 + C\,\frac{a_k}{a_i} .$$

Since b_i/b_k and a_k/a_i remain bounded, the last quantity in (2.6) can be made arbitrarily small by choosing δ sufficiently small.

Using the first Borel-Cantelli theorem, it is seen that this lemma (applied to $I_{-\varepsilon}$) has as immediate consequence the

COROLLARY. *The first assertion in part* (a) *of Theorem* 3 *is valid.*

We proceed to construct a special sequence $\{n_k\}$, and to avoid illegible double subscripts, we put generally $\nu_k = n_{k-1} + 1$. Consider $\delta > 0$ as given. Put $\nu_1 = 1$. Proceeding by induction we define n_k as the *last* subscript for which

$$a_{n_k} - a_{\nu_k} \leqq \delta b_{\nu_k} . \tag{2.7}$$

The possibility $n_k = \nu_k$ is not excluded.

The sequence is tight, but it need not be separated. However, omitting every second term in $\{n_k\}$ produces separated sequences. Indeed, for those k for which $a_{\nu_k} \leqq 2a_{n_{k-1}}$ we have $b_{\nu_k} \geqq \tau_2 b_{n_{k-1}}$ (see (1.7)), and hence

$$a_{n_{k+1}} - a_{n_{k-1}} \geqq a_{\nu_{k+1}} - a_{\nu_k} \geqq \delta b_{\nu_k} \geqq \delta\tau_2 b_{n_{k-1}} .$$

For the remaining k we have $a_{\nu_k} > 2a_{n_{k-1}}$ and hence

$$a_{n_{k+1}} - a_{n_{k-1}} \geqq a_{\nu_k} - a_{n_{k-1}} > a_{n_{k-1}} \geqq b_{n_{k-1}} .$$

It follows that the two sequences $\{n_{2k}\}$ and $\{n_{2k+1}\}$ are separated with a mesh $\geqq \min[1, \delta\tau_2]$.

The following lemma refers to the sequence $\{n_k\}$ just constructed.

LEMMA 2. *For arbitrary* $\varepsilon > 0$ *and all sufficiently small* δ

$$\Sigma(I) < \infty \quad \textit{implies} \quad \sum P\{S^*_{n_k} \in I_{-\varepsilon}\} < \infty. \tag{2.8}$$

$$\Sigma(I) = \infty \quad \textit{implies} \quad \sum P\{S^*_{n_k} \in I_{\varepsilon}\} = \infty. \tag{2.9}$$

PROOF. From Lemma 1 we know that for sufficiently small δ

$$P\{S^*_j \in I\} \leqq 2P\{S^*_{n_k} \in I_\varepsilon\} , \qquad \nu_k \leqq j \leqq n_k . \tag{2.10}$$

The same lemma applied to $I_{-\varepsilon}$ (with i, k replaced by ν_k, j) shows that

$$P\{S^*_j \in I\} \geqq \frac{1}{2}P\{S^*_{\nu_k} \in I_{-\varepsilon}\} , \qquad \nu_k \leqq j \leqq n_k . \tag{2.11}$$

We now partition the series $\Sigma(I)$ into blocks (ν_k, n_k) and put for abbreviation

$$A_k = \sum_{j=\nu_k}^{n_k} \min\left[1, \frac{a_{j+1} - a_j}{b_j}\right] . \tag{2.12}$$

We overestimate $\Sigma(I)$ if we multiply the right side in (2.10) by A_k and sum over k. To prove (2.9) it suffices therefore to show that A_k remains bounded. In view of condition (1.6) we may assume that all the ratios b_j/b_i within a block $\nu_k \leqq i < j \leqq n_k$ remain within a fixed neighborhood of 1, say $b_i/2 \leqq b_j \leqq 2b_i$. For $\nu_k \leqq j \leqq n_k - 1$ we have then $a_{j+1} - a_j < \delta b_{\nu_k}$ and hence

$$A_k \leqq 1 + \frac{2}{b_{\nu_k}} \sum_{j=\nu_k}^{n_k-1} (a_{j+1} - a_j) = 1 + \frac{2}{b_{\nu_k}}(a_{n_k} - a_{\nu_k}) \leqq 1 + 2\delta .$$

The proof of (2.8) is less straightforward. The A_k remain bounded away from 0 because either $a_{j+1} - a_j \geqq b_j$ for some j, in which case $A_k \geqq 1$, or else

$$A_k \geqq \frac{1}{2b_{\nu_k}} \sum_{j=\nu_k}^{n_k} (a_{j+1} - a_j) = \frac{1}{2b_{\nu_k}}(a_{\nu_{k+1}} - a_{\nu_k}) \geqq \frac{1}{2}\delta$$

by the definition of ν_{k+1}. This together with (2.11) proves that $\Sigma(I) < \infty$ implies

$$\sum P\{S^*_{\nu_k} \in I_{-\varepsilon}\} < \infty \ , \tag{2.13}$$

but it remains herein to replace ν_k by n_k. To do this, we split the series into two parts. Let $\mathcal{K}$ be the class of all k such that

$$a_{\nu_{k+1}} - a_{n_k} \leqq h b_{n_k} \ . \tag{2.14}$$

Applying Lemma 1 to $I_{-\varepsilon}$ and the pair of integers n_k and $n_k + 1 = \nu_{k+1}$ we see that if h is chosen sufficiently small we have for $k \in \mathcal{K}$

$$P\{S^*_{n_k} \in I_{-2\varepsilon}\} \leqq 2P\{S^*_{\nu_{k+1}} \in I_{-\varepsilon}\} \ . \tag{2.15}$$

Thus

$$\sum_{k \in \mathcal{K}} P\{S^*_{n_k} \in I_{-2\varepsilon}\} < \infty \ . \tag{2.16}$$

But when (2.14) does not hold, $P\{S^*_{n_k} \in I\}$ appears in the series $\Sigma(I)$ with a coefficient $\geqq \min [1, h]$, and so $\Sigma(I) < \infty$ implies

$$\sum_{k \notin \mathcal{K}} P\{S^*_{n_k} \in I_{-2\varepsilon}\} \leqq \sum_{k \notin \mathcal{K}} P\{S^*_{n_k} \in I\} < \infty \ . \tag{2.17}$$

This proves (2.8) except for the notational change from ε to 2ε.

Outline of the proof. We have already seen that Lemma 1 implies the first assertion in Theorem 3(a). If $\Sigma(I) < \infty$ there exist tight sequences $\{n_k\}$ with arbitrarily small mesh for which $\sum P\{S^*_n \in I_{-\varepsilon}\} < \infty$. By Theorem 3(a) this implies $p(I_{-2\varepsilon}) = 0$, and as ε is arbitrary this proves part (a) of Theorem 1. Furthermore, when there are no tight sequences with the stated properties, then $\Sigma(I)$ necessarily diverges, and the second assertion in Theorem 3(a) merely restates part (b) of Theorem 1.

We saw that the two subsequences $\{n_{2k}\}$ and $\{n_{2k+1}\}$ are separated, and (2.9) shows that when $\Sigma(I) = \infty$ at least one of the series

$$\sum P\{S^*_{n_{2k}} \in I_\varepsilon\} \qquad \textit{and} \qquad \sum P\{S^*_{n_{2k+1}} \in I_\varepsilon\}$$

diverges. Part (b) of Theorem 1 is therefore contained in part (b) of Theorem 3, and we now turn to its lengthy proof.

3. The framework

From now on we shall deal principally with one fixed subsequence $\{S_{n_k}\}$ and, to avoid the nuisance of many subscripts, we shall denote it by $\{S'_k\}$. More generally, given a symbol with subscripts i, j, ν, etc., *a dash will signify that these subscripts are replaced* by n_i, n_j, n_ν, etc. Thus

(3.1) $$a'_j = a_{n_j}\ , \qquad S'_{j,k} = S_{n_k} - S_{n_j}\ , \qquad S'^*_j = S^*_{n_j}\ .$$

Throughout the next two sections $I = (L_1, L_2)$ stands for an interval with $0 \leqq L_1 < L_2 < \infty$. For abbreviation we put

(3.2) $$\Gamma_j = \{S'^*_j \in I\}\ , \qquad \gamma_\varepsilon = P\{S'^*_j \in I_\varepsilon \ i.o.\}$$

The stronger assertion of Theorem 3(b) may now be restated succinctly as follows. *If the sequence* $\{n_k\}$ *is separated and*

(3.3) $$\sum P\{\Gamma_k\} = \infty\ ,$$

then $\gamma_\varepsilon = 1$ *for every* $\varepsilon > 0$.

Here the mesh of $\{n_k\}$ is arbitrary. The next lemma states essentially that there is no loss of generality in assuming that the mesh is arbitrarily large.

LEMMA 3. *Every separated sequence* $\{n_k\}$ *satisfying* (3.3) *contains a subsequence with arbitrarily large mesh such that the corresponding subseries of* (3.3) *diverges.*

PROOF. Consider the N arithmetic sequences of the form $m + Nk$ and the corresponding subseries of (3.3). At least one of them diverges and we may suppose that

(3.4) $$\sum \Gamma(\Gamma_{Nk}) = \infty\ .$$

We show that by choosing N sufficiently large the mesh of the subsequence $\{n_{Nk}\}$ may be made to exceed a pre-assigned number t. If h is a mesh of $\{n_k\}$ then

(3.5) $$a'_{N(k+1)} - a'_{Nk} \geqq h(b'_{Nk} + \cdots + b'_{Nk+N-1})\ .$$

The assumption (1.7) concerning $\{b_k\}$ assures us that either the left side exceeds ta'_{Nk}, or else all b'_j appearing on the right in (3.5) exceed $\tau_t b'_{Nk}$, and then the right side exceeds $Nh\tau_t b'_{Nk}$. Since $a'_n \geqq b'_n$, this implies that $\{n_{Nk}\}$ has a mesh $>t$ provided only $Nh\tau_t > t$.

Suppose then that for a given separated sequence $\{n_k\}$ the series (3.3) diverges and $\gamma_\varepsilon < 1$. In passing from $\{n_k\}$ to the subsequence $\{n_{Nk}\}$ the probabilities corresponding to (3.2) cannot increase, and so $\{n_{Nk}\}$ shares the stated properties of $\{n_k\}$ but its mesh exceeds t. From now on we shall deal exclusively with the subsequence $\{n_{Nk}\}$, and to avoid overburdening the notation we shall treat it as if it were the given sequence; in other words, we write n_k instead of n_{Nk}.

To recapitulate, we have shown that if Theorem 3(b) were false, there would exist two numbers ε and $\gamma < 1$ for which the following would be true.

For each $t > 0$ there exists a separated sequence with mesh $>t$ such that the series (3.3) *diverges while the probability γ_ε of* (3.2) *satisfies $\gamma_\varepsilon < \gamma < 1$.* It remains to show that this is impossible.

To achieve this we introduce

$$\mu_p = P\{S'_j \in I_\varepsilon \ \textit{for some} \ j > p\} . \tag{3.6}$$

We shall prove the following

PROPOSITION. *Given the interval I and the positive numbers ε and η, there exists t such that for any separated sequence with mesh $>t$*

$$P\{\textstyle\bigcup_{j=p}^{q} \Gamma_j\} \geqq \sum_{j=p}^{q} P\{\Gamma_j\}\cdot[1 - \eta - \mu_p] \tag{3.7}$$

for any pair of integers $p < q$ with p sufficiently large.

For a sequence $\{n_k\}$ with the properties stated above in italics and $\eta = (1-\gamma)/2$ the quantity within brackets would be $>(1-\gamma)/2$ for all p sufficiently large. Then, because of the assumed divergence of (3.3), one can always choose q so large that the right side exceeds any preassigned number. But the left side remains ≤ 1, and this contradiction shows that no sequence $\{n_k\}$ can have the stated properties. Thus *our remaining task reduces to proving the proposition.*

To estimate the probability on the left in (3.7) we rewrite the set as the union of the non-overlapping sets

$$\Gamma_j^\# = \Gamma_j - \Gamma_j \textstyle\bigcup_{k=j+1}^{q} \Gamma_k , \qquad p \leqq j \leqq q . \tag{3.8}$$

Clearly

$$P\{\textstyle\bigcup_{j=p}^{q} \Gamma_j\} = \sum_{j=p}^{q} P\{\Gamma_j^\#\} . \tag{3.9}$$

Unfortunately no sufficiently sharp estimates can be obtained from the individual components $P\{\Gamma_j\Gamma_k\}$, but it is necessary to partition the Γ_k into blocks. These will depend on j, but for better legibility the dependence on j will not be emphasized. For the time being consider j as fixed.

Let $\lambda > 1$ be the number occurring in the condition A on $\{a_n\}$. For $m = 0, 1, \cdots$, put

$$K_m = \{k : \lambda^m a'_j < a'_k \leqq \lambda^{m+1} a'_j\} \tag{3.10}$$

and

$$\Lambda_m = \textstyle\bigcup_{k \in K_m} \Gamma_k . \tag{3.11}$$

Some of these sets may be empty. We require upper estimates for $P\{\Gamma_j \Lambda_m\}$.

4. Estimates

In this section we are dealing with a sequence $\{n_k\}$ such that with the notation introduced in (3.1)

$$a'_{k+1} - a'_k > tb'_k \,. \tag{4.1}$$

$I = (L_1, L_2)$ is an interval with $0 \leqq L_1 < L_2 < \infty$.

LEMMA 4. *If $t > 2L_2$, then for $m = 0, 1, \cdots$*

$$P\{\Gamma_j \Lambda_m\} \leqq P\{\Gamma_j\} \cdot P\left\{S'_{j,k} > \frac{1}{2}(a'_k - a'_j) \quad \textit{some} \quad k \in K_m\right\} \tag{4.2}$$

PROOF. Obviously for $k > j$

$$\Gamma_j \Gamma_k \subset \Gamma_j \{S'_{j,k} \geqq a'_k + L_1 b'_k - a'_j - L_2 b'_j\} \tag{4.3}$$

and the two events on the right are independent. But

$$L_2 b'_j < L_2 t^{-1}(a'_{j+1} - a'_j) \leqq L_2 t^{-1}(a'_k - a'_j); \tag{4.4}$$

when $L_2 t^{-1} < 1/2$ it follows that (4.3) implies (4.2).

LEMMA 5. *For t sufficiently large*

$$P\{\Gamma_j \Lambda_0\} \leqq \frac{1}{2}\eta P\{\Gamma_j\} \,. \tag{4.5}$$

(More precisely, (4.5) holds if t satisfies (4.15).)

PROOF. There is nothing to be proved if K_0 is vacuous. Otherwise put

$$Y_{\rho-j} = \sum\nolimits_{h=j+1}^{\rho} \frac{S'_{h-1,h}}{a'_h - a'_j} \,, \qquad \rho \in K_0 \tag{4.6}$$

and let $Y_0 = 0$. Then $S'_{\rho-1,\rho}$ is a linear combination of $Y_{\rho-j}$ and $Y_{\rho-j-1}$ and hence for $k > j$

$$\begin{aligned} S'_{j,k} &= \sum\nolimits_{\rho=j+1}^{k} (a'_\rho - a'_j)(Y_{\rho-j} - Y_{\rho-j-1}) \\ &= (a'_k - a'_j) Y_{k-j} - \sum\nolimits_{h=j+1}^{k} (a'_{h+1} - a'_h) Y_{h-j} \,. \end{aligned} \tag{4.7}$$

Thus

$$| S'_{j,k} | \leqq 2(a'_k - a'_j) \max\nolimits_{\nu \leqq k-j} | Y_\nu | \,. \tag{4.8}$$

Now the Y_ν are the partial sums of a sequence of independent random variables with zero expectation and so we conclude from (4.8) and Kolmogorov's inequality that

$$\begin{aligned} &P\left\{| S'_{j,k} | > \frac{1}{2}(a'_k - a'_j) \quad \textit{some} \quad k \in K_0\right\} \\ &\qquad \leqq P\left\{\max\nolimits_{k \in K_0} | Y_{k-j} | > \frac{1}{4}\right\} \leqq 16E(Y^2_{r-j}) \end{aligned} \tag{4.9}$$

where r is the greatest integer in K_0. Now

(4.10) $$E(Y_{r-j}^2) = \sum_{h=j+1}^{r} \frac{s_h'^2 - s_{h-1}'^2}{(a_h' - a_j')^2} .$$

By the definition of K_0 we have $a_r' \leqq \lambda a_j'$ and from the assumed property (1.4) of $\{\alpha_n\}$ we get therefore for $h \leqq r$

(4.11) $$s_h'^2 - s_{h-1}'^2 \leqq 2Cs_h' \frac{a_h' - a_{h-1}'}{a_{h-1}'} \leqq \frac{2Cs_r'^2}{a_j'}(a_h' - a_{h-1}') .$$

Thus

(4.12) $$E(Y_{r-j}^2) \leqq \frac{2Cs_r'^2}{a_j'} \sum_{h=j+1}^{r} \frac{a_h' - a_{h-1}'}{(a_h' - a_j')^2} .$$

Now for $h \geqq j + 2$

$$\frac{a_h' - a_{h-1}'}{(a_h' - a_j')^2} \leqq \frac{1}{a_{h-1}' - a_j'} - \frac{1}{a_h' - a_j'}$$

and hence by telescoping the sum from $h = j + 2$ to r

(4.13) $$\begin{aligned} E(Y_{r-j}^2) &\leqq \frac{4Cs_r'^2}{a_j'} \cdot \frac{1}{a_{j+1}' - a_j'} \\ &\leqq \frac{4Cs_r'^2}{ta_j'b_j'} \leqq \frac{4C}{t\omega} \cdot \left(\frac{s_r'}{s_j'}\right)^2 . \end{aligned}$$

Again from (1.4)

(4.14) $$\frac{s_r'}{s_j'} \leqq 1 + C\frac{a_r'}{a_j'} \leqq 1 + C\lambda$$

and hence finally

(4.14) $$P\left\{ |S_{j,k}'| > \frac{1}{2}(a_k' - a_j') \ \ some \ \ k \in K_0 \right\} \leqq \frac{64C}{t\omega}(1 + C\lambda)^2 .$$

In view of Lemma 4 this implies the truth of (4.5) for all t such that

(4.15) $$t > \frac{128}{\eta\omega} C(1 + C\lambda)^2 \ \ and \ \ t > 2L_2 .$$

LEMMA 6. *Let $m \geqq 1$ and denote by r_m the largest integer in K_m (if any). There exists a constant A_λ depending only on λ such that*

(4.16) $$P\{\Gamma_j \Lambda_m\} \leqq \frac{A_\lambda}{\alpha_{r_m}'^2} P\{\Gamma_j\} .$$

If K_m contains an integer k such that

(4.17) $$a_k' < a_j' \alpha_k'^4$$

then also

(4.18) $$P\{\Gamma_j \Lambda_m\} \leqq B_\lambda \lambda^{-m/2} P\{\Gamma_j\}$$

with B_λ depending only on λ.

PROOF. By the definition of K_m we have for $k \in K_m$

$$a'_k - a'_j \geqq (\lambda^{-1} - \lambda^{-m-1})a'_{r_m} \geqq \frac{\lambda-1}{\lambda^2}a'_{r_m} \tag{4.19}$$

and hence by Kolmogorov's inequality

$$P\left\{S'_{j,k} > \frac{1}{2}(a'_k - a'_j) \quad some \quad k \in K_m\right\} \leqq \left(\frac{2\lambda^2}{\lambda-1} \cdot \frac{s'_{r_m}}{a'_{r_m}}\right)^2 . \tag{4.20}$$

Introducing this into (4.2) we get (4.16).

For $k \in K_m$ we have $a'_{r_m} < \lambda a'_k$ and hence from (1.4)

$$\frac{\alpha'_k}{\alpha'_{r_m}} = \frac{a'_k}{a'_{r_m}} \cdot \frac{s'_{r_m}}{s'_k} \leqq (1 + C\lambda) . \tag{4.21}$$

If K_m contains a k satisfying (4.17) we get therefore

$$\frac{1}{\alpha'^2_{r_m}} \leqq \frac{(1+C\lambda)^2}{\alpha'^2_k} \leqq (1+C\lambda)^2\left(\frac{a'_j}{a'_k}\right)^{1/2} \leqq (1+C\lambda)^2\lambda^{-m/2} \tag{4.22}$$

and so (4.18) holds with $B_\lambda = A_\lambda(1 + C\lambda)^2$.

LEMMA 7. *Let M be the class of all integers $k > j$ such that*

$$a'_k \geqq a'_j \alpha'^4_k , \tag{4.23}$$

and put

$$\Lambda = \bigcup_{k \in M} \Gamma_k . \tag{4.24}$$

Then for any $\eta > 0$ and $\varepsilon > 0$

$$P\{\Gamma_j \Lambda\} \leqq \left(1 + \frac{1}{4}\eta\right) P\{\Gamma_j\} \cdot P\{S'^*_k \in I_\varepsilon \quad some \quad k \in M\} , \tag{4.25}$$

provided j is sufficiently large (satisfies (4.30)).

PROOF. Suppose that for some $k > j$

$$\frac{1}{2}\varepsilon b'_k > (1 + L_2)a'_j . \tag{4.26}$$

Remembering that $b'_j \leqq a'_j$ one verifies easily that this implies

$$\Gamma_j \Gamma_k \subset \Gamma_j\{S'_{j,k} \in a'_k + b'_k I_{\varepsilon/2}\} \tag{4.27}$$

as well as

$$\{S'_k \in a'_k + b'_k I_\varepsilon\} \supset \{S'_{j,k} \in a'_k + b'_k I_{\varepsilon/2}\}\{|S'_j| < a'_j\} . \tag{4.28}$$

For $k \in M$ we have

$$a'_j \leqq \frac{a'_k}{\alpha'^4_k} = \frac{b'_k}{\alpha'^3_k \beta'_k} \leqq \frac{b'_k}{\omega \alpha'^2_k} . \tag{4.29}$$

We have assumed $\alpha_n \to \infty$ and hence we may assume j to be so large that

(4.30) $$\frac{1}{\alpha_j'^2} < \frac{\eta}{\eta + 4} \quad and \quad \frac{1 + L_2}{\omega \alpha_k'^2} < \frac{1}{2}\varepsilon \quad for \quad k > j .$$

From (4.29) it follows then that (4.26) holds for all $k \in M$, and the last event in (4.28) has a probability $\geqq 4/(4 + \eta)$.

The events on the right being independent we get then from (4.27)

(4.31) $$P\{\Gamma_j \Lambda\} \leqq P\{\Gamma_j\} P\{S'_{j,k} \in a'_k + b'_k I_{\varepsilon/2} \quad some \quad k \in M\} .$$

To pass from this to (4.25) it suffices to use (4.28) where the events on the right are independent and the probability of the last exceeds $(1 + \eta/4)^{-1}$.

5. Proof of the proposition

We are finally in a position to finish the proofs by verifying the validity of the proposition in § 3.

We start from (3.9). For given j every Γ_k with $k > j$ is a subset of one among the following (not mutually exclusive) sets:

(i) The set Λ_0.

(ii) The union of Λ_m with $1 \leqq m \leqq \nu$, where ν is any integer.

(iii) The union of all those Λ_m with $m \geqq 1$ for which K_m contains an integer k such that (4.17) holds.

(iv) The set Λ defined in (4.24).

If t is large enough for Lemma 5 to apply, and j as stipulated in Lemma 7, then Lemmas 5–7 show that

(5.1) $$P\{\Gamma_j \textstyle\bigcup_{k>j} \Gamma_k\} < \zeta P\{\Gamma_j\}$$

where we put for abbreviation

(5.2) $$\zeta = \frac{1}{2}\eta + A_\lambda \sum_{m=1}^{\nu} \frac{1}{\alpha_{r_m}'^2} + B_\lambda \sum_{m=\nu+1}^{\infty} \lambda^{-m/2} + \left(1 + \frac{\eta}{4}\right) P\{S_k'^* \in I_\varepsilon \quad some \quad k \in M\} .$$

Since $\lambda > 1$, the third term can be made $<\eta/8$ by choosing ν sufficiently large. Since $\alpha_n \to \infty$ the second term will be $<\eta/8$ for all j sufficiently large. Finally, since M contains only integers $>j$ we get for sufficiently large j

(5.3) $$\zeta < \eta + P\{S_k'^* \in I_\varepsilon \quad some \quad k > j\} .$$

Substituting from (5.1) into (3.9) we find

(5.4) $$P\{\Gamma_j^*\} \geqq P\{\Gamma_j\} \cdot [1 - \eta - P\{S_k'^* \in I_\varepsilon \quad some \quad k \geqq j\}] .$$

Summing over $p \leqq j \leqq q$ we get the assertion (3.8) of the proposition.

6. Normal variables

Let X_n be normal with $E(X_n) = 0$ and $E(X_n^2) = \sigma_n^2 > 0$. Assume $s_n^2 \to \infty$, and put

$$d_n = \min\left[1, \frac{s_{n+1} - s_n}{s_n}\right], \qquad D_n = d_1 + \cdots + d_n . \tag{6.1}$$

The series with terms $s_{n+1} - s_n$ diverges, and by the Abel-Dini theorem [5, p. 290] the same is therefore true of the series with general term $(s_{n+1} - s_n)/s_n$. One concludes easily that also $\sum d_n = \infty$, and the theorem of Abel-Dini assures us that

$$\sum \frac{d_n}{D_n^\lambda} \begin{matrix} < \infty & \quad if\ \lambda > 1 \\ = \infty & \quad if\ \lambda \leq 1 . \end{matrix} \tag{6.2}$$

This leads us to[7]

Theorem 5. *With probability one,*

$$\limsup \frac{S_n}{s_n\sqrt{2 \log D_n}} = 1 . \tag{6.3}$$

Proof. Let $\alpha_n^2 = \beta_n^2 = 2 \log D_n$. Then

$$\begin{aligned} P\{S_n > a_n + \eta b_n\} &= P\left\{\frac{S_n}{s_n} > \alpha_n(1 + \eta)\right\} \\ &\sim \frac{1}{\sqrt{2\pi}(1 + \eta)\alpha_n} \cdot e^{-(1+\eta)^2 \log D_n} . \end{aligned} \tag{6.4}$$

In the present case $(a_{n+1} - a_n)/b_n \sim (s_{n+1} - s_n)/s_n$, and it is obvious that the series $\Sigma(I)$ corresponding to $I = (\eta, \infty)$ converges or diverges at the same time as

$$\sum \min\left[1, \frac{s_{n+1} - s_n}{s_n}\right] \cdot P\{S_n > a_n + \eta b_n\} \quad or \quad \sum \frac{d_n}{\alpha_n D_n^{(1+\eta)^2}} . \tag{6.5}$$

From (6.2) it follows that the last series converges for $\eta > 0$ and diverges for $\eta < 0$. The assertion now follows from Theorem 1.

For example, if σ_n/s_n remains bounded away from 0 then $\log D_n \sim \log n$, which may be a far cry from $\log \log s_n$. In this way we obtain a variety of forms replacing the $\log \log s_n$ of the classical law. Of course, more precise results can be obtained from Theorem 2 with $\alpha_n^2 = 2 \log D_n$ and $\beta_n = 1/\alpha_n$.

7. A counter-example

We show that part (b) of Theorem 1 does not apply to semi-infinite intervals. For that purpose put for $r \geq 3$

[7] Special cases were treated by Hartman and Wintner [4], and by Hartman [3].

(7.1) $$\nu_r = r^r , \qquad p_r = \frac{1}{r \log^2 r} .$$

We define $\{X_n\}$ as follows. If n is *not* a term of the sequence $\{\nu_r\}$ we let X_n be a binomial variable assuming the values ± 1 each with probability 1/2. If, however, $n = \nu_r$ then X_n assumes the values $\pm\sqrt{\nu_r/p_r}$ each with probability $(1/2)p_r$, and the value 0 with probability $1 - p_r$. The variance $\sigma_n^2 = E(X_n^2)$ equals either 1 or ν_r and hence

(7.2) $$n \leqq s_n^2 \leqq 3n .$$

Put

(7.3) $$\alpha_n^2 = \beta_n^2 = 9 \log\log n .$$

From $\sum p_r < \infty$ it follows that with probability one only finitely many among the X_{ν_r} will be different from zero, and so $\{X_n\}$ is essentially equivalent to a sequence of binomial variables. In particular, by the law of the iterated logarithm

(7.4) $$P\{S_n > s_n\alpha_n \ \ i.o.\} = 0 .$$

We show that nevertheless for $I = (1, \infty)$ we have $\Sigma(I) = \infty$.

Since $\alpha_n = \beta_n$ increases monotonically

$$\frac{a_{n+1} - a_n}{b_n} \geqq \frac{s_{n+1} - s_n}{s_n} \geqq \frac{\sigma_{n+1}^2}{2s_{n+1}^2} \geq \frac{1}{12n}$$

and hence

(7.5) $$\Sigma(I) \geqq \frac{1}{12} \sum \frac{1}{n} P\{S_n > 2s_n\alpha_n\} .$$

From Chebyshev's inequality we get for $n > \nu_r$

(7.6) $$P\{S_n < 2s_n\alpha_n\} \geqq P\{X_{\nu_r} > 3s_n\alpha_n\}\left(1 - \frac{1}{\alpha_n^2}\right) .$$

But for $\nu_r < n < \nu_{r+1}$ we have

(7.7) $$\begin{aligned} 3s_n\alpha_n &< 20(\nu_{r+1} \log\log \nu_{r+1})^{1/2} < 100(\nu_r r \log r)^{1/2} \\ &= \frac{100}{(\log r)^{1/2}} \left(\frac{\nu_r}{p_r}\right)^{1/2} . \end{aligned}$$

When $\log r > 10^4$ and $\alpha_n^2 > 2$ it follows that

$$\begin{aligned} \sum_{n=\nu_r+1}^{\nu_{r+1}} \frac{1}{n} P\{S_n > 2s_n\alpha_n\} &\geqq \frac{1}{2} \sum_{n=\nu_r+1}^{\nu_{r+1}} \frac{1}{n} P\left\{X_{\nu_r} = \left(\frac{\nu_r}{p_r}\right)^{1/2}\right\} \\ &= \frac{1}{4r \log^2 r} \sum_{n=\nu_r+1}^{\nu_{r+1}} \frac{1}{n} \sim \frac{1}{4r \log^2 r} \log \frac{\nu_{r+1}}{\nu_r} \sim \frac{1}{4r \log r} \end{aligned}$$

Summing over r one sees that the series in (7.5) diverges.

PRINCETON UNIVERSITY

REFERENCES

[1] W. FELLER, *The general form of the so-called law of the iterated logarithm*, Trans. Amer. Math. Soc. **54** (1943), 373-402.

[2] ———, *Limit theorems for probabilities of large deviations*. To appear in the Zs. Wahrscheinlichkeitsrechnung. For a summary see Proc. Natl. Acad. Sci. **61** (1968), 1224-1227.

[3] P. HARTMAN, *Normal distributions and the law of the iterated logarithm*, Amer. J. Math. **63** (1941), 584-588.

[4] ——— and A. WINTNER, *On the law of the iterated logarithm*, Amer. J. Math. **63** (1941), 169-176.

[5] K. KNOPP, Theory and Applications of Infinite Series, Engl. Transl. (London-Glasgow) 1928.

[6] P. LÉVY, Théorie de l'addition des variables aléatoires, (p. 147 ff.) 2nd ed., Paris 1954.

(Received April 14, 1969)

Kolibri 360 GmbH
Ferdinand-Jühlke-Straße 7
99095 Erfurt, Deutschland
produktsicherheit@kolibri360.de